SMITHSONIAN

TIMELINES of SCIENCE

DK SMITHSONIAN

TIMELINES *of* SCIENCE

·ARISMETRICA·
·GEOMETRIA·
·MVSICA·
·ASTRO

LONDON, NEW YORK, MELBOURNE,
MUNICH, AND DELHI

DK LONDON

Senior Art Editor
Ina Stradins

Project Art Editors
Alison Gardner, Clare Joyce, Francis Wong

Senior Preproduction Producer
Ben Marcus

Producer
Vivienne Yong

Creative Technical Support
Adam Brackenbury

Jacket Designer
Mark Cavanagh

Picture Researcher
Liz Moore

New Photography
Gary Ombler

New Illustrations
Peter Bull

Jacket Design Development Manager
Sophia MTT

Managing Art Editor
Michelle Baxter

Art Director
Philip Ormerod

Senior Editors
Peter Frances, Janet Mohun

US Senior Editor
Rebecca Warren

US Editor
Jill Hamilton

Project Editors
Jemima Dunne, Joanna Edwards, Lara Maiklem, David Summers, Miezan van Zyl, Laura Wheadon

Editors
Ann Baggaley, Martyn Page, Carron Brown

Editorial Assistant
Kaiya Shang

Jacket Editor
Manisha Majithia

Indexer
Jane Parker

Managing Editor
Angeles Gavira Guerrero

Publisher
Sarah Larter

Associate Publishing Director
Liz Wheeler

Publishing Director
Jonathan Metcalf

DK INDIA

Deputy Managing Art Editor
Sudakshina Basu

Senior Art Editor
Devika Dwarakadas

Art Editors
Suhita Dharamjit, Amit Malhotra

Assistant Art Editor
Vanya Mittal

Production Manager
Pankaj Sharma

Managing Editor
Rohan Sinha

Senior Editor
Anita Kakar

Editors
Dharini Ganesh, Himani Khatreja, Priyaneet Singh

DTP Manager
Balwant Singh

Senior DTP Designer
Jagtar Singh

DTP Designers
Nand Kishor Acharya, Sachin Gupta

SMITHSONIAN ENTERPRISES

Senior Vice President Carol LeBlanc

Director of Licensing Brigid Ferraro

Licensing Manager and Project Coordinator Ellen Nanney

Product Development Coordinator Kealy Wilson

First American edition, 2013 Published by DK Publishing, 4th Floor, 345 Hudson Street, New York 10014

13 14 15 10 9 8 7 6 5 4 3 2 1

184801 – 001 – Oct/2013

Published in Great Britain by Dorling Kindersley Limited

A catalog record for this book is available from the Library of Congress.

ISBN 978-1-4654-1434-2

DK books are available at special discounts when purchased in bulk for sales promotions, premiums, fund-raising, or educational use. For details, contact: DK Publishing Special Markets, 345 Hudson Street, New York, NY 10014 or SpecialSales@dk.com

Color reproduction by Alta Images, London

Printed and bound in China by Hung Hing

Discover more at
www.dk.com

LOGIA· ·LOGICA· ·RETHORICA· ·GRAMMATICA·

CONTRIBUTORS

Jack Challoner
Science writer and communicator with a background in physics. He contributed to DK's *Science* and has written more than 30 other books on science and technology, for readers of all ages.

Derek Harvey
Naturalist and science writer for titles including DK's *Science* and *The Natural History Book*.

John Farndon
Popular science writer, specializing in Earth science and the history of ideas.

Philip Parker
Historian and writer whose books include DK's *Eyewitness Companion: World History*, *Timelines of History*, and *Engineers*.

Marcus Weeks
Writer on history, economics, and popular science. He has contributed to DK's *Science*, *Engineers*, and *Help Your Kids with Math*.

Giles Sparrow
Popular science writer, specializing in astronomy and space science.

Mary Gribbin
Science writer for young readers and a Visiting Fellow at the University of Sussex.

GLOSSARY

Richard Beatty
Edinburgh-based science writer, editor, and scientific lexicographer.

EDITOR-IN-CHIEF

Professor Robert Winston
Robert Winston is Professor of Science and Society and Emeritus Professor of Fertility Studies at Imperial

CONSULTANTS

John Gribbin
Science writer, astrophysicist, and Visiting Fellow in Astronomy at the University of Sussex. He is the

SMITHSONIAN INSTITUTION

Smithsonian contributors include historians and museum specialists from:

National Air and Space Museum

Foreword

"The past is never dead. It's not even past."
William Faulkner, *Requiem for a Nun*, 1950

Modern science carries multiple traces of its historical origins: we encounter its past every day. Even the most sophisticated clocks mark off time in sixties, a survival from Babylonian numbering systems used many thousands of years ago. Scientific heroes are celebrated in units of measurement—Volts, Curies, Richters—and in parts of our body, such as the Eustachian tubes in our ears. Discarded scientific theories live on in language: "melancholic" and "sanguine" originated in ancient Greek medicine, while "mesmerizing" refers to an 18th-century French therapy based on magnets. Plants and animals still bear the Latin names of Carl Linnaeus's classification system, introduced in Sweden long before Charles Darwin's evolutionary theories made sense of life's complicated variety—and rainbows have seven colors because Isaac Newton believed they should follow the mathematics of musical scales worked out by Pythagoras.

Technological science now permeates society, inseparable from political, commercial, military, and industrial projects, yet the word "scientist" was invented only in 1833. Despite that apparently late start, science has ancient roots. Long before universities and laboratories were created, stargazers studied the heavens to calculate the dates of religious festivals, while scholars attached to mosques and monasteries deciphered God's designs by interpreting the natural world. Uneducated men and women were building up the practical expertise that later provided the foundation of scientific disciplines—how to distil medicines from herbs, smelt ores to produce metals, navigate by the stars, detect the signs of bad weather, mix chemicals to make soap.

From the earliest attempts to make fires, pots, and tools, people have always experimented to find out how the world works and how they can make their lives more comfortable. These twin goals of scientific research were spelled out in the early 17th century by philosopher Francis Bacon. "Knowledge is power," he declared, and the rate of change accelerated as governments increasingly recognized the advantages to be gained from investment in scientific projects. Expanding exponentially, technological science rapidly came to dominate the world, uniting it in an international web of instantaneous electronic communication.

Science has uncovered many of nature's secrets, but it has also unleashed some genies—atomic energy, global warming, genetic modification—that may ultimately destroy us. As citizens of a scientific global community, we need to understand the past in order to control our own future.

PATRICIA FARA
Chief Editorial Consultant

Extremophile habitat
Vivid colors in the Grand Prismatic Spring in Yellowstone National Park, US, result from a film of pigmented bacteria around the edge of the hot spring. Different species of microbes flourish in specific temperatures and contain pigments suited to their environments.

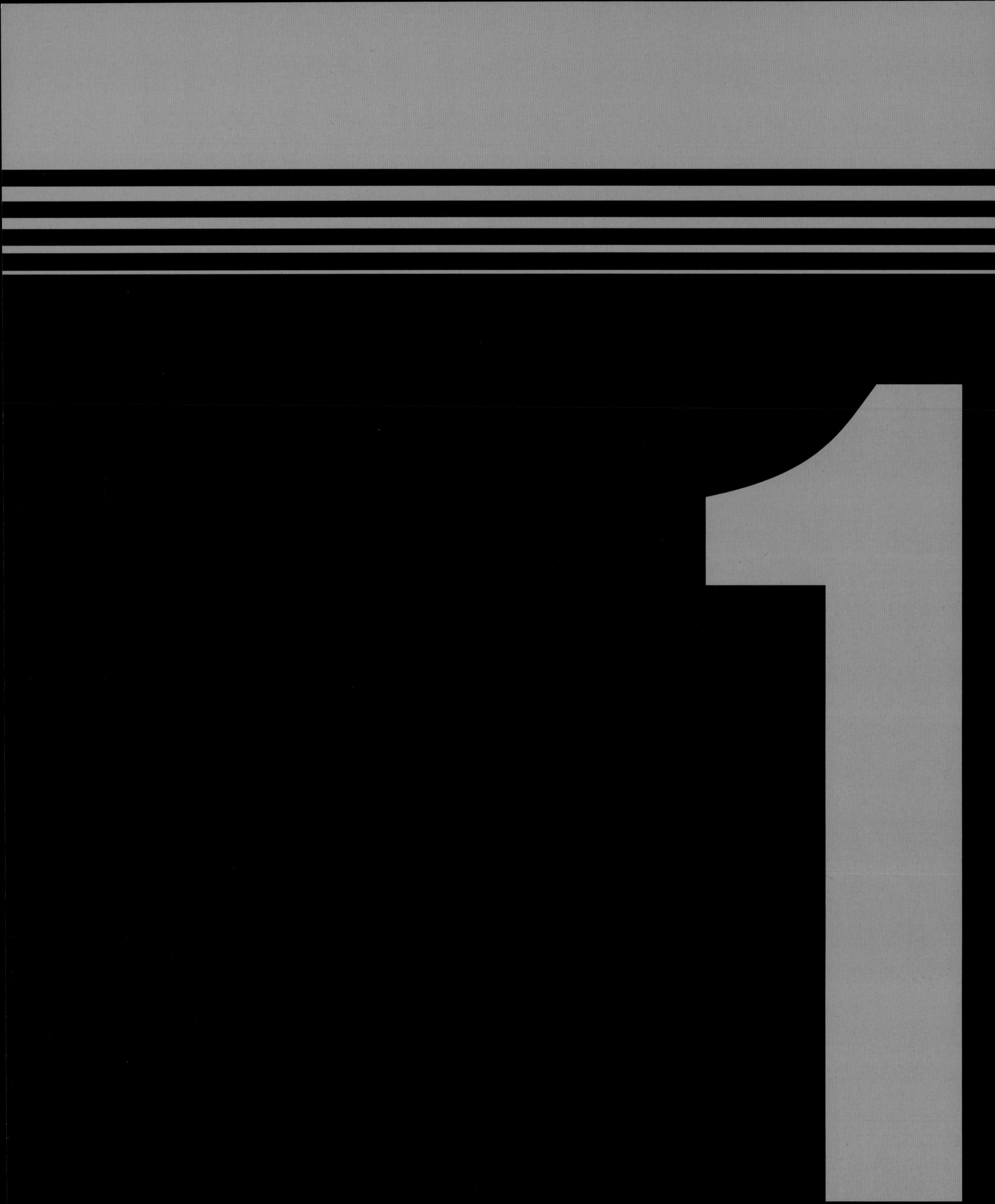

BEFORE SCIENCE BEGAN 2.5 MYA–799 CE

Starting with early experiments to make tools and use fire, humans gradually learned how to control, explore, and understand their surroundings by developing techniques in astronomy, medicine, and mathematics.

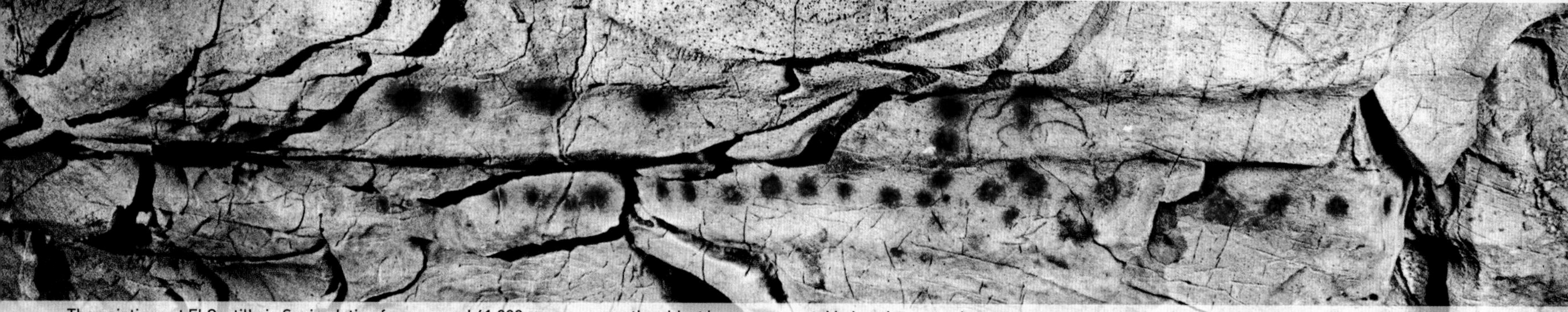

The paintings at El Castillo in Spain, dating from around 41,000 YA, are among the oldest known cave art. Made using natural pigments, the paintings include depictions of horses and bison, although the very earliest are abstract disks and dots.

THE FIRST SIGNIFICANT SCIENTIFIC ADVANCE was the production of stone tools. Around 2.5 million years ago (MYA), early hominids (either *Homo habilis* or *Australopithecus*) began to modify cobbles by striking them with another stone, thus removing flakes of stone and creating a sharp edge—a method known as **hard-hammer percussion**. These early pebble tools, or choppers, are known as **Oldowan tools**. They were used for dismembering killed animals, cracking bones for the marrow, and scraping hides. Oldowan technology spread throughout Africa, where it lasted until around 1.7 MYA.

Early hominids must have seen and understood the power of fire by observing wildfires caused by lightning strikes. They may have lit branches from these fires to use as weapons against predators or to provide light and heat. There is possible **evidence for sporadic controlled use of fire** from around 1 MYA, with evidence of regular use from around 400,000 YA. Finds at Gesher Benot Ya'aqov in Israel (790,000 YA) show signs of the **active use of fire**.

Early humans were able to use devices such as fire plows or fire drills to produce their own fire with friction. Fire was important for warmth and protection, splitting stones, hardening the points of wooden tools, and cooking. **Heating food breaks down proteins**, which makes it easier to digest. It also protects food from decay and extends the range of edible resources to include plants containing toxins that can be broken down by heat. The earliest evidence of cooking comes from sites such as Gesher Benot Ya'aqov in Israel (790,000 YA), where concentrations of burnt seeds and wood were found.

sharp edge where stone flake struck off

Oldowan tool
Choppers like this were the earliest stone tools. They were suitable for tasks such as cutting animal hides.

SURFACE 1
SURFACE 2
upper and lower blocks rubbed together
surface of both blocks heat up

GENERATING HEAT FROM FRICTION

Rubbing two surfaces together causes the kinetic movement energy of the rubbing motion to be transferred to the atoms in the surfaces. This process, known as friction, causes the atoms to heat up. The smoother the surfaces, the more heat is generated; in extreme cases this can cause nearby material to catch fire.

flame generated in kindling such as twigs

Making fire
Early humans probably made fire using a fire drill or fire plow, which generates heat by rubbing two pieces of wood together. The heat causes wood dust to ignite and this can then be used to light larger kindling.

Around 1.76 MYA, more advanced stone tools began to appear. Unlike Oldowan tools, **Acheulian tools**, particularly the multipurpose handaxe, were deliberately shaped. **Hard-hammer percussion** (striking off flakes with a hammerstone) was used to rough out the tool's shape. It was then refined by removing smaller flakes using a soft hammer of bone or antler.

Mousterian tools are particularly associated with Neanderthals and occurred from *c.*300,000 YA. They include sharp-edged Levallois flakes that were struck off a prepared core (see panel, opposite), and a wide range of flake tools—such as knives, spear points, and scrapers—shaped for different purposes.

In the late Middle and Upper Paleolithic (*c.*35,000–10,000 YA), a new technique, indirect percussion, allowed for many blades to be struck from a single core. The final stage of tool development first appeared *c.*70,000 YA and became widespread post-glacially from about 10,000 YA. It involved **microliths**—tiny flakes and blades for use in composite tools.

The earliest weapons were rocks or handaxes, but by about 400,000 BCE early people had adapted **sticks for use as spears**. At first, these had sharpened wooden ends, but by around 200,000 BCE stone points started being attached to create more effective weapons. **The** bow was probably first developed around 64,000 BCE, but the earliest examples found

Upper Paleolithic leaf point
This skillfully crafted tool was made by flaking small pieces off a larger core using a sharp piece of bone or antler to apply pressure.

c.2.5 MYA Earliest Oldowan stone tools are made

c.1.76 MYA Earliest Acheulian handaxes

c.790,000 YA First controlled use of fire, found at Benot Ya'aqov, Israel

c.500,000 BCE Earliest antler tools, found at Boxgrove, England

c.400,000 BCE Oldest wooden spears, found at Gröningen, Germany

125,000 BCE Mousterian flake tools made using Levallois technique become predominant in Europe

c.64,000 BCE Bow is first developed

c.39,000 BCE Earliest known cave paintings, at El Castillo, Spain

c.30,000 BCE Dogs first domesticated

c.30,000 BCE Earliest bone needles, found in Europe

Einkorn is the ancestor of modern wheat and still occurs naturally throughout southwest Asia. It has a higher protein content than its domesticated descendent.

date from around 9000 BCE. The arrows of this period show evidence of fletching—attaching feathers to the shaft to improve flight and accuracy.

The first deliberate use of fire to harden clay dates from around 24,000 BCE, with the manufacture of ceramic Venus figurines found at Dolni Vestonice in the Czech Republic. Examples of the **first pottery vessels**, from around 18,000 BCE, were found in Xianrendong Cave in China, but the earliest ceramic vessels to have survived in any quantity are Jomon pots **from Japan**. These date from about 14,000 BCE and were probably used for cooking food. The growing stability of settlements probably played a role in the spread of pottery vessels, which were useful for storing food and also for cooking. Early pottery was generally formed by pinching (shaping the wet clay by hand) or gradually coiling rolls of clay up and into the shape of a pot. These pots were **fired in pit-kilns**, or bonfire kilns, which were shallow pits dug in the ground and lined with fuel.

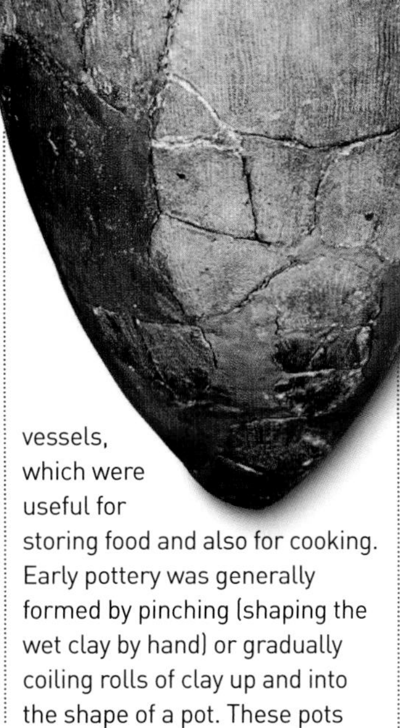

Jomon pot
This style of pottery was produced in Japan for over 10,000 years. The earlier examples generally have pointed bottoms.

In western Asia, unbaked clay was initially used for **making bricks**. The first containers were made from gypsum and lime plaster, which was made by burning chalk. It was not until around 6900 BCE that ceramic pottery appeared at sites such as Çayönü in Turkey.

The **earliest bone needles** date from around 30,000 BCE and come from Europe. They may have been **used to join skins** together, using threads of gut or sinew, and to thread pierced objects, such as shells or beads.

Ancient clay impressions of textiles date the **first woven cloth** to around 27,000 BCE. **Cordage**—the twisting together of fibers to increase the strength of the threads—appeared around 18,000 BCE, when **three-ply cord** was in use in the Lascaux caves of southern France.

Until at least 13,000 YA, early humans were hunter-gatherers or foragers. The first evidence of **plant domestication** (the deliberate selection and manipulation of plants for cultivation) is of wild rye seeds that were sown and harvested around the settlement of Abu Hureyra in Iraq around 10,500 BCE. About a thousand years later, a group of **wild cereals**—notably einkorn (*Triticum boeoticum*) and emmer (*Triticum dicoccoides*), both varieties of wheat, and wild barley (*Hordeum vulgare*)—were domesticated. Cultivation of these cereals was widely distributed in southwest Asia, particularly in a fertile crescent of land that stretched from the Persian Gulf to the coastlands of the Near East. By 7000 BCE, **barley had also been domesticated on the Indian subcontinent**. In China, however, a different set of plants, notably millet and rice, was domesticated beginning in the 8th millennium BCE.

Bone shuttle
Needles and shuttles of bone were the first means of binding materials together, using animal gut or vegetable fibers such as flax.

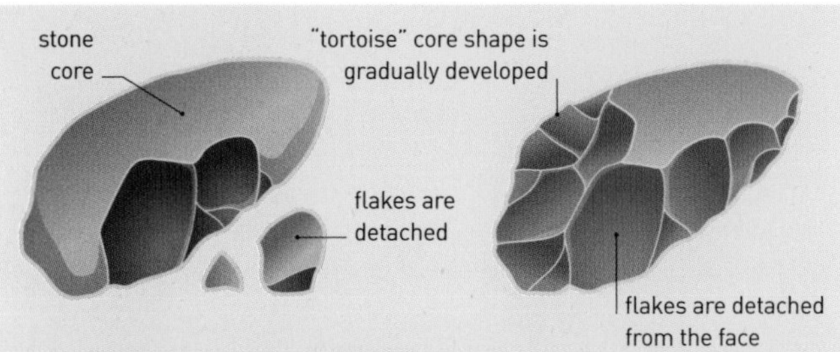

stone core
"tortoise" core shape is gradually developed
flakes are detached
flakes are detached from the face

THE LEVALLOIS TECHNIQUE

This technique involves shaping a "tortoise" core using hard and soft percussion. Flakes are struck from the edges and one face to produce the desired shape of the final flake, which is then detached from the core. The resulting flake has a sharp edge on all sides and can be used without further modification.

DEVELOPMENT OF AGRICULTURE

The upland areas of the Fertile Crescent, an area of relatively fertile land in southwest Asia, were home to wild cereals, sheep, and goats. Around 10,000 BCE the climate cooled, leading to a contraction of the range of wild cereals to areas with higher rainfall. Perhaps due to the greater difficulty in gathering the seeds of these plants, communities began cultivating them next to their villages. Sheep and goats were also domesticated for their meat. More productive sources of food led to increased population densities, while the demand in time and labor needed for agriculture led to settlements becoming both larger and more sedentary.

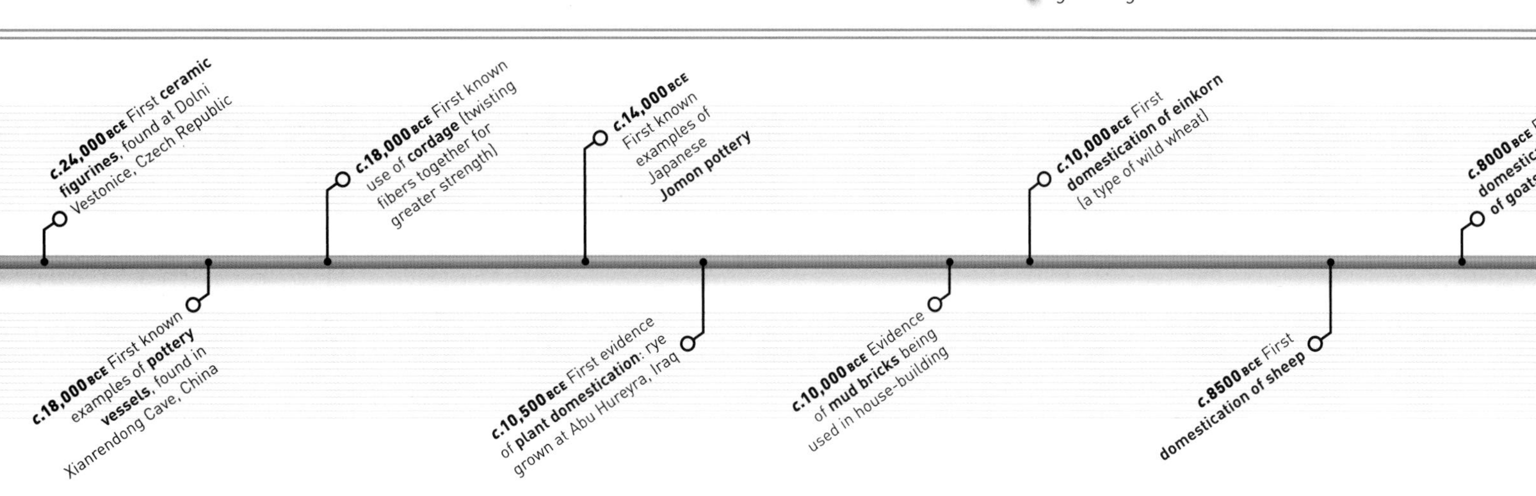

c.24,000 BCE First ceramic figurines, found at Dolni Vestonice, Czech Republic
c.18,000 BCE First known examples of pottery vessels, found in Xianrendong Cave, China
c.18,000 BCE First known use of cordage (twisting fibers together for greater strength)
c.14,000 BCE First known examples of Japanese Jomon pottery
c.10,500 BCE First evidence of plant domestication: rye grown at Abu Hureyra, Iraq
c.10,000 BCE Evidence of mud bricks being used in house-building
c.10,000 BCE First domestication of einkorn (a type of wild wheat)
c.8500 BCE First domestication of sheep
c.8000 BCE First domestication of goats

15 THOUSAND THE MAXIMUM NUMBER OF SOAY SHEEP ALIVE TODAY

The Soay sheep is native to a small island off the west coast of Scotland. It is a primitive breed, very similar to the first domesticated sheep in Europe.

Farming hillsides
The use of terraces to allow hilly areas to be farmed began in Yemen in about 4000 BCE but was also widely practiced in China and in mountainous areas of Peru.

ONE OF THE FIRST ANIMALS to be domesticated by humans, around 30,000 BCE, was the dog, which was selectively bred from domesticated wolves and was used for hunting. Around 8500 BCE, people in southwest Asia began to domesticate other animals, beginning with sheep and goats. **Cattle and pigs were domesticated** around 7000 BCE in many places across the world, and by 3000 BCE a number of other animals had been domesticated including, in the Americas, the guinea pig (around 5000 BCE) and the llama (about 4500 BCE).

The first large-scale construction of **stone buildings** began around 9000 BCE, with the building of a ritual structure at Göbekli Tepe in southeast Anatolia (in modern Turkey). It consisted of a number of free-standing T-shaped pillars within a low circular enclosure wall. In around 8000 BCE the **first settlement wall was built** at Jericho in Palestine. Made of stone, the wall was about 16 ft (5 m) high with a circumference of 1,970 ft (600 m). **Architectural techniques became more sophisticated**, with the use of corbelling (overlapping stone to create a type of vaulted roof) in northwest Europe by 4000 BCE, and buttresses to strengthen walls in Mesopotamia by around 3400 BCE. From about 5000 BCE, the practice of building large structures using massive stones—**megaliths**—spread throughout western Europe, resulting in structures such as the Carnac stones in Brittany, France (dating from around 4500 BCE), Newgrange Passage tomb in Ireland (around 3400 BCE), and Stonehenge in England (from 2500 BCE).

By around 6500 BCE the people of Mehrgarh (in modern Pakistan) were making bitumen, a sticky liquid that seeps from crude oil deposits, to **make reed baskets waterproof**, and around 2600 BCE the people of the Indus Civilization were using it to create a watertight coating for brick-built basins. In Mesopotamia from the 4th millennium BCE, bitumen was mixed with sand to create a mortar for building and as a **tar for caulking ships.**

Where there was insufficient rainfall for agriculture, farmers developed irrigation to transport water to their fields. At Choga Mami, in eastern Iraq, water channels from the Tigris River were constructed from around 6000 BCE, and by the 4th millennium BCE, **dams and dikes** were used to store water in reservoirs in parts of western Asia. In Egypt, the annual flooding of the Nile River inundated fields naturally, but from at least as early as 3000 BCE, excess water was diverted for storage. **Terrace agriculture**, in which flat, cultivable areas are cut into a hillside and irrigated by water channels, was developed in Yemen in around 4000 BCE. In China, **networks of banks and ditches** were built to flood and drain wet-rice cultivation fields (paddies).

Cold-working (beating or hammering) of naturally occurring metals, such as gold and copper, was practiced as early as 8000 BCE. **Smelting**—heating metallic ores with a reducing agent to extract the pure metal (see 1800–700 BCE)—appeared as early as 6500 BCE in Çatal Höyük in Turkey. The technique spread widely: it was being used from southeast Europe to

ALLOYS

The combination of two or more metals produces an alloy, which may have different characteristics from the original metals. In the mid to late 5th millennium BCE, it was discovered that smelting a small amount of arsenic with copper produced arsenical bronze, which is harder and stronger than copper alone. By around 3200 BCE, true bronze was being produced in southwest Asia by using tin instead of arsenic in the smelting process, and objects such as this early 2nd century bronze figurine were being made. By the late 3rd millennium BCE, it had also been discovered that copper could be alloyed with zinc, forming brass.

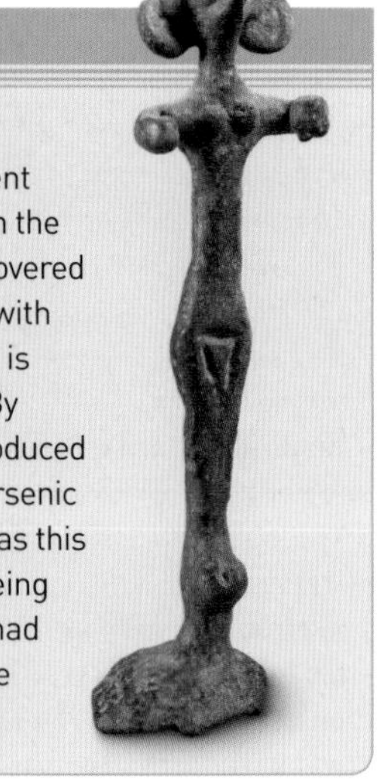

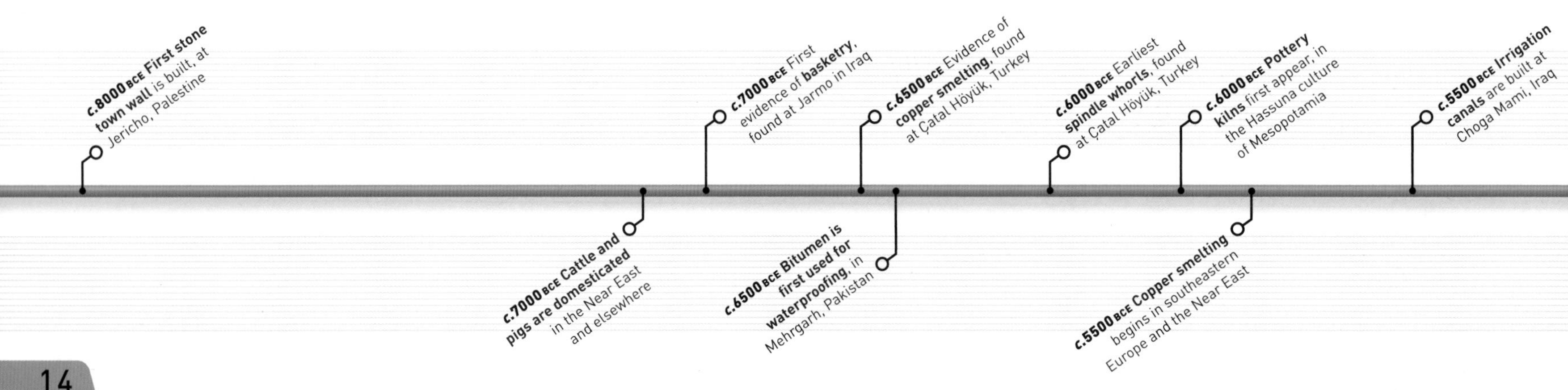

c.8000 BCE First stone town wall is built, at Jericho, Palestine
c.7000 BCE Cattle and pigs are domesticated in the Near East and elsewhere
c.7000 BCE First evidence of **basketry**, found at Jarmo in Iraq
c.6500 BCE Evidence of **copper smelting**, found at Çatal Höyük, Turkey
c.6500 BCE Bitumen is first used for waterproofing, in Mehrgarh, Pakistan
c.6000 BCE Earliest **spindle whorls**, found at Çatal Höyük, Turkey
c.6000 BCE Pottery kilns first appear, in the Hassuna culture of Mesopotamia
c.5500 BCE Copper smelting begins in southeastern Europe and the Near East
c.5500 BCE Irrigation canals are built at Choga Mami, Iraq

The Carnac stones in Brittany, France, are a series of more than 3,000 upright megaliths. The oldest stones date from around 4500 BCE.

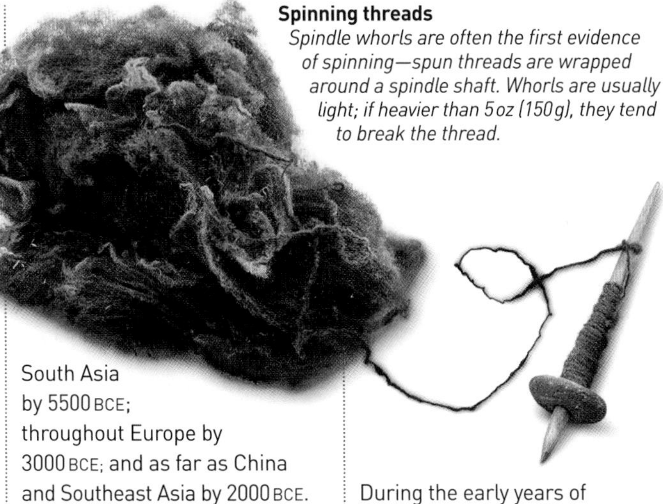

Spinning threads

Spindle whorls are often the first evidence of spinning—spun threads are wrapped around a spindle shaft. Whorls are usually light; if heavier than 5 oz (150 g), they tend to break the thread.

South Asia by 5500 BCE; throughout Europe by 3000 BCE; and as far as China and Southeast Asia by 2000 BCE. **Casting metal objects** with a mold developed in the 5th millennium BCE. The first known cast metal object comes from Mesopotamia and dates from about 3200 BCE. **Spinning raw fibers** to make a thread may have begun as early as the 7th millennium BCE, which is the date of spindle whorls found at Çatal Höyük, Turkey. **Weaving** may have arisen from the late Paleolithic skill of making nets and baskets. The loom—a frame or brace to keep one set of threads (the warp) tense while another (the weft) is interwoven with it—appeared in the form of warp beams (simple sticks) and backstrap looms (the warp beams were held taut by a strap around the user's back) in West Asia and Egypt by the 4th millennium BCE.

During the early years of agriculture, ground for sowing had to be cultivated using **handheld digging sticks** or hoes. The use of cattle as draft animals made the eventual use of **the ard, or scratch plow**, possible. This primitive wooden plow, sometimes with a metal tip, cut shallow furrows. The earliest evidence of its use comes from the 4th millennium BCE, and it spread widely in Egypt, West Asia, and Europe.

The quality of ceramics was improved by the invention, in around 6000 BCE, of **kilns**—specially built chambers in which pottery could be fired. **Two-chamber, updraft kilns** (in which the fire is in the lower chamber) appeared in the Hassuna culture of Mesopotamia in about 6000 BCE. Around 5000 BCE, the method of "coiling" pots was improved by a simple turntable (tournette) beneath the pot. By 3500 BCE, the tournette had been replaced in southern Mesopotamia by a **true potter's wheel**, consisting of a heavy stone wheel that could be turned rapidly and continuously. This allowed the potter to throw the pot by placing a lump in the center of the device and shaping it as the wheel spun around.

> **BARLEY** IS **THRESHED** FOR YOU, **WHEAT** IS **REAPED** FOR YOU, YOUR **MONTHLY FEASTS** ARE MADE WITH IT, YOUR **HALF-MONTHLY FEASTS** ARE MADE WITH IT.
>
> **Ancient Egyptian pyramid text,** *c.*2400–2300 BCE

For many millennia, all transportation on land was by foot. The first artificial aids were sleds, which have been found in Finland dating from 6800 BCE, and skis, in use in Russia around 6300 BCE. The **invention of the wheel** revolutionized transportation.

Four-wheeled wagons appeared in Poland and the Balkans around 3500 BCE and soon afterward in Mesopotamia. At first, wheels were solid disks connected to the wagon by a wooden axle, but around 2000 BCE, **spoked wheels were developed**, which made lighter, more mobile vehicles possible. Around 3100 BCE, efficient harnesses for attaching draft animals to wagons were developed in Mesopotamia, allowing greater loads and distances to be attained.

As commercial transactions grew more complex, accurate measurements of goods became essential. **Standardized weight and length measures** were introduced in Mesopotamia, Egypt, and the Indus Valley in the late 4th millennium BCE. The earliest weights were often based on grains of wheat or barley, which have a uniform weight. The standard unit of length, the cubit, was based on the length of a man's forearm.

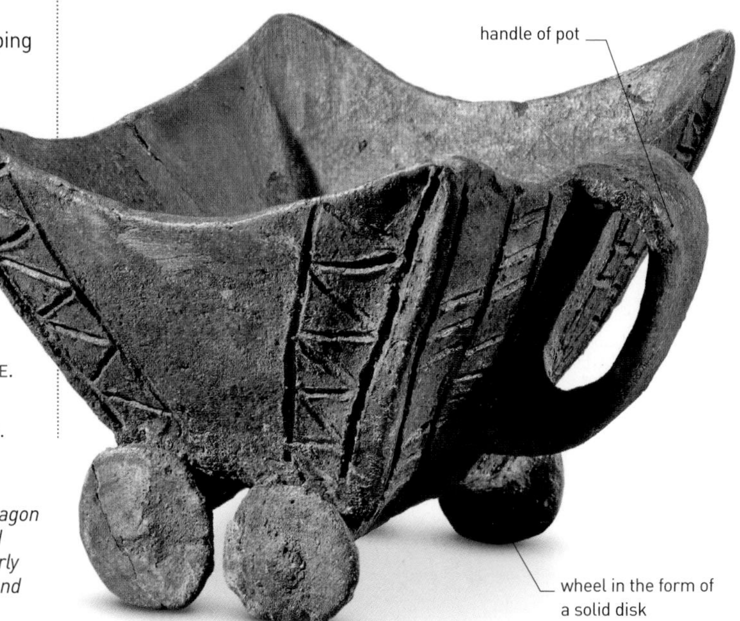
handle of pot
wheel in the form of a solid disk

Wagon in clay

This clay pot in the shape of a wagon dates from around 3000 BCE and shows the typical features of early wheeled vehicles from central and southern Europe.

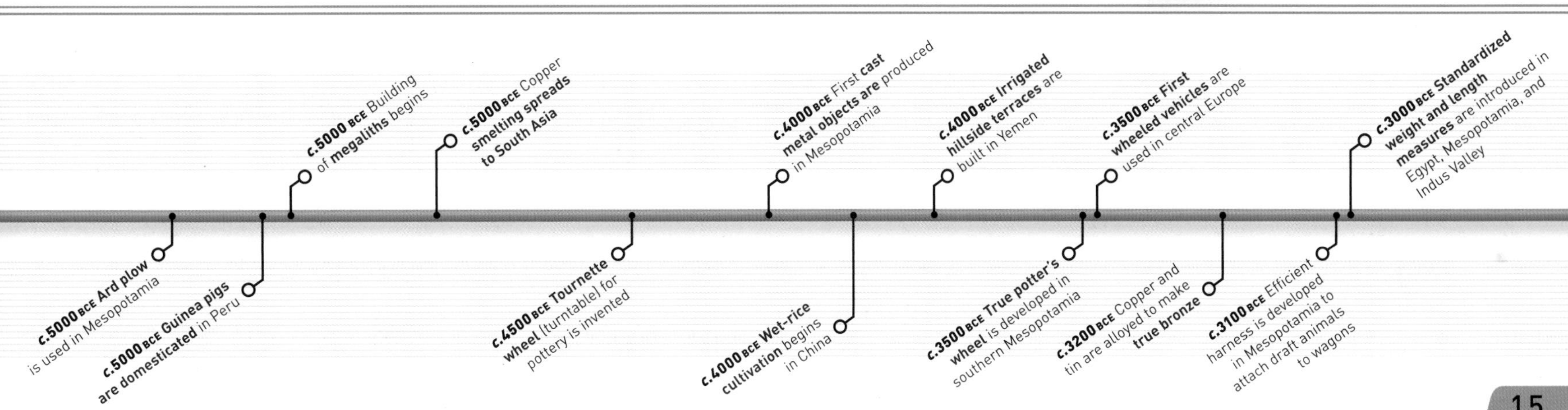
c.5000 BCE Ard plow is used in Mesopotamia
c.5000 BCE Guinea pigs are domesticated in Peru
c.5000 BCE Building **of megaliths** begins
c.5000 BCE Copper **smelting spreads to South Asia**
c.4500 BCE Tournette wheel (turntable) for pottery is invented
c.4000 BCE First **cast metal objects are** produced in Mesopotamia
c.4000 BCE Wet-rice cultivation begins in China
c.4000 BCE Irrigated hillside terraces are built in Yemen
c.3500 BCE True potter's wheel is developed in southern Mesopotamia
c.3500 BCE First wheeled vehicles are used in central Europe
c.3200 BCE Copper and tin are alloyed to make **true bronze**
c.3100 BCE Efficient harness is developed in Mesopotamia to attach draft animals to wagons
c.3000 BCE Standardized weight and length measures are introduced in Egypt, Mesopotamia, and Indus Valley

curved blade adapted for harvesting grain

Reaping hook
Date unknown
By the Iron Age, metal harvesting sickles had replaced flint-bladed ones since the metal was readily available and easier to sharpen and mend than flint.

Metal shears
Date unknown
These iron shears from Italy are similar to those used by later sheep-shearers. They were found in Riva del Garda, in the Italian province of Trento.

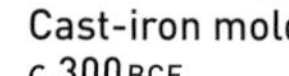

Cast-iron mold
*c.*300 BCE
The Chinese had invented high-temperature furnaces capable of melting iron as early as 500 BCE. This enabled them to produce cast iron by pouring molten metal into molds such as this one, used to make agricultural tools.

sharp tip for piercing

Bronze sword
*c.*1200 BCE
Sword blades could be created using bronze, an alloy of copper and tin. Bronze Age swords, such as this one from France, were carried only by the rich.

flat pommel

Iron sword
*c.*500–700
The Anglo-Saxons used the pattern-welding technique to make swords, in which rods of iron were twisted together and forged to form the core. An edge was then added.

blade with rounded tip

EARLY METALLURGY

ANCIENT METALLURGISTS PRODUCED A VARIETY OF OBJECTS—FROM LETHAL WEAPONRY TO STUNNING JEWELRY

The development of metallurgy, from around 6500 BCE, made possible the production of ornamental objects of great beauty as well as tools and weapons that were more durable and effective than those made of wood.

The earliest metalworking was cold-hammering—the beating of naturally occurring metals. After smelting (heating ore to extract metal) was developed, techniques became more sophisticated. Metal casting began around 5000 BCE, and alloys were developed in the 5th millennium BCE. By the end of the ancient period, techniques such as gilding and inlaying had been developed and metalworking had spread across much of the world.

Chariot decoration
*c.*100 BCE–100 CE
Enameling, or the fusing of molten glass with metal, was invented around 1200 BCE. The use of red glass in enameling became especially popular in the late Iron Age, as seen in this Celtic chariot decoration.

red enamel

Bronze Celtic brooch
*c.*800 BCE
This ornate brooch was created by Hallstatt craftsmen in Austria. The spiral pattern was part of the Celtic artistic repertoire for over 1,500 years.

Bronze pin
*c.*1200
Pins with flattened heads were a common decorative item used for fastening clothes in Bronze-Age Europe.

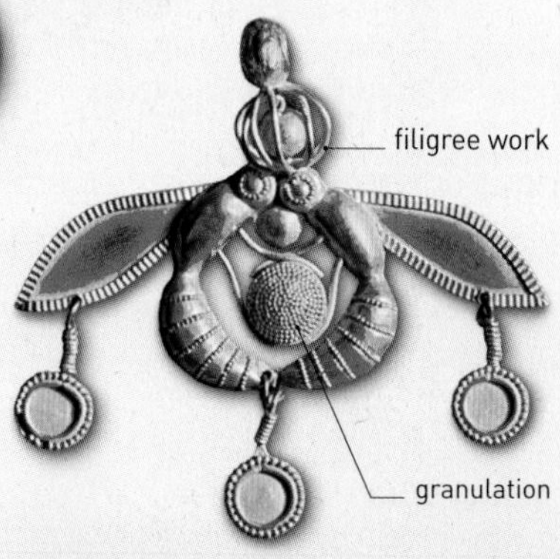
filigree work
granulation

Gold Minoan pendant
*c.*1700–1550 BCE
This pendant, depicting bees depositing honey on a honeycomb, exhibits granulation (minute balls of gold soldered onto the surface) and filigree (fine threads of metal).

bird's head in profile
writhing snake pattern

Anglo-Saxon belt buckle
*c.*620
This gold belt buckle features an intertwined pattern of snakes and beasts, highlighted in black niello—an enamel-like substance formed from an alloy of silver, copper, lead, and sulfur.

hemispherical iron cap
rigid face mask, riveted to cap
neck guard

Anglo-Saxon helmet (reconstruction)
*c.*620
Found in a ship burial at Sutton Hoo, UK, the original helmet was made of iron and covered with tinned bronze sheets. It was decorated with silver wire and garnets.

Corinthian helmet
*c.*700 BCE
This helmet is made from a single piece of bronze, giving it extra strength. Such helmets were popular in Greece from the 8th to the 6th centuries BCE.

decorative roundel
red glass inlay

Ceremonial shield cover
*c.*350–50 BCE
Made from a bronze sheet, this shield cover displays the repoussé technique of hammering the reverse side to create a raised design on the front.

Silver plaque
*c.*300–200 BCE
This plaque depicts the figures of the Greek goddess Aphrodite, her son Eros, and a girl attendant, was made by repoussé. Other decorative incisions have been highlighted with gilding.

stamped design

Lydian coins
*c.*700 BCE
The earliest coinage was produced in Lydia (now in Turkey). It was made from electrum—a naturally occurring alloy of silver and gold, which was once believed to be a metal in its own right.

Bronze figure
*c.*1000 BCE
This statuette of a Canaanite god was made with a technique called cire-perdue casting, which uses a single-use mold, and plated with silver using a direct application technique.

leg shaped as dragon

Bronze Age vessel
*c.*800 BCE
This animal-shaped ritual vessel, known as *yi*, was used in Late Western Zhou China for washing hands before making a sacrifice.

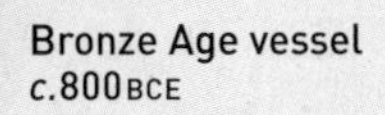

turquoise eye

Copper mask
*c.*250
Found in the tomb of a nobleman from the Peruvian Moche culture, this mask shows mastery of metal sculpture. Both eyes were originally inset with turquoise.

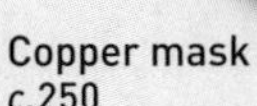

3000–1800 BCE

The ruins of Uruk, the world's oldest city, are in present-day Iraq. The site of Uruk was first settled around 4800 BCE and became a town around 4000 BCE.

> "CLIMB UPON **THE WALL** OF **URUK,** WALK ALONG IT, I SAY; REGARD THE **FOUNDATION TERRACE** AND EXAMINE THE MASONRY…"
>
> ***Epic of Gilgamesh,*** Tablet I, *c.*2000 BCE

IRRIGATION TECHNIQUES BECAME MORE COMPLEX during the 3rd millennium BCE. The **shadoof was developed** in Mesopotamia in around 2400 BCE. It consisted of an upright frame with a pole suspended from it; on one end of the pole was a bucket for scooping up water, while on the other was a counterweight. By 1350 BCE, the shadoof had spread to Egypt. There, devices called **nilometers had already been developed** to measure the rise and fall of the river, which predicted how good the harvest would be.

In the period 4000–3000 BCE farming communities in Mesopotamia had coalesced to form the **world's first cities**, such as Uruk *c.*3400 BCE. By 3100 BCE, cities had begun to appear in Egypt, beginning with Hierakonpolis. By 2600 BCE, Mohenjo Daro and Harappa, great cities of the Indus Civilization, had been built.

As towns and cities developed, the first **true writing emerged** in Mesopotamia around 3300 BCE, probably prompted by the need to keep detailed records. Originally largely **pictographic**, with signs looking like the things they represented, they were written using a stylus that produced wedge-shaped marks. Cuneiform script developed as the curved outlines of these early signs changed into a series of wedge-shaped lines that gradually became more stylized over time. These symbols were impressed into soft clay, which then hardened to create durable documents. At around the same time, another writing system developed in Egypt. Known as **hieroglyphic**, this system was initially primarily pictographic. The earliest known examples are clay labels from Abydos, *c.*3300 BCE. Writing also developed in the Indus Valley around 2600 BCE, in China by at least 1400 BCE, and in Mesoamerica around 600 BCE.

Soda-lime glass was first developed in Mesopotamia around 3500 BCE. It was made by firing silica (sand), soda ash, and lime in a furnace, but initially was suitable only for small objects. In Egypt, **faience became common** from around 3000 BCE. Consisting of a mixture of crushed quartz, calcite lime, and soda lime, which when vitrified produced a blue-turquoise glaze, faience was used by the Egyptians on small sculptures and beads.

The early 3rd millennium BCE saw the **spread of true bronze**, created by alloying copper with tin, which became the most commonly used metal in Mesopotamia between 3000 and 2500 BCE. Clay **crucible furnaces for smelting** appeared there in around 3000 BCE. Mesopotamian metallurgists also invented the technique of **gold granulation** around 2500 BCE. This produced tiny gold balls, which were used to decorate jewelry.

EARLY ASTRONOMY

Evidence of interest in astronomical phenomena dates from Neolithic times in Europe, when many megaliths were laid out in an orientation that indicated particular lunar or solar events. Some of the stones at Stonehenge (first erected around 2500 BCE) were aligned to indicate the times of year at which the winter and summer solstices occurred. Other features may have been connected with lunar events.

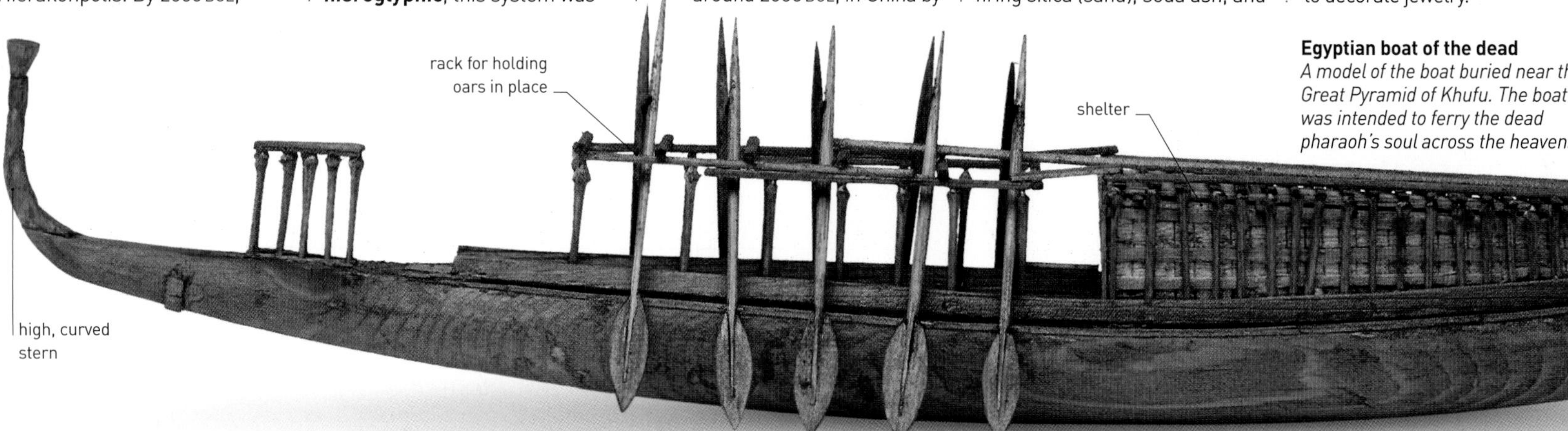

rack for holding oars in place
shelter
high, curved stern

Egyptian boat of the dead
A model of the boat buried near the Great Pyramid of Khufu. The boat was intended to ferry the dead pharaoh's soul across the heavens.

***c.*3000 BCE Cities begin to appear** in Egypt

***c.*3000 BCE** Use of **faience** becomes common in Egypt

***c.*3000 BCE** Clay **crucible furnaces** for smelting appear in Mesopotamia

***c.*3000 BCE** Egyptians develop **sewn-plank ships**

***c.*2900 BCE Mastaba tombs** are built in Egypt

***c.*2625 BCE Step Pyramid of Djoser** is built in Egypt

***c.*2600 BCE Cities of Mohenjo Daro and Harappa** are established by the Indus Civilization

2550–2472 BCE Great Pyramids of Giza are built in Egypt

1ST CENTURY THE DATE BY WHICH **CUNEIFORM SCRIPT** HAD BECOME **EXTINCT**

Cuneiform (wedge-shaped) script developed from the earliest writing, invented in Mesopotamia around 3300 BCE. It was used for a wide range of ancient Near Eastern languages, including Sumerian and Akkadian.

Boat-building also developed significantly around the 3rd millennium BCE. Early humans had probably been using some form of boat from as long ago as 50,000 BCE, although the earliest surviving water craft is a dugout canoe that dates from around 7200 BCE. In the Gulf region, boats were being made of bitumen-coated reeds as early as 5000 BCE.

By around 3000 BCE, more sophisticated vessels made of wooden planks that were sewn together were being built in Egypt. Early boats were powered solely by oars. **Sailing boats, with square-rigged sails**, appeared in Egypt in around 3100 BCE, supplementing muscle power with wind power. By 3000 BCE, large steering oars had been developed in Egypt, and by about 2500 BCE, pairs of **side oars and tillers** had been introduced.

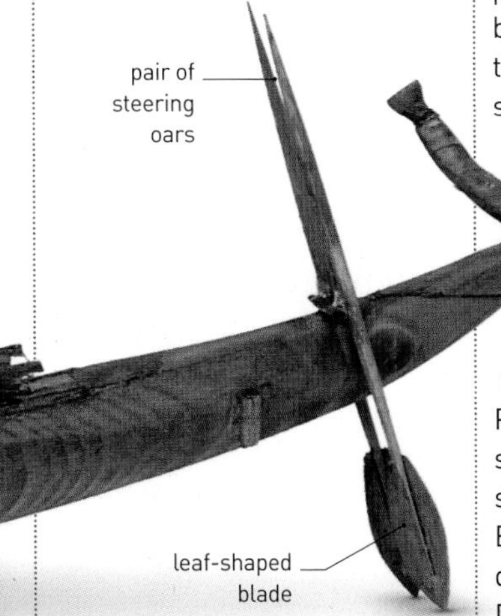
pair of steering oars
leaf-shaped blade

Before 3000 BCE, few monumental structures were residential. The practice of building some temples on platforms began before 4000 BCE, the platform rising with each rebuilding. After 2900 BCE, temple platforms in Sumerian cities such as Ur and Kish reached a considerable height, leading to the **development of ziggurats**—initially three-tiered structures with a shrine on the top platform. Largely made of mud bricks, with a baked-brick facing, these monumental structures suggest growing sophistication in structural engineering.

In Egypt, **most architecture was religious (temples) or funerary (tombs)**. The tombs of the nobility and rulers of the Early Dynastic Period (around 2900 BCE) were simple mud-brick rectangular structures known as **mastabas**. Between 2630 and 2611 BCE, during the reign of the pharaoh Djoser, a huge mastaba was modified by building six stepped platforms to create a **step pyramid**. By the reigns of Khufu, Khafre, and Menkaure in mid-3rd millennium BCE, the creation of **smooth-sided stone pyramids** had been perfected, and each of these pharaohs erected a huge pyramid tomb for himself at Giza. Collectively known as the **Great Pyramids**, each was oriented and built with great precision, which suggests that sophisticated surveying techniques were in use.

Egyptian faience
This Middle Kingdom (1975–1640 BCE) statue of a woman with a tattooed body shows the deep blue color typical of much Egyptian faience work.

An interest in observational astronomy arose early in Mesopotamia, culminating in the **Venus tablet of Ammi-Saduqa** (dating from around 1650 BCE). It contained the rising and setting times of the planet Venus over a period of 21 years. A carved piece of mammoth tusk found in Germany, dating from about 32,500 BCE, may possibly represent the constellation Orion, but **systematic division of the sky into constellations** dates from Babylonian manuscripts *c.*1595 BCE.

With the growing administrative demands of cities in the 3rd millennium BCE, the development of an **accurate calendar** became vital. The first known version is in the Umma calendar of Shulgi, a Sumerian document dating from about 2100 BCE that contains **12 lunar months of either 29 or 30 days**. When this 354-day year became too out of phase with the real 365.25-day year, an extra month was added by royal decree. The ancient Egyptians had a similar calendar, but five days were added each year to give a 365-day year.

There have been claims that some prehistoric carvings represent **topographical maps**, but **true cartography** and real maps were not developed until the 3rd millennium BCE. The Akkadian **Ga-Sur tablet** (dating from about 2500 BCE) shows the size and location of a plot of land between two hills and was probably part of a land transaction. Fragments of a statue of Gudea of Lagash, from around 2125 BCE, show a plan of a temple. The **first real street map** discovered to date shows a scale plan of the Sumerian town of Nippur (in present-day Iraq) and dates from about 1500 BCE. The first surviving attempt to map the entire known world is the Babylonian "world map" from about 600 BCE, which shows the regions surrounding Babylon (see 700–400 BCE).

Monumentally tall
Built around 2560 BCE, the Great Pyramid of Khufu was 482 ft (147 m) tall and remained the world's tallest building for nearly 4,000 years.

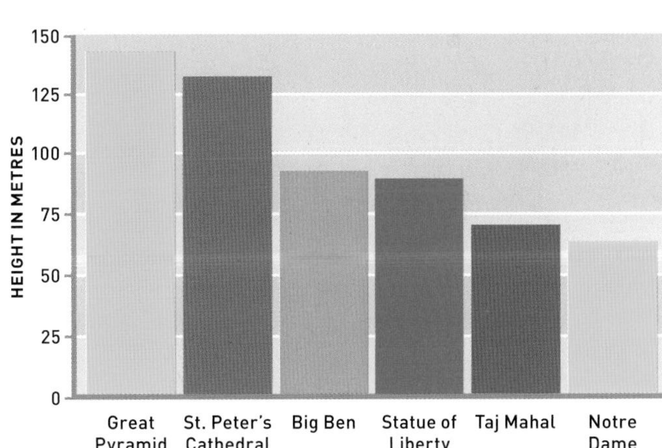
HEIGHT IN METRES
150
125
100
75
50
25
0
Great Pyramid
St. Peter's Cathedral
Big Ben
Statue of Liberty
Taj Mahal
Notre Dame

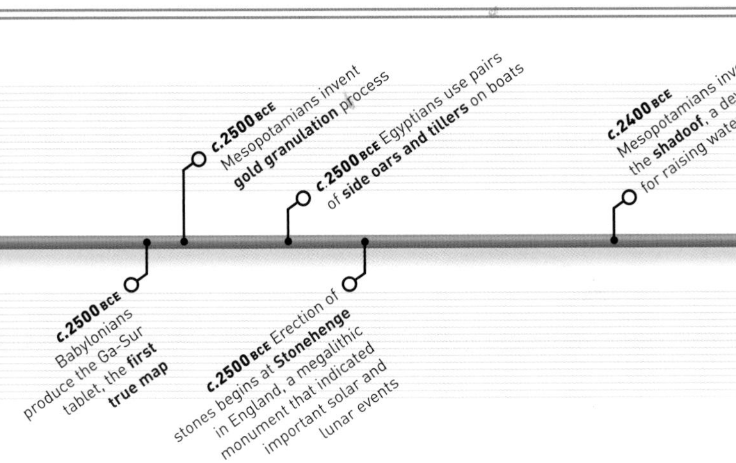
c.2500 BCE Mesopotamians invent gold granulation process
c.2500 BCE Egyptians use pairs of side oars and tillers on boats
c.2500 BCE Babylonians produce the Ga-Sur tablet, the first true map
c.2500 BCE Erection of stones begins at Stonehenge in England, a megalithic monument that indicated important solar and lunar events

c.2400 BCE Mesopotamians invent the **shadoof**, a device for raising water

c.2200 BCE First **ziggurats** are built in Mesopotamia

c.2100 BCE Sumerians produce the **earliest known calendar**, the Umma calendar of Shulgi

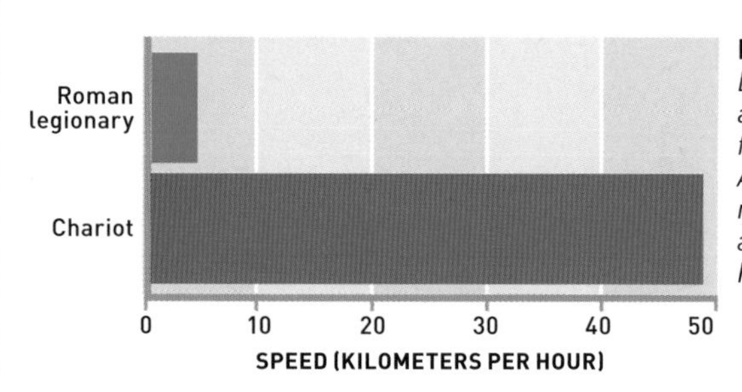
Roman legionary
Chariot
0
10
20
30
40
50
SPEED (KILOMETERS PER HOUR)

Faster armies
Lightweight war chariots enabled armies to maneuver much faster than an infantry ever could. A Bronze Age chariot (c.1200 BCE) moved more than 10 times as fast as the marching pace of a Roman legionary.

Egyptian chariot
Around 1600 BCE, the Egyptians developed lightweight war chariots that had spoked wheels and a thin wooden semicircular frame. The platform could accommodate two people, one to maneuver at high speed, and another armed with a bow.

leather bindings connect shaft to chariot body

footboard made of sycamore wood

wood bent into V-shape to make spokes

hub or nave

cattle intestines fasten spokes to hub

> “…BUT LET THE **LEFT-HAND HORSE** KEEP SO CLOSE IN THAT THE NAVE [HUB] OF THE **WHEEL** SHALL ALMOST **GRAZE THE POST.**”

Homer, Greek poet, from *Iliad Book XXIII*, first description of chariot race, *c.*750 BCE

Neolithic period
Logroller
Neolithic people place loads on rollers made from logs. These logs, however, are not always smooth and the difficulties in keeping them aligned make this an inefficient method.

Early logrollers

3500 BCE
Potter's wheel
In southern Mesopotamia, potters become the first to use wheels to mechanize an industrial process—that of making pottery. They use a heavy, rapidly turning stone wheel to shape clay on.

Egyptian potter

c.2500 BCE
Disk wheel
The first true transportation wheels—disks of wood connected by axles—are developed in the Balkans and Mesopotamia. The Sumerians used these on battle wagons.

Disk wheels on the Standard of Ur

c.1323 BCE
Spoked wheel
Wheels with spokes are lighter than disk wheels, and allow a cart or war chariot to be pulled by a lighter animal, such as a horse. First developed in the steppes of central Asia a little after 2000 BCE, these wheels spread to Egypt by 1600 BCE.

c.750 BCE
Iron-rimmed wheel
The Celts add iron rims to the wooden wheels of chariots to improve their durability on rough surfaces. They do so first by nailing the metal to the rim and later, by applying strips of hot iron, which shrink to fit as they cool.

Celtic chariot

c.300 BCE
Water wheel
The Greeks invent water wheels as a means of harnessing the power of running water. They use water wheels either to raise water in buckets to a higher level for irrigation, or to drive around a shaft that operates a milling machine.

yoke to attach horses to shaft

THE STORY OF THE WHEEL

THIS SIMPLE INNOVATION HAS MOVED ARMIES, CARRIED LOADS, AND POWERED INDUSTRIES

One of the most important inventions in history, the wheel allowed the transportation of loads over long distances, revolutionized early warfare, and made the development of the first mechanized processes possible. It opened up the globe to human exploration and revolutionized industry.

movement
box is easier to move
wheels turn around static axles
friction gives outside edges of wheels grip on the road
small contact area, so friction is less

WHEELS AND FRICTION

The force needed to pull a load pressing down directly on the ground is increased by the friction or "rolling resistance" between the load and the ground. The use of wheels resolves this problem. Since only a small part of the wheel is in contact with the ground at any one point in time, the rest of it can rotate freely, without being impeded by friction. The little friction that remains allows the wheel to grip the ground without sliding. Wheels are mounted on sturdy shafts, called axles, which facilitate the rolling motion.

The earliest wheel, the logroller, was used by neolithic people to transport heavy weights, such as large stones used in the construction of megaliths. By 3500 BCE, the logroller was adapted to create the first true wheels—solid disks of wood connected by an axle. These wheels, however, were very heavy. Lighter, spoked wheels were invented around 1600 BCE. The more hard-wearing, iron-rimmed wheels came around 800 years later, making for faster, more durable vehicles suitable for battle and long-distance transportation. Wheels steadily evolved, using materials such as iron and steel as they were developed. Modern wheels use high-tech alloys of titanium or aluminum that are light and allow vehicles to move faster, using much less power.

Spoked wheel construction
The spokes of a wheel distribute the force applied to a vehicle evenly around its rim. As the wheel rotates, each spoke shortens slightly.

outer rim of wheel
spokes radiating from central hub

THE WHEEL IN INDUSTRY

Beginning with the potter's wheel around 4500 BCE, the wheel was also adapted for use in industrial processes. By 300 BCE, watermills were employed in Greece to harness the power of water, via a turbine, for use in milling. By the time of the Industrial Revolution, the wheel appeared in one form or another in almost all industrial machinery. Gears (toothed wheels) and cogs were used in the Antikythera mechanism—an astronomical calculating machine created in Greece around 100 BCE—but it is possible they were used earlier in China. Gears and cogs eventually became common components of machines as diverse as clocks and automobiles. Yet there were some cultures where the wheel did not feature as prominently. Some ancient civilizations of Central America and Peru did not develop wheels, or, as in the case of the Aztecs of Mexico, used them only in children's toys.

c.100 BCE
Wheelbarrow
The Chinese create a wheelbarrow with a large central wheel, which makes all the weight fall on the axle. Easy to push, each wheelbarrow can carry up to six men.

"Wooden ox" wheelbarrow

Chinese spinner

c.1035
Spinning wheel
In China, a hand-crank-operated driving wheel is added to a hand-spindle, automating it and allowing multiple spindles to be operated simultaneously.

1845
Vulcanized rubber tire
Robert Thomson uses vulcanized rubber—invented by Charles Goodyear—to make pneumatic (air-filled) tires, which are lighter and harder to wear out.

1848
Mansell wheel
The quieter and more resilient Mansell railroad wheel has a steel central boss (hub), surrounded by a solid disk of 16 teak segments.

Gazelle steam engine

1910
Early automobile spoked wheel
Earliest automobile wheels have wooden spokes, which are more suitable for narrow tires, but tend to warp and crack.

Ford Model T

1915
Radial tire
Patented by Arthur Savage, radial ply tires are made of rubber-coated steel or polyester cords. They are now the standard tire for almost all cars.

1960s Mini

2010
Modern wheel types
Ultra-lightweight racing bicycles use composite carbon spokes, while car wheels are made of magnesium, titanium, or aluminum alloys.

High-tech racing bike

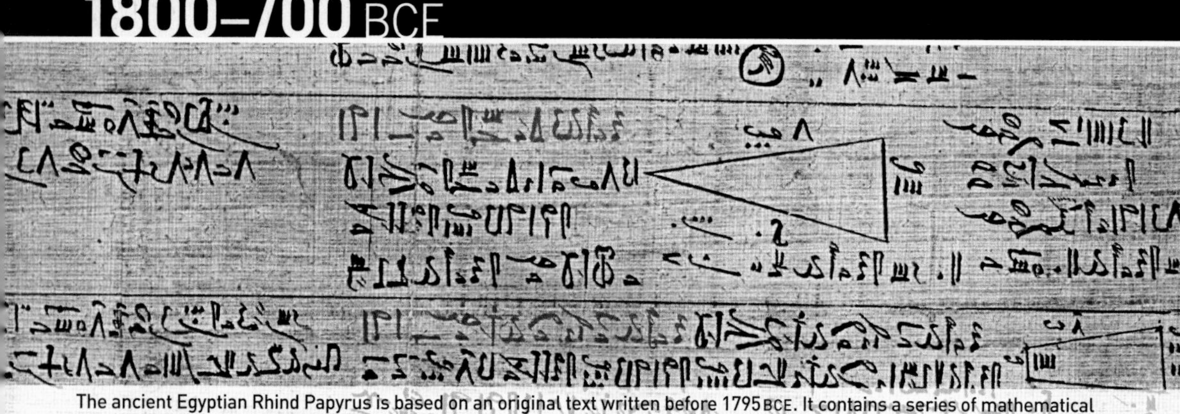

The ancient Egyptian Rhind Papyrus is based on an original text written before 1795 BCE. It contains a series of mathematical problems and their solutions, including calculations of the areas and volumes of geometrical figures.

IN THE EARLY 2ND MILLENNIUM BCE, the composite bow was developed, probably in the steppes of Central Asia. Unlike self bows made of a single piece of wood, the **composite bow** was made of laminated strips of horn, wood, and sinew, which together provided greater range and penetration, and allowed the bow to be smaller and easier to use on horseback. The bow was **further modified to become recurved**, with the ends curving forward, which added even more strength. Composite bows spread from the Steppes to China, where they were used during the Shang (1766–1126 BCE) and Zhou (1126–256 BCE) dynasties, and west into Egypt and Mesopotamia.

> “ANOTHER **REMEDY** FOR SUFFERING IN HALF THE HEAD. THE SKULL OF A **CATFISH,** FRIED IN OIL. **ANOINT** THE HEAD THEREWITH.”
>
> **Ebers Papyrus 250,** Egyptian medical treatise, *c.*1555 BCE

There is evidence that **doctors existed in Egypt** during the Old Kingdom (*c.*2700–2200 BCE) and depictions of surgery have been found on temple walls, but most knowledge of ancient Egyptian medicine comes from papyri written around 1550 BCE. These show that medicine had moved beyond a belief that disease was a divine punishment. The Edwin Smith Papyrus (*c.*1600 BCE) contains **details of human anatomy**, shows awareness of the link between the pulse and heartbeat, and also gives instructions for the diagnosis and **treatment of a range of ailments and injuries**. The Ebers Papyrus (*c.*1555 BCE), dating from about the same time, includes descriptions of diseases, tumors, and even of mental disorders such as depression.

The earliest intentional **production of iron** was in Anatolia in Turkey, which was exporting small quantities of iron by the 19th century BCE. At first, iron was smelted only on a small scale, but by 700 BCE production was widespread in Europe. Smelting also developed independently in a number of places, including Africa and India, where the earliest evidence of ironworking is thought to date from around 1300 BCE.

Until medieval times, smelting in the West produced only bloom that needed to be hammered to remove impurities. It was only in China that furnaces capable of melting iron were developed and iron could be cast. **Evidence of cast iron** production in China dates from the 9th century BCE.

In mathematics, the Babylonians had made major advances by 1800 BCE, producing **tables of reciprocals, squares, and cubes** and using them to solve algebraic problems, such as quadratic equations. Several tablets are thought to show an awareness of Pythagoras's theorem (see 700–400 BCE). The Babylonians also estimated pi to be about 3.125, close to the actual value of about 3.142.

Most of what is known of ancient Egyptian mathematics comes from **mathematical texts such as the Rhind Papyrus**. It is based on a text written before 1795 BCE and consists of a series of problems and solutions. It shows the use of unit fractions ($1/n$), **solutions for linear equations**, and methods for calculating the areas of triangles, rectangles, and circles. It also shows the volumes of cylinders and pyramids.

The earliest boats recovered date from before 6000 BCE, but **early navigation was not sophisticated**. The most effective navigators of this period were the Lapita people of the Pacific (ancestors of the Polynesians), who from 1200 BCE expanded eastward to Vanuatu, New Caledonia, Samoa, and Fiji. Their voyage to Fiji involved a 530 miles (850 km) journey across open sea. To accomplish

Stick chart
Made by Marshall Islanders in Micronesia, this chart uses sticks to represent currents and waves, a technique that may have been passed down from ancient Polynesians.

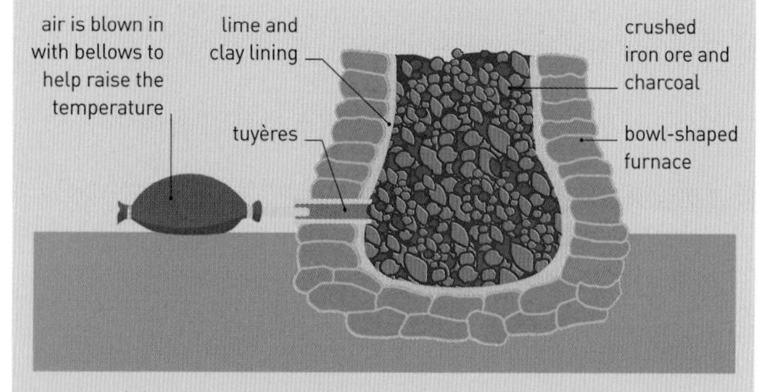
air is blown in with bellows to help raise the temperature
lime and clay lining
tuyères
crushed iron ore and charcoal
bowl-shaped furnace

SMELTING

Pure iron melts at 2800°F (1540°C), higher than early technology could achieve, so instead it was smelted by reducing iron ore with charcoal at around 2200°F (1200°C). The ore was packed with charcoal in bowl furnaces, and tuyères (clay nozzles) were used to blow air in to raise the temperature. The resulting molten metal was cooled to form a "bloom," a solid lump containing iron and various impurities, which was then hammered repeatedly to remove the impurities and extract the iron.

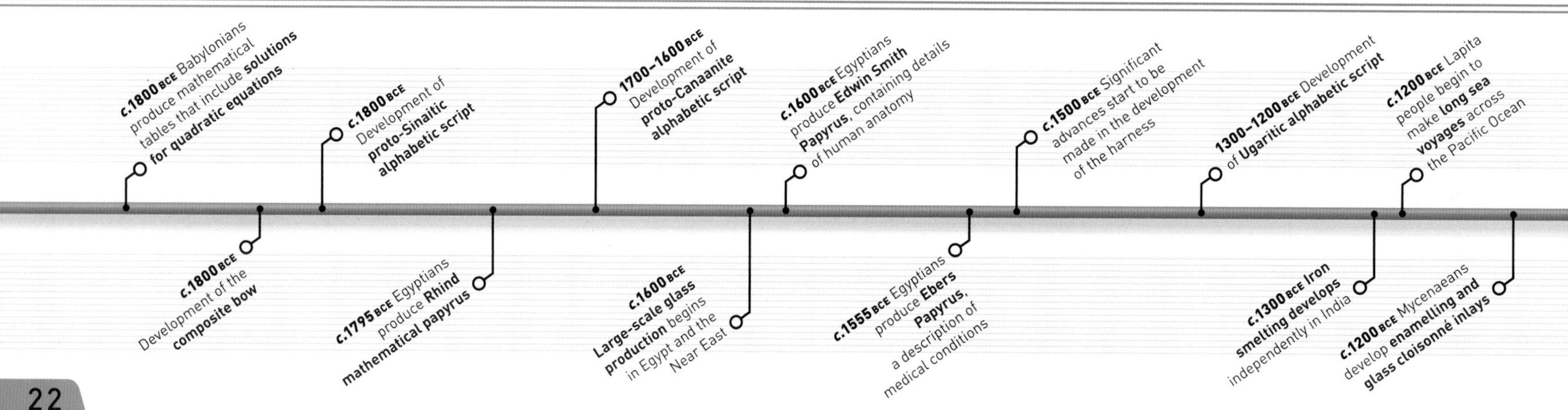
c.1800 BCE Babylonians produce mathematical tables that include solutions for quadratic equations
c.1800 BCE Development of the composite bow
c.1800 BCE Development of proto-Sinaitic alphabetic script
c.1795 BCE Egyptians produce Rhind mathematical papyrus
1700–1600 BCE Development of proto-Canaanite alphabetic script
c.1600 BCE Large-scale glass production begins in Egypt and the Near East
c.1600 BCE Egyptians produce Edwin Smith Papyrus, containing details of human anatomy
c.1555 BCE Egyptians produce Ebers Papyrus, a description of medical conditions
c.1500 BCE Significant advances start to be made in the development of the harness
1300–1200 BCE Development of Ugaritic alphabetic script
c.1300 BCE Iron smelting develops independently in India
c.1200 BCE Lapita people begin to make long sea voyages across the Pacific Ocean
c.1200 BCE Mycenaeans develop enamelling and glass cloisonné inlays

An Assyrian bronze relief of the mid-9th century BCE shows war chariots carrying soldiers to assault the city of Khazazu (present-day Azaz, in Syria).

this, the Lapitan sailors must have used knowledge of winds, stars, and currents. They may also have created **stick maps**, like those later used by the Polynesians who settled as far as Easter Island, Hawaii, and (by 1000–1200 CE) New Zealand.

In Egypt and the Near East, glass began to be made in significant amounts from about 1600 BCE. In the late 2nd millennium BCE, the technique of **bonding glass to ceramics to produce glazes** was discovered. Glass cloisonné inlays and enameling (fusing glass to metal surfaces) were developed by the Mycenaeans in Greece around 1200 BCE. **Casting glass** (by pouring molten glass into a mold) was discovered in Mesopotamia around 800 BCE. Around 100 years later, the Phoenicians had developed **clear glass**.

crouching lion cub

poppy pods in crown

clay figure glazed with quartz and metal oxides

Snake goddess
Faience reached its peak in the Minoan civilization, with works such as this goddess statuette (c.1700 BCE), but with more available glass, faience was replaced by glass-glazed ceramics.

The invention of the spoked wooden wheel around 2000 BCE, along with the **domestication of the horse**, opened up new possibilities in land transport, permitting lighter vehicles and eventually the use of adaptable riding animals. Although harnesses were in use from the 3rd millennium BCE, significant advances began to be made from c.1500 BCE. The **halter yoke**, with flat straps across the neck and chest of the animal, made horses more efficient at **pulling light chariots**. Weighing as little as 66 lb (30 kg), these chariots could carry two warriors and became crucial to many Near Eastern armies.

The **preservation of corpses** had its origins in the natural process of drying out and preserving bodies in the desert sand. These corpses were wrapped in linen bandages dipped in resin, which also helped to prevent the bodies from decaying. By around 2700 BCE, the Egyptians had discovered that natron (a mixture of salts) **desiccated flesh and could be used to mummify bodies**. They gradually refined this process until it reached a peak of sophistication around 1000 BCE. Bodies were mummified by removing the internal organs (apart from the heart), washing out the body cavity, and packing it with natron for 40 days to dry it out. The natron was removed and replaced with clean packets of natron and linen soaked in resin to restore the body's shape before being coated in resin and bandaged in linen.

METALS	MELTING POINT (°C)
Iron	1539
Copper	1083
Gold	1064
Bronze	950
Lead	328
Tin	232

Melting point of metals
Iron melts at a far higher temperature than other metals used in early metallurgy. China was first to master the technology to melt iron.

EARLY SCRIPTS

The transition from symbolic scripts to an alphabetic one, where each individual sign represents a sound in the language, seems to have first taken place among miners in the Sinai desert of Egypt around 1800 BCE. The signs appear to derive from Egyptian hieratic script (a cursive script that developed alongside the hieroglyphic system), but there are few inscriptions in this proto-Sinaitic alphabet and it is not certain whether slightly later alphabets in the region, such as proto-Canaanite (17th century BCE) and Ugaritic (13th century BCE), derived from it or developed separately. By 1050 BCE, proto-Canaanite had evolved into the Phoenician script that is the ancestor of Greek and other European scripts.

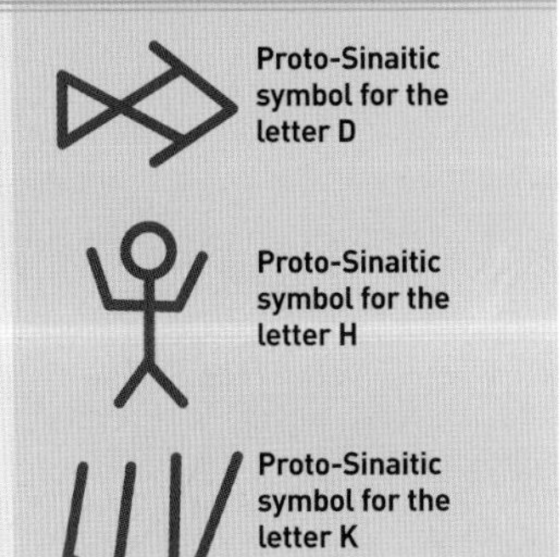
Proto-Sinaitic symbol for the letter D
Proto-Sinaitic symbol for the letter H
Proto-Sinaitic symbol for the letter K

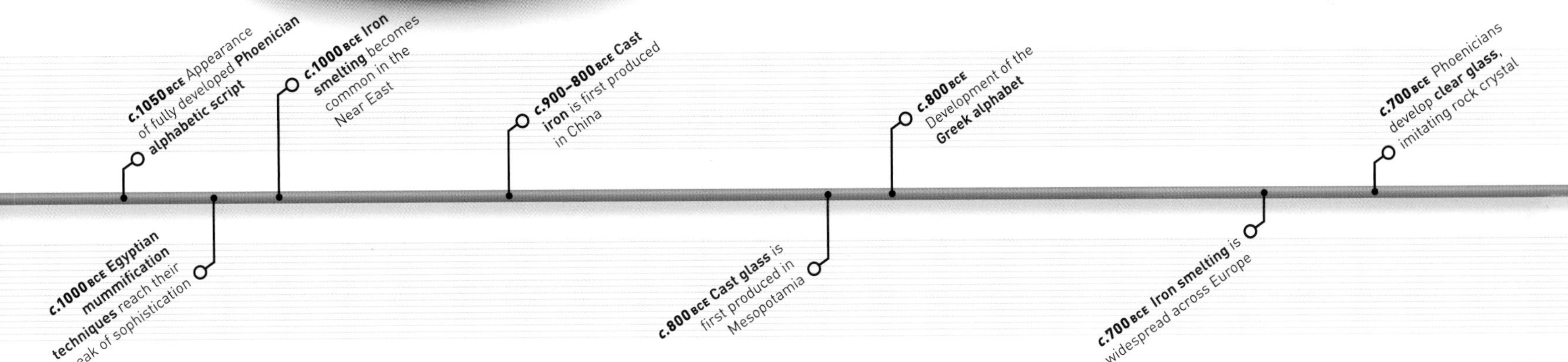
c.1050 BCE Appearance of fully developed Phoenician alphabetic script
c.1000 BCE Egyptian mummification techniques reach their peak of sophistication
c.1000 BCE Iron smelting becomes common in the Near East
c.900–800 BCE Cast iron is first produced in China
c.800 BCE Cast glass is first produced in Mesopotamia
c.800 BCE Development of the Greek alphabet
c.700 BCE Iron smelting is widespread across Europe
c.700 BCE Phoenicians develop clear glass, imitating rock crystal

700–400 BCE

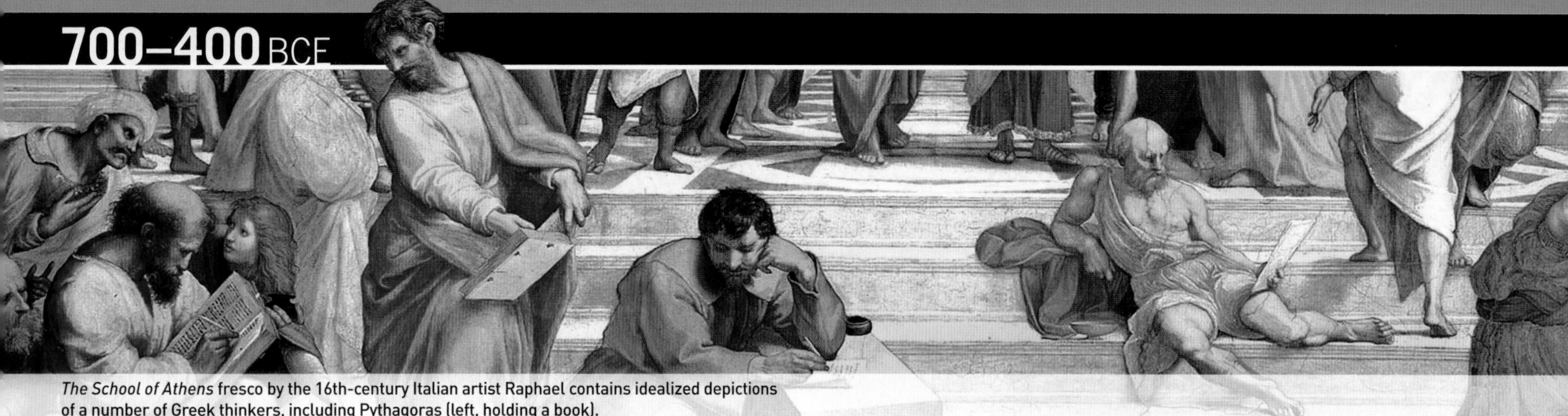

The School of Athens fresco by the 16th-century Italian artist Raphael contains idealized depictions of a number of Greek thinkers, including Pythagoras (left, holding a book).

AS EARLY AS 2300 BCE, the Babylonians had developed a sexagesimal number system (based on **writing numbers** in multiples of 60) and the principle of position (where numbers in different positions represent different orders of magnitude). By 700 BCE they sometimes used a **marker to indicate a null value (zero)**.

The screw pump (or Archimedes Screw) is a cylindrical pump with a central shaft surrounded by inner blades in the shape of a spiral and encased in wood. As the shaft is rotated, **water is pulled up the spiral**, transferring it from a lower to a higher level. The invention of the pump is traditionally ascribed to the ancient Greek mathematician Archimedes (287–212 BCE) in around 250 BCE, but it was probably first invented much earlier, in the 7th century BCE under the rule of King Sennacherib of Assyria to water his palace gardens at Nineveh.

By the 1st millennium BCE, the Babylonians had begun to make maps of larger areas. By around 600 BCE they had produced a **"world map,"** which showed the **city of Babylon** in relation to eight surrounding regions. The first known **Chinese map**, found on an engraved bronze plaque in the tomb of King Cuo of Zhongshan, was a plan of the king's proposed necropolis.

The ancient Greek cartographic tradition began in Ionia in the 6th century BCE. **Anaximander** (*c.*611–546 BCE) is said to have drawn the **first world map** that showed Earth surrounded by a great ocean. Hecataeus of Miletus (*c.*550–480 BCE) also drew a map to accompany his *Survey of the World* that showed three great continents, Libya (Africa), Asia, and Europe.

The first evidence of scientific (as opposed to supernatural) thinking about the nature of the world came from **ancient Greek philosophers** in the 6th and 5th centuries BCE. Thales of Miletus (b. *c.*620 BCE) believed that **water was the fundamental material of the universe**, and that earthquakes happened when the surface of Earth rocked on the watery surface on which it floated. In contrast, **Anaximander**, who was also from Miletus, believed that the prime material of the universe was apeiron, a substance that preceded air, fire, and water. He also put forward an **early evolutionary theory**, suggesting that humans had developed from a type of fish.

The first atomic theory was also proposed by a Greek, the philosopher Democritus of

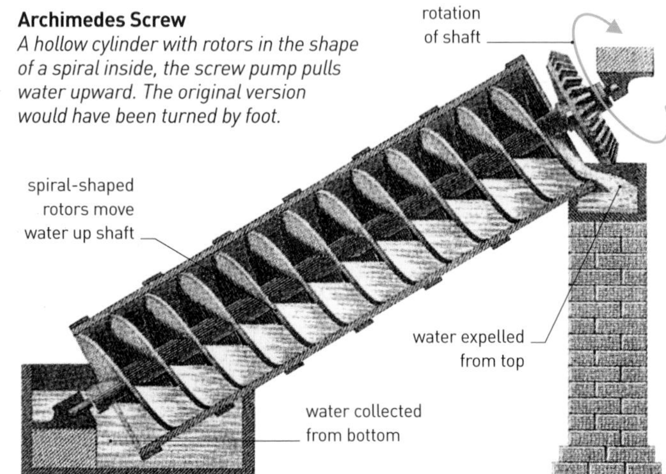

rotation of shaft
spiral-shaped rotors move water up shaft
water expelled from top
water collected from bottom

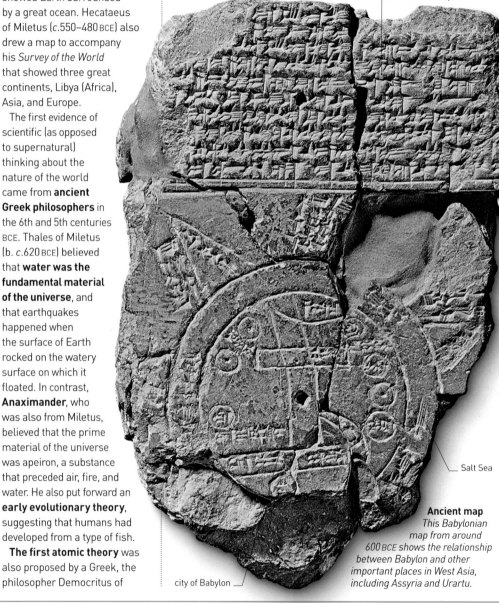

Archimedes Screw
A hollow cylinder with rotors in the shape of a spiral inside, the screw pump pulls water upward. The original version would have been turned by foot.

cuneiform inscriptions
Salt Sea
city of Babylon

Ancient map
This Babylonian map from around 600 BCE shows the relationship between Babylon and other important places in West Asia, including Assyria and Urartu.

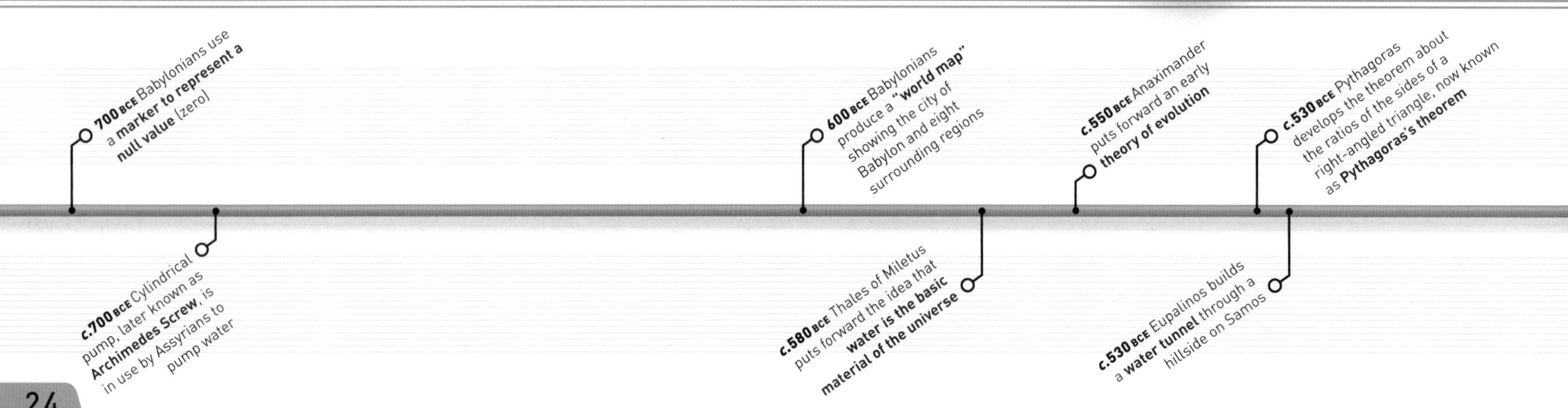

700 BCE Babylonians use a **marker to represent a null value** (zero)
c.700 BCE Cylindrical pump, later known as **Archimedes Screw**, is in use by Assyrians to pump water
600 BCE Babylonians produce a **"world map"** showing the city of Babylon and eight surrounding regions
c.580 BCE Thales of Miletus puts forward the idea that **water is the basic material of the universe**
c.550 BCE Anaximander puts forward an early **theory of evolution**
c.530 BCE Pythagoras develops the theorem about the ratios of the sides of a right-angled triangle, now known as **Pythagoras's theorem**
c.530 BCE Eupalinos builds a **water tunnel** through a hillside on Samos

0.65 MILES THE LENGTH OF THE TUNNEL OF EUPALINOS AT SAMOS

The Tunnel of Eupalinos, built in the 6th century BCE, may have been excavated accurately by surveying a series of right-angled triangles above ground.

Abdera (460–370 BCE), who postulated that matter was made up of an infinite number of minute, indivisible particles.

The most famous mathematician of the ancient world was the Greek, **Pythagoras of Samos** (*c.*580–500 BCE). He established a school that promoted the **mystical powers of numbers** and particularly of the *tetraktys*, the perfect arrangement of 10 as a triangle of four rows. He is best known for the theorem bearing his name (see panel, below), but he also firmly believed in the **transmigration of souls** and his followers lived by a strict set of rules, including a prohibition on eating beans.

The oldest major Chinese mathematical treatise, the *Zhou Bi Suan Jing* (some parts of which date to as early as 500 BCE), contain proof of Pythagoras's theorem. At about the same time, or possibly earlier, **Chinese mathematicians also invented magic squares**—square grids of numbers in which the numbers in all rows, all columns, and both diagonals add up to the same total.

By around 530 BCE, Greek surveying expertise had advanced sufficiently to allow the engineer **Eupalinos of Samos** to excavate a **water channel** 0.65 miles (1.04 km) through a hillside by digging tunnels from each end. The two tunnels met almost perfectly in the middle. Eupalinos may have used Pythagoras's theorem to survey right-angled triangles above ground to determine the path of the channel.

Indian astronomy is thought to have its roots in the Indus Civilization. Ancient Hindu sacred scriptures called the *Vedas*—completed *c.*500 BCE—contain references to using astronomical observations for calculating the dates of religious ceremonies and identifies **28 star patterns in the night sky** to help track the movements of the Moon.

In the 5th century BCE, Greek thinkers moved away from simple cosmological theories toward more sophisticated ideas about **the nature of the universe**. Heraclitus (*c.*535–475 BCE) sought to explain phenomena in terms of flux and change. He also believed in the unity of opposites, saying "the road is the same both up and down." **Empedocles of Acragas** (494–434 BCE) believed that all matter consisted of varying proportions of earth, air, fire, and water. This **theory of four elements** remained influential for many centuries.

The Maya of Mesoamerica developed a complex **calendrical system** based on a series of cycles based on the number 20, which may initially have been developed by the Olmecs (the first major civilization in Mexico) before the 5th century BCE. The Mayan Haab (year) had 18 months of 20 days plus one of five days—**one of the two elements of the Calendar Round cycle**. Mayan astronomers also oriented monuments to sunset positions at the equinoxes and solstices, and were able to predict eclipses.

The cities of the Indus Valley were laid out in a grid pattern around 2600 BCE, but the first person to theorize **urban planning** was Hippodamus of Miletus (493–408 BCE). He is said to have devised an ideal city for 10,000 citizens, laid out on a grid. Using his "Hippodamian grid," he also laid out Piraeus, the harbour town of Athens, and Thurii in Italy.

PLATO (424–348 BCE)

One of the most influential of the ancient Greek philosophers, Plato proposed a type of ideal society ruled by philosopher-kings, and espoused the importance of ethics as a guide to a just life. In his many works, he set out a theory of ideal "forms," of which the material world is only a reflection. Most of his books are cast in the form of dialogues by his teacher Socrates.

PYTHAGORAS'S THEOREM

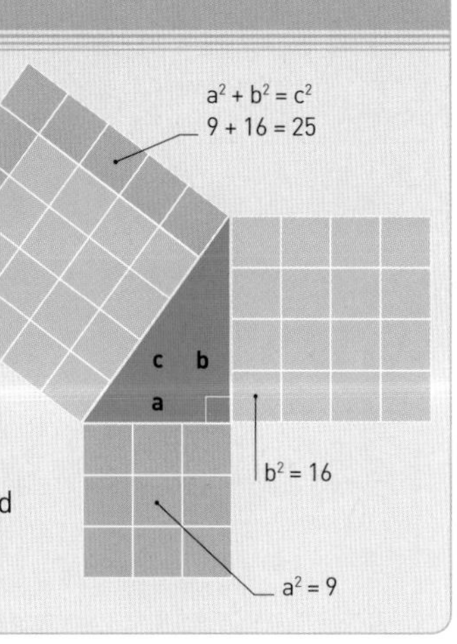

a² + b² = c²
9 + 16 = 25
c
b
a
b² = 16
a² = 9

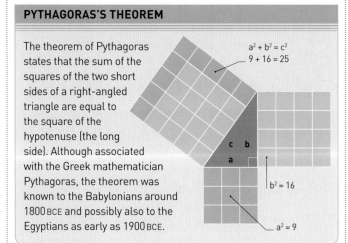

The theorem of Pythagoras states that the sum of the squares of the two short sides of a right-angled triangle are equal to the square of the hypotenuse (the long side). Although associated with the Greek mathematician Pythagoras, the theorem was known to the Babylonians around 1800 BCE and possibly also to the Egyptians as early as 1900 BCE.

Mesoamerican calendrics
This Zapotec stele from Monte Albán in Mexico dates from 500–400 BCE and contains some of the earliest calendary glyphs from Mesoamerica.

glyph for Zapotec year "Four Serpent"

glyph for Zapotec day "Eight Water"

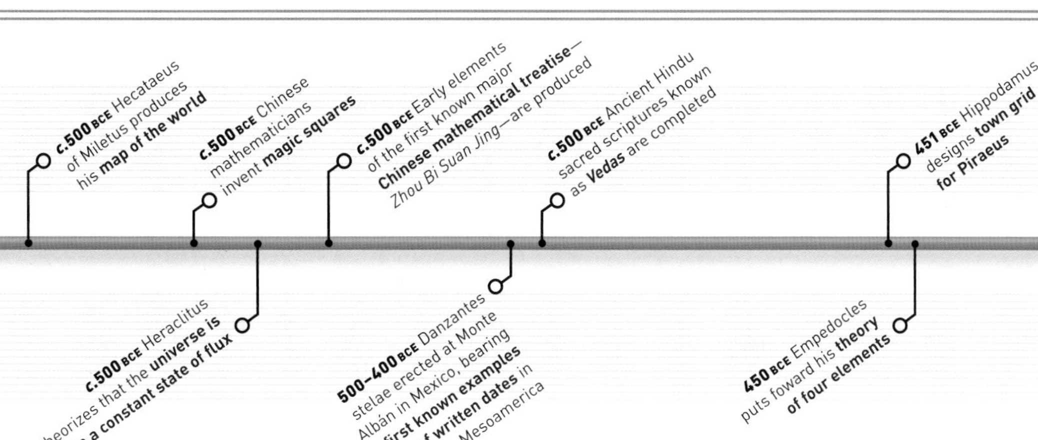

c.500 BCE Hecataeus of Miletus produces his **map of the world**
c.500 BCE Heraclitus theorizes that the **universe is in a constant state of flux**
c.500 BCE Chinese mathematicians invent **magic squares**
c.500 BCE Early elements of the first known major **Chinese mathematical treatise**—*Zhou Bi Suan Jing*—are produced
500–400 BCE Danzantes stelae erected at Monte Albán in Mexico, bearing **first known examples of written dates** in Mesoamerica
c.500 BCE Ancient Hindu sacred scriptures known as **Vedas** are completed

451 BCE Hippodamus designs **town grid for Piraeus**

450 BCE Empedocles puts forward his **theory of four elements**

c.420 BCE Democritus of Abdera proposes the **first atomic theory**

THE STORY OF GEOMETRY

ONE OF THE OLDEST BRANCHES OF MATHEMATICS, GEOMETRY IS EXPANDING INTO NEW AREAS

The term "geometry" derives from ancient Greek words meaning "Earth measurement," but this branch of mathematics encompasses more than map-making. It is about relationships between size, shape, and dimension—and also about the nature of numbers and mathematics itself.

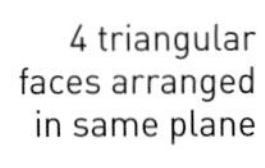

4 triangular faces arranged in same plane
6 edges

Tetrahedron

Octahedron

12 edges

8 triangular faces

Geometry first arose as a series of ad hoc rules and formulas used in planning, construction, and mathematical problem-solving across the ancient world. Greek philosophers such as Thales, Pythagoras, and Plato were the first to recognize geometry's fundamental relationship to the nature of space, and to establish it as a field of mathematics worthy of study in its own right. Euclid, probably a student of Plato and a teacher at Alexandria, summed up early Greek geometry in his great work *The Elements*, written around 300 BCE, and established fundamental mathematical and scientific principles through complex geometrical models developed from a handful of simple rules or axioms.

Axioms of geometry
Euclid's approach to geometry had a huge and lasting influence on later mathematicians.

BREAKTHROUGHS IN UNDERSTANDING

Throughout medieval times, philosophers and mathematicians from various cultures continued to use geometry in their models of the Universe, but the next major breakthrough did not come until the 17th century, with the work of French mathematician and philosopher René Descartes. His invention of coordinate systems to describe the positions of points in two-dimensional and three-dimensional space gave rise to the field of analytical geometry, which used the new tools of mathematical algebra to describe and solve geometrical problems.

Descartes's work led to more exotic forms of geometry. Mathematicians had long known that there were regions, such as the surface of a sphere, where the axioms of Euclidian geometry did not hold. Investigation of such non-Euclidian geometries revealed even more fundamental principles linking geometry and number, and in 1899 allowed German mathematician David Hilbert to produce a new, more generalized, set of axioms. Throughout the 20th century, and into the 21st, these have been applied to a huge variety of mathematical scenarios.

c.2500 BCE
Practical geometry
Early geometry is driven by the need to solve problems such as working out the volume of material required to build a pyramid.

Pyramids at Giza

c.500 BCE
Pythagoras
The Greek philosopher lends his name to the formula for calculating the hypotenuse (long side) of a right-angled triangle from the lengths of its other two sides.

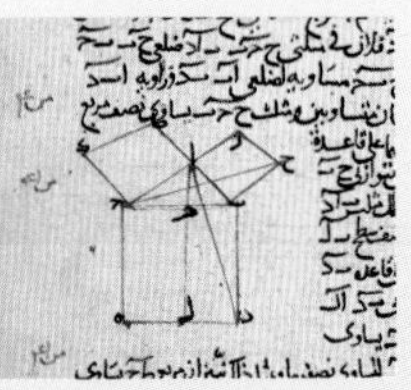

Theorem of Pythagoras

4th century BCE
Geometric tools
The hugely influential philosopher Plato argues that the tools of a true geometrician should be restricted to the compass and straight edge, and so helps establish geometry as a science rather than a practical craft.

Pair of compasses

360 BCE
Platonic solids
These five regular, convex polyhedra (solids with several sides) are long known, but Plato now links them to ideas about the structure of matter. They comprise five shapes that can be formed by the joining together of identical faces along their edges.

c.400 CE
"Archimedean" solids
Greek mathematician Pappus describes 13 convex polyhedra, comprising regular polygons of two or more types meeting in identical vertices or corners.

9th century
Islamic geometry
Mathematicians and astronomers of the Islamic world explore the possibilities of spherical geometry; geometric patterns used in Islamic decoration at this time show similarities to modern fractal geometry.

Mosaic at Alhambra

1619
Kepler's polyhedra
German mathematician Johannes Kepler discovers a new class of polyhedra known as star polyhedra.

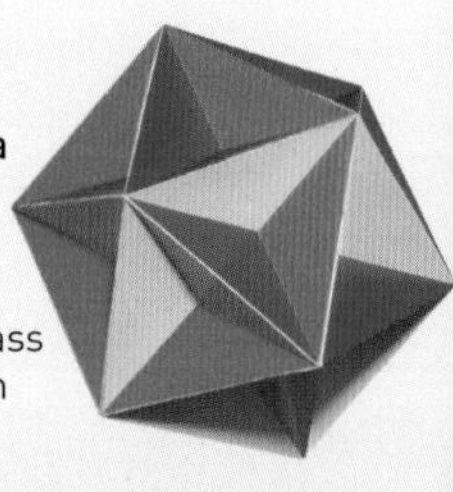

Platonic solids

There are only five convex polyhedra (solids having several sides) that can be formed by joining identical polygons (shapes with three or more sides). Known as the Platonic solids, they are the cube (hexahedron), tetrahedron, octahedron, dodecahedron, and icosahedron.

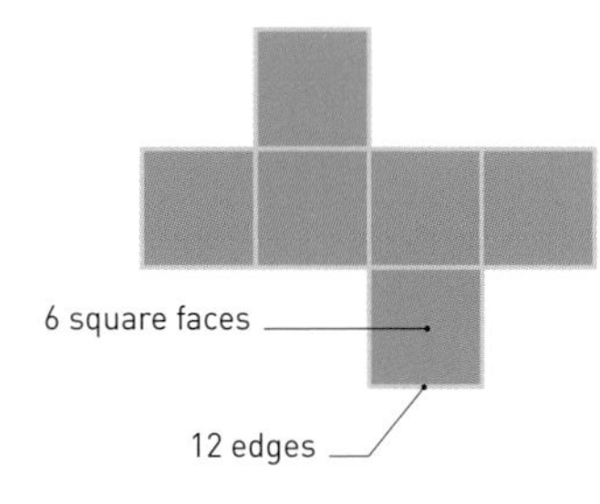
6 square faces
12 edges

Hexahedron (cube)

Dodecahedron

Icosahedron

SPHERICAL GEOMETRY

So-called "spherical geometry" allows the calculation of angles and areas on spherical surfaces, such as points on a map or the positions of stars and planets on the imaginary celestial sphere used by astronomers. This system does not follow all Euclidean rules. In spherical geometry, the three angles in a triangle sum to more than 180 degrees and parallel lines eventually intersect.

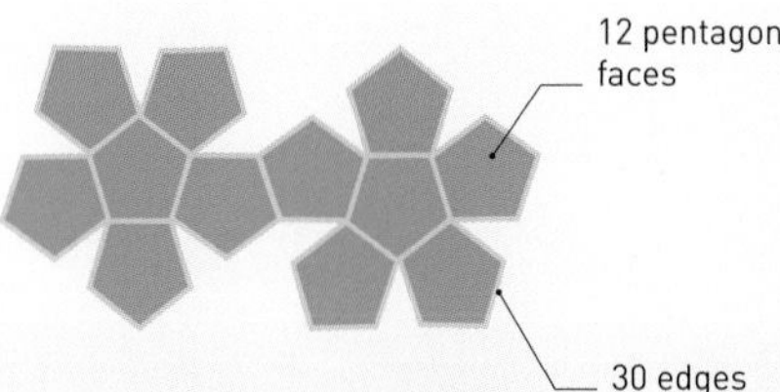
12 pentagonal faces
30 edges

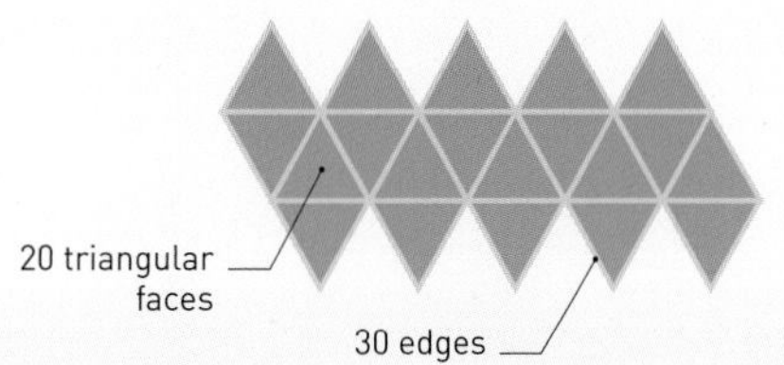
20 triangular faces
30 edges

> " LET NO ONE **DESTITUTE OF GEOMETRY** COME UNDER MY ROOF. "
>
> **Plato, Greek philosopher and mathematician,** *c.*427–347 BCE

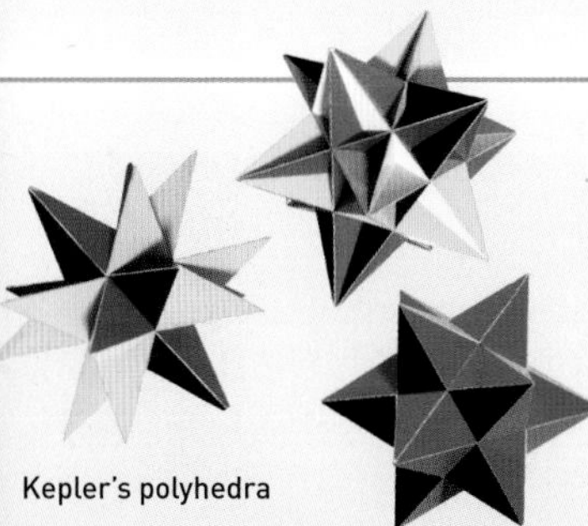

Kepler's polyhedra

1637
Analytic geometry
René Descartes's influential work *La Géometrie* introduces the idea that points in space can be measured with coordinate systems, and that geometrical structures can be described by equations—a field known as analytic geometry.

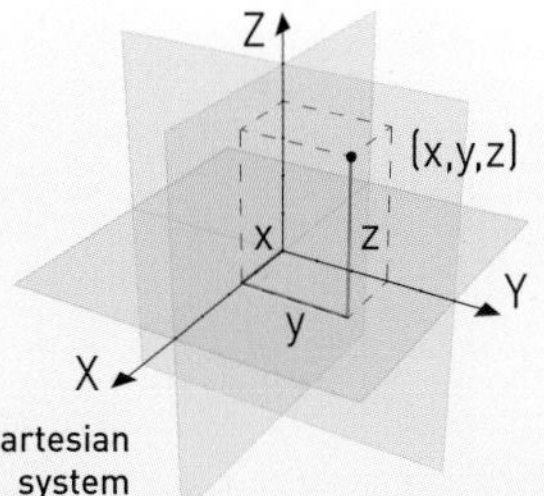
Z
(x,y,z)
x
z
y
Y
X

Cartesian system

20th century
Fractal geometry
Computing power allows fractals—equations in which detailed patterns repeat on varying scales—to be illustrated in graphical form, producing iconic images such as the famous Mandelbrot set.

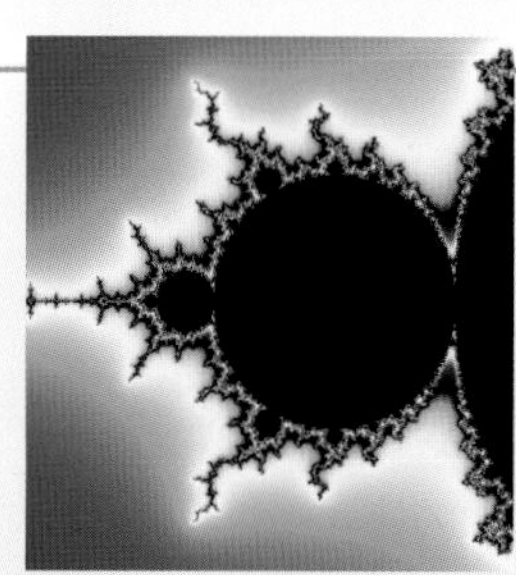

Mandelbrot fractal

Möbius strip

1858
Topology
Mathematicians become fascinated by topology—edges and surfaces, rather than specific shapes. The iconic Möbius strip is an object with a single surface and a single continuous edge.

1882
Klein discovery
Investigating geometries with more than three dimensions, German scholar Felix Klein discovers a construct with no surface boundaries.

Modern Klein bottle

Present day
Computerized proofs
Computer power solves problems such as the four-color theorem (only four colors are needed to distinguish between regions of even complex maps).

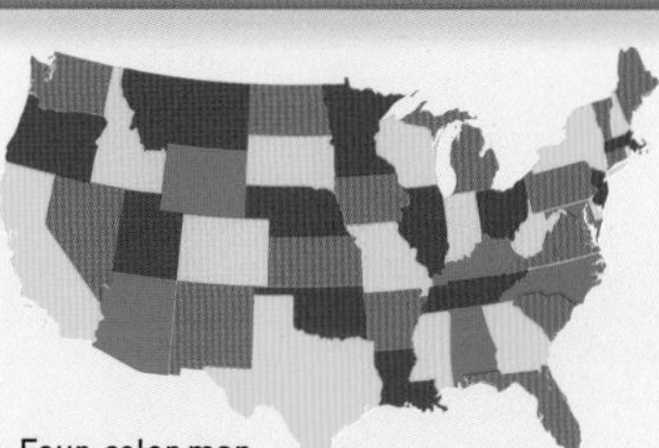

Four-color map

Euclid's *Elements* is one of the most important mathematical texts from the ancient world. It consists of 13 books and was originally written in Greek.

Healing hands *A marble frieze showing Hippocrates treating a sick woman. He advocated careful examination to determine the underlying disease.*

> “ IF YOU CUT OPEN THE **HEAD,** YOU WILL FIND THE **BRAIN HUMID,** FULL OF **SWEAT** AND HAVING A **BAD SMELL...** ”
>
> **Hippocrates,** from *On the Sacred Disease,* 400 BCE

ASTRONOMERS IN GREECE were interested in predicting the location of celestial bodies. This led the Greek astronomer Eudoxus of Cnidus (*c.*408–355 BCE) to develop a geometrical **model of the heavens**, in which the Sun, Moon, and planets moved in a series of 27 concentric spheres. He also made an accurate estimate of the length of the year at 365.25 days. At the time, most Greek astronomers believed Earth was stationary at the center of the Solar System, but Heraclides of Pontus (388–312 BCE) offered a variation on this theory. He claimed that **Earth rotated on an axis**, which explained the changing seasons.

Greek medicine moved in a more scientific direction when Alcmaeon of Croton began to teach that health is achieved by balancing the elements in the body. Hippocrates of Cos (460–370 BCE), who valued clinical observation, including taking a patient's pulse, applied this theory, teaching that **imbalances in the body** and impurities in the air could cause disease. In the mid-5th century BCE, Euryphon of Cnidus, who was from a rival school, taught that diseases were caused by residues building up in the body and advised that these be neutralized.

The Greek polymath, Aristotle, **refined the theory of the four elements**—earth, air, fire, and water—to include a fifth—*aither*—which caused the stars and planets to move in a circular motion. Aristotle modified Eudoxus's theory to explain anomalies, adding additional spheres to a total of 55. He also began the study of dynamics by theorizing that speed could be directly proportional to the weight of the body, the force applied, and the density of the medium in which the body moved.

The **foundations of geometry** were laid in the mid-4th century BCE by the Greek mathematician and father of geometry, **Euclid of Alexandria** (325–265 BCE), in his 13-book work called *Elements*. In it he puts forward a set of five "geometrical postulates" and nine "common notions" (or axioms). From these he deduced a set of theorems, including Pythagoras's theorem, and that the sum of angles in a triangle is always 180 degrees. *Elements* also included pioneering work on number theory, including an algorithm for the greatest common divisor.

MOTION OF THE SPHERES

Greek astronomers explained irregularities in planetary motions by theorizing that the Sun, Moon, and planets each sat in a series of concentric spheres. The circular motion (at differing speeds) of each sphere generated the planet's orbits.

In the early 2nd century, the astronomer Ptolemy replaced the spheres with circles in his model of the Solar System.

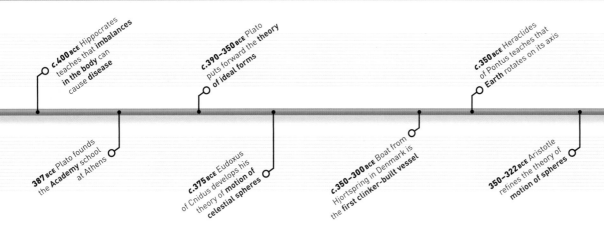

c.400 BCE Hippocrates teaches that **imbalances in the body** can cause **disease**

387 BCE Plato founds the **Academy** school at Athens

c.390–350 BCE Plato puts forward the **theory of ideal forms**

c.375 BCE Eudoxus of Cnidus develops his theory of **motion of celestial spheres**

c.350–300 BCE Boat from Hjortspring in Denmark is the **first clinker-built vessel**

c.350 BCE Heraclides of Pontus teaches that **Earth** rotates on its axis

350–322 BCE Aristotle refines the theory of **motion of spheres**

334–300 BCE

14 THOUSAND

THE NUMBER OF **SPECTATORS** THAT CAN BE SEATED AT THE **THEATER** IN **EPIDAURUS**

The acoustic properties of the theater at Epidaurus in Greece, built by Polycleitus the Younger in the 4th century BCE, allowed the actors to be heard perfectly up to 197 ft (60 m) from the stage.

GREEK MEDICINE MADE SIGNIFICANT ADVANCES in the 4th century BCE after the dissection of human bodies was pioneered by Diocles of Carystus, who wrote the first book devoted to anatomy. The **foundation of the Museum**, a scientific academy set up by Ptolemy I of Egypt (367–283 BCE), helped give rise to an Alexandrian school of medicine. One member, Herophilus of Chalcedon (335–280 BCE), identified the brain as the **seat of the nervous system** and made a distinction between arteries and veins.

Greek understanding of physics also progressed under Strato of Lampsacus (c.335–269 BCE). He rejected the idea of a force pushing light objects, such as air, upward to counter the force that pulls heavy objects down. He argued for the **existence of a vacuum** and showed that, because air can be compressed, voids must exist between the particles of which it is made up.

Via Appia
The first major Roman road, the Via Appia, originally ran from Rome to Capua. It was gravelled; paving stones were added in 295 BCE.

In Europe, wooden trackways had been used to traverse wet and marshy ground since Neolithic times, but proper roads needed a strong, centralized political authority to build and maintain them. In 312 BCE, the Romans began to construct **a vast network of roads** that bound their empire together. The first road they built, which ran from Rome to Capua, was called Via Appia. Roman roads were 10–26 ft (3–8 m) wide and were laid out on solid clay beds or timber frameworks, filled with loose flint or gravel. Sometimes they were **bound together with lime mortar** and topped with paving stones, or cobblestones in cities.

The Pharos of Alexandria was commissioned c.300 BCE by the ruler of Egypt, Ptolemy I. It was the tallest lighthouse in the ancient world at 410–492 ft (125–150 m) high. **Innovative engineering** was used in the hydraulic machinery needed to raise fuel to the fire that burned on top by night. During the day, a mirror of polished metal or glass reflected the Sun to create a warning beacon for ships.

Pythagoras had experimented with acoustics in the 6th century BCE. Aristotle advanced his work further in the 4th century BCE by theorizing that **sound consisted of contractions and expansions in the air**. The Greek theater at Epidaurus used stepped rows of seats to filter out low-frequency background noise, which allowed actors to be heard perfectly in the back row.

Compiled before 300 BCE, the Chinese text *Huang Di Nei Jing* explains human physiology and pathology in terms of the balancing forces of the universe: the opposing, but mutually dependent, principles of yin and yang; the five elements (earth, fire, wood, water, and metal); and qi, the essence of which everything is composed. **Ill health was thought to be caused by an imbalance** of yin and yang, in the patient's qi, and in the five elements that had their counterparts in the organs of the body and the environment.

Pharos of Alexandria
The Pharos of Alexandria was one of the Seven Wonders of the World. It was destroyed by an earthquake in the 14th century.

ARISTOTLE (384–322 BCE)

A founding figure in Western philosophy, Aristotle was a pupil at Plato's Academy in Athens. During his career he wrote more than 150 treatises on almost every aspect of Greek philosophy and science. He taught an empirical approach, that knowledge is gained from experience, and that all matter consists of a changeable form and an unchangeable substance.

5

THE NUMBER OF **PLATONIC** SOLIDS (REGULAR POLYHEDRALS) IN **EUCLIDEAN GEOMETRY**

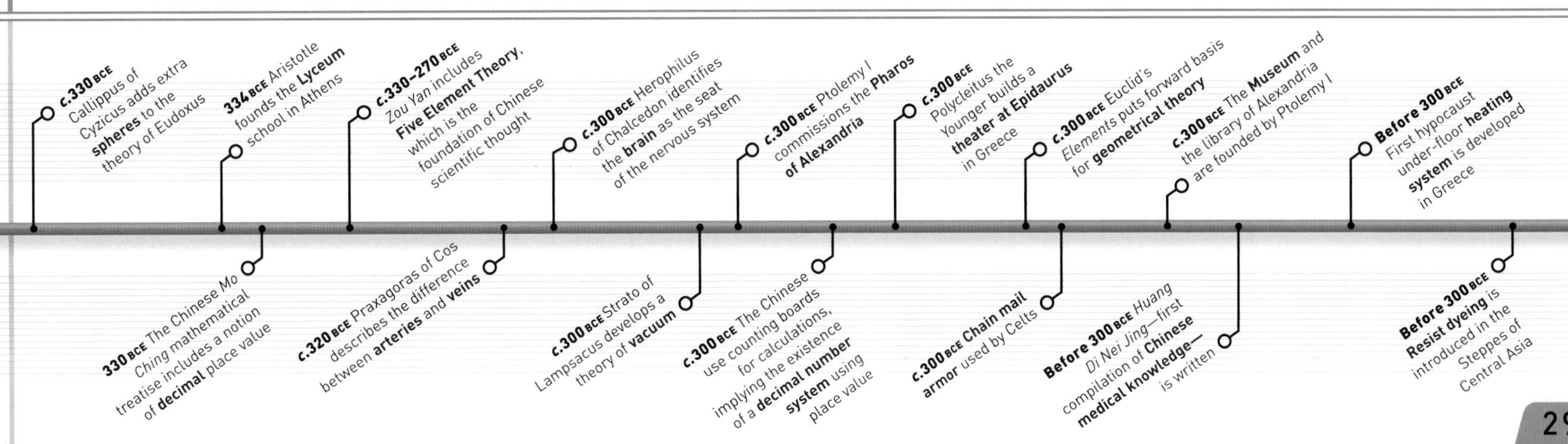

c.330 BCE Callippus of Cyzicus adds extra spheres to the theory of Eudoxus
330 BCE The Chinese Mo Ching mathematical treatise includes a notion of decimal place value
334 BCE Aristotle founds the Lyceum school in Athens
c.330–270 BCE Zou Yan includes Five Element Theory, which is the foundation of Chinese scientific thought
c.320 BCE Praxagoras of Cos describes the difference between arteries and veins
c.300 BCE Herophilus of Chalcedon identifies the brain as the seat of the nervous system
c.300 BCE Strato of Lampsacus develops a theory of vacuum
c.300 BCE Ptolemy I commissions the Pharos of Alexandria
c.300 BCE The Chinese use counting boards for calculations, implying the existence of a decimal number system using place value
c.300 BCE Polycleitus the Younger builds a theater at Epidaurus in Greece
c.300 BCE Euclid's Elements puts forward basis for geometrical theory
c.300 BCE Chain mail armor used by Celts
c.300 BCE The Museum and the library of Alexandria are founded by Ptolemy I
Before 300 BCE Huang Di Nei Jing—first compilation of Chinese medical knowledge—is written
Before 300 BCE First hypocaust under-floor heating system is developed in Greece
Before 300 BCE Resist dyeing is introduced in the Steppes of Central Asia

300–250 BCE

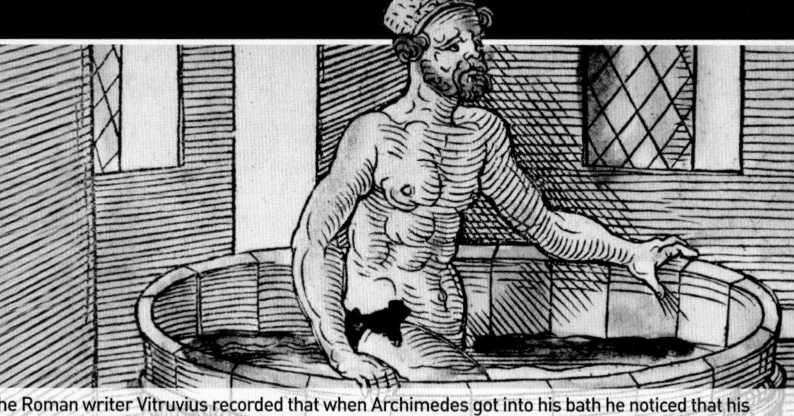

“EUREKA! I HAVE FOUND IT.”

Attributed to Archimedes, Greek inventor and philosopher *c.*287–*c.*212 BCE

The Roman writer Vitruvius recorded that when Archimedes got into his bath he noticed that his body displaced a certain amount of water. This gave him the idea for the Archimedes Principle.

MAGNETIC IRON LODESTONES were described in Chinese literature of the 3rd century BCE. By *c.*83 the Chinese text *Lun-heng* (*Discourses Weighed in the Balance*) had mentioned the electrostatic qualities of amber, which becomes charged when rubbed.

At around this time, Chinese diviners may also have discovered that iron, when rubbed against a lodestone, **becomes magnetized** and will point in a particular direction. The first primitive compasses were iron ladles set on divining boards that pointed south.

In Greece, **Theophrastus of Lesbos** (*c.*370–287 BCE), a pupil of Aristotle and also his successor as head of the Lyceum school in Athens, **extended Aristotle's work**, particularly in botany. He wrote *Enquiry into Plants* and *On the Causes of Plants*, which **classified plants into trees, shrubs, and herbs**. He also began the study of plant reproduction and discussed the best methods of cultivation for agriculture and companion planting to combat pests.

In astronomy, **Aristarchus of Samos** (*c.*310–230 BCE) rejected the prevailing view among early Greek astronomers that Earth was at the center of the Solar System. He believed that **Earth rotated in orbit around the Sun**; whether he thought the other planets also orbited the Sun is unclear. Aristarchus estimated the **comparative sizes** of the Sun and Earth at a ratio of about 20:1, and calculated the distance between Earth and the Sun to be 499 times the radius of Earth.

The science of **pneumatics** was founded by **Ctesibius of Alexandria** in the early 3rd century BCE. It is said that one of his first inventions was an adjustable-height mirror for his father's barber shop that used air compressed by counterweights to move up and down. He developed this idea to produce the **Ctesibian device**, a two-chamber force pump that used pistons attached to a rocker to create pressure. With the chambers of the device immersed in water, the rocker was moved up and down, alternately sucking water into one chamber and forcing it out of the other.

Chinese compass
A Han-era compass in the form of a magnetized ladle set on a bronze plate, featuring a diviner's representation of the cosmos.

Ctesibian pump
The rocker arm pushes the piston down on one side, creating pressure that closes the inlet valve and forces water through the outflow tube. Reduced pressure on the opposite side opens the valve to let more water in.

pivot
rocker arm moves pistons
water forced up and out
piston goes up
piston goes down
chamber fills with water
pressure pushes outlet valve open
reduced pressure opens inlet valve
water sucked in
reduced pressure shuts outlet valve
pressure forces inlet valve shut

Another inventor and philosopher, **Archimedes** (287–212 BCE) was also one of the greatest mathematicians of Ancient Greece. In *On the Measurement of a Circle* he presented a method for **calculating the area and circumference of a circle**. He also produced methods for calculating the **volumes of solids**, proving that the volume of a sphere inside a circumscribed cylinder is two-thirds that of the cylinder. Archimedes was the founder of hydrostatics (the science of fluids at rest). He showed that objects placed in water will displace a quantity of liquid equal to their buoyancy. He also developed a systematic **theory of statics**, showing how two weights balance each other at distances proportional to their relative magnitude. His aptitude for practical applications led him to develop the Archimedes screw (see 700–400 BCE) to pump out the bilges of a huge ship he built for the ruler of Syracuse. During

c.300 BCE Theophrastus of Lesbos starts to **classify plants**

c.250 BCE Aristarchus of Samos develops **heliocentric theory**

c.260 BCE Archimedes works on calculating the **circumference and area of a circle**

c.250 BCE Ctesibius of Alexandria develops the **Ctesibian pump**

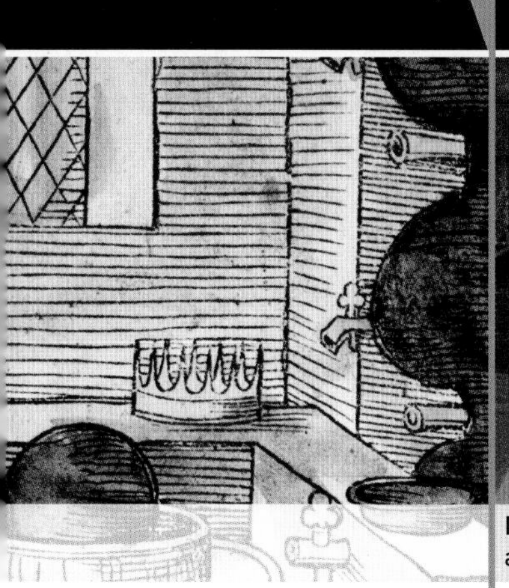

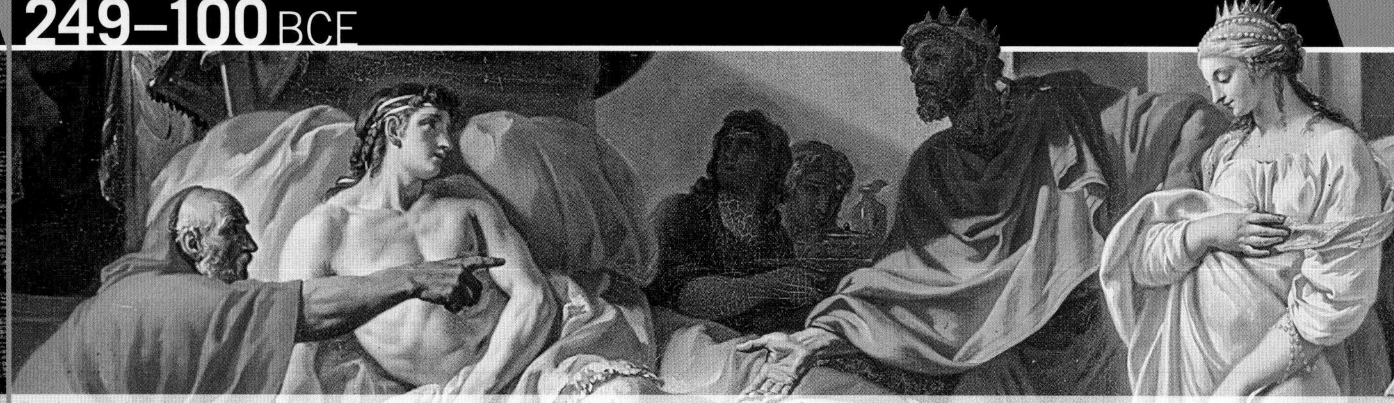

Erasistratus is said to have cured Antiochus, the son of Seleucus I of Syria, who was gravely ill. He identified the disease as love-sickness for his stepmother Stratonice, one of the first diagnoses of a psychosomatic illness.

the Roman conquest of Sicily, in 214 BCE, he was employed by the state to build various machines to defend Syracuse from attack. This included the **Claw of Archimedes**—a type of crane with a huge grappling hook that could capsize enemy ships.

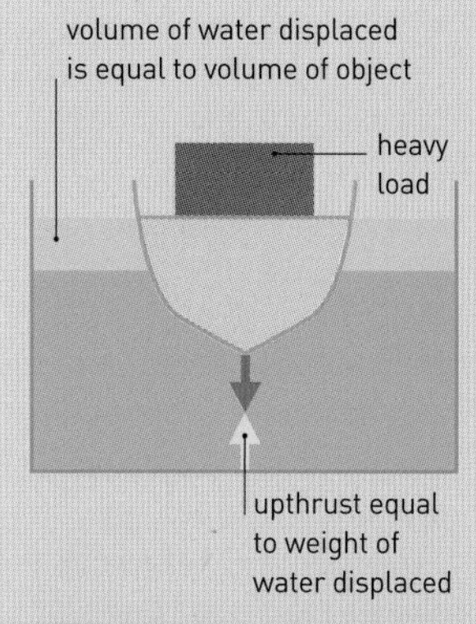

volume of water displaced is equal to volume of object
heavy load
upthrust equal to weight of water displaced

ARCHIMEDES PRINCIPLE

This states that a solid object, partly or wholly immersed in a liquid, has a buoyant force acting on it that is equal to the weight of the fluid it displaces. The relative density of the object can be worked out by dividing the weight of the object by the weight of the displaced liquid. The boat above can support a heavy load because it displaces a lot of water; therefore, the buoyant force supporting it is equally great.

ANATOMY ADVANCED CONSIDERABLY IN GREECE with the work of **Erasistratus of Cos** (*c.*304–250 BCE). He developed a theory of **vascular circulation**, in which he said that blood passed through the body in veins, while arteries distributed *pneuma* (air) to vital organs. He also gave an **accurate description of the brain**, including the cerebellum, and distinguished sensory from motor nerves.

Eratosthenes of Cyrene (*c.*275–195 BCE) made the first map of the world that featured lines of longitude and latitude in around 240 BCE. He also calculated the **dimensions of Earth** by comparing the angles of shadows at noon at Alexandria and Syene in Egypt, which are on roughly the same longitude. He yielded a figure of 250,000 stades—about 29,870 miles (48,070 km)—which is within one percent of the true figure. Eratosthenes also worked out a simple method of finding prime numbers, known as the **Sieve of Eratosthenes** (see panel, right).

Basilica Maxentius
This early 4th-century concrete Basilica was the largest building in Rome at the time.

Greek geometry advanced further in the late 3rd century BCE with the work of **Apollonius of Perga** (*c.*262–190 BCE), whose major work was entitled *On Conics*. In it he described the properties of the three fundamental **types of conic** section—the ellipse, parabola, and hyperbola. He also developed the **theory of epicycles**—circular orbits rotating around a larger circumference—to refine the theory of the motion of the spheres (see 400–335 BCE).

The Romans found a way of **bonding small stones** to produce **concrete** in the late 2nd century BCE. By adding pozzolana stone (ash from prehistoric volcanoes) to lime, they produced a strong binding mortar. This enabled them to build **stronger and cheaper** monumental buildings. The first structure built using concrete was the Porticus Aemilia in Rome in 193 BCE.

circled numbers are primes — crossed-out numbers are non-primes

	(2)	(3)	~~4~~	(5)	~~6~~	(7)	~~8~~	~~9~~	~~10~~
(11)	~~12~~	(13)	~~14~~	~~15~~	~~16~~	(17)	~~18~~	(19)	~~20~~
~~21~~	~~22~~	(23)	~~24~~	~~25~~	~~26~~	~~27~~	~~28~~	(29)	~~30~~
(31)	~~32~~	~~33~~	~~34~~	~~35~~	~~36~~	(37)	~~38~~	~~39~~	~~40~~
(41)	~~42~~	(43)	~~44~~	~~45~~	~~46~~	(47)	~~48~~	~~49~~	~~50~~

SIEVE OF ERATOSTHENES

This is a simple algorithm for finding prime numbers. Starting at 2 without striking it out, strike out all multiples of 2 to the end of the series. Return to the next non-struck out number (3) and without striking it out, strike out every multiple of 3 to the end. Repeat the process; eventually all the non-struck out numbers will be prime.

Observational astronomy was revolutionized by **Hipparchus of Nicaea** (c.190–120 BCE), who made a **new map of the heavens** that catalogued 850 stars. He invented a new astronomical sighting tool and surveying instrument called the dioptra that was in use until it was replaced by the armillary sphere. Using the dioptra, he discovered the phenomenon of **precession**, by which stars appear to move gradually in relation to the equinoxes.

Hipparchus also calculated the length of the year to be 365.2467 days—very close to the true value.

At this time, the Chinese were busy refining the **production of paper**. The process of soaking and pulping textile rags then drying them out on a screen to produce a fibrous mat for writing on, probably dates from the late 3rd century BCE. Although the **invention of paper** is often ascribed to Cai Lun (50–121), he probably just refined this process and introduced new pulp materials, such as tree bark.

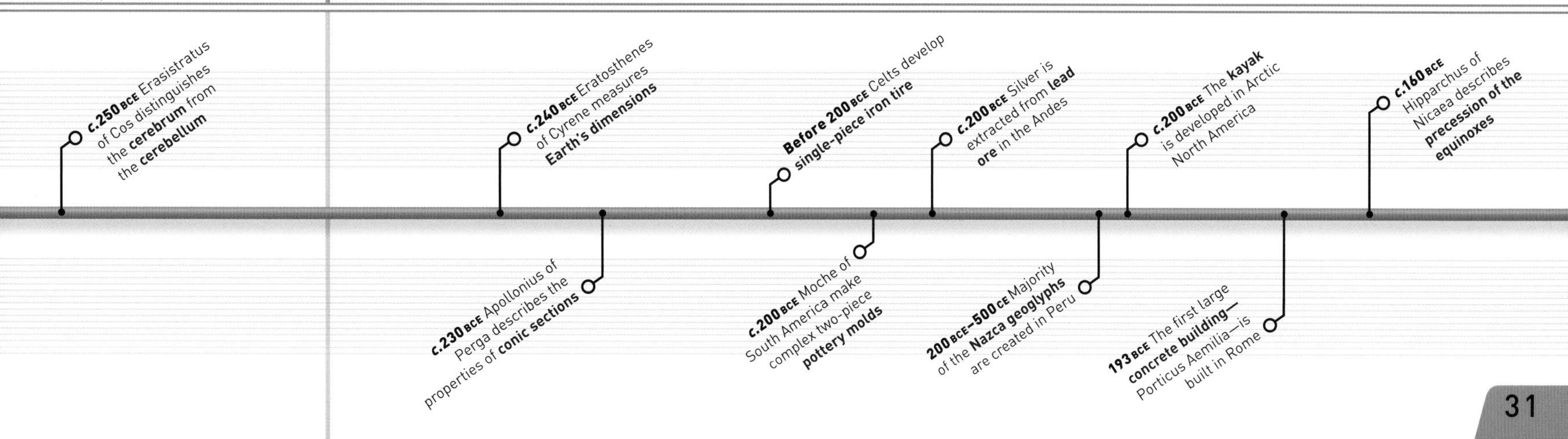

c.250 BCE Erasistratus of Cos distinguishes the **cerebrum** from the **cerebellum**
c.230 BCE Apollonius of Perga describes the properties of **conic sections**
c.240 BCE Eratosthenes of Cyrene measures **Earth's dimensions**
Before 200 BCE Celts develop **single-piece iron tire**
c.200 BCE Moche of South America make complex two-piece **pottery molds**
c.200 BCE Silver is extracted from **lead ore** in the Andes
200 BCE–500 CE Majority of the **Nazca geoglyphs** are created in Peru
c.200 BCE The **kayak** is developed in Arctic North America
193 BCE The first large **concrete building**—Porticus Aemilia—is built in Rome
c.160 BCE Hipparchus of Nicaea describes **precession of the equinoxes**

100 BCE–50 CE

A medieval depiction of a Vitruvian undershot waterwheel. Operated with a hand lever, the buckets fill with water as the wheel rotates and the buckets dip into a water source. The water is deposited at the top.

> “ THE **LAWS OF MECHANICS** ARE FOUNDED ON THOSE OF **NATURE,** AND ARE **ILLUSTRATED** BY STUDYING THE MASTER-MOVEMENTS OF **THE UNIVERSE ITSELF.** ”
>
> **Marcus Vitruvius Pollio, Roman architect and engineer,** from *Ten Books on Architecture, c.*15 BCE

THE ANTIKYTHERA MECHANISM IS A COMPLEX DEVICE that shows the earliest understanding of gears. Dating from around 80 BCE, it was recovered in 1900 from a shipwreck off the Greek island of Antikythera. Made up of a series of bronze toothed dials and at least 30 gears, it is thought to have been used to **predict solar and lunar eclipses** and to track other time cycles, such as the 19-year Metonic cycle—the basis for the ancient Greek calendar.

By the 1st century BCE the **Maya calendar** had developed a 5,125-year era known as the **Long Count**. Twenty *tun* (years) made a *katun*, 20 katun were a *baktun*, and 13 of these completed the whole era. The earliest known date inscribed in the Long Count system is December 9, 36 BCE; this is found on a stele at Chiapa de Corzo in Mexico. The Maya also used a 52-year Calendar Round, with two elements working in combination—the 260-day Tzolk'in calendar and the 365-day Haab.

Around 90 BCE **Posidonius of Apamea** (*c.*135–50 BCE) used the relative position of the star Canopus, seen from Alexandria and Rhodes, to **calculate the size of Earth**. His calculation was 240,000 stades, only slightly smaller than the estimate of **Eratosthenes of Cyrene** (see 250–100 BCE). Posidonius also calculated the size of the Moon and made a study of tides, relating them to the phases of the Moon.

Around this time, the Greek physician Asclepiades of Bithynia (*c.*129–40 BCE) put forward his idea of the **brain being the seat of sensation**. He developed a theory of disease based on the flow of atoms in the body, a doctrine he derived from the atomic theory of the 5th-century-BCE philosopher Democritus. His treatment methods were very subtle, prescribing baths and exercises. Perhaps less humane were the practices of his follower Themison of Laodicea, who was the first recorded physician to use **leeches to bleed patients**.

The Roman writer Celsus (*c.*25 BCE–50 CE) produced one of the most important **texts on medicine**, *De Medicina*, an encyclopedic summary of medical knowledge of the time. In it, he gave accounts of the use of opiates for calming patients and laxatives to purge them. He also detailed many **surgical techniques**, including the removal of kidney stones and how to operate on cataracts (clouding of the lens in the eye).

The Romans also advanced **engineering** in this period. The architect Vitruvius (*c.*84–15 BCE) was the first to explain the use of siphons to lessen hydraulic pressure in pumps. He also described the **Vitruvian turning wheel**. When the wheel was turned, buckets emptied water into a channel at the top and filled up from a water source at the bottom. This type of "undershot" waterwheel had probably been invented earlier, but Vitruvius may have refined it to make it more effective.

Glassblowing was developed around 50 BCE in Roman-controlled Syria. **Glassmakers** obtained a more even flow by blowing **molten glass** through a tube (either freely or into a mold), rather than just pouring it. The **higher-quality glassware** that resulted led to the establishment of glassworks throughout the Roman Empire.

Roman glass *The strong colors of this 1st-century CE vase from Lebanon are typical of the early Imperial period.*

GEARS

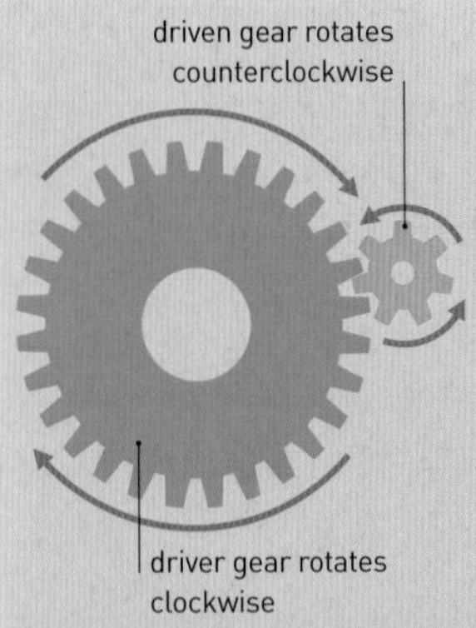
driven gear rotates counterclockwise
driver gear rotates clockwise

Mentioned *c.*330 BCE by Aristotle, the Romans brought gears into common use during this period in waterwheels and hoists. Gears are made up of sets of interlocking toothed wheels. They work when a larger wheel engages with a small wheel and alters the speed of a driving mechanism.

ROMAN VETERINARY SCIENCE

Roman interest in veterinary science sprang from the needs of farmers and also of the army, which had large cavalry units. In the army, specialists called *mulomedicus* cared for military donkeys and horses. Around 45 CE the Roman writer Columella wrote extensively on the care and diseases of farm animals.

early terracotta horse head

3 MILLION
THE **DIAMETER** OF THE **SUN** IN **STADES,** AS PER **POSIDONIUS**

- ***c.*90 BCE** Evidence of use of **acupuncture** in China
- ***c.*90 BCE** Posidonius of Apamea calculates the **size of Earth and Moon**
- ***c.*80 BCE** Construction of the **Antikythera mechanism**
- ***c.*75 BCE** Asclepiades of Bithynia develops the atomist **theory of disease**
- ***c.*50 BCE** **Glassblowing** is developed by the Syrians in the Levant
- ***c.*45** Columella writes a treatise covering **animal diseases**
- **36 BCE** Earliest inscription containing **Maya Long Count** date
- ***c.*15 BCE** Roman architect Vitruvius describes use of the **force pump**
- ***c.*15 BCE** Vitruvius describes methods of surveying and building **aqueducts**
- ***c.*1 CE** Chinese are the first to **extract salt** by brine mining and boiling
- ***c.*25–50 CE** Celsus produces the medical encyclopedia ***De Medicina***

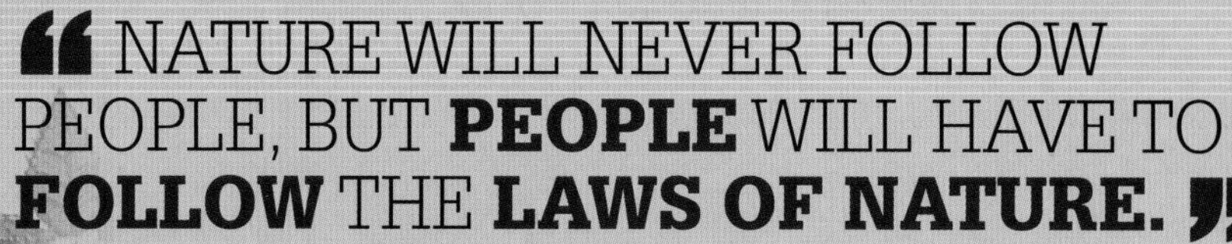

❝ NATURE WILL NEVER FOLLOW PEOPLE, BUT **PEOPLE** WILL HAVE TO **FOLLOW** THE **LAWS OF NATURE. ❞**

Dioscorides, Greek physician and botanist, from *De Materia Medica, c.*50–70

An illustration of the common bilberry, traditionally used for circulatory problems, from a 6th-century manuscript of Dioscorides's *Materia Medica* (*Regarding Medical Materials*).

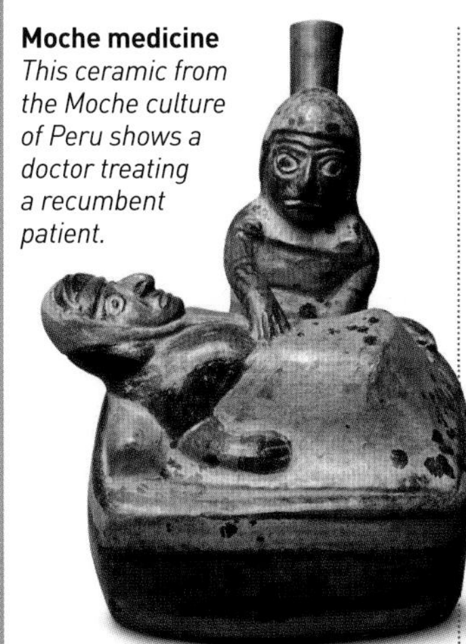

Moche medicine
This ceramic from the Moche culture of Peru shows a doctor treating a recumbent patient.

INDIAN MEDICINE HAD ITS ROOTS IN THE VEDIC PERIOD before 1000 BCE, but in the period 100 BCE–100 CE, the *Caraka Samhitā* (*Compendium of Caraka*) appeared as one of the earliest **Indian medical texts**. The book highlights the importance of clinical examination and the use of careful regimens of drugs or diets to cure illnesses. Traditional Indian, or **ayurvedic**, medicine came to stress the importance of balancing humors in the body and ensuring *srotas* (channels) in the body transport fluids correctly.

Much of what is known about **medicine in ancient South America** comes from examination of the ceramics of the Moche people from the late 1st century CE onward. These depict a variety of injured patients, including some with facial paralysis, and also show the use of crutches, and primitive prosthetic legs for amputees.

The first **pharmacopeia** (compilation of medicinal plants) was compiled by Dioscorides in Greece. In it he described **over 600 plants**, including their physical properties and effects on patients. Hugely influential, it was used by physicians throughout the Middle Ages.

The *Huainanzi* (*Master Huainan*) is a **compilation of Chinese knowledge** composed before 122 BCE. It touches on a range of subjects, including philosophy, metaphysics, natural science, and geography. It is notable for its analysis of mathematical and musical harmonies, including a description of the traditional 12-tone Chinese scale.

The Greek geometer and inventor Hero of Alexandria (*c.*10–70 BCE) described a variety of cranes including the *barulkos*, which operated using a toothed worm-gear that could not reverse and which prevented loads from slipping. He provided the **first description of a lathe** for the precision cutting of screws, and was also the first to describe the use of a wind wheel, in which the rotating vanes operated pistons that made the pipes of a water organ sound. Hero is perhaps most well known for his studies into the properties of steam. He used his knowledge to build an **aeolipile**. This is a primitive form of steam engine that uses steam to spin a hollow sphere.

sphere is spun by steam power
steam forced through pipe and into sphere
bung blocks steam from exiting the cauldron
bent pipe allows steam to escape, which pushes the sphere around
cauldron filled with water
cauldron stand
fuel for fire

Hero's aeolipile
The aeolipile is the only known ancient machine operated by steam. It makes the sphere spin by channeling steam from a cauldron into a hollow sphere and out of the bent pipes that are attached to it.

600 THE NUMBER OF **PLANTS** DESCRIBED IN DIOSCORIDES'S ***DE MATERIA MEDICA***

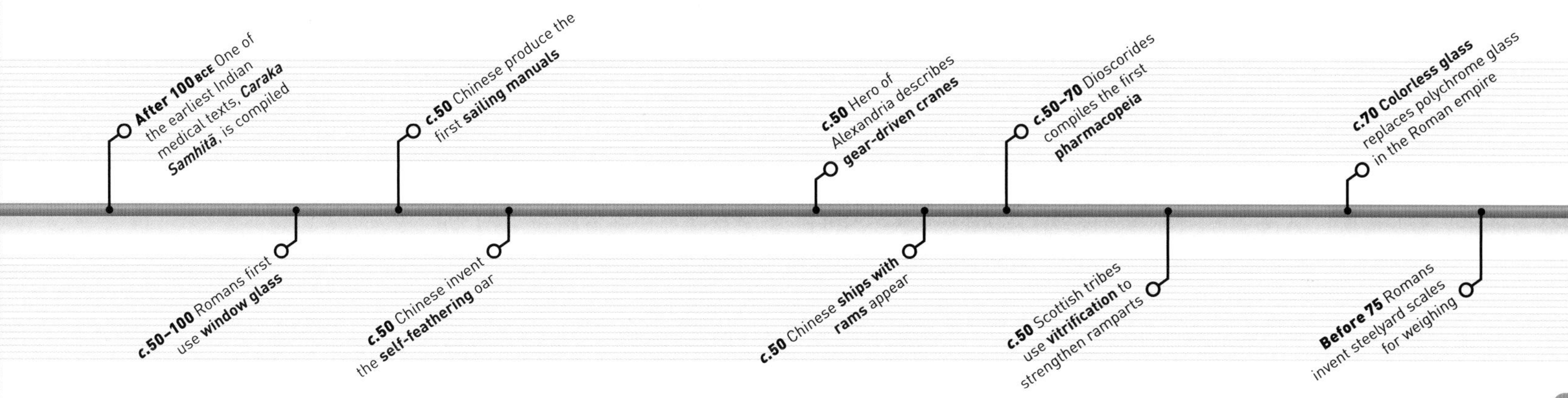

After 100 BCE One of the earliest Indian medical texts, **Caraka Samhitā**, is compiled
c.50–100 Romans first use **window glass**
c.50 Chinese produce the first **sailing manuals**
c.50 Chinese invent the **self-feathering** oar
c.50 Hero of Alexandria describes **gear-driven cranes**
c.50 Chinese **ships with rams** appear
c.50–70 Dioscorides compiles the first **pharmacopeia**
c.50 Scottish tribes use **vitrification** to strengthen ramparts
c.70 Colorless glass replaces polychrome glass in the Roman empire
Before 75 Romans invent steelyard scales for weighing

UNDERSTANDING SIMPLE MACHINES

DEVICES THAT CHANGE THE SIZE AND DIRECTION OF FORCES HAVE BEEN USED SINCE ANCIENT TIMES

Mechanical devices are composed of different working parts. Among the most important are six basic components called simple machines, which mathematicians and engineers have studied since ancient times: the wheel and axle, the inclined plane, the lever, the pulley, the wedge, and the screw.

Greek engineer Hero of Alexandria (1st century CE) was the first person to bring together the simple machines, in his book *Mechanica*, although the inclined plane was not included in his account. Hero illustrated and explained various devices for lifting heavy objects. Others before him had studied why these devices work—most notably, Archimedes of Syracuse (3rd century BCE), who studied levers. Archimedes worked out that the ratio between the input force (the effort) and the output force (the load) is equal to the ratio between the distances from the pivot at which those forces act. So, to gain a very large "mechanical advantage" (multiplication of force)—and move a heavy load—a very long lever should be used, but the load needs to be close to the pivot point. What the ancient engineers didn't realize is that there is always a pay-off between force and the distance—to gain a large mechanical advantage, the long end of the lever moves through a large distance, while the load moves only a small way. Similarly, using pulleys to lift a heavy load, the length of rope you must pull is much greater than the distance the load moves. The amount of "work" done by the effort is the same as the amount of work done by the load (neglecting friction).

HERO
Hero (or Heron) of Alexandria was one of the most prolific engineers of ancient Greece. He is seen here demonstrating his aeolipile, an early example of the use of steam power.

INCLINED PLANE

People have used simple ramps (inclined planes) to gain a mechanical advantage since prehistory. A person raising an object by pushing it up a ramp pushes with a lesser force than if the object were being lifted directly; however, the object must be pushed along the ramp's length, while the load moves, vertically, a much shorter distance.

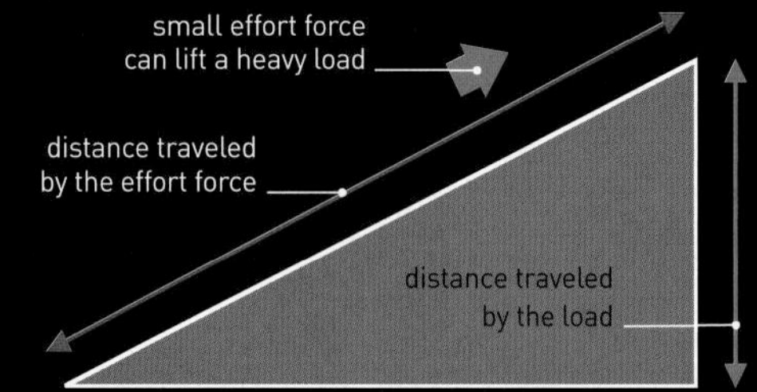
small effort force can lift a heavy load
distance traveled by the effort force
distance traveled by the load

RAMP
The simplest example of an inclined plane is a ramp. A heavy load can be pushed up a ramp in a continuous motion that requires a smaller force than would be required to lift the load straight up.

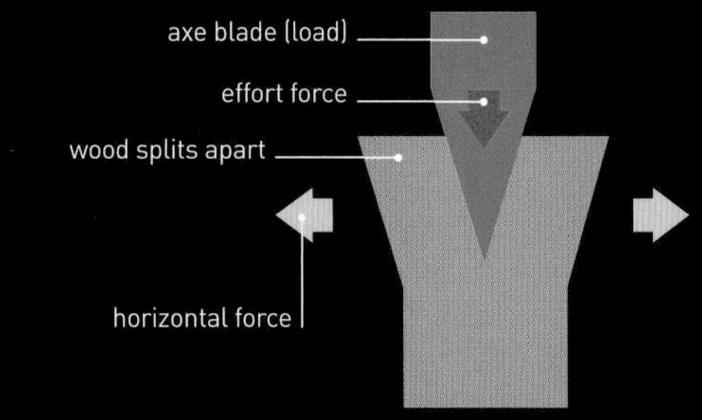
axe blade (load)
effort force
wood splits apart
horizontal force

WEDGE
Two inclined planes back to back make a wedge. An ax blade is a wedge, which, forced vertically into a block of wood, produces a strong horizontal force. The force splits the wood—but the two pieces move only a small distance apart.

WHEEL AND AXLE

The wheel was invented in Mesopotamia around 3500 BCE. When a wheel is fixed to an axle, the two turn together; ancient engineers used wheels in devices such as the windlass by winding ropes around the axle. The mechanical advantage of a windlass is the ratio of the crank wheel's radius to the axle's radius—if the crank wheel has twice the radius of the axle, the effort force will be doubled. Door handles and bicycle cranks are modern examples of the wheel and axle. Gears are interlocking wheels without axles; the mechanical advantage is the ratio of diameters between one gear and the next.

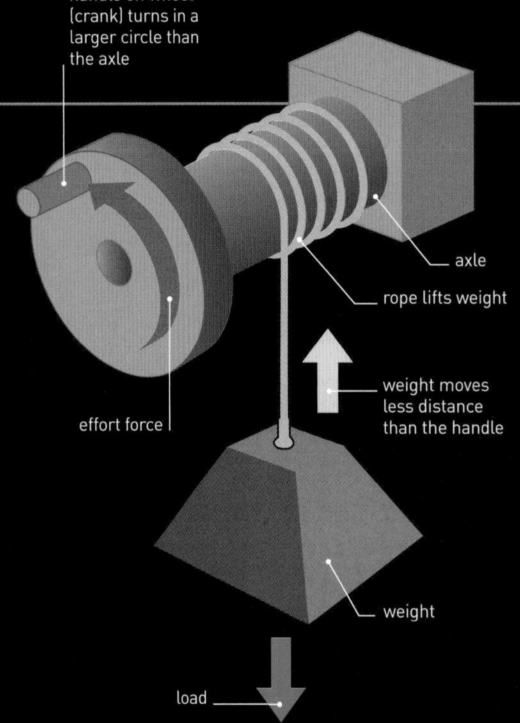
handle on wheel (crank) turns in a larger circle than the axle
axle
rope lifts weight
weight moves less distance than the handle
effort force
weight
load

TURNING FORCE
A rope is pulled by an axle turned by a wheel. By making a wheel much larger than the axle, it is possible to gain a large mechanical advantage – but the handle moves through a much greater distance than the weight.

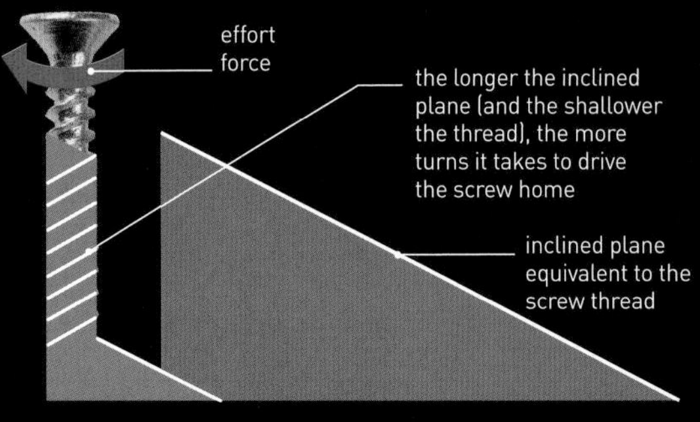
effort force
the longer the inclined plane (and the shallower the thread), the more turns it takes to drive the screw home
inclined plane equivalent to the screw thread

SCREW
A screw thread is equivalent to an inclined plane wrapped around a shaft. Turning a screw inside a material pulls it inward. Screws are also used to move water, grain, and other bulk materials in screw conveyors.

PULLEYS

A simple pulley—a rope passed over a free-moving wheel—has no mechanical advantage, because the rope is continuous. But by passing the rope underneath a pulley, the load is shared between two sections of the rope, and the effort is reduced by half; in that case, the load moves half as far as the end of the rope is pulled. By combining two or more pulley blocks, the mechanical advantage can be increased further.

SINGLE PULLEY

A rope passed over one pulley can raise a weight attached to the end of the rope. This set-up has no mechanical advantage, but it does change the direction of the force—and it can be more convenient than simply lifting the weight.

fixed pulley block

pulley wheel around which the rope moves

weight rises

end of the rope moves the same distance as the weight

effort is equal to load

load is the weight of the object

effort is half the load

movable pulley block

load is raised half as far as rope is pulled

HALF THE EFFORT

A single pulley block can be used to create a mechanical advantage of two. If the rope is slung under the pulley wheel, the force is shared between the two sections of rope either side of the pulley.

BLOCK AND TACKLE

An arrangement of two pulley blocks, one fixed and one moving, is called a block and tackle. The mechanical advantage is still two, because the load is pulled by two ropes—but pulling the rope downward is more convenient.

fixed pulley block

movable pulley block

effort force is half the load

load is the weight of the object

EASY TO LIFT

A block and tackle with more pulley wheels gives an increased mechanical advantage. In this example, the job of lifting the load is shared between four sections of rope, so the mechanical advantage is four.

fixed pulley block with two wheels

rope must be pulled four times as far as the weight rises

movable pulley block with two wheels

effort force is one quarter the load

load is the weight of the object

LEVERS

The mechanical advantage of a lever is the ratio of distances from the fulcrum (pivot) to the effort and the load. The ratio can be equal to one, or greater than or less than one. There are three types of lever, distinguished by the positions of the effort and load relative to the fulcrum.

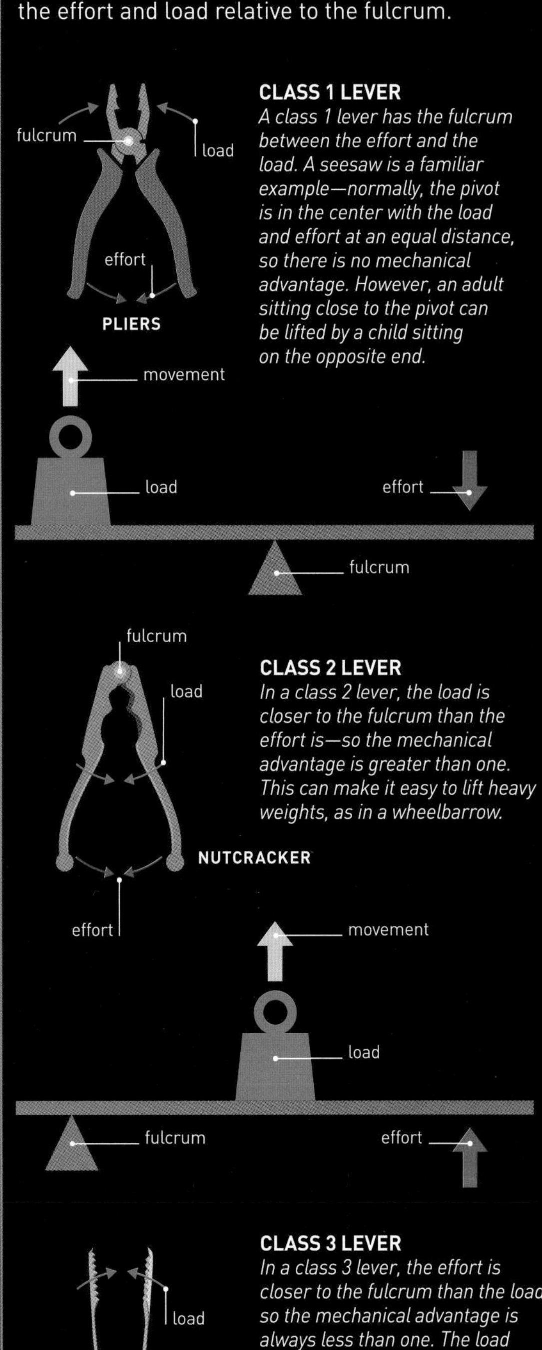
fulcrum
load
effort
PLIERS
movement
load
effort
fulcrum
fulcrum
load
NUTCRACKER
effort
movement
load
fulcrum
effort

CLASS 1 LEVER

A class 1 lever has the fulcrum between the effort and the load. A seesaw is a familiar example—normally, the pivot is in the center with the load and effort at an equal distance, so there is no mechanical advantage. However, an adult sitting close to the pivot can be lifted by a child sitting on the opposite end.

CLASS 2 LEVER

In a class 2 lever, the load is closer to the fulcrum than the effort is—so the mechanical advantage is greater than one. This can make it easy to lift heavy weights, as in a wheelbarrow.

CLASS 3 LEVER

In a class 3 lever, the effort is closer to the fulcrum than the load, so the mechanical advantage is always less than one. The load moves farther (and faster) than the effort; a golf club benefits from this effect.

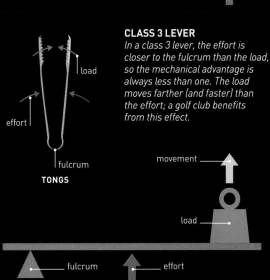

load

effort

fulcrum

TONGS

movement

load

fulcrum

effort

75–250

“ **BEARS** WHEN **FIRST BORN** ARE SHAPELESS MASSES OF WHITE FLESH **A LITTLE LARGER THAN MICE,** THEIR **CLAWS** ALONE BEING **PROMINENT.** ”

Pliny the Elder, Roman historian and philosopher from *Natural History*, Book VIII, 77

Pliny the Elder holds a pair of surveyor's dividers in this medieval frontispiece of his book entitled *Natural History*.

THE ROMAN HISTORIAN AND PHILOSOPHER PLINY THE ELDER (23–79) compiled *Natural History*, a 37-volume summary of ancient knowledge, which he completed in 77. It contains much of what we know about Greek and Roman science, covering mineralogy, astronomy, mathematics, geography, and ethnography, as well as including detailed sections on botany and zoology. *Natural History* is also significant because it contains the only references we have to the **work of earlier scientists**.

During this period three Greek physicians published notable works on **anatomy and diseases**. In the late 1st century, Aretaeus of Cappadocia wrote *The Causes and Signs of Acute and Chronic Diseases*, describing a vast range of diseases, their diagnosis, causes, and treatment. He was the **first physician to describe both diabetes and celiac disease**. Among the other conditions he dealt with were pleurisy, pneumonia, asthma, cholera, and phthisis (tuberculosis), for which he prescribed trips to the seaside.

In 100, Greek physician Rufus of Ephesus wrote *On the Names of the Parts of the Human Body*, summarizing the **Roman knowledge of anatomy**. He gave a detailed description of the eye, and was the **first to identify the optic chiasma**, where the optic nerves partially cross in the brain. He was also the first to name the pancreas and made a detailed study of melancholia (depression).

3:10 Infant mortality rate in Rome
Despite medical advances, in the 1st and 2nd centuries, the infant mortality rate in Rome was still roughly 30 percent.

In the early 2nd century, the Greek physician Soranus of Ephesus produced ***On the Diseases of Women***. This was the most comprehensive work on the subject from the ancient world. In it, he described the appropriate training for midwives and gave instructions for managing childbirth, such as the use of the obstetric chair or birthing stool and how to administer intrauterine injections, and explained the use of the speculum mirror for internal examinations, as well as giving a detailed description of specific gynecological conditions. Soranus of Ephesus also pioneered the **science of pediatrics**. His work contained advice on the early care of infants, including the making of artificial teats for feeding, and accounts of childhood afflictions such as tonsillitis, a variety of fevers, and heatstroke.

Zhang Heng (78–139) was a polymath whose work in China in the early 2nd century included calculating a value for pi, the identification of 124 constellations in the sky, and the construction of an armillary sphere with moving parts to show the rotation of the planets. He is best known for the **construction of the earliest seismograph**, which he completed in 132. It consisted of a bronze urn with a pendulum inside. When a tremor occurred, the pendulum swung in the direction of one of eight dragons' heads, which opened and released a ball into the mouth of a bronze frog below, indicating the direction of the earthquake. In 138, Zhang Heng used the seismograph to **successfully detect an earthquake** that had happened more than 400 miles (640 km) from the Chinese court, where he was demonstrating it.

In the 3rd century BCE, the Romans discovered the principle

Zhang Heng's seismograph
Earth's vibrations caused a pendulum in the seismograph to move, which released a ball from a dragon's teeth into a frog's mouth, indicating the direction of the earthquake.

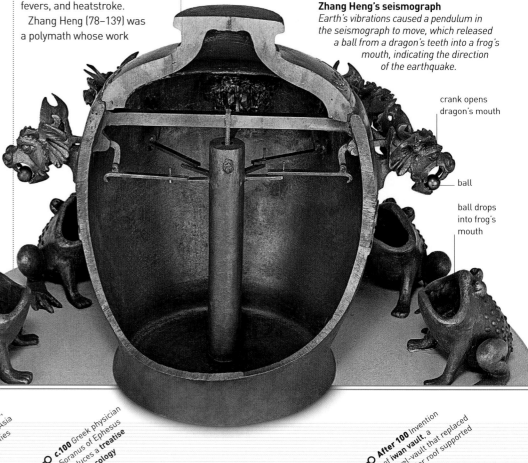

crank opens dragon's mouth
ball
ball drops into frog's mouth

c.75 Demosthenes Philalithes writes a textbook on **ophthalmology**

77 Roman historian Pliny the Elder completes his ***Natural History***

100 Rufus of Ephesus writes a **treatise on anatomy**

c.100 Herodotus, a physician of Asia Minor, identifies **smallpox**

c.100 Greek physician Soranus of Ephesus produces a **treatise on gynecology**

c.100 Aretaeus of Cappadocia describes **diabetes**

c.100 Menelaus of Alexandria produces work on **trigonometry**, including proof of Menelaus's theorem of spherical triangles

After 100 Invention of **iwan vault**, a barrel-vault that replaced an earlier roof supported by columns

After 100 The **Treadle loom is** developed in China

After 100 Archigenes sets down 10 criteria for **measuring the pulse**

One of the most complete of the original Roman bridges, the Pons Aelius was built by Emperor Hadrian to provide a processional route to his mausoleum (now the Castel Sant'Angelo). In its original form the bridge had eight arches.

Ptolemy's map
The coordinates and topographic lists in Ptolemy's Almagest *enabled maps to be composed of his view of the world. This map dates from 1492.*

of the weight-supporting arch, and used it in bridge-building. In around 104, the engineer Apollodorus of Damascus had **constructed a great bridge across the Danube** to facilitate the Emperor Trajan's invasion of Dacia (modern Romania).

Trajan's bridge was destroyed in *c.*120 by his successor, Hadrian, who himself had several great bridges built, including the Pons Aelius in Rome in *c.*134.

The best-known works of Greek–Roman astronomer **Ptolemy of Alexandria** (*c.*90–168) are on mathematical geography and astronomy. In *Geography*, he gave a description of the known world, including **coordinates for longitude and latitude** (the latter derived from the length of the longest day) and gave instructions for the creation of a world map. In *Mathematical Compendium,* also known as *Almagest*, Ptolemy presented a star catalog with over 1,000 listed stars and 48 constellations. He refined the **theory of the celestial spheres**, introducing additional epicycles to explain irregularities in the motion of the Sun and the Moon and the apparent retrograde motion of certain planets, when they appear to orbit in a contrary direction to other bodies in the Solar System. He was the first astronomer to **convert observational data into a mathematical model** to back up his theories, using spherical trigonometry to do so. His model of the Solar System remained the basis of astronomical theory until the Renaissance.

In 169, **Claudius Galen** became personal physician to the Roman emperor Marcus Aurelius. Galen specialized in anatomy and had earlier worked as a surgeon to a gladiatorial school, where he gained valuable knowledge of human physiology and surgery. He championed the theory that the **body had four basic humors** (see panel, below).

The Bakshali manuscript, found in what is now Pakistan, dates from around 200 and contains instructions for the computation of square roots. It is probably the **earliest document to use a specific sign for zero in the decimal system**, making it the first complete decimal notation with a single sign for each number value. This system spread westward through the Arabic world, acquiring the popular name of Arabic numerals.

Chinese mathematics had made significant advances by the time ***Jiuzhang Suanshu*** (*Nine Chapters on the Mathematical Art*) was in existence in 179. It included rules for **calculating the area of arcs of circles** and the volume of solid figures such as cones, and for the treatment of vulgar fractions (written in the form x/y). It contained instructions for the **calculation of linear equations**, including the earliest appearance of equations with negative numbers.

CLAUDIUS GALEN (*c.*130–*c.*210)

Born in the ancient Greek city of Pergamum, Claudius Galen consolidated the works of his predecessors to create a single scientific framework. His insistence on direct observation of the body cut across his view that each of the body's organs functioned according to a divinely ordained scheme. He wrote 350 medical works.

THE FOUR HUMORS

The theory of the four humors stated that the body is composed of four substances: blood, phlegm, yellow bile, and black bile. In blood, the four elements of the universe (fire, air, earth, and water) are mixed equally, while in the other humors, one element predominates. An excess of one humor was believed to cause disease. Too much yellow bile led to jaundice, too much black bile to leprosy, and too much phlegm to pneumonia.

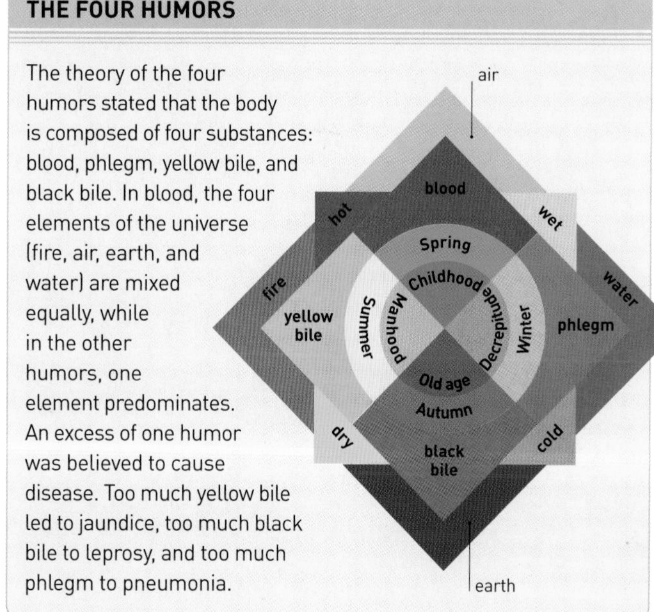
air
blood
hot
wet
Spring
fire
water
Childhood
yellow bile
Summer
Manhood
Decrepitude
Winter
phlegm
Old age
Autumn
dry
cold
black bile
earth

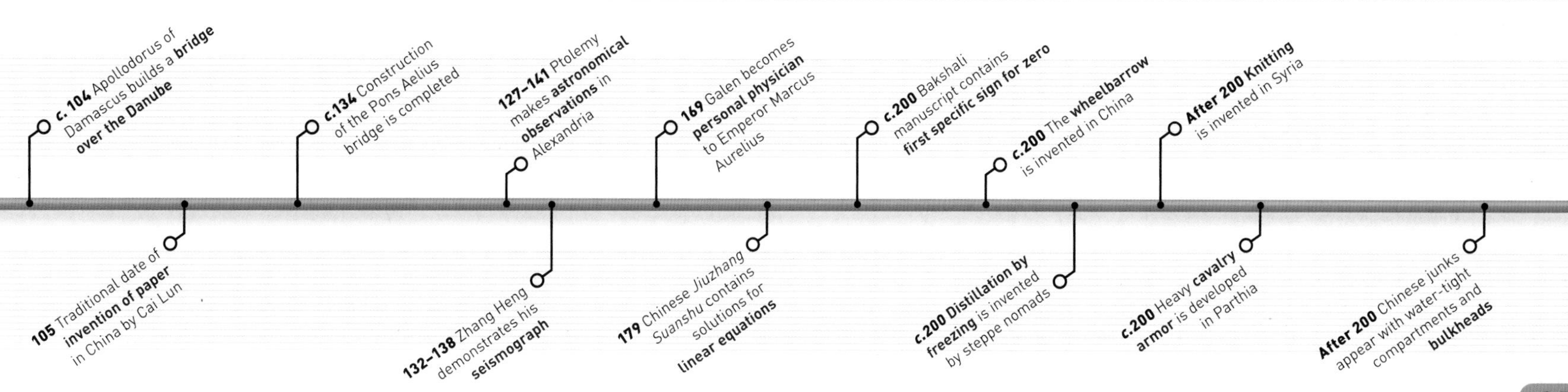
c. 104 Apollodorus of Damascus builds a bridge over the Danube
c.134 Construction of the Pons Aelius bridge is completed
127–141 Ptolemy makes astronomical observations in Alexandria
169 Galen becomes personal physician to Emperor Marcus Aurelius
c.200 Bakshali manuscript contains first specific sign for zero
c.200 The wheelbarrow is invented in China
After 200 Knitting is invented in Syria
105 Traditional date of invention of paper in China by Cai Lun
132–138 Zhang Heng demonstrates his seismograph
179 Chinese Jiuzhang Suanshu contains solutions for linear equations
c.200 Distillation by freezing is invented by steppe nomads
c.200 Heavy cavalry armor is developed in Parthia
After 200 Chinese junks appear with water-tight compartments and bulkheads

250–500

·ARISMETRICA·
·GEOMETRIA·
·MVSICA·
·ASTRO LOGIA·
·LOGICA·

This 15th-century painting depicts the seven liberal arts, core subjects such as arithmetic, music, astronomy, rhetoric, and grammar that the 5th-century writer Martianus Capella established as the basis of early medieval European education.

DIOPHANTUS OF ALEXANDRIA (*c.*200–*c.*284) founded the mathematical discipline of algebra around 250 by introducing a systematic notation to indicate an unknown quantity and its power; for example, in the equation $x^2 - 3 = 6$, x^2 represents an unknown number raised to the power of 2 (or squared). In his *Arithmetica*, Diophantus provided solutions for linear equations (in which no variable in the equation is raised to a power greater than 1—as in $ax + b = 0$) and quadratic equations (in which at least one of the variables is squared—as in $ax^2 + bx + c = 0$). Diophantus also made a particular study of **indeterminate equations**, proposing a method of solving them that is now known as Diophantine analysis. Fermat's Last Theorem (see 1635–37) is probably the most famous example of such an equation.

Roman surgical instruments
Ancient Roman physicians used a wide variety of surgical instruments, including spatulas and hooks (right), specula for internal examinations, and saws.

In around 320, **Pappus of Alexandria** (*c.*290–*c.*350) compiled *Collections*, an eight-volume work that contained the major results of the great mathematicians who preceded him and also introduced novel concepts. Among these new ideas were work on the **centers of gravity** and the volumes created by plane figures revolving. He also proposed what is now known as **Pappus's hexagon theorem**, which states that the intersections of three collinear points (points along the same line) with three similar points along a similar line will themselves be collinear.

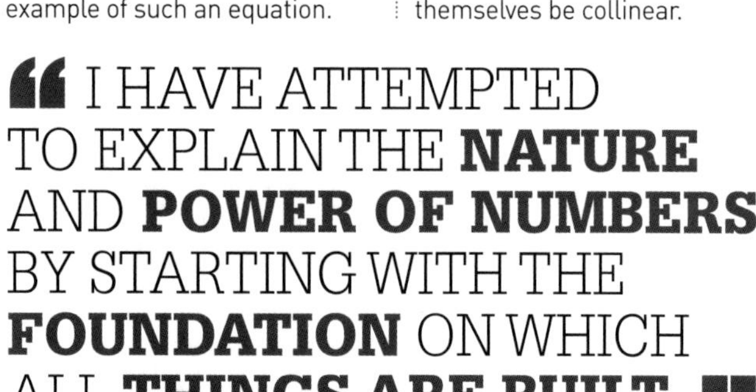

" I HAVE ATTEMPTED TO EXPLAIN THE NATURE AND POWER OF NUMBERS BY STARTING WITH THE FOUNDATION ON WHICH ALL THINGS ARE BUILT. "

Diophantus of Alexandria, Greek mathematician, from *Arithmetica, c.*250

In the 3rd century, Plotinus (*c.*205–270) created a modified form of Plato's teachings (see 700–400 BCE) known as **Neoplatonism**, which remained influential into the Middle Ages. Plotinus taught that there is a transcendent being (the "One"), which cannot be described, from which emanated a series of other beings. These included the "Divine Mind" and the "World Soul," from which human souls are derived. Plotinus's follower Iamblichus of Apamea (*c.*245–*c.*325) developed these ideas, adding number symbolism derived from Pythagoras (see 700–400 BCE). Iamblichus believed that **mathematical theorems applied to the whole universe**, including divine beings, and that numbers themselves had a form of concrete existence.

In line with the general trend in the 3rd and 4th centuries for gathering together the work of earlier scientists, **Oribasius of Pergamum** (*c.*323–400) produced *Collections*, a set of 70 volumes that **brought together the works of Galen and other earlier medical writers**. Only 20 of these volumes survive, of which four, collectively titled *Euporista*, give advice on food, drink, and diets. Oribasius also described a sling to bind a fractured jaw, which he attributes to the 1st-century physician Heraklas. Oribasius became personal physician to the Roman emperor Julian, but failed to save his patron when he was struck by a spear during a battle in Persia in 363.

In China, mathematicians continued to make advances. *Hai Tao Suan Ching*, which dates from 263, contains a discussion

Raised fields
The Maya cut drainage channels through swamps, heaping up the fertile silt to create raised fields, similar to the ones seen here.

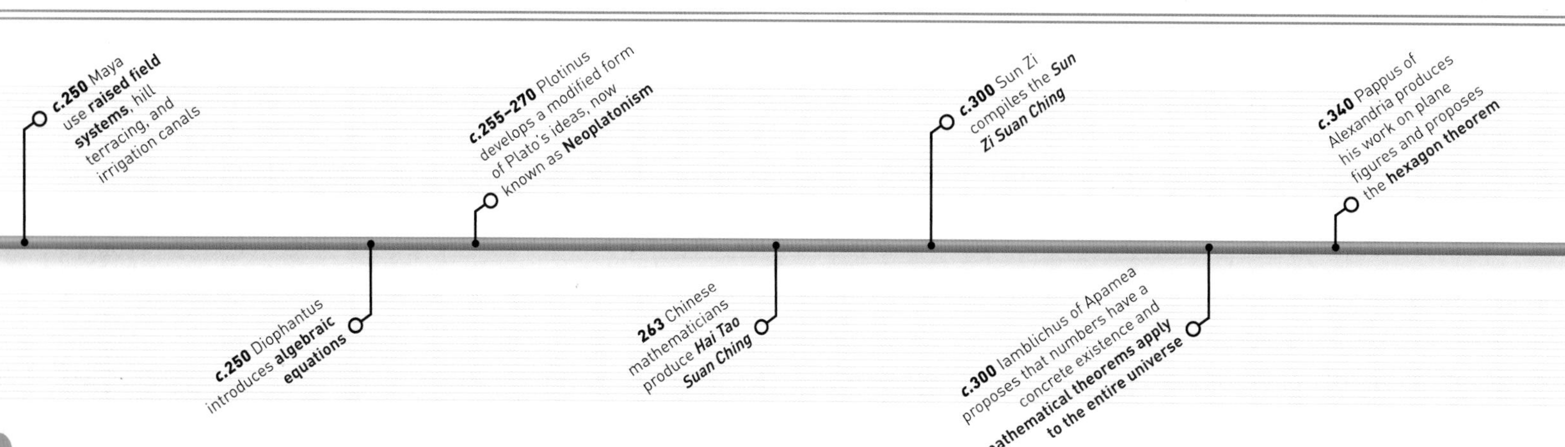

c.250 Maya use raised field systems, hill terracing, and irrigation canals
c.250 Diophantus introduces algebraic equations
c.255–270 Plotinus develops a modified form of Plato's ideas, now known as Neoplatonism
263 Chinese mathematicians produce Hai Tao Suan Ching
c.300 Sun Zi compiles the Sun Zi Suan Ching
c.300 Iamblichus of Apamea proposes that numbers have a concrete existence and mathematical theorems apply to the entire universe
c.340 Pappus of Alexandria produces his work on plane figures and proposes the hexagon theorem

·RETHORICA·
·GRAMMATICA

“ THE **SHAPE** OF THE **EARTH** IS NOT **FLAT,** AS SOME SUPPOSE WHO IMAGINE IT TO BE LIKE AN **EXPANDED DISK... ”**

Martianus Capella, from *On the Marriage of Philology and Mercury,* 410–439

7 THE NUMBER OF **LIBERAL ARTS** IDENTIFIED BY **ROMAN** WRITER **CAPELLA**

of right-angled triangles and in around 300, **Sun Zi** compiled the *Sun Zi Suan Ching*, which includes an analysis of indeterminate equations. It also contains what is now known as the **Chinese remainder theorem**, which provides a method of finding solutions to problems in modular arithmetic (also called clock arithmetic, because numbers are arranged in a circle, rather than along the number line). In the 5th century, **Zu Chongzh**i (429–500) wrote *Zhui Shu* (*Method of Interpolation*), in which he **calculated pi** to be $^{355}/_{113}$. He refined this to produce a value for pi that was accurate to seven decimal places (see panel, below), a figure that was not improved upon until the 16th century.

Martianus Capella, from Madaura in North Africa, established the basic structure of early medieval European education. In his *On the Marriage of Philology and Mercury* (410–439), he presented a **compendium of knowledge**, which he divided into the trivium (grammar, dialectic, and rhetoric) and the quadrivium (geometry, arithmetic, astronomy, and music). In this work, he stated that **Mercury and Venus orbit around the Sun**, a view that Copernicus used to support his heliocentric view of the solar system (see 1543).

Mathematics progressed only slowly during the later Roman empire. In about 450, the Neoplatonist philosopher **Proclus** (*c.*410–485) produced his *Commentary on Euclid*, in which he preserved the work of earlier mathematicians. Proclus's contemporary, Domninus of Larissa (*c.*410–480) wrote *Manual of Introductory Arithmetic*, which included a summary of number theory.

By the 5th century, the Maya had devised a **sophisticated calendrical system** and a notation system for numbers that could express any number using only three symbols: a dot for 1, a bar for 5, and a shell for 0. Maya astronomers were particularly concerned with lunar cycles, the Sun, eclipses, and movements of the planet Venus.

There is also evidence that the Maya were practicing **raised-field agriculture** from as early as the mid-3rd century to utilize fertile land that would otherwise have been too waterlogged for agricultural use.

THE VALUE OF PI

Pi, the ratio of the circumference of a circle to its diameter, was estimated at 3.125 by the Babylonians. The Greeks discovered a method of calculating it by using the sides of a polygon inside a circle to approximate the circumference, and Archimedes used this method to give a figure of $^{22}/_{7}$. In about 475, Zu Chongzhi calculated pi as 3.1415926—accurate to seven decimal places. Computers have now calculated this value to trillions of decimal places.

Astronomical codex
A section from the Dresden Codex, *a 9th-century Maya astronomical work that includes detailed tables for movements of the planet Venus.*

c.360 Oribasius of Pergamum writes on **dietetics**

c.400 Theon of Alexandria introduces a method of **calculating square roots by approximation**

410–439 Martianus Capella asserts that **Mercury and Venus orbit around the Sun**

c.450 Greek philosopher Proclus writes his **Commentary on Euclid**

mid- to late 400s Domninus of Larissa writes his **Manual of Introductory Arithmetic**

c.475 Zu Chongzhi calculates **pi to seven decimal places**

c.499 Indian mathematician Aryabhata (476–550) **estimates pi** to be 3.1416

500–540

“ YET IT SEEMS NOT TO REST UPON SOLID MASONARY, BUT TO COVER THE SPACE WITH ITS GOLDEN DOME SUSPENDED FROM HEAVEN. ”

Procopius, Byzantine scholar, from *The Buildings Book, c.*500–65

The dome of Hagia Sophia was completed in 537 and collapsed in an earthquake in 558. It was rebuilt by Isidore the Younger, who raised it by about 20ft (6m) to make it more stable.

MUCH ANCIENT KNOWLEDGE reached the Middle Ages through the efforts of Roman nobleman **Boethius** (*c.*480–*c.*524). He acted as a link in the **transmission of Greek and Roman science** to scholars of his time. He translated sections of Aristotle's *Logic*, produced an adaption of Greco-Roman mathematician Nicomachus's (*c.*60–*c.*120) *Arithmetike Eisagoge* (*Introduction to Arithmetic*), and **compiled manuals of the liberal arts**, including accounts of Euclidean geometry and Ptolemaic astronomy. Without his work, much ancient knowledge might have been lost in western Europe.

Flavius Cassiodorus (*c.*480–*c.*575), who succeeded Boethius as the leading Roman nobleman at the court of the Ostrogothic kings of Italy, retired around 540 to a monastery he founded at Vivarium in southern Italy. There, he composed ***Institutiones Divinarum et Humanarum Lectionum*** (*An Introduction to Divine and Human Readings*). This handbook on monastic life included a **compilation of secular knowledge**, divided according to the seven liberal arts (see 250–500). Cassiodorus also **established a library** in which many ancient scientific and philosophical treatises were collected. He instituted the **practice of copying manuscripts**, thus ensuring that important works survived into the later Middle Ages.

Great minds
Boethius is shown here calculating with written numbers in a competition against Pythagoras, who is using a counting board.

Before the 6th century, scholars had largely accepted Aristotle's view that motion was inherent in a body or caused by the medium through which it traveled (such as air). Greek philosopher **John Philoponus** (*c.*480–*c.*570) opposed this view, arguing that the medium actually resisted the body's movement. He proposed that **motion is caused externally** through energy impressed upon it by the person or thing moving it. This was the first expression of the theory of **impetus and inertia**.

Around 500, **Li Tao-Yuan recorded fossil animals** in his *Commentary on the Waterways Classic*. He called these fossils stone oysters, or stone swallows; they were said to emerge from the rock and fly around during thunderstorms. By the mid-7th century, such fossils were being dissolved in vinegar for use as medicine in China.

In early 6th century, Chinese mathematician Zhang Qiujian gave the first example of the **modern method of division**—inversing the divisor and multiplying. He also gave examples of problems involving **arithmetical progressions** (where the difference between successive terms is constant).

Around 532–37, Byzantine architects **Anthemius of Tralles** (*c.*474–*c.*534) and **Isidore of Miletus** succeeded in setting a **round dome over a square room using pendentives**. The dome of Hagia Sophia (in Istanbul, Turkey) remained the largest in the world for nearly a thousand years.

BUILDING WITH PENDENTIVES

Pendentives, such as those employed in the church of Hagia Sophia in Constantinople, are curved, concave sections of masonry that are used to join a square lower section of a building to the circular base of a domed top section. They allow the weight of the dome to be equally distributed onto square supporting walls or piers, which allows far larger domes to be built.

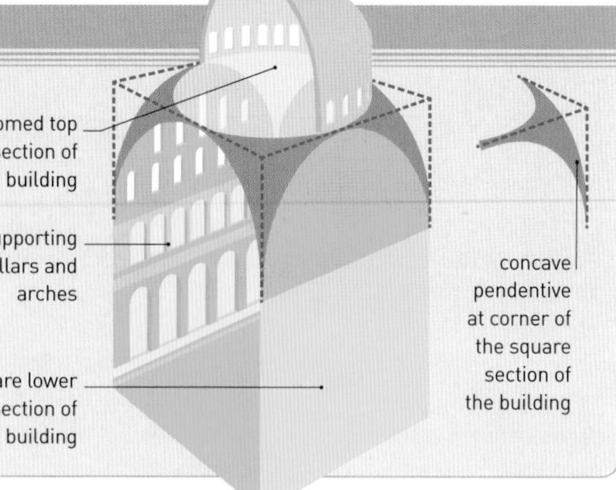
domed top section of building
supporting pillars and arches
square lower section of building
concave pendentive at corner of the square section of the building

grooved, shell-like "bird's wing"

Brachiopod fossil
Resembling bird's wings, these became known as "stone swallows" in China.

c.500 Cotton gin with **worm gearing** used in India

c.500 First use of **wooden blocks** for printing in China

c.500 Zhang Qiujian describes the method of division using **reciprocal of divisor**

c.500 Li Tao-Yuan records **fossil brachiopods** in China

Xu Yue writes *Memoirs on the Mathematical Art*

510–24 Boethius writes books on **arithmetic and geometry**, and translates Aristotle's *Logic*

531–79 Nahrawan canal built in Persia

532–37 First **round dome**—in Hagia Sophia church, Turkey—is set over a square room

540 Cassiodorus retires to Vivarium and composes ***Institutiones Divinarum et Humanarum Lectionum***

c.540 Cosmas Indicopleustes composes *Topographia Christiana*

2 MILLION

THE ESTIMATED **NUMBER** OF INDIVIDUAL **MOSAIC TILES** ORIGINALLY USED TO **CREATE** THE **MADABA MAP**

The Madaba map is a mosaic showing Palestine and lower Egypt, with particular focus on towns and other sites of Biblical importance. This part of the map shows Jerusalem.

THE FIRST DESCRIPTION of the **bubonic plague** was given by Roman historian Procopius (*c.*500–*c.*565). He was present in Constantinople (now Istanbul) when the disease struck the Byzantine empire in 542. He described the characteristic swellings (or buboes) under the arms and around the groin, and a type of delirium brought on by septicemia (blood poisoning) that caused sufferers to run around screaming.

By the 6th century, the cartographic tradition inspired by Ptolemy was waning, to be replaced by a religiously inspired view of Earth. The **Madaba map**, thought to be the oldest surviving map of Biblical cities, dates back to this time. In around 550, **Cosmas Indicopleustes**, a merchant from Alexandria, composed the *Topographia Christiana* (*Christian Topography*), which controversially presented the **world as a flat space** dividing the heavens from the underworld, and in which Jerusalem occupied a central position. Cosmas located Paradise just beyond the ocean that surrounded Earth.

7:10 **Bubonic plague death toll** *At its peak, the bubonic plague, which struck the Byzantine empire in 542, killed 10,000 people a day in Constantinople alone.*

One of the leading medical practitioners at the time of the Byzantine emperor Justinian was Alexander of Tralles (*c.*525–*c.*605). His *Twelve Books on Medicine* **described a range of diseases** including those caused by intestinal parasites. He was the first physician to **identify melancholy (depression)** as a cause of suicidal tendencies.

Around 570, Chinese mathematician **Chen Luan** mentioned the abacus for the first time in a commentary on an earlier work of the 2nd century. He described 14 methods of arithmetical calculation, one of which he referred to as "ball arithmetic," in which a series of wires were suspended on a wooden frame, with four balls strung on the lower half of each wire representing a unit each, and a ball on the upper half representing five units.

Although the Chinese had a **long tradition of canal building**, their greatest project was the cutting of the **Grand Canal from Changan to Loyang**, under the Sui dynasty, which joined up earlier, smaller canals. Its main section, the Pien Chu canal, which was 621 miles (1,000 km) long was completed in 605 and was said to have taken five million laborers to build.

By the early 7th century, Chinese engineers had worked out that bridges did not need semicircular arches. In 605, Chinese engineer Li Chun completed construction of the **Anji Bridge** in Hebei. The arch was flattened by two smaller arches in its spandrels (the triangular area bounded by the outer curve of an arch and adjacent wall), which spread the weight more evenly and meant that only one main arch was needed to span the river.

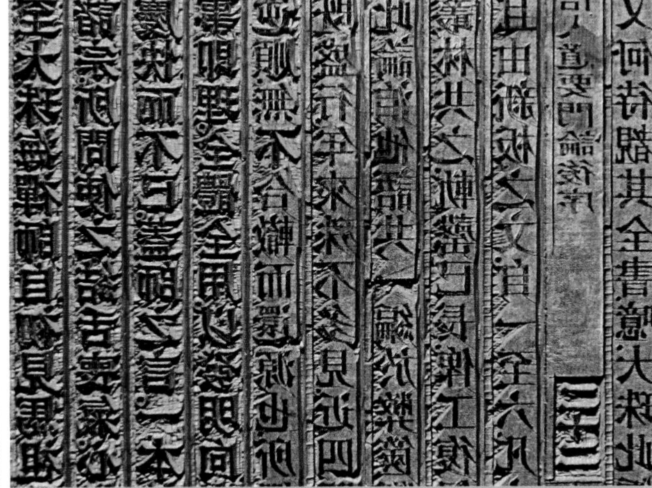

CHINESE BLOCK PRINTING

Printing using wooden blocks was probably invented in China in the 6th century, although the first complete surviving printed book dates to 868. A manuscript was prepared on waxed paper, which was rubbed against a wooden block to transfer a mirror image of the characters onto it. The block was then carved and used for printing.

Rainbow bridge *This bridge over a side section of the Grand Canal at Wuxi, China, arches in a dramatic fashion, which gave this type of construction the nickname "rainbow bridge."*

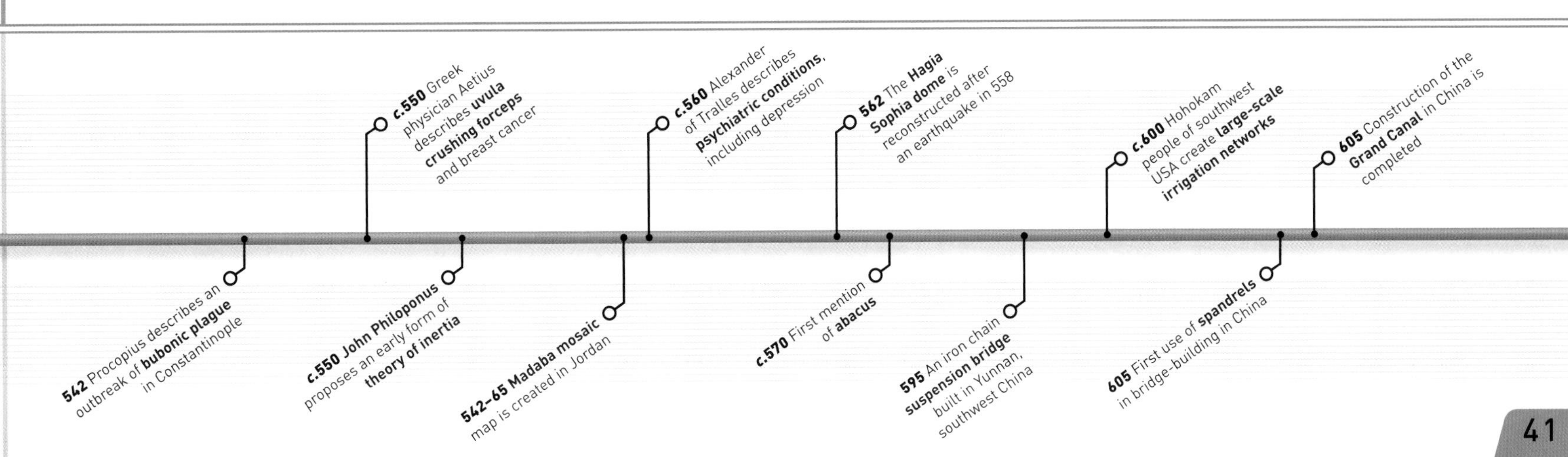

542 Procopius describes an outbreak of **bubonic plague** in Constantinople
c.550 Greek physician Aetius describes **uvula crushing forceps** and breast cancer
c.550 John Philoponus proposes an early form of **theory of inertia**
542–65 Madaba mosaic map is created in Jordan
c.560 Alexander of Tralles describes **psychiatric conditions**, including depression
562 The **Hagia Sophia dome** is reconstructed after an earthquake in 558
c.570 First mention of **abacus**
595 An iron chain **suspension bridge** built in Yunnan, southwest China
c.600 Hohokam people of southwest USA create **large-scale irrigation networks**
605 First use of **spandrels** in bridge-building in China
605 Construction of the **Grand Canal** in China is completed

610–700

This illustration from a 12th-century manuscript graphically depicts the use of Greek Fire. Flames are being projected from a handheld tube onto a fleet of invading soldiers.

IN CHINA, in the year 610 court physician **Chao Yuanfang** (550–630) compiled the first **comprehensive Chinese treatise on diseases**. One of the diseases he described was smallpox; he explained that lesions with purple or black coloration were far more deadly than those that contained white pus. He also recommended brushing teeth daily and proposed a routine of rinsing and gargling then gnashing the teeth seven times.

> “AS THE **SUN ECLIPSES THE STARS** BY ITS BRILLIANCY, SO THE **MAN OF KNOWLEDGE** WILL ECLIPSE THE FAME OF OTHERS IN ASSEMBLIES OF THE PEOPLE IF HE PROPOSES **ALGEBRAIC PROBLEMS,** AND STILL MORE IF HE **SOLVES THEM.**”
>
> **Brahmagupta, Indian mathematician,** from *Brahmasphutasiddhanta* (*The Revised System of Brahma*), 628

Before 644, **windmills had been developed in Persia**. They used wind to drive wooden vanes set in a circle around a windshaft. This generated rotational energy, which could be used to grind wheat. The earliest windmills had **vertical windshafts**, unlike the more familiar horizontal types that were later developed in Europe.

Spanish bishop **Isidore of Seville** was a prolific author who wrote books on cosmology and arithmetic. In the 7th century, he compiled a **20-volume manuscript of contemporary knowledge**, entitled *Etymologiae*, using the work of earlier encyclopedists such as Roman author Marcus Terentius Varro (116–27 BCE). It helped disseminate classical knowledge in the Middle Ages.

In the field of surgery, Greek physician **Paul of Aegina** (*c.*625–*c.*690) compiled *The Epitome of Medicine*—a digest of **medical treatises by ancient authorities** such as Galen. It also contained descriptions of new surgical procedures, such as tracheotomy (surgery to the windpipe) and sterilizing wounds through cauterization.

Chinese mathematician **Wang Xiaotong** (*c.*580–*c.*640) was the first to provide **solutions for cubic equations** (of the form $a^3+ba^2+ca=n$). It was a technique that European mathematicians did not master until Fibonacci (see 1220–49) in the 13th century.

In India, one of the greatest early mathematicians was **Brahmagupta** (598–*c.*668). His *Brahmasphutasiddhanta* (*The Revised System of Brahma*) contained **rules** for using **negative numbers** in arithmetic and also first stated the rule that two negative numbers multiplied together yield a positive number.

In the late 7th century, a new incendiary weapon was developed in the Byzantine empire. Known as **Greek Fire**, it was discharged by tubes and burned even in contact with water. Its exact composition is still unknown, but it was probably a compound of naphtha (a hydrocarbon mixture).

Vertical windmills
Because the area around Nishtafun in Persia (Iran) experienced high winds, but had little water, windmills were a very useful adaptation.

ISIDORE OF SEVILLE
(*c.*560–636)

The Bishop of Seville for more than 30 years, Isidore wrote several important texts, including the encyclopedic *Etymologiae*, a dictionary of synonyms, and a manual of basic physics. He also established a system of seminaries to promote ecclesiastical education. He was canonized in 1598 by Pope Clement VIII.

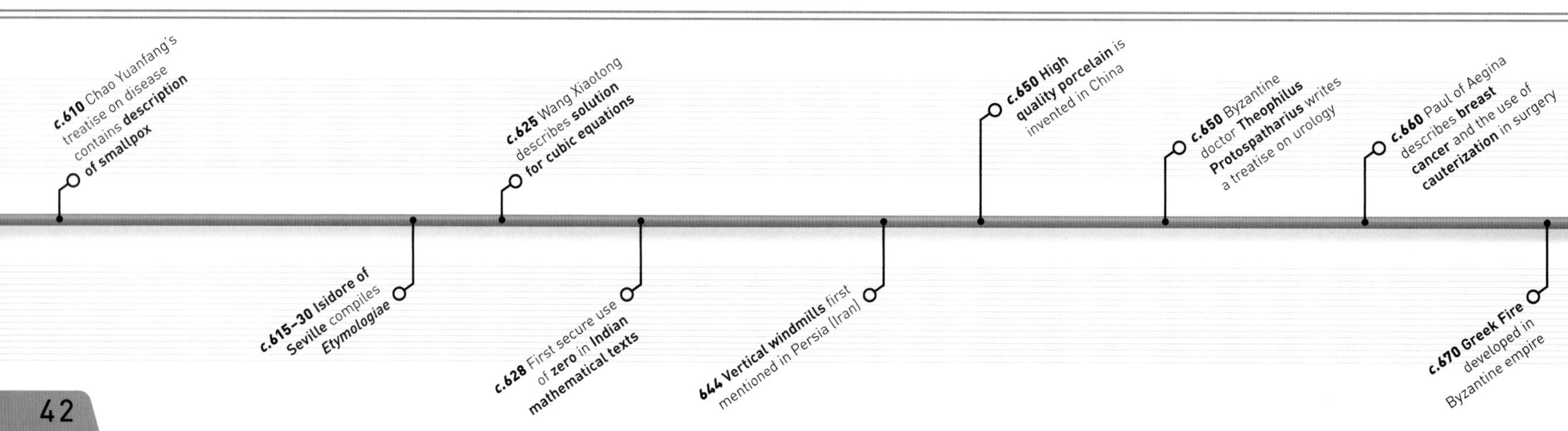

c.610 Chao Yuanfang's treatise on disease contains description of smallpox
c.615–30 Isidore of Seville compiles Etymologiae
c.625 Wang Xiaotong describes solution for cubic equations
c.628 First secure use of zero in Indian mathematical texts
644 Vertical windmills first mentioned in Persia (Iran)
c.650 High quality porcelain is invented in China
c.650 Byzantine doctor Theophilus Protospatharius writes a treatise on urology
c.660 Paul of Aegina describes breast cancer and the use of cauterization in surgery
c.670 Greek Fire developed in Byzantine empire

700–800

This image depicts Jabir ibn-Hayyan giving a lecture on alchemy in his home town, Edessa, modern Turkey. The town played an important role in the transmission of Greek science into the Islamic world.

THE ISLAMIC WORLD'S FIRST MAJOR TREATISES on zoology were produced by al-Asmai, a philologist from Basra, Iraq. His ***Kitab al-Khail*** (*Book of the Horse*) and ***Kitab al-Ibil*** (*Book of the Camel*) described in detail the physiology of these animals. He also wrote books on sheep and wild animals, as well as a **book on human anatomy**.

As knowledge of Greek astronomy spread to the Islamic world, Ibrahim al-Fazari (d. *c.*796), an astronomer from Baghdad, wrote the **first Islamic treatise on the astrolabe**—a device that transferred observations of the celestial sphere onto flat plates and helped predict the location of celestial bodies.

Astrolabe
A Greek invention refined by Arab astronomers, the astrolabe helped perform complex astronomical calculations.

pivoted sighting rule

climate plate with coordinates to locate user's latitude

plate with star map

13 THE NUMBER OF **TEST SITES** SET UP BY **YI XING** FOR HIS **ASTRONOMICAL SURVEY**

In China, around 725, engineer and astronomer Yi Xing (683–727) invented **the first escapement** for a mechanical clock. The device was attached to an armillary sphere (a model of the celestial sphere) that was powered by water. It used a toothed gear to transfer energy to the moving parts of the sphere and to regulate their movement. Yi Xing also carried out a **major astronomical survey** to help predict solar eclipses more accurately and reform the calendar.

In India, mathematician and astrologer Lalla (*c.*720–790) became the first to **describe a perpetuum mobile**, a machine that once set in motion would carry on moving forever. His *Sisyadhivrddhidatantra* (*Treatise for Increasing the Intelligence of Students*) also gave details of **planetary movements, conjunctions, and eclipses**, although he rejected the idea that Earth rotated.

A few years later, in 762, the **city of Baghdad** was founded by the caliph al-Mansur. The **first planned city in the Islamic world**, its perfectly round shape was laid out by al-Naubakht, a Persian astrologer. His son, al-Fadl ibn Naubakht, founded the House of Wisdom in Baghdad, which became a major Islamic center for the study of science.

Jabir ibn-Hayyan (*c.*722–804) was an early Islamic alchemist who has become known as the father of Arab chemistry. He invented the alembic, an enclosed flask for heating liquids, established the **classification of substances** into metals and nonmetals, and identified the **properties of acids and alkalis**.

Jurjish ibn Bakhtishu was the first of a dynasty of **Islamic physicians** who served the Abbasid caliphs at Baghdad. He rose to prominence when he cured the caliph al-Mansur of a stomach complaint in 765. His grandson Jibril **founded the first hospital in Baghdad** some time after 805.

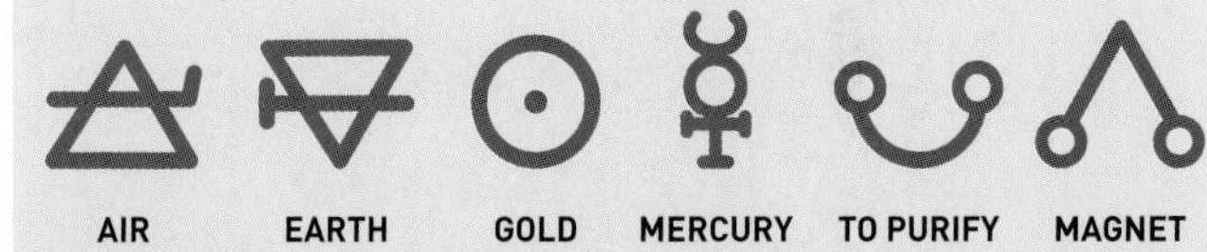

AIR
EARTH
GOLD
MERCURY
TO PURIFY
MAGNET

ALCHEMY

First developed in Hellenistic Egypt (4th–1st century BCE) by scholars such as Zosimos of Panopolis, alchemy was advanced further by Arab practitioners such as ibn-Hayyan and al-Razi in the 8th–9th centuries. It was concerned mainly with the transmutation of base metals, such as lead, into noble metals, such as gold, through the use of the "philosopher's stone." It led to the development of many practical chemical processes, such as distillation and fermentation.

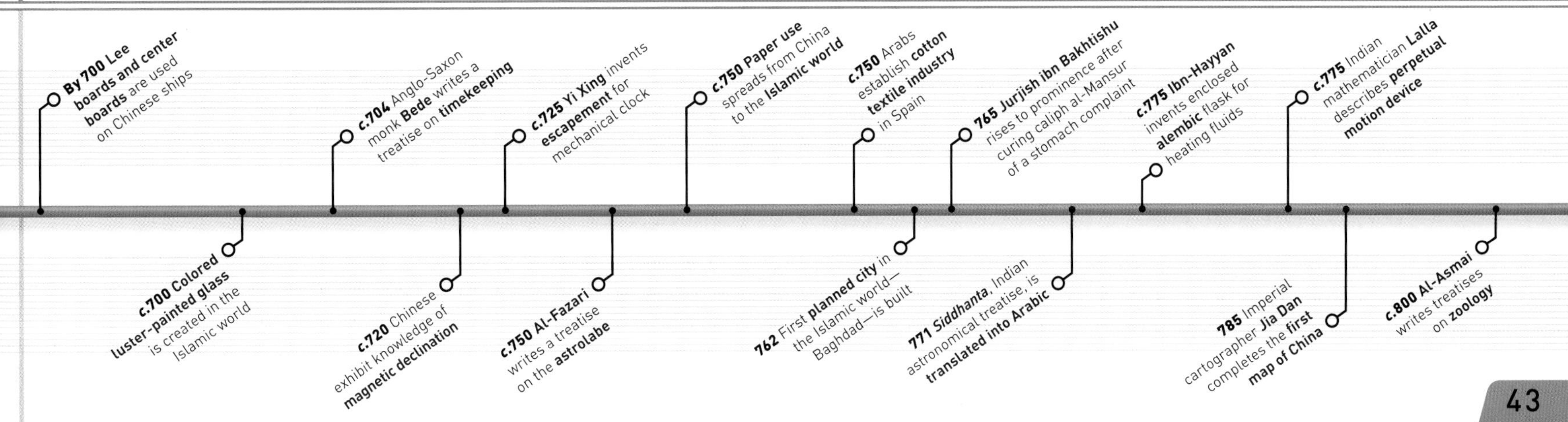

By 700 Lee boards and center boards are used on Chinese ships
***c.*700** Colored luster-painted glass is created in the Islamic world
***c.*704** Anglo-Saxon monk **Bede** writes a treatise on **timekeeping**
***c.*720** Chinese exhibit knowledge of **magnetic declination**
***c.*725** Yi Xing invents **escapement** for mechanical clock
***c.*750** Al-Fazari writes a treatise on the **astrolabe**
***c.*750** Paper use spreads from China to the **Islamic world**
***c.*750** Arabs establish **cotton textile industry** in Spain
762 First **planned city** in the Islamic world—Baghdad—is built
765 Jurjish ibn Bakhtishu rises to prominence after curing caliph al-Mansur of a stomach complaint
771 *Siddhanta*, Indian astronomical treatise, is **translated into Arabic**
***c.*775** Ibn-Hayyan invents enclosed **alembic** flask for heating fluids
***c.*775** Indian mathematician Lalla describes **perpetual motion device**
785 Imperial cartographer **Jia Dan** completes the **first map of China**
***c.*800** Al-Asmai writes treatises on **zoology**

2

THE EUROPEAN AND ISLAMIC RENAISSANCE

800–1542

Classical knowledge was revived and expanded by Islamic scholars attached to the mosques and the courts. Subsequently translated into Latin, their Arabic texts circulated through Western Europe and formed the basis of modern science

The House of Wisdom in Baghdad was a major center of Islamic scholarship, attracting the foremost thinkers from across the Islamic world.

THE ARABIC AND PERSIAN EMPIRES had a long tradition of scholarship, and this continued after the birth of the Islamic religion. Islam encouraged scientific and philosophical pursuits, which were not seen as incompatible with theology. **Libraries and other centers of learning** were established in many Islamic cities during the **Islamic "Golden Age."** Perhaps the greatest of these was the **House of Wisdom** (Bayt al-Hikma), founded in Baghdad at the beginning of the 9th century. As well as housing thousands of books, the House of Wisdom **encouraged research** and the **translation** of mathematical, scientific, and philosophical texts from ancient Greece.

Persian mathematician and astronomer Muhammad ibn Musa **al-Khwarizmi** (*c.*780–850) was one of the most important scholars at the House of Wisdom, studying both Greek and Indian scientific treatises. In around 820, he **described the use of the astrolabe**—an instrument used to observe the position of stars. Although not the first to produce a work on the astrolabe, al-Khwarizmi's contribution was significant, especially in the Islamic world, where the astrolabe could be used to calculate the time of daily prayers.

The Chinese were pioneers in the technology of **printing**, largely due to their invention of paper—possibly as early as the 2nd century BCE—which lent itself better to printing than the papyrus and parchment used elsewhere. Developing a form of **woodblock printing on silk** that had appeared around 200, they applied the technique to paper and used it for the mass production of books. By the 9th century, it was being used to **print promissory notes** that were in effect a form of paper money issued by the Chinese government.

400 THOUSAND THE NUMBER OF **BOOKS** IN THE **HOUSE OF WISDOM**

"FONDNESS FOR SCIENCE... HAS ENCOURAGED ME TO COMPOSE A SHORT **WORK ON...** WHAT IS **EASIEST** AND **MOST USEFUL** IN **ARITHMETIC."**

Al-Khwarizmi, Persian mathematician, *c.*780–850

ONE OF AL-KHWARIZMI'S MAJOR ACHIEVEMENTS was a treatise on mathematics entitled *The Compendious Book on Calculation by Completion and Balancing*, published around 830. It contained a description of the branch of mathematics now known as **algebra**. Although he drew on sources such as Greek and Indian texts (see 250–500), he is considered to be the inventor of algebra. In his book, al-Khwarizmi explained the process of **balancing both sides of an equation** (*al-jabr* in Arabic, hence the modern term algebra), and gave a systematic way of **solving quadratic equations**, which had been described almost 500 years earlier by Greek mathematician Diophantus of Alexandria. Central to his method was the principle of balancing an equation by transposing terms from one side to the other and canceling out terms that appear on both sides.

Another prominent scholar at the House of Wisdom was the polymath **Abu Yusuf Ya'qub ibn 'Ishaq al-Kindi** (also known as al-Kindi), who in the mid-9th century wrote a large number of **treatises on various scientific subjects**, ranging from mathematics, astronomy, and optics, to medicine and geography. A scholar of theology and philosophy, he was also responsible for the **translation of many classic Greek texts** and their incorporation into Islamic thinking. It is largely through al-Kindi's translations and commentaries on Indian texts that **Indian numerals were introduced to the Islamic world**, and subsequently became the **basis for the modern system of numbers,** although zero was probably "discovered" later (see 861–99).

Al-Kindi was very **sceptical about alchemy**, refuting one of its central ideas—the

$$ax^2 + bx + c = 0$$

ALGEBRA

Algebra is a branch of mathematics that uses letters to represent unknown quantities (called variables), and symbols for operations such as addition and subtraction. These can be combined in an algebraic statement known as an expression, such as "a + 3". A mathematical statement, such as "a + 3 = 7", is known as an equation. Equations in which the highest power of an unknown quantity is two are known as quadratic equations (as above), and those in which the highest power is three are called cubic equations.

AL-KINDI (*c.*801–873)

Born and educated in Kufa, near Baghdad, Al-Kindi was one of the first major scholars of the newly founded House of Wisdom. He translated Greek scientific and philosophical texts into Arabic, and incorporated Hellenistic ideas into Islamic scholarship. He wrote treatises on many subjects, including medicine, chemistry, astronomy, and mathematics.

c.810 The **Bayt al-Hikma** (House of Wisdom) is **founded** in Baghdad

812 The Chinese government issues a form of **paper money**

c.820 Al-Khwarizmi describes the **astrolabe**

c.830 Al-Khwarizmi describes **algebra** in his Compendious Book on Calculation by Completion and Balancing

c.850 Al-Kindi writes treatises on optics, perspective, medicine, and cryptography

855 Chinese alchemists describe the **discovery of gunpowder**

This statue of Al-Khwarizmi stands in Khiva, Uzbekistan, his birthplace.

The Chinese edition of the Buddhist *Diamond Sutra*, printed using woodblocks on a scroll of paper, is the earliest surviving printed book.

transmutation of metals. However, alchemy was at the root of another discovery—this time in China. In the early 9th century, Chinese alchemists were experimenting with various mixtures of substances to find the "elixir" of life. One of the by-products of this quest was the **discovery**, in about 855, of **gunpowder**—the first man-made explosive. It consisted of a mixture of sulfur, carbon (in the form of charcoal), and saltpeter (potassium nitrate) – all of which occur naturally as minerals. The mixture's explosive properties meant that it was initially used in the **manufacture of fireworks**, but gunpowder later came to fuel rockets, and was eventually used in the development of firearms.

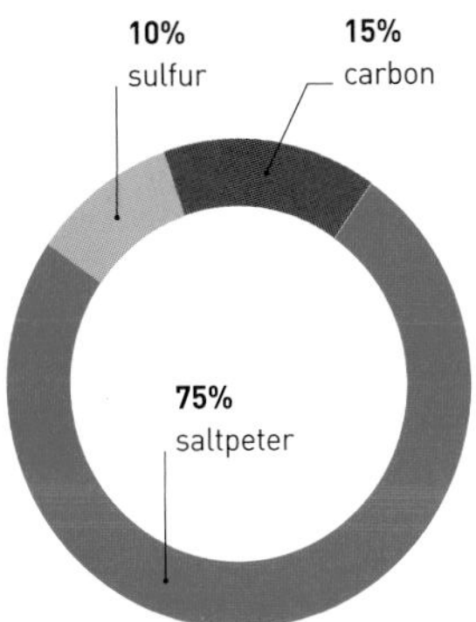

10%
sulfur
15%
carbon
75%
saltpeter

Composition of gunpowder
Sulfur, carbon, and saltpeter, while quite innocuous individually, become highly explosive when mixed in the correct proportions.

A CHINESE EDITION OF THE BUDDHIST TEXT, *Diamond Sutra*, was discovered in 1907 in Dunhuang, northwest China. Although it is probably not the first example of a **woodblock printed book**, it is the earliest known one, and bears the date May 11, 868. The text and illustrations of *Diamond Sutra* exhibit a great deal of sophistication, suggesting that the techniques of printing on paper were well known in China by this time. An inscription at the end of the manuscript indicates that this was one of a number of copies printed for distribution.

An inscription on a stone in Gwalior, India, dated 876, contains one of the earliest known uses of the **symbol for zero—"0"**. Prior to the appearance of a specific symbol, a space was used to indicate zero, which led to ambiguity and prevented the development of a place value system of numbers (a system in which the position of the numeral indicates its value). The introduction of a symbol for zero in Indian mathematics was a vital step in the development of the **decimal system of notation** we use today. This decimal system came to Europe through the influence of Islamic mathematicians, and eventually replaced the use of cumbersome Roman numerals.

Alchemist Jabir ibn-Hayyan at work
Alchemy in the Islamic world involved much experimentation, and led to the development of many processes that were later used in chemistry.

Toward the end of the century, Arab alchemists developed the process of **distillation**—a method of separating the ingredients of a liquid mixture. **Muhammad ibn Zakariya al-Razi** (*c.*854–925/35), along with other alchemists, perfected the technique and was successful in extracting a form of alcohol—ethanol or ethyl alcohol—by distilling wine. The word alcohol derives from the Arabic *al kuhl*, originally used to describe a powder extracted from a mineral, but which later came to mean the essence or "spirit" of a liquid. The apparatus developed by al-Razi for distillation has remained fundamentally unchanged to the present day.

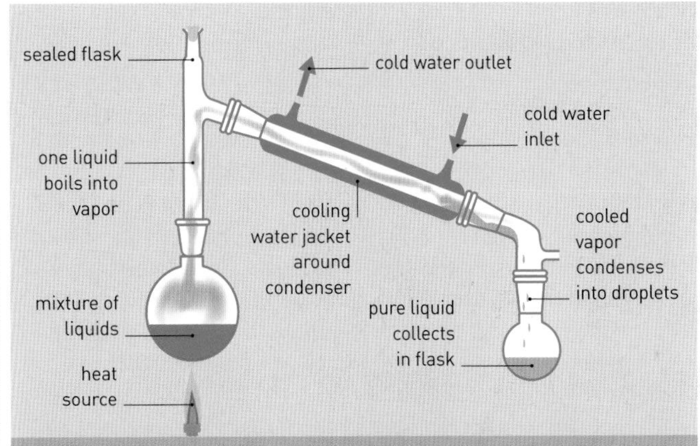

sealed flask
cold water outlet
cold water inlet
one liquid boils into vapor
cooling water jacket around condenser
cooled vapor condenses into droplets
mixture of liquids
pure liquid collects in flask
heat source

DISTILLATION

Distillation is a method of separating the components of a liquid mixture. The liquid mixture is converted into vapor by heating. As the components of the mixture have different boiling points, they vaporize at different rates. The vapor is then cooled so that it condenses back into a liquid, which can be collected separately. Distillation can be used to extract liquids such as alcohol and gasoline, and also to purify liquids, such as salt water.

868 The earliest surviving printed book, the ***Diamond Sutra***, is printed in China

876 Indian mathematicians use a **symbol for zero**

c.890 Al-Razi distils alcohol from wine

“ TRUTH IN MEDICINE IS AN UNATTAINABLE GOAL, AND THE ART AS DESCRIBED IN BOOKS IS FAR BENEATH THE KNOWLEDGE OF AN EXPERIENCED AND THOUGHTFUL PHYSICIAN. ”

Al-Razi, Arab physician, 10th century

Arab doctor and chemist al-Razi's belief in practical experimentation on substances led him to propose an early classification of elements.

MUHAMMAD IBN ZAKARIYA AL-RAZI (Rhazes) was one of the greatest physicians of the Arab world. Around 900, he wrote *Al-Shukuk ala Jalinus* (*Doubts About Galen*), in which he **criticized Galen's theory of the four humors** (see 75–250). He rejected the notion that a balance of these humors was necessary for the health of the patient, and dismissed the idea that body temperature is automatically raised or lowered when a patient drinks warm or cold fluids. His clinical practices were advanced for this time; he ran a psychiatric ward, and he wrote a treatise attacking untrained physicians. The *Kitab al-Hawi* (*Comprehensive Book*), a collection of his clinical notes, ran to 23 volumes, and contained **medical diagnoses**, including the first description of **hay fever** (or rose-cold). He also wrote a monograph, *Kitab al-Judwar wal Hasba* (*Treatise on the Smallpox and Measles*), which was the first work to detail the symptoms of smallpox, although his explanation—that the disease was caused by the impurities from menstrual blood that stay in the fetus during pregnancy and then bubble up to the skin in later life—betrayed a belief in sympathetic magic. He was particularly concerned with **preventing blindness caused by smallpox pustules**, and advocated regularly bathing the eyes in rose-water.

An alchemist as well as a physician, al-Razi devised a **classification of elements** into spirits and metals and minerals. He divided the latter into stones, vitriols, boraces, salts, and other substances, and gave a detailed account of the behavior of each under various processes, such as melting and extraction. He described the **distillation of kerosene and petroleum** from crude oil and gave recipes for preparing **hydrochloric and sulfuric acids**.

Around 920, Arab astronomer and mathematician **al-Battani** (*c.*858–929) proffered greater insights into the working of the **planispheric astrolabe**—a device with a number of overlapping plates for making astronomical observations. Although al-Fazari was the first to describe it in the 8th century, al-Battani worked out the mathematics underlying the instrument. He presented formulas in spherical trigonometry, replacing Ptolemy's geometrical methods.

AL-RAZI (*c.*865–925)

Born in Rayy, Mesopotamia (now in Iran), al-Razi was a physician and philosopher, as well as an alchemist. He encouraged experimentation as a means of discovery and his clinical notes became a key medieval medical text. He headed a hospital in Rayy, and then two in Baghdad. Among his innovations was the first recorded clinical trial—on patients with meningitis.

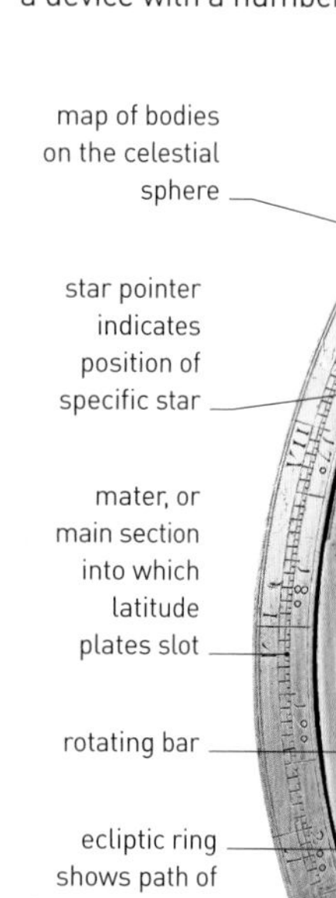
map of bodies on the celestial sphere
star pointer indicates position of specific star
mater, or main section into which latitude plates slot
rotating bar
ecliptic ring shows path of Sun through sky

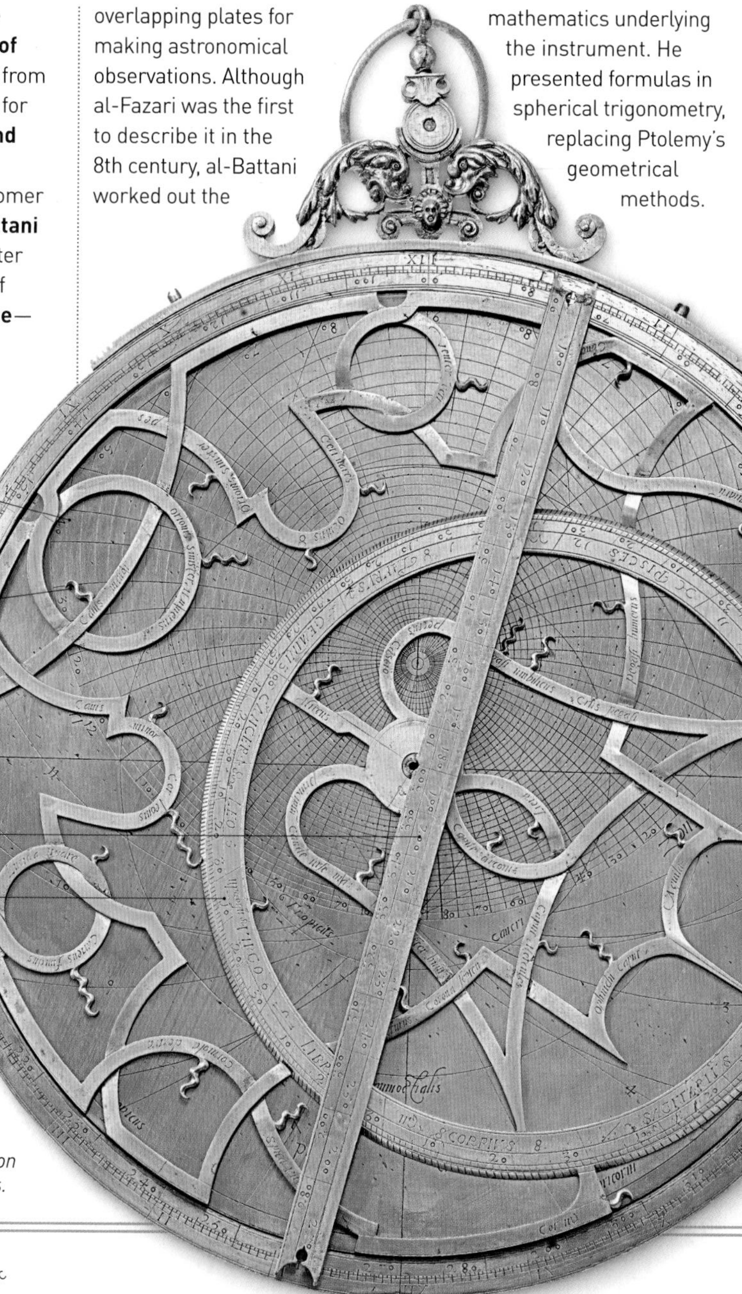

Astrolabe
The user of an astrolabe adjusted its moveable parts to indicate a specific date or time, and the markings on the plates would then indicate the position of the various heavenly bodies.

c.900 Islamic mathematician Abu Kamil develops al-Khwarizmi's **algebra**, dealing with **powers greater than 2**

c.912 Qusta ibn Luqa writes a **treatise on numbness**

900–30 Al-Razi describes symptoms of **smallpox**

900–30 Al-Razi criticizes Galenic **humor theory**

c.920 Al-Battani discovers **mathematical principles** underlying the astrolabe

c.925 Al-Farabi writes treatise on **music therapy**

c.927 Earliest dated surviving **astrolabe** is built by Islamic astronomer Nastulus

200 THE APPROXIMATE NUMBER OF **NEW SURGICAL INSTRUMENTS** INTRODUCED BY AL-ZAHRAWI

A 14th-century manuscript shows two innovative surgical instruments introduced by Spanish–Arab physician al-Zahrawi (Albucasis).

THE MODERN NUMERALS for expressing the **decimal system** first appeared in Europe in 976 in a treatise written by the monk Vigila in the northern Spanish convent of Albelda (although he included only the symbols for 1 to 9 and not zero). This number system, now known as Hindu–Arabic, had originated in the numerical notation of the Brahmi script used in India in the mid-3rd century BCE, which then spread westward after the Arabs came in contact with India in the early 8th century CE.

Abacus
This is a modern example of an abacus, a counting device that appeared in Mesopotamia around 2700 BCE. It was introduced to medieval Europe by Gerbert in around 990 CE.

Although **mechanical armillary spheres and mechanical escapements for clocks** had been devised in China by Zhang Heng in the 2nd century and Yi Xing in the 8th century, a superior version was constructed by the astronomer **Zhang Sixun** in 979. It was powered by a waterwheel with scoops that deposited the liquid into a clepsydra (a device that measures time by the flow of liquid through a small hole) as it turned around, which in turn regulated the measurement of the hours. In order to avoid the liquid freezing in winter—a problem that had afflicted earlier such clocks—Yi Xing had substituted **mercury** for water. Zhang Sixun's improved version made one complete revolution each day, with each quarter-hour and hour sounded out by mechanical jacks that emerged to strike bells and drums, or to display the time on a tablet. The clock also showed the **position of the Sun, Moon, and five planets on a celestial globe**, and was said to be so advanced that after Sixun's death no one could keep it in working order.

In 984, Persian mathematician **Ibn Sahl** (*c.*940–1000) wrote *On the Burning Instruments*, a treatise in which he examined the bending of light by lenses and curved mirrors. He was the first to express a **geometric theory of refraction**. He suggested that the amount of light that is deflected when it enters another medium (such as glass) varies, depending on the refractive index (see 1621–24) of the substance.

Christian monastic scholar **Gerbert** (c.943–1003), who became Pope in 999, was one of the **first Western European mathematicians** of the Middle Ages. He sought to recover mathematical and astronomical treatises by ancient scholars such as Boethius, studied the work of Islamic mathematicians, and **introduced the abacus to Europe**, giving instructions for its use in multiplication and division.

The greatest Arab surgeon of medieval times was **Abu al-Qasim al-Zahrawi** (c.936–1013), also known as Albucasis. He was court surgeon to al-Hakam, the Umayyad Caliph of Cordoba. His *Kitab al-Tasrif* (*The Method of Medicine*)—containing detailed descriptions of **human anatomy** and the **pathology of diseases**—became the main textbook for medieval European physicians.

> “...HE WHO DEVOTES HIMSELF TO **SURGERY** MUST BE VERSED IN... **ANATOMY.**”
>
> **Al-Zahrawi (Albucasis)**, in *Kitab al-Tasrif*, *c.*990

Babylonian	Ancient Egyptian	Ancient Greek	Ancient Roman	Ancient Chinese	Mayan	Modern Hindu-Arabic
𒁹	|	α	I	一	•	1
𒈫	||	β	II	二	••	2
𒐈	|||	γ	III	三	•••	3
𒐉	||||	δ	IV	四	••••	4
𒐊	|||||	ε	V	五	—	5
𒐋	||||||	ϛ	VI	六	• —	6
𒑂	|||||||	ζ	VII	七	•• —	7
𒑄	||||||||	η	VIII	八	••• —	8
𒑆	|||||||||	θ	IX	九	•••• —	9
𒌋	∩	ι	X	十	=	10

DEVELOPMENT OF NUMBERS

Many early number systems, such as the Egyptian, were additive—the value of the number symbol did not depend on its position; to make 20, the symbol for 10 would be written twice. Around 2000 BCE, the Babylonians began to use a partly positional system—where the order of magnitude depends on the position in which the symbol appears. Positional systems using 10 as the base developed in India, and gradually evolved into the modern Hindu–Arabic numerals.

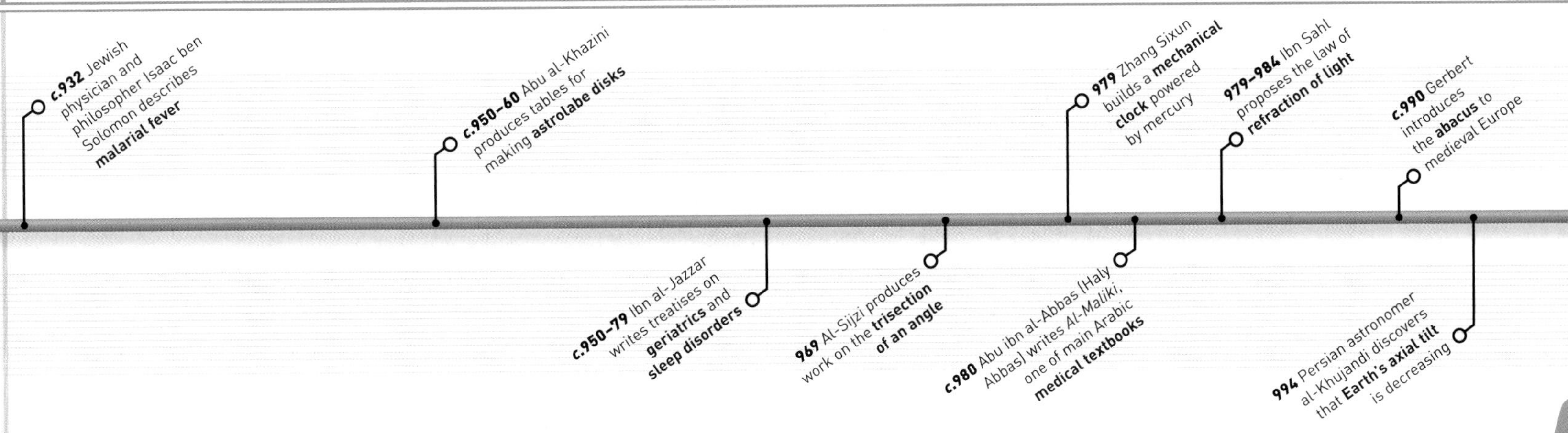

c.932 Jewish physician and philosopher Isaac ben Solomon describes **malarial fever**
c.950–60 Abu al-Khazini produces tables for making **astrolabe disks**
c.950–79 Ibn al-Jazzar writes treatises on **geriatrics and sleep disorders**
969 Al-Sijzi produces work on the **trisection of an angle**
979 Zhang Sixun builds a **mechanical clock** powered by mercury
c.980 Abu ibn al-Abbas (Haly Abbas) writes *Al-Maliki*, one of main Arabic **medical textbooks**
979–984 Ibn Sahl proposes the law of **refraction of light**
c.990 Gerbert introduces the **abacus** to medieval Europe
994 Persian astronomer al-Khujandi discovers that **Earth's axial tilt** is decreasing

1000–1029

> **“** NOW IT IS **ESTABLISHED** IN **THE SCIENCES** THAT NO **KNOWLEDGE** IS ACQUIRED SAVE THROUGH THE **STUDY** OF ITS **CAUSES** AND **BEGINNINGS. ”**
>
> **Ibn Sina (Avicenna), Arab polymath,** from *Canon of Medicine*, c.1005

A page from Avicenna's *Canon of Medicine* shows the heart and skull as part of an illustration explaining the theory of the four humors.

AROUND 1005, THE ARAB MEDICAL SCHOLAR and polymath **Ibn Sina** (known as Avicenna in Europe) wrote the *Canon of Medicine*, a major compendium that sought to provide a systematic understanding of medical knowledge of the time. Avicenna tried to **reconcile theories** of four humors (blood, yellow bile, black bile, and phlegm; see 100–250) **with Aristotle's idea** of three life forces (psychic, natural, and human). Avicenna's careful and comprehensive account in **five volumes** of physiology, diagnosis, therapy, the pathology of diseases, and pharmacology made it an **extremely valuable medical handbook**. It was commented on by many subsequent Arabic physicians and was printed as Latin translations 36 times.

The Persian astronomer and mathematician **Abu Sahl al-Quhi** (*c.*940–1000) was **head of the observatory** founded by Sharaf al-Dawla in Baghdad in 988, but he was particularly noted for his work in **solving equations of greater than the second degree** (in which the highest power of a variable is more than two). He employed a geometrical method of intersecting curved lines to achieve this. Around 1000, he wrote *On the Construction of an Equilateral Pentagon in a Known Square*, in which he demonstrated the solution by solving an equation to the fourth degree.

In 1005, the Fatimid caliph **al-Hakim** founded the **House of Knowledge (Dar al-'ilm)** in Cairo. Equipped with a vast library covering subjects ranging from Islamic philosophy and law to physics and astronomy, it became **a center for philosophers and theologians**. At first, the House of Knowledge hosted a series of public lectures, but these ended in 1015 after fears that religious dissidents were establishing a presence there.

The Arabic sage **Abu 'Ali ibn al-Haytham** (*c.*965–1039), also known as **Alhazen**, is best known for his *Kitab al-Munazir* (*Book of Optics*), which he wrote between 1011 and 1021. He proposed that the blinding effect of bright light and the existence of after-images proved that **vision was caused by light coming into the eye**. He also developed a new theory of the eye's physiology, describing it as being made up of various **humors** and separated into sections by **spherical sheaths**.

HOW LENSES WORK

A convex lens is thicker in the middle than at the sides. When light rays strike the lens they are diffracted and converge behind the lens at a single point called the principal focus. Convex lenses are used to treat far-sightedness because they bring objects that are closer, into focus. A concave lens is thinner in the middle than at the sides. Light rays diverge and seem to focus in front of the lens. These lenses address near-sightedness.

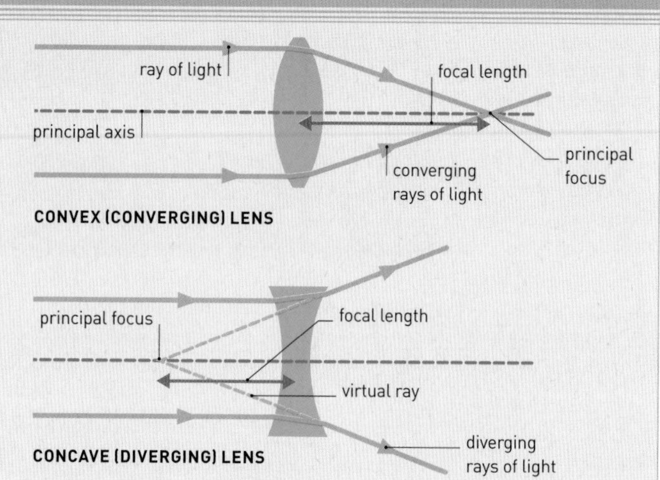

ray of light
focal length
principal axis
converging rays of light
principal focus
CONVEX (CONVERGING) LENS
principal focus
focal length
virtual ray
diverging rays of light
CONCAVE (DIVERGING) LENS

Alhazen's eye
Shown here is a diagrammatic eye from a 1575 Latin translation of Alhazen's Book of Optics.

IBN SINA (980–1037)

Born near Bukhara, Uzbekistan, Ibn Sina (Avicenna) was a medical prodigy. He claimed to have successfully treated patients by age 16. He served the Samanid rulers of Bukhara, but their overthrow in 999 led to his exile. He ended up at the court of Shams al-Dawla of Hamadan, where he wrote his great *Canon of Medicine*.

1000 Al-Quhi describes the **solution of equations beyond the second degree**

1004 Caliph al-Hakim founds the **House of Knowledge** in Cairo

1005 Ibn Sina (Avicenna) composes the Canon of Medicine, **a medical compendium**

1006 Muslim astronomer Ali ibn Ridwan gives **first description of a supernova**

1011–21 Alhazen proposes his theory of vision in ***Book of Optics***

1015 Arab physician Maswijah al-Marindi **prescribes the gem electuary**—a paste made of ground gems—to cure melancholia

The earliest movable type in China was made of clay, and later of wood. Metal movable type such as these blocks did not become common until the Ming dynasty in the 17th century.

IN THE EARLY 11TH CENTURY, Spanish–Arab astronomer Abu **Abdallah ibn Mu'adh al-Jayyani** (989–1079) carried out work integrating trigonometry and optics. His *Book of Unknown Arcs of a Sphere* was the first comprehensive work on spherical trigonometry. Around 1030, al-Jayyani used this work in his *Book on Twilight* to calculate the angle of the Sun below the horizon at the end of evening twilight to be 18 degrees. By taking this as the lowest angle at which the Sun's rays can meet the upper edge of the atmosphere, he worked out the height of Earth's atmosphere as 64 miles (103 km).

Printing using carved wooden blocks had appeared in China around the 6th century, but the process was cumbersome, requiring a new block to be carved for each individual page. Around 1040, a commoner named **Pi Sheng** developed **a form of movable type** by creating thin strips of clay, each impressed with a single character, which he baked in a fire. He then placed these on an iron tray to compose the page to be printed. The clay letters could be rearranged as desired to create a new page. The method fell into disuse after Pi Sheng's death until its revival in the mid-13th century. By then, far **more durable type made of iron** had been invented in Korea, where it was **first used in 1234**.

The Chinese had understood the properties of **magnetic lodestones** in transferring polarity to a needle several centuries earlier (see 300–250 BCE), but no real application was made. In 1044, the first mention is made of a "south-pointing carriage" used to find directions on land during gloomy weather using an "iron fish." The needle of **this early compass** probably floated on top of a bowl of water and the technique was **later adapted for navigation at sea**. One document referring to the period around 1086 tells of a "south-pointing needle" used for finding bearings at night. In 1123, an account of a diplomatic mission to South Korea describes the sailors' use of the compass. It would be another 67 years, however, before such knowledge spread to Europe.

Crossbows had made an appearance in China as early as the 8th century BCE and are recorded in Greece in the early 3rd century BCE. **Hand-held crossbows** came into use in France in the 10th century, but their power was limited by the ability of the user to pull back the bowstring by hand. By the mid-11th century, **a stirrup** was placed at the end of the stock, so that the user could push against this with his legs while pulling the string back. **Mechanical cranks** were also invented that could be turned to tighten the string. By the early 13th century, **complex windlasses** (contraptions used to move heavy objects) were devised, which imparted high tensile strength to the crossbow bolt.

NUMBER OR ARROWS PER MINUTE

12, 10, 8, 6, 4, 2, 0

Longbow Crossbow

WEAPON

The deadliest weapon?
Medieval crossbowmen fired at as little as a tenth the rate of longbow archers, although their bolts had more power.

Anatomy of a crossbow
This 16th-century German crossbow could not be used without a crannequin—a toothed wheel attached to a crank—which was used to bend the crossbow.

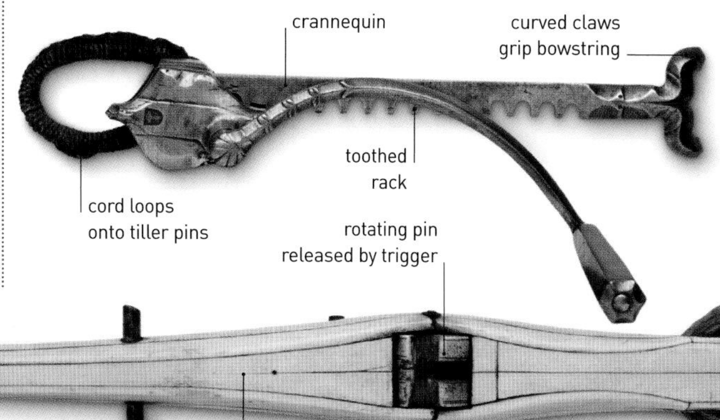
crannequin
curved claws grip bowstring
toothed rack
cord loops onto tiller pins
rotating pin released by trigger

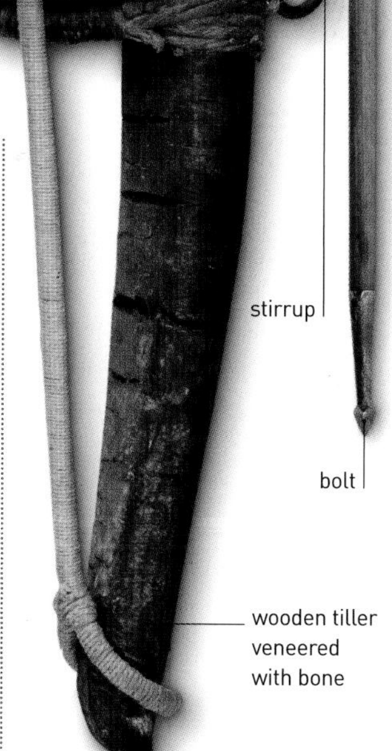
flight
steel pin to engage spanning mechanism
composite lathe of bone, sinew, and wood
stirrup
bolt
wooden tiller veneered with bone

c.1030 Persian astronomer Al-Biruni suggests that **Earth may rotate around the Sun**, but says he cannot prove it

c.1030 Guido d'Arezzo devises **new theory of musical notation** and develops system of hexachords

c.1030 Al-Nasawi summarizes Euclid's *Elements* and describes method for extraction of **cube roots**

c.1040 Pi Sheng invents **movable type for printing** using clay blocks

1044 First mention of **magnetic compass used for navigation**, in China

c.1050 Ibn Butlan writes *Taqwim al-sihha* treatise emphasizing **importance of good diet and hygiene**

“THEN WAS ALL OVER ENGLAND **SUCH A TOKEN** SEEN AS **NO MAN** EVER **SAW BEFORE...**”

From *The Anglo-Saxon Chronicle*, describing the comet of 1066

IN THE 11TH CENTURY, Chinese mathematician **Jia Xian** described a method of **calculating square and cubic roots** using numbers arranged in rows. Each row contained one more number than the row above it, to form a triangle in which each number is the sum of the two directly above it. Known as the **Jia Xian triangle**, it is also often referred to in the West as Pascal's triangle, after French mathematician Blaise Pascal, who described it 600 years later.

In 1054, the massive explosion of a **supernova** (which formed what we now know as the Crab Nebula) was visible from Earth. It was **observed by Arabic and Chinese astronomers**, who described it as a "guest star," but its significance was not realized by observers in Europe.

In 1066, the comet now called **Halley's Comet** made one of its regular 76-year periodic appearances and was described by European astronomers. Astrologers viewed the comet as an omen, and found it especially significant in the year of the Norman invasion of England.

22 THE NUMBER OF MONTHS THAT THE **SUPERNOVA** WAS **VISIBLE** FROM **1054** TO **1055**

HAROLD

The Bayeux tapestry
This embroidered record of the events surrounding the Battle of Hastings in 1066 shows the appearance of Halley's Comet.

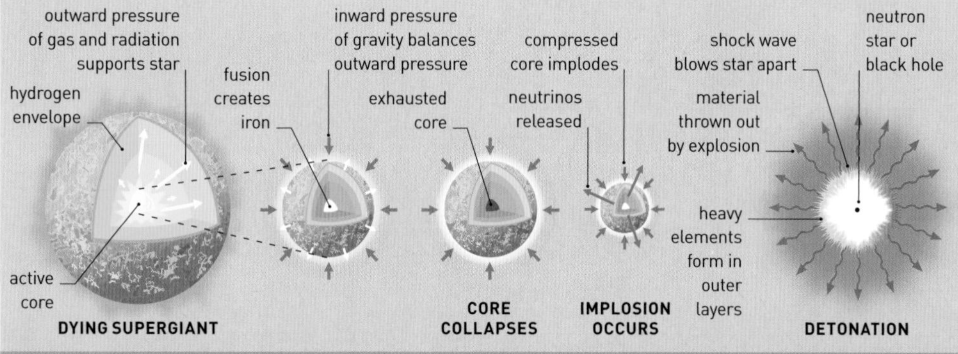
outward pressure of gas and radiation supports star
hydrogen envelope
fusion creates iron
inward pressure of gravity balances outward pressure
exhausted core
compressed core implodes
neutrinos released
shock wave blows star apart
neutron star or black hole
material thrown out by explosion
heavy elements form in outer layers
active core
DYING SUPERGIANT
CORE COLLAPSES
IMPLOSION OCCURS
DETONATION

FORMATION OF A SUPERNOVA

A supernova is an explosion of a massive supergiant star at the last stage of its life. Over a long period of time, a star builds up a core of iron, which eventually collapses in on itself as the star runs out of fuel for fusion. This results in an implosion that rapidly reheats the star and restarts the process of fusion. Subatomic particles called neutrinos are released as implosion occurs. Now out of control, the star explodes with a huge amount of energy—billions of times more than the Sun, which is also a star—shining brighter than other stars and scattering debris in all directions, over vast distances.

c.1050 Jia Xian describes the **Jia Xian triangle**, later known as the Pascal triangle

1054 Chinese and Arab astronomers **observe the supernova** that forms the Crab Nebula

1066 The comet later known as **Halley's Comet** is sighted

> “BY THE **HELP OF GOD** AND WITH HIS PRECIOUS ASSISTANCE, I SAY THAT **ALGEBRA** IS A **SCIENTIFIC ART.**”

Omar Khayyam, from *Treatise on Demonstration of Problems of Algebra*, 1070

This manuscript is one of the many treatises that Omar Khayyam wrote on mathematics, astronomy, mechanics, and philosophy.

PERSIAN MATHEMATICIAN AND ASTRONOMER Omar Khayyam began work on his ***Treatise on Demonstration of Problems of Algebra*** in 1070, the year that he moved to Samarkand, Uzbekistan, and devoted himself to study and writing. In it, he gave a complete **classification of the types of cubic equation** (an equation involving a term to the power of three, such as $x+y^3=15$) and described for the first time a general theory for solving them using geometry. The method he used involved the use of conic sections and curves. He realized that some equations, such as quadratic (involving a squared term) and cubic equations, had more than one solution.

Khayyam was also an accomplished astronomer: from 1073 he worked at the observatory in Isfahan, Iran. Much of his work was concerned with the compilation of **astronomical tables**, but he also helped improve the accuracy of the calendar.

At Isfahan, he also worked on his poetry, later collected in ***The Rubaiyat of Omar Khayyam***. In 1079, he calculated the **length of a year** as 365.24219858156 days—a greater degree of precision than ever before, and remarkably close to the modern measurement of 365.242190 days. This led to the introduction of a **new calendar in the Islamic world**, which was more accurate than the Julian calendar used in Europe at the time.

Meanwhile, in China, the polymath **Shen Kuo** retired from a successful career as a civil servant and military leader in the court of the Song dynasty, and devoted his time to study. He wrote an extraordinarily wide-ranging collection of essays on subjects as diverse as politics, divination, music, and the sciences. The *Dream Pool Essays*, named after his garden estate, were finished in 1088, and included an overview of the sciences of the time, as well as some innovative ideas. For example, Shen was the first to give a description of the **magnetic compass needle**. He explained how it could be used in navigation to determine the direction of North. He also contributed to the fields of paleontology and geology. Describing the discovery of the remains of **marine creatures** in the strata of a cliff hundreds of miles from the coast, he suggested that these must have been covered by silt over a long period of time—which would have been later eroded—and also proposed that the cliff must have at some time been a coastal area. He described **fossilized bamboo** unearthed by a landslide, in an area where bamboo does not grow, and came to the conclusion that this was the remains of an ancient forest from a time when the climate of the area had been significantly different.

OMAR KHAYYAM (1048–1131)

Born in Persia (now Iran), Omar Khayyam showed a talent for astronomy and mathematics at an early age; he wrote many of his treatises before he was 25 years old. In 1073, he was invited by Sultan Malik-Shah to set up an observatory in Isfahan. Here, he worked on calendar reform and astronomical tables, before returning to his home town.

Bamboo
Shen's discovery of fossilized bamboo in a cool, dry area led him to conclude that the region would have been warm and humid in the past.

> “UNDER THE GROUND, A **FOREST OF BAMBOO SHOOTS** WAS REVEALED… THESE WERE SEVERAL DOZENS OF FEET **BELOW** THE **PRESENT SURFACE** OF THE **GROUND.**”

Shen Kuo, from *Dream Pool Essays*, 1088

1070 Omar Khayyam begins writing ***Treatise on Demonstration of Problems of Algebra***

1079 Omar Khayyam calculates the **length of a year** and enables **calendar reform**

1088 Shen Kuo completes ***Dream Pool Essays***

UNDERSTANDING STARS

MASSIVE BALLS OF HOT, IONIZED GAS, STARS ARE POWERED BY NUCLEAR REACTIONS

Our galaxy contains hundreds of billions of stars—and there are hundreds of billions of galaxies, each containing similar numbers of these huge balls of plasma (hot, ionized gas). A star glows because it is hot, and most of the heat is generated by nuclear reactions in the star's core.

HANS BETHE
In the 1930s, German-born physicist Hans Bethe (1906–2005) worked out how nuclear fusion builds elements inside stars, for which he was awarded the 1967 Nobel Prize in Physics.

Around 6,000 stars are visible to the naked eye in the night sky. Apart from the Sun, they are so far away that, despite their enormous size, they appear only as tiny points of light, even through powerful telescopes.

THE SUN IS A STAR

The Sun is by far the closest star: the light and other radiation it produces takes eight minutes to reach Earth, compared with over four years from the next nearest star. Like other stars, the Sun is composed mostly of hydrogen and helium, with small amounts of other elements. Its luminous surface (photosphere) is white hot, with a temperature of about 10,000°F (5,500°C), and its outer atmosphere, the corona, is much hotter. The Sun is about 5 billion years old, and is about halfway through its life cycle.

109

THE NUMBER OF **TIMES GREATER** THE SUN'S DIAMETER IS **COMPARED** WITH THE **EARTH'S**

LIFE CYCLES OF STARS

Stars form in huge masses of gas and dust called molecular clouds. Gravity causes matter in denser regions of these clouds to clump together to form protostars. This gravitational collapse produces heat, which causes atoms to lose electrons, becoming ions, so the matter in the protostar becomes plasma—a mixture of ions and electrons. At the protostar's center, the high temperature and pressure cause nuclei of hydrogen atoms to fuse together to form nuclei of helium and some heavier elements. This nuclear fusion reaction releases energy, which heats the protostar further: a star is born. When the hydrogen runs out, nuclear fusion ends, and the star cools and collapses under its own gravity. A star's final destiny depends upon its mass; the most massive stars end up as black holes (see opposite).

STAR BIRTH
The molecular cloud in the Carina Nebula (part of which is shown in this image from the Hubble Space Telescope) is one of the largest known regions of star birth in our galaxy, the Milky Way.

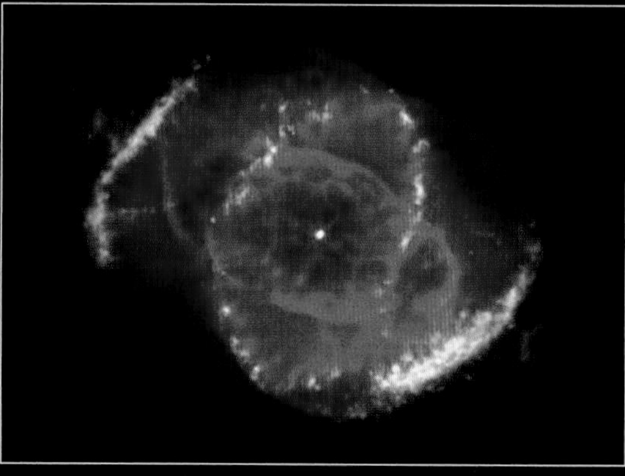

STAR DEATH
As stars of low to intermediate mass near the ends of their lives, they eject haloes of hot gas, forming objects known as planetary nebulae. At the center of each such nebula is a small remnant of the once much larger star, called a white dwarf.

STAR SIZES

Stars come in a huge variety of sizes. Supergiants, among the largest stars, can be over 1,500 times bigger than the Sun. The Sun itself has a diameter of about 870,000 miles (about 1.4 million km)—roughly average for a star in the main part of its life. The smallest stars, neutron stars, are only about 12.5 miles (20 km) across.

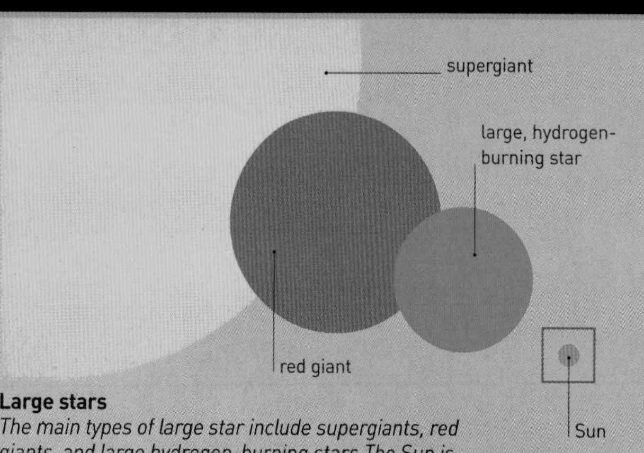
supergiant
large, hydrogen-burning star
red giant
Sun

Large stars
The main types of large star include supergiants, red giants, and large hydrogen-burning stars. The Sun is an average-sized hydrogen-burning star.

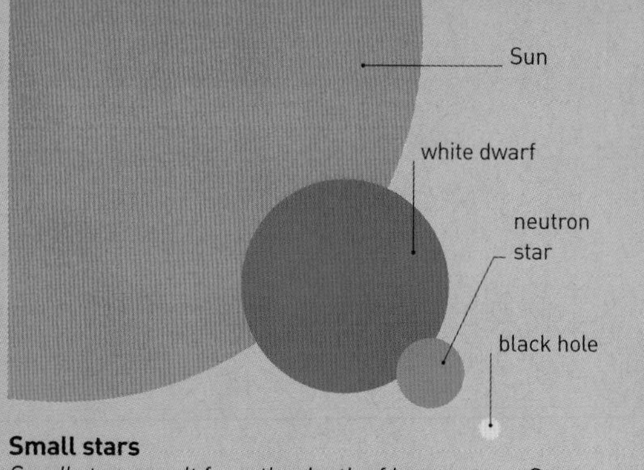
Sun
white dwarf
neutron star
black hole

Small stars
Small stars result from the death of larger stars. Stars like the Sun become white dwarfs, while more massive stars become tiny neutron stars or even black holes.

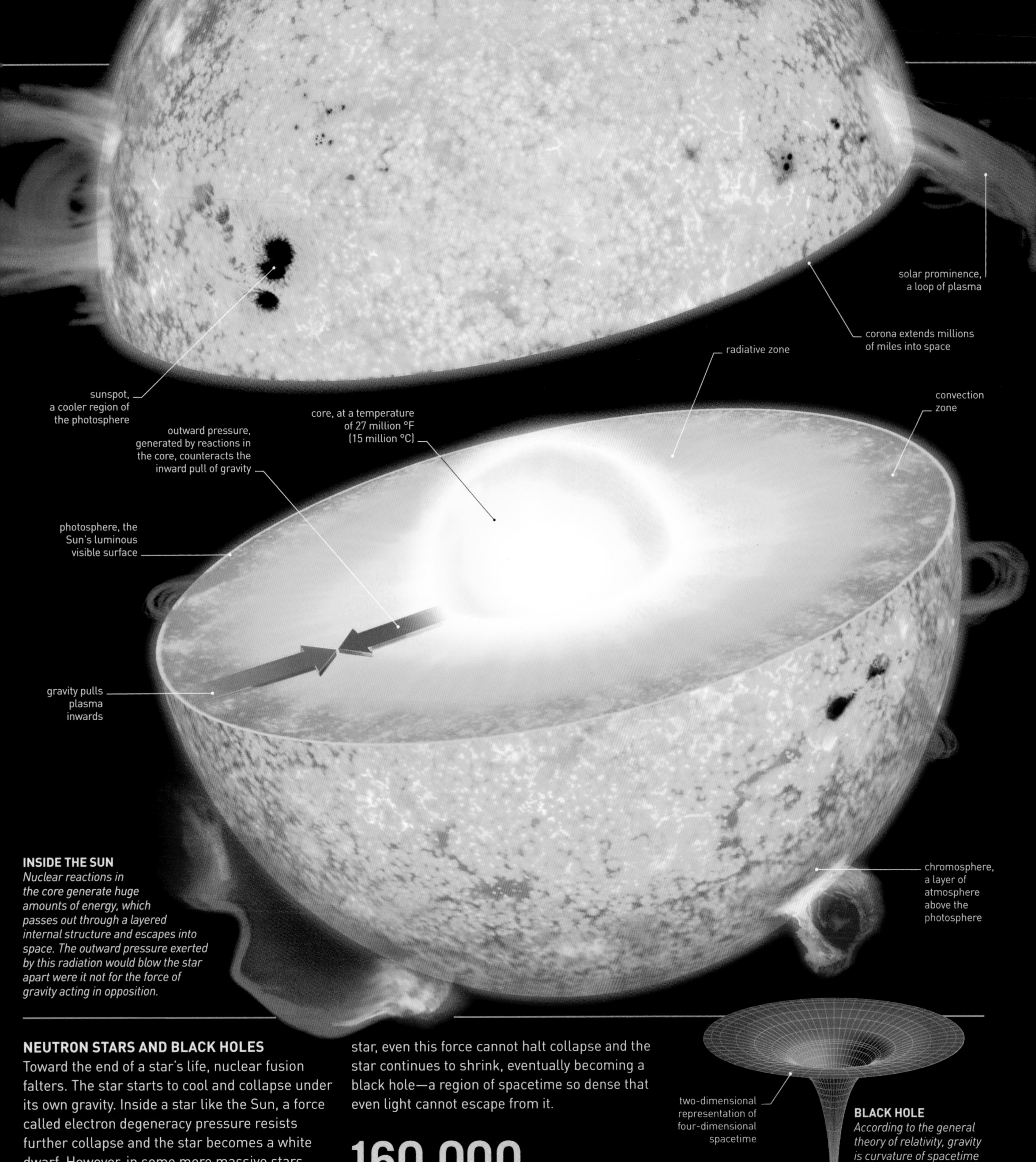
solar prominence,
a loop of plasma
corona extends millions
of miles into space
radiative zone
sunspot,
a cooler region of
the photosphere
convection
zone
core, at a temperature
of 27 million °F
(15 million °C)
outward pressure,
generated by reactions in
the core, counteracts the
inward pull of gravity
photosphere, the
Sun's luminous
visible surface
gravity pulls
plasma
inwards
chromosphere,
a layer of
atmosphere
above the
photosphere

INSIDE THE SUN
Nuclear reactions in the core generate huge amounts of energy, which passes out through a layered internal structure and escapes into space. The outward pressure exerted by this radiation would blow the star apart were it not for the force of gravity acting in opposition.

NEUTRON STARS AND BLACK HOLES

Toward the end of a star's life, nuclear fusion falters. The star starts to cool and collapse under its own gravity. Inside a star like the Sun, a force called electron degeneracy pressure resists further collapse and the star becomes a white dwarf. However, in some more massive stars, gravitational collapse overcomes this force and pushes electrons and protons together to form neutrons. The result is a neutron star, which is prevented from further collapse by a force called neutron degeneracy pressure. In a very massive star, even this force cannot halt collapse and the star continues to shrink, eventually becoming a black hole—a region of spacetime so dense that even light cannot escape from it.

160,000

THE NUMBER OF **LIGHT-YEARS** FROM EARTH TO THE NEAREST **BLACK HOLE**

two-dimensional
representation of
four-dimensional
spacetime
steep-sided
gravitational well
singularity

BLACK HOLE
According to the general theory of relativity, gravity is curvature of spacetime due to mass (see 1916). A black hole is a region of spacetime with a central point of infinite density—a singularity—that produces an infinitely deep well in spacetime.

16,000

THE NUMBER OF **WORKERS** ASSEMBLING **SAILING SHIPS** AT THE **VENICE** ARSENALE IN THE 17TH CENTURY

A 17th-century painting shows workers at Venice's Arsenale. Innovative construction techniques enabled the Venetians to dominate the seaways for centuries.

AROUND 1104, the city authorities in Venice ordered the construction of the Arsenale, a state shipyard and armory, which would employ 16,000 workers by the 17th century. The Arsenale **pioneered new production techniques**, producing prefabricated parts and using a method of frame-building for ships that made it possible to **construct a vessel in a day**.

The Chinese had started to print on silk using stencils before the end of the Han Dynasty (220 CE), but they began to use **multiple colors** in the wooden block printing of pictures *c.*1107. By 1340, the technique was applied to an edition of the *Diamond Sutra* (see 861–99), in which the main text is in black, and the prayers are in red.

> “BECAUSE OF THE **FREQUENCY** OF THE **EXPERIENCE,** THESE JUDGMENTS MAY BE REGARDED AS CERTAIN, EVEN WITHOUT OUR **KNOWING THE REASON.**”
>
> **Abu'l Barakat al-Baghdadi**, in *Kitab al-Mu'tabar*, early 12th century

In the 11th century, Avicenna had theorized that the **motion of a projectile** continues because of the *mail* (inclination, or motive power) imparted to it by the projector, but said that only one such force could exist in a body at any time. This was later confirmed by French priest Jean Buridan (see 1350–62). Around 1120, Baghdad philosopher **Abu l'Barakat** (*c.*1080–1165) suggested that more than one *mail* could exist in a projectile. As it fell, the *mail* pushing it forward weakened, and another *mail* took over, causing it to accelerate downward. These *mail* forces caused acceleration. In this way, he expressed the idea of the relationship between force and acceleration.

TRANSLATING ANCIENT MANUSCRIPTS

The works of many classical philosophers had been lost in the Christian West, but they were preserved through translations made into Arabic in the 8th and 9th centuries. These manuscripts in turn became available in Europe from the 12th century, where they were translated into Latin by scholars such as Gerard of Cremona.

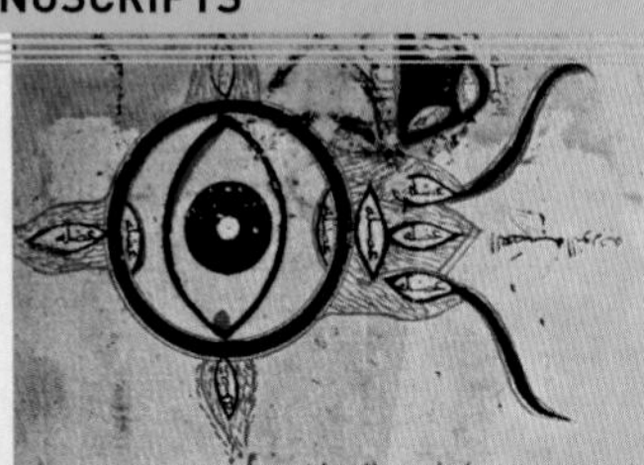

Raymond of Toledo
Archbishop Raymond is seen standing before King Alfonso VII at his coronation in 1135, a demonstration of the importance of royal patronage.

Around 1121, in the Persian city of Merv, **al-Khazini** wrote *Book of the Balance of Wisdom* in which he put forward a **theory of centers of gravity**. He suggested gravity varies according to the distance from the center of the world—the farther the objects are, the heavier they seem.

English philosopher **Adelard of Bath** (1080–1152) spent seven years in Salerno and Sicily, where he learned Arabic. His extensive knowledge of Arabic culture and language led him in 1126 to **translate al-Khwarizmi's astronomical work**, the *Sindhind Zij*, (*Astronomical Tables of Sindhind*) into Latin, bringing his work to a wider audience.

Raymond, Archbishop of Toledo in Spain (1126–52), **encouraged the translation of books from Arabic into Latin**. The first translators were succeeded in 1167 by **Gerard of Cremona** (1114–87), who translated more than 80 Arabic works.

INDIAN MATHEMATICIAN AND ASTRONOMER Bhaskara II (1114–85) described **a perpetual motion machine**, one that would, once a force was imparted to it, continue to work indefinitely. Bhaskara's device was **a wheel whose spokes were filled with mercury**. He theorized that mercury was sufficiently heavy so that as the wheels turned it would flow to the edge of the spokes and impel the machine around another part-turn.

Bhaskara II was better known for his **astronomical and mathematical works**, which made him one of the most respected Indian mathematicians of the Middle Ages. In *Lilavati* (named after his daughter), his

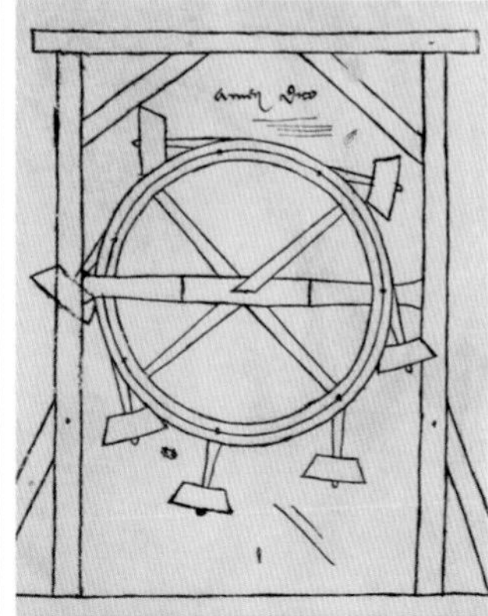

In perpetual motion
This 13th century version of a perpetual motion machine is an overbalanced wheel with hinged mallets around its rim.

1104 Construction of the **Venice Arsenale** begins

c.1107 Chinese use **multiple colors** in woodblock printing

1120 Abu l'Barakat expresses notion of the **relationship between force and acceleration**

1121 Persian scholar al-Khazini proposes an early form of **theory of gravitation**

1125 Chinese fleet uses **magnetic floating compass** on voyage to Korea

1126 Adelard of Bath **translates Euclid's *Elements*** from Arabic into Latin

1126–51 Raymond of Toledo orders **translations of many classical works** from Arabic into Latin

1145 Spanish Jewish mathematician Savasorda produces work on **quadratic equations**

1150 Bhaskara II demonstrates that a number has **two square roots**, one positive and the other negative

1150 Trotula, one of Europe's **first known female physicians**, practices at Salerno, Italy

The philosopher Ibn Rushd (Averroës) is banished from the court of the Almohads, after their overthrow of the Almoravids, whom he served as court physician.

> **“KNOWLEDGE** IS THE CONFORMITY OF THE **OBJECT** AND THE **INTELLECT.”**
>
> **Ibn Rushd (Averroës)**, from *Commentaries on the Physics*, late 12th century

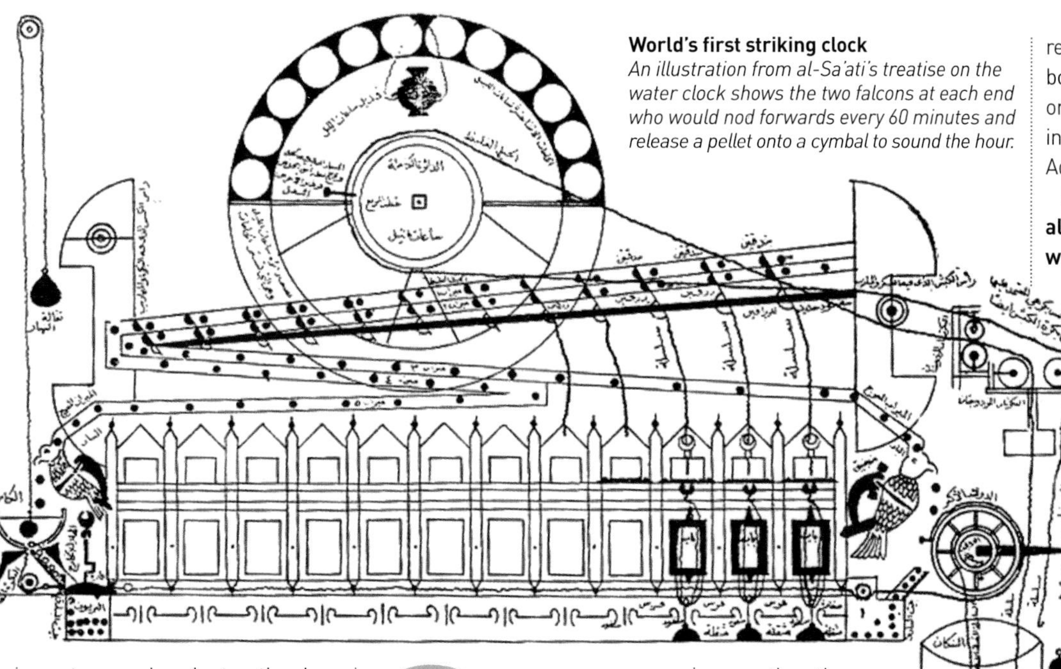

World's first striking clock
An illustration from al-Sa'ati's treatise on the water clock shows the two falcons at each end who would nod forwards every 60 minutes and release a pellet onto a cymbal to sound the hour.

most comprehensive treatise, he discussed fractions, algebra and algorithms, permutations and combinations, and the geometry of triangles and quadrilaterals. He also introduced the idea of negative quantities in geometry. In his *Bija-Ganita* (*Seed Counting*), he concluded that the **division of a number by zero would produce infinity**. He also became the first mathematician to realize that there are **two square roots of a number**, one positive and one negative. In his astronomical work of 1150, the *Siddhanta-siromani* (*Head Jewel of Accuracy*), Bhaskara II performed calculations on small increments

2
THE NUMBER OF **SQUARE ROOTS** OF ANY NUMBER

of motion that came close to **an idea of differential calculus**, which studies the rates at which quantities change. However, his ideas were of much narrower scope than those developed by Isaac Newton or Gottfried Leibniz five centuries later.

The Spanish-born philosopher **Ibn Rushd** (1126–98), known as Averroës in Europe, commented extensively on Aristotle's work in the 4th century BCE, seeking to integrate his ideas with Islamic theology. Around 1154, in his work on Aristotle's theory of motion, Averroës made a distinction for the first time between the **motive force of an object** (its weight) **and the inherent resistance of a body to motion** (its mass), although he restricted this analysis to celestial bodies. Its extension to bodies on Earth would be made only in the 13th century by Thomas Aquinas (*c.*1224–74).

In 1154, Arab engineer **al-Kaysarani** constructed **the world's first striking clock**, near the Umayyad mosque in Damascus. It was powered by water and was described by al-Kaysarani's son Ridwan al Sa'ati in his 1203 treatise *On the Construction of Clocks and their Use*. Islamic water clocks became so sophisticated that in 1235 one was built in Baghdad that told people the times of prayer, day and night.

The advanced state of both cartography and printing in China are indicated by **the first printed map**, which dates from around 1155 (at least three centuries before its first European counterpart, in 1475). Contained in the *Liu Ching Tu* (*Illustrations of Objects mentioned in the Six Classics*), it depicted parts of western China with **rivers and provincial names given**, and showed the line of the Great Wall.

A more grandiose cartographic creation of the Chinese Sung dynasty was the *Yu Ji Tu*, an 1137 **map of the country carved in stone**, which included grid lines and an indication of the scale of the map.

While waterwheels had long been used in Europe for the grinding of grain, around 1180 the idea was adapted to the use of **windpower**. Unlike earlier Persian windmills, which were horizontal, the **European vertical mills used a post design, with sails** mounted on a vertical tower that itself was free to rotate as the wind varied. By the 1190s, windmills had become so commonplace that Pope Celestine III imposed a tax on them.

Vane power
This German windmill shows the typical arrangement of four sails attached to a vertical post, but unlike earlier post-mills only the cap of the mill rotates to face the wind.

1151–58 Abbess Hildegard of Bingen writes treatises on **medicine and botany**

1154 Ibn Rushd's (Averroës's) **theory of kinetics** expresses the idea that objects have mass that resists motive force

1154 Al-Kaysarani builds world's **first striking clock** in Damascus

c.1155 French theologian Thierry of Chartres writes works reconciling Biblical account of creation with Platonic cosmology

1155 First printed map, of part of western China

1167 Italian scholar Gerard of Cremona becomes head of Toledo translation school; **translates nearly 90 works** into Latin

c.1180 Horizontal **windmills** appear in northwestern Europe

c.1180 Italian scholar Burgundio of Pisa **translates works of Galen** into Latin

1191 Salerno is sacked by Emperor Henry VI, leading to **decline of medical school**

1198 Jewish physician Moses Maimonides writes treatise on **poisons and their antidotes**

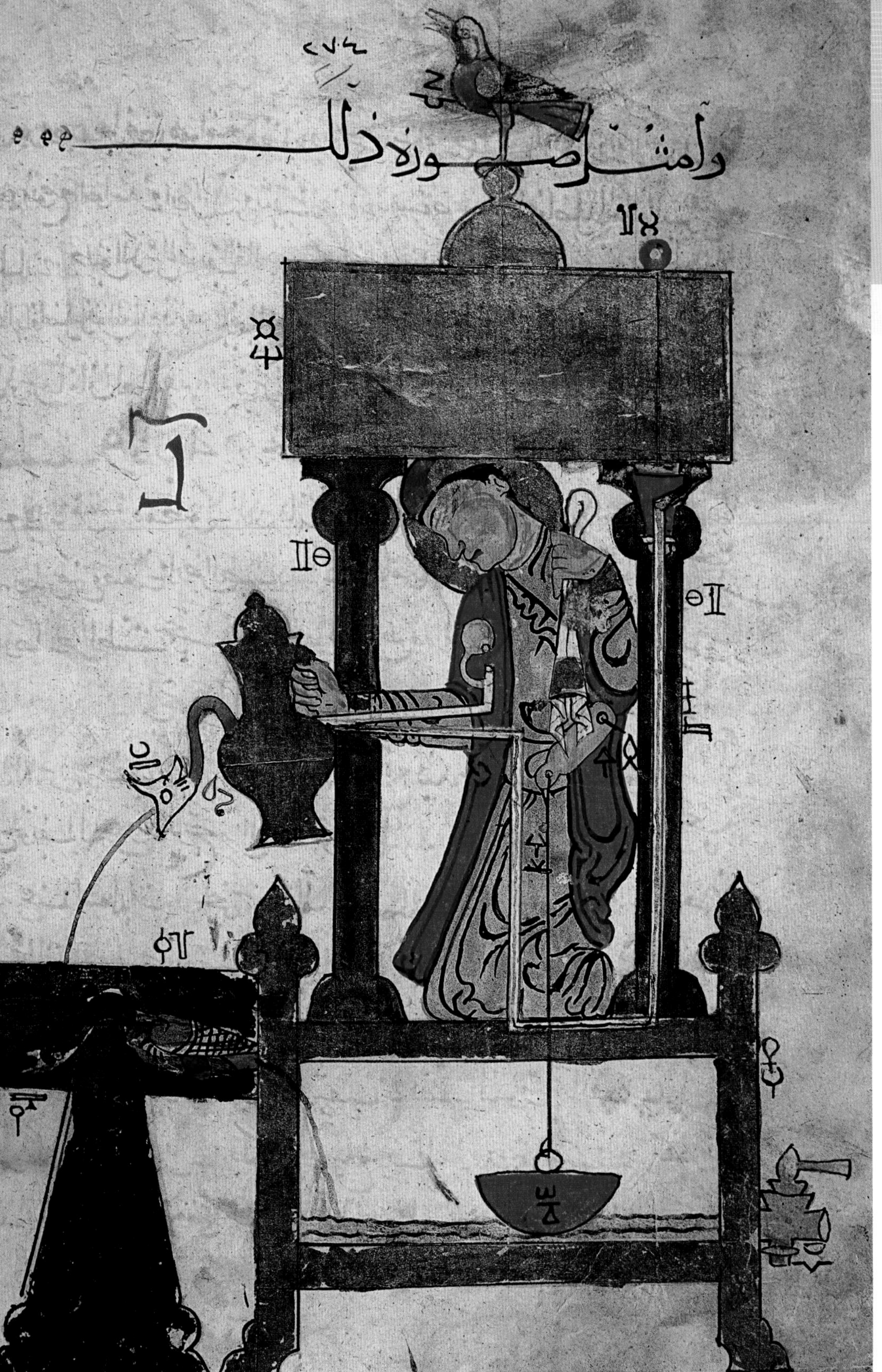

Ingenious devices
This illustration of one of al-Jazari's mechanical devices shows an automaton that pours water from a pot, then returns to a chamber where it is scooped back up again.

> **" THE IMPACT OF AL-JAZARI'S INVENTIONS IS STILL FELT IN MODERN CONTEMPORARY MECHANICAL ENGINEERING. "**
>
> **Donald Hill,** from *Studies in Medieval Islamic Technology*, 1998

ITALIAN MATHEMATICIAN Leonardo Pisano (Fibonacci) published the *Liber Abaci* (*Book of Calculations*) in 1202, the first major western European work **popularizing the use of Hindu-Arabic numerals** and **place notation** (see 861–99). The book also presented **rules for algebra**, which he probably derived from al-Khwarizmi (see 821–60), as well as solutions for finding square and cube roots. Fibonacci described techniques that were useful for the Pisan merchants of his day, including a method for multiplication using a grid, advice on the barter of goods, and the use of alloys to make coins. The Fibonacci sequence (below) is derived from a problem that concerned the growth of a rabbit population.

In 1206, Arab engineer **Ibn Isma'il al-Jazari** published the *Book of Knowledge of Ingenious and Mechanical Devices*, detailing 50 machines, including the **first descriptions of crankshafts and camshafts**. The most spectacular of these was a 2 m- (6.5 ft-) high water clock in the form of an elephant with a phoenix that marked half-hours.

sequence starts with 1

each number in sequence is sum of two numbers before it

sequence continues in same way indefinitely

1+1   1+2   2+3   3+5

1, 1, 2, 3, 5, 8...

FIBONACCI SEQUENCE

The Fibonacci sequence is a series of numbers in which each successive number is the sum of the two numbers preceding it. Any number in the series is known as a Fibonacci number. These occur surprisingly often in nature, with the number of petals of many flowers being Fibonacci numbers (daisies have 13, 21, or 34), while the arrangement of leaves on a plant stem is also determined according to a ratio connected to this sequence.

1202 Fibonacci publishes the *Liber Abaci* expounding **use of Hindu-Arabic numerals**

1206 Al-Jazari's *Book of Knowledge of Ingenious and Mechanical Devices* describes **crankshafts and camshafts**

1210 A church synod **bans the study of the works of Aristotle** at Paris University

1214 Italian physician Hugh of Lucca uses **wine as an antiseptic** and observes role of pus in infection

1217 Scottish scholar Michael Scot translates al-Bitruji's ***Planetary Theory*** into Latin

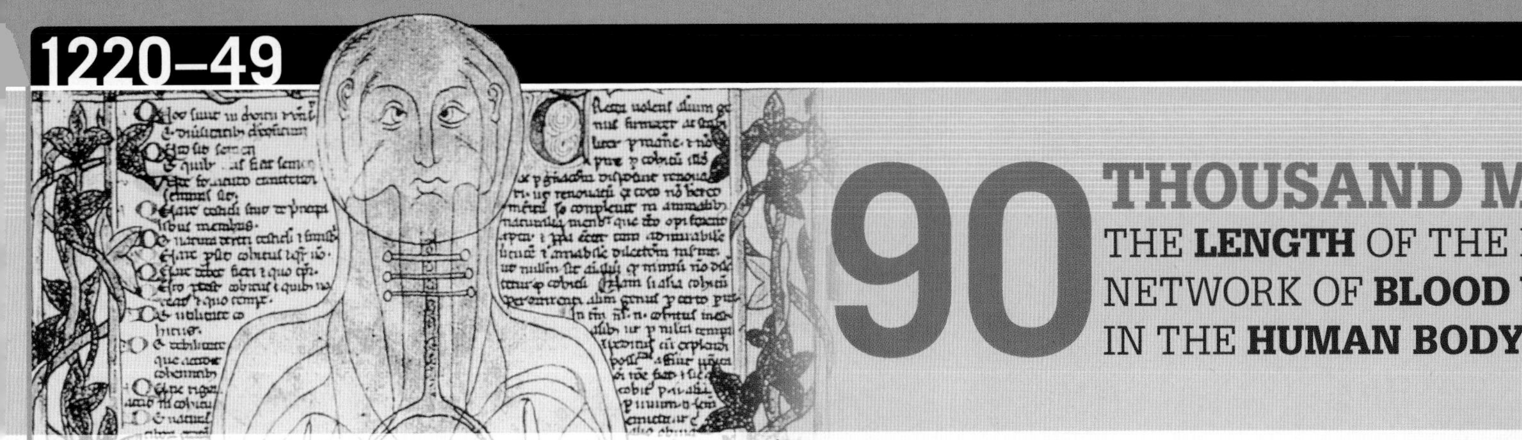

90 THOUSAND MILES

THE **LENGTH** OF THE ENTIRE NETWORK OF **BLOOD VESSELS** IN THE **HUMAN BODY**

A 13th-century illustration shows the flow of blood through the body by means of veins. The top of the heart can just be seen.

EARLY CHINESE GUNPOWDER WEAPONRY had been relatively low powered, in the form of hand-hurled grenades or fire-arrows, with a small charge attached to the shaft. In 1231, faced with a Mongol invasion, the Chinese defenders of Ho Chung deployed the "Heaven-Shaking Thunder Crash Bomb," which contained **gunpowder rich enough in saltpeter to burst an iron casing**. The resulting explosion could be heard 31 miles (50 km) away, and the shrapnel from the explosion was said to have torn iron armor to pieces. In 1232, the **Chinese also used an early form of rocket** consisting of a spear with a bamboo container packed with gunpowder attached to it. When lit, these "fire-spears" were propeled forward by the explosive charge. Another form of the weapon was used as **an early form of flamethrower**, which could shoot fire up to 6.5 ft (2 m) towards an enemy, causing appalling injuries.

Rockets in the making
Shown here is an early Chinese rocket of the kind used at the siege of Kaifeng in 1232. The soldier is about to light the fuse on the bamboo gunpowder container.

Robert Grosseteste (*c.*1168–1253), Bishop of Lincoln, played a key role in **reconciling Aristotelian philosophy and scientific method with Christian thinking** through his commentary on Aristotle's *Posterior Analytics*, published from 1220 to 1235. His logical method was rigorous, a process he called **"resolution and composition,"** which involved the testing, by experiment if possible, of hypotheses, and the rejection of any conclusions that were not based on observation. His theory that all changes were caused by **the action of forces acting through a medium** led him to study optics and to write treatises on rainbows and astronomy.

Around 1230, European mathematician **Jordanus de Nemore** produced **a new theory of levers** in his *Elementa Super Demonstrationem Ponderum* (*Elements on the Demonstration of Weight*). Building on Aristotle's axiom that equal weights at equal distances from a fulcrum are in equilibrium (see pp.34–35), Jordanus introduced the idea of **virtual displacement** (which looks at the effects of infinitesimal changes on a mechanical system) into the science of mechanics. His *De Ratione Ponderis* (*On the Theory of Weight*) also investigated the problem of **downward forces acting along the trajectory of a moving body**. He demonstrated that the more oblique the object's trajectory, the smaller the downward forces (later understood as positional gravity). Jordanus also developed proof to show the point at which weights supported by angled (or bent) **levers on a fulcrum will be in equilibrium (balanced)**.

Syrian polymath and anatomist **Ibn al-Nafis** (1213–88) produced **a major medical compendium** the *Sharh Tashrish al-Qanun* (*Commentary on Anatomy in Ibn Sina's Canon*). It contained a host of anatomical discoveries, but al-Nafis's major breakthrough was his discovery of how **blood circulated between the heart and lungs**. He showed that blood circulates from the right-hand side of the heart to the left through the lungs, in contrast to the traditional view of Galen (see 100–250), who held that blood seeped in from the right ventricle to the left through pores in the wall that separated the two chambers. However, he did not explain how the blood then returned from the heart's left ventricle to the right. A full theory of blood's circulation would not be formulated until William Harvey in the 17th century (see 1628–30).

> " THE CONSIDERATION OF **LINES, ANGLES,** AND **FIGURES** IS OF THE GREATEST UTILITY SINCE IT IS IMPOSSIBLE FOR **NATURAL PHILOSOPHY** TO BE KNOWN WITHOUT THEM. "
>
> **Robert Grosseteste, English philosopher and theologian,** in *On Lines, Angles, and Figures*, *c.*1235

FIBONACCI (*c.*1170–1250)

Leonardo Pisano (Fibonacci) was born into a wealthy merchant family in Pisa, Italy. His father was in charge of the Pisan trading colony in Bugia in Tunisia, and there Fibonacci came into contact with Arabic mathematical ideas. Aged 32, he published the *Liber Abaci*, which brought him great fame and he gave a mathematical demonstration to King Frederick II of Sicily.

1220 Robert Grosseteste **publishes commentary** on Aristotle's *Posterior Analytics*

c.1230 Jordanus de Nemore develops theory of **equilibrium of levers** and the idea of virtual displacement

1231 Lift of ban on the study of Aristotle at Paris University

1231 Al-Dakhwar bequeaths money for establishment of **first school exclusively for teaching of medicine**, in Damascus

c.1230–40 Chemical encyclopedia of Bartholomew the Englishman published

1245–46 Syrian physician Ibn Abi Usaybi'ah publishes *Sources of Information on the Classes of Physicians*, **first history of Arabic medicine**

1247 Song Ci produces **first ever work on forensic medicine**

c.1248 Islamic physician Ibn al-Baytar compiles the *Comprehensive Book on Simple Drugs*, the **most influential medieval Arab pharmacopeia**

> TO CURE **MELANCHOLY,** CUT A CROSS-SHAPED **HOLE** IN THE… **SKULL…** THE **PATIENT** IS TO BE **HELD IN CHAINS.**

Roger Frugardi, Italian surgeon, from *Chirurgia*, late 12th century

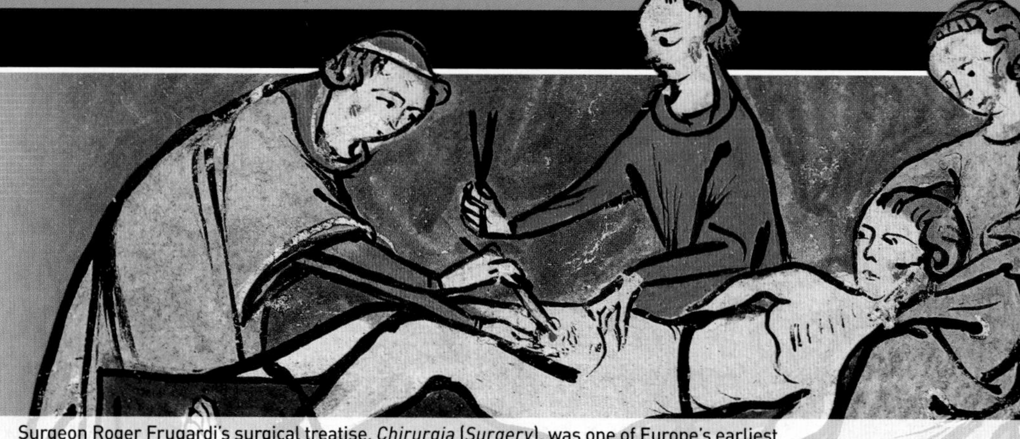

Surgeon Roger Frugardi's surgical treatise, *Chirurgia* (*Surgery*), was one of Europe's earliest books on surgery. This illustration shows a hernia operation in progress.

AROUND 1250, PRIEST AND PHYSICIAN Gilbert the **Englishman** completed the *Compendium Medicinae* (*Compendium of Medicine*). It became one of the most **widely used medical works** of the Middle Ages and was translated from Latin into German, Hebrew, Catalan, and English. The work had separate volumes devoted to the head, heart, respiratory organs, fevers, and women's diseases. In the books, Gilbert also wrote about the **diagnosis of leprosy** by its numbing effect on skin.

In 1266, English friar and scholar **Roger Bacon** completed his ***Opus Majus*** (*Greater Work*). Ostensibly a plea for church reform, the work included large sections on experimental observation and natural sciences, intended to convince the Church of the **virtues of new learning**. It also contained the first description of **gunpowder** in western Europe, and ideas for flying machines and steamships. The section on **optics** was particularly important. In it, Bacon agreed with Arab scholar Alhazen's view that vision is made possible when rays emanating from the object viewed enter into the eye (see 1000–29). He examined the properties of differently shaped lenses, and described the use and mathematics behind magnifying lenses—although he did not, as commonly supposed, actually invent eye glasses.

> EXPERIMENTAL SCIENCE IS THE **QUEEN OF SCIENCES** AND THE **GOAL OF ALL SPECULATION.**

Roger Bacon, in *Opus Tertium*, 1267

Italian surgeons **Hugo** (c.1180–1258) and **Teodorico Borgognoni** (1205–98) came from a family of doctors who practiced in Bologna, a leading center of medicine. By the 1260s, Teodorico was advocating the **cleansing of wounds with wine** and their rapid closing up. This practice was in contrast to most contemporary medical practitioners who went along with Greek physician Galen's insistence that pus be allowed to form in wounds. The Borgognonis also advocated using **dry bandaging for wounds**—discarding the salves and poultices used at the time. They also used an **early form of anesthesia** by holding sponges soaked with narcotics such as opium or hemlock near the noses of patients who were about to undergo surgery.

ROGER BACON (1220–92)

Educated at Oxford University, England, Roger Bacon travelled to Paris, where he lectured on Aristotle. In 1247, he gave up his post to research privately. He joined the Franciscan Order in 1257 in order to continue his studies. He was commissioned by Pope Clement IV to produce a work on church reform, which led to his *Opus Majus* in 1267.

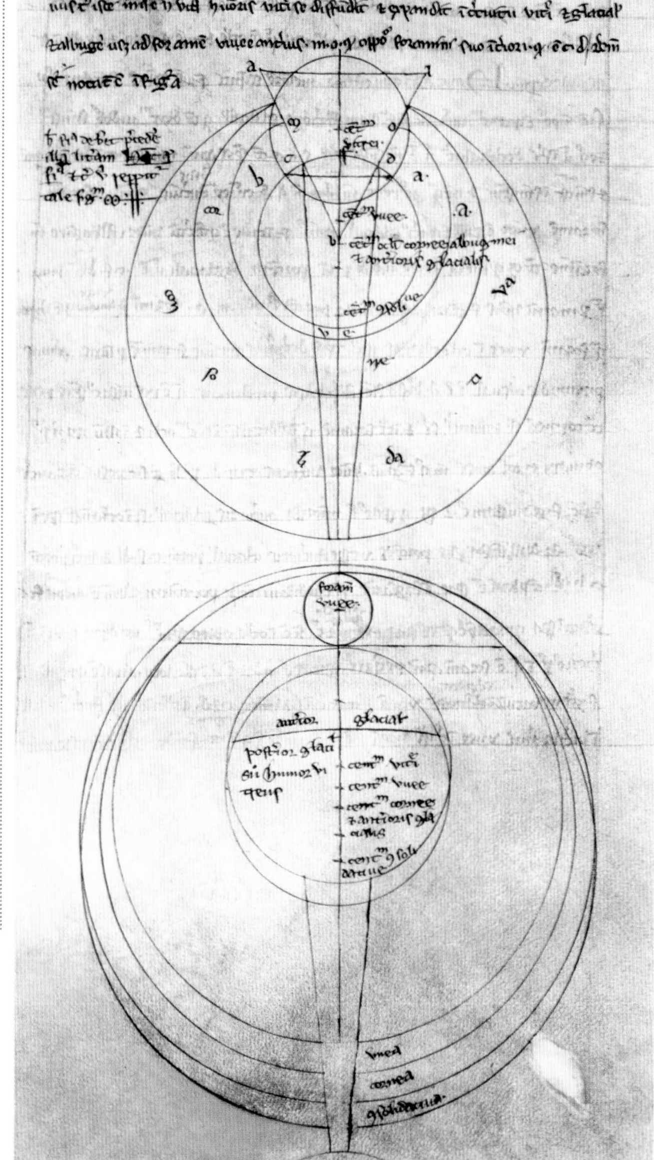

Opus Majus
This diagram from Roger Bacon's major work shows the structure of the eye, the curvature of its lens, and how light rays striking the lens produce vision.

1250 Gilbert the Englishman's *Compendium of Medicine* includes **diagnosis of leprosy**

c.1260 Teodorico Borgognoni advocates cleansing of wounds with wine and use of **anaesthetic narcotics** in operations

1260 Albertus Magnus writes that **volcanoes** are caused by subterranean winds

1267 Roger Bacon describes the **structure of the eye** and the properties of magnifying lenses

1269–79

“ YOU WILL BE ABLE TO **DIRECT YOUR STEPS** TO CITIES AND ISLANDS AND TO **ANY PLACE** WHATEVER IN THE **WORLD.** **”**

Pierre de Maricourt, French scholar, describing the compass in *Epistola de Magnete*, 1269

Gaocheng observatory
One of its two towers contained an armillary sphere. Between them lay a "sky-measuring scale" to measure the shadow of the 39-foot gnomon.

IN 1269, FRENCH SCHOLAR PIERRE DE MARICOURT wrote *Epistola de magnete* (*Letter on the Magnet*), the first work to describe the **properties of magnets**. In it, he set out the laws of magnetic attraction and repulsion, and explained how to identify the poles of compasses. Maricourt's work led to the construction of **better magnetic compasses**, which became invaluable aids in sea navigation. He also described the operation of a **perpetual motion machine**, which worked using magnetism.

In 1276, the Mongol ruler of China, Kublai Khan, asked mathematician and engineer **Guo Shoujing** (1231–1316) to reform the calendar. To perform this task, Guo first had a series of astronomical instruments constructed. This included a vast equatorial armillary sphere—calibrated with a ring representing equatorial coordinates, a system not used in Europe until the time of Tycho Brahe (see 1565–74) three centuries later. He then established **astronomical observatories** at Peking (Beijing) and Gaocheng, near Loyang, China, between 1279 and 1280. At the latter, a 39 ft- (12 m-) high gnomon (shaft on a sundial) sat on top of a pyramid, casting shadows, which were measured at the time of the Sun's solstices to help determine the **length of the year**. Guo used advanced trigonometry to calculate the length of a year.

1280–99

Figures wearing eyeglasses soon found their way into religious art, as is shown here in this 1491 detail from the *Betrayal by Judas* in Notre Dame, Paris, France.

IN 1280, GUO SHOUJING finally completed his **calendar**. According to his calculations, a year had 365.2425 days.

The **earliest surviving cannon** is from China and was made *c.*1288. Before cannons were strengthened by the use of cast iron barrels, the Chinese had probably used bronze tubes to eject projectiles using gunpowder explosives. Manuscripts of 1274 and 1277, however, refer to *huo pa'o*—explosive weapons used by the Mongols to demolish the ramparts of Chinese cities—so the invention may have occurred a little earlier.

26 SECONDS THE **DIFFERENCE** BETWEEN **GOU'S CALCULATION** AND **ACTUAL DAYS IN A YEAR**

Although the magnifying properties of glass lenses had been studied by English bishop Robert Grosseteste (1175–1253) and Roger Bacon earlier in the 13th century, the **first description of eyeglasses** was given by a Dominican friar **Giordano da Pisa** (*c.*1260–1310), who wrote that he had seen them in 1286. Early glasses were convex (curving out) to correct far-sightedness. Concave lenses (curving in) for near-sightedness did not appear for more than a century.

In 1290, French astronomer **William of St. Cloud** gave an account of a **solar eclipse** witnessed by him five years earlier. Many of those who had observed the eclipse had damaged their eyes by viewing the Sun directly. In order to prevent this, William used a **camera obscura**—a type of pinhole camera in which light goes into a dark chamber and is projected through a tiny aperture onto another surface, such as a card, opposite it. The technique had been used by earlier astronomers, such as Alhazen in the 11th century, to prove that intersecting rays do not interfere with each other. William, however, was the **first to explain its use in solar observation**. He also calculated an accurate value by which the Earth tilts on its axis, by observing the Sun's position at the solstices. In addition, he produced an almanac with detailed positions of the Sun, Moon, and planets at various dates between 1292 and 1312.

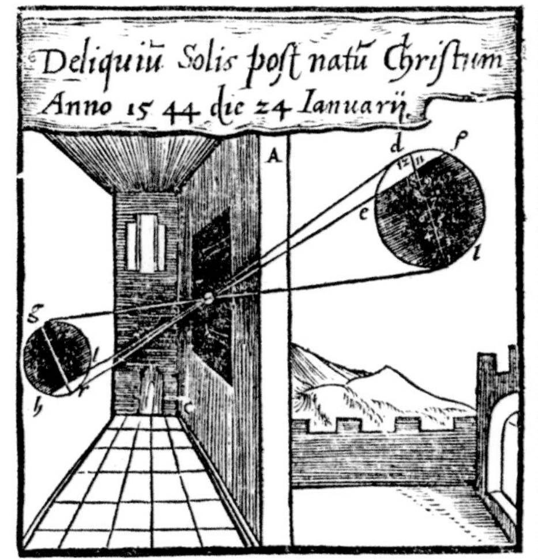
Deliquiū Solis post natū Christum
Anno 1544 die 24 Januarij.

Camera obscura
This 16th-century illustration of a camera obscura shows how an image of the Sun is reversed after light passes through an aperture onto a surface in a darkened room.

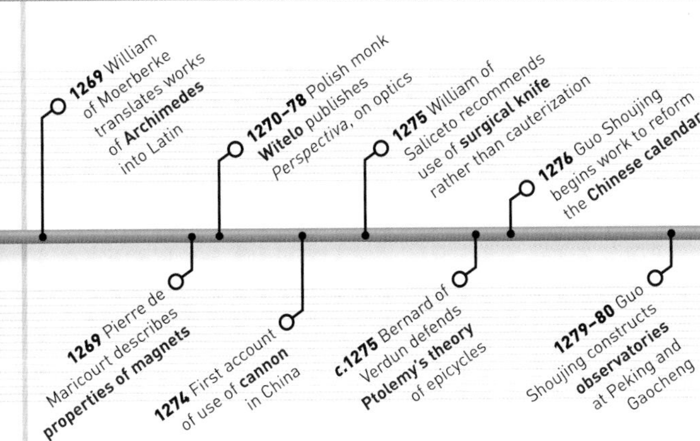
1269 William of Moerberke translates works of **Archimedes** into Latin
1269 Pierre de Maricourt describes **properties of magnets**
1270–78 Polish monk **Witelo** publishes *Perspectiva*, on optics
1274 First account of use of **cannon** in China
1275 William of Saliceto recommends use of **surgical knife** rather than cauterization
c.1275 Bernard of Verdun defends **Ptolemy's theory** of epicycles
1276 Guo Shoujing begins work to reform the **Chinese calendar**
1279–80 Guo Shoujing constructs **observatories** at Peking and Gaocheng

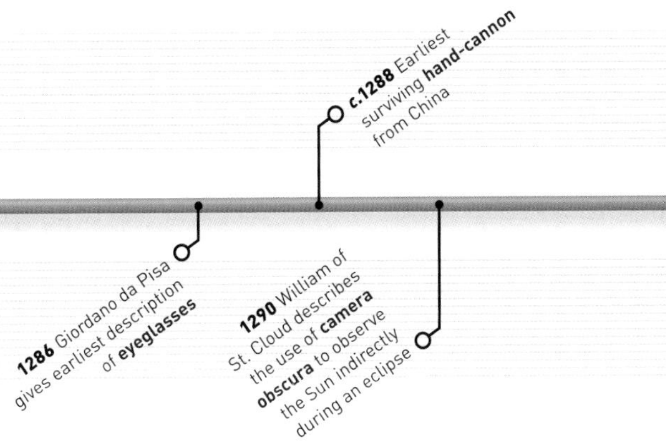
1286 Giordano da Pisa gives earliest description of **eyeglasses**
c.1288 Earliest surviving **hand-cannon** from China
1290 William of St. Cloud describes the use of **camera obscura** to observe the Sun indirectly during an eclipse
1296 Use of *Alfonsine Tables*—astronomical tables—spreads to **Paris**

"programmable" gears can be set to steer the car in a particular direction

coiled springs beneath chassis store and release energy

tiller alters direction of front wheel for manual steering

like a child's toy car, spinning the wheels in reverse stores energy in the spring

brake mechanism keeps the car stationary until released

30 THE NUMBER OF **GEAR WHEELS** FOUND IN THE REMAINS OF THE **ANTIKYTHERA MECHANISM** RECOVERED FROM A MEDITERRANEAN **SHIPWRECK**

c.230 BCE
Early gears
The Chinese south-pointing chariot probably uses a gear arrangement that ensures the figure at the front points to the south as the wheels turn.

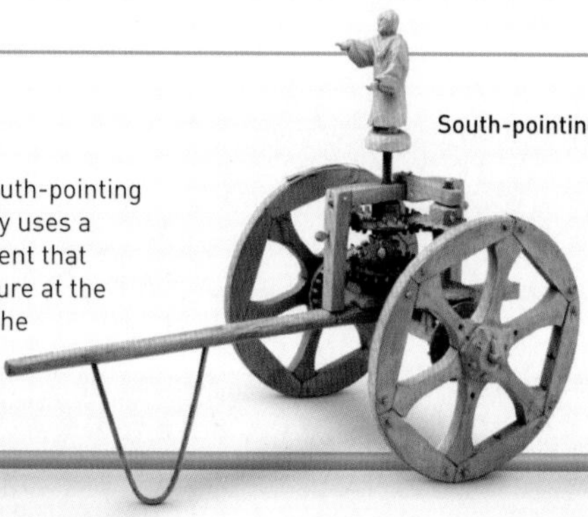

South-pointing chariot

c.125 BCE
Zhang Heng's armillary sphere
Chinese scholar Zhang Heng builds an armillary sphere driven by gears and water. His model, showing the motions of the Sun, Moon, and stars, would go on to influence not only Chinese gear technology, but also later clockmakers (see 700–799).

Clock at Salisbury Cathedral

13th century
Mechanical clocks in Europe
The first mechanical clocks using gears to drive the rotation of pointers and to control the striking of chimes are invented. They are powered by the controlled drop of a weight attached to a drive chain.

Chinese watermill

c.200 BCE
Watermills
Greek watermills that use gears to harness hydropower begin to spread throughout the Graeco-Roman world. The Chinese develop their own water wheel technology about 200 years later, with gearing mechanisms to drive various motions.

Windmill gear

c.7th century
Persian windmills
The first functional windmills, developed in Persia, have horizontal sails that drive the rotation of a vertical shaft.

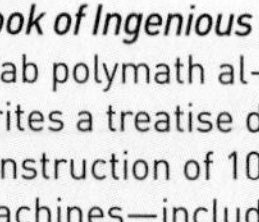

1206
Book of Ingenious Devices
Arab polymath al-Jazari writes a treatise describing the construction of 100 remarkable machines—including the crankshaft—many of which rely on gears.

Al-Jazari's treatise

THE STORY OF GEARS

SIMPLE BUT EFFECTIVE MECHANISMS FOR TRANSMITTING FORCES OF ROTATION, GEARS HAVE A LONG AND COMPLEX HERITAGE

The ability to alter the direction of a force, transmit it from one axis of rotation to another, or trade force with movement, is a vital aspect of many modern machines. Yet, such mechanical functions often rely on gearing techniques that are centuries old.

A gear is a wheel mounted on a central rotating axis, with a series of teeth or cogs around its outer edge that can engage with the cogs of another gear. The teeth allow the gear to transmit its angular motion to its neighbor, forming a pairing known as a transmission. The ratio of teeth between the two wheels determines the speed and force with which the second gear rotates, providing a so-called "mechanical advantage"; a smaller secondary gear rotates more rapidly, but with less torque or rotational force.

The earliest specific devices known to have used gears were the Chinese south-pointing chariots—direction-finding devices used in the 3rd century BCE.

Leonardo da Vinci's vehicle
This model of self-propelled automobile was built from a Leonardo da Vinci sketch. The force from the expansion of two wound springs is transmitted through ingenious gearing to drive the rear wheels.

In ancient Greece, gear technology reached its apex with devices such as the Antikythera mechanism—a complex astronomical calculator recovered from a Mediterranean shipwreck around 1900.

PRACTICAL APPLICATIONS

More immediate practical applications of gears, such as their use in harnessing power from flowing water and wind, gradually spread throughout the ancient world. Treadmills powered by animals—or even humans—became common. Gears found various applications in mills—flour mills are perhaps the most familiar, but sawmills also used gears to turn rotating cutting blades, and hammer mills used gearing to lift and drop heavy hammers for beating metal or minting coins.

New advances led to the development of traditional clockwork in Europe in around the 13th century, and the Industrial Revolution saw the development of further ingenious transmissions to harness the power of steam engines. The use of gears has continued to the present day in modern machines ranging from automobiles to inkjet printers.

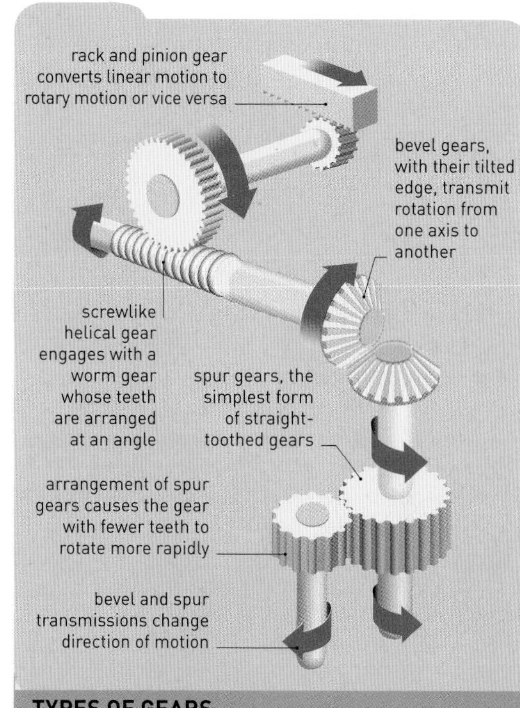
rack and pinion gear converts linear motion to rotary motion or vice versa
bevel gears, with their tilted edge, transmit rotation from one axis to another
screwlike helical gear engages with a worm gear whose teeth are arranged at an angle
spur gears, the simplest form of straight-toothed gears
arrangement of spur gears causes the gear with fewer teeth to rotate more rapidly
bevel and spur transmissions change direction of motion

TYPES OF GEARS

Gears can be designed or arranged in a variety of ways to transmit motion from one direction to another. Complex transmission assemblies can take the motive power from a single rotating drive shaft and apply it to drive a range of linear movements or to propel further rotating shafts at any speed required.

1480
Leonardo da Vinci's work on gears
Italian polymath Leonardo da Vinci utilizes complex gear assemblies in many of his inventions, such as lens-grinding and metal-rolling devices, and shows a deep theoretical understanding of their function.

18th century
Industrial Revolution
The rise of steam power during the Industrial Revolution drives advances in gear technology. The linear motion of steam pistons is applied to the rotation of locomotive wheels.

Steam locomotive

1781
Murdoch's gearing
Scottish engineer William Murdoch's sun and planet gear transmission converts vertical motion, such as that of a steam-driven beam, into the rotational motion of a driveshaft.

Sun and planet gearing

19th century
Development of bicycles
Through the 19th century, bicycles gradually develop from the scooterlike velocipedes that were invented around 1817 to pedal-powered machines that use gears and a drive chain.

"Safety" bicycle

1835
Gear hobbing proccess
British engineer Joseph Whitworth invents hobbing—the first process for the production of high precision gears, on an industrial scale.

Plastic gear

1950s
Plastic gears
Gears made from new plastics materials are introduced from the 1950s. They lack the strength of properly machined metal gears, but are far more easily and cheaply manufactured.

1990s
Nanotechnology
Machines created on a nanoscale often rely on the same gearing principles as larger devices, except that the gear wheels are mere micrometers across.

1300–10

The search for an explanation of the colors of the rainbow preoccupied medieval scholars, including Roger Bacon and Theodoric of Freiberg.

0.45 KILOGRAMS THE **MODERN** EQUIVALENT OF THE **AVOIRDUPOIS POUND**

EARLY MEDIEVAL EUROPE USED A SYSTEM OF WEIGHTS based on the Roman pound, which had 12 *unciae* (or ounces) and was mainly used for weighing pharmaceuticals and coins. A new set of measures suitable for bulky goods, such as wool, was introduced in England around 1303 (when it was mentioned in a charter). It was called **avoirdupois**, from the Norman French *Habur de Peyse* meaning "goods having weight" and was **based on a 16-ounce pound**—a measure that would be used for the next 700 years, and still is in parts of the world.

The avoirdupois pound probably originated in Florence, where an almost identical unit was in use for the weighing of wool. Soon supplementary weights were added, including the hundredweight (112 pounds), defined in an ordinance of 1309.

Weighty issues
This is one of a set of standard avoirdupois weights that were issued by Elizabeth I of England in 1582. These weights were to remain the standard measure until the 1820s.

The properties of the rainbow had fascinated philosophers from Aristotle to Roger Bacon, who thought their color was due to the reflection of light from spherical raindrops in a cloud.

Around 1310, the Dominican friar **Theodoric of Freiberg** (*c.*1250–1311) carried out scientific experiments to determine the **origins of rainbows** by using glass balls filled with water through which he passed light, which was then projected onto a screen. He concluded that the rainbow was caused by light striking spherical raindrops, which was first refracted, then reflected internally on the inner surface of the drop, and then refracted again. Theodoric also properly described the **color spectrum**. He discovered that the light that projected out of his glass balls produced the same range of colors as a rainbow, and in the same order (red, yellow, green, and blue).

1311–16

The title page of Mondino de Luzzi's *Anatomia* shows a corpse on the dissecting table; it has already been opened up and had organs removed.

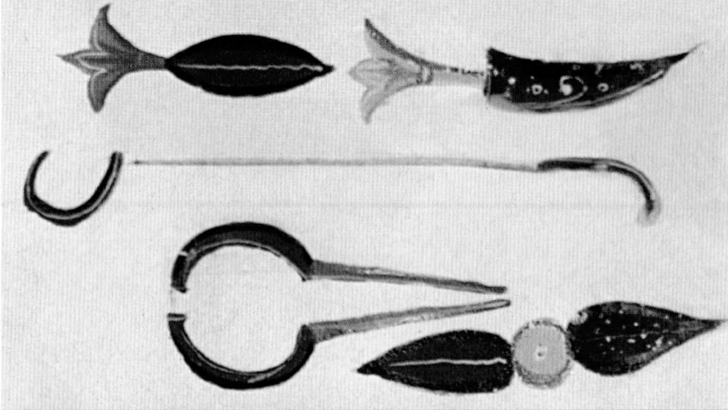

EARLY SURGERY AND DISSECTION

Surgery took a long time to become established as a separate discipline, but written records since 1170 indicate a growing medical sophistication. By 1200, operations for bladder-stones, hernias, and fractures were routine. By the 14th century, surgeons were aware of the need to avoid infection, sometimes cleaning wounds with wine and closing them up as soon as possible.

THE EARLIEST WRITTEN RECORDS OF WEIGHT-DRIVEN CLOCKS feature in the Italian writer Dante Alighieri's book *Paradiso* (*Paradise*) (*c.*1313–21), although such clocks probably first appeared decades earlier. Weight-driven clocks use a weight to act as an energy storage device so that the clock can run for a certain period of time (such as a day or a week). Winding such clocks pulls on a cord that lifts the weight, which is affected by gravity and falls; the clock uses the potential energy as the weight falls to drive the clock's mechanism. The first clock faces were probably divided up according to the canonical hours (seven regulated times of prayer), which punctuated the church day. **Clocks showing 12 equal hours** were first recorded in 1330.

Among the first major surgical writers was **Henri de Mondeville** (*c.*1260–1316). A former military surgeon who came to teach medicine at Montpellier, de Mondeville had, by 1308, **begun to use anatomical charts and a model of a skull as aids to his teaching**. Around 1312, he

1303 Avoirdupois system of weights introduced in England

1305–07 English physician John of Gaddesden describes the **pelican**, an instrument for extracting teeth

1309 Hundredweight is added to avoirdupois system

c.1310 Theodoric of Freiberg carries out experiments on the **color spectrum** and rainbow

1312 French surgeon **Henri de Mondeville** publishes a major manual on surgery entitled *Surgery*

1313–21 Italian writer Dante Alighieri mentions a **weight-driven clock** in his book *Paradise*

“ IT IS **FUTILE** TO DO WITH **MORE THINGS** THAT WHICH CAN BE DONE WITH **FEWER.** ”

William of Ockham, Franciscan friar, from *Summa Totius Logicae, c.*1323

“ **GOD** HIMSELF WAS A **PRACTICING SURGEON** WHEN HE **MADE** THE FIRST **MAN** FROM MUD AND **EVE** FROM ONE OF HIS RIBS. ”

Henri de Mondeville, French surgeon, from *Cyrurgia, c.*1312

produced his *Cyrurgia* (*Surgery*), a manual based in part on his observations of dissected corpses, although his definitions were not always accurate.

The **practice of dissection was revived** by Italian medical professor **Mondino de Luzzi** (*c.*1275–1326), who taught at Bologna, and who performed **a public dissection** in 1315. Dissections featured regularly in his teaching. In 1316, Mondino completed *Anatomia*, **the first textbook specifically concerned with anatomy** (rather than surgery).

An atlas of the body
This anatomical drawing from Henri de Mondeville's Cyrurgia *shows the lower part of the torso cut away, revealing the internal organs.*

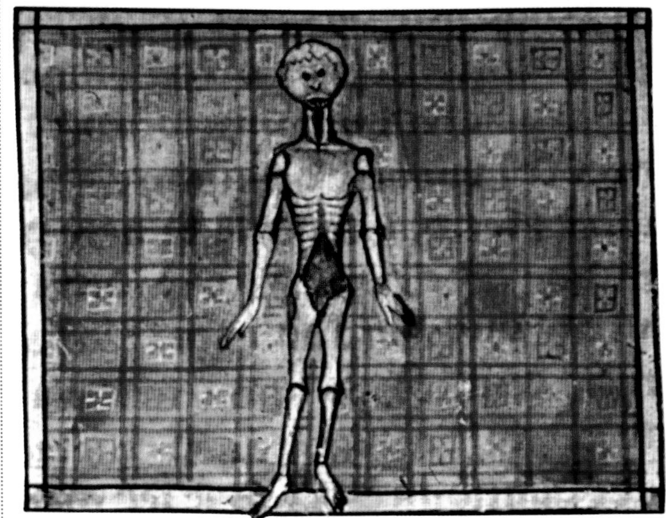

IN 1323, WILLIAM OF OCKHAM produced one of the greatest works of logic of the Middle Ages—*Summa Logicae* (*The Logic Handbook*)—in which he radically diverged from traditional Christian philosophy. Most notable in William's ideas is the idea of economy, that if a cause or factor is unnecessary to prove an argument, then it should be discarded, a principle that came to be called **Ockham's razor**. He promoted the idea that individual perception is the foundation of all knowledge about the world, attacking long-held metaphysical explanations for the order of the Universe. He also advocated the separation of secular and ecclesiastical power.

Windmills had been used to grind flour since the 12th century in Europe, but in 1345 **windmills** are first recorded as being used to operate water-pumps **to drain land in the Netherlands**. The resulting reclaimed land, or polders, ultimately came to make up a fifth of the country, which is still protected from the sea by a system of dikes.

Around 1349, the French scholar **Nicholas of Oresme** (*c.*1320–82) expounded a **system using graphs to represent the growth of a function** (such as the velocity of an object), which was a great aid to mathematical analysis. Later, in 1377, he proposed the idea in his *Livre du Ciel et du Monde* (*Treatise of the Sky and the World*) that the Earth was not immobile at the center of the Universe, as traditional cosmology held, but that it **rotated on its axis**. He met objections from those who said birds would simply fly off it, by affirming that the oceans were included in the rotation.

WILLIAM OF OCKHAM (*c.*1285–1349)

Franciscan friar William of Ockham studied at Oxford and by 1315 he was lecturing on the Bible. His theories of logic were seen by many as attacking Christian tenets, and he was summoned to the Papal court at Avignon to answer charges of erroneous teaching. He fled before the enquiry was concluded and spent the rest of his life at the court of the Holy Roman Emperor Louis IV of Bavaria.

20 PERCENT
THE PROPORTION OF THE **NETHERLANDS** THAT IS **RECLAIMED LAND**

1315 Mondino de Luzzi performs the **first public dissection** in Bologna

1316 De Luzzi publishes ***Anatomia***, the first textbook specifically on anatomy

1319 First cannon appears in Europe

1323 William of Ockham's *Summa Logicae* describes **Ockham's razor**

c.1340 Englishman John of Dumbleton suggests that no parts of a substance are eliminated during condensation, paving way for a **theory of molecules**

1343 First **Public Health Commission** established in Venice

1345 Windmills used for **reclaiming land** through drainage in the Netherlands

1348 Italian clockmaker Giovanni de Dondi begins construction of his **planetarium**

1348 English mathematician Richard Swineshead proves equations concerning movement of bodies with uniform velocity

c.1349 Nicholas of Oresme expounds a **system of graphical representation**

A catapult in the fortress of Edessa hurls a ball at a siege tower. According to Jean Buridan's theory, the catapult has imparted impetus to the projectile.

> “ A PROJECTILE WOULD BE **MOVED BY AN IMPETUS...** BY THE THROWER AND WOULD **CONTINUE TO BE MOVED** AS LONG AS THE **IMPETUS REMAINED STRONGER** THAN THE **RESISTANCE.** ”
>
> **Jean Buridan, French priest,** in *Questions on the Physics of Aristotle, c.*1357

IN THE LATE 1340S, A NEW AND TERRIBLE DISEASE STRUCK Europe, the Middle East, and North Africa. The Black Death was an **epidemic of bubonic plague** that infected humans via rat fleas. It was discovered later that the cause was a bacterium called *Yersinia pestis*. The plague reached Constantinople in 1347 and spread by ship throughout the Mediterranean, arriving in France and England in 1348. The disease began with swellings, or buboes, in the groin and armpit, followed by the spread of black spots over the body and high fever. It caused millions of deaths across Europe.

Contemporary physicians, who **had no cure for the plague**, believed that it was caused by putrefaction in the air brought on by humidity or rotting corpses. Remedies included controlling the body's heat by avoiding "putrefying" foods, such as meat and fish, fumigating rooms, and wearing pomanders infused with spices close to the nose.

Although doctors had failed to control the plague, after its end a renewed energy was given to medical science. By 1351, Padova had 12 medical professors (as against three in 1349). **Measures to promote public health** were also enacted. In 1377, the Republic of Ragusa (Dubrovnik) ordered **a quarantine** of 30 days for anyone coming from plague-infected areas, as did Marseilles in 1383. By 1450 Milan would establish **a permanent board of health**, and **health passports** were introduced in Italy in 1480.

Plague victims
This 15th-century Swiss manuscript shows victims of the plague with the characteristic swellings, or buboes, covering large parts of their bodies.

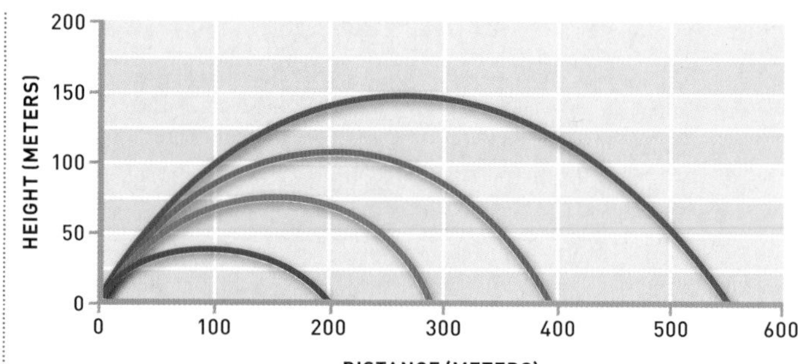

HEIGHT (METERS)
200
150
100
50
0
0
100
200
300
400
500
600
DISTANCE (METERS)

Free fall
The impetus provided by throwing an object to a greater height means it travels farther before downward forces pulls it back to the ground.

Aristotle's explanation of projectiles in motion had long puzzled scholars. In 1357, the French priest **Jean Buridan** (*c.*1300–58) published *Questions on the Physics of Aristotle*. He pointed out that a thrown stone continues to move even out of contact with the thrower. He theorized that the person throwing an object imparts a force to it, which he called **impetus**, and this causes it to continue to move, so long as the resistance of the air does not stop it. He believed the amount of impetus in an object depended on the amount of matter in it, so that feathers would not move quickly when thrown, whereas heavier objects would.

50 PERCENT

THE **PROBABLE DEATH RATE** IN EUROPE DURING THE PEAK OF **THE BLACK DEATH** IN THE MID-14TH CENTURY

1351 Black Death epidemic peters out. At its peak, it had killed about 50 percent of Europe's population

1357 French priest Jean Buridan publishes his **theory of impetus**

> **A SURGEON** WHO DOES NOT KNOW HIS **ANATOMY** IS **LIKE A BLIND MAN** CARVING A LOG.
>
> **Guy de Chauliac, French physician,** in *Great Surgery*, 1363

This illustration from Guy de Chauliac's *Chirurgia Magna* shows patients with a variety of injuries, including a broken arm and an eye wound, visiting a surgeon.

FRENCH PHYSICIAN GUY DE CHAULIAC (*c.*1300–68) was personal surgeon to three Popes. He remained in Avignon during the outbreak of the Black Death in 1348, an experience that led him to **distinguish for the first time between pneumonic** (affecting the lungs) **and bubonic plague**.

His *Chirurgia Magna* (*Great Surgery*, 1363), became one of the medieval world's **most important surgical textbooks**. In its seven volumes, he gave advice on the treatment of fractures, advising extending broken limbs with pulleys and weights, and noting the loss of cerebrospinal fluid in skull fractures. He outlined procedures such as tracheotomies (cutting open the windpipe) and the replacement of lost teeth by ox bone. But his over-reliance on the work of Galen (see 100–250 BCE) led him to some retrograde steps, such as **abandoning antiseptic treatment of wounds** and encouraging pus to form as part of the healing process.

Accurate marking of time
This reconstruction of de Dondi's 1364 astrarium shows three of its seven dials, as well as the balance wheel and weights that regulated its movement.

In 1364, the Italian clockmaker **Giovanni de Dondi** (1318–89) published his *Planetarium*, a description of the **complex astronomical clock (astrarium)** that he had just completed after 16 years of work. This 1 m- (3.3 ft-) high, weight-driven clock with an escapement and balance wheel was the most advanced of its time. Its **seven dials showed the celestial movements of the Sun, Moon, and five calendars**, and it acted as a perpetual calendar, including showing the date of Easter. The clock was regulated by a balance that swung 1,800 times an hour; and the addition or removal of small weights enabled corrections to be made if the device ran too fast or slow.

The **first recorded use of rockets** in Europe as a military weapon came in 1380 at the Battle of Chioggia, fought between the fleets of Venice and Genoa. Rockets need an ignition that provides continued and regular thrust as the projectile flies through the air (unlike cannon balls). Since gunpowder packed into a tube burns unevenly and mostly at the surface, military technologists had to devise new techniques. They left **a conical hole in the centre of the tube**, which encouraged an even burn (and sufficient thrust), and **made the rocket airtight**, except for a small opening at the rear. These methods were discussed by the German military engineer **Konrad Kyeser** (1366–after 1405) in his *Bellifortis* (*War Fortifications*) in 1405. Kyeser also advised the adding of feathers (like the fletching of an arrow) or weights to the rear of the rocket to make its trajectory more even and to enable more accurate aiming.

Making gunpowder
German legend attributed the invention of gunpowder to alchemist Barthold Schwartz. This woodcut depicts him stirring together the ingredients for gunpowder.

- **1363** French physician Guy de Chauliac finishes his ***Great Surgery***
- **1364** Italian clockmaker Giovanni de Dondi completes his **astrarium clock** after 16 years
- **1368 Fellowship of Surgeons** recorded in London, England
- **1370** English surgeon John Arderne describes a **new type of syringe**
- **1377** Republic of Ragusa enacts **quarantine laws**
- **1377 First effective use of cannons** in European siege warfare at Oudenaarde
- **1377** French philosopher Nicholas Oresme proposes idea that the **Earth rotates on its axis**
- **1380 First recorded use of rockets** in Europe, at the Battle of Chioggia
- **1383** Marseilles introduces **quarantine regulations**
- **1391 First dissection** recorded in Spain

1400–21

Italian artist Masolino da Panicale was an early master of visual perspective. He made good use of the technique in *St. Peter Curing a Cripple and the Raising of Tabitha*, painted for the Brancacci Chapel , Florence.

THE PRINCIPLES OF LINEAR PERSPECTIVE were known to the ancient Greeks, particularly **Euclid** who wrote of it in his *Elements*, but knowledge of these was lost after the fall of the Roman Empire.

Although Italian artist **Giotto** (1266–1337) had attempted to use algebraic formulas to create perspective, he had only partially succeeded. Renewed efforts to achieve true linear perspective included works from 1377 to 1397 by Italian mathematician **Biagio Pelacani** (*c.*1347–1416), who showed how **mirrors** could be used as **aids to view objects at a distance**. In 1415–16, Italian architect **Filippo Brunelleschi** (1377–1446) demonstrated in public for the first time the use of mirrors, by **reflecting an image of the Florence Baptistery** onto a 12 in- (30 cm-) canvas, which could then be drawn in perspective.

138 FEET THE WIDTH OF THE DOME OF FLORENCE CATHEDRAL

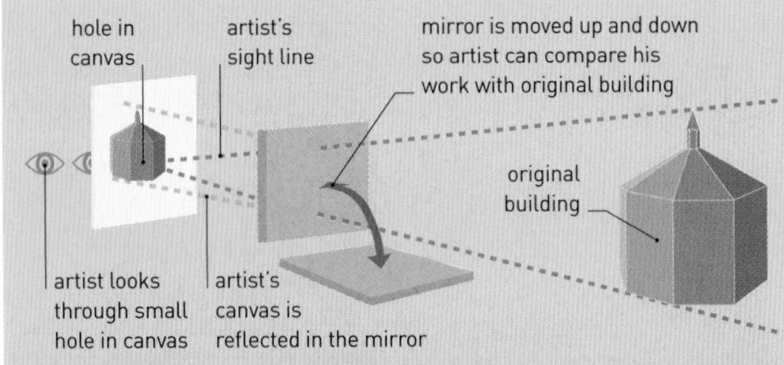
hole in canvas
artist's sight line
mirror is moved up and down so artist can compare his work with original building
original building
artist looks through small hole in canvas
artist's canvas is reflected in the mirror

BRUNELLESCHI'S PERSPECTIVE

Florence architect and artist Filippo Brunelleschi used mirrors to recreate an accurate depiction of Florence's Baptistry on canvas. He realized that linear perspective could be used to give an accurate impression of a three-dimensional object on a two-dimensional surface. Using a single perspective point (a hole in the canvas) and a mirror, he produced a painting that was identical to the original.

Brunelleschi may have been taught the technique by Florentine physician **Paola Toscanelli** (1397–1482), but he did not publish his theories until 1460. The first full account of the **application of linear perspective to painting**, including the creation of a grid to organize the placement of objects in a picture and of the principles of the **vanishing point and horizon line**, was set out by **Leon Battista Alberti** (1404–72), another of Toscanelli's pupils, in his *On Painting* in 1436.

As well as being an innovator in drawing, Brunelleschi devised advanced machinery for the construction of his many building projects in Florence. Among these was a *colla grande* (great crane), **a massive barge-based hoist** that could lift weights of more than one ton, had three different lifting velocities, and could operate in reverse without unhitching the load. In 1421, the authorities in Florence granted him **the first recorded monopoly patent**. The Venetian government would go on to regularize the process of granting patents, giving inventors 10 years' monopoly rights, as long as the invention was properly registered.

1422–49

The dome of Santa Maria del Fiore cathedral, Florence, is 137.8 ft (42 m) in diameter and 171 ft (54 m) high and took its architect, Brunelleschi, 16 years to complete.

IN 1420, THE MONGOL RULER ULUGH BEG (1411–49) had established a scientific institute at Samarkand, Uzbekistan; in 1424, he **started to build an observatory** there. It had a huge sextant that had a radius of 131 ft (40 m). Among the astronomers recruited was **Jamshid al-Kashi** (*c.*1380–1429), who produced **a mathematical encyclopedia** with a section on astronomical calculations, calculated the **value of pi** to 17 decimal places, and helped produce an extremely **accurate set of trigonometric tables**. In 1437, the astronomers at the observatory published the *Zij-i-Sultani*, **a star catalog** showing the position of 1,018 fixed stars.

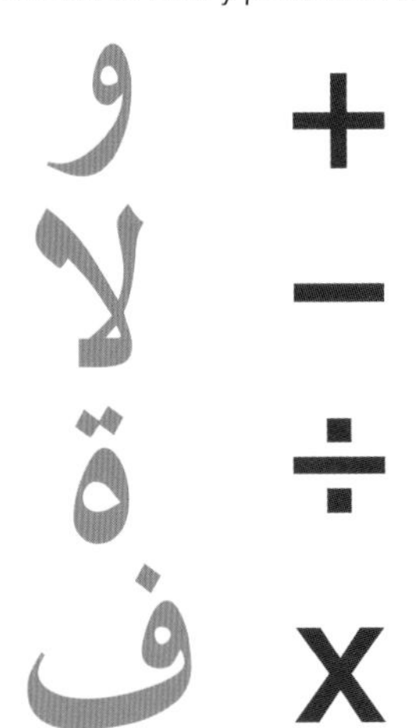
و
لا
ة
ف
+
−
÷
x
ARABIC CHARACTER
MATHEMATICAL SYMBOL

A new mathematical language
The mathematician al-Qalasadi used short Arabic words for algebraic operations, such as wa ("and") for addition and ala ("over") for division.

Between 1430 and 1440, **Ibn Ali al-Qalasadi** (1412–86), a Spanish Muslim mathematician, published a work in which he used a series of **short words and abbreviations** to stand for **arithmetical operations in algebraic equations**. He was not the first to do so—such Arabic abbreviations had appeared a century earlier in North Africa, and Diophantus had devised a form of algebraic notation—but al-Qalasadi's widely diffused works were responsible for popularizing the system.

In 1436, Brunelleschi finally completed the dome of Florence Cathedral after 16 years work. The dome was **the largest unsupported structure yet built**, and Brunelleschi solved the problem of its weight by building a lighter inner shell, on which was built a tougher outer dome. He used a ring and rib pattern of stone and timber supports between the two shells, and devised a herringbone pattern for the bricks, both of which helped diffuse the weight of the structure.

Nicholas of Cusa (1401–64), a German philosopher, wrote a number of treatises such as *De Docta Ignorantia* (*On Learned Ignorance*), which included advanced astronomical and

1415–16 Brunelleschi demonstrates the use of **perspective** in Florence

1420 Ulugh Beg establishes **scientific institute** in Samarkand

1421 First recorded **patent granted to Brunelleschi** in Florence

1424 Ulugh Beg starts to build **observatory** at Samarkand

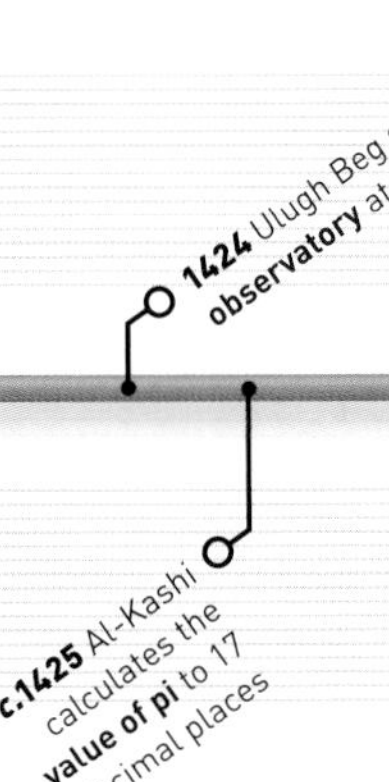

c.1425 Al-Kashi calculates the **value of pi** to 17 decimal places

“ WE **OUGHT NOT TO SAY** THAT BECAUSE THE **EARTH IS SMALLER THAN THE SUN** AND IS INFLUENCED BY THE SUN, IT IS **MORE LOWLY. ”**

Nicholas of Cusa, German philosopher, in *De Docta Ignorantia*, 1440

An early 16th-century woodcut shows Nicholas of Cusa caught between a group advocating church reform (as Nicholas did) and conservative Papal supporters.

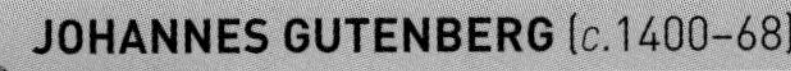

JOHANNES GUTENBERG (*c.*1400–68)

Born in Mainz, Johannes Gutenberg later moved to Strasbourg where he engaged in a mysterious venture he called “adventure and art”—perhaps his first experiments in printing. By 1448, he was back in Mainz, where, by 1450, his printing press was in operation. The venture did not prosper and by 1459 Gutenberg was bankrupt.

cosmological ideas. He held the radical view that **Earth rotates on its own axis and orbits around the Sun**, prefiguring Copernicus's theory 100 years later.

Around 1440, **Johannes Gutenberg** began experiments with **printing using movable type**. The blocks could be moved as required and later reused. By 1450, he had established **a printing press** that produced the earliest extant printed work in Europe, an edition of the *Ars Grammatica*. Gutenberg's printing techniques grew more sophisticated and, in 1454, he published an edition of the Bible.

Printing for the masses
This replica of Gutenberg's press shows the type of machinery that he employed in Mainz for the production of his 42-line Bible, so called because of its 42-line columns.

lever to tighten plates together to impress ink

plate for placing paper

ink transferred to printing block

plate for laying movable type in strips

“ IT IS A **PRESS…** FROM WHICH SHALL FLOW IN INEXHAUSTIBLE STREAMS… **LIKE A NEW STAR IT SHALL SCATTER** THE DARKNESS OF **IGNORANCE. ”**

Johannes Gutenberg, German printer, *c.*1450

1430–40 Al-Qalasadi publishes work using **symbols for algebraic operations**

1436 Leon Battista Alberti's *On Painting* describes **mathematical principles of perspective** in painting

1436 Brunelleschi completes the **dome of Florence Cathedral**

1437 Astronomers at Samarkand observatory publish ***Zij-i-Sultani*** star catalog

1440 Nicholas of Cusa proposes that **Earth moves around the Sun**, and that other stars may have inhabited planets

***c.*1440** Gutenberg begins experiments with printing using **movable type**

1450–67

An illuminated page from Gutenberg's 42-line Bible. The 48 surviving copies are among the most valuable books in the world.

IN 1454, GERMAN PRINTER JOHANNES GUTENBERG completed his edition of the Bible printed with 42 lines on each page. It was the first substantial book printed in Europe and its 180 or so copies sold out almost immediately. The Gutenberg Bible was soon followed by hundreds of works by Gutenberg and other printers, allowing the much **more rapid dissemination of scientific ideas**.

In 1464, German mathematician **Johannes Müller**, also known as Regiomontanus (1436–76), completed his *De Triangulis Omnimodis (On Triangles)*, a systematic **textbook for trigonometry**. One of his fundamental propositions was that two triangles that have sides in similar proportions will also have similar angles. He relied on the work of Arabic mathematicians.

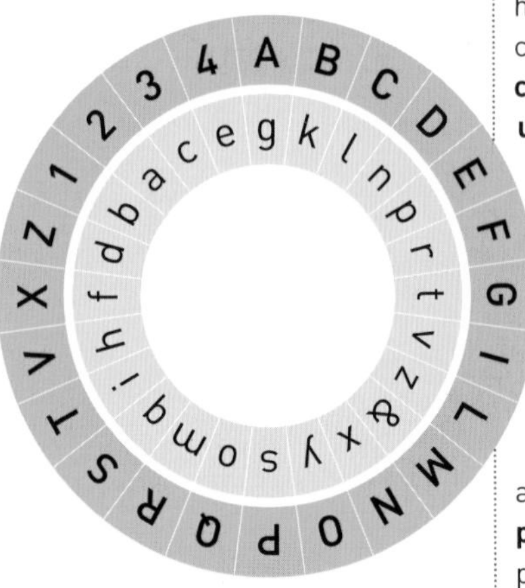
1 2 3 4 A B C D E F G I L M N O P Q R S T V X Z
a c e g k l n p r t v & x y s o m q i h f d b
Cracking the code
Alberti's disk operated by rotating the inner ring so that an agreed character (such as "g") lined up with the A of the outer ring.

30 FLORINS
THE ORIGINAL COST OF THE GUTENBERG BIBLE

The first work on cryptography had been written in the 13th century. By the 15th century, **cyphers were in widespread use for diplomatic correspondence**. Codes relied on **monoalphabetic substitution**, in which each letter is transformed into the same encoded letter.

In 1466, Italian painter and philosopher **Leon Battista Alberti** (1404–72) devised a cypher disk that made **polyalphabetic substitution** possible. Each new rotation of the disk generated an entirely new alphabet for coding.

1468–82

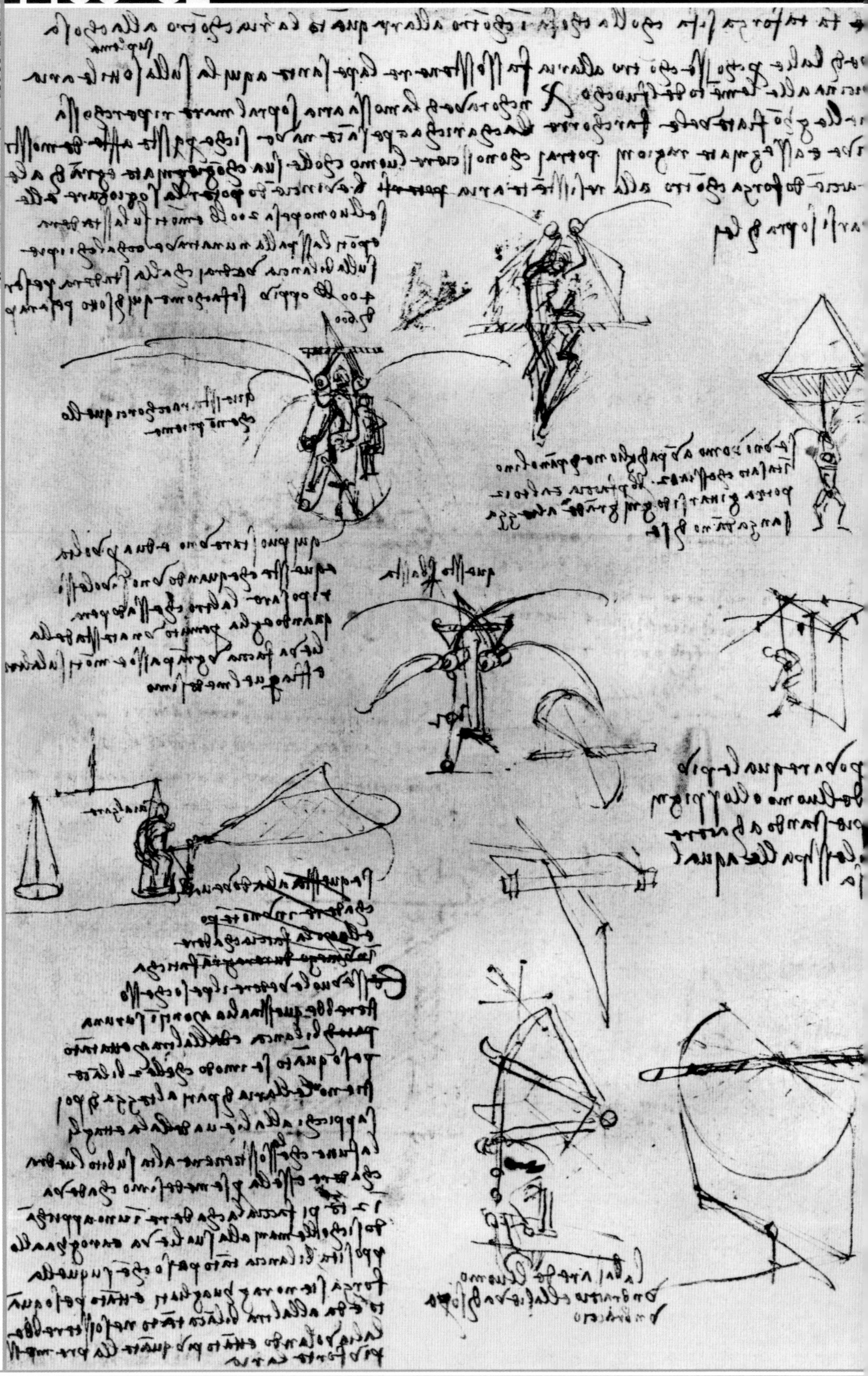

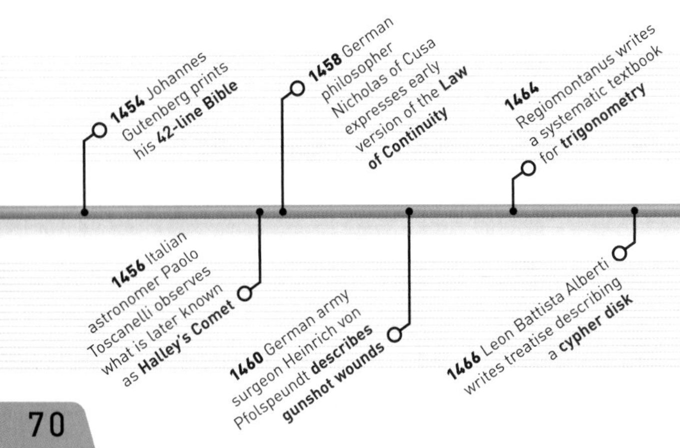
1454 Johannes Gutenberg prints his **42-line Bible**
1456 Italian astronomer Paolo Toscanelli observes what is later known as **Halley's Comet**
1458 German philosopher Nicholas of Cusa expresses early version of the **Law of Continuity**
1460 German army surgeon Heinrich von Pfolspeundt **describes gunshot wounds**
1464 Regiomontanus writes a systematic textbook for **trigonometry**
1466 Leon Battista Alberti writes treatise describing a **cypher disk**

1472 Austrian astronomer Georg Peurbach publishes *New Theories of the Planets*—the **first widely circulated printed astronomical work**

1478 Publication of the ***Treviso Arithmetic***, the first printed general work on mathematics

> **“** ALTHOUGH **NATURE COMMENCES** WITH **REASON** AND ENDS IN EXPERIENCE IT IS NECESSARY **FOR US** TO **DO THE OPPOSITE… ”**

Leonardo da Vinci, Italian painter, architect, and engineer, in *Notebooks*

This woodcut shows Columbus's three ships, the *Niña*, *Pinta*, and *Santa Maria*, on their five-week crossing of the Atlantic Ocean, which ended in the discovery of the Americas.

THE LATE 15TH CENTURY saw the production of the first printed practical mathematical textbooks. The *Treviso Arithmetic*, printed in 1478, demonstrated techniques of **addition** and **subtraction**, as well as **division** and five different ways of performing **multiplication**, including cross-multiplication and a "chessboard" technique similar to modern practice. It also dealt with the rule of mixtures (for instance, showing the proportions of a precious metal in alloys) and **methods of calculating the Golden Number** (see 1723–24).

In 1483, it was followed by a German counterpart, the *Bamberger Rechenbuch* (*Bamberger Arithmetic*), which, as well as setting out five procedures for multiplication, gave **rules for the summing of geometrical and arithmetical progressions**.

The prolific scientific interests of the Italian painter, architect, and engineer Leonardo da Vinci led him to his studies on the mechanisms of flight. He considered that "the bird is an instrument operating through mathematical laws" and **he worked on designs for flying machines** using birdlike wings.

In 1481, he also devised a **parachute,** with a sealed linen cloth supported by wooden poles that made a pyramidal shape, that would reduce the rate of acceleration in a fall and cushion the wearer's impact. There is, however, no evidence that Leonardo actually built any of these fantastic machines.

LEONARDO DA VINCI
(1452–1519)

Born in Tuscany, Leonardo da Vinci was the most creative mind of the Renaissance. He became an apprentice sculptor before moving to Milan to work for the ruling Sforza family. A talented artist, his paintings include *The Last Supper* (1495–98) and *Mona Lisa* (c.1503). His scientific interests were vast and he produced 13,000 pages of notebooks.

Leonardo da Vinci's notebooks
This page from Leonardo's notebooks shows his sketches of flying machines. He used mirror-writing to make notes, although it is uncertain why he did so.

THE DEVELOPMENT OF MATHEMATICAL NOTATION progressed rapidly during the late 15th century. The *Dresden Manuscript* (1461) gave **special symbols for the first four powers of x**, and, around 1489, the German mathematician Johannes Widman (1462–98) wrote the first work to use the signs "+" and "–" to represent the operations of addition and subtraction. He also used a long line to represent "equals."

Around 1489, Leonardo **began a study of human anatomy**, using dissections of animals and human corpses (he claimed to have dissected 10). He recorded his findings in notebooks between 1489 and 1507. In these, he made the most detailed anatomical drawings yet seen.

In 1490, Leonardo was **the first person to describe capillary action**, the ability of water in tiny spaces to "crawl upward," acting contrary to other natural forces (such as gravity).

In October 1492, the Genoese explorer Christopher Columbus landed at San Salvador in the Bahamas, **the first European to reach the Americas** since the Vikings in the 11th century. His voyage led to an exchange of population, food crops, and diseases as well as the discovery of large numbers of hitherto unseen species, such as the llama and armadillo.

> **“ I… STEERED FOR THE CANARY ISLANDS…** THENCE TO TAKE MY DEPARTURE AND PROCEED **TILL I ARRIVED** AT THE **INDIES. ”**

Christopher Columbus, from *Journal of the First Voyage*, 1492

NAVIGATING AND MAPPING THE WORLD

The translation of Ptolemy's *Geography* from Greek to Latin in 1409 and the Portuguese voyages down the west coast of Africa gave an impetus to map-making techniques.

Maps of the 15th century, such as this 1540 map (left) by Venetian monk and mapmaker Fra Mauro, combined a knowledge derived from Ptolemy with information sourced from mariners' charts, but did not use a projection that portrayed distances accurately. It wasn't until 1569 and Flemish geographer and cartographer Gerardus Mercator's world map that maps really helped sailors determine routes at sea more easily.

1481 Leonardo da Vinci produces a **design for a parachute**

1489 A treatise by Johannes Widman is first to use **"+" and "–" signs**

1489 Leonardo da Vinci begins series of **anatomical drawings** based on dissections

1490 Leonardo describes the principle of **capillary action**

1492 Christopher Columbus lands in the Bahamas

1492 Italian scholar Ermolao Barbaro's ***Castigationes Plinianae*** suggests thousands of **corrections** to Pliny's *Natural History*

1494 Italian mathematician Luca Pacioli's ***Algebra***—the first comprehensive work on the subject—is printed

1496 Regiomontanus's *Epitome of Ptolemy's Almagest* is published, making his **astronomical theories more accessible**

1500–16

Martin Waldseemüller's 1507 world map was the first to name America, although much of the coastline of North and South America remained unknown.

THE EXPLORATION OF NEW LANDS IN THE LATE 15TH CENTURY, and in particular, Columbus's discovery of the Americas in 1492 and Vasco da Gama's circumnavigation of Africa en route to India in 1497–98, provided **much new material for map-makers**.

In 1504, a letter written by Amerigo Vespucci (1454–1512) detailing his third voyage to America came into the hands of a group based in St.-Dié in Lorraine (in modern-day France). One of them, **Martin Waldseemüller** (*c.*1470–1522), produced **a globe and world map in 1507 in which he suggested that the new-found continent be called America**, the first occurrence of the term.

In 1508, Waldseemüller wrote a treatise on surveying, in which **he described the theodolite** (which he called polimetrum) for the first time. Using a theodolite, surveyors and cartographers could now measure angles of up to 360 degrees.

4 YEARS
THE AVERAGE LIFESPAN OF A SWISS MILITARY WATCH BATTERY

40 hours The average lifespan of a 16th-century wind-up pocket watch

Batteries versus springs
Henlein's first watch would have lasted less than two days before needing rewinding, but this was a major achievement at the time.

In the late 15th century, clockmakers learned how to **construct spring-driven clocks** in which the gradual uncoiling of the spring operates the mechanism. The Nuremberg clockmaker **Peter Henlein** (1485–1542) applied this system to portable clocks ("watches"). In 1512, Henlein was recorded as having **made a watch that went for 40 hours and could be carried in a pocket**.

In 1513, Polish astronomer **Nicolaus Copernicus** (1473–1543) wrote his *Commentariolus* (*Little Commentary*), a preliminary outline of **his revolutionary view that Earth revolved in orbit around the Sun**. Feeling dissatisfied with the old planetary theory of Ptolemy, with its multiplicity of celestial spheres, geocentric view, and its anomalies (such as the apparent retrograde motion of some planets), Copernicus explained how a planet's periodicity varied in proportion to its distance from the Sun. Fearing the reaction of the Church, Copernicus kept his findings to himself for 30 years.

Around 1500, **gunsmiths devised the wheel-lock mechanism** for firearms. It used a serrated metal wheel that rotated rapidly, striking against a lump of the mineral pyrite, and creating sparks that lit the gunpowder charge.

First pocket-watch
The compact workings of Peter Henlein's portable clock (c.1512) were driven by a slowly uncoiling spring. It was the first timepiece small enough to be carried in the user's pocket.

1507 Martin Waldseemüller produces first map that uses the name **America**

1508 Waldseemüller describes the **theodolite**

1512 Peter Henlein makes the first **pocket watch**

1513 Nicolaus Copernicus writes his *Commentariolus*, an outline of his **heliocentric theory** of the solar system

1515 The first **wheel-lock pistols** appear in Germany

1517–35

"FRACASTORO… DECLARED… THAT **FOSSIL SHELLS** HAD ALL BELONGED TO **LIVING ANIMALS."**

Charles Lyell, Scottish geologist,
from *Principles of Geology*, 1830–33

THE MEDIEVAL 13TH-CENTURY SCHOLAR Albertus Magnus had described stones with the "figures of animals," but Arab and medieval scholars believed they were produced by Earth itself or were the remains of animals drowned in the Biblical flood. In a debate in 1517, Italian physician **Girolamo Fracastoro** (1478–1553) was the first to publicly express the view that **fossils are organic matter**, originally animals, that has been ossified over time.

FAMOSO·DOCTOR PARESEL

Portrait of Paracelsus
Paracelsus was both physician and chemist, and stressed the importance of using chemical techniques in the production of medicines.

In August 1522, the 18 survivors of Ferdinand Magellan's expedition arrived in Spain, having completed the **first circumnavigation of Earth**. The voyage had taken three years, and more than 230 crew (including Magellan) perished. However, it did definitively prove the size of **Earth's circumference** to be about 24,800 miles (40,000 km).

In 1525, German artist **Albrecht Dürer** (1471–1528) published his *Instructions for Measuring with Compass and Ruler*, **one of the first works on applied mathematics**, which contained detailed accounts of the properties of curves, spirals, and regular and semiregular polygons and solids and their use as **an aid for artists in producing scientifically accurate images**.

German chemist and physician **Theophrastus von Hohenheim** (1493–1541), known as Paracelsus, devised a new classification of chemical substances, rejecting Aristotle and Galen's four humors. In his *De Mineralibus* (*On Minerals*), he divided them instead using the three principal units of sulfur, mercury, and salt. Paracelsus spurned the study of anatomy and promoted the idea that the

1517 Frascatoro expresses his view that **fossils** were originally organic life forms

1522 Survivors of Ferdinand Magellan's expedition complete first **circumnavigation** of Earth

1525 Christoph Rudolff produces first **German algebraic manual** and introduces modern symbol for square root

1525 Albrecht Dürer publishes first work on **applied mathematics** for use by artists

Girolamo Fracastoro was one of the first to believe that fossilized shells, such as this fossil of the Archimedes species, had once been animals.

“ TO KEEP **ALIVE** THE **MEMORY OF OLD KINGDOMS** AND EVENTS AND... MAKE KNOWN TO **COMING GENERATIONS OUR TIME.** ”

Imperial charter describing Gerardus Mercator’s terrestrial globe, *c.*1535

body (microcosm) must be in balance with nature (macrocosm). His interest in distilling chemicals led him to use apparently noxious substances such as sulfuric acid (which he employed against gout), mercury, and arsenic as medicines. Some time before 1529 he began to use a **pain reliever he called laudanum**.

In 1533, Flemish cartographer **Gemma Frisius** (1508–55) gave the **first full description of the method of triangulation**, by which a large area could be surveyed from an accurately measured base line. He went on in 1547 to suggest a new way of calculating longitude, by using a portable clock set to the time of the point of departure that could be compared with a clock showing the time at the point of arrival. The imprecision of clocks, however, rendered the technique of limited practical value.

In 1521 Italian surgeon **Berengario da Carpi** (*c.*1460–1530), who lectured on anatomy at the University of Bologna, wrote about the importance of the anatomy of things that can be observed, which included the **use of dissection of human corpses**. He used this as the basis for his *Anatomia Carpi* (*The Anatomy of Carpi*), which was the first anatomical work to use printed figures to illustrate the text.

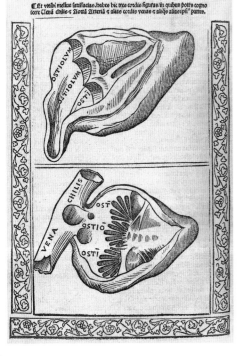

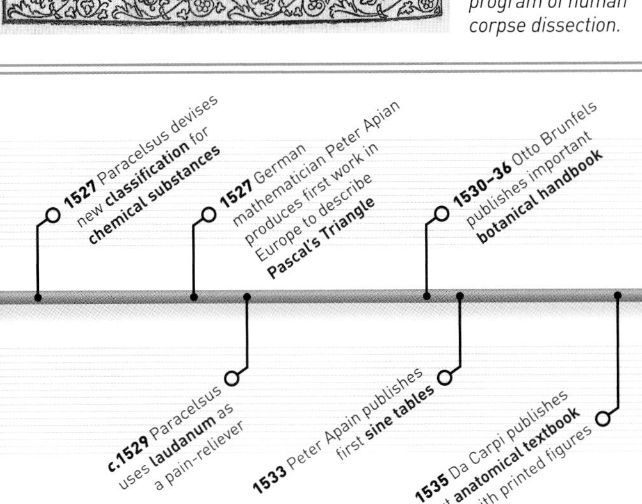
DE ANATOMIA
Et vtibi melius satisfacias.habes hic tres cordis figuras/in quibus potes cogno scere Venã chilis & Aortã Arteriã & alias cordis venas & aliq̃s alias ipsi' partes.
OSTIOLVM
OSTIOLVM
OSTI
CHILIS
OST
OSTIO
VENA
OSTI

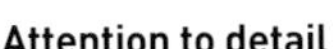

Attention to detail
Drawings from Anatomia Carpi *show the veins leading into the heart. These examples show how accurately da Carpi derived his drawings from his program of human corpse dissection.*

GERARDUS MERCATOR
(1512–94)

Born in Flanders, Gerardus Mercator embarked on a career making mathematical instruments. He began producing maps in 1537 and published his first world map in 1538. In 1569, he compiled another map of the world, this time using a projection that showed constant lines of course as straight lines, which came to bear his name.

AN IMPETUS TO BOTANY IN THE RENAISSANCE had been provided by the desire to illustrate plants found in classical authors’ texts, such as those of Roman naturalist Pliny, and the possibilities provided by printed illustrations. Between 1530 and 1536, German Carthusian monk **Otto Brunfels** (*c.*1488–1534) published his *Herbarum Vivae Iconis* (*Living Pictures of Herbs*) with 260 woodcuts of plants, whose accurate detail set an exacting **standard for botanical drawing**.

In 1530, Gemma Frisius had compiled a manual explaining how to construct a globe showing Earth’s geography (a terrestrial globe). In 1541, Flemish cartographer Gerardus Mercator produced what would become the **first surviving terrestrial globe**. He also included a selection of stars superimposed on the globe, as well as rhumb lines (which showed the straightest course between two points on the same latitude); both were invaluable aids at sea.

In 1542, German botanist **Leonhard Fuchs** (1501–66) published his *De Historia Stirpium* (*The History of Plants*), in which he **described around 550 plants** (mainly medicinal ones), providing their names and therapeutic virtues. Its drawings were so clear that it became the **first botanical work to be widely used by laymen**.

Botanical drawing
Accurate and attractive illustrations, such as this drawing of a borage, or starflower, made Fuchs’s De Historia Stirpium *a valued botanical handbook.*

1527 Paracelsus devises new **classification for chemical substances**

1527 German mathematician Peter Apian produces first work in Europe to describe **Pascal’s Triangle**

c.1529 Paracelsus uses **laudanum** as a pain-reliever

1530–36 Otto Brunfels publishes important **botanical handbook**

1533 Peter Apain publishes first **sine tables**

1535 Da Carpi publishes first **anatomical textbook** with printed figures

1537 Italian mathematician Niccolò Tartaglia issues his *Nova Scientia*, first major modern **work on ballistics**

1539 German botanist Jerome Bock (Tragus) classifies plants into **herbs, grasses and trees, or shrubs**

1541 Mercator produces his first **celestial globe**

1542 Leonard Fuchs’s *De Historia Stirpium* accurately **describes 550 plants**

3

THE AGE OF DISCOVERY 1543–1788

As international travel increased, greater emphasis was placed on first-hand observations and accurate instruments. Instead of relying on written authority, natural philosophers devised experiments to construct testable theories about the Universe.

“ AT REST, HOWEVER, IN THE **MIDDLE OF EVERYTHING** IS **THE SUN.** ”

Nicolaus Copernicus, Polish astronomer, from *De Revolutionibus Orbium Coelestium*, 1543

Copernicus's idea that all the planets revolve around the Sun, rather than Earth being at the center of the Universe, was a departure from conventional astronomy and challenged the authority of the Church.

A CENTURY AFTER GUTENBERG REVOLUTIONIZED PRINTING with the invention of **movable type** (see 1450–67), scientists were able to publish their work for a mass readership, giving new ideas wider influence. The year 1543 was a milestone in **scientific publishing**, when several important books first appeared. Two books stood out—Nicolaus Copernicus's *De Revolutionibus Orbium Coelestium (On the Revolutions of Celestial Bodies)* and Andreas Vesalius's *De Humani Corporis Fabrica (On the Structure of the Human Body)*. They are often seen as marking the beginning of a **new scientific age**, as they called into question the conventional authorities on **astronomy** and **anatomy**.

Up to that time, most astronomers believed that Earth was at the center of the Universe—a view put forward by **Ptolemy** in the 2nd century. **Copernicus**, however, calculated that **Earth, and all the planets, revolved around the Sun**. He had been working on this idea since about 1510, and by the 1530s had put together mathematical calculations to support his argument. However, he was reluctant to publish his theory because it **challenged convention** and went against the Church. He was persuaded to publish *De Revolutionibus* by Georg Rheticus, an Austrian mathematician who had come to study with him. It is said that Copernicus was presented with its first edition on his deathbed.

An expensive book, *De Revolutionibus* sold only a few hundred copies and did not have an immediate impact. However, Copernicus's mathematical arguments for a heliocentric (sun-centered) Universe were soon accepted by most astronomers, leading to a rift between them and the Church.

In contrast, **Vesalius** was 28 years old when he published his comprehensive seven-volume study of human anatomy, *De Humani Corporis Fabrica*. This was the **first book of human anatomy to be fully illustrated**, showing in detail what Vesalius had discovered in his dissections of human bodies. Unlike Copernicus's work, *De Humani* sold well, and Vesalius published a single-volume summary of the book later in 1543.

This year saw groundbreaking publications in the field of mathematics as well. Italian engineer and mathematician **Niccolò Fontana Tartaglia** published his translation of Euclid's *Elements* into Italian, the first translation of that work into a modern European language.

Welsh mathematician **Robert Recorde** published ***The Ground of Artes***, the first printed book on mathematics in English. It was to remain a standard textbook for more than a century.

400 THE FIRST **PRINT RUN** OF COPERNICUS'S *DE REVOLUTIONIBUS ORBIUM COELESTIUM*

The heliocentric Universe
In De Revolutionibus Orbium Coelestium, *Copernicus used mathematics and astronomical observations to show how Earth and the other five planets move in circular orbits around the Sun.*

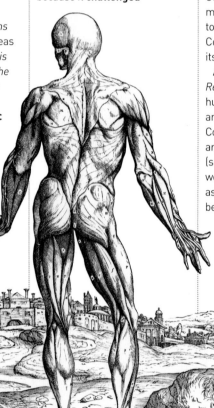

De Humani Corporis Fabrica
Vesalius's treatise on human anatomy was lavishly illustrated with detailed drawings of various stages of dissection. The figures are drawn in poses similar to the allegorical paintings of the time.

NICOLAUS COPERNICUS (1473–1543)

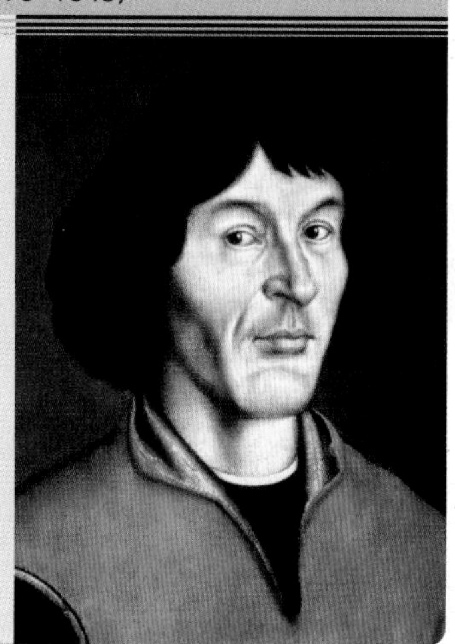

Born in Torun, Poland, into a German family, Nicolaus Copernicus was brought up by his uncle after his father's death. He studied law in Bologna and medicine in Padua. Copernicus lectured in mathematics in Rome before returning to Poland to work as a physician. He developed the idea of a heliocentric Universe, but had only just published his work when he died in 1543.

May Nicolaus Copernicus suggests the **Sun is at the center of the Universe**

Andreas Vesalius publishes *De Humani Corporis Fabrica*, a pioneering book on human anatomy

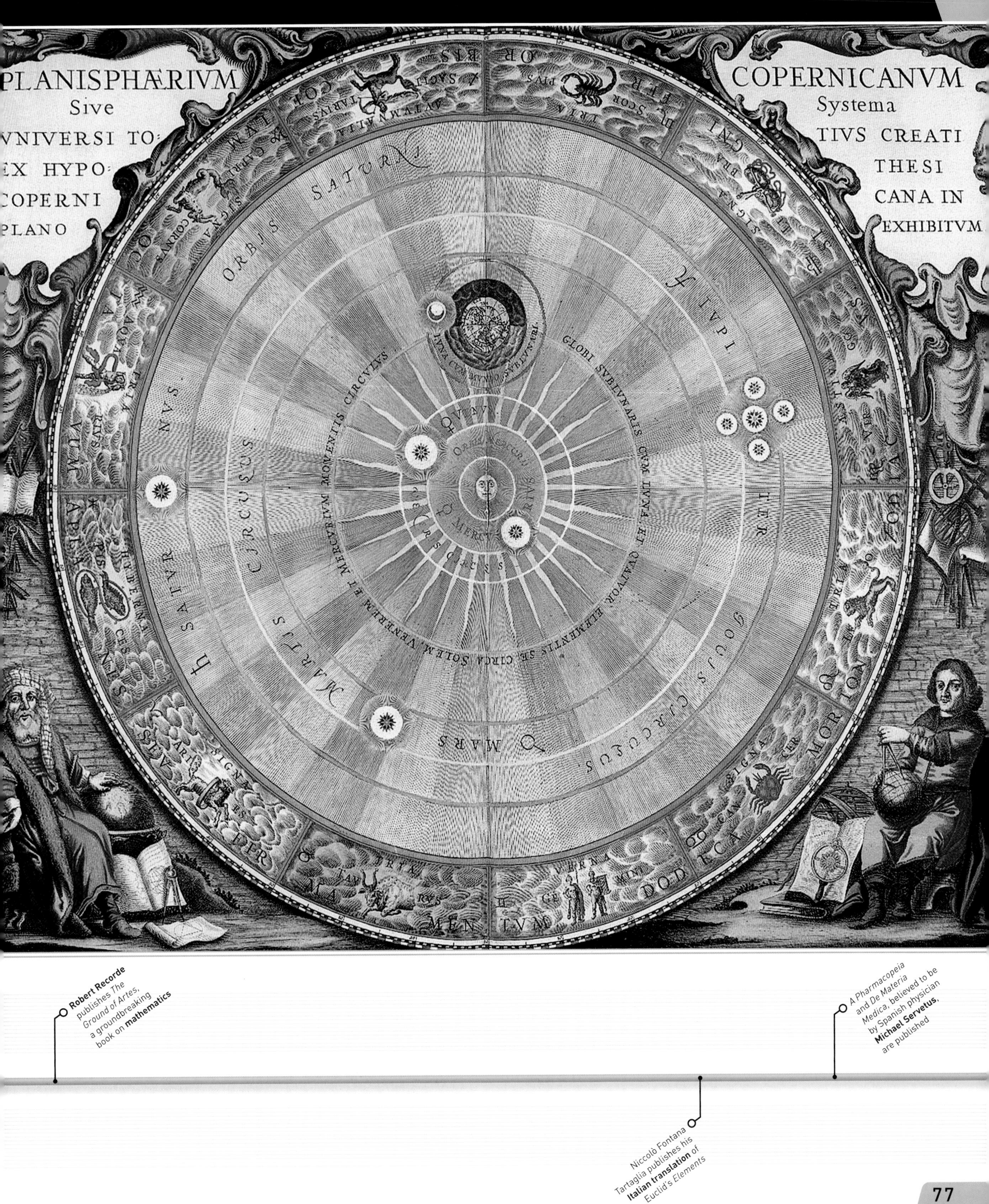

PLANISPHÆRIVM
Sive
VNIVERSI TO:
EX HYPO:
COPERNI
PLANO
COPERNICANVM
Systema
TIVS CREATI
THESI
CANA IN
EXHIBITVM

Robert Recorde publishes *The Ground of Artes*, a groundbreaking book on **mathematics**

Niccolò Fontana Tartaglia publishes his **Italian translation** of Euclid's *Elements*

A Pharmacopeia and *De Materia Medica*, believed to be by Spanish physician **Michael Servetus**, are published

Anatomy of the shoulder
Leonardo da Vinci brought both artistic aptitude and scientific enquiry to his studies of anatomy. He often worked in collaboration with an anatomist to ensure the accuracy of his work.

deltoid (shoulder muscle)

biceps brachii (biceps)

trapezius (neck-shoulder muscle)

scapula (shoulder blade) connects the humerus with the clavicle

clavicle (collarbone)

humerus (upper arm bone)

pectoralis (pectoral muscle)

ribcage

sternum (breast bone)

1600 BCE
Mummification
In ancient Egypt, bodies are mummified; the internal organs are removed for religious reasons and to help preservation, and are stored in canopic jars.

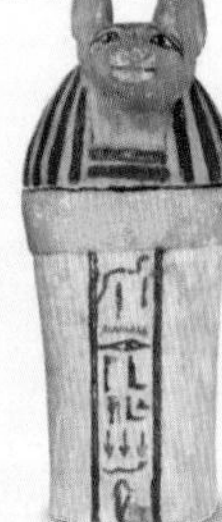
Canopic jar

12th century
Islamic physicians refute Galenic wisdom
There is no prohibition of human dissection in the medieval Islamic world, where physicians such as Ibn Zuhr (Avenzoar) perform routine autopsies. Ibn Zuhr corrects some of Galen's human anatomy—much of it based on the Barbary ape.

Late 15th century
New observations
With physicians refuting Galen, Leonardo da Vinci begins his study of human anatomy. Italian physician Jacopo da Carpi's *Anatomia Carpi* introduces a new age of original observation.

500 BCE
Early Greek anatomy
The Greek physician Hippocrates promotes animal dissection as a way of learning about the human body.

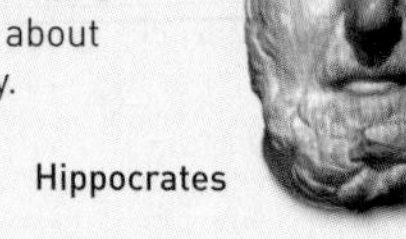
Hippocrates

180 BCE
Galenic circulation
Greek-born physician Galen concludes that blood is continually made in the body, an idea not corrected until the 17th century.

Galen's anatomy

1300s
Mondino de Liuzzi
Italian physician Mondino de Liuzzi performs the first public human dissections c.1315, but his catalog of anatomy perpetuates many erroneous ideas of antiquity.

Liuzzi's *Anathomia*

THE STORY OF ANATOMY

THE SECRETS OF THE LIVING BODY HAVE LONG FASCINATED BOTH SCIENTISTS AND ARTISTS

The exploration of biological structure—anatomy—is the basis for understanding how bodies work. Early anatomists had to dissect cadavers to find answers to even simple questions; later, technologies such as the microscope helped physicians chart the body in greater detail.

In the ancient world, anatomists dissected the bodies of animals but were forbidden to open human cadavers, which were considered sacred. As a result, Galen (129–200 CE), Rome's most celebrated physician, circulated erroneous ideas about the human body that were based on animal anatomy. When human dissection was sanctioned, Galen's ideas were corrected through direct observation. By the Renaissance, artists such as Leonardo da Vinci (see 1468–82) were illustrating bodies with exquisite realism, and each new anatomical publication charted and named new structures. Flemish-born anatomist Andreas Vesalius (see 1543) dominated the scene with his illustrated *De Humani Corporis Fabrica*.

Anatomical waxwork
Three-dimensional figures, such as this 19th-century wax fetus, were important tools for teaching medicine.

DELVING DEEPER

With the invention of microscopes in the 1600s, anatomists could see that organs were made up of cellular tissues. By the 1900s, the discovery of X-rays heralded new directions for anatomy. Today, powerful electron microscopes can probe the detailed structure of cells, and new imaging techniques reveal internal structures in 3-D without the need to cut the body open.

PRESERVING ANATOMICAL SPECIMENS

Dead body parts decay quickly. Preservation in alcohol prolongs opportunities for study, but also dehydrates specimens, causing distortion. Formalin is commonly used as a fixative to avoid this. Some of the most sophisticated modern methods preserve bodies in a dry state, for example, by replacing water and fat with plastic.

> “THE **HUMAN FOOT** IS A **MASTERPIECE** OF **ENGINEERING** AND A **WORK** OF **ART.**”

Leonardo da Vinci, Italian polymath, from his notebooks, 1508–18

1543
Father of anatomy
Artists attend the dissections of Flemish-born Andreas Vesalius to draw accurate illustrations for his *De Humani Corporis Fabrica*.

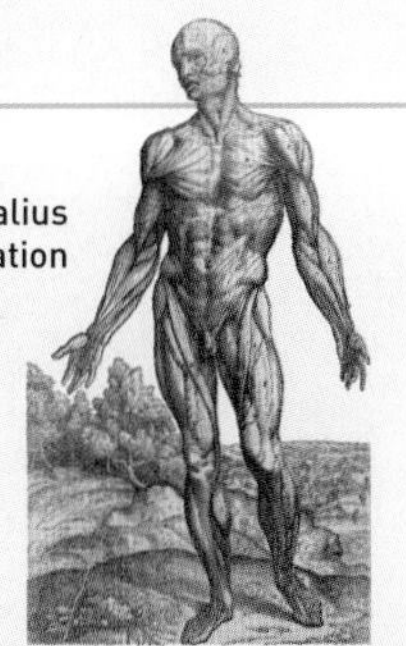

Vesalius illustration

1665
Compound microscopes
Anatomists such as Marcello Malpighi, Jan Swammerdam, and Robert Hooke use sophisticated microscopes to record the structure of cells, capillaries, and tissues.

Hooke's microscope

1770
Microtome
The microtome is invented for cutting tissue into extremely thin, almost transparent sections. This enables samples to be examined under high-power light microscopes.

Microtome

Mid-19th century
Comparative anatomy
Directed by Charles Darwin's evolutionary theory of 1859, many anatomists seek evidence of common descent among species.

Chimpanzee skeleton

1895
X-rays
German physicist Wilhelm Röntgen uses his newly discovered X-rays to reveal the bones of his wife's hand, and shows a way of examining internal bony structures without the need to dissect.

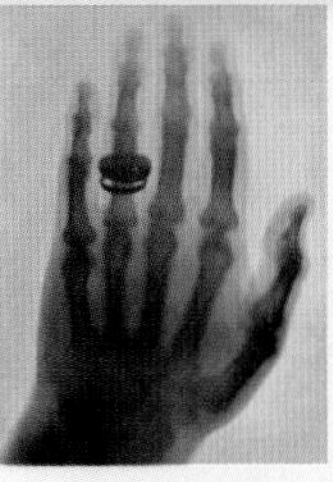

X-ray of hand

1940s–1950
MRI scanning
In 1946, American physicists find a way of detecting signals from atoms that enable scientists to obtain images of the soft internal structures of living bodies.

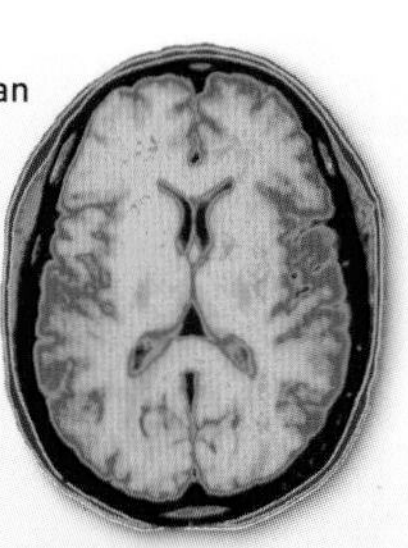

MRI scan

1544–45

The Orto Botanico di Padova (Padua Botanical Garden), the oldest existing botanical garden in Europe, continues to be a major center for research in botany and pharmacology even today.

GERMAN CLERGYMAN AND INSTRUMENT-MAKER GEORG HARTMANN (1489–1564), was the first to notice and describe, in 1544, the phenomenon of magnetic inclination. Also known as **magnetic dip**, this is the phenomenon whereby the needle of a compass follows the line of Earth's magnetic field, which curves around Earth's surface. As a result, the north-pointing end of a compass needle tends to point slightly downward in the Northern Hemisphere and upward in the Southern Hemisphere. Hartmann's discovery was not widely known until centuries later. In 1581, English instrument-maker Robert Norman published his own account of the phenomenon.

> **“** MATHEMATICS IS… ITS OWN EXPLANATION… FOR THE **RECOGNITION** THAT A **FACT** IS SO, IS THE **CAUSE UPON WHICH WE BASE** THE **PROOF. ”**
>
> **Gerolamo Cardano, Italian mathematician**, in *De Vita Propria Liber*

MICHAEL SERVETUS (1511–53)

Spanish scientist Miguel Servet, also known as Michael Servetus, wrote several treatises on medicine and human anatomy. He was the first European to correctly explain pulmonary circulation in *Christianismi Restitutio* (*The Restoration of Christianity*). His theological works were considered heretical, and he was burned at the stake in Geneva for his views.

In 1545, Italian mathematician **Girolamo Cardano** (1501–76) published ***Ars Magna*** (*The Great Art*), an important book on algebra. He presented **solutions to cubic and quartic equations**, involving unknown quantities to the power of three and four respectively. He drew on his own ideas as well as those of compatriots such as **Niccolò Fontana Tartaglia** (1499–1557), who translated works of Euclid and Archimedes. *Ars Magna* made reference for the first time to **imaginary numbers**—those numbers that are a multiple of the square root of −1.

Italy was also becoming a center for botanical research, with the opening of the **botanical garden** in Padua, in 1545. The first of its kind and a model for subsequent botanical gardens, it was established by the Senate of the Venetian Republic. It was used for **growing and studying medicinal plants** and comprised a circular plot of land, symbolizing the world, surrounded by water. The **first custodian** of the gardens was **Luigi Squalermo** (1512–70), also known as Anguillara. He cultivated around **1,800 species of medicinal herbs**, making a significant contribution to the modern scientific studies of botany, medicine, and pharmacology.

1546–50

> **“ CLOTHES, LINEN, ETC…** NOT THEMSELVES CORRUPT, **CAN… FOSTER** THE ESSENTIAL SEEDS OF THE **CONTAGION** AND THUS CAUSE INFECTION. **”**
>
> **Girolamo Fracastoro, Italian physician, poet, and geologist,** 1546

PHYSICIAN, GEOLOGIST, AND POET GIROLAMO FRACASTORO (1478–1553) published his most important work, ***On Contagion and Contagious Diseases***, in Italy in 1546. Best known at the time for his poem *Syphilis, or the French Disease* in 1530, he covered his examination of diseases in greater depth in *On Contagion*, offering an early **explanation** of the **mechanism by which diseases are spread**. His theory was that each disease is caused by very small bodies, or "spores," which are carried in the body, skin, and clothing of the person affected. These minute bodies, he believed, could multiply rapidly, and be transmitted from person to person by physical contact by handling unwashed clothes, or even through the air. Despite being initially accepted by the medical establishment, his ideas had little effect on the treatment and prevention of disease until his theory was proven right by Louis Pasteur (see 1857–58) and others centuries later.

Fracastoro also took an interest in the

3.4 BILLION YEARS

THE AGE OF THE **EARLIEST FOSSILS,** OF SINGLE CELLS, DISCOVERED IN AUSTRALIA

Ammonite fossil

Early scholars believed that fossils were the remains of animals laid down in the biblical flood. By the 16th century, people were considering other theories.

1544 Georg Hartmann discovers **magnetic dip**

1545 Gerolamo Cardano publishes his treatise on algebra, ***Ars magna***

1545 The **botanical garden in Padua** is opened

1546 Girolamo Fracastoro publishes **On Contagion and Contagious Diseases**

1546 In a letter to a colleague, **Gerardus Mercator** mentions that **magnetic and geographic North poles** are different

1546 **Georgius Agricola** publishes **On the Nature of Fossils**

Fracastoro made major advances in understanding the spread of disease.

Konrad von Gesner published several volumes of *Historia Animalium*, illustrated with dramatic pictures and accurately detailed drawings.

“ LET THERE BE FOR **EVERY PULSE A THANKSGIVING,** AND FOR **EVERY BREATH A SONG. ”**

Konrad Von Gesner, German naturalist, 1550s

emerging study of geology. After examining the fossils of marine creatures found by building workers who were excavating a site in Verona, he expressed the **controversial idea** that they may be the **fossilized remains** of animals that had lived there many years before.

This, however, was not the view held by other geologists of the time. **German scholar** Georg Pawer, known as **Georgius Agricola** (1494–1555), dismissed the idea. He maintained that these were organic shapes created by the action of heat on "fatty matter" within the rocks. Despite this erroneous opinion, Agricola was among the first to lay a **scientific foundation** for the **study of geology**. In his 1546 publication *De Veteribus et Novis Metallis*, better known as ***De Natura Fossilium***, he attempted to categorize various minerals and rocks according to their characteristics. This, along with his earlier text ***De Re Metallica***, provided a comprehensive **overview of mineralogy and geology**, and was a practical guide to various mining techniques and machinery used at the time. It also showed the inadequacy of contemporary theories, which had not changed since the time of the Romans.

ENGLISH SURVEYOR LEONARD DIGGES (1520–59) made measurement of distance more accurate with the invention, in 1551, of the **theodolite**.

In the same year, **German naturalist Konrad von Gesner** (1516–65) published the first of five volumes of his ***Historia Animalium*** (*The History of Animals*). This book attempted to present a comprehensive **catalog** of the real and mythical **creatures of the world** and included illustrations and engravings. More importantly, it introduced exotic and recently discovered animals to European readers. Despite its popularity in northern Europe, the series was banned by the Catholic Church because of von Gesner's Protestant beliefs.

The anatomical drawings of Italian physician **Bartolomeo Eustachi** (*c.*1520–74) , completed in 1552, were not published until 1714 because he feared excommunication from the Catholic Church. He studied human teeth and was the first to describe **adrenal glands**, but he is more commonly known for his research into the workings of the ear, specifically the tube now known as the **Eustachian tube**.

Michael Servetus (see panel, opposite) published his ***Christianismi Restitutio*** (*The Restoration of Christianity*) in 1553, but fell foul of both the Catholic and Protestant authorities in doing so. In it, he included the first correct description of **pulmonary circulation**.

Also controversial was the theory proposed by **Giambattista Benedetti** (1530–90) concerning "bodies" (objects) in free fall. In his book published in 1554, he stated that bodies of the same material fell at the same speed, no matter what their weight, contradicting the law proposed by Aristotle. In a second edition of the work, he modified his theory to account for air resistance (friction), but maintained that different-sized bodies would fall at the same speed in a vacuum.

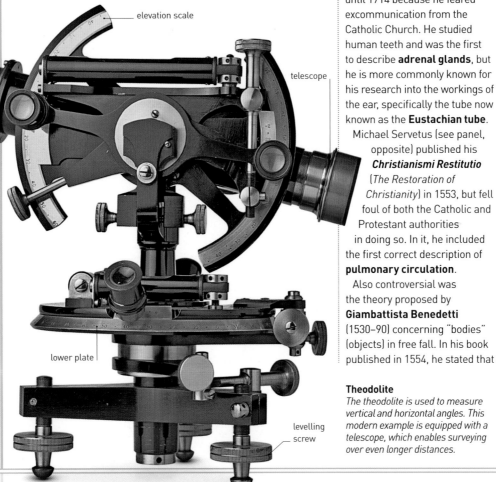

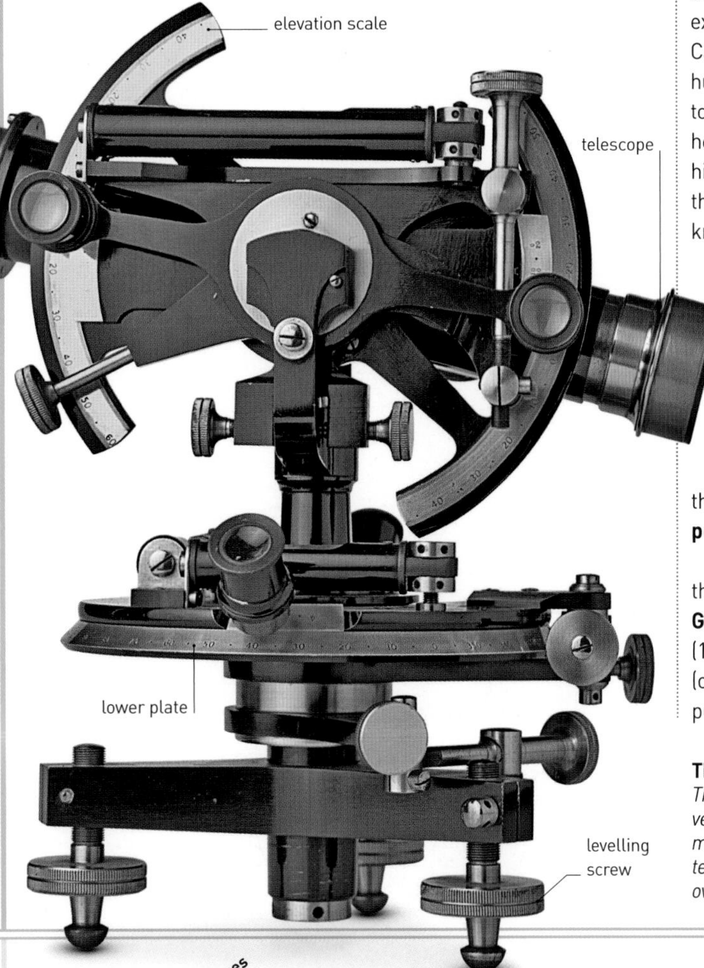
elevation scale
telescope
lower plate
levelling screw

Theodolite
The theodolite is used to measure vertical and horizontal angles. This modern example is equipped with a telescope, which enables surveying over even longer distances.

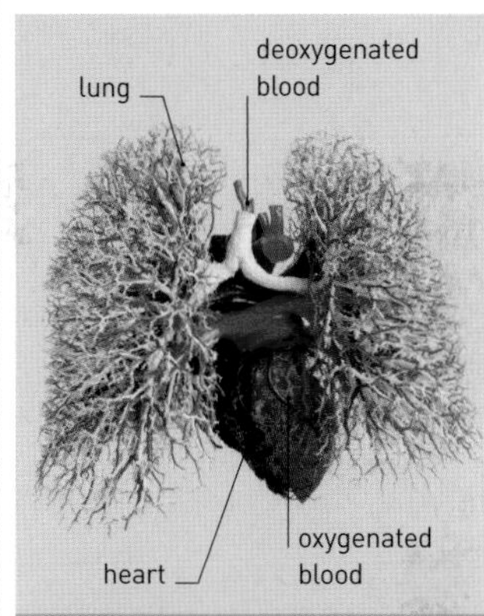
lung
deoxygenated blood
heart
oxygenated blood

PULMONARY CIRCULATION

The heart pumps deoxygenated blood through the pulmonary artery to small capillaries around the lungs, where carbon dioxide is replaced with oxygen. The pulmonary vein returns this oxygen-rich blood to the heart. Michael Servetus was the first to describe this system in 1553, but it had little influence at the time.

1550 Gerolamo Cardano publishes a survey of the natural sciences, ***De Subtilitate Rerum***

1550s Taqi al-Din designs programmable weight-driven **astronomical clock**

1551 Leonard Digges invents **theodolite**

1551 Konrad von Gesner publishes first volume of ***Historia Animalium***

1552 Bartolomeo Eustachi finishes his ***Anatomical Engravings***

1553 Spanish physician **Michael Servetus** publishes ***Christianismi Restitutio***

1554 Giambattista Benedetti proposes theory of bodies in free fall

> **“** TO **AVOID** TEDIOUS REPETITION OF… **‘IS EQUAL TO’,** I WILL SETTLE… ON A **PAIR OF PARALLELS** OF ONE LENGTH… **”**

Robert Recorde, Welsh physician and mathematician, 1557

Robert Recorde was the author of the first books on algebra in English.

De Re Metallica *Georgius Agricola's lavishly illustrated book on mining techniques describes the formation of ores in the ground, and how metal can be extracted from them.*

GEORGIUS AGRICOLA'S BOOK ***De Re Metallica*** (*On the Nature of Metals*) was published posthumously in 1556. In it, he described various techniques for mining minerals, and the machines, especially water mills, used for raising them from the mines. This **classic text of mining engineering** included descriptions of the veins of ores found in rock, and how metals could be extracted from them, as well as a comprehensive catalog of the minerals known at that time. Sometimes referred to as "the father of mineralogy," Agricola made contributions to the emerging fields of geology, metallurgy, and chemistry.

After his books *The Ground of Artes* (1543) on arithmetic, and *The Pathway to Knowledge* (1551) on geometry, the **Welsh mathematician Robert Recorde** published a companion volume, ***The Whetstone of Witte***, in 1557, probably the **first book on algebra in English**. As well as presenting the principles of algebra, this book established usage of the symbols + (plus) and – (minus), which had previously been used only occasionally by some German mathematicians. It introduced a symbol he invented: **=, the equals sign**. Best known for popularizing mathematics in Britain, Recorde had originally studied medicine and worked as a physician to the Royal family, and for a time was controller of the Royal Mint supervising the manufacture of coins. Despite fame and standing, he died in a debtors' prison a year after publishing *The Whetstone*.

3214°F THE **MELTING POINT** OF **PLATINUM**

Nugget of platinum
A rare metal, platinum is one of the least reactive elements.

Although **platinum** had already been used by the indigenous people of Central and South America to make jewelry and ornaments, it was unknown in Europe until the 16th century. The **first written reference** to the metal came in 1557 in the writings of **Julius Caesar Scaliger** (1484–1558), an Italian scholar. He described how Spanish explorers came across an unknown element, with an **unusually high melting point and resistance to corrosion**. Originally known as "white gold," platinum was later recognized as an element which occurs naturally in both pure and alloy (combined with another element) forms in South America, Russia, and South Africa.

Smoking became a fashionable habit in 16th-century Europe.

FRENCH DIPLOMAT JEAN NICOT, (1530–1600) while ambassador in Lisbon, Portugal, was introduced to **tobacco, brought by Spanish explorers from America**. Native Americans smoked it in religious rituals, and ingested or made poultices with its leaves for medicinal purposes. Nicot sent tobacco plants and snuff to the **royal court in France**, where smoking and snuff-taking soon became fashionable. The tobacco plant *Nicotiana*, and the chemical nicotine, are named after him.

In the same year, Italian anatomist and surgeon **Realdo Colombo** (*c.*1516–59) published

Title page of *De Re Anatomica*
Although Realdo Colombo published only one book on anatomy, his discoveries rivaled those of Andreas Vesalius and Gabriele Falloppio.

1556 Georg Pawer's *De Re Metallica* is published in Basel under the Latinized form of his name, Georgius Agricola

1557 Robert Recorde introduces the symbol =, for "equals," in *The Whetstone of Witte*

1557 First European reference to platinum in writings of Julius Caeser Scaliger

1560–64

Ambroise Paré developed new surgical techniques while working as a battlefield surgeon, and demonstrated his ideas to his students.

GABRIELE FALLOPPIO (1523–62)

Born in Modena, Italy, Falloppio studied medicine in Ferrara, and went on to teach anatomy and surgery at the Universities of Ferrara and Padua. He also served as superintendent of the botanical garden at Padua.

He is credited with important discoveries in the anatomy of the human head and reproductive systems. He died, aged only 39, in Padua.

his treatise ***De Re Anatomica*** (*On Things Anatomical*). Colombo's practical background in surgery led to a sometimes acrimonious rivalry with his more academic contemporary Andreas Vesalius. However, he is credited with **advances in anatomy**, including work on **pulmonary circulation**.

> **"** IT IS GOOD FOR NOTHING BUT TO **CHOKE A MAN**, AND FILL HIM FULL OF **SMOKE AND EMBERS. "**
>
> **Ben Jonson, English playwright,** in *Every Man in His Humor*, 1598

ITALIAN POLYMATH PLAYWRIGHT GIAMBATTISTA DELLA PORTA (*c.*1535–1615), particularly interested in the sciences, founded a group of like-minded thinkers in Naples, nicknamed the Otiosi, or men of leisure, who met to "uncover the secrets of nature." Their more formal title was the **Academia Secretorum Naturae** (*Academy of Secrets of Nature*), considered to be the **first scientific society**. Its membership was open to anyone who could demonstrate having made a new discovery in one of the natural sciences, and meetings were held at della Porta's home until, thanks to his interest in occult philosophy, Pope Paul V ordered the Academy to disband in 1578. Della Porta went on to encourage the founding of another society, the **Accademia dei Lincei** (Academy of Lynxes) in 1603.

Gabriele Falloppio, who had succeeded Realdo Colombo as chair of anatomy and surgery at Padua University in 1551, published his major work, ***Observationes Anatomicae (Anatomical Observations)***, in 1561 at the height of what in retrospect was a golden age of **anatomical discovery**. Sometimes known by the Latinized form of his name, Fallopius, he made valuable contributions to the **study of the ears, eyes, and nose, as well as human reproduction and sexuality**. A nerve canal in the face (the *aquaeductus Fallopii*) and the Fallopian tube connecting the ovaries with the uterus are named after him. Falloppio was a respected physician and skillful surgeon as well as an anatomist. He wrote a number of treatises on surgery, medication, and treatments of various kinds, although only one, *Anatomy*, was published in his lifetime. His work complemented the writings of his compatriots Vesalius and Colombo, and often quietly corrected their misconceptions.

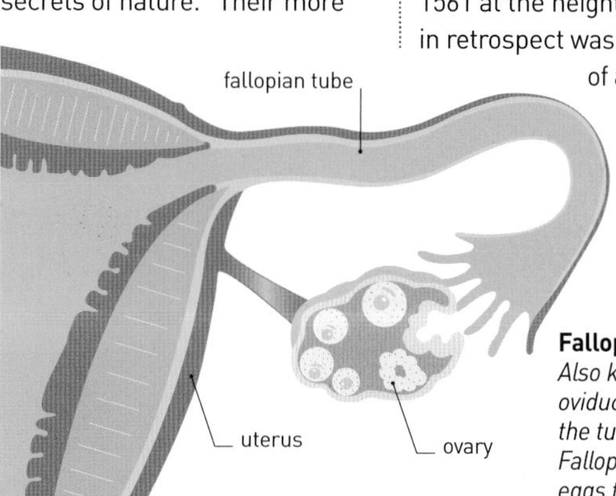

fallopian tube
uterus
ovary

Fallopian tube
Also known as the oviducts or salpinges, the tubes named after Falloppio allow the eggs to pass from the ovaries to the uterus.

Meanwhile, Frenchman **Ambroise Paré** (1510–90) wrote one of the **first manuals of modern surgery**. His ***Treatise***, written in French rather than Latin, was based on his experience as a surgeon on the battlefield. As well as describing surgical procedures, some of his own invention, Paré's book proposed the idea of surgery as a **restorative procedure**, which should involve minimum suffering. It stated that pain relief, healing, and even compassion were essential to successful surgery. This came from his experience of treating amputated limbs with balms and ointments rather than cauterizing the wounds with boiling oil, which often harmed the very tissues the surgeon was attempting to mend. Paré's innovative scientific approach based on **empirical observation** did much to improve the status of the "barber-surgeon," which was previously considered inferior to the medical physician.

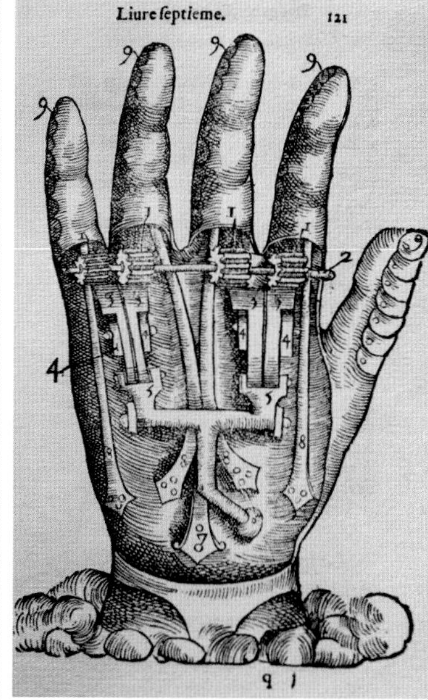

Liure septieme.
121

Prosthetic hand by Paré
Paré designed sophisticated prostheses to replace missing limbs, such as the hand shown in this 1585 drawing.

1559 Jean Nicot introduces **tobacco** to France

1559 Realdo Colombo publishes his treatise ***De Re Anatomica***

1560 The **first scientific society** is founded in Naples by Giambattista della Porta

1561 Gabriele Falloppio describes the human **ovaries, the uterus, and the tubes** connecting them

1564 Ambroise **Paré** publishes his ***Treatise*** on surgery

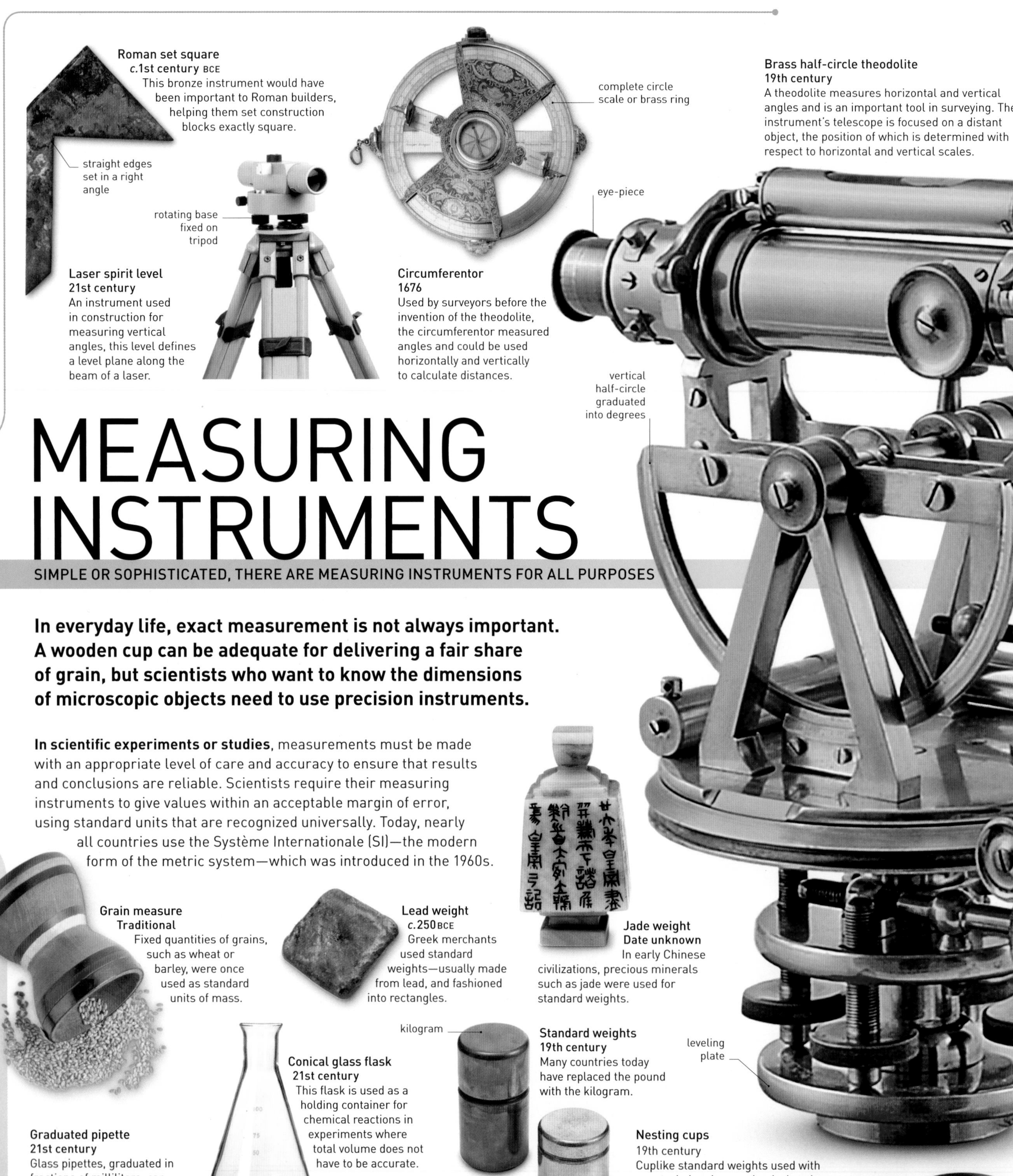
straight edges set in a right angle
rotating base fixed on tripod
complete circle scale or brass ring
eye-piece
vertical half-circle graduated into degrees
leveling plate
kilogram
pound
milliliter graduations

MEASURING INSTRUMENTS

SIMPLE OR SOPHISTICATED, THERE ARE MEASURING INSTRUMENTS FOR ALL PURPOSES

In everyday life, exact measurement is not always important. A wooden cup can be adequate for delivering a fair share of grain, but scientists who want to know the dimensions of microscopic objects need to use precision instruments.

In scientific experiments or studies, measurements must be made with an appropriate level of care and accuracy to ensure that results and conclusions are reliable. Scientists require their measuring instruments to give values within an acceptable margin of error, using standard units that are recognized universally. Today, nearly all countries use the Système Internationale (SI)—the modern form of the metric system—which was introduced in the 1960s.

Roman set square
*c.*1st century BCE
This bronze instrument would have been important to Roman builders, helping them set construction blocks exactly square.

Laser spirit level
21st century
An instrument used in construction for measuring vertical angles, this level defines a level plane along the beam of a laser.

Circumferentor
1676
Used by surveyors before the invention of the theodolite, the circumferentor measured angles and could be used horizontally and vertically to calculate distances.

Brass half-circle theodolite
19th century
A theodolite measures horizontal and vertical angles and is an important tool in surveying. The instrument's telescope is focused on a distant object, the position of which is determined with respect to horizontal and vertical scales.

Grain measure
Traditional
Fixed quantities of grains, such as wheat or barley, were once used as standard units of mass.

Lead weight
*c.*250 BCE
Greek merchants used standard weights—usually made from lead, and fashioned into rectangles.

Jade weight
Date unknown
In early Chinese civilizations, precious minerals such as jade were used for standard weights.

Standard weights
19th century
Many countries today have replaced the pound with the kilogram.

Conical glass flask
21st century
This flask is used as a holding container for chemical reactions in experiments where total volume does not have to be accurate.

Graduated pipette
21st century
Glass pipettes, graduated in fractions of milliliters, can measure liquid volumes precisely, drop by drop.

Nesting cups
19th century
Cuplike standard weights used with counterbalancing mechanical scales could be nested together in multiples.

measurement read from point where screw touches scale
measuring rod
dial around movable screw
telescope for viewing
calipers for measuring internal dimension
sliding scale
MITUTOYO
199.826
calipers for measuring external dimension
pointer
cross beam is horizontal when weights are equal
horizontal full circle graduated into degrees with Vernier scale
suspended weight
central pivot
coin
links at fixed-distance intervals

Yardstick
18th century
The Imperial yard (3 ft or 0.9 m) has long been a popular unit of measurement for construction work. Most yardsticks can be used as rulers.

Brass micrometer
Early 19th century
The first micrometers opened up the field of precision engineering; these adjustable screwlike devices enable accurate measurement of small distances.

Modern micrometer
21st century
Most modern micrometers work as calipers that move by tiny distances as they close around an object.

Laser distance meter
21st century
This shoots a laser pulse at a distant object and measures the time taken for the pulse to be reflected back.

Vernier calipers
20th century
In 1631, Paul Vernier invented a sliding scale for taking small measurements with great accuracy. The principle of the Vernier scale remained in use for modern instruments.

Spring balance
18th century
Originating in the 18th century, spring balances rely on stretching a spring in proportion to an applied force—the weight. The dial can be calibrated in units of mass (for example, kilograms) or force (Newtons).

Cased balance
18th century
Beam-balances were used in science and medicine, but small portable balances were also used for such purposes as measuring coins.

Surveyor's chain
19th century
Land surveyors began using chains in the 1600s. This example is about 66 ft (20 m) long, and is divided into 100 links.

Weighing scales
18th century
Scales determine an unknown weight by counterbalancing it with known weights until equilibrium is achieved.

Analytical balance
21st century
The most sophisticated modern digital balances can weigh minute fractions of a gram, and are so sensitive that they must be protected from vibrations, dust, and air movements.

1565–69

The Exeter Ship Canal reconnected the English inland port of Exeter to the sea, bypassing the section of the Exe River that was no longer navigable.

BEGUN IN 1564, the Exeter Ship Canal was completed in 1566 or 1567. A man-made **navigable channel**, it reestablished the town of Exeter in England as a port after centuries of blockages on the Exe River by weirs constructed to power watermills. Probably the UK's first artificial waterway, the Exeter Canal was a forerunner of the prolific canal-building that came with the Industrial Revolution in the 18th century.

> **“** HE DOES SMILE HIS FACE INTO MORE LINES THAN IS **IN THE NEW MAP** WITH THE AUGMENTATION OF THE INDIES. **”**
>
> **William Shakespeare, English playwright,** *Twelfth Night, c.*1602

Maps became increasingly important during the 16th century, as traders and explorers made voyages around the world. While cartographers could make accurate globes showing oceans and continents, these were not convenient for navigation. The problem was how to represent the three-dimensional Earth in two dimensions. In 1569, Flemish cartographer **Gerardus Mercator** (1512–94), already famous for his globes and maps of Europe, devised a new way to represent the curved surface of Earth when designing his paper **map of the world**. Now known as **Mercator's projection**, this method shows the lines of longitude as equally spaced, parallel, vertical straight lines, and the lines of latitude as perpendicular to them—as if they were projected onto a cylinder enclosing a globe. Although this projection distorted the shapes and sizes of large land masses and oceans, it was particularly useful for navigation, since compass courses could be shown as straight lines.

GERARDI MERCATORIS RVPELMVNDANI EFFIGIEM ANNOR. DVORVM ET SEX – AGINTA, SVI ERGA IPSVM STVDII CAVSA DEPINGI CVRABAT FRANC. HOG. CIↃ. IↃ. LXXIV.

Gerardus Mercator
Mercator's projection, a method of showing the spherical world as a two-dimensional map, enabled accurate nautical navigation and still remains in use.

Although the German–Swiss alchemist and physician **Paracelsus** (1493–1541) was a prolific writer and notorious self-publicist, few of his works were published during his lifetime. One of the most influential, *Archidoxa*, was published posthumously in Kraków in 1569. This **rejected the magical elements of alchemy** and was important in the development of modern chemistry and medicine.

1570–71

HISPANIA NOVA
ARCHIPELAGO DI
SAN LAZARO
OCCIDENS
NOVA GVINEA
The first modern world atlas, the *Theatrum Orbis Terrarum*, showed the extent of voyages of discovery in the 16th century.

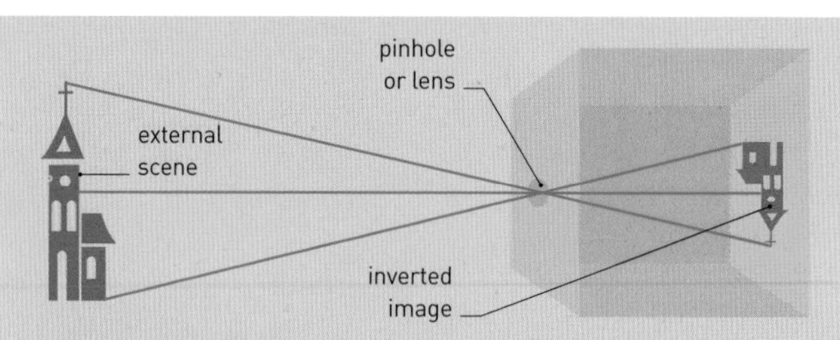
pinhole
or lens
external
scene
inverted
image

CAMERA OBSCURA

A camera obscura (Latin for "dark chamber") is a simple room or box with a small hole in one of its walls. Light from outside passes through the hole and falls onto the inside of the opposite wall, projecting an image of the scene outside. The image, which appears upside down, is sharper with a smaller pinhole, or brighter with a larger hole, and can be focused by adding a lens.

ENCOURAGED BY HIS FRIEND AND COLLEAGUE MERCATOR, the Flemish cartographer **Abraham Ortelius** (1527-98) published his *Theatrum Orbis Terrarum* (*Theater of the World*) in 1570. He had previously made a large map of the world on eight sheets, and several separate maps of various parts of the world, but this collection of 53 maps and accompanying text in book form was the **first modern world atlas**—a term suggested by Mercator. After its original publication in Latin, it was translated into several other languages, and in these later editions Ortelius added several more maps and corrected inaccuracies.

Italian polymath **Giambattista della Porta** (*c.*1535–1615) first published his book *Magiae Naturalis* (*Natural Magic*) with descriptions and observations of all the sciences in 1558, but it proved so popular that it was revised and expanded in a

53
THE **NUMBER OF MAPS** IN THE **FIRST WORLD ATLAS**

1566/67 The **Exeter Ship Canal** is completed

1569 Gerardus Mercator publishes a complete **map of the world**

1569 Paracelsus's treatise on **alchemy and medicine**, *Archidoxa*, is published posthumously

1570 Abraham Ortelius publishes the **first modern atlas**, the *Theatrum Orbis Terrarum*

c.1570 Gianbattista della Porta describes an improved **camera obscura**

A supernova is a massive exploding star that shines very brightly, giving the impression that a new star has appeared. Tycho Brahe first described the phenomenon in 1573.

number of editions, eventually becoming a 20-volume work. In one of the later editions of around 1570, he included a description of the **camera obscura**. The principles of this device date back to China and Greece around 2,000 years before, but della Porta was the **first to suggest the use of a convex lens** rather than a pinhole to focus the image on the screen inside, allowing more light through the larger aperture without a subsequent loss of clarity. He also used this camera obscura with a lens, an innovation that was of great value to Kepler in the following century in his studies of the workings of the human eye (see 1598–1604).

Although a successful attorney in Paris, **François Viète** (1540–1603), also known as Franciscus Vieta, was a talented mathematician who devoted much of his spare time to the subject. One of his first achievements in the field was a set of **trigonometric tables** to aid calculation, the *Universalium Inspectionum ad Canonem Mathematicum Liber Singularis*, which he started to publish in 1571.

COMPOSING A BOOK ON ALGEBRA that was both comprehensive and intelligible to the nonmathematician, **Rafael Bombelli** (1526–72) published his treatise simply titled *Algebra* shortly before his death in 1572. He explained in everyday language the algebra known at that time and tackled a problem barely understood by his contemporaries, **imaginary numbers**: numbers whose **square is less than zero**. These, he explained, cannot be dealt with in the same way as other numbers, but are essential in solving equations involving powers of two, three, or four. Bombelli effectively laid down the rules for using these imaginary numbers for the first time. Despite his pioneering work, imaginary numbers were not accepted in mathematics until almost 200 years later.

Tycho Brahe had observed what appeared to be a bright new star in the **constellation of Cassiopeia** in 1572, and the following year published his account of it, *De Nova Stella* (*On the New Star*). It was in fact a **massive stellar explosion**, not a new star, but the Latin word "nova" later came to be applied to what is now known as a **supernova**, as well as to other stars that brighten abruptly. Brahe realized that his "star" was very distant from Earth, certainly beyond the orbit of the Moon. Conventional wisdom since the time of Aristotle had maintained that anything outside the immediate vicinity of Earth was unchanging, including the stars of the celestial realm, but Brahe's observations **contradicted the idea of the immutability of the stars**.

> " IT WAS NOT JUST THE **CHURCH** THAT **RESISTED** THE **HELIOCENTRISM** OF **COPERNICUS.** "
>
> **Tycho Brahe, Danish astronomer,** 1587

animated figures
staircase

Astronomical clock
This astronomical clock, built between 1547 and 1574, stood in Strasbourg Cathedral until the 19th century. An 1840s replica now exists in its place.

One of the most elaborate **astronomical clocks** of the period was built in the Cathedral of Notre Dame in Strasbourg to replace the existing 200-year-old clock, which had stopped working. The new clock was designed by mathematician **Christian Herlin** in the 1540s, but only the preliminary building had been accomplished in 1547 when work was interrupted by Herlin's death. Political problems further delayed work until the 1570s, when Herlin's pupil, mathematician **Conrad Dasypodius** (1532–1600), took over the project. The clock was finally built by Isaac and Josias Habrecht. It incorporated many of the latest ideas in mathematics and astronomy—as well as clockmaking—into its design, which included a celestial globe, an astrolabe, a calendar dial, and automata.

TYCHO BRAHE
(1546–1601)

Born in Scania, now in Sweden but then part of Denmark, Tycho Brahe became interested in astronomy while studying law in Copenhagen. Under the patronage of King Frederick II of Denmark he established an observatory equipped with the finest astronomical instruments.

1571 François Viète begins publication of his **trigonometric tables**

1572 Rafael Bombelli publishes *Algebra*

1573 Danish astronomer **Tycho Brahe** publishes *De Nova Stella*

1574 The **astronomical clock** in Strasbourg Cathedral is completed

1575–77

Designed by Taqi al-Din, the observatory at Istanbul was equipped with the latest technology and attracted the finest astronomers of the Ottoman Empire.

AN INHABITANT OF SICILY, Greek mathematician and astronomer **Francesco Maurolico** (1494–1575) published several treatises on mathematics. In *Arithmeticorum Libri Duo* (*Two Books on Arithmetic*), published in 1575, he was the first mathematician known to prove a mathematical statement explicitly using **mathematical induction**. This is a method of proof using a series of successive logical steps.

To persuade astronomer **Tycho Brahe** (see 1572–74) to return to his native Denmark, King Frederick II offered him land and funding to **establish an observatory** in Hven (an island now belonging to Sweden). Work on the building, known as **Uraniborg**, began in 1576. However, the completed structure was not considered steady enough for accurate observations. A second complex, **Stjerneborg**, was built close by in 1584 to house the delicate equipment. Together, these two complexes formed a **major centre for astronomical and scientific research**.

Italian polymath **Gerolamo Cardano** trained in medicine and was a respected physician. He wrote several treatises, including the **first description of typhoid fever** in 1576. He was the first to recognize the disease's distinctive symptoms.

Stjerneborg observatory
The complex of buildings that replaced Tycho Brahe's Uraniborg observatory was equipped with the latest astronomical instruments.

There were **many advances in botany** in the second half of the 16th century. Emphasis moved from the study of plants for their medicinal properties to a more comprehensive study and classification of plant life. Botanist **Charles de l'Écluse** (1526–1609), also known as Carolus Clusius, published the first of his studies of the flora of Spain in 1576. He went on to found the botanical garden at the University of Leiden in Holland, where his work helped lay the foundation of the Dutch tulip industry.

While Brahe was sponsored by the King of Denmark, the Ottoman Turkish engineer and astronomer **Taqi al-Din** persuaded Sultan Murad III to fund an equally prestigious observatory in Istanbul. Built in 1577, this was **designed to be the major observatory of the Islamic world**. However, it existed only briefly: after a mistaken astrological prediction of Ottoman victories in battle, the Sultan had the observatory destroyed in 1580.

1578–81

IN 1579, HIERONYMUS FABRICIUS (1537–1619), professor of anatomy and surgery at the University of Padua, noticed in his dissections folds of tissue on the inside of veins. He described these folds as valves, but did not propose any function for them. It was only later that they were found to prevent the backflow of blood as it returns to the heart. Fabricius's treatise on the subject, *De Venarum Ostiolis* (*On the Valves of the Veins*), particularly influenced one of his later students, William Harvey (see 1628–30).

Although trained in medicine, Venetian physician **Prospero Alpini** (1553–*c*.1616) was more interested in botany. In 1580, he took up a post as physician to the Venetian consul in Cairo, Egypt, where he studied the plant life. He also worked as the manager of a date palm plantation in Egypt. While there, he observed that the **pollination of flowers was necessary to produce fruit**, and deduced that there were two sexes of plants. Alpini's study of plants in Egypt inspired him to write several books on exotic plants, including *De Plantis Aegypti Liber* (*Book of Egyptian Plants*), published in Venice in 1592, and *De Plantis Exoticis* (*Of Exotic Plants*), published posthumously in 1629. He is also credited with introducing the banana and baobab to Europe.

1575 Francesco Maurolico makes the first known use of **mathematical induction**

1576 Gerolamo Cardano gives the **first description of typhoid fever**

1576 Tycho Brahe builds an observatory called Uraniborg on the island of Hven, Denmark

1576 Carolus Clusius publishes the **first of his books on botany**

1577 Taqi al-Din **builds an astronomical observatory** in Galata, Istanbul

1579 Hieronymus Fabricius **describes valves in veins**

While managing date palm plantations in Egypt, Prospero Alpini observed the difference between the sexes of plants.

> “THE **MARVELLOUS PROPERTY** OF THE **PENDULUM** IS THAT IT MAKES ALL ITS **VIBRATIONS… IN EQUAL TIMES.**”

Galileo Galilei, Italian astronomer and physicist

English explorer and navigator **Steven Borough** (1525–84) had previously organized the English translation of the standard textbook on navigation of the time: *Breve Compendio* (*Brief Summary*) or *Arte de Navigar* (*Art of Sailing*) by Martín Cortés de Albacar. In 1581, Borough published his own treatise, dealing with the properties of magnetism and its effects on a compass needle. The treatise, reflecting his experience as a seaman, contributed considerably to the understanding and practical use of the magnetic compass in navigation and cartography.

In 1581, on his father's insistence, **Galileo Galilei** (see 1611–13) was studying medicine at Pisa, Italy. However, he already had a fascination for mathematics and physics. Observing a chandelier swaying in Pisa Cathedral, he noted that each swing took the same amount of time, regardless of how far it traveled. He then **experimented with pendulums** and found that the rate of swing was constant, no matter how wide the swing, and that two pendulums of the same length would swing in unison even if their sweeps were different. Galileo later published his observations on the constancy of pendulum swing.

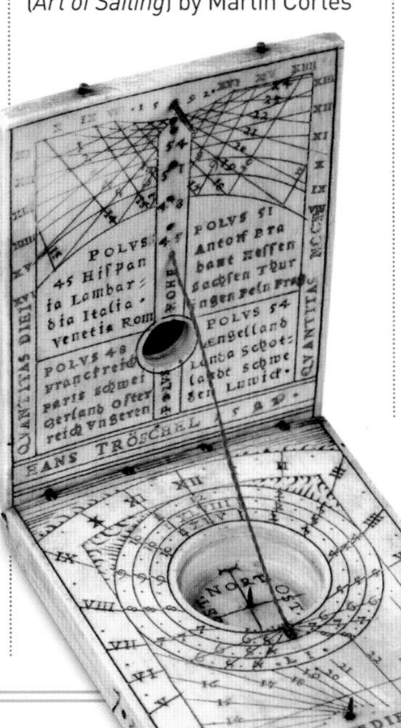
16th-century magnetic sundial
This portable sundial has a magnetic compass that is used to align it in different locations. The gnomon (diagonal string) must be set north–south.

Galileo and the pendulum
Galileo first discovered the constancy of a pendulum swing by observing a swaying chandelier and timing its motion against his pulse.

> “…THE **FORM AND COLOUR** OF EXTERNAL OBJECTS… [AND] **LIGHT** ENTER THE **EYE** THROUGH… THE **PUPIL** AND ARE PROJECTED ON **[THE OPTIC NERVE]** BY THE **LENS.**”

Felix Platter, Swiss physician, 1583

BY 1582, THE JULIAN CALENDAR, which had been in use in Europe since Roman times, had become out of step with the times of the equinoxes by about ten days, so Pope Gregory XIII issued a decree introducing a new calendar. The Julian calendar had approximated the year, the time between successive spring equinoxes, as 365.25 days. This led to a discrepancy of about three days in 400 years. The reformed calendar, which came to be known as the **Gregorian calendar**, worked from a more accurate measurement of the time between the spring equinoxes. It was adopted first by the Catholic countries and gradually elsewhere.

Colonization of the New World gained pace towards the end of the 16th century. Writer **Richard Hakluyt** (1552–1616) helped to promote the English settlement of North America. In his 1582 publication *Divers Voyages Touching the Discoverie of America* and other later books, he pointed out the advantages of colonization, citing the possibility of establishing plantations for foods and tobacco.

768
THE NUMBER OF PLANTS IN **CESALPINO'S HERBARIUM**

Italian physician and botanist **Andrea Cesalpino** (1519–1603), who had been director of the botanical garden connected to the University of Pisa, developed the first scientific method of botanical classification in his *De Plantis Libri XVI* (*The Book of Plants XVI*), published in 1583. Cesalpino classified flowering plants according to their fruit, seeds, and roots, rather than by their medicinal properties.

Andrea Cesalpino
One of the foremost botanists of the 16th century, Cesalpino revolutionized the classification of plants.

ANDREA CESALPINO

1580 Sultan Murad III orders the **Istanbul observatory to be destroyed**

c.1580 Prospero Alpini describes **two different sexes in plants**

1581 Steven Borough publishes *A Discourse of the Variation of the Compas*

1581 Galileo Galilei observes **constancy of pendulum swing** in Pisa Cathedral

1582 Catholic countries, including Spain and Italy, adopt the **Gregorian Calendar**

1582 Richard Hakluyt publishes *Divers Voyages Touching the Discoverie of America*, promoting **English settlement of America**

1583 Andrea Cesalpino **classifies plants** according to their fruit, seeds, and roots in *De Plantis Libri XVI*

1583 Felix Platter (or Plater) suggests that the **retina is stimulated by light**

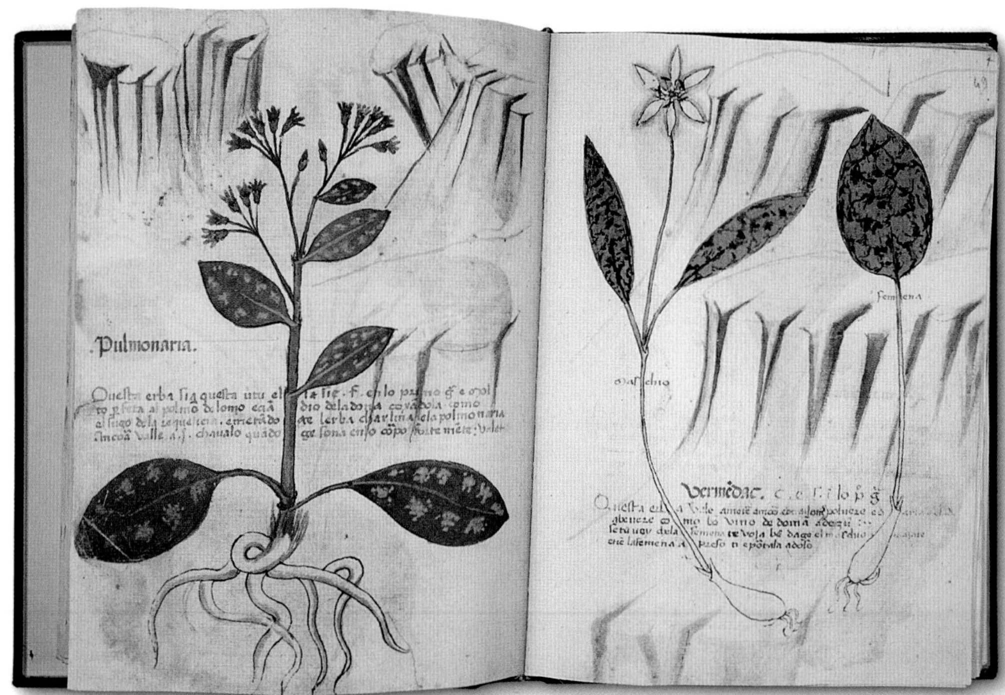
Pulmonaria.

Book of herbs
16th century
Early texts called "herbals" classified plants according to their medicinal properties—or supposed magical powers.

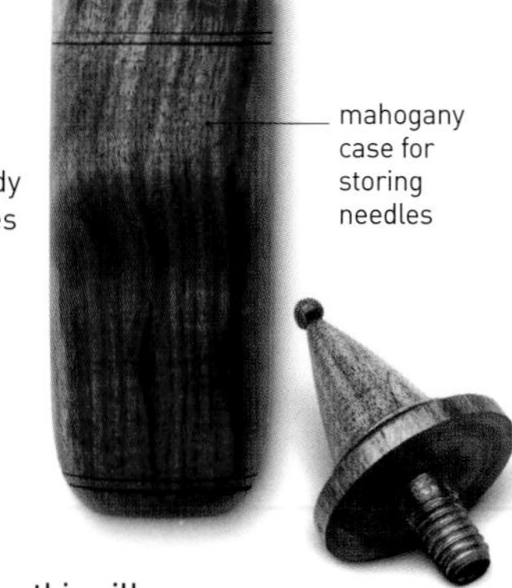
steel needles
mahogany case for storing needles

Acupuncture needles
19th century
The earliest evidence of acupuncture dates back to 3000 BCE. Practitioners produced maps of the body to show where the needles would be most effective.

Homeopathic pills
19th century
Homeopathic ideas—that small amounts of a substance that cause symptoms in healthy people can cure similar symptoms caused by disease—originated in Ancient Greece, but German physician Samuel Hahnermann started the formal practice in the 1790s.

MEDICINE

MEDICAL CUSTOMS AND TRADITIONS HAVE GRADUALLY BEEN SUPPLANTED BY A SCIENTIFIC APPROACH

The history of medicine is as old as humanity itself. For thousands of years, breakthroughs in understanding the human body and innovations in technology have improved the way disease is diagnosed and treated.

Evolving from its origins in herbalism and shamanism, medicine flourished in the ancient classical world—where the first physicians assessed patients using scientific judgment. The early study of anatomy, physiology, and diseases, together with its drugs, vaccines, and new instruments, turned medicine into a complex, multifaceted discipline.

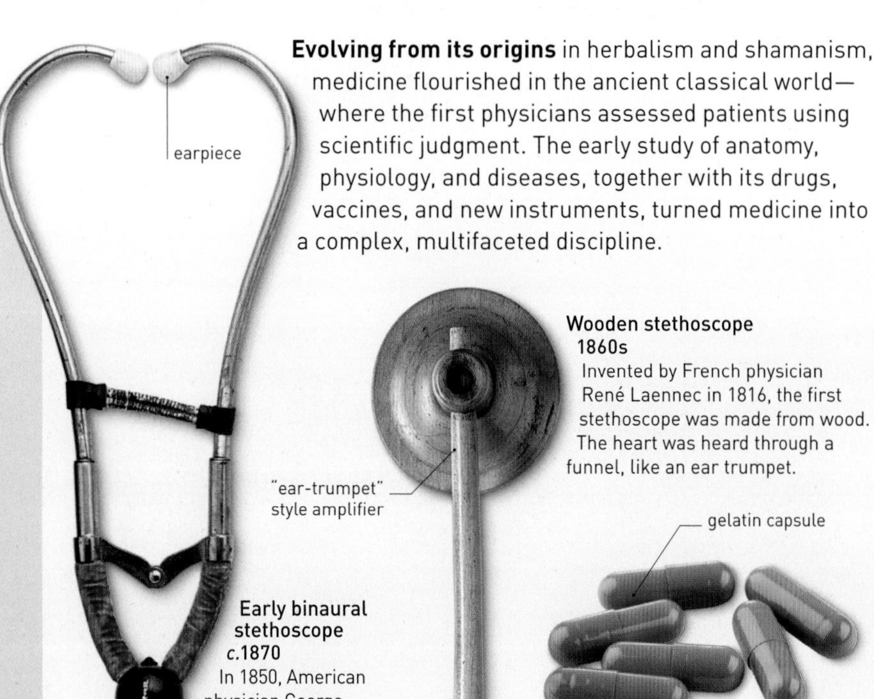
earpiece
"ear-trumpet" style amplifier
gelatin capsule

Early binaural stethoscope
*c.*1870
In 1850, American physician George Camman incorporated rubber into his binaural (two-earpiece) stethoscope to make it easier to use. It was the first commercially successful stethoscope.

Wooden stethoscope
1860s
Invented by French physician René Laennec in 1816, the first stethoscope was made from wood. The heart was heard through a funnel, like an ear trumpet.

Pills
20th century
Pills accurately deliver a small drug dose. They were first made by encasing the active ingredient in hardened glucose syrup.

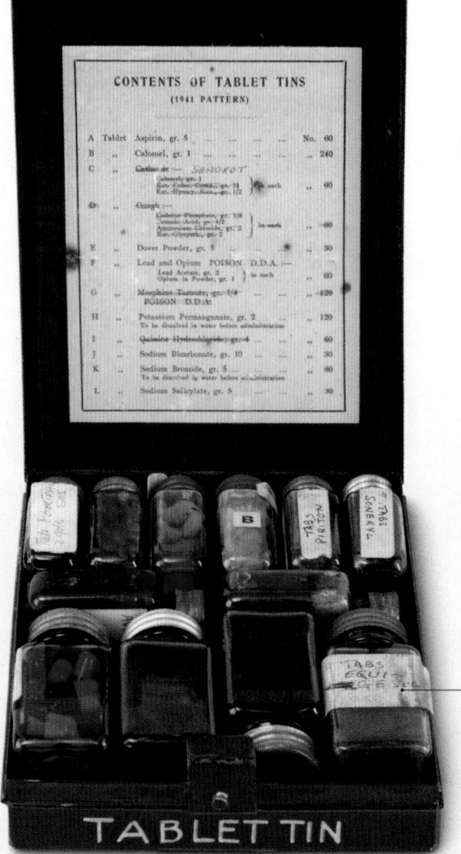
CONTENTS OF TABLET TINS
(1941 PATTERN)
TABLET TIN
bottle of painkilling tablets

Military tablet tin
*c.*1942
An armament of drugs was used on the battlefield to help deal with injuries and illness to hasten return to duty. Medical officers had first-aid tablet tins that contained painkillers, sedatives, and antiseptics.

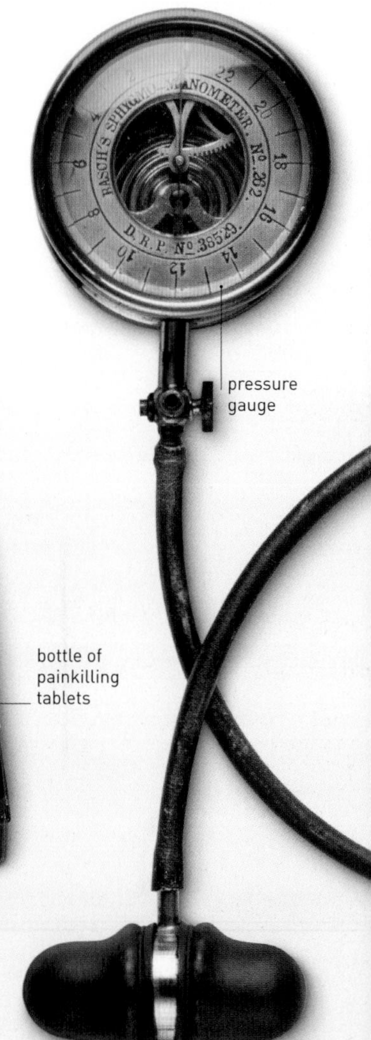
pressure gauge

hypodermic needle
metallic barrel
glass barrel
funnel for concentrating light
candle
"piston-type" plunger
viewing lens
hollow needle
protective casing
LONDON
N
M
Night.
reservoir of mercury
digital temperature display
rubber tubing
metal plunger

Disposable syringe
21st century
The modern disposable syringe—made from plastic—reduces the chance of cross-infection. It was patented by New Zealand pharmacist Colin Murdoch in 1956.

Ophthalmoscope
***c.*1875**
The first optical devices for examining the back of the eye were made in the 1840s and 1850s. Early models came with a selection of interchangeable lenses.

Brass endoscope
19th century
Invented by German physician Phillip Bozzini in 1805, the first rudimentary endoscope was illuminated using a candle. An endoscope is used to "see inside" a patient.

Mechanical syringe
18th century
Piston-type syringes have been used since antiquity, but metallic non-plunger mechanical syringes were developed in the 1600s and 1700s, and used for extracting fluids.

Clinical glass thermometer
18th century
German physician Hermann Boerhaave began using glass thermometers in the 1700s. In 1866, British physician Thomas Allbutt designed a portable, 6 in (15 cm) clinical model.

Sphygmomanometer
1883
The first attempts to measure blood pressure were inaccurate, until Austrian physician Samuel Ritter von Basch invented the sphygmomanometer in 1876. In early models, pressure was measured by a water-filled bulb applied to the skin, but later ones used an inflatable cuff.

Digital thermometer
21st century
Invented in the 1950s, electronic thermometers measure body temperature with a far greater degree of accuracy and use an unambiguous digital display.

X-ray
20th century
German physicist Wilhem Röntgen produced the first X-ray photograph—of his wife's hand—in 1895. Today, various scanning techniques are used to examine the interior of the body.

Glass syringe
1940s
The invention of fine needles in 1853 meant syringes could be used to inject drugs. The first precision-bore glass syringes in 1946 were easier to sterilize en masse because barrels and plungers were interchangeable.

1585–89

Simon Stevin wrote in Dutch, which he felt was better suited to technical subjects.

> **"[THE DECIMAL FRACTION]** TEACHES… THE EASY PERFORMANCE OF ALL **RECKONINGS, COMPUTATIONS, AND ACCOUNTS,** WITHOUT BROKEN NUMBERS. **"**
>
> **Simon Stevin, Flemish mathematician and engineer,** from *De Thiende* (*The Tenth*), 1585

FLEMISH MATHEMATICIAN AND ENGINEER SIMON STEVIN (1548–1620) published the booklet *De Thiende* (*The Tenth*) in 1585. It promoted the use of decimal fractions and predicted the adoption of a decimal system of weights and measures. While Islamic mathematicians had used decimal fractions centuries before, Stevin presented a comprehensive case for their use, citing ease of calculation. His notation was awkward and different from that used today.

The following year, Stevin published two works on water and "statics," in which he showed that because of its weight, **the pressure of water increases with depth**. His ideas became the foundation for a field of engineering called hydrostatics.

In 1588, Danish astronomer **Tycho Brahe** published further works, including the second part of his *Astronomiae Instauratae Progymnasmata* (*Introduction to New Astronomy*). He described the observation of comets and the instruments he used, and also included a catalog of stars, and described a **geo-heliocentric universe**, in which most of the planets orbited the Sun, and the Sun and Moon orbited the Earth.

In 1589, English inventor **William Lee** (1563–1614) designed the **stocking frame machine**, which mimicked the action of hand-knitters. Although it had the potential to revolutionize the textile industry, fear of upsetting the hand-knitters kept Lee from obtaining a patent in England, and he moved to France.

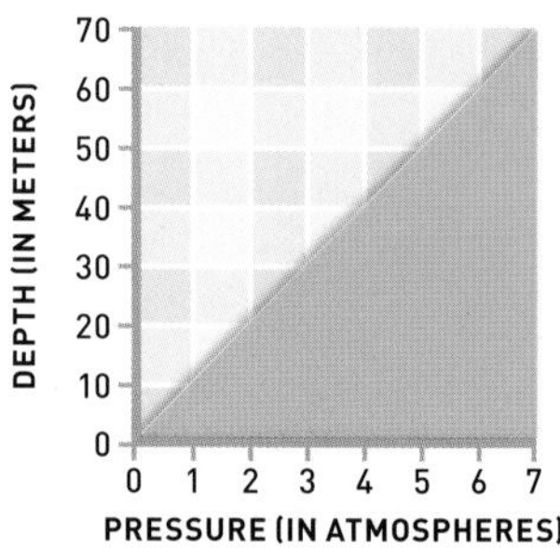

DEPTH (IN METERS)
70
60
50
40
30
20
10
0
0 1 2 3 4 5 6 7
PRESSURE (IN ATMOSPHERES)

Water pressure
Directly proportional to depth, water pressure increases by 1 atmosphere (atm) for every 10 m (33 ft) in depth.

WATER PRESSURE

As an object descends into water, the weight of the water above it exerts pressure on it. As a result, water pressure increases with depth. At a depth of about 33 ft (10 m), water pressure is double the atmospheric pressure at the surface. Water pressure at ocean floors can be as much as 1,000 atmospheres; 1 atm equals 14 lb/in^2 (1 kg/cm^2).

spring maintains tension
needles
knitted material
wool or silk yarn

Knitting machine
English inventor William Lee improved on his original knitting machine with more needles per inch, which enabled production of fine silk fabrics as well as wool.

1590–93

Galileo's thermoscope was an early device for measuring temperature.

DUTCH LENS-MAKER ZACHARIAS JANSEN (1580–1638) is believed to have invented the microscope, initially using a single magnifying lens. Around 1590, he combined two lenses to form the **first compound optical microscope**, which was capable of magnifying images about nine times. Jansen is also associated with the development of the telescope, an invention credited to his rival Hans Lippershey in 1608.

With the publication of the ten-part *In Artem Analyticien Isagoge* (*Introduction to the Art of Analysis*) in 1591, French mathematician **François Viète**, also known as Vieta, laid the **foundations for modern algebra**. One of the key innovations of his system of analysis—known as "new algebra"—was the use of letters of the alphabet for parameters and unknowns in equations. Viète thereby created a symbolic algebra to replace the Classical and Islamic rhetorical algebra, which relied on explanation rather than signs and symbols.

In 1592, Italian mathematician **Galileo Galilei** invented the **thermoscope**, a tube in which liquid rises and falls with changes in temperature. This was the forerunner of the liquid thermometer, which was developed later by adding a scale to the tube.

1585 Simon Stevin publishes *De Thiende*, proposing a form of **decimal fraction**

1586 Stevin describes how **water pressure** increases with depth

1588 Tycho Brahe publishes **two works on astronomy**, including a **catalog of stars**

1589 William Lee builds a **knitting machine**

1590 Zacharias Jansen builds a **compound microscope**

1590 Galileo writes ***De Motu***, an unfinished treatise on motion

1591 François Viète introduces **"new algebra"**

1592 Galileo invents the **thermoscope**

1594–95

Designed by Hieronymus Fabricius, the dissection theater in Padua, Italy, offered public demonstrations of anatomical dissections.

THE UNIVERSITY OF PADUA had been at the forefront of the "golden age" of anatomy since Andreas Vesalius became professor of surgery and anatomy there in 1537. The university attracted students from all over Europe, and the department was led by a succession of distinguished surgeons and anatomists.

Hieronymus Fabricius was appointed to the post in 1565, and became well known for demonstrating **the dissection of both humans and animals**, and instituting a new style of investigative anatomy. In order to make these demonstrations available to a wider audience, he designed a theater for dissections. The **theater was built** in 1594 with funds provided by the Senate of the Republic of Venice. Although some public dissections had been performed before, this was the first **permanent structure designed and built especially for such demonstrations**. He was succeeded by his students, Julius Casserius, and later Adriaan van der Spiegel, who continued the tradition of public demonstrations of anatomical dissections.

In the same year, Simon Stevin wrote his treatise, ***Arithmetic***. This book dealt with, among other things, the **solution of quadratic equations**—equations involving a squared quantity—and important concepts in the field of **number theory**.

HIERONYMUS FABRICIUS (1537–1619)

Born in Acquapendente, Italy, Hieronymus Fabricius studied at the University of Padua, where he eventually became Professor of Anatomy in 1562 and Professor of Surgery in 1565. Famous for his public demonstrations of anatomical dissections, he is best known as a pioneer in the field of embryology and for describing the valves in veins.

1596–97

> **“** I MUCH **PREFER** THE **SHARPEST CRITICISM** OF **A SINGLE INTELLIGENT MAN** TO THE **THOUGHTLESS APPROVAL** OF THE **MASSES... ”**
>
> **Johannes Kepler, German astronomer**

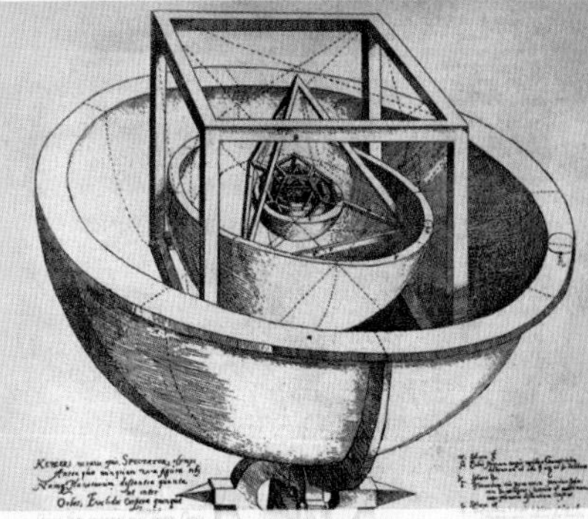
An illustration of Kepler's planetary model from *Mysterium Cosmographicum*.

IN 1596, GERMAN ASTRONOMER JOHANNES KEPLER (1571–1630) published his first important work on astronomy, *Mysterium Cosmographicum* (*The Cosmographic Mystery*). As well as defending the heliocentric model of the universe proposed by Copernicus (see 1543), Kepler **explained the orbits of the known planets around the Sun in geometric terms** in an attempt to unravel "God's mysterious plan of the universe." To do this, he drew upon the classical notion of "the harmony of the spheres," which he linked to the five Platonic solids—octahedron, icosahedron, dodecahedron, tetrahedron, and cube. These, when inscribed in spheres and nested inside one another in order, corresponded to the orbits of the planets Mercury, Venus, Earth, Mars, Jupiter, and Saturn.

In 1596, Flemish cartographer **Abraham Ortelius** noted that the coastlines on either side of the Atlantic Ocean seemed to fit like pieces of a puzzle. He was the first to suggest that Africa, Europe, and the Americas may once have been connected. Although he attributed their separation to a major cataclysm, his ideas anticipated the modern theory of continental drift (see 1914–15).

Also in 1596, English author **Sir John Harington** (1561–1612) published *A New Discourse upon a Stale Subject: The Metamorphosis of Ajax*. This text was part political satire and part description of his invention, a **rudimentary flush toilet** called "Ajax." The name was a pun on "a jakes," contemporary slang for toilet. The invention was a major step toward modern sanitation.

Water closet
John Harington's "Ajax," the prototype of the modern flush toilet, was invented with the aim of eliminating disease.

One of the most important textbooks of alchemy, ***Alchemia***, was published in 1597 by German metallurgist **Andreas Libavius**. Unlike previous books on alchemy, *Alchemia* stressed the importance of systematic laboratory procedures. It also contained a catalog of various medicaments and metals, and included the first description of the properties of zinc.

Zinc
Known in China and India since the 14th century, zinc was first described in Europe by 16th-century alchemist Andreas Libavius.

1594 Hieronymus Fabricius opens the first permanent theater for public **anatomical dissections** in Padua

1594 Simon Stevin publishes ***Arithmetic***

1596 Johannes Kepler publishes *Mysterium Cosmographicum*, defending the **Copernican heliocentric system** of planetary motion

1596 Abraham Ortelius compares the **coastlines of continents** and suggests that they **may have been joined** at one time

1596 John Harington describes his invention of an early **flush toilet**, the "Ajax"

1596 Austrian astronomer and mathematician Georg Rheticus's **trigonometric tables** are published posthumously

1597 Andreas Libavius publishes ***Alchemia***

1598

1599–1600

Giordano Bruno, arrested by the Roman Inquisition on charges of heresy, was imprisoned from 1562 until his execution in 1600.

1,004 THE NUMBER OF STARS CATALOGUED IN TYCHO BRAHE'S *ASTRONOMIAE INSTAURATAE MECHANICA*

IN 1598, DANISH ASTRONOMER TYCHO BRAHE published *Astronomiae Instauratae Mechanica* (*Instruments for the Restoration of Astronomy*)—a star catalog listing **the positions of more than a thousand stars** he had observed. Brahe had recently left his observatories at Hven, Denmark, after falling out with his patron, King Christian IV, who did not share his predecessor's enthusiasm for astronomy. Before leaving, Brahe wrote detailed descriptions with illustrations of the instruments and equipment he had used at his observatories. He included these in *Astronomiae* as well. In the following year, Brahe found a new sponsor, the Holy Roman Emperor Rudolf II, and moved to Prague (now in the Czech Republic).

In September, Dutch sailors landed on the island of Mauritius in the Indian Ocean, claiming it for the Netherlands. They were **the first to describe the dodo**, a flightless bird related to the pigeon and unique to Mauritius. Within less than a century, the bird had been **driven to extinction**, hunted by the settlers and preyed upon by imported animals.

Extinct species
The dodo is one of the first recorded examples of a species' extinction due to human interference. Its last known sighting was in 1662.

While much ship design was geared toward voyages of discovery and trade, in 1598, Admiral **Yi Sun-sin** of Korea turned his attention to the **design of warships**. He improved on the traditional Korean "turtle ship" by adding metal armor. These first iron-clad warships were protected by iron plates covered with spikes.

mast
rigging
spiked iron plates

Korean ship
Admiral Yi Sun-sin's redesign of the Korean "turtle ship" was a precursor of the iron-clad steam warships of the 19th century.

ITALIAN NATURALIST ULISSE ALDROVANDI (1522–1605), who had founded the botanical garden of Bologna in 1568, published the first of three volumes of *Ornithologiae*—a treatise on birds—in 1599. As well as designing and managing the botanical garden, Aldrovandi organized expeditions to collect plants for his herbarium and established a large collection of flora and fauna specimens. He also wrote many books covering all aspects of natural history, helping to lay the **foundations for the modern study of botany and zoology**.

In 1600, English physician and scientist **William Gilbert** (1544–1603) published *De Magnete, Magneticisque Corporibus, et de Magno Magnete Tellure* (*On the Magnet, Magnetic Bodies, and the Great Magnet of the Earth*). In this treatise, he described his experiments with magnets, many of which used small magnetic spheres called "terrellae," to model Earth's behavior. He concluded that **Earth behaves as a giant magnet**, making compass needles point north, and that **the center of Earth is made of iron**.

31–34 MILES / YEAR THE SPEED AT WHICH THE MAGNETIC NORTH POLE IS MOVING

EARTH'S MAGNETIC POLES

The Earth's core—composed of an iron alloy—behaves like a gigantic bar magnet. Magnetic compass needles are attracted to the two poles of the Earth's magnetic core, also known as the magnetic poles. These coincide roughly with the geographical north and south poles. Because the core is fluid, the magnetic poles can shift position.

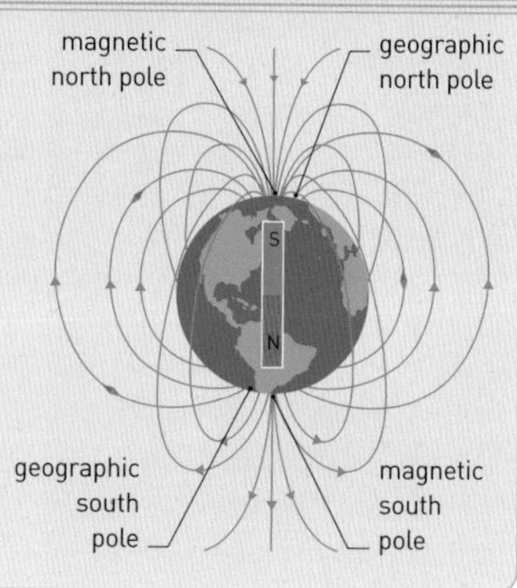

magnetic north pole
geographic north pole
S
N
geographic south pole
magnetic south pole

Tycho Brahe publishes the star catalog ***Astronomiae Instauratae Mechanica***

Dutch sailors discover the **dodo** on the island of Mauritius

Korean Admiral Yi Sun-sin develops an improved **"turtle ship"**

1599 Italian naturalist **Ulisse Aldrovandi** publishes the first volume of ***Ornithologiae***

1599 Tycho Brahe builds an **observatory at Benátky nad Jizerou**

1600 William Gilbert publishes ***De Magnete, Magneticisque Corporibus, et de Magno Magnete Tellure***

1601–04

> **“** THE DISCUSSION OF **NATURAL PROBLEMS** OUGHT TO BEGIN… WITH **EXPERIMENTS AND DEMONSTRATIONS. ”**

Galileo Galilei, Italian astronomer and physicist, from *The Authority of Scripture in Philosophical Controversies*

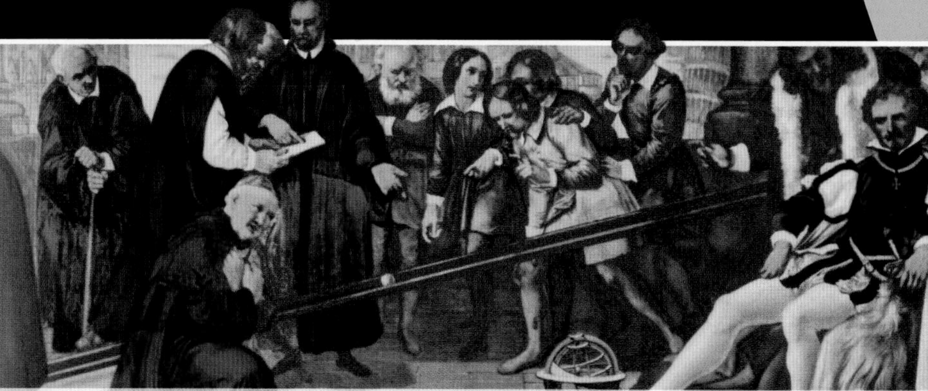

Galileo demonstrated his "law of falling bodies" by rolling a ball down an inclined plane and measuring its rate of acceleration.

De Magnete also claimed that **magnetism and electricity are two distinct kinds of force**. To show the **properties of static electricity**, Gilbert created a **versorium—the first electroscope**, comprising a freely rotating unmagnetized needle on a stand. The versorium's needle was attracted to static-charged amber, as if it were a compass needle moved by magnetism. Gilbert incorrectly inferred that gravity was a magnetic force, and that Earth's magnetism held the Moon in its orbit.

In the same year, Italian friar and astronomer **Giordano Bruno** (1548–1600) was **burned at the stake** by the Roman Inquisition on charges of heresy. It is possible that he was originally arrested purely for his unconventional theological beliefs, but it is more likely that his **scientific views** were the real reason for the Inquisition's wrath. Bruno's theory of the cosmos went a step further than Copernicus's (see 1543), and was potentially more threatening to the Church's authority: he believed that **the Sun**, far from being the center of the Universe, **was just a star like any other**, and that it was possible that **Earth was not the only world inhabited** by intelligent life.

ENGLISH ASTRONOMER AND MATHEMATICIAN THOMAS HARRIOT (1560–1621) was fascinated by the behavior of light. In 1602, he studied the relationship between the different angles produced as light is refracted, or bent, when passing from one medium to another, such as from air to water. Now known as the **law of refraction**, this phenomenon had first been discovered by **Persian mathematician** Ibn Sahl in 984. Unfortunately, Harriot did not publish his findings, and the principle is now known as Snell's Law, after Willebrord Snellius (see 1621–24), who rediscovered the idea around 20 years later.

The following year, naturalist **Federico Cesi** (1585–1630) founded a scientific society called **Accademia dei Lincei** (Academy of the Lynx-eyed) in Rome—a successor to the Academia Secretorum Naturae (Academy of the Secrets of Nature), which had been founded in 1560 but forced to disband. The Accademia dei Lincei later became the national academy of Italy.

The story of **Galileo Galilei** (see 1611–13) dropping balls of different weights from the top of the Leaning Tower of Pisa, Italy, to ascertain the rate of their fall may or may not have been true; however, it is known that in 1604 he hypothesized for the first time that **bodies made of the same material and falling through the same medium would fall at the same speed**, regardless of their mass. This idea contradicted the prevalent Aristotelian theory that the heavier the object, the faster it would fall. Galileo published the final version of his **law of falling bodies** in 1638.

JOHANNES KEPLER (1571–1630)

Born in Germany, Johannes Kepler studied at the University of Tübingen, where he encountered the ideas of Copernicus. He worked as a teacher in Graz, Austria, before moving to Prague to study with Tycho Brahe in 1600. He remained there as Imperial Astronomer after Brahe's death, until political and family problems forced him to leave 12 years later.

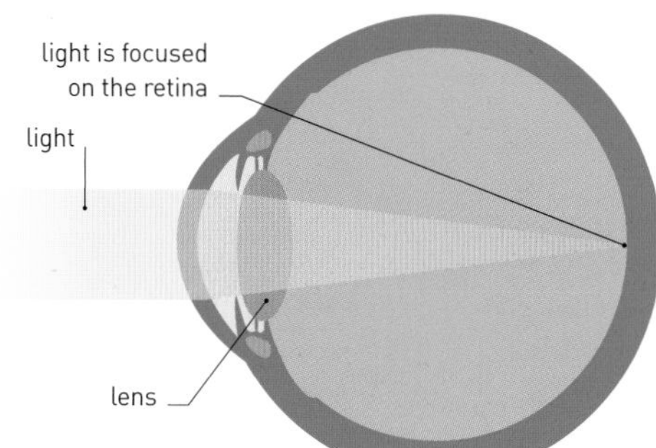

light is focused on the retina
light
lens

Human eye
The eye sees by allowing light in through a lens near the front. The lens projects the light onto the retina at the back, and focuses it to give a clear though inverted image.

Best known as an astronomer, **Johannes Kepler** was also a pioneer in the field of optics, publishing *Astronomiae Pars Optica* (*The Optical Part of Astronomy*) in 1604. In addition to describing astronomical instruments, he devoted much of the text to optical theory, including explanations of **parallax** (the apparent change in position of a heavenly body when viewed from different points), **reflections in flat and curved mirrors**, and **the principle of the pinhole camera**. He examined the **optics of the human eye**, describing how the lens reverses and inverts the image projected onto the retina, and suggested that this is corrected in the brain.

In 1604, Italian surgeon and anatomist **Hieronymus Fabricius** (see 1594–95) published the results of his dissections of various animals' fetuses, establishing **embryology** as a new field of study. He showed various stages of fetal development, and combined these studies with his work on the circulation of blood to produce one of the first studies of **embryonic circulation**.

> **“** DISCOVER **THE FORCE OF THE HEAVENS,** O MEN: ONCE RECOGNIZED, **IT CAN BE PUT TO USE. ”**

Johannes Kepler, German astronomer, from *De Fundamentis*, 1601

1600 Giordano Bruno is **burned at the stake** for his heretical views

1600 Johannes Kepler begins work as **Tycho Brahe's assistant** in Prague

1602 Thomas Harriot arrives at the **law of refraction**

1603 The **Accademia dei Lincei** is founded in Rome

1604 Galileo Galilei begins to formulate his **law of falling bodies**

1604 Johannes Kepler describes how the eye focuses light in ***Astronomiae Pars Optica***

1604 Hieronymus Fabricius begins his study of **blood circulation in embryos**

> **“** IF A MAN WILL **BEGIN WITH CERTAINTIES,** HE SHALL **END IN DOUBTS;** BUT IF HE WILL BE CONTENT TO **BEGIN WITH DOUBTS,** HE SHALL **END IN CERTAINTIES. ”**

Francis Bacon, English philosopher, from *The Advancement of Learning,* 1605

Around 1,344 light-years away and situated in the constellation of Orion, the Orion nebula is one of the brightest and closest nebulae to Earth.

AT THE BEGINNING OF THE 17TH CENTURY, SCIENCE was still known as "Natural Philosophy." In 1605, however, English philosopher **Francis Bacon** (1561–1626) published his first work, *The Advancement of Learning*, setting out arguments for using **induction**—a process of drawing conclusions from data accumulated by observation—as the basis for scientific knowledge. Later known as the **Baconian method**, or "scientific method," induction became important in modern experimental science.

German astronomer **Johannes Kepler** observed the appearance of a comet (now known as **Comet Halley**) in 1607, and noted its position and path across the night sky. He realized that it was traveling well outside the orbit of the Moon. His observations later influenced his laws of planetary motion (see pp.100–101).

The invention of the first two-lens telescope is generally attributed to Dutch inventor **Hans Lippershey** (*c.*1570–1619) in 1608. Unlike later reflecting telescopes that used mirrors, Lippershey's refracting telescope had a lens at each end. Although Lippershey could not obtain a patent for it, his invention earned him money and recognition for its military and commercial uses.

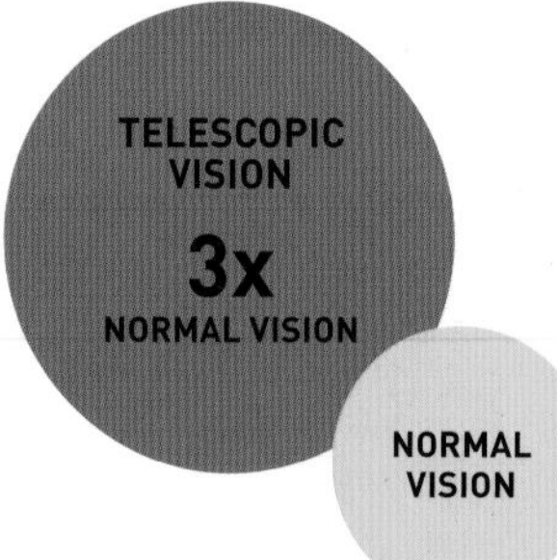
TELESCOPIC VISION
3x
NORMAL VISION
NORMAL VISION

Magnification of early telescopes
The telescopes built by Lippershey and his contemporaries were capable of magnifying an image to about three times its size.

Another invention that was to revolutionize military campaigns was the **flintlock mechanism** for firearms. Probably the work of gunsmith and violin-maker **Marin le Bourgeoys** (*c.*1550–1634), the flintlock appeared around 1608 in France. Quicker and more efficient than previous mechanisms, the flintlock was also safer because it could be locked into position during reloading. It remained in use for more than 200 years.

Lippershey in his workshop
In creating his refracting telescope, Hans Lippershey may have used a combination of concave and convex lenses, or two convex lenses.

IN 1609, GERMAN ASTRONOMER JOHANNES KEPLER'S book *Astronomia Nova* (*New Astronomy*) was published, describing his observations of the motion of the planet Mars. His detailed measurements and calculations confirmed the theory that the **planets revolved around the Sun**, and went further in suggesting that they did so in **elliptical, rather than circular, orbits**. He also pointed out that the **speed** at which they orbited the Sun did not remain constant but **changed according to their position on the orbit**.

These principles formed the basis of the first two of the three laws that are now known as **Kepler's laws of planetary motion** (see pp.100–101). The first law states that each planet has an elliptical orbit with the Sun as one of the focuses; the second that the speed of a planet is inversely proportional to its distance from the Sun—that is, a planet moves fastest when it is closest to the Sun.

News of the invention of the refracting telescope reached Italy in 1609, and Galileo Galilei set about building one for himself. **Galileo's telescope** allowed him to make detailed astronomical observations. His early telescopes had a magnification of about eight times normal vision, but he later improved the design to around 30 times magnification. English astronomer **Thomas Herriot** (1560–1621) had used a telescope to study the Moon in 1609, and had produced the first drawings of its surface. The following year, Galileo used his superior telescope and artistic training to produce **detailed maps of the lunar landscape**, clearly showing the irregularities to be craters and mountains. So accurate were his maps that he was even able to estimate the height of the mountains on the surface of the Moon.

4
THE **NUMBER OF MOONS OF JUPITER** OBSERVED BY GALILEO

Galileo was also able to **examine other planets**, and in 1610 turned his attention to **Jupiter**. He noticed three

1605 Francis Bacon publishes ***The Advancement of Learning***

1607 Johannes Kepler records the appearance and motion of a comet, later known as Comet Halley

1608 Hans Lippershey builds a **refracting telescope** with two lenses

1608 The **flintlock mechanism** for firearms first appears

1609 Galileo Galilei demonstrates one of his **early telescopes** in Venice

1609 In *Astronomia Nova*, Johannes Kepler states his first two **laws of planetary motion**

1609 Dutch inventor Cornelis Drebbel invents a **thermostat**

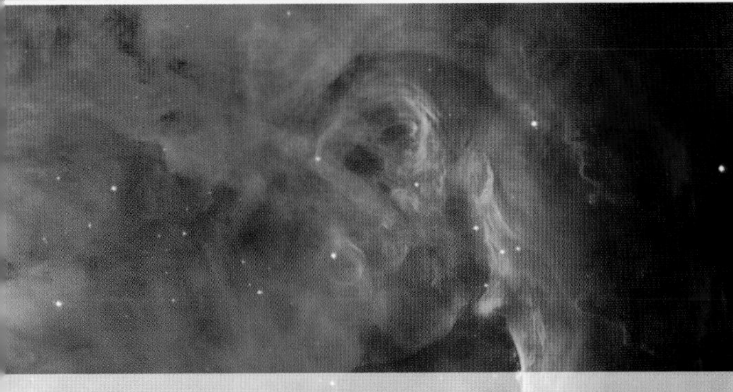

> **“** IN **QUESTIONS OF SCIENCE,** THE **AUTHORITY OF A THOUSAND** IS **NOT WORTH** THE HUMBLE **REASONING OF A SINGLE INDIVIDUAL. ”**

Galileo Galilei, Italian mathematician and astronomer, 1632

previously undetected "stars" close to Jupiter. However, their behavior indicated that they were, in fact, not stars but moons or satellites, orbiting the planet—a theory confirmed when one of them disappeared behind Jupiter. In further observations, he discovered a fourth satellite following a similar orbit. These **Galilean satellites**, as they were later called, were the four largest moons of Jupiter, now known as Io, Europa, Ganymede, and Callisto, their names being associated with classical myths.

The telescope also led to new discoveries elsewhere: French astronomer **Nicolas-Claude Fabri de Peiresc** (1580–1637) acquired one in 1610 and saw the Galilean satellites for himself. Later that year, he became the **first person to observe the Orion Nebula**.

Galileo's moon map
Although not the first maps of the Moon, Galileo's detailed charts were the first to show the distinctive craters and mountains on its surface.

GALILEO CONTINUED HIS ASTRONOMICAL DISCOVERIES in 1611, describing temporary dark areas seen on the surface of the Sun—now called **sunspots**. Although he claimed to be the first to have observed these, others may have done so beforehand. The importance of their discovery lay in the fact that the periodic appearance of sunspots was yet another **challenge to the Aristotelean notion** of the perfect immutability of the heavens.

In 1611, **Kepler** published a treatise on optics, *Dioptrice*, in which he explained the **workings of the microscope** and the **refracting telescope**. He also explored the effects of using lenses of different shapes and focal lengths. He explained the workings of the Galilean telescope, with its convex and concave lenses, and also suggested a way of improving Galileo's design using two convex lenses to achieve greater magnification.

In the same year, Kepler also wrote an extraordinary "thought experiment" entitled ***Somnium* (*The Dream*)**, which was published posthumously. In it, he described a form of interplanetary travel, and attempted to explain a model of the universe from a perspective that is not geocentric (Earth-centred).

Florentine priest and chemist, **Antonio Neri** (1576–1614) devoted much of his time to the study of glassmaking. In 1612, he published a comprehensive book ***L'Arte Vetraria* (*The Art of Glass*)** on the manufacture and uses of glass, which remained a standard textbook until the 19th century.

Galileo was interested not only in astronomy, but also in many other fields. In 1613, he studied the **concept of motion**, and put forward his **principle of inertia**, which states that "a body moving on a level surface will continue in the same direction at constant speed unless disturbed." It explained that moving objects retain their velocity unless a force, such as friction, acts upon them, a principle later important for Isaac Newton's First Law of Motion (see pp.120–21).

GALILEO GALILEI (1564–1642)

Born in Pisa, Italy, Galileo Galilei studied medicine and mathematics at university. In 1592, he took up professorship at Padua. His interests included astronomy and science of motion.

Galileo's scientific views were seen as heretical by the Catholic Church, and he was placed under house arrest in 1633, where he remained until his death.

Galileo's telescope
Based only on vague descriptions of Lippershey's telescope, Galileo made a telescope with a combination of convex and concave lenses.

- **1610** Galileo describes mountains and craters on the surface of the **Moon**
- **1610** Galileo identifies four of **Jupiter's moons**
- **1610** Nicolas-Claude Fabri de Peiresc discovers the **Orion nebula**
- **1611** Galileo observes and describes **sunspots**
- **1611** Kepler publishes *Dioptrice*, on the **optics of the telescope**, and the fantasy *Somnium*
- **1612** Antonio Neri publishes the first textbook on **glassmaking**, *L'Arte Vetraria*
- **1613** Galileo first describes his **principle of inertia**

1614–17

Napier's Bones, a set of rods inscribed with numbers, provided a quick and simple way of multiplying, dividing, and finding square and cube roots.

IN 1614, SCOTTISH MATHEMATICIAN JOHN NAPIER (1550–1617) published a description of logarithms (a logarithm is the power to which a base, such as 10, must be raised to produce a given number), showing how they could resolve "the tedium of lengthy multiplications and divisions, the finding of ratios, and... the extraction of square and cube roots." The logarithm of the product of two numbers multiplied together is equal to the sum of each number's logarithms. Using tables that Napier included in his book, it was possible to find the product of two numbers by looking up their logarithms, adding them, and then looking up the result in a table of antilogarithms.

Santorio Santorio (1561–1636), also known as Sanctorius, Professor of Anatomy in Padua, Italy, described his experiments in the **study of metabolism** in *De Statica Medicina* (*On Medical Measurement*). Over 30 years, he maintained a record of his weight, and the weights of everything he ate and drank, and of all feces and urine that he passed, and discovered a discrepancy that he attributed to "insensible perspiration."

3,959 MILES

THE MEAN **RADIUS** OF **EARTH**

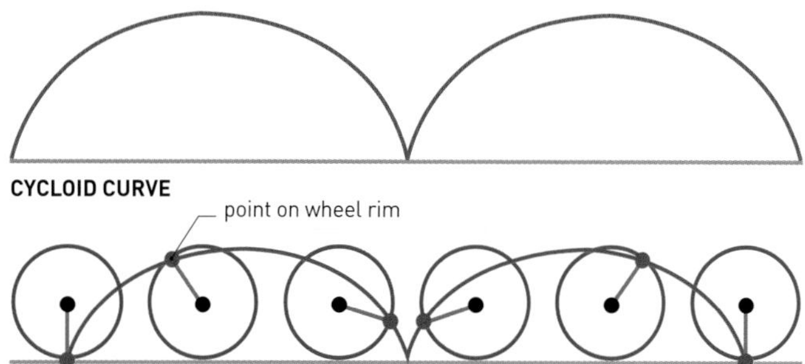
CYCLOID CURVE
point on wheel rim
WHEEL ROLLING IN STRAIGHT LINE

Cycloid
The curve described by a point on the rim of a circular wheel as it rolls on a flat surface, known as a cycloid, fascinated mathematicians in the 17th century.

In 1615, French mathematician and theologian, **Marin Mersenne** (1588–1648), was the first to properly define the cycloid curve traced by a point on the rim of a wheel. He also made an unsuccessful attempt to calculate the area under the curve, and so posed a problem that several 17th-century mathematicians tried to solve.

In April 1616, for the first of his annual lectures at the Royal College of Physicians in London, England, **William Harvey** (1578–1657) spoke on the **circulation of blood**. He was the first to explain the way the heart pumps oxygenated blood around the body. During this seven-year lecture series he expounded on his theory, but he did not publish a full account until 1628.

In 1617, Dutch astronomer and mathematician **Willebrord Snellius** (1580–1626) published his work *Eratosthenes Batavus* (*The Dutch Eratosthenes*), in which he described a new method for **measuring Earth's radius**, by first finding, using triangulation, the distance between two points separated by just one degree of latitude. His work is seen as the foundation for modern geodesy—surveying and measuring Earth.

This year, Napier presented another aid to calculation in his book *Rabdologiae* (meaning measuring rods). This was a set of rods inscribed with figures derived from multiplication tables, which became known as **Napier's Bones**.

1618–20

This reconstruction of Drebbel's submarine, the first navigable underwater craft, is complete with fins, a rudder, and watertight portholes for oars.

"MY **METHOD,** THOUGH HARD TO PRACTICE, IS **EASY TO EXPLAIN...** I PROPOSE TO ESTABLISH PROGRESSIVE STAGES OF **CERTAINTY.**"

Francis Bacon, English philosopher, from *Novum Organum,* 1620

IN 1619, German astronomer **Johannes Kepler** (see 1601–04) published *Harmonices Mundi* (*The Harmony of the World*). In this book, he explained the structure and proportions of the universe in terms of geometric shapes and musical harmonies, in much the same way as the ancient philosophers Pythagoras and Ptolemy had done before him. Much of Kepler's thesis concerned the **harmony of the spheres**—the idea that each planet produces a unique sound based on its orbit. It also discussed the relationships of astrological aspects (the angles between planets) and musical tones. More influentially, the thesis contained in its final section a statement of **Kepler's third law of planetary motion**, describing the relationship between a planet's distance from the Sun and the time taken to orbit around it, and the speed of the planet at any time in that orbit (see pp.100–01).

Dutch inventor **Cornelis Drebbel** (1572–1633) had moved to England around 1604. In 1620, he invented the **first navigable submarine**, while in the employment of the British Royal Navy. It was based on a design suggested by British writer

FRANCIS BACON
(1561–1626)

Born into an aristocratic family, Francis Bacon studied at Trinity College, Cambridge, UK, from age 12. A lawyer and Member of Parliament, he was knighted by James I, who appointed him Attorney General (1613) and Lord Chancellor (1618). In 1621, Bacon was found guilty of corruption. He devoted the rest of his life to writing.

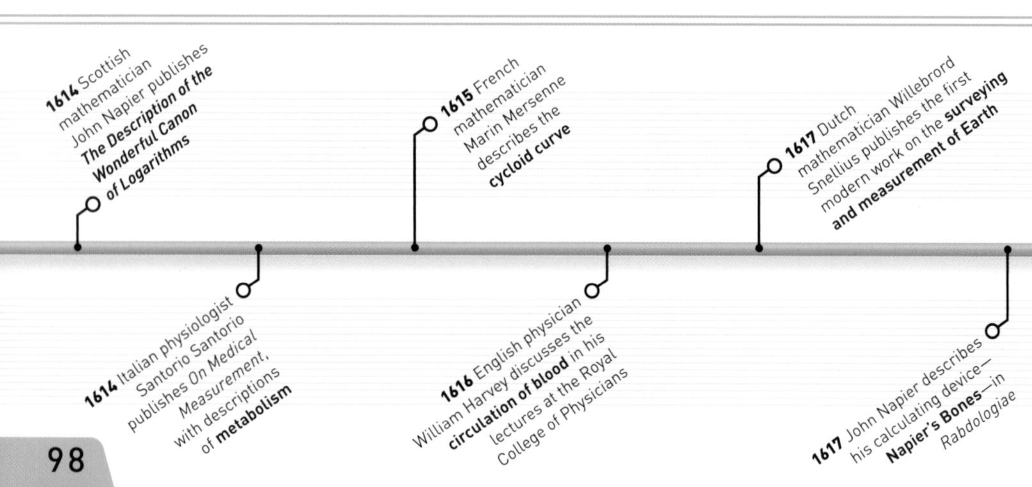
1614 Scottish mathematician John Napier publishes ***The Description of the Wonderful Canon of Logarithms***
1614 Italian physiologist Santorio Santorio publishes *On Medical Measurement*, with descriptions of **metabolism**
1615 French mathematician Marin Mersenne describes the **cycloid curve**
1616 English physician William Harvey discusses the **circulation of blood** in his lectures at the Royal College of Physicians
1617 Dutch mathematician Willebrord Snellius publishes the first modern work on the **surveying and measurement of Earth**
1617 John Napier describes his calculating device—**Napier's Bones**—in *Rabdologiae*
1619 German astronomer Johannes Kepler proposes his **third law of planetary motion** in *The Harmony of the World*

“ MELANCHOLY IS A HABIT, A SERIOUS AILMENT, A SETTLED HUMOUR, NOT ERRANT, BUT FIXED... ”

Robert Burton, English scholar, *The Anatomy of Melancholy,* 1621

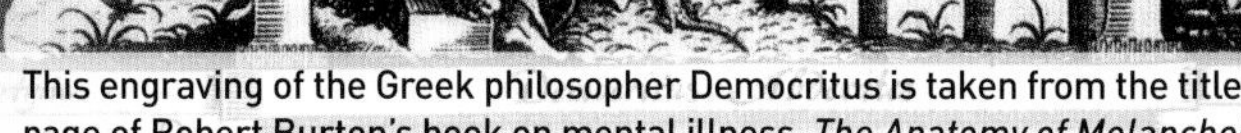

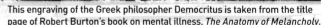

This engraving of the Greek philosopher Democritus is taken from the title page of Robert Burton’s book on mental illness, *The Anatomy of Melancholy.*

William Bourne (*c.*1535–82) in 1578, and consisted of a wooden frame covered with leather and powered by oars. Drebbel subsequently built two **larger submarines, capable of carrying a number of passengers**, which were demonstrated on the Thames River in England. In tests, the final version of Drebbel’s submarine managed to stay submerged for over three hours, suggesting that it had some means of providing oxygenated air for the occupants—although there are no records to explain how Drebbel could have achieved this. Despite the success of the submarine, the Royal Navy had no interest in using it.

In 1605, in the book *The Advancement of Learning,* English philosopher **Francis Bacon** had advocated **inductive reasoning for scientific investigation**. He wrote on the subject again in 1620, in a major treatise on logic called *Novum Organum Scientiarum* (*New Instrument of Science*). Bacon also advocated a process of reduction, which involved explaining the nature of things in terms of the relationships of their parts.

WITH HIS 1621 TREATISE ON THE CIRCLE, *Cyclometricus*, Dutchman Willibrord Snellius became the first European to publish the **law of refraction** describing the relationship between the angles of incidence and refraction of light passing through two different transparent substances, such as air and glass. Although known as **Snell’s Law** (see panel, right), this principle had been mentioned by English mathematician Thomas Harriot about 20 years earlier, and was originally described by Persian mathematician Ibn Sahl in 984.

The best known book by English scholar **Robert Burton** (1577–1640), *The Anatomy of Melancholy*, appeared in 1621. It attempted to describe various forms of **mental disorder** and their symptoms, and suggested possible medical causes and remedies. Although the book was written in the style of a medical textbook, it was more a literary work than a scientific one. Nonetheless, it was a forerunner of later scientific studies of the psychology and psychiatry of mental disorders.

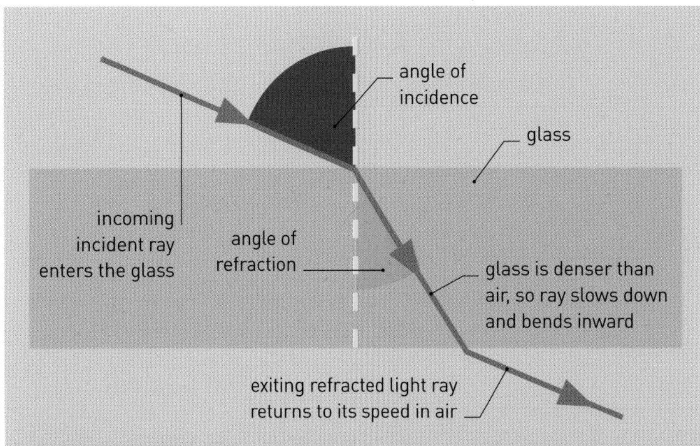

angle of incidence
glass
incoming incident ray enters the glass
angle of refraction
glass is denser than air, so ray slows down and bends inward
exiting refracted light ray returns to its speed in air

SNELL’S LAW

The law of refraction, or Snell’s Law, concerns the relationship between the angle of incidence (the angle at which light approaches the surface of a transparent medium) and the angle of refraction (the angle the light takes as it changes speed through the medium). The relationship is constant for all angles of incidence and refraction, but varies from substance to substance.

Following from Napier’s discovery of logarithms, English mathematician **Edmund Gunter** (1581–1626) devised **logarithmic scales** that could be engraved on a ruler to help seamen make calculations for navigation using a pair of compasses or dividers.

In 1622, English mathematician, **William Oughtred** (1574–1660), discovered that multiplication and division could be done by sliding two of Gunter’s scales against each other and reading the result—the principle of the slide rule. Oughtred experimented with several designs for his slide rule, starting with a circular shape, but eventually settling on the familiar straight ruler with a sliding middle section, which remained in use until the invention of the pocket calculator some 300 years later.

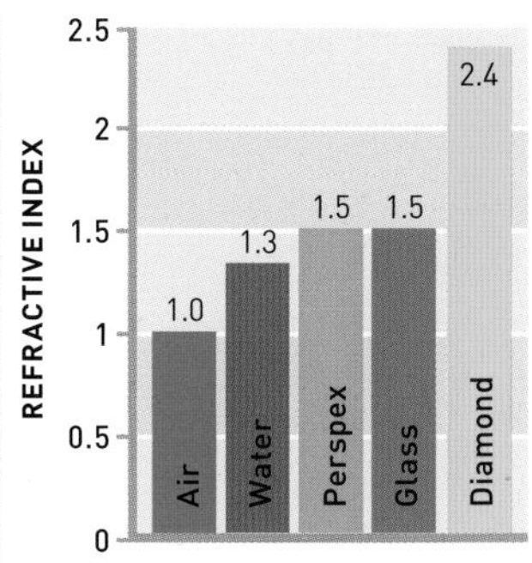

REFRACTIVE INDEX
2.5
2
1.5
1
0.5
0
1.0
1.3
1.5
1.5
2.4
Air
Water
Perspex
Glass
Diamond
SUBSTANCE

Refractive indices
The refractive index of a substance compares the speed of light when it passes through the substance to the speed of light in a vacuum.

Modern slide rule
Complex calculations could be done rapidly by lining up the different logarithmic scales inscribed on the rulers of the slide rule and reading the result using the cursor.

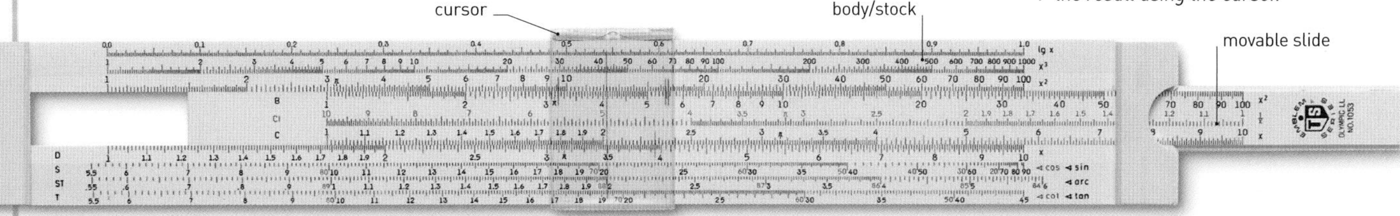

cursor
body/stock
movable slide

1620 Dutch inventor Cornelis Drebbel builds the **first navigable submarine**

1620 English philosopher Francis Bacon explains his method of **inductive reasoning** in *New Instrument of Science*

1621 Dutchman Willebrord Snellius rediscovers the **law of refraction**

1621 English scholar Robert Burton publishes ***The Anatomy of Melancholy***

1622 English mathematician William Oughtred invents the **slide rule**

UNDERSTANDING PLANETARY ORBITS

THE MOVEMENTS OF THE PLANETS CAN BE DESCRIBED WITH THREE LAWS AND EXPLAINED BY GRAVITY

The eight planets of the Solar System, as well as millions of smaller bodies such as comets and asteroids, travel around the Sun in closed loops called orbits. What keeps these objects on their curved trajectories is the same force that makes things fall to the ground on Earth: gravity.

For centuries it was generally believed that the Earth was the center of the Universe, with the Sun, Moon, planets, and stars rotating around it. However, this geocentric model could not satisfactorily account for planetary orbits, and in 1543 Danish astronomer Nicolaus Copernicus (1473–1543) proposed his heliocentric (Sun-centered) model, with the planets moving around the Sun in circular orbits (see 1543).

KEPLER'S LAWS

In the early 1600s, German astronomer Johannes Kepler (1571–1630) used observations of planetary movements to try to prove Copernicus right. However, he could make the observations fit a heliocentric system only if the orbits were not circles but ellipses, with the Sun at one focus (see below left). This fact became the first of Kepler's three laws of planetary motion. His second law (see below right) relates to the way a planet's speed changes during its orbit, and his third law (see opposite) concerns the relationship between a planet's distance from the Sun and how long it takes to complete each orbit (its orbital period).

GRAVITATIONAL FORCE

Kepler had no idea why orbits should be elliptical. The answer came after his death from English scientist Isaac Newton (1642–1727), who suggested that the same force that makes objects fall to the ground on Earth—gravity—might also be keeping the Moon in orbit around our planet. Newton realized that the force of gravity is weaker the further you are from the center of the Earth, and he proposed that gravity weakened in direct proportion to the square of the distance. When he applied this to the Moon, he was able to work out the Moon's orbital period. This allowed him to formulate his universal law of gravitation (see 1687) and to realize that gravity must also be responsible for keeping the planets in orbit around the Sun.

JOHANNES KEPLER
Kepler was an assistant to the great Danish astronomer Tycho Brahe. He used Brahe's observations of the planets when formulating his laws of motion.

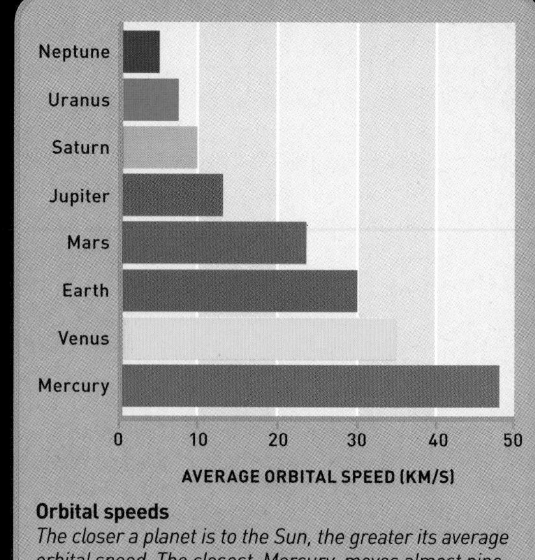
Neptune
Uranus
Saturn
Jupiter
Mars
Earth
Venus
Mercury
0
10
20
30
40
50
AVERAGE ORBITAL SPEED (KM/S)

Orbital speeds
The closer a planet is to the Sun, the greater its average orbital speed. The closest, Mercury, moves almost nine times faster than Neptune, the farthest.

92,955,778

THE AVERAGE **DISTANCE** IN MILES **BETWEEN** THE **SUN** AND THE **EARTH**

ELLIPTICAL ORBITS
Kepler's first law states that every planet's orbit is an ellipse with the Sun at one focus. An ellipse has two foci; they are the points from which two lines meeting any point on the ellipse always have the same total length.

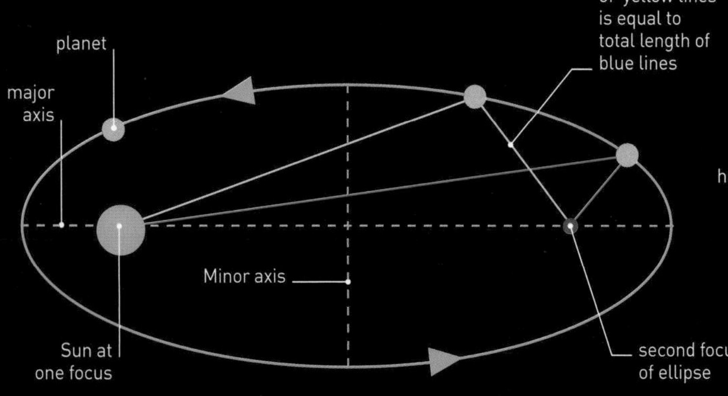
total length of yellow lines is equal to total length of blue lines
planet
major axis
Minor axis
Sun at one focus
second focus of ellipse

SPEED AND DISTANCE
The second of Kepler's laws states that an imaginary line joining a planet to the Sun sweeps across equal areas in equal times. This takes into account the fact that a planet moves faster when it is closer to the Sun and slower when it is farther away.

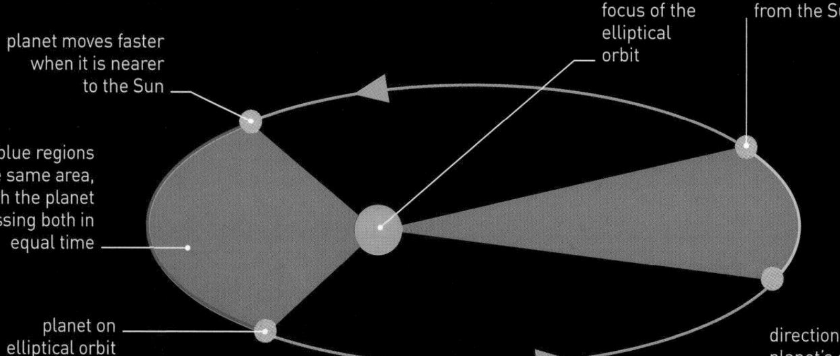
planet moves more slowly when it is farther away from the Sun
Sun at one focus of the elliptical orbit
planet moves faster when it is nearer to the Sun
both blue regions have the same area, with the planet crossing both in equal time
planet on elliptical orbit
direction of planet's orbit

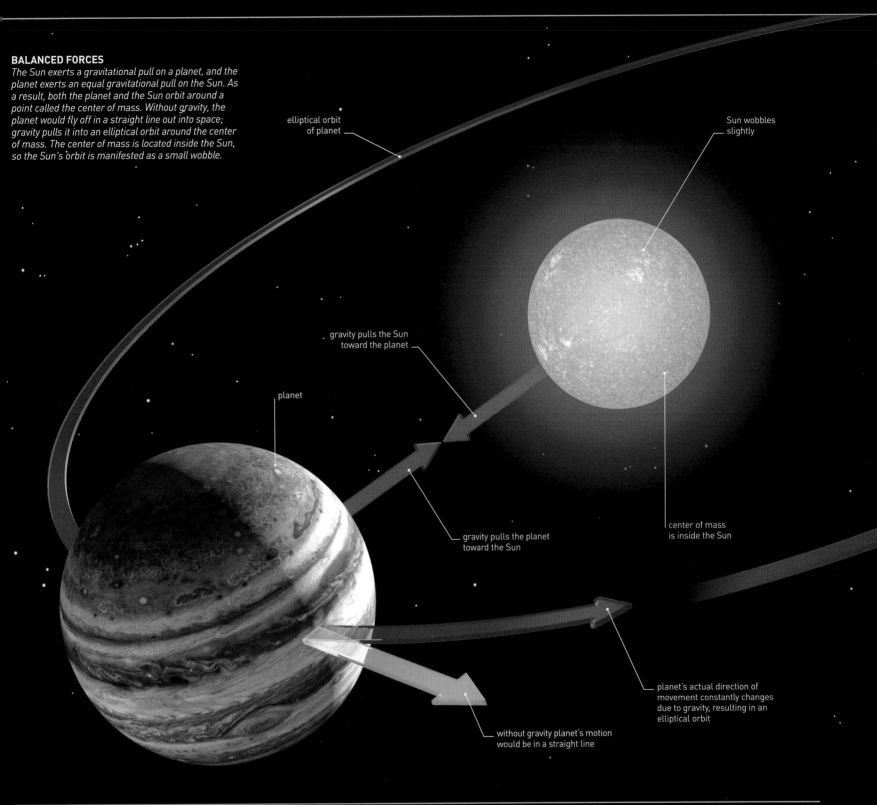

BALANCED FORCES

The Sun exerts a gravitational pull on a planet, and the planet exerts an equal gravitational pull on the Sun. As a result, both the planet and the Sun orbit around a point called the center of mass. Without gravity, the planet would fly off in a straight line out into space; gravity pulls it into an elliptical orbit around the center of mass. The center of mass is located inside the Sun, so the Sun's orbit is manifested as a small wobble.

elliptical orbit of planet

Sun wobbles slightly

gravity pulls the Sun toward the planet

planet

gravity pulls the planet toward the Sun

center of mass is inside the Sun

planet's actual direction of movement constantly changes due to gravity, resulting in an elliptical orbit

without gravity planet's motion would be in a straight line

ORBITAL PERIODS

Kepler's third law gives a mathematical relationship between a planet's average distance from the Sun and its orbital period (the time to complete each orbit). Specifically, it states that the square of the orbital period is proportional to the cube of the semimajor axis (half the diameter of an ellipse at its widest point). This makes it possible to quantify the increase in the orbital period with increasing distance from the Sun. Although Kepler's third law is not as simple as the second law, it enabled Newton to develop his universal law of gravitation.

PLANETARY YEARS

The length of a planet's "year," or orbital period, depends on its average distance from the Sun. The innermost planet, Mercury, has the shortest year at just 88 Earth days. Neptune's is the longest: 60,190 Earth days (164.8 Earth years). The diagram on the right (which is not to scale), shows the planets' orbital periods in Earth years.

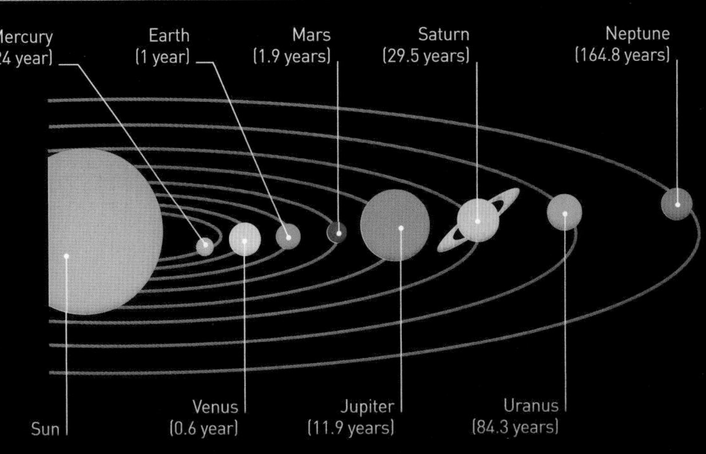
Mercury (0.24 year)
Earth (1 year)
Mars (1.9 years)
Saturn (29.5 years)
Neptune (164.8 years)
Sun
Venus (0.6 year)
Jupiter (11.9 years)
Uranus (84.3 years)

1625–27

Sodium sulfate crystals, known as Glauber's salt up to the late 18th century, are mostly sourced from natural minerals.

IN 1625, YOUNG DUTCH-GERMAN CHEMIST JOHANN GLAUBER (1604–70) recovered from a stomach bug after drinking from a spring. The following year, he succeeded in crystallizing *sal mirabile* (miraculous salt) from the spring's water. This became known as **Glauber's salt** and is in fact **sodium sulfate**, which has laxative properties. For nearly 300 years physicians would use it as a purgative.

In 1626, while traveling through icy London, English philosopher **Francis Bacon**—champion of the idea that theories must be built up from empirical evidence—acted on an impulse. He wanted to see if he could preserve meat by stuffing a chicken carcass with snow. While the experiment was a success, Bacon contracted **pneumonia** and never recovered.

In 1627, the most accurate **catalog of astronomical measurements** was published since Nicolaus Copernicus had suggested that the Sun was at the center of the Solar System (see 1543). Much of the data had been collected by Danish astronomer **Tycho Brahe**, but he died before he could publish the work. It fell to his collaborator, German astronomer **Johannes Kepler**, to complete the catalog, which was named ***Rudolphine Tables*** after the Holy Roman Emperor Rudolph II. This work contains data on the positions of nearly 1,500 stars, and the planets known at the time. Kepler financed the book's printing and dedicated it to Brahe.

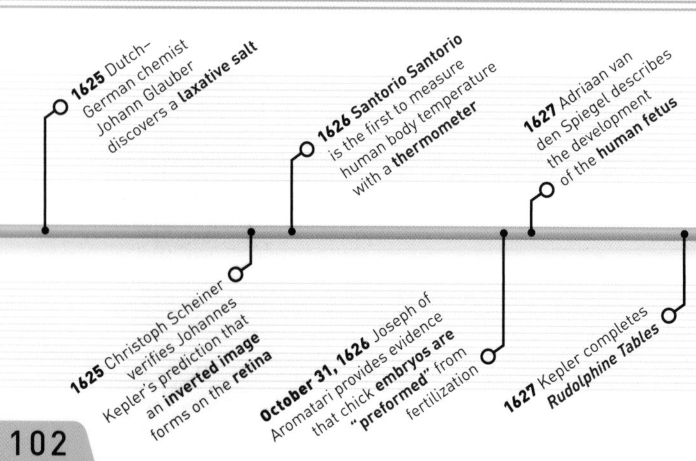

Frontispiece of *Rudolphine Tables*
This depicts an imaginary monument celebrating the achievement of generations of astronomers, including Hipparchus, Ptolemy, Nicolaus Copernicus, and Tycho Brahe.

1628–30

“ THE **HEART OF ANIMALS** IS THE FOUNDATION OF THEIR LIFE… THE **SUN OF THEIR MICROCOSM,** UPON WHICH ALL GROWTH DEPENDS, FROM WHICH **ALL POWER PROCEEDS. ”**

William Harvey, English physician, from *An Anatomical Essay,* 1628

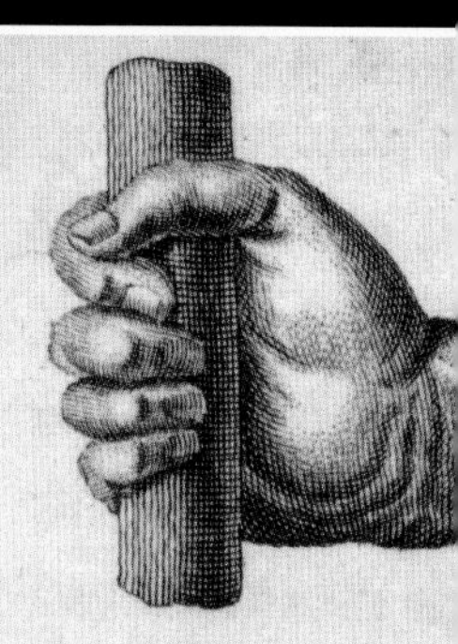

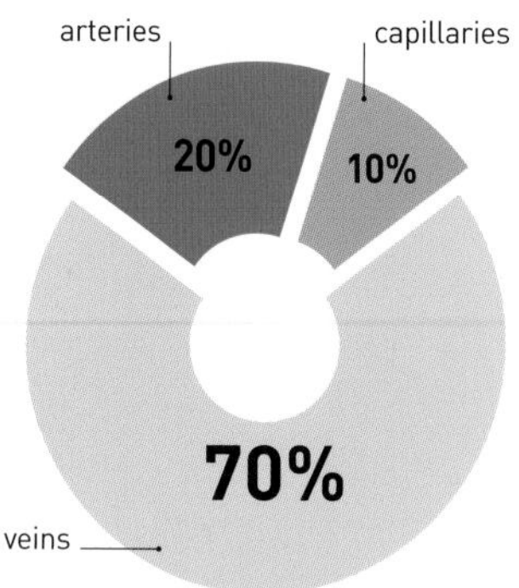

arteries
capillaries
20%
10%
70%
veins

Distribution of blood
Most blood in the systemic circulation is carried in the veins. The blood in these thin-walled vessels is at a very low pressure, so it effectively pools there.

IN 1628, ENGLISH PHYSICIAN WILLIAM HARVEY published his most celebrated work: *An Anatomical Essay Concerning the Movement of the Heart and Blood in Animals*. Harvey was a great believer in the need for science to progress by experimentation, and had closely studied the **blood systems of animals**. In the previous century, Italian physician Matteo Colombo had demonstrated that the heart worked as a pump, and not by suction, as the Ancient Greeks had thought. But the traditional view persisted—originating with Ancient Greek surgeon, and philosopher Galen—that blood was continuously made in the liver. After assessing the pumping effect of the heart, Harvey doubted this was true. His experiments had shown that so much blood was pumped by the heart that continuous production was improbable. Instead, he deduced that the volume of blood is fixed and this is continuously circulated in the body. High-pressure blood, coming from the heart, is distributed around the body through arteries, while low pressure blood returns through veins. He also theorized about a specific **circulation for the lungs**.

In 1629, Italian inventor **Giovanni Branca** published

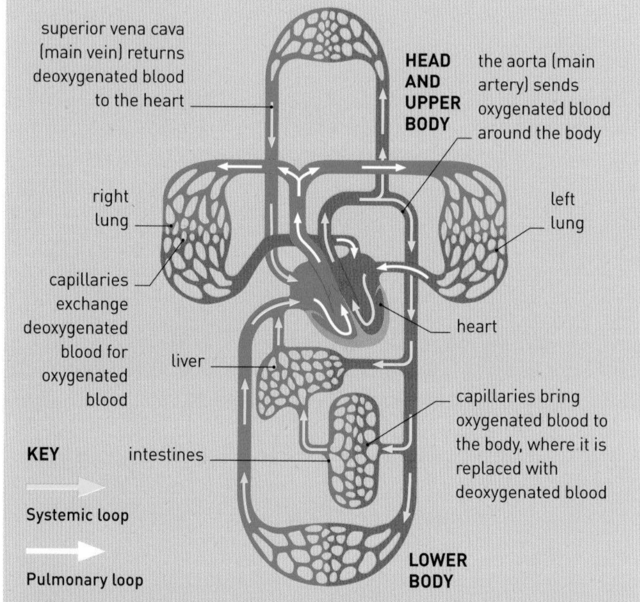

superior vena cava (main vein) returns deoxygenated blood to the heart
HEAD AND UPPER BODY
the aorta (main artery) sends oxygenated blood around the body
right lung
left lung
capillaries exchange deoxygenated blood for oxygenated blood
heart
liver
capillaries bring oxygenated blood to the body, where it is replaced with deoxygenated blood
KEY
intestines
Systemic loop
Pulmonary loop
LOWER BODY

CIRCULATION OF BLOOD

A double blood circulation system facilitates the exchange of oxygen and carbon dioxide, and ensures maximum pressure of blood in the lungs and around the body. Blood that has been oxygenated in the lungs (red) is pumped around the body by the left side of the heart. Deoxygenated blood from body tissues (blue) is pumped back to the lungs by the heart's right side.

1625 Dutch-German chemist Johann Glauber discovers a **laxative salt**

1625 Christoph Scheiner verifies Johannes Kepler's prediction that an **inverted image** forms on the **retina**

1626 Santorio Santorio is the first to measure human body temperature with a **thermometer**

October 31, 1626 Joseph of Aromatari provides evidence that chick **embryos are "preformed"** from fertilization

1627 Adriaan van den Spiegel describes the development of the **human fetus**

1627 Kepler completes ***Rudolphine Tables***

1628 William Harvey publishes *An Anatomical Essay Concerning the Movement of the Heart and Blood in Animals*

1629 Giovanni Branca publishes a collection of inventions that includes an early **steam engine**

1629 John Parkinson publishes *Park-in-Sun's Terrestrial Paradise*, said to be the **first gardening book**

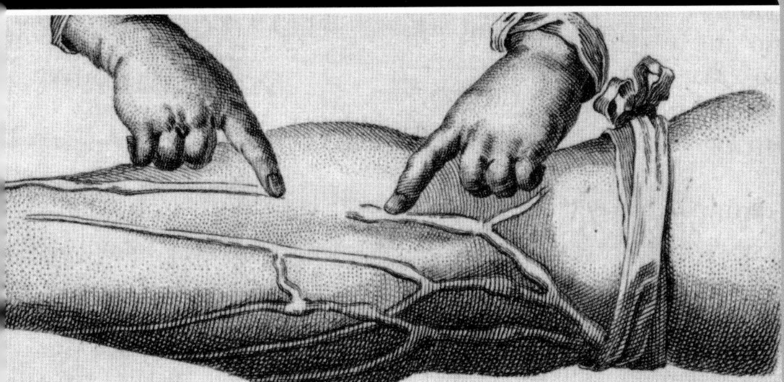

William Harvey demonstrated that one-way valves in veins stop blood from flowing back to the hand.

WILLIAM HARVEY (1578–1657)

Born in England, William Harvey studied at the universities of Cambridge and then Padua, in Italy, before a career in England devoted to studying blood and circulation; later, he also investigated reproduction and development. He served as physician to both King James I and King Charles I, and treated victims of the English Civil War.

a collection of machine designs that included an **early steam engine**. The steam-blasting vessel blew through a pipe that was directed at the vanes of a paddle-wheel, causing the wheel to revolve. Branca came up with several uses for his machine: lifting water, and grinding stone or gunpowder. In reality, however, it would have limited practical use. It was also entirely unrelated to later, more successful steam engine designs.

Englishman **John Parkinson** (1567–1650) was a herbalist and apothecary to the king. He was also a plantsman caught between the ancient herbalists and a new generation of botanists. His first major horticultural book—wryly entitled *Park-in-Sun's Terrestrial Paradise*—became an important text on the cultivation of plants, both for their esthetic and medicinal qualities. Although widely acknowledged as the **first gardening book**, it made limited impact on the scientific understanding of plants.

800

THE APPROXIMATE NUMBER OF **PLANTS ILLUSTRATED** IN PARKINSON'S 1629 WORK

1631–34

Galileo Galilei was tried by the Inquisition and forced to retract his heliocentric views. He was placed under house arrest, where he remained till he died.

“AND YET IT **MOVES.**”

Galileo Galilei, Italian astronomer, supposedly after his forced recantation of the theory that Earth moves around the Sun, 1633

IN 1631, FRENCH MATHEMATICIAN PIERRE VERNIER (1580–1637) described a device that assisted in the accurate **measurement of length**. It was based on an earlier idea by German mathematician Christopher Clavius. The original instrument had a sliding scale along the edge of a quadrant, which meant the user could measure a fractional part of the smallest division on the scale. To this day, the **Vernier scale** remains one of the best mechanical devices for accurate measurement.

In the same year, English mathematician **William Oughtred**—inventor of the slide rule—published a text that would have a lasting influence on many other mathematicians, including Isaac Newton. Oughtred's ***The Key of the Mathematicks*** introduced some fundamental algebraic symbols: the multiplication sign (x), and the proportion sign (:). For years it was described as the **most influential mathematical publication** in England.

Early in 1632, Italian astronomer **Galileo Galilei** published *Dialogue Concerning Two Chief World Systems*. In it he defended the heliocentric model of Copernicus that Earth orbited the Sun, against the classical view of Ptolemy, who said that Earth was at the center of the Solar System. As a result of this heresy, Galileo was **tried by the Inquisition and convicted**. He was forced to recant his views.

In the early 1630s, Italy faced a deadly natural threat. Malaria was spreading northward into swampy, low-lying regions; it had already claimed the lives of several popes and countless Roman citizens. **Agostino Salumbrino** (1561–1642) had worked as an apothecary in Peru, where the bark of the cinchona tree was used to control the disease. He sent the bark to Europe, where demand for it escalated. Its active ingredient, **quinine**, would be used to **treat malaria** for more than 300 years.

measuring jaw

retainer

Vernier scale

main scale

Vernier scale
Two adjoining scales that slide against one another are used to make accurate measurements. This device helps subdivide the smallest of divisions.

1629 Neil O'Glacan publishes a **treatise on plague**, after tending its victims

1631 Pierre Vernier first describes a **Vernier scale**

1631 English mathematician William Oughtred introduces **algebraic symbols**

1631 Peruvian Agostino Salumbrino sends **quinine-containing bark** to Rome to treat **malaria**

February 22, 1632 **Galileo Galilei** publishes *Dialogue Concerning the Two Chief World Systems*

1632 Italian surgeon Marco Severino publishes first textbook on **surgical pathology**

1635–37

René Descartes asserted that knowledge had to be distinct and precise.

> **EACH PROBLEM THAT I SOLVED BECAME A RULE** WHICH SERVED AFTERWARD TO SOLVE **OTHER PROBLEMS.**
>
> **René Descartes, French philosopher and mathematician,** from *Discourse on Method*, 1637

1638–40

In 1639, Jeremiah Horrocks was the first person known to record a transit of Venus as the planet moved across the surface of the Sun.

IN 1636, THE FRENCH MATHEMATICIAN MARIN MERSENNE published a treatise on the **mathematical analysis of musical sound**, in which he described laws to explain a stretched string's **frequency of vibration**. He stated that the frequency was lower in longer strings but increased with more stretching force.

Following Italian physicist Galileo Galilei's conviction for heresy in 1633, French philosopher and mathematician **René Descartes** delayed the release of *The World*—a bold account of his scientific views, including an agreement with Galileo's theory of Earth revolving around the Sun. Part of the text appeared in 1637, in *Discourse on the Method of Properly Conducting One's Reason and of Seeking the Truth in the Sciences*, which included essays on astronomy, geometry, and optics. In the book's appendix, *Geometry*, Descartes explained how **algebra and geometry were connected**. Two quantities, x and y, could be represented on two intersecting coordinate lines, the x and y axes, in a graph, and the relationship between the two could be represented in an algebraic equation. Another French mathematician, **Pierre de Fermat** (1601–65), had independently devised this method in 1629, but it was given Descartes' name, and was called the **Cartesian coordinate system** (see panel, below).

Fermat became better known for his **"Last Theorem,"** which stated that no positive whole numbers fit the equation $a^n + b^n = c^n$, where n is greater than 2. He wrote the theorem in the margin of an old textbook, claiming that he had proof for the theorem but no room to write it. Independent proof for the theorem was not found until 1995.

PIERRE DE FERMAT (1601–65)

Trained as a lawyer, Pierre de Fermat influenced several branches of mathematics. While he considered himself an amateur, and often refrained from providing proof for his discoveries, his work in geometry anticipated that of René Descartes. In 1654, Fermat corresponded with Blaise Pascal and helped develop probability theory.

CARTESIAN COORDINATE SYSTEM

When Descartes and Fermat realized how coordinates link algebra and geometry, it was a major breakthrough for science. Coordinates consist of two intersecting axis lines—x (horizontal) and y (vertical). It is possible to define the position of any point within this two-dimensional space by stating the values of x and y, that is, the x and y coordinates.

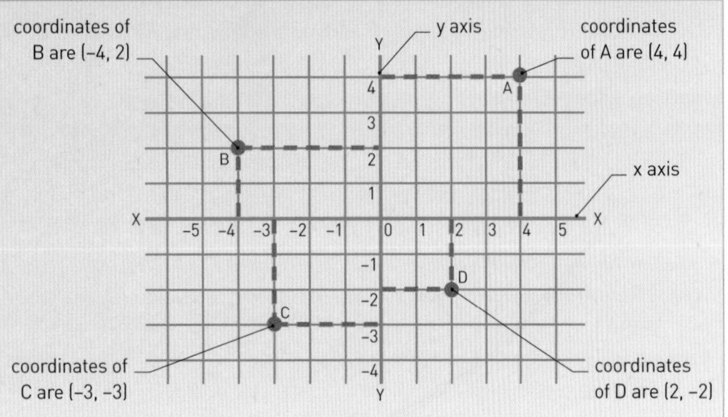

coordinates of B are (−4, 2)
y axis
coordinates of A are (4, 4)
x axis
coordinates of C are (−3, −3)
coordinates of D are (2, −2)
Y
X
A
B
C
D
−5 −4 −3 −2 −1 0 1 2 3 4 5
4 3 2 1 −1 −2 −3 −4

IN 1638, GALILEO GALILEI published his final word on physics: *Discourses and Mathematical Demonstrations Relating to Two New Sciences*, which dealt with the **strength of materials and kinematics**—the study of the motion of bodies without reference to mass or force. The Inquisition had banned publication of any work by Galileo after his trial in 1633. However, *Discourses* was published in Leiden, Netherlands, where the Inquisition had little influence.

English astronomer **Jeremiah Horrocks** (1618–41) had been studying Venus and estimated that the planet would pass the Sun on December 4, 1639. His prediction was based on the understanding that transits of Venus occur in pairs, eight years apart, and each paired transit occurs more than 100 years apart. When the time came, Horrocks focused the Sun's image onto paper and spotted the shadow of Venus only 15 minutes later than his prediction. Horrocks went on to calculate **Venus's size and distance** more accurately than ever before.

In 1640, sixteen-year-old French prodigy **Blaise Pascal** (1623–62) published *Essay on Conics*, in which he described the geometric relationship that occurs when a hexagon is drawn within a circle. In doing so he completed a mathematical theorem so advanced that at first many, including René Descartes, refused to believe that the young mathematician had done it.

In 1640, English botanist **John Parkinson** (1566–1650) published a plant catalog called *Theatrum Botanicum* (*Theater of Plants*). This was the most comprehensive work of its kind at the time, and remained a popular guide for many years.

7,522 MILES
THE **DIAMETER** OF **VENUS**

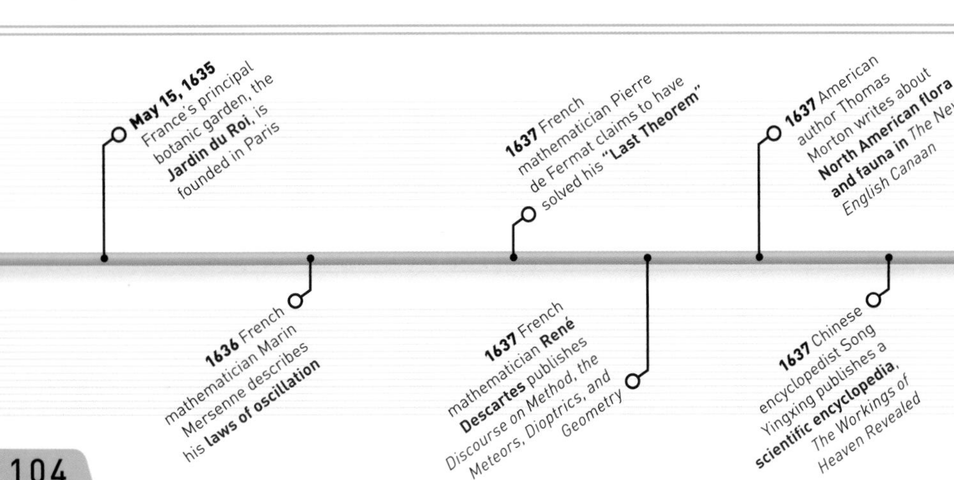

May 15, 1635 France's principal botanic garden, the **Jardin du Roi**, is founded in Paris
1636 French mathematician Marin Mersenne describes his **laws of oscillation**
1637 French mathematician Pierre de Fermat claims to have solved his **"Last Theorem"**
1637 French mathematician **René Descartes** publishes *Discourse on Method, the Meteors, Dioptrics, and Geometry*
1637 American author Thomas Morton writes about **North American flora and fauna** in *The New English Canaan*
1637 Chinese encyclopedist Song Yingxing publishes a **scientific encyclopedia**, *The Workings of Heaven Revealed*

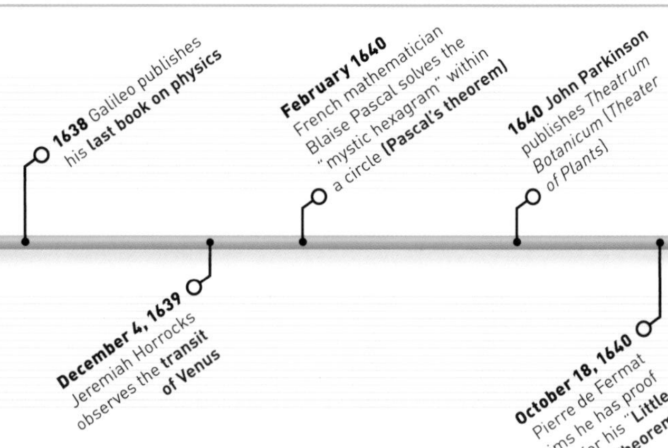

1638 Galileo publishes his **last book on physics**
December 4, 1639 Jeremiah Horrocks observes the **transit of Venus**
February 1640 French mathematician Blaise Pascal solves the "mystic hexagram" within a circle **(Pascal's theorem)**
1640 John Parkinson publishes *Theatrum Botanicum (Theater of Plants)*
October 18, 1640 Pierre de Fermat claims he has proof for his **"Little Theorem"**

1641–42

Although the symptoms of cholera were first cataloged in the 1600s, the cholera bacterium, seen here, was not isolated until the 19th century.

THE GRAND DUKE OF TUSCANY, Italy, Ferdinand II (1610–70) invented **the sealed glass thermometer** in 1641. Working with Italian physicist Evangelista Torricelli, he improved on Galileo's thermoscope (see 1590–93) by sealing the liquid column in a glass capillary, and using wine, which did not freeze as easily as water.

A year later, Blaise Pascal invented a **mechanical calculating machine** to help his father with his work in government taxes. Known as the Pascaline, this device worked by a system of wheels and gears and could perform routine arithmetic functions of addition, subtraction, division, and multiplication.

Dutch physician **Jacobus Bontius** (1591–1631) had traveled to the tropical East Indies for the Dutch East India Company in 1627. In 1642, his medical treatise—*De Medicina Indorum* (*Indian Medicine*) was published posthumously. This book contained some of the **earliest descriptions of tropical diseases**, including beriberi and cholera.

Another Dutchman, explorer and merchant **Abel Tasman** (1603–59), became the first European to reach Van Dieman's Land (now called Tasmania in his honor). He went on to visit New Zealand and islands of the Southwest Pacific. On his voyages, he recorded the earliest **European observations of Australasian fauna and flora**.

display window shows number input and answers to calculations
dial for inputting numbers

Pascal's calculating machine
This first Pascaline was primarily used by accountants, and its dials were calibrated in accordance with the French currency.

1643–44

30 INCHES
THE STANDARD HEIGHT OF **MERCURY** IN A **BAROMETER** AT SEA LEVEL

IN THE EARLY 1640S, EVANGELISTA TORRICELLI was investigating the practical problems associated with pumping water from deep wells. He imitated the action of a suction pump in a small tube and used a denser substance, mercury, instead of water, to study the effects. Torricelli discovered that mercury would rise into the sealed tube to a fixed height of 30 in (76 cm), and leave a gap at the top, which later became known as the **Torricelli vacuum**. He deduced that pressure from the atmosphere was forcing the mercury up the tube, and its height was determined by the value of this pressure (see p.106). Later, it was discovered that the pressure value varied according to altitude and weather, and small changes in atmospheric pressure signaled impending changes in weather. Torricelli's instrument therefore came to be adopted as **the first barometer**.

In 1644, René Descartes published *Principles of Philosophy*, in which he proposed an entirely **mechanical basis for the Universe**. He proposed that the Universe was filled with small, unobservable particles of matter that were set in motion by God, and that all aspects of science could ultimately be explained in accordance with this mechanical principle.

Torricellian tube
Evangelista Torricelli's device consisted of an evacuated glass tube containing a mercury column—the height of which was determined by atmospheric pressure.

1641 Ferdinand II invents a **sealed liquid thermometer**

1641 Dutch scientist Franciscus Sylvius describes the **deep cleft that separates lobes of the brain**

March 2, 1642 German anatomist Johann Georg Wirsung discovers the **pancreatic duct**

1642 Blaise Pascal makes a **wooden calculating machine**

1642 Dutch physician Jacobus Bontius publishes the **first book on tropical medicine**

November 24, 1642 Dutch explorer Abel Tasman becomes the first European to see the land now called **Tasmania**

1643 Evangelista Torricelli develops a **forerunner of the mercury barometer**

1644 Torricelli publishes *Opera Geometrica*, and **applies Galileo's laws of motion to fluids**

1644 Italian mathematician Pietro Mengoli poses the **Basel Problem**: the precise value of a particular number series

1644 René Descartes publishes *Principles of Philosophy*, describing a **mechanical universe**

1644 Italian astronomer Giovanni Odierna describes the microscopic appearance of a **fly's eye** in the first book on microscopic life

1645–48

Florin Périer, Blaise Pascal's brother-in-law, climbed France's Puy-de-Dôme volcano to measure changes in atmospheric pressure at higher altitudes with a rudimentary barometer.

IN THE 1640S, FRENCH MATHEMATICIAN BLAISE PASCAL started to investigate hydraulics—the mechanical properties of liquids. He found that unlike gas, liquid cannot be compressed, so when a force is applied, it is transmitted through the liquid. Pascal's studies led to the **invention of the hydraulic press** and the **syringe**. By 1646, Pascal had confirmed Italian physicist Evangelista Torricelli's observation that a fluid would rise in a glass column because of air pressure bearing downward (see 1643–44). Pascal also predicted that this pressure would diminish at higher altitudes. He asked his brother-in-law **Florin Périer**, who lived near a mountain, to test the idea. Receiving proof, Pascal suggested that air would thin out into a vacuum at still greater altitudes.

Polish astronomer **Johannes Hevelius's** (1611–87) greatest achievement came in 1647, when he published *Selenography* (*Description of the Moon*). The **first atlas of the Moon's surface**, it became a standard reference for years to come.

In 1648, a collection of essays written by Flemish chemist **Jan Baptist van Helmont** (1580–1644) was posthumously published by his son. Helmont had articulated an early version of the law of conservation of matter by describing a **five-year experiment in growing a willow tree**. By weighing both the plant and soil, he had deduced that material for the tree's growth had come from water. More than a century later, experimenters found that an even greater quantity came from air, in the form of carbon dioxide.

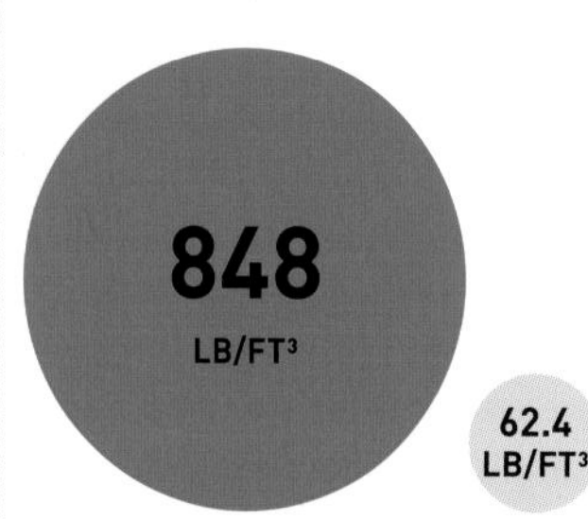

848
LB/FT³
62.4
LB/FT³
MERCURY
WATER

Comparative densities
Nearly 14 times denser than water, mercury rises short measurable distances in capillary tubes, making it useful in barometers.

MEASURING ATMOSPHERIC PRESSURE

For centuries, it was believed that air had no weight. But in fact it exerts a measurable force per surface area of the Earth. Blaise Pascal demonstrated atmospheric pressure by inverting a mercury-filled glass tube over a mercury reservoir. The tube's mercury falls to create an airless space (a vacuum), but atmospheric pressure pushes down on the reservoir to maintain a column: the bigger the pressure, the taller the column.

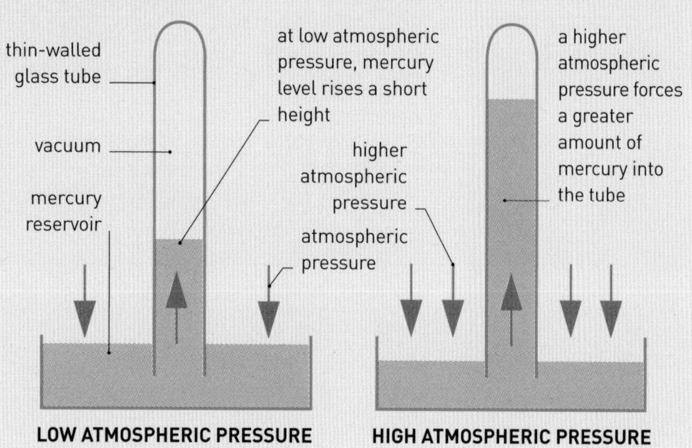

thin-walled glass tube
vacuum
mercury reservoir
at low atmospheric pressure, mercury level rises a short height
atmospheric pressure
higher atmospheric pressure
a higher atmospheric pressure forces a greater amount of mercury into the tube
LOW ATMOSPHERIC PRESSURE
HIGH ATMOSPHERIC PRESSURE

1649–51

> **“** WHILE **[THE ATOMS]** ARE **MOVING...** MEETING, INTERWEAVING, INTERMINGLING, UNROLLING, UNITING AND BEING **FITTED TOGETHER, MOLECULES...** ARE **CREATED. ”**

Pierre Gassendi, French philosopher, from *Syntagma Philosophiae Epicuri*, 1649

IN 1644, FRENCH PHILOSOPHER AND MATHMETICIAN RENÉ DESCARTES (1596–1650) had described a mechanical universe that was filled with particles of matter, within which a vacuum (a space devoid of matter) was an impossibility. In 1649, French priest, experimenter, and philosopher **Pierre Gassendi** (1592–1655) rejected the notion that everything could be explained in purely mechanical terms, and proposed an alternative. He suggested that the properties of matter were determined by the **shapes of the atoms**, and that atoms joined together to make bigger molecules. Gassendi accepted the existence of vacuums and even proposed that most matter consisted of "void." Gassendi's views anticipated later ideas concerning the bonding of atomic elements and the idea that an atom's mass is almost entirely concentrated at its nucleus (see 1911).

German physicist **Otto von Guericke** (1602–86) performed many experiments to prove that the vacuum existed. Around 1650, he invented a **piston-operated vacuum pump** with a valve system that could remove the air from a container by

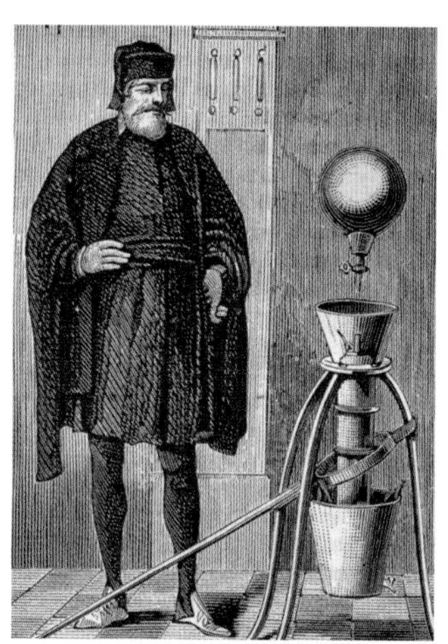
Piston-operated vacuum pump
Using an elaborate piston system, Otto von Guericke created a vacuum inside two joined hemispheres, called the hemispheres of Magdeburg, after his hometown in Germany.

> **“** THE **AIR... FLOWS ALL AROUND** US. JUST AS IT **PRESSES FROM ABOVE** ON THE HEAD, IT LIKEWISE **PRESSES ON THE SOLES** OF THE FEET... AND... **ON ALL PARTS OF THE BODY** FROM ALL DIRECTIONS. **”**

Otto von Guericke, German physicist, from *Experimenta Nova*, 1672

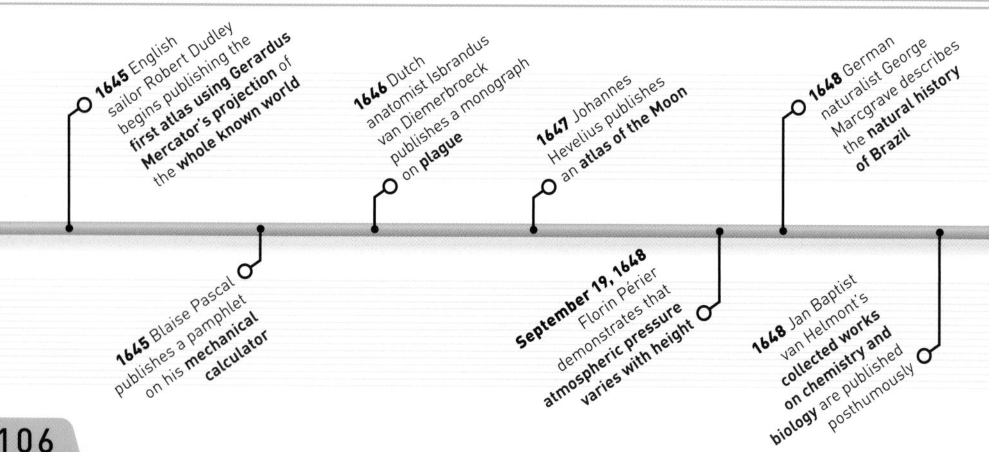

1645 English sailor Robert Dudley begins publishing the **first atlas using Gerardus Mercator's projection** of the whole known world
1645 Blaise Pascal publishes a pamphlet on his **mechanical calculator**
1646 Dutch anatomist Isbrandus van Diemerbroeck publishes a monograph on **plague**
1647 Johannes Hevelius publishes an **atlas of the Moon**
September 19, 1648 Florin Périer demonstrates that **atmospheric pressure varies with height**
1648 German naturalist George Marcgrave describes the **natural history of Brazil**
1648 Jan Baptist van Helmont's **collected works on chemistry and biology** are published posthumously

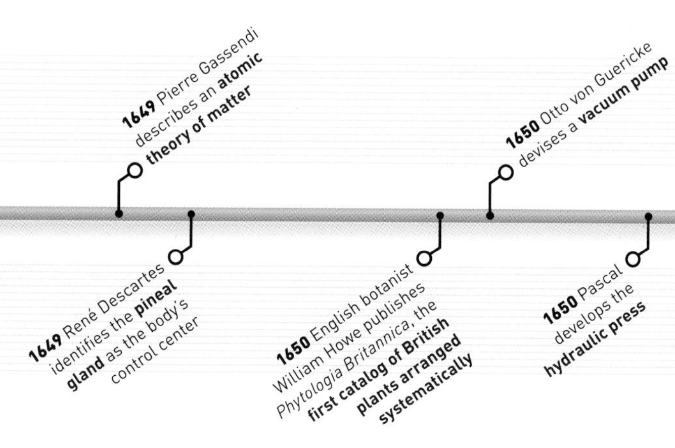

1649 Pierre Gassendi describes an **atomic theory of matter**
1649 René Descartes identifies the **pineal gland** as the body's control center
1650 English botanist William Howe publishes *Phytologia Britannica*, the **first catalog of British plants arranged systematically**
1650 Otto von Guericke devises a **vacuum pump**
1650 Pascal develops the **hydraulic press**

Pierre Gassendi's "atomic" theory was ahead of its time.

Fig. III.
Otto von Guericke's evacuated copper hemispheres were sealed together by nothing more than a smear of grease, yet two teams of eight horses were unable to pull them apart.

pumping, rather than suction. Von Guericke published descriptions of his experiments in 1672 in *Experimenta Nova* (*New Experiments*), which also contained illustrations of his vacuum pump design.

While physical investigators debated the nature of matter, **biologists questioned the origins of life**. Many took the view that life could arise spontaneously. In 1651, English physician **William Harvey**, who had previously described the circulation of blood (see 1628–30), maintained that **animals could only originate from eggs**. After studying chickens, he set out to find the mammalian "egg." As royal physician, Harvey was granted access to the king's fallow deer for his studies. He examined pregnant animals killed ever closer to the point of copulation in the hope of tracing the source of the egg. Harvey did not know that embryonic development in deer is naturally delayed for up to eight weeks after fertilization—so he wrongly concluded that the egg arose spontaneously in the womb. Mammalian eggs were not found until the 1800s, when ovaries were examined with microscopes.

ENGLISH PHYSICIAN NICHOLAS CULPEPER (1616–54) combined interests in medicine and botany to publish *The English Physician* in 1652—a book that integrated herbal medicine with astrology—followed by *Complete Herbal* in 1653, which catalogued the **medicinal uses of plants**. Many of the plants he listed are still in use in medicine today—for example, foxglove (*Digitalis*) for heart conditions. Culpeper's work included descriptions of many remedies that had been kept a secret up to that time.

In 1653, Blaise Pascal published the results of his exhaustive studies in the physics of liquids in *Treatise on the Equilibrium of Liquids*. This publication contained the idea that later came to be known as **Pascal's Law**: an incompressible liquid's pressure in a small closed system is equal in all directions.

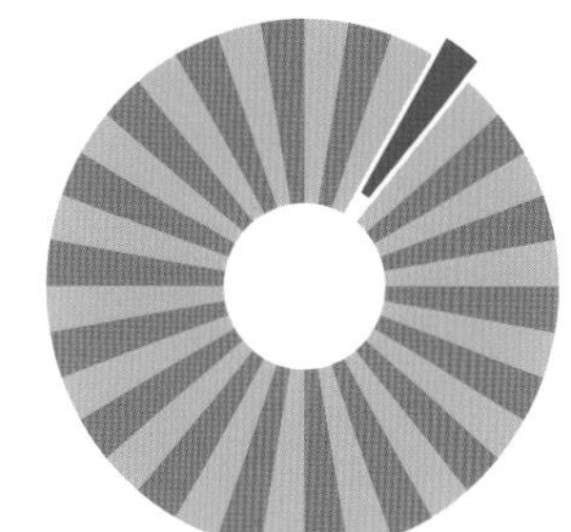
Probability
As the Pascal–Fermat theory tells a gambler, the probability of throwing a double 6 with two dice is 1 in 36.

> “ THE **HERBS** OUGHT TO BE **DISTILLED** WHEN THEY ARE IN THEIR **GREATEST VIGOR…** SO OUGHT THE **FLOWERS.** ”
>
> **Nicholas Culpeper, English botanist,** from *The English Physician,* 1652

Herb Robert
Marsh Pennywort
White Horehound
Complete Herbal
This illustration of medicinal plants is from an 1850 edition of English physician Nicholas Culpeper's Complete Herbal.

In 1654, four years after creating a vacuum pump to extract air from two sealed hemispheres, **Otto von Guericke** staged one of the most **dramatic public experiments of all time** in front of an audience at his hometown Magdeburg, Germany. Guericke sealed two copper hemispheres with grease, evacuated them using the pump, and suspended them between two teams of eight horses. Such was the strength of the air pressing in on the two copper hemispheres, that the horses could not pull them apart despite their best efforts. This astounded the assembled audience, and helped Guericke prove the **power of the vacuum**.

Meanwhile, the gambling habits of French nobleman Antoine Gombaud (1607–84) were about to help open a new field of mathematics. Gombaud questioned the profitability of a certain strategy in a game of dice, so he called upon the assistance of mathematician **Blaise Pascal** to explore the subject. Pascal initiated correspondence with his contemporary **Pierre de Fermat** (see 1635–37) to solve the problem. This collaboration resulted in the **formalization of the principles of probability**. After hearing about the Pascal–Fermat exchange, Dutch scholar Christiaan Huygens published the first book on probability in 1657. Because of the widespread enthusiasm for gambling, probability theory became popular among those who took the trouble to understand it.

BLAISE PASCAL
(1623–62)

Born in Clermont-Ferrand in France, Blaise Pascal was a child prodigy who was tutored by his tax-collector father. While still in his teens, he solved a complex mathematical problem and invented a mechanical calculating machine. He helped form the basis of probability theory and the physics of hydraulics.

1651 William Harvey asserts that **all animals develop from eggs**

1652 Nicholas Culpeper publishes ***The English Physician***

December 1652 Danish physician Thomas Bartholin names and describes the **lymphatic system**

1653 Culpeper publishes ***Complete Herbal***

1653 Pascal publishes studies on **hydrodynamics and hydraulics**

1653 Francis Glisson describes the **anatomy of the liver**

1653 Pascal publishes a treatise on an arithmetical triangle, later known as **Pascal's Triangle**

May 8, 1654 Otto von Guericke demonstrates the power of the vacuum in the **Magdeburg demonstration**

July 1654 Frenchmen Pierre de Fermat and Blaise Pascal develop **probability theory**

THE STORY OF MEASURING TIME

THE ACCURATE MEASUREMENT OF TIME IS VITAL TO MANY ASPECTS OF THE MODERN WORLD

The modern conception of time of a standardized quantity is shared across the world. It combines knowledge of astronomical calendars and clocks based on the apparent motions of stars and planets with recent technologies for measuring and recording relatively short intervals of time.

Humans have probably been aware of the passage of time from the dawn of consciousness, but a proper understanding of the seasons and changing length of days throughout the year only became important with the beginning of settled agriculture around 8000 BCE. Prehistoric monuments from around the world, including Stonehenge in Britain, show a clear ability to track seasons from the rising and setting of the Sun.

The need to measure smaller time intervals arose only with the advanced civilizations of ancient Mesopotamia around 2000 BCE, probably driven by religious, ritual, and administrative requirements. Sundials were used to roughly track the time of day, while shorter time intervals were measured by tracking the dripping of water, or later the flow of sand, through a narrow aperture.

CLOCK TIME

The earliest weight-driven mechanical clocks probably originated in Europe early in the 2nd millennium CE. A single clock on a public building such as a church sufficed for an entire community. Mechanical clocks became portable with the introduction of the spring drive around 1500, and their accuracy was greatly improved in the late 17th century. The Industrial Revolution, bringing with it faster travel and telegraph communication, eventually forced a standardization of timekeeping across widespread areas.

zodiac ring shows the constellations

main hand indicates local solar time

the Sun moves through the zodiac constellations over the course of a year

Astronomical clock, Prague, Old Town Square, Czech Republic

> **“...TRUE** AND MATHEMATICAL TIME... **FLOWS UNIFORMLY** AND... IS CALLED DURATION.”

Isaac Newton, from *Principia*, 1687

2000 BCE
First calendars
Ancient Babylonians develop the earliest known calendars. The year is split into 12 months based on the lunar cycle and an extra month added to bring the lunar and solar cycles back into line. Other civilizations develop similar calendars.

Mayan calendar

1600 BCE
Water clocks
Although probably developed in Mesopotamia, water clocks (clepsydra) become popular in Greece and Rome. Typically, a graduated marker is used to track the level of water in a container with a small hole on the base.

Greek clepsydra

1500 BCE
Early sundials
Developed in both Babylon and Egypt, the first sundials track time through the shadow cast by an upright rod called a gnomon.

Ancient Egyptian sundial

520 CE
Time candles
The first reference to time candles—slow-burning wax candles or sticks of incense, which can roughly reveal the time even at night—is made in a Chinese poem.

800 CE
Hourglasses
The first definitive references to this sand-based equivalent of the water clock date from the 14th century, but the sandglass was probably invented in Europe, or at least introduced there, in the early 9th century.

1088
Su Song's clock tower
Chinese scholar Su Song builds a water clock that uses a complex series of gears to keep track of astronomical cycles, anticipating advances in European clockmaking technology.

Clock tower

the Moon circles the sky roughly every 29.5 days; ball rotates to represent the lunar phase

star, representing local sidereal (star) time, moves as the Sun shifts against the background sky

24-hour dials in ancient Czech, Roman, and Arabic numerals

start and end of ancient Czech day

shaded areas separate day, night, and twilight

Astronomical clockface
Installed in Prague's old City Hall in 1410, this clock combines a 24-hour clockface with mechanisms to show the directions of the Sun and Moon among the stars, and the lunar phases.

TIME ZONES

Until the early 19th century, towns kept their own local time based on the Sun's position at noon. The advent of rail travel—which reduced travel times from days to hours—made the time difference between locations problematic. Railroad companies drove the move to adopt agreed "mean times" that would be applicable across broad regions or even countries. Near-instantaneous telegraph communication drove a similar revolution later in the century, with many territories in the British Empire adopting time zones that were a set number of hours behind or ahead of Greenwich Mean Time, as measured at London's Royal Greenwich Observatory. By 1929, this system was adopted almost universally.

13th century
Weight-driven mechanical clocks
The earliest mechanical clocks, known from English cathedrals such as Salisbury and Norwich, use a falling weight on a chain to power the rotation of the gears, which is regulated by an escapement-and-oscillator mechanism.

1430
Spring-driven clock
Harnessing the force from an uncoiling spring helps reduce the size of clocks and watches. German clockmaker Peter Henlein uses this technique to make the first pocket watches.

Henlein's pocket watch

1656
Huygens' pendulum clock
Dutch inventor Christiaan Huygens harnesses the regular oscillations of a weighted pendulum to build clocks that keep time accurately to within a few seconds each day.

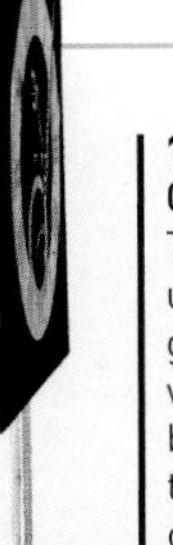
Huygens' pendulum clock

1759
Marine chronometer
English clockmaker John Harrison perfects a spring-driven timepiece that can keep time accurately over long periods at sea, permitting the exact calculation of longitude on board a ship for the first time.

1927
Quartz clock
The first electronic clock using the natural electricity generated by a rapidly vibrating quartz crystal is built. It measures time with the accuracy of a fraction of a second per day.

Quartz clock

1947
Atomic clock
These instruments use rapid transitions in the internal structure of elements such as cesium to measure time with tremendous accuracy.

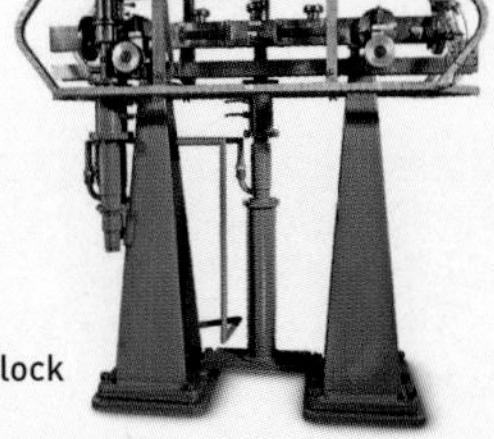
Atomic clock

1967
The second defined
A second is redefined as the duration of 9,192,631,770 cycles of transition between two energy levels in a cesium atom.

1970s
Digital timekeeping
The use of liquid crystals to display changing digits in digital devices revolutionizes the way time is represented.

Casio watch

1655–59

Dutch mathematician Christiaan Huygens was the first person to see Saturn's rings, and he suggested they were composed of solid particles.

IN 1655, JOHN WALLIS, AN ENGLISH MATHEMATICIAN, helped develop a way of finding the tangential lines to a curve—a fundamental aspect of the study of infinitesimal changes known as calculus. He devised a new mathematical **symbol to denote infinity** (a quantity larger than any number): **∞**. Four years later, Swiss mathematician Johann Rahn would invent the **symbol for division**: ÷.

Christiaan Huygens, a Dutch mathematician and instrument-maker, invented new kinds of timepieces and telescopes. Early in 1655, he **discovered Titan**—Saturn's biggest moon—using a telescope he had made with his brother. By the end of 1656, he had noticed that Saturn's crescents cast a shadow on the surface, suggesting that **these rings were made of solid material** not directly connected to the planet. In the same year, Huygens invented an **accurate pendulum clock**. Until the early 1600s, clocks could lose up to 15 minutes a day. Huygens' clock was a hundred times more accurate. By 1657, he was back to mathematics, collaborating with Pierre de Fermat and Blaise Pascal to publish the **first major textbook of probability theory**. The mechanism of Huygens' pendulum clock was further improved by the invention of the **anchor escapement** in 1657. Widely attributed to English inventor **Robert Hooke**, it allowed pendulum clocks to work with smaller swings and longer pendulums with heavier weights. Later, in 1658, Hooke devised the **balance spring for watches** as part of an improved escapement.

crutch

toothed wheel

pendulum bob

Pendulum clock
Huygens' clock kept better time because the period of the pendulum remained the same, regardless of the amplitude of its swing.

1% white blood cells and platelets

45% red blood cells

54% plasma

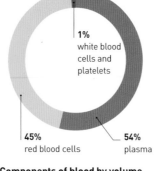

Components of blood by volume
The scarcity of white blood cells, along with inadequate microscopy, meant that 17th-century microscopists were able to record only red blood cells.

In 1657, a new scientific society was established in Florence, Italy. **Accademia del Cimento** (The Academy of Experiment) aimed to further enquiry by experimentation. Its prospectus became a popular laboratory manual in the 1700s.

Jan Swammerdam, a Dutch biologist, spent much of his career studying anatomy and insects using a microscope. In 1658—before his university training—he was, purportedly, the **first person to observe red blood cells**.

1660–61

Gresham College in London, England, was the original home of the Royal Society; the college was founded by financier Sir Thomas Gresham.

THE ROYAL SOCIETY, one of the oldest learned scientific societies, was founded in London in November 1660. The first meeting of 12 natural philosophers took place at Gresham College and included English architect **Christopher Wren** and **Robert Boyle**. The society met weekly to discuss "natural knowledge" and watch experiments; the first Curator of Experiments was Robert Hooke.

A year later, **Robert Boyle** published *The Sceptical Chymist*, which established his reputation as the **father of chemistry**. In it, he criticized the old alchemy and described a new scientific way of studying chemistry that advanced by experimentation. He replaced old ideas about nature's elements with the modern concept of an element as a pure substance that cannot be degraded into simpler forms.

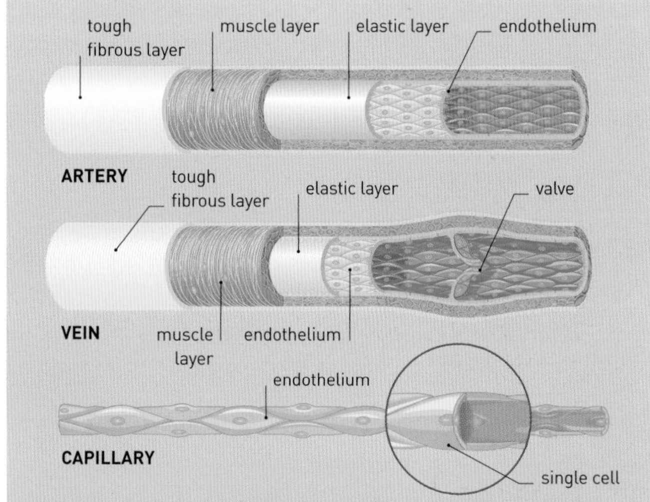

tough fibrous layer
muscle layer
elastic layer
endothelium
ARTERY
tough fibrous layer
elastic layer
valve
VEIN
muscle layer
endothelium
endothelium
CAPILLARY
single cell

TYPES OF BLOOD VESSELS

Thick-walled arteries have the most elastic fibers to help sustain the high pressure of blood from the heart. Thin-walled veins transport low-pressure blood and have valves to stop backflow as the blood returns to the heart. Between them are microscopic capillaries composed of endothelium (a single-celled layer) only, which facilitates the transfer of food and oxygen into tissues.

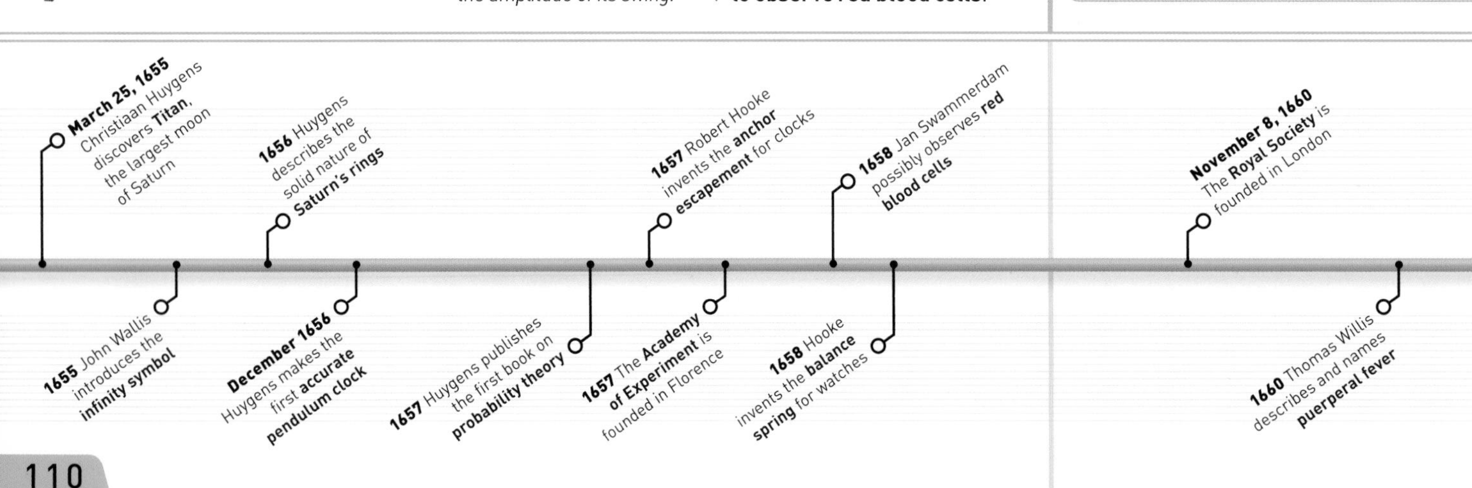

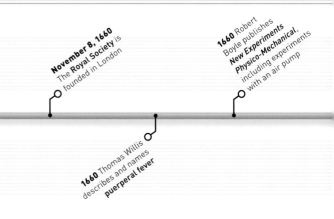

March 25, 1655 Christiaan Huygens discovers Titan, the largest moon of Saturn
1655 John Wallis introduces the infinity symbol
1656 Huygens describes the solid nature of Saturn's rings
December 1656 Huygens makes the first accurate pendulum clock
1657 Huygens publishes the first book on probability theory
1657 Robert Hooke invents the anchor escapement for clocks
1657 The Academy of Experiment is founded in Florence
1658 Jan Swammerdam possibly observes red blood cells
1658 Hooke invents the balance spring for watches
November 8, 1660 The Royal Society is founded in London
1660 Thomas Willis describes and names puerperal fever
1660 Robert Boyle publishes New Experiments Physico-Mechanical, including experiments with an air pump

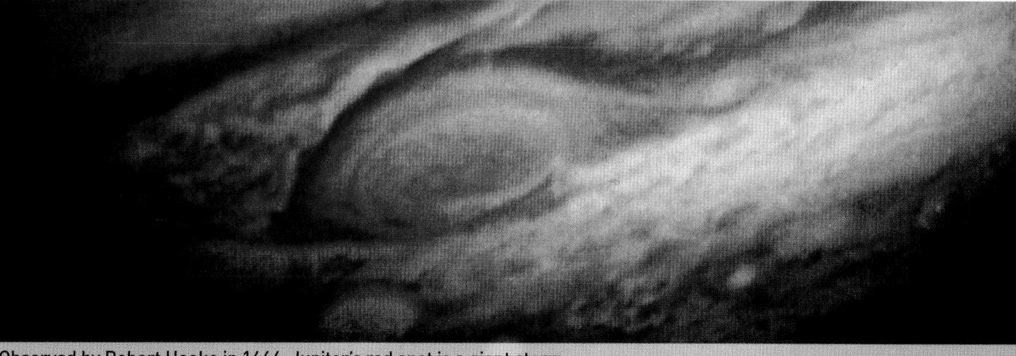

Observed by Robert Hooke in 1664, Jupiter's red spot is a giant storm, large enough to engulf three Earths.

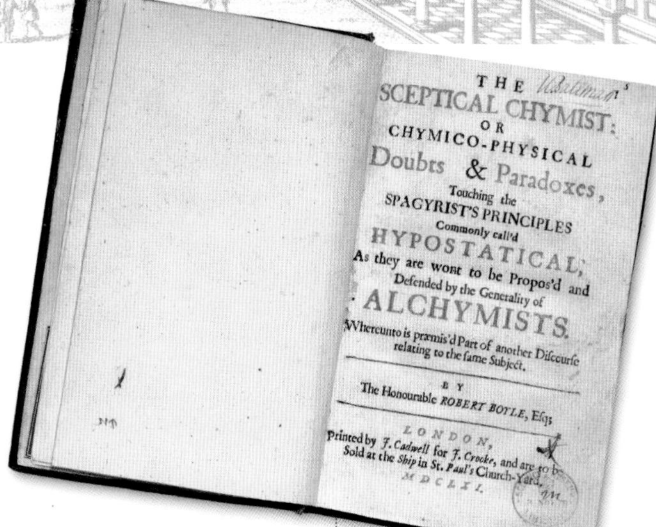

THE SCEPTICAL CHYMIST: OR CHYMICO-PHYSICAL Doubts & Paradoxes, Touching the SPAGYRIST'S PRINCIPLES Commonly call'd HYPOSTATICAL; As they are wont to be Propos'd and Defended by the Generality of ALCHYMISTS. Whereunto is præmis'd Part of another Discourse relating to the same Subject.

BY The Honourable ROBERT BOYLE, Esq;

LONDON, Printed by J. Cadwell for J. Crooke, and are to be Sold at the Ship in St. Paul's Church-Yard. MDCLXI.

The Sceptical Chymist
Robert Boyle's book is a dialog between fictitious supporters of alchemy and the "voice of reason" extolling a science based on atoms, definable elements, and experiments.

More than 30 years earlier, in his great treatise on blood circulation, English physician William Harvey had suggested that the body contained minute blood vessels that connected arteries with veins—and thereby completed the circuit. In 1661, Italian physician and biologist **Marcello Malpighi** used his microscope and discovered these **blood capillaries**. Malpighi devoted much of his career to the microscopic study of anatomy. He would go on to make important discoveries about the kidneys, embryos, insects, and even plants.

> **"** I NOW MEAN BY **ELEMENTS,** CERTAIN **PRIMITIVE OR SIMPLE,** OR **PERFECTLY UNMINGLED BODIES... "**
>
> **Robert Boyle, English scientist and inventor,** from *The Sceptical Chymist*, 1661

TWENTY YEARS after it was proved that air pressure decreases with height, English meteorologist **Richard Towneley** noted that a fixed amount of **trapped air expanded** in volume **at high altitude**. Robert Hooke subsequently confirmed these observations by experimentation. Robert Boyle published what he called "Towneley's hypothesis" in 1662, but later it became known as **Boyle's law**.

In the same year, at a time when an outbreak of the bubonic plague in London was imminent, English shopkeeper **John Graunt** published his analysis of *Bills of Mortality*. Although not a scholar, Graunt used these records to work out **population trends**. Impressed with Graunt's efforts, Charles II ordered the Royal Society to admit him as a fellow. Today, Graunt's **life tables** mark the foundation of the statistical study of populations: **demography**.

In 1663, Scottish astronomer **James Gregory** proposed a design for a **reflecting telescope**. It incorporated mirrors as a way of avoiding the color aberrations that arose when lenses refracted (bent) different wavelengths of light. However, it was Isaac Newton who was able to get the first reflecting telescope made (see 1667–68).

Although astronomers had studied **Jupiter** earlier, they did not record its **Great Red Spot** until the 1660s. This may have been because of inadequate telescopes or because, until then, it was not there at all. The spot, which is a giant storm, probably started only around 1600. **Robert Hooke** observed it in 1664, but Italian astronomer Giovanni Cassini may have seen it as early as 1655.

ROBERT BOYLE (1627–91)

English physicist and inventor, Robert Boyle was a pioneer of chemistry as well. He pursued science through experiment and reasoning, and he was inspired by Galileo's work (see 1611–13). Boyle made an air pump and used it to study the behaviour of gases. One of the first fellows of the Royal Society, he came up with the modern concept of chemical elements.

BOYLE'S LAW

Unlike liquids, gases are compressible. Physicist Robert Boyle formalized a law describing the relationship between the pressure of a gas and its volume. As long as temperature stays the same, the pressure and volume of a gas are inversely proportional. In quantitative terms this means that if pressure is doubled, volume is halved and vice versa.

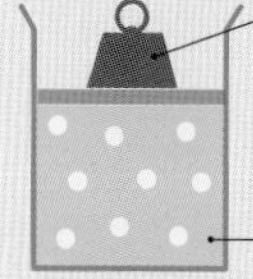

one weight produces pressure in the container
molecules spread evenly
DIFFUSION

two weights produce double the pressure in the container
high pressure squeezes molecules into half the original volume
PRESSURE

March 15, 1661 Marcello Malpighi publishes a treatise on the **lung**, describing its **alveoli** (air sacs) and network of **capillaries**

March 1661 Malpighi makes the first unambiguous reference to **blood cells**

1661 Niels Stensen discovers the duct of the largest salivary gland, the **parotid gland**

1661 Robert Boyle publishes ***The Sceptical Chymist***

1662 Lorenzo Bellini describes the **anatomy of the kidneys**

1662 Robert Boyle describes the relationship between volume and pressure of a gas, later called **Boyle's Law**

1662 **John Graunt** publishes *Natural and Political Observations Made Upon Bills of Mortality*

1663 Nicholas Steno recognizes that the **heart** is made of **muscle**

1663 James Gregory describes the design of a **reflecting telescope**

May 9, 1664 Robert Hooke observes **Jupiter's Great Red Spot**

1664 **Thomas Willis** publishes *Anatomy of the Brain, with a Description of the Nerves and their Functions*

1665–66

“...THERE IS A **NEW VISIBLE WORLD DISCOVERED** TO THE UNDERSTANDING.”

Robert Hooke, English inventor, from *Micrographia*, 1665

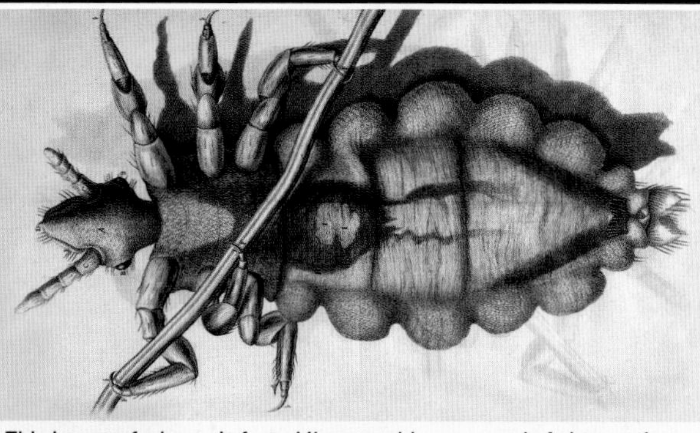

This image of a louse is from *Micrographia*—a record of observations Robert Hooke made using a microscope.

ENGLISH INVENTOR ROBERT HOOKE, Curator of Experiments at the Royal Society, London, had turned his attention to **microscopy**. In 1665, he published the Society's first monograph, *Micrographia*, with exquisite illustrations of miniature life, including the **first depiction of a microorganism**—in this case, a mould. The book contained the first published reference to a **biological cell, which Hooke named after** looking at cork tissue.

In this year, Cambridge University, England, was closed as a precaution against the plague. One of its students, physicist and mathematician **Isaac Newton** (1642–1727), used his freedom to make extraordinary discoveries. Within two years, he invented **calculus**, had his first insight into **gravity**, and used **prisms** to study the colors of a rainbow.

In 1666, Italian astronomer **Giovanni Cassini** (1625–1712) was the first to observe that the planet Mars had a polar cap. He also calculated the planet's rotational period to be about 24 hours, 40 minutes. In the previous two years, he had determined the rotational periods of the planets Jupiter and Venus as well.

Ice cap on Mars
Giovanni Cassini observed an ice cap on Mars, although it was centuries later that images such as this helped reveal its makeup.

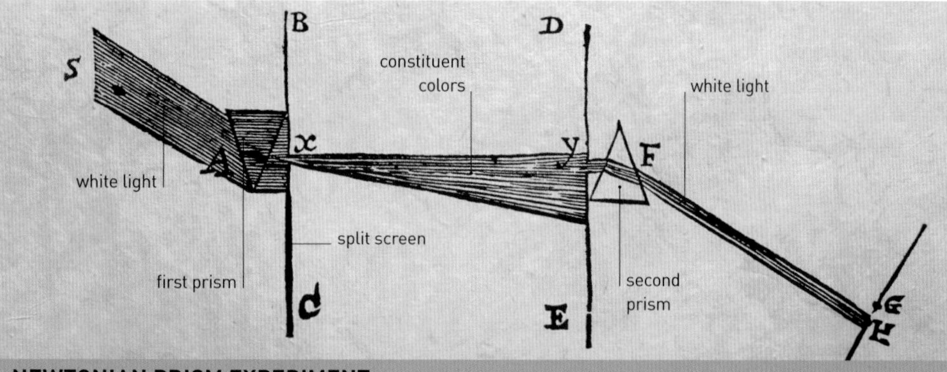

S
B
D
constituent colors
white light
x
y
F
white light
split screen
first prism
d
E
second prism
G
H

NEWTONIAN PRISM EXPERIMENT

Although other scientists had shone white light through a prism to produce a rainbow of colors, Isaac Newton had the novel idea that the colors were constituents of white light, which were separated by the prism. He proved this by placing a second prism upside-down in front of the first. The first prism splits white light into seven colors, each made up of light with a different wavelength. Splitting happens because light with the longest wavelength (red) bends less than light with the shortest wavelength (violet). The second prism bends them again—and so recombines them.

1667–68

11 PINTS THE AVERAGE **VOLUME OF BLOOD** IN THE **ADULT HUMAN BODY**

IN 1666, THE FIRST BLOOD TRANSFUSIONS—dog to dog—had been demonstrated before the Royal Society. In 1667, animal-to-human "therapeutic" blood transfusions were attempted; animal blood was regarded as less likely "to be rendered impure by passion or vice." Independently of one another, English physician **Richard Lower** (1631–91) and French physician **Jean-Baptiste Denis** (1643–1704) transfused small quantities of **lamb's blood into their patients**. Those patients lucky enough to survive doubtless did so because the allergic reaction was minimal.

1¼ INCHES THE DIAMETER OF THE **OBJECTIVE MIRROR** IN **NEWTON'S TELESCOPE**

However, the procedure was eventually banned in France following a number of fatalities.

In 1668, **Isaac Newton** built the **first reflecting telescope**. By using mirrors, the design avoided the lens aberration associated with refracting telescopes. Scottish astronomer James Gregory had described such an instrument five years earlier, but had no means of producing it.

horizontal eyepiece

Although the book *Experiments on the Generation of Insects* appeared as an obscure publication in 1668, its author, Italian physician **Francesco Redi** (1626–97), had described potentially ground-breaking experiments in it. Redi was testing the idea that life—specifically maggots—could form spontaneously, as was the prevailing wisdom. He placed pieces of meat in jars, sealing some with gauze and leaving others open. Maggots appeared only in the open jars—evidence they could not form on their own. However, his debunking of **spontaneous generation** had little impact on the progress of biological thought

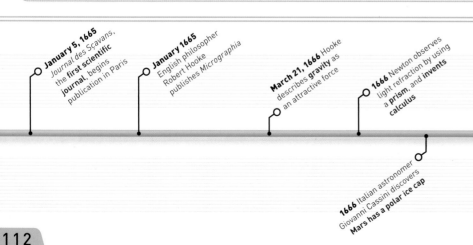

January 5, 1665 *Journal des Sçavans*, the **first scientific journal**, begins publication in Paris

January 1665 English philosopher Robert Hooke publishes *Micrographia*

March 21, 1666 Hooke describes **gravity** as an attractive force

1666 Newton observes light refraction by using a **prism**, and **invents calculus**

1666 Italian astronomer Giovanni Cassini discovers **Mars has a polar ice cap**

June 15, 1667 French physician Jean-Baptiste Denys performs the first recorded **human blood transfusion**

October 1667 Hooke shows heart function depends on inflation of the **lungs**

1667 Danish biologist Nicolas Steno produces the **earliest geological treatise**

Richard Lower was part of a transfusion "craze" that spread across Europe. Here, he is seen transfusing blood from a lamb to a man.

A 19th-century engraving depicts Dutch naturalist Jan Swammerdam being plagued by a swarm of bees after removing their queen from the hive.

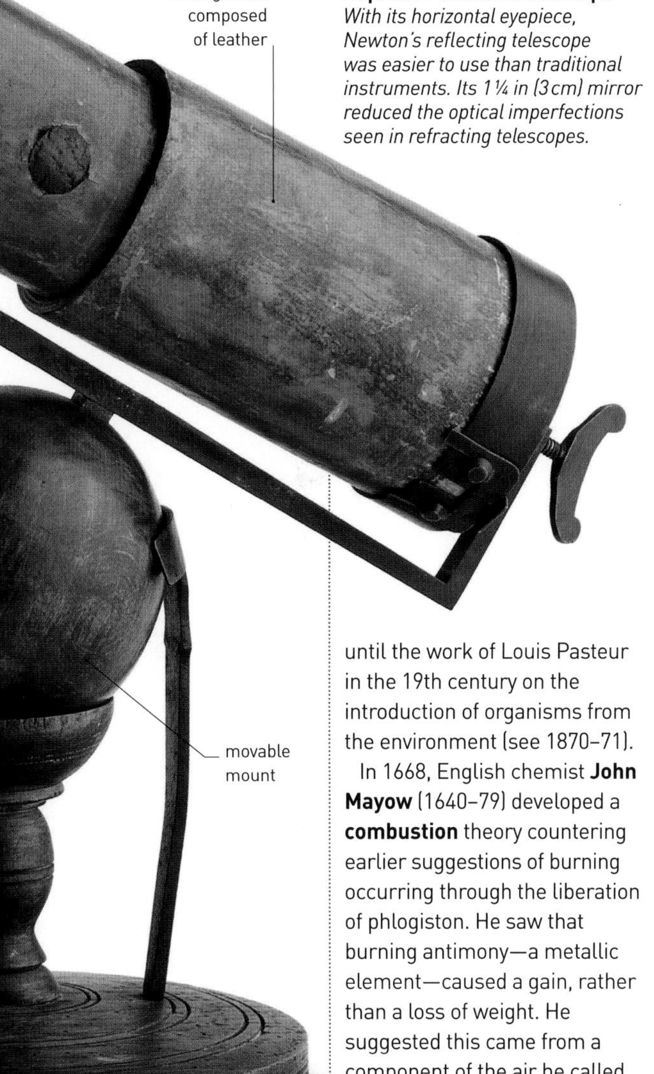
sliding focus composed of leather
movable mount
Replica of Newton's telescope
With its horizontal eyepiece, Newton's reflecting telescope was easier to use than traditional instruments. Its 1¼ in (3 cm) mirror reduced the optical imperfections seen in refracting telescopes.

until the work of Louis Pasteur in the 19th century on the introduction of organisms from the environment (see 1870–71).

In 1668, English chemist **John Mayow** (1640–79) developed a **combustion** theory countering earlier suggestions of burning occurring through the liberation of phlogiston. He saw that burning antimony—a metallic element—caused a gain, rather than a loss of weight. He suggested this came from a component of the air he called *spiritus igneo-aereus*. This idea anticipated the **discovery of oxygen** a century later.

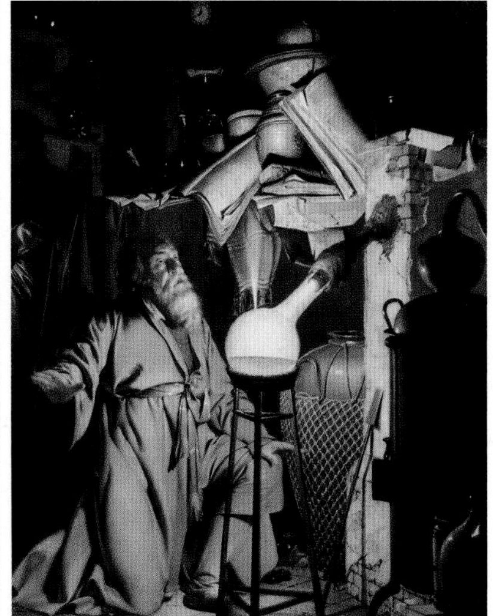
Joseph Wright's *The Alchymist*
This highly dramatic painting of 1771 is a fanciful recreation, rather than a true portrait, of Hennig Brand's accidental discovery of phosphorus

GERMAN ALCHEMIST HENNIG BRAND was searching for the "philosopher's stone" that supposedly changed base metal into gold and, in 1669, he thought he had found it. But the glowing substance Brand discovered was **phosphorus**.

That year, Danish biologist and geologist **Nicolas Steno** (1638–86) explained that as sediment layers accumulated, old rock strata were overlain by newer ones. This meant that **fossils**—mineralized remains of extinct organisms—found in these strata could be **sorted by age**.

The scientific study of insects arguably had its foundations in **Jan Swammerdam's** 1669 book *General History of Insects*. In this, the Dutch microscopist described the larval and pupal stages of **insect life histories**.

In 1670, English chemist **Robert Boyle** poured acid onto a metal and obtained inflammable air. Boyle had isolated **hydrogen**.

Cassini's work on **astronomical dimensions** in 1671 included his computation of the Earth–Mars distance, which gave the first indications of the Solar System's size. His 1672 calculation of the Earth–Sun distance is close to current estimates.

> " IT IS **UNWORTHY OF** EXCELLENT **MEN TO LOSE HOURS** LIKE SLAVES IN THE **LABOR** OF CALCULATION. "
>
> **Gottfried Leibniz, German philosopher and mathematician,** 1685

In 1672, **Isaac Newton** presented a paper to the Royal Society on his observations of the **rainbow of colors that make up white light** (see panel, opposite). He was subsequently elected a fellow of the Society, but Curator of Experiments Robert Hooke criticized Newton's paper, triggering an ongoing dispute between the two men.

In 1673, German mathematician **Gottfried Leibniz** created a **calculating machine** and presented it to the Royal Society. In the same year, Dutch astronomer **Christiaan Huygens**, inventor of the pendulum clock, published the **mathematical analysis of pendulum motion**, showing how length and weight affect swing.

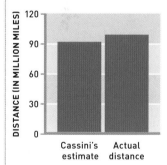

	Distance (in million miles)
Cassini's estimate	~92
Actual distance	~99

Earth–Sun distance
Cassini's religious faith made him resist the idea of a Sun-centered Universe, but his views changed as he computed astronomical distances.

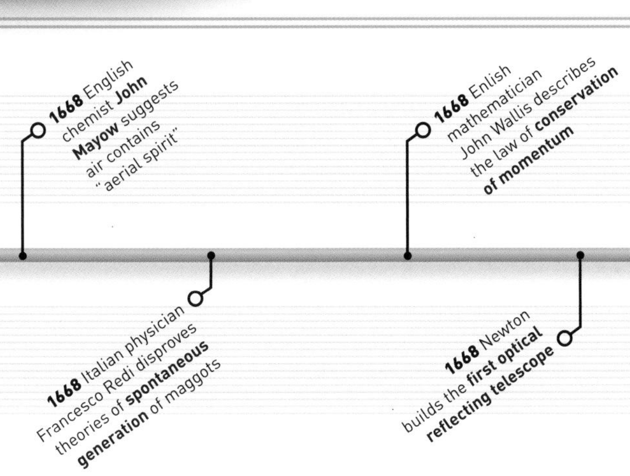
1668 English chemist **John Mayow** suggests air contains "aerial spirit"
1668 Italian physician Francesco Redi disproves theories of **spontaneous generation** of maggots
1668 Enlish mathematician John Wallis describes the law of **conservation of momentum**
1668 Newton builds the **first optical reflecting telescope**

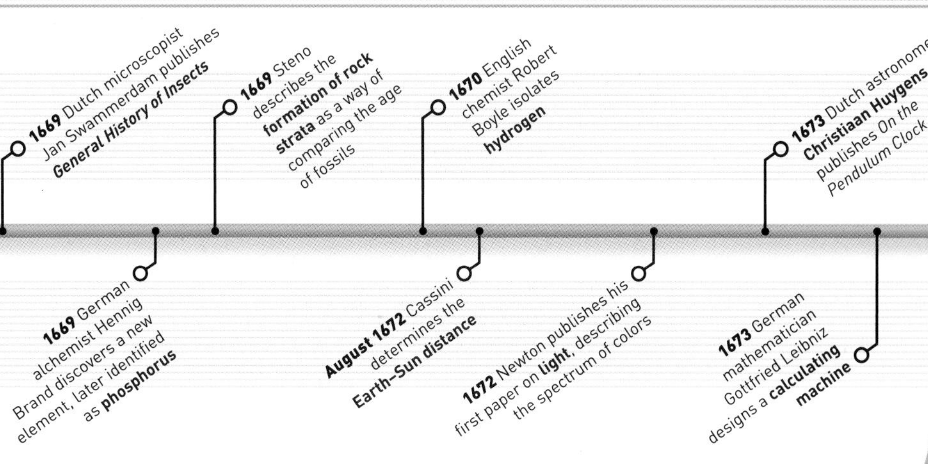
1669 Dutch microscopist Jan Swammerdam publishes ***General History of Insects***
1669 German alchemist Hennig Brand discovers a new element, later identified as **phosphorus**
1669 Steno describes the **formation of rock strata** as a way of comparing the age of fossils
1670 English chemist Robert Boyle isolates **hydrogen**
August 1672 Cassini determines the **Earth–Sun distance**
1672 Newton publishes his first paper on **light**, describing the spectrum of colors
1673 Dutch astronomer **Christiaan Huygens** publishes *On the Pendulum Clock*
1673 German mathematician Gottfried Leibniz designs a **calculating machine**

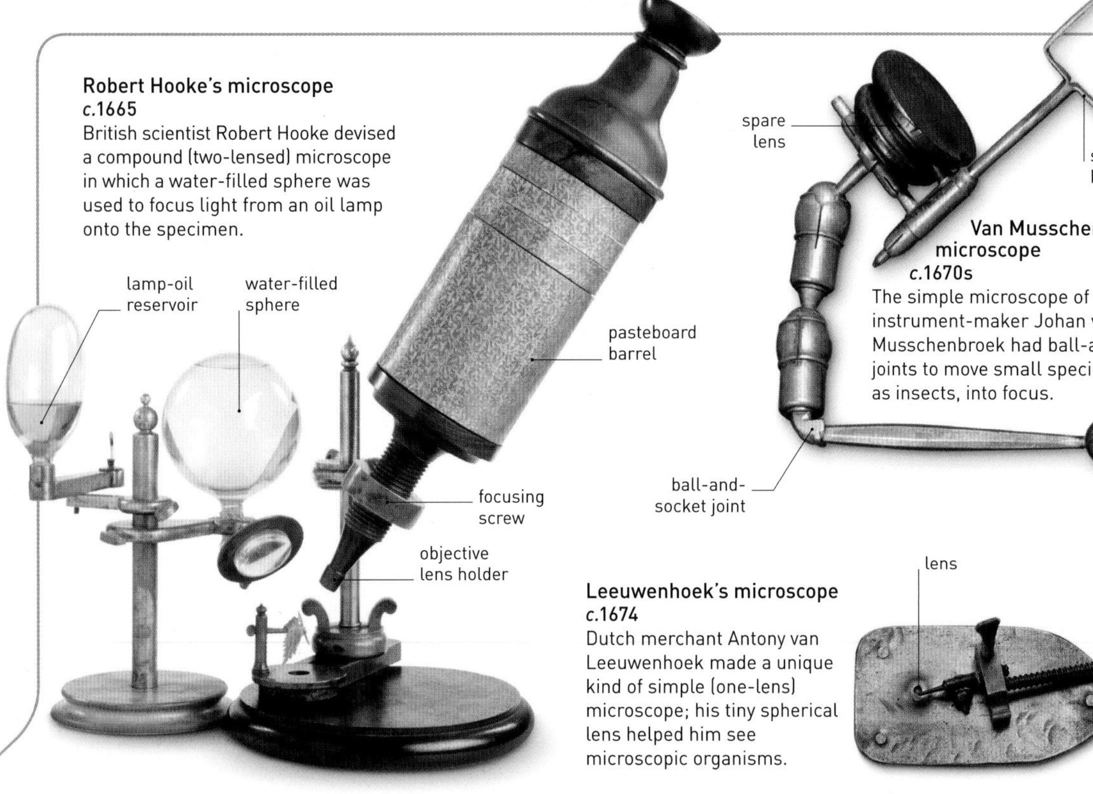
lamp-oil reservoir
water-filled sphere
pasteboard barrel
focusing screw
objective lens holder
spare lens
specimen holder
lens
ball-and-socket joint
lens
screw moves specimen up or down

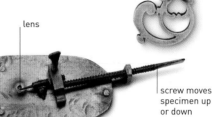

Robert Hooke's microscope
*c.*1665
British scientist Robert Hooke devised a compound (two-lensed) microscope in which a water-filled sphere was used to focus light from an oil lamp onto the specimen.

Van Musschenbroek's microscope
*c.*1670s
The simple microscope of Dutch instrument-maker Johan van Musschenbroek had ball-and-socket joints to move small specimens, such as insects, into focus.

Leeuwenhoek's microscope
*c.*1674
Dutch merchant Antony van Leeuwenhoek made a unique kind of simple (one-lens) microscope; his tiny spherical lens helped him see microscopic organisms.

MICROSCOPES

VIEWING THE MICROSCOPIC WORLD BEYOND THE NAKED EYE

By opening up a world in miniature, microscopes have helped scientists to understand the building blocks of the world around us—from the cells of living things down to individual molecules and even atoms.

In the early 1600s, Dutch eyeglass-makers made the first microscopes by fixing two lenses together in a tube to create a magnifying power greater than that of a single lens used on its own. As lenses were refined, so the quality of the magnified image improved. Then, in the 20th century, breakthroughs in atomic physics led to the invention of the electron microscope, which—instead of light rays—used electron beams with shorter wavelengths to reveal even tinier particles.

barrel
coarse focus
mirror to focus light on specimen
stage (holds specimen)
mirror
instruction booklet
interchangeable objective lenses

Tulley & Sons achromatic microscope
*c.*1835
Designed by British scientist Joseph Lister, this microscope was made with new achromatic lenses that focused different colors accurately together, yielding better images.

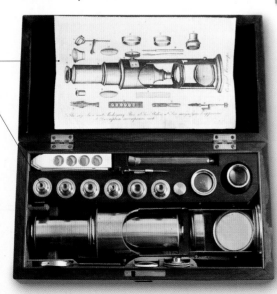

Compound drum microscope
*c.*1850
The popular drum microscope focused on a specimen mounted on a basal stage, using a sliding body tube that contained the lenses. This design made it easy to transport the microscope and lenses.

Culpeper compound microscope
*c.*1740
British instrument-maker Edmund Culpeper produced inexpensive tripod-style microscopes; early models were made partly of wood. However, the fixed upright design and crude focus made them difficult and uncomfortable to use.

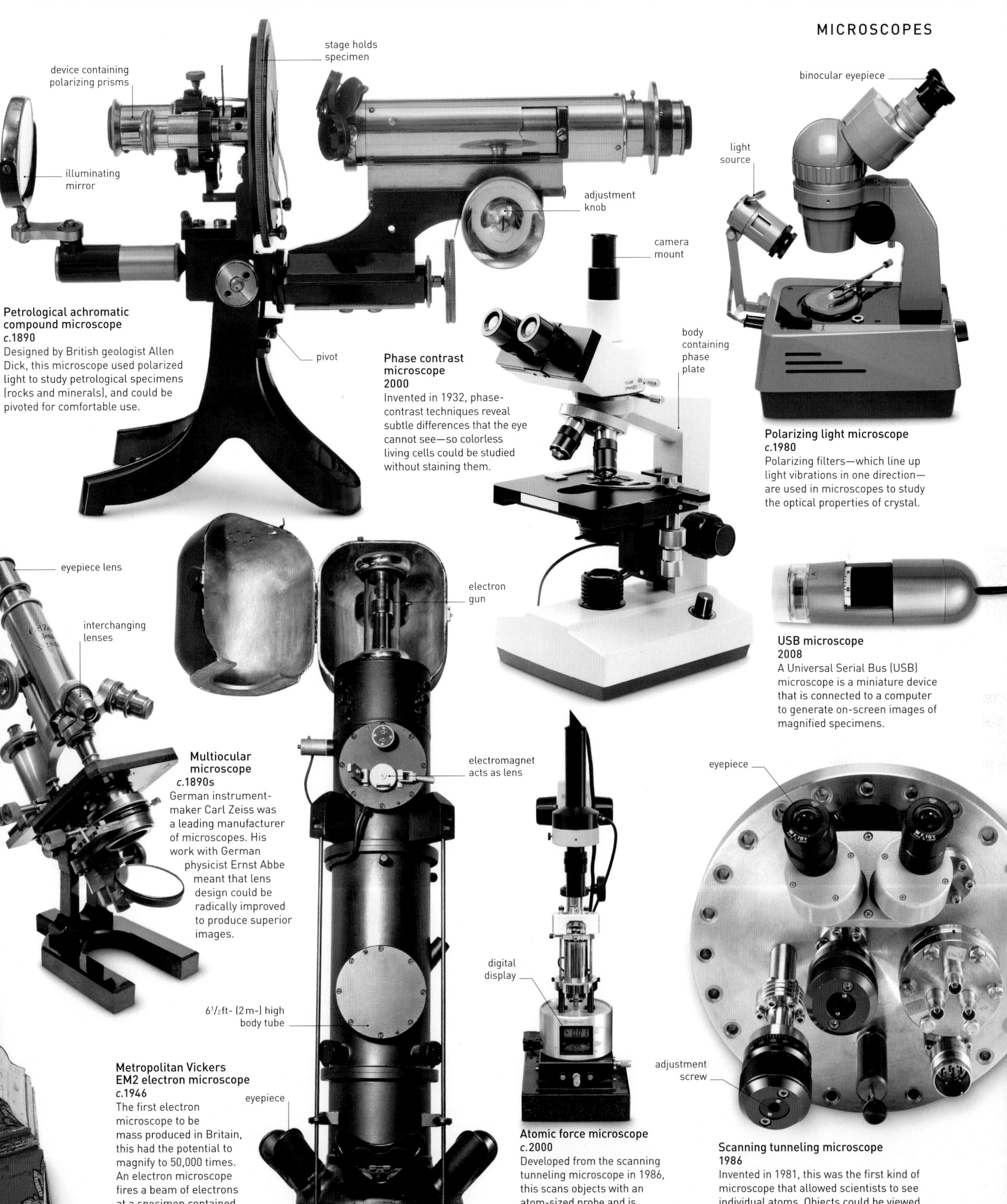

stage holds specimen
device containing polarizing prisms
illuminating mirror
adjustment knob
pivot
binocular eyepiece
light source
camera mount
body containing phase plate
eyepiece lens
interchanging lenses
electron gun
electromagnet acts as lens
6½ft- (2m-) high body tube
eyepiece
digital display
eyepiece
adjustment screw

Petrological achromatic compound microscope
*c.*1890
Designed by British geologist Allen Dick, this microscope used polarized light to study petrological specimens (rocks and minerals), and could be pivoted for comfortable use.

Phase contrast microscope
2000
Invented in 1932, phase-contrast techniques reveal subtle differences that the eye cannot see—so colorless living cells could be studied without staining them.

Polarizing light microscope
*c.*1980
Polarizing filters—which line up light vibrations in one direction—are used in microscopes to study the optical properties of crystal.

USB microscope
2008
A Universal Serial Bus (USB) microscope is a miniature device that is connected to a computer to generate on-screen images of magnified specimens.

Multiocular microscope
*c.*1890s
German instrument-maker Carl Zeiss was a leading manufacturer of microscopes. His work with German physicist Ernst Abbe meant that lens design could be radically improved to produce superior images.

Metropolitan Vickers EM2 electron microscope
*c.*1946
The first electron microscope to be mass produced in Britain, this had the potential to magnify to 50,000 times. An electron microscope fires a beam of electrons at a specimen contained in a vacuum and uses electromagnetic "lenses."

Atomic force microscope
*c.*2000
Developed from the scanning tunneling microscope in 1986, this scans objects with an atom-sized probe and is one of the most powerful microscopes available today.

Scanning tunneling microscope
1986
Invented in 1981, this was the first kind of microscope that allowed scientists to see individual atoms. Objects could be viewed to a resolution of a nanometer (one millionth of a millimeter).

1675

2,920

MILES

THE GAP BETWEEN RINGS A AND B OF **SATURN,** KNOWN AS THE **CASSINI DIVISION**

ISAAC NEWTON PUBLISHED HIS HYPOTHESIS OF LIGHT in 1675, suggesting that light was made up of particles that he called corpuscles. Physicists had long debated the nature of light: some, like Newton, favored particles, others the theory that light traveled like waves. The corpuscular theory prevailed until the 1800s, when British physicist Thomas Young proved that light was wavelike (see 1801).

In March, Charles II appointed British astronomer **John Flamsteed** (1646–1719) as the first Astronomer Royal for a new observatory at Greenwich, London. The **Royal Observatory** was built to improve ways of measuring longitude for sea navigation: it marked what later became the prime meridian between east and west. Many years later, through international agreement, it would mark an official starting point of each day—at the stroke of midnight **Greenwich Mean Time (GMT)**.

Italian microscopist **Marcello Malpihgi** (1628–94) published his principal work *Anatome Plantarum* (*Anatomy of Plants*) on the fine structure of plant tissues, naming the outer layer of a leaf the epidermis and its tiny breathing pores, stomata.

Another Italian, astronomer **Giovanni Cassini**, noticed that Saturn's distinctive ring was divided. The dark gap became known as the **Cassini division**. Scientists now know that this gap comprises small particles at low density.

Royal Greenwich Observatory
Home of the prime meridian and Greenwich Mean Time, the Royal Observatory was made a World Heritage Site by UNESCO in 1997.

1676–78

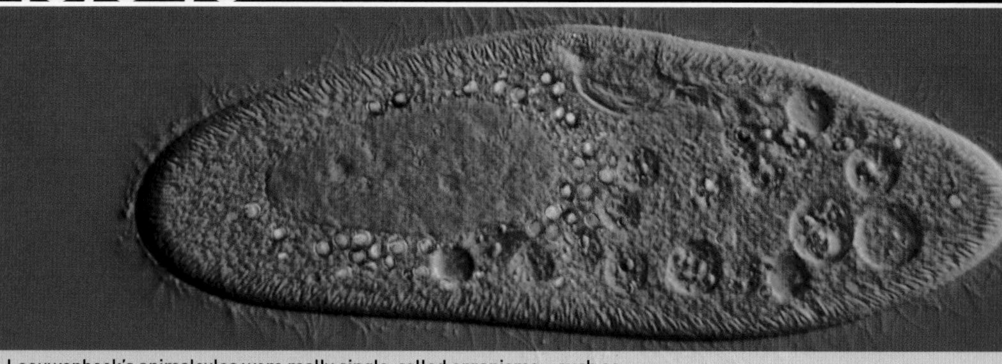

Leeuwenhoek's animalcules were really single-celled organisms—such as this *Paramecium*—many of which reproduce rapidly in standing water.

IN 1676, DUTCH ASTRONOMER OLE RÖMER (1644–1710) used astronomical measurements to deduce that **light has a fixed speed**—something that was not readily accepted until the mid-1700s.

In 1668, Dutch textile merchant **Antony van Leeuwenhoek** (1632–1723) had traveled to London and been impressed by Englishman Robert Hooke's publication on microscopic life: ***Micrographia***. On his return home, Leeuwenhoek designed his own **microscopes**—with small spherical lenses fashioned by drawing out a thread of glass, which was rounded off at the tip (see p.114). The magnifications rivaled those of any microscopes then in use—and he set about exploring miniature worlds. When he saw the microscopic taste buds on an ox's tongue, he was curious to study taste. This led him to soak pepper and spices in water. One of his pepper infusions ended up teeming with tiny living beings, which Leeuwenhoek referred to as **animalcules**. Many of these organisms were likely to have been the microbes that were later referred to as protozoans.

In 1676, Leeuwenhoek wrote the first of many letters to the Royal Society describing what he saw—initially provoking scepticism. The following year, he became the first person to see **human spermatozoa** and, as he persisted, Leeuwenhoek's scientific reputation improved.

In 1677, English astronomer **Edmond Halley** suggested it was possible to calculate the **distance between Earth and the Sun**—later known as the astronomical unit—by making geometric measurements during a **transit of Venus** as the planet passed in front of the Sun. Halley could not test this theory in his lifetime, but at the next transit, in 1761, his technique was used to produce a value very close to modern estimates.

In England, **Robert Hooke** had been turning his attention to the physics of elastic **clock springs**. He formalized an everyday observation—that the force applied is proportional to the amount of stretch—into what became known as **Hooke's Law**, which he published in 1678.

HOOKE'S LAW

Hooke originally applied his law of elasticity to a clock spring, but it applies to any elastic material—a solid that can change shape but then return to its original form.

As applied force (F) increases, so does the stretch length (X): doubling the force stretches the elastic twice as much. The law applies up to a certain elastic limit, beyond which the material does not recover and may snap.

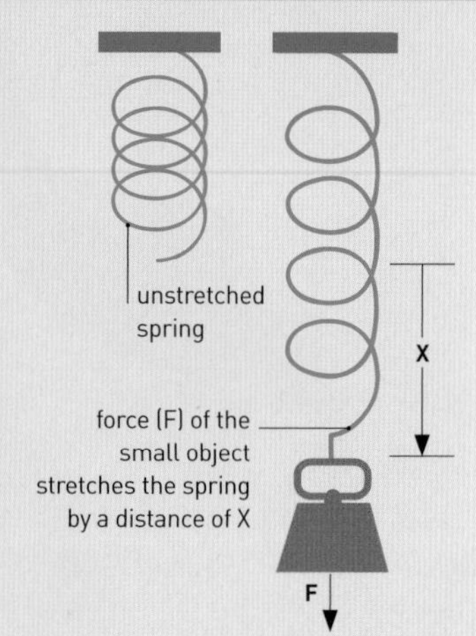
unstretched spring
X
force (F) of the small object stretches the spring by a distance of X
F

Antony van Leeuwenhoek
This merchant's experience of using a magnifying glass in the textile trade led him to make microscopes, and then microbiological discoveries.

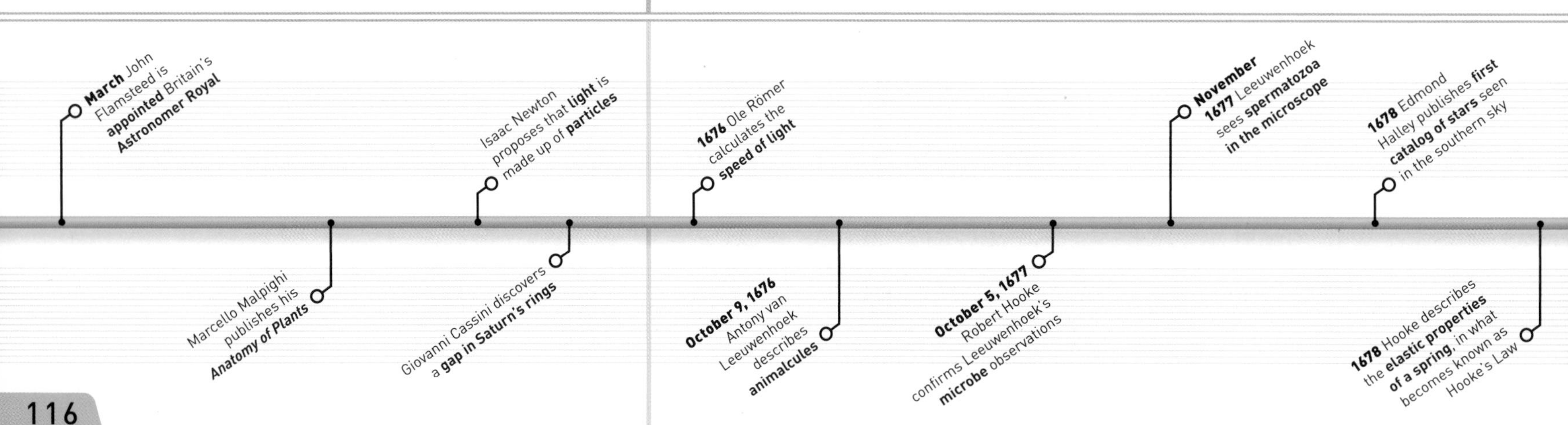
March John Flamsteed is **appointed** Britain's **Astronomer Royal**
Marcello Malpighi publishes his ***Anatomy of Plants***
Isaac Newton proposes that **light** is made up of **particles**
Giovanni Cassini discovers a **gap in Saturn's rings**
1676 Ole Römer calculates the **speed of light**
October 9, 1676 Antony van Leeuwenhoek describes **animalcules**
October 5, 1677 Robert Hooke confirms Leeuwenhoek's **microbe** observations
November 1677 Leeuwenhoek sees **spermatozoa in the microscope**
1678 Edmond Halley publishes **first catalog of stars** seen in the southern sky
1678 Hooke describes the **elastic properties of a spring**, in what becomes known as Hooke's Law

“ IF THE SPINE OF A STEVEDORE [DOCKWORKER]… SUPPORTS A LOAD OF 120 POUNDS… THE FORCE OF NATURE EXERTED IN THE DISK AND… MUSCLES OF THE SPINE IS EQUAL TO 25,585 POUNDS. ”

Giovanni Borelli, Italian physiologist, 1670s

screw
safety valve lever
weight
vessel

GERMAN MATHEMATICIAN Gottfried Leibniz had investigated a **binary number system** in which numbers are represented by just two symbols: 0 and 1. In 1679, he suggested using such a system as the **basis for a computing machine**.

This year, French inventor **Denis Papin** (1647–1712) collaborated with his Anglo-Irish counterpart Robert Boyle on a **steam digester**. This cooking device using high-pressure steam made it possible to extract fat from bones, and led to the development of the steam engine and pressure cooker.

Italian physiologist and physicist **Giovanni Borelli** (1608–79) spent much of his career studying the movements of animals. He realized that **muscle contraction** relied on chemical processes and nervous stimulation. His pioneering work in this new field of biomechanics was published in 1680, a year after his death.

Papin's steam digester
The safety valve that Papin invented for his steam digester was an important technological advance for the use of steam as a motive power.

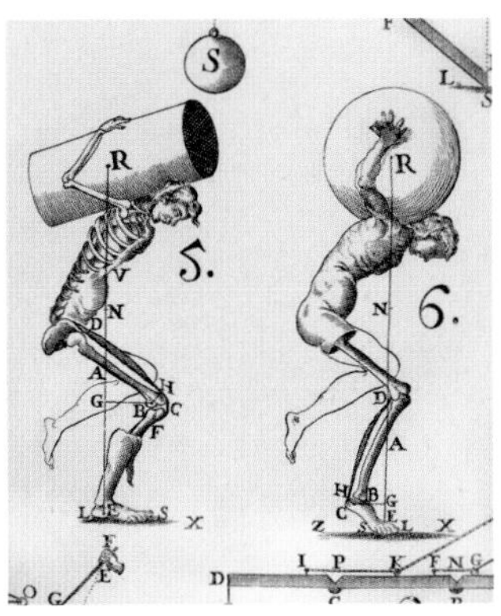

On the Motion of Animals
In this book, Giovanni Borelli applied the physical principles of mechanics to describe how the living body works and moves.

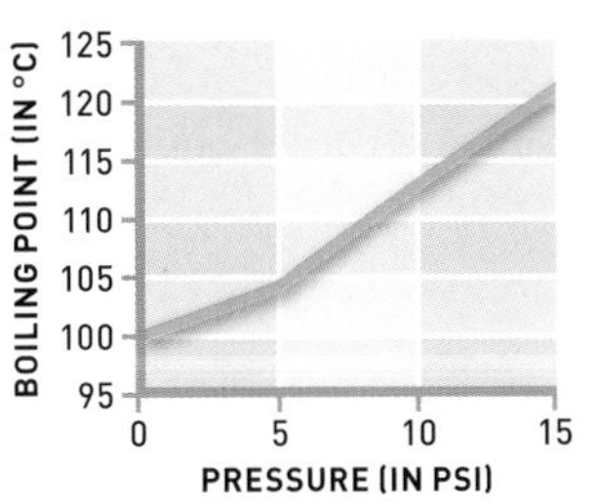

125
120
115
110
105
100
95
BOILING POINT (IN °C)
0
5
10
15
PRESSURE (IN PSI)

Boiling points
Water boils at higher temperatures as pressure is raised. As a result, food cooks at higher temperatures in boiling water under pressure.

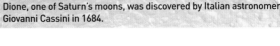

Dione, one of Saturn's moons, was discovered by Italian astronomer Giovanni Cassini in 1684.

IN 1682, AS ENGLISH ASTRONOMER EDMOND HALLEY plotted the orbit of a comet, he realized that its characteristics matched those of comets recorded in 1531 and 1607. He deduced that they were all the same comet—which today bears his name.

Also this year, English botanist **Nehemiah Grew** (1641–1712) published his book *The Anatomy of Plants*, one of the earliest comprehensive texts on **plant biology**. Grew often collaborated with the Italian **Marcello Malpighi** on **microscopic anatomy**—Grew concentrated on plants, Malpighi on animals. Previously, Grew had extracted the green plant pigment—called **chlorophyll** today—and may have made some of the earliest observations on chloroplasts. He also asserted that plants reproduce sexually (in other words, have male and female parts), and found that pollen grains had distinctive surface sculpturing (see 1916–17).

Giovanni Cassini had been studying the planet Saturn, and by 1684 discovered four of its moons. He called them *Sidera Lodoicea*—the **Stars of Louis**—in honor of Louis XIV, patron of the Paris Observatory. Individually, the moons are now named Iapetus, Rhea, Dione, and Tethys.

Robert Hooke and Edmond Halley had been collaborating on trying to explain observed **planetary motions** based on mathematical laws that had been described by German astronomer **Johannes Kepler** at the start of the century. When they were unable to do so, in 1684, Halley visited English physician and mathematician **Isaac Newton** in Cambridge to gauge his opinion, only to be told that Newton had already resolved the issue. Encouraged by Halley, Newton went on to explain the **elliptical orbits of planets**, which he eventually incorporated in *Principia* (see 1687–89).

“ ALL NATURE IS AS ONE GREAT ENGINE, MADE BY, AND HELD IN HIS HAND. ”

Nehemiah Grew, English botanist, in *The Anatomy of Plants*, 1682

1679 Gottfried Leibniz describes a system of **binary arithmetic**

1679 Denis Papin demonstrates a **steam digester**

1680 Giovanni Borelli explains the **mechanics of the living body**

September 1682 **Halley** plots the course of the comet that will later carry his name

1682 Nehemiah Grew publishes ***Anatomy of Plants***

1684 Cassini discovers four of **Saturn's moons**

December 10, 1684 Isaac Newton **explains planetary motion** based on **Kepler's laws**

1685–86

> **"** IN A WORD THE CORRUPTION AND **WANT OF [TEETH]** IS AS GREAT A **DEFORMITY,** AND OF AS MUCH **PREJUDICE** TO ONE, AS ANYTHING WHATSOEVER CAN BE. **"**

Charles Allen, British dental practitioner, from *The Operator for the Teeth,* 1685

Painted by Dutch artist Gerrit von Honthorst, *The Tooth Extractor* illustrates the crudities of 17th-century dentistry.

IN 1685, EARLY DENTAL PRACTITIONER CHARLES ALLEN published the first book written in English on dental procedures, *The Operator for the Teeth*. Dentistry had been attempted—with varying degrees of success—since the ancient civilizations, but specific "operators for the teeth" emerged only in the 17th century. These **early dentists gave advice on dental hygiene**, made artificial teeth, and also performed extractions, without anesthetic, using a "pelican"—an instrument so-called because of its resemblance to the bird's bill.

This latter part of the 17th century saw important advances in the classification of life's diversity. Naturalists **catalogued and classified animals and plants** based on their structure, often performing painstaking dissections of specimens to do so. Prominent among them was English naturalist **John Ray** (1627–1705), who published the first volume of his treatise *The History of Plants* in 1686, a work relying heavily on his travels in Europe. Ray created a system of classification to organize his catalog and, significantly, **formalized the idea of a species**. He emphasized the importance of reproduction: that seeds sprouting from the same parent plant belong to the same species, even though they may exhibit accidental variations. Ray's concept was to be adopted by generations of naturalists.

Another English naturalist, **Francis Willughby** (1635–72), had studied at Cambridge University, England, under John Ray with whom he collaborated in much of their work on classification. Following Willughby's premature death in 1672, Ray published his studies posthumously. Willughby's treatise *Ornithology*, which had appeared in 1676, was the first book to take a **scientific approach to the study of birds**. *The History of Fishes*, published in 1686, was another **ground-breaking achievement in natural history** but sold poorly, which meant that its publisher, the Royal Society, could not afford to fund Isaac Newton's *Principia* a year later.

English mathematician **Edmond Halley**, already known for his astronomical discoveries, also studied the terrestrial atmosphere. In 1686, he suggested that surface winds occurred because of a pattern of atmospheric circulation that was ultimately driven by heat from the Sun. Tropical warmth at the equator makes air there rise, causing more air to rush in to the region of low pressure. This phenomenon provided the basis for Halley's explanation of the behavior of trade winds and monsoons. At this time he also revisited observations made by other researchers 40 years earlier: that **atmospheric pressure decreases with altitude** (see 1645–54). Halley searched for the quantitative relationship between pressure and altitude, and so established routine use of the **barometer** in practical surveying.

Edmond Halley
Although perhaps best known for his work on astronomy, Halley was also a mathematician and geophysicist, and became professor of geometry at Oxford University in England.

Historia Plantarum
John Ray's three-volume treatise appeared from 1686 to 1704. He classed plants as either herbs or trees, and distinguished between spore- and seed-bearing plants.

> **"** IN ORDER THAT AN **INVENTORY OF PLANTS** MAY BE BEGUN… WE MUST… DISCOVER **CRITERIA…** FOR **DISTINGUISHING** WHAT ARE CALLED **'SPECIES.'** **"**

John Ray, from *Historia Plantarum,* 1686

1687–89

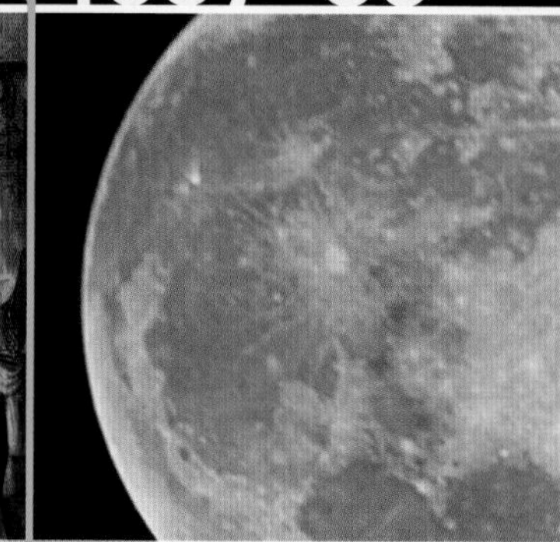

Newton argued that the Moon is subject to the force of Earth's gravity.

IN THE SUMMER OF 1687, THE ROYAL SOCIETY IN LONDON AUTHORIZED PUBLICATION of Isaac Newton's *Philosophiae Naturalis Principia Mathematica* (*Mathematical Principles of Natural Philosophy*). In this celebrated book (usually referred to simply as *Principia*), regarded by some as the most important scientific work ever produced,

ISAAC NEWTON
(1642–1727)

Arguably the greatest of all mathematicians, Newton founded classical mechanics, invented calculus, and made breakthrough discoveries about gravity and light. He studied at Cambridge University, England, where he became a professor of mathematics. After reforming the coinage of the Royal Mint, he was elected president of the Royal Society in 1703 and knighted in 1705.

1685 Englishman Charles Allen publishes ***The Operator for the Teeth***, an early work on dentistry

1686 Francis Willughby's ***History of Fishes*** is published posthumously

1686 John Ray publishes the first volume of ***History of Plants***, containing the first definition of the term "species"

1686 Edmond Halley explains atmospheric circulation and the link between **atmospheric pressure and altitude**

July 5, 1687 Isaac Newton publishes his ***Principia***, the foundation for classical mechanics

> **“** THE **LAW OF GRAVITATION** IS RENDERED **PROBABLE,** THAT **EVERY PARTICLE ATTRACTS** EVERY OTHER PARTICLE WITH **A FORCE WHICH VARIES INVERSELY** AS THE SQUARE OF THE DISTANCE. **”**

Isaac Newton, English mathematician, from *Principia,* 1687

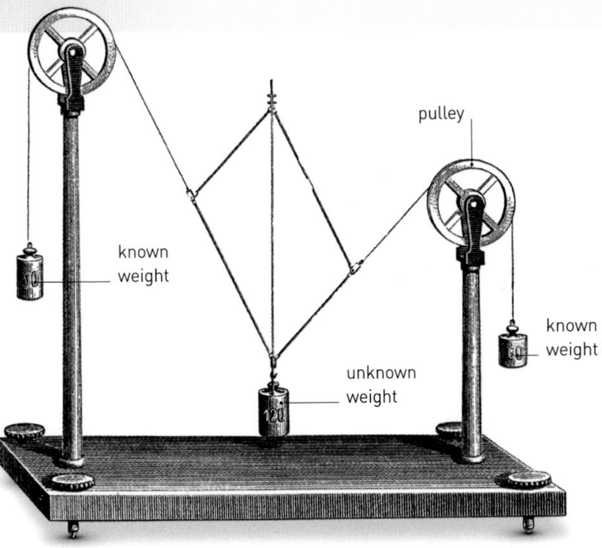
pulley
known weight
known weight
unknown weight

First law of motion
According to Newton, these weights stay still because no net force acts upon them. Unknown weight can be calculated if the forces acting to keep it still are known.

Newton described the laws of motion and universal gravitation that became the foundation of physical science. The appearance of *Principia* was due in part to the efforts of Edmond Halley. At a time when the Royal Society had already spent its annual publishing budget, Halley stepped in to finance its production. He had even been responsible for Newton starting work on it in the first place.

Three years earlier, three members of the Royal Society—**Christopher Wren, Robert Hooke**, and **Halley**—were debating **mathematical laws that govern the orbits of planets**, and Halley asked Newton for help in resolving a technical matter. Newton's response was a manuscript on planetary motion; impressed, Halley asked Newton to prepare a more exhaustive text for the Royal Society. For more than a year, Newton immersed himself in a study of physical laws, the result of which was his three-part masterpiece. In *Principia*, Newton describes his **three laws of motion** (see pp.120–21) and the **universal law of gravitation** (see panel, right), the basis of the branch of physics dealing with forces and motion: mechanics.

32.2 NEWTONS
THE GRAVITATIONAL FORCE PULLING ON **1 LB MASS** ON EARTH

Just before his death, Polish astronomer **Johannes Hevelius** (1611–87) completed the most **comprehensive celestial atlas** and star catalog of the time, in which he identified several new constellations, including Triangulum Minus. His work was published a few years later. In 1688, German astronomer **Gottfried Kirch** (1639–1710), director of the Berlin Observatory, described **another new constellation**, named Sceptrum Brandenburgicum in honor of the royal Prussian province. Today, its stars are considered part of the constellation Eridanus.

Naturalists continued to chart the diversity of the living world. In France, the botanist **Pierre Magnol** (1638–1715) had just become curator of France's biggest botanical garden at Montpellier. Magnol corresponded with English naturalist **John Ray**, who had embarked on his own survey of plant groups. Both men followed the principle of **classifying species according to anatomical similarities**. Their work implied underlying affinities within plant groups, although the evolutionary implications were not fully recognized for nearly two centuries. Magnol published his work in 1689 and, remarkably, many of his plant families are still recognized.

One of the earliest books on **pediatric medicine** appeared in 1689, published by English physician Walter Harris (1647–1732). This treatise on the diseases of children became a standard text on the subject.

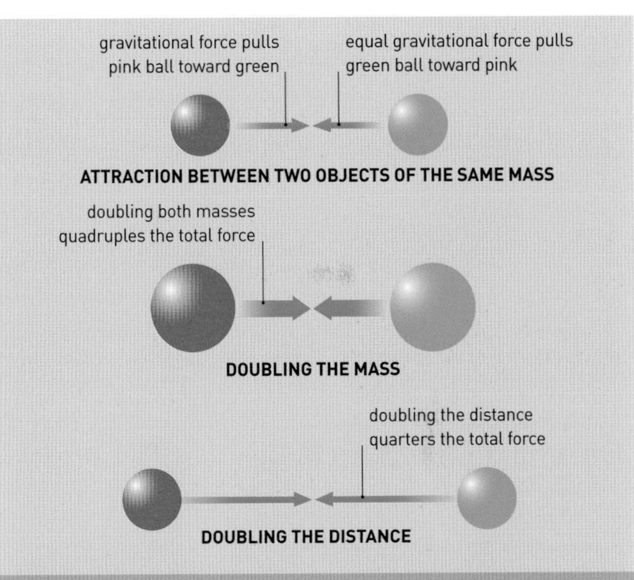
gravitational force pulls pink ball toward green
equal gravitational force pulls green ball toward pink
ATTRACTION BETWEEN TWO OBJECTS OF THE SAME MASS
doubling both masses quadruples the total force
DOUBLING THE MASS
doubling the distance quarters the total force
DOUBLING THE DISTANCE

UNIVERSAL LAW

Newton applied the physics of planetary interactions to create a Universal Law of Gravitation. Gravity is the force of attraction between bodies: stronger for more massive objects, weaker for a bigger distance apart. But whereas force and mass have a simple relationship, that between force and distance follows an inverse-square rule—doubling the distance reduces force by a quarter.

1687 Johannes Hevelius completes his **star atlas** shortly before his death.

1688 Gottfried Kirch describes the **new constellation** Sceptrum Brandenburgicum

1689 Pierre Magnon publishes his **classification of plants**, identifying families

1689 Walter Harris publishes one of the **earliest works on pediatric medicine**

UNDERSTANDING
NEWTON'S LAWS OF MOTION

THREE STRAIGHTFORWARD RULES DESCRIBE AND PREDICT HOW THINGS MOVE

In the late 17th century, English physicist and mathematician Isaac Newton established the science of mechanics—the study of forces and motion—with three simple but revolutionary scientific laws that are still used today.

When Newton was a student, scholarly understanding of forces and motion was based on the ideas of ancient Greek philosopher Aristotle (384–322 BCE), who believed that an object moves only as long as a force acts on it. For example, according to Aristotle, projectiles in free motion are pushed along by following air currents. Thinkers in the Middle Ages expanded on this idea with the "impetus" theory that suggests that the force with which an object is thrown is stored in the object, and gradually runs out. Italian mathematician and physicist Galileo Galilei (1564–1642) overturned these ideas, realizing that an object continues to move at the same speed and in the same direction unless a force—such as gravity or air resistance—acts upon it. Newton adopted this idea as the first of his three laws, which he expressed in mathematical form in his book *Philosophiae Naturalis Principia Mathematica* (1687). Newton's laws accurately describe and predict the motion of objects in most situations. At very high speeds or in strong gravitational fields, they are not accurate because of effects explained by Einstein's theories of relativity (see pp.244–45).

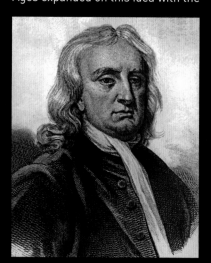

ISAAC NEWTON
Newton was the most influential thinker and experimentalist of the 17th and 18th centuries. He made enormous contributions to the study of gravity, light, astronomy, and mathematics.

31,000

THE SPEED, IN MILES PER HOUR, AT WHICH THE VOYAGER 1 SPACECRAFT IS LEAVING THE SOLAR SYSTEM. **VOYAGER** KEEPS **MOVING** THROUGH SPACE BECAUSE **NO AIR RESISTANCE** ACTS ON IT.

rocket remains stationary until a force acts on it
tanks contain fuel and oxygen that will produce a force when ignited
liquid fuel
liquid oxygen
rocket's weight is the force of gravity
reaction force balances rocket's weight

ROCKET AT REST
A rocket stands on a launch pad. Its enormous weight is the result of gravity pulling it downward toward Earth. The launch pad produces an upward reaction force that exactly balances the weight, and the rocket does not move.

FIRST LAW
Newton's first law states that an object remains at rest or continues moving in a straight line unless a force acts upon it. Most objects have many different forces acting on them at all times, but often the forces balance. A book lying on a table, for example, is being pulled downward by gravity—but the table pushes upward on the book with a force of exactly the same magnitude (see third law). Since the forces balance, the book remains stationary.

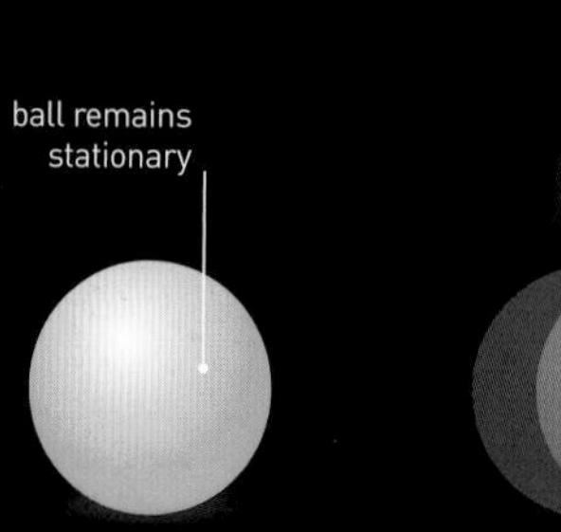

ball remains stationary

AT REST
A ball remains stationary until a force acts upon it. The ball's weight pulls it downward, but the ground exerts an upward reaction force of the same magnitude, so the net force is zero.

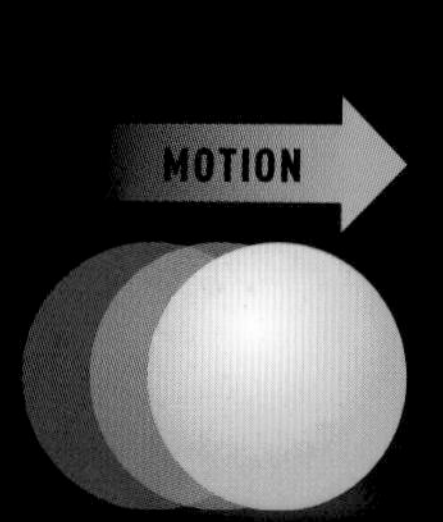

MOTION

IN MOTION
Once the ball is in motion, its velocity—or particular combination of speed and direction—continues. In reality, friction between the ball and the surface would slow it down.

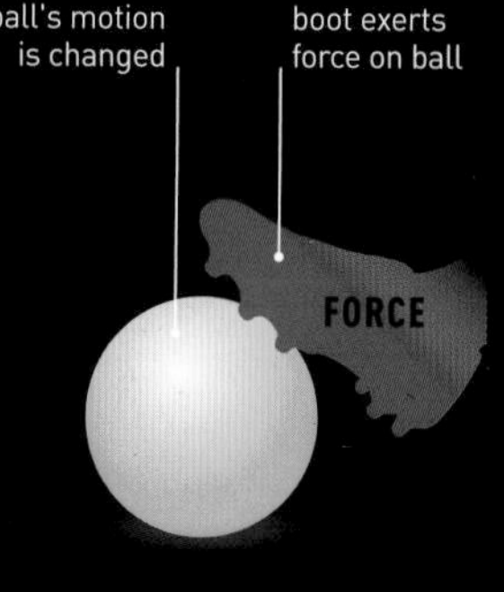

ball's motion is changed
boot exerts force on ball
FORCE

FORCE APPLIED
A force, such as a kick from a boot, alters the ball's velocity, a change termed acceleration. The ball either slows down, speeds up, or changes its direction with or without changing its speed.

ROCKET ENGINE NOZZLES
Causing a rocket to accelerate upward requires enormous forces, not least to overcome the gravity pulling the rocket downward. The force is generated inside the engine by expanding exhaust gases, which escape through these nozzles.

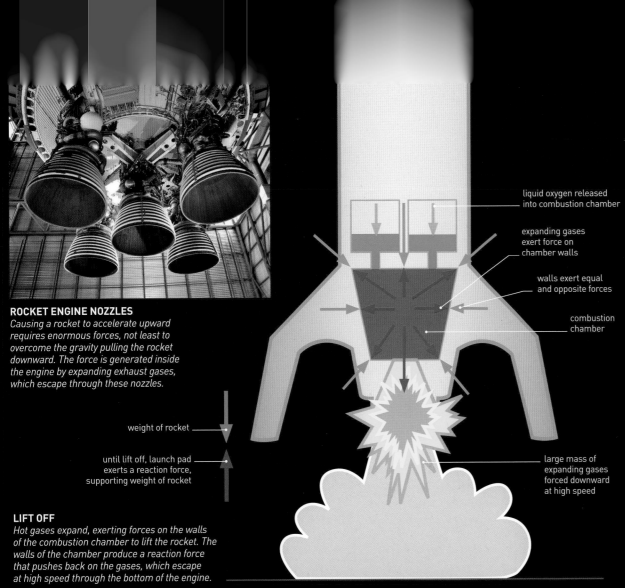

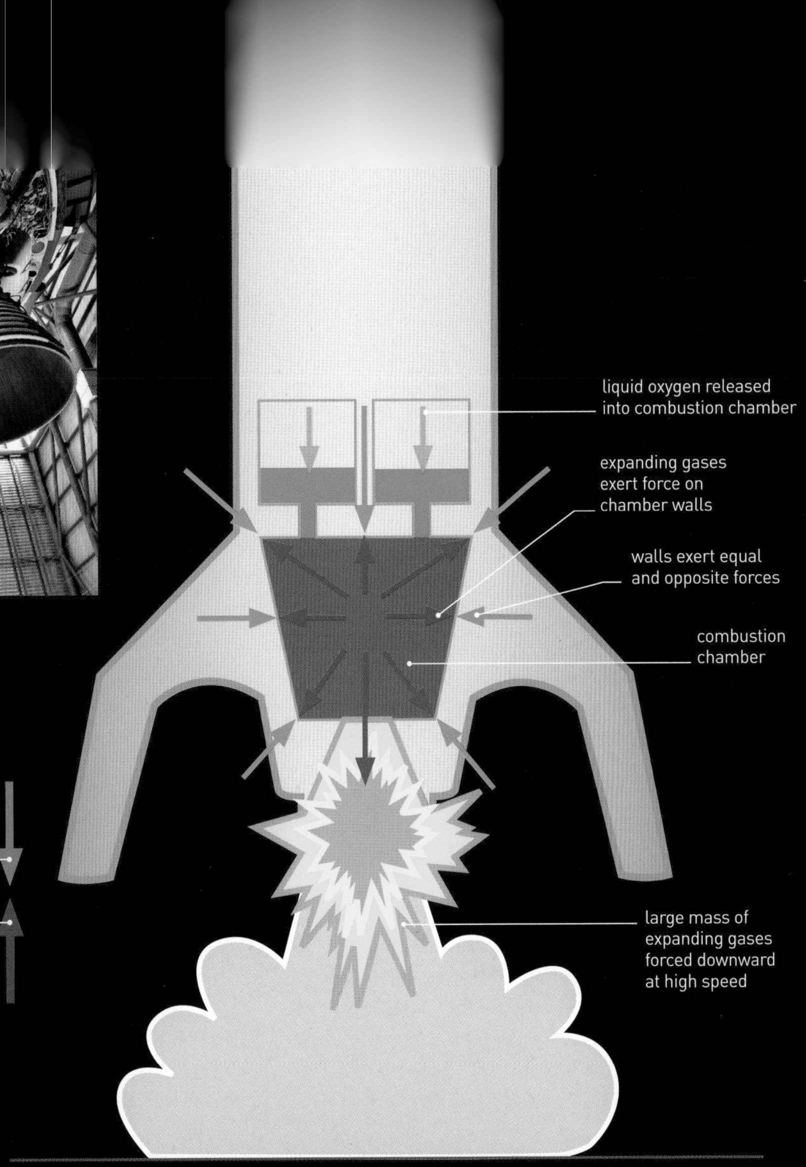

liquid oxygen released into combustion chamber

expanding gases exert force on chamber walls

walls exert equal and opposite forces

combustion chamber

weight of rocket

until lift off, launch pad exerts a reaction force, supporting weight of rocket

large mass of expanding gases forced downward at high speed

LIFT OFF
Hot gases expand, exerting forces on the walls of the combustion chamber to lift the rocket. The walls of the chamber produce a reaction force that pushes back on the gases, which escape at high speed through the bottom of the engine.

SATURN V
The Saturn V rocket, used during NASA's Apollo missions of the 1960s and 1970s, had a weight of 28 million newtons, and an engine thrust of 34 million newtons.

SECOND LAW

Newton's second law involves momentum: an object's mass multiplied by its velocity. The law states that the change in momentum is proportional to the force exerted. So, a force doubled will accelerate an object twice as much; but the same force applied to an object with twice the mass will produce only half the acceleration. The second law is often summarized with a simple equation: $a = F/m$, in which a is the acceleration, F is the force, and m is the mass of the object.

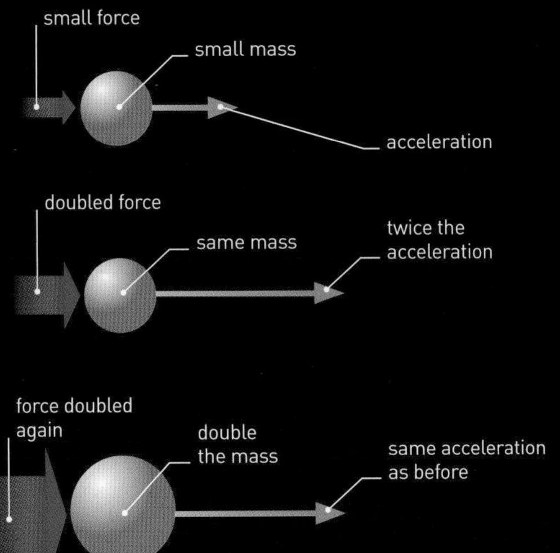

THIRD LAW

Newton's third law states that forces exist in pairs. When one object exerts a force on another, the second object exerts an equal and opposite force on the first. If one of the objects is immobile, then the other object will move; push against a wall on a ice rink, and the wall pushes back on you—which makes you slide on the ice. If both objects can move, then the object with less mass will accelerate more than the other; for example, a heavy gun recoils slightly as the bullet shoots out at high speed.

SMALL MASS, SMALL FORCE
An applied force causes an object to accelerate. The acceleration —change in velocity per second— depends upon the size of the force, but also on the mass of the object.

SMALL MASS, DOUBLE FORCE
Since $a = F/m$, doubling the force but keeping the same mass will cause the object to accelerate at twice the rate.

DOUBLE MASS, DOUBLE FORCE
Doubling the force again (to four times the original value) but also doubling the mass produces the same acceleration as before.

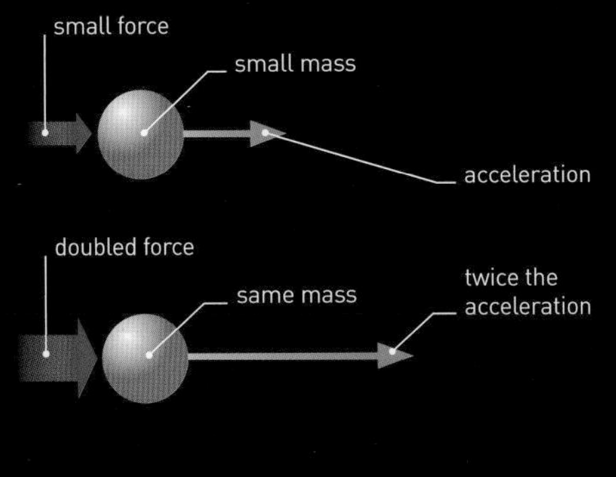
small force
small mass
acceleration
doubled force
same mass
twice the acceleration
force doubled again
double the mass
same acceleration as before

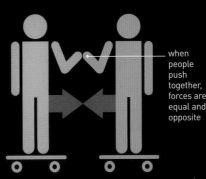

when people push together, forces are equal and opposite

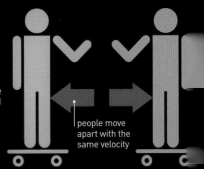

people move apart with the same velocity

EQUAL AND OPPOSITE
Two people on skateboards pushing against one other will move apart. Even if only one person does the pushing, the other person's body will produce a reaction force of equal strength in the opposite direction.

EQUAL MASSES
If the two people have the same mass, they will accelerate equally; but if one has much less mass, he or she will move away more quickly, since the same force will produce a greater acceleration.

“ ONE MAY CONCEIVE **LIGHT** TO **SPREAD** SUCCESSIVELY, BY **SPHERICAL WAVES. ”**

Christiaan Huygens, Dutch physicist, from *Treatise on Light*, 1690

mandible (jawbone)
vertebral column (spine)
femur (thighbone)
tibia (shinbone)

IN 1690, a decade after making an early pressure cooker that produced high-pressure steam, French inventor **Denis Papin** modified his original design by incorporating a piston, producing the **first working "atmospheric" engine**. Boiling water in a cylinder created steam, which pushed the piston up; as the steam condensed, it created a vacuum in the cylinder and atmospheric pressure plunged the piston back down. This invention marked the beginning of steam engine development.

Papin received advice on his designs from Dutch astronomer **Christiaan Huygens** (1629–95) who, also in 1690, made a significant contribution to other areas of knowledge with his *Treatise on Light*. Based on the observation that light beams could cross without bouncing, he deduced that **light is composed of waves**—supporting a theory proposed by French philosopher René Descartes in the 1630s and English inventor Robert Hooke in the 1660s. But Isaac Newton's idea that light was made from particles (see 1675–84) was to predominate over the wave theory for more than 100 years.

Clopton Havers (1657–1702), an English physician, was the first to study the **detailed anatomy of bones**—including marrow and cartilage. He published his results in 1691, describing the microscopic pores and cavities running through a bone's structure. Havers surmised that they carried oil, but it is now known that these so-called **Haversian canals** contain blood and lymphatic vessels, and provide bone cells with oxygen and nourishment.

Permeated bones
The bones of the body are permeated by tiny organic channels, named after Clopton Havers, the physician who discovered them.

70 PERCENT THE AMOUNT **OF BONE** MADE OF **NONLIVING** MINERAL

The location of fossil fish and other marine creatures far inland gave rise to conflicting theories among naturalists and theologians.

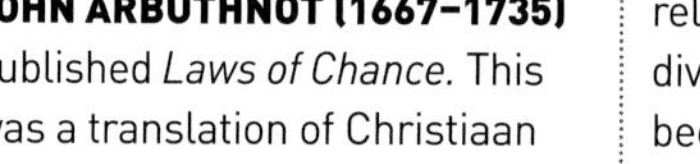

John Ray
Philosopher and theologian, John Ray is also regarded as the founding father of English natural history.

IN 1692, SCOTTISH PHYSICIAN JOHN ARBUTHNOT (1667–1735) published *Laws of Chance*. This was a translation of Christiaan Huygen's 1657 classic work on **probability theory**, and the first publication in English devoted to the subject.

English naturalist **John Ray** had written extensively on the diversity of plants since the 1660s, but by the 1690s he was also active in paleontology (the study of fossil organisms) and zoology. His accurate descriptions of fossils supported the idea that they were the remains of once-living species. Ray also **tried to explain the locations of fossils**. A popular concept was that the Biblical flood had been responsible for the forming of fossils, but Ray saw that a deluge would not have created the specific layers that could be observed in geological deposits. He suggested that in an ancient world covered by sea, land rose by volcanic activity—which would explain the occurrence of fossilized marine animals on land. However, Ray's theological leanings meant that he was reluctant to take the view that divinely created species could become extinct. He proposed that organisms so far known only as fossils would one day be found living in remote areas.

In 1693, Ray published one of his most important works of zoology, *Synopsis of Quadruped Animals and Serpents*. Based on anatomical features, it provided

“ NEVER... DID I EXPECT TO PRODUCE A **HISTORY** OF **QUADRUPEDS. ”**

John Ray, from *Synopsis of Quadruped Animals*, 1693

1690 French inventor Denis Papin develops a **piston steam engine**

1690 Dutch astronomer Christiaan Huygens describes a **wave theory** in his ***Treatise on Light***

1691 English physician Clopton Havers publishes a **treatise on bone anatomy**

1692 Scottish physician John Arbuthnot publishes the **first English work** on **probability theory**

> **“** MANY **SPECIES** OF ANIMALS HAVE BEEN **LOST** OUT OF THE WORLD, WHICH PHILOSOPHERS AND DIVINES ARE **UNWILLING TO ADMIT…** **”**
>
> **John Ray, English naturalist,** from *Three physico-theological discourses, concerning the primitive Chaos, and creation of the world*, 1713

By the end of the 17th century, the sexual function of flowers was recognized. *Clematis marmoria*, seen here, has separate male and female plants.

the **first scientific classification of animals**. He identified mammals as viviparous (giving birth to live young) quadrupeds, and placed them into groups according to structures such as feet and teeth.

In the same year, Belgian physician **Philip Verheyen** (1648–1711) published his illustrated *Anatomy of the Human Body*, which would become a standard textbook on the subject in European universities. In this work, Verheyen introduced the term Achilles tendon for the structure at the back of the leg, named for the Greek hero killed by an arrow wound in his heel. Opportunistically, Verheyen had been able to dissect his own left leg, which had been amputated because of illness nearly 20 years before. Verheyen had insisted on preserving the limb so he could study it. Based on personal experience, he was one of the first physicians to report **phantom limb phenomenon**: the sensation that an amputated limb is still attached to the body.

Also in 1693, English astronomer and mathematician **Edmond Halley** was the first person to draw up **life annuity charts** based on mortality tables. Thirty years earlier, a shopkeeper called John Graunt had produced “life-tables” as part of a scheme to monitor the advance of bubonic plague, but Halley had the mathematical skills to carry out a more sophisticated analysis. Using data on births and deaths from the European city of Breslau, he estimated the city’s population size and the probabilities of its citizens surviving to particular ages. The study became a model for future demographic investigations.

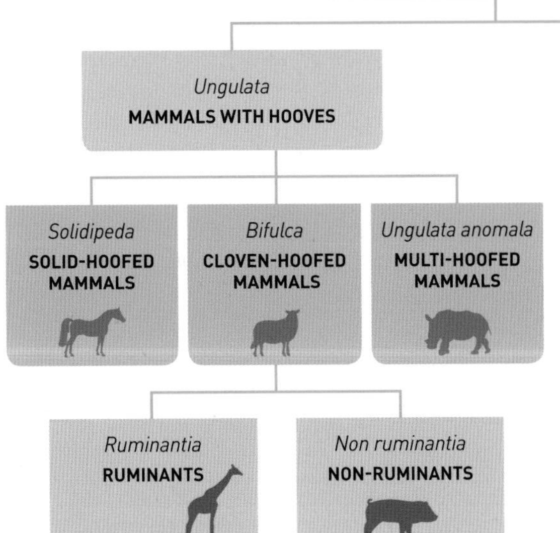

Animalium viviparorum quadrupeda
MAMMALS
Ungulata
MAMMALS WITH HOOVES
Unguiculata
MAMMALS WITH CLAWS OR NAILS
Solidipeda
SOLID-HOOFED MAMMALS
Bifulca
CLOVEN-HOOFED MAMMALS
Ungulata anomala
MULTI-HOOFED MAMMALS
Ruminantia
RUMINANTS
Non ruminantia
NON-RUMINANTS

Ray’s classification of mammals
John Ray grouped mammals according to whether they had hooves, claws, or nails. The division he called Unguiculata *is no longer valid, but his hoofed (ungulate) groups are partly supported by modern biology. He also recognized ruminants: cud-chewing herbivores with multichambered stomachs.*

BY THE END OF THE 17TH CENTURY THE WORK OF SEVERAL NATURALISTS had revealed some of the secrets of flowering plants. In 1694, French botanist **Joseph Pitton de Tournefort** (1656–1708) published a classification system that was based on the structure of flowers and fruits, as well as leaves and roots. Although Tournefort’s conclusions were often misguided, his work had a lasting influence because of the clarity of his species-level accounts. He was also one of the first botanists to use the **genus as a taxonomic category** that included similar species: a forerunner of the binomial system of naming formalized by Linnaeus in the 1700s (see 1733–39).

German botanist **Rudolf Camerarius** (1665–1721) went much further in studying flowers. His 1694 paper on the reproduction of plants provided experimental evidence for the notion that not only did **plants have sexual organs**, but that pollen was the agent of male fertilization. Camerarius observed that female shoots separated from male shoots often failed to set seed, and that when pollen-producing stamens were removed, no seeds were produced at all. But he was frustrated by the fact that he could not probe deeper into the minute functions of flowers. It would not be until better microscopes opened up the world of cells more than a century later that the microscopic basis of plant reproduction could be properly explored.

English instrument maker **Daniel Quare** (1649–1724) is credited with a number of innovations in horology (the study of time), including the invention of repeating watches and the introduction of the minute hand. By 1694, he had also produced the **first portable barometer**, which he patented the following year. Until then, the system of tubes associated with a barometer was not easily moved, but a portable instrument could allow experimentalists to measure atmospheric pressure in places such as mines or mountains.

Portable barometer
The construction of Daniel Quare’s barometer ostensibly allowed free movement of the instrument without letting air in or spilling its mercury.

1693 English naturalist John Ray makes the first scientific **classification of animals**

1693 Belgian physician Philip Verheyen publishes a **book on human anatomy**

1693 English astronomer Edmond Halley creates the **first life annuity charts**

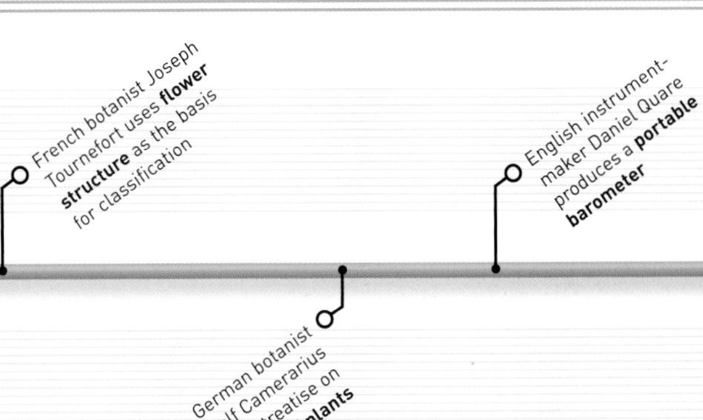

French botanist Joseph Tournefort uses **flower structure** as the basis for classification
German botanist Rudolf Camerarius publishes a treatise on the **sexuality of plants**
English instrument-maker Daniel Quare produces a **portable barometer**

1695–97

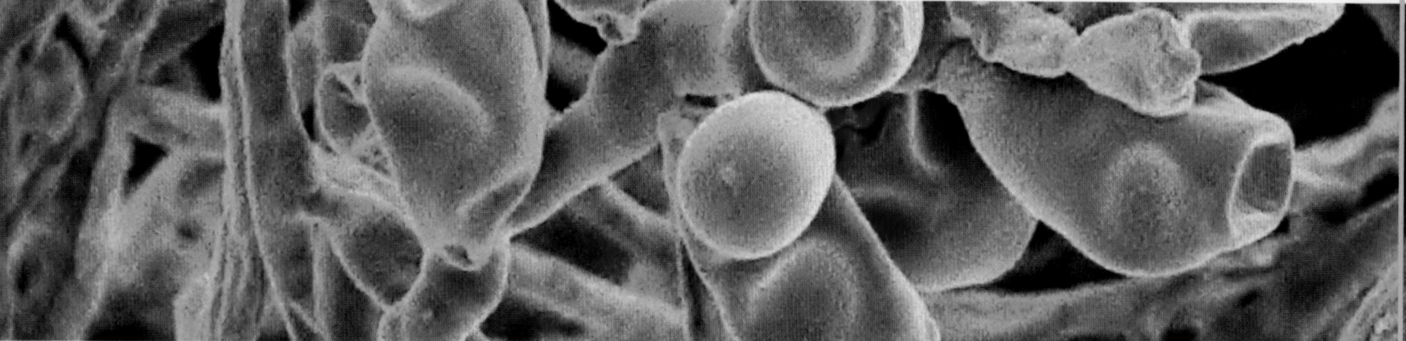
Dutchman Antony van Leeuwenhoek was one of the first observers of microscopic organisms such as these mold spore capsules.

SOME 20 YEARS AFTER HE MADE HIS FIRST OBSERVATIONS of miniature life (see 1675–84), Dutch microscopist **Antony van Leeuwenhoek** published a compilation of his work, *Arcana Naturae* (*Secrets of Nature*) in 1695. As well as describing and illustrating a range of biological curiosities—from tadpoles to red blood cells—the book contained descriptions of the techniques Leeuwenhoek had used to carry out his studies. Many of these, including his microscope, were his own inventions.

In the same year, English theologian and mathematician **William Whiston** (1667–1752) published his *New Theory of the Earth*, which was a combination of religious and scientific thought. He supported the idea of divine creation, and his work was praised by many, including Isaac Newton. Whiston suggested that the global catastrophe of the Biblical flood had been caused by a comet. He would succeed Newton as the third Lucasian Professor of Mathematics at Cambridge University in England.

In 1697, many decades before the discovery of oxygen, German chemist **Georg Stahl** (1660–1734), proposed **a theory to explain combustion**. He suggested that metals and minerals contained two components—one being the **calx** (ashy residue), and the other being a substance called **phlogiston**, which was given off when something burned. Stahl thought that the amount of phlogiston varied: there was a great deal in coal, which diminished to ashes during combustion, but very little in iron, which did little more than rust. The phlogiston theory had its roots with Stahl's mentor, German alchemist Johann Becher, who conceived phlogiston as *terra pinguis* (oily matter), one of the classical elements. Later, the French chemist Antoine Lavoisier argued that **combustion happened by oxidation**: reaction of the substance with oxygen in the air. Stahl's idea of the calx was equivalent to the modern idea of oxide.

In 1697, Swiss mathematician **Johann Bernoulli**, prompted by a dispute with his brother (who was often his bitter rival), **solved a trajectory problem**. He described the path followed by a particle moving under gravity. By studying the rates of movement along this curve, Bernoulli's work had important implications for the development of calculus: the mathematics of infinitesimal changes.

Also in 1697, English explorer **William Dampier** (1651–1715) published an account of his first voyage, *A New Voyage Around the World*—containing descriptions of the Americas and East Indies. The British Admiralty granted him the command for another trip, and Dampier eventually circumnavigated the globe three times. His **work on navigation influenced explorer James Cook**, while his studies of natural history would be used by biologists such as Alexander von Humboldt and Charles Darwin (see 1859).

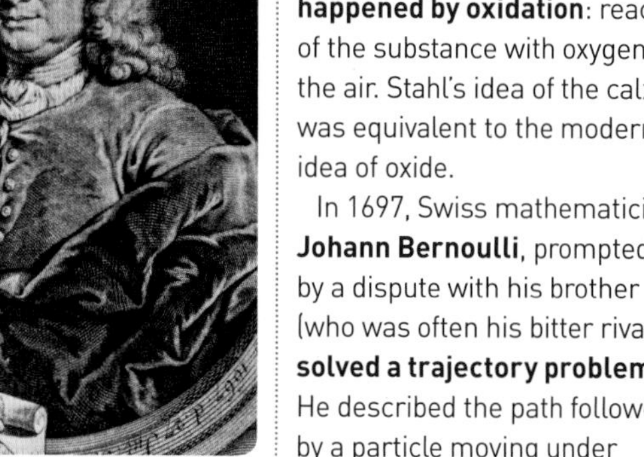

main screw
body plate

Simple microscope
In his 1695 study Arcana Naturae, *Antony von Leeuwenhoek explained the use of the microscope he had designed himself.*

JOHANN BERNOULLI (1667–1748)

Bernoulli was born into a prominent mathematical family and had professorships in Groningen, the Netherlands, and Basel, Switzerland. His work included studying the mathematical trajectories of curves and investigating the reflection and refraction of light. Together with his brother Jacob, he helped Newton and Leibniz develop calculus.

1698–99

The first dissection of a chimpanzee, in 1698, revealed a humanlike brain.

ISAAC NEWTON HAD ALREADY ESTABLISHED that sound moves as longitudinal compression waves, not by transverse waves (oscillation at right angles to the direction of travel), as previously thought. In 1698, he went on to calculate the **speed of sound in air**, which he determined to be 979 ft (298 m) per second. (The modern value is 1,125 ft per second.)

Dutch astronomer Christiaan Huygens died in 1695, but his final book, *Cosmothereos,* appeared in 1698. He had delayed publication because he feared offending religious sensibilities—he had conjectured upon the possibility that life existed on other planets with habitable conditions.

British physician **Edward Tyson** (1650–1708) was governor of the Bethlem Hospital for psychiatric patients, in London. He routinely performed

1,125 FEET PER SECOND THE **SPEED** OF **SOUND**

1695 Antony van Leeuwenhoek publishes ***Arcana Naturae***

1695 William Whiston publishes ***New Theory of the Earth***

1697 Georg Stahl proposes the ***phlogiston theory*** of combustion

1697 Johann Bernoulli solves a **trajectory** problem

1697 William Dampier **publishes** an account of his **first world voyage**

1698 English mathematician Isaac Newton **measures the speed of sound**

> **‘OUR PYGMIE’** IS **NO MAN**, NOR YET THE **COMMON APE;** BUT A SORT OF **ANIMAL** THAT BELONGED **BETWEEN MAN AND THE APES.**

Edward Tyson, British physician, in *Orang-Outang, sive Homo Sylvestris* (*Orang-Outang, or Man of the Woods*), 1699

STEAM POWER

When water is heated to boiling point it creates gaseous steam; if this steam is then trapped in a sealed container and cooled, it condenses back into water. As the quantity of gas drops, so does its pressure, creating a partial vacuum. Force is then generated when the atmosphere is let in to fill the void. The idea of harnessing this force in an "atmosphere engine" originated in the 1690s and would be fully realized in the steam engines of the next century.

autopsies in an effort to understand the causes of mental illness. But he dissected animals too, so becoming the **father of comparative anatomy**. In 1699, he published his **study of the chimpanzee** (which he called an "orang-outang"), concluding that it had more in common with humans than it did with monkeys.

That year, an English inventor **Thomas Savery** (1650–1715) demonstrated his latest creation to the Royal Society: "an engine to raise water by fire." Patented the year before, it exploited the recently discovered power of gas pressure, which could generate considerable force when gas rushed in to fill a vacuum. **Savery's steam pump** consisted of a boiler to produce steam that was directed into a vessel below a cold-water shower. This created a vacuum in the vessel as the steam condensed, which sucked up water from below. The sequence was controlled by a system of taps. Savery claimed his pump could be used to pull water up from mines, but it had a working height limit of about 25 ft (7.5 m). It was also vulnerable to explosion.

Again in 1699, Welsh naturalist **Edward Lhyud** (1660–1709) published a **catalog of fossils**. This included one of the earliest unambiguous specimens—a tooth—later identified as that of a dinosaur. Lhyud had fanciful notions about his specimens, suggesting that fossils grew in rocks from vaporous spawn that came from the sea.

French physicist Guillaume Amontons (1663–1705) was **an accomplished instrument maker** and perfected thermometers and barometers for measuring temperature and pressure. He was also the first experimenter to discuss the **idea of an absolute zero for temperature**. In 1699, Amontons turned to mechanics, describing how friction force depended upon load. **Amontons' friction law** had a prestigious history, being based on experiments first performed by Leonardo da Vinci.

cold water shower
tap
funnel for filling with water
steam boiler
vessel for trapping steam
suction pipe

Savery's steam pump
Having understood the principles of atmospheric force, Savery created a steam generator to pump water vertically.

1699 Edward Tyson publishes his study on **chimpanzees**

1699 Thomas Savery demonstrates his **steam pump**

1699 Edward Lhyud publishes a catalog of **fossils**

1699 Guillaume Amontons describes his **friction law**

1700–1701

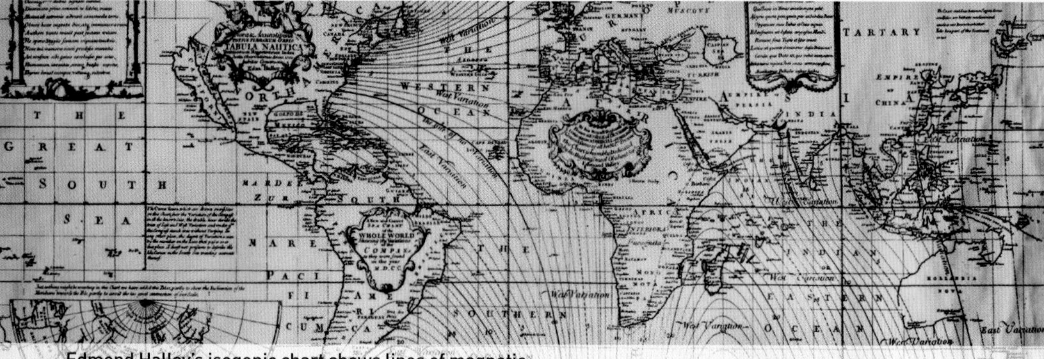
Edmond Halley's isogenic chart shows lines of magnetic variation from true magnetic north.

AS THE 18TH CENTURY DAWNED, British astronomer **Edmond Halley** sailed the Atlantic on his third **voyage of discovery**. In January 1700, he made the first observation of the **Antarctic convergence**, where icy Antarctic waters come up against warmer Atlantic waters in a ring around Antarctica. On February 1, he made the first recorded sighting of **tabular icebergs**, which have steep sides and a flat top. Halley also showed that **Earth's magnetism fluctuates** too much for compasses to be used to find longitude at sea. He confirmed that **magnetic north** does not correspond with **true north**, a phenomenon known as **magnetic declination** (see1598–1604).

Also in 1700, French physician **Nicolas Andry** (1658–1742) suggested that **smallpox** was caused by **tiny microorganisms** or "worms" that he had seen through a microscope.

In 1701, English agriculturalist **Jethro Tull** helped modernize farming practices when he invented the **mechanical seed drill**—a machine that automatically planted seeds in neat, evenly spaced rows. The adoption of Tull's method increased crop yields by as much as 900 percent.

JETHRO TULL (1674–1741)

Born in Berkshire, England, Jethro Tull intended to enter politics in London, but ill health kept him at home, farming. Noticing that hand-sown seeds were scattered chaotically, he developed the mechanical drill to sow seeds in even rows. He became a key figure in the agricultural reforms that swept through England in the 1700s and then around the world.

Sowing seeds
Jethro Tull's 1701 seed drill planted seeds in uniform, equally spaced rows. By giving seeds enough space to grow, it increased yields and reduced waste during sowing.

1702–1703

Isaac Newton, here seen speaking at a meeting of the Royal Society, was elected its president in 1703.

ISAAC NEWTON'S SCIENTIFIC ACHIEVEMENTS became increasingly well known in the 18th century. A key moment in the growth of his fame was the 1702 publication of *Astronomiae Physicae et Geometricae Elementa* (*Elements of Astronomy, Physics, and Geometry*) by Scottish mathematician David Gregory. One of the first popular accounts of Newton's theories, this work discussed his ideas on gravity and the movement of the planets. Newton was **elected president of the Royal Society** in London in 1703, a post he held until his death in 1727.

In 1703, German chemist **Georg Stahl** developed Johann Becher's 1667 idea that an element called *terra pinguis* is released from substances such as wood when they burn. Stahl called the element phlogiston, and the **phlogiston theory of combustion** came to dominate 18th-century chemistry until finally disproved by Antoine Lavoisier (see 1789) later in the century.

By the 18th century, scholars were starting to consider natural events, such as earthquakes, as phenomena to be investigated scientifically rather than as acts of God. In 1703, French priest and inventor **Abbé Jean de Hautefeuille** (1647–1724) described a **seismometer** for measuring the severity of earthquakes. De Hautefeuille's device, a simple balanced pendulum whose swing responded to ground movement, was one of the earliest seismometers used in Europe.

500,000
THE **NUMBER OF EARTHQUAKES** THAT OCCUR **EACH YEAR**

Botanists, meanwhile, were beginning to embark on voyages of exploration to study the rich variety of unknown plants in newly discovered parts of the world. After three plant-hunting voyages to the West Indies, French botanist **Charles Plumier** published *Nova Plantarum Americanarum Genera*, a huge and groundbreaking work on **plant classification**. In it, he described the plants fuchsia and magnolia for the first time.

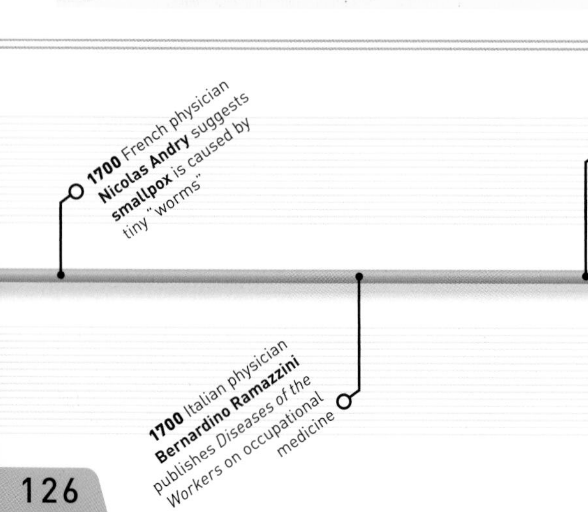
1700 French physician **Nicolas Andry** suggests **smallpox** is caused by tiny "worms"
1700 Italian physician **Bernardino Ramazzini** publishes *Diseases of the Workers* on occupational medicine

January 1700 English astronomer **Edmond Halley** observes the **Atlantic convergence**

1701 Jethro Tull develops the **seed drill**

1703 German chemist **Georg Stahl** devises **phlogiston theory**

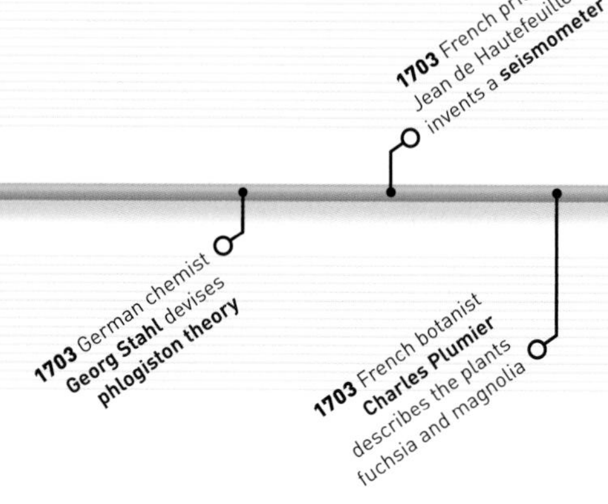
1703 French priest Abbé Jean de Hautefeuille invents a **seismometer**
1703 French botanist **Charles Plumier** describes the plants fuchsia and magnolia

Edmond Halley correctly surmised that comets seen at regular 76-year intervals had been the same comet. It later became known as Comet Halley.

> **“** IN… **1456… A COMET WAS SEEN** PASSING **RETROGRADE BETWEEN** THE **EARTH** AND THE **SUN… HENCE** I DARE VENTURE **TO FORETELL,** THAT IT WILL **RETURN** AGAIN IN… **1758. ”**

Edmond Halley, from *A Synopsis of the Astronomy of Comets,* 1705

> **“** THE **CHANGING** OF BODIES INTO **LIGHT,** AND LIGHT INTO **BODIES,** IS VERY CONFORMABLE TO THE **COURSE OF NATURE,** WHICH SEEMS **DELIGHTED** WITH **TRANSMUTATIONS. ”**

Isaac Newton, from *Opticks,* 1704

ISAAC NEWTON PUBLISHED HIS SECOND GREAT SCIENTIFIC BOOK, entitled *Opticks,* in 1704. The experiments he described in this book proved that the spectrum of brilliant colors produced when sunlight shines through a prism is not an effect of the glass (see 1665–66). Instead, as Newton showed, the colors are all contained in "white" sunlight and are simply separated when each color of light is bent, or refracted, differently as it enters the prism and slows down slightly. He also suggested that **light is a stream of tiny particles**, or "corpuscles," traveling at great speed. The theory ignited a debate that lasted more than 200 years about whether light is indeed formed of particles or, as suggested by Newton's Dutch rival Christiaan Huygens, waves.

That same year, English instrument maker and experimenter **Francis Hauksbee** (1660–1713) began a series of demonstrations at the **Royal Society** in London on the effects of **static electricity**. In 1704, Hauksbee thrilled witnesses with a demonstration of **"barometric light"**—the sparks of light that appear when mercury in the vacuum at the top of a mercury barometer is shaken. Two years later, Hauksbee built the **first electrical machine**, which he called the "influence machine," in which a hand-turned spindle rubbed wool against amber inside a glass vacuum globe to generate a glowing **static charge**. It was a forerunner of electric light.

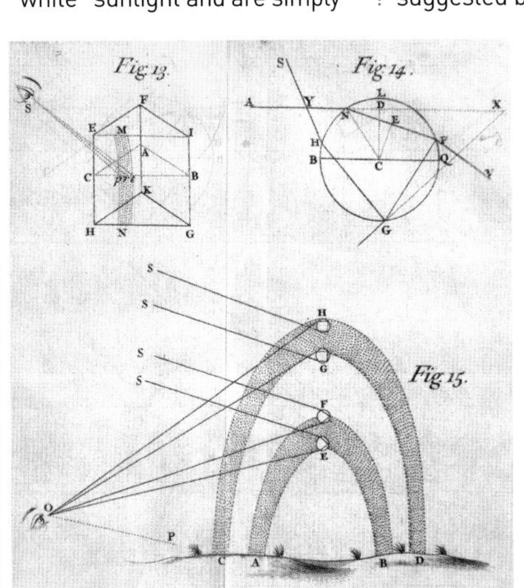
Fig 13.
Fig 14.
Fig 15.

Splitting light
Isaac Newton's findings, published in Opticks *in 1704, showed that "white" sunlight contained all the colors of the rainbow.*

In 1703, Dutch mathematician and astronomer **Christiaan Huygens** had published details of the gearing needed to drive a clockwork model of the Solar System that would precisely represent how the Sun and planets move in a year of 365.242 days. By 1704, English clockmakers **George Graham** (1764–51) and **Thomas Tompion** (1639–1713) had built a **clockwork mechanism**, based on Huygens' calculation, to show how Earth and the Moon move around the Sun. The pair were asked to make another mechanism for English nobleman Charles Boyle, 4th Duke of Orrery. Such devices subsequently became known as orreries.

In 1705, English astronomer **Edmond Halley** explained how **comets are on a great elliptical journey** around the Sun, and appear periodically when their journey brings them near to the Sun and Earth. He argued that comets seen in 1456, 1531, 1607, and 1682 were a single comet—now known as Halley's Comet—and predicted, correctly, that it would return in 1758.

In 1706, Welsh mathematician **William Jones** (1675–1749) proposed the **Greek letter pi (π)** to describe the ratio of the circumference of a circle to its diameter—approximately equal to the number 3.14159. Also this year, English inventor **Thomas Newcomen** (1663–1729) built a prototype for his steam engine that was to kickstart the Industrial Revolution in Europe (see 1712–13).

Earth and Moon
This orrery, made by George Graham and Thomas Tompion, shows how Earth and the Moon move around the Sun. Later orries included the movement of all the planets.

1704 English astronomer, mathematician, and physicist **Isaac Newton** publishes *Opticks*

1704 English clockmakers **George Graham and Thomas Tompion** invent the **orrery**

1704 English scientist **Francis Hauksbee** demonstrates **barometric light**

1705 English astronomer **Edmond Halley** predicts the return of **Halley's Comet**

1705 German naturalist **Maria Sibylla Merian** publishes an important **study of butterflies**

1706 English scientist **Francis Hauksbee** invents the static electric **influence machine**

1706 English ironmonger **Thomas Newcomen** invents a prototype of the **steam engine**

1.707 BILLION TONS

THE **AMOUNT** OF **IRON PRODUCED** EACH YEAR AROUND THE WORLD

It was in Coalbrookdale in Shropshire, England, that engineer Abraham Darby built the first coke-fired blast furnace to cast iron.

Coral reefs may look like plants but really they are colonies of animals.

THE HUMAN PULSE WAS KNOWN AS AN INDICATOR OF HEALTH more than 2,500 years ago. But it was not until English physician **John Floyer** invented his **pulse watch** in 1707 that western physicians began to measure the pulse in terms of heartbeats per minute. Floyer's timepiece was a watch that ran for exactly a minute while the physician counted pulses.

The following year, Dutch botanist and physician **Herman Boerhaave** developed a systematic approach to **diagnosis** that involved considering the patient's history, conducting a physical examination at the bedside, taking the pulse, and studying excretions.

Also in 1708, German physician, mathematician, and experimenter **Ehrenfried Walther von Tschirnhaus** (1651–1708) discovered that he could make **porcelain** with a paste mixed from clay, alabaster, and calcium sulfate. Although the Chinese had been making fine porcelain for centuries, the technology had eluded the west until this time.

In 1709, English experimenter **Francis Hauksbee** published *Physico-Mechanical Experiments on Various Subjects*, in which he described his celebrated **experiments with static electricity**. Hauksbee discovered that by rubbing glass, he could create static electricity and produce astounding electrical effects, such as "electric light" (the glow inside a rotating evacuated glass sphere when rubbed), electric wind (the prickling sensation when rubbed glass is brought near the face), and electric repulsion and attraction.

Alcohol thermometer
Gabriel Fahrenheit's 1709 thermometer was the first compact device of its kind. It showed temperature by the expansion of colored alcohol. Later, versions using mercury were popular.

English engineer **Abraham Darby** (1678–1717) revolutionized ironmaking in 1709 by **producing cast iron** in a coke-fueled blast furnace at Coalbrookdale in England. For the first time, iron could be cast in very large shapes, paving the way for the machines and engineering feats of the **Industrial Revolution**.

In Amsterdam, also in 1709, Polish-Dutch physicist **Gabriel Daniel Fahrenheit** (1686–1736) constructed an **alcohol-filled thermometer**. It was the first compact, modern-style thermometer with graduated markings, and it was similar to today's devices. The Fahrenheit temperature scale (see 1740–1742) was named after him.

In Lisbon, meanwhile, Brazilian-born priest and naturalist **Bartholome de Gusmão** (1685–1724) sent a ball to the roof using **hot air** and designed a **hot-air airship**. Although the first recorded manned flight in a hot-air balloon would not happen for another 74 years, de Gusmão's experiment anticipated future developments in aviation.

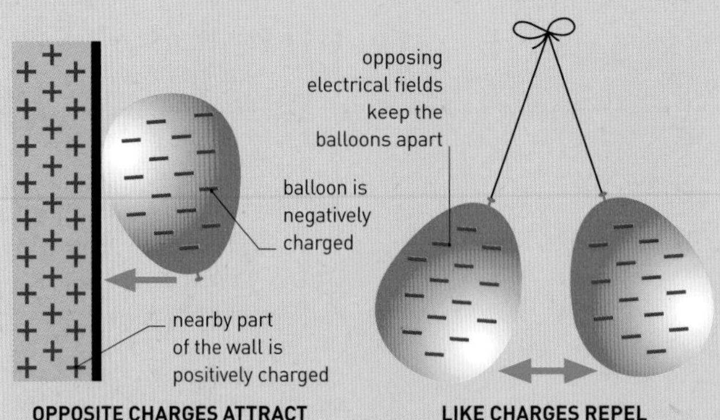
opposing electrical fields keep the balloons apart
balloon is negatively charged
nearby part of the wall is positively charged
OPPOSITE CHARGES ATTRACT
LIKE CHARGES REPEL

STATIC ELECTRICITY

Static electricity is the build-up or deficiency of electrons (particles contained in atoms). Surfaces charged with excess electrons are attracted to surfaces that have lost electrons. Experiments to create static electricity were widely practiced in the 18th century, often with striking results. Some of the most important investigations were carried out by English experimenter Francis Hauksbee.

IN 1710, GERMAN PAINTER JACOB CHRISTOPH LE BLON (1667–1741) found he could print pictures in a range of colors with just three different-colored inks. Paint of almost any color could be created by mixing three primary colors, but Le Blon realized that the colors did not have to be mixed. Instead, they could be **printed one on top of the other** in three layers. He started off in 1710 with three colors: red, blue, and yellow. Later, he discovered that even better results could be achieved with four colors: **black (K)** and the three primary colors used in printing—now known as **cyan (C), magenta (M), and yellow (Y)**, today called the CMYK system.

Also in this year, French entomologist **René de Réaumur** (1683–1757) set out to investigate whether spiders can make silk like silkworms. He showed that

17 CUBIC INCHES

THE TOTAL **CAPACITY** OF THE **HUMAN HEART**

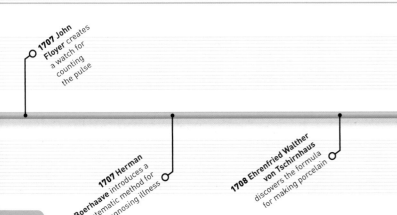

1707 John Floyer creates a watch for counting the pulse

1707 Herman Boerhaave introduces a systematic method for diagnosing illness

1708 Ehrenfried Walther von Tschirnhaus discovers the formula for making porcelain

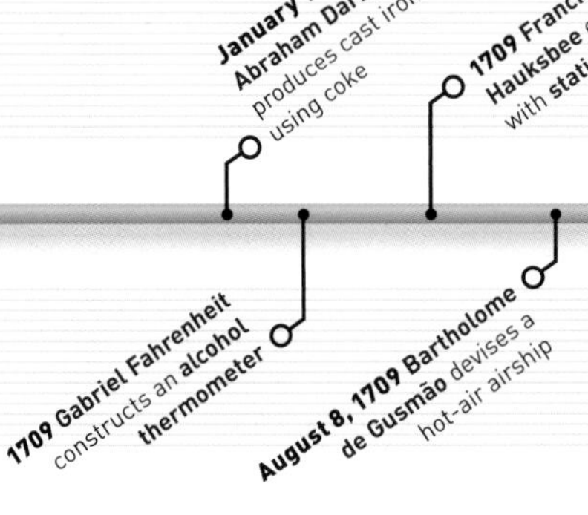

January 10, 1709 Abraham Darby produces cast iron using coke

1709 Gabriel Fahrenheit constructs an **alcohol thermometer**

1709 Francis Hauksbee experiments with **static electricity**

August 8, 1709 Bartholome de Gusmão devises a hot-air airship

1710 Jacob Le Blon invents the three-color printing method

1710 René de Réaumur shows that spiders make silk

English astronomer John Flamsteed's meticulous observations of the night sky formed the basis of the first modern star catalog.

athough **spiders do make silk**, it is much thinner thread, and he argued that spiders were too aggressive to use commercially.

Mathematician John Keill (1671–1721) published a paper claiming that Gottfried Liebniz, a German mathematician, stole the idea of calculus from British mathematician and physicist Isaac Newton. It is now thought that both men independently developed the **basis of calculus**.

The following year in Italy, Bolognese nobleman **Luigi Fernando Marsili** asserted that **corals** are plants, not animals. His mistaken view prevailed at the time, although others had realized that corals are animals.

Spider web
In 1710, Frenchman René de Réaumur showed that spiders produce silk. Spiders use the thread to make webs to catch prey or as cocoons for their young.

THE FIRST FULL-SCALE STEAM ENGINE, built by English engineer **Thomas Savery** in 1698, proved too dangerous for general use because high pressure in its boiler tended to cause explosions. But English ironmonger **Thomas Newcomen** (1663–1729) overcame the danger in 1712 to create the world's **first practical steam engine**. Newcomen's solution was to boil water in an isolated chamber and send the steam into a cylinder with a piston at low pressure. When steam flowed into the cylinder, it pushed the piston up. A valve closed, cold water was sprayed in, and the steam condensed, creating a vacuum that pulled the piston down, moving the engine's beam. Newcomen's steam engine was so successful that soon thousands of them were installed in mines across Britain and Europe to pump out floodwater.

In London this year, **Isaac Newton** and astronomer **Edmond Halley** enraged British astronomer **John Flamsteed** by publishing a catalog of more than 3,000 stars based on Flamsteed's observations made over 40 years at the Royal Greenwich Observatory. Newton and Halley believed the data should be published, but Flamsteed felt it was not thorough enough. He was so incensed by the publication that he gathered and burned 300 of the 400 printed copies.

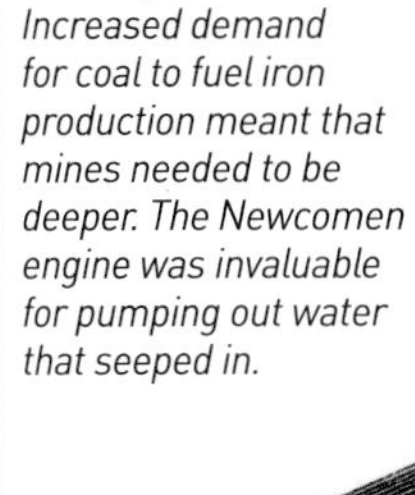

10 GALLONS
THE AMOUNT OF **WATER PUMPED** EACH MINUTE BY NEWCOMEN'S **FIRST ENGINE**

Swiss mathematician **Jacob Bernoulli's** book *Ars Conjectandi* (*The Art of Conjecture*) was published seven years after his death, in 1713. It introduced the **Law of Large Numbers**, which says that the more times you perform an experiment, the closer the average result tends to be to the average of a large number of experiments. That year, Bernoulli's nephew, **Nicolas Bernoulli**, devised the **St. Petersburg paradox** familiar to probability theorists today. It is based on a theoretical lottery game that seems to allow an infinite win, yet it is one that nobody with any sense would enter.

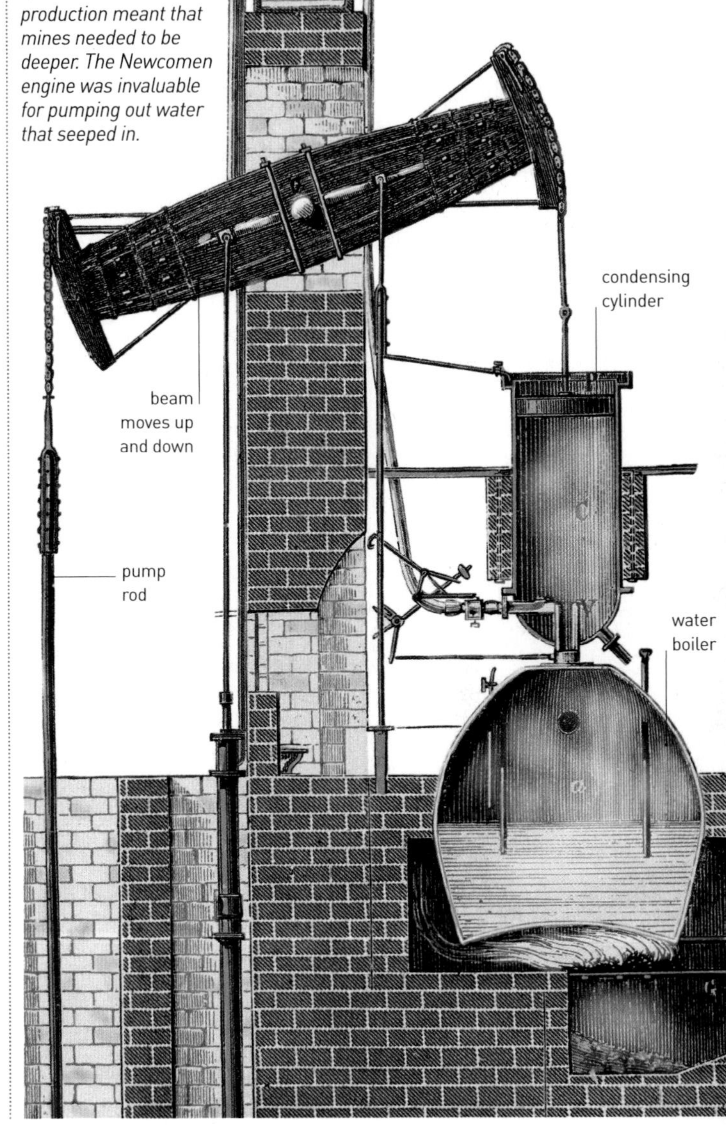
water tank
condensing cylinder
beam moves up and down
pump rod
water boiler
Steam power
Increased demand for coal to fuel iron production meant that mines needed to be deeper. The Newcomen engine was invaluable for pumping out water that seeped in.

1711 Luigi Fernando Marsili insists, incorrectly, that corals are plants

1711 English clergyman **Stephen Hales** measures animal **blood pressure**

1712 Thomas Newcomen invents the first **practical steam engine**

1712 John Flamsteed's star catalogue is published

1713 Johann Bernoulli's *Ars Conjectandi* is published posthumously

September 9, 1713 Swiss mathematician Nicolas Bernoulli outlines the **St. Petersburg paradox**

5 BILLION THE NUMBER OF YEARS BEFORE OUR SUN BECOMES A PLANETARY NEBULA

The Horsehead Nebula is a cloud of interstellar gas and dust. Edmond Halley was the first to suggest that indistinct objects in space could be nebulae.

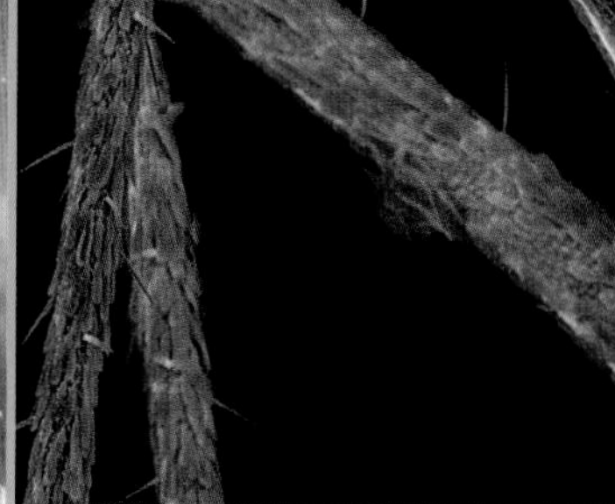

Giovanni Lancisi was first to realize that malaria is spread by mosquitos.

right-angled corner

scale in degrees

17th-century quadrant
This quadrant, designed by mathematician Edward Gunter in 1605, showed latitude. But there was still no way to be sure of longitude.

BY THE EARLY 18TH CENTURY, BRITAIN WAS SENDING OUT thousands of ships over the oceans to serve its growing overseas empire. But every ship's captain had the same problem—of not knowing where the ship was when out of sight of land. A good navigator could work out his **latitude**—how far north or south he was—from the altitude angle of the Sun and the North Star. The problem was to calculate **longitude**—how far east or west. The technique of **dead reckoning**, or estimating how far he had sailed from his average speed, gave a clue. But miscalculation meant that many ships were lost at sea. In 1714, the British parliament launched a competition with a prize of £20,000—a huge amount at the time—for the person who found a way to determine longitude accurately. Similar competitions were held in France and Holland.

One reason why longitude calculation was tricky was that the clocks of the day were wildly inaccurate. So, in 1715, English inventor George Graham's development of the **deadbeat escapement** was a great breakthrough. This mechanism eliminated recoil when a clock's time gear moved around a notch, enabling clocks to keep time within a second per day—a huge improvement. Deadbeat escapement clocks were preferred for scientific observation for the next 200 years because of their accuracy.

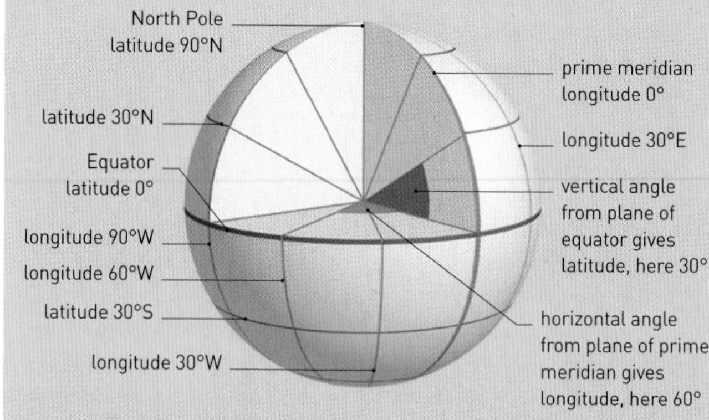
North Pole latitude 90°N
prime meridian longitude 0°
latitude 30°N
longitude 30°E
Equator latitude 0°
vertical angle from plane of equator gives latitude, here 30°
longitude 90°W
longitude 60°W
latitude 30°S
horizontal angle from plane of prime meridian gives longitude, here 60°
longitude 30°W

GEOGRAPHICAL COORDINATES

Solving the problem of how to find longitude at sea—how far east or west a ship is—was a priority in the 1700s. Lines of longitude, or meridians, run north-south around the world, dividing it like the segments of an orange. Zero degrees longitude is the Prime Meridian which passes through Greenwich, London, and a position's longitude is its angle east or west of this in degrees.

In 1715, English astronomer **Edmond Halley** suggested that the **age of Earth** could be determined by the **salinity of the oceans**, since the salt content would build up steadily as salt is washed in from the land. But his theory was impossible to prove, and, in fact, the salinity is too variable to be a measure of this. Halley was also the first astronomer to argue, correctly, that nebulae, which are seen as pale fuzzy shapes in the night sky, could comprise clouds of dust and gas.

IN 1716, ITALIAN PHYSICIAN GIOVANNI MARIA LANCISI (1654–1720) was the first to recognize the source of malaria. This often fatal disease, then common in Europe, was known as "ague" or "marsh fever" because it tended to occur near marshes, such as those around Rome. People believed it was caused by fumes from the damp ground—*mal'aria* is Italian for "bad air." But Lancisi realized that **malaria was caused by bites from swamp-inhabiting mosquitoes**. Few listened to

July 1714 British Parliament offers prize for a **method of determining longitude**

1715 Edmond Halley identifies five **nebulae**

1715 George Graham creates **deadbeat escapement** for clocks

1716 Italian physician Giovanni Lansisi identifies **mosquitoes as carriers of malaria**

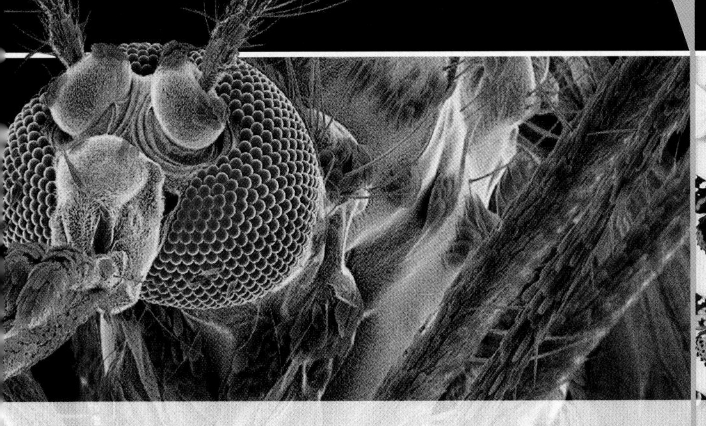

The 18th century saw the creation of the first scientific collections of butterflies, like this collection of British butterflies, first named by English naturalist James Petiver.

him, but we now know the disease is caused by a parasite spread by female *Anopheles* mosquitoes (see 1893–94).

In England, astronomer **Edmond Halley** made the first safe and practical **diving bell**—a bell-shaped diving chamber that enabled a person to go under water, breathing the air trapped inside. The idea of the diving bell dated back to the age of Aristotle, and in the 1600s, less sophisticated bells were used to recover goods from shipwrecks. But Halley, who studied the problem over two decades, realized that air is compressed by water pressure at depth, which is why a simple air tube to the surface did not work. Halley's ingenious solution was to continually replenish the air in the bell with air pressurized in weighted barrels lowered beside the bell. He also added a weighted tray to keep the bell upright, and a glass window to let in light.

bell continually replenished with pressurized air

Halley's diving bell
This engraving of Edmund Halley's diving bell illustrates the weighted platform at the base and the separate barrel that replenished the air to the side.

SMALLPOX WAS A DEADLY DISEASE IN THE 18TH CENTURY. Millions of people, many of them children, died from the illness, and even those who recovered were left with faces permanently disfigured by the scars. Yet long ago, the Chinese had noticed that once people had survived smallpox, they never caught it again, no matter how much they were exposed to the disease. Chinese physicians began deliberately rubbing infected material into a scratch on healthy people. Some died quickly from the infection this caused, but most survived, and seemed to gain immunity to the disease. The **practice of "variolation,"** as it became known, spread across Asia to Turkey, where it was noticed by Greek physician **Giacomo Pylari** (1659–1718), and then the young wife of the British ambassador to Constantinople (Istanbul), **Lady Mary Wortley Montagu** (1689–1762). Montagu was so impressed that she wrote a famous series of letters home advocating its use. She had her own children inoculated, and campaigned ardently to introduce the practice to the British upper classes. Her pioneering efforts led Edward Jenner to discover vaccination (see 1796).

LADY MARY MONTAGU
(1689–1762)

Mary Montagu's campaign to introduce smallpox inoculation in Britain helped to establish the idea that disease could be prevented through immunity. She had had smallpox herself as a young woman. Besides her pioneering work on disease, she was a celebrated writer, much admired by some of the leading figures of the day.

In 1717, London apothecary **James Petiver** (1685–1718) published *Papilionum Brittaniae Icones* (*Images of British Butterflies*). It was one of the **first great catalogs of butterflies**, based on Petiver's collection of species, now in London's Natural History Museum.

In 1718, English inventor James Puckle (1667–1724) was working on the design of a forerunner of the machine gun. The **Puckle gun** was a flintlock rifle mounted on a tripod with a revolving cylinder holding 11 shots that could be turned by a handle to fire 63 shots in 7 minutes—three times as fast as the best musketman.

17,500
WORLD
3,700
PERU
56
Britain

Butterfly species
James Petiver described 48 species of British butterflies. Now 56 are known (out of 17, 500 around the world), but species are vanishing with habitat loss.

In 1720, English instrument-maker **Jonathan Sisson** (1690–1747) **added a telescopic sight to the theodolite**, paving the way for the first accurate regional surveys and maps. The first theodolite had been invented by Leonard Digges in 1554 (see 1551–54) but theodolites equipped with a telescope could be used to measure angles over long distances. It meant that the height and position of every feature in the landscape could be surveyed by the method of triangulation, which uses simple trigonometry.

1717 James Petiver publishes his catalog of British butterflies

15 May 1718 James Puckle invents the **Puckle Gun**

1720 Jonathan Sisson invents the **telescopic theodolite**

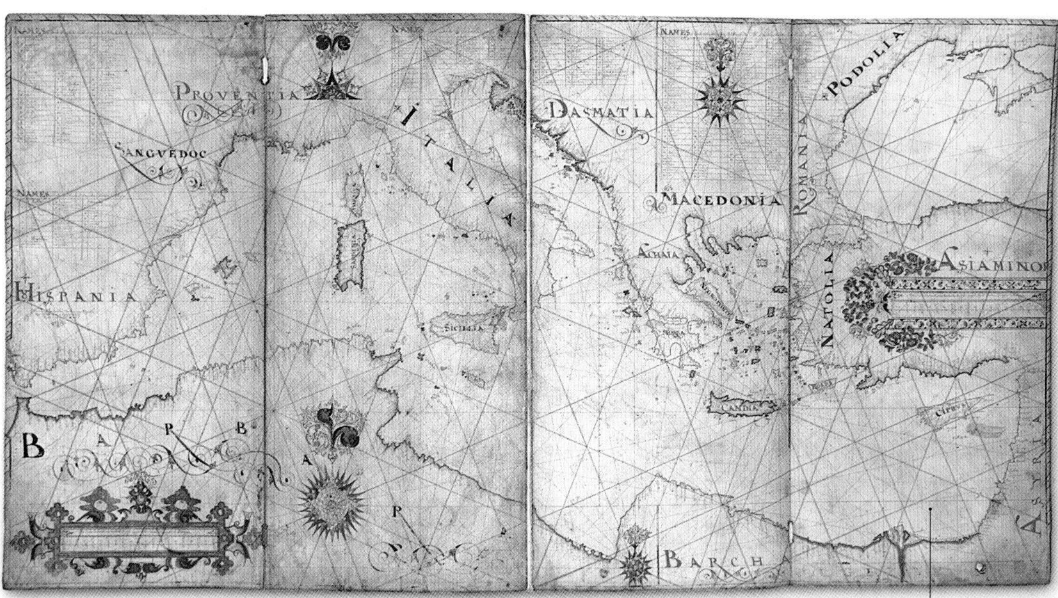

line indicating compass bearing

Portolan map
Date unknown
From the 13th century on, sailors relied on portolan charts—maps showing compass bearings—to guide them between ports. This early chart of the Mediterranean depicts the navigational lines between hundreds of ports.

alidade, for sighting stars

star pointer

Astrolabe
Late 15th century
Developed over 2,000 years ago to sight stars and make astronomical calculations, astrolabes were later simplified to find latitude at sea by measuring the height of the Sun and stars.

NAVIGATION TOOLS

IMPROVEMENTS IN INSTRUMENT DESIGN HAVE MADE NAVIGATION INCREASINGLY PRECISE

Ancient navigators relied on the position in the sky of the Sun and the stars to determine their location and chart their course. Later, a compass and an accurate timepiece could be used to work out direction and location.

For much of history, sailors found their latitude with tools such as sextants, astrolabes, and quadrants that indicated the angle of the Sun and stars above the horizon. From about a thousand years ago, compasses gave them a direction to sail in—a bearing. And from the 1700s, chronometers finally enabled them to work out their longitude. For most modern navigators, these instruments have been replaced by satellite systems.

Navigator's compass
*c.*1860
From the 13th century onward, navigators used a magnetic compass with a wire lozenge or metal needle, mounted to swing freely, to find north.

magnetic mineral

Lodestone
*c.*1550–1600
Chinese sailors used swinging lodestones—magnetic stones that turn to align with Earth's magnetic field—to gauge direction in overcast conditions.

sliding cover for viewing window

binnacle

iron sphere compensates for magnetism of ship's iron hull

Binnacle compass
*c.*1930
From the mid-18th century, compasses were mounted inside cases called binnacles on "gimbals"—pivots to keep the needle level however much the ship pitched and rolled.

Marine chronometer
*c.*1893
High-precision clocks, chronometers provided the accurate timekeeping necessary to keep track of longitude (distance east or west) on a long voyage.

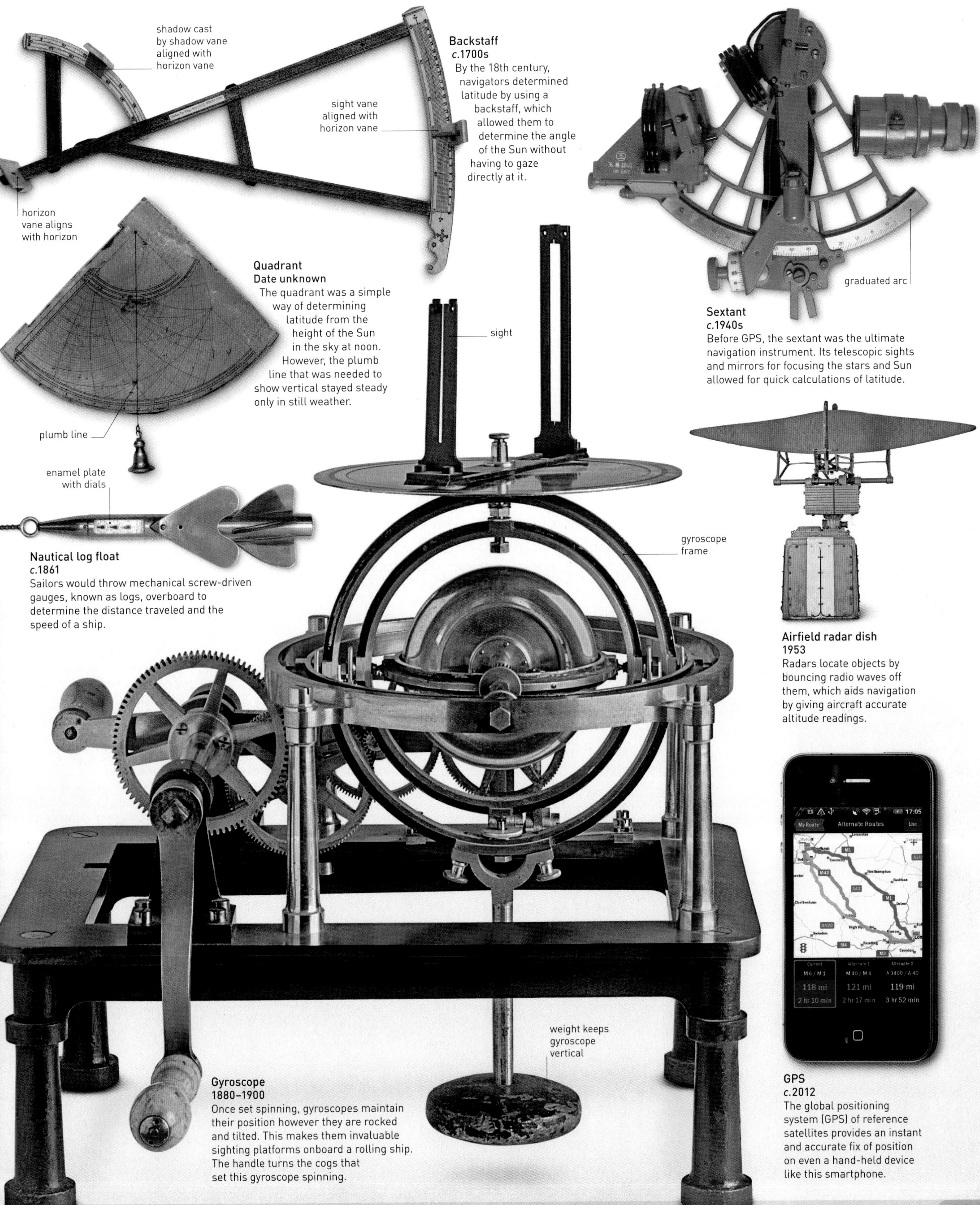
shadow cast by shadow vane aligned with horizon vane
sight vane aligned with horizon vane
horizon vane aligns with horizon
plumb line
sight
graduated arc
enamel plate with dials
gyroscope frame
weight keeps gyroscope vertical
My Route
Alternate Routes
List
17:05
M6 / M1
M40 / M4
A3400 / A40
118 mi
121 mi
119 mi
2 hr 10 min
2 hr 17 min
3 hr 52 min

Backstaff
*c.*1700s
By the 18th century, navigators determined latitude by using a backstaff, which allowed them to determine the angle of the Sun without having to gaze directly at it.

Quadrant
Date unknown
The quadrant was a simple way of determining latitude from the height of the Sun in the sky at noon. However, the plumb line that was needed to show vertical stayed steady only in still weather.

Sextant
*c.*1940s
Before GPS, the sextant was the ultimate navigation instrument. Its telescopic sights and mirrors for focusing the stars and Sun allowed for quick calculations of latitude.

Nautical log float
*c.*1861
Sailors would throw mechanical screw-driven gauges, known as logs, overboard to determine the distance traveled and the speed of a ship.

Airfield radar dish
1953
Radars locate objects by bouncing radio waves off them, which aids navigation by giving aircraft accurate altitude readings.

Gyroscope
1880–1900
Once set spinning, gyroscopes maintain their position however they are rocked and tilted. This makes them invaluable sighting platforms onboard a rolling ship. The handle turns the cogs that set this gyroscope spinning.

GPS
*c.*2012
The global positioning system (GPS) of reference satellites provides an instant and accurate fix of position on even a hand-held device like this smartphone.

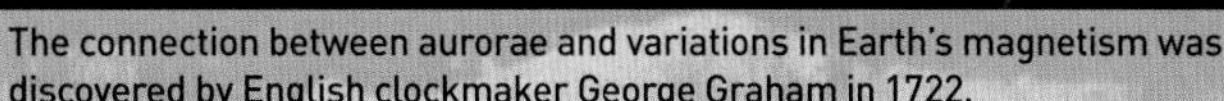

The connection between aurorae and variations in Earth's magnetism was discovered by English clockmaker George Graham in 1722.

The Russian Academy of Sciences was founded in St Petersburg in 1724.

WITH SO MANY BUILDINGS MADE MOSTLY OF WOOD, fire was a major hazard in 18th-century cities. In the 1600s, the Dutch had rushed **water pumps** mounted on handcarts to fires, but they delivered little more than a trickle of water. The breakthrough came when London buttonmaker **Richard Newsham** (d.1743) patented a pump in North America in 1721. Newsham's **fire pump cart** was the forerunner of today's fire engines. It had a 169-gallon (640-liter) watertank, and its pump operated by long handles and foot treadles extending either side could squirt 100 gallons (380 liters) of water per minute.

In 1722, clockmaker George Graham (1674–1751) noticed the link between aurorae (natural light displays in the sky) and **Earth's magnetism**. He observed that magnetic "storms" that made a compass needle fluctuate significantly coincided with sightings of the aurorae. Graham's discovery followed a particularly dramatic display of the **aurora borealis**, or northern lights, in 1716 that had fascinated people at the time and was seen as far south as London. Graham also improved the accuracy of **pendulum clocks** by replacing the solid lead weight with a flask of liquid mercury. This eliminated the variations in the length and swing of a solid weight caused by expansion and contraction due to temperature change.

In France, polymath **René Réaumur** (1683–1757) was experimenting with **iron and steel**. He realized that the difference between the metals was caused by their differing **sulfur and salt contents**. Steel, produced by smelting iron, was more brittle than pure iron because it contained sulfur, while cast iron was even more brittle because it contained still higher levels of sulfur. Réaumur discovered that the brittleness of cast iron could be reduced by burying it in lime to draw sulfur out. He believed this method was too expensive to be practical but it later became widely used.

Accurate timekeeping
The mercury pendulum helped eliminate inaccuracies in timekeeping caused by temperature variations with solid weights.

RENÉ ANTOINE FERCHAULT DE RÉAUMUR (1683–1757)

Born in La Rochelle, France, René Réaumur was a naturalist who made contributions to many different fields of science, from the study of insects to ceramics and metallurgy. Elected to the French *Academie des Sciences* (Academy of Science) aged just 24, his greatest work was in natural history, where he showed that some crustaceans can regenerate lost limbs.

620 MILES
THE HEIGHT OF SOME **NORTHERN LIGHT DISPLAYS**

IN PARIS IN 1723, Italian astronomer **Giacomo Filippo Maraldi** (1665–1729) noticed that there was a bright spot in the center of the shadow of any disk. This phenomenon, later called the **Arago spot**, is caused by interference between waves of light coming around the edge of the object. Maraldi's observation later became proof of the theory that **light travels in waves**, not particles, because only waves can produce an interference pattern (see 1801).

Also in Paris that year, naturalist **Antoine de Jussieu** (1686–1758) compared stones called **ceraunia**, thought to be natural, to the stone tools of Native Americans. The likeness proved that ceraunia were ancient axes and arrowheads.

In 1724, Russian emperor **Peter the Great** (1672–1725) founded the St. Petersburg Academy of Sciences and installed Swiss mathematician **Daniel Bernoulli**

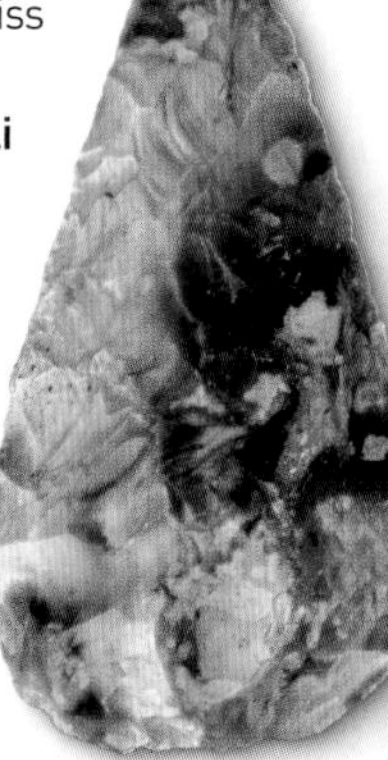

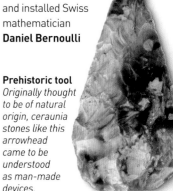

Prehistoric tool
Originally thought to be of natural origin, ceraunia stones like this arrowhead came to be understood as man-made devices.

1721 London buttonmaker **Richard Newsham** patents a **fire pump** in North America

1722 British clockmaker George Graham shows the link between **aurorae and Earth's magnetism**

1722 George Graham introduces the **mercury pendulum**

1722 French polymath René Réaumur shows the importance of sulfur content in iron and steel

1723 Italian astronomer **Giacomo Maraldi** observes the Arago spot

William Ged's stereotype printing process involved making a copy of the typeset page from a mold, so the copy could be used again and again for reprints.

(1700–82) as professor. That year, Bernoulli linked two ancient concepts: the **golden number,** which the Ancient Greeks believed gave perfect artistic proportions, and the **Fibonacci sequence** (see 1200–19). The golden number (approximately 1.618) is the ratio of a rectangle divided in two so that the ratio of the larger piece to the smaller is the same as the ratio of the whole rectangle to the larger. In the Fibonacci sequence, each number is the sum of the previous two numbers. Bernoulli showed that the golden number is in fact the ratio of any Fibonacci number to the previous number in the sequence.

nautilus shell

Golden spiral
In a nautilus shell, the growth factor by which each spiral section increases in size is the golden number.

FROM THE 16TH CENTURY, Lyon in France had been the center of European **silkmaking**. It was here in 1725 that silkmaker **Basile Bouchon** invented a system for setting up the cords on the silk loom. Normally, this was a long and laborious job, but by arranging for threading needles to be raised or not according to holes on a moving roll of paper, Bouchon could **partially automate** the machine. This reduced mistakes, speeding up the process. Bouchon's paper roll paved the way for all programmable machines, including, ultimately, today's computers.

Although by 1700 it was widely accepted that **Earth** is not fixed in position but **moves around the Sun**, it was hard to actually prove. Then, in 1725, English astronomer **James Bradley** (1693–1762) observed the star Gamma Draconis moving in the opposite direction to the way it usually did. This was difficult to explain, but it is said that while sailing on the Thames River Bradley realized the weather vane on the mast sometimes changed direction not because the wind's direction was changing but because the boat was changing its course. In the same way, Bradley surmized, the mysterious change in the direction of the stars, now known as **stellar aberration**, must be caused by the changing motion of Earth. In London this year, Scottish printer **William Ged** (1699–1749) invented the **stereotype**—a copy of an original typeset page made using a mold. This meant that limitless copies could be made from the stereotype without the trouble of laboriously resetting the type.

Meanwhile in China, the *Gujin Tushu Jicheng* (*Collection of Pictures and Writings*) was being printed. It was a vast encyclopedia overseen by the Qing Dynasty emperors Kangxi and Yongzheng. Only 64 copies were ever printed, but it consisted of 10,000 volumes, 800,000 pages, and 100 million Chinese characters.

10 THOUSAND THE **NUMBER** OF VOLUMES IN THE ***GUJIN TUSHU JICHENG***

In 1726, English clergyman Stephen Hales (1677–1761) described how he made the **first measurements of blood pressure** by observing how far blood rose up a tube inserted in the artery of a horse. He measured the **heart's capacity and output** in various animals, and the **speed and resistance of blood flow** in the arteries.

Blood pressure measurement
English clergyman Stephen Hales inserted an 11½ft- (3.5m-) long glass tube into the neck artery of a horse, and held it vertically to see how far the blood rose up the tube.

1723 French naturalist **Antoine de Jussieu** shows that ceraunia stones are ancient tools

1724 St. Petersburg Academy of Sciences is founded

1724 Swiss mathematician **Daniel Bernoulli** links the **golden number and Fibonacci sequence**

1725 British astronomer **James Bradley** observes **stellar aberration**

1725 French silkmaker **Basile Bouchon** builds the first **semiautomated machine**

1725 The *Gujin Tushu Jicheng* (*Collection of Pictures and Writings*) is printed in China

1725 William Ged invents **stereotype printing**

1726 English clergyman **Stephen Hales** describes the first **blood pressure measurements**

1727–28

The Cyclopaedia summarized human knowledge, reflecting the growing belief that people could learn about the world by studying it scientifically.

IN 18TH CENTURY INDIA there was no better symbol of power and enlightenment than knowledge of the heavens, which may be why the **Maharajah of Amber Jai Singh II** had five massive observatories built across his kingdom. The greatest of them was the Jantar Mantar at Jaipur, begun in 1727, which still stands today. **Jantar Mantar means "calculation instrument"** and this site contains the world's largest sundial, the Samrat Yantra, which is accurate to within two seconds. Its significance is as much astrological and religious as it is scientific.

In the same year, English clergyman and naturalist **Stephen Hales** wrote about his experiments on plant physiology in *Vegetable Staticks*. He noticed how plants drew water up through their stems due to **root pressure and transpiration** (the evaporation of water through the leaves). He also suggested that plants absorb food from air using energy from sunlight—an idea eventually leading to our understanding of **photosynthesis** (see 1787–88).

Jaipur's Jantar Mantar
The Samrat Yantra in the Jantar Mantar is the world's biggest sundial, at over 88 ft (27 m), and the Sun's shadow can be seen visibly moving over ½ in every 10 seconds.

In 1728, English physicist **James Bradley** looked at the stars to make one of the first **accurate measurements of the speed of light**. He used **stellar aberration**, the apparent movement of stars caused by Earth's motion, which he had discovered in 1722. Bradley measured the stellar aberration of starlight from a star in the constellation of Draco and calculated the speed of light to be 987,532,800 ft/s (301,000,000 m/s), remarkably close to today's estimate of 983,571,056 ft/s (299,792,458 m/s).

In Paris, French physician **Pierre Fauchard** launched **modern dentistry** in his book *Le Chirurgien Dentiste* (*The Surgeon Dentist*). He introduced fillings and advocated cutting down on sugar to avoid tooth decay.

In London, English writer **Ephraim Chambers** published *The Cyclopaedia* or *A Universal Dictionary of Arts and Sciences*, one of the first great **encyclopedias of knowledge** written in English.

90 FEET
THE HEIGHT OF THE **SAMRAT YANTRA SUNDIAL** NEEDLE AT **JANTAR MANTAR**

1729–30

> " ...AS THE TUBE COMMUNICATED A **LIGHT TO BODIES...** MIGHT [IT] NOT AT THE SAME TIME **COMMUNICATE ELECTRICITY TO THEM...** "
>
> **Stephen Gray, English experimenter,** in *A Letter to Cromwell Mortimer Containing Several Experiments Concerning Electricity*, 1731

FOR MUCH OF HIS LIFE, ENGLISHMAN STEPHEN GRAY worked in the family trade as a dyer, and appears to have been largely self-educated. When he retired in the 1720s he began experimenting with electrical effects. It was the simplicity of Gray's experiments that introduced many people to the phenomenon of electricity. Most significantly, he demonstrated how an **electric charge** could be **transmitted over distances** by showing that it could be conducted through a damp silk thread for hundreds of yards.

In France, the prodigious mathematician and astronomer **Pierre Bouguer** was making key discoveries about the transmission of light. Appointed a professor and lecturer in physics and mathematics when he was just 15 years of age, Bouguer began to study how **light is absorbed by transparent substances** such as the atmosphere. He found that light does not decrease in intensity arithmetically (uniformly) as it passes through the air but geometrically (at an ever-increasing rate).

It is not just the atmosphere that distorts starlight. **Telescopes** of the day suffered from **chromatic aberration**—the blurring and color fringing caused by the fact that a simple, conventional lens cannot focus all the different wavelengths of light at the same point. British inventor **Chester Moor Hall** solved this problem by producing the first **achromatic lenses**.

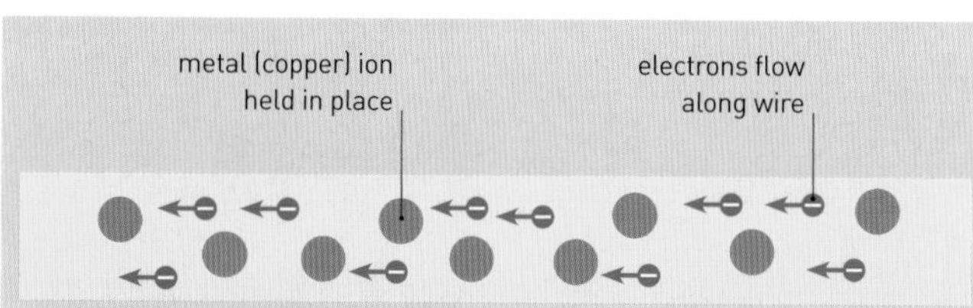
metal (copper) ion held in place
electrons flow along wire

ELECTRICAL CONDUCTION

Electrical conduction is the movement of electrical charge. It is essentially a relay race of electrons (discovered later, in 1897). Electrons are normally attached to atoms, but can sometimes break free. The more easily electrons can break free, the better a substance can conduct electricity, which is why metals such as copper are good conductors.

- **1727** Maharaja Jai Singh II begins construction of **Jantar Mantar observatory**
- **1727 Stephen Hales** publishes *Vegetable Staticks*
- **1728** Pierre Fauchard pioneers **dentistry**
- **July 14, 1728** Danish navigator Vitus Bering enters the Bering Strait
- **1728** English physicist **James Bradley** uses stellar aberration to calculate the speed of light
- **1728** English writer **Ephraim Chambers** publishes *Cyclopaedia*

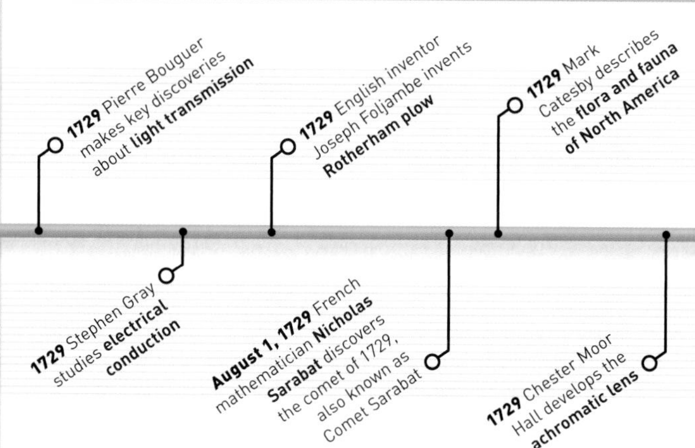

- **1729** Pierre Bouguer makes key discoveries about **light transmission**
- **1729** Stephen Gray studies **electrical conduction**
- **1729** English inventor Joseph Foljambe invents **Rotherham plow**
- **August 1, 1729** French mathematician **Nicholas Sarabat** discovers the comet of 1729, also known as Comet Sarabat
- **1729** Mark Catesby describes the **flora and fauna of North America**
- **1729** Chester Moor Hall develops the **achromatic lens**

Stephen Gray transmitted electricity along damp silken thread.

French engineer Henri Pitot devised the Pitot tube to measure the speed of flow beneath the bridges crossing the Seine River in Paris, France.

Catesby's account of flora and fauna
One bird described by Catesby was the Ivory-billed Woodpecker: one of the world's largest woodpeckers and now probably extinct.

These were fused lenses designed to bring different wavelengths of light to focus together.

In the same year, **Joseph Foljambe** developed a fast, light plow that came to be the standard for the next 180 years. It was called the **Rotherham swing plow** and could be driven by just one man and two horses. The design became so popular it was the first plow ever to be made in a factory.

In North America, British naturalist **Mark Catesby** began to publish the first account of the continent's flora and fauna.

SEVERAL INVENTIONS IN 1731 highlighted the growing scientific interest in measuring the natural world. Italian inventor Nicholas Cirillo created the **first modern seismograph** for measuring the intensity of earthquakes. It consisted of a sensitively balanced pendulum that drew lines on paper as it swung with each tremor, so that the size of the swings recorded their intensity.

Englishman **John Hadley** and American **Thomas Godfrey** independently invented the **octant** to measure **the angle of the stars and Sun at sea** by lining up their image in a mirror with the horizon. The addition of a telescope in 1759 was important and the octant was widely used for navigation.

In that same year, agricultural changes gained impetus with Jethro Tull's book on horse-hoeing husbandry, which showed that crops could be sown every year without a fallow period.

Dutch scientist and physician **Herman Boerhaave** put 18th-century chemistry on a firm footing in his book *Elementa Chemiae* (*Elements of Chemistry*), published in 1732. He emphasized meticulous measurement and helped turn chemistry into a science based on principles. Boerhaave also **founded the science of biochemistry** with his brilliant demonstrations on the chemistry of natural substances such as urine and milk.

In 1732, French hydraulic engineer **Henri Pitot** created the Pitot tube for measuring how fast a river flowed. This right-angled tube could be immersed in a river, pointing into the flow so that the height of the water in the upright of the tube indicated the speed of flow. Pitot tubes are now widely used to **measure airspeed on aircraft**.

5 FT/S
THE FLOW **VELOCITY** OF THE **SEINE** THROUGH **PARIS**

LAURA BASSI (1711–78)

Born to a wealthy Bolognese family, Laura Bassi was patronized in her scientific work by Cardinal Lambertini, the future Pope Benedict XIV. She was appointed professor of anatomy at the University of Bologna in 1731 and professor of philosophy in 1732. Her work introduced Newtonian physics to Italy and broke the ground for many women in science.

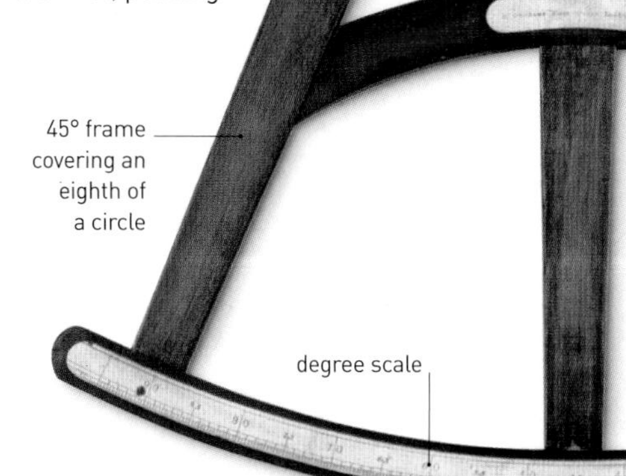
reflector
pivoting sight
45° frame covering an eighth of a circle
degree scale
Octant used in navigation
The octant enabled the angle of the Sun and stars to be measured easily at sea by lining up their reflection in a mirror with the horizon.

1730 French astronomer **Grandjean de Fauchy** shows analemma, the figure-of-eight path of the Sun and satellites

1731 Swiss mathematician **Leonhard Euler** introduces "e" as **base for natural logs**

1731 Italian inventor **Nicholas Cirillo** creates first modern **seismograph**

1731 Dutch scientist **Herman Boerhaave** publishes seminal work on biochemistry

1731 English agriculturalist **Jethro Tull** publishes book on horse-hoeing husbandry

1731 Italian **Laura Bassi** becomes first woman to teach science at a university

1731 John Hadley and Thomas Godfrey independently invent the **reflecting octant**

1732 French hydraulic engineer Henri Pitot develops the **Pitot tube** for **measuring the flow of liquids**

1732 Russians Mikhail Gvozdev and Ivan Fyodorov cross the **Bering Strait** from Russia to Alaska

1733–34

John Kay's flying shuttle was the first of many devices that transformed textile making and led to the Industrial Revolution.

THE MACHINE AGE HAD ITS TRUE BEGINNINGS IN 1733, when English inventor **John Kay** (1704–*c.*1779) designed a **machine to weave cotton**, a material soon cheap enough for the mass market. Kay's semiautomatic loom swiftly wove an important new cloth called broadloom. The machine was christened the **"flying shuttle"** because of its operating speed.

In Paris that year, wealthy experimenter **Charles François de Cisternay du Fay** (1689–1739) was researching electricity by conducting a series of experiments. He observed the difference between **substances that conduct** electricity or heat and those that **insulate**. He also proposed that there are two kinds of electricity—one created by rubbing glass (which he called vitreous electricity) and the other by rubbing resin (resinous electricity). These terms were replaced 15 years later with "positive" and "negative." Du Fay also found that like-charged objects repel and unlike-charged objects attract.

At around the same time, another French aristocrat, **René Antoine Ferchault de Réaumur** (1683–1757), was beginning his great **study of insects**, *Mémoires pour servir à l'histoire des insectes* (*Memoirs Serving as a Natural History of Insects*). This work contained accurate descriptions of the **life and habitats of nearly all insects then known**, and laid the groundwork for the science of entomology.

Philosophers across Europe were questioning established theories. In 1734, English philosopher **Bishop George Berkeley** (1685–1753) criticized **calculus** for the way it never solved the problem of pinning down movement at a single instance—preferring to fudge it instead by calculating it over an infinitely small distance, between what are known as limits.

Swedish philosopher **Emanuel Swedenborg** proposed the idea that the **Solar System formed from a cloud of gas and dust** that collapsed due to gravity, and then began spinning to conserve angular momentum.

925,000

THE NUMBER OF **INSECT SPECIES** KNOWN TODAY

1735–36

" GOD CREATED, **LINNAEUS ORGANIZED.** "

Carl Linnaeus, Swedish botanist

IN 1735, THE GREAT SCIENTIFIC QUESTION was determining the **dimensions of Earth**. **Isaac Newton** held that Earth is not quite spherical: it is fatter around the equator than around the poles, because of Earth's rotation. French astronomer Jacques Cassini insisted it was fatter from pole to pole. With national pride at stake, King Louis XV of France sent off two expeditions to measure an arc (the distance between two points of the same longitude) near the equator and the North Pole. The polar expedition, led by mathematician and biologist **Pierre Maupertuis** (1698–1759), set off to Lapland while the equatorial team, led by naturalist

CAROLI LINNÆI REGNU[M]

I. QUADRUPEDIA. Corpus hirsutum. Pedes quatuor. Feminæ viviparæ, lactiferæ.

II. AVES. Corpus plumosum. Alæ duæ. Pedes duo. Rostrum osseum. Feminæ oviparæ.

III. AMPHIBIA. Corpus nudum, vel squamosum. Dentes molares nulli: reliqui semper. Pinnæ nullæ.

IV. PISCES. Corpus apodum, pinnis veris instructum, nudum, vel squamosum.

PARADOXA

1733 Charles du Fay discovers **electrical charges are of two kinds**

1733 John Kay creates the **"flying shuttle"**

1734 George Berkeley **questions calculus**

1734 Emanuel Swedenborg proposes the **nebular hypothesis**

1735 Swedish naturalist Carl Linnaeus publishes ***Systema Naturae***

May 22, 1735 George Hadley explains the **trade winds**

1735 German mineralogist George Brandt **discovers cobalt**

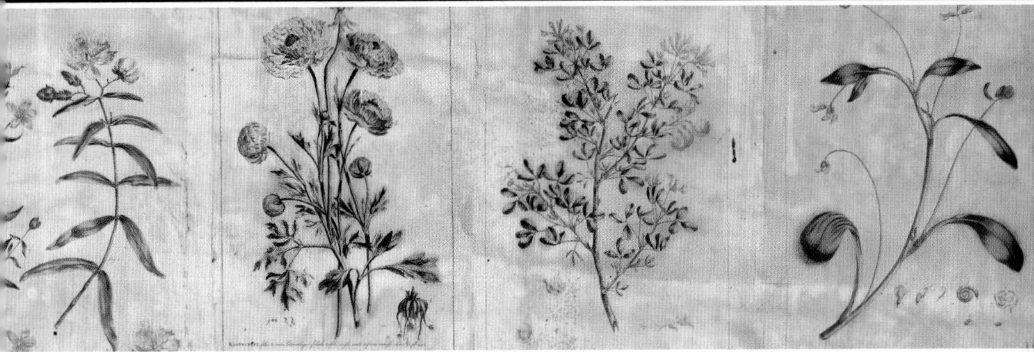
The botanical wallpaper in Carl Linnaeus's former home, now a museum, in Uppsala, Sweden, is a modern replication of the original 18th-century wall covering.

and explorer **Charles Marie de La Condamine** (1701–74), went to Peru and Ecuador. When the teams reported their findings, they proved that Newton, not Cassini, was right—**Earth is fatter at the equator**. Also in 1735, as the French exploratory teams set sail, English meteorologist George Hadley (1685–1768) had a key insight into the **trade winds** that drive ships across the Atlantic: **these winds blow east-west, not straight toward the equator** because they are deflected by Earth's rotation.

This year, too, English clockmaker **John Harrison** (1693–1776) completed the first version of his **marine chronometer**, a clock that could keep time accurately enough at sea to allow **longitude to be calculated**. By 1759, Harrison had produced a fourth, pocket-sized version known as H4 that was of even greater accuracy (see 1759–64).

Three years before Maupertius and his team went to Lapland, Swedish naturalist **Carl Linnaeus** had traveled there to collect plant and bird specimens. It was this trip that planted the seeds for his great scheme for **classifying life**, the *Systema Naturae* (*System of Nature*), first published in 1735. **Linnaeus divided the natural world** into three kingdoms—animal, plant, and mineral—and **subdivided each into class, order, genus, and species**. He introduced the now internationally recognized Latin **binomial** (two-part name) classification system which indicates first the genus and then the species.

NIMALE.

VI. VERMES.

Linnaeus's animal kingdom
In this table from Systema Naturae, *Carl Linnaeus sets out his six subdivisions of the animal kingdom: mammals, birds, amphibians, fish, insects, and worms.*

HADLEY CELL

It is now known that there are three great bands of vertical air circulation, or "cells," on either side of the equator, including the tropical cell named after the meteorologist George Hadley. The east-west deflection of these cells caused by Earth's rotation creates a corkscrew pattern that accounts for prevailing winds at different latitudes.

direction of Earth's rotation
Hadley cell

“ THOSE, WHO PRETEND TO DISCOVER ANYTHING NEW... INSINUATE PRAISES OF THEIR OWN SYSTEMS, BY DECRYING ALL THOSE... ADVANCED BEFORE THEM. ”

David Hume, Scottish philosopher, from *A Treatise on Human Nature*, 1739

LATE IN THE EVENING OF MAY 28, 1737, English physician and astronomer **John Bevis** (1695–1771) witnessed a rare event through a telescope at the Royal Greenwich Observatory in London: a **planetary occultation**—in which one celestial body passes in front of another, temporarily hiding it from view. What Bevis watched was Venus passing in front of Mercury, the only planetary occultation ever recorded.

In Switzerland, mathematician **Daniel Bernoulli** (1700–82) published *Hydrodynamics*, a **study of the flow of water**, based on his work in St. Petersburg, Russia. Bernoulli noted that as the speed of moving fluid increases, the pressure within it decreases—a phenomenon now known as **Bernoulli's principle**. Also in St. Petersburg, **French astronomer Joseph-Nicolas Delisle** (1688–1768) established a method for **tracking sunspots** as they moved across the Sun.

In 1739, French explorer **Jean-Baptiste Charles Bouvet de Lozier** (1705–86) found the **world's most remote island**, now called Bouvet Island, in the South Atlantic Ocean. In France, physicist **Émilie du Châtelet** (1706–49) published her 1739 paper on combustion, in which she predicted the existence of what is now recognized as infrared radiation. In Anjou, France, young Scottish philosopher **David Hume** completed his ***A Treatise on Human Nature***, in which he tried to create a complete psychological profile of man.

CARL LINNAEUS
(1707–78)

Born in Rashult, Sweden, Carl Linneaus was one of the greatest naturalists of his time. A practicing physician, he spent most of his time classifying plants. His students traveled the world, sending back samples and spreading Linnaean theories. He became Professor of Botany at the University of Uppsala in 1741.

1736 Charles Marie de la Condamine **discovers rubber**

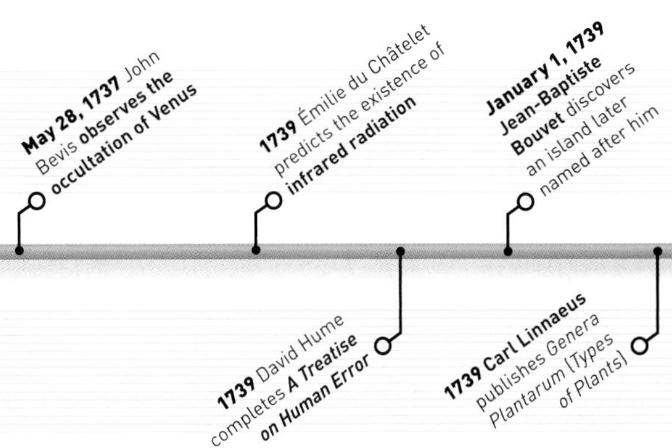
May 28, 1737 John Bevis **observes the occultation of Venus**
1739 Émilie du Châtelet predicts the existence of **infrared radiation**
January 1, 1739 Jean-Baptiste Bouvet discovers an island later named after him
1739 David Hume completes ***A Treatise on Human Error***
1739 Carl Linnaeus publishes *Genera Plantarum* (*Types of Plants*)

1740–42

> **“ THE ANIMAL** NEVER COMES OUT ON SHORE… ITS SKIN **IS BLACK** AND THICK… ITS HEAD… **IS SMALL, IT HAS NO TEETH,** BUT ONLY TWO FLAT WHITE BONES. **”**

Georg Steller, German zoologist, 1740

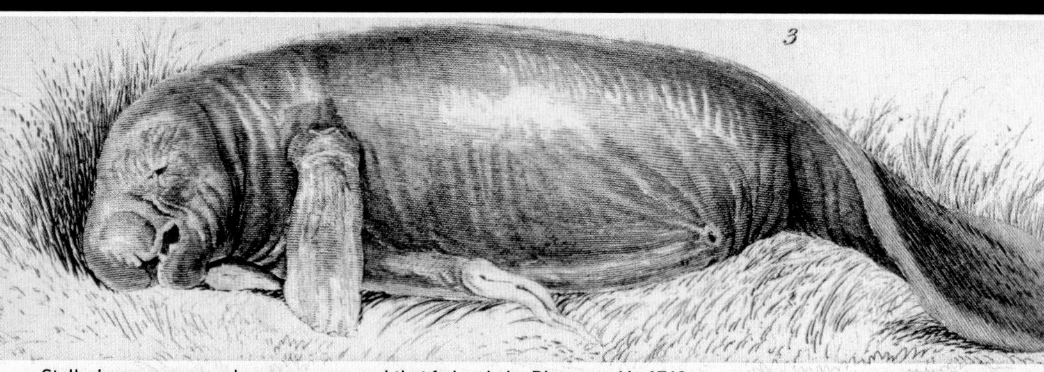

Steller's sea cow was a large sea mammal that fed on kelp. Discovered in 1740 by German naturalist Georg Steller, it was extinct by 1767.

TOUGH AND RESISTANT TO CORROSION, STEEL is a practical metal for construction and machinery. But for thousands of years it was so hard to make reliably in any quantity that it was used only for blades. Then, in 1740, English clockmaker **Benjamin Huntsman** (1704–76) perfected his **"crucible" method of making steel** in Sheffield, England. It involved heating steel to 2912°F (1600°C) in a coke-fired furnace in clay pots or "crucibles" to make ingots of tough steel large and pure enough to cast into many shapes. Huntsman's crucible revolutionized steel making and over the next century, Sheffield's steel production rose from 200 to 80,000 tons per year. This was almost half of Europe's steel.

On June 4, 1740, Danish explorer **Vitus Bering** (1681–1741) launched an expedition to **map the remote Arctic coast of Siberia**. He sailed from Kamchatka in eastern Russia aboard the *St. Peter*, while fellow explorer Aleksey Chirikov (1703–48) sailed aboard the *St. Paul*. The ships became separated and **Bering discovered the Alaskan Peninsula** while **Chirikov found some of the Aleutian Islands**. After Bering fell ill with scurvy, his ship was wrecked on the Aleutians and he died there. Some of his crew built a small boat and returned to Russia with news of fur-trading possibilities that would make Russia rich. Among the survivors was **German naturalist Georg Steller** (1709–46), who had collected **specimens of hitherto unknown species** of wildlife during the expedition. Steller's sea cow, Steller's jay, Steller's sea eagle, and Steller's eider all bear his name. Within 27 years of its discovery, Steller's sea cow had been hunted to extinction.

Developed in the early 17th century by innovators such as the physician **Robert Fludd** and astronomer **Galileo Galilei**, the **thermometer** displayed temperature by showing the level of liquid in a glass tube as it expanded or contracted. But a century on, it had still not been agreed how to calibrate it. Among other suggestions, English physicist and mathematician Isaac Newton proposed a scale with the melting point of snow at one end and the boiling point of water at the other, with points divided by 33 degrees in between. In the end, the winner turned out to be the **temperature scale invented by Swedish astronomer Anders Celsius** (1701–44) in 1742, which developed into the **modern Celsius scale** (known until 1948 as Centigrade). In Celsius's scale, the two end points were set 100 degrees apart, with 100 degrees signifying the freezing point of water, while 0 degrees signified boiling point at standard atmospheric pressure. Two years later, Swedish naturalist **Carl Linnaeus** adopted Celsius's scale for his greenhouse thermometers, but reversed the scale so that 100 degrees was the boiling point of water.

ANDERS CELSIUS (1701–41)

Born in Uppsala in Sweden, Anders Celsius succeeded his father as professor of astronomy at Uppsala University in 1730. He is most famous for devising the temperature scale that now bears his name, but he also helped discover the link between magnetic storms in the Sun and the aurora phenomenon on Earth.

6 THE **NUMBER** OF SPECIES **GEORG STELLER** DISCOVERED DURING THE 1740 VOYAGE

In 1742, French mathematician **Jean Le Rond D'Alembert** (1717–83) found another way to consider Newton's Second Law of Motion, by introducing a fictitious balancing or "equilibrium" force. Known as **D'Alembert's principle,** this **made calculations about dynamic or changing forces simpler** by reducing them to static calculations. Also this year, American inventor and statesman **Benjamin Franklin designed a cast-iron stove** that could be set in the middle of a room to **maximize its heating effect**. Cast-iron stoves soon became immensely popular.

TEMPERATURE SCALES

In the 1700s, many temperature scales were used, including Réaumur's. Now just three are common: the Kelvin (K, introduced in 1848), Celsius (C), and Fahrenheit (F) scales. Each just shows degrees between fixed points. The Kelvin scale starts at absolute zero. One Kelvin is equal to one degree Celsius, so 273.15 K is 0°C, the melting point of water, and 373.15 K is 100°C, the boiling point of water.

	KELVIN	CELSIUS	FAHRENHEIT
	373K	100°C	212°F
	300K	27°C	81°F
	273K	0°C	32°F
	255K	-18°C	0°F
	200K	-73°C	-99°F
	100K	-173°C	-279°F
Absolute zero	0K	-273°C	-460°F

1740 British clockmaker Benjamin Huntsman develops **crucible steel making**

May 1741 Danish explorer Vitus Bering **maps the coast of Alaska and Arctic Siberia**

July 1741 Russian explorer Aleksey Chirikov **discovers the Aleutian Islands**

July 20, 1741 German naturalist George Steller lands in Alaska

1742 Swedish astronomer Anders Celsius proposes a **centigrade temperature scale**

1743–44

In March 1744, six unusual tails were seen shooting above the horizon from the amazingly bright Great Comet of 1744.

IN SPRING 1744, NIGHT SKIES around the world were illuminated by **one of the brightest ever comets**. It was spotted through a telescope late in 1743 by German astronomers **Jan de Munck** and **Dirk Klinkenberg**, and Swiss astronomer Jean-Philippe de Chéseaux—it later became known as the Comet de Klinkenberg-Chéseaux. By the next spring, this rare **"Great Comet"** was so bright that it outshone Venus at night and, for a few weeks in March 1744, it could even be seen by day.

Thanks to new surveying equipment, it became possible to make accurate maps using triangulation—a technique that establishes positions through measuring angles. In 1744, French mapmaker **César-François Cassini de Thury** (1714–84), also known as Cassini III, began a huge project to make the **first accurate map of all France** at the scale of 1:84,600. The project was a landmark in mapmaking.

LA MANCHE
OCEAN
MER MEDITERRANEE

Cassini's map of France
In making the first detailed, accurate map of France, Cassini set out: "To measure distances by triangulation and thus establish the exact position of the settlements."

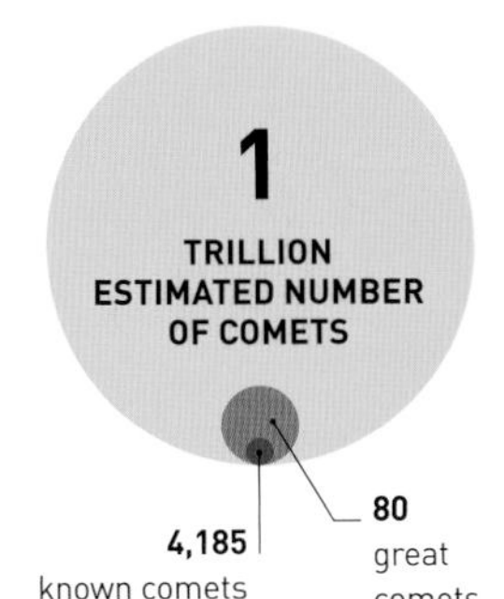

1
TRILLION
ESTIMATED NUMBER
OF COMETS
4,185
known comets
80
great
comets

Comets
Great comets—comets that are exceptionally bright—are seen less frequently than other comets. There are trillions of comets that remain undiscovered to this day.

During this year, Swiss mathematician **Leonhard Euler** (1707–83) was working in Berlin on optics. The clarity of his papers helped ensure that **Huygens' theory that light travels in waves** prevailed over Isaac Newton's theory of light "corpuscles" or particles (see 1675).

1745–46

> **“** THE **SPECIES... TODAY** ARE... **SMALLEST PART** OF WHAT BLIND **DESTINY HAS PRODUCED... ”**
>
> **Pierre-Louis Moreau de Maupertuis,** in *Essai de cosmologie,* 1750

French philosopher Pierre Maupertuis's ideas hinted at later evolutionary theory.

IN 1745, SWISS NATURALIST CHARLES BONNET (1720–93) wrote a **key study on insects**, *Traité d'Insectologie* (*Treatise on Insectology*). In it he noted that **caterpillars breathe through pores**, and that **aphids reproduce by parthogenesis** (without the need for mating).

In France that year, the mathematician and philosopher **Pierre Louis Maupertuis** (1698–1759) was writing *Vénus Physique,* in which he hinted at **ideas that emerged later in evolutionary theory**. He suggested that only those animals made in a way that best met their needs would survive, while those lacking appropriate characteristics would not. Maupertius also suggested that **all life descends from a common ancestor**.

Also this year, French mathematician, **César-François Cassini de Thury** developed the **Cassini map projection**. All map projections are accurate but they distort in various ways. The Cassini projection was accurate at right angles to a central meridian, so was good for local grid-based maps. For this reason, the Cassini projection was used for the well-known **Ordnance Survey maps** of the UK.

In 1746, French mineralogist **Jean-Étienne Guettard** (1715–86) was pioneering another kind of map, showing all the country's surface minerals. It was perhaps the **first ever major geological map**.

> **“** I HAVE FOUND OUT SO MUCH ABOUT **ELECTRICITY** THAT... I **UNDERSTAND NOTHING** AND CAN **EXPLAIN NOTHING. ”**
>
> **Pieter van Musschenbroek, Dutch physicist,** 1746

In the city of Leyden in Holland, German cleric Ewald Georg von Kleist (1700–48) and Dutch physicist Pieter van Musschenbroek (1692–1761)—quite independently of one another—**developed the first device for storing electricity. The Leyden jar stored a static electric charge** between two electrodes on the inside and outside of a glass jar. It was not a battery, because it did not produce a charge itself—instead it stored a static charge built up by friction generators. However, it was a compact way of storing electricity and provided a useful and ready source of charge.

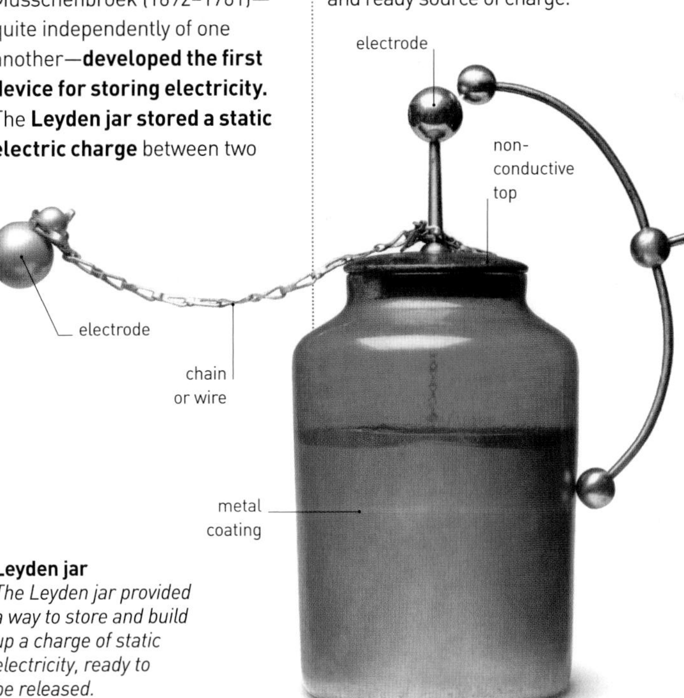

electrode
non-conductive top
electrode
chain or wire
metal coating

Leyden jar
The Leyden jar provided a way to store and build up a charge of static electricity, ready to be released.

1743 German astronomers Jan de Munck and Dirk Klinkenberg discover the **Great Comet**

1744 French mathematician César-François Cassini de Thury begins mapping France

1744 Swiss mathematician Leonhard Euler publishes important papers on **optics**

1745 Swiss naturalist Charles Bonnet publishes a **treatise on insectology**

1745 The **Leyden jar** is developed by two separate inventors, Ewald Georg von Kleist and Pieter van Musschenbroek

1745 French mathematician César-François Cassini de Thury creates the **Cassini map projection**

1745 French philosopher **Pierre Louis Maupertuis** proposes that **all life is descended from a common ancestor**

1746 English pharmacist **William Cookworthy discovers kaolin** in Cornwall, England

1747–48

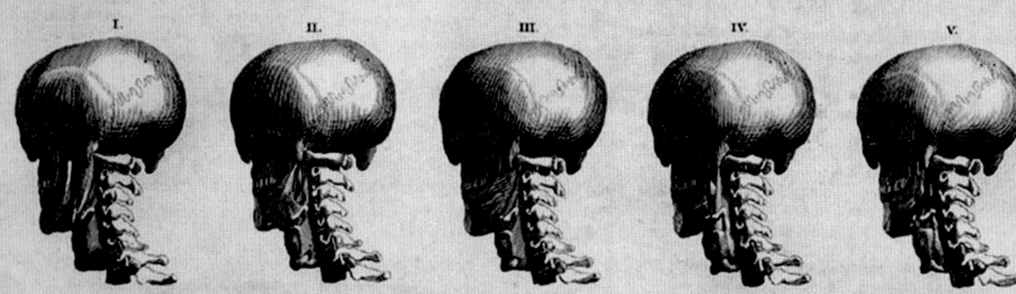

I.
II.
III.
IV.
V.

Bernhard Siegfried Albinus's *Tabulae Sceleti et Musculorum Corporis Humani* contained anatomical drawings of unprecedented accuracy.

AFTER ENGLISH PHYSICIST ISAAC NEWTON'S DISCOVERY of the law of gravitation (see 1687–89) many others became interested in the **gravitational effect of the Moon**. In 1747, French mathematician **Jean le Rond d'Alembert** argued that winds are caused by "tides" in the atmosphere that are driven by the Moon, just like tides in the sea. He was mistaken—winds are really driven by variations in the way the air is warmed by the sun, as warm air rises and cold air rushes in to take its place. However, his work did introduce **partial differential equations (PDEs)**, complex equations involving several variables. Later developed by Swiss mathematician **Leonhard Euler**, PDEs are now used for **calculating how fast one variable changes** when others are held constant, and they are central to calculations about the movement of sound, heat, electricity, and fluids.

Countless sailors were **dying from scurvy** on long voyages. Nobody knew at the time that the illness was caused by a **vitamin C deficiency**, but some people suspected it might be prevented by eating lemons and limes. In 1747, British naval surgeon **James Lind** (1716–94) carried out an experiment to test the effect of different dietary supplements on six pairs of sailors suffering from scurvy. Only the pair fed limes recovered, and we now know that eating **citrus fruit prevents scurvy** because it contains vitamin C.

In 1748, Dutch anatomist **Bernhard Siegfried Albinus** (1697–1770) published an important **study of human anatomy**, entitled *Tabulae Sceleti et Musculorum Corporis Humani* (*Drawings of the Skeleton and Muscles of the Human Body*). The drawings were plotted on grids to ensure their accuracy.

Also in 1748, English physicist **James Bradley** explained an astronomical effect he had been studying for 20 years. This was **nutation**—the way Earth's axis nods slightly with a period of 18.6 years. The Moon's orbit does not lie exactly in the plane of the ecliptic (Earth's orbit around the Sun), so its changing unsymmetrical gravitational pull unbalances Earth's rotation.

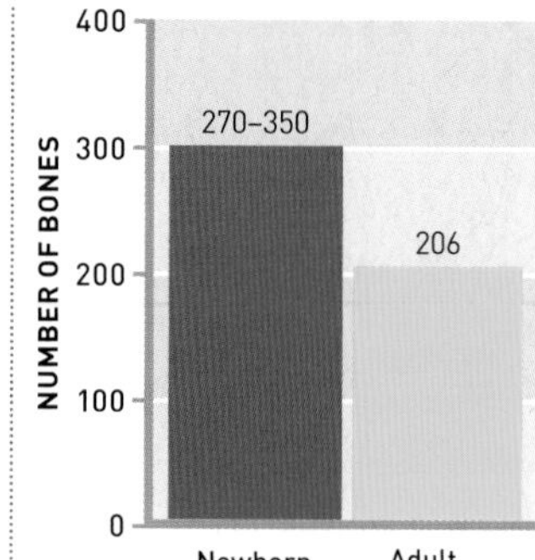
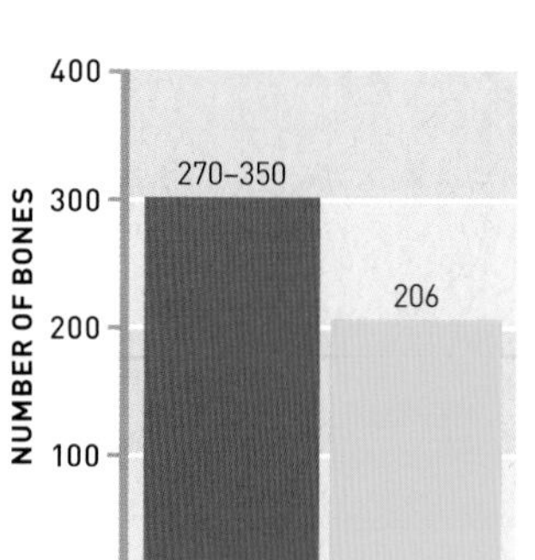

400
300
200
100
0
NUMBER OF BONES
270–350
206
Newborn
Adult

Human bones
Babies have more bones than adults, and some adult bones result from the fusion of bones that are separate in newborns.

Treating scurvy
James Lind's 1747 study proved that citrus fruits prevented scurvy, but it was many years before his ideas were put into practice.

1749–50

English astronomer Thomas Wright described the Milky Way as being shaped like a disk.

IN 1749, FRENCH NATURALIST GEORGES BUFFON (1707–88) began the publication of his study *Histoire Naturelle* (*Natural History*), a 44-volume **study of animals and minerals**. He was one of the first to recognize that the world is very ancient and that many species have come and gone since it was formed. This laid the groundwork for Darwin's theory of evolution a century later (see pp.204–205).

Also this year, Swiss mathematician **Leonhard Euler** turned his attention to the stability of ships in his book *Scientia Navalis* (*Naval Science*). He analysed their **three-dimensional motion** at sea with such mathematical precision that he had to add a third axis to graphs to show

Buffon's *Natural History*
This turkey is one of the many accurately observed drawings in Georges Buffon's important study, Histoire Naturelle, *a work translated into several languages.*

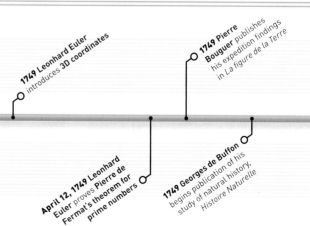

1747 Jean le Rond d'Alembert develops **partial differential equations**

1747 James Lind suggests that **citrus fruit cures scurvy**

1748 English physician **John Fothergill** describes the illness **diphtheria**

1748 James Bradley publishes his explanation of **Earth's nutation**

1748 Bernhard Siegfried Albinus publishes his study of human anatomy

1749 Leonhard Euler introduces **3D coordinates**

April 12, 1749 Leonhard Euler proves **Pierre de Fermat's theorem for prime numbers**

1749 Pierre Bouguer publishes his expedition findings in *La figure de la Terre*

1749 Georges de Buffon begins publication of his study of natural history, *Histoire Naturelle*

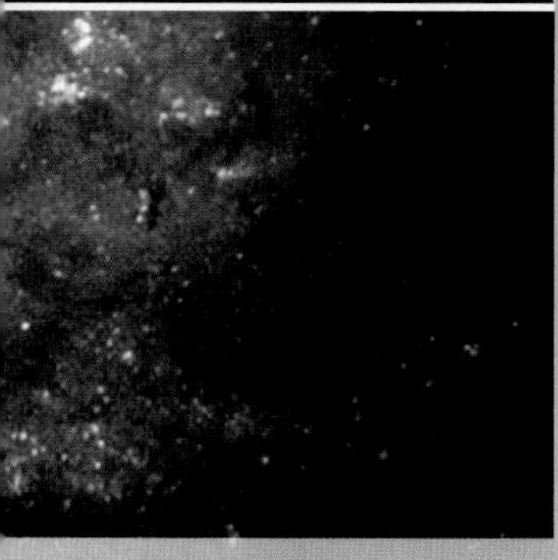

“ THE GOAL OF ENCYCLOPÉDIE IS TO ASSEMBLE ALL THE KNOWLEDGE... OF THE EARTH... SO THAT THE WORK OF CENTURIES PAST IS NOT USELESS. ”

Denis Diderot, French philosopher, in *Encyclopédie*, 1751

This illustration of a state-of-the-art scientific laboratory of the mid-18th century is from Denis Diderot's and Jean d'Alembert's *Encyclopédie*.

variations in depth as well as length and breadth. The positions or coordinates on these three axes, known as x, y, and z, are now central to trigonometry (see 1635–37). Also this year, Euler proved French mathematician Pierre de Fermat's theorem that certain **prime numbers**—numbers that are divisible only by themselves and the number one—can be expressed as the **sum of two square numbers**.

Meanwhile, French hydrographer **Pierre Bouguer** (1698–1758) was embroiled in a dispute concerning the shape of Earth. The French expedition to South America led by Bouguer and Charles de la Condamine in the 1730s (see 1733–39) had helped prove that **Earth's circumference is flattened at the poles**—but the pair disagreed bitterly on the exact results. Bouguer published his claim in *La figure de la terre* (*The Shape of the Earth*), in 1749. De la Condamine published his counterclaim two years later.

In 1750, English astronomer **Thomas Wright** (1711–86) began to think about the shape of the Milky Way, not then recognized as a galaxy. Wright speculated correctly that although we cannot see it because we are in the middle of it, the Milky Way is shaped like a flat disk.

Lightning charge
In Philadelphia in June 1752, experimenter and statesman Benjamin Franklin risked death as he proved lightning is electricity by flying a kite into a thundercloud to draw down the charge.

FRENCH MATHEMATICIAN PIERRE LOUIS MAUPERTUIS (1698–1759) wrote *Systeme de la Nature* (*The System of Nature*) in 1751. In it he discussed how characteristics are passed on from animals to their offspring, later the basis of the science of genetics. His ideas also foreshadowed naturalist Charles Darwin's once discredited ideas on pangenesis, an early theory of heredity now receiving renewed attention.

Also in this year, French philosophers **Denis Diderot** (1713–84) and **Jean d'Alembert** started work on their book that attempted to summarize all knowledge of the time, *Encyclopédie*. It was the first encyclopedia to include work from a variety of named contributors, and it aimed to collate the world's knowledge in one place.

In Edinburgh in 1751, Scottish physician **Robert Whytt** (1714-1766) discovered how the pupil of the eye **automatically opens or closes** in response to levels of light. His **pupil reflex** was the first discovery of a bodily reflex, an automatic response to a stimulus.

Future American statesman **Benjamin Franklin** (1706–90) was intrigued by the similarity between lightning and the sparks in his home demonstrations of electrical phenomena. Franklin became convinced that **lightning is natural electricity** and in *Experiments and Observations in Electricity*, published in London in 1751, he described an experiment to prove his theory. This involved drawing lightning down to a spike on a sentry box.

In May 1752, Frenchman **Jean-Francois d'Alibard** (1703-99) tried Franklin's experiment in France and found that it worked. The following month, Franklin, not yet aware of d'Alibard's success, went out in a summer storm in Philadelphia to fly a **kite** under the clouds to **draw electrical charge** down the line to a key, insulated from the experimenter by a silk ribbon. As sparks streamed off the key, Franklin, like d'Alibard, could see that the **cloud was electrified**.

“ HE SNATCHED THE LIGHTNING FROM THE SKY AND THE SCEPTER FROM TYRANTS. ”

Anne-Robert Jacques Turgot, French economist and statesman, on Benjamin Franklin, in a letter to Samuel P. du Pont, 1778

Also this year, physicist **Thomas Melvill** (1726-53) discovered that when he set different substances aflame, the flame gave **differently colored spectra** when shone through a prism. Salt gave a spectra dominated by bright yellow, for instance. This was the beginning of **spectroscopy,** by which substances are identified by the color of light they emit.

BENJAMIN FRANKLIN
(1706–90)

Born in Boston, Franklin lived most of his life in Philadelphia, where he ran a printing business. He was one of the founding fathers of American independence and became famous for his investigations into the nature of electricity. He also invented the lightning rod and a type of cast-iron stove, and he made studies of the Gulf Stream.

1750 Thomas Wright describes the **Milky Way** as **shaped like a disk**

1751 Robert Whytt discovers the **pupil reflex**

1751 Pierre Maupertuis suggests the theory of **pangenesis**

1751 Denis Diderot and Jean d'Alembert begin work on their *Encyclopédie*

1752 Benjamin Franklin proves lightning is electric

1752 English chemist Thomas Melville pioneers **spectroscopy**

1753–54

K
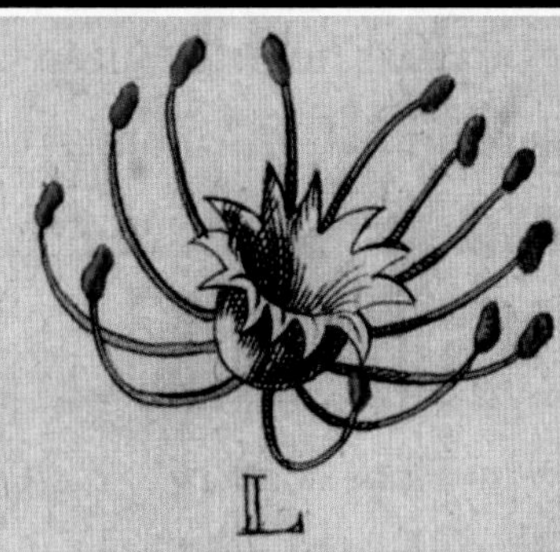
L
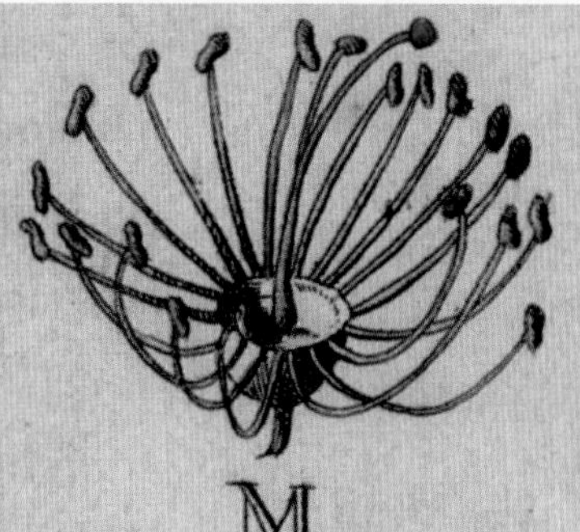
M

Carl Linnaeus's classification of plants focused on their sexual organs, the pistils and stamens, as illustrated here by botanical artist Georg Ehret, who worked with Linnaeus in the 1730s.

AMERICAN INVENTOR AND POLYMATH BENJAMIN FRANKLIN (1705–90) had proved in 1751 that lightning is natural electricity. Some two years later, he demonstrated with his new invention, the **lightning rod**, how buildings could be protected against this hazard. The simple device, still used today, comprised a metal rod placed on top of a building to draw down lightning and conduct it harmlessly to the ground through a metal wire, saving the building from damage. Although many argued that drawing lightning only increased the likelihood of a strike, the idea caught on quickly. But Czech inventor **Prokop Diviš** (1698–1765) independently invented a similar device around the same time, and it was Diviš's design that became more widely used.

British naval surgeon **James Lind** published in 1753 the first edition of ***A Treatise of the Scurvy***, the result of six years research into the benefits of citrus fruit as a preventive of the dread shipboard disease. It was several decades before anyone acted fully on Lind's theory.

Lightning conductor
Franklin's lightning rod initially met with worried opposition, but soon many buildings were sprouting these "property protectors."

Swiss mathematician **Leonhard Euler** (1707–83) addressed the question of how three heavenly bodies, such as the Sun, Moon, and Earth, interact. He approached what is known as the three-body problem in his book ***Theoria Motus Lunae (A Theory of Lunar Motion)***, and eventually found a solution in 1760. Euler also pioneered studies into how the gravitational pull between the Moon and Earth drives tides on Earth. Understanding of such forces was still in its infancy in 1754, when Dutch dike supervisor **Albert Brahms** (1692–1758) began the first scientific recordings of tide levels.

Ruđer Josip Bošković (1711–87) of Dubrovnik, who made significant contributions to at least half a dozen scientific fields, claimed that the Moon has no atmosphere. In fact, we now know that the Moon does have a sparse atmosphere, although not enough to support life as known on Earth. But Bošković's almost-correct theory was a key step in the process of understanding other worlds beside our own.

Scottish chemist **Joseph Black** (1728–99) also found out something about Earth's atmosphere with his discovery of **carbon dioxide**, which he called "fixed air." Black learned that this gas is heavier than air, does not allow flames to burn, can cause asphyxiation, and is exhaled by animals in breath.

During this period, Swedish naturalist **Carl Linnaeus** (see 1737–39) produced his work on classifying plants, ***Species Plantarum (The Species of Plants)***, which covered some 6,000 plants and gave each one a binomial (two-part) Latin name indicating genus and species. This system provides the basis for plant nomenclature used by botanists today.

30,000
ESTIMATED NUMBER OF **AMPS** CARRIED IN A **BOLT OF LIGHTNING**

1755–56

> " ...SYSTEMS OF MANY **STARS,** WHOSE DISTANCE PRESENTS THEM **IN SUCH A NARROW SPACE** THAT THE **LIGHT... REACHES US,** IN A UNIFORM PALE GLIMMER... "

Immanuel Kant, German philosopher in *Universal Natural History and Theory of the Heavens,* 1755

AROUND MIDMORNING ON NOVEMBER 1, 1755, the city of Lisbon in Portugal was devastated by an earthquake now estimated at magnitude 8.5 on the Richter scale (see 1935). After studying the after-effects, British geologist **John Michell** (1724–93) correctly suggested that earthquakes travel as **seismic waves**, which alternately compress and stretch the ground. Michell worked out that the quake's epicenter—the point where the waves that shook Lisbon originated—was in the eastern Atlantic between the Azores and Gibraltar.

British engineer **John Smeaton** (1724–92) improved the stability of buildings with his pioneering use of hydraulic lime, a cement that sets underwater and is resilient to deterioration when wet. Smeaton used hydraulic lime to build the third Eddystone lighthouse off the coast of southwest England.

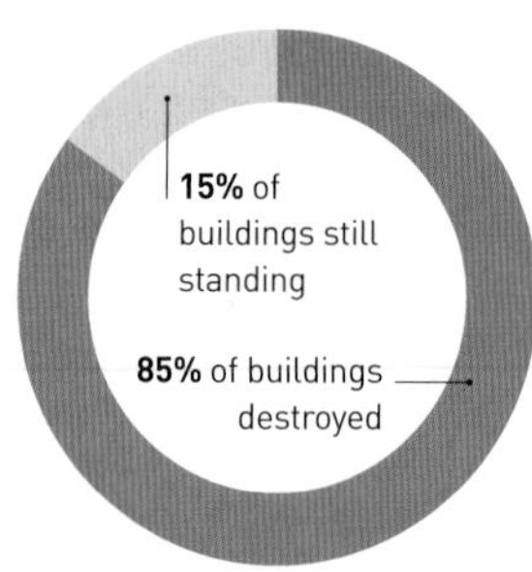
15% of buildings still standing
85% of buildings destroyed

Great quake
Many major buildings and around 12,000 dwellings were destroyed by Lisbon's earthquake of 1755.

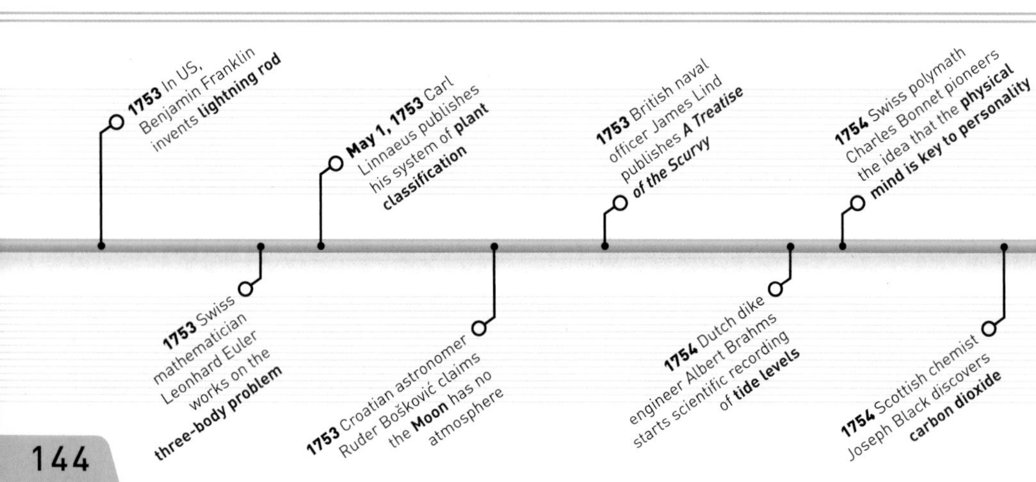
1753 In US, Benjamin Franklin invents **lightning rod**
1753 Swiss mathematician Leonhard Euler works on the **three-body problem**
May 1, 1753 Carl Linnaeus publishes his system of **plant classification**
1753 Croatian astronomer Ruđer Bošković claims the **Moon** has no atmosphere
1753 British naval officer James Lind publishes ***A Treatise of the Scurvy***
1754 Dutch dike engineer Albert Brahms starts scientific recording of **tide levels**
1754 Swiss polymath Charles Bonnet pioneers the idea that the **physical mind is key to personality**
1754 Scottish chemist Joseph Black discovers **carbon dioxide**
1755 German philosopher Immanuel Kant develops **nebular theory** for formation of the Solar System
November 1, 1755 An earthquake devastates Lisbon, and prompts John Michell to develop theory of **seismic waves**

Immanuel Kant believed, as do many scientists today, that the Solar System originated as a cloud of dust between the stars.

“A PARTICLE... OF POINTS QUITE HOMOGENEOUS, **SUBJECT TO A LAW OF FORCES...** MAY EITHER **ATTRACT, REPEL, OR HAVE NO EFFECT...** ON ANOTHER PARTICLE... **”**

Ruder Boškovi´c, Croatian scientist, in *Theory of Natural Philosophy,* 1758

In Russia, polymath **Mikhail Lomonosov** proved the law of **conservation of matter** by showing that when lead plates are heated inside a jar the collective weight of jar and contents stays the same, although the materials have altered. His findings predated by nearly 30 years the similar law formulated by French chemist Antoine Lavoisier (see 1781–82).

German philosopher **Immanuel Kant** (1724–1804) developed the idea of the nebular hypothesis, first suggested by Swedish thinker and visionary **Emanuel Swedenborg** (1688–1772) in the 1730s. This theory postulates that the Solar System originated in a rotating gas cloud that collapsed inward under its own gravity, the matter forming into the Sun and planets.

MIKHAIL LOMONOSOV (1711–65)

Little known in the West, Lomonosov was born a peasant and went on to become a pioneer in physics, chemistry, and astronomy, a poet, and a key thinker in the Russian Enlightenment. His achievements include the discovery of planet Venus's atmosphere, and devising theories for light waves and iceberg formation.

Destruction of a city
Over 30,000 people were killed by the massive earthquake that rocked the Portuguese city of Lisbon in November 1755.

THE IDEA THAT THINGS ARE MADE FROM ATOMS had been important since the early 17th century. But **Ruđer Bošković**, living in Venice at the time, went further and developed his own atomic theory, which he explained in his book ***Theoria Philosophiae (Theory of Natural Philosophy)***. Bošković suggested that matter is built from pointlike particles interacting in pairs.

In France, astronomer **Alexis Claude Clairaut** (1713–65), who had earlier made a name for himself with his theory on why Earth must be flattened at the poles, developed a **theory about comets**. He suggested that Halley's comet, due to reappear in 1759, might be subject to unknown gravitational forces, such as another comet. Clairaut also compiled lunar tables, but they were not as important or as accurate as those made in Göttingen, in Germany, by astronomer **Tobias Mayer** (1723–62) and used to calculate longitude at sea. Also working on longitude calculations at about this time was English clockmaker **John Harrison** (1693–1776), whose H3 chronometer proved precise enough to be used for longitude calculations under all conditions.

In Switzerland, the three engineering **Grubenmann brothers**—Jakob (1694–1758), Johannes (1707–71), and Hans (1709–83)—erected the world's longest road bridges, including a 220 ft- (67 m-) long bridge at Reichenau over a tributary of the Rhine.

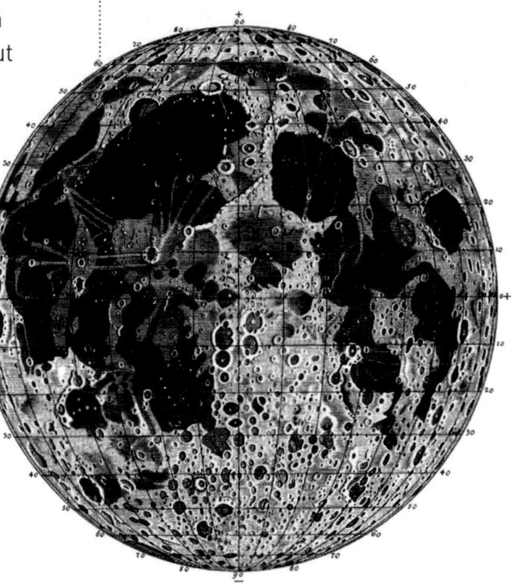
Moon map
German astronomer Tobias Mayer's close study of the Moon resulted in the first maps to show accurately the positions of the craters on the lunar surface.

1756 Russian polymath Mikhail Lomonosov proves the law of **conservation of matter**

1756 British engineer John Smeaton develops **watersetting concrete**

1757 English clockmaker John Harrison completes his H3 **chronometer**

1757 German astronomer Tobias Mayer presents **accurate tables** of the Moon's motion

1758 French astronomer **Alexis Claude Clairaut** suggests Halley's Comet may be steered by another comet

1758 Croatian scientist Ruder Bošković publishes his **atomic theory**

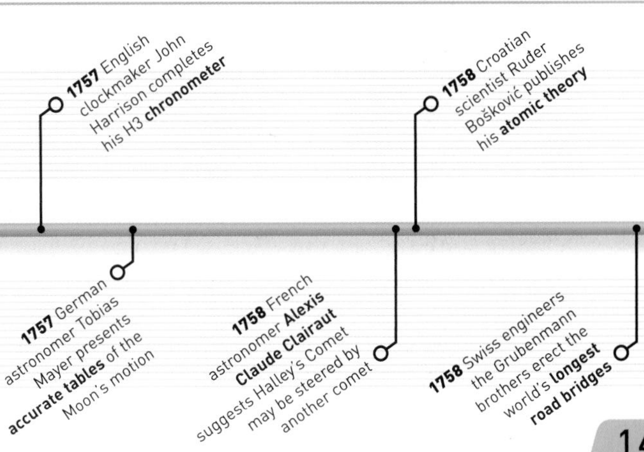

1758 Swiss engineers the Grubenmann brothers erect the world's **longest road bridges**

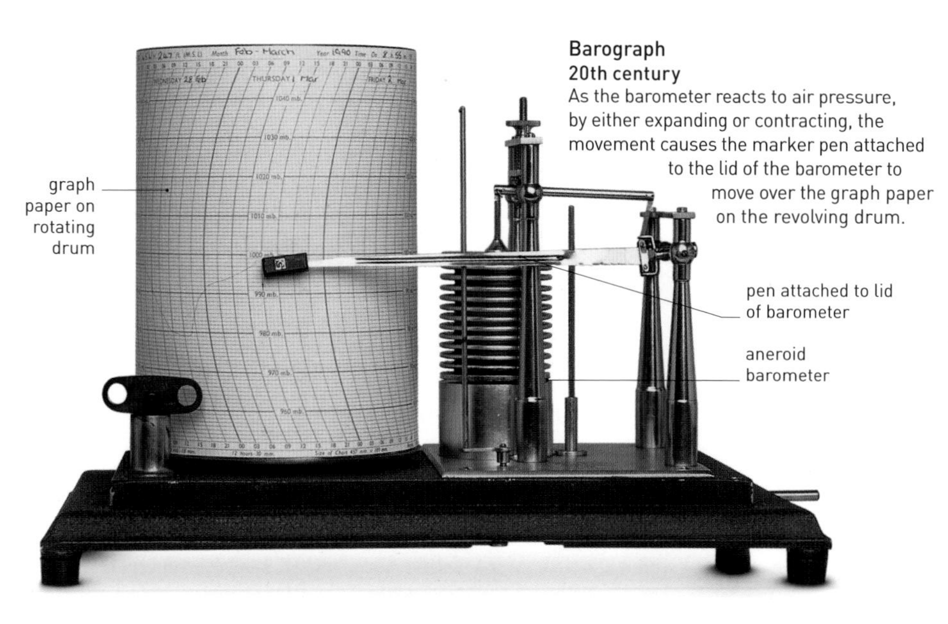
graph paper on rotating drum
pen attached to lid of barometer
aneroid barometer

Barograph
20th century
As the barometer reacts to air pressure, by either expanding or contracting, the movement causes the marker pen attached to the lid of the barometer to move over the graph paper on the revolving drum.

vane veers to show wind direction

rotors turn vane into the wind

METEOROLOGICAL INSTRUMENTS

ACCURATE METEOROLOGICAL MEASUREMENTS PAVED THE WAY FOR WEATHER FORECASTING

People have always sought ways to understand and predict the changes to the weather around them, and this led to the development of devices to investigate the properties of air, such as its temperature and pressure.

The first meteorological instruments were made in 17th-century Italy. At first, tools were created simply to learn about the atmosphere. Thermometers measured changes in temperature; barometers revealed variations in air pressure; anemometers registered wind strength; and hygrometers showed humidity. Gradually, it was realized that these measurements could be used to help predict the weather, and now countless readings from weather stations all around the world are fed into powerful computers to build up weather forecasts.

reflective surface of glass ball focuses Sun's rays
focused rays scorched onto strip of card held here

Parheliometer
1881
Hours of sunshine can be recorded on a parheliometer, in which a glass ball focuses the Sun's rays onto card so that the Sun's passage leaves a scorch trace.

drum held paper on which wind speed was recorded

mercury thermometer

Aneroid barometer
20th century
Air pressure changes, shown on the dial of an aneroid barometer, are a good indicator of weather to come. Falling pressure suggests stormy weather, and steady high pressure heralds clear weather.

air pressure in millibars

Spinning-cup anemograph
20th century
The spinning-cup anemometer was invented in 1846 by Irish astronomer John Robinson to measure wind speed. This "anemograph" records wind speed continuously on a cylindrical chart.

Aneroid barometer/thermometer
20th century
In the 19th century, before broadcast weather reports, a combined thermometer and aneroid barometer, typically housed in a case shaped like a banjo, helped householders make their own weather predictions.

wind speed indicated by how fast cup spins
collection funnel
water runs down funnel and collects in cylinder
dry bulb
wet cloth keeps bulb moist
scale shows humidity
strands of human hair
antenna
propeller wind vane measures speed and direction of wind
sensor measures air temperature
vane to orient buoy into the wind
sensor measures temperature of sea's surface

Soil thermometers
1990s
Right-angled thermometers are used to measure soil temperature at varying depths. This is done to see how deeply frost has penetrated the ground.

Weather calculator
1920s
British meteorologist Lewis Richardson helped develop numeric weather prediction by creating special calculators.

Glass thermometer
1700s
This beautiful thermometer was made by Italian glass blowers. It is filled with alcohol that expands and climbs the spiral when the temperature rises.

Thermometer
1990s
Maximum and minimum temperatures reached each day are recorded on either arm of a double-ended thermometer.

Cotton-reel thermometer
c.1855–77
This desk-top combination instrument features a mercury thermometer and a compass.

Rain gauge
1980s
Rainfall can be recorded simply by the depth of water funnelled down inside a rain gauge, typically mounted 8 in (20 cm) above the ground to avoid splashes.

Sea thermometer
c.1870
This thermometer for measuring sea temperature was used on the HMS *Challenger* oceanographic expedition of 1872–76.

Ocean weather station
1980s
Since the 1970s, floating weather buoys have been used to monitor weather conditions at sea. They move freely with ocean currents and beam back continual measurements via satellite links.

Hair hygrometer
c.1830
This simple way of measuring humidity depended on the ability of a human hair to stretch in moist air and shrink in dry air in a regular and predictable way.

Hygrometer
1836
Evaporation causes cooling, so the temperature difference between two bulbs of a thermometer, one kept wet, one dry, can be compared to calculate humidity.

30,000

THE NUMBER OF **PLANT SPECIES** IN THE LIVING COLLECTIONS AT **KEW GARDENS**

In 1760, the botanic gardens at Kew in London were enlarged to accommodate the many exotic plants brought back from distant lands. This is Kew's great Palm House of 1848.

H4 chronometer
The first practical device for calculating longitude at sea, Harrison's H4 chronometer was like a large pocket watch, 5 in (13 cm) across and weighing 3.2 lb (1.45 kg).

THE PROBLEM OF CALCULATING LONGITUDE AT SEA WAS finally solved in 1759 with a highly accurate clock, or chronometer. Most people had assumed that such a clock would be large and complex. Between 1730 and 1759, English clockmaker **John Harrison** had built three chronometers, all very accurate but not accurate enough. Then Harrison realized the clock did not have to be big. In 1760, he built a **chronometer** the size of a pocket watch. Called H4, it worked astonishingly well, losing just 5.1 seconds in a two-month journey across the Atlantic in 1761.

As mariners sailed farther, they brought **exotic plants** to Europe from across the globe. These were planted in newly created botanical gardens, such as **Kew Gardens** in London, which was greatly enlarged in 1760 by Augusta of Saxe-Coburg, dowager Princess of Wales.

Mariners sailing through the Southern Ocean brought back tales of **giant icebergs**. Russian polymath **Mikhail Lomonosov** suggested that they must have formed on dry land, on a **continent as yet undiscovered**, which later proved to be **Antarctica**. He also suggested that some rocks were much older than others and that the **history of Earth's landscapes** must be long and complex, not simply the result of a few brief catastrophes.

One year previously, Italian geologist **Giovanni Arduino** (1714–95) suggested that the **geological history of Earth** could be divided into four periods: Primitive, Secondary, Tertiary, and Volcanic or Quaternary.

> **"** THE **THEATER OF THE MIND** COULD BE **GENERATED** BY THE **MACHINERY OF THE BRAIN. "**
>
> **Charles Bonnet, Swiss scientist,** from *Essai Analytique sur les Facultés de L'âme* (*Analytical Essay on the Powers of the Soul*), 1760

In 1760, **Charles Bonnet** described what came to be known as **Charles Bonnet syndrome**, a condition in which people with poor eyesight are afflicted with hallucinations. He observed the symptoms in his grandfather and suggested that the visualizations that the mind sees are generated by the physical brain.

CHARLES BONNET (1720–93)

Born near Geneva, Charles Bonnet lived all his life in his hometown. His studies included research on parthenogenesis in insects (reproduction without sex) and the discovery that caterpillars breathe through pores. A naturalist as well as a philosopher, he also pioneered the idea that the mind is the product of the physical brain.

1759 Giovanni Arduino proposes dividing the **geological history of Earth** into four periods

1760 Kew Gardens, London, enlarged by Augusta of Saxe-Coburg

1760 English geologist John Michell suggests **cause of earthquakes**

1760 John Harrison completes **H4 chronometer**

1760 Mikhail Lomonosov suggests that **Earth's landscape** is created by long, **slow natural processes**

1761–62

> **“** THE **HEAT** WHICH **DISAPPEARS** IN THE **CONVERSION OF WATER INTO VAPOR,** IS **NOT LOST. ”**
>
> **Joseph Black, Scottish chemist,** from his *Lectures on Elements of Chemistry*, 1960

Joseph Black became famous for his groundbreaking lectures on the study of heat at Glasgow University, Scotland.

ONE OF THE MILESTONES on the road to the Industrial Revolution was the establishment in 1761 of the **Soho Manufactory** in Birmingham, England. The brainchild of entrepreneur Matthew Boulton, the Soho factory pioneered the **assembly line**, with mass-production of cheap items, such as buttons, buckles, and boxes for the general public. It was here that British engineer **James Watt's steam engines** (see 1765–66) were built a few years later.

An early patron of James Watt, Scottish chemist **Joseph Black** (1728–99) discovered one of the properties of heat. He found that much more heat is needed to melt ice than to warm ice-cold water, just as it takes extra heat to evaporate water. The discovery of this **latent heat**, highlighting the difference between heat and temperature, was fundamental to the modern understanding of energy. It also helped Watt turn the steam engine from an inefficient contraption to the powerhouse of the Industrial Age.

In Sweden, chemist Johan **Gottschalk Wallerius** (1709–85) showed how science could be applied to farming as well as to industry. His pioneering work on **agricultural chemistry**, *Agriculturae Fundamenta Chemica* (*The Natural and Chemical Elements of Agriculture*), discussed the chemical components most conducive to plant growth.

Meanwhile, another Swede, Johan Carl Wilcke (1732–96) designed a **dissectible condenser**—a device to generate static electricity.

In Switzerland, mathematician **Johann Heinrich Lambert** (1728–77) published a treatise in which he put forward his version of the **nebular hypothesis**, the theory that the Solar System developed from a cloud of interstellar dust. English astronomer Thomas Wright and German philosopher Immanuel Kant independently proposed a similar theory. Lambert also suggested rightly that the Sun and nearby stars travel together as a group through the Milky Way.

LATENT HEAT

Substances absorb or release heat when their physical state undergoes change. For a solid to melt to a liquid, it must absorb heat energy; and when a liquid freezes to a solid, it loses heat. This energy is called latent heat. Much more heat energy is needed to change solids to liquids and liquids to gases than to simply raise the temperature of a substance.

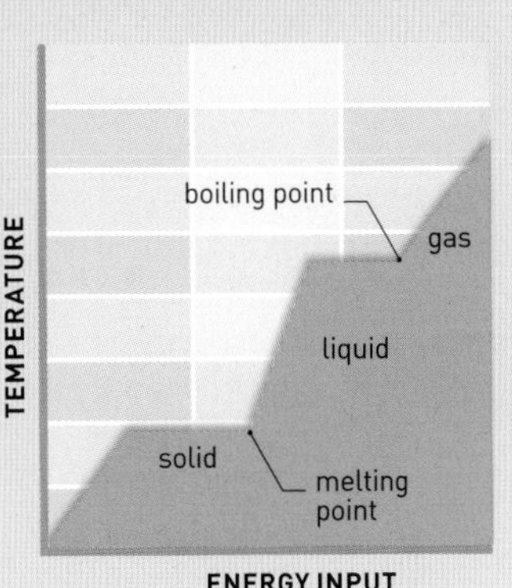
TEMPERATURE
boiling point
gas
liquid
solid
melting point
ENERGY INPUT

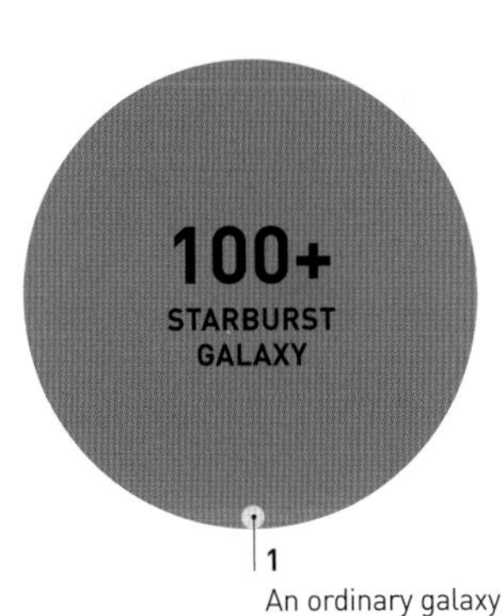
100+
STARBURST GALAXY
1
An ordinary galaxy

Stars made by galaxies per year
The Milky Way, the only known galaxy in the 1760s, makes one new star a year, compared to hundreds in the "starburst" galaxies now recognized.

1763–64

Edward Stone's discovery of the medicinal properties of willow bark was a great breakthrough in palliative medicine.

IN 1763, BRITISH CLERGYMAN EDWARD STONE (1702–68) discovered the **medicinal properties of willow bark**. Stone found, after careful testing, that willow bark dramatically reduced ague, a fever with symptoms similar to malaria. It was later found that the active ingredient in willow bark is salicylic acid, which forms the basis of aspirin (see 1897).

One of the most influential inventions of 1764 was English carpenter and weaver **James Hargreaves'** (1721–78) **spinning jenny**, which could spin cotton eight times faster than a manual worker. The spinning jenny was hand-driven and required some skill to operate, but it was a key step toward automatic, powered machines that could turn out cloth in vast quantities.

In 1764, Italian-born French mathematician **Joseph Louis Lagrange** (1736–1813) showed why the **Moon oscillates, or librates**, continually. He also explained why the same lunar face is always turned toward Earth. This explanation later became the basis of his equations of motion, which provide a simple way of calculating the movement within a system.

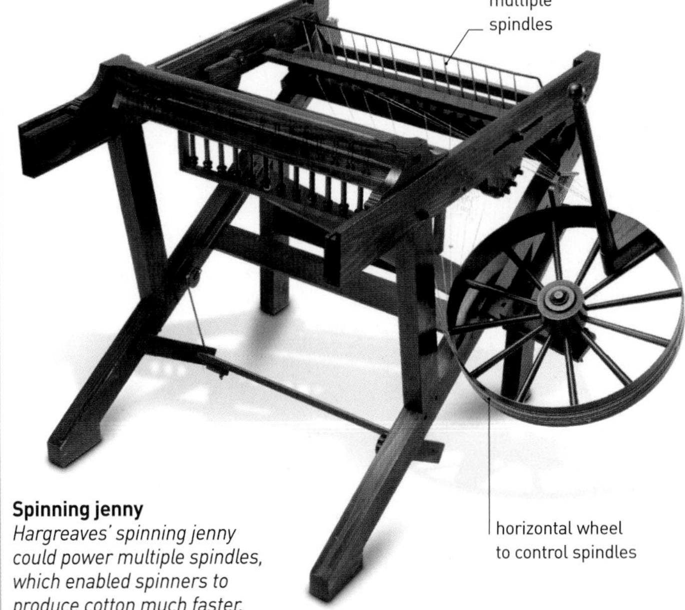
multiple spindles
horizontal wheel to control spindles

Spinning jenny
Hargreaves' spinning jenny could power multiple spindles, which enabled spinners to produce cotton much faster.

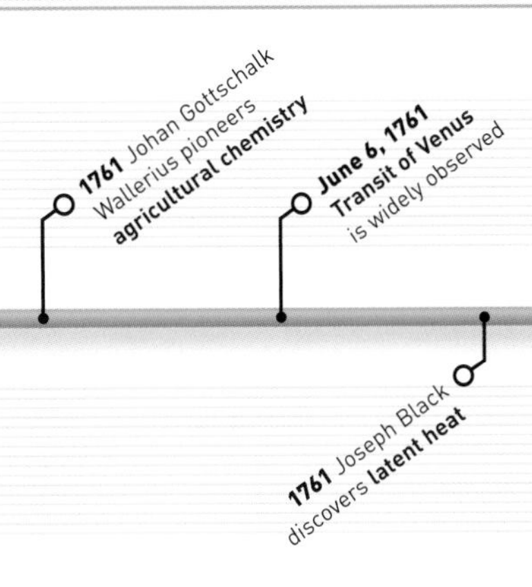

1761 Johan Gottschalk Wallerius pioneers **agricultural chemistry**

June 6, 1761 **Transit of Venus** is widely observed

1761 Joseph Black discovers **latent heat**

1762 Johan Carl Wilcke designs **electrostatic generator**

1764 James Hargreaves invents the **spinning jenny**

1764 Joseph Louis Lagrange shows the **libration of the Moon**

2.02 THE PROPORTION BY WHICH **GANYMEDE'S** MASS IS GREATER THAN THAT OF **EARTH'S MOON**

Mathematician Joseph Louis Lagrange calculated the motions of the four moons of Jupiter known to 18th-century science: Io, Europa, Ganymede, and Callisto.

IN 1765, SECLUDED IN HIS WORKSHOP IN THE UNIVERSITY OF GLASGOW, young Scottish engineer **James Watt** (1736–1819) was working on improving Thomas **Newcomen's steam engine**, which was still only in small-scale use. In Newcomen's engine, steam was let into the cylinder to push the piston up, then cold water was sprayed in and the steam condensed, creating a vacuum that pulled the piston down. The cold water wasted a great deal of heat, so Watt added a **separate component to condense steam outside the cylinder** and avoid continuous chilling. When Watt teamed up in the 1770s with Birmingham industrialist **Matthew Boulton** (1728–1809) to put his improvements into practice, Watt's invention transformed the steam engine from a pump of limited use to the source of power that drove the Industrial Revolution.

Boulton established the Lunar Society of Birmingham, a small group of pioneering thinkers that included Erasmus Darwin, Josiah Wedgwood, and **Joseph Priestley** (1733–1804). Later Watt, too, joined them. Priestley and chemists Joseph Black and **Henry Cavendish** (1731–1810) became known as the "pneumatic chemists" for their work on air and gases. In 1765, Cavendish discovered that **hydrogen**, which he called "inflammable air," is an element, and can be made by dissolving metals in acid.

In 1766, Swiss mathematician **Leonhard Euler** accepted a new post at the Academy of Sciences in St. Petersburg, where he had spent much of his earlier career. During this year he developed key **equations for the movement of rigid bodies**—that is, any objects that keep their shape, such as a planet. Italian-born French mathematician **Joseph Louis Lagrange** (1736–1813) took over Euler's position at the Prussian Academy of Sciences in Berlin, writing a paper on the **motion of Jupiter's moons**, only four of which were then known. It was a century later that they were given the names Io, Europa, Ganymede, and Callisto.

In Kurume, Japan, **Arima Yoriyuki** (1714–83) **calculated pi** to 29 decimal places.

beam
piston
cylinder
flywheel

Industrial power
James Watt's revolutionary engine, incorporating a double-acting cylinder that allowed steam to escape to a separate condenser, was a key development for industry.

JAMES WATT
(1736–1819)

Born on January 19, 1736 in Greenock, Scotland, engineer James Watt was one of the greatest-ever inventors. His improvements to the steam engine transformed it from erratic mine pump to factory powerhouse. His many other inventions included a machine for duplicating sculptures and the world's first copying machine.

1765 British engineer James Watt develops his **steam engine** with external condenser

1765 Swiss mathematician Leonhard Euler develops his theory on the **motions of rigid bodies**

1766 French mathematician Joseph Louis Lagrange analyzes the **movements of Jupiter's moons**

1766 Japanese mathematician Arima Yoriyuki **calculates pi** to 29 decimal places

1766 English chemist Henry Cavendish discovers **hydrogen**

1767–68

“ THE ELECTRIC SUBSTANCE WHICH SEPARATES THE **TWO CONDUCTORS,** POSSESSING THE TWO OPPOSITE KINDS OF ELECTRICITY, IS SAID TO BE **CHARGED.** **”**

Joseph Priestley, English clergyman and scientist, from *The History and Present State of Electricity,* 1767

WHILE A CLERGYMAN IN LEEDS, FROM 1667–1773, JOSEPH PRIESTLEY experimented with chemistry and electricity, and published a highly successful book, ***The History and Present State of Electricity***, in 1767. Priestley noticed that **carbon dioxide**, or "fixed air" as it was then called, was a by-product of the fermenting beer vats of the local brewery. Experimenting with making fixed air himself, Priestley discovered that water impregnated with bubbles of carbon dioxide was pleasantly tangy. He passed on his discovery in the false hope that carbonated water could be used medicinally to prevent scurvy among sailors. In 1783, Swiss watchmaker Johann Jacob Schweppe (1740–1821) used the idea to launch the world's carbonated drinks industry.

Italian naturalist **Lazzaro Spallanzani** (1729–99), the first to suggest that food might be preserved in airtight containers, discovered in 1767 that microbes are ever-present in the air and multiply naturally, instead of bursting into life spontaneously as was commonly believed. Most of the credit for this discovery would go to Louis Pasteur (see 1857–58), who demonstrated the same thing a century later.

On a more theoretical level, Leonhard Euler made the insightful suggestion that the **color of light is determined by its wavelength**.

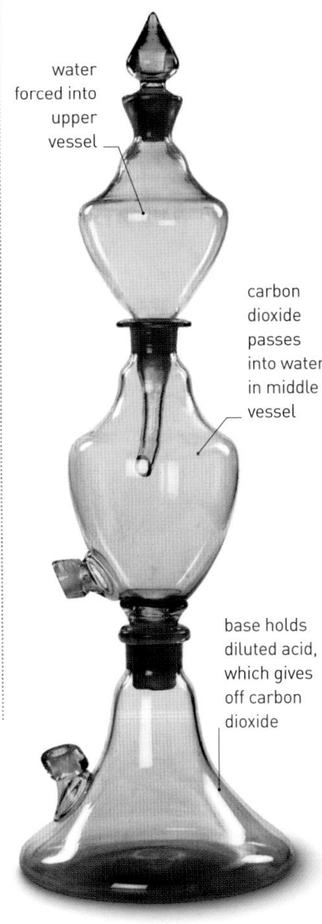
water forced into upper vessel
carbon dioxide passes into water in middle vessel
base holds diluted acid, which gives off carbon dioxide

Nooth's apparatus
This apparatus was built by John Mervyn Nooth (1737–1828) in 1774 to make carbonated water for medicinal purposes in the way suggested by Priestley.

1769–70

French engineer Nicolas-Joseph Cugnot built his steam carriage to haul guns, but it was far too heavy to be practical, and may have crashed into a wall as this old print shows.

30,000

THE NUMBER OF PLANT SPECIMENS **JOSEPH BANKS** DISCOVERED IN **BOTANY BAY**

THE YEAR 1769 SAW THE CREATION OF MACHINES that would lead to the development of two key technologies. One could be called the forerunner of the automobile: the **steam-carriage** built by **French military engineer Nicolas-Joseph Cugnot** (1725–1804). The second, full-scale version of Cugnot's carriage, built in 1770, had three wheels and a large copper boiler hanging over the front wheel. This machine could run for only 20 minutes and was so heavy and so impossible to steer or stop that it is said to have run out of control and demolished a wall.

The other revolutionary breakthrough was the **powered factory machine**, developed by **Richard Arkwright** (1732–92). It is often said that Arkwright's **spinning machine**, which he commissioned clockmaker John Kay to build in 1769, was his greatest invention. This could automatically spin strong cotton yarn from frail English thread. Arguably Arkwright's most inspired idea was to install scores of these machines in a specially built factory, where they were driven not by manpower but by water wheels—hence they were known as **water frames**. Arkwright's revolutionary water-frame mill opened at Cromford on the Derwent River in Derbyshire, England, in 1771.

Across the other side of the world, British naval officer **Captain James Cook** (1728–79) was commanding HMS *Endeavour* on the first of his great voyages of discovery. He and his crew became the first Europeans to encounter the eastern coastline of Australia. On board was botanist **Joseph Banks**, who found 30,000 specimens of plant life, 1,600 of them unknown to European science, when they landed in a harbor later called Botany Bay. On June 3, 1769, Cook and his crew observed the **transit of Venus**—the passage of the planet in front of the Sun—from Tahiti. The event was also witnessed by astronomers all over the world. That year, French astronomer **Charles Messier** (1730–1817) made the first record of the vast cosmic cloud known as the **Orion Nebula**, a region in the constellation of Orion where new stars are formed.

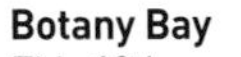

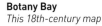

Botany Bay
This 18th-century map of Botany Bay was engraved from charts made by James Cook during his exploration of Australia's east coast.

861,030 miles
DIAMETER OF THE SUN

7,521 miles
diameter of Venus

Comparative sizes
Venus is tiny compared with the Sun, so when it is observed in transit the planet appears as just a small dark dot crossing the face of the Sun.

1767 English chemist Joseph Priestley invents **carbonated water**

1767 Italian naturalist **Lazzaro Spallanzani** challenges view that microbes generate spontaneously

1768 Leonhard Euler proposes that the color of light depends on wavelength

March 4, 1769 French astronomer Charles Messier first records the **Orion Nebula**

June 3, 1769 Transit of Venus observed worldwide, and seen by James Cook in Tahiti

1769 British inventor Richard Arkwright's **water-powered spinning frame** built by John Kay

1769 French engineer **Nicolas-Joseph Cugnot** builds his steam carriage

1770 James Cook's expedition lands in Botany Bay, Australia

> **“** I HAVE... DISCOVERED AN **AIR** FIVE OR SIX TIMES **AS GOOD** AS **COMMON AIR. ”**

Joseph Priestley, English chemist, on his discovery of nitrous oxide, in *Experiments and Observations on Different Kinds of Air,* 1775

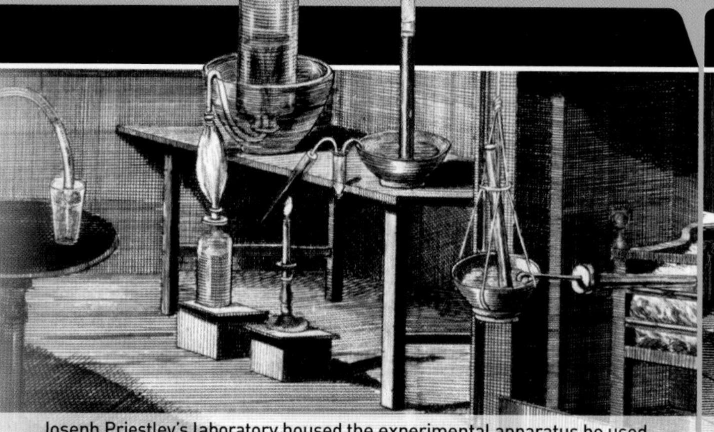

Joseph Priestley's laboratory housed the experimental apparatus he used for the investigation of gases.

Swiss physicist Georges-Louis LeSage u[ses] his pioneering electric telegraph.

IN 1771, GERMAN NATURALIST PETER SIMON PALLAS (1741–1811) started sending reports back from his six-year expedition to eastern Russia and Siberia, then almost as unknown to Europeans as South America. Pallas identified many **new species of plants and animals**, including what is now known as Pallas's cat.

In 1772, Swiss physicist and mathematician **Leonhard Euler** showed that the number 2,147,483,647 is a **Mersenne prime number**—a number that is one less than twice any prime number. It is named after Marin Mersenne, a French monk who had studied prime numbers in the 17th century. It was the **largest prime number** that had been discovered for over a century. The proof involved 372 hand-calculated divisions.

Euler's fame across Europe was confirmed with the 1772 publication of his *Letters to a German Princess*. Apparently written at a rate of two per week between April 1760 and May 1763, the letters were **lessons in elementary science** for the young Prussian Princess Louise Henriette Wilhelmine of Anhalt Dessau. This 800-page work covered a wide range of scientific topics including light, color, gravity, astronomy, magnetism, optics, and more.

Also this year, Leonhard Euler's successor at the Berlin Academy, Italian-born French mathematician **Joseph Louis Lagrange,** began an important work on mechanics, entitled *Mécanique Analytique* (*Analytical Mechanics*). From 1766 onward, Lagrange also produced a series of works on how astronomical movements could be calculated using mathematics. In particular, he looked at the **three-body problem**—how three bodies moving in space, such as the Sun, Earth, and Moon, affect each other gravitationally. This was a problem that had fascinated mathematicians for 90 years. Lagrange discovered that there are five places where a small body (such as the Moon) can remain in equilibrium with respect to two larger bodies (such as the Sun and Earth). These are now known as Lagrangian points.

In England this year, chemist **Joseph Priestley** discovered **nitrous oxide**, which he called "phlogisticated nitrous air." It was later used as a sedative, and became known as laughing gas because of the euphoric effects of inhaling it. The same year, Scottish physician **Daniel Rutherford** (1749–1819) **isolated nitrogen gas** in air for the first time. He called it "noxious air" because he found that mice could not survive when confined to a space filled with the gas. He also discovered that the gas did not support combustion.

Wild cat
Pallas's cat is a small, now-endangered wild cat that lives in the mountains of central Asia. It was one of the species first identified by German naturalist Peter Pallas.

LEONHARD EULER (1707–83)

Born in Basel in Switzerland, Leonhard Euler spent most of his life in St. Petersburg and Berlin. He is credited with laying the basis of modern mathematical notation and made great advances in calculus and graph theory. His output in all areas of science was prodigious. Although in later life he was totally blind, he continued to perform calculations mentally.

> **“** ...**NOTHING AT ALL** TAKES PLACE IN THE **UNIVERSE** IN WHICH SOME RULE OF **MAXIMUM** OR **MINIMUM** DOES NOT APPEAR. **”**

Leonhard Euler, Swiss mathematician, from *Methodus Inveniendi Lineas Curvas* (*Method for Finding Curved Lines*), 1744

SCIENCE HISTORY IS BESET WITH DISPUTES over who discovered things first, and one of these disputes involves the discovery of oxygen. Because of the part it played in **combustion,** and in particular the widely held phlogiston theory of combustion (see 1702–03), many chemists were investigating it in the 1770s. The first to announce its discovery was **Joseph Priestley**, who in 1775 described an experiment he conducted on August 1,1774 in which he **created oxygen gas** (which he called "dephlogisticated air") by focusing sunlight on mercuric oxide inside a glass tube. But in 1777, Swedish pharmacist **Carl Wilhelm Scheele** (1742–86) described how he had already produced oxygen gas by heating

DISCOVERY OF OXYGEN

Oxygen is, with nitrogen, one of two main parts of air, vital for human life and for combustion. Priestley showed that without oxygen candles will not burn and mice cannot breathe. Later, in 1777, French chemist Antoine Lavoisier went on to prove that it was oxygen involved in combustion, not phlogiston as then widely thought.

- **1771** German naturalist **Peter Simon Pallas** reports on Siberian wildlife
- **1772** Swedish chemist **Carl Wilhelm Scheele** discovers **oxygen gas**
- **1772** French mathematician **Joseph Louis Lagrange** describes the **three-body problem**
- **1772** English chemist **Joseph Priestley** synthesizes **nitrous oxide**
- **1772** Scottish physician **Daniel Rutherford** identifies **nitrogen**
- **1772** Swiss mathematician **Leonhard Euler** finds the largest then known **prime number**
- **1772 Leonhard Euler** publishes *Letters to a German Princess*
- **1773** English Captain **James Cook** crosses the Antarctic Circle
- **1773** French astronomer **Charles Messier** discovered a number of nebulae, drawing attention to non-stellar objects in the sky

26 THE NUMBER OF **INSULATED WIRES** IN LESAGE'S **TELEGRAPH**

Scottish chemist James Keir realized that the strange columns of Northern Ireland's Giant's Causeway are made from molten lava.

mercuric oxide and various nitrates in 1772. French chemist **Antoine Lavoisier** (1743–94) later claimed to have first discovered oxygen and gave it its name—but both Priestley and Scheele had already told him about their discoveries.

American inventor **David Bushnell** (1742–1824) built an underwater explosive device in 1773 to help the American Revolutionary War effort. In 1775, he held trials of the vessel intended to deliver this device, the world's first working submarine. Known as the Turtle, but shaped like a lemon, it had an airtight hatch at the top for the submariner and was maneuvered by hand-cranked propellers and rudder.

The following year, **Carl Wilhelm Scheele** discovered a gas for which he is credited—**chlorine**. He called his discovery "dephlogisticated muriatic acid air" because it came from muriatic acid, now known as hydrochloric acid.

In Geneva, Switzerland, in 1774, physicist **Georges-Louis LeSage** (1724–1803) created an early form of electric telegraph. It had a separate wire for each of the 26 letters of the alphabet and could send messages between rooms.

vertical propeller for submerging
ventilation pipe
hatch

First submarine
Bushnell's Turtle submarine (shown here as a model) had a trial run in 1775. It successfully laid explosives in the Hudson River.

BY 1775, ELECTRICITY was the subject of numerous experiments and this year Italian physicist **Alessandro Volta** (1745–1827) developed the **electrophorus**. It consisted of a metal disk with an insulating handle that could be given a **static charge** by holding it against a disk of **resin** that had already been given a charge by being rubbed with fur or wool. It was a simple and effective way of magnifying and accumulating an **electric charge**.

Also this year, Swedish naturalist **Peter Forsskål's** (1732–63) studies of Middle Eastern fauna were published posthumously; Forsskål had died of malaria while specimen hunting in Yemen.

In France, future engineer **Pierre-Simon Girard** (1765–1836), just ten years old, invented a **water turbine**. He went on to complete important work on fluid mechanics.

Another key area of research was investigating **how Earth's rocks and landscapes were formed**. In 1776, German geologist Abraham Werner (1750–1817) insisted, incorrectly, that all rocks settled out of a ocean that once covered the Earth—a theory called Neptunism. Plutonists, however, believed that rocks were formed by **volcanic processes**. Plutonist **James Keir** (1735–1820), a Scottish chemist, realized that the interlocking basalt columns of formations like the Giant's Causeway in Northern Ireland were formed by the **crystallization of molten lava**.

40,000 THE NUMBER OF **BASALT COLUMNS** THAT MAKE UP THE **GIANT'S CAUSEWAY**

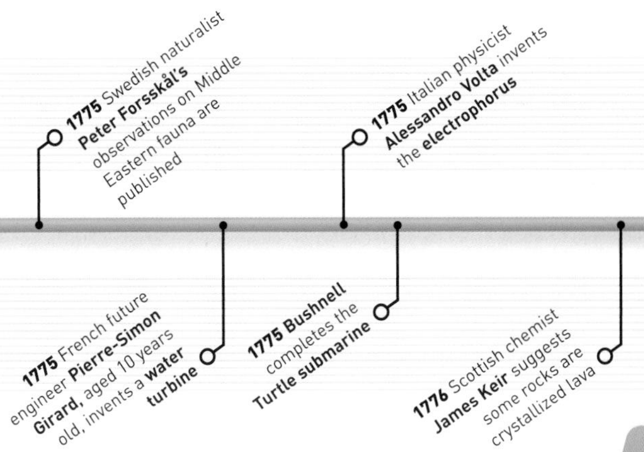
insulating handle
metal disk
wax or resin dish

Electrophorus
Alessandro Volta's device was a simple and convenient way to magnify and accumulate electric charge.

1773 American inventor **David Bushnell** builds an explosive device

1774 Swedish chemist **Carl Wilhelm Scheele** discovers **chlorine, manganese, and barium**

1774 Swiss physicist **Georges-Louis LeSage** creates the **first electric telegraph**

August 1, 1774 English chemist **Joseph Priestley isolates oxygen**

1775 Swedish naturalist **Peter Forsskål's** observations on Middle Eastern fauna are published

1775 French future engineer **Pierre-Simon Girard**, aged 10 years old, invents a **water turbine**

1775 Italian physicist **Alessandro Volta** invents **the electrophorus**

1775 Bushnell completes the **Turtle submarine**

1776 Scottish chemist **James Keir** suggests some rocks are crystallized lava

1777–78

1779–80

> **"** MY **SIGHT BECOMES DISORDERED...** LIKE A PERSON WHO HAS LOOKED AT THE SUN... THE **PAIN IN MY HEAD** COMMENCES... WITH **GREAT SEVERITY... "**

Samuel-Auguste Tissot, Swiss physician, record of a patient's symptoms from *Treatise on the Nerves and Nervous Disorders*, 1783

Designed by Thomas Pritchard and built by Abraham Derby III, the bridge at Ironbridge, which crosses the Severn River in Shropshire, England, was fabricated entirely from precast pieces of iron.

MATHEMATICIANS HAD LONG BEEN BAFFLED BY THE PROBLEM OF finding the roots of negative numbers, calling them "imaginary numbers." Then, in 1777, Swiss mathematician **Leonhard Euler** introduced the imaginary unit, the **symbol *i***, which gives –1 when squared. Euler's insight meant the square root of any negative number could be included in equations as *i* times the square root of the number.

In 1777, London clockmaker **John Arnold** (1736–99) created a **watch of unprecedented accuracy** for calculating longitude at sea, improving on Harrison's H4 of 1759. Arnold named such timepieces "chronometers."

Swiss physician **Samuel-Auguste Tissot** (1728–97) described **migraine**. Although Tissot wrongly thought migraines began in the stomach, he described the symptoms very accurately—the severity of the pain, its recurrence, the suddenness of onset, the effect on vision, and vomiting.

Scottish surgeon **John Hunter** (1728–93) wrote an important **study of teeth**. He also advocated transplanting good teeth from donors to replace rotten ones, but transplanted teeth were rejected by the recipient's immune system.

Tooth transplant
This cartoon satirizes John Hunter's practice of buying healthy teeth from the poor to implant in place of lost teeth in the mouths of the rich.

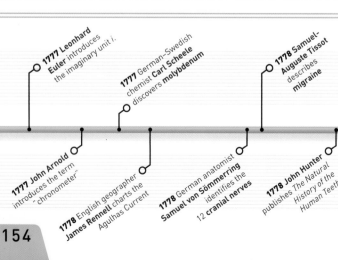

WITH A SIMPLE EXPERIMENT IN 1779, Dutchborn physician **Jan Ingenhousz** (1730–99) discovered the essence of **photosynthesis**, the chemical reaction by which plants make food from sunlight (see 1783–88). When Ingenhousz set plants under water in a glass container, he saw gas bubbles form on the undersides of leaves, which he showed was oxygen. But the bubbles formed only in sunlight, not in darkness. Ingenhousz found that **plants need sunlight for respiration**—when they take in gases from the air to make glucose for energy—and oxygen is released as a waste product.

Factory production of cotton was accelerated in 1779 when English inventor **Samuel Crompton** (1753–1827) created an **ingenious hybrid, or mule**, combining a spinning machine with a weaving machine to produce finished cloth from raw thread. Also in 1779, large-scale construction using iron began with the **bridge at Ironbridge** in England, designed by **Thomas Pritchard** (*c.*1723–77) and built by Abraham Derby III (1750–91).

In 1780, in the memoir *On Combustion in General*, French chemist **Antoine Lavoisier** finally demolished the theory that burning materials lose a substance called phlogiston. By meticulous weighing before and after burning, Lavoisier showed that substances can gain weight when they burn—by combining with oxygen in the air—not lose it. Together with **Pierre-Simon Laplace** (1749–1827), a French mathematician, Lavoisier invented the **ice calorimeter** for measuring the heat produced by chemical changes. This was the beginning of the science of thermochemistry. With his ice calorimeter, Lavoisier demonstrated that animals produce heat without losing weight, so disproving the

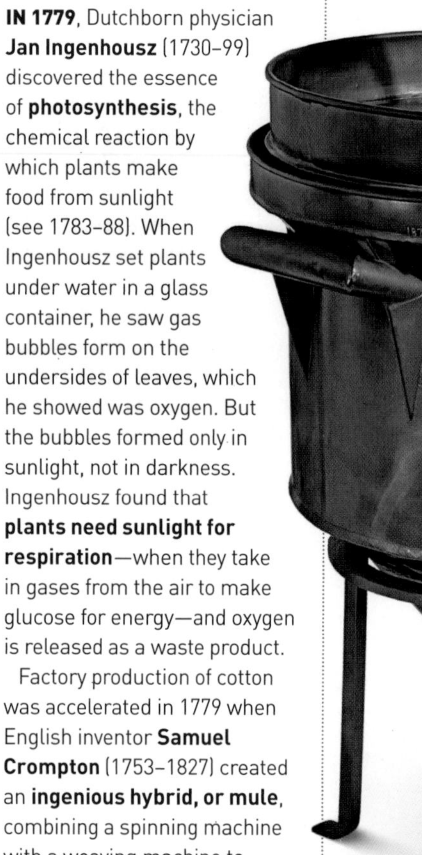
mouse placed in inner chamber
ice placed in outer chambers
outlet for meltwater

Lavoisier–Laplace calorimeter
Heat produced by a mouse placed in the inner chamber was measured by the volume of water produced by melted ice in the outer chambers.

1777 Leonhard Euler introduces the imaginary unit *i*.

1777 John Arnold introduces the term "chronometer"

1777 German-Swedish chemist **Carl Scheele** discovers **molybdenum**

1778 English geographer **James Rennell** charts the Agulhas Current

1778 German anatomist **Samuel von Sömmerring** identifies the 12 **cranial nerves**

1778 Samuel-Auguste Tissot describes **migraine**

1778 John Hunter publishes *The Natural History of the Human Teeth*

1779 Jan Ingenhousz discovers plant **photosynthesis**

1779 Construction of the **first all-iron bridge** begins at Ironbridge, Shropshire, in England

1779 Samuel Crompton invents the spinning **mule**

1780 Antoine Lavoisier demonstrates the role of oxygen in combustion

1780 Lavoisier and Pierre-Simon Laplace invent the **ice calorimeter**

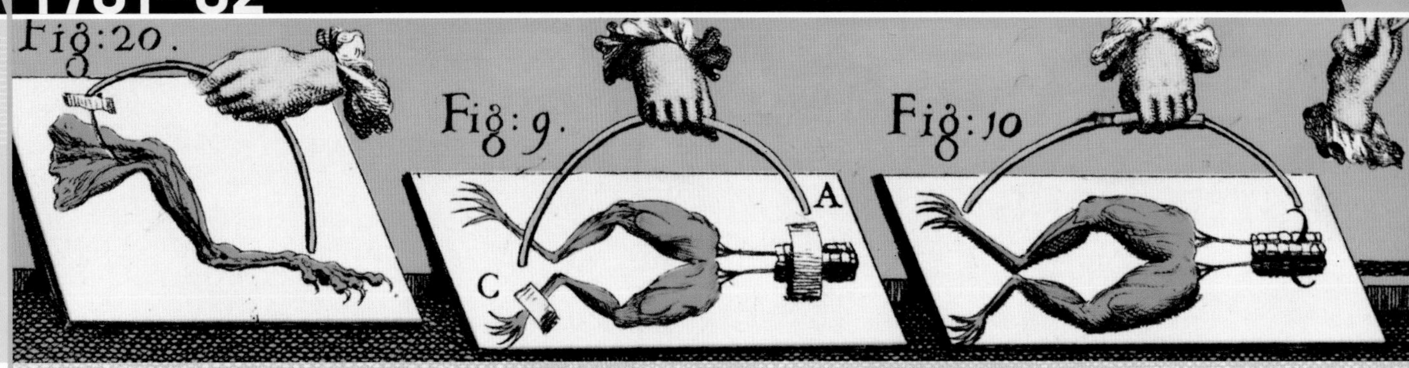
Fig:20.
Fig:9.
A
C
Fig:10

Macabre by today's standards, Galvani's experiments in generating movement in the leg muscles of dead frogs were a key step in the understanding of electricity.

418 TONS

THE **AMOUNT OF IRON** USED TO MAKE THE **BRIDGE**

JAN INGENHOUSZ
(1730–99)

Born in Breda in the Netherlands, Ingenhousz was a physician who had wide-ranging interests in other scientific fields, including the study of electricity. He was one of the pioneers of vaccination against smallpox and successfully inoculated the Austrian Empress Maria Theresa and her family.

previously held theory that body heat comes from phlogiston. In fact, animals make heat by combustion using oxygen.

Swiss physicist **Almé Argand** (1750–1803) revolutionized home lighting with the **Argand oil lamp**, 10 times brighter than a candle. The lamp had a circular wick and a tall glass chimney to improve the airflow.

THE SCIENTIFIC SENSATION OF 1781 WAS THE DISCOVERY OF A SIXTH PLANET by German-born British astronomer **William Herschel** (1738–1822). When Herschel first saw the object through the telescope in his garden in Bath in England on March 13, he assumed it was a comet. But its brightness and elliptical orbit soon left astronomers in no doubt. Herschel named the new planet George's Star in honor of British King George III, but astronomers eventually decided on **Uranus**, after Ouranos, the Greek god of the sky.

French astronomer **Charles Messier** (1730–1817) published the final version of his *Catalogue des Nébuleuses et des Amas d'Étoiles* (***Catalog of Nebulae and Star Clusters***) in 1781. This list, which Messier had been compiling since the 1770s, recorded 103 vague, distant objects in the night sky, all still known to modern astronomers by the numbers Messier gave them. Also in 1781, another astronomer, the Czech **Christian Mayer** (1719–83), published a catalog listing 80 pairs of stars, later known as **binaries**, that John Michell had argued in 1767 revolve together and which were observed by astronomers such as Herschel.

The same year, Italian physician **Luigi Galvani** (1737–98) began carrying out experiments showing that **static electricity** could cause the muscles of dissected frogs' legs to twitch. In April, he connected the nerves of a dead frog to a metal wire pointed skyward in a thunderstorm. In this experiment, which inspired Mary Shelley's novel *Frankenstein*, the frog's legs jumped with each flash of lightning. In September, the legs twitched when just hung on a brass hook over a metal rail. A fierce argument sprang up between Galvani and fellow Italian experimenter **Alessandro Volta** over whether the electrical effect was intrinsic to muscles and life (animal electricity), as Galvani claimed, or whether it was a chemical reaction, as Volta believed. Animal and chemical electricity were later shown to be the same.

In 1782, Lavoisier formulated what is perhaps now regarded as the most **important law in chemistry**—that of the conservation of matter. This law shows that in any chemical reaction no matter—not even the smallest part—is ever lost; it is simply rearranged in different combinations.

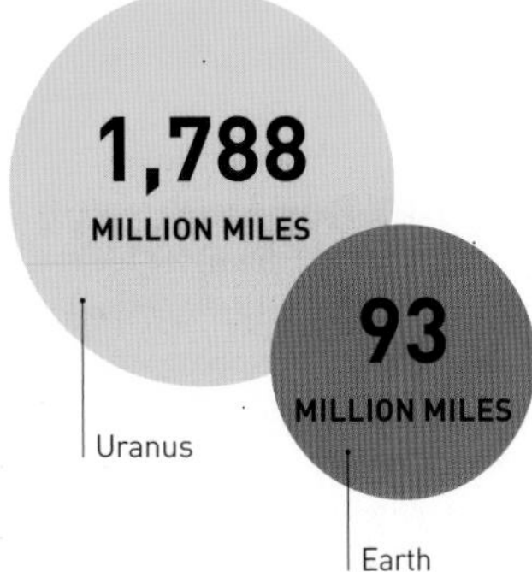
1,788
MILLION MILES
Uranus
93
MILLION MILES
Earth

Distance from the Sun
Uranus orbits the Sun at a distance of 1,788 million miles, almost 20 times the distance of Earth from the Sun. Each orbit takes 84 Earth years.

103

THE NUMBER OF **ASTRONOMICAL OBJECTS** IN MESSIER'S LIST

Messier's objects
Many of the objects in Messier's list have been revealed to be distant galaxies far from our own Milky Way, which in Messier's time was thought to comprise the entire Universe.

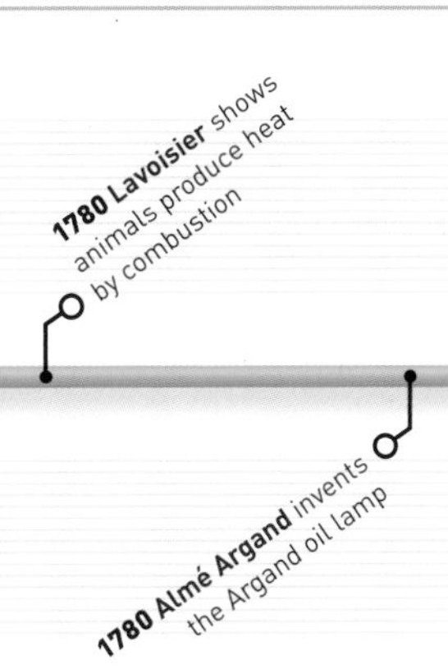
1780 Lavoisier shows animals produce heat by combustion
1780 Almé Argand invents the Argand oil lamp

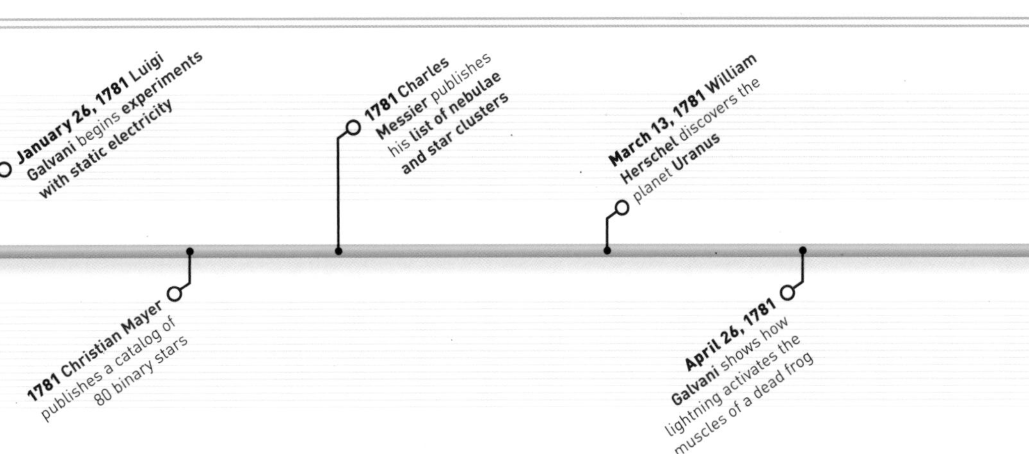
January 26, 1781 Luigi Galvani begins **experiments with static electricity**
1781 Christian Mayer publishes a catalog of 80 binary stars
1781 Charles Messier publishes his **list of nebulae and star clusters**
March 13, 1781 William Herschel discovers the planet **Uranus**
April 26, 1781 Galvani shows how lightning activates the muscles of a dead frog
1782 Lavoisier formulates the law of the **conservation of matter**

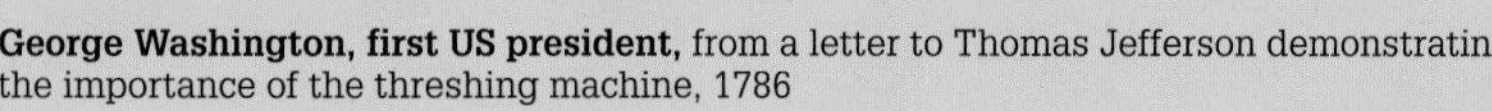

“ IF YOU CAN BRING A **MOVABLE THRESHING MACHINE…** IT WILL BE AMONG THE **MOST VALUABLE INSTITUTIONS** IN THIS COUNTRY. ”

George Washington, first US president, from a letter to Thomas Jefferson demonstrating the importance of the threshing machine, 1786

THE SENSATION OF 1783 was the first manned flight, in a hot-air balloon on November 21, in Paris. Piloted by Pilâtre de Rozier (1754–85) and the Marquis d'Arlandes, the balloon was the invention of French papermakers the **Montgolfier brothers**, Joseph-Michel (1740–1810) and Étienne (1745–99). The Montgolfiers had previously discovered that when hot air is trapped inside a bag it causes the bag to float upward, since hot air is less dense than cold air.

Ten days later, French physicist **Jacques Charles** (1746–1823) and craftsman **Noël Robert** (1760–1820) flew over Paris in a **balloon filled with hydrogen gas**. They achieved much longer, more controlled flights than with hot air (which quickly cooled).

Also in 1783, Spanish brothers **José** (1754–96) and **Fausto Elhuyar** (1755–1833) isolated a new metal element by reacting charcoal with tungstic acid. It was later named tungsten, one of the toughest of all metals.

2,953 FEET
THE **HEIGHT** REACHED BY **MONTGOLFIER BALLOON** IN 1783

In 1784, English chemist **Henry Cavendish** (1731–1810) showed that water is composed of two gases, hydrogen and oxygen. Astronomer **John Michell** predicted the **existence of black holes** when he observed that a star of sufficient mass would have such a strong gravitational field that light could not escape.

In London, inventor **Joseph Bramah** (1748–1814) patented his **unpickable security lock**, which opened only when the key moved a unique combination of sliders into grooves.

The **biggest question for geologists** in the 1780s was how marine fossils ended up in rocks high in mountains. German geologist Abraham Werner (1749–1817) believed that a single catastrophe, such as an almighty flood, had reshaped the world dramatically (see 1775–76). But Scottish geologist **James Hutton** (1726–97) did not agree. In his **groundbreaking *Theory of the Earth***, published in 1785, he suggested that landscapes are shaped slowly and continually over long periods by repeating cycles of erosion, sedimentation, and uplift. Hutton's theory (now proved correct) meant that Earth must be millions of years old—not just thousands, as was thought at the time.

In France, physicist **Charles de Coulomb** (1736–1806) established an important law

expansion chamber for gases

brass casing

inlet for gas

Henry Cavendish's eudiometer
This eudiometer is a replica of one used by Cavendish to measure the volume of gases, such as oxygen, in air, produced in a reaction.

First balloon ascent
In this early 19th-century print, Pilâtre de Rozier and the Marquis d'Arlandes acknowledge spectators as they ascend in their Montgolfier hot-air balloon in November 1783.

November 21, 1783 First manned flight in the **Montgolfier hot-air balloon** takes place in Paris

December 1, 1783 First flight in a **hydrogen gas balloon** takes place in Paris

January 15, 1784 Henry Cavendish shows that **water comprises hydrogen and oxygen**

August 21, 1784 Joseph Bramah patents his **security lock**

1785 Charles de Coulomb establishes his **Law of electrical charges**

1785 Andrew Meikle invents the **threshing machine**

1785 James Hutton publishes ***Theory of the Earth***, suggesting slow shaping of landscapes

1785 Jean Blanchard invents the **parachute**

1785 Edmund Cartwright develops the **power loom**

Andrew Meikle's threshing machine brought to an end the age-old practice of separating grain from stalks and husks with a hand flail.

Swiss naturalist Jean Senebier was one of the first to realize that plants respire. Gases move in and out of leaves through tiny surface pores (stomata), seen here under an electron microscope.

about electrical charges, showing that the force of attraction or repulsion between two charges varies in inverse proportion to the square of the distance between them. **The SI unit of charge, the coulomb, is named after him**.

Also in France in 1785, aviator **Jean Blanchard** (1753–1809) invented the **first folded silk parachute**. Previous parachute designs usually had a heavy wooden frame and failed to work. Blanchard used his parachute successfully to drop a dog in a basket from a balloon high in the air. With American **John Jeffries** (1744–1819), Blanchard made the first airborne crossing of the English Channel, in a gas balloon. In danger of descending too early, the pair shed all of their ballast—and most of their clothes—to keep the balloon in the air.

> **“...LITTLE CAUSES** ARE CONSIDERED AS BRINGING ABOUT THE **GREATEST CHANGES.”**
>
> **James Hutton, Scottish geologist,** from *Theory of the Earth*, 1795

In the same year, Scottish engineer **Andrew Meikle** (1719–1811) invented a **threshing machine**, and English inventor **Edmund Cartwright** (1743–1823) developed a **steam-powered loom**. Both machines dramatically reduced the need for manual labor, causing protests from laid-off workers.

JAMES HUTTON (1726–97)

After inheriting his father's farm in Berwickshire, Scotland, in 1753, Edinburgh-born James Hutton became fascinated by geology. He studied rock formations and noticed breaks in the sequence, which suggested to him, correctly, that the landscape is shaped in repeated cycles over an immensely long time.

GERMAN-BORN BRITISH ASTRONOMER WILLIAM HERSCHEL was the greatest astronomer of his age. He conducted many deep-sky surveys of objects in space with the powerful telescopes he designed himself, and made many thousands of discoveries. On January 11, 1787, he observed that **Uranus**, the planet he had discovered just six years earlier (see 1781–82), **had two moons**, which were named Oberon and Titania, after the fairy king and queen in Shakespeare's *A Midsummer Night's Dream*.

Swiss naturalist **Jean Senebier** (1742–1809) built on the discoveries made about photosynthesis by Jan Ingenhousz in 1779. Senebier showed that when plants are exposed to light, they take up carbon dioxide from the air and release oxygen (see panel, right).

Silicon is the second most abundant element in Earth's crust after oxygen, but it was not identified until 1787. In that year, French chemist **Antoine Lavoisier** understood that sand is an oxide (a chemical combination with oxygen) of a **hitherto unknown element, which he named silicon**.

Also in this year, various inventors were trying to **harness steam power to propel boats**.

PHOTOSYNTHESIS

Plants take carbon dioxide from the air in the process called photosynthesis. Sunlight, which is absorbed by chlorophyll, the green pigment in leaves, fuels the reaction between carbon dioxide and water in the leaves to make glucose, the plant's energy food. Oxygen, a waste byproduct, is exuded through pores (stomata) on the underside of leaves.

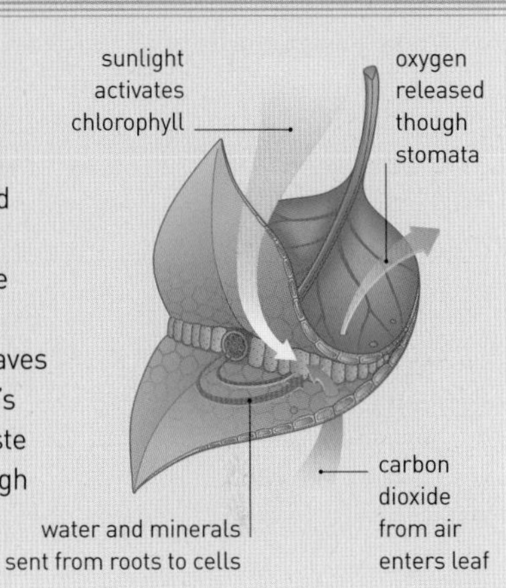
sunlight activates chlorophyll
oxygen released though stomata
carbon dioxide from air enters leaf
water and minerals sent from roots to cells

American inventor **John Fitch** (1743–98) launched the *Perseverance,* a **boat rowed by steam-driven oars**, on the Delaware River. Scottish engineer William Symington (1764–1831) built a **paddle steamer**, which had its first trial on Dalswinton Loch in Scotland. In West Virginia, on the Potomac River, American engineer James Rumsey (1743–92) experimented with a boat propelled by a jet of water forced out by a steam pump.

In 1788, French **mathematician Joseph Louis Lagrange** (1736–1813) published his *Mécanique Analytique* (*Analytic Mechanics*), perhaps the greatest work about mechanics (the science of forces and movement) since Isaac Newton's *Principia* (see 1685–89). In this book, Lagrange used his own **new method of calculus** to reduce mechanics to a few basic formulas from which everything else could be calculated.

5 MILES PER HOUR THE SPEED OF **SYMINGTON'S PADDLE STEAMER**

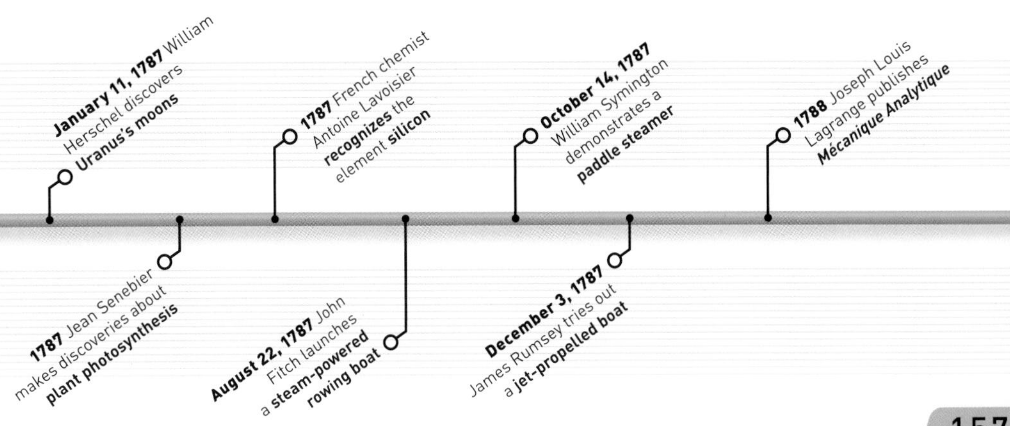
1786 American scientist Benjamin Franklin's 1770 **map of the Gulf Stream** is published in the US
January 11, 1787 William Herschel discovers **Uranus's moons**
1787 Jean Senebier makes discoveries about **plant photosynthesis**
1787 French chemist Antoine Lavoisier **recognizes** the element **silicon**
August 22, 1787 John Fitch launches a **steam-powered rowing boat**
October 14, 1787 William Symington demonstrates a **paddle steamer**
December 3, 1787 James Rumsey tries out a **jet-propelled boat**
1788 Joseph Louis Lagrange publishes ***Mécanique Analytique***

4

THE AGE OF REVOLUTIONS
1789–1894

While the drive for greater efficiency and power stimulated technological invention and industrial growth, the search to uncover nature's secrets overturned older beliefs about Earth and its inhabitants. Scientists came to understand that life emerged after an extremely long process of evolutionary development.

1789

“ …WE MUST ADMIT, **AS ELEMENTS,** ALL THE **SUBSTANCES INTO WHICH WE ARE CAPABLE,** BY ANY MEANS, TO REDUCE BODIES **BY DECOMPOSITION. ”**

Antoine Lavoisier, French chemist, from *Traité élémentaire de chimie* (*Elements of Chemistry*), 1789

AGAINST THE BACKDROP OF THE FRENCH REVOLUTION, a scientific revolution was underway in 1789. French chemist **Antoine Lavoisier** completed his *Elements of Chemistry*, a book that laid the foundations of **chemistry as a science**. Lavoisier created the first table of elements, which also included heat and light. He introduced international chemical symbols and introduced modern names, such as oxygen and hydrogen for gases and sulphates for compounds.

On 28 August, German-born British astronomer **William Herschel** looked through his **largest telescope** for the first time. He had already built several telescopes, but this 40 ft (12 m) reflecting telescope was the biggest. That night, Herschel discovered **a new moon of Saturn, Enceladus**, and three weeks later he spotted another, **Mimas**.

French botanist **Antoine Laurent de Jussieu** published a system for classifying flowering plants. It used Linnaeus's two-part Latin naming system (see 1754) and divided flowering plants into classes according to the number of stamens and pistils. This system is still used today.

In Canada, British explorer **Alexander Mackenzie** (1764–1820) set out in his canoe, following the uncharted 1,100 mile (1,770 km) river—now known as Mackenzie River—to the Arctic Ocean.

Herschel's telescope
Astronomer William Herschel took four years to build this giant telescope in his garden in Slough, England. It was paid for by King George III, and dismantled in 1839.

ANTOINE LAVOISIER
(1743–94)

Born in Paris, France, Antoine Lavoisier is sometimes said to be the father of chemistry—he proved the role of oxygen in combustion, which laid the foundation for modern chemistry. Lavoisier established that matter is neither created nor destroyed during chemical changes. A target for revolutionaries, he was guillotined on May 8, 1794.

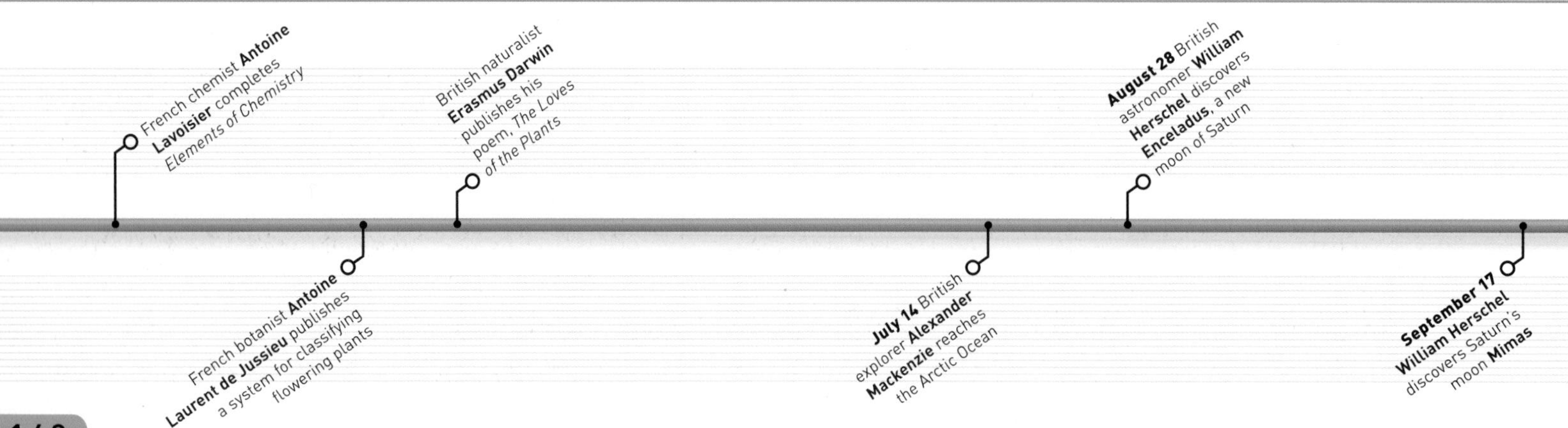
French chemist **Antoine Lavoisier** completes *Elements of Chemistry*
French botanist **Antoine Laurent de Jussieu** publishes a system for classifying flowering plants
British naturalist **Erasmus Darwin** publishes his poem, *The Loves of the Plants*
July 14 British explorer **Alexander Mackenzie** reaches the Arctic Ocean
August 28 British astronomer **William Herschel** discovers **Enceladus**, a new moon of Saturn
September 17 **William Herschel** discovers Saturn's moon **Mimas**

1790–91

The Armagh Observatory, the oldest scientific institution in Northern Ireland, was built in 1790 and was one of the largest and most advanced observatories of its day.

IN PHILADELPHIA, AMERICAN INVENTOR OLIVER EVANS (1755–1819) developed an idea that would make powered transport a reality. The low-pressure Watt steam engines of the day (see 1765–66) were too big and heavy for land vehicles. But Evans realized that if the steam is much hotter, the piston can be driven by steam pressure alone—so the condensing process needed to create a vacuum in low-pressure engines could be dispensed with. This revelation led to Evans' invention of **high-pressure steam engines**, which could give the same power as Watt engines that were 10 times larger.

In Warwickshire, England, inventor **John Barber** (1734–1801) was developing an even more revolutionary engine—the **gas turbine**. His idea was to compress gas made from wood or coal, then burn it explosively to turn the vanes of a paddle wheel. Barber never built a prototype, but his ideas re-emerged centuries later when the jet engine was developed.

Another discovery, which is now vital for modern aircraft, was **titanium**. The metal was isolated from the mineral ilmenite in Cornwall, England, by British geologist **Reverend William Gregor** (1761–1817). It was independently discovered a few months later by German chemist **Martin Klaproth (1743–1817)**, who named it titanium.

9:1 **Lightning injuries**
Nine of every ten people struck by lightning survive. James Parkinson proposed that survivors suffer from a form of muscle paralysis and skin burns.

There were several developments in the field of pure science too. Swiss physicist **Pierre Prévost (1751–1839)** proved that **all bodies**—no matter how hot or cold—**radiate heat**. German physiologist Franz Joseph Gall (1758–1828) showed that the **nervous system** is actually **made from masses of nerve fibres**, or ganglia. British surgeon **James Parkinson** (1755–1824), who lent his name to a devastating disease of the nervous system, gave the first medical description of injuries to people struck by lightning.

Stargazing continued to attract funding from wealthy patrons, not just because of people's fascination with it but also because of its potential value to shipping. Archbishop of Armagh, **Richard Robinson**, built an expensive **observatory in Armagh**, Northern Ireland, which continues to be one of the leading scientific research establishments today.

titanium

Titanium
The metal titanium has several ores, including ilmenite and rutile, and it also occurs in trace amounts in rocks, water bodies, soils, and all living things.

1792

A cartoon from 1807 shows the initial mixed reaction in London to the new gas streetlighting, which would soon transform cities at night.

THIS YEAR SAW A DISCOVERY THAT WOULD ONE DAY HELP technology transform cities across the world—**Alessandro Volta's** discovery of **chemical electricity**. Previously, static electricity had been generated through friction. Volta found that by bringing metals into contact, he could create chemical reactions that generated current electricity. He went on to produce the first battery (see 1800).

Another notable development in the field of communication was French inventor **Claude Chappe's telegraph**—a line of towers that could relay messages visually by showing panels at different angles. It could send a message across 137 miles (220 km)—from Lille to Paris—in an hour.

Scottish engineer **William Murdoch** (1754–1839) invented **gas lighting**; his home in Cornwall, England, was the **first gas-lit house**. Soon many houses, factories, and city streets would be lit by gas.

The **first iron-cased rockets**—forerunners of modern missiles—were used in India by **Tipu Sultan**, ruler of Mysore, against the British.

One discovery that was little appreciated at the time was British scientist **Thomas Wedgwood's** (1771–1805) observation that all **materials glow** the **same color when heated** to the same temperature. For example, we know the Sun's surface is 5,800 K (9,980°F) because of its color.

pulley for swiveling telegraph arm
movable signaling panels
pivot
receiving platform
weight for balancing signals

Chappe telegraph
This telegraph tower shows Chappe's semaphore system, with its mechanism for changing the panels' angle to pass on messages.

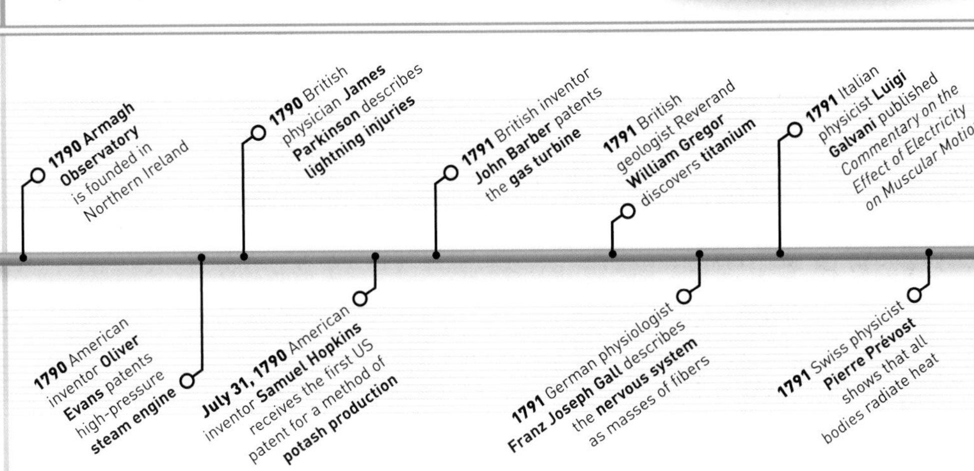

1790 Armagh Observatory is founded in Northern Ireland
1790 American inventor **Oliver Evans** patents high-pressure **steam engine**
1790 British physician **James Parkinson** describes **lightning injuries**
July 31, 1790 American inventor **Samuel Hopkins** receives the first US patent for a method of **potash production**
1791 British inventor **John Barber** patents the **gas turbine**
1791 British geologist Reverend **William Gregor** discovers **titanium**
1791 German physiologist **Franz Joseph Gall** describes the **nervous system** as masses of fibers
1791 Italian physicist **Luigi Galvani** published *Commentary on the Effect of Electricity on Muscular Motion*
1791 Swiss physicist **Pierre Prévost** shows that all bodies radiate heat

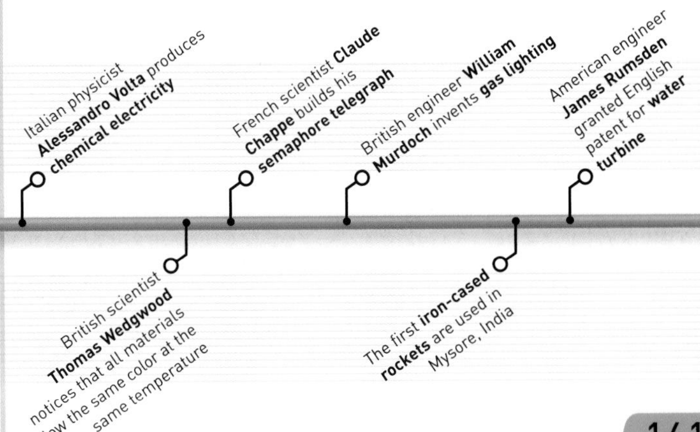

Italian physicist **Alessandro Volta** produces **chemical electricity**
British scientist **Thomas Wedgwood** notices that all materials glow the same color at the same temperature
French scientist **Claude Chappe** builds his **semaphore telegraph**
British engineer **William Murdoch** invents **gas lighting**
The first **iron-cased rockets** are used in Mysore, India
American engineer **James Rumsden** granted English patent for **water turbine**

1793

“ THE **OBJECT OF THIS WORK** IS TO EXPLAIN... THE **CHANGES OF STRUCTURE** ARISING FROM MORBID ACTIONS IN SOME OF THE MOST **IMPORTANT PARTS** OF THE **HUMAN BODY. ”**

Matthew Baillie, British physician, in Preface to *The Morbid Anatomy of Some of the Most Important Parts of the Human Body,* 1793

IN REVOLUTIONARY PARIS, the first natural history museum in France, the Muséum national d'Histoire naturelle (National Museum of Natural History), was founded on June 10. It was built in the former Jardin Royal des Plantes Médicinales (Royal Garden of Medicinal Plants), created by King Louis XIII in 1635. It became the Jardin des Plantes—a botanical garden and home to **one of the world's first public zoos**, founded in 1795.

On July 22, British explorer **Alexander Mackenzie** (1764–1820) became the first known man to **cross the continent of North America** north of Mexico, when his Peace River Expedition finally reached the end of its journey just west of Bella Coola in British Columbia, Canada.

A lead mine near the British village Strontian became the site of the **discovery of a new element, strontium**. In 1790, Irish chemist Adair Crawford (1748–95) and his British colleague, chemist William Cruikshank (d. *c.*1810) noticed that some local lead ores had an unusual make-up. Various scientists, such as German chemist Martin Klaproth (1743–1817), began investigating them and in 1793 British chemist **Thomas Charles Hope** (1766–1844) **named them strontites** after the village in the Scottish highlands where the mine was located. In 1808, British chemist Humphry Davy **isolated a new element from these ores**, a soft metal called strontium.

John Dalton (1766–1844), the British scientist who proposed an atomic theory in 1803, published *Meteorological Observations and Essays* in 1793. In it, Dalton wrote on everything from barometers to cloud formation, explaining how moisture in the air turns to rain when air cools.

British physician Matthew Baillie (1761–1823) published *The Morbid Anatomy of Some of the Most Important Parts of the Human Body.* This was the **first modern scientific study of disease**, and showed how diseases could be better understood by studying their effect on organs after death.

400 YARDS
THE DISTANCE COVERED BY **AGUILERA'S GLIDER**

Muséum national d'Histoire naturelle
The world's first great natural history museum opened in Paris just a few months after Louis XVI of France was guillotined nearby.

1794–95

The limestone rocks of the Jura Mountains on the French-Swiss border formed the basis of the identification of the Jurassic Age.

199 MILLION YEARS
THE AGE OF THE **JURA MOUNTAINS**

IN 1794, AMERICAN INVENTOR ELI WHITNEY (1765–1825) devised a machine for separating the seeds from cotton to obtain clean fibers for spinning into cloth. His gin, and others like it, was soon in use throughout the American South, leading to a huge **increase in the production of raw cotton** after 1800.

The increase in cotton production was a massive boost to the growing factory towns of northwest England, including Manchester, where in 1794 **John Dalton** presented his **pioneering studies on color blindness.** Dalton was blind to green, and he believed it was blue fluid in his

Day light
1 Brown

- British chemist Thomas Charles Hope **names "strontites"**
- British physician Matthew Baillie writes about **pathology**
- **May 15** Spanish inventor Diego Marín Aguilera **flies a glider** about 1,180ft (360m)
- **June 10** The **Muséum national d'Histoire naturelle** is founded in Paris
- **July 22** British explorer Alexander Mackenzie reaches the **Pacific Ocean**
- British scientist John **Dalton writes about meteorology**
- **1794** The Jardin des Plantes in Paris becomes home to one of the **first public zoos**
- **1794** British scientist John Dalton writes about **color blindness**
- **1794** American inventor Eli Whitney devises **cotton gin**
- **1794** German physicist Ernst Chladni suggests **meteorites come from space**

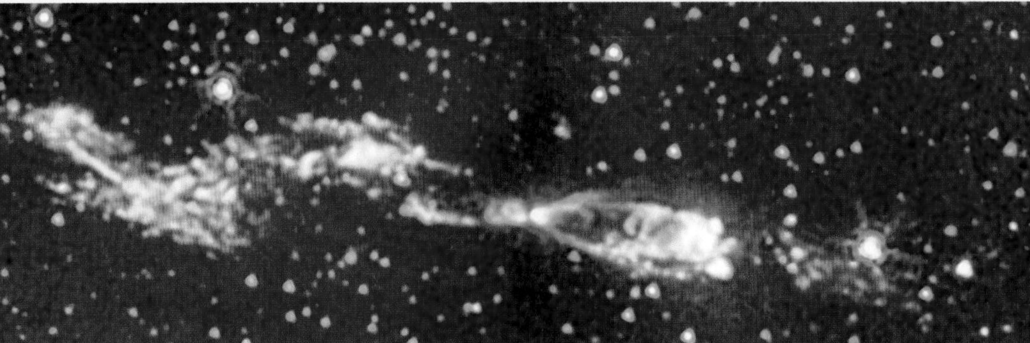

An image from the Spitzer space telescope reveals a distant star and planets forming from a spinning gas cloud, in the same way that our Solar System was created billions of years ago.

eyes that was robbing him of the ability to see green or red. He requested that his eyes be examined after his death to prove this thesis, but the fluid was normal and clear. A DNA test on his preserved eyes in the 1990s showed that he had a common genetic deficiency that reduced his sensitivity to green, proving **he did indeed have color blindness**.

Up to this point in time, meteorites were thought to come from volcanoes, but in 1794, German physicist **Ernst Chladni** (1756–1827) made the suggestion that **they came from space**. The following year a large meteorite fell to earth at Wold Newton in Yorkshire, England, which was far from any volcano and backed up Chladni's theory.

In 1795, the French government was championing a **new decimal system of measures**. They decreed that a gram was "the absolute weight of a volume of water equal to the cube of the hundredth part of the meter, at the temperature of melting ice."

Dalton's color blindness test
Dalton used this book of colored threads to test his own color blindness. He suffered from inherited red-green color blindness.

CARL FRIEDRICH GAUSS (1777–1855)

Born in the German Duchy of Brunswick, Gauss was a mathematical genius who astonished people with his brilliance at an early age. He grew up to be one of the greatest mathematicians ever, making important contributions to the theory of numbers, non-Euclidean geometry, planetary astronomy, probability, and the theory of functions.

Also in France, the naturalist **George Cuvier** (1769–1832) introduced the idea that **Earth's past held many species of animal now extinct**. This was shocking for some, since many people in Europe still believed all animals were created at the time of the Creation.

Fossil hunters found more fossils in the Jurakalk rock formation in the French Jura Mountains. Prussian geographer **Alexander von Humboldt** (1769–1859) identified the age of these limestone formations, thus setting a **basis for geologists to identify the Jurassic age**—the age of dinosaurs.

FRENCH MATHEMATICIAN PIERRE-SIMON LAPLACE (1749–1827) published his *Exposition du Système du Monde* (*The System of the World*), which included his theories on orbits and tides, and the **first fully scientific exposition of the nebular hypothesis**, first suggested by Immanuel Kant in 1755, which proposes that the Solar System began as a giant gas cloud (nebula) spinning around the Sun. It cooled and contracted to form planets.

For 19-year-old German mathematician, Carl Gauss, 1796 was a year of breakthroughs. On March 30, he proved that it was **possible to construct certain regular polygons**, including the 17-sided heptadecagon, with just a compass and a ruler—a problem that had eluded mathematicians since the time of Pythagoras. On April 8, he made key contributions to solving quadratic equations. On May 31, he created the **prime number theorem** to show how prime numbers are spread. And, July 10, he discovered that **every positive integer is representable as a sum of, at most, three triangular numbers**.

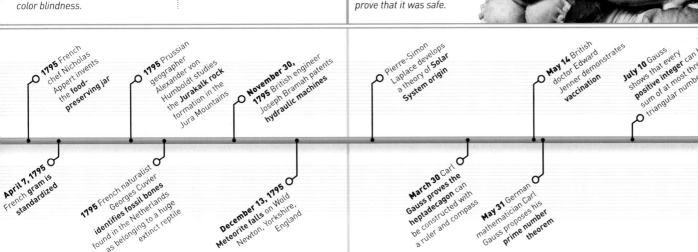

Edward Jenner vaccinating his son
Many people remained unconvinced about vaccination, even after it had been successfully performed in 1796. In 1797, Jenner vaccinated his 11-month-old son to prove that it was safe.

Since the 1720s, many had been protected against the deadly disease smallpox by inoculation, which involved deliberate infection with smallpox germs. But the inoculation had risks and could kill. In 1796, British country doctor Edward Jenner (1748–1823) noticed that dairymaids often failed to contract smallpox and began to wonder if they gained their immunity through exposure to a similar, milder disease known as cowpox. He thought it possible that **"vaccination" with cowpox would be safer than inoculation** with smallpox, and deliberately infected his gardener's son with cowpox. Vaccination originally meant inoculation with cowpox, but the term is now used for any inoculation by weak or killed germs. A few weeks later he attempted to infect the boy with smallpox, who proved to be immune and remained so all his life. Vaccination using a harmless form of a dangerous disease is now one of the most effective methods of disease prevention.

1795 French chef Nicholas Appert invents the **food-preserving jar**

April 7, 1795 French **gram is standardized**

1795 Prussian geographer Alexander von Humboldt studies the **Jurakalk rock** formation in the Jura Mountains

1795 French naturalist Georges Cuvier **identifies fossil bones** found in the Netherlands as belonging to a huge extinct reptile

November 30, 1795 British engineer Joseph Bramah patents **hydraulic machines**

December 13, 1795 Meteorite falls on Wold Newton, Yorkshire, England

Pierre-Simon Laplace develops a theory of **Solar System origin**

March 30 Carl **Gauss proves the heptadecagon** can be constructed with a ruler and compass

May 14 British doctor Edward Jenner demonstrates **vaccination**

May 31 German mathematician Carl Gauss proposes his **prime number theorem**

July 10 Gauss shows that every **positive integer** can be sum of at most three triangular numbers

First **cast iron aqueducts** completed in Britain

Cyanobacteria
PRECAMBRIAN: 4,000–542 MYA
Cyanobacteria are among the oldest and simplest organisms alive today. Sediment has built up in round clusters around a cyanobacteria nucleus.

Clavilithes shell
PLIOCENE: 5.3–1.8 MYA
Clavilithes is related to the modern whelk. Shell fossils may be found far from the sea, leading Greeks to deduce that land was once covered by water.

Stromatolites
PRECAMBRIAN: 4,000–542 MYA
Stromatolites are layers of sediment that were deposited by microorganisms in the coastal shallows of ancient seas. Living stromatolites still exist today.

Calymene trilobite
SILURIAN: 444–416 MYA
Early scholars called trilobite fossils "petrified insects." They were identified as arthropods in the eighteenth century.

Harpoceras ammonite
LOWER JURASSIC: 199.6–175.6 MYA
The Roman naturalist Pliny named the ammonite fossil after the ram horns of the Egyptian god Ammon.

Pentacrinites crinoid
PALEOGENE: 65–23 MYA
Delicate animals such as crinoids could be beautifully preserved when the soft mud in which they lived turned into rock.

FOSSILS

SINCE THE EARLIEST TIMES, FOSSILS HAVE HELPED US UNDERSTAND THE HISTORY OF LIFE ON EARTH

Fossils have been known to man since prehistory, but early Christian scholars believed they were the remains of creatures that died in the Biblical flood. We now know they are evidence of evolution on a changing planet.

The scientific study of fossils has its roots in the Age of the Enlightenment, when naturalists began to describe and classify their finds. By the 1800s, geologists recognized not only that rock layers built up over time, but also that each layer had a distinctive set of fossil types. Supporting evidence was recovered from this fossil record to support Charles Darwin's theory of evolution. Today, paleontology routinely uses technology that can determine the age of fossils and even study their DNA.

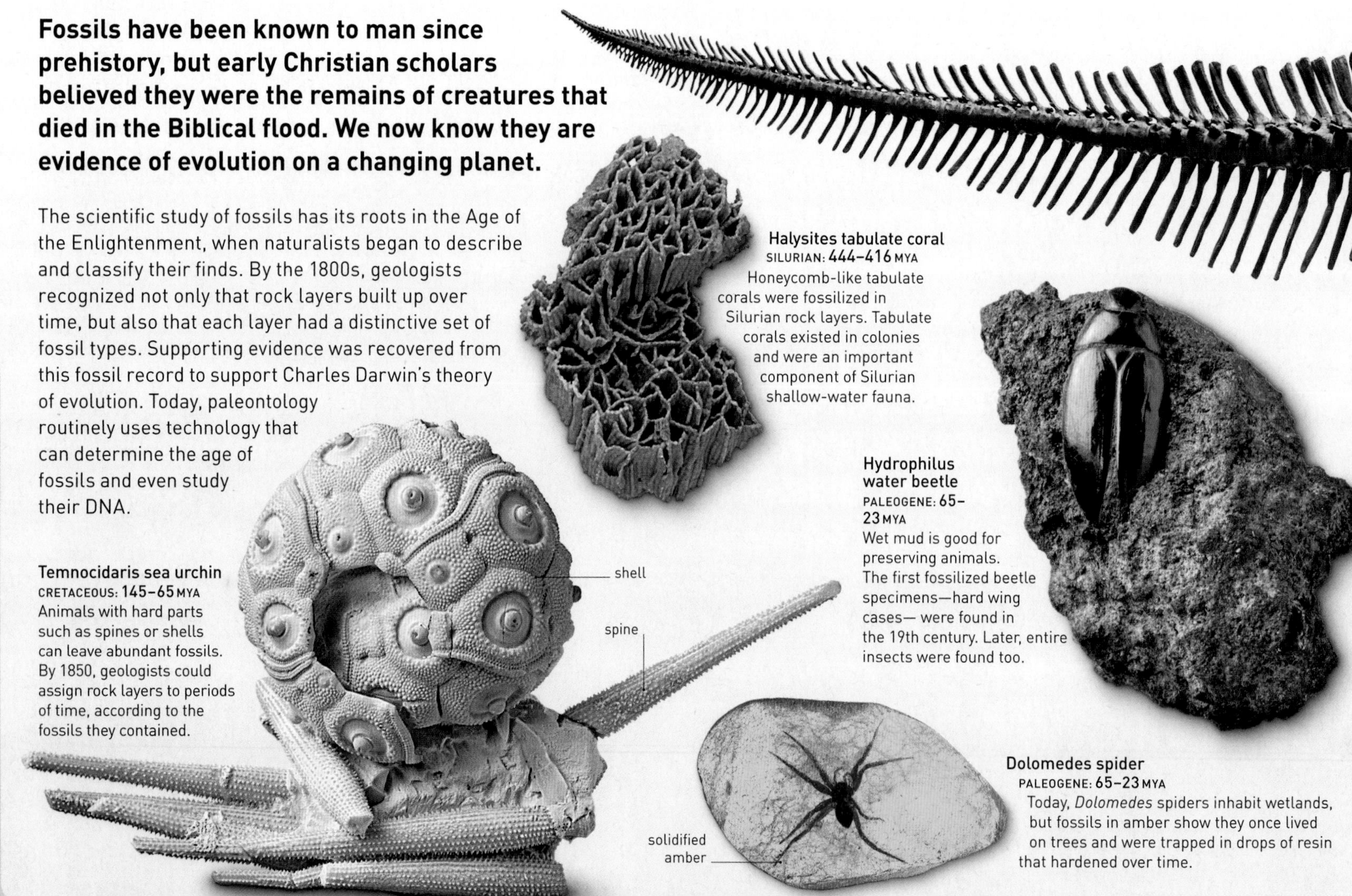
shell
spine
solidified amber

Halysites tabulate coral
SILURIAN: 444–416 MYA
Honeycomb-like tabulate corals were fossilized in Silurian rock layers. Tabulate corals existed in colonies and were an important component of Silurian shallow-water fauna.

Hydrophilus water beetle
PALEOGENE: 65–23 MYA
Wet mud is good for preserving animals. The first fossilized beetle specimens—hard wing cases— were found in the 19th century. Later, entire insects were found too.

Temnocidaris sea urchin
CRETACEOUS: 145–65 MYA
Animals with hard parts such as spines or shells can leave abundant fossils. By 1850, geologists could assign rock layers to periods of time, according to the fossils they contained.

Dolomedes spider
PALEOGENE: 65–23 MYA
Today, *Dolomedes* spiders inhabit wetlands, but fossils in amber show they once lived on trees and were trapped in drops of resin that hardened over time.

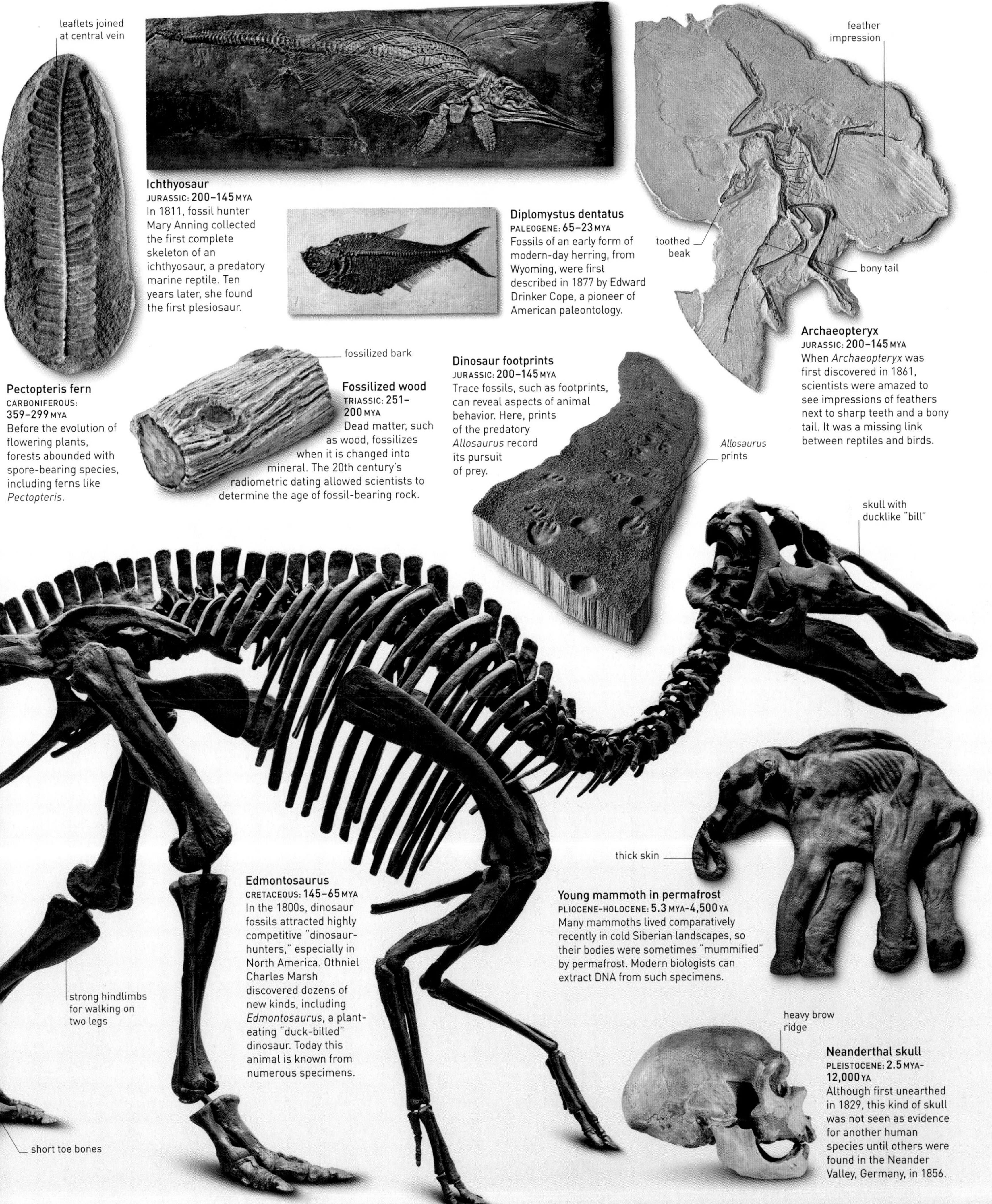
leaflets joined
at central vein
feather
impression
toothed
beak
bony tail
fossilized bark
Allosaurus
prints
skull with
ducklike "bill"
thick skin
strong hindlimbs
for walking on
two legs
heavy brow
ridge
short toe bones

Ichthyosaur
JURASSIC: **200–145** MYA
In 1811, fossil hunter Mary Anning collected the first complete skeleton of an ichthyosaur, a predatory marine reptile. Ten years later, she found the first plesiosaur.

Diplomystus dentatus
PALEOGENE: **65–23** MYA
Fossils of an early form of modern-day herring, from Wyoming, were first described in 1877 by Edward Drinker Cope, a pioneer of American paleontology.

Archaeopteryx
JURASSIC: **200–145** MYA
When *Archaeopteryx* was first discovered in 1861, scientists were amazed to see impressions of feathers next to sharp teeth and a bony tail. It was a missing link between reptiles and birds.

Pectopteris fern
CARBONIFEROUS: **359–299** MYA
Before the evolution of flowering plants, forests abounded with spore-bearing species, including ferns like *Pectopteris*.

Fossilized wood
TRIASSIC: **251–200** MYA
Dead matter, such as wood, fossilizes when it is changed into mineral. The 20th century's radiometric dating allowed scientists to determine the age of fossil-bearing rock.

Dinosaur footprints
JURASSIC: **200–145** MYA
Trace fossils, such as footprints, can reveal aspects of animal behavior. Here, prints of the predatory *Allosaurus* record its pursuit of prey.

Edmontosaurus
CRETACEOUS: **145–65** MYA
In the 1800s, dinosaur fossils attracted highly competitive "dinosaur-hunters," especially in North America. Othniel Charles Marsh discovered dozens of new kinds, including *Edmontosaurus*, a plant-eating "duck-billed" dinosaur. Today this animal is known from numerous specimens.

Young mammoth in permafrost
PLIOCENE-HOLOCENE: **5.3** MYA-**4,500** YA
Many mammoths lived comparatively recently in cold Siberian landscapes, so their bodies were sometimes "mummified" by permafrost. Modern biologists can extract DNA from such specimens.

Neanderthal skull
PLEISTOCENE: **2.5** MYA-**12,000** YA
Although first unearthed in 1829, this kind of skull was not seen as evidence for another human species until others were found in the Neander Valley, Germany, in 1856.

1797

A wood engraving of a magpie from the first volume of Thomas Bewick's *History of British Birds*.

CARBON EXISTS IN THREE FORMS—as graphite, diamond, and charcoal. It was only in the late 18th century, however, that it was established that they are all the same substance. In 1772, French chemist **Antoine Lavoisier** had burned graphite in oxygen and proved that the only product was carbon dioxide. In 1797, British chemist **Smithson Tennant** (1761–1851) repeated the experiment, using diamond instead of graphite. The product was carbon dioxide once again, proving that diamond is an allotrope (form) of carbon.

In this year, amid the turmoil of post-revolutionary France, Italian-French mathematician **Joseph Louis Lagrange** (1736–1813) published *Théorie des Fonctions Analytiques* (*Theory of Analytic Functions*), setting out a new approach to **calculus**. Although not widely appreciated at the time, his ideas proved invaluable in 20th-century science, especially in the study of **quantum mechanics**.

Thomas Jefferson (1743–1826), vice-president of the US at the time, presented a paper to the American Philosophical Society, Philadelphia, in which he described the fossil remains of a creature he named *Megalonyx* (giant claw). The fossil was that of the **extinct ground sloth**, now known as ***Megalonyx jeffersonii***.

British artist and ornithologist **Thomas Bewick** (1753–1828) published the first volume of his ***History of British Birds***. He illustrated these volumes using the wood-engraving technique—an advancement in the printing of illustrated books.

CRYSTALLINE CARBON

A crystalline form of carbon, diamonds are formed at high pressure within the Earth and brought to the surface by tectonic activity. They are found in up to 3-billion-year-old rock strata. Carbon atoms are arranged in a crystal structure called a diamond lattice (a face-centered cube). This rigid structure is why diamonds are hard and also transparent.

carbon atom
UNCUT DIAMOND IN ROCK
ATOMIC STRUCTURE OF DIAMOND

1798–99

760 POUNDS
THE **WEIGHT** OF **ROSETTA STONE**

The Rosetta Stone is a decree passed on behalf of Egyptian King Ptolemy V in 196 BCE. It carries the same text in three scripts.

FRENCH PHARMACIST AND CHEMIST LOUIS VAUQUELIN (1763–1829) discovered **beryllium**, a metal that melts at 2,349°F (1,287°C), in 1798. He used a process of chemical extraction to obtain beryllium from a variety of emerald known as beryl.

British economist and demographer **Thomas Robert Malthus** (1766–1834), then an Anglican priest in Surrey, England, anonymously published the first edition of *An Essay on the Principle of Population*. The book argued that **population growth** would ultimately outstrip the carrying capacity of Earth. It had a profound influence on Charles Darwin in his development of the idea of a struggle for survival (see pp.204–05).

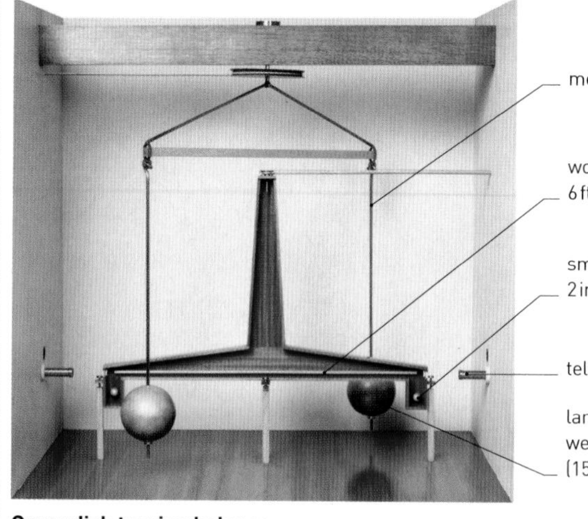
metal wire
wooden torsion rod, 6ft (2m) long
small lead sphere, 2in (5cm) wide
telescope
large lead sphere weighing 350lb (159kg)

Cavendish torsion balance
This model of Cavendish's torsion balance experiment shows the arrangement of small and large lead spheres attached to a wooden rod.

One of the most important experiments in the history of science was completed in 1798 by reclusive British physicist **Henry Cavendish** (1731–1810). He used a **torsion balance** apparatus, designed by British geologist and astronomer John Michell (1724–93), to measure the mass of Earth. The apparatus had a wooden rod suspended from a wire, with a lead sphere attached to each end, and two smaller lead spheres. He measured the force or **gravitational attraction** between the large and small lead spheres, and calculated the gravitational force of Earth on the small ball by weighing it. The ratio of these two forces allowed him to **calculate the mass and density of Earth**. He found the density to be 5.48 times that of water—the modern estimate is 5.52 times. Cavendish's careful calculation of possible errors and the painstaking detail of his work arguably make this the first modern physics experiment.

Meticulous experimentation was also a feature of the work of American-born British physicist **Benjamin Thompson** (1753–1814). While working in Bavaria, Germany, he investigated the way **heat was produced by friction** when cannons were being bored, and disproved the idea that heat was a kind of fluid, called caloric. Thompson published these findings in a paper in 1798.

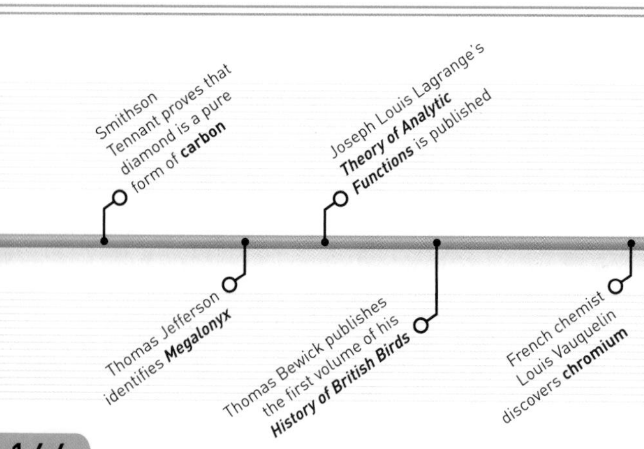
Smithson Tennant proves that diamond is a pure form of **carbon**
Thomas Jefferson identifies **Megalonyx**
Joseph Louis Lagrange's **Theory of Analytic Functions** is published
Thomas Bewick publishes the first volume of his **History of British Birds**
French chemist Louis Vauquelin discovers **chromium**

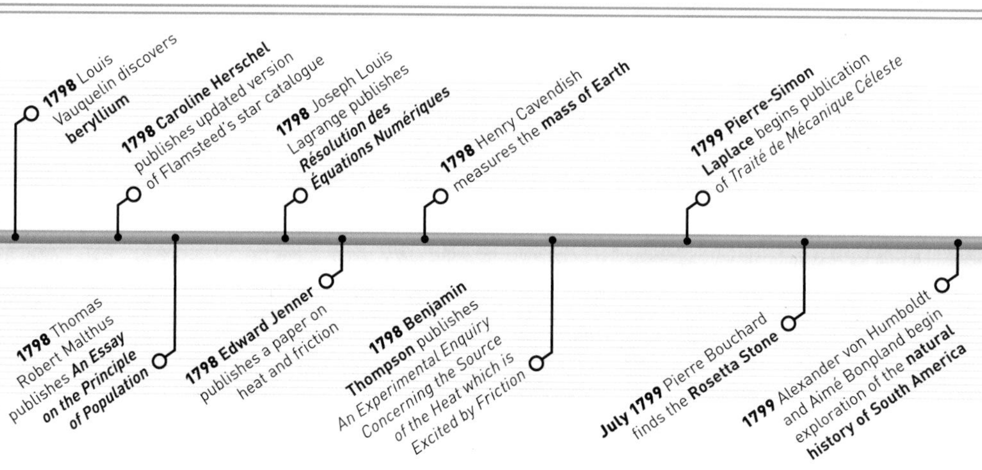
1798 Louis Vauquelin discovers **beryllium**
1798 Thomas Robert Malthus publishes **An Essay on the Principle of Population**
1798 Caroline Herschel publishes updated version of Flamsteed's star catalogue
1798 Edward Jenner publishes a paper on heat and friction
1798 Joseph Louis Lagrange publishes **Résolution des Équations Numériques**
1798 Henry Cavendish measures the **mass of Earth**
1798 Benjamin Thompson publishes *An Experimental Enquiry Concerning the Source of the Heat which is Excited by Friction*
1799 Pierre-Simon Laplace begins publication of *Traité de Mécanique Céleste*
July 1799 Pierre Bouchard finds the **Rosetta Stone**
1799 Alexander von Humboldt and Aimé Bonpland begin exploration of the **natural history of South America**

1800

The duck-billed, egg-laying, web-footed platypus, first described by British zoologist George Shaw, is amphibious—capable of living both on land and in water.

A year later, a specially made platinum cylinder with a mass equal to 1.000025 liters of water, was legally declared the **official prototype of the kilogram**. The modern kilogram is derived from this prototype.

In July 1799, **Pierre Bouchard** (1772–1832), an officer in Napoleon's French army, found a black granite stone at Rosetta, Egypt. The stone carried the **same text in three scripts**—hieroglyphs, Egyptian demotic script, and ancient Greek. This provided the key to translating hieroglyphs. The **Rosetta Stone**, was among the items captured by the British in 1801.

In the same year, French astronomer **Pierre-Simon Laplace** (1749–1827) published the first part of his five-volume *Traité de Mécanique Céleste* (*Treatise on Celestial Mechanics*). Described as "an encyclopaedia of calculations about the six known planets and their satellites, their shapes, and tides in Earth's oceans," it proved that the Solar System is stable on timescales that are relevant to humankind.

In 1799, physician **Thomas Beddoes** (1760–1808), established the **Pneumatic Institution** in Bristol, UK, to research the medical implications of newly discovered gases. It was here that British chemist Humphry Davy served his apprenticeship.

THIS YEAR, ITALIAN SCIENTIST ALESSANDRO VOLTA (1745–1827) wrote to the Royal Society, London, to report the battery he had made, which became known as a **voltaic pile**. His invention was actually a **"wet cell" battery**. In 1799, Volta had found that stacking disks of zinc, copper, and cardboard soaked in brine in multiple layers could generate an electric current. Adding more disks increased the amount of electricity generated.

The year also produced a puzzle for scientists, when British zoologist **George Shaw** (1751–1813) published the first scientific description of a **platypus** (*Ornithorhynchus anatinus Shaw*), based on preserved samples and sketches sent from Australia. He considered the creature so bizzare that, at first, he thought it might be a hoax perpetrated by an accomplished taxidermist.

Another surprise came when British–German astronomer **William Herschel** studied the heat associated with the **different colors of sunlight, as split into a spectrum by a prism**. Herschel placed thermometers in different parts of the spectrum and found that the temperatures of the colors increased from the violet part of the spectrum to the red. He then placed a thermometer just beyond the red and observed that this had the highest temperature of all. This invisible radiation is now called **infrared** (see pp.234–35).

In industry, an important step toward the era of precision engineering came when British engineer and inventor **Henry Maudslay** (1771–1831) developed the first practical **screw-cutting lathe**—a machine for accurately cutting screw threads.

In 1800, the **Royal Institution of Great Britain**—based in London—was granted its charter. Its aim was to provide a more popular forum for science than the established Royal Society. Henry Cavendish and Benjamin Thompson were instrumental in founding this institution, and Humphry Davy soon became its director. It became a **major research center** and a center for the popularizing of science throughout the 19th and into the 20th century.

HOW BATTERIES WORK

Batteries contain one or more electrochemical cells. Chemical reactions encourage electrons to move to the cathode end of the battery, building up negative charge there. Positive charge collects at the anode. When the cathode and anode are connected by an external wire, the electrons flow along the wire to produce an electric current. This is a simple "wet cell" battery.

−
cathode
silver plate
zinc plate
blotting paper
individual element
anode
+

porous cardboard soaked in electrolyte
zinc disk
copper disk

Volta's battery
This "wet cell" battery, invented by Italian scientist Alessandro Volta, was made of alternating layers of zinc and copper separated by cardboard disks soaked in salt water.

- **1799** Thomas Beddoes establishes the **Pneumatic Institution** in Bristol
- **1799** Humphry Davy argues that **heat** is not a fluid called caloric but that it is **"motion"**
- **1799 Prototype kilogram** is manufactured
- **Royal Institution of Great Britain** is granted its charter
- Alessandro Volta describes the **voltaic cell** in a letter to the Royal Society
- William Nicholson and Anthony Carlisle discover **electrolysis**
- Yeast is used as a fermenting agent in **beer** for the first time
- George Shaw describes the **platypus**
- William Herschel discovers **infrared radiation**
- Henry Maudslay develops the first practical **screw-cutting lathe**

“ THE **SUPERIORITY** OF THE **HIGH-PRESSURE ENGINE** WILL SERVE TO... **HONOR** THROUGH ALL TIMES... **THE NAME** OF **TREVITHICK. ”**

Michael Williams, Member of Parliament for West Cornwall, 1853

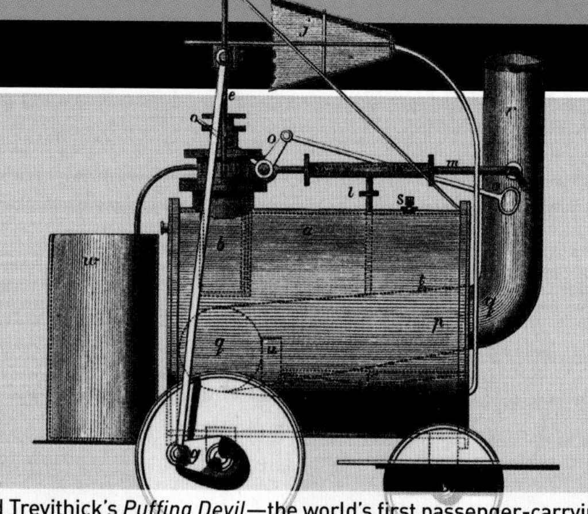

Richard Trevithick's *Puffing Devil*—the world's first passenger-carrying steam road locomotive—ran on the road, not on rails.

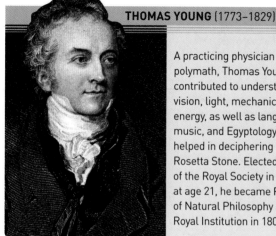

THOMAS YOUNG (1773–1829)

A practicing physician and polymath, Thomas Young contributed to understanding vision, light, mechanics, and energy, as well as language, music, and Egyptology. He helped in deciphering the Rosetta Stone. Elected fellow of the Royal Society in London at age 21, he became Professor of Natural Philosophy at the Royal Institution in 1801.

AT THE START OF THE YEAR, Italian astronomer **Giuseppe Piazzi** (1746–1826) **discovered Ceres**, a celestial body made of rock and ice that orbits the Sun every 1,679.819 days. Thought at the time to be a planet, it is now known to be the archetype of a class of smaller objects known as dwarf planets. Ceres is sometimes wrongly referred to as an asteroid because its orbit between Mars and Jupiter roughly coincides with that of a band of rocky rubble known as the asteroid belt.

The 19th century witnessed a growing understanding of atoms and molecules. In 1801, British chemist **John Dalton** developed his **law of partial pressures**, which stated that the pressure of a gas mixture is the sum of its individual components' partial pressures—the pressure each component gas would exert if it occupied the same space alone. So, in a mixture of nitrogen and oxygen, the total pressure is equal to the partial pressure of oxygen plus the partial pressure of nitrogen.

Since the time of English physicist and mathematician **Isaac Newton** (see 1687–89), it had been widely believed that light was a stream of particles. In May 1801, British polymath Thomas Young (1773–1829) performed the **double slit experiment** (see right), which **demonstrated the wavelike properties of light**. It is now known that light sometimes behaves like waves and at other times like particles. The same year, German mathematician **Johann Georg von Soldner** (1776–1833) predicted how **light** traveling as a stream of particles would be **deflected by gravity** if it passed near the Sun. According to his calculation, a light ray would bend by 0.84 seconds of arc (a unit of angular measurement). More than 100 years later, German-American physicist Albert Einstein (1879–1955) would make a different prediction using his general theory of relativity (see pp.244–245).

British scientist **William Hyde Wollaston** (1766–1828) showed that electricity produced by friction (static electricity) and **galvanic electricity** produced by what is now referred to as a battery, are exactly the same.

French biologist **Jean-Baptiste Lamarck** (1744–1829) published his book *Système des Animaux Sans Vertèbres* (*System of Invertebrate Animals*), in which he coined the term **invertebrate** and developed a system of classification for this group of animals.

different parts of light wavefront travel through parallel slits

light diffracts again, and waves interfere with each other

screen with pattern of light and dark bands caused by interference

light passes through first slit and diffracts

double slit panel

single slit panel

light source

Double slit experiment
British scientist Thomas Young showed that light can exhibit wavelike properties of interference and diffraction, like ripples in a pond.

380

MILES

THE DIAMETER OF **CERES**, THE **FIRST** KNOWN **DWARF PLANET**

The field of engineering was developing rapidly. Designed by Irish-American engineer **James Finley** (1756–1828), the **first suspension bridge hung from wrought-iron chains** was completed at Jacob's Creek, Pennsylvania. It cost $600.

French inventor **Joseph-Marie Jacquard** (1752–1834) developed a system of controlling **textile looms** using punched cards. This was the **beginning of the era of programmable machines**.

On Christmas Eve, English engineer **Richard Trevithick** (1771–1833) successfully tested the **first successful steam road locomotive**. It carried several men up a hill in Camborne in Cornwall, UK, and traveled at approximately 4 mph (6.4 kmph).

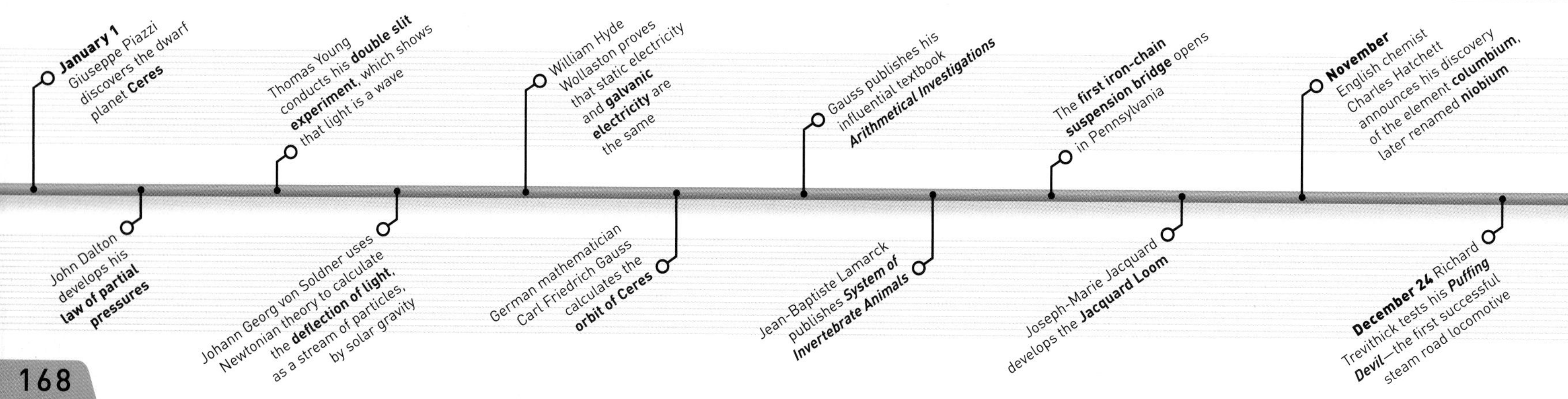
January 1 Giuseppe Piazzi discovers the dwarf planet **Ceres**
John Dalton develops his **law of partial pressures**
Thomas Young conducts his **double slit experiment**, which shows that light is a wave
Johann Georg von Soldner uses Newtonian theory to calculate the **deflection of light**, as a stream of particles, by solar gravity
William Hyde Wollaston proves that static electricity and **galvanic electricity** are the same
German mathematician Carl Friedrich Gauss calculates the **orbit of Ceres**
Gauss publishes his influential textbook ***Arithmetical Investigations***
Jean-Baptiste Lamarck publishes ***System of Invertebrate Animals***
The **first iron-chain suspension bridge** opens in Pennsylvania
Joseph-Marie Jacquard develops the **Jacquard Loom**
November English chemist Charles Hatchett announces his discovery of the element **columbium**, later renamed **niobium**
December 24 Richard Trevithick tests his ***Puffing Devil***—the first successful steam road locomotive

800 THE NUMBER OF BINARY STAR SYSTEMS DISCOVERED BY WILLIAM HERSCHEL

This artist's impression shows an unusual binary star system known as J0806, 1,600 light-years away from Earth. In this rare example, two dense white dwarf stars orbit one another.

ON MARCH 28, GERMAN ASTRONOMER WILHELM OLBERS (1758–1840) discovered the asteroid **Pallas** in an orbit similar to that of Ceres (see 1801). He incorrectly thought that Pallas, Ceres, and other asteroids discovered later were remnants of an exploded planet.

JEAN-BAPTISTE LAMARCK (1744–1829)

Originally a bank worker and army officer, Jean-Baptiste Lamarck became interested in botany and published the popular book *French Flora* in 1773. He also wrote a history and classification of invertebrates and became a highly regarded taxonomist. Lamarck is best known for his theory that acquired traits can be inherited by future generations.

While carrying out experiments with glass prisms, **William Hyde Wollaston** noticed **dark lines in the spectrum of the Sun**. Although it was not known at the time, these lines indicate the absence of particular colors in sunlight. They are usually called **Fraunhofer lines**, after German physicist Joseph von Fraunhofer (1787–1826), who independently rediscovered the lines and studied them in more detail in 1814. In the 1860s, German scientists Gustave Robert Kirchhoff (1824–87) and Robert Bunsen (1811–99) showed how such lines functioned as fingerprints of different elements.

The study of evolution continued at a steady pace. **Jean-Baptiste Lamarck** became one of the first to use the word **biology** in its modern sense. German naturalist and botanist Gottfried Reinhold Treviranus (1776–1837) also used this term, independently of Lamarck, in *Biologie oder Philosophie der Natur Lebenden* (*Biology or Philosophy of Natural Living*). Both Lamarck and Treviranus came up with the idea that evolution occurs through the **inheritance of acquired characteristics**—a concept that came to be known as **Lamarckism**. This concept was often demonstrated by citing the example of the giraffe: a giraffe that stretches its neck to reach high branches in the search for food will bear offspring that have a longer neck as a consequence (see 1809). Although incorrect, this concept was a step toward understanding evolution.

British inventor **Thomas Wedgwood** (1771–1805) was responsible for an important development in the field of photography. Using silver nitrate, he made the first **permanent images**—photographs in the form recognized today. British chemist Humphry Davy (1778–1829) described Wedgwood's work in the *Journal of the Royal Institution* in 1802.

The term **binary stars** was coined by German-born British astronomer **William Herschel** (1738–1822) in this year. A binary is a system where two stars—referred to as binary stars—orbit around each other. These types of stars are distinct from double stars that, despite being very far apart, lie almost along the same line of sight.

As investigations into astronomy gathered pace, French chemist **Joseph Gay-Lussac** (1778–1850) described a law that concerned the **behavior of gases**. According to this law, the pressure of a fixed amount of gas at a fixed volume is directly proportional to its temperature (on the scale that is now known as the Kelvin scale). At first, this law was named Gay-Lussac's Law, but because it follows directly from earlier work by French chemist Jacques Charles and Anglo-Irish chemist Robert Boyle, it is now called the **Third Gas Law** (see panel, above). Today, there is another, quite separate, law known as Gay-Lussac's Law.

one weight produces pressure in the beaker
cool particles move slowly
two weights produce double the pressure in the beaker
hot molecules move faster, increasing the pressure while maintaining the volume
heat

THIRD GAS LAW

Described by French chemist Joseph Gay-Lussac, this law states that for a particular mass of gas at a constant volume, the pressure and temperature are proportional to each other. In the example shown here, when the pressure on the gas in the beaker is doubled, the temperature reached by the gas as it holds up the weights increases proportionately.

Devil's toenail
French biologist Jean-Baptiste Lamarck named numerous species in his lifetime. One of these was the Mesozoic fossil oyster Gryphaea arcuata, *commonly known as Devil's Toenail.*

- Wollaston discovers dark lines in the **solar spectrum**
- Lamarck and Gottfried Reinhold Treviranus promote the idea of **inheritance of acquired characteristics**
- Humphry Davy describes Thomas Wedgwood's method of creating **photographs** using silver nitrate
- **March 28** Wilhelm Olbers discovers the asteroid **Pallas**
- **April 10** The **Great Trigonometric Survey of India** begins with the measurement of a baseline near Madras
- **June** The **first pediatric hospital** opens in Paris, France, on the site of a former orphanage
- British mathematician John Playfair publishes ***Illustrations of the Huttonian Theory of the Earth***
- William Herschel uses the term **binary star** for the first time to refer to a star that revolves around another star
- Joseph Gay-Lussac demonstrates the **Third Gas Law**

THE STORY OF THE ENGINE

THE DRIVING FORCE BEHIND INDUSTRY, ENGINES POWER AN ARRAY OF MACHINES RANGING FROM CARS TO ROCKETS

Engines burn fuel to create hot gases, which expand powerfully to create the mechanical force needed to make things move. As they have developed through the centuries, engines have taken on different forms, from steam engines to rotary engines and gas turbines.

Greek thinkers noticed 2,400 years ago that heat could make things move. In the 1st century CE, they created the aeolipile, in which steam jetted out from a metal sphere to make it spin on a pivot. It was another 1,600 years before the first practical steam engine was built. The breakthrough came in the 1670s, with the discovery of the power of vacuums. French inventor Denis Papin realized that if steam is trapped in a cylinder, it will shrink dramatically as it condenses to create a partial vacuum powerful enough to move things. In 1698, English inventor Thomas Savery built the first full-scale steam engine using this principle.

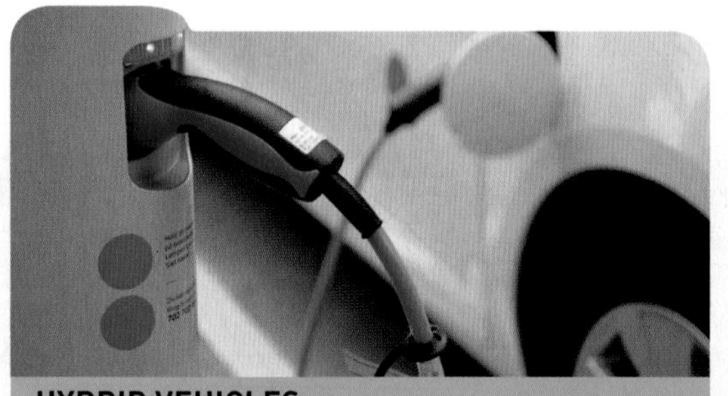

HYBRID VEHICLES

Heat engines burn a lot of fuel and produce waste gases. Fuel shortage and environmental concerns have led to the development of hybrid vehicles that combine different power sources, such as internal combustion engine and an electric motor, to provide a greener, more cost-effective compromise.

FROM CARS TO MARS

For 150 years, engines all depended on steam. They drove the Industrial Revolution, providing power for everything from machines to ships and locomotives. In the mid-19th century, engineers began to develop internal combustion engines based on the rapid expansion of gases burning inside a cylinder. More compact, these engines used gasoline—a more concentrated source of fuel, which could be drawn in automatically, unlike coal, which had to be manually added. Internal combustion engines were key to the development of automobiles, which transformed mobility in the 20th century. The development of jet and rocket engines helped flying machines achieve previously unimaginable speeds, and eventually, propelled spacecraft to the Moon and beyond.

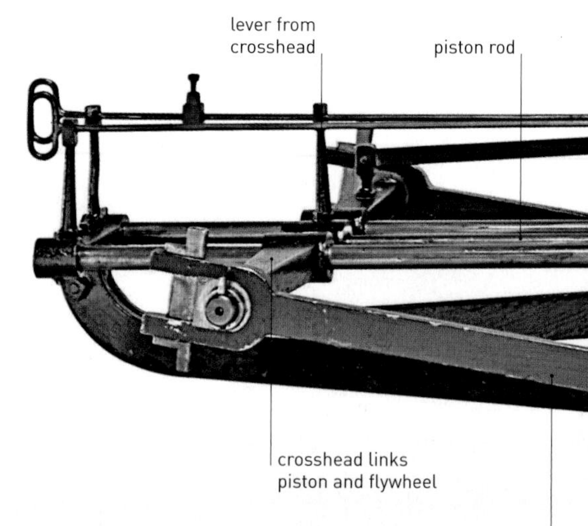
lever from crosshead
piston rod
crosshead links piston and flywheel
connecting rod joins crosshead and crank

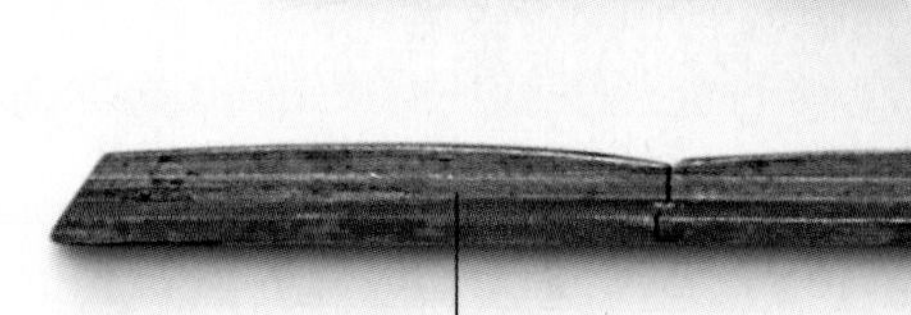

1st century CE
Aeolipile
Alexandrian scholar Hero designs a device in which a sphere is spun by steam jets. A scientific curiosity, it has no apparent practical purpose.

Aeolipile

1679
Papin's steam digester
French inventor Denis Papin invents the steam digester, which traps steam inside a cylinder. This creates a powerful vacuum as the steam cools, condenses, and shrinks.

Steam digester

1698
Savery engine
Thomas Savery builds the first steam engine, to pump water out of mines. It is, however, prone to exploding.

Savery engine

1712
Newcomen engine
British inventor Thomas Newcomen builds a steam engine that avoids the danger of explosion by boiling water separately, and sending steam at low pressure into a cylinder with a piston.

Newcomen engine

1774
Watt engine
Scottish engineer James Watt produces an improved steam engine, which has a separate condensation chamber, and is more efficient.

Watt engine

1791
Barber gas turbine
English inventor John Barber patents a gas turbine, intended to propel a "horseless carriage." In this device, fuel is mixed with air and ignited to produce hot gases, which expand and spin a turbine.

1804
Trevithick engine
Low-pressure vacuum-based engines are big and heavy. English engineer Richard Trevithick's develops a compact and powerful high-pressure steam engine.

smokestack releases exhaust gases

flywheel regulates speed

large gearwheel

cylindrical boiler

crank

gearwheel connects driving wheel and axle

driving wheel

track

> **“ JAMES WATT** [SAID]… I DESERVED **HANGING** FOR BRINGING INTO USE THE **HIGH-PRESSURE ENGINE. ”**
>
> **Richard Trevithick, English engineer,** on Watt's perception of the high-pressure engine as potentially explosive, 1833

First steam locomotive
Richard Trevithick's Penydarren *locomotive exhibited the effectiveness of his high-pressure steam engine when it made its first journey on February 21, 1804 in Wales. The locomotive's engine could also be used as a stationary engine.*

1816
Closed-cycle steam engine
Scottish engineer Robert Stirling invents a steam engine in which gases remain within the system, so there is no exhaust and little noise from explosions.

1860
Gas engine
Invented by Belgian engineer Étienne Lenoir, the first successful internal combustion engine—the gas engine—generates power by burning gas and air inside a cylinder.

Étienne Lenoir's gas engine

1876
Four-stroke engine
German engineer Nikolaus Otto's powerful, four-stroke cycle engine fires four cylinders in turn, so fuel ignited in each pushes down the piston on every fourth stroke.

Four-stroke Daimler motorcycle

1897
Diesel engine
The first diesel engine is built by French–German engineer Rudolf Diesel. Despite being heavier, his invention is more efficient than gasoline engines, and uses the heat of compression, rather than a spark, to ignite fuel.

Diesel engine

1926
Liquid fuel rocket
American engineer Robert Goddard's invents the rocket engine. Thrust is achieved for flight by burning liquid fuel.

1937
Turbojet engine
English engineer Frank Whittle and German engineer Hans von Ohain independently develop and test engines that burn fuel and use a fan to fire out a continuous jet of hot air to thrust a plane forward.

W2/700 Turbojet engine

1956
Rotary engine
German engineer Felix Wankel creates a rotary engine that has a triangular rotor inside an oblong cylinder instead of pistons.

Wankel's rotary engine

> **CONVEX** OR **CONICAL HEAPS,** INCREASING UPWARD FROM A **HORIZONTAL BASE.**

Luke Howard, British meteorologist, describing cumulus clouds in *Essay on the Modifications*, 1803

British pharmacist and meteorologist Luke Howard derived the term cumulus from the Latin word for "heap." As seen in this image, cumulus clouds appear cottonlike and tend to have flat bases.

ON MARCH 28, 1803, THE CHARLOTTE DUNDAS, designed by British engineer **William Symington** (1764–1831), became the **first practical steamboat** when it towed two laden barges through the Forth and Clyde Canal in Scotland.

The following year, another British engineer, **Thomas Telford** (1757–1834), began work on the **Caledonian Canal** in Scotland. When completed in 1822, this canal was 60 miles (100 km) long and 101 ft (30.5 m) wide, with 28 locks—stretches of water enclosed by gates—that were the largest in the world at the time.

In October 1803, British chemist **John Dalton** presented his **atomic theory** to an audience in Manchester, UK. He suggested that: all matter is composed of atoms; atoms cannot be made or destroyed; all atoms of the same element are identical; and different elements have different types of atoms. Dalton also stated that chemical reactions occur when atoms are rearranged, and that compounds are formed from atoms of the constituent elements.

Also in 1803, British scientist **William Hyde Wollaston** added to the number of known elements by discovering **palladium** and **rhodium**.

British pharmacist and meteorologist **Luke Howard** (1772–1864) published a **description of clouds**. He used Latin names to classify them into three simple categories—cirrus, cumulus, and stratus—that are still used.

On February 21, 1804, 25 years before Stephenson's *Rocket* (see 1829), a **steam locomotive** designed by British engineer **Richard Trevithick** (1771–1833) hauled 70 passengers, 11.2 tons (10 tonnes) of iron, and five wagons from the ironworks at Penydarren to the Merthyr-Cardiff Canal in Wales, a distance of 9 miles (14 km). It reached a speed of nearly 5 mph (8 kmph).

Swiss chemist Nicolas-**Théodore de Saussure** (1767–1845) outlined the process of **photosynthesis** (see 1787–88) and proved that both water and carbon dioxide are absorbed by plants as they grow. Saussure later analyzed the ashes of plants and showed that their mineral composition differed from that of soil, indicating that plants absorbed nutrients selectively.

German pharmacist **Friedrich Sertürner** (1783–1841) became the first person to isolate the active ingredient of a medicinal plant. In experiments starting in 1803 and published in 1805, he isolated **morphine** from opium. This substance would later become invaluable in surgery.

Japanese surgeon **Hanaoka Seishū** (1760–1835) performed the first successful surgery using **general anesthesia** in October. The anesthetic was an orally administered herbal concoction.

ELEMENTS

	W.t		W.t
Hydrogen	1	Strontian	46
Azote	5	Barytes	68
Carbon	54	Iron	50
Oxygen	7	Zinc	56
Phosphorus	9	Copper	56
Sulphur	13	Lead	90
Magnesia	20	Silver	190
Lime	24	Gold	190
Soda	28	Platina	190
Potash	42	Mercury	167

Dalton's table of elements
John Dalton was the first to use symbols for elements, and to calculate their atomic weights. This table shows his symbols and atomic weights for 20 elements.

12 PERCENT THE AMOUNT OF **MORPHINE** THAT **OPIUM** CAN CONTAIN

Surgical breakthrough
Hanaoka Seishū carried out the first successful surgery using general anesthesia on a 60-year-old woman who suffered from breast cancer.

JOHN DALTON (1766–1844)

A Quaker schoolteacher, John Dalton was secretary of the Manchester Literary and Philosophical Society, UK, from 1800. Best remembered for his theory of atoms, Dalton also contributed to meteorology and studied color blindness, from which he suffered. He made meteorological observations and published scientific papers well into his seventies.

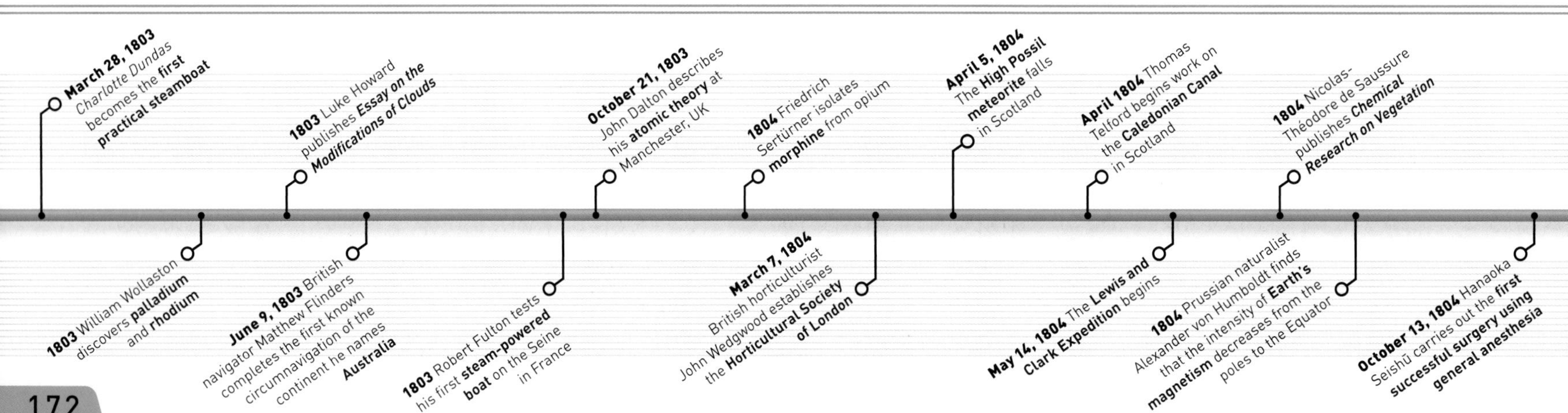
March 28, 1803 Charlotte Dundas becomes the **first practical steamboat**
1803 William Wollaston discovers **palladium** and **rhodium**
1803 Luke Howard publishes **Essay on the Modifications of Clouds**
June 9, 1803 British navigator Matthew Flinders completes the first known circumnavigation of the continent he names **Australia**
1803 Robert Fulton tests his first **steam-powered boat** on the Seine in France
October 21, 1803 John Dalton describes his **atomic theory** at Manchester, UK
1804 Friedrich Sertürner isolates **morphine** from opium
March 7, 1804 British horticulturist John Wedgwood establishes the **Horticultural Society of London**
April 5, 1804 The **High Possil meteorite** falls in Scotland
April 1804 Thomas Telford begins work on the **Caledonian Canal** in Scotland
May 14, 1804 The **Lewis and Clark Expedition** begins
1804 Nicolas-Théodore de Saussure publishes **Chemical Research on Vegetation**
1804 Prussian naturalist Alexander von Humboldt finds that the intensity of **Earth's magnetism** decreases from the poles to the Equator
October 13, 1804 Hanaoka Seishū carries out the **first successful surgery using general anesthesia**

> **"** ...IT WOULD BE **USELESS** TO ATTEMPT TO **SPECULATE** ON THE **REMOTE CAUSE** OF THE **ELECTRICAL ENERGY... "**
>
> **Humphry Davy, British chemist,** from *On Some Chemical Agencies of Electricity*, 1806

A 1920 illustration of the American Lewis and Clark Expedition.

This engraving shows British chemist Humphry Davy conducting experiments with metals, such as magnesium and barium.

FRENCH CHEMISTS NICOLAS-LOUIS VAUQUELIN (1763–1829) and **Pierre-Jean Robiquet** (1780–1840) isolated **asparagine** from asparagus in 1805. This was the first amino acid (the building blocks of protein) to be identified.

The next year, British inventor **Ralph Wedgwood** (1766–1837) was granted a patent for a type of **carbon paper**. He had originally intended it to help visually impaired people to write, but he later realized it could be used to make duplicates of letters.

In September, the **Lewis and Clark Expedition**, commanded by Captain **Meriwether Lewis** (1774–1809) and Lieutenant **William Clark** (1770–1838) of the US military, reached the Pacific Coast of North America. It had been commissioned by US President Thomas Jefferson to explore the Missouri River after the Louisiana Purchase—the US purchase of 828,000 sq miles (2,100,000 sq km) of French territory—in 1803. The expedition discovered several new species of plants and animals.

In November, British chemist **Humphry Davy** presented his work on the **electrolysis of water**—breaking down water into hydrogen and oxygen by passing an electric current through it (see 1834)—at the Royal Society in London, UK.

IN 1807, THE NORTH RIVER STEAMBOAT, later renamed *Clermont*, carried passengers from New York City to Albany on the Hudson River in New York, making it the **first commercially successful steamboat**. Designed by American engineer Robert Fulton (1765–1815), the boat completed the 150 mile (240 km) journey in just over 30 hours.

In Britain, chemist **Humphry Davy** used electrolysis to isolate pure forms of many metals, including magnesium, sodium, barium, and calcium. The first metal separated in this way was **potassium**, in 1807.

William Hyde Wollaston patented the **camera lucida** (light room), a drawing aid for artists, in 1807. This device uses a four-sided prism to project an image of the scene to be drawn onto a drawing surface, allowing the artist to trace it.

Another patent was granted in the same year to French inventors and brothers **Nicéphore** (1765–1833) and **Claude Niépce** (1763–1828) for their invention of an internal combustion engine called the **pyréolophore** (from the Greek words for fire, wind, and bearer). A boat powered by the engine, which burned fine powders such as crushed coal dust, was tested on the Seine River in France.

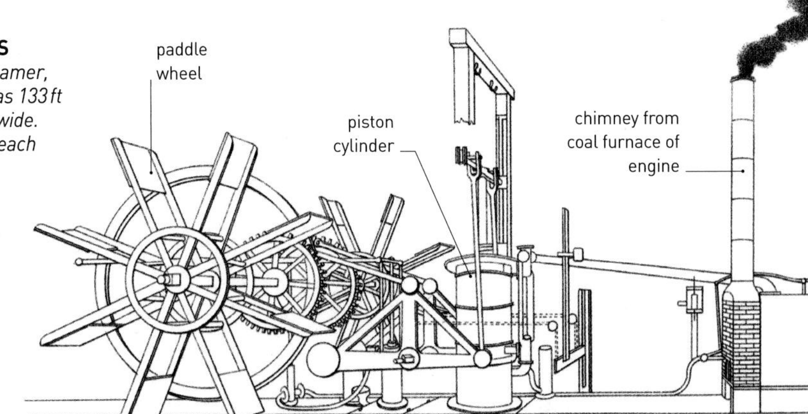
paddle wheel
piston cylinder
chimney from coal furnace of engine

First steamboat in the US
Robert Fulton's paddle steamer, later named Clermont, *was 133 ft (41 m) long and 12 ft (4 m) wide. It had two paddle wheels, each 15 ft (5 m) in diameter.*

Swiss engineer **François Isaac de Rivaz** (1752–1828) was also working on an engine design and received the patent for a **hydrogen-powered internal combustion engine** in 1807. This early engine used only two strokes. The four-stroke engine (see panel, below) was not developed until 1876.

Unaware of earlier work in the field in Europe, Irish-born American mathematician **Robert Adrain** (1775–1843) published his version of the **method of least squares** in 1808. This statistical technique minimizes the sum of the squares of errors made in a data set, and is used to fit curves (graphs) to data.

INTERNAL COMBUSTION ENGINE

Power is generated in an internal combustion engine by burning fuel within a cylinder. Four-stroke engines operate on a four-stroke cycle. First, a valve lets fuel and air into the cylinder. A piston then moves upward in the cylinder, compressing the air–fuel mixture. A spark plug ignites the mixture, making it explode and pushing the piston down. The piston then pushes the exhaust gases out.

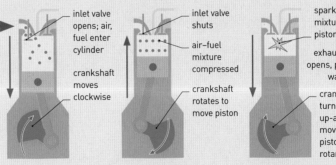

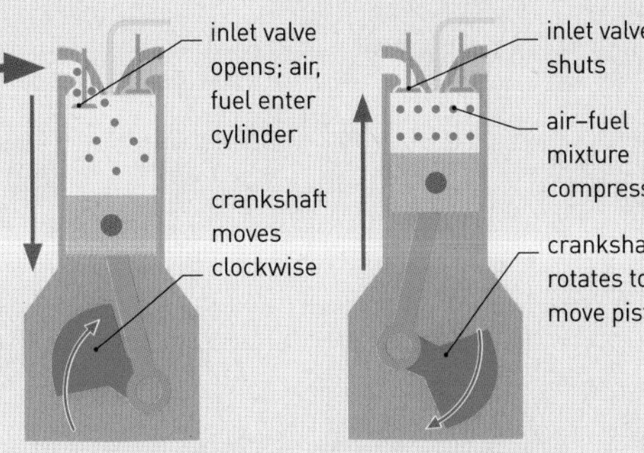
inlet valve opens; air, fuel enter cylinder
crankshaft moves clockwise
INTAKE STROKE
inlet valve shuts
air–fuel mixture compressed
crankshaft rotates to move piston
COMPRESSION STROKE

spark plug ignites mixture, pushing piston down
exhaust valve opens, pushing waste out
crankshaft turns up-and-down movement of piston into rotary motion
motion of crankshaft makes wheels spin
IGNITION STROKE
EXHAUST STROKE

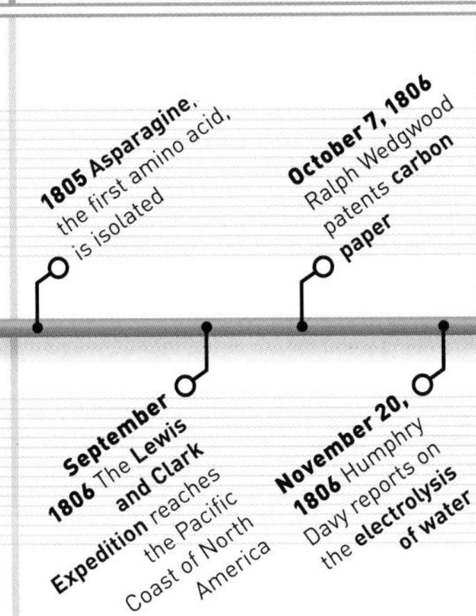
1805 Asparagine, the first amino acid, is isolated
October 7, 1806 Ralph Wedgwood patents carbon paper
September 1806 The Lewis and Clark Expedition reaches the Pacific Coast of North America
November 20, 1806 Humphry Davy reports on the electrolysis of water

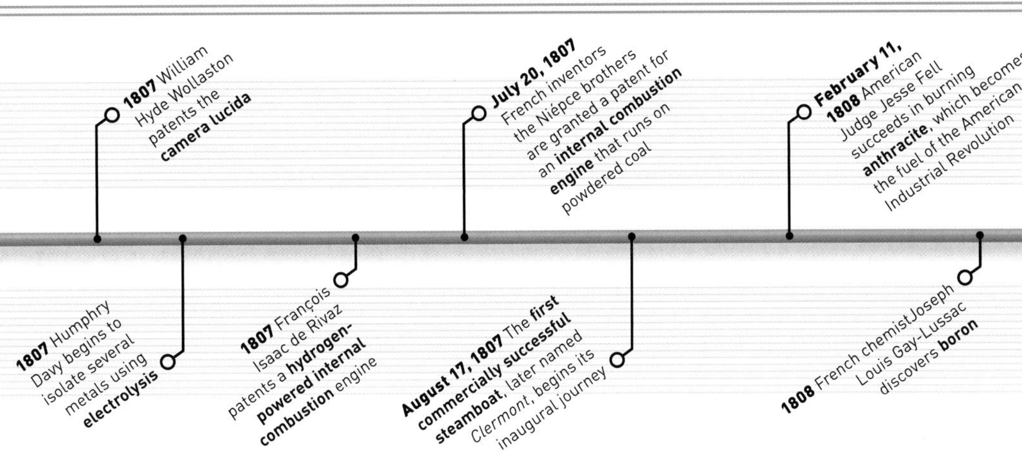
1807 William Hyde Wollaston patents the camera lucida
1807 Humphry Davy begins to isolate several metals using electrolysis
1807 François Isaac de Rivaz patents a hydrogen-powered internal combustion engine
July 20, 1807 French inventors the Niépce brothers are granted a patent for an internal combustion engine that runs on powdered coal
August 17, 1807 The first commercially successful steamboat, later named Clermont, begins its inaugural journey
February 11, 1808 American Judge Jesse Fell succeeds in burning anthracite, which becomes the fuel of the American Industrial Revolution
1808 French chemist Joseph Louis Gay-Lussac discovers boron
1808 Robert Adrain publishes his version of the method of least squares

UNDERSTANDING COMPOUNDS AND REACTIONS

SUBSTANCES CAN CHANGE THEIR FORM IN CHEMICAL REACTIONS

POTASSIUM IN WATER
When potassium reacts with water, hydrogen gas is released. The reaction also produces heat, which ignites the hydrogen.

A chemical compound is a substance composed of two or more types of atom held together by chemical bonds. Water, for example, is made of hydrogen atoms bonded to oxygen atoms. Chemical reactions involve the breaking or forming of chemical bonds, resulting in new substances.

Most solids, liquids, and gases are mixtures of compounds or elements (an element is a substance made of only one kind of atom). Air, for example, is a mixture composed mainly of the elements nitrogen and oxygen, with most of the rest consisting of the element argon and the compounds water, carbon dioxide, and methane. In some compounds, the atoms bond by sharing electrons to form molecules (see right). This kind of bond is called a covalent bond. In other types of compound, the atoms have lost or gained electrons and are in the form of electrically charged ions. These ions are held together by the electrical forces between them: ionic bonds.

1,000

MILLION MILLION MILLION—ROUGHLY THE NUMBER OF **MOLECULES** IN **ONE DROP** OF WATER

COMPOUNDS

Any sample of a particular compound always has the same ratio of the elements of which it is composed. For example, if the compound methane was broken down into its constituent atoms and the atoms were counted, the carbon (C) and hydrogen (H) atoms would always be in the ratio of 1:4. As a result, every compound has a chemical formula—methane's is CH_4.

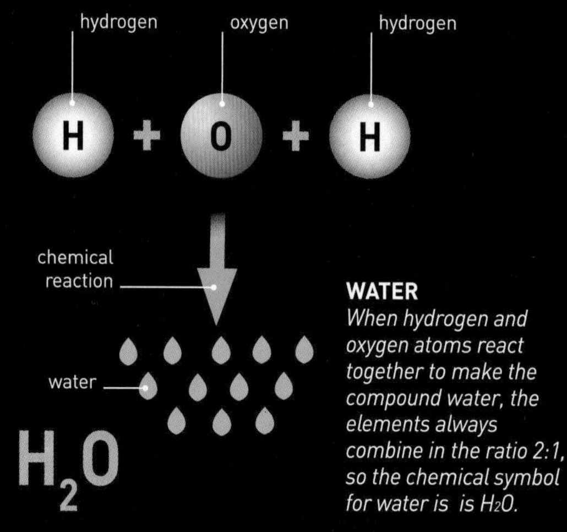

hydrogen
oxygen
hydrogen
H + O + H
chemical reaction
water
H_2O

WATER
When hydrogen and oxygen atoms react together to make the compound water, the elements always combine in the ratio 2:1, so the chemical symbol for water is is H_2O.

MOLECULES

The atoms that make up a molecule are joined together by one or more covalent bonds, rather than the ionic bonds that hold atoms together in nonmolecular compounds such as sodium chloride (common salt). The smallest molecules are composed of just two atoms, but some are much larger; proteins, for example, may consist of tens of thousands of atoms. Some elements can also exist as molecules. For instance, pure hydrogen and oxygen are typically composed of two-atom (diatomic) molecules—H_2 and O_2.

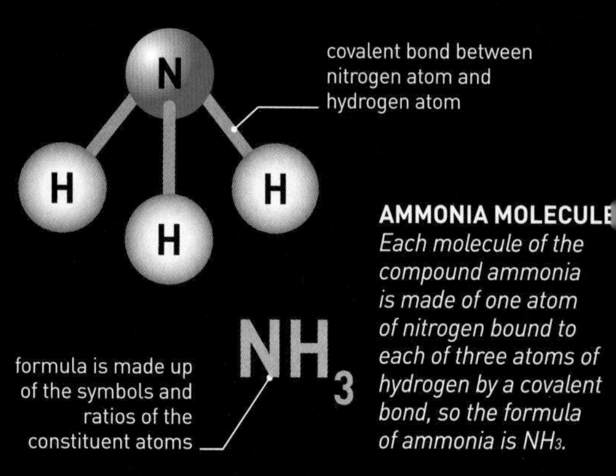

covalent bond between nitrogen atom and hydrogen atom
N
H
H
H
formula is made up of the symbols and ratios of the constituent atoms
NH_3

AMMONIA MOLECULE
Each molecule of the compound ammonia is made of one atom of nitrogen bound to each of three atoms of hydrogen by a covalent bond, so the formula of ammonia is NH_3.

REACTIONS

The elements or compounds that take part in a reaction are called reactants. During a reaction, the bonds of the reactants break and new bonds may form, producing one or more different substances—known as the products. For example, in the reaction shown on the right, the atoms of two reactants combine to form a single compound as the product. The atoms involved in a reaction do not go out of or come into existence—they are just rearranged, so the total mass of the products is the same

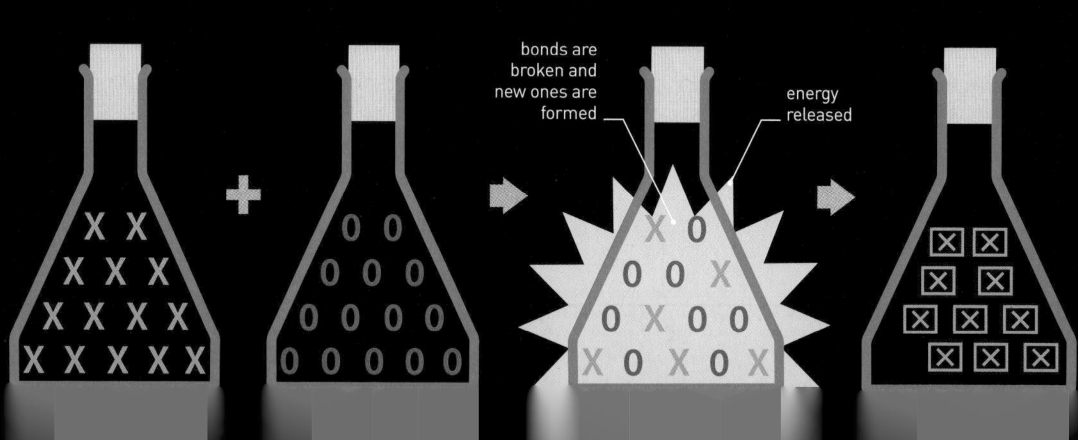

bonds are broken and new ones are formed
energy released

ENERGETIC REACTION
Two reactants may react spontaneously when mixed, forming a new compound. In

TYPES OF REACTION

There are many different types of reaction, including, for example, electrolysis (in which an electric current splits a compound into its constituent parts) and acid-base reactions (in which an acid and base, or alkali, react together). In general, however, reactions can be categorized into three main types according to what happens to the substances involved: decomposition, synthesis, and displacement (or replacement) reactions. In a decomposition reaction, a compound breaks up into smaller parts. Synthesis is the opposite: two or more compounds combine to form a single product. In a displacement reaction, part of one compound breaks away and becomes part of another.

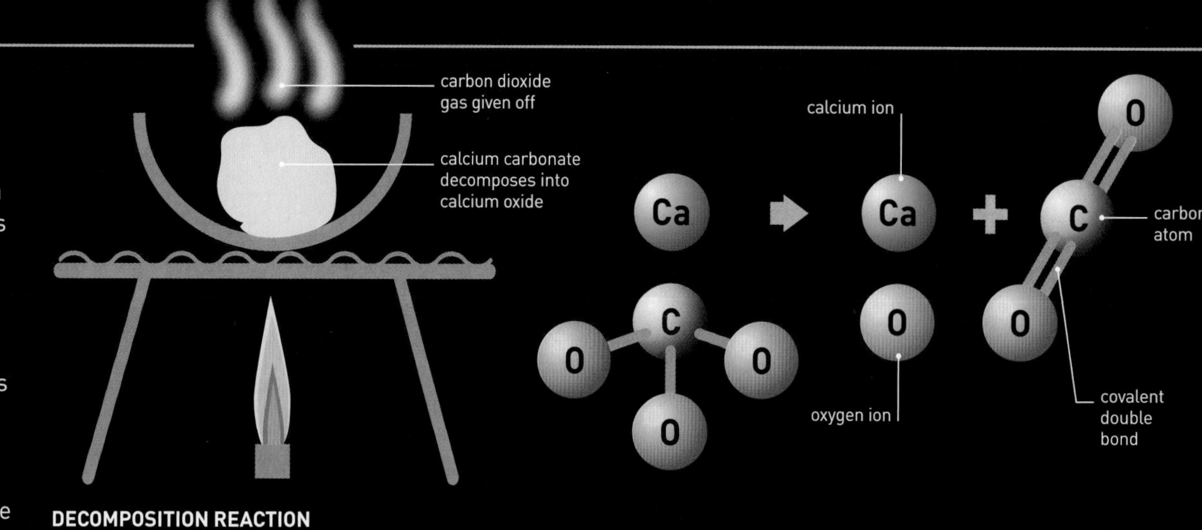
carbon dioxide gas given off
calcium carbonate decomposes into calcium oxide
calcium ion
carbon atom
oxygen ion
covalent double bond
Ca
C
O
$CaCO_3 \rightarrow CaO + CO_2$
calcium carbonate
calcium oxide
carbon dioxide

DECOMPOSITION REACTION
Heating the mineral limestone (calcium carbonate) causes it to decompose into calcium, oxygen, and carbon dioxide. The calcium and oxygen are in the form of ions (charged particles); they form the ionic solid calcium oxide. The carbon dioxide is a gas composed of covalently bonded molecules.

> THE WORLD OF **CHEMICAL REACTIONS** IS LIKE A STAGE, ON WHICH SCENE AFTER SCENE IS CEASELESSLY PLAYED. THE ACTORS ON IT ARE THE **ELEMENTS.**

Clemens Alexander Winkler, German chemist, 1887

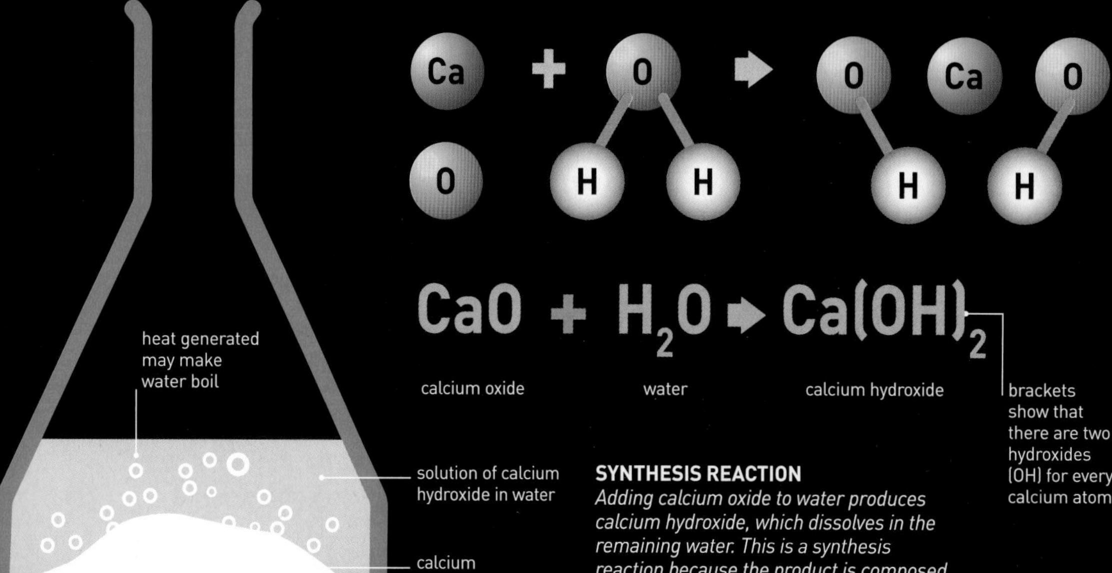
Ca
O
H
heat generated may make water boil
solution of calcium hydroxide in water
calcium oxide
$CaO + H_2O \rightarrow Ca(OH)_2$
calcium oxide
water
calcium hydroxide
brackets show that there are two hydroxides (OH) for every calcium atom

SYNTHESIS REACTION
Adding calcium oxide to water produces calcium hydroxide, which dissolves in the remaining water. This is a synthesis reaction because the product is composed of all the atoms of the two ingredients.

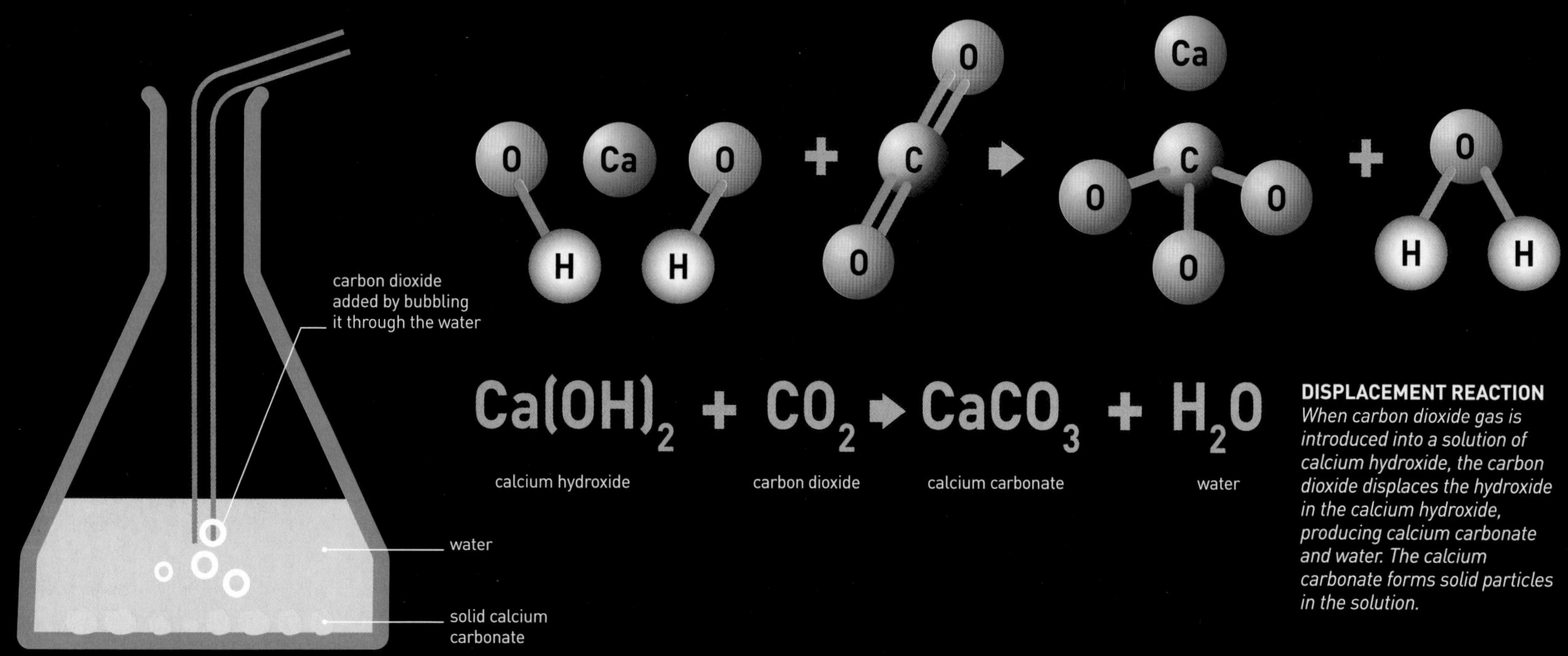
carbon dioxide added by bubbling it through the water
water
solid calcium carbonate
Ca
C
O
H
$Ca(OH)_2 + CO_2 \rightarrow CaCO_3 + H_2O$
calcium hydroxide
carbon dioxide
calcium carbonate
water

DISPLACEMENT REACTION
When carbon dioxide gas is introduced into a solution of calcium hydroxide, the carbon dioxide displaces the hydroxide in the calcium hydroxide, producing calcium carbonate and water. The calcium carbonate forms solid particles in the solution.

1809

Later commentators often depicted Lamarckian evolution with the idea that giraffes acquired long necks by stretching to reach high branches.

IN 1809, French biologist **Jean de Monet, Chevalier de Lamarck** (1744–1829) came up with one of the first systematic theories about the **evolution of life**. Lamarck argued that life evolved gradually, from the simplest to the most complex. He suggested that a change in the environment can provoke a change in an organism, and that these **changes can also be inherited**. According to him, useful characteristics develop further over the generations, and those that are not useful fall into disuse and may disappear. Unlike Charles Darwin (see 1859), Lamarck had no mechanism to explain how these changes occur. One of his ideas was that an **organism changes during its life to adapt to its environment**, and these changes are then passed on to its offspring. This idea, called **Lamarckian inheritance**, was largely ridiculed by followers of Darwin, but there has been renewed interest with recent discoveries that the environment can alter genes and their expression—the study of this is know as **epigenetics**.

In Germany, mathematician **Carl Friedrich Gauss** (1777–1855) laid the foundations of astronomical mathematics with his **gravitational constant**. Newton had shown that there is a single universal figure, or constant, for the power of gravitational attraction. Gauss's insight was to devise a simple set of three measurements for calculating **gravity's effects** in which masses are measured in solar masses (the mass of the Sun), distance is measured in terms of the longest diameter of Earth's orbit, and time is measured in days. From these measurements he found a constant for gravity of 0.01720209895. This number was fed into calculations to work out **planetary orbits**. We now know that the measurements Gauss used were not as invariable as originally thought, but despite this his work has been of great value to astronomers.

0.01720209895

THE **UNIVERSAL CONSTANT FOR GRAVITY,** ACCORDING TO CARL FRIEDRICH GAUSS

1810–11

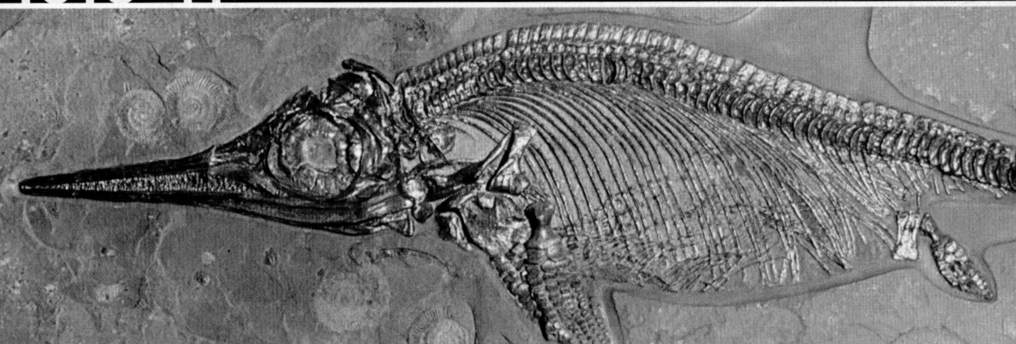

Mary Anning's discovery of a fossil, later called ichthyosaur, confirmed that the oceans too once held strange creatures that are no longer alive today.

BRITISH CHEMIST HUMPHRY DAVY (1778–1829) amazed audiences at his London science demonstrations with the glow from the **first electric lamp**, the arc lamp, in which high voltage was shot across the gap between two carbon electrodes. Although bright, the arc lamp was not practical to use for everyday lighting. Everyday electric light arrived only when American inventor Thomas Edison (1847–1931) and British physicist Joseph Swan (1828–1914) developed the incandescent lamp (see 1878–79). Davy also proved that **chlorine is an element**, and that muriatic acid is a compound of hydrogen and chlorine (now known as **hydrochloric acid**), disproving French chemist Antoine-Laurent Lavoisier's theory that every acid contained oxygen.

> “THE **CEREBRUM** I CONSIDER AS THE **GRAND ORGAN** BY WHICH THE **MIND** IS **UNITED** TO THE **BODY.**”
>
> **Sir Charles Bell,** from *Idea of a New Anatomy of the Brain,* 1811

In 1811, Italian chemist **Amedeo Avogadro** (1776–1856) reconciled John Dalton's atomic theory of elements (see 1803–04) with Gay-Lussac's law of 1808, which said that when two gases react, the volumes of the reactants and products are in simple whole number ratios. Avogadro realized the difference between atoms and molecules. So, simple gases such as hydrogen and oxygen are made of molecules of two or more atoms joined together. From this, Avogadro deduced his hypothesis that **any gas at the same temperature and pressure** always contains the **same number of molecules**.

Another milestone in chemistry was the system of **chemical symbols and formulae** proposed by Swedish chemist **Jöns Jacob Berzelius** (1779–1848) in 1811 that is still used today. He suggested that every element be identified simply by its initial letter as a capital. Where two elements begin with the same letter, he added a second letter or consonant of the name. To show the number of atoms of an element in a compound, he added a figure to the symbol. So the formula for water is H_2O, indicating there are two atoms of hydrogen for each one of oxygen.

Meanwhile, on the south coast of England, 11-year-old **Mary Anning** made the first of her many key fossil finds. It was of an **ichthyosaur**—a marine reptile

MARY ANNING (1799–1847)

The daughter of a poor cabinet maker in the British coastal town of Lyme Regis, Mary Anning became the greatest fossil hunter of the age. Among her key finds were the almost complete skeletons of marine reptiles such as the ichthyosaur and the plesiosaur. At the time, Anning was one of the foremost experts on the anatomy of these fossil creatures.

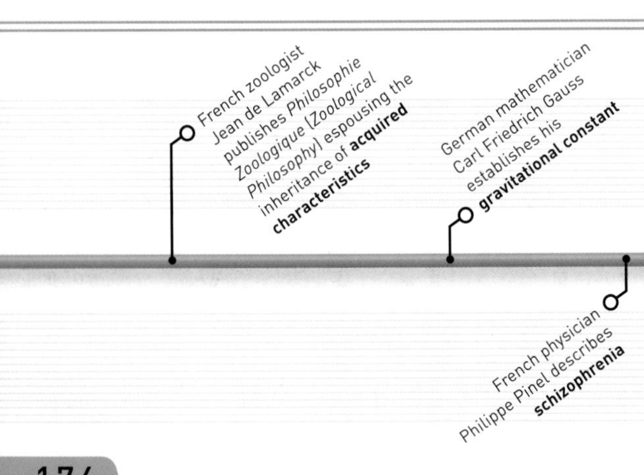
French zoologist Jean de Lamarck publishes Philosophie Zoologique (Zoological Philosophy) espousing the inheritance of acquired characteristics
German mathematician Carl Friedrich Gauss establishes his gravitational constant
French physician Philippe Pinel describes schizophrenia

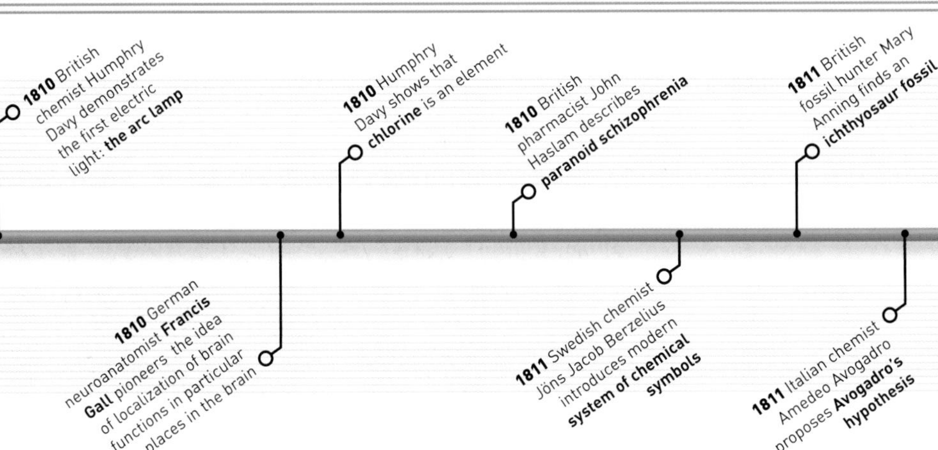
1810 British chemist Humphry Davy demonstrates the first electric light: the arc lamp
1810 German neuroanatomist Francis Gall pioneers the idea of localization of brain functions in particular places in the brain
1810 Humphry Davy shows that chlorine is an element
1810 British pharmacist John Haslam describes paranoid schizophrenia
1811 Swedish chemist Jöns Jacob Berzelius introduces modern system of chemical symbols
1811 British fossil hunter Mary Anning finds an ichthyosaur fossil
1811 Italian chemist Amedeo Avogadro proposes Avogadro's hypothesis

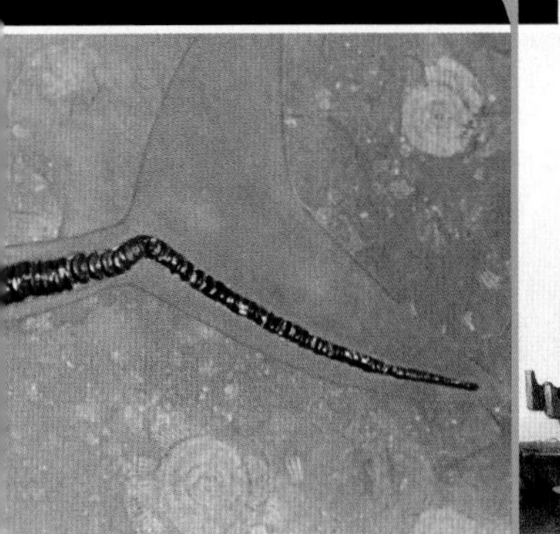

A coal train is hauled by steam locomotive *Salamanca* at the Middleton Colliery Railway, UK, in 1814.

> **“ ...RAGE FOR RAILROADS** IS SO GREAT THAT **MANY WILL BE LAID** IN PARTS WHERE THEY WILL NOT PAY. **”**
>
> **George Stephenson, British civil engineer,** in a letter to Joseph Sandars, December 1824

atomic number | atomic weight

26 55.845

Fe

IRON

name | chemical symbol

CHEMICAL SYMBOLS

The system of symbols devised by Berzelius is used by chemists even today. Each element is indicated by the initial letter or two of its Latin name—the symbol for iron, Fe, comes from its Latin name, *ferrum*. The box in the periodic table for each element shows its atomic number, its atomic weight, and the number of protons in the nucleus of each atom.

shaped like a dolphin—that lived at the time of the dinosaurs.

Also in Britain, anatomist **Charles Bell** published *Idea of a New Anatomy of the Brain*, which distinguished between the **sensory and motor nerves** of the brain.

THE FRENCH REVOLUTION'S introduction of the **metric system** in the 1790s created chaos, as many people insisted on continuing to use the local units of measure that existed in different towns. So, in 1812, French emperor **Napoléon Bonaparte** introduced the *mesures usuelles* (standard measures), which combined basic metric units—such as the meter and kilogram—and old familiar measures. This system was finally replaced in 1840 with the full metric system.

New metric system
This engraving satirizes the confusion French people had in adopting the metric system, which is why Napoleon introduced the compromise—mesures usuelles.

In this year, German geologist **Friedrich Mohs** (1773–1839) devised a system to **identify minerals**. It was based on the physical properties, such as hardness, color, and shape. Mohs noticed that hard minerals could scratch softer ones. He developed a scratch test to determine the hardness of each mineral and a scale of 10 standard minerals—now known as the **Mohs scale**—on which to place each mineral. French paleontologist Georges Cuvier (1769–1832) published the *Discours Préliminaire* (*Preliminary Discourse*), an introduction to his essays on fossil quadrupeds (animals with four legs), in which he argued that many more species lived on Earth in the past and that every rock bed contains fossils from a different time in Earth's past. In line with geologists who believed that the **landscapes of the world** were **shaped by a series of catastrophes**, Cuvier argued that the world had been overcome by past catastrophes or revolutions, which had swept away a large number of species.

Meanwhile, the **first steamboat service** in Europe opened with the paddlesteamer PS *Comet* plying on the River Clyde in Scotland. In Middleton, West Yorkshire, UK, steam locomotives were used to **haul trains** successfully for the first time, on an adapted track initially built in the 1750s to enable horses to pull wagons full of coal from the Middleton mine.

In the 1790s, Frenchman **Nicholas Appert** had developed sealed **glass jars to preserve food**, but the glass was breakable. Then in 1810, British merchant **Peter Durand** patented the tin can, which was made of iron coated with tin to prevent rusting. In 1812, American engraver **Thomas Kensett** (1786–1829) established the **first food preservation factory** in New York, for preserving oysters, meat, fruit, and vegetables in glass jars. In 1825, Kensett set up the first US canning factory.

James Barry (*c.*1792–1865)—born and raised as Margaret Ann Bulkley—chose to live as a man so she could be accepted into university. In 1812, she became the **first woman to qualify as a medical doctor**, graduating from the University of Edinburgh, Scotland. She went on to become a distinguished surgeon.

MOHS SCALE

The scale rates mineral hardness from 1 to 10 in terms of standard minerals. Geologists use scratch tests to identify a mineral—for example, one that scratches apatite but is scratched by quartz is a 6 on the scale.

1 TALC

2 GYPSUM
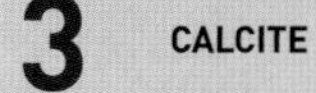

3 CALCITE

4 FLUORITE
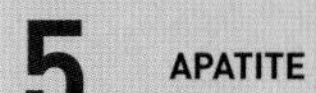

5 APATITE

6 ORTHOCLASE
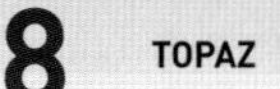

7 QUARTZ

8 TOPAZ

9 CORUNDUM

10 DIAMOND

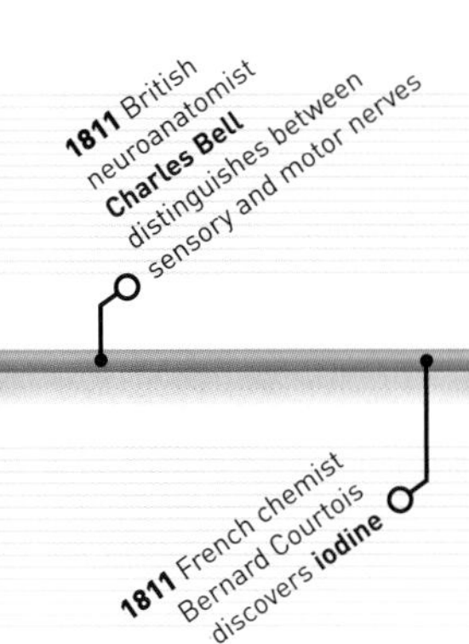

1811 British neuroanatomist **Charles Bell** distinguishes between sensory and motor nerves

1811 French chemist Bernard Courtois discovers **iodine**

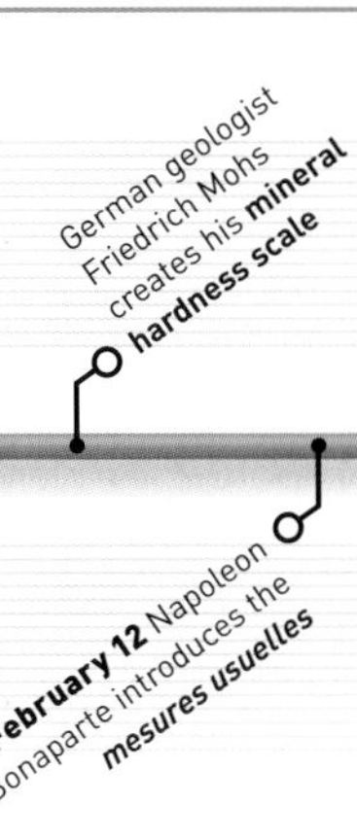

German geologist Friedrich Mohs creates his **mineral hardness scale**

February 12 Napoleon Bonaparte introduces the ***mesures usuelles***

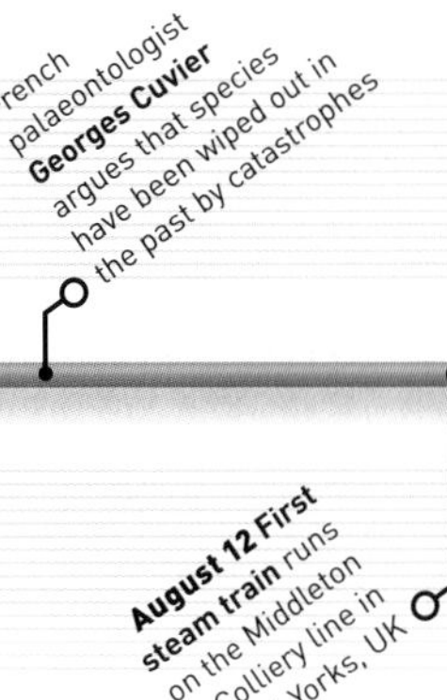

French palaeontologist **Georges Cuvier** argues that species have been wiped out in the past by catastrophes

August 12 First steam train runs on the Middleton Colliery line in West Yorks, UK

August 15 First steamboat service opens on the River Clyde in Scotland

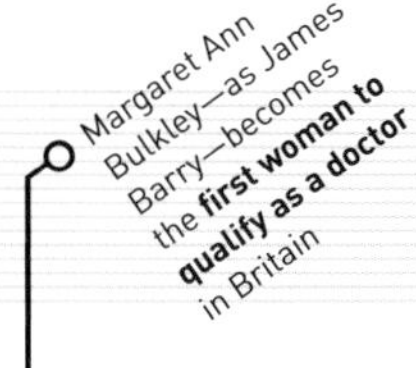

Margaret Ann Bulkley—as James Barry—becomes the **first woman to qualify as a doctor** in Britain

American engraver Thomas Kensett sets up the US's first **food preservation** factory

1813–14

Fraunhofer lines are dark lines in a light spectrum created by the absorption of certain wavelengths by gases. The pattern reveals the identity of the gas.

ON MARCH 13, 1813, BRITISH ENGINEER William Hedley (1779–1843) patented a design for a steam locomotive known as **Puffing Billy**. It began hauling coal trucks in Northumberland, England, in 1814 and is **the world's oldest surviving steam locomotive**. The greatest pioneer of steam railroads, **George Stephenson** (1781–1848), also built his first steam locomotive in the north of England; it first ran on July 25, 1814.

In London, in 1813, American physician **William Wells** (1757–1817) read a paper to the Royal Society in which he explained racial differences on the basis of a **process of evolution involving natural selection**.

In Bavaria, German optician **Joseph von Fraunhofer** was making fine optical glass for separating light into different colors. When sunlight passed through the glass, Fraunhofer noticed **dark lines (Fraunhofer Lines) where color in the light spectrum was missing**. Fraunhofer was not the first to notice these lines, but in 1814 he was the first to start an extensive study of them, and in doing so provided a basis for the science of spectroscopy (see 1884–85).

> “ HE [WELLS] DISTINCTLY RECOGNIZES THE PRINCIPLE OF **NATURAL SELECTION...** THIS IS THE **FIRST RECOGNITION...** BUT HE APPLIES IT ONLY TO MAN. ”
>
> **Charles Darwin, British naturalist,** *The Origin of Species by Means of Natural Selection,* 4th edition, 1866

Elsewhere in 1814, news was delivered with the help of **steam-powered presses** at *The Times* newspaper in London.

In Connecticut, inventor **Eli Terry** (1772–1852) developed a groundbreaking design for a **mass-produced clock** that could be made by machines instead of being hand-assembled by skilled clockmakers. This made clocks more affordable.

1815

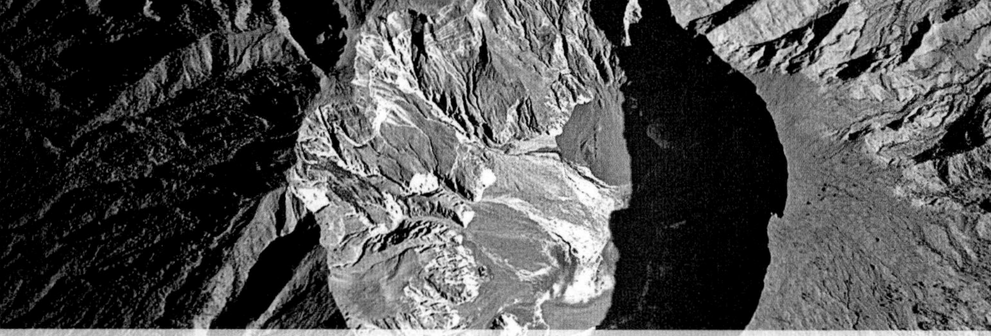

Mount Tambora on Sumbara Island in Indonesia was the site the largest volcanic events in recorded history. The eruption took 4,593 ft (1,400 m) in height off the cone of the volcano.

MOUNT TAMBORA ON THE INDONESIAN ISLAND of Sumbawa erupted on April 5. It was the most **powerful volcanic eruption** in recorded history—the explosions could be heard as far away as 1,616 miles (2,600 km).

The role of fossils in studying Earth's history was revealed by British geologist **William Smith** (1769–1839). Working as a surveyor, overseeing the digging of canals, Smith noticed that widely separated outcrops of the same **rock strata could be identified by the fossils they contain**. He used this to create the first geological map in 1799, and in 1815 he published a geological map of Britain. His map became the model for all geological maps.

The canals Smith helped to build were essential to the accelerating Industrial Revolution in Britain, as were the mines that provided coal for fires and steam engines. Mining was dangerous work, however, and miners lived in constant fear of hitting pockets of methane or other flammable gases, known as firedamp, which could explode if they reached the naked flame of their candles. British scientist **Humphry Davy invented the miners' safety lamp**. The lamp's flame was wrapped in wire mesh that reduced the chance of it igniting gases.

In chemistry, the atomic theory of elements was gaining supporters. British scientist **William Prout** (1785–1850) concluded from studying tables of atomic weights (see 1803–04) that every weight is a multiple of the weight of a hydrogen atom, and that the **hydrogen atom is the only fundamental particle** from which all other elements are made up. He was not right, but a century later, in 1920, Ernest Rutherford (1871–1937) named the proton partly in honour of Prout.

In France, scientist **Jean-Baptiste Biot** (1774–1862) was **experimenting with polarized light**—light vibrating in just a single plane (see panel, opposite). On October 23, he shone a beam of polarized light through a tube of turpentine and noticed how the plane of polarization was rotated. Other organic liquids, such as lemon juice, produced the same effect. This rotation is at the heart of the Liquid Crystal Displays (LCD) now widely used in display screens.

wire mesh prevents flame from igniting mine gasses

Davy lamp
This miner's safety lamp consisted of a cylinder of wire gauze containing a wick attached to an oil reservoir.

William Smith's geological map
This pioneering map of Britain showed the geological make-up of the country and set a precedent for geological maps in the future.

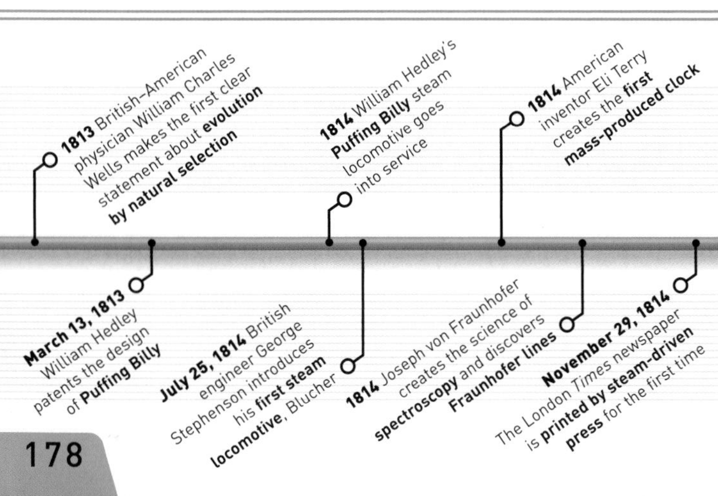

1813 British-American physician William Charles Wells makes the first clear statement about **evolution by natural selection**
March 13, 1813 William Hedley patents the design of **Puffing Billy**
1814 William Hedley's **Puffing Billy** steam locomotive goes into service
July 25, 1814 British engineer George Stephenson introduces his **first steam locomotive**, Blucher
1814 Joseph von Fraunhofer creates the science of **spectroscopy** and discovers **Fraunhofer lines**
1814 American inventor Eli Terry creates the **first mass-produced clock**
November 29, 1814 The London *Times* newspaper is **printed by steam-driven press** for the first time

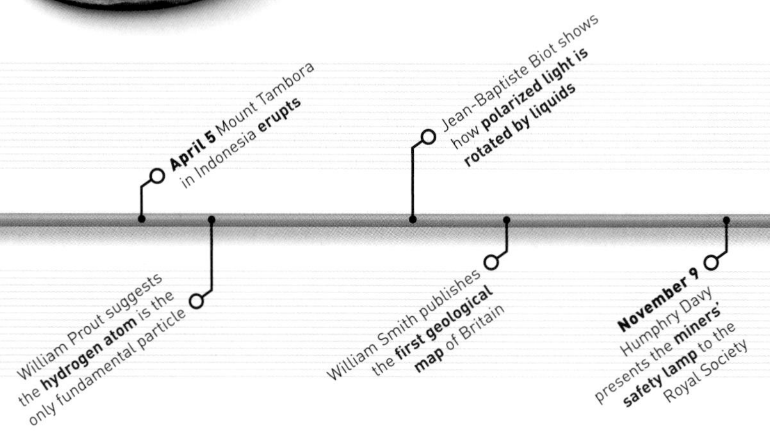

April 5 Mount Tambora in Indonesia **erupts**
William Prout suggests the **hydrogen atom** is the only fundamental particle
Jean-Baptiste Biot shows how **polarized light is rotated by liquids**
William Smith publishes the **first geological map** of Britain
November 9 Humphry Davy presents the **miners' safety lamp** to the Royal Society

10 MILES PER HOUR THE AVERAGE SPEED OF DRAIS'S LAUFMASCHINE

Karl von Drais's Laufmaschine (running machine) was a forerunner of the bicycle and the first two-wheeled form of personal transportation.

each ring refracts light at a different angle
stepped lens focuses the beam

Fresnel lighthouse lens
His experiments in light and optics led Fresnel to create a special lens for use in lighthouses; it was also sometimes used in theater lights. It focuses the light using multiply stepped glass, rather than a single thick lens.

AUGUSTIN-JEAN FRESNEL (1788–1827)

Fresnel worked as an engineer during the Napoleonic Wars of 1803 to 1815. Afterward, he began to research light and optics, making key contributions to understanding the nature of light waves, diffraction, and polarization. He is best known for his invention of the stepped glass Fresnel lens, which is commonly used in lighthouses.

THE THEORY THAT LIGHT TRAVELS IN WAVES (see 1801) was backed up in 1816 by a series of precise experiments with diffraction—the way light spills around objects into shadows—by French engineer **Augustin-Jean Fresnel**. When Fresnel shone a light through slits, he detected tiny fringes of light that could only be produced by interference between waves. Fresnel backed this up with detailed calculations of **how a light wave might move and produce diffraction.**

Working with French physicist **François Arago** (1786–1853), in 1817 Fresnel began to explore polarization, which was then thought irreconcilable with wave theory. **Polarized light is reflected in just one plane**; Fresnel found that beams of polarized light do not create interference fringes if they are polarized in different planes.

In 1816, British physicist **David Brewster** (1781–1868) **calculated Brewster's Angle**—the angle at which light must strike an object at for maximum polarization.

In 1817, **three new elements were discovered**: cadmium by German chemist **Friedrich Stromeyer** (1776–1835); lithium by Swedish chemist Johann Arfwedson (1792–1841); and selenium by Swedish chemist Jöns Berzelius (1779–1848).

In Germany, Baron **Karl von Drais** (1785–1851) **invented an early bicycle** known as the Laufmaschine—propelled by feet rather than pedals. His first public ride took place on June 12, 1817.

POLARIZATION

Augustin-Jean Fresnel worked out that light moves forward in waves that vibrate transversely—perpendicularly to the direction in which they are traveling.

Ordinary light vibrates at every angle or plane, but when it is polarized—by passing through a polarizing filter—the vibrations are reduced to a single plane.

beam of ordinary light vibrating on numerous planes

polarizing light vibrating in one plane

second filter blocks polarized light

polarizing filter

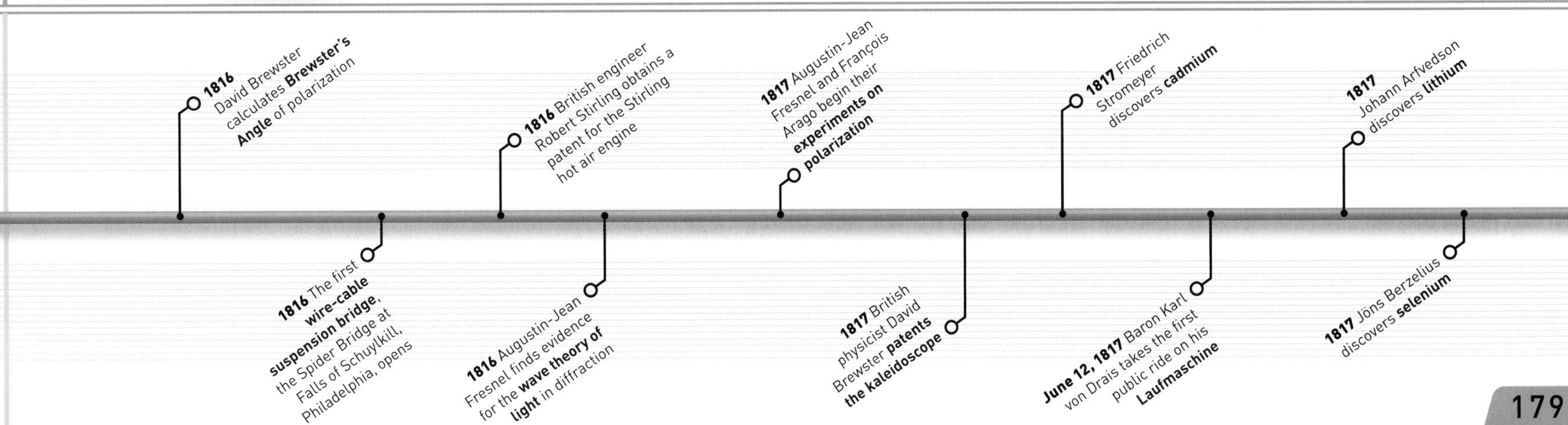

1816 David Brewster calculates **Brewster's Angle** of polarization
1816 The first **wire-cable suspension bridge**, the Spider Bridge at Falls of Schuylkill, Philadelphia, opens
1816 British engineer Robert Stirling obtains a patent for the Stirling hot air engine
1816 Augustin-Jean Fresnel finds evidence for the **wave theory of light** in diffraction
1817 Augustin-Jean Fresnel and François Arago begin their **experiments on polarization**
1817 British physicist David Brewster **patents the kaleidoscope**
1817 Friedrich Stromeyer discovers **cadmium**
June 12, 1817 Baron Karl von Drais takes the first public ride on his **Laufmaschine**
1817 Johann Arfvedson discovers **lithium**
1817 Jöns Berzelius discovers **selenium**

4 FLUID OUNCES THE AMOUNT OF BLOOD BLUNDELL EXTRACTED

To perform the first successful blood transfusion, James Blundell took blood from the arm of his assistant and injected it into the patient.

THE BRITISH PHYSICIST MICHAEL FARADAY (1791–1867), later famous for his work on electromagnetism, spent the initial years of his career concentrating on chemistry. Together with the utensil maker James Stodart, Faraday began to experiment with different **steel alloys** to incorporate rare metals such as platinum. These new alloys were too expensive to be commercial, but they showed the value of a scientific approach.

The demands of industry accelerated progress in technology, and the increased status of engineers was reflected in the founding of the Institute of Civil Engineers in London. Steam locomotives were becoming more than curiosities, but they were still expensive to run and liable to explode. British engineer **Robert Stirling** invented a **heat engine** that was intended as an alternative. His engine worked by continually **compressing and expanding air** or another gas in a closed space. The Stirling engine didn't catch on at the time although recently it has excited interest as a simple and low maintenance power source for everything from use in the Third World to space exploration.

At the same time, the natural world continued to hold fascination for many people. French naturalist **Georges Cuvier** studied fossils in the possession of the British clergyman William Buckland, found near Stonesfield in England a few years earlier. Cuvier confirmed that these fossils belonged to a gigantic extinct lizard.

This period also saw the growing professionalism of surgeons and doctors. London doctor **James Blundell** saved a mother from bleeding to death after giving birth with the **first successful blood transfusion**. He used a syringe to extract blood from the arm of a donor and he injected it into the arm of the patient. This was before doctors realized what caused blood to clot or knew about blood groups (see 1901).

gas is heated in the hot cylinder, building pressure to move the piston
cool cylinder where the hot gas is cooled, lowering its pressure
conducting pipe carries heat to the hot cylinder
cooling pipes draw heat from gas in the cool cylinder
piston rod driven by pressure changes in the hot and cool cylinders
wheel is driven by piston
power meter

Stirling hot air engine
The Stirling engine could alternately compress and expand hot air, making it a far quieter and more efficient machine.

William Parry's ships *HMS Hecla* and HMS *Griper*, trapped in the Arctic ice.

hole placed on patient's body

earpiece

Stethoscope
Leaennec's stethoscope allowed doctors to hear murmurs inside a patient's chest.

ANOTHER MEDICAL ADVANCE was the **simple stethoscope**, invented by Parisian physician **René Laennec** (1781–1826) in 1816 to listen to the heartbeart and breathing patterns. It was first described in his 1819 book on diagnosis, *De l'Auscultation Mediate* (*On Mediate Auscultation*). His stethoscope allowed physicians to **diagnose diseases sooner and far more accurately**.

Also in Paris, French physicists **Alexis Petit** (1791–1820) and **Pierre Dulong** (1785–1838) found a way to **verify the atomic weights of elements**. In 1803, British chemist **John Dalton** had put forward an atomic theory, stating that each element is made of atoms of a particular weight, but establishing the

London doctor James Blundell carries out the first successful **blood transfusion**

British engineer Robert Stirling invents the **Stirling engine**

British scientist Michael Faraday experiments with **steel alloys**

French chemist Louis Jacques Thénard discovers **hydrogen peroxide**

French physician René Laennec describes his invention of the **stethoscope** in *De l'Auscultation Mediate*

> **“ THE EXPERIMENTAL INVESTIGATION** BY WHICH **AMPÈRE ESTABLISHED** THE LAW OF THE **MECHANICAL ACTION** BETWEEN **ELECTRIC CURRENTS** IS ONE OF THE MOST **BRILLIANT ACHIEVEMENTS** IN SCIENCE. **”**
>
> **James Clerk Maxwell, British theoretical physicist,** from *A Treatise on Electricity and Magnetism,* 1873

112°51’W

THE **LONGITUDE** IN THE **NORTHWEST PASSAGE** REACHED BY **HMS *HECLA*** AND **HMS *GRIPER***

weights proved to be a difficult task. Petit and Dulong found that an element’s **specific heat** (see 1761–62)—the amount of heat required to raise the temperature by one degree Celsius—is **inversely proportional** to its **atomic weight**. By measuring the specific heat of an element, Petit and Dulong were able to make an estimate of its atomic weight.

During this period, finding a sailing route to the Pacific through the Arctic was an attractive proposition for commercial interests in Europe, because the routes to the south were long and stormy. British naval officer William Parry led an expedition to find the **Northwest Passage** in 1819 and succeeded. He reached Melville Island in the Arctic and won the prize offered by Parliament for crossing a longitude of 110° West. Parry’s ships, HMS *Hecla* and HMS *Griper*, were trapped by frozen sea until the spring of 1820 when the ice finally melted.

Another ship, the SS *Savannah,* became the **first ship to cross the Atlantic using steam engines**, sailing from Savannah in the USA on May 22 and arriving in Liverpool in England 18 days later. This feat was not repeated for over 20 years. Despite being equipped with luxury cabins, however, no passengers could be persuaded to join the voyage because of the ship’s revolutionary design: a sailing ship that also contained a steam engine operated by paddle wheels.

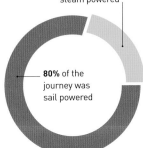

20% of the journey was steam powered

80% of the journey was sail powered

Voyage of the SS *Savannah*
While the SS Savannah *was the first ship to cross the Atlantic using steam power, the ship used its steam engines for only 41½ hours during the 207-hour voyage.*

DANISH PHYSICIST HANS CHRISTIAN ØRSTED (1777–1851) revealed the **link between electricity and magnetism**. At a public lecture in Copenhagen, he astonished the audience by showing a compass needle move as he brought it near a wire conducting electricity. Influenced by Ørsted’s discovery, French physicist **André-Marie Ampère** created a theory of **electromagnetism**. This showed that electric currents flowing in opposite directions create magnetic fields that cause the wires to be attracted, while currents flowing the same way lead to the wires being repelled.

British physicist **John Herapath** (1790–1868) explained how temperature and pressure in gas are created by moving molecules, an early version of the **kinetic theory of gases**.

ANDRÉ-MARIE AMPÈRE (1775–1836)

Born near Lyon in France, André-Marie Ampère was a talented mathematician and teacher. He laid the foundations of electromagnetic theory and discovered that the magnetic interaction of two electrical wires is proportional to their length and the strength of the current flowing through them. This is known as Ampère’s Law.

In Paris, French naturalist **Georges Cuvier** ridiculed the ideas of fellow naturalist **Jean-Baptiste Lamarck**, who argued that **species have transformed, or evolved**, through time. This idea is now an accepted part of the theory of evolution.

Also in Paris, chemists **Pierre-Joseph Pelletier** (1788–1842) and **Joseph Bienaimé Caventou** (1795–1877) worked on **isolating medically active ingredients from plants**. In 1820, they isolated quinine from cinchona bark, which later became important in treating malaria.

ELECTROMAGNETS

An electric current creates its own magnetic field, and this is the basis of an electromagnet. These are powerful magnets that can be switched on and off with an electric current. A solenoid (coil of wire) is a common form of electromagnet. The more coils of wire, the stronger the magnetic field. Electromagnets are vital to the operation of everything from the speaker in a phone to electric motors.

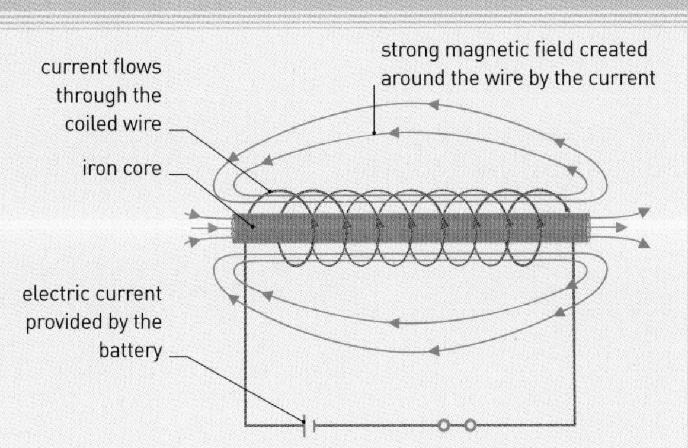
current flows through the coiled wire
iron core
electric current provided by the battery
strong magnetic field created around the wire by the current

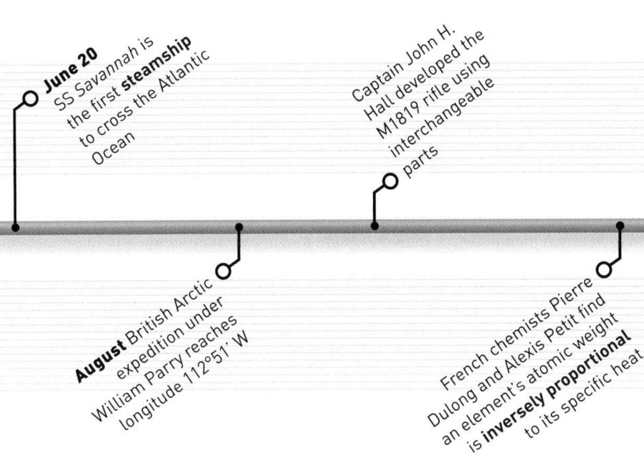
June 20 SS *Savannah* is the first **steamship** to cross the Atlantic Ocean
August British Arctic expedition under William Parry reaches longitude 112°51’ W
Captain John H. Hall developed the M1819 rifle using interchangeable parts
French chemists Pierre Dulong and Alexis Petit find an element’s atomic weight is **inversely proportional** to its specific heat

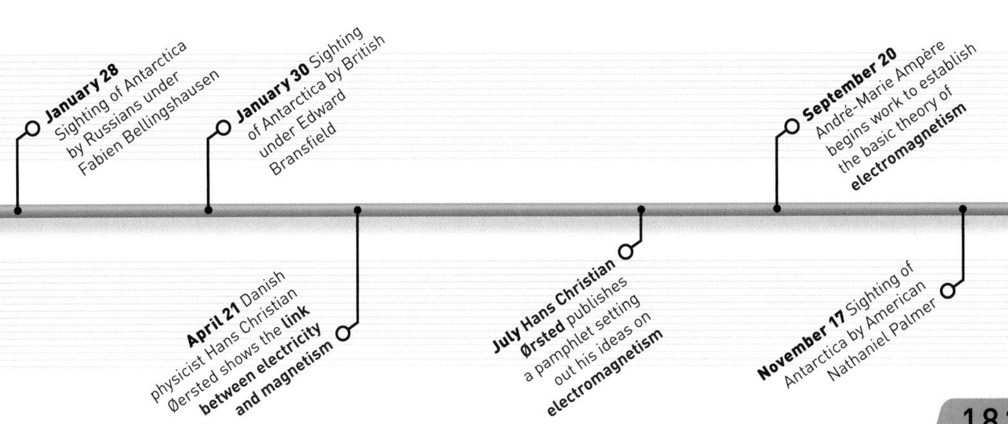
January 28 Sighting of Antarctica by Russians under Fabien Bellingshausen
January 30 Sighting of Antarctica by British under Edward Bransfield
April 21 Danish physicist Hans Christian Øersted shows the **link between electricity and magnetism**
July Hans Christian Ørsted publishes a pamphlet setting out his ideas on **electromagnetism**
September 20 André-Marie Ampère begins work to establish the basic theory of **electromagnetism**
November 17 Sighting of Antarctica by American Nathaniel Palmer

1821–22

“I… FOUND **A PORTION** OF THE **LUNG** AS LARGE AS A TURKEY’S EGG, **PROTRUDING THROUGH** THE **EXTERNAL WOUND, LACERATED AND BURNT.**”

William Beaumont, US army surgeon, from *Experiments and Observations on the Gastric Juice, and the Physiology of Digestion,* 1833

US army surgeon William Beaumont inserted a tube into the stomach of his colleague Alexis St. Martin, who had been injured by a gunshot.

BUILDING ON DANISH PHYSICIST HANS CHRISTIAN ØRSTED’S discovery in 1820 that an electric current makes a magnet move, British scientist **Michael Faraday** (see 1837) showed that a wire carrying an electric current moved in a circle around a fixed magnet, and that a suspended magnet moved in a circle around a fixed wire carrying a current. He had discovered **electromagnetic rotation—the principle of the electric motor**.

German-Estonian scientist **Thomas Seebeck** (1770–1831) observed that a compass needle wavers when it is close to a loop of two different metals that are cooled in one place and heated in another. This is because the slightly different movement of heat through each of the two metals disturbs atoms to generate an electric current, a phenomenom called the Seebeck or **thermoelectric effect**.

In the field of geology, Swiss geologist **Ignatz Venetz** (1788–1859) suggested that in the past, during an ice age, the world was colder and Europe was covered in glaciers that shaped much of its landscape.

On the English coast, near Lyme Regis in Dorset, fossil collector **Mary Anning** (see 1810–11) found the first fossil *plesiosaur*—a huge marine reptile that lived 65–195 million years ago. The following year, another fossil hunter, **Gideon Mantell** (1790–1852) found teeth of a huge reptile he called an *iguanadon,* which was later identified as a dinosaur. In Yorkshire, British naturalist **William Buckland** (1784–1856) discovered ancient remains of a hyena’s den, with bones of rhinoceroses, elephants, and lions, showing that wildlife in the British Isles was once very different.

freely pivoting wire

magnet

water completes the circuit

Faraday’s experiment
A replica of the apparatus Michael Faraday used to demonstrate the principle of electromagnetic rotation, which is the basis of electric motors.

Geologists were starting to identify the different ages in Earth’s past from fossils found in rocks. In 1822, British geologists **William Phillips** (1775–1828) and **William Conybeare** (1787–1857) made the **first identification of a geological period**. They named it the **Carboniferous period** after the carboniferous (coal-bearing) strata of northern England .

The same year, British computer pioneer **Charles Babbage** (1791–1871) proposed his ingenious idea for a **Difference Engine**: a calculating machine built from cogs and rods that would work automatically and eliminate human error.

American army surgeon **William Beaumont** (1785–1853) was the first person to observe human digestion in the stomach. He performed experiments on a soldier who had been shot in the abdomen, and pioneered **gastric endoscopy**—inserting a tube to look inside the stomach.

GEOLOGICAL PERIODS

Layers of rock form one on top of the other, with the oldest at the bottom, unless they have been disturbed. The sequence forms the geological column and is the basis for dividing Earth’s history into geological periods, identified originally by the fossils found in each rock layer. The oldest, deepest layer is the Cambrian Period 542–488 million years ago—the first era when life left enough fossils to date it.

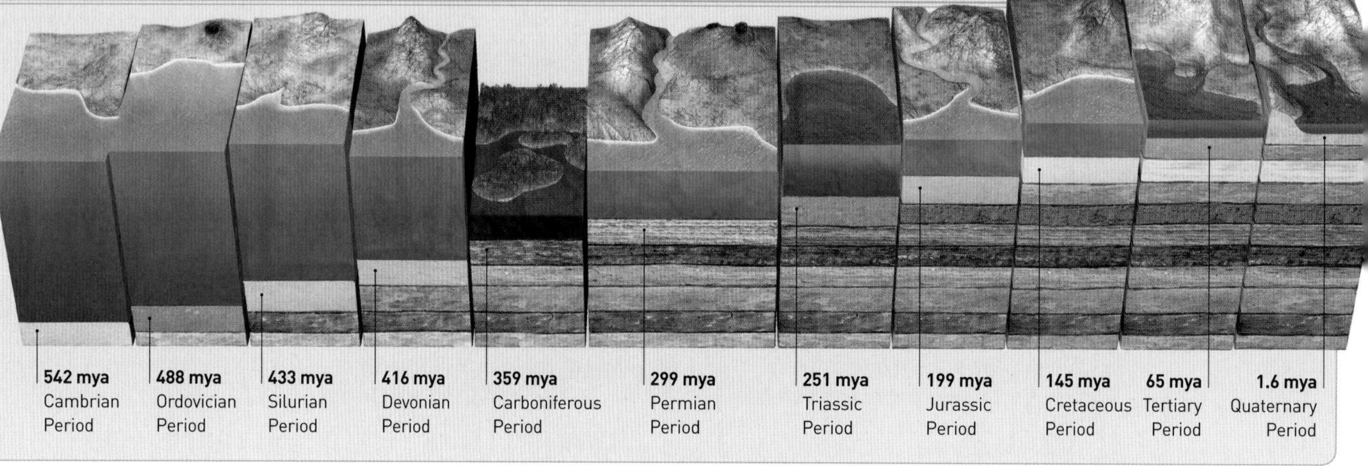

542 mya Cambrian Period
488 mya Ordovician Period
433 mya Silurian Period
416 mya Devonian Period
359 mya Carboniferous Period
299 mya Permian Period
251 mya Triassic Period
199 mya Jurassic Period
145 mya Cretaceous Period
65 mya Tertiary Period
1.6 mya Quaternary Period

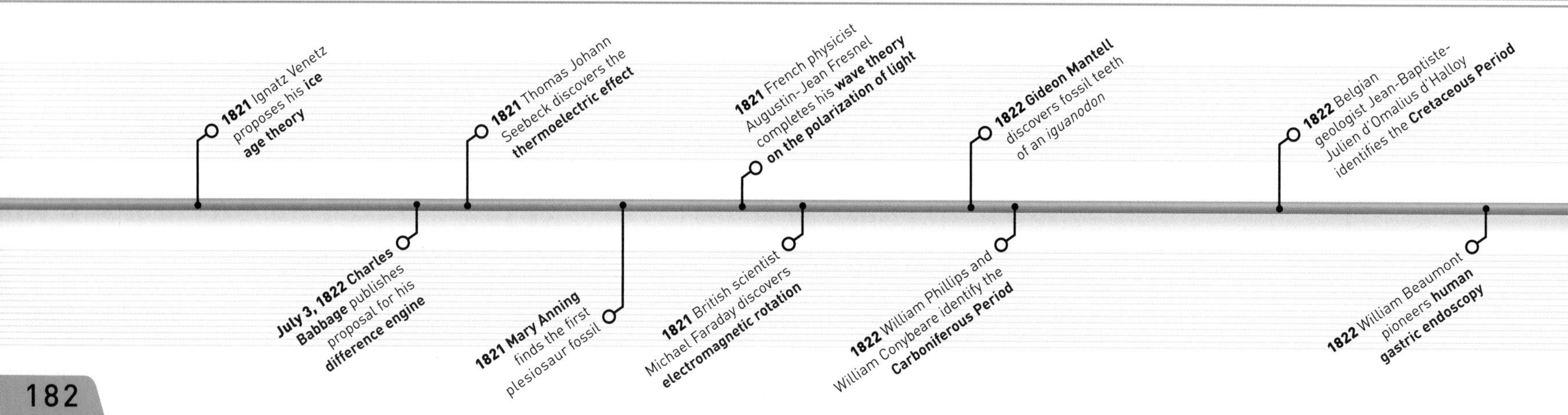

1821 Ignatz Venetz proposes his ice age theory
July 3, 1822 Charles Babbage publishes proposal for his difference engine
1821 Thomas Johann Seebeck discovers the thermoelectric effect
1821 Mary Anning finds the first plesiosaur fossil
1821 French physicist Augustin-Jean Fresnel completes his wave theory on the polarization of light
1821 British scientist Michael Faraday discovers electromagnetic rotation
1822 Gideon Mantell discovers fossil teeth of an iguanodon
1822 William Phillips and William Conybeare identify the Carboniferous Period
1822 Belgian geologist Jean-Baptiste-Julien d’Omalius d’Halloy identifies the Cretaceous Period
1822 William Beaumont pioneers human gastric endoscopy

A B C D E F G H I J

K L M N O P Q R S T

> **" BRAILLE** IS **KNOWLEDGE;** KNOWLEDGE IS **POWER. "**
>
> **Louis Braille, French inventor of Braille writing**

A simple invention by a blind French boy to help him read, Braille has become a window into the world of books for millions of visually impaired people.

DURING 1823–24, SCIENTISTS WERE STUDYING THE NIGHT sky as well as the history of Earth. Bavarian astronomer **Franz von Gruithuisen** (1774–1852) realized craters on the Moon were formed by past meteorite impacts. Another German astronomer, **Heinrich Olbers** (1758–1840), asked why the night sky is dark. Surely, if there is an infinite number of stars, then it should be possible to see a star in every direction and, as a result, the night sky should be bright. Olbers was not the first scientist to ask this question, but it has become known as **Olbers' Paradox**. Today, this paradox is known to be the result of space expanding, which diminishes the apparent brightness of distant stars in many directions, causing the sky to appear dark.

British naturalist **William Buckland** made two momentous contributions to the field of geology. The first was his **discovery**, in a cave on the coast of Wales, UK, **of the first fossilized human remains ever found**. Buckland wrongly identified them as being those of a Roman woman. Carbon-dating (see 1955) has since confirmed they are, in fact, of a 33,000-year-old man. Buckland's second contribution, in 1824, was when he gave the first scientific description of a dinosaur (although the term was not coined until 1842)—he identified some fossils as a giant extinct lizard called **megalosaurus**.

French mathematician **Joseph Fourier**, calculated that Earth is too far from the Sun to be warmed to the temperature it is by solar radiation alone. He suggested that heat is trapped by Earth's atmosphere. This was the first identification of what later became known as **the greenhouse effect**.

JOSEPH FOURIER
(1768–1830)

Joseph Fourier was a brilliant mathematician who went with Napoleon to Egypt in 1798 to decode hieroglyphs. He studied heat transfer and identified the greenhouse effect. His studies of waves led to Fourier analysis, the mathematical analysis of wave forms, now used in everything from touch screens to understanding brain function.

Megalosaurus bones
Drawings of megalosaurus bones from William Buckland's 1824 paper, which contained the first scientific description of a dinosaur.

Hungarian mathematician **János Bolyai** (1802–60) pioneered a new form of geometry—**non-Euclidean geometry**. It breaks away from Euclid's definition of parallel lines on a flat, two-dimensional surface (see 400–335 BCE), and frees mathematicians to contemplate abstract multidimensional ideas, such as the **curved nature of space, time, and the Universe**, and parallel lines that can actually cross.

A blind 15-year-old French boy called **Louis Braille** (1809–52) invented the **six-dot code** later known as **Braille**. This writing system enables blind or partially sighted people to read and is now used in virtually every country.

Also in France, engineer **Nicolas Sadi Carnot** (1796–1832) published *Reflections on the Motive Power of Fire and on Machines Fitted to Develop that Power*. This book contained the **first successful theory of heat engines**, which is now known as the **Carnot cycle**. All heat engines are inefficient because they lose heat every time hot gases are released before the next cycle. The Carnot cycle shows the maximum theoretical efficiency for all engines. Carnot laid the foundations for the **science of thermodynamics** (see 1847–48).

700 THE NUMBER OF **DINOSAUR SPECIES** THAT HAVE BEEN **IDENTIFIED**

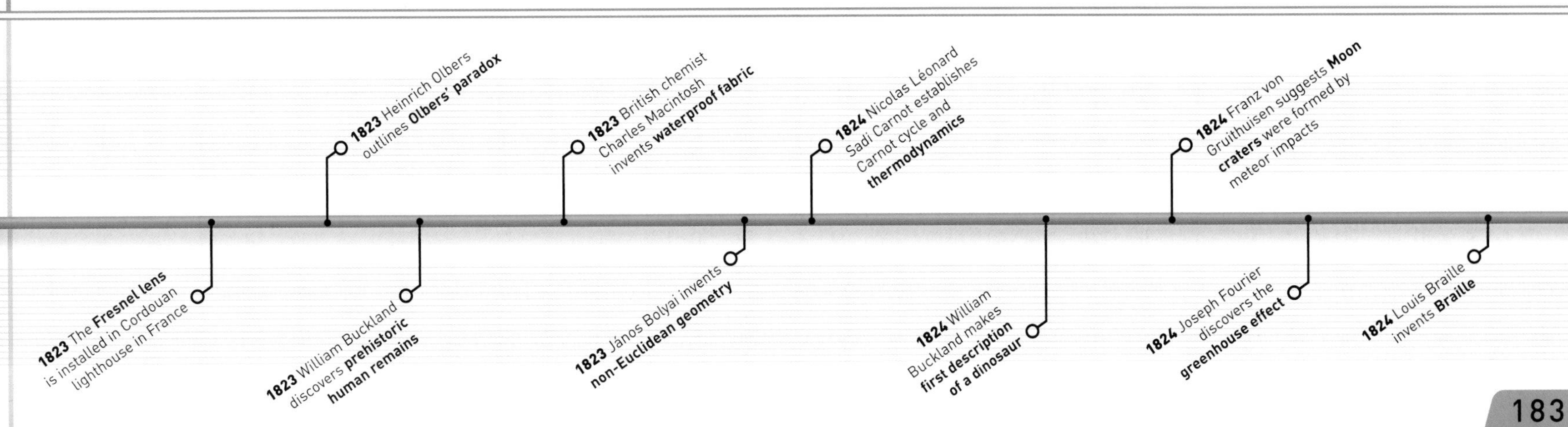
1823 The **Fresnel lens** is installed in Cordouan lighthouse in France
1823 Heinrich Olbers outlines **Olbers' paradox**
1823 William Buckland discovers **prehistoric human remains**
1823 British chemist Charles Macintosh invents **waterproof fabric**
1823 János Bolyai invents **non-Euclidean geometry**
1824 Nicolas Léonard Sadi Carnot establishes Carnot cycle and **thermodynamics**
1824 William Buckland makes **first description of a dinosaur**
1824 Franz von Gruithuisen suggests **Moon craters** were formed by meteor impacts
1824 Joseph Fourier discovers the **greenhouse effect**
1824 Louis Braille invents **Braille**

Abacus

c.2700 BCE
First abacus
The abacus is invented in Sumer (present-day Iraq) and is soon widespread.

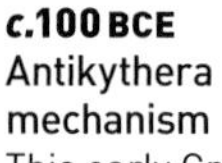

Antikythera mechanism

c.100 BCE
Antikythera mechanism
This early Greek device uses bronze gears to calculate astronomical positions.

1617
Napier's Bones
Invented by John Napier, this set of inscribed rods or "bones" provides a quick way of multiplying or dividing large numbers.

Napier's Bones

1630
Slide rule
Invented by English mathematician William Oughtred, the slide rule is used to multiply, divide, and calculate roots and logarithms.

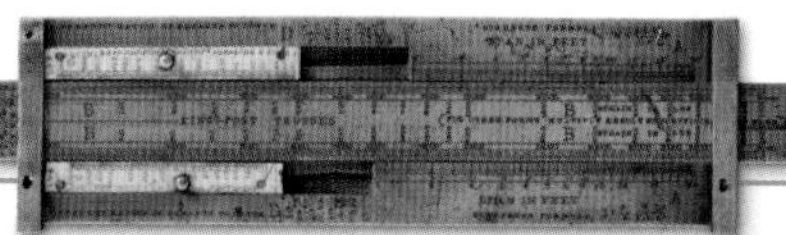
Slide rule

1642
Pascaline
Blaise Pascal invents one of the first mechanical calculators capable of simple arithmetic.

Pascaline

1801
Mechanical loom
Joseph Marie Jacquard's loom is controlled by cards with punched holes, a system later adopted by early computers.

Jacquard loom

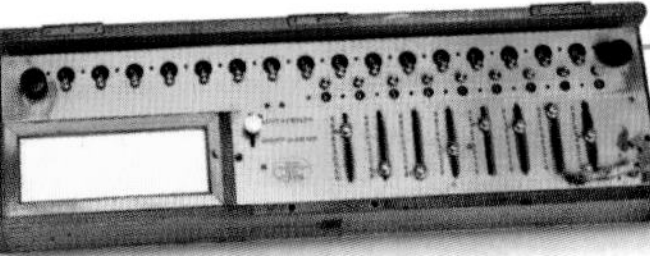
Arithmometer

1820
Arithmometer
Thomas de Colmar builds the first commercially successful mechanical calculating machine.

THE STORY OF CALCULATING MACHINES

CALCULATING HAS BEEN IMPORTANT FOR SCIENCE, INDUSTRY, AND COMMERCE FROM THE EARLIEST TIMES TO THE PRESENT DAY

The word calculate is derived from the Latin *calculus*, or "little stone," referring to the ancient practice of using stones to perform calculations. Since then, increasingly sophisticated devices have been invented to perform the complex calculations demanded by advances in the sciences.

The first calculating device, the abacus, evolved when counting stones were arranged on a frame, and remained the most widely used means of calculating until the 17th century. A breakthrough came with Scottish mathematician John Napier's discovery of logarithms (see 1614–17), and his invention of the calculating device known as Napier's Bones. The first mechanical calculators also appeared in the 17th century, prompted by the need for accurate astronomical tables.

PROGRAMMABLE MACHINES

During the Industrial Revolution, a French weaver, Joseph Marie Jacquard, used punched cards to control the working of his looms. The idea of a calculating machine capable of carrying out different programmable functions was originated by British mathematician Ada Lovelace. The same idea was taken up by British inventor Charles Babbage in his concept of an "Analytical Engine."

The first electrical computers began to appear in the 1930s, and with the introduction of integrated circuits, or chips, smaller and more powerful computers and calculators eventually became viable. By the mid-1970s, entire processing units on silicon chips—microprocessors—enabled the production of personal computers.

BINARY NUMBERS

Unlike the decimal system, which uses the numbers 0 to 9, the binary numeral system uses only two symbols, 0 and 1. In this system, 1 is represented by 1, 2 by 10, 3 by 11, 4 by 100, and so on. Because there are only two symbols, the binary system is ideal for use in digital electronic computers, where the two possible states of an electronic circuit, off or on, correspond to the digits 0 and 1.

> " ALL WHICH IS **BEAUTIFUL** AND **NOBLE** IS THE RESULT OF REASON AND **CALCULATION.** "

Charles Baudelaire, French poet (1821–67)

Hollerith tabulator

1889
Hollerith tabulator
The electrical tabulating machine invented by Herman Hollerith in the US is the first device to use punched cards to store data rather than control a process.

1960s
Electronic calculator
Electronic desktop calculators emerge in the 1960s with the invention of the transistor. Pocket calculators soon follow.

Pocket calculator

1970s
Microprocessor chips
Integrated circuits containing thousands of transistors become commercially available for use in computers.

Microchip

1822
Babbage's Difference Engine
Charles Babbage begins work on the design of a machine capable of complex calculations.

1939
The Bombe
Based on a Polish design, this electromechanical device is built in Britain during World War II to decipher codes.

1980s–present
Personal computers
The so-called "microcomputer revolution" takes off in the 1980s as personal computers become smaller, more powerful, and more affordable.

Early Apple computer

interconnected gear wheels
dial on final column shows result of calculation
figures engraved on wheels
brass framework

Babbage's Difference Engine
Charles Babbage designed the first of his calculating machines in the 1820s to overcome human error in compiling numerical tables. He improved on the design with a second Difference Engine in 1847–49. This demonstration model of Babbage's first design was built by his son Henry.

1825

Marc Seguin's bridge across the Rhône between Tournon-sur-Rhône and Tain-l'Hermitage used a pioneering wire cable suspension system.

THE OPENING OF THE STOCKTON AND DARLINGTON railway on September 27 in the north of England marked the **beginning of the railroad age**. Smaller steam railways had existed before, but this project involved massive bridges and viaducts. The railway was engineered by **George Stephenson**, who also designed its first locomotive, *Locomotion No.1*.

One early passenger was French engineer **Marc Seguin** (1786–1875), whose experience inspired him to create his own steam railways in France. In August, Seguin also opened **Europe's first large wire cable suspension bridge** between Tournon-sur-Rhône and Tain-l'Hermitage, which spanned almost 298 ft (91 m).

Another technological first was an **electromagnet that was capable of supporting more than its own weight**. Made by British electrical engineer **William Sturgeon** (1783–1850), the 7 oz (200 g) magnet could lift 9 lb (4 kg).

The Danish physicist who discovered electromagnetism in 1820, **Hans Ørsted**, created **aluminium** in a chemical reaction in 1825.

British scientist **Michael Faraday**, who was another pioneer of electromagnetism, **discovered benzene** in the oil residue that was created from making coal gas for gaslights. Benzene is a key ingredient of petroleum and is used to make plastic.

Meanwhile, French naturalist **Georges Cuvier** published his idea (first proposed in 1812) that large groups of animals had been wiped out by past catastrophes, in the book *Discourse on the Revolutions of the Surface of the Globe*. German geologist **Christian von Buch** (1774–1853) argued that natural variations between animals led to **separate species**.

8 PERCENT
THE AMOUNT OF **ALUMINUM** PRESENT IN **EARTH'S CRUST**

Stockton and Darlington railway
The opening of George Stephenson's Stockton and Darlington Railway was a major event, attracting over 40,000 spectators and worldwide attention.

1826–27

500
THE NUMBER OF **BIRD SPECIES** PAINTED BY JOHN JAMES AUDUBON

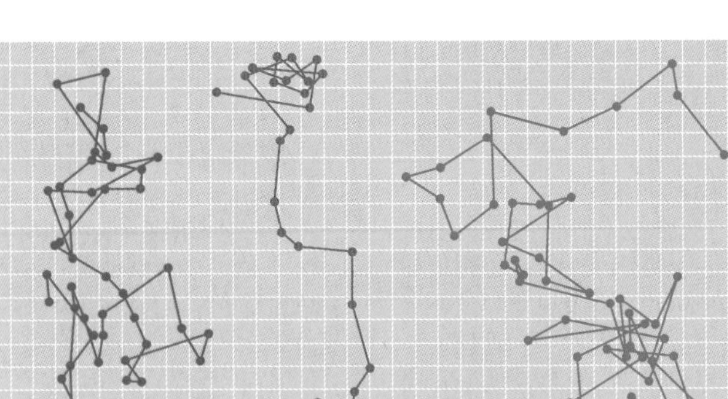
BROWNIAN MOTION

In 1827, when Robert Brown observed pollen grains in water through a microscope, he noticed they moved around at random, but could offer no explanation for it. In 1905 Albert Einstein (1879–1955) showed that the pollen grains were being knocked by water molecules, which caused them to move. This map of Brownian motion shows the random path of individual grains.

IN THE NATURAL WORLD, Russian naturalist **Karl von Baer** (1792–1876) discovered in 1826 that mammals start life as eggs or ova, and Scottish biologist **Robert Grant** (1793–1874), German naturalist **August Schweigger** (1782–1821), and German anatomist **Friedrich Tiedemann** (1781–1861) all argued that both **plants and animals have a common origin**.

Another Russian, mathematician **Nikolai Lobachevsky** (1792–1856), presented his system of **hyperbolic geometry** in February 1826, involving imaginary surfaces and lines.

Two important developments in engineering also occurred in 1826. American inventor **Samuel Morey** (1762–1843) patented an **early version of the internal combustion engine**.

In the summer of 1826, French inventor **Joseph Nicéphore Niépce** (1765–1833) used a camera obscura to take the world's oldest-known **photograph** on a light-sensitive, bitumen-coated, pewter plate.

In nearby Montpellier, French chemist **Antoine Balard**

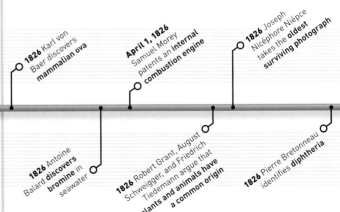

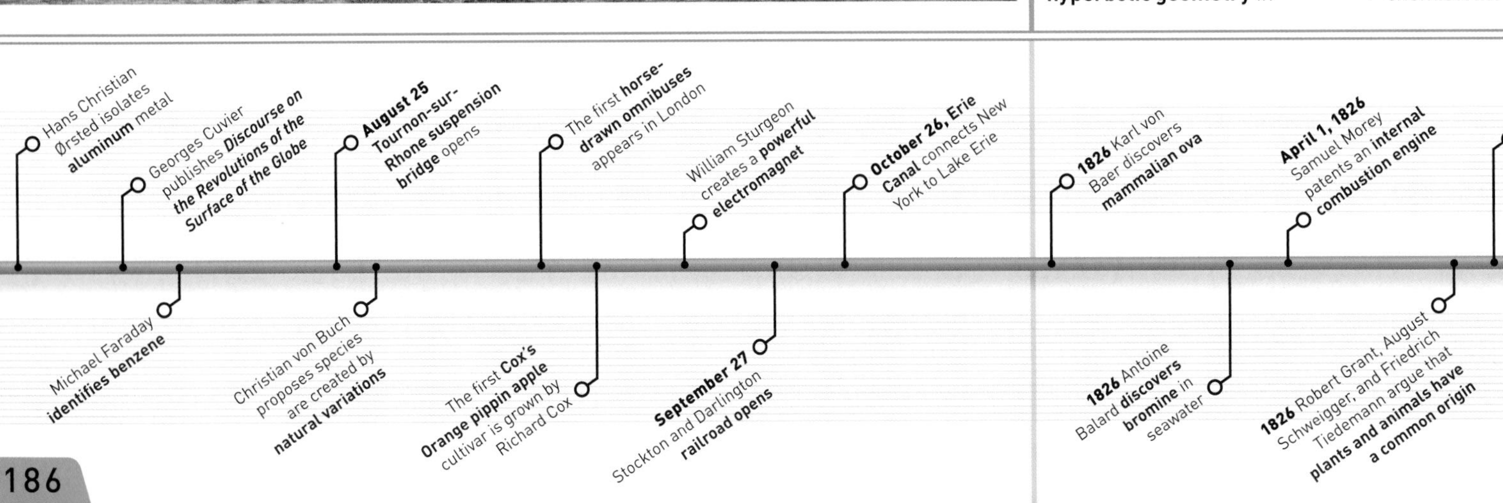
Hans Christian Ørsted isolates **aluminum metal**
Georges Cuvier publishes ***Discourse on the Revolutions of the Surface of the Globe***
Michael Faraday **identifies benzene**
August 25 Tournon-sur-Rhone suspension bridge opens
Christian von Buch proposes species are created by **natural variations**
The first **horse-drawn omnibuses** appears in London
The first **Cox's Orange pippin apple** cultivar is grown by Richard Cox
William Sturgeon creates a **powerful electromagnet**
September 27 Stockton and Darlington **railroad opens**
October 26, Erie Canal connects New York to Lake Erie
1826 Karl von Baer discovers **mammalian ova**
1826 Antoine Balard **discovers bromine** in seawater
April 1, 1826 Samuel Morey patents an **internal combustion engine**
1826 Robert Grant, August Schweigger, and Friedrich Tiedemann argue that **plants and animals have a common origin**
1826 Joseph Nicéphore Niépce takes the **oldest surviving photograph**
1826 Pierre Bretonneau identifies **diphtheria**

An illustration from Audubon's book, *The Birds of America*.

(1802–76) **discovered bromine in seawater**. And in Tours, French doctor **Pierre Bretonneau** (1778–1862) identified **diphtheria**.

In 1827, English chemist **William Prout** (1785–1850) **classified food into the three main divisions** known today: carbohydrates, fats, and proteins.

Scottish naturalist **Robert Brown** (1773–1858) observed the phenomenon now known as **Brownian motion** (see panel, opposite). French-American naturalist **John James Audubon** (1785–1851) sold the first prints of his book, *The Birds of America, a 13-year project that culminated in 1838 with a total of 435 plates.*

> “ THE **OBJECTS** APPEAR WITH **ASTONISHING SHARPNESS…** DOWN TO THE SMALLEST DETAILS… THE **EFFECT** IS DOWNRIGHT **MAGICAL.** ”
>
> **Joseph Nicéphore Niépce, French inventor,** describing an earlier photographic experiment to his brother, September 16, 1824

1828

Originally set up as a place for scientific study, the Gardens of the London Zoological Society (London Zoo) was eventually opened to the public in 1847.

AS THE WESTERN WORLD BECAME INCREASINGLY URBANIZED, interest turned toward the natural world and **botanic gardens and zoos** were opened to display exotic plants and animals. The first zoo for scientific study was London Zoo in England, which opened on April 27, 1828.

The **foundations of embryology** (the science of embryo development) were laid by Estonian naturalist **Karl Ernst von Baer** (1792–1876), who showed that different animal species could appear similar at early stages of development.

English fossil collector **Mary Anning** (see 1810–11) had made a number of important prehistoric discoveries on the English coast, but in 1828 she found the **fossil of a pterosaur**, a huge flying reptile. It was only the third such fossil to be found and the first to be identified. Anning's pterosaur was named *Dimorphodon* by paleontologist Richard Owen (1804–92) in 1859.

In Berlin, German chemist **Friedrich Wöhler** (1800–82) **pioneered organic chemistry**—the chemistry of living things—with his discovery of what is now called **Wöhler synthesis**. This is a chemical reaction that produces the organic chemical urea. Its discovery showed that organic chemicals were not only made by living things, as Wöhler's former tutor Swedish chemist **Jön Jakob Berzelius** (1779–1848) had insisted. Berzelius also made a discovery in 1828 when he **isolated the radioactive element thorium**, a dense metal found in a black mineral discovered by the Norweigian geologist Morten Esmark (1801–82).

The world's **first electric motor** was made in Budapest by Hungarian inventor and Benedictine monk **Ányos Jedlik** (1800–95), and in Nottingham, England, self-taught mathematician **George Green** (1793–1841) published an essay in which he outlined the first **mathematical theory of electricity and magnetism**. This was later built on by James Clerk Maxwell (see 1861–64).

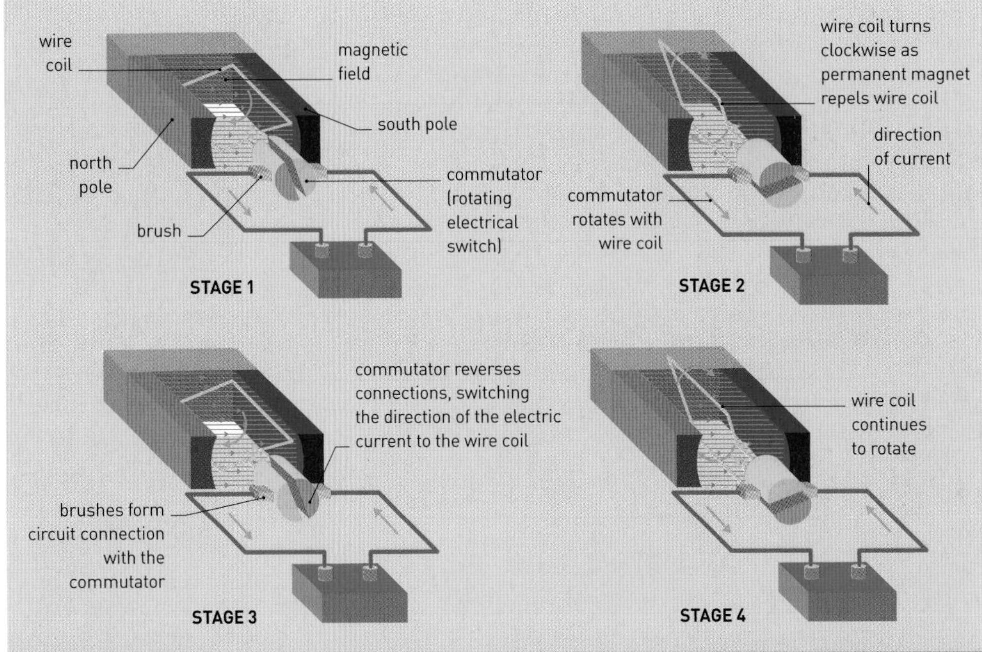
wire coil
magnetic field
south pole
north pole
commutator (rotating electrical switch)
brush
STAGE 1
wire coil turns clockwise as permanent magnet repels wire coil
direction of current
commutator rotates with wire coil
STAGE 2
commutator reverses connections, switching the direction of the electric current to the wire coil
brushes form circuit connection with the commutator
STAGE 3
wire coil continues to rotate
STAGE 4

ELECTRIC MOTORS

Hungarian inventor Ányos Jedlik showed how an electric motor could be driven by the repulsion between the poles of a permanent magnet and an electromagnetic coil. A half turn moves the like poles of the coil away from each other, which effectively means that there would be no repulsion to drive the motor. The coil is therefore connected to the circuit by contacts called commutators that allow the circuit to swap direction as each half of the coil spins past. This swaps the coil's polarity too, so it continually presents like (repelling) poles to the permanent magnet.

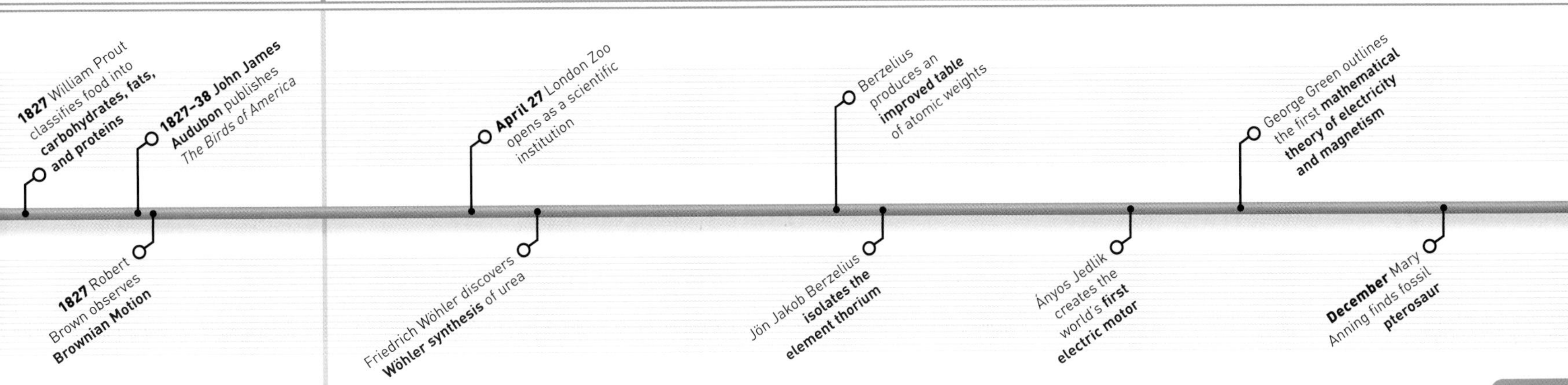
1827 William Prout classifies food into carbohydrates, fats, and proteins
1827–38 John James Audubon publishes The Birds of America
1827 Robert Brown observes Brownian Motion
April 27 London Zoo opens as a scientific institution
Friedrich Wöhler discovers Wöhler synthesis of urea
Berzelius produces an improved table of atomic weights
Jön Jakob Berzelius isolates the element thorium
Ányos Jedlik creates the world's first electric motor
George Green outlines the first mathematical theory of electricity and magnetism
December Mary Anning finds fossil pterosaur

29mph

THE **MAXIMUM SPEED** REACHED BY THE ***ROCKET*** DURING THE **LOCOMOTIVE TRIALS** AT **RAINHILL**, LANCASHIRE, UK

Civil engineer Robert Stephenson designed the *Rocket* steam locomotive, which gained worldwide fame.

IN NEW YORK, AMERICAN SCIENTIST JOSEPH HENRY (1797–1878) was exploring the power of electromagnetism (see 1820). He found that by carefully insulating the wires and winding them closely and in several layers, he could make very strong electromagnets. In December 1830, Henry finally demonstrated the first powerful electromagnet, able to hold up 750 lb (340 kg) of iron.

In October, the directors of the pioneering Liverpool and Manchester Railway (L&MR) held **locomotive trials at Rainhill** in Lancashire, UK. The public was yet to be convinced of the merits of steam locomotives, so the competition was held to generate publicity as well as to choose which experimental locomotive would pull L&MR's trains. The trials were a huge success. Although only one of the five competing locomotives completed the event—**Robert Stephenson's *Rocket***—the steam age had arrived.

As steam engines developed, engineers became interested in extracting maximum efficiency from them, and theoretical scientists began to **explore the concept of energy.** French scientist **Gustave-Gaspard Coriolis** (1792–1843) published *Calcul de l'Effet des Machines* (*Calculation of the Effect of Machines*), in which he looked at the relationship between energy and work (the effect of energy), and introduced **the concept of kinetic energy**—energy that is produced when an object is in motion (see 1847–48).

smokestack
multitube boiler
driving wheel

Stephenson's *Rocket*
The Rocket *was the first locomotive to have a multitube boiler, with 25 tubes to carry hot exhaust gas from the firebox where fuel was burned. This helped generate more steam power.*

The **identification of geological periods** gathered pace. While studying sediments in the Seine valley in France, geologist **Jules Desnoyers** (1800–87) coined **the term Quaternary** to describe the most recent geological period, when loose material (gravel, sand, and clay deposits) was laid down on top of solid rocks.

In a cave in Engis, Belgium, Dutch-Belgian prehistorian **Philippe-Charles Schmerling** (1791–1836) found a fragment of a small child's skull. This was only the **second discovery of a human fossil**; British geologist William Buckland had discovered the first in 1823. Schmerling's find later proved to be 30,000 to 70,000 years old—the **first Neanderthal remains ever discovered**.

ELECTROMAGNETIC INDUCTION

This demonstration of electromagnetic induction involves moving a bar magnet in and out of a coil of wire to produce an electric current. The magnet's field pulls the electrons in the wire and produces a voltage. If the coil is connected to a circuit, an electric current will flow. If the magnet is moved in the other direction, the flow is reversed.

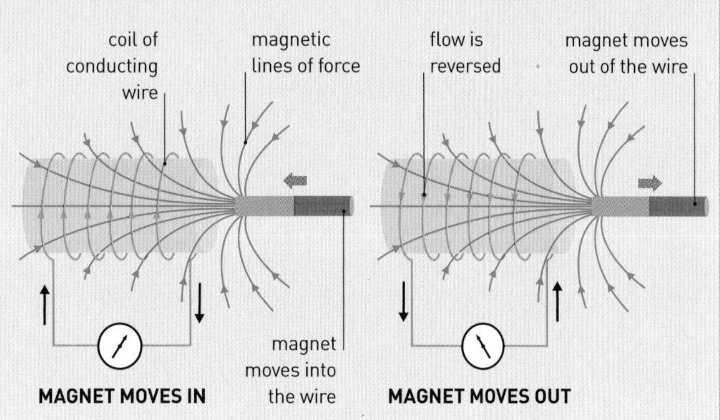
coil of conducting wire
magnetic lines of force
flow is reversed
magnet moves out of the wire
magnet moves into the wire
MAGNET MOVES IN
MAGNET MOVES OUT

Metamorphic rocks, such as these, were first identified by Charles Lyell.

IN THE EARLY 19TH CENTURY, GEOLOGISTS WERE DIVIDED into two groups. The **catastrophists** claimed that the surface of Earth was shaped by a few huge and brief catastrophes, such as floods and earthquakes (see 1812). The **uniformitarians**, meanwhile, believed that Earth's landscapes were shaped and reshaped gradually over very long periods of time by steady processes, such as river erosion.

More evidence was being found confirming the uniformitarian school of thought. Scottish geologist **Charles Lyell** (1797–1875) summarized these findings, insisting that change was continuous and gradual, in his monumental book *Principles of Geology,* which was published in three parts between 1830 and 1833. Lyell's text was so authoritative and convincing that after its publication, few doubted that the **Earth had gone through many geological ages** over millions of years. It was this picture of Earth's vast geological history that paved the way for Charles Darwin's theory of evolution (see 1859), which was partly influenced by Lyell's work.

The same year, German astronomer **Johann von Mädler** (1794–1874) began to make **drawings of the surface of Mars**. These were later

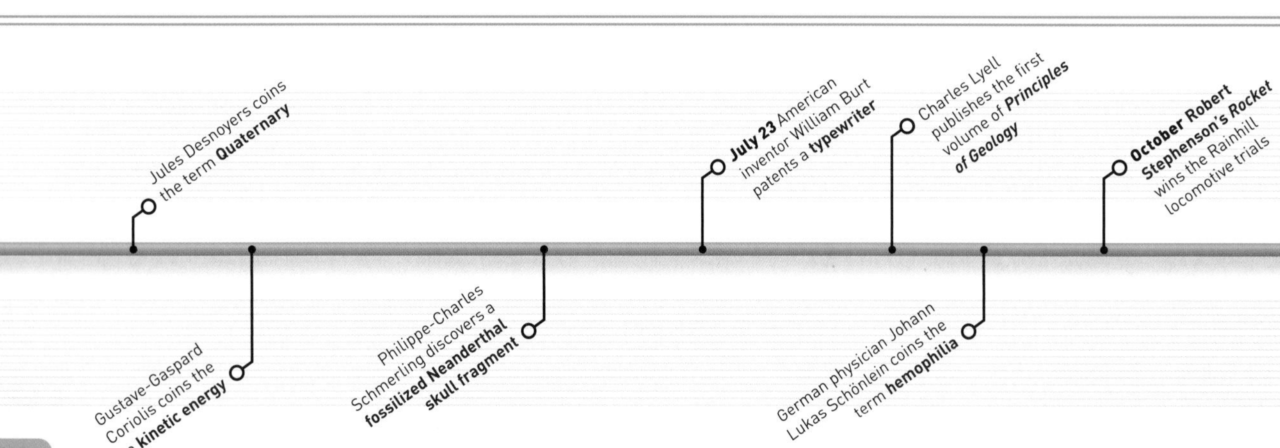
Jules Desnoyers coins the term **Quaternary**
Gustave-Gaspard Coriolis coins the term **kinetic energy**
Philippe-Charles Schmerling discovers a **fossilized Neanderthal skull fragment**
July 23 American inventor William Burt patents a **typewriter**
Charles Lyell publishes the first volume of ***Principles of Geology***
German physician Johann Lukas Schönlein coins the term **hemophilia**
October Robert Stephenson's *Rocket* wins the Rainhill locomotive trials

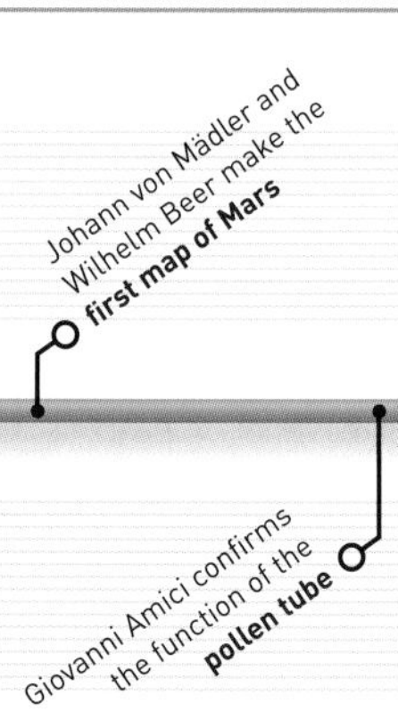
Johann von Mädler and Wilhelm Beer make the **first map of Mars**
Giovanni Amici confirms the function of the **pollen tube**

1831–32

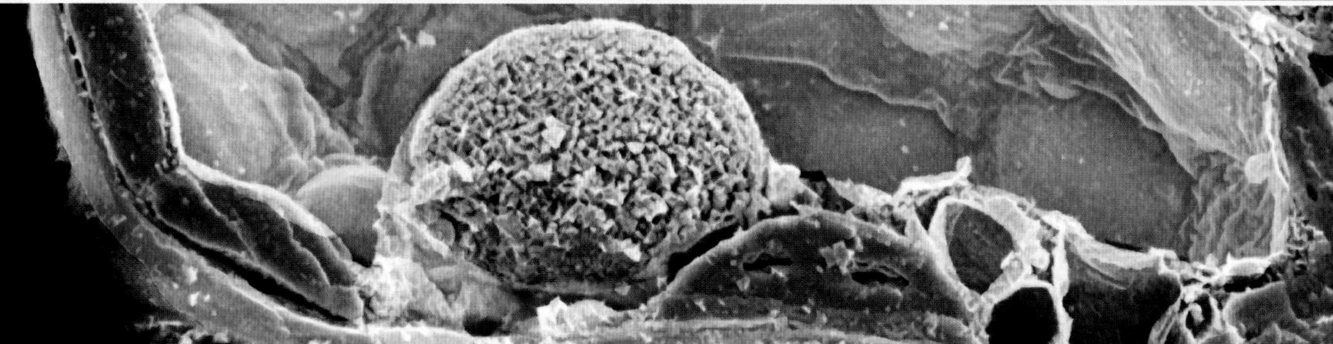

British botanist Robert Brown was the first to observe and appreciate the significance of a plant cell's nucleus, orange in this modern colored scanning electron microscope image.

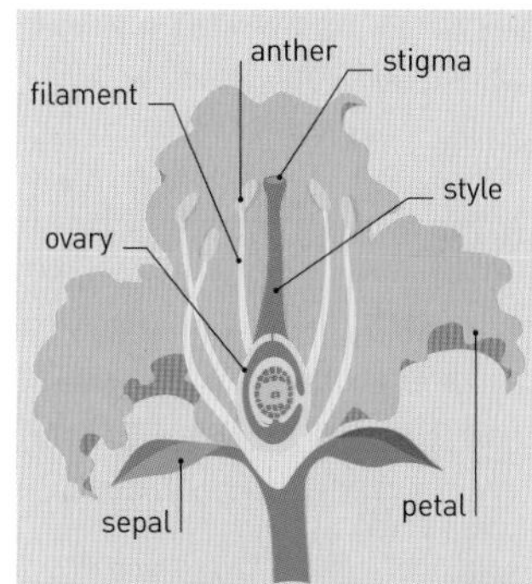

anther
stigma
filament
style
ovary
sepal
petal

PLANT REPRODUCTION

Flowering plants reproduce sexually, and have both male (a stamen, made up of an anther and filament) and female parts (a pistil, made up of a stigma, style, and ovary). Fertilization begins when pollen from an anther lands on the stigma. A tube grows down the style to the ovary to deliver the male sex cells to the ovule (female sex cell).

regarded as the first true maps of the planet.

Italian microscopist and astronomer **Giovanni Battista Amici** (1786–1863) also studied flowers. He first noticed the **pollen tube**—a single cell tube that transports male sex cells to the plant's ovule—in 1824. In 1830, Amici used a microscope to observe the process whereby the pollen tube is formed.

THE STUDY OF ELECTRICITY AND MAGNETISM was proceeding at such a speed in the early 1830s that there were often disputes over who had made a particular discovery first. In 1831, British scientist **Michael Faraday** and his American counterpart **Joseph Henry** independently found that moving a magnetic field near a wire induced an electric current, thereby discovering the principle of **electromagnetic induction**. In time, this led to the development of machines that could generate large quantities of electricity, enabling advances such as electric lighting.

Less controversially, German mineralogist **Franz Ernst Neumann** (1798–1895) extended Pierre Dulong and Alexis Petit's discovery of the **inverse relationship of specific heat to the atomic weight of elements** (see 1819) to include molecules. Building up the picture of the relationship between atoms and molecules and the energy they carry, Neumann showed that the molecular heat of a compound is equal to the sum of the atomic heat of its constituents. This came to be known as **Neumann's Law**.

Swedish chemist **Jöns Jacob Berzelius** had already published a list of atomic weights of 43 elements in 1825. In 1831, he **introduced the term isomer** for different compounds with the same chemical composition.

In the same year, British botanist **Robert Brown** used the word **nucleus for the first time in biology**, to describe the central globule he saw through a microscope when he was observing the cells of orchids. Other scientists had seen the nucleus before, but Brown linked it to reproduction. In Germany, astronomer **Heinrich Schwabe** (1789–1875) made **the first detailed drawing of Jupiter's Great Red Spot** (see 1662–64).

In 1832, French instrument maker **Hippolyte Pixii** (1808–35) invented a magnet-electric machine, which was the **first direct current generator**, and British physician **Thomas Hodgkin** (1798–1866) described **Hodgkin's Lymphoma** for the first time.

copper disk
belt
driving wheel
horseshoe magnet
contact touching rim of copper disk
disk drives belt

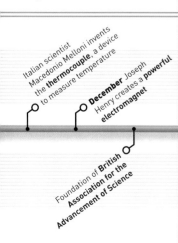

Faraday disk
Also known as the homopolar generator, the Faraday disk was developed by Michael Faraday in 1831. As the disk spins within the magnet's field, a weak current flows.

Italian scientist Macedonio Melloni invents the **thermocouple**, a device to measure temperature

December Joseph Henry creates a **powerful electromagnet**

Foundation of **British Association for the Advancement of Science**

1831 Michael Faraday and Joseph Henry independently discover the phenomenon of **electromagnetic induction**

1831 Robert Brown discovers the cell **nucleus**

1831 Jöns Jacob Berzelius coins the term **isomer**

1831 Franz Ernst Neumann **extends the Dulong–Petit law** to include molecules and compounds

1831 Heinrich Schwabe draws **Jupiter's Great Red Spot**

December 27, 1831 British naturalist **Charles Darwin** embarks on his voyage in HMS **Beagle**

1832 British physician Thomas Hodgkin describes **Hodgkin's lymphoma**

1832 French instrument maker **Hippolyte Pixii** invents one of the first machines to generate electricity

1833

Radiates:

Louis Agassiz's work on fossils gave impetus to the study of extinct life.

BIOCHEMISTRY, THE STUDY OF THE CHEMISTRY OF LIVING THINGS, can be said to have begun in 1833 with the **discovery and isolation of the enzyme** diastase by French chemist **Anselme Payen** (1795–1871). Enzymes are produced by living organisms and act as a catalyst to bring about biochemical reactions (see 1893–94). Diastase is the enzyme in beer mash that helps the starch in barley seed change into soluble sugars.

The term "scientist" was also coined this year by the polymath **William Whewell** (1794–1866). Until this time the only terms in use were "natural philosopher" and "man of science."

British physician and physiologist **Marshall Hall** (1790–1857) **discovered the reflex arc**—the primitive part of the body's nervous system. It takes time for the brain to receive sense signals, then process and act on them. Reflex arcs provide a rapid automatic response by short-circuiting the brain. When the hand, for example, touches something hot, the sense signal only goes as far as the spinal cord before a message is sent back to move the hand.

German mathematician **Carl Friedrich Gauss** and physicist **Wilhelm Weber** (1804–91) developed the **first practical electromagnetic telegraph**.

1834

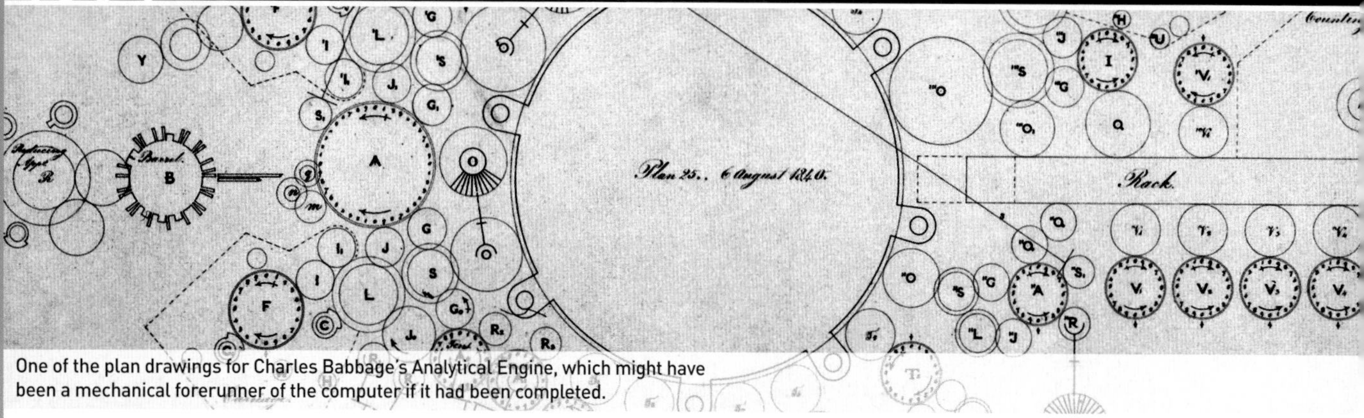

One of the plan drawings for Charles Babbage's Analytical Engine, which might have been a mechanical forerunner of the computer if it had been completed.

BRITISH SCIENTIST MICHAEL FARADAY had conducted some brilliant work in 1833 on electrolysis—the chemical reactions that occur when an electric current passes through a liquid (see panel, opposite). In 1834, he published his two laws relating to it. Faraday's **first law of electrolysis** states that the amount of **chemical change varies in proportion to the current**. The second law states that the amount of material deposited on the electrodes by the reaction is proportional to the mass of the material involved in the reactions.

Another law relating to electricity was developed by Russian physicist **Heinrich Lenz** (1804–65). **Lenz's law** stated that an electric current, induced by an electromagnetic field for example, always creates a magnetic force counter (opposite) to the force inducing it.

French engineer **Émile Clapeyron** (1799–1864) began to formulate a third key scientific law. Clapeyron took the work of fellow French physicist Nicolas Léonard Sadi Carnot (see 1823–24) on heat engines and presented it in graphic form. It clarified Carnot's observation that in energy exchanges, the energy available to make things happen (potential energy) must always gradually run down unless more energy is put in. Burning fuel is an irreversible process, which is why a car needs continual refueling. This became the **basis of the second law of thermodynamics** (see 1849–51).

Also in 1834, after producing his mechanical calculator, or Difference Engine (see 1822), British inventor Charles Babbage (1791–1871) began to work on plans for an **Analytical Engine**.

memory racks

metal frame

calculating wheels

Charles Babbage's Analytical Engine
Charles Babbage simplified the design of his Analytical Engine and managed to have a small part of it built before he died in 1871.

Anselme Payen discovers the **first enzyme**, diastase, in beer malt

Marshall Hall discovers the **reflex arc**

The term **"scientist"** is coined

Carl Friedrich Gauss and William Weber develop the **first electric telegraph**

Michael Faraday publishes the **laws of electrolysis**

March 14 British astronomer **John Herschel** discovers the NGC 3630 star cluster

Henry Lenz establishes **Lenz's law for electromotive force**

German physician **Herman Helmholtz** wrongly suggests that the Sun's heat comes from gravitational contraction

June 21 American inventor Cyrus Cormick patents his **corn reaper**

Émile Clapeyron presents Nicolas Léonard Sadi Carnot's work on **heat engines**

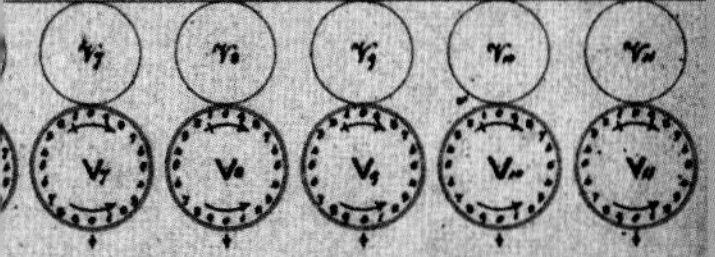

1835–36

> **“** WE FOUND THE VERY **SUPERIOR SPECIES OF THE VESPERTILIO-HOMO...** THEY... APPEARED IN OUR EYES SCARCELY **LESS LOVELY THAN** THE **REPRESENTATIONS OF ANGELS** BY... PAINTERS. **”**
>
> A description of the inhabitants of the moon in the *New York Sun*, 1835

The Great Moon Hoax in the *New York Sun* caused a sensation and fooled many with illustrations of people on the Moon.

If it had been completed it would have been a genuine forerunner to the computer (see pp.184–85), because it would have been programmable, had a memory, and could have been programmed to perform many tasks besides simple calculation.

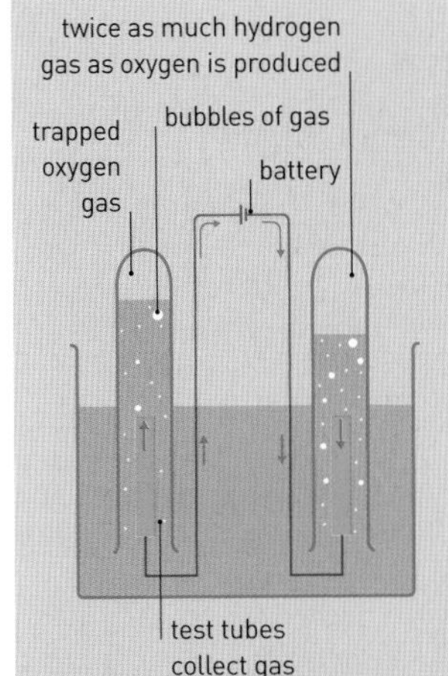
twice as much hydrogen gas as oxygen is produced
trapped oxygen gas
bubbles of gas
battery
test tubes collect gas

ELECTROLYSIS OF WATER

Electrolysis splits water into its component elements (hydrogen and oxygen) by passing an electric current through it. Hydrogen gas gathers in the test tube above the negatively charged electrode and oxygen gas gathers in the test tube above the positively charged electrode. Twice as much hydrogen is released.

A POPULAR SCIENTIFIC STORY OF 1835 TURNED OUT TO BE A HOAX. The Great Moon Hoax was a series of stories in the *New York Sun* newspaper that covered the "discoveries" of the famous British astronomer **John Herschel** (1792–1871). The illustrated articles claimed there was **life, and even civilizations, on the Moon**. It was several weeks before the joke was revealed.

In 1835, French engineer **Gaspard-Gustave Coriolis** (1792–1843) identified what is now known as the **Coriolis effect**—the effect of Earth's rotation on both wind and water moving across Earth's surface. Winds do not follow a straight path, but veer to the east in the Northern Hemisphere and to the west in the Southern Hemisphere, sometimes curling around in loops in clockwise and counterclockwise circulations. The same is true of ocean currents.

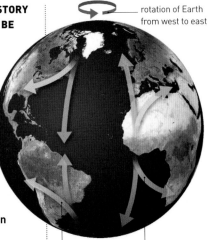

rotation of Earth from west to east

air moving toward the equator is deflected from east to west

air moving away from the equator is deflected from west to east

Coriolis effect
Earth's rotation is the cause of the Coriolis effect. One of the main impacts of it is the deflection of winds and currents in the ocean.

Knowledge of the history of Earth took a step forward when **two geological periods were identified** (see 1821–22). English geologist **Adam Sedgwick** (1785–1873) proposed the **Cambrian period**, naming it after Cambria, the Latin name for Wales, where Britain's Cambrian rocks are best exposed. Scottish geologist, Roderick Murchison (1792–1871), proposed the **Silurian period** in the same year, naming it after an ancient Celtic tribe, the Silures, and the two presented their findings the same year in a joint paper.

On September 16, 1835, **Charles Darwin** (1809–82) landed on the **Galapagos Islands** for the first time. His visit to the islands had a significant impact on his theory of evolution (see 1857–58).

In England, inventor **Henry Fox Talbot** (1800–77) produced the **world's first photographic negative** (dark and light reversed). Although French physicist Louis Daguerre (1787–1851) went public with his process known as daguerreotype first (see 1837), this process produced a one-off positive photograph. Talbot's process, called the **calotype**, produced a negative photograph from which many positive prints could be made.

Talbot exposes photo negatives
Fox Talbot's calotype process recorded a negative image, from which a large number of positive prints could be made.

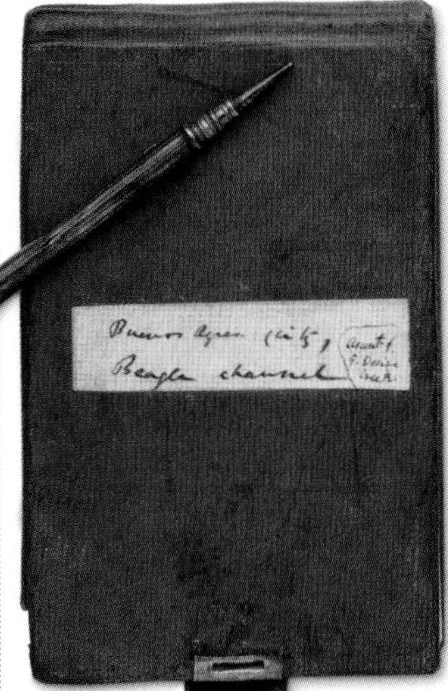

Darwin's notebook
Darwin recorded his observations on the Beagle *voyage (1831–36) in notebooks. His records of this trip helped him develop his theory of evolution over the next 20 years.*

On the other side of the Atlantic, in 1836, American inventor **Samuel Morse** (1791–1872) developed a code (later called **Morse code**) for sending messages by telegraph, with each letter of the alphabet represented by a unique combination of short pulses (dots) and long pulses (dashes).

Another American, **Samuel Colt** (1814–62), was granted a US patent for the revolver. This gun could fire six shots in quick succession by using a revolving cylinder that automatically moved a new bullet into the firing position after each shot.

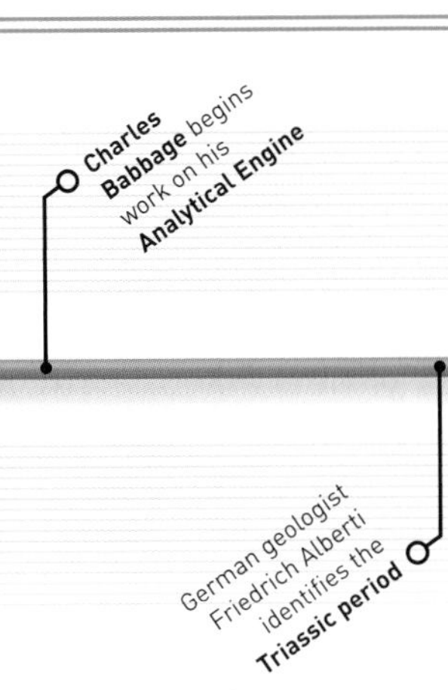

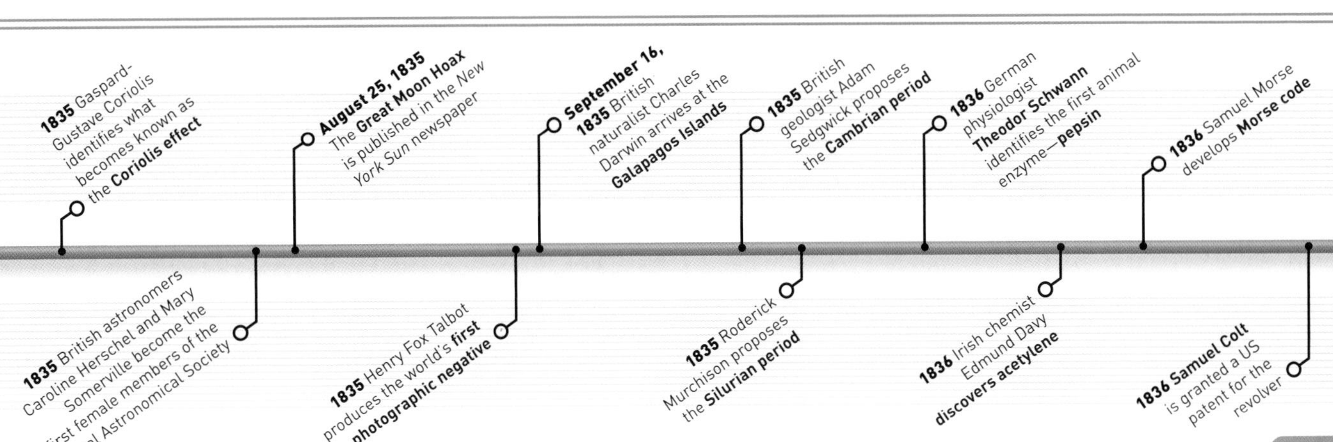
Charles Babbage begins work on his **Analytical Engine**
German geologist Friedrich Alberti identifies the **Triassic period**
1835 Gaspard-Gustave Coriolis identifies what becomes known as the **Coriolis effect**
1835 British astronomers Caroline Herschel and Mary Somerville become the first female members of the Royal Astronomical Society
August 25, 1835 The **Great Moon Hoax** is published in the New York Sun newspaper
1835 Henry Fox Talbot produces the world's **first photographic negative**
September 16, 1835 British naturalist Charles Darwin arrives at the **Galapagos Islands**
1835 British geologist Adam Sedgwick proposes the **Cambrian period**
1835 Roderick Murchison proposes the **Silurian period**
1836 German physiologist **Theodor Schwann** identifies the first animal enzyme—**pepsin**
1836 Irish chemist Edmund Davy **discovers acetylene**
1836 Samuel Morse develops **Morse code**
1836 Samuel Colt is granted a US patent for the revolver

“ WHAT WAS THE **USE OF THIS GREAT ENGINE** SET AT WORK AGES AGO TO… **FURROW AND KNEAD OVER…** THE SURFACE OF THE EARTH? THE GLACIER WAS **GOD’S GREAT PLOW.** ”

Louis Agassiz, Swiss geologist, in *Geological Sketches*, 1866

The Marjerie Glacier in Glacier Bay, Alaska, USA, flows out from the mountains to the sea and extends a distance of 21 miles (34 km).

THIS WAS THE YEAR in which telecommunications really began. Three men had already been working on the idea of an electric telegraph—**Samuel Morse** in the US and **William Cooke** (1806–79) and **Charles Wheatstone** (1802–75) in England. In 1837, it became a reality. Cooke and Wheatstone were the first to succeed with their 1.2 mile- (2 km-) long telegraph from Euston to Camden in London. In the end, the simplicity of the Morse telegraph's single wire system and dot-dash code meant it was adopted as the standard telegraph and the Wheatstone-Cooke model was discarded. An equally momentous breakthrough was the development of the **first successful photographic process** by French painter **Louis-Jacques-Mandé Daguerre** (1787–1851). Daguerre had long hunted for a means to fix the fleeting images he saw in his artist's camera obscura. He finally succeeded with his invention of the process known as **daguerreotyping**, which involved capturing a one-off photograph in chemicals coated on a silver-plated copper sheet.

Daguerreotype camera
Dating from the 1840s, this is one of the first cameras made for taking photographs. The plate for each photo is slotted into the back.

His first daguerreotype, taken in 1837, was blurry, but within a few years daguerreotypes were recording photographs of astonishing clarity.

The most extraordinary scientific insight of this year was understood by no one at the time but its creator, British scientist **Michael Faraday**. This was the idea of **fields of force**—regions around magnets and electric currents in which their effects are felt. Faraday argued that in these fields, electric charges are pushed by the movement of invisible **lines of force**, which can be revealed by the pattern iron filings make around a magnet. This is why compasses swivel in magnetic fields and magnets move charged particles to create electric currents.

French mathematician **Siméon Denis Poisson** (1781–1840) developed the valuable statistical idea of **Poisson distribution**: the probability of a given number of events occurring in a particular time if they occur at an average rate.

In Paris, following the discovery of the **green pigment chlorophyll** in plant leaves in 1817, French physiologist **Henri Dutrochet** (1776–1847) argued that chlorophyll is the key to **photosynthesis**, by which plants fix oxygen from the air using sunlight (see 1787–88).

Meanwhile, in Switzerland, geologist **Louis Agassiz** (1807–73) published *Études sur les Glaciers* (*Studies of Glaciers*), arguing that the **Earth was once subject to an ice age** and that traces of ice erosion and deposition by vast glaciers and ice sheets are still evident in the landscape today.

MICHAEL FARADAY (1791–1867)

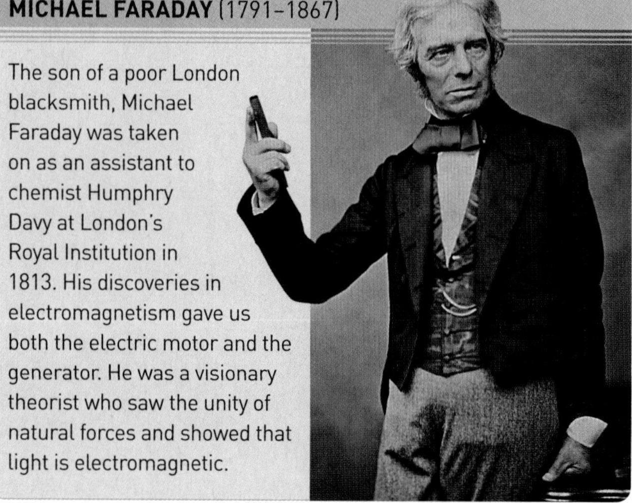

The son of a poor London blacksmith, Michael Faraday was taken on as an assistant to chemist Humphry Davy at London's Royal Institution in 1813. His discoveries in electromagnetism gave us both the electric motor and the generator. He was a visionary theorist who saw the unity of natural forces and showed that light is electromagnetic.

GLACIATION

GLACIER
cirque
ice-fall, where glacier flows over steep gradient
crevasse
terminus (snout)
glacial lake

GLACIAL-ERODED LANDSCAPE
cirque
hanging valley
U-shaped valley
river
moraine-dammed lake
tarn (small lake in scooped-out base of cirque)

Glaciers accumulate such a weight of ice over the years that they possess an immense power to shape the landscape. They can carve out vast U-shaped valleys, cut-off hillsides, and leave behind piles of rock debris. This process of modifying the landscape is known as glaciation and the landscapes left behind by the glaciers of the ice age are unmistakable to geologists today.

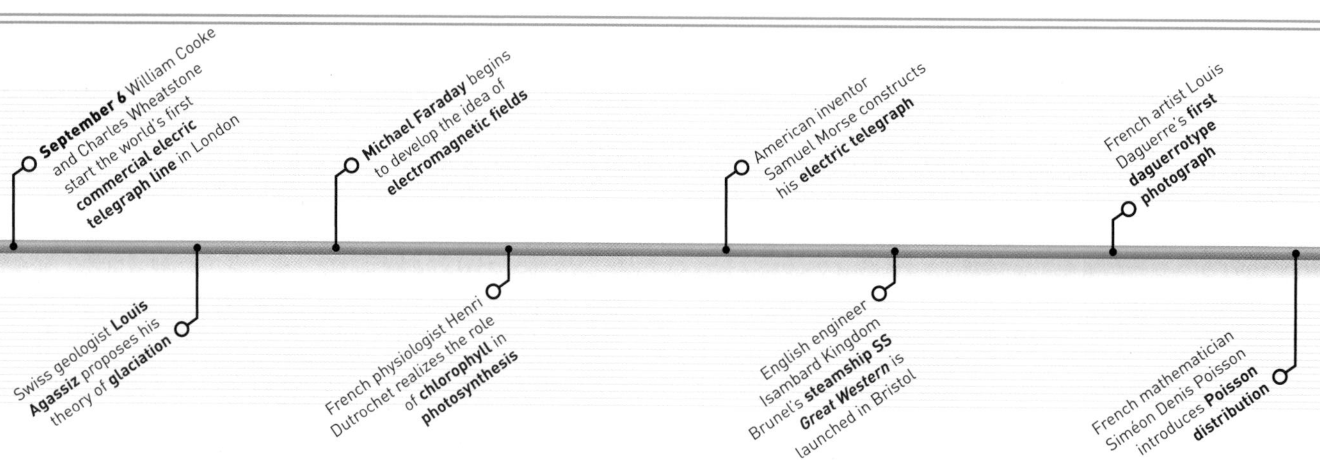
September 6 William Cooke and Charles Wheatstone start the world's first **commercial elecric telegraph line** in London
Swiss geologist **Louis Agassiz** proposes his theory of **glaciation**
Michael Faraday begins to develop the idea of **electromagnetic fields**
French physiologist Henri Dutrochet realizes the role of **chlorophyll** in **photosynthesis**
American inventor Samuel Morse constructs his **electric telegraph**
English engineer Isambard Kingdom Brunel's **steamship SS *Great Western*** is launched in Bristol
French artist Louis Daguerre's **first daguerrotype photograph**
French mathematician Siméon Denis Poisson introduces **Poisson distribution**

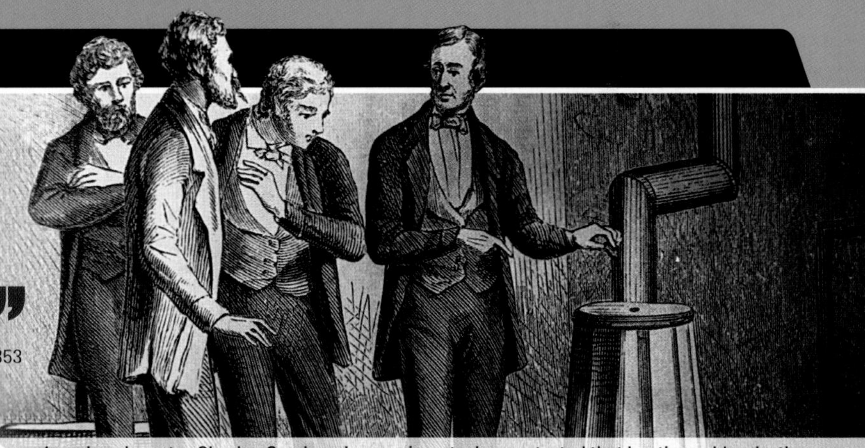

“ THE INVENTOR MADE SOME EXPERIMENTS TO ASCERTAIN THE EFFECT OF HEAT... THE SPECIMEN, BEING... BROUGHT INTO CONTACT WITH A HOT STOVE, CHARRED LIKE LEATHER. ”

Charles Goodyear, American inventor, in *The Applications and Uses of Vulcanized Gum-Elastic,* 1853

American inventor Charles Goodyear’s experiments demonstrated that heating rubber by the right amount and adding sulfur could toughen or “vulcanize” it.

EARLY MICROSCOPES had been plagued by color blurring, or chromatic aberration. By 1838, achromatic microscopes—solving this problem—gave scientists a clearer view of living cells. As German physiologist **Theodor Schwann** (1810–82) studied plant and animal cells through his microscope, he realized that all living things are made of cells and cell products, and the **cell is the basic unit of life**.

Dutch chemist **Gerardus Mulder** (1802–80) came to an equally important conclusion about the basic material that cells are made of. After experimenting with heating albuminous substances, such as egg white, blood, milk solids, and plant gluten with lye (a strong alkali solution), he always ended up with the same material. Mulder believed this material was composed of a **single large molecule common to all living things**. It was the Swedish chemist **Jöns Jacob Berzelius** (1779–1848) who suggested the name **protein** for this substance. It is now known there are numerous proteins and they are the basic chemicals of life.

The same year, French physicist **Claude Pouillet** (1791–1868) made the first accurate measurements of the **solar constant** (the amount of solar heat received at Earth).

German astronomer **Friedrich Bessel** (1784–1846) made the first accurate estimate of the distance to a star using the **parallax method**, which depends on slight shifts in the star’s apparent position due to the movement of the Earth.

German–Swiss chemist **Christian Friedrich Schönbein** (1799–1868) developed the idea of a fuel cell that **converts chemical energy from a fuel**, such as hydrogen, into **electricity**. In 1839, British physicist William Grove (1811–96) made the **first fuel cell**. He knew electricity could split water into hydrogen and oxygen; his fuel cell reversed this process and made electricity by combining the two gases to produce water. At the time, though, Grove was much better known for the battery he invented that same year, known as a **Grove cell**.

American inventor **Charles Goodyear** (1800–60) developed a technique for vulcanizing rubber, a process that toughened it to make it suitable for use such as in tires.

In 1840, **Jacob Berzelius** suggested the word allotrope to describe **different forms of the same element**. Allotropes differ from each other as the result of different bonding between atoms results in different chemical and physical properties. Also in 1840, **Christian Friedrich Schönbein** discovered an allotrope of oxygen, which he gave the name “ozone.”

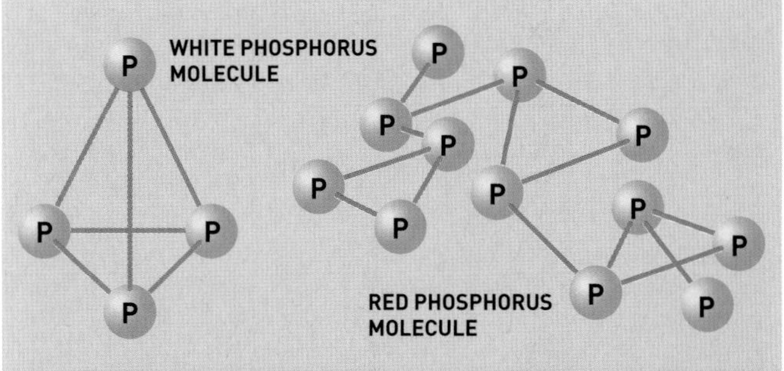

WHITE PHOSPHORUS MOLECULE
RED PHOSPHORUS MOLECULE
P

ALLOTROPY

Some elements come in a number of different physical forms, called allotropes. Each form is made from the same type of atom, but the atoms link up in different ways. Carbon has eight allotropes, including diamond, graphite, and fullerenes. Phosphorus has at least twelve, the most common of which are white and red solids.

plate electrodes of zinc and platinum

cells filled with acid

Grove cell
In this battery invented by William Grove, the charge is generated on zinc and platinum electrodes immersed in acid.

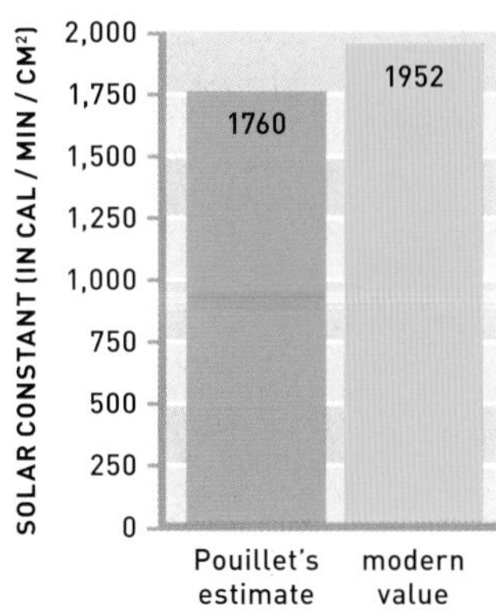

SOLAR CONSTANT (IN CAL / MIN / CM²)
2,000
1,750
1,500
1,250
1,000
750
500
250
0
1760
1952
Pouillet’s estimate
modern value

Solar constant
Claude Pouillet’s 1839 estimate of solar heat radiation, made using a device called a pyrheliometer, was very close to today’s value.

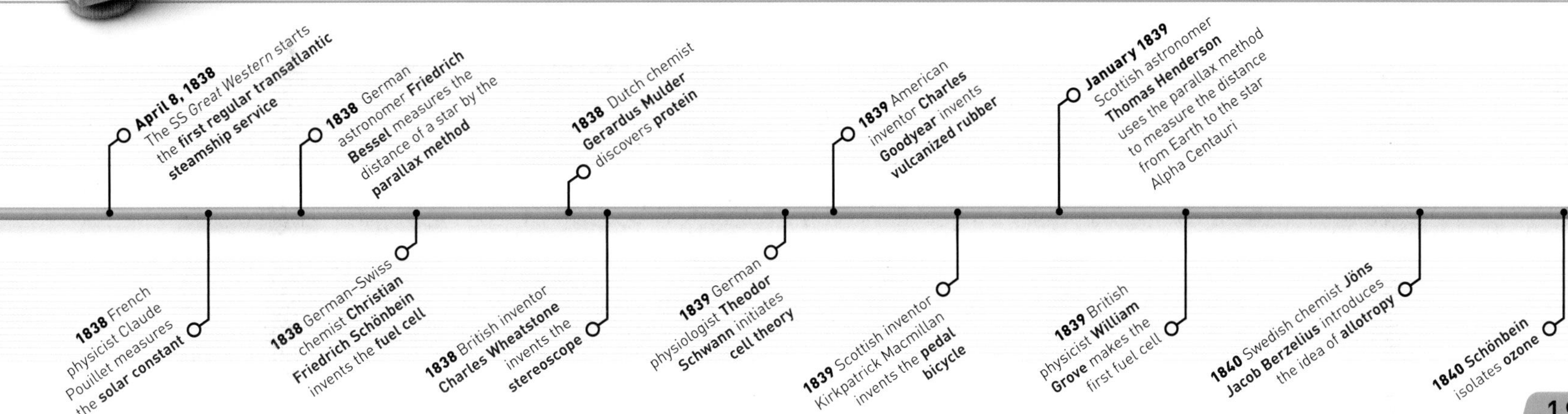

April 8, 1838 The SS Great Western starts the **first regular transatlantic steamship service**
1838 French physicist Claude Pouillet measures the **solar constant**
1838 German astronomer **Friedrich Bessel** measures the distance of a star by the **parallax method**
1838 German–Swiss chemist **Christian Friedrich Schönbein** invents the **fuel cell**
1838 British inventor **Charles Wheatstone** invents the **stereoscope**
1838 Dutch chemist **Gerardus Mulder** discovers **protein**
1839 German physiologist **Theodor Schwann** initiates **cell theory**
1839 American inventor **Charles Goodyear** invents **vulcanized rubber**
1839 Scottish inventor Kirkpatrick Macmillan invents the **pedal bicycle**
January 1839 Scottish astronomer **Thomas Henderson** uses the parallax method to measure the distance from Earth to the star Alpha Centauri
1839 British physicist **William Grove** makes the first fuel cell
1840 Swedish chemist **Jöns Jacob Berzelius** introduces the idea of **allotropy**
1840 Schönbein isolates **ozone**

UNDERSTANDING CELLS

A CELL HAS COMPLEXITY ON A MINISCULE SCALE AND IS THE SMALLEST THING ALIVE

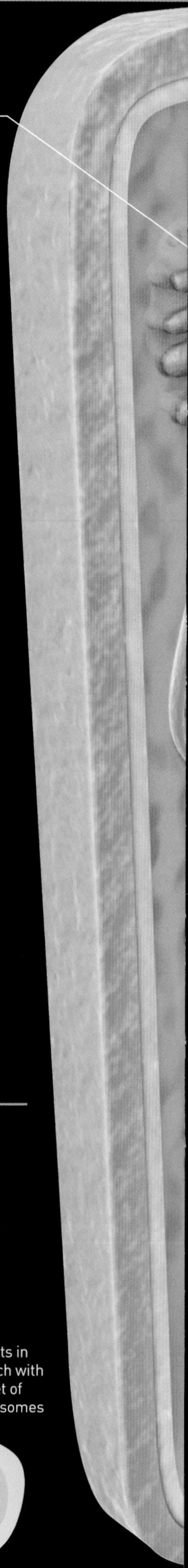

A plant or animal body contains more cells than people who have ever lived—hundreds could fit on a pinhead. Within its outer structure, a cell has chemistry of unrivaled intricacy—to manage its growth, reproduction, and nourishment.

Golgi apparatus—used for sorting and packaging substances, including those for export

THEODOR SCHWANN
German scientist Schwann, a founding father of Cell Theory, insisted that cells could be understood in chemical terms with no mysterious "life force."

The first cells were seen in 1663 when English scientist Robert Hooke found cork cells with his microscope. But it wasn't until the nineteenth century that their significance was better appreciated and German biologists developed a "Cell Theory." They suggested that cells were the units of all organisms and could arise only from other cells. In other words, cellular life could not form spontaneously. By 1900, scientists began to see how this reproductive ability was linked to the cell's nucleus and its chromosomes. This work would culminate with the discovery that a self-copying chemical inside the nucleus—called DNA—lay at the heart of the process.

While some cells, such as those of bacteria, are structurally simple, those of animals and plants contain even smaller compartments for specific roles. These so-called organelles enclose the particular mixture of chemicals needed for a specific job.

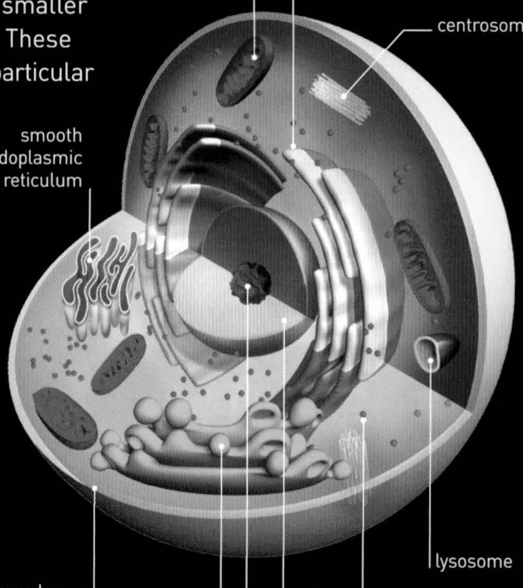

ANIMAL CELL
Most animal cells are smaller than plant cells because they lack a rigid supporting cell wall. This also makes animal cells less angular in shape. Many organelles found in plant cells (right) are also found in animal cells.

mitochondrion
rough endoplasmic reticulum
centrosome
smooth endoplasmic reticulum
cell membrane
Golgi apparatus
nucleus
nucleolus
ribosome
lysosome

60 TRILLION
THE POSSIBLE NUMBER OF **CELLS** THAT MAKE UP **THE HUMAN BODY**

CELL DIVISION
By the late 1800s, microscopes were good enough to scrutinize dividing cells. Scientists saw threadlike chromosomes moving around in precise ways—chromosomes are bundles of the DNA that carry the information for producing new cells. Just before cell division, the cell's DNA doubles up by self-copying—so when body cells divide during growth (mitosis), each daughter cell ends up with a copied set of each chromosome, or DNA bundle. During sexual reproduction, a special kind of division (meiosis) halves the chromosome number to make sperm and eggs; the normal number is restored when the egg and sperm cells combine during fertilization.

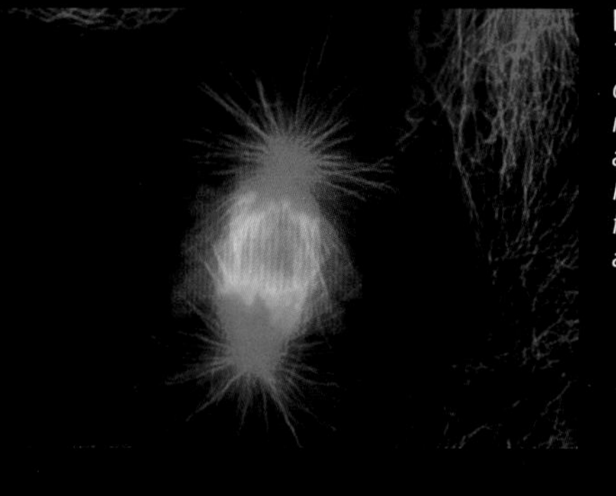

IMMUNOFLUORESCENT MICROGRAPH
Today, scientists can reveal aspects of the cell that are invisible with conventional microscopy by using fluorescently dyed antibodies that stick to specific structures. Here, the antibodies illuminate the spindle fibers (green) that pull chromosomes (blue) around during cell division.

MITOSIS
During growth, this multistage division keeps all cells genetically identical. A system of protein cables called the spindle pulls chromosomes in such a way that daughter cells end up with the same chromosome number as the parental cell.

nucleus
chromosomes (four shown as examples)
EARLY PROPHASE

chromatids are copies of the same chromosome joined together
LATE PROPHASE

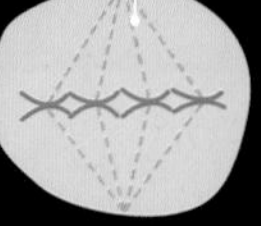

spindle fibers grow out from poles to attach to chromosomes
METAPHASE

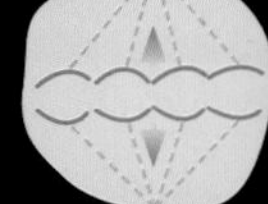

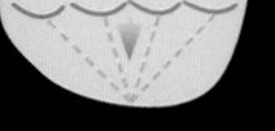

spindle fibers shorten to pull apart chromosomes
ANAPHASE

cell membrane forms across the cell
TELOPHASE

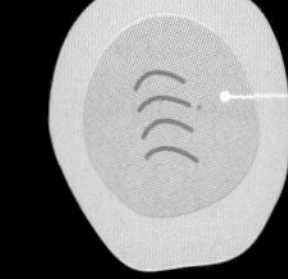

cell splits in two, each with a full set of chromosomes
nucleus forms around the chromosomes in each cell
CYTOKINESIS

chloroplast—the site of food-producing reactions of photosynthesis

cytoplasm—fluid between the nucleus and the cell membrane

cell membrane—controls what enters and leaves the cell

cell wall—a rigid layer of cellulose fibers that supports the cell

vacuole—contains stored substances, pigment, or poisons

vesicle—a fluid-filled sac that stores cellular substances

mitochondrion—generates energy for the cell

ribosome—site where proteins are made

rough endoplasmic reticulum (RER) has ribosomes for making and transporting proteins

nucleus—stores genetic material

nucleolus—used for producing ribosomes

smooth endoplasmic reticulum (SER) used for making and transporting lipids

PLANT CELL
The cell membrane of a plant cell is overlain by a porous cell wall. Inside the cell are membrane-bound organelles that package the chemicals needed for biological processes, such as mitochondria for respiration.

MOVEMENT ACROSS CELL MEMBRANES

Cells and their organelles are bound by oily membranes that separate watery mixtures of chemicals on either side. Small molecules, or substances that can blend with oil, can penetrate the membrane—and tend to move from areas of high to low concentration by diffusion. Other particles can only get across by special molecular "pumps" in a process called active transport. The cell has to use up energy to do this. The energy comes from chemical processes of respiration (see below). Cells take up oxygen and excrete carbon dioxide by diffusion, but need active transport for moving salts and big molecules.

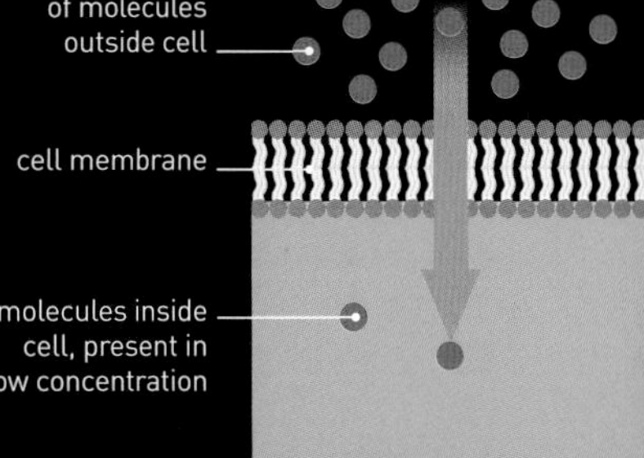
high concentration of molecules outside cell
cell membrane
molecules inside cell, present in low concentration

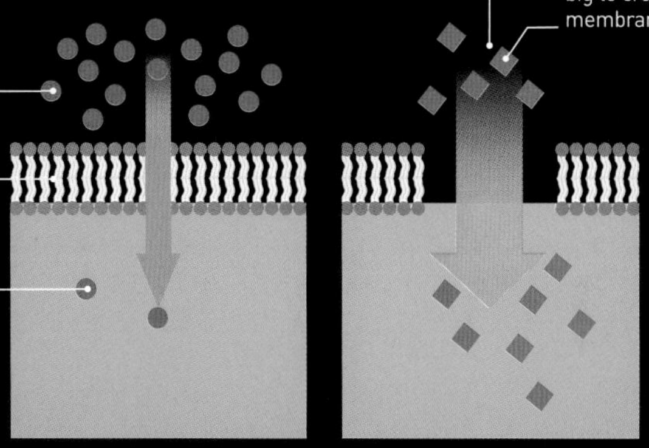

DIFFUSION
This moves substances from high to low concentration. The bigger the difference (concentration "gradient"), the more they move.

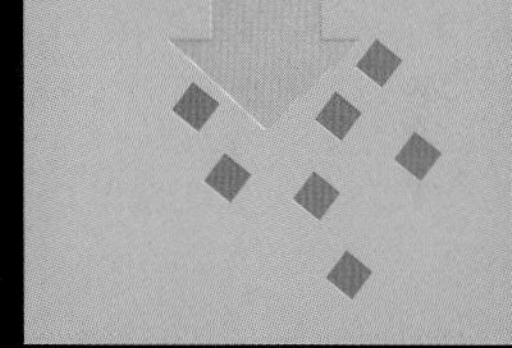
energy is needed to pump molecules into cell
molecules too big to cross membrane

ACTIVE TRANSPORT
Here, substances move from low to high concentration, giving a cell the means to accumulate a substance in the cytoplasm or in an organelle.

RELEASING ENERGY

Cells are driven by energy, which comes from food. Plants make their food by photosynthesis during which carbon dioxide and water react with light and chlorophyll. Almost all cells obtain energy in the same way: by breaking down a sugar called glucose. The process begins in the cell's cytoplasm and finishes in organelles called mitochondria. These "powerhouses" use oxygen to extract energy in a particularly efficient way, generating a great deal of an energy-rich compound called ATP, which powers cellular activities involved in building or movement.

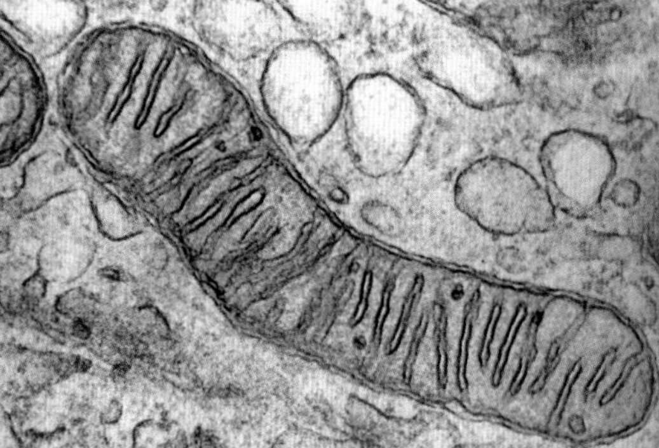

MITOCHONDRION
Chemical reactions that release most of the cell's energy happen on a mitochondrion's internal membranes; the most active of cells have the most membrane-packed mitochondria.

1841

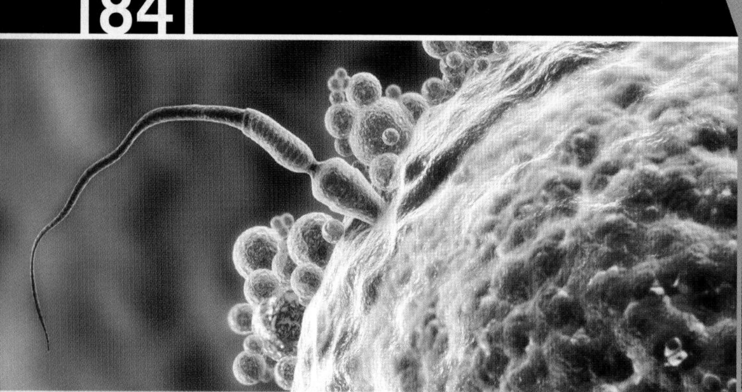
Swiss physiologist Albert von Kölliker showed that, like every other living cell, each sperm is a single cell with its own nucleus.

1842–1843

British engineer James Nasmyth's steam hammer was able to forge large pieces of wrought iron and was developed in response to the increased demands of 19th-century engineering.

1-3 MILLIMETERS PER MINUTE
THE **SPEED** A **SPERMATOZOAN** CAN TRAVEL

IN 1841, GERMAN PHYSICIAN JULIUS VON MAYER (1814–78) first proposed that **energy could never be created or destroyed**. This is now known as the first law of thermodynamics (see 1847–48). Von Mayer also proposed that work and heat are equivalent—a certain amount of work will always produce a certain amount of heat—as English physicist James Joule (1818–89) discovered independently two years later. However, it was some time before either von Mayer or Joule's work was acknowledged.

In contrast, improvements made by Swiss physiologist **Albert von Kölliker** (1817–1905) to microscope techniques, such as the staining of samples, were soon acknowledged and adopted. Von Kölliker also confirmed that **each sperm and egg is a cell with a nucleus**, adding to the emerging science of **histology**—the science of living cells.

In the UK, engineer **Joseph Whitworth** (1803–87) identified and solved a basic problem with assembling precision machines—the variation in screws. Whitworth devised a system of **standardization for screw threads and pitches**. When several railroad companies decided to adopt it, other organizations quickly followed suit. This system is now known as the BSW (British Standard Whitworth) system.

In Antarctica, British explorer **James Clark Ross** (1800–62) discovered and named the Victoria Barrier, later known as the Ross Ice Shelf.

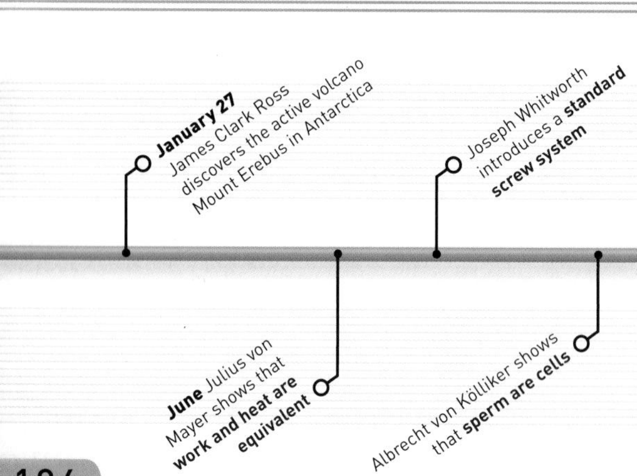
crest
pitch
thread root

Screw threads
Joseph Whitworth set a standard for screw threads with a fixed angle of 55° for the thread.

THE DOPPLER EFFECT

The sound of a police siren approaching rises in pitch as the sound waves are squashed in front of the moving car. As the car moves past and away from the listener, the pitch drops as sound waves stretch out in its wake. This occurs because, as the police car approaches, each successive sound wave begins closer to the car. As the police car moves away, each successive sound wave starts farther away from the car.

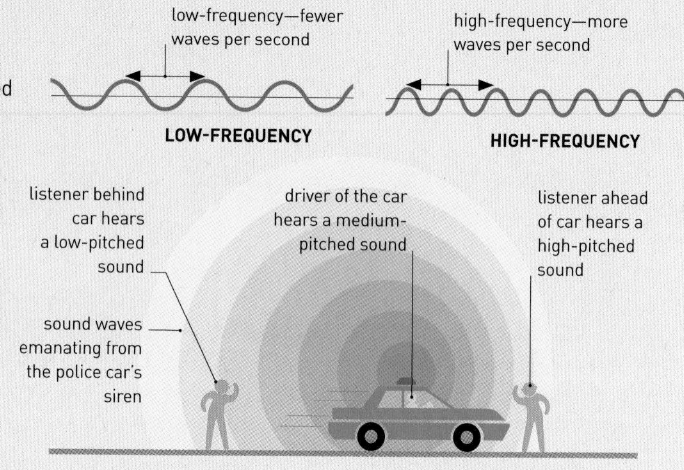
low-frequency—fewer waves per second
high-frequency—more waves per second
LOW-FREQUENCY
HIGH-FREQUENCY
listener behind car hears a low-pitched sound
driver of the car hears a medium-pitched sound
listener ahead of car hears a high-pitched sound
sound waves emanating from the police car's siren

IN 1842, 25 YEARS AFTER THE FIRST RECOGNIZED FOSSIL was found, British biologist **Richard Owen** (1804–92) first used the word dinosauria to describe these "terrible reptiles." Ironically, Owen had been one of the first people to wrongly pour scorn on British geologist Gideon Mantell's (1790–1852) idea that the Iguanadon fossils he had found belonged to an **extinct giant reptile**.

Meanwhile, a revolution in manufacturing was taking place with the help of the steam hammer, patented in June 1842 by British engineer **James Nasmyth** (1808–90). Previously, iron foundries had shaped iron weakly and inaccurately with a pivoting tilt-hammer that was lifted mechanically then allowed to drop. In contrast, Nasmyth's **vertical steam hammer** was forced down with great power, but could still stop short with enough precision to crack an egg in a wine glass. The power and precision of Nasmyth's steam hammer meant that such things as railroad wheels and the first steel hulls for ships could be pounded out of solid steel for the first time, thus revolutionizing the manufacturing process.

British inventors **John Stringfellow** (1799–1883) and **William Henson** (1812–88) worked together on ideas

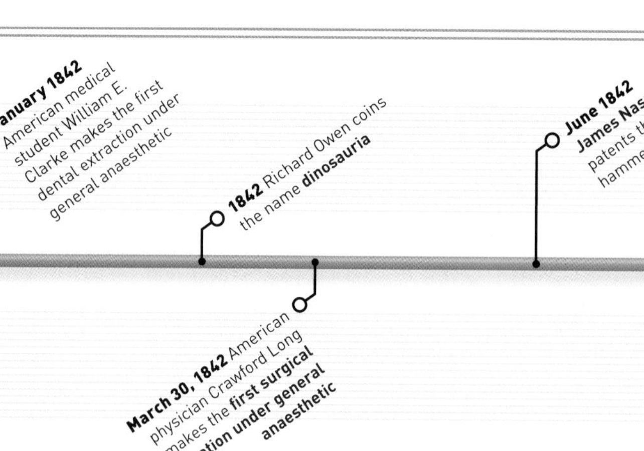

Skeleton of an iguanodon
This species of dinosaur was discovered by British geologist Gideon Mantell in 1822, although the term dinosaur was coined only in 1842.

January 27 James Clark Ross discovers the active volcano Mount Erebus in Antarctica

June Julius von Mayer shows that **work and heat are equivalent**

Joseph Whitworth introduces a **standard screw system**

Albrecht von Kölliker shows that **sperm are cells**

January 1842 American medical student William E. Clarke makes the first dental extraction under general anaesthetic

1842 Richard Owen coins the name **dinosauria**

March 30, 1842 American physician Crawford Long makes the **first surgical operation under general anaesthetic**

June 1842 James Nasmyth patents the steam hammer

1842 Christopher Doppler discovers what became known as **the Doppler effect**

28 TONS THE MAXIMUM WEIGHT OF NASMYTH'S STEAM HAMMER

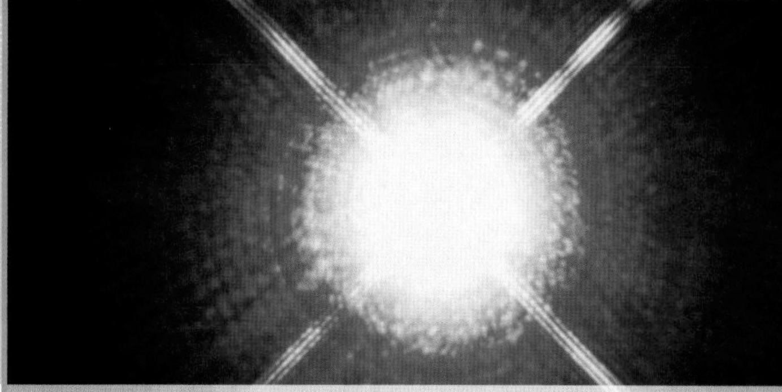

The star Sirius A and its almost invisible companion, the white dwarf star Sirius B, whose existence was deduced in 1844 by Bessel.

for powered flight and, in 1842, they designed a **large, steam-powered, passenger plane**. They received a patent for it in 1843 and launched the Aerial Steam Transit Company, which they advertised as flying to exotic locations such as the pyramids, but the idea was eventually abandoned.

In 1842, Austrian scientist **Christian Doppler** (1803–53) suggested how the frequency of sound and light waves vary as objects approach and move away from an observer (see panel, opposite). This is known as **the Doppler effect**. It is these so-called Doppler shifts that later helped reveal that **the Universe is expanding** (see 1929–1930).

ribs

femur (thighbone)

The world's first "computer program" was written in 1843 by British mathematician **Ada Lovelace**, daughter of the poet Lord Byron. Lovelace worked with British inventor Charles Babbage (1791–1871) on his Analytical Engine (see 1834), which would have been the world's first computer had it ever been built. Between 1842 and 1843 she worked on translating an article about the Analytical Engine by Italian mathematician Luigi Menabrea (1809–96). As she worked through the article she added a note that included an encoded algorithm, designed to be processed by a machine. If the machine had been built, this algorithm would have been the **first computer program**.

ADA LOVELACE (1815–52)

The daughter of poet Lord Byron, Ada Lovelace was a brilliant mathematician. In 1843, she began to publicize British inventor Charles Babbage's ideas for his Analytical Engine, and wrote what is called the world's first computer program for it. She foresaw that Babbage's machine would have uses far beyond mere calculation.

> **“** CREATURES FAR SURPASSING IN SIZE THE **LARGEST** OF **EXISTING REPTILES...** I WOULD PROPOSE THE NAME OF **DINOSAURIA. ”**
>
> **Richard Owen, British Biologist,** in *Report on British Fossil Reptiles*, 1842

20 ASTRONOMICAL UNITS THE APPROXIMATE DISTANCE BETWEEN SIRIUS A AND SIRIUS B

IN 1844, GERMAN ASTRONOMER FRIEDRICH BESSEL (1784–1846) spotted a wobble in the movement of the stars Sirius and Procyon. Isaac Newton's laws of gravity (see pp.120–21) allow astronomers to calculate the motions of distant stars with such precision that discrepancies can be revealed. Bessel deduced that these **stars have dark companion stars**, now known as Sirius B and Procyon B.

Another German astronomer, **Heinrich Schwabe** (1789–1875), observed how **sunspots vary cyclically** over a period of 10 years—later, this cycle was shown in fact to be 11 years.

Meanwhile, on May 24, American inventor **Samuel Morse** sent the **US's first long-distance telegraph** message down a new line from Washington to Baltimore. It was a Biblical message written in his own Morse code, "What Hath God wrought?"—instant communication had arrived.

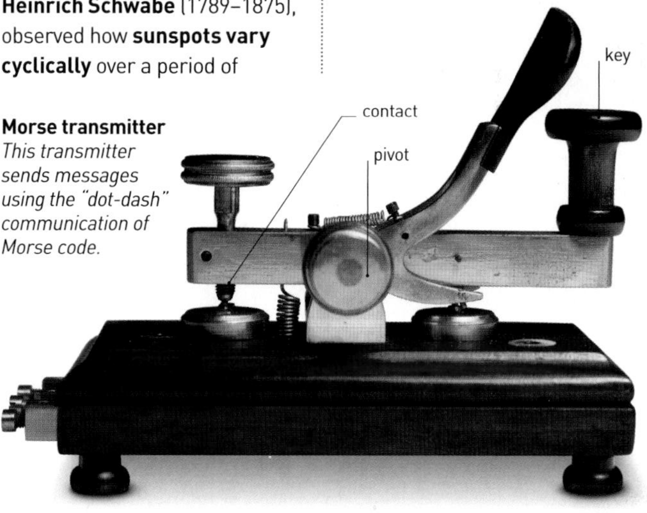
key
contact
pivot

Morse transmitter
This transmitter sends messages using the "dot-dash" communication of Morse code.

- **1843** Henson and Stringfellow patent their **steam-powered passenger plane**
- **March 25, 1843** Completion of British engineer Marc Isambard Brunel's Thames Tunnel in London, the first bored underwater tunnel
- **July 19, 1843** Launch of Brunel's pioneering, iron-hulled steamship *SS Great Britain*
- **1843** Ada Lovelace writes the **world's first computer algorithm**
- Friedrich Bessel deduces that Sirius and Procyon have **dark companion stars**
- **May 24** Samuel Morse sends the **first message using Morse code**
- German astronomer Heinrich Schwabe observes cyclic variation in sunspots

1845

John Franklin's doomed expedition to find the elusive Northwest Passage through the Arctic Ocean. His two ships HMS *Terror* and HMS *Erebus* vanished after entering Baffin Bay.

ONE OF THE KEY SEARCHES IN SCIENCE TODAY is to discover the underlying unity of forces and matter. In 1845, British physicist **Michael Faraday** made an early contribution to this with a series of experiments in which he showed that a magnetic field could alter the polarization of light as it traveled through heavy lead glass. This finding revealed a previously unsuspected **link between light, magnetism, and electricity**, paving the way for the discovery of the complete spectrum of electromagnetic radiation (see pp.234–35).

At the same time, astronomers were continuing to scrutinize the night sky. An English astronomer, the **3rd Earl of Rosse**, constructed a huge reflecting telescope in Ireland dubbed the Leviathan of Parsonstown. Through its 6 ft (1.8 m) aperture, Rosse observed that the galaxy M51—now known as the Whirlpool Galaxy—had a spiral structure. This was the **first galaxy observed to be spiral**.

340 TONS

THE **WEIGHT** OF THE THREE-STORY-HIGH **STEAM ENGINE** IN THE SS *GREAT BRITAIN*

Other astronomers questioned why Uranus kept appearing in places that it should not appear according to Kepler's laws (pp.100–101) and Newton's laws (pp.120–21). British mathematician **John Couch Adams** (1819–1892) suggested that there was another planet beyond Uranus disturbing its orbit. French astronomer **Urbain Le Verrier** (1811–77) observed disturbances in Uranus's orbit to predict where this planet might be.

Another mystery was the disappearance of **British explorer John Franklin** and his ships HMS *Erebus* and HMS *Terror*, in Baffin Bay during an expedition to find the Northwest Passage in the Arctic.

The SS *Great Britain* successfully sailed from Liverpool to New York, becoming the first iron steamship driven by a screw-propeller rather than paddle wheels to make this journey across the Atlantic.

Lord Rosse's reflecting telescope *The Leviathan of Parsonstown, constructed in County Offaly, Ireland in 1845, was the biggest telescope in the world for more than 70 years.*

1846

The planet Neptune rising over its moon Triton in an image constructed from pictures taken by the Voyager space probes. Both were discovered in 1846.

GENERAL ANESTHESIA

The introduction of general anesthetics (substances used to render a person unconscious) transformed surgery. It allowed surgeons to perform all kinds of operations painlessly. The term anesthesia means "loss of sensation": anesthetics work by blocking signals that pass along nerves to the brain. The earliest anesthetics were ether, laughing gas (nitrous oxide), and chloroform.

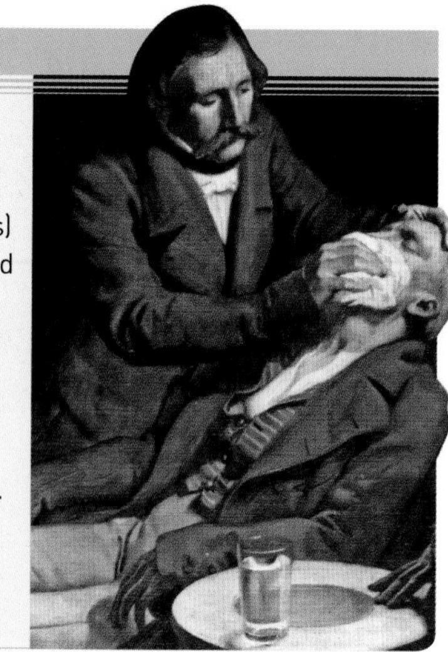

ON SEPTEMBER 23, 1846, German astronomer **Johann Gottfried Galle** (1812–1910) received a letter from French astronomer **Urbain Le Verrier**. It contained instructions telling him where to look to find the solar system's eighth planet, soon to be called **Neptune**. A long dispute followed over who deserved the credit for Neptune's discovery—Urbain Le Verrier or **John Couch Adams**, who had also predicted its existence the previous year. Most authorities credit Le Verrier because he was the one who calculated its position accurately enough for Galle to find it immediately. Within 17 days, English astronomer **William Lassell** (1799–1880) found that Neptune had a moon. It was named Triton a century later.

> "THIS SIGNIFIES **INSENSIBILITY... TO OBJECTS OF TOUCH.** THE ADJECTIVE WILL BE **ANESTHETIC.** THUS WE MIGHT SAY, THE **'STATE OF ANESTHESIA'.**"
>
> **Oliver Wendell Holmes, US physician**, in a letter to Dr. William Morton, 1846

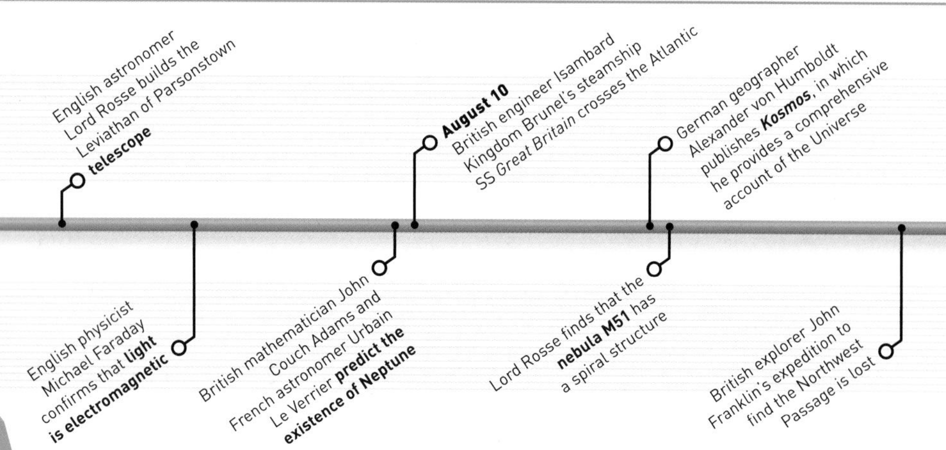

English astronomer Lord Rosse builds the Leviathan of Parsonstown **telescope**
English physicist Michael Faraday confirms that **light is electromagnetic**
August 10 British engineer Isambard Kingdom Brunel's steamship SS *Great Britain* crosses the Atlantic
British mathematician John Couch Adams and French astronomer Urbain Le Verrier **predict the existence of Neptune**
German geographer Alexander von Humboldt publishes **Kosmos**, in which he provides a comprehensive account of the Universe
Lord Rosse finds that the **nebula M51** has a spiral structure
British explorer John Franklin's expedition to find the Northwest Passage is lost

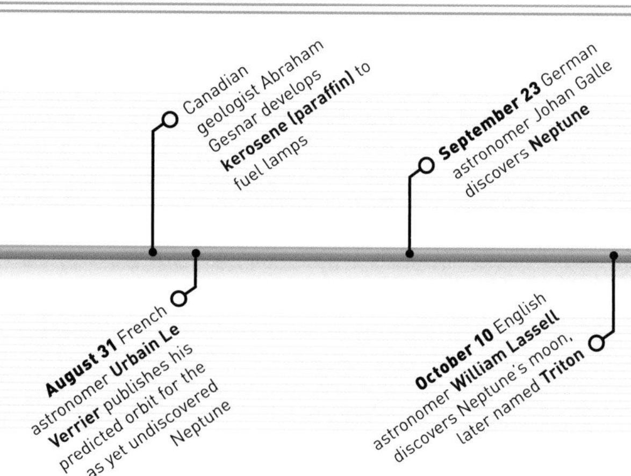

Canadian geologist Abraham Gesnar develops **kerosene (paraffin)** to fuel lamps
August 31 French astronomer **Urbain Le Verrier** publishes his predicted orbit for the as yet undiscovered Neptune
September 23 German astronomer Johan Galle discovers **Neptune**
October 10 English astronomer **William Lassell** discovers Neptune's moon, later named **Triton**

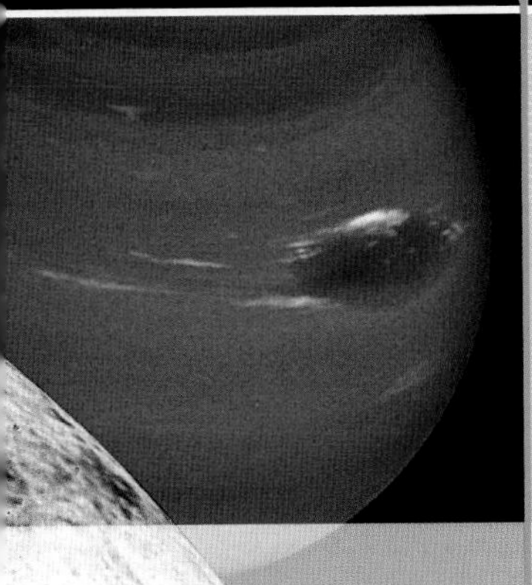

> **"** THE **EQUILIBRIUM…** BETWEEN HIS INTELLECTUAL FACULTIES AND ANIMAL PROPENSITIES… [HAS] BEEN **DESTROYED. "**
>
> **John Martyn Harlow, American physician,** in *Recovery from the Passage of an Iron Bar through the Head,* 1868

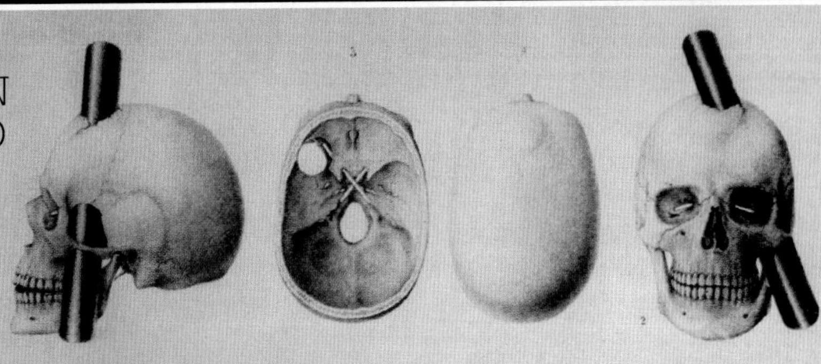

Following an iron bar penetrating Phineas Gage's skull and brain, studies on how his personality changed after this accident revealed much about brain function.

Advances in anesthesia were made by surgeons on both sides of the Atlantic. On October 16, at the Massachusetts General Hospital in Boston, American surgeon **William Morton** (1819–68) anesthetized his patient, Gilbert Young, **sending him to sleep with fumes from ether** while Morton cut out a tumor from his neck. Half an hour later Young woke up, unaware the operation had been done. Others had used anesthetics before, such as American surgeon **Crawford Long** (see 1842–43) and American dentist **Horace Wells**, but it was Morton's demonstration that made an impact. Two months later, Scottish surgeon **Robert Liston** (1794–1847) performed a leg amputation under anesthetic in London.

Another important discovery was the production of **kerosene**, or paraffin, **from coal or oil**, by Canadian geologist **Abraham Pineo Gesnar** (1797–1864). In 1846, Gesnar began experimenting with methods for distilling coal and oil. By 1853, he had perfected a process to produce a new fuel he named kerosene that was used in lamps. Up until this time most lamps were fuelled by whale oil, but kerosene was much cheaper, so people could afford to burn lights brighter and longer.

IN 1847 Hungarian physician **Ignaz Semmelweis**. made an important discovery for medicine. He realized that if **doctors washed their hands** this could **reduce infection** by puerperal fever: a disease responsible for the deaths of many woman during childbirth. His procedures were not adapted until many years later.

Scottish surgeon **James Simpson** realized that neither ether nor laughing gas could keep a patient unconscious long enough for a long operation, and introduced **chloroform** as an anesthetic.

German physicist **Hermann von Helmholtz** outlined the **law of the conservation of energy**, which was first put forward by **Julius von Mayer** (see 1841). This stated that energy cannot be created or destroyed (now known as the first law of thermodynamics).

The following year, Scottish physicist **William Thomson** (Lord Kelvin) formulated the third law of thermodynamics with his **idea of absolute zero**. He realized there must be a temperature at which all molecular movement ceases and calculated it to be –459.67°F (–273.15°C). Thomson used this as the starting point for a new temperature scale—the **Kelvin scale** (see 1740–42).

By 1848, increasingly powerful telescopes were revealing more about the solar system. Astronomers discovered **planets** often had **more than one moon**. **William Lassell** and American astronomer **William Bond** (1789–1859) discovered Saturn's eighth moon, Hyperion.

English inventors **John Stringfellow** (1799–1883) and **William Henson** (1812–88) flew a model of their steam-powered aircraft, the Aerial Steam Carriage, for 33 ft (10 m). It was the **first ever powered flight**, but attempts to fly a larger model were unsuccessful.

Vermont railroad worker Phineas Gage survived a 3.3 ft (1 m) iron rod being driven through his head, resulting in intellectual and personality changes. This was the first record of how damage to the frontal lobe of the brain affected function.

outlet pipe to patient's face mask

chloroform holder

Chloroform inhaler
The chloroform inhaler was developed in 1848. It enabled surgeons to quickly anesthetize patients with fumes of vaporized chloroform.

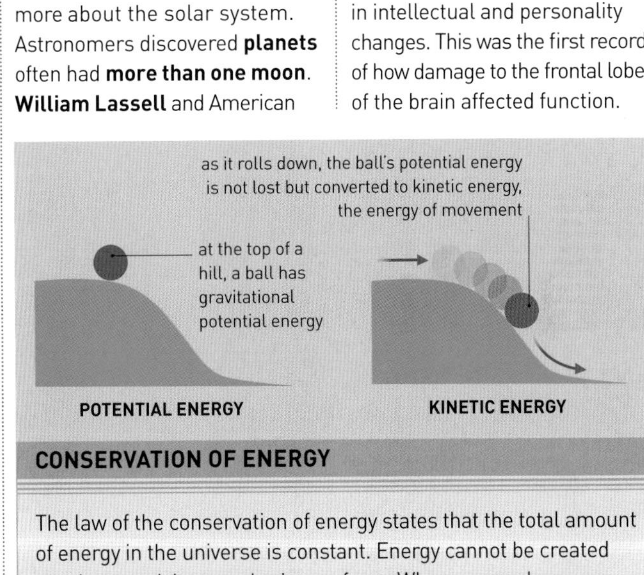
as it rolls down, the ball's potential energy is not lost but converted to kinetic energy, the energy of movement
at the top of a hill, a ball has gravitational potential energy
POTENTIAL ENERGY
KINETIC ENERGY

CONSERVATION OF ENERGY

The law of the conservation of energy states that the total amount of energy in the universe is constant. Energy cannot be created nor destroyed, it can only change form. When any work occurs, energy is converted from one form to another or transferred from one object to another. For example, potential energy (in a static object) can be converted to kinetic energy (motion energy).

October 16 American surgeon William Morton performs an operation under **general anesthesia**

December 21 Scottish surgeon Robert Liston performs a leg amputation under **general anesthetic**

1847 German physicist Hermann von Helmholtz formally states the **law of conservation of energy**

1847 Hungarian doctor Ignaz Semmelweis recommends **washing hands to reduce childbirth infections**

October 1, 1847 American astronomer Maria Mitchell discovers **comet 1847 VI**

November 4–8, 1847 Scottish surgeon James Simpson uses **chloroform** for an **operation**

1848 Scottish physicist William Thomson calculates the **absolute zero point of temperature**

June 1848 English inventors John Stringfellow and William Henson make the **first powered flight**

September 13, 1848 American railroad worker **Phineas Gage** survives an iron rod penetrating his head

September 20, 1848 The American **Association for the Advancement of Science** is established

In 1986, the Voyager 2 space probe took these close-ups of Uranus's five largest moons, including Ariel and Umbriel, discovered in 1851, and the smallest, called Miranda, which was discovered in 1948.

IN 1849, FRENCH ASTRONOMER EDOUARD ROCHE (1820–83) explained why Saturn has rings as well as moons. If planets and moons get too close to Saturn they are **ripped apart by tidal forces**—the changing pull of gravity created as they all rotate. There is a limit to how close planets and moons can get without being pulled apart; this became known as the **Roche limit**. If the planets and moons have identical densities, the Roche limit is 2.446 times the radius of the planet. If our Moon ever strayed closer to Earth than about 11,476 miles (18,470 km), it too would be shredded into rings.

Two other French scientists, physicists **Hippolyte Fizeau** (1819–96) and **Jean Foucault** (1819–68), **measured the speed of light** in 1849 by bouncing a beam of light off a mirror, through slots on a rapidly rotating wheel, and onto another mirror approximately 22 miles (35 km) away. The beam was reflected through the wheel and onto the first mirror. By the time the beam reached the rotating wheel again, the wheel had moved on to another slot. By measuring how fast the wheel was rotating, the spacing between the slots, and the distance from the more distant mirror to the wheel, the two scientists could calculate the speed of light.

It was difficult to get an accurate figure for the speed of light using this method, so in 1850 Foucault replaced the spinning wheel with a rotating mirror. As the mirror swiveled, it reflected the returning beam to a slightly different position. This difference revealed the speed clearly. Foucault got a very **close calculation for the speed of light** at 185,168 miles per second (298,000 km per second).

Further advances in science include the publication in Germany of physicist **Rudolf Clausius's** (1822–88) 1850 paper on the movement of heat. He laid **foundations for the science of thermodynamics** with two basic laws. The first is conservation of energy—energy is never lost, but simply redistributed. The second is that heat can never move from a cold place to a hot one, only the reverse.

In 1850, British chemist **James Young** (1811–83) patented a method of **distilling paraffin** from coal, which gradually replaced whale oil in domestic oil lamps. Scotch-born American inventor **John Gorrie** (1803–55) also **introduced refrigeration**, when he created a machine for making ice using circulating liquid to draw out heat (see 1872–73).

In 1851, British astronomer **William Lassell** (1799–1880) **discovered two more moons of Uranus**, Ariel and Umbriel, and British sculptor **Frederick Scott Archer** (1813–57) introduced the **wet plate photographic process**. This involved coating a photographic plate with a sticky liquid called collodion in darkness before each photo was taken. It had the fine detail of daguerreotypes (see 1837) and the repeatability of Fox Talbot's calotype (see 1835).

mirror
light source
outward beam
returning beam
toothed wheel
wheel rotates as beam travels out and returns
half-silver beam
to observer

Fizeau's apparatus
Fizeau measured the speed of light by reflecting a beam of light through the slots in a toothed wheel, onto a mirror 22 miles (35 km) away, and back.

lens
bellows for adjusting focus
camera front

Wet plate camera
Working in a portable darkroom, the photographer had under 10 minutes to take a picture and process it before the plate dried.

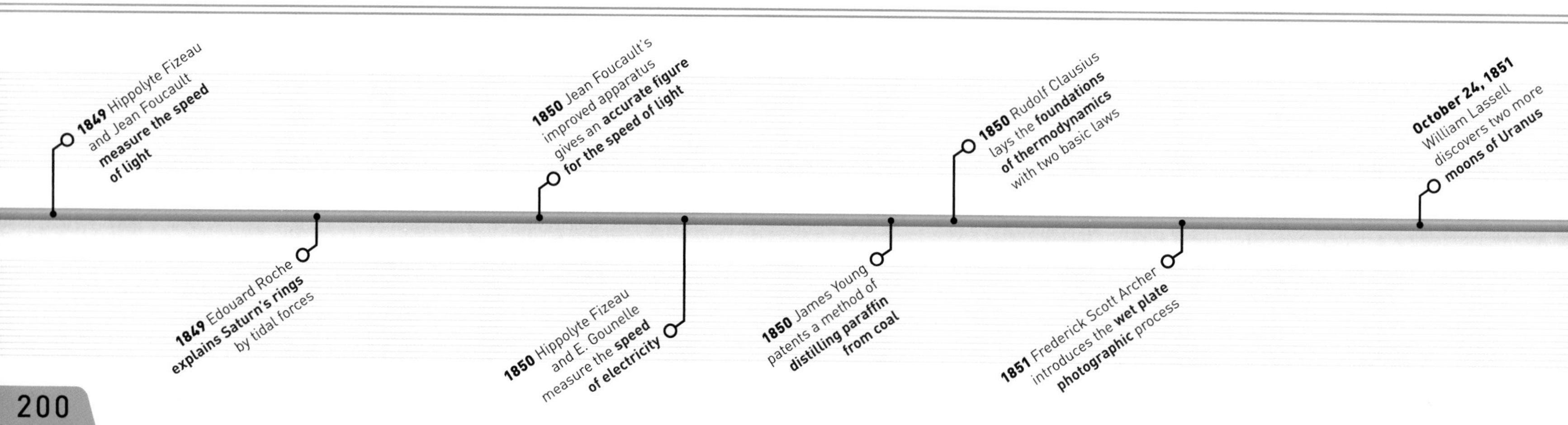
1849 Hippolyte Fizeau and Jean Foucault **measure the speed of light**
1849 Edouard Roche **explains Saturn's rings** by tidal forces
1850 Jean Foucault's improved apparatus gives an **accurate figure for the speed of light**
1850 Hippolyte Fizeau and E. Gounelle **measure the speed of electricity**
1850 James Young patents a method of **distilling paraffin from coal**
1850 Rudolf Clausius lays the **foundations of thermodynamics** with two basic laws
1851 Frederick Scott Archer introduces the **wet plate photographic** process
October 24, 1851 William Lassell discovers two more **moons of Uranus**

1852–53

A replica of George Cayley's glider, which made the world's first fixed-wing flight in 1853 in Yorkshire, England.

> **"** I AM WELL CONVINCED THAT **AERIAL NAVIGATION** WILL FORM A MOST **PROMINENT FEATURE** IN THE **PROGRESS OF CIVILIZATION. "**
>
> **George Cayley, British aeronautical engineer**, 1804

GREAT AUKS HAD BEEN EASY TARGETS FOR HUNTERS in areas around the North Atlantic for many years, and by the 1840s they were practically extinct. **The last bird was spotted** off the coast of Newfoundland in Canada in 1852.

British physicists **James Joule** (1818–89) and **William Thomson** (1824–1907) discovered the Joule–Thomson effect in 1852. This explains **the way that gases and liquids cool and expand** after flowing through a restriction or throttle. The Joule–Thompson effect is central to the way refrigerators and air conditioning systems work.

The age of aviation began on September 24, 1852 when French engineer **Henri Giffard** (1825–82) made the **first powered and controlled fight**, flying 17 miles (27 km) from Paris to Trappes in France. The flight was made in a powered airship, which consisted of a cigar-shaped, hydrogen-filled balloon that provided the lift, and a steam-driven propeller that moved it through the air.

In 1853 there was another aviation first, the **first flight in a full-sized aircraft**—a glider built by the British engineer **George Cayley** (1773–1857), who had pioneered understanding of the theory of flight. Details of the flight, across Brompton Dale in Yorkshire, England, are not clear; some reports say Cayley's butler was the pilot, while others say it was his footman. Nonetheless, it was a historic achievement.

Great Auk
The last Great Auk, one of the largest birds of the North Atlantic at 31.5 in (80 cm) tall, was last seen in 1853.

1853 was also an important year for medicine and physiology. French physiologist **Claude Bernard** (1813–78) discovered that **glucose sugar, the body's energy food, is stored temporarily in the liver** in the form of a starchlike substance called glycogen, ready to be released into the blood as glucose when energy is needed.

French physician **Antoine Desormeaux** (1815–82) **developed the endoscope** for surgical operations. This was a long metal tube that could be inserted into the body to make examinations, using light from a paraffin-fueled lamp reflected in a mirror.

Another Frenchman, surgeon **Charles Pravaz** (1791–1853), and British physician **Alexander Wood** (1817–84) independently **invented a practical hypodermic syringe** with a hollow metal needle that could be inserted into the body to deliver drugs directly into veins for much faster effect than taking the drugs by mouth.

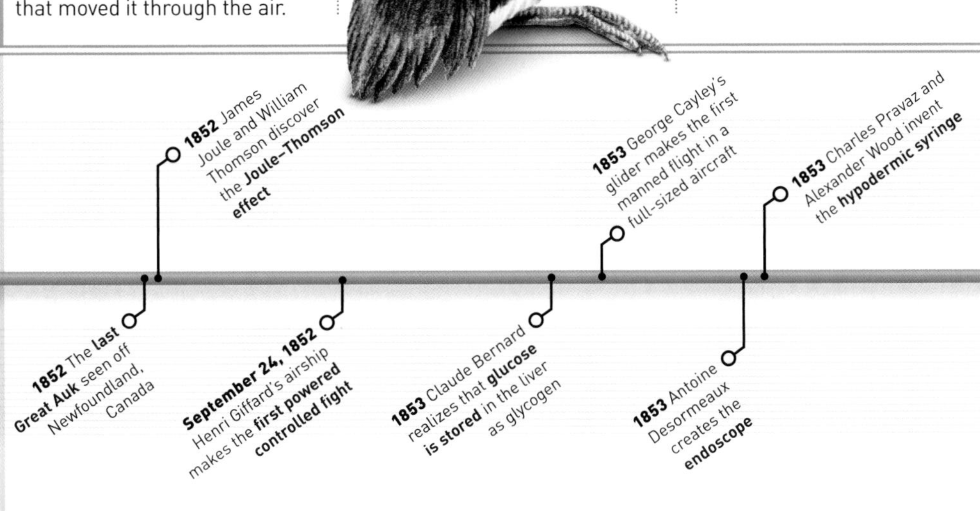

1852 James Joule and William Thomson discover the Joule–Thomson effect
1853 George Cayley's glider makes the first manned flight in a full-sized aircraft
1853 Charles Pravaz and Alexander Wood invent the hypodermic syringe
1852 The last Great Auk seen off Newfoundland, Canada
September 24, 1852 Henri Giffard's airship makes the first powered controlled fight
1853 Claude Bernard realizes that glucose is stored in the liver as glycogen
1853 Antoine Desormeaux creates the endoscope

1854

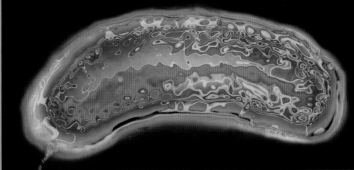

This color-enhanced micrograph shows the cholera bacterium, now known to be responsible for cholera.

FOLLOWING A SUGGESTION BY BRITISH PHYSICIST William Thomson, German physicist **Hermann von Helmholtz** (1821–94) and British engineer **William Rankine** (1820–72) developed Rudolf Clausius's theory that heat never flows from a cold to a hot place. This implies that heat will ultimately spread evenly through the Universe and once this happens energy will not be able to move and the Universe will come to a stop in what is called **"heat death of the Universe."**

British astronomer **George Airy** (1801–92) calculated the **density of Earth** by measuring the swing of a pendulum on Earth's surface and 1,257 ft (383 m) down a coal mine. Different measurements revealed slight variations in the effects of gravity, from which he obtained the figure 3.795 oz/in^3 (6.566 g/cm^3). Today's accepted figure is 3.19 oz/in^3 (5.52 g/cm^3).

Two new types of mathematics were introduced in 1854. One was the non-Euclidean geometry of German mathematician **Bernhard Riemann** (1826–66). Euclidean geometry applies only to flat surfaces; **Riemannian geometry is the geometry of curved surfaces**, important because the surface of Earth is curved. In Euclidean geometry, the angles in a triangle add up to two right angles and the shortest distance between two points is a straight line, in Riemannian geometry angles in a triangle add up to more than two right angles and there is no such thing as a straight line along a surface.

The new algebra of British mathematician **George Boole** (1815–64) was intended to make logic mathematical, not philosophical. Boole argued that any proposition could be reduced to just "and," "or," and "not," and worked through to a conclusion. Today, **Boolean logic** combines with the binary system of numbers to shape all computer programs.

> **3.19 OZ/IN^3**
> THE AVERAGE **DENSITY** OF **EARTH**

In August, there was an outbreak of cholera in London's Soho. British physician **John Snow** (1813–58) traced the source to a single water pump, thus validating his theory that **cholera is water-borne**.

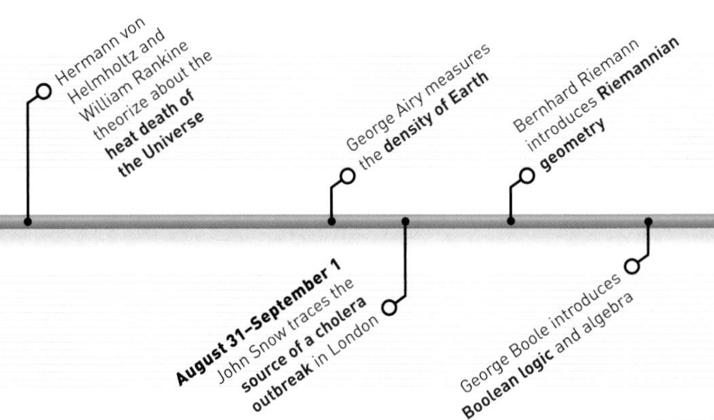

Hermann von Helmholtz and William Rankine theorize about the heat death of the Universe
George Airy measures the density of Earth
Bernhard Riemann introduces Riemannian geometry
August 31–September 1 John Snow traces the source of a cholera outbreak in London
George Boole introduces Boolean logic and algebra

The Bessemer process revolutionized engineering by making steel production cheaper and more efficient.

Montgomerie surveyed the Karakoram range as part of the Great Trigonometric Survey of India. It includes K2, the second-highest mountain in the world.

AT THE FOREFRONT OF SCIENCE in 1855 was the search for the relationship between **atoms, light, and electromagnetism**. British mathematician **James Clerk Maxwell** (1831–79) began working on a theory to unify electricity, light, and magnetism, while other scientists conducted practical experiments to find out how **atoms emitted light**. They had already worked out that each kind of atom emits and absorbs a particular range of colors, or spectrum, with dark lines (gaps) at some wavelengths and bright lines (peaks) at others. Swedish physicist **Anders Ångström** (1814–74) and American scientist David Alter (1807–81) independently described the **spectrum of hydrogen gas**, which would prove crucial in the understanding of the link between **light and atoms**.

Geissler tube
These gas-filled tubes came in an assortment of elaborate shapes and glowed in a variety of colors.

spiral electric discharge tube

German physicist **Julius Plücker** (1801–79) investigated spectra by studying the glow (undistorted by air) from **electric sparks**. To do this he commissioned instrument maker **Heinrich Geissler** (1814–79) to create a sealed glass tube, with a near-perfect vacuum and electric terminals at either end. When switched on, the **electric charge traveled** through the tube, between the terminals, creating a bright glow.

French chemist Charles Gerhardt (1816–56) had suggested in 1853 that four basic types of organic chemicals are created by carbon linking with hydrogen, hydrogen chloride, water, or ammonia molecules. In 1855, English chemist **William Odling** (1829–1921) added a fifth type, based on methane. This led German chemist Friedrich Kekulé (1829–96) and Scottish chemist Archibald Couper (1831–1892) to begin developing a **structural theory of molecules**.

> **"ATOMS** WERE GAMBOLLING **BEFORE MY EYES..."**
>
> **Frederich Kekulé, German chemist,** describing a daydream that led to his structural theory, 1855

The technological breakthrough of the year was the development of a special furnace, developed by the British engineer Henry Bessemer (1813–98), which allowed **steel to be made cheaply** and in quantity from pig iron (a crude form of iron).

In Germany, an unusual fossil was found at Riedenburg. It was thought to be a flying reptile until 1970 when it was finally shown to have feathers, identifying it as the **first *Archeopteryx*** ever discovered. It is evidence that birds evolved from the dinosaurs.

THE SCIENCE OF GENETICS BEGAN WITH THE WORK of the Austrian monk **Gregor Mendel** (1822–84), but his work was not fully appreciated until years later. Mendel began to experiment with pea plants in his monastery garden in 1856. He laid the foundations of genetics as he showed how **characteristics are passed on** from generation to generation through what he called "factors," which later became known as genes, that are inherited from both parents.

Another discovery that was not fully appreciated until much later, was the discovery of the first recognized **fossil of a human ancestor**. In August

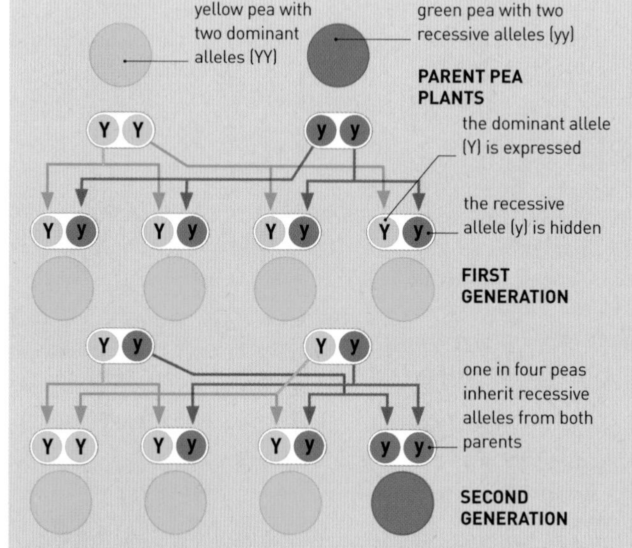
yellow pea with two dominant alleles (YY)
green pea with two recessive alleles (yy)
PARENT PEA PLANTS
the dominant allele (Y) is expressed
the recessive allele (y) is hidden
FIRST GENERATION
one in four peas inherit recessive alleles from both parents
SECOND GENERATION

DOMINANT AND RECESSIVE GENES

Mendel's work on peas showed that inherited characteristics, such as color, are determined by particles, later called genes. Genes come in different forms, called alleles, that combine in different ways in offspring. A dominant Y allele determines yellow pea color and a single one of these being present is sufficient to make a yellow pea. A recessive y allele determines green pea color and two of these must come together to make a green pea.

- Swedish physicist Anders Ångström publishes **spectrum of hydrogen**
- **March** British chemist William Perkin discovers **aniline dye** mauveine
- American scientist **David Alter** describes the spectra of hydrogen and other gases
- **May 10** German chemist Robert Bunsen invents the **Bunsen burner**
- German chemist Friedrich Kekulé starts to develop the **structural theory** of organic molecules
- British mathematician **James Clerk Maxwell** begins work on his theory to unite light, electricity, and magnetism
- **October 17** Henry Bessemer patents the Bessemer process of **steelmaking**
- The first ***Archeopteryx*** fossil was discovered at Riedenburg in Germany
- Austrian monk **Gregor Mendel** starts his research on plant inheritance
- American **meteorologist** William Ferrel explains wind circulation
- **August** German teacher John Fuhlrott identifies fossil bones of **Neanderthal man**
- British surveyor Thomas Montgomerie names the **Himalayan mountain peaks** K1 and K2

17 THE NUMBER OF **HOURS** TAKEN TO **TRANSMIT** THE **FIRST MESSAGE** ON THE 1858 CABLE

Workers inpecting the transatlantic telegraph cable. The first message sent was "Glory to God in the highest; on earth, peace and good will toward men."

1856, quarry workers found some bones in a cave and local teacher **Johan Fuhlrott** (1803–77) identified them as humanlike. Named after the Neander Valley in Germany where the bones were found, this human ancestor is now known as **Neanderthal man** (*Homo neandethalis*). Neanderthal man is thought to have lived in Europe between 300,000 and 30,000 years ago, but it was some time before many would accept that any other humanlike creatures had ever lived.

In America, meteorologist **William Ferrel** (1817–91) explained how **rising warm air** and Earth's rotation creates spiralling circulations of air in the mid-latitudes, called **Ferrel cells**. These cells drive the stormy, westerly winds that are characteristic of these latitudes.

In India, British surveyor **Thomas Montgomerie** (1830–78) began his **survey of the Karakoram Range** as part of the Great Trigonometric Survey of India that had started in 1802.

British inventor Alexander Parkes (1830–90) patented the **first plastic**, Parkesine, which was made from cellulose treated with acid and a solvent.

IN FRANCE, IN 1857, CHEMIST AND MICROBIOLOGIST LOUIS PASTEUR (1822–95) published his seminal results on **fermentation** and yeast multiplication in 1857. He found that when beer and wine ferment it is not chemicals that are responsible, but **tiny microbes** known as yeast. Pasteur later developed a process known as pasteurization that involved killing the yeast microbes with heat in order to prolong the life of certain foods.

On February 13, 1858, British explorers **Richard Francis Burton** (1821–90) and **John Hanning Speke** (1827–64) became the first Europeans to see **Lake Tanganyika** in Africa, the second largest freshwater lake in the world. Speke continued alone to discover Lake Victoria.

In the cities, **fast-rising populations** were putting new demands on engineers and builders. In New York, **Elisha Otis** (1811–61), inventor of a special safety device that prevented lifts from falling if the cables failed, installed his **first elevator** at 488 Broadway on March 23, 1857. In Germany, Friedrich Hoffmann played his part in speeding up urbanization by patenting the Hoffmann kiln in 1858, which was capable of **firing bricks non-stop**. Cities also began to communicate across oceans. The first undersea **transatlantic telegraph cable**, laid between western Ireland and Newfoundland in Canada, went into service in August 1858.

On July 1, 1858, a scientific paper was delivered to the Linnean Society in London. It combined the ideas of the British naturalists Charles Darwin and Alfred Russel Wallace in a new theory—the theory of **evolution by natural selection**. The idea that species evolved through time was not new, but Darwin and Wallace argued that all species on Earth had evolved gradually due to a process of change known as natural selection—members of the species less well suited to the environment either fail to reproduce or die early, thus failing to pass on their inferior traits. Wallace had written to Darwin from Indonesia in June 1858 with an outline of his idea, but, unbeknown to him, Darwin had already spent two decades developing the theory.

The most widely used book in medical history was published in 1858. British surgeon Henry Gray's (1827–61) *Anatomy: Descriptive and Surgical* has been in publication ever since and is now known simply as ***Gray's Anatomy***.

Gray's Anatomy
The anatomy book now known as Gray's Anatomy *was first published by British surgeon Henry Gray in 1858. There have been over 40 editions.*

ALFRED WALLACE
(1823–1913)

Alfred Wallace is best known for independently conceiving a theory of evolution by natural selection. He also pioneered biogeography, the study of species in a geographical area. His work in Indonesia between 1854 and 1862 led to the Wallace Line, which divides Asian and Australian species (see pp.204–205).

420 MILES THE **LENGTH** OF **LAKE TANGANYIKA** IN AFRICA

British inventor Alexander Parkes patents the **first plastic**, Parkesine

1857 French chemist Louis Pasteur shows that **fermentation** is caused by living microbes

March 23, 1857 Elisha Otis's **first elevator** is installed in New York

February 13, 1858 British explorers Richard Burton and John Speke reach **Lake Tanganyika** in Africa

May 14, 1858 British missionary David Livingstone's African **Zambesi expedition** begins

June 1858 Wallace writes to Darwin from Indonesia

July 1, 1858 British naturalists Darwin and Wallace's **theory of evolution** by natural selection is presented

August 3, 1858 John Speke reaches **Lake Victoria** in Africa

August 16, 1858 First transatlantic **telegraph cable** message is sent

1858 First publication of *Gray's Anatomy*

1858 First Hoffmann **brick kiln**, capable of firing bricks non-stop, patented in Germany

UNDERSTANDING EVOLUTION

THE SOURCE OF BIOLOGICAL DIVERSITY, EVOLUTION ALSO PROVIDES A LINK TO OUR PREHISTORIC ANCESTORS

Fossils reveal how prehistoric life forms were different from those of today, and studies of different species suggest that all life originated from a single, simple ancestor that lived billions of years ago. Today, scientists understand the biological and genetic processes behind the evolutionary changes that gave rise to the diversity of life.

In the early 1800s, the French naturalist Jean-Baptiste Lamarck suggested, erroneously, that features acquired during an organism's lifetime could be passed on to its offspring. Later, Charles Darwin (see 1859) proposed that individuals are born with variations that make some of them "fitter"—more likely to survive and pass on their characteristics. This is called evolution by natural selection. It is now known that characteristics are determined by genes and that random gene mutations cause variation (see pp.284–85). But only natural selection can explain how some variations are better adaptations to the environment and come to predominate in organisms.

CONVERGENT EVOLUTION

Physical resemblance can indicate common ancestry but sometimes completely unrelated species evolve independently to become alike. Known as convergent evolution, this often happens when species have comparable roles in the same environment, so natural selection works on them in similar ways.

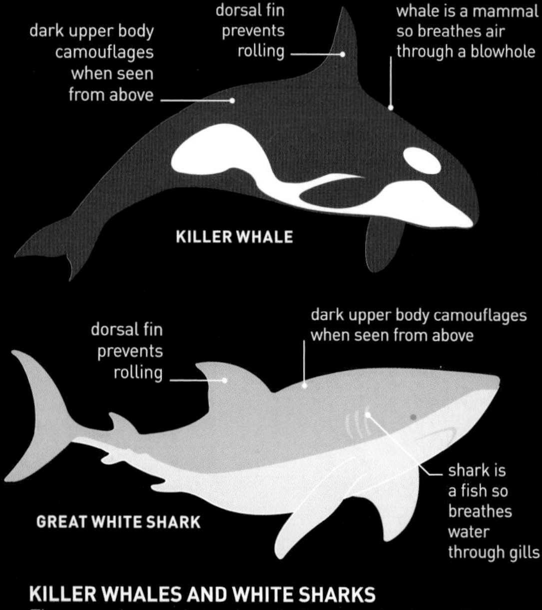
dark upper body camouflages when seen from above
dorsal fin prevents rolling
whale is a mammal so breathes air through a blowhole
KILLER WHALE
dorsal fin prevents rolling
dark upper body camouflages when seen from above
shark is a fish so breathes water through gills
GREAT WHITE SHARK

KILLER WHALES AND WHITE SHARKS
These marine predators have both evolved a streamlined shape for greater speed when chasing prey and have a darker upper body and lighter lower body to provide camouflage.

ADAPTIVE RADIATION

When descendents of a common ancestor adapt to different circumstances and diversify, it is known as adaptive radiation. It tends to happen most rapidly in habitats where there are many opportunities for exploiting new ways of life. For example, on newly formed islands there may be few competitors and new food sources, so over many generations a pioneering population of one species can diverge to produce many new species, each adapted to a slightly different role.

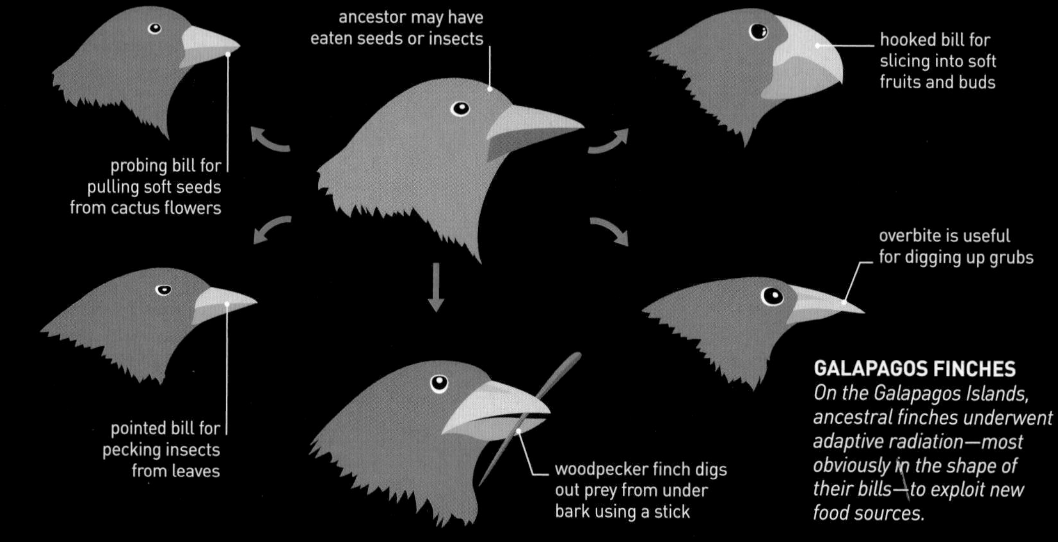
ancestor may have eaten seeds or insects
probing bill for pulling soft seeds from cactus flowers
hooked bill for slicing into soft fruits and buds
overbite is useful for digging up grubs
pointed bill for pecking insects from leaves
woodpecker finch digs out prey from under bark using a stick

GALAPAGOS FINCHES
On the Galapagos Islands, ancestral finches underwent adaptive radiation—most obviously in the shape of their bills—to exploit new food sources.

SEXUAL SELECTION

Not all adaptations increase an individual's ability to survive. Sometimes, the selective advantage that drives evolution comes from being better able to attract a mate. For instance, showy plumage can make male birds more vulnerable to predators but it also makes them more successful at courting. As a result, they father more offspring and the genes for showy plumage are passed on.

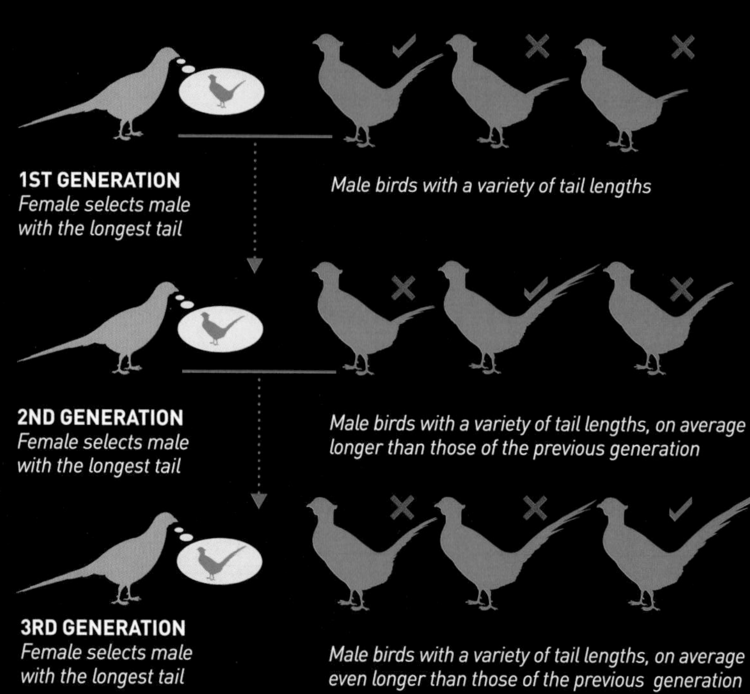
1ST GENERATION
Female selects male with the longest tail
Male birds with a variety of tail lengths
2ND GENERATION
Female selects male with the longest tail
Male birds with a variety of tail lengths, on average longer than those of the previous generation
3RD GENERATION
Female selects male with the longest tail
Male birds with a variety of tail lengths, on average even longer than those of the previous generation

PHEASANT'S TAIL
Male pheasants with longer tails are more attractive to females, so the genes for a long tail get passed on and male tail length increases with successive generations.

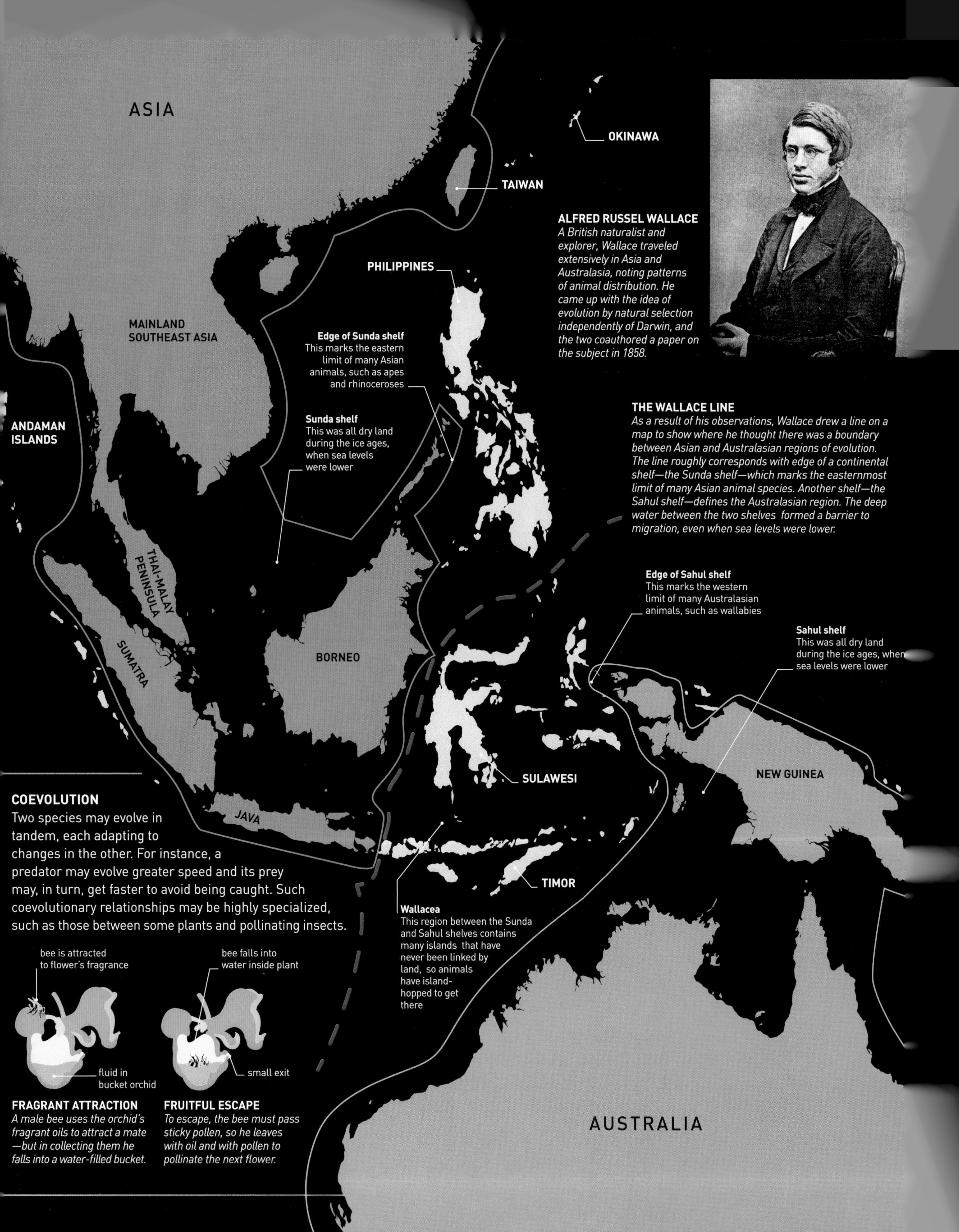
ASIA
OKINAWA
TAIWAN
ALFRED RUSSEL WALLACE
A British naturalist and explorer, Wallace traveled extensively in Asia and Australasia, noting patterns of animal distribution. He came up with the idea of evolution by natural selection independently of Darwin, and the two coauthored a paper on the subject in 1858.
PHILIPPINES
MAINLAND SOUTHEAST ASIA
Edge of Sunda shelf
This marks the eastern limit of many Asian animals, such as apes and rhinoceroses
Sunda shelf
This was all dry land during the ice ages, when sea levels were lower
ANDAMAN ISLANDS
THE WALLACE LINE
As a result of his observations, Wallace drew a line on a map to show where he thought there was a boundary between Asian and Australasian regions of evolution. The line roughly corresponds with edge of a continental shelf—the Sunda shelf—which marks the easternmost limit of many Asian animal species. Another shelf—the Sahul shelf—defines the Australasian region. The deep water between the two shelves formed a barrier to migration, even when sea levels were lower.
THAI-MALAY PENINSULA
Edge of Sahul shelf
This marks the western limit of many Australasian animals, such as wallabies
Sahul shelf
This was all dry land during the ice ages, when sea levels were lower
BORNEO
SUMATRA
SULAWESI
NEW GUINEA
COEVOLUTION
Two species may evolve in tandem, each adapting to changes in the other. For instance, a predator may evolve greater speed and its prey may, in turn, get faster to avoid being caught. Such coevolutionary relationships may be highly specialized, such as those between some plants and pollinating insects.
JAVA
TIMOR
Wallacea
This region between the Sunda and Sahul shelves contains many islands that have never been linked by land, so animals have island-hopped to get there
bee is attracted to flower's fragrance
bee falls into water inside plant
fluid in bucket orchid
small exit
FRAGRANT ATTRACTION
A male bee uses the orchid's fragrant oils to attract a mate—but in collecting them he falls into a water-filled bucket.
FRUITFUL ESCAPE
To escape, the bee must pass sticky pollen, so he leaves with oil and with pollen to pollinate the next flower.
AUSTRALIA

Darwin visited the Galapagos Islands in 1835; the animals and plants he observed there are said to have contributed to his becoming an evolutionist, but it wasn't until 1859 that Darwin consolidated his observations into a book—*The Origin of Species*.

ON NOVEMBER 24, 1859, CHARLES DARWIN PUBLISHED his epoch-making book *On the Origin of Species*. In it he explained in detail his theory of evolution, first introduced in 1858. Darwin's theory was that species change and develop automatically through a **process of natural selection**; this idea was neatly summed up by the philosopher Herbert Spencer (1820–1903) as **"survival of the fittest."** In his book, Darwin explained how, occasionally, a chance mutation at birth may equip some organisms with a trait that gives them a better chance of survival, which means they are more likely to pass the trait onto their offspring. Different mutations will suit (or not suit) particular conditions, so species gradually diversify and become **adapted to suit specific habitats.** If habitats change, however, these special advantages may become weaknesses and as a result the species could die out.

What made Darwin's theory such a turning point was that the mechanism he proposed worked for all of life, and asserted that every organism is **descended from a common ancestor**. Although many people accepted the force of Darwin's arguments, some were bitterly critical and debate flared up.

In April, two British archaeologists, **John Evans** (1823–1908) and **Joseph Prestwich** (1812–96), made a startling discovery that **pushed human origins further back into prehistory**. In St. Acheul in northern France, they found a stone ax in layers that also contained fossils of extinct creatures, including mammoths. If mammoths and humans had existed together, then human life must date back tens of thousands of years.

CHARLES DARWIN (1809–82)

Darwin trained as a doctor before embarking, as a naturalist, on a round-the-world voyage in 1831–36 aboard HMS *Beagle*. The voyage sowed the seeds for his theory of evolution by natural selection, which he finally revealed in 1858. He also applied his theory to humans in *The Descent of Man*, published in 1871.

Number of species on Earth
Naturalists have identified 1.25 million living species on Earth today, and estimate there may be over 8.7 million in total.

MILLIONS
0 1 2 3 4 5 6 7 8 9
Estimated species
Identified species

eyepiece for viewing spectrum
telescope
telescope with eyepiece replaced by collimator
diffraction grating to split light into spectrum
spectroscope mount

Spectroscope
The first instrument for analyzing spectra was built from old telescopes. This specially built spectroscope dates from slightly later.

Earlier chemists had already realized that the range of colors (spectrum) in the glow of light from heated chemicals could help identify them. In autumn 1859, German scientists **Robert Bunsen** (1811–99) and **Gustav Robert Kirchhoff** (1824–87) began to study spectra systematically, heating chemicals in the special gas burner devised by Bunsen in 1855. They found that **every element has its own unique spectrum**, and realized that spectra can be used to show the presence of even tiny traces of chemicals. By passing sunlight through a sodium flame, Kirchhoff also found that the flame absorbed spectral lines in sunlight in a mirror image to those emitted by sodium—showing that the Sun contains sodium.

German scientists Robert Bunsen and Gustav Kirchhoff study systematically the **glows of elements**

Using a spectroscope, Gustav Kirchhoff discovers that the **Sun contains sodium**

French engineer Étienne Lenoir produces the first **two-stroke gasoline** engine

April British **archaeologists** John Evans and Joseph Prestwich discover ancient stone tools

Russian astronomer Marian Kowalski develops a way to work out the **rotation of the Milky Way**

August 28–September 2 The largest geomagnetic solar storm on record, known as the Carrington Event

French mathematician Urbain Le Verrier proposes the existence of an **undiscovered planet,** Vulcan, to explain Mercury's orbit

November 24 The first **iron-hulled** battleship, **the Gloire**, launched in France

November 24 Charles Darwin's ***On the Origin of Species*** is published

> **"** SEVERAL MINUTES… BEFORE **TOTALITY,** I… FOUND THAT THE **SUN'S IMAGE** MAY BE **VIEWED WITHOUT** THE SLIGHTEST **INCONVENIENCE. "**
>
> **Warren de la Rue, British Astronomer**

Warren de la Rue developed a special camera that allowed him to photograph a total eclipse of the sun as seen from Spain on July 18, 1860.

THE STIR CAUSED BY THE PUBLICATION OF DARWIN'S *Origin of Species* came to a head in fierce public debate in June at the Oxford House Museum in Oxford, UK. Opposing Darwin from a religious perspective was Bishop **Samuel Wilberforce** (1805–73); supporting Darwin and science was the British biologist **Thomas Huxley** (1825–95). The **debate focused on whether humans are descended from apes**, even though Darwin had not actually suggested they were. Huxley is generally considered to have won the debate.

Following their breakthrough in 1859, Bunsen and Kirchhoff made further progress in the field of spectroscopy. Bunsen **discovered two new elements from the spectra of light** absorbed by drops of mineral water. Each was named after the brightest colors in their spectra: cesium, from the Latin for "sky blue," and rubidium, meaning "dark red." Kirchhoff learned more about the sun's composition from the spectrum of sunlight, discovering more than 16 different elements in it.

Photographs taken of a solar eclipse by British astronomer **Warren de la Rue** (1815–89) proved that the flames, now known as prominences, that sometimes appear around the Moon during an eclipse come from the Sun's surface.

The **oldest known recording of a human voice** was made in April 1860 using a phonautograph made by French bookseller **Édouard-Léon Scott de Martinville** (1817–79) that used sound vibrations to draw on a carbon-coated cylinder. The phonautograph was not designed to play back sound, it was just used for turning sound into a graphic. The recording has only recently been converted to sound using computer technology.

British scientist **Joseph Wilson Swan** (1828–1914) demonstrated the **first working incandescent light bulb**. The light was created by passing an electric current through a thin carbon filament inside a vacuum-filled glass bulb, which heated the filament until it glowed. With only a partial vacuum in the bulb, however, the filament quickly burned out. He made improvements and patented it in 1878 (see 1878–79).

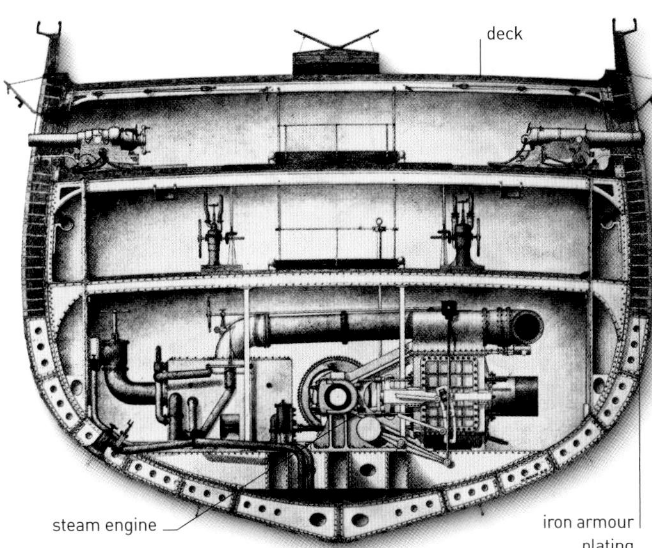

deck
steam engine
iron armour plating

HMS *Warrior*
With its armor-plated iron hull, HMS Warrior *was the first modern battleship. It was armed with 26 muzzle-loading cannons that were able to fire 68 lb (31 kg) shot over 8,858 ft (2,700 m).*

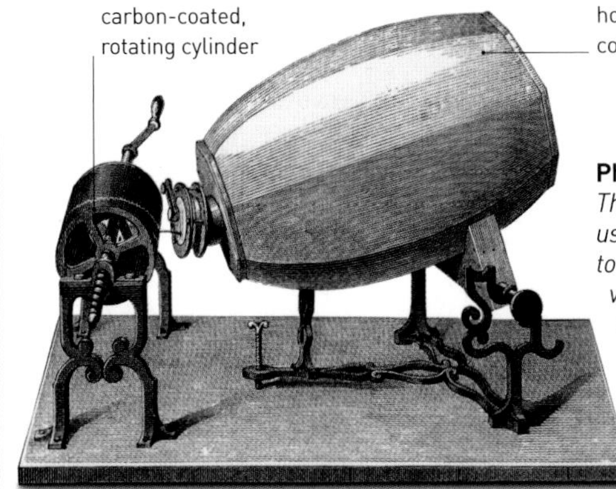

carbon-coated, rotating cylinder
horn for collecting sound

Phonautograph
The phonautograph used a horn attached to a diaphragm that vibrated a stiff bristle to inscribe an image on a lamp-black (carbon) coated, hand-cranked cylinder.

400,000
THE APPROXIMATE NUMBER OF KNOWN PLANT SPECIES IN THE WORLD

1,500
The number of plant species collected on the Antarctic voyages of Joseph Hooker aboard *Erebus* and *Terror*

4½
INCHES
THE **WIDTH** OF **HMS *WARRIOR'S*** BELT OF WROUGHT **IRON**

The launch of HMS *Warrior* on December 29 on the Thames in London was a landmark in naval technology. *Warrior* was only the second **armor-plated, iron-hulled warship** to be built, after the French ship *La Gloire* of 1859, and it was on a vastly different scale from anything that had gone before, at over 417 ft (127 m) long and weighing almost 10,000 tons.

More peacefully, British botanist **Joseph Hooker** (1817–1911) concluded his **account of the many previously unknown plants** he had discovered on his voyages to Antarctica, between 1839 and 1843 aboard the naval ships *Erebus* and *Terror*.

Number of plant species
New species of plants are being identified all the time; 400,000 are now known. Joseph Hooker added 1,500 on his Antarctic voyages.

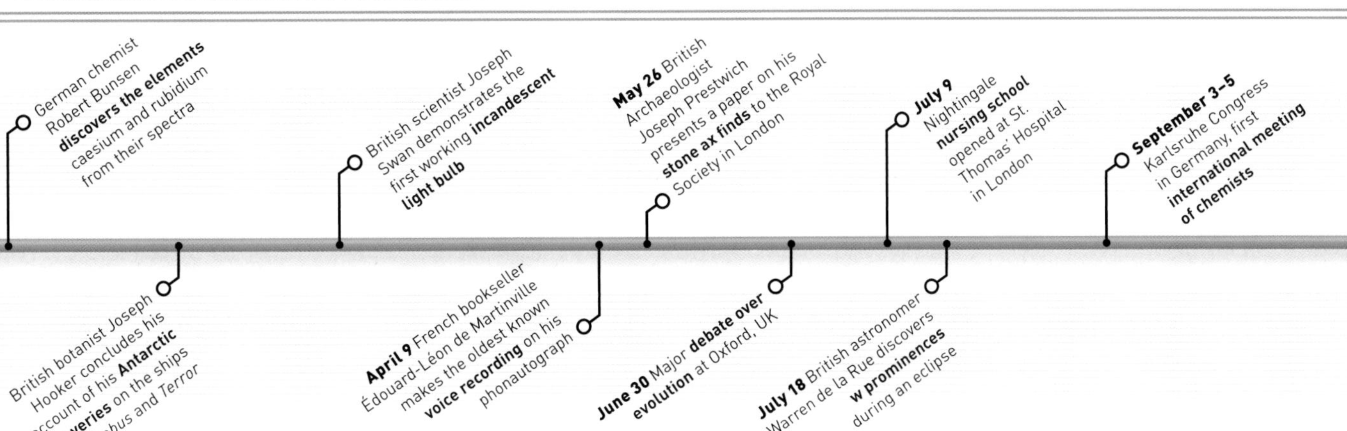

German chemist Robert Bunsen **discovers the elements** caesium and rubidium from their spectra
British botanist Joseph Hooker concludes his account of his **Antarctic discoveries** on the ships *Erebus* and *Terror*
British scientist Joseph Swan demonstrates the first working **incandescent light bulb**
April 9 French bookseller Édouard-Léon de Martinville makes the oldest known **voice recording** on his phonautograph
May 26 British Archaeologist Joseph Prestwich presents a paper on his **stone ax finds** to the Royal Society in London
June 30 Major **debate over evolution** at Oxford, UK
July 9 Nightingale **nursing school** opened at St. Thomas' Hospital in London
July 18 British astronomer Warren de la Rue discovers **w prominences** during an eclipse
September 3–5 Karlsruhe Congress in Germany, first **international meeting of chemists**
December 29 The **iron battleship**, HMS *Warrior*, launched

The Yosemite Grant, signed by President Abraham Lincoln on June 30, 1864, preserved the natural grandeur of the Yosemite region of California for the public.

Red blood cells get their red color from the protein hemoglobin. Felix Hoppe-Seyler discovered its vital role in oxygen transport in 1864.

BRITISH PHYSICIST JAMES CLERK MAXWELL REVEALED the pivotal advances he had made in the science of electricity and magnetism in two books: *On the Lines of Physical Force* (1861) and *A Dynamical Theory of the Electromagnetic Field* (1864). He began by explaining how an **electromagnetic field is created by waves** that radiate outward. He proved that these waves radiate at exactly the same speed as light, which showed that light is also an electromagnetic wave. Finally, he summed up his findings in four equations, now known as **Maxwell's equations**, which underpin all calculations relating to electricity and magnetism, in the same way that Isaac Newton's equations underpin all studies of motion.

In 1861, Maxwell also took the **first color photograph**. He had already proved that we see colors as varying intensities of three colors (see panel, right), but to demonstrate this he asked photographer Thomas Sutton (1819–75) to take three black and white photographs of a tartan ribbon, each one through a different color filter—red, green, and blue. He then projected all three images together, through filters of the same colors. The three color images mixed to recreate the picture in full color.

The colors found in the spectrum of sunlight by the Swedish physicist **Anders Ångström** (1814–74) in 1861, showed that the **Sun contains hydrogen gas**.

In the same year, French physician Paul Broca (1824–80) discovered a **key area of the brain for speech** in a man who had suffered a brain injury that rendered him unable to talk but able to understand. When the man died Broca performed an autopsy that revealed damage to a region of the brain now known as **Broca's area**.

After fossils of the *Glossopteris* fern were found in Africa, India, and South America, Austrian geologist **Edward Suess** (1831–1914) theorized in 1861 that these three continents were once joined by land bridges to create

The Berlin *Archaeopteryx*
The Berlin fossil of Archaeopteryx, *found in 1874, is the most complete. It plainly shows the toothed, dinosaur-like beak and feathered, birdlike wings.*

THREE-COLOR SYSTEM

Mixing three colors of light—red, green, and blue—in varying proportions makes every color in the rainbow. Images can be reproduced in full color by registering and then displaying how much light there is of each of these three colors in each part of the picture. Color was created this way in everything from the first color photograph to modern phone displays.

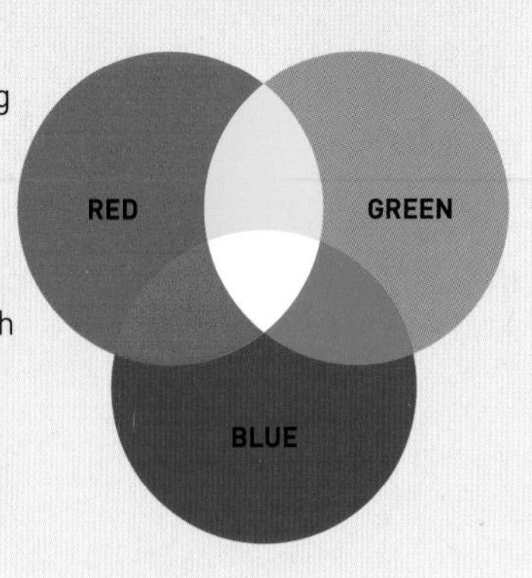
RED
GREEN
BLUE

> **"...LIGHT CONSISTS** IN THE TRANSVERSE UNDULATIONS OF THE **SAME MEDIUM** WHICH IS THE CAUSE OF **ELECTRIC AND MAGNETIC PHENOMENA."**
>
> **James Clerk Maxwell, British physicist,** January 1862

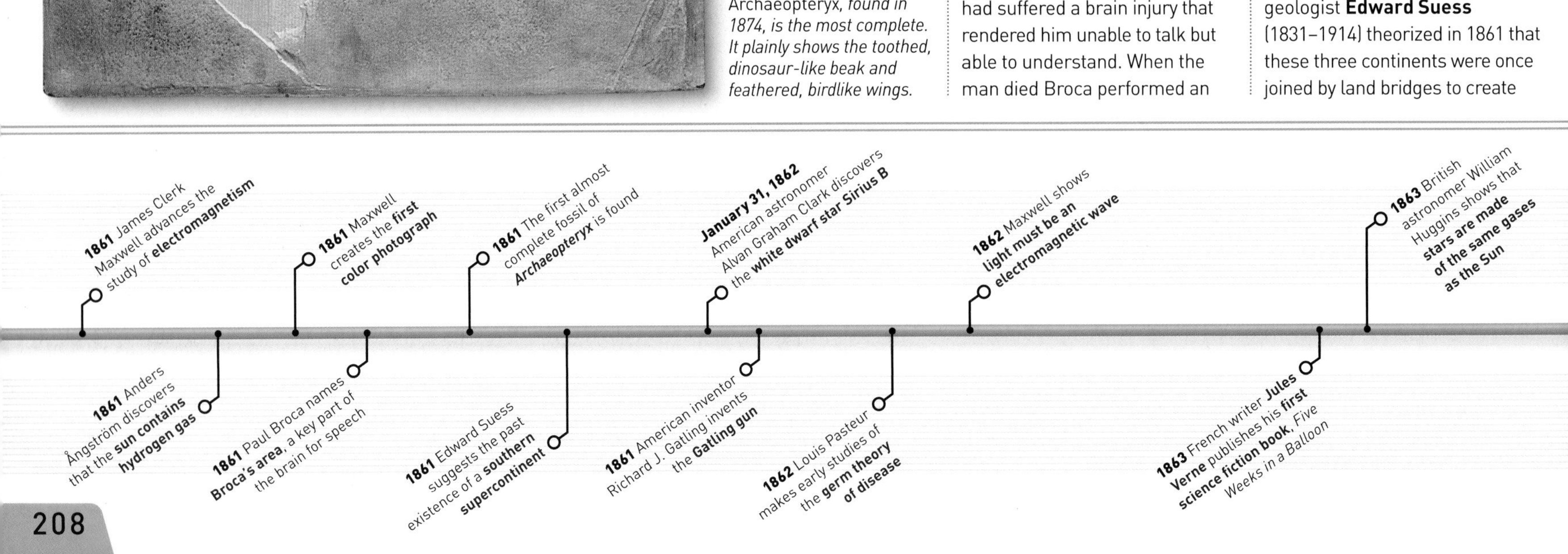
1861 James Clerk Maxwell advances the study of electromagnetism
1861 Anders Ångström discovers that the sun contains hydrogen gas
1861 Maxwell creates the first color photograph
1861 Paul Broca names Broca's area, a key part of the brain for speech
1861 The first almost complete fossil of Archaeopteryx is found
1861 Edward Suess suggests the past existence of a southern supercontinent
January 31, 1862 American astronomer Alvan Graham Clark discovers the white dwarf star Sirius B
1861 American inventor Richard J. Gatling invents the Gatling gun
1862 Louis Pasteur makes early studies of the germ theory of disease
1862 Maxwell shows light must be an electromagnetic wave
1863 French writer Jules Verne publishes his first science fiction book, Five Weeks in a Balloon
1863 British astronomer William Huggins shows that stars are made of the same gases as the Sun

15 GRAMS THE AMOUNT OF HEMOGLOBIN PER LITER OF HEALTHY HUMAN BLOOD

one **giant continent**, which he named Gondwanaland, before the seas rose and they were separated. He was right about the giant continent, but we now know that they separated through continental drift (see 1915) and not because the seas rose.

At Langenaltheim in Germany, the **first almost complete fossil of *Archaeopteryx*** was found. This winged, feathered creature with reptilian teeth, which lived 150 million years ago, shows the transition between dinosaurs and birds. The discovery was a major piece of evidence supporting Darwin's theory that one species evolves gradually into another. But Darwin's theory received a major setback in 1862 when British physicist **William Thomson** calculated **the age of Earth** from how fast it probably cooled since the time it was formed. The figure he came up with was no more than 400 million years, and possibly as short as 20 million years ago. Even 400 million years was not long enough for Darwin's gradual evolution to happen. Earth is now known to be closer to 4.5 billion years old.

In 1862, French chemist **Louis Pasteur** (1822–95) came closer to establishing that **microbes are responsible for many infectious diseases** with his studies of puerperal fever, an infection often caught by women during childbirth. Medicine was also advanced in 1864 when German physiologist **Felix Hoppe-Seyler** (1825–95) identified the **role of the iron-containing protein hemoglobin** in binding oxygen to red blood cells for transport through the blood. This was also the year that US president Abraham Lincoln signed the Yosemite Grant, which was the first step in creating California's now famous National Park in 1890.

JAMES CLERK MAXWELL
(1831–79)

Born in Edinburgh, Scotland, British physicist James Clerk Maxwell laid the cornerstones of electromagnetic theory. With his brilliant math, he showed that electricity, magnetism, and light are all forms of electromagnetic fields. His four equations are the basis of all classical electromagnetic science.

1865

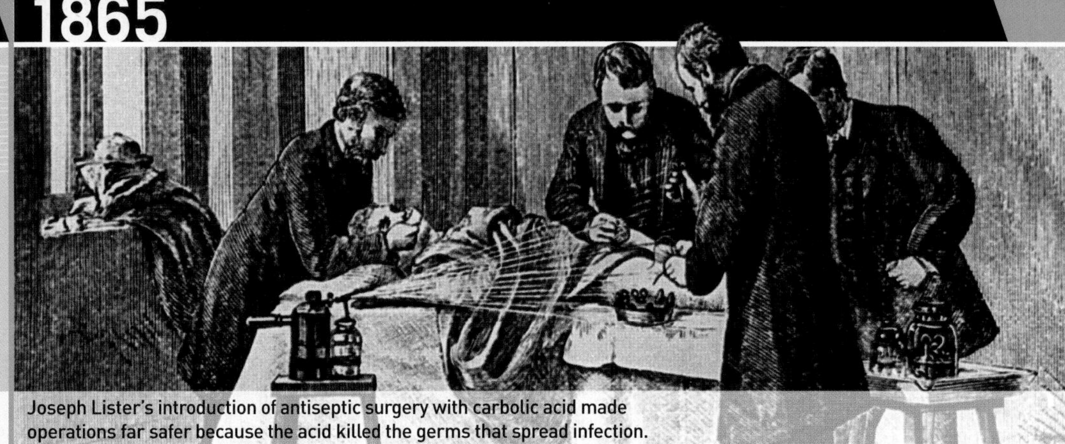

Joseph Lister's introduction of antiseptic surgery with carbolic acid made operations far safer because the acid killed the germs that spread infection.

IN 1865, GERMAN PHYSICIST RUDOLF CLAUSIUS made an assertion about heat that was more significant than most people appreciated at the time. Starting from the second law of thermodynamics (heat flows only from hot places to cold), he came up with the **concept of entropy**. Entropy is a mathematical measure of the disorder in any system. To create order, heat needs to be concentrated, and this requires energy. So in any system, whether it is the human body or the entire Universe, entropy and disorder will increase **unless there is a continuous input of energy from outside** to maintain the concentration of heat. For example, life on Earth depends on the input of energy from the Sun; once the Sun's fuel has run out, this input will cease.

> “THE **ENERGY** OF THE UNIVERSE IS **CONSTANT.** THE **ENTROPY** OF THE UNIVERSE TENDS TO A **MAXIMUM.**”
>
> **Rudolf Clausius, German physicist,** from *On Various Forms of the Laws of Thermodynamics that are Convenient for Applications,* 1865

Another German, **Otto Friedrich Karl Deiters** (1834–63), observed the **basic features of nerve cells** for the first time under a microscope. He noted that each nerve cell has a main cell body, a long tail fiber (axon), and a series of branching treelike fibers (dendrites).

In medicine, British surgeon **Joseph Lister** (1827–1912) pioneered the **use of carbolic acid as an antiseptic** during surgery, to clean instruments and wounds in order to reduce the chance of infection.

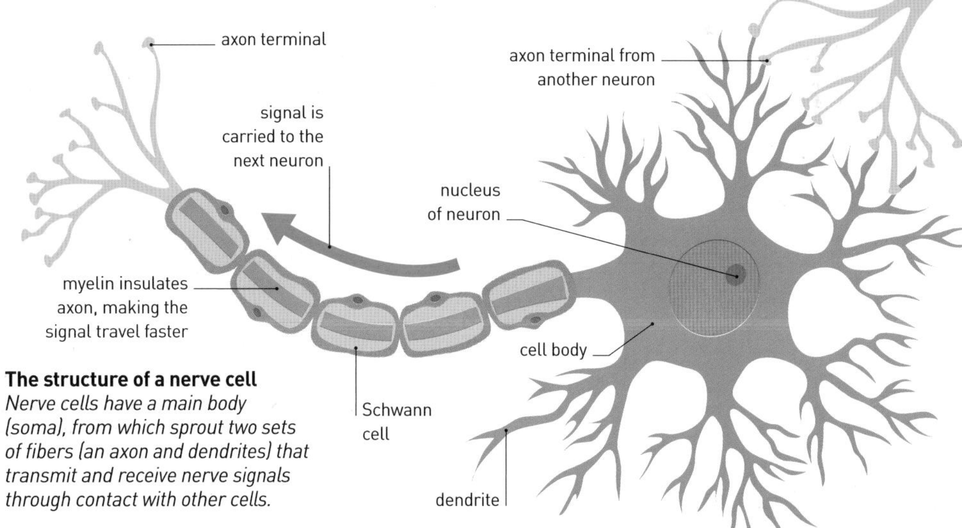

axon terminal
axon terminal from another neuron
signal is carried to the next neuron
nucleus of neuron
myelin insulates axon, making the signal travel faster
cell body
Schwann cell
dendrite

The structure of a nerve cell
Nerve cells have a main body (soma), from which sprout two sets of fibers (an axon and dendrites) that transmit and receive nerve signals through contact with other cells.

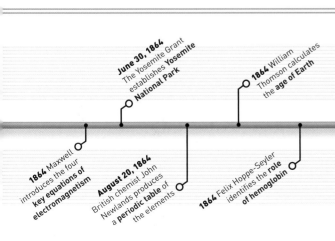

1864 Maxwell introduces the four **key equations of electromagnetism**

June 30, 1864 The Yosemite Grant establishes **Yosemite National Park**

August 20, 1864 British chemist John Newlands produces a **periodic table** of the elements

1864 William Thomson calculates the **age of Earth**

1864 Felix Hoppe-Seyler identifies the **role of hemoglobin**

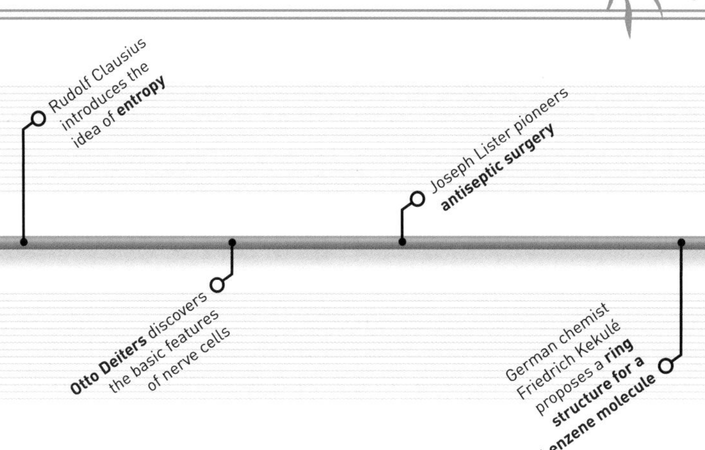

Rudolf Clausius introduces the idea of **entropy**
Otto Deiters discovers the basic features of nerve cells
Joseph Lister pioneers **antiseptic surgery**
German chemist Friedrich Kekulé proposes a **ring structure for a benzene molecule**

1866–67

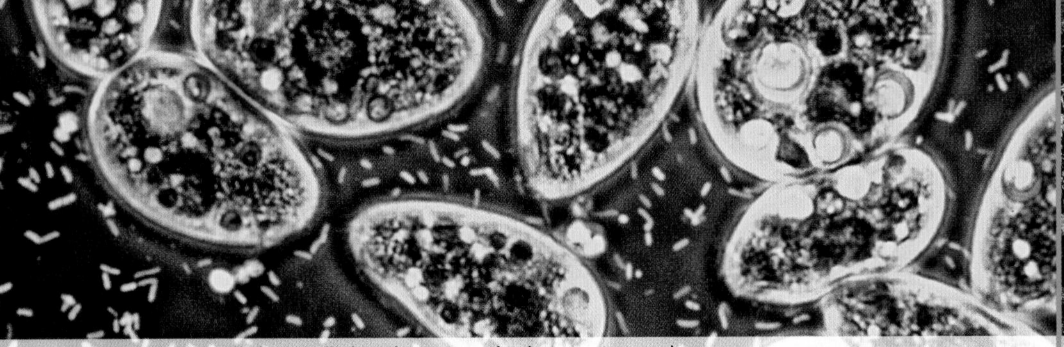

Ernst Haeckel identified one-celled protists as completely separate organisms, although the exact definition is still being debated.

THE INITIAL RELIGIOUS DEBATE OVER DARWIN'S THEORY OF EVOLUTION died down (see 1860), but scientific doubts persisted. One doubt was the question of how traits persist. In 1867, British engineering professor **Fleeming Jenkin** (1833–85) argued that adaptations would eventually blend in with the general population, **getting lost over the generations through what he called a "swamping effect."**

Unknown to both Darwin and Jenkin, Austrian monk **Gregor Mendel** (1822–84) answered this question with his work on peas (see 1856), which he completed in 1866. Mendel showed that **inherited characteristics persist** as they are passed on through "factors," now known as genes. It was not until the early 20th century that scientists fully appreciated the importance of this in relation to evolution.

Another commentator on evolution was German naturalist **Ernst Haeckel** (1834–1919), who argued, wrongly, in 1866 that stages in embryonic development retell evolutionary history. He made drawings to show **similarities between fish and human embryos**. Haeckel also proposed in 1866 that the single-celled organisms called **protists should have a kingdom of their own**, separate from plants and animals.

Also in 1866, German microscopist **Max Schultze** (1825-1974) undertook the most important **early study on the structure of the retina**—the light-sensitive tissue on the inside of the eye. He identified the layers of the retina and drew detailed illustrations of its cellular makeup, including the individual structure of rods and cones—the two types of light-receiving cells inside the back of the eye that react to light and colors (see 1935).

In technology, British engineer **Robert Whitehead** (1823–44) developed the **first self-propelled torpedo** in 1866, a device that later proved devastatingly effective in both World Wars. A year later, Swedish chemist **Alfred Nobel** (1833–96), invented the **explosive dynamite**.

In 1867 Parisian blacksmith **Pierre Michaux** (1813–83) developed one of the **first practical bicycles**, or velocipedes, with pedals and a chain. He also invented one of the **earliest motorcycles**, powered by a tiny steam engine.

small steam engine
iron bicycle frame

Early motorcycle
Dating from between 1867 and 1871, Pierre Michaux's bicycle was powered by a small steam engine. It is thought to be the world's first motorcycle.

1866 Gregor Mendel publishes his research on inheritance

1866 Ernst Haeckel identifies **protists** as separate from plant and animals

1866 Robert Whitehead develops a **self-propelling torpedo**

1867 Pierre Michaux invents the **first proper bicycle**

March 16, 1867 British surgeon Joseph Lister **describes antiseptic surgery**

May 7, 1867 Alfred Nobel patents **dynamite**

1867 Fleeming Jenkin talks about the problem of **blending inheritance** with Darwin's theory

1867 French engineer Louis-Guillaume Perreaux attaches a small steam engine to Michaux's bike to create the **first motorcycle**

1868–69

The Suez Canal was finally opened on November 17, 1869, after 10 years of labor involving tens of thousands of men.

WHILE HE WAS PREPARING A TEXTBOOK on chemistry in 1868, Russian chemist **Dmitri Mendeleyev** (1834–1907) began to wonder if chemical elements could be arranged in a table that **linked their atomic weights and properties**. Laying out the 60 elements known at the time in weight order, he saw that certain properties were repeated periodically and realized he could organize the elements into eight groups or "periods" of elements. Tellingly, the elements that appear in particular positions in each period have similar properties.

Mendeleyev presented his idea, now known as the **Periodic Table**, to a meeting of the Russian Chemical Society on March 6, 1869, and it has been a key reference ever since. What made it so powerful was that Mendeleyev could use it to predict the existence of three as-yet-undiscovered elements to fill gaps in the table. Over the following 16 years, all three missing elements—gallium, scandium, and germanium—were found, and since then **more than 50 further elements have been identified** and discovered.

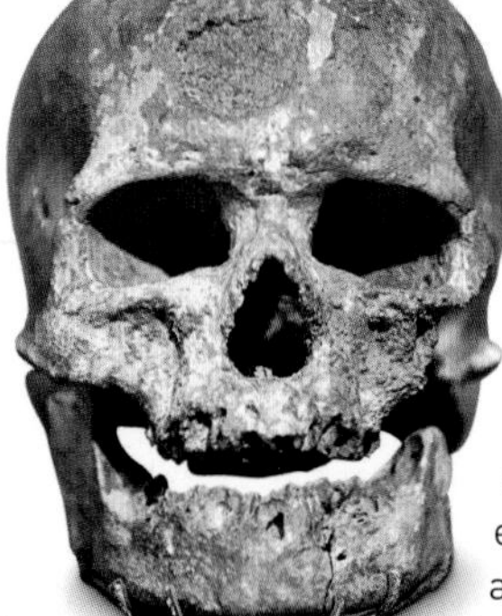

Another new element was discovered in 1868 using the knowledge that every substance glows with its own particular spectrum, or light signature (see 1884–85). While studying the spectra of light from the edge of the Sun during a total eclipse, British astronomer **Norman Lockyer** (1836–1920) and French astronomer **Jules Janssen**

Homo sapien's skull
This European Early Modern Human skull is from Cro-Magnon cave in France. Its discovery provided evidence that humans evolved.

17 MILES
THE DISTANCE **GIFFARD'S AIRSHIP** TRAVELED

January 30, 1868 British naturalist Charles Darwin publishes his **theory of pangenesis**

March 1868 French geologist Louis Lartet discovers the first identified skeletons of **Cro-Magnon man**

1868 Swedish physicist Anders Ångström **maps the solar spectrum**

1868 British inventor J. P. Knight **invents traffic lights**

102 MILES
THE LENGTH OF THE SUEZ CANAL WHEN IT WAS FIRST BUILT

(1824–1907) both noticed a bright yellow line at a wavelength that did not match any known substance. Lockyer and Janssen suggested that this indicated an unknown element in the Sun, which Lockyer **called helium, after the Greek "helios"** for Sun. That year, Swedish physicist Anders Ångström (1814–74) also **mapped the complete spectrum of light from the Sun**, identifying a thousand spectral lines in units that became known as angstroms in his honor.

The debate about evolution continued. Darwin described **how traits might be passed on from generation to generation through a process called pangenesis**. He proposed that in the body there are countless particles, or gemmules, which are like seeds that can reproduce the whole organism. Some claim this has similarities with DNA, the genetic material found in every body cell, which was coincidentally identified in cell nuclei the following year by Swiss biology student Friedrich Miescher (1844–95). It is now known that only DNA in the sex, or germ, cells (eggs and sperm) is ever used to make a new organism.

In France, geologist **Louis Lartet** (1840–99) added weight to the idea that humans also evolved, with his discovery of the first identified skeletons of what came to be known as **Cro-Magnon man**. They were named after the cave near Les Eyzies in France where the remains were found. They are now more generally known as European Early Modern Humans.

More controversially, in 1869 Darwin's half-cousin **Francis Galton** (1822–1911) used Darwin's theories to suggest a hereditary basis for human intelligence, an idea that was to lead him to develop the **science of eugenics**.

Technological change gathered pace as British inventor **John Peake Knight** (1828–86) invented **traffic lights**, and in the US American engineer **George Westinghouse** (1846–1914) invented **air brakes**. Another American, inventor **John Hyatt** (1837–1920), developed **celluloid**.

The **Suez Canal opened** to shipping in November 1869, linking the Red Sea to the Mediterranean. When it first opened the canal was 102 miles (164 km) long. It took 10 years of construction work to complete.

DMITRI MENDELEEV
(1834–07)

Born in 1834 in Tobolsk, Siberia, Dmitri Mendeleyev hiked to St. Petersburg to enroll in the university. Despite suffering from tuberculosis, he became Russia's leading chemist and developed the Periodic Table, which earned him worldwide acclaim and enabled him to predict the discovery of the elements gallium, scandium, and germanium.

ПЕРИОДИЧЕСКАЯ СИСТЕМА ЭЛЕМЕНТОВ
Д.И.МЕНДЕЛЕЕВА

Table of elements
This Russian periodic table is based on Mendeleyev's original table. It includes his predicted elements—gallium, scandium, and germanium.

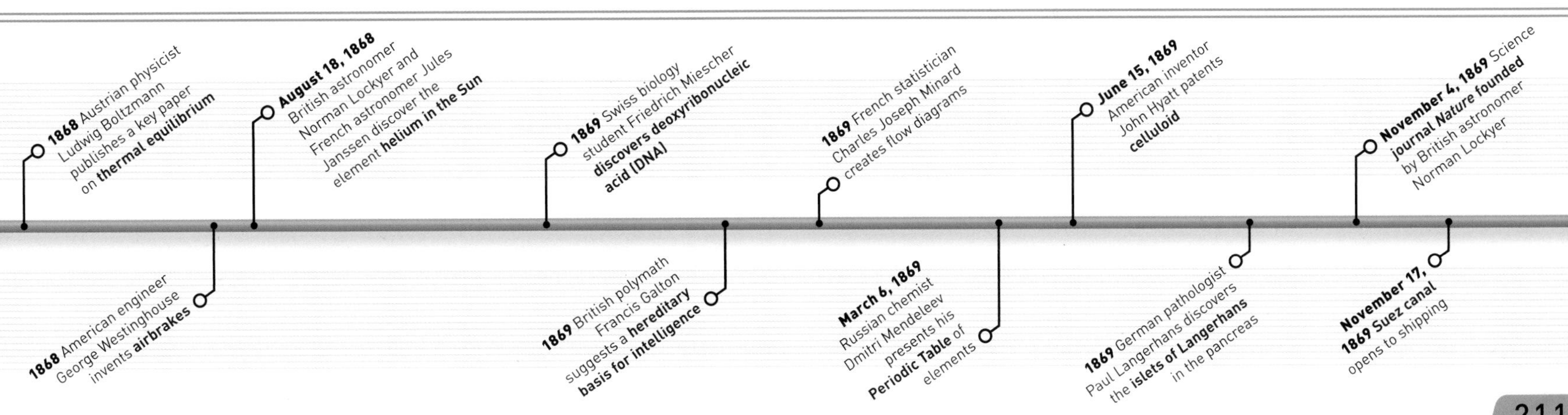
1868 Austrian physicist Ludwig Boltzmann publishes a key paper on thermal equilibrium
1868 American engineer George Westinghouse invents airbrakes
August 18, 1868 British astronomer Norman Lockyer and French astronomer Jules Janssen discover the element helium in the Sun
1869 Swiss biology student Friedrich Miescher discovers deoxyribonucleic acid (DNA)
1869 British polymath Francis Galton suggests a hereditary basis for intelligence
1869 French statistician Charles Joseph Minard creates flow diagrams
March 6, 1869 Russian chemist Dmitri Mendeleev presents his Periodic Table of elements
June 15, 1869 American inventor John Hyatt patents celluloid
1869 German pathologist Paul Langerhans discovers the islets of Langerhans in the pancreas
November 4, 1869 Science journal Nature founded by British astronomer Norman Lockyer
November 17, 1869 Suez canal opens to shipping

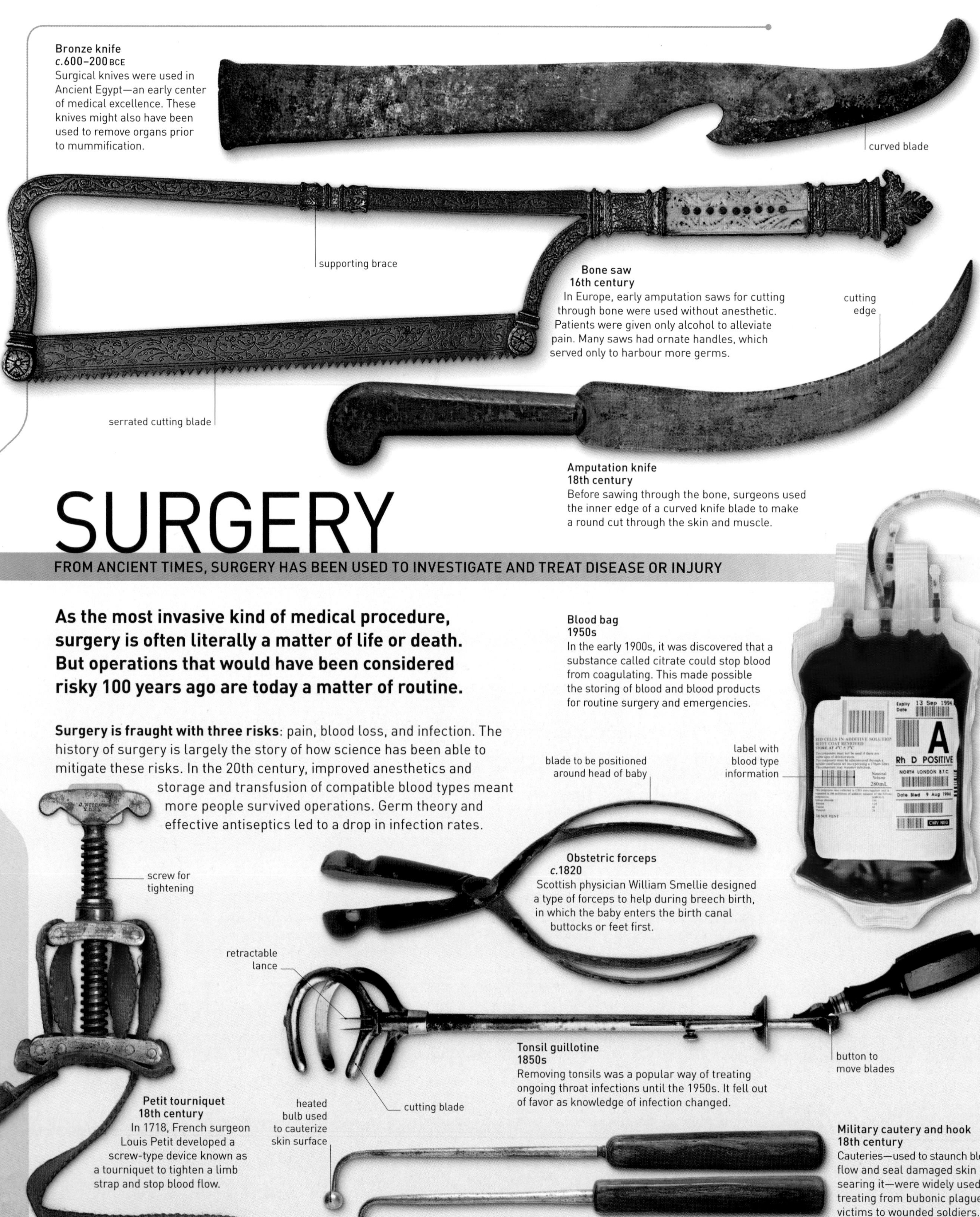
curved blade
supporting brace
serrated cutting blade
cutting edge
label with blood type information
blade to be positioned around head of baby
screw for tightening
retractable lance
cutting blade
button to move blades
heated bulb used to cauterize skin surface
Rh D POSITIVE
A

Bronze knife
*c.*600–200 BCE
Surgical knives were used in Ancient Egypt—an early center of medical excellence. These knives might also have been used to remove organs prior to mummification.

Bone saw
16th century
In Europe, early amputation saws for cutting through bone were used without anesthetic. Patients were given only alcohol to alleviate pain. Many saws had ornate handles, which served only to harbour more germs.

Amputation knife
18th century
Before sawing through the bone, surgeons used the inner edge of a curved knife blade to make a round cut through the skin and muscle.

SURGERY

FROM ANCIENT TIMES, SURGERY HAS BEEN USED TO INVESTIGATE AND TREAT DISEASE OR INJURY

As the most invasive kind of medical procedure, surgery is often literally a matter of life or death. But operations that would have been considered risky 100 years ago are today a matter of routine.

Surgery is fraught with three risks: pain, blood loss, and infection. The history of surgery is largely the story of how science has been able to mitigate these risks. In the 20th century, improved anesthetics and storage and transfusion of compatible blood types meant more people survived operations. Germ theory and effective antiseptics led to a drop in infection rates.

Blood bag
1950s
In the early 1900s, it was discovered that a substance called citrate could stop blood from coagulating. This made possible the storing of blood and blood products for routine surgery and emergencies.

Obstetric forceps
*c.*1820
Scottish physician William Smellie designed a type of forceps to help during breech birth, in which the baby enters the birth canal buttocks or feet first.

Petit tourniquet
18th century
In 1718, French surgeon Louis Petit developed a screw-type device known as a tourniquet to tighten a limb strap and stop blood flow.

Tonsil guillotine
1850s
Removing tonsils was a popular way of treating ongoing throat infections until the 1950s. It fell out of favor as knowledge of infection changed.

Military cautery and hook
18th century
Cauteries—used to staunch blood flow and seal damaged skin by searing it—were widely used, treating from bubonic plague victims to wounded soldiers.

nozzle

water chamber

Carbolic steam spray
1860s
British surgeon Joseph Lister developed a device to spray carbolic acid in the operating room. The pungent chemical acted as an antiseptic, reducing the chance of wounds becoming infected.

reservoir of carbolic acid

Surgical sterilizing equipment
1860s
On battlefields, a hot flame from a portable alcohol burner was used to sterilize surgical instruments supported on a brass stand.

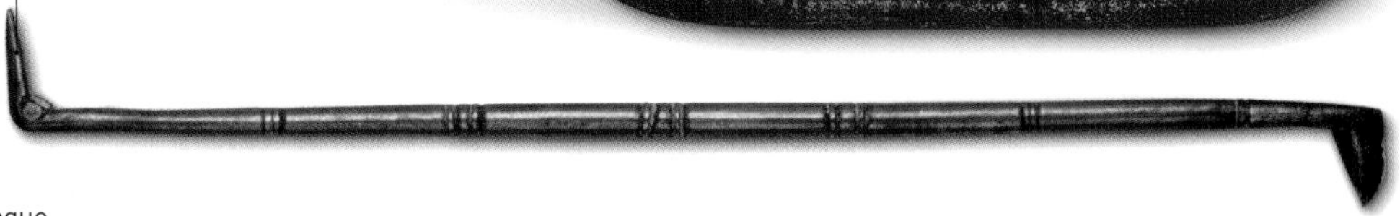

scraping blade

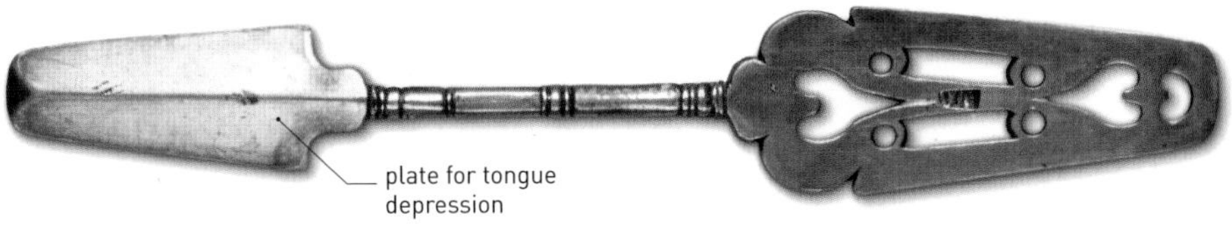

Barber-Surgeon instruments
Late 1860s
The scraper and tongue depressor shown here once belonged to a member of the Barber-Surgeon's Company (a guild created by Henry VIII in 1540).

plate for tongue depression

dropping tube

Esmarch chloroform-ether dropper
c.1890
English surgeon James Simpson used chloroform as an alternative anesthetic to ether in 1847. Although glass dropping bottles delivered more precise doses, the danger of overdosing persisted until safer anesthetics were developed in the 1950s.

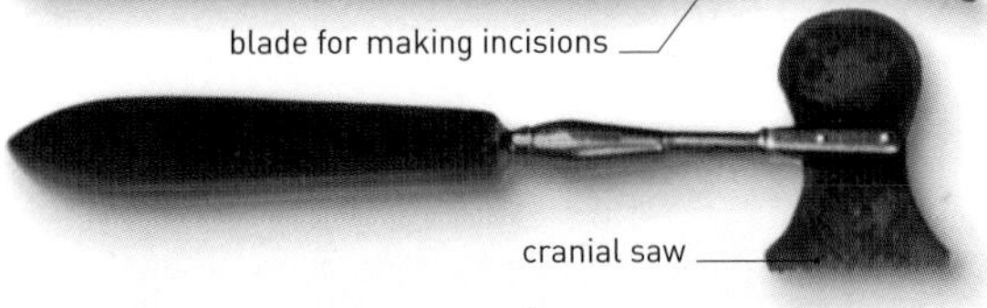

sharp blade for cutting muscle

American Civil War surgical equipment
1860s
During the American Civil War, two soldiers died from disease and infection for every one killed in battle. Surgical instruments were used by medical attendees with limited training.

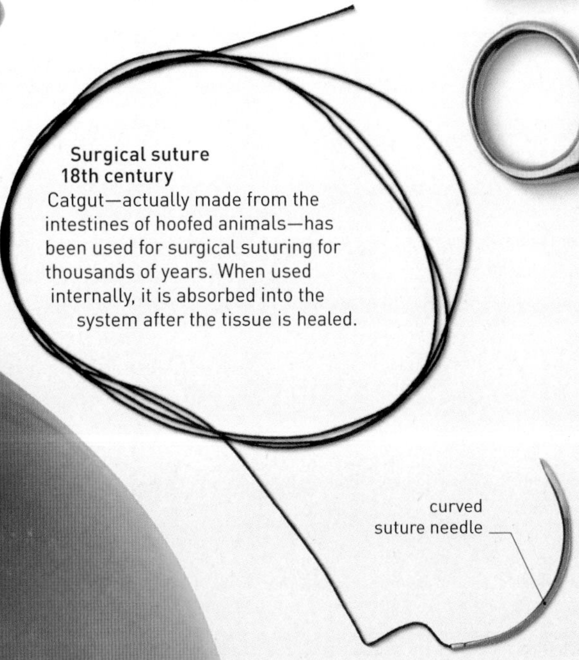

blade for making incisions

cranial saw

locking catch

Surgical suture
18th century
Catgut—actually made from the intestines of hoofed animals—has been used for surgical suturing for thousands of years. When used internally, it is absorbed into the system after the tissue is healed.

scalpel with interchangeable blade

curved suture needle

graduations on side of bottle

surgical forceps

Stainless-steel instruments
20th century
Stainless steel surgical instruments became available in 1930. They were corrosion-free and easily sterilized surgical instruments could be made. Additional modifications have helped make surgical stainless steel especially smooth and scratch-resistant.

retractor with blunt hook

1870–71

“ THE **LIGHT** WHICH WE RECEIVE FROM THE **CLEAR SKY** IS DUE… TO **SMALL SUSPENDED PARTICLES** WHICH **DIVERT** THE LIGHT FROM ITS **REGULAR COURSE. ”**

Lord Rayleigh, English scientist, from *On the Light from the Sky, its Polarization and Color*, 1871

The sky appears blue because of Rayleigh scattering—the scattering of sunlight by molecules in the air.

AROUND 1870, ITALIAN PHYSICIAN CAMILLO GOLGI (1843–1926) developed a technique to stain brain and other tissue to be able to observe them under the microscope. He used this method to identify **neurons** (nerve cells) which process and transmit information between the brain and other parts of the body. For his work in this field, Golgi was awarded the Nobel Prize in Physiology or Medicine in 1906.

In a related discovery, German scientists **Gustav Theodor Fritsch** (1838–1927) and **Eduard Hitzig** (1839–1907) demonstrated the link between **electricity and brain function**. They showed that an electric stimulus applied to different parts of a dog's brain would cause distinct muscular contractions.

The same year, English mathematician **William Kingdon Clifford** (1845–79) suggested that energy and matter are caused by the **curvature of space**. Clifford died young, and was unable to develop the theory further; his ideas resurfaced in German-born physicist Albert Einstein's general theory of relativity (see 1914–15).

British astronomer **Joseph Norman Lockyer** (1836–1920) and his French colleague **Pierre Jules César Janssen** (1824–1907) independently suggested that certain lines in the Sun's spectrum were produced by a previously unknown element (see 1868–69). In 1870, Lockyer named this element **helium**, after the Greek Sun god—Helios.

Also in 1870, French microbiologist **Louis Pasteur** (1822–95) published a book documenting the mysterious disease killing silkworms, tracing it back to microbes. This, together with the discovery of the anthrax bacterium (see 1876–77) led to the development of **germ theory**.

In 1871, English scientist **Lord Rayleigh** (1842–1919) discovered that when light bounces off small particles, it scatters—now called **Rayleigh scattering**. He stated that for visible light to scatter, the particles must be smaller than 400 to 700 nanometers, that is, smaller than the wavelength of the light being scattered.

“ WITH SAVAGES, THE WEAK IN BODY OR MIND ARE **SOON ELIMINATED…** WE **CIVILIZED MEN,** ON THE OTHER HAND, DO OUR UTMOST TO **CHECK THE PROCESS** OF **ELIMINATION. ”**

Charles Darwin, British naturalist, from *The Descent of Man*, 1871

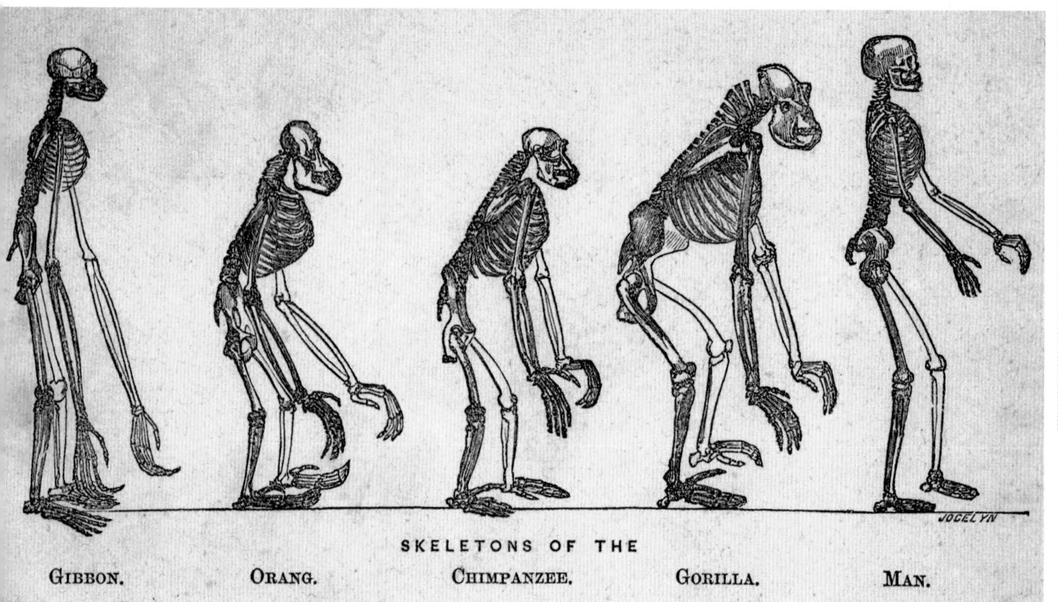
SKELETONS OF THE
GIBBON.
ORANG.
CHIMPANZEE.
GORILLA.
MAN.

Ape and man
This illustration from an 1863 book by English biologist Thomas Henry Huxley—an advocate of Charles Darwin's theories—shows similarities between humans and living apes.

LOUIS PASTEUR
(1822–95)

While working at Lille University, France, Louis Pasteur investigated the problem of beer and wine (and later milk) going sour. He found that this was caused by bacteria, which could be killed by boiling (pasteurization). These studies helped Pasteur develop the germ theory of disease, and the prevention of disease by vaccination.

This year, English naturalist **Charles Darwin** published a book on **human evolution**—*The Descent of Man, and Selection in Relation to Sex*. He had been wary of publishing on the topic before, expecting the sort of furor that had accompanied the publication of *On the Origin of Species* in 1859.

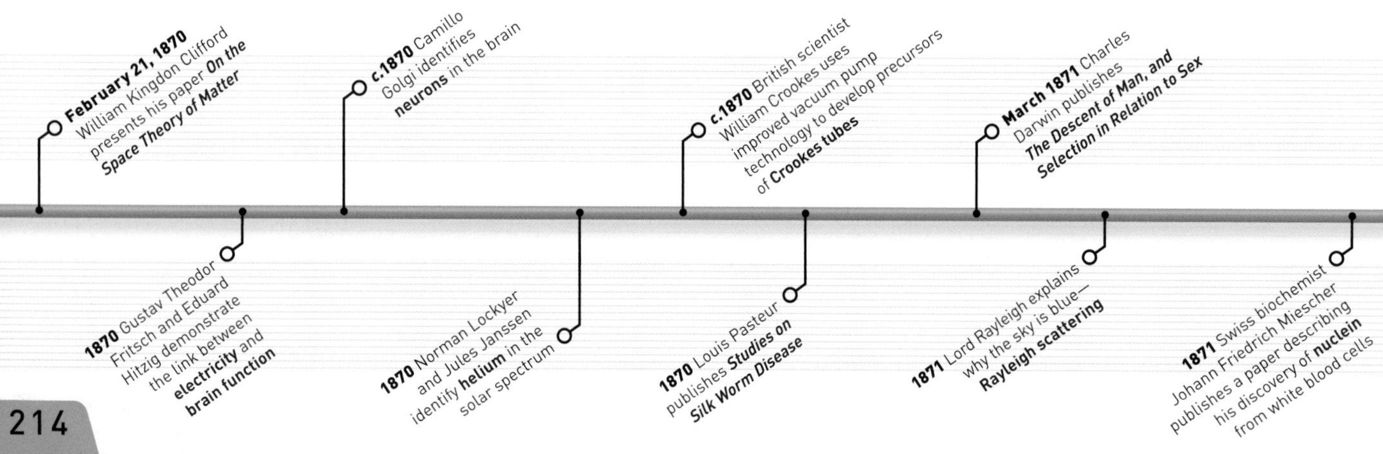
February 21, 1870 William Kingdon Clifford presents his paper On the Space Theory of Matter
1870 Gustav Theodor Fritsch and Eduard Hitzig demonstrate the link between electricity and brain function
c.1870 Camillo Golgi identifies neurons in the brain
1870 Norman Lockyer and Jules Janssen identify helium in the solar spectrum
c.1870 British scientist William Crookes uses improved vacuum pump technology to develop precursors of Crookes tubes
1870 Louis Pasteur publishes Studies on Silk Worm Disease
March 1871 Charles Darwin publishes The Descent of Man, and Selection in Relation to Sex
1871 Lord Rayleigh explains why the sky is blue—Rayleigh scattering
1871 Swiss biochemist Johann Friedrich Miescher publishes a paper describing his discovery of nuclein from white blood cells

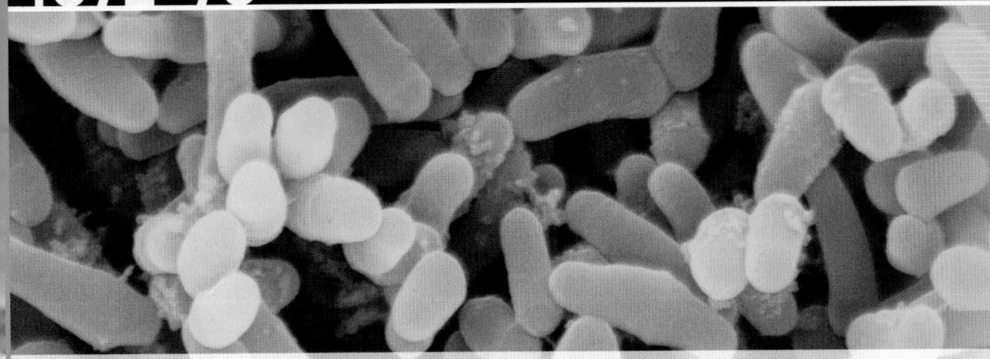

> **“** I THINK **LEPROSY** TO BE **INOCULABLE;** I, MOREOVER, THINK THAT LEPROSY **IN MOST CASES IS TRANSFERRED BY INOCULATION. ”**

Gerhard Hansen, Norwegian physician, from a letter to British social reformer William Tebb, *c.*1889

This micrographic image shows rod-shaped cells of the leprosy bacterium *Mycobacterium leprae*—the first bacterium to be identified as a cause of disease in humans.

IN 1872, AUSTRIAN PHYSICIST LUDWIG EDUARD BOLTZMANN (1844–1906) developed an equation that described the **behavior of a fluid** (gas or liquid) by applying probability distributions (see 1652–54) to the interactions of large numbers of atoms or molecules. This equation gave a **mathematical foundation** to the **second law of thermodynamics,** which states that systems tend toward a state of equilibrium.

In December 1872, a Royal Society expedition set out on board **HMS *Challenger*** from Portsmouth, UK. Over the next four years, it discovered many large features of the planet, including the mid-Atlantic ridge in the Atlantic Ocean and the Marianas Trench in the western Pacific, where it took samples from the Challenger Deep—a point near the deepest recorded point on Earth. The expedition also catalogued approximately 4,700 previously unknown species of animals and plants.

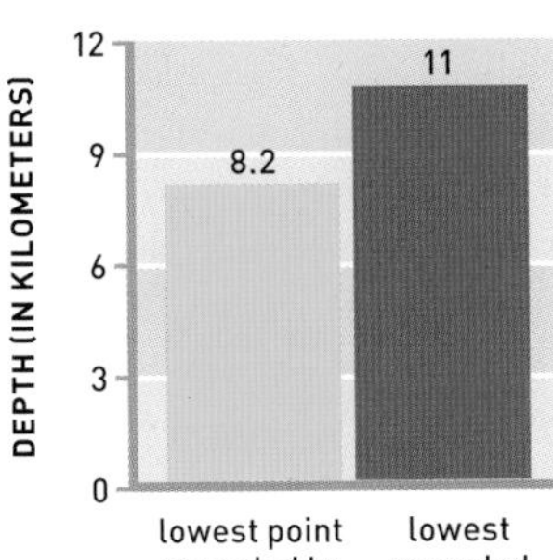

DEPTH (IN KILOMETERS)
12
9
6
3
0
8.2
11
lowest point recorded by Challenger
lowest recorded point on Earth

Challenger Deep
HMS Challenger *estimated the depth of Earth's lowest point—Challenger Deep—at 5.1 miles (8.2 km). Recent estimates suggest 6.8 miles (11 km).*

The following year, Dutch physicist **Johannes Diderik van der Waals** (1837–1923) derived an "equation of state," which described how the liquid and gas states of a substance merge into each other. He assumed that molecules exist, are of finite size, and attract each other by a weak force—now called a **van der Waals force**. Together, these ideas helped provide a better understanding of atoms.

In 1873, Scottish physicist **James Clerk Maxwell** published *A Treatise on Electricity and Magnetism,* in which he described his theory of **electromagnetism** (see pp.234–35). The theory predicted the existence of radio waves (see 1886), and was a major influence on 20th-century science.

William Crookes invented his **radiometer** while investigating the nature of light as a form of electromagnetic radiation. This device is a partially evacuated, airtight glass bulb that contains a set of vanes mounted on a spindle, like a horizontal windmill. One side of each vane is painted white, while the other is black. When the vanes are exposed to light, the "light mill" rotates, with the white side leading the way. This occurs because the dark side of the paddle absorbs more radiant energy and gets hotter than the white side; some of the energy is transferred to the molecules hitting the surface, giving the paddle a kick. The same thing happens on the white side, but to a lesser degree.

HMS *Challenger*
Launched on February 13, 1858, HMS Challenger *was primarily a sailing ship, but was also fitted with an auxiliary steam engine.*

The understanding of disease took a step forward in 1873, when Norwegian physician **Gerhard Hansen** (1841–1912) discovered the **leprosy** bacterium—*Mycobacterium leprae*. It was previously thought that leprosy was either inherited or spread by "bad air" known as miasmas.

Meanwhile, the mechanism of heredity was starting to be understood, partly through the work of German zoologist **Anton Schneider** (1831–90). He gave the first accurate description of **mitosis**—the process whereby a dividing cell provides identical copies of its chromosomes (then known as nuclear filaments) to each of the daughter cells (see pp.194–95).

compressor
evaporator

First practical refrigerator
Von Linde improved on his 1873 design by using glycerine to seal the compressor and using ammonia as a refrigerant.

The **first modern refrigeration system** was designed by German engineer **Carl von Linde** (1842–1934) and built by the mechanical engineering company Maschinenfabrik Augsburg for a brewery in Munich, Germany. Three years later, Linde designed a more reliable system—the first practical compressed-ammonia refrigerator. The huge commercial success of this invention enabled von Linde to focus on research, becoming the first person to **liquify air,** and **separate oxygen and nitrogen from it**.

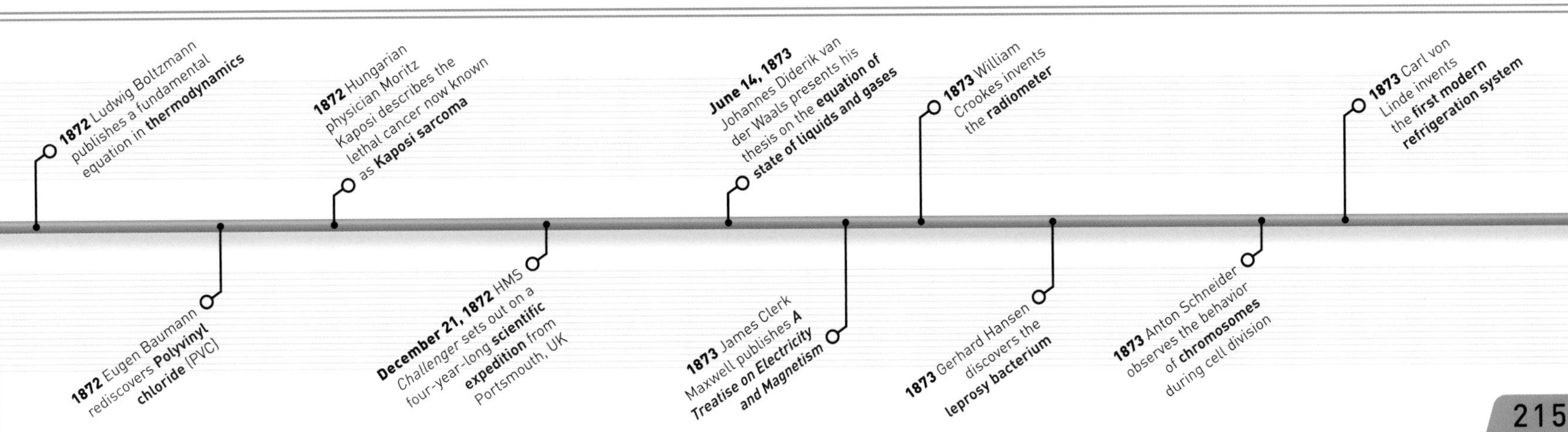

1872 Ludwig Boltzmann publishes a fundamental equation in **thermodynamics**
1872 Eugen Baumann rediscovers **Polyvinyl chloride** (PVC)
1872 Hungarian physician Moritz Kaposi describes the lethal cancer now known as **Kaposi sarcoma**
December 21, 1872 HMS Challenger sets out on a four-year-long **scientific expedition** from Portsmouth, UK
June 14, 1873 Johannes Diderik van der Waals presents his thesis on the **equation of state of liquids and gases**
1873 James Clerk Maxwell publishes **A Treatise on Electricity and Magnetism**
1873 William Crookes invents the **radiometer**
1873 Gerhard Hansen discovers the **leprosy bacterium**
1873 Anton Schneider observes the behavior of **chromosomes** during cell division
1873 Carl von Linde invents the **first modern refrigeration system**

“ **NATURE** IS ALL THAT A MAN BRINGS WITH HIMSELF **INTO THE WORLD; NURTURE** IS EVERY INFLUENCE WITHOUT THAT AFFECTS HIM **AFTER HIS BIRTH.** ”

Francis Galton, from *English Men of Science: Their Nature and Nurture*, 1874

British scientist Francis Galton's contributions to science include studies of twins and meteorology.

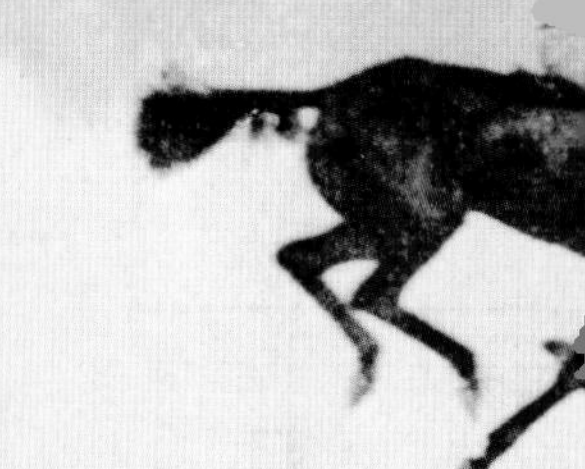

GERMAN MATHEMATICIAN GEORG CANTOR (1845–1918) published *On a Characteristic Property of all Real Algebraic Numbers* in 1874. This paper laid the foundations of **set theory**, and introduced the idea of different kinds of **infinity**.

The same year, Russian mathematician **Sofia Vasilyevna Kovalevskaya** (1850–91) became the **first woman to be awarded a doctorate in mathematics**, which she received from the University of Göttingen, Germany. In 1889, she became the **first woman to be appointed professor of mathematics**, at Stockholm University, Sweden.

Meanwhile, British economist **William Stanley Jevons** (1835–82) published *The Principles of Science: A Treatise on Logic and Scientific Method*, which criticized the method of **induction** as a source of new scientific ideas and instead recommended **random hypotheses**.

Austrian chemist Othmar Zeidler (1849–1911) synthesized **DDT** (dichlorodiphenyltrichloroethane) as a doctoral student at the University of Strasbourg, France, in 1874. The importance of the substance as a powerful insecticide was not realized until the late 1930s.

French chemist **Paul-Émile Lecoq de Boisbaudran** (1838–1912) plugged a gap in Mendeleyev's periodic table of elements (see 1868–69) by identifying the metal **gallium** spectroscopically in 1875. Before the end of the year he had synthesized pure gallium by the electrolysis of a solution of gallium hydroxide in potassium hydroxide.

Silvery white metal
Gallium is one of the few metals that occur in liquid form at near-room temperatures. Like water, this metal expands on solidifying.

Austrian geologist **Eduard Suess** (1831–1914) **coined the term biosphere** (see panel, left) in 1875, defining it as "the place on Earth's surface where life dwells." The concept was intended to complement the **three geological zones**—the lithosphere (the rocky outer layer of the planet), the hydrosphere (the layer of water at Earth's surface), and the atmosphere (the gas envelope surrounding Earth). However, the theory made little impact on the scientific community until it was developed by Russian geochemist Vladimir Vernadsky (1863–1945) in his 1926 book, *La Biosphere*.

The scientific study of twins began in 1875, when British scientist **Francis Galton** (1822–1911) published his landmark paper *The History of Twins, as a Criterion of the Relative Powers of Nature and Nurture*. While he paved the way for further scientific enquiry into this subject, Galton was unaware of the distinction between monozygotic (identical) twins, born from one fertilized egg, and dizygotic twins born from two. These studies would only be carried out in the 20th century.

BIOSPHERE

The term biosphere is used to refer to the sum of all ecosystems. A closed, self-regulating system, the biosphere shows that living things do not exist in isolation from the nonliving world. Devised by Eduard Suess in 1875, the idea was developed by James Lovelock (b.1919) and Lynn Margulis (1938–2011) as Gaia theory (see 1979).

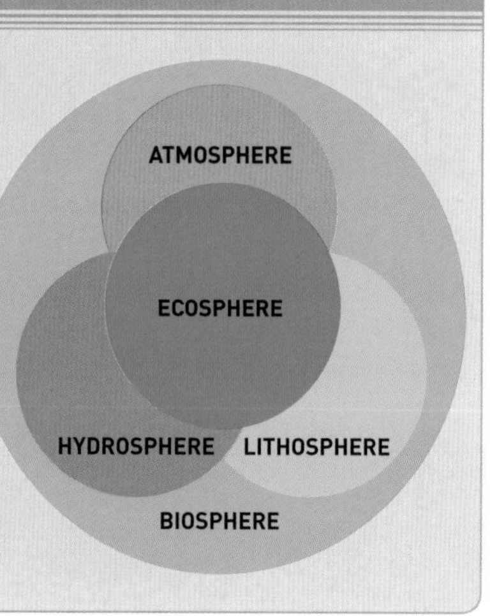
ATMOSPHERE
ECOSPHERE
HYDROSPHERE
LITHOSPHERE
BIOSPHERE

0.4 PERCENT THE PROPORTION OF **BABIES** BORN AS **IDENTICAL TWINS**

BRITISH NATURALIST Alfred Russel Wallace (see 1855–58) published *The Geographical Distribution of Animals* in 1876. Co-founder of natural selection theory with Charles Darwin (see 1859–60), Wallace also made contributions to **biogeography**, including the concept of warning coloration in animals and the way barriers to hybridization, such as mountains, contribute to the evolution of new species.

American inventors Elisha Gray (1835–1901) and **Alexander Graham Bell** (see panel, opposite) independently came up with the design of a working **telephone**. However, Bell won the race to patent the device on March 7, 1876. Three days later, he spoke the famous sentence: "Mr. Watson, come here, I want to see you." into the machine, summoning his assistant from the next room.

Just about two months later, German inventor **Nikolaus Otto** (1832–91) completed building the first practical **four-stroke piston internal combustion engine** (see 1807–09).

In 1876, German neurologist **Karl Wernicke** (1848–1905) found that damage to a specific part of the brain—now called **Wernicke's area**—resulted in language disorders. This area of the brain is connected to Broca's area (see 1861–64) by a nerve

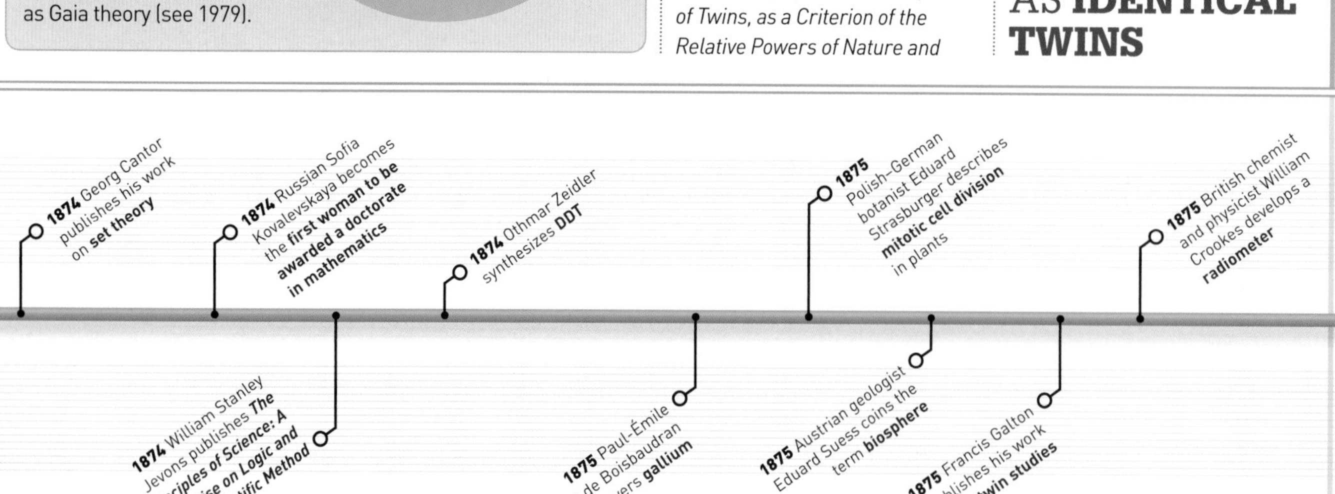
1874 Georg Cantor publishes his work on set theory
1874 Russian Sofia Kovalevskaya becomes the first woman to be awarded a doctorate in mathematics
1874 Othmar Zeidler synthesizes DDT
1875 Polish-German botanist Eduard Strasburger describes mitotic cell division in plants
1875 British chemist and physicist William Crookes develops a radiometer
1876 Alfred Russel Wallace publishes The Geographical Distribution of Animals
1874 William Stanley Jevons publishes The Principles of Science: A Treatise on Logic and Scientific Method
1875 Paul-Émile Lecoq de Boisbaudran discovers gallium
1875 Austrian geologist Eduard Suess coins the term biosphere
1875 Francis Galton publishes his work on twin studies
March 7, 1876 Alexander Graham Bell is granted a patent for the telephone

English photographer Eadweard Muybridge's pictures showed the exact position of a trotting horse's legs for the first time. The original pictures shot in 1877 did not survive. The ones seen here are from 1878.

ALEXANDER GRAHAM BELL (1847–1922)

Alexander Graham Bell was born in 1847 in Edinburgh, Scotland. In 1870, he first moved to Ontario, Canada, and then Boston, US, where he began his career as an inventor. Both Bell's mother and wife suffered hearing impairments, inspiring his interest in speech and hearing. This interest led him to invent the microphone and the telephone.

fiber bundle called the arcuate fasciculus. Damage to these fibers can leave people able to understand language, but with speech that makes no sense.

German physiologist **Wilhelm Kühne** (1837–1900) discovered and named the pancreatic enzyme **trypsin** in 1876. He also coined the term **enzyme**, which refers to large biological molecules that act as catalysts and control the rate of chemical reactions in the cells of living things (see 1893–94).

In 1877, German bacteriologist **Robert Koch** (1843–1910) grew *Bacillus anthracis*—the organism that causes the infectious disease **anthrax**—in the laboratory and injected it into animals to induce the disease. This was the first bacterium shown to be the cause of a disease.

In the world of psychology, **Charles Darwin** published ***Biographical Sketch of an Infant***, a 10-page book based on his observations of his newborn son, from 37 years earlier. Darwin speculated that his son went through the same stages of learning that his ancestors had for centuries, and that "each individual somehow retains rudiments of the long evolutionary past."

In 1877, English photographer **Eadweard Muybridge** (1830–1904) made a **camera** with a **shutter speed of only one-thousandth of a second**. With this and an ultrasensitive photographic plate he was able to "freeze" the motion of a trotting horse in a picture. The picture caused a sensation, proving that all four of the horse's legs left the ground at once.

274

THE NUMBER OF **AMINO ACIDS** IN **TRYPSIN**

American astronomer **Asaph Hall** (1829–1907) discovered the two moons of Mars in 1877. He spotted **Deimos**, the smaller moon, on August 12, and **Phobos** on August 18. Hall also worked out the mass of Mars, the orbits of moons, and measured the rotation of Saturn.

Italian astronomer **Giovanni Schiaparelli** (1835–1910) was also studying Mars in 1877. He reported seeing ***canali*** on the planet. This word, simply meaning "channels," was mistranslated to "canals," giving rise to a frenzy of speculation that Mars was inhabited by intelligent beings. It was later discovered that these markings were optical illusions.

horseshoe magnet
diaphragm
trumpetlike mouthpiece
terminal for external connection
wire
coil of copper wire

Bell's electric telephone
Used as both transmitter and receiver, this early telephone converted sound vibrations into electrical signals (and vice versa).

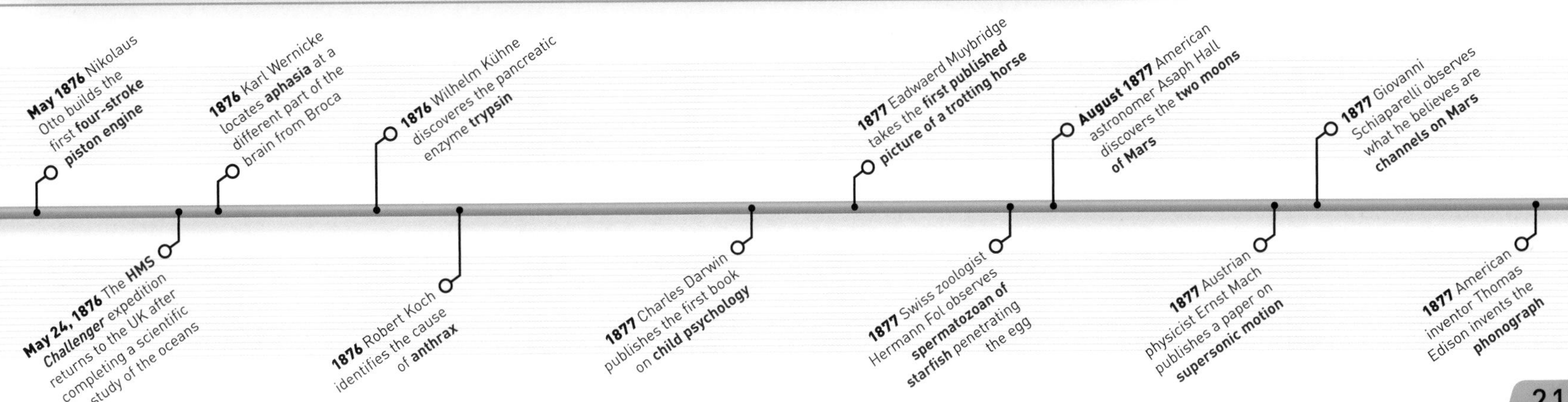

May 1876 Nikolaus Otto builds the first **four-stroke piston engine**
May 24, 1876 The HMS *Challenger* expedition returns to the UK after completing a scientific study of the oceans
1876 Karl Wernicke locates **aphasia** at a different part of the brain from Broca
1876 Wilhelm Kühne discovers the pancreatic enzyme **trypsin**
1876 Robert Koch identifies the cause of **anthrax**
1877 Charles Darwin publishes the first book on **child psychology**
1877 Eadwaerd Muybridge takes the **first published picture of a trotting horse**
1877 Swiss zoologist Hermann Fol observes **spermatozoan of starfish** penetrating the egg
August 1877 American astronomer Asaph Hall discovers the **two moons of Mars**
1877 Austrian physicist Ernst Mach publishes a paper on **supersonic motion**
1877 Giovanni Schiaparelli observes what he believes are **channels on Mars**
1877 American inventor Thomas Edison invents the **phonograph**

1815
Multicylinder music box
First produced in Switzerland in 1815, these contain a rotating cylinder covered in spikes that pluck the teeth of a steel comb. In 1862, a system that allows cylinders to be changed to play different melodies is invented.

Multicylinder music box

1888
Gramophone player
Invented by Emile Berliner, the gramophone uses disks of shellac. They can be copied numerous times from a brass master without any loss of quality.

Gramophone

1857
Phonautograph
Édouard-Léon Scott invents the first device able to record sound but it is unable to play the sound back.

1876
Player piano
The automatic-playing piano becomes popular when it is shown at an exhibition. It operates by an electromagnet and contains a paper music roll.

Player piano

1877
Edison's phonograph
Thomas Edison's phonograph is the first device able to play sounds back as well as record them. The vibrations of sound are funneled through a horn and recorded on a cylinder covered in tin foil.

THE STORY OF SOUND RECORDING

horn concentrates sound for recording and amplifies it for playback

RECORDING SOUND WAS AN ANCIENT DREAM OF MANKIND BUT IT ONLY BECAME A REALITY DURING THE 19TH CENTURY

Little more than a century ago, the only music most people heard was performed live. The advent of technology to record sound and play it back has not only transformed the way we listen to music but has also had other applications, including broadcasting, film-making, and sound archiving.

Frenchman Édouard-Léon Scott's phonautograph of 1857 was the first device capable of recording sound, using a moving needle to trace a line on a carbon-coated surface. In 1877, American Thomas Edison invented the phonograph—the first device that could record sound and play it back. Early sound recording machines worked mechanically—they captured sound by playing into a horn to make the sound vibrations move a needle to engrave scratches on a disk or cylinder.

In the 1920s, sound recording entered the electrical era with the invention of microphones. Sound was soon being reproduced electrically using electromagnet-driven loudspeakers with improved sound quality and volume.

After 1945, sound recording of music took off with vinyl records that played back at 33 or 45 revolutions per minute (rpm) (early records played back at 78 rpm). Magnetic devices that recorded sound as varying magnetic patterns on tape rather than physical grooves on a disk were also developed.

Edison and his phonograph
The phonograph was invented by Thomas Edison in 1877. It recorded sounds in grooves embossed in tin foil wrapped around a cylinder. Later cylinders were made from wax-coated cardboard.

DIGITAL

The next big breakthrough was digital recording (see panel, right), which made for much more robust, practical systems. These include the first compact disks, as well as digital audio formats such as MP3, which have made it possible to store vast amounts of music on small devices, and download limitless music from the internet.

> "...I CAN TRANSPORT YOU TO THE **REALMS OF MUSIC.**"

The first promotional message recorded on the Edison phonograph, 1877

Microphone

1925
Microphone
The horn is replaced by electric microphones, in which vibrations move electromagnets to create a varying electrical signal that moves the needle to make the grooves.

1931
Reel-to-reel tape recorder
German Fritz Pfleumer invents the reel-to-reel tape recorder, in which the fluctuations in the electrical signal are recorded in the magnetic coating of moving tape. The AEG company develops it into the Magnetophone.

Magnetophone

1978
Personal stereo
German–Brazilian Andreas Pavel's Stereobelt of 1972 plays cassette tapes through ear phones on a tiny, portable, battery-driven player. In 1978, Sony launches a portable music player—the Sony Walkman.

SONY
WALKMAN

Sony Walkman

1898
Telegraphone
Danish engineer Valdemar Poulsen's telegraphone is the first device to record and play back sound magnetically, using a wire wrapped around a cylinder to record the varying magnetic fields created by sound vibrations.

Telegraphone

1948
Record player
Long-playing vinyl records, playing at 33 and 45 revolutions per minute (rpm) instead of 78 rpm, are introduced. They give much longer playing times and the sound quality is better.

1950s record player

1982
Compact disks
CDs digitally store large amounts of sound data that can be read by a laser. They quickly replace large, easily scratched vinyl records.

1999
MP3 players
These use digital recordings stored as computer data, enabling music to be downloaded, or swapped from computer to personal player instantly.

Ipod *nano*

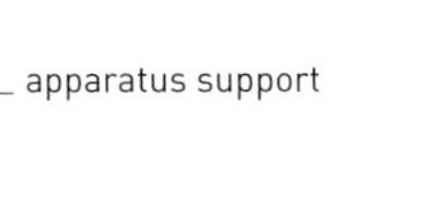
original sound wave rises and falls continuously
analog sound wave follows original sound wave
original sound wave
analog sound wave

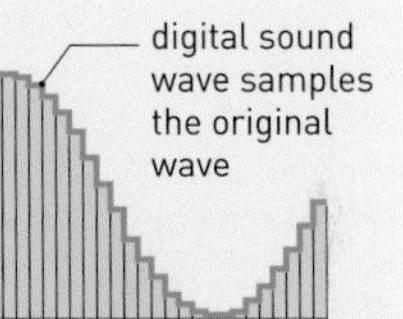
digital sound wave samples the original wave
digital sound wave

SOUND WAVES

Sound is created by vibrations in the air that create a rise and fall of air pressure. It can be plotted as a simple wave. An analog sound wave follows this original sound wave form exactly, with the intensity of the signal going up and down continuously with the wave. Digital sound waves are made by repeatedly taking samples of the original wave at a very high rate and converting them into a series of numbers. They are not continuous and have a "staircase" shape.

apparatus support
recording stylus, or needle
diaphragm turns sound waves into physical vibrations (and vice versa)
wax cylinder with grooves for recording
cylinder shaft
sound recording box
TRADE
Thomas A. Edison
MARK
Edison
cylinder shaft crank
crank handle

Edison Fireside phonograph, 1909
By 1909, Edison phonographs such as this one were a feature of many homes, and there was a wide range of professional wax-cylinder recordings that could be inserted in the machine and played by turning the handle. Unfortunately, wax-cylinders could only be played a few times before the playback quality started to deteriorate.

8 MILES PER HOUR
THE **SPEED** OF THE **SIEMENS LOCOMOTIVE**

This photograph shows people traveling on carriages pulled by the Werner-Siemens locomotive, which was demonstrated at the Berlin Trade Fair in 1879. This locomotive was the first to be powered by electricity.

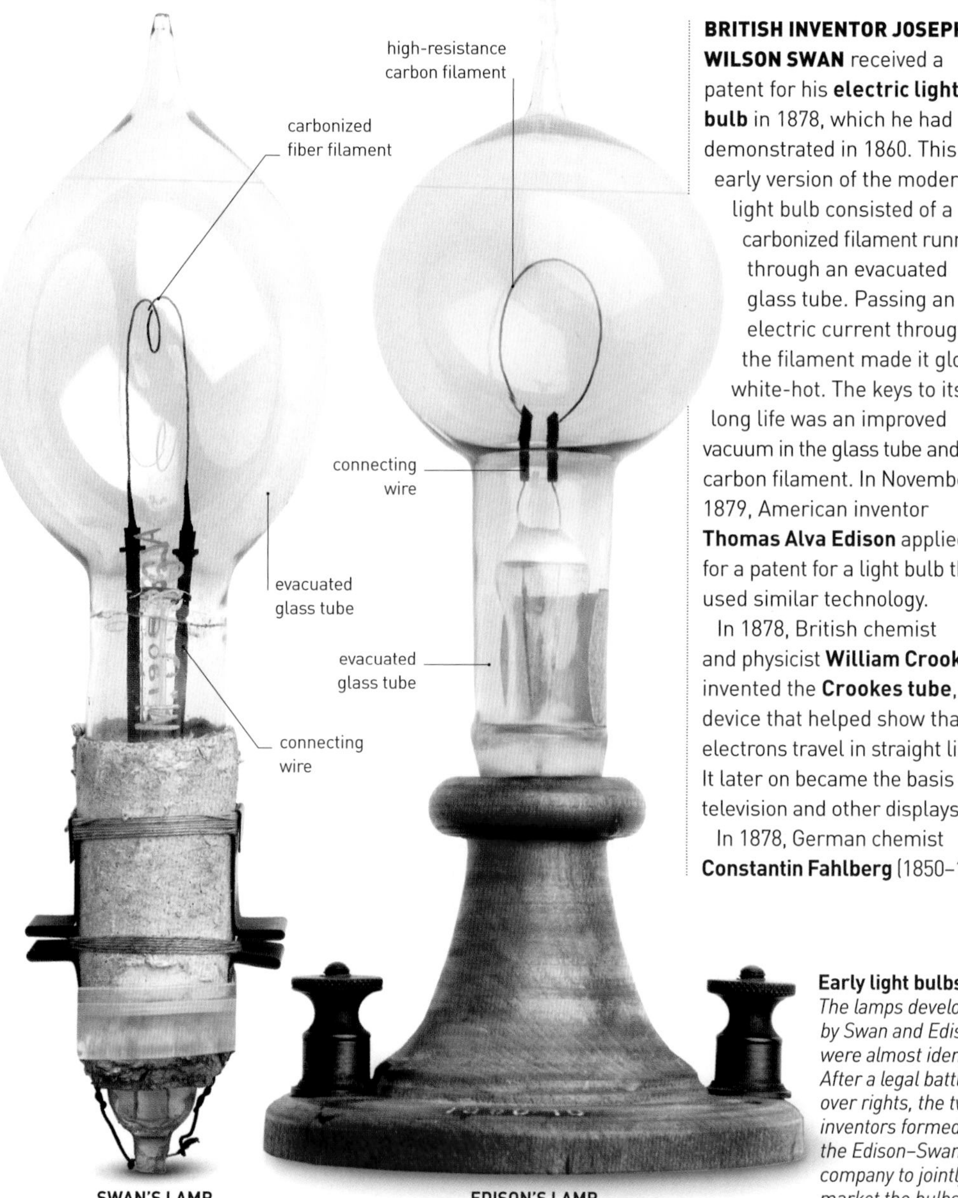
high-resistance carbon filament
carbonized fiber filament
connecting wire
evacuated glass tube
evacuated glass tube
connecting wire
SWAN'S LAMP
EDISON'S LAMP

Early light bulbs
The lamps developed by Swan and Edison were almost identical. After a legal battle over rights, the two inventors formed the Edison-Swan company to jointly market the bulbs.

BRITISH INVENTOR JOSEPH WILSON SWAN received a patent for his **electric light bulb** in 1878, which he had demonstrated in 1860. This early version of the modern light bulb consisted of a carbonized filament running through an evacuated glass tube. Passing an electric current through the filament made it glow white-hot. The keys to its long life was an improved vacuum in the glass tube and the carbon filament. In November 1879, American inventor **Thomas Alva Edison** applied for a patent for a light bulb that used similar technology.

In 1878, British chemist and physicist **William Crookes** invented the **Crookes tube**, a device that helped show that electrons travel in straight lines. It later on became the basis of television and other displays.

In 1878, German chemist **Constantin Fahlberg** (1850–1910) accidentally discovered a natural sweetener, later called **saccharin**, while working with coal tar in the US. Saccharin is about 200 times sweeter than sugar.

German engineer **Karl Benz** (1844–1929) developed a one-cylinder **two-stroke gas engine**, demonstrated for the first time on December 31, 1879.

In his 1879 book, *Cartography of Russian Soils*, Russian geologist **Vasily Dokuchaev** (1846–1905) introduced the concept of **pedology**, the scientific study of soil.

American physicist **Edwin Hall** (1855–1938) discovered that a magnetic field created at a right angle to the flow of electric current would create a voltage difference across the conductor.

cathode
anode
object

Crookes tube
In this Crookes tube, electrons stream past a cross-shaped object to cast a shadow on the fluorescent glass beyond.

Now known as the **Hall effect**, this phenomenon is important in semiconductor technology and magnetic sensors.

Austrian physicist **Josef Stefan** (1835–1893) formulated the equation now known as the **Stefan–Boltzmann law**, which concerned the calculation of radiation emitted from black bodies—surfaces that absorb all the electromagnetic radiation that strikes them. In 1884, Stefan's colleague Ludwig Boltzmann (1844–1906) explained the law using thermodynamics.

American scientist **Albert Abraham Michelson** (1852–1931) measured the **speed of light**

> “WE WILL MAKE **ELECTRICITY** SO **CHEAP** THAT ONLY THE **RICH WILL BURN CANDLES.**”
>
> **Thomas Edison, American inventor,** 1879

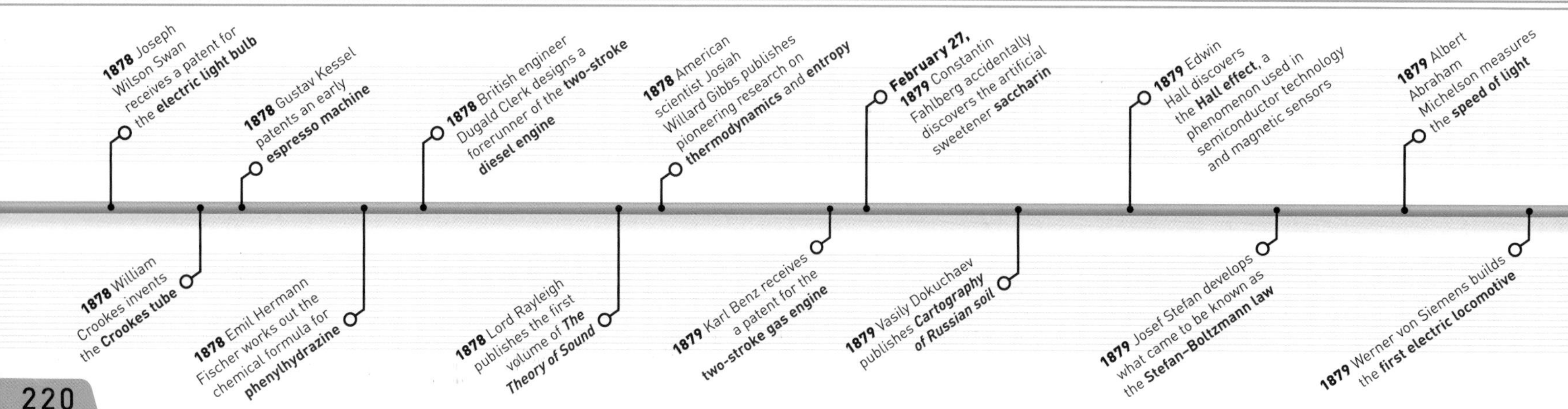
1878 Joseph Wilson Swan receives a patent for the **electric light bulb**
1878 William Crookes invents the **Crookes tube**
1878 Gustav Kessel patents an early **espresso machine**
1878 Emil Hermann Fischer works out the chemical formula for **phenylhydrazine**
1878 British engineer Dugald Clerk designs a forerunner of the **two-stroke diesel engine**
1878 Lord Rayleigh publishes the first volume of ***The Theory of Sound***
1878 American scientist Josiah Willard Gibbs publishes pioneering research on **thermodynamics** and **entropy**
1879 Karl Benz receives a patent for the **two-stroke gas engine**
February 27, 1879 Constantin Fahlberg accidentally discovers the artificial sweetener **saccharin**
1879 Vasily Dokuchaev publishes ***Cartography of Russian soil***
1879 Edwin Hall discovers the **Hall effect**, a phenomenon used in semiconductor technology and magnetic sensors
1879 Josef Stefan develops what came to be known as the **Stefan–Boltzmann law**
1879 Albert Abraham Michelson measures the **speed of light**
1879 Werner von Siemens builds the **first electric locomotive**

> **"** ...THE **ANTHRAX VACCINE** THAT **FIRST SPREAD** THROUGH THE PUBLIC MIND **FAITH IN** THE **SCIENCE OF MICROBES. "**

Emile Duclaux, French chemist and microbiologist, from *Pasteur: The History of a Mind*, 1896

This engraving depicts French bacteriologist Louis Pasteur vaccinating sheep against anthrax at Pouilly-le-Fort, France, in 1881.

in air to be 186,327 miles per second (299,864 kmps). This estimate matched the prediction of British physicist James Clerk Maxwell (see 1872–73).

German engineer **Werner von Siemens** (1816–92) demonstrated the **first electric locomotive** using an external power source at the 1879 Berlin trade fair.

THOMAS ALVA EDISON
(1847–1931)

American inventor Thomas Edison is credited with many inventions, particularly in the field of telecommunications. He filed more than a thousand patents in the US, and others around the world. Edison applied the idea of mass production and teamwork to science and developed the world's first industrial research laboratory—his greatest invention.

IN 1880, BRITISH LOGICIAN AND PHILOSOPHER John Venn (1834–1923) developed the concept of the **Venn diagram**, which represents sets of things as circles, with overlapping circles indicating the subsets they have in common. Venn wrote about this concept in *Symbolic Logic*, which was published in 1881.

In February 1880, **Thomas Alva Edison** rediscovered a phenomenon that had previously been observed by others. He noticed that electricity would flow from a hot filament to a cool metal plate in an evacuated bulb. Edison patented this concept, which came to be known as the **Edison effect**. The electricity could flow only one way, so the setup acted like a valve to control the flow of current in the same way that a valve in a pipeline controls the flow of water. This concept became the basis of the valves used to amplify electrical signals in television and radio before the invention of the transistor.

The understanding of electricity was also aided by the **discovery of the piezoelectric effect** by French scientists **Pierre Curie** (1859–1906) and **Paul-Jacques Curie** (1856–1941). They discovered that voltage can be produced by applying pressure to a suitable material. Conversely, applying an electric potential to crystals can make them vibrate at a very precise frequency. This effect has many applications, including driving the vibrating crystals in quartz clocks and watches.

Known as the father of modern seismology, British geologist **John Milne** (1850–1913), was teaching science and engineering in Japan when he became interested in earthquakes. In 1880 he was instrumental **in inventing the seismograph**—an instrument used to measure earthquakes. The instrument, developed in collaboration with British engineer Thomas Gray (1850–1908) and properly known as the Milne-Gray instrument, was based on a horizontal pendulum design.

metal wire bears pendulum

metal coil

roll of paper

Seismograph
This seismograph—designed in conjunction with John Milne—was made by James White in 1885. It records earthquake vibrations on a roll of paper.

The year 1881 saw two key advances in the application of electricity. In May, the **first electric tramway** was opened in a suburb in Berlin, Germany. Later, Godalming, UK, became the first town to have its **streets lit by electricity**.

In September 1881, German gynecologist **Ferdinand Adolf Kehrer** (1837–1914) carried out the **first modern cesarian section operation**. Both mother and baby survived.

French microbiologist **Louis Pasteur made a vaccine for anthrax** by using the oxidizing agent potassium dichromate to weaken its bacteria. He began using British surgeon Edward Jenner's term "vaccine" to refer to all such artificially weakened disease organisms. Jenner had coined the term in reference to smallpox (see 1796).

German scientist **Paul Ehrlich** (1854–1915)—whose cousin, pathologist Karl Weigert, (1845–1904) had been the first person to stain bacteria with dyes in the 1870s—found a more effective dye, **methylene blue**. This made it easier to identify and investigate bacteria, and was used by German physician Heinrich Koch to discover the bacteria that causes tuberculosis (see 1882–83).

First electric tramway
Developer of the first electric train, Werner von Siemens also worked on the first electric tram—the Gross-Lichterfelde in Germany.

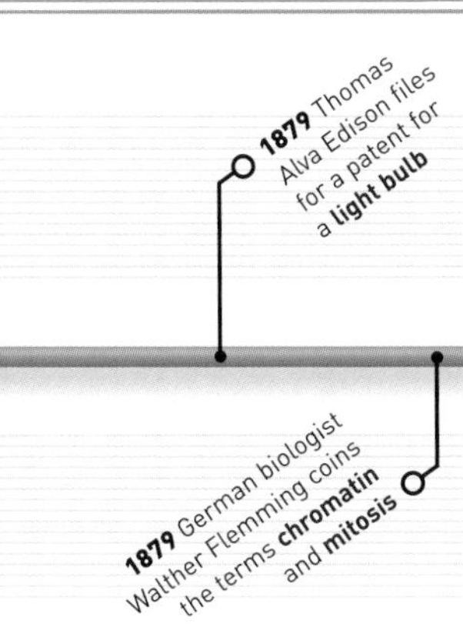

1879 Thomas Alva Edison files for a patent for a **light bulb**

1879 German biologist Walther Flemming coins the terms **chromatin** and **mitosis**

1880 John Milne invents the **seismograph**

February 13, 1880 Thomas Edison rediscovers the **Edison effect**, also known as thermionic emission

1880 Pierre and Paul-Jacques Curie discover **piezoelectricity**—electricity generated due to pressure

1881 John Venn publishes *Symbolic Logic*, introducing the concept of **Venn diagrams**

May 16, 1881 The world's **first electric tramway** opens in Germany

1881 Alfred Russel Wallace proposes that the **Cambrian period** began about 28 million years ago

1881 Louis Pasteur develops a vaccine for **anthrax**

September 1881 Godalming, UK, becomes the first town to have its streets illuminated by **electric light**

September 25, 1881 Ferdinand Kehrer performs the first modern **cesarean section**

1881 Paul Ehrlich uses **methylene blue** to stain bacteria

1,595 FEET THE SPAN OF THE BROOKLYN BRIDGE

The Brooklyn Bridge spans the East River between Brooklyn and the island of Manhattan in New York City. It was the longest bridge in the world when it was built.

BUILDING ON the work of British surgeon Joseph Lister and French microbiologist Louis Pasteur (see 1870–71), German physician **Robert Koch** (1843–1910) **isolated the organism that causes tuberculosis** (TB) in 1882. He discovered that TB was transmitted in water droplets and could quickly spread in overcrowded slums.

Another medical landmark was achieved when Russian biologist **Élie Metchnikoff** (1845–1916) discovered **phagocytosis**—the process used by the immune system to remove bacterial invaders. In phagocytosis, one cell engulfs another and, in effect, eats it. This discovery led to an **improved understanding of the immune system**.

> “I CONCLUDE THAT THESE **TUBERCLE BACILLI** OCCUR IN ALL TUBERCULOUS DISORDERS, AND THAT THEY ARE **DISTINGUISHABLE** FROM ALL OTHER **MICROORGANISMS.**”
>
> **Robert Koch, German physician,** from *The Etiology of Tuberculosis,* 1882

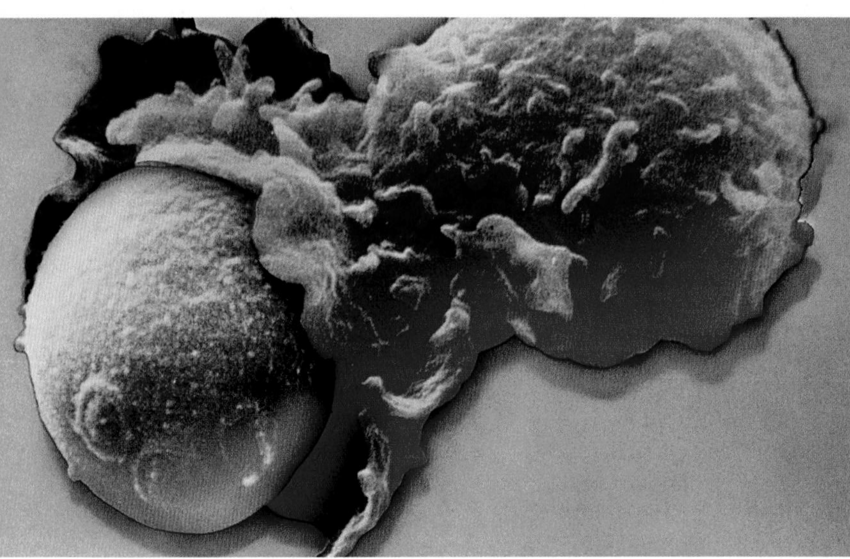
Phagocytosis
In this color-enhanced micrograph, a lymphocyte (white blood cell) engulfs a yeast cell in the process known as phagocytosis.

German botanist **Eduard Strasburger** (1844–1912) also contributed to the understanding of how cells work. He coined the terms **cytoplasm** for the jellylike outer region of a cell, and **nucleoplasm** for the compact material of the cell nucleus.

In 1882, Italian volcanologist and meteorologist **Luigi Palmieri** (1807–96) made the **first observation of the element helium** on Earth by conducting spectral analysis (see panel, opposite) of lava during an eruption at Mount Vesuvius, Italy. Previously, the element had only been identified by analysis of light from the Sun.

The following year, English polymath **Francis Galton** came up with the **controversial concept of eugenics**. His theory aimed to improve the human race by selective breeding, and was later misused by the Nazis as an excuse for their attempted extermination of the Jews. The scientific basis for eugenics drew in part on the germ line theory developed by **August Weismann** (1834–1914), which stated that characteristics were only passed on by egg and sperm cells, and were not affected by other cells of the body (somatic cells). So, for example, a bodybuilder who develops muscles through exercise will not pass on the muscles to his children. This **marked the end of Lamarckism**, according to which acquired characteristics could be passed on to children (see 1809).

Quagga
A close relative of the zebra, the quagga had distinctive stripes only on the front of its body.

A more practical contribution to human society was made by **Osborne Reynolds** (1842–1912), an Irish-born engineer who studied fluids. In 1883, he came up with what is now known as the **Reynolds number**, which characterizes the way fluids flow. Today, Reynold's work is important in the designing of pipes to carry different fluids, and in shipbuilding, where the behavior of full-size vessels must be estimated from models tested in water tanks.

On May 24, 1883, the longest bridge of its time, the **Brooklyn Bridge**, was opened in New York City. It was the world's first steel suspension bridge.

A female **quagga**, the last surviving member of its species, died in a zoo in Amsterdam in 1883. This southern African species had been extinct in the wild since the late 1870s.

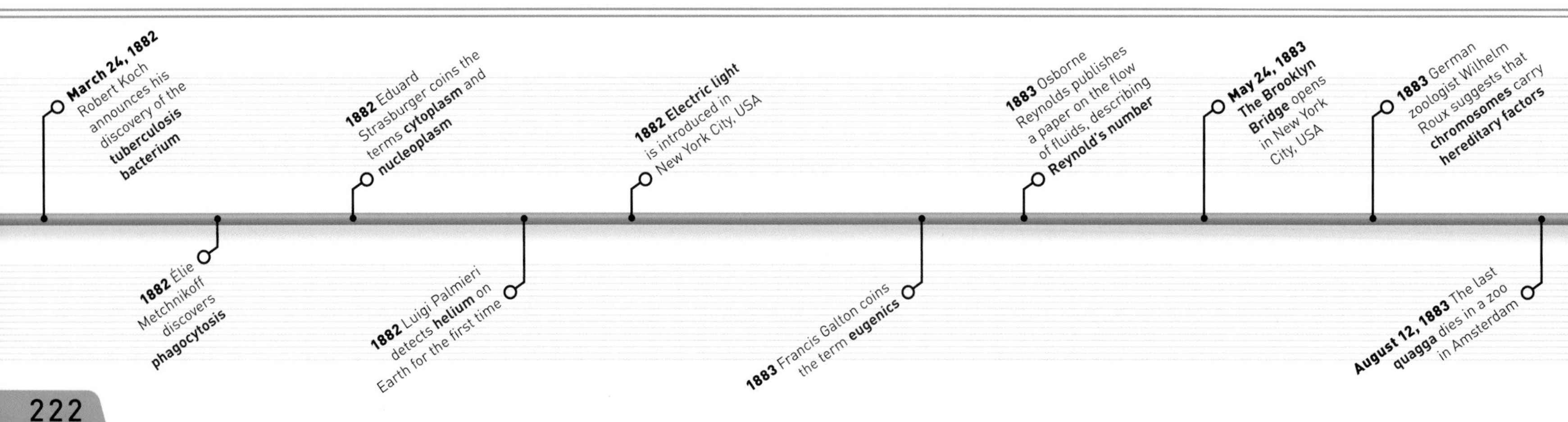
March 24, 1882 Robert Koch announces his discovery of the **tuberculosis bacterium**
1882 Élie Metchnikoff discovers **phagocytosis**
1882 Eduard Strasburger coins the terms **cytoplasm** and **nucleoplasm**
1882 Luigi Palmieri detects **helium** on Earth for the first time
1882 Electric light is introduced in New York City, USA
1883 Francis Galton coins the term **eugenics**
1883 Osborne Reynolds publishes a paper on the flow of fluids, describing **Reynold's number**
May 24, 1883 The Brooklyn Bridge opens in New York City, USA
1883 German zoologist Wilhelm Roux suggests that **chromosomes** carry **hereditary factors**
August 12, 1883 The last **quagga** dies in a zoo in Amsterdam

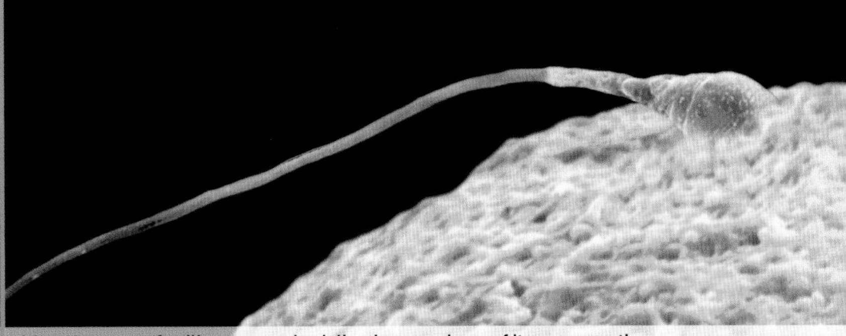

A human sperm fertilizes an egg by delivering a package of its own genetic material (germ line DNA) to combine with the genes of the egg.

> **“ NEW CELL NUCLEI** CAN ONLY ARISE FROM THE **DIVISION** OF **OTHER NUCLEI. ”**
>
> **Eduard Strasburger, Polish–German botanist,** from *Über Zellbildung und Zelltheilung* (*On Cell Formation and Cell Division*), 1880

FRENCH CHEMIST HILAIRE DE CHARDONNET (1839–1924) received a patent for **artificial silk** in 1884. He discovered the substance accidentally in 1878, when he knocked over a bottle of nitrocellulose—a highly flammable compound. When he started to clean it up, strands of nitrocellulose stuck to his cleaning cloth in thin, silklike fibers. It was not until the 20th century that this substance was developed in the form of the material known as rayon.

The 19th-century understanding of disease developed further in 1884, when German physicist **Friedrich Löffler** (1852–1915), Robert Koch's colleague, isolated the diptheria-causing bacterium *Corynebacterium diphtheriae*.

Koch and Löffler also formulated Koch's postulates in 1884, which set out the criteria for establishing whether an organism is responsible for a disease. Koch published their findings in 1890, and in doing so he dramatically refined the science of microbiology.

Starting in 1884, **Eduard Strasburger**, German zoologist **Wilhelm Hertwig** (1849–1922), and Swiss anatomist **Rudolf von Kölliker** (1817–1905) each separately identified the cell nucleus as the **origin of heredity**. Hertwig stated that, from the biological point of view, sex is merely a union of two cells (strictly speaking, two nuclei).

Austrian ophthalmologist **Carl Koller** (1857–1944) ushered in the era of **local anesthesia** when he used cocaine as a surface anesthetic in an eye operation in 1884 (see 1846). While looking into whether cocaine could be used to wean patients off morphine—at the request of his colleague at Vienna General Hospital, **Sigmund Freud**—Koller discovered the tissue-numbing properties of cocaine.

On July 6, 1885, French chemist **Louis Pasteur** used the **rabies vaccine** for the first time on a 9-year-old boy who had been bitten by a rabid dog. The success of the treatment paved the way for the widespread use of vaccines.

On August 20, 1885, German astronomer **Ernst Hartwig** (1851–1923) observed a **bright new star** in the Andromeda nebula. A belief that this object was similar to the novas seen in the Milky Way encouraged the idea that the nebula, too, was part of the Milky Way. In the 20th century, it was discovered that the Andromeda nebula is a galaxy (see 1924), far beyond the Milky Way, and that Hartwig's star is a supernova, much brighter than a nova.

In a key discovery in the fields of astronomy and atomic physics, Swiss mathematician **Johann Balmer** (1825–98), developed a mathematical formula to describe the positions of **lines in the spectrum of hydrogen**—the Balmer series. Using his formula, Balmer predicted the wavelengths of lines that were discovered later.

In the field of psychology, German psychologist **Hermann Ebbinghaus** (1850–1909) pioneered the experimental study of memory and developed the concept of the **Forgetting Curve**. He published *Memory: A Contribution to Experimental Psychology* in 1885.

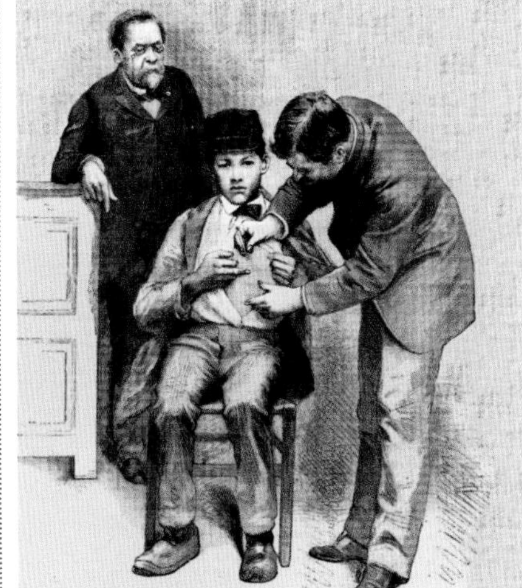

Rabies vaccine
This 1885 engraving shows Louis Pasteur watching as his assistant inoculates Joseph Meister, a shepherd boy who had been bitten by a rabid dog.

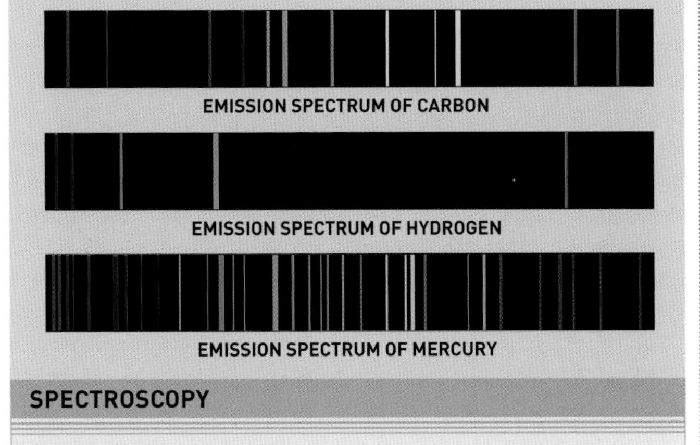
EMISSION SPECTRUM OF CARBON
EMISSION SPECTRUM OF HYDROGEN
EMISSION SPECTRUM OF MERCURY

SPECTROSCOPY

When hot, each chemical element produces a distinctive set of bright spectral lines, like a barcode, that can identify the element. Cold gases absorb light in exactly the same wavelengths, producing dark spectral lines. Analyzing spectra makes it possible to determine the composition of substances in the laboratory and also to measure the composition of stars.

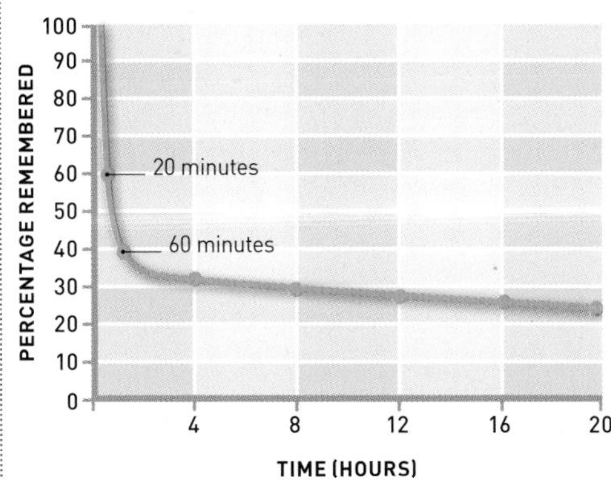
PERCENTAGE REMEMBERED
100
90
80
70
60
50
40
30
20
10
0
20 minutes
60 minutes
4
8
12
16
20
TIME (HOURS)

Forgetting Curve
Ebbinghaus gave people a list of nonsense 3-letter words and measured how long they could remember them. This gave him the data for his Forgetting Curve.

1884 Hilaire de Chardonnet patents **artificial silk**

1884 Friedrich Löffler isolates the **diptheria bacterium**

1884 The meridian through the Royal Greenwich Observatory in London, UK, is established as the **Prime Meridian**

1884 Karl Koller uses cocaine as a **local anaesthetic** for the first time

1884 Austrian physicist Ludwig Boltzmann refines the **Stefan–Boltzmann law**

January 4, 1885 American physician William W. Grant performs the first **successful appendix removal**

July 6, 1885 Louis Pasteur uses the **rabies vaccine** for the first time

August 20, 1885 Ernst Hartwig observes a **supernova** in the Andromeda galaxy

1885 Johann Balmer develops a formula representing the wavelengths of hydrogen's spectral lines

1885 German mechanical engineer Karl Benz manufactures the **first car**

1885 Theodor Escherich discovers the ***Escherichia coli*** bacteria

1885 Hermann Ebbinghaus publishes ***Memory: A Contribution to Experimental Psychology***

1886

“…MAESTRO MAXWELL WAS RIGHT… THESE MYSTERIOUS **ELECTROMAGNETIC WAVES** THAT WE CANNOT SEE WITH THE NAKED EYE. **BUT THEY ARE THERE.”**

Heinrich Hertz, German physicist, 1887

IN 1886, AMERICAN CHEMIST CHARLES MARTIN HALL (1863–1914) and French scientist **Paul-Louis-Toussaint Héroult** (1863–1914) independently developed a technique for converting alumina—the powdery white oxide of aluminum—into **aluminum** using electrolysis.

The same year, German-born American inventor **Ottmar Mergenthaler** (1854–99) revolutionized the world of publishing when he developed the **linotype line casting machine**. This device could set an entire line of type at a time, reducing the costs and production time of printed material. He was dubbed the “second Gutenberg” (see 1450) for his invention.

The linotype and other machines would soon run on alternating current or AC electricty (see panel, below), thanks to American physicist **William Stanley Jr** (1858–1916). He demonstrated the **first full AC power generating system** on March 20, using it to light the town of Great Barrington, Massachusetts.

Meanwhile, German physicist Heinrich Hertz confirmed the existence of **long-wave electromagnetic radiation** (see pp.234–35)—a kind of invisible light now referred to as radio waves—that had been predicted by Scottish physicist James Clerk Maxwell in 1867.

American physicist **Henry Augustus Rowland** (1848–1901) analysed sunlight using **diffraction gratings**—glass plates or mirrors with a number of parallel lines etched onto the surface to diffract light—that he had made himself.

ALTERNATING CURRENT

When a loop of wire is rotated between the poles of a magnet, alternating electrical current is generated. The current flowing in the wire reverses repeatedly (alternates) as it turns. Main, or three-phase electricity, is generated using three coils oriented at 120 degrees to each other. Domestic power supplies most commonly alternate 50 or 60 times a second.

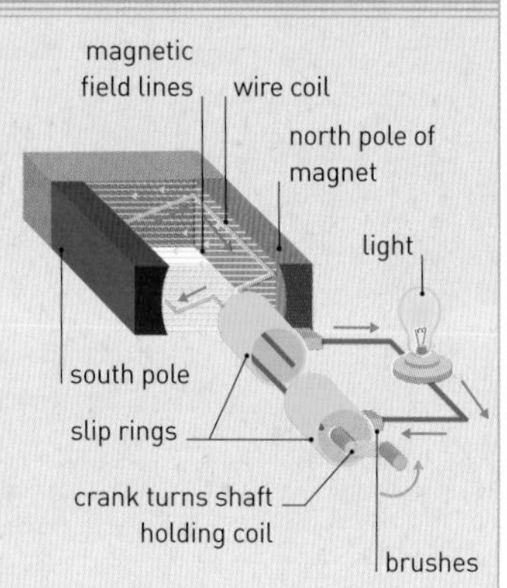
magnetic field lines
wire coil
north pole of magnet
light
south pole
slip rings
crank turns shaft holding coil
brushes

1887–88

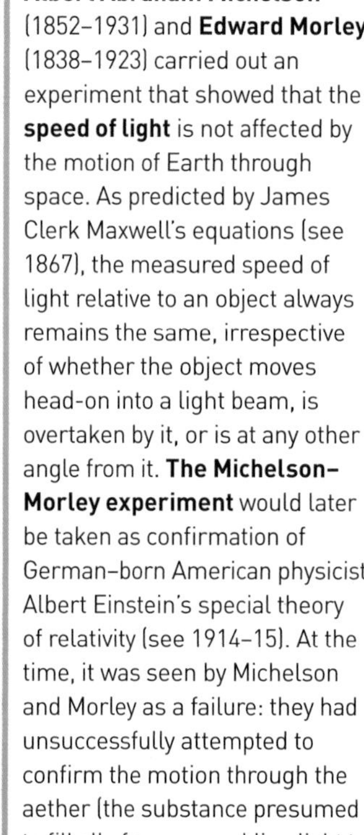

Lick Observatory, on Mount Hamilton, near San Jose, in California, was the first permanently occupied mountain-top observatory in the world.

THE STUDY OF LIGHT continued in 1887 as American scientists **Albert Abraham Michelson** (1852–1931) and **Edward Morley** (1838–1923) carried out an experiment that showed that the **speed of light** is not affected by the motion of Earth through space. As predicted by James Clerk Maxwell’s equations (see 1867), the measured speed of light relative to an object always remains the same, irrespective of whether the object moves head-on into a light beam, is overtaken by it, or is at any other angle from it. **The Michelson–Morley experiment** would later be taken as confirmation of German-born American physicist Albert Einstein’s special theory of relativity (see 1914–15). At the time, it was seen by Michelson and Morley as a failure: they had unsuccessfully attempted to confirm the motion through the aether (the substance presumed to fill all of space, enabling light to travel through a vacuum).

Heinrich Hertz’s work with radio waves led him to discover the **photoelectric effect**. He observed that a transmitter’s radio waves generated sparks between two small metal spheres that were almost touching. We now know that this occurs because electromagnetic radiation knocks electrons out of a metal surface. Hertz published his observations of the phenomenon in the journal *Annalen der Physik* (*Annals of Physics*) in 1887.

The same year, American inventor **Herman Hollerith** (1860–1929) received a patent for his **punched card tabulating machine**, which helped tabulate census statistics. This machine was a forerunner of the electronic computer.

On the last day of the year, a refracting telescope with a 36-in (91-cm) diameter lens was completed at Lick Observatory in California. It was first used on January 3, 1888. At the time, it was the **biggest in the world**.

HEINRICH HERTZ (1857–94)

German physicist Heinrich Hertz is best known for the series of experiments he carried out to test James Clerk Maxwell’s theories of electromagnetism. These included transmission and detection of radio waves and proving that light is a form of electromagnetic vibration. The unit of frequency—cycles per second—is called the hertz (Hz) in his honor.

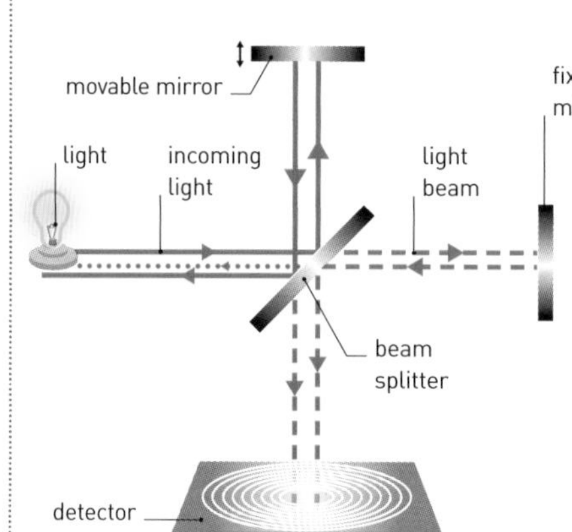
movable mirror
fixed mirror
light
incoming light
light beam
beam splitter
detector

Michelson–Morley interferometer
Michelson and Morley built a device consisting of a light source, two mirrors, and a detector. They used it to study interference between beams of light moving with Earth and at right angles to Earth’s motion.

Charles Martin Hall and Paul-Louis-Toussaint Héroult independently discover how to convert alumina into **aluminum**

Ottmar Mergenthaler invents the **linotype line casting machine**

March 20 William Stanley, Jr. builds the first complete system for **transmission of high-voltage alternating current**

July 3 American newspaper *New York Tribune* becomes the **first to use the linotype**

Austro-German psychiatrist Richard vin Krafft-Ebing’s ***Sexual Psychopathy: A Clinical-Forensic Study*** is published

Henry Rowland studies sunlight using **diffraction gratings**

November 11 Heinrich Hertz confirms the existence of **electromagnetic waves** predicted by Maxwell

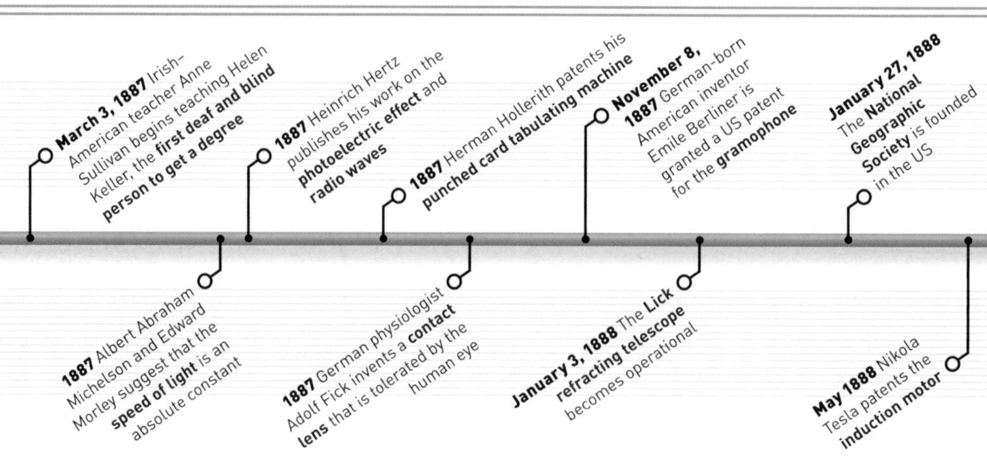
March 3, 1887 Irish-American teacher Anne Sullivan begins teaching Helen Keller, the first deaf and blind person to get a degree
1887 Albert Abraham Michelson and Edward Morley suggest that the speed of light is an absolute constant
1887 Heinrich Hertz publishes his work on the photoelectric effect and radio waves
1887 German physiologist Adolf Fick invents a contact lens that is tolerated by the human eye
1887 Herman Hollerith patents his punched card tabulating machine
January 3, 1888 The Lick refracting telescope becomes operational
November 8, 1887 German-born American inventor Emile Berliner is granted a US patent for the gramophone
May 1888 Nikola Tesla patents the induction motor
January 27, 1888 The National Geographic Society is founded in the US

36 INCHES

THE SIZE OF THE **LICK TELESCOPE LENS,** THE **LARGEST REFRACTOR** AT THE TIME

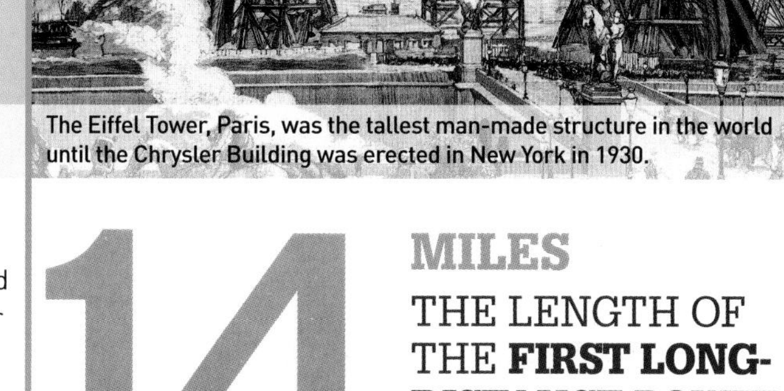
The Eiffel Tower, Paris, was the tallest man-made structure in the world until the Chrysler Building was erected in New York in 1930.

Applications of alternating current electricity were being developed by Serbian–American inventor **Nikola Tesla** (1856–1943). In 1888, he patented the **induction motor**. This is a two-phase machine that uses two alternating currents to produce the rotating magnetic field that makes the rotor turn. The patent was bought by the Westinghouse Electric Company and used to develop motors, which were widely used in industry and household appliances around the world.

Scottish inventor **John Boyd Dunlop** (1840–1921) developed a **pneumatic bicycle tire** (a tire filled with air) in 1887 and patented it in England in1888. The patent was later declared invalid because the priniciple had been patented earlier by another Scotsman, Robert Thomson (1822–73), in France in 1846 and the US in 1847. However, Dunlop's were the **first practical pneumatic tires**.

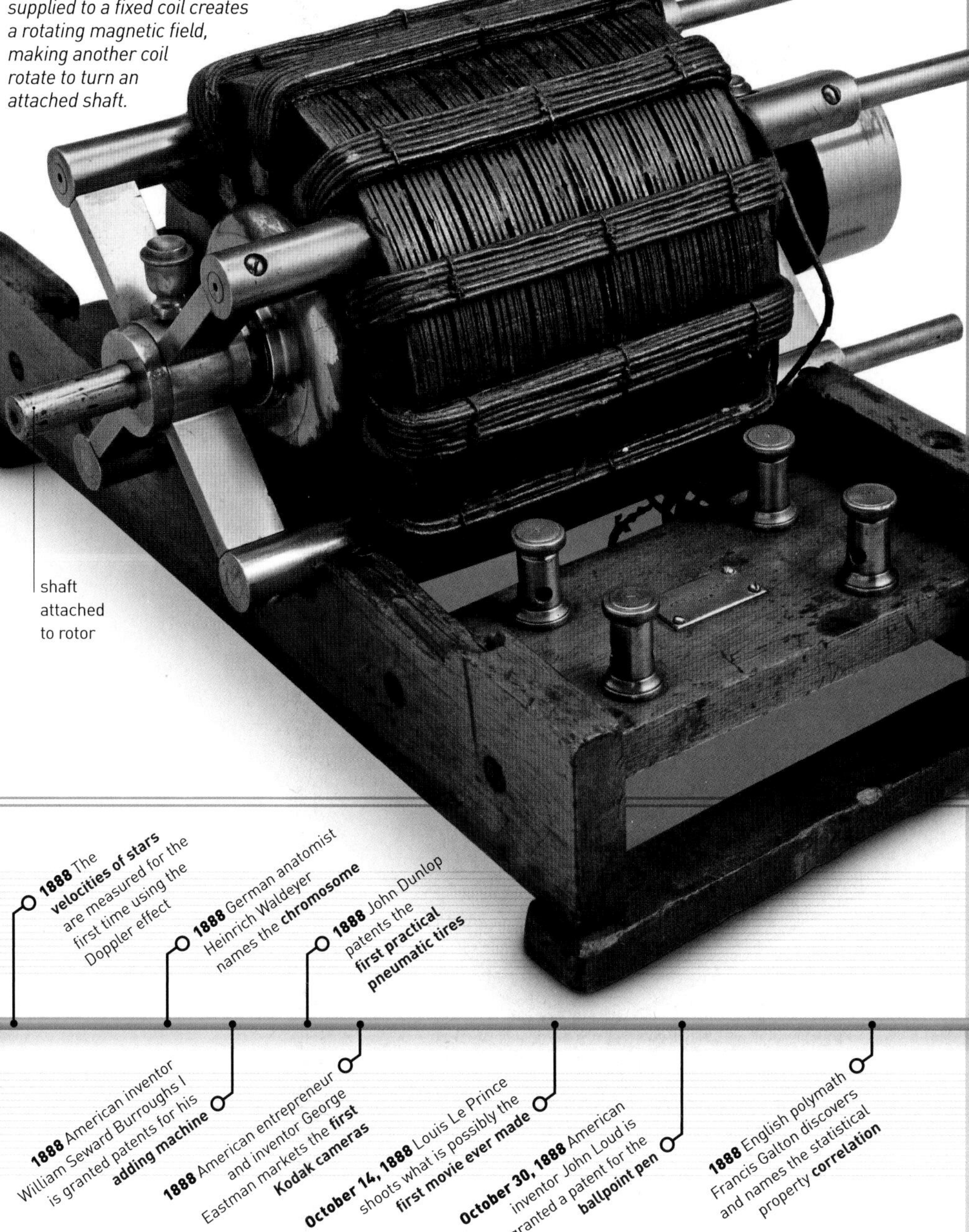
fixed coil (stator)
shaft attached to rotor

Tesla's induction motor
In this motor, alternating current supplied to a fixed coil creates a rotating magnetic field, making another coil rotate to turn an attached shaft.

14 MILES

THE LENGTH OF THE **FIRST LONG-DISTANCE POWER TRANSMISSION LINE** IN THE **US**

GERMAN PHYSIOLOGIST Oskar **Minkowski** (1858–1931) and German physician **Joseph von Mering** (1849–1908) showed that the pancreas produces a substance (later identified as insulin) that regulates sugar glucose in the body, and that **diabetes** occurs when this organ malfunctions.

English chemist **Frederick Abel** (1827–1902) and Scottish chemist **James Dewar** (1842–1923) patented **cordite**—an explosive designed to burn vigorously and produce high-pressure gases that could propel bullets and shells.

Designed by French engineer Gustave Eiffel, the **Eiffel Tower** in Paris, France, was opened on March 31. At 984 ft (300 m), the tower was the tallest building in the world at the time.

On June 3, **the first long-distance power transmission line in the US** was completed. It was installed in Oregon, between a string of lights in Portland and a generator at Willamette Falls.

Russian physiologist **Ivan Petrovich Pavlov** (1849–1936) began studying **conditioning in dogs** in 1889. He had noticed that dogs would begin salivating when they saw the lab technician who fed them. Pavlov began to signal their feeding with the sound of a metronome; soon, the dogs began to salivate each time they heard the metronome (see 1907).

Irish physicist **George FitzGerald** (1851–1901) published a paper suggesting that if all moving objects shrank in the direction of their motion, the results of the Michelson–Morley experiment could be explained. This speculation was based on the idea that electromagnetic forces would squeeze the moving objects. Dutch physicist **Hendrik Lorentz** (1853–1928) came up with a similar idea as well; this shrinking emerges naturally from special theory of relativity (see 1905).

- **1888** The **velocities of stars** are measured for the first time using the Doppler effect
- **1888** German anatomist Heinrich Waldeyer names the **chromosome**
- **1888** American inventor William Seward Burroughs I is granted patents for his **adding machine**
- **1888** John Dunlop patents the **first practical pneumatic tires**
- **1888** American entrepreneur and inventor George Eastman markets the **first Kodak cameras**
- **October 14, 1888** Louis Le Prince shoots what is possibly the **first movie ever made**
- **October 30, 1888** American inventor John Loud is granted a patent for the **ballpoint pen**
- **1888** English polymath Francis Galton discovers and names the statistical property **correlation**
- Oskar Minkowski and Joseph von Mering discover the role of the pancreas in **diabetes**
- **March 12** American inventor Almon Strowger files a patent for an **automatic telephone exchange**
- **March 31** The **Eiffel Tower** is opened to the public
- **June 3 US's first long-distance power transmission line** is completed
- British naturalist Alfred Russel Wallace publishes his book on **natural selection**—*Darwinism*
- George FitzGerald publishes ***The Ether and the Earth's Atmosphere***
- Frederick Abel and James Dewar patent **cordite**
- Ivan Pavlov begins his studies of **conditioning in dogs**

1890

164 FEET THE DISTANCE TRAVELED BY THE ADER ÉOLE

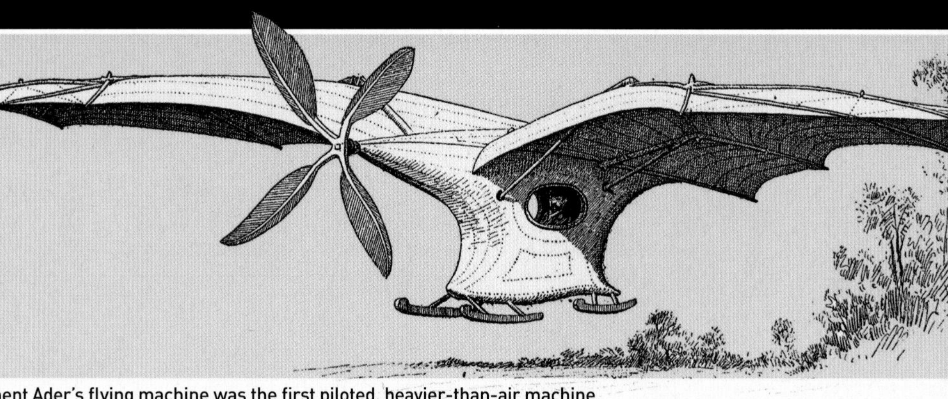

Clément Ader's flying machine was the first piloted, heavier-than-air machine to take off, literally under its own steam, on October 9, 1890.

ON OCTOBER 1, THE US CONGRESS PASSED AN ACT that founded the **Yosemite National Park in California**. This brought the park, which had existed since 1872, under federal control.

The process of **meiosis**—a stage in cell division responsible for the production of gametes—had first been described by German biologist Oscar Hertwig (1849–1922) during his study of sea urchin eggs in 1876. The full significance of meiosis in reproduction and inheritance was appreciated only with the work of German biologist **August Weismann** (1834–1914) in 1890. He realized that two cell divisions are necessary to transform one diploid cell (with two sets of chromosomes) into four haploid cells (with one set of chromosomes each) to maintain the number of chromosomes (see panel, below).

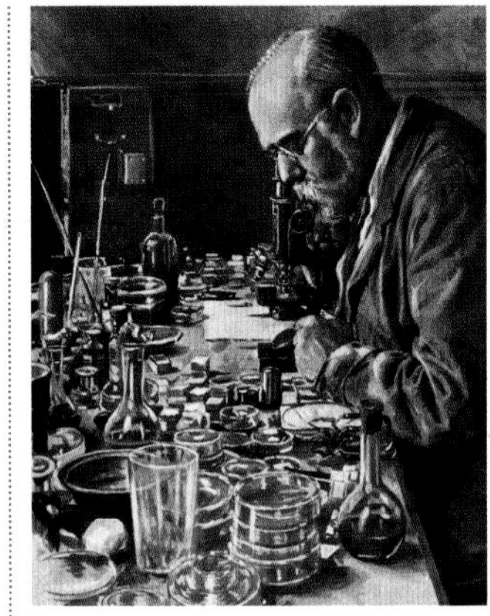

Robert Koch
This lithograph—copied from an 1890s photograph—shows the German bacteriologist working on the Rinderpest virus in his laboratory.

German bacteriologist **Emil von Behring** (1854–1917) and his Japanese counterpart **Kitasato Shibasaburō** (1853–1931) discovered that injecting dead or weakened **disease-causing bacteria**, such as diptheria or tetanus, into an animal caused its blood to **produce antibodies** that gave immunity against the disease. This provided a practical counterpoint to German physician Robert Koch's (1843–1910) theories about the relationship between microbes and disease, which were published in the same year. German bacteriologist Friedrich Löffler (1852–1915) had worked with Koch to develop these ideas.

Arguably, the most important book in the history of psychology, ***The Principles of Psychology***, by American philosopher and psychologist **William James** (1842–1910), was published in 1890. Comprising two volumes and 1,200 pages, the book took James 12 years to write and covered everything in the field known at the time.

> **"** AN AIRPLANE-CARRYING VESSEL IS **INDISPENSABLE...** IT WILL LOOK LIKE A **LANDING FIELD. "**
>
> **Clément Ader, French inventor,** from *L'Aviation Militaire*, 1909

Contrary to popular belief, the Wright brothers did not make the **first manned flight** of a **heavier-than-air flying machine**. This accolade goes to the French inventor **Clément Ader** (1841–1926), who achieved this feat 13 years before the Wrights. His machine—the Ader Éole—had a batlike design and a wingspan of 46 ft (14 m). It was powered by a lightweight, four-cylinder steam engine with 20 horsepower—that weighed only 112 pounds (51 kg). On October 9, the aircraft took off—with Ader on board—and reached a height of 8 in (20 cm). It flew uncontrolled for roughly 165 ft (50 m).

MEIOSIS

Meiosis, the division of cells for sexual reproduction, produces gametes, such as sperm and egg cells. Chromosomes from two parents undergo "recombination," which shuffles the genes to produce a different genetic combination in each gamete. Meiosis produces four genetically unique cells, each with a single set of chromosomes.

set of chromosomes from mother (blue) and father (red); four shown for example

nucleus

chromosomes duplicate, forming chromatid pairs, and some of their genes are mixed up

pairs line up in middle of cell, after nuclear membrane dissolves

microtubules form at the cell's poles and attach to the pairs

microtubules pull chromosome pairs to either end of the cell

cell membrane forms across the cell

each of the two daughter (new) cells has a unique genetic make-up, different from each other and the parent cells

further divisions known as meiosis II result in four separate daughter cells

MEIOSIS II

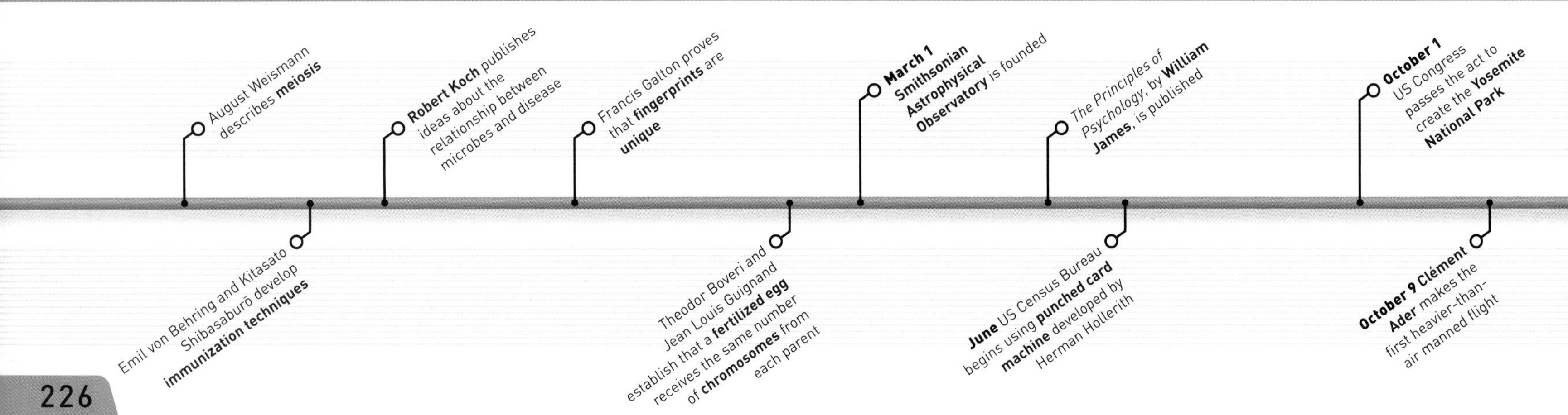
August Weismann describes **meiosis**
Emil von Behring and Kitasato Shibasaburō develop **immunization techniques**
Robert Koch publishes ideas about the relationship between microbes and disease
Francis Galton proves that **fingerprints** are **unique**
Theodor Boveri and Jean Louis Guignand establish that a **fertilized egg** receives the same number of **chromosomes** from each parent
March 1 Smithsonian Astrophysical Observatory is founded
The Principles of Psychology, by **William James**, is published
June US Census Bureau begins using **punched card machine** developed by Herman Hollerith
October 1 US Congress passes the act to create the **Yosemite National Park**
October 9 Clément Ader makes the first heavier-than-air manned flight

86 BILLION THE AVERAGE **NUMBER** OF **NEURONS** IN THE **HUMAN BRAIN**

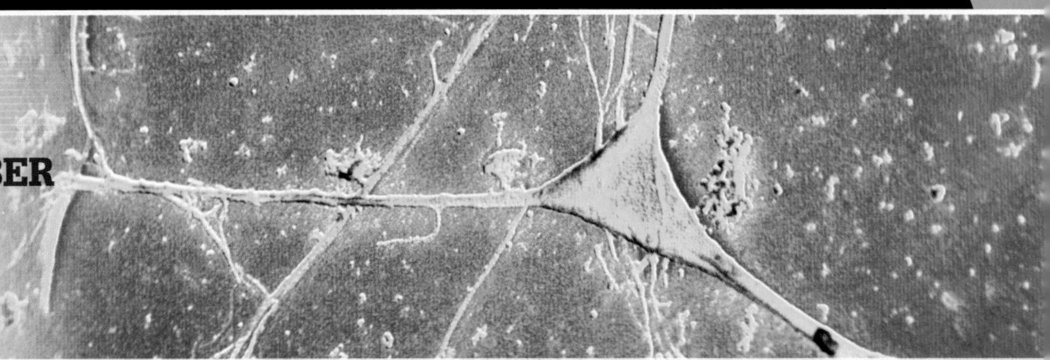

This color-enhanced scanning electron micrograph shows a roughly triangular neuron cell body from the human cerebral cortex—the outer gray matter of the brain.

IN 1891, SOME "KITCHEN SINK" DISCOVERIES were reported in the journal *Nature*. In Germany, **Agnes Pockels** (1862–1935), who had not been able to go to university because she was a woman, had been investigating the effect of different substances on water **surface tension**—a result of observations she had made while washing up. Pockels sent a letter to the British physicist Lord Rayleigh (1842–1919) describing her discoveries and he had her letter translated and published in *Nature*. Pockels went on to publish another 15 scientific papers.

Around this time, the Dutch paleoanthropologist **Eugene Dubois** (1858–1940) made a profound discovery in East Java, Indonesia. He found fragments of what he called "a species in between humans and apes," and gave it the name *Pithecanthropus erectus*, meaning upright ape-man. It is now known as ***Homo erectus***. His interpretation of the finds was controversial, but is now recognized as a step toward understanding human evolution.

Water walker
This insect does not sink in water because its low weight is supported by surface tension—a phenomenon studied by Agnes Pockels.

Panhard car
The 1891 Panhard car ushered in the era of the modern automobile. Panhard cars went on to win several races and established many records.

In the same year, German anatomist **Heinrich Wilhelm Gottfried von Waldeyer-Hartz** (1836–1921) introduced the term **neuron** to describe cells that transmit nerve impulses.

In France, the brothers André (1859–1940) and Édouard Michelin (1859–1940) patented the **removable pneumatic tire**. Their tires were used the same year to win the world's first long-distance cycle race, from Paris to Brest and back.

The **first front-engine, rear-wheel-drive car** was produced by Panhard et Levassor, a French car manufacturing company. This car design, known as **Système Panhard**, became the standard layout for cars for decades. French inventor Amédee Bollée (1844–1917) had, however, used the same layout for steam-powered cars in 1878.

In 1892, French–German engineer **Rudolf Diesel** (1858–1913) received a patent for a forerunner of the engine named after him. The actual **diesel engine** itself was **patented** two years later.

Science was also using new technology in 1891. For the first time, German astronomer **Max Wolf** (1863–1932) used a photographic machine—the **Bruce double-astrograph**, a device for comparing two star fields to see if objects have moved—to find an asteroid. He named the asteroid 323 Brucia, after American philanthropist Catherine Bruce, who paid for the astrograph.

In 1892, French mathematician **Henri Poincaré** (1854–1912) published the first volume of *New Methods of Celestial Mechanics*, which introduced the many techniques used in calculating orbits.

Dutch physicist **Hendrik Lorentz** (1853–1928) applied new ideas on electromagnetism to a theory of the electron as a charged particle. The name "electron" had been proposed by Irish physicist George Johnstone Stoney (1826–1911). Lorentz implied that **electrons are part of atoms**, and that atoms are not indivisible.

British scientist **James Dewar** (1842–1921) invented the **vacuum flask**, or Dewar flask; and French engineer **François Hennebique** (1842–1921) patented the **reinforced concrete technique**, which transformed building technology.

HENRI POINCARÉ
(1844–1912)

Known for not always following through with his many bright ideas, Henri Poincaré did complete the three-volume epic *New Methods of Celestial Mechanics*. In this work, he elaborated on celestial mechanics—a branch of astronomy that deals with orbits and other motions, especially under the influence of gravity.

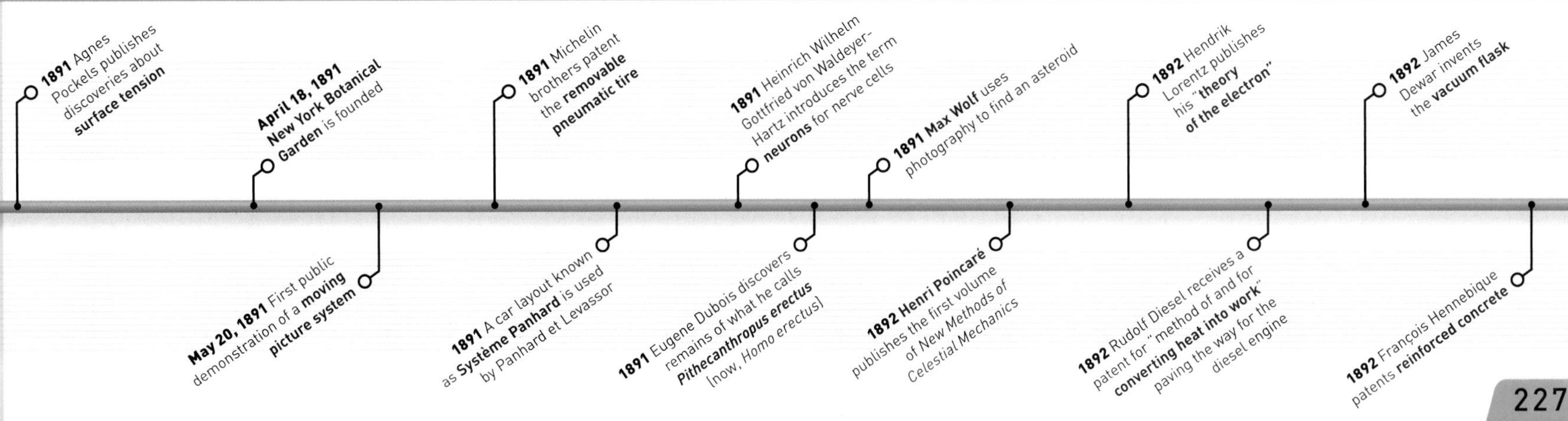
1891 Agnes Pockels publishes discoveries about surface tension
April 18, 1891 New York Botanical Garden is founded
May 20, 1891 First public demonstration of a moving picture system
1891 Michelin brothers patent the removable pneumatic tire
1891 A car layout known as Système Panhard is used by Panhard et Levassor
1891 Heinrich Wilhelm Gottfried von Waldeyer-Hartz introduces the term neurons for nerve cells
1891 Eugene Dubois discovers remains of what he calls Pithecanthropus erectus (now, Homo erectus)
1891 Max Wolf uses photography to find an asteroid
1892 Henri Poincaré publishes the first volume of New Methods of Celestial Mechanics
1892 Hendrik Lorentz publishes his "theory of the electron"
1892 Rudolf Diesel receives a patent for "method of and for converting heat into work" paving the way for the diesel engine
1892 James Dewar invents the vacuum flask
1892 François Hennebique patents reinforced concrete

> **“** I WILL **FORCE** UPON **POLITICIANS** THE **RECOGNITION** OF **ANTHROPOLOGY** IF I HAVE TO DO IT WITH THE **STAKE** AND **THUMBSCREW. ”**

Mary Kingsley, British anthropologist, in a letter to anthropologist E.B. Teylor

Mary Kingsley—shown here traveling on the Ogowe River in Africa—wrote extensively about the African continent and its people.

DECADES AFTER ESPOUSING THE IDEA that there had once been a great continent in the Southern Hemisphere, which he dubbed Gondwana (see 1861–64), Austrian geologist **Eduard Suess** came up with a new theory. In 1893, he suggested that this southern continent had been separated from its northern counterpart, Laurasia, by an inland sea he named **Tethys**, after the Greek goddess of the sea. A modern approach based on plate tectonics suggests the existence of a larger version of this, the Tethys Ocean, in the Mesozoic era, 251–65.5 million years ago.

On February 1, 1893, American inventor **Thomas Edison** and his team completed the **first studio** for the production of movies. Officially called the Kinetographic Theater, it was also known as **The Black Maria**—a slang term for police wagons—because both were small, cramped, and dark.

The **first Ferris wheel**, designed by American engineer **George Washington Gale Ferris, Jr.** (1859–96) opened in Chicago, Illinois, on June 1, 1893, and it operated until November 6, the same year.

In July this year in Japan, inventor **Kokichi Mikimoto** (1858–1954) produced the first perfect pearl at his farm. Although Swedish botanist Carl Linnaeus had cultivated freshwater pearls in Europe in the 18th century, Mikimoto was the first person to **cultivate pearls commercially**.

On August 17, 1893, pioneering English anthropologist **Mary Kingsley** (1862–1900) arrived in Sierra Leone on her first trip to Africa. She drew on her experiences of living with the indigenous people to give lectures and write books that helped **debunk the stereotype** of Africans being "savages," and raised questions on the benefits of colonialism.

In 1891, German neurologist **Arnold Pick** (1851–1924) had introduced the term **dementia praecox** (premature dementia) to refer to a psychotic disorder beginning in the late teens. In 1893, German psychiatrist **Emil Kraepelin** (1856–1926) gave a detailed textbook description of this condition, later reinterpreted and renamed **schizophrenia**.

This was also the year in which Austrian psychoanalyst **Sigmund Freud** (1856–1939) and Austrian physician **Josef Breuer** (1842–1925) published their paper *Über Den Psychischen Mechanismus Hysterischer Phänomene* (*On the Psychical Mechanism of Hysterical Phenomena*), which marked the **beginning of psychoanalysis**. The paper was based on Breuer's work with the patient, Anna O. Freud and Breuer also elaborated their ideas in a book, *Studien über Hysterie* (*Studies on Hysteria*), first published in 1895.

American inventor **Edward Goodrich Acheson** (1856–1931) patented a process to manufacture the industrial abrasive carborundum, essential in the manufacture of precision-ground, interchangeable metal parts, in 1893. In 1926, the US Patent Office would include **carborundum** in the list of 22 patents most responsible for the industrial age.

In 1893, using the technique of **interferometry**, Albert Abraham Michelson, an American scientist,

First motion picture studio
The Black Maria, as Thomas Edison's first motion picture studio was called, operated in West Orange, New Jersey, from December 1893 to 1901.

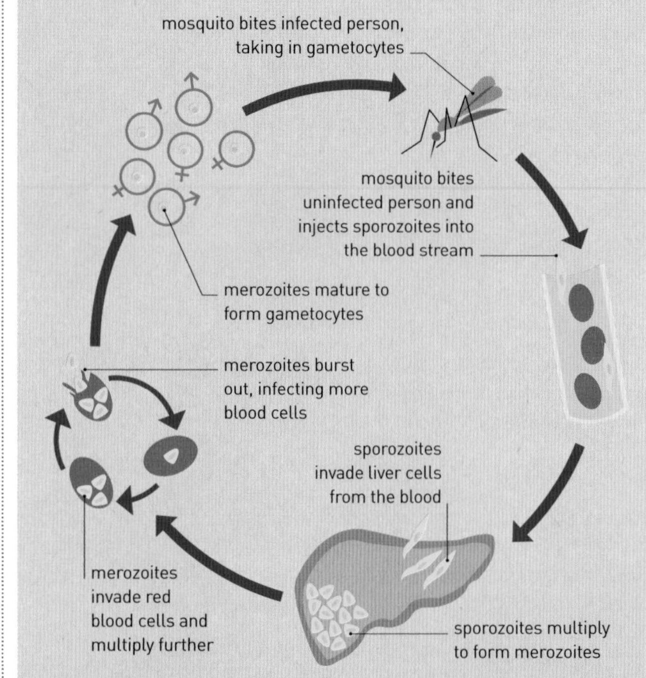

mosquito bites infected person, taking in gametocytes
mosquito bites uninfected person and injects sporozoites into the blood stream
merozoites mature to form gametocytes
merozoites burst out, infecting more blood cells
sporozoites invade liver cells from the blood
merozoites invade red blood cells and multiply further
sporozoites multiply to form merozoites

MALARIA LIFE CYCLE

A malaria-carrying female *Anopheles* mosquito feeds on a human and injects parasites in the form of *sporozoites* into the bloodstream. These multiply in the liver cells and produce *merozoites*, which reproduce in the red blood cells. Some of the infected cells produce *gametocytes* that are ingested by other feeding mosquitoes, which then become carriers of the disease.

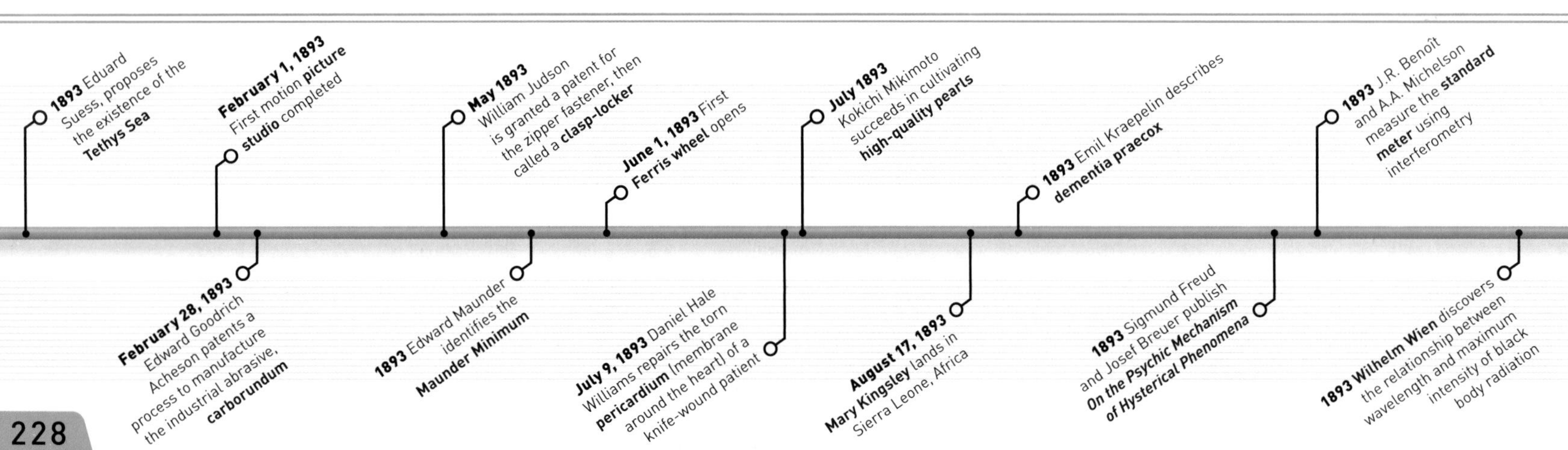

1893 Eduard Suess, proposes the existence of the **Tethys Sea**
February 1, 1893 First motion **picture studio** completed
February 28, 1893 Edward Goodrich Acheson patents a process to manufacture the industrial abrasive, **carborundum**
May 1893 William Judson is granted a patent for the zipper fastener, then called a **clasp-locker**
1893 Edward Maunder identifies the **Maunder Minimum**
June 1, 1893 First **Ferris wheel** opens
July 9, 1893 Daniel Hale Williams repairs the torn **pericardium** (membrane around the heart) of a knife-wound patient
July 1893 Kokichi Mikimoto succeeds in cultivating **high-quality pearls**
August 17, 1893 **Mary Kingsley** lands in Sierra Leone, Africa
1893 Emil Kraepelin describes **dementia praecox**
1893 Sigmund Freud and Josef Breuer publish ***On the Psychic Mechanism of Hysterical Phenomena***
1893 J.R. Benoît and A.A. Michelson measure the **standard meter** using interferometry
1893 Wilhelm Wien discovers the relationship between wavelength and maximum intensity of black body radiation

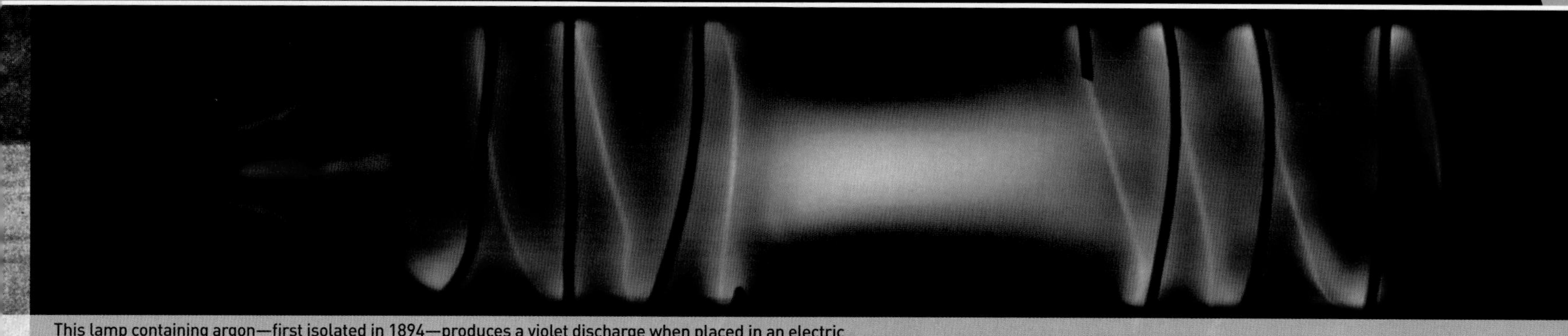
This lamp containing argon—first isolated in 1894—produces a violet discharge when placed in an electric field produced from a high voltage transformer to produce neon lighting.

RAMÓN Y CAJAL (1852–1934)

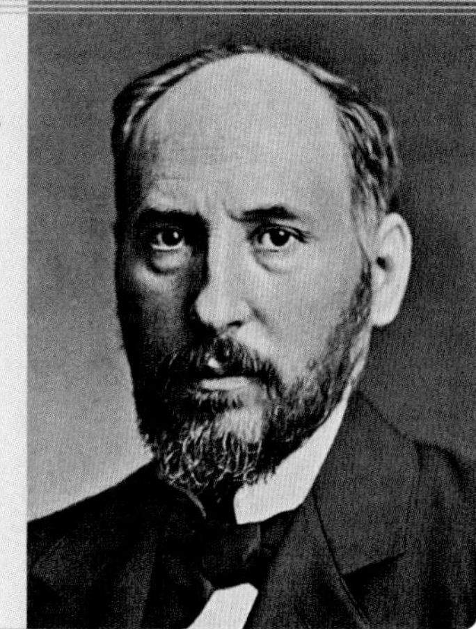

Spanish pathologist and histologist, Ramón y Cajal, was responsible for identifying a type of cell that controls the slow waves of contraction that move food along the intestine. He was also a neuroscientist and an expert in hypnotism, which he used to help his wife during labor. In 1906, he was awarded the Nobel Prize in recognition of his work on the nervous system.

and J.R. Benoît, Director of the International Bureau of Weights and Measures, decided to use wavelengths of light to redefine a standard of distance. They measured **the meter**—a prototype platinum–iridium bar of which was kept in Paris, France—in terms of the wavelength of the red light emitted by heated cadmium.

Edward Maunder (1851–1928), a British astronomer working at the Royal Greenwich Observatory in London, was investigating the historical records of sunspots in 1893 when he discovered that very few spots had been observed between 1645 and 1715. This interval, now known as the **Maunder Minimum**, coincided with the coldest part of the **Little Ice Age** (*c.*1500–1800), a time when Earth cooled considerably.

In 1894, British parasitologist **Patrick Manson** (1844–1922) developed the idea that **malaria is spread by mosquitoes**. In November, Manson mentioned the hypothesis to British doctor **Ronald Ross** (1857–1932), who received the Nobel Prize in 1902 for working out the details of the process.

In 1784, British physicist Henry Cavendish had discovered that air contains a small proportion of a substance less reactive than nitrogen, but he was unable to isolate it. In August 1894, following a suggestion by British scientist Lord Rayleigh, British chemist **William Ramsay** (1852–1916) reported that he had isolated this gas, which he named **argon**. It was the first of the so-called **noble gases** to be isolated.

French engineer Édouard Branly (1844–1940) had developed an **early radio signal detector** at the beginning of the 1890s. In 1894, in lectures at the Royal Institution, London, British physicist **Oliver Lodge** (1851–1940) dubbed this the **coherer**. Lodge used this invention in his work, which became an important part of Italian physicist Guglielmo Marconi's system of wireless telegraphy.

0.93
THE PERCENTAGE OF **ARGON** IN THE **ATMOSPHERE**

In 1894, Spanish histologist (histology is the study of tissues and cells) **Ramón y Cajal**, known as the father of modern neuroscience, theorized that **memories** do not involve growing new neurons (nerve cells), but making new connections between existing neurons. The connections between neurons came to be known as **synapses**.

Also in 1894, British physiologist **Edward Sharpey-Schafer** (1850–1935) and English physician **George Oliver** (1841–1915) found that an extract from the adrenal gland caused a rise in blood pressure. This led them to identify the hormone **epineprine**.

German chemist **Emil Fischer** (1852–1919) in 1894 came up with the **lock and key theory** that explains how **enzymes target specific molecules** and function so efficiently.

> THE **BRAIN** IS A **WORLD...** [WITH] A NUMBER OF **UNEXPLORED CONTINENTS** AND **STRETCHES** OF **UNKNOWN TERRITORY.**
>
> **Ramón y Cajal, Spanish pathologist and histologist,** 1906

HOW ENZYMES WORK

Enzymes are proteins that act as catalysts to increase the rate of specific chemical reactions. They fold into complex shapes that allow smaller molecules to fit into them. The active site where these molecules fit may either encourage molecules to join together or split apart. However, the enzyme remains unchanged and can repeat the process indefinitely.

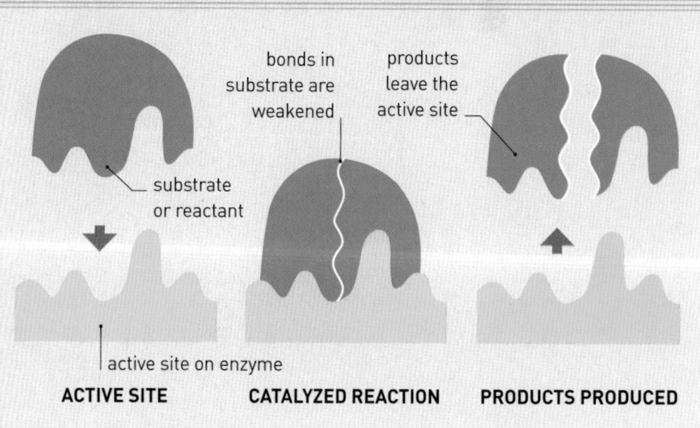
substrate
or reactant
active site on enzyme
bonds in
substrate are
weakened
products
leave the
active site
ACTIVE SITE
CATALYZED REACTION
PRODUCTS PRODUCED

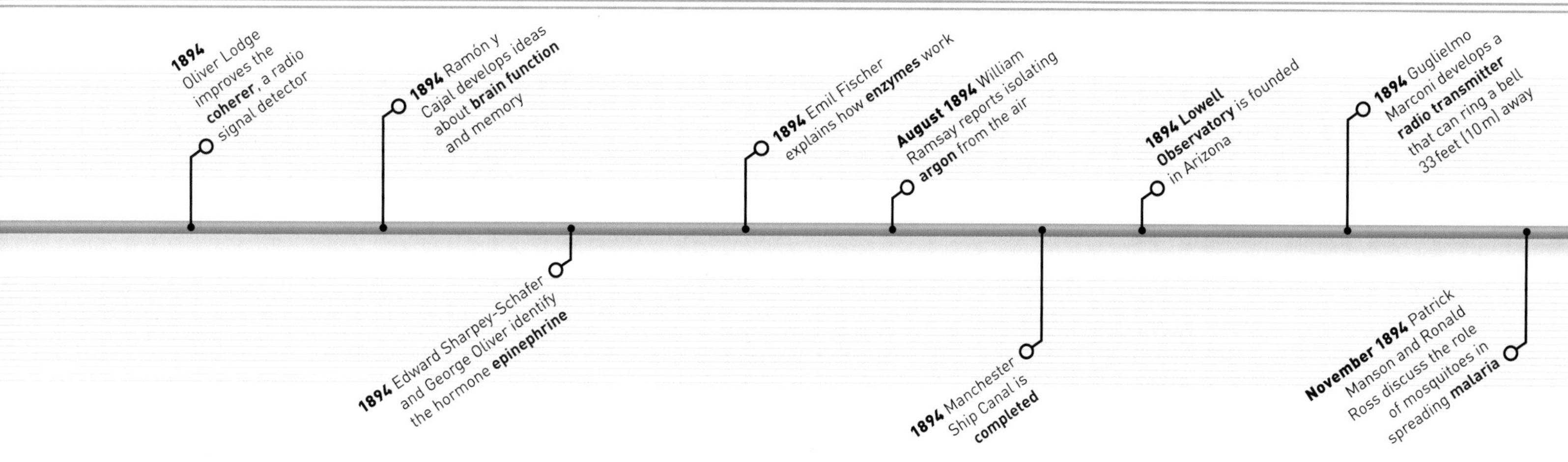
1894 Oliver Lodge improves the coherer, a radio signal detector
1894 Ramón y Cajal develops ideas about brain function and memory
1894 Edward Sharpey-Schafer and George Oliver identify the hormone epinephrine
1894 Emil Fischer explains how enzymes work
August 1894 William Ramsay reports isolating argon from the air
1894 Manchester Ship Canal is completed
1894 Lowell Observatory is founded in Arizona
1894 Guglielmo Marconi develops a radio transmitter that can ring a bell 33 feet (10 m) away
November 1894 Patrick Manson and Ronald Ross discuss the role of mosquitoes in spreading malaria

5

THE ATOMIC AGE 1895–1945

The unanticipated discovery of radioactivity revealed that massive amounts of energy are hidden inside atoms, available to be unleashed. New and surprising theories of relativity and quantum mechanics described a Universe of four-dimensional space-time containing interchangeable waves and particles that, at the subatomic level, can never be pinned down with absolute certainty.

20th-century botanists classified plant formations, such as vegetation on crumbling sand dunes (shown here), by their ecological characteristics.

THE SCIENCE OF PLANT ECOLOGY reached a milestone when Dutch botanist **Eugenius Warming** (1841–1924) and German botanist **Andreas Schimper** (1856–1901) published their books on the subject at the end of the century. Together, they showed how **vegetation could be classified** into different formations based on climate and soil conditions.

In Britain, physicists **Lord Rayleigh** (1842–1919) and **William Ramsay** (1852–1916) discovered the gas **argon**. Rayleigh realized that air must contain an unknown chemical component, since the density of atmospheric nitrogen was different from that of pure nitrogen made in a laboratory. He found that atmospheric nitrogen contained traces of argon and other unreactive elements later dubbed **noble gases**.

In November, German physicist **Wilhelm Röntgen** (1845–1923) discovered that electrically charged vacuum tubes emitted rays that made a fluorescent screen glow; he called them **X-rays**. He found that they went through human skin and exposed photographic plates. This led to the development of **medical radiography**. By 1896, scientists knew that X-rays could ionize (charge up) air. Some physicians even tried firing X-rays at tumors—**radiotherapy**—to try to cure cancer.

Inspired by Röntgen, French physicist **Henri Becquerel** (1852–1908) studied whether **phosphorescent substances**, such as uranium salts, produced X-rays. He expected radiation to be emitted only after exposure to sunlight, but found that the salts could fog a photographic plate even in darkness. He had discovered a new phenomenon: **radioactivity**.

> “I HAVE **SEEN MY DEATH.**”
>
> **Anna Röntgen**, wife of Wilhelm, on seeing her hand X-ray, 1895

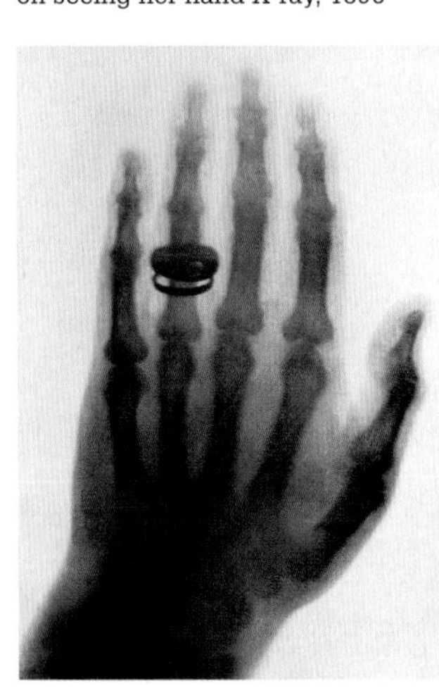

X-ray of Anna Röntgen's hand
Röntgen's X-ray of his wife's hand shows that the rays penetrated her skin and muscle, but were impeded by denser bones—and her ring.

> “EVERY DAY SEES **HUMANITY** MORE **VICTORIOUS** IN THE STRUGGLE WITH **SPACE AND TIME.**”
>
> **Guglielmo Marconi, Italian inventor**

IN APRIL, BRITISH PHYSICIST J.J. THOMSON (1856–1940) was studying **cathode rays**. These rays are produced by the negative electrode (cathode) of an electrically charged vacuum tube, and are attracted to the positive electrode (anode). They cause the glass at the far end of the tube to glow. Thomson demonstrated that the rays were composed of particles much lighter than the smallest atoms. He concluded that these "corpuscles," as he called them, were negatively charged components present in all atoms; Thomson had discovered the **first subatomic particles**. They were later called **electrons**.

In May of the same year, the first radio communication over water was made across Britain's Bristol Channel. Italian inventor **Guglielmo Marconi** (1874–1937) had been experimenting with **wireless technology**, and in 1897, his team of scientists succeeded in sending a **Morse code signal** from Flat Holm Island to a receiver on the Welsh coast. Later, German physicist Karl Braun (1850–1918) improved the technology to increase broadcasting range. Marconi and Braun went on to share the Nobel Prize 12 years later for their work on wireless radio.

In 1897, Braun was also working on vacuum tubes. He **modified cathode ray tubes** so that the rays struck a surface to produce images. Braun's tube was the first **oscilloscope**—a device that made graphical presentations of electrical signals. It paved the way for the invention of the television and the development

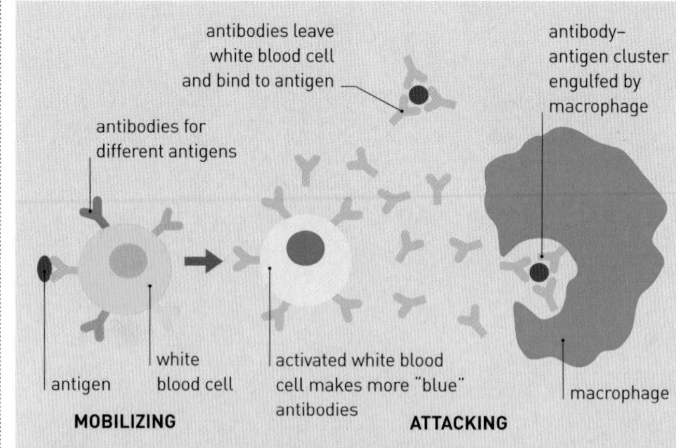

antibodies leave white blood cell and bind to antigen
antibody–antigen cluster engulfed by macrophage
antibodies for different antigens
antigen
white blood cell
activated white blood cell makes more "blue" antibodies
macrophage
MOBILIZING
ATTACKING

ANTIGEN–ANTIBODY INTERACTION

Paul Ehrlich explained how the immune system could be mobilized to destroy infection. White blood cells carry side-chain antibodies, which bind themselves to foreign particles called antigens. As they are bound together, the white cells are prompted to produce more antibodies. These then cluster around the antigens and enable macrophages—other immune system cells—to destroy them.

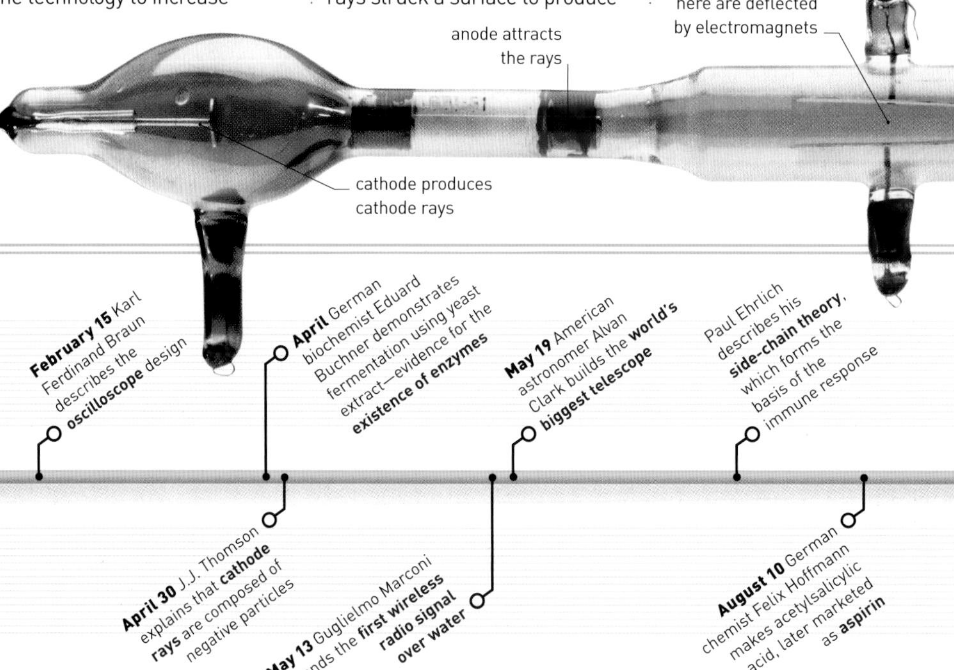

rays that reach here are deflected by electromagnets
anode attracts the rays
cathode produces cathode rays

November 8, 1895 Wilhelm Röntgen discovers **X-rays**

1895 Eugenius Warming publishes *Plantesamfund*, classifying **vegetation formations**

January 12, 1896 American physician Emil Grubbe describes treating **breast cancer with X-rays**

March 1, 1896 Henri Becquerel accidentally discovers **radioactivity**

March 1896 J.J. Thomson describes how **X-rays** can **ionize** (charge up) gas

May 1896 Ronald Ross correctly suggests that **malaria** is transmitted by **mosquito bites**

February 15 Karl Ferdinand Braun describes the **oscilloscope** design

April 30 J.J. Thomson explains that **cathode rays** are composed of negative particles

April German biochemist Eduard Buchner demonstrates fermentation using yeast extract—evidence for the **existence of enzymes**

May 13 Guglielmo Marconi sends the **first wireless radio signal over water**

May 19 American astronomer Alvan Clark builds the **world's biggest telescope**

August 10 German chemist Felix Hoffmann makes acetylsalicylic acid, later marketed as **aspirin**

Paul Ehrlich describes his **side-chain theory**, which forms the basis of the immune response

Marconi formed his Wireless Telegraph and Signal company in 1897.

Malaria "oocysts" (blue) in the mosquito stomach lining, first observed by Ronald Ross, burst into cells that infect the insect's salivary glands—and thereby, its bite.

of medical technology, such as the electrocardiogram used for monitoring heartbeat.

In August, German chemist **Felix Hoffmann** (1868–1946), working at the Bayer pharmaceutical laboratories, produced a painkilling substance called acetylsalicylic acid. It was modeled on a related ingredient derived from certain medicinal plants, such as willow and meadowsweet, which had been known since Ancient Greece. The new painkiller was later marketed as **aspirin** (see 1899).

Another medical breakthrough was made by German physician **Paul Ehrlich**. He developed the **side-chain theory**, which explained how the **immune system** could attack specific infections. It remains the basis of immunological theory to this day.

Cathode ray tube
Glass vacuum tubes proved useful to scientists in the discovery of X-rays and electrons. Rays pass along the tube and the pattern formed at the end shows they are negatively charged particles (electrons).

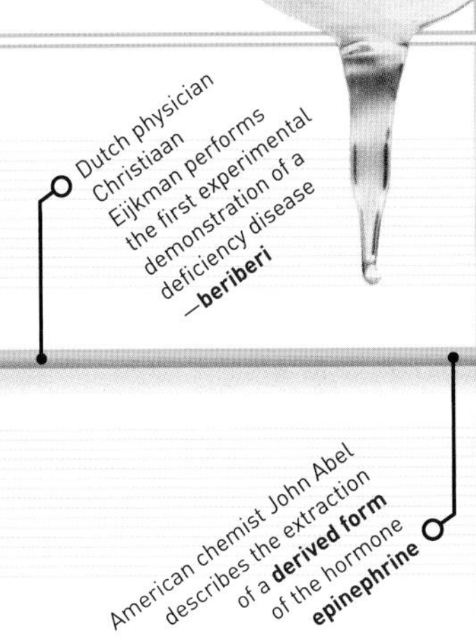
rays cause deflection pattern

AFTER HENRI BECQUEREL DISCOVERED RADIOACTIVITY, Polish–French physicist **Marie Curie** and her husband, French physicist **Pierre Curie** (1859–1906), embarked on a lifelong career studying radioactivity at the Becquerel Laboratory. Becquerel had found that pure uranium emissions could cause air to conduct electricity. The Curies discovered that a uranium ore—called pitchblende—was 300 times stronger in this respect, and deduced that a new element present in the ore must be responsible. They named the element **polonium**, after Marie's native country Poland, and coined the term **radioactive** at the same time. Later that year, the Curies discovered another radioactive element, **radium**, and managed to purify quantities of both for further study.

By the 1890s, scientists had discovered two **unreactive noble gases**, helium and argon, but William Ramsay suggested that gaps in the periodic table and laboratory analysis of air pointed to the existence of other elements. In 1898, working with British chemist Morris Travers (1872–1961), he discovered three more noble gases: **krypton**, **xenon**, and **neon**.

In July, the Annual Meeting of the British Medical Association reported

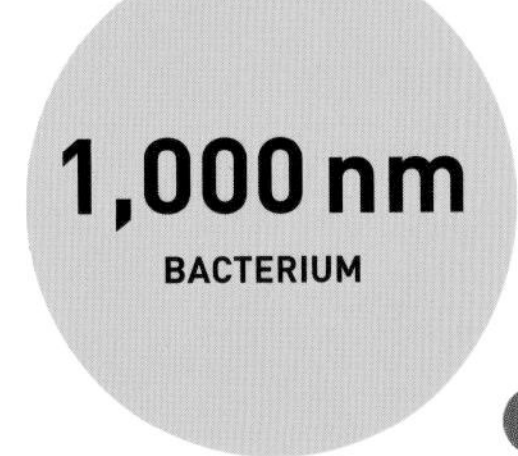
1,000 nm
BACTERIUM
Virus 20–40 nm

Size differential between virus and bacterium
Viruses are measured in nanometers (nm). They lack the cellular structure of bacteria, being just particles of protein and genetic material.

on key discoveries about a deadly disease—malaria. British physician **Ronald Ross** (1857–1932), working in India, had proved that mosquitoes spread the **malaria parasite** through their bite. The previous year—after painstakingly dissecting mosquito guts—he had found malaria parasites lodged in the stomach walls of these insects. His work demonstrated that the life cycles of the malaria parasite (see 1893–94) and certain kinds of mosquitoes were linked.

Meanwhile, Dutch biologist **Martinus Beijerinck** (1851–1931) made another breakthrough. He found that a plant disease called **tobacco mosaic** could be spread even when infected plant extract was passed through a filter that held back bacteria. He deduced that the contagious particles were smaller than bacteria, and called them viruses. Beijerink's tobacco mosaic virus would not be isolated until the 1930s.

In Germany, physicist **Wilhelm Wien** (1864–1928), experimenting with the positively charged rays produced in certain types of vacuum tubes, laid the foundations of a new area of analytical science, **mass spectrometry**. This was a technique used to determine the make-up of molecules by vaporizing them into ions (charged particles). Wien invented a way of separating the different kinds of ions in electromagnetic fields according to their mass and charge. The accuracy of mass spectrometry has resulted in it being used in fields as diverse as the medical testing of blood and urine to analyzing atmospheric samples in space exploration.

MARIE CURIE (1867–1934)

Born in Poland, Marie married Pierre Curie in France in 1895. The couple shared the Nobel Prize in 1903 with Henri Becquerel for their work on radioactivity. Marie received a second Nobel Prize in 1911 for the discovery of polonium and radium. She donated all her medals to the World War II effort. She died of leukemia caused by radiation exposure.

" **NOTHING** IN LIFE IS TO BE **FEARED**, IT IS ONLY **TO BE UNDERSTOOD**. NOW IS THE TIME TO UNDERSTAND **MORE**, SO WE… **FEAR LESS** "

Marie Curie, Polish–French physicist

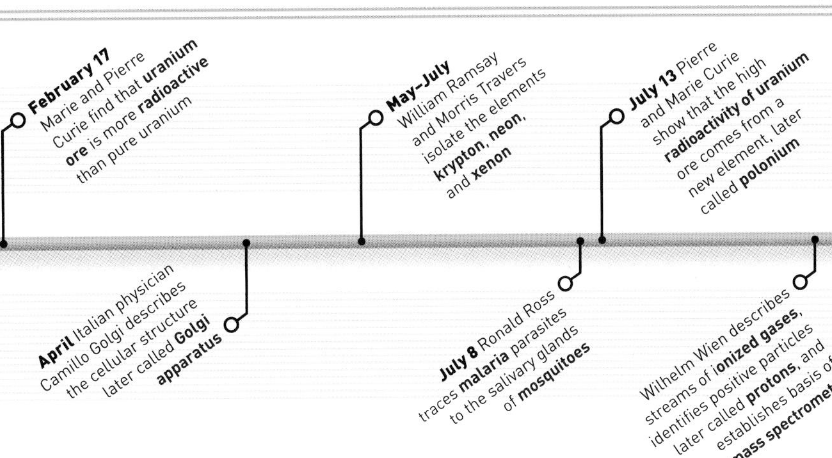

Dutch physician Christiaan Eijkman performs the first experimental demonstration of a deficiency disease—**beriberi**

American chemist John Abel describes the extraction of a **derived form** of the hormone **epinephrine**

February 17 Marie and Pierre Curie find that **uranium ore** is more **radioactive** than pure uranium

April Italian physician Camillo Golgi describes the cellular structure later called **Golgi apparatus**

May–July William Ramsay and Morris Travers isolate the elements **krypton**, **neon**, and **xenon**

July 8 Ronald Ross traces **malaria** parasites to the salivary glands of **mosquitoes**

July 13 Pierre and Marie Curie show that the high **radioactivity of uranium** ore comes from a new element, later called **polonium**

Wilhelm Wien describes streams of **ionized gases**, identifies positive particles later called **protons**, and establishes basis of **mass spectrometry**

December 26 Pierre and Marie Curie discover a new radioactive element, later called **radium**

December German microbiologist Carl Benda coins the term **mitochondria**

UNDERSTANDING ELECTROMAGNETIC RADIATION

DISCOVERIES IN THE 19TH CENTURY LED TO A NEW UNDERSTANDING OF THE NATURE OF RADIATION

Light, infrared and ultraviolet radiation, X-rays, gamma rays, microwaves, and radio waves all propagate through space at extremely high speed. They are all different forms of electromagnetic radiation, which can be understood as waves, but also as particles called photons.

By the 19th century, evidence suggested that light travels as waves and that wavelength determines the light waves' color. Two invisible forms of "light"—longer-wavelength infrared radiation (IR) and shorter-wavelength ultraviolet radiation (UV)—had also been discovered.

ELECTRIC AND MAGNETIC FIELDS

In the 1860s, British physicist James Clerk Maxwell formulated a set of equations that describes how electric fields produce magnetic fields and how magnetic fields produce electric fields. Maxwell realized that his formula was a "wave equation," describing wave motion. The speed of the waves described by the equation exactly matched the speed of light. Maxwell concluded that light is an "electromagnetic wave," and went on to predict the existence of other as yet unknown forms of radiation. Within 20 years, German physicist Heinrich Hertz had produced radio waves—electromagnetic waves with much longer wavelengths than light (see 1887).

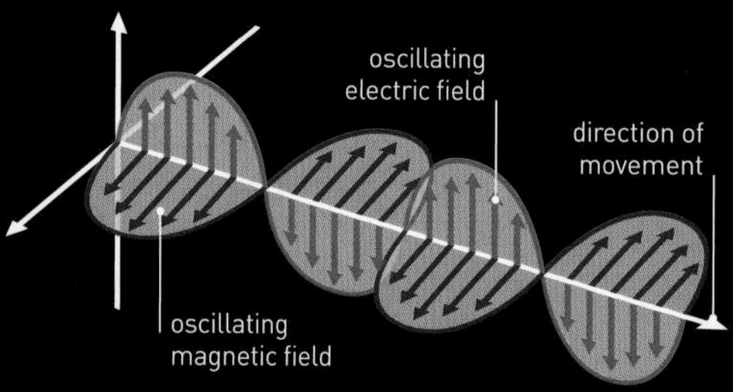
oscillating electric field
direction of movement
oscillating magnetic field

ELECTROMAGNETIC WAVE
In these self-propagating waves, oscillating electric and magnetic fields travel in the same direction but in perpendicular planes.

JAMES CLERK MAXWELL
As well as theorizing the existence of electromagnetic waves, James Clerk Maxwell played a key role in interpreting the emerging science of thermodynamics, and took the first color photograph *(see 1861).*

2,050 miles
THE **DISTANCE** COVERED BY GUGLIEMO MARCONI'S FIRST **TRANSATLANTIC RADIO** SIGNAL IN 1901

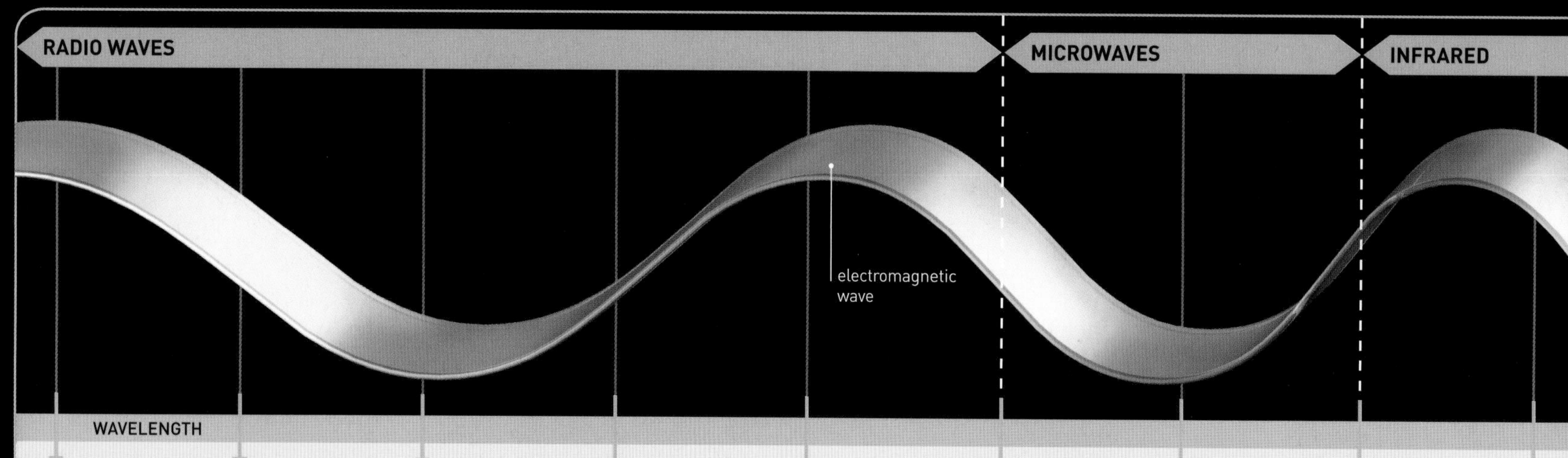
RADIO WAVES
MICROWAVES
INFRARED
electromagnetic wave
WAVELENGTH
1km (0.6 miles)
100m (330ft)
10m (33ft)
1m (3ft 3in)
10cm (4in)
1cm (2.5in)
1mm
100µm
10µm

Electromagnetic spectrum
Visible light represents a tiny part of the whole spectrum of electromagnetic radiation. Every part of the spectrum features in some way in the modern world; a few representative examples are shown here.

Radio telescope
Large dish telescopes that detect radio waves provide essential information about distant space.

Microwave oven
Produced by magnetrons, short-wavelength radio waves called microwaves are used to heat food.

Remote controller
Most remote controllers use coded infrared radiation to send instructions to electrical devices.

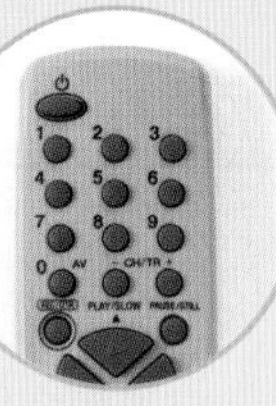

WAVES AND PARTICLES

Scientists had long debated whether light travels through space as streams of particles or as waves. The wave theory was in favor in the 19th century, even before Maxwell's discovery. However, there were phenomena the wave theory could not explain, including the "photoelectric effect."

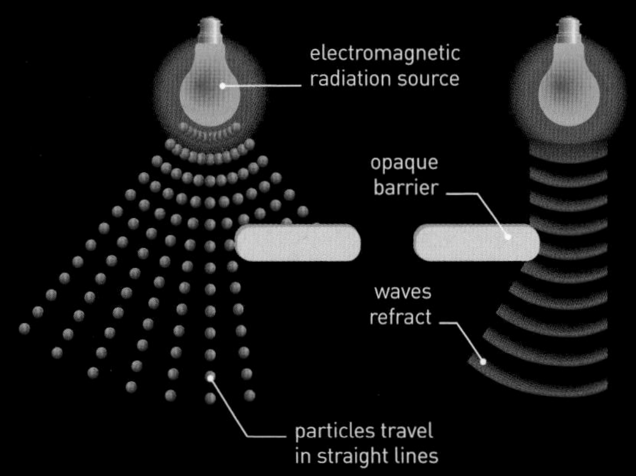
electromagnetic radiation source
opaque barrier
waves refract
particles travel in straight lines

WAVE-PARTICLE PARADOX
All waves "diffract" or spread out as they pass the edges of stationary objects. Water waves do this as they enter harbors, for example. The diffraction of light is hard to explain if light is understood to be a stream of particles.

In 1887, Hertz attached two electrodes to a battery and set them a small distance apart in a vacuum tube. When he shone light onto them, an electric current fired between them, but above a certain wavelength it stopped, however intense the light. Albert Einstein explained this effect by proving that electromagnetic radiation exists as particles (photons), and that different colors of light, and different forms of radiation, differ in the amount of energy their photons carry.

> “**X-RAYS** WILL PROVE TO BE A **HOAX.**”
>
> **British physicist William Thomson (Lord Kelvin)**, 1899

USING ELECTROMAGNETIC RADIATION

In addition to ultraviolet, infrared, and radio waves, scientists discovered two new forms of electromagnetic radiation in the 1890s, both with very short wavelengths (or high-energy photons): X-rays and gamma rays.

Devices that can produce or detect the various forms of electromagnetic radiation have many, and varied, applications. For example, different types of radio waves are used to carry television, radio, and telephone signals. In medicine, penetrating X-rays are used to produce images of the inside of the body, while gamma rays are used in radiation therapy.

INFARED RADIATION
Cameras sensitive to infrared radiation can produce color-coded images that reveal temperature differences—allowing engineers to detect heat loss from houses, for example.

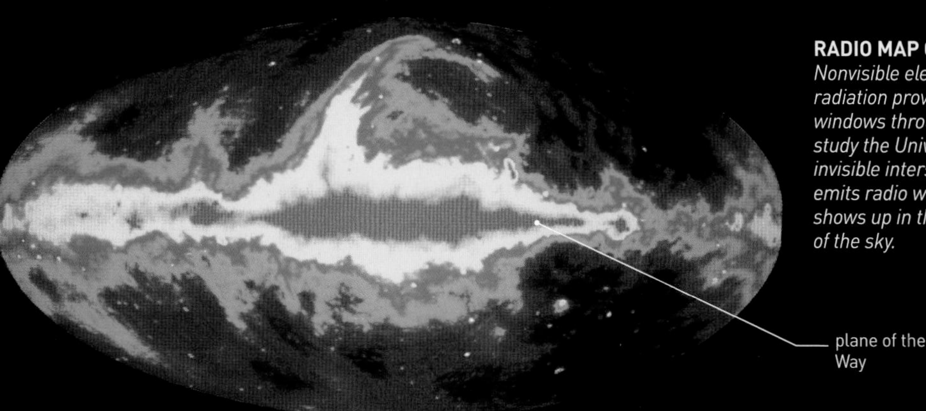
plane of the Milky Way

RADIO MAP OF THE SKY
Nonvisible electromagnetic radiation provides new windows through which to study the Universe. Normally invisible interstellar dust emits radio waves, so it shows up in this radio map of the sky.

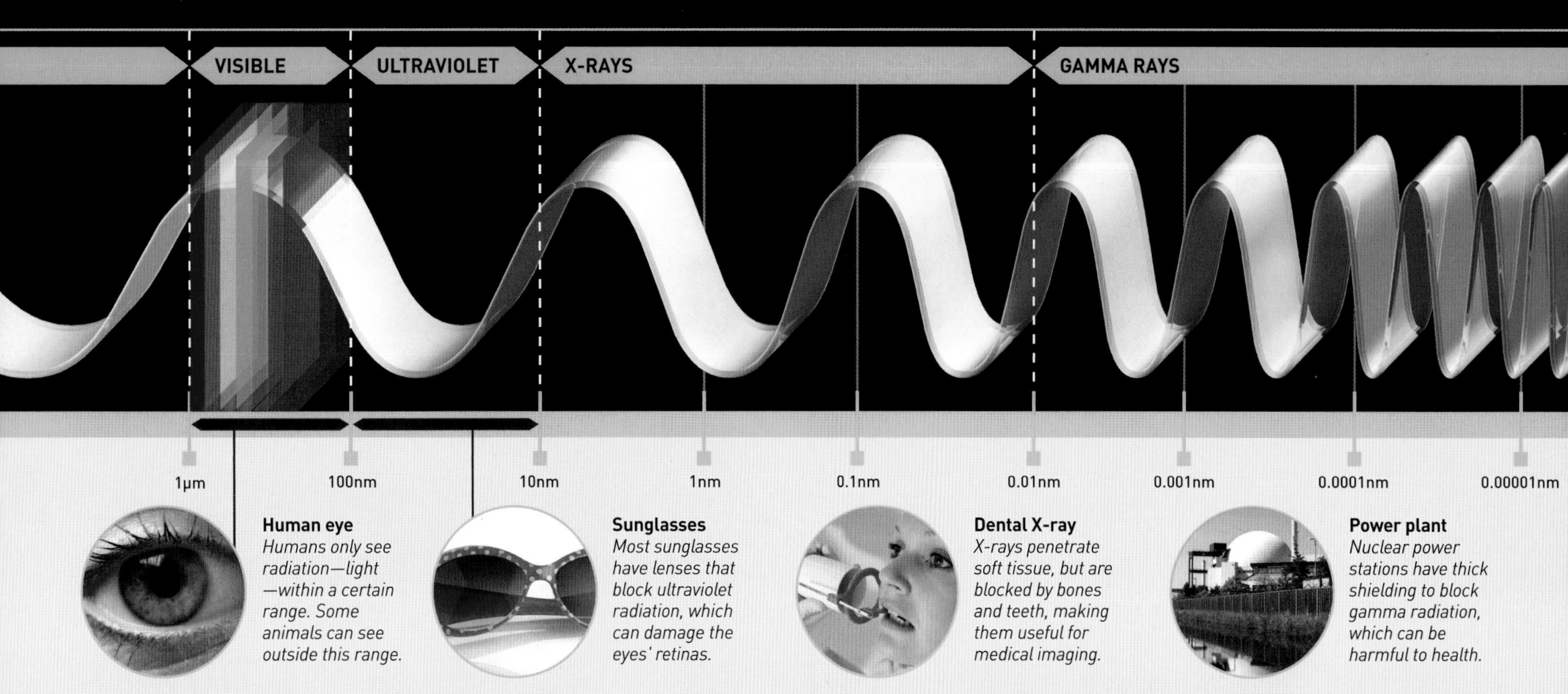
VISIBLE
ULTRAVIOLET
X-RAYS
GAMMA RAYS
1µm
100nm
10nm
1nm
0.1nm
0.01nm
0.001nm
0.0001nm
0.00001nm

Human eye
Humans only see radiation—light—within a certain range. Some animals can see outside this range.

Sunglasses
Most sunglasses have lenses that block ultraviolet radiation, which can damage the eyes' retinas.

Dental X-ray
X-rays penetrate soft tissue, but are blocked by bones and teeth, making them useful for medical imaging.

Power plant
Nuclear power stations have thick shielding to block gamma radiation, which can be harmful to health.

1899

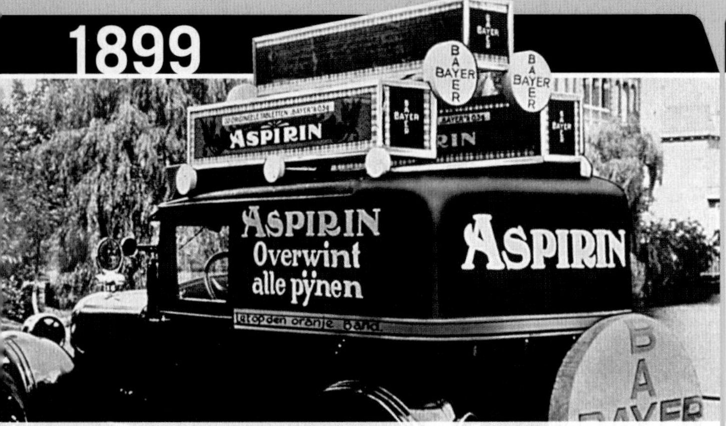
ASPIRIN
Overwint
alle pynen
ASPIRIN
BAYER

This car from the 1920s advertises Bayer's drug bearing a Dutch slogan, which translates as "Overcomes all sufferings."

NEW ZEALAND-BORN PHYSICIST ERNEST RUTHERFORD (1871–1937) was studying the radiation given off by uranium salts—first discovered by Henri Becquerel (see 1896). Rutherford was interested in the way radioactivity caused gases to be able to conduct electricity. This happens because the gas becomes ionized: radiation knocks out one or more electrons (negative particles), leaving positive charges behind. Rutherford also discovered that uranium emitted **two types of radiation, which he named alpha and beta**. His alpha rays were later identified as particles that are the nuclei of helium atoms, while beta rays were found to be streams of electrons. Both were the by-products of radioactive decay.

In March, the Imperial Patent Office in Berlin **trademarked a new drug for Bayer**, a German pharmaceutical company. The drug was **aspirin**, a painkiller that had been developed by Bayer's scientists two years earlier. Aspirin would become **the world's best-selling drug**.

45,000 TONS THE QUANTITY OF **ASPIRIN CONSUMED GLOBALLY** EACH YEAR

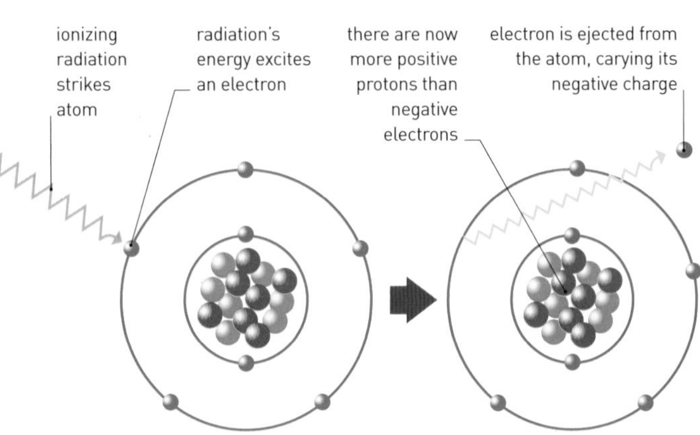
ionizing radiation strikes atom
radiation's energy excites an electron
there are now more positive protons than negative electrons
electron is ejected from the atom, carying its negative charge
ATOM
POSITIVE ION

Effect of ionization
Ionizing radiation (alpha and beta particles, gamma rays, and X-rays) carries enough energy to create ions (charged particles) from atoms.

1900

420 FEET THE **LENGTH** OF THE **FIRST ZEPPELIN**

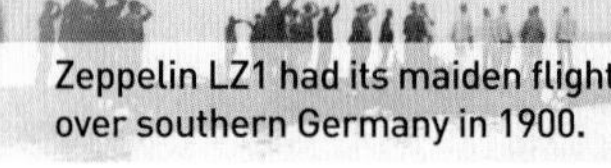

Zeppelin LZ1 had its maiden flight over southern Germany in 1900.

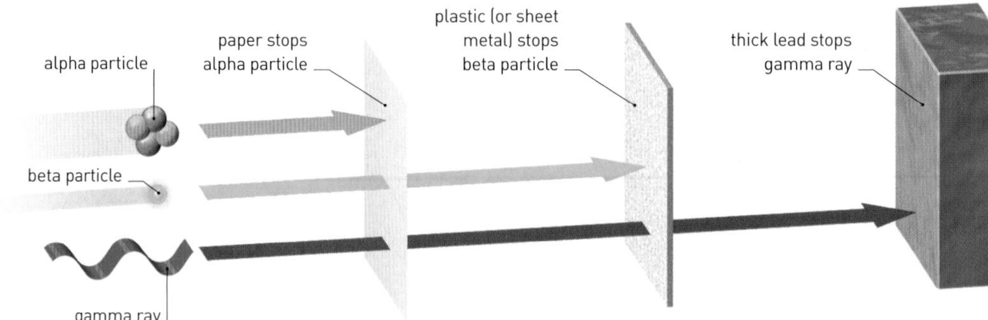
alpha particle
paper stops alpha particle
plastic (or sheet metal) stops beta particle
thick lead stops gamma ray
beta particle
gamma ray

Penetration of radiation
Alpha particles cannot penetrate paper, unlike smaller beta particles. Gamma radiation is not made up of particles and its high-energy rays are stopped only by lead.

THE FRENCH CHEMIST PAUL VILLARD (1860–1934) announced that he had found a **third type of radiation** only a year after the discovery of alpha- and beta-radiation. Villard's rays, emitted by radium salts, were far more penetrating: they were similar to X-rays but had shorter wavelengths and high energy. **Rutherford later called them gamma rays.**

July saw the **first flight of the rigid airship** named after the German **Count Ferdinand von Zeppelin** (1838–1917). Its light-alloy framework—buoyed by an internal system of hydrogen balloons—proved difficult to control. It hailed the start of a period of commercial airship success. The program was scrapped after the fatal Hindenburg crash of 1937.

MAX PLANCK (1858–1947)

Physicist Max Planck studied in Munich and Berlin, and became a professor at Kiel, then Berlin. He helped organize the first **Solvay Conference for Physics** in 1911, when scientists met to discuss quantum theory. He was awarded the Nobel Prize in 1918. Unlike many scientists, Planck remained in Germany during the Nazi government.

In October, German physicist Max Planck had a **theory that proposed a new way of looking at physics**. He was interested in the science behind an everyday phenomenon—darker objects are warmed more by light than paler ones. The theoretically darkest object, a so-called black body, absorbs all electromagnetic radiation, including visible light—and then is a perfect emitter of this radiation. Planck reasoned that there were discrete vibrations of atoms in a body, equivalent to "packages" of energy—which when added together give the total amount of energy emitted. The idea that radiation, such as light, comes in packages of energy later called quanta was **the foundation of quantum physics**.

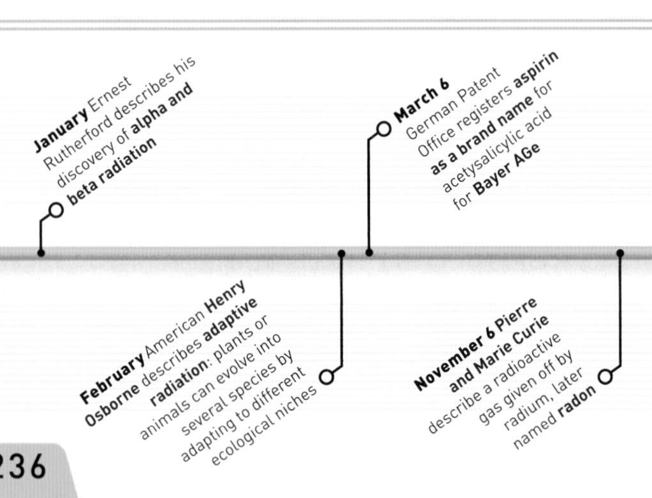
January Ernest Rutherford describes his discovery of **alpha and beta radiation**
February American **Henry Osborne** describes **adaptive radiation**: plants or animals can evolve into several species by adapting to different ecological niches
March 6 German Patent Office registers **aspirin as a brand name** for acetysalicylic acid for **Bayer AGe**
November 6 Pierre and Marie Curie describe a radioactive gas given off by radium, later named **radon**

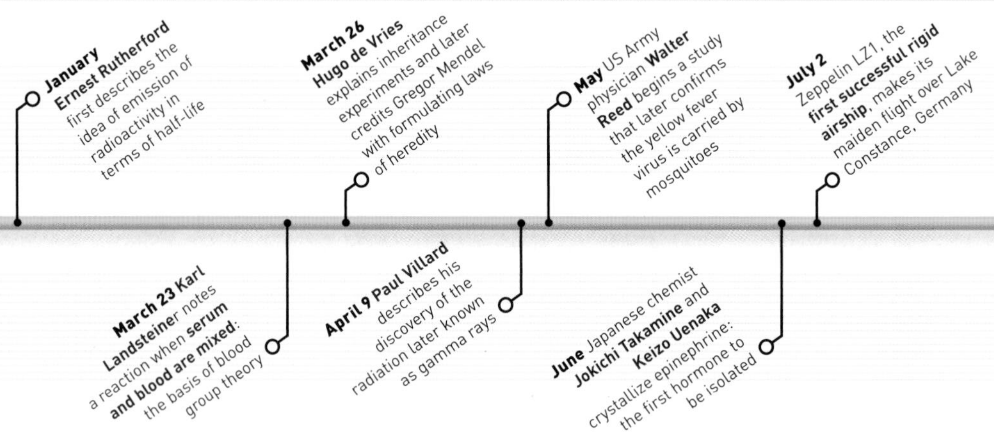
January Ernest Rutherford first describes the idea of emission of radioactivity in terms of half-life
March 23 Karl Landsteiner notes a reaction when **serum and blood are mixed**: the basis of blood group theory
March 26 Hugo de Vries explains inheritance experiments and later credits Gregor Mendel with formulating laws of heredity
April 9 Paul Villard describes his discovery of the radiation later known as gamma rays
May US Army physician **Walter Reed** begins a study that later confirms the yellow fever virus is carried by mosquitoes
June Japanese chemist **Jokichi Takamine** and **Keizo Uenaka** crystallize epinephrine: the first hormone to be isolated
July 2 Zeppelin LZ1, the **first successful rigid airship**, makes its maiden flight over Lake Constance, Germany

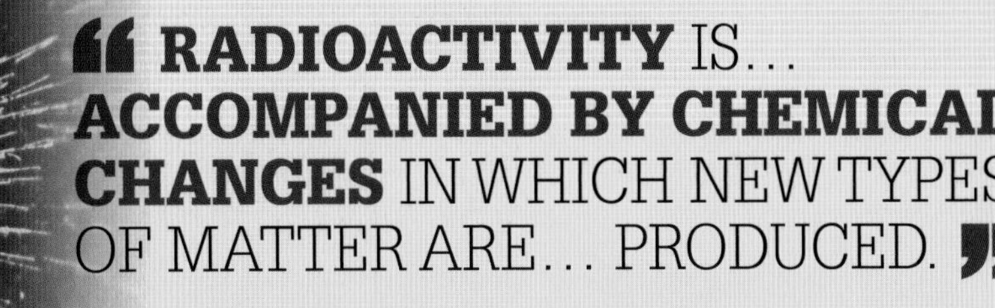

" RADIOACTIVITY IS... **ACCOMPANIED BY CHEMICAL CHANGES** IN WHICH NEW TYPES OF MATTER ARE... PRODUCED. **"**

Ernest Rutherford, *Philosphical Magazine*, 1902

Radium salt left on photographic plate shows strong radioactivity after development—the yellow tracks are emitting alpha particles.

This year also saw the origins of a **revolution in biology**: several biologists rediscovered the **laws of inheritance** that had been established by Gregor Mendel (see 1866). Dutch botanist **Hugo de Vries** (1848–1935) found that the inheritance of characteristics in plants followed rules dictated by particles he called **pangenes (later changed to genes)**.

Austrian biologist **Karl Landsteiner** (1868–1943) proposed a theory about **blood group compatibility** in a footnote of a scientific paper. He had found that if the serum (the liquid part of blood) from one person was mixed with another's entire blood, it could cause clumping of the red blood cells. This explained why some blood transfusions were fatal.

De Vries in his garden
Hugo de Vries experimented with breeding plants—as Gregor Mendel had done years before. Although he retired in 1918, he continued his research until his death.

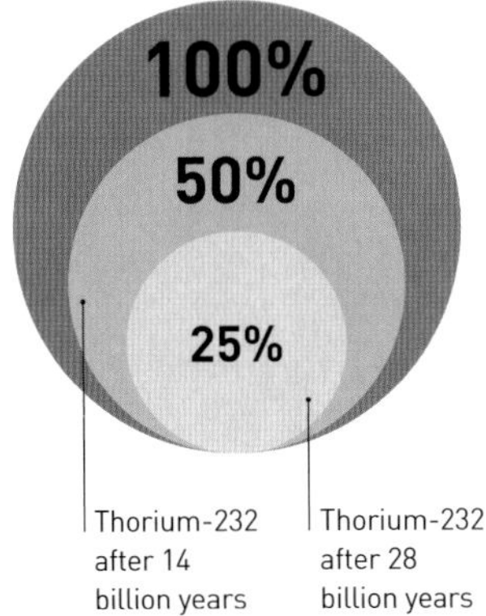
100%
50%
25%
Thorium-232 after 14 billion years
Thorium-232 after 28 billion years

Radioactive decay
The half-life of a radioactive element is the time taken for a particular element to decay into another form. Thorium-232 has a half-life of 14 billion years.

IN FEBRUARY AND AUGUST, BRITISH ENGINEER Hubert Cecil Booth (1871–1955) filed patents for a device that sucked air through a filter system. His invention was the **first powered vacuum cleaner**. In November, American electrical engineer **Miller Reese Hutchinson**, (1876–1944), inspired by a friend rendered deaf by scarlet fever, **patented the first electrical hearing aid**. In a modification of Alexander Bell's telephone technology, Reese Hutchinson's device transmitted sound from a microphone to the ear via a set of headphones.

Ernest Rutherford and British physicist Frederick Soddy (1877–1956) found that **radioactive elements changed into other forms when they emitted radioactivity**. This transmutation, as they later called it (see 1916), always happened in the same way: for instance, thorium changes into radium. Rutherford identified the time it takes for **half the radioactive material to decay into another form**, which he later **named its half-life**. Soddy went on to demonstrate that some elements had variants, known as isotopes, which may or may not be radioactive. Half-lives of elements and their isotopes vary: for some isotopes of beryllium it is a fraction of a second, but for the element bismuth-209, it is a billion times longer than the estimated age of the Universe.

In biology, **Karl Landsteiner elaborated on his theory of blood compatibility**. On November 14, he announced that he had identified **three different blood groups, A, B, and O**, on the basis of compatibility patterns. Another rarer blood group, AB, was discovered later.

A meeting of the Zoological Society of London reported on **the discovery of a spectacular new large mammal** from the forests of Africa. **The okapi** had been **discovered by explorer Harry Johnston** (1858–1927) and was described on the basis of examination of its skin and skull.

In November, German psychiatrist **Alois Alzheimer** (1864–1915) examined a woman exhibiting signs of a **severe form of dementia**—and described symptoms of the disease that eventually carried his name. Following her death in 1906, Alzeimer examined her brain and observed the **abnormal plaques** that are characteristic of **Alzheimer's disease**.

On December 12, Italian inventor **Guglielmo Marconi** was reported to have **sent the first radio signal across the Atlantic Ocean**—from Porthcurno, the most southwesterly tip of England, to Newfoundland, in North America. Although some people suggested that it was nothing more than interference, others described it as a deliberate Morse code signal.

BLOOD GROUPS

Red blood cells from some people carry surface components called antigens. Karl Landsteiner identified two types—A and B—which can trigger the fatal clumping (agglutination) if blood is given to someone with appropriately sensitive antibodies. Blood group O has no antigens so it can be donated to anyone. Blood from group AB has both antigens, and can only be given to a person with AB blood.

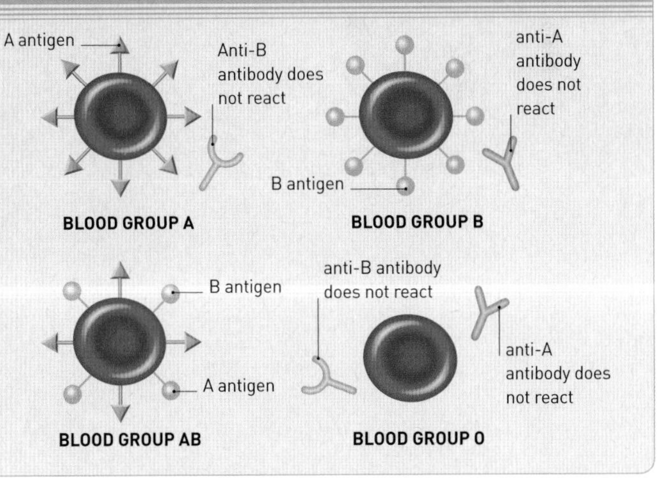
A antigen
Anti-B antibody does not react
BLOOD GROUP A
anti-A antibody does not react
B antigen
BLOOD GROUP B
B antigen
A antigen
BLOOD GROUP AB
anti-B antibody does not react
anti-A antibody does not react
BLOOD GROUP O

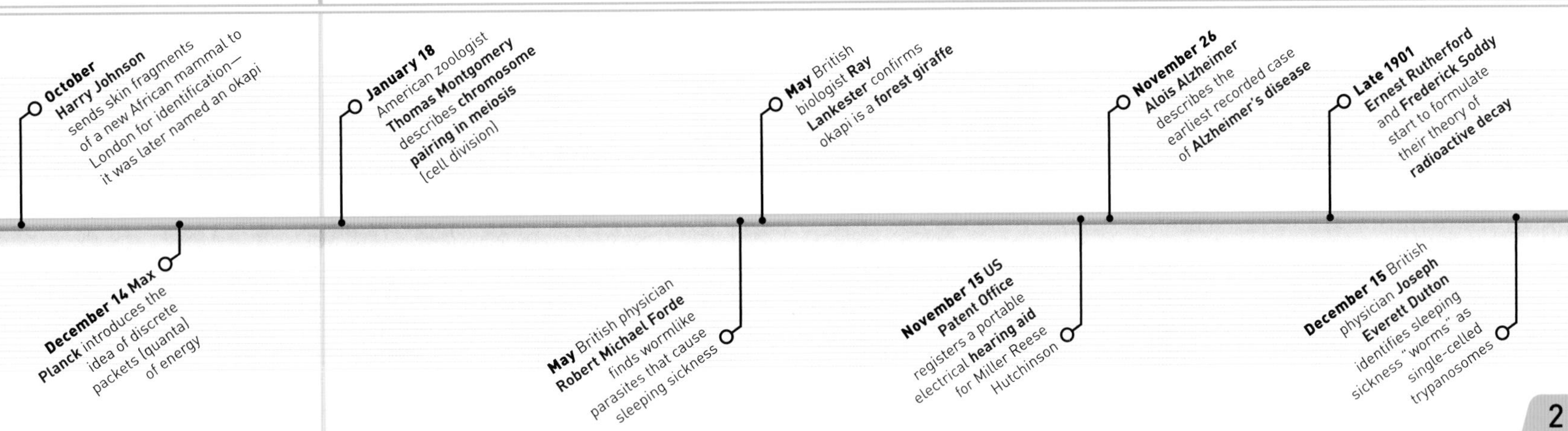
October Harry Johnson sends skin fragments of a new African mammal to London for identification—it was later named an okapi
December 14 Max Planck introduces the idea of discrete packets (quanta) of energy
January 18 American zoologist **Thomas Montgomery** describes **chromosome pairing in meiosis** (cell division)
May British physician **Robert Michael Forde** finds wormlike parasites that cause sleeping sickness
May British biologist **Ray Lankester** confirms okapi is a **forest giraffe**
November 15 US **Patent Office** registers a portable electrical **hearing aid** for Miller Reese Hutchinson
November 26 Alois Alzheimer describes the earliest recorded case of **Alzheimer's disease**
Late 1901 Ernest Rutherford and **Frederick Soddy** start to formulate their theory of **radioactive decay**
December 15 British physician **Joseph Everett Dutton** identifies sleeping sickness "worms" as single-celled trypanosomes

1902

Close to the Earth's surface is the orange-red glow of the troposphere, which contains breathable air and our weather systems. The brownish layer, the tropopause, marks the transition to the gray-blue stratosphere beyond.

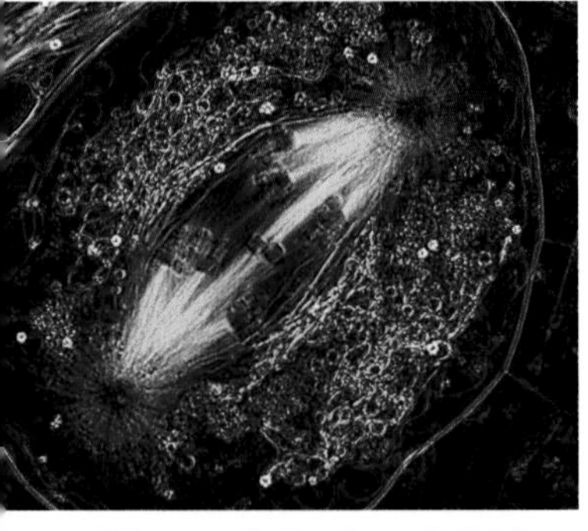

Micrograph showing cell division
As a cell divides, chromosomes are pulled apart, and the genes they carry are passed on to the two new cells formed.

ON JANUARY 1, NATHAN STUBBLEFIELD (1860–1928), a Kentucky farmer and inventor, showed off an electrical device that could send voice and music wirelessly over a distance of half a mile. Although it provoked scientific discussion, this wireless technology did not survive because it relied on disturbances generated by electromagnetic induction, rather than radio signals, and was vulnerable to interference.

American and German biologists **Walter Sutton** (1877–1916) and **Theodor Boveri** (1862–1915) independently **identified chromosomes as the carriers of genetic material**. Nearly 40 years earlier, Gregor Mendel had shown that inherited characteristics were the result of particles (see 1866). Sutton looked at sperm-forming cells of grasshoppers and saw that moving chromosomes mirrored Mendel's particles of inheritance. Boveri saw that sea urchin embryos needed an **intact chromosome set** to develop properly.

> INDIVIDUAL CHROMOSOMES POSSESS **DIFFERENT QUALITIES.**
>
> **Theodor Boveri**, *On Multipolar Mitosis as a Means of the Cell Analysis*, February 17, 1902

UP TO
621
MILES:
THERMOSPHERE

up to 53 miles: mesosphere
up to 31 miles: stratosphere
up to 10 miles: troposphere
sea level

Layers of Earth's atmosphere
The atmosphere is made up of four layers. The gases are concentrated in the thin troposphere.

In April, meteorologist **Léon Teisserenc de Bort** (1855–1913) reported to the French Academy of Sciences on his **investigation of the atmosphere**. Over a period of 10 years he had sent up more than 200 specially equipped hydrogen balloons. He found that weather systems occurred in a layer that extended at least 6 miles (9 km) above the Earth's surface. Beyond this layer the air was thinner and conditions calmer. Later, de Bort called the lower layer the troposphere, and the upper one, the stratosphere.

On February 17, the Stanley Motor Carriage Company was founded in the US for the **production of a steam-driven car** first built in 1897; the factory closed in 1924.

The Stanley Steamer
Early Stanley cars had boilers under the seats that generated steam. Known as Stanley Steamers, they were fuelled by a gasoline burner and started by a crank.

1903

The Wright brothers' 1903 Flyer first flew in Kitty Hawk, North Carolina

THE FIRST DESCRIPTION OF A PROCEDURE that would eventually become known as **chromatography** was presented at the Proceedings of the Warsaw Society of Naturalists in 1903. Russian botanist **Mikhail Tsvet** (1872–1919) had managed to separate the **chemical components of plant pigments**. He first let the mixture dissolve in petroleum ether and then ran it through a column of finely ground calcium carbonate. The orange, yellow, and green pigments separated into different bands: the ones that dissolved better in the solvent travelled faster and further. Tsvet's technique would later be adopted as an important analytical tool for separating mixtures of substances.

Humankind's attempts at powered flight reached an important breakthrough on **December 17**, when American inventor brothers **Orville** (1871–1948) and **Wilbur Wright** (1867–1912) achieved the **first controlled man-carrying flight powered by an engine**. For many years pioneers of aviation had tried hot-air balloons and gliders—with varying degrees of success. The

January 1 Nathan Stubblefield demonstrates a new device that sends sound **wirelessly**

February 17 Theodor Boveri shows that an intact set of chromosomes is needed for normal development

February 17 Stanley Motor Carriage Company is established

April 28 Léon Teisserenc de Bort describes stratosphere

American paleontologist **Barnum Brown** documented first remains of ***Tyrannosaurus rex***

Austrian biologists **Alfred von Decastello** and **Adriano Sturli** describe blood group AB

October 17 Walter Sutton suggests that chromosomes are the physical basis of Mendel's laws of inheritance

September Ernest Rutherford and Frederick Soddy set out their theory that decay of atoms changes an element into another form

December Robert Scott's Antarctic expedition arrives at the southernmost point reached by humans

February Ernest Rutherford calls **Paul Villard's radiation** gamma rays

March 21 Mikhail Tsvet introduces chromatography

June 21 Dutch physiologist **Willem Einthoven** describes first electrocardiograph

Russian scientist **Konstantin Tsiolkovsky** describes how rockets could study outer space

" I NOW FELT… CERTAIN THAT THE DAY WOULD COME WHEN MANKIND WOULD BE ABLE TO SEND MESSAGES WITHOUT WIRES ACROSS THE ATLANTIC. "

Guglielmo Marconi, Italian inventor, in *Messages without Wires,* 1901

The equipment at the Marconi wireless station in Massachusetts, US, which transmitted across the Atlantic to Cornwall, England.

Wright brothers experimented with glider design to maximize lift and added a lightweight aluminium gasoline engine to provide the power. **Orville made the first 12-second flight** in their aircraft, the 1903 Flyer, and flew 40 yd (37 m). The same day, Wilbur managed 284 yd (260 m) in 59 seconds.

Orville and Wilbur Wright
The Wright brothers ran a bicycle shop, but were inspired by early attempts at aviation. By 1908, they had managed an hour-long flight, and carried a passenger.

The Wright brother's first aircraft was designed to minimize load and maximize flexibility. It had a spruce ash wooden frame covered in muslin. The engine gave it enough speed for the wings to generate more lift than the weight of the machine: the **principle of flight**.

AFTER THREE YEARS of sending transatlantic radio signals—including coherent messages—Italian inventor **Guglielmo Marconi** (1874–1937) set up the **first commercial transatlantic radio service**. By 1907 it had become a regular service.

The March issue of the science journal the *Philosophical Magazine* contained a piece written by British physicist **J.J. Thomson** (1856–1940) in which he described a **new way of looking at the atom** that accommodated the newly discovered electron, known as the **plum pudding model**. Thomson thought of an atom as a "pudding" of positive charge, in which the negatively charged electron "plums" were embedded. Later in the year, the Japanese physicist **Hantaro Nagaoka** (1865–1950) **rejected the idea** that positive and negative charges could intermingle in this way, and suggested his **Saturnian model**.

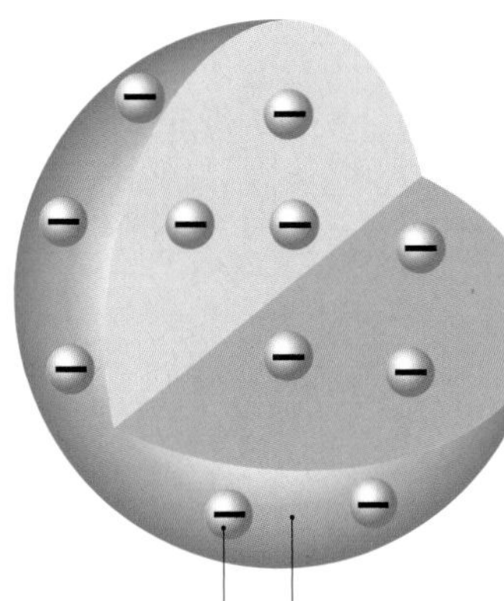
static electron, or plum
cloudlike body —the pudding

The plum pudding model
Thomson's model for atomic structure was an early attempt at suggesting how charged particles could coexist in a neutral atom.

His model had a large positive core, which was orbited by negative electrons, rather like the rings around planet Saturn. Within a few years, experiments in the UK would show that atoms have a dense positive nucleus with encircling electrons—more like Nagaoka's model.

7,000,000 MILES
THE **DISTANCE** OF **HIMALIA** FROM **JUPITER**

In July, Piero Conti (1865–1939), an Italian businessman from the volcanic region of Larderello, Tuscany, **demonstrated a steam engine that ran on geothermal power**. Conti succeeded in producing enough electricity from a dynamo to illuminate five light bulbs. Conti's legacy is that the Larderello region now produces 10 percent of the world's geothermal electricity.

In November, British physicist **John Ambrose Fleming** (1849–1945) **filed a patent for the first vacuum diode**—designed by adding a positive electrode (anode) to an Edison light bulb. Electrons from the hot filament bulb flowed through the bulb's vacuum to the cold anode—converting alternating current (AC) signals to direct current (DC). This marked the **start of an era of electronics,** and for decades improved versions of Fleming's diodes were used in many devices, from radios to the first computers.

In early December at the Lick Observatory in California, American astronomer **Charles Dillon Perrine** (1867–1951) **discovered a moon of Jupiter** that was later **named Himalia**—after a mythical Greek nymph. It is the sixth largest of Jupiter's moons and is thought to have been formed from an asteroid captured in the planet's orbit.

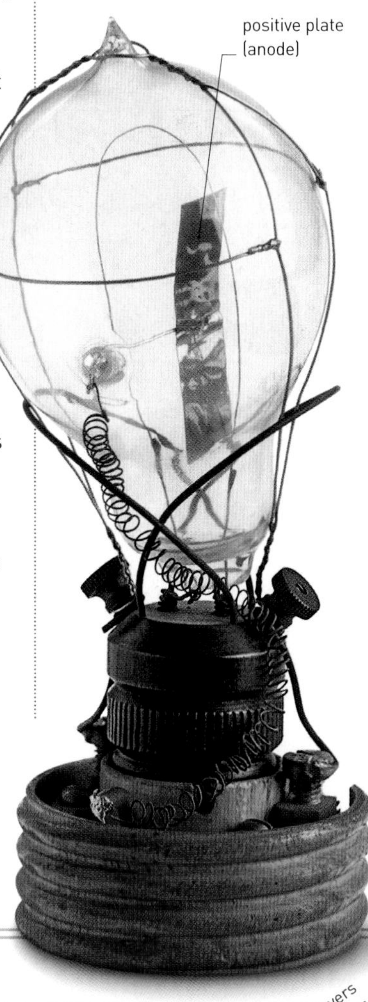
positive plate (anode)

Fleming's vacuum diode
The design features a metal plate that acts as an anode (positively charged electrode) to attract electrons from the bulb filament, thereby creating direct current.

September 24 George Darwin suggests the **Earth is heated by radioactivity**

October 17 David Bruce reports that **sleeping sickness** is spread by tsetse flies

Dutch scientist **Cornelius Koning** suggests that fungi help decomposition of organic matter to form humus

March J.J. Thomson describes plum pudding model of atomic structure

June British astronomer **Edward Maunder** describes patterns of sunspot distribution

German biologist **Friedrich Meves** describes finding mitochondria in plant cells

July 4 Piero Conti tests the first generator powered by geothermal energy

November 16 British Patent Office registers the thermionic diode valve for John Fleming

December 3 Dillon Perrine discovers the 10th moon of Jupiter

December 5 Hantaro Nagaoka describes a **Saturnian model** of atomic structure

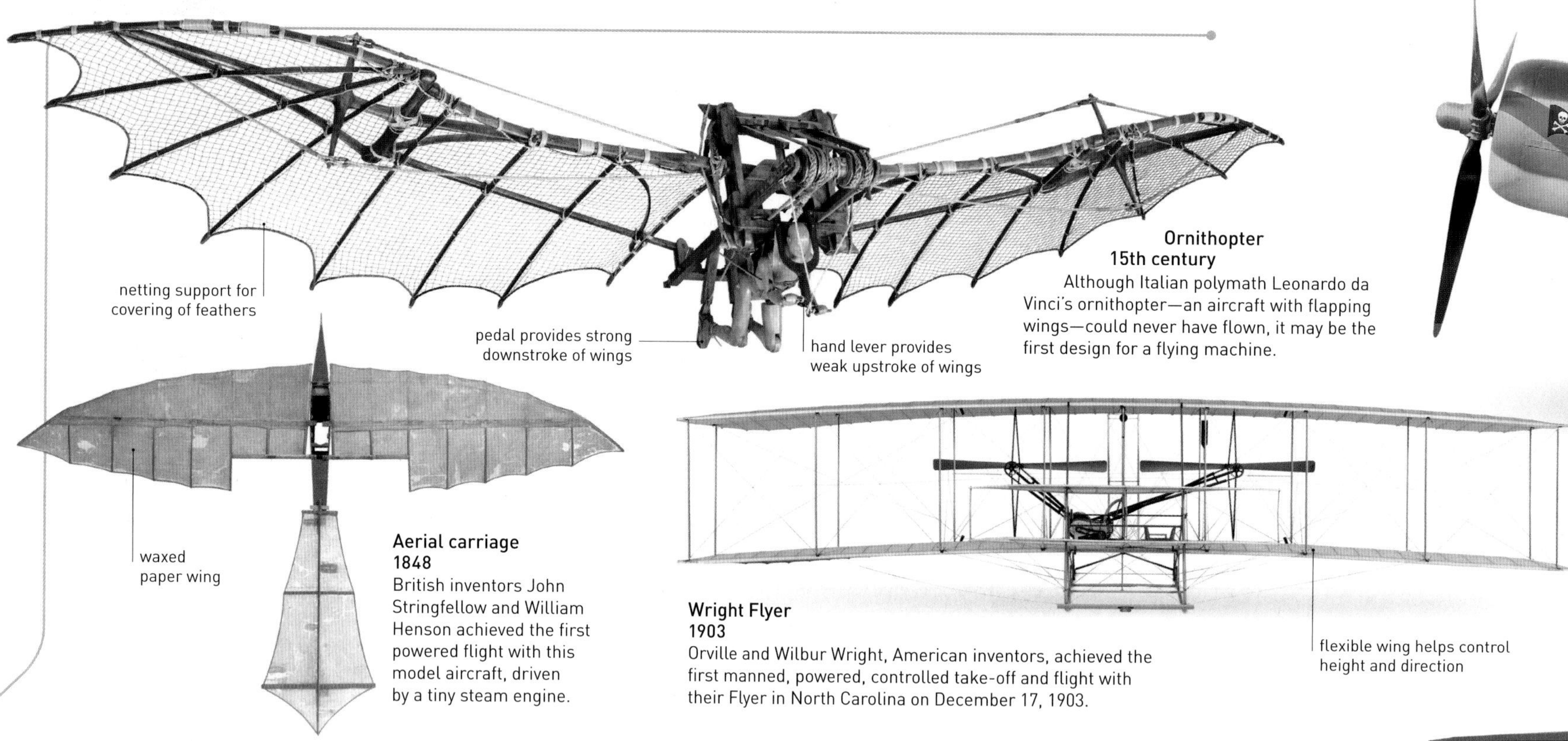
netting support for covering of feathers
pedal provides strong downstroke of wings
hand lever provides weak upstroke of wings
waxed paper wing
flexible wing helps control height and direction

Ornithopter
15th century
Although Italian polymath Leonardo da Vinci's ornithopter—an aircraft with flapping wings—could never have flown, it may be the first design for a flying machine.

Aerial carriage
1848
British inventors John Stringfellow and William Henson achieved the first powered flight with this model aircraft, driven by a tiny steam engine.

Wright Flyer
1903
Orville and Wilbur Wright, American inventors, achieved the first manned, powered, controlled take-off and flight with their Flyer in North Carolina on December 17, 1903.

FLYING MACHINES

THE DEVELOPMENT OF POWERED FLYING MACHINES LED TO A MILITARY AND TRANSPORT REVOLUTION

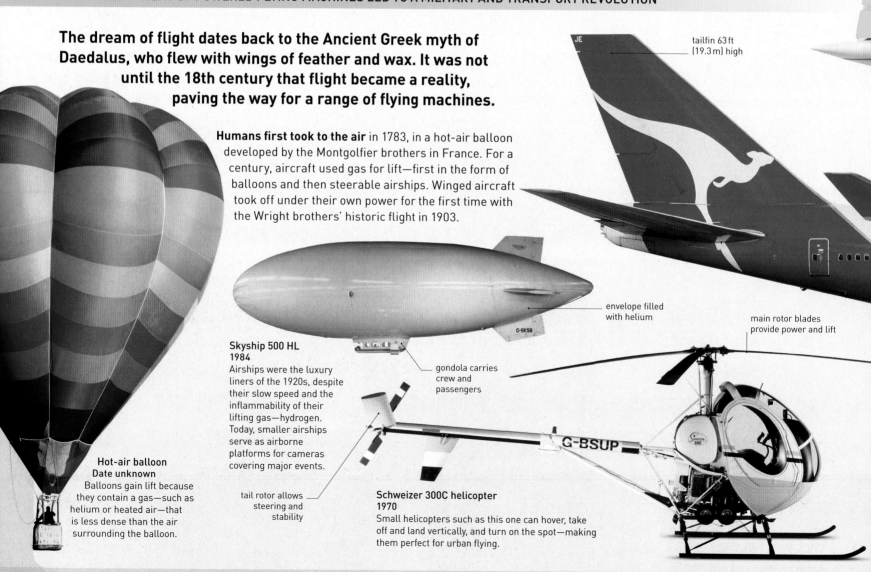

The dream of flight dates back to the Ancient Greek myth of Daedalus, who flew with wings of feather and wax. It was not until the 18th century that flight became a reality, paving the way for a range of flying machines.

Humans first took to the air in 1783, in a hot-air balloon developed by the Montgolfier brothers in France. For a century, aircraft used gas for lift—first in the form of balloons and then steerable airships. Winged aircraft took off under their own power for the first time with the Wright brothers' historic flight in 1903.

tailfin 63 ft (19.3 m) high

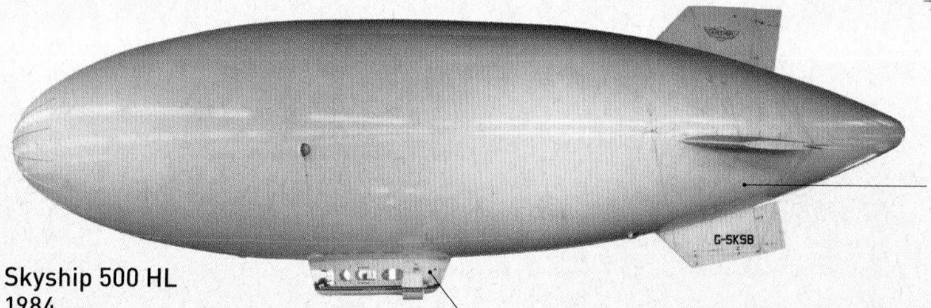

envelope filled with helium

gondola carries crew and passengers

Skyship 500 HL
1984
Airships were the luxury liners of the 1920s, despite their slow speed and the inflammability of their lifting gas—hydrogen. Today, smaller airships serve as airborne platforms for cameras covering major events.

main rotor blades provide power and lift

tail rotor allows steering and stability

Hot-air balloon
Date unknown
Balloons gain lift because they contain a gas—such as helium or heated air—that is less dense than the air surrounding the balloon.

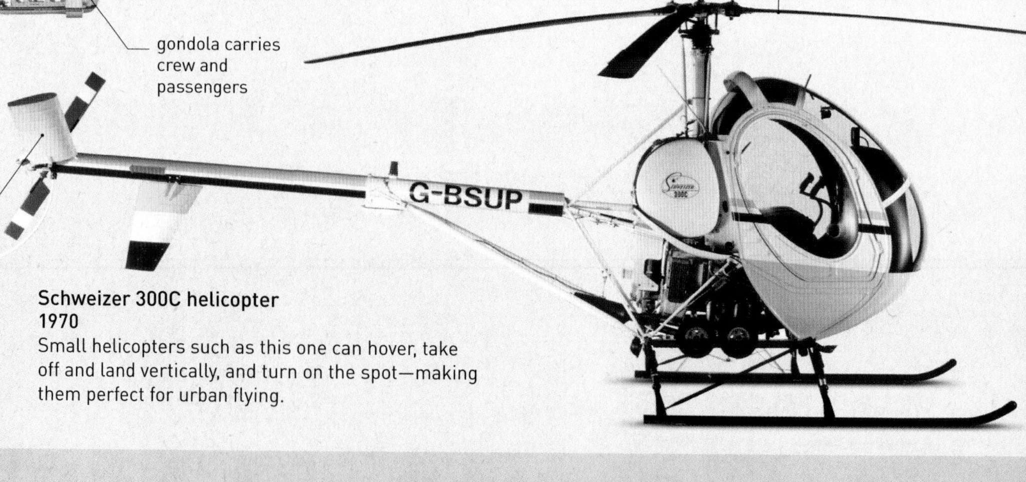

Schweizer 300C helicopter
1970
Small helicopters such as this one can hover, take off and land vertically, and turn on the spot—making them perfect for urban flying.

8
fully retractable landing gear
light aluminum body
foldable wings
metal body
wing float
faceted body shape
narrow, streamlined fuselage
drooping nose cone allows runway visibility
triangular wings, roundly tapered at the ends
wide body
QANTAS
THE SPIRIT OF AUSTRALIA
LONGREACH
VH-OJE
high-power, low-fuel-consumption turbofan engine
tailfin provides directional stability
hold can be opened in space to release payload
NASA
United States
main engine
long, thin wing reduces drag

Vought F4U Corsair
1943
Fast and highly maneuverable, single-engine fighter-bomber planes such as the Corsair were in great demand during World War II.

Supermarine Walrus
1935
Used for survey missions during World War II, the Walrus could be catapulted into the air from the deck of a ship. It could also land on water and be lifted back aboard.

Lockheed F-117A Nighthawk
1982
Stealth aircraft such as the Nighthawk are designed to avoid detection by enemy radar. They are shaped and colored to diffract and reflect radar waves indistinctly.

Concorde
1976
A British–French engineering collaboration and powered by turbojet engines, the Concorde was one of two supersonic airliners to enter commercial service. It could fly from New York, US, to London, UK, in under three hours.

Boeing 747
1970
Until it was surpassed by the A380 in 2007, the Boeing 747 was the world's biggest airliner, with the capacity to carry up to 680 passengers.

US Space Shuttle
1982
Although it had to be launched by rockets, the US Space Shuttle was the first spacecraft to land on a runway and to be used for further flights like a conventional aircraft.

Schleicher ASK13
1966
Made using strong, ultralight modern materials, gliders can fly far and high, using rising air currents for lift.

1905

> **"** BODIES OF **MICROSCOPICALLY VISIBLE SIZE** SUSPENDED IN A LIQUID WILL PERFORM **MOVEMENTS** OF SUCH A MAGNITUDE THAT THEY CAN BE… OBSERVED IN A **MICROSCOPE. "**

Albert Einstein, on Brownian motion, July 18, 1905

ALBERT EINSTEIN PUBLISHED FOUR REVOLUTIONARY PAPERS in what became known as his Annus Mirabilis (Miracle Year). His first expanded Max Planck's idea from 1900 that energy existed in minute packets of energy. Einstein proposed that **packets of light energy** could help explain **the photoelectric effect**, in which wavelength (not intensity) of light provides the energy needed to eject electrons from a metal surface—later proven experimentally (see 1921).

Einstein's second theory explained how the random movement of tiny particles in gas or water was caused by the motion of molecules bombarding the particles—Brownian motion. In September, **Einstein published his theory of special relativity** (see pp.244–45). In it he reconciled the constancy of the speed of light with the principle of relativity: the idea that mechanical processes happened in the same way whether at rest or moving. Light had previously been regarded as an exception to this principle—but Einstein used an analysis of time and space to show that it was compatible. Finally in November, he **published a conclusion** that arose from his work on relativity. He suggested that when an object emitted energy, it lost mass too, so energy and mass were interchangeable. This relationship was expressed in a **simple equation: $E = mc^2$.**

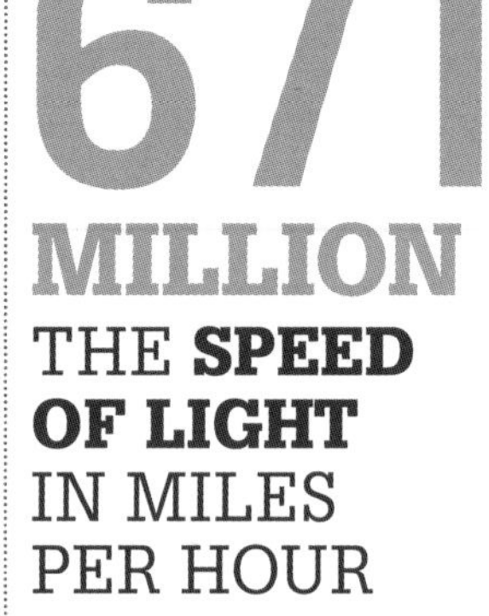
671
MILLION
THE SPEED OF LIGHT
IN MILES PER HOUR

British biologist **William Bateson** (1861–1926) was among a number of scientists interested in the study of inheritance. In a letter of April this year he called it **genetics**.

Growing populations needed more food, so the demand for crop fertilizer increased. German chemists devised a way to make ammonia, the compound that provided nitrogen for plant growth. Ammonia is formed from hydrogen and nitrogen. **Fritz Haber** (1868–1934) described a key reaction between atmospheric nitrogen and hydrogen. During World War I, when natural sources of nitrate came under Allied control, **Carl Bosch** (1874–1940) used Haber's principle to produce industrial quantities of ammonia.

German chemist **Alfred Einhorn** (1856–1917) succeeded in making the local anesthetic **procaine, later traded as Novocaine**. The drug would become a standard painkiller.

ALBERT EINSTEIN (1879–1955)

Born in Germany, Einstein took Swiss citizenship before finding work at the Bern patent office in 1903. Here he wrote his ground-breaking papers and was awarded a doctorate by the University of Zurich. He completed his general theory of relativity in 1915 and was awarded the Nobel Prize in 1922. In later years, Einstein became a US citizen.

1906

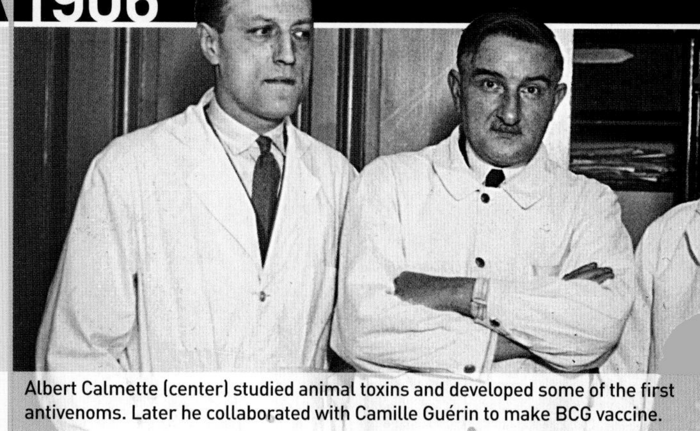

Albert Calmette (center) studied animal toxins and developed some of the first antivenoms. Later he collaborated with Camille Guérin to make BCG vaccine.

BRITISH GEOLOGIST RICHARD OLDHAM (1858–1936) studied the seismic shocks that went through Earth following earthquakes. At the end of the previous century Oldham had identified two kinds of waves: fast-moving, longitudinal, primary P-waves and slower, transverse, secondary S-waves. He found that below a certain depth within Earth, **seismic waves travel much more slowly—and S-waves stop** altogether. In February, Oldham suggested this phenomenon could be explained by the **presence of a core inside Earth** that was composed of a different kind of material. Later studies would show that Earth has a core located at a depth of 1,802 miles (2,900 km), and confirmed that the outer core is fluid.

On Christmas Eve, a Canadian inventor, working at the US Weather Bureau, **broadcast the first radio program**. Reginald Fessenden (1866–1932), a rival of Guglielmo Marconi, transmitted his program from Brant Rock, Massachusetts. It included a voice message and music was directed to ships in the Atlantic Ocean more used to receiving messages in Morse code.

At the Institut Pasteur, France, work began on a program that led to the development of a **new vaccine** that would protect millions of peoples from a **deadly disease: tuberculosis (TB)**. French scientists **Albert Calmette** (1863–1933) and **Camille Guérin** (1872–1961) had been inspired by the historical uses of the harmless cowpox as a way to vaccinate humans against dangerous smallpox (see 1796). They thought that a similar process could be tried

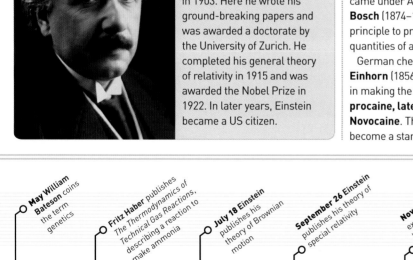

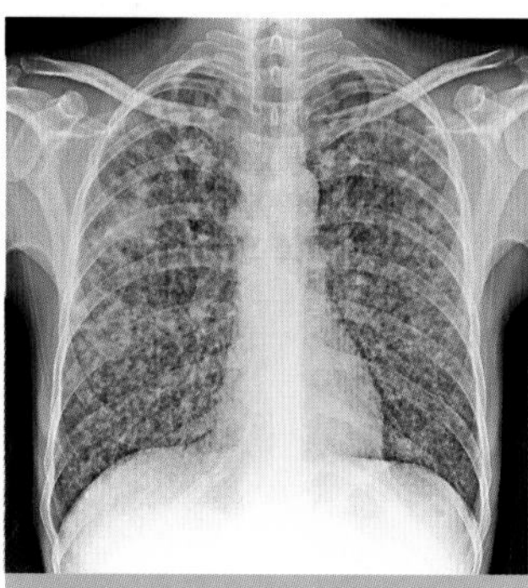

TUBERCULOSIS (TB)

German physician Robert Koch (1843–1910) identified the bacteria that caused TB, and was awarded the Nobel Prize in 1905. Formerly called consumption, TB was often fatal. It infects the lungs—causing lesions, or tubercles—and is spread through droplets dispersed by coughs and sneezes. The BCG vaccine was the first protection against TB.

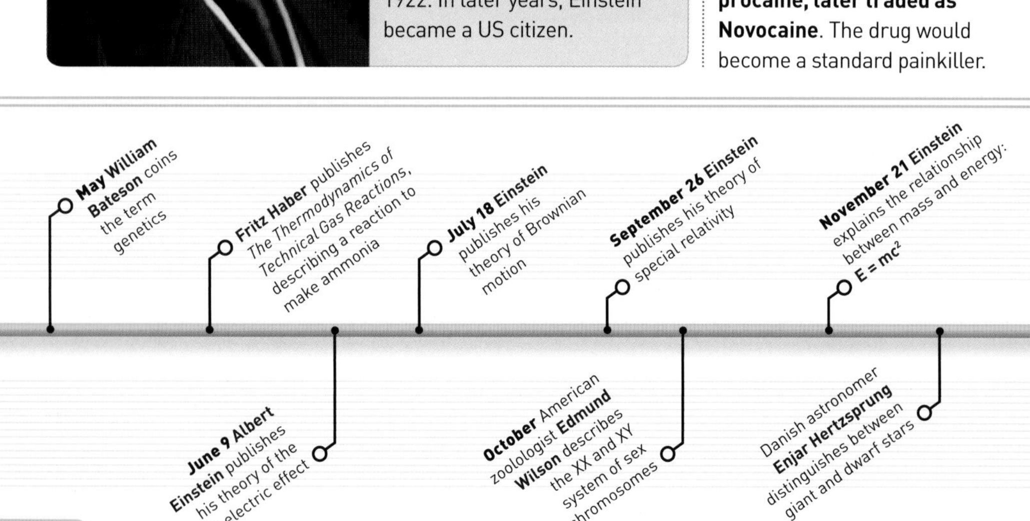
May William Bateson coins the term genetics
Fritz Haber publishes *The Thermodynamics of Technical Gas Reactions*, describing a reaction to make ammonia
June 9 Albert Einstein publishes his theory of the photoelectric effect
July 18 Einstein publishes his theory of Brownian motion
September 26 Einstein publishes his theory of special relativity
October American zoologist **Edmund Wilson** describes the XX and XY system of sex chromosomes
November 21 Einstein explains the relationship between mass and energy: $E = mc^2$
Danish astronomer **Enjar Hertzsprung** distinguishes between giant and dwarf stars

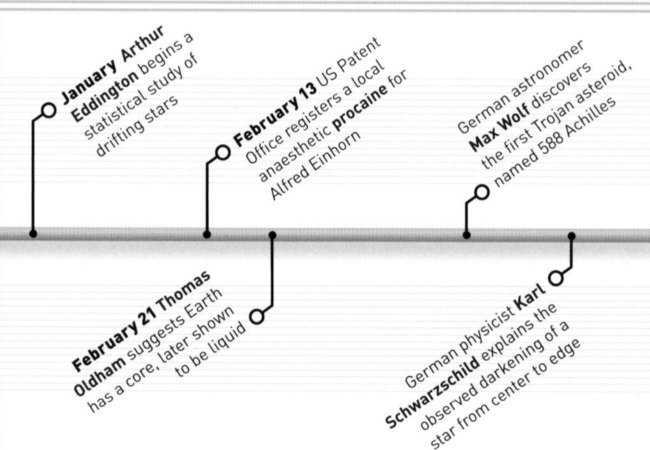
January Arthur Eddington begins a statistical study of drifting stars
February 13 US Patent Office registers a local anaesthetic **procaine** for Alfred Einhorn
February 21 Thomas Oldham suggests Earth has a core, later shown to be liquid
German astronomer **Max Wolf** discovers the first Trojan asteroid, named 588 Achilles
German physicist **Karl Schwarzschild** explains the observed darkening of a star from center to edge

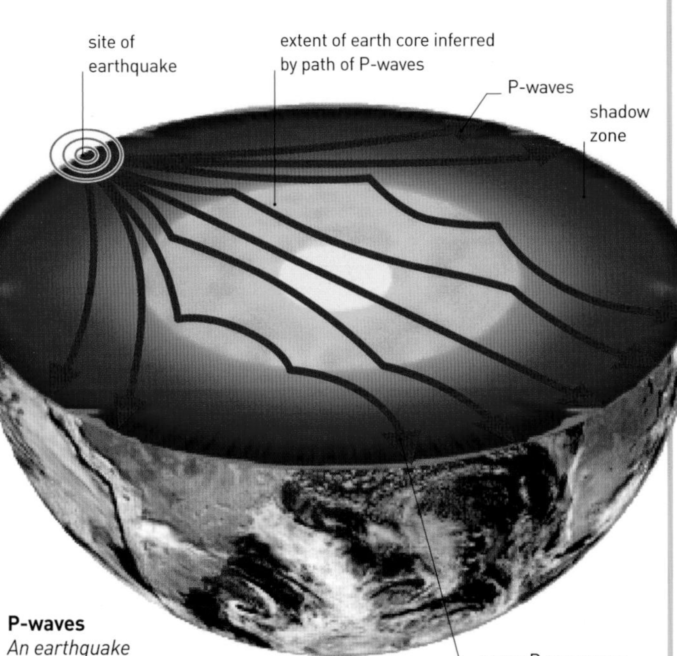

site of earthquake
extent of earth core inferred by path of P-waves
P-waves
shadow zone
some P-waves are refracted by the core

P-waves
An earthquake produces fast-moving waves that slow down and bend as they pass through Earth's core, creating a "shadow" zone where a seismograph cannot detect them.

out for TB. Calmette and Guerin wanted to use a **bovine (cattle-borne) form of TB to develop a vaccine** against the human form of TB. They set up cultures of bovine TB on potato slices that had been soaked with ox gall (obtained from cow's liver) and set out to continue the culture line until it was safe to use. It was 10 years before their **BCG (Bacille Calmette Guérin) vaccine** was ready for use on animals—it was not tried on humans until 1921.

9,800°F

THE **TEMPERATURE** OF **EARTH'S INNER CORE**

By 1907, Nobel laureate Ivan Pavlov, shown here with his staff and one of the dogs, had received much acclaim for his experiments with reflex responses in animals.

AMERICAN INVENTOR LEE DE FOREST PATENTED THE TRIODE, which could be used to amplify electrical signals and act as a switch. Until the invention of the transistor years later (see 1947), diode and triode valves were used in circuits in radios, televisions, and computers.

The Belgian-born chemist **Leo Baekeland** (1863–1944) produced the **first plastic made from synthetic materials**. Baekeland's new product, later named Bakelite, was heat-resistant and non-conductive. It was used for electrical insulation as well as to make domestic utensils and children's toys.

French filmmakers **Auguste** (1862–1954) and **Louis Lumière** (1864–1948) began marketing the **first commercial color photographic method**. This **autochrome** (self-coloring) process involved using a negative plate that was coated with transparent grains of dye-colored starch, which filtered light before it hit a layer of photographic emulsion.

Following J.J. Thomson's demonstration that electrons were particles (see 1896), American physicist **Robert Millikan** (1868–1953) began experiments to calculate their electrical charge. He found that the charge on falling droplets of oil was always a multiple of a tiny value: **the charge on a single electron**. Millikan published his first results in 1910.

> “THE **MATERIAL** OF **1,000 USES.**”
>
> **Leo Baekeland,** on Bakelite

Following the discovery that radioactive materials decay at fixed rates (see 1901), American scientist **Bertram Boltwood** (1849–1936) used this **to calculate the age of rocks**. He found that ores of uranium (uraninite) contained a proportion of lead, and the older the rock, the more lead it contained. Lead was a product of a known rate of uranium decay, so it accumulated over time. The study of radiometric dating was later taken up by British geologist Arthur Holmes (see 1913).

The German chemist **Emil Fischer** (1852–1919) succeeded in **linking together the building blocks of proteins**. These units—called **amino acids**—came in a variety of types. Fischer identified many of them and showed how they bonded in protein chains. His work was the foundation of protein chemistry.

Russian physiologist **Ivan Pavlov** (1849–1936) converted his laboratory so that he could concentrate on his **study of animal behavior**. Using dogs, Pavlov had shown that animals could be taught to salivate when they heard a ringing bell, known as the **conditioned, or learned, reflex** (see 1889).

Early color photoraph
This autochrome image shows Doug, niece of the Lumière brothers, out in her pram with her nurse. It was taken between 1906 and 1912.

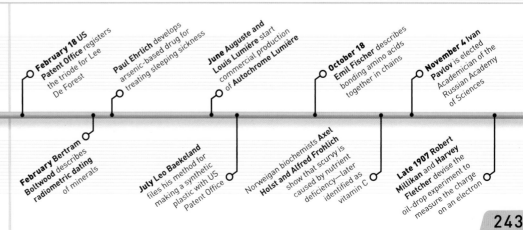

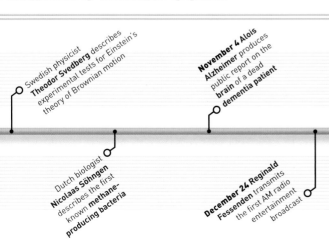

- Swedish physicist **Theodor Svedberg** describes experimental tests for Einstein's theory of Brownian motion
- Dutch biologist **Nicolaas Söhngen** describes the first known **methane-producing bacteria**
- **November 4 Alois Alzheimer** produces public report on the **brain** of a dead **dementia patient**
- **December 24 Reginald Fessenden** transmits the first AM radio entertainment broadcast
- **February 18 US Patent Office** registers the triode for Lee De Forest
- **February Bertram Boltwood** describes **radiometric dating** of minerals
- **Paul Ehrlich** develops arsenic-based drug for treating sleeping sickness
- **June Auguste and Louis Lumière** start commercial production of **Autochrome Lumière**
- **July Leo Baekeland** files his method for making a synthetic plastic with US Patent Office
- **October 18 Emil Fischer** describes bonding amino acids together in chains
- Norwegian biochemists **Axel Holst and Alfred Frohlich** show that scurvy is caused by nutrient deficiency—later identified as vitamin C
- **November 4 Ivan Pavlov** is elected Academician of the Russian Academy of Sciences
- **Late 1907 Robert Millikan** and **Harvey Fletcher** devise the oil-drop experiment to measure the charge on an electron

UNDERSTANDING RELATIVITY

EINSTEIN'S GROUND-BREAKING THEORIES REVEALED THAT SPACE AND TIME ARE INTIMATELY LINKED

In the early 20th century, German-born physicist Albert Einstein published two theories that revolutionized our understanding of space, time, energy, and gravity. The first, known as the special theory of relativity, only applies in certain circumstances; the second is the general theory of relativity.

ALBERT EINSTEIN
When Einstein published his theory of special relativity, he was working as a clerk at the patent office in Bern, Switzerland.

In the 19th century, physicists thought that "empty" space was actually filled with a substance, which they called ether, and that light travels through the ether at a fixed speed. Since our planet is moving, they predicted that the measured speed of light would differ from its actual speed—just as a passing car appears to move faster or slower than its actual speed if you are also moving. To test this idea—and in an attempt to determine Earth's actual, or "absolute" speed through space—they measured the speed of light in different directions and at different times of the year. But in every case, the speed of light was always exactly the same.

SPECIAL THEORY OF RELATIVITY

The fact that the speed of light is constant was perplexing, and it challenged common-sense assumptions about the nature of time and space. In his theory of special relativity Einstein showed that time and space are indeed "relative" quantities: that two observers in relative motion both measuring the distance between the same two points in space, or the time between the same two events, would come up with different answers. Einstein's relativity also did away with the "stationary" ether and showed that there is no absolute reference point in space or time.

OBSERVER WITHIN REFERENCE FRAME

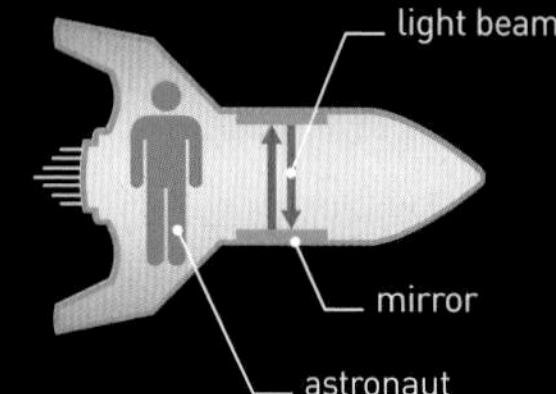
light beam
mirror
astronaut

OBSERVER OUTSIDE REFERENCE FRAME

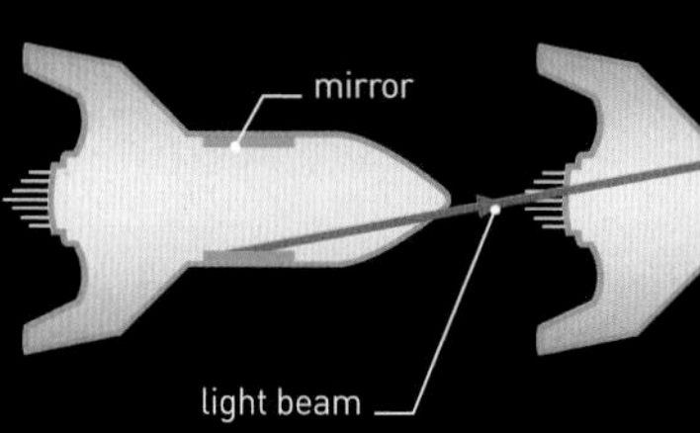
mirror
light beam

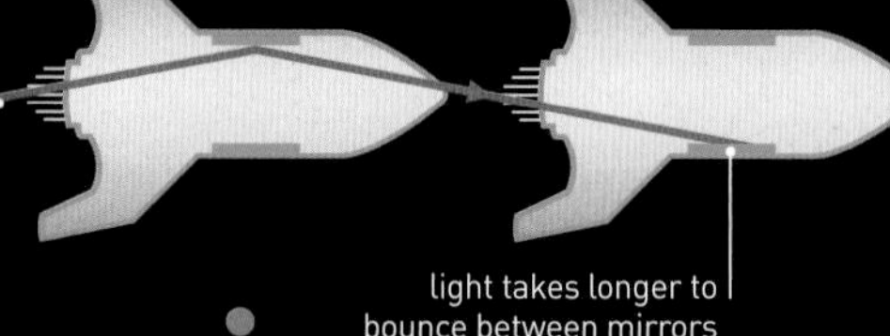
light takes longer to bounce between mirrors
observer on Earth

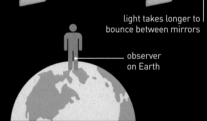

TIME DILATION
A beam of light bounces between mirrors inside a spacecraft that is moving past Earth. An astronaut inside the craft perceives only vertical movement in the light. Viewed from Earth, it appears that the light travels farther between the mirrors, and takes longer to make the same journey. So, time runs slowly in the "moving" frame of reference.

> “ IT FOLLOWED FROM THE **SPECIAL THEORY OF RELATIVITY** THAT **MASS AND ENERGY** ARE ... MANIFESTATIONS OF **THE SAME THING.** ”

Albert Einstein, in the film *Atomic Physics* (1948)

MASS AND ENERGY

In working through the mathematical equations in his theory of relativity, Einstein encountered a surprising result: the mass of an object increases as it gets faster, and at the speed of light an object would have infinite mass. Einstein realized that the speed of light must therefore be the Universe's ultimate speed limit. His equations suggested that mass and energy are equivalent to each other, and he defined a new quantity: "mass-energy." The equivalence of mass and energy is expressed in Einstein's most famous equation, $E = mc^2$, in which E stands for energy, m for mass, and c^2 for the speed of light squared.

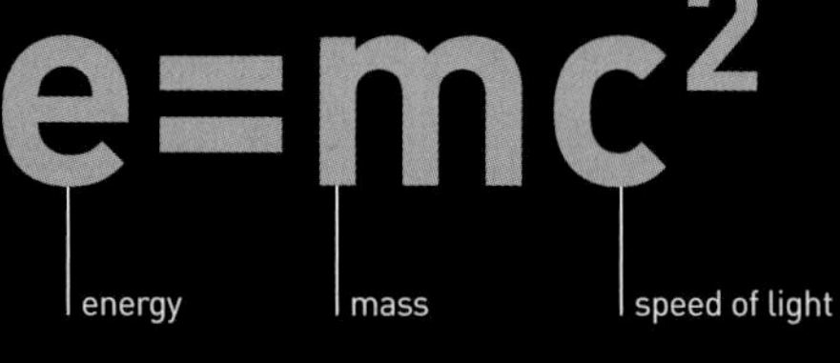
e=mc²
energy
mass
speed of light

PARTICLE ACCELERATOR
Physicists working with particle accelerators routinely use Einstein's theory to predict the increase in mass of high-speed particles, and to work out how much longer it takes for them to decay as a result of time dilation (see above).

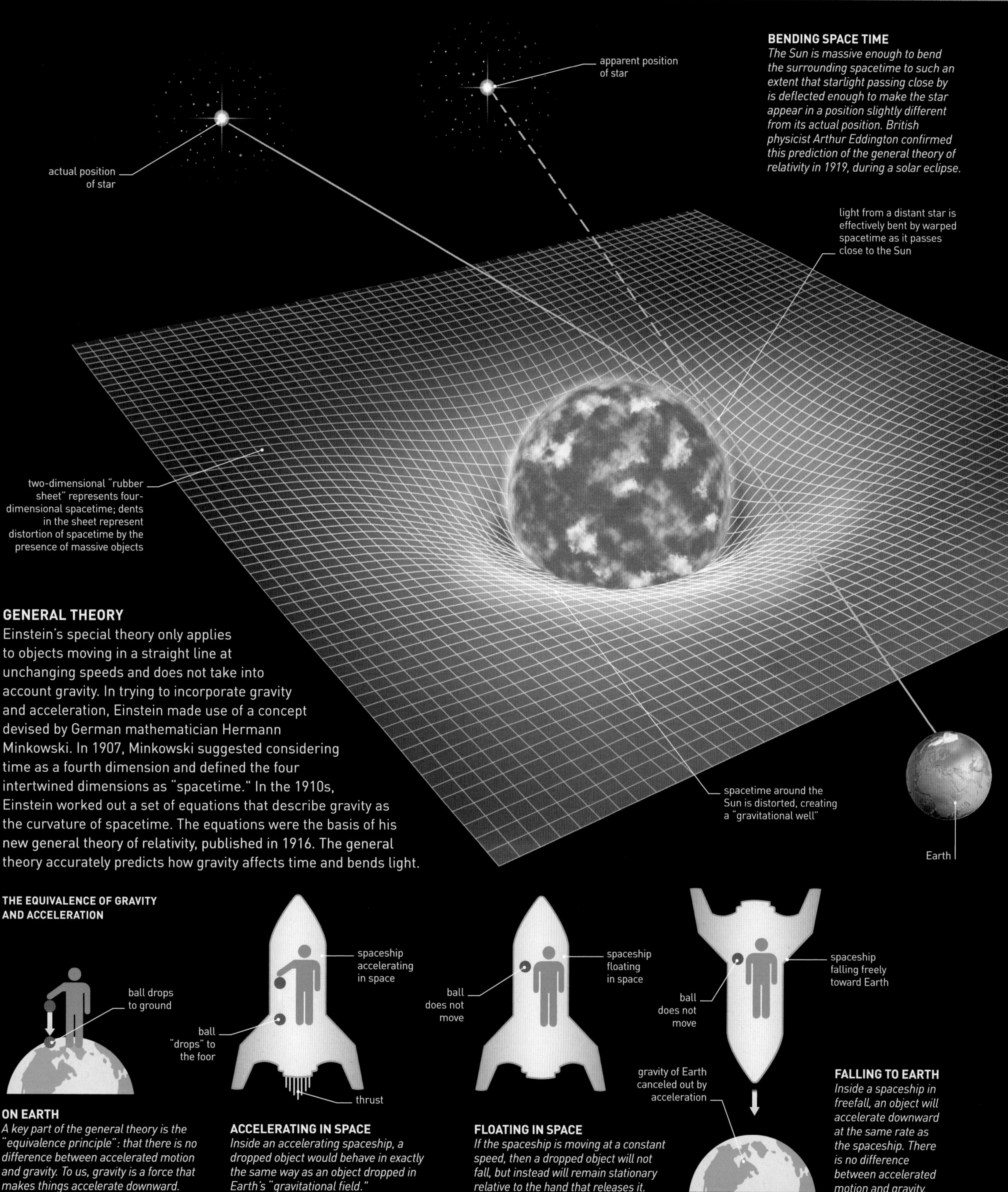

BENDING SPACE TIME
The Sun is massive enough to bend the surrounding spacetime to such an extent that starlight passing close by is deflected enough to make the star appear in a position slightly different from its actual position. British physicist Arthur Eddington confirmed this prediction of the general theory of relativity in 1919, during a solar eclipse.
apparent position of star
actual position of star
light from a distant star is effectively bent by warped spacetime as it passes close to the Sun
two-dimensional "rubber sheet" represents four-dimensional spacetime; dents in the sheet represent distortion of spacetime by the presence of massive objects
spacetime around the Sun is distorted, creating a "gravitational well"
Earth
THE EQUIVALENCE OF GRAVITY AND ACCELERATION
ball drops to ground
spaceship accelerating in space
ball "drops" to the foor
thrust
ball does not move
spaceship floating in space
ball does not move
spaceship falling freely toward Earth
gravity of Earth canceled out by acceleration
ON EARTH
A key part of the general theory is the "equivalence principle": that there is no difference between accelerated motion and gravity. To us, gravity is a force that makes things accelerate downward.
ACCELERATING IN SPACE
Inside an accelerating spaceship, a dropped object would behave in exactly the same way as an object dropped in Earth's "gravitational field."
FLOATING IN SPACE
If the spaceship is moving at a constant speed, then a dropped object will not fall, but instead will remain stationary relative to the hand that releases it.
FALLING TO EARTH
Inside a spaceship in freefall, an object will accelerate downward at the same rate as the spaceship. There is no difference between accelerated motion and gravity.

GENERAL THEORY

Einstein's special theory only applies to objects moving in a straight line at unchanging speeds and does not take into account gravity. In trying to incorporate gravity and acceleration, Einstein made use of a concept devised by German mathematician Hermann Minkowski. In 1907, Minkowski suggested considering time as a fourth dimension and defined the four intertwined dimensions as "spacetime." In the 1910s, Einstein worked out a set of equations that describe gravity as the curvature of spacetime. The equations were the basis of his new general theory of relativity, published in 1916. The general theory accurately predicts how gravity affects time and bends light.

1908

According to some, the inspiration for cellophane came after its inventor, Jacques Brandenberger, saw wine spilled on a tablecloth. But the film proved more useful in waterproof packaging than for stain protection.

SWISS CHEMIST JACQUES E. BRANDENBERGER (1872–1954) invented a way of producing sheets of thin waterproof film made from wood cellulose. The **film** came to be called **cellophane** (for cellulose and *diaphane*, French for transparent). Brandenberger's original idea was to spray liquefied cellulose onto fabric as a stain-repellent, but he found that he could pull away a dry film that was far more useful.

New Zealand-born physicist Ernest Rutherford's **half-life explanation of radioactivity** still commanded attention. He and German physicist Hans Geiger (1882–1945) had devised **a way of measuring radioactivity** by scintillation—counting the flashes of light when the rays struck a zinc sulfide screen. They carried out experiments firing radiation through barriers, enlisting the help of a student called Ernest Marsden (1889–1970). Geiger and Marsden studied the effects of metallic foils on a stream of alpha particles. Marsden's **work using gold foil threw up unexpected results**. According to the atomic structure theory of the time (the plum pudding model, see 1904) alpha particles should have passed straight through the gold, but instead a few bounced back. Rutherford later said it was, "as though you fired a bullet at tissue paper and it bounced back." His analysis of Marsden's results suggested that the deflected particles were striking **a very dense nucleus at the core of each atom**.

In August, five years after their historic maiden flights, the **Wright brothers were in the air again**, but this time they had an audience. In an atmosphere of scepticism, Wilbur had traveled to France to show off their manned, powered aircraft. Over several days he demonstrated his mastery of flying and the assembled crowds grew daily: the Wright brothers became **aviation celebrities**.

> " IT WAS QUITE THE **MOST INCREDIBLE EVENT** THAT HAS EVER HAPPENED TO ME **IN MY LIFE.** "
>
> **Ernest Rutherford, New Zealand-born physicist,** from his lecture *The Development of the Theory of Atomic Structure*, 1936

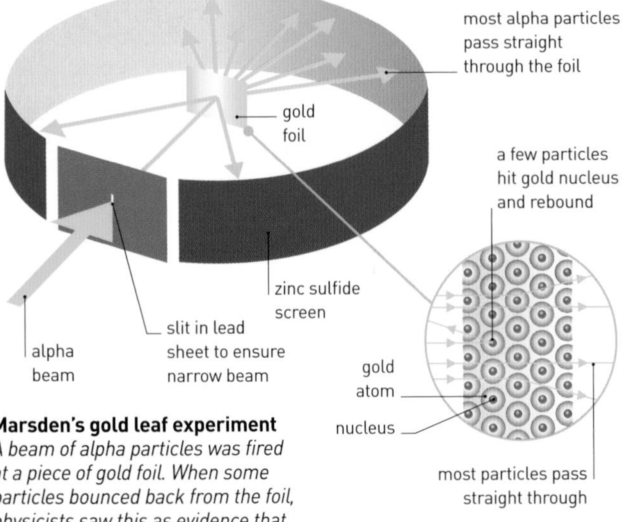
most alpha particles pass straight through the foil
gold foil
a few particles hit gold nucleus and rebound
zinc sulfide screen
slit in lead sheet to ensure narrow beam
alpha beam
gold atom
nucleus
most particles pass straight through
DETAIL OF GOLD FOIL

Marsden's gold leaf experiment
A beam of alpha particles was fired at a piece of gold foil. When some particles bounced back from the foil, physicists saw this as evidence that atoms had very dense nuclei.

1909

A remarkable repository of some of the oldest animal fossils, Canada's Burgess Shale is a record of Cambrian ocean life half a billion years ago.

THE GERMAN DRUG COMPANY Bayer—pioneer of aspirin—was granted a **patent on a sulfur-based drug**. The drug was a derivative of sulfonamides, a class of bacterial agents that would rise to prominence in 1932 and dominate the preantibiotic period. Sulfonamides marked a significant step in **chemotherapy**—the scientific development of drugs that cured disease using sound pharmaceutical principles—but Bayer was unaware of their importance. At the same time, German physician Paul Ehrlich and Japanese biologist **Sahachiro Hata** (1873–1978) were working on their own **custom-made drug** to treat the sexually transmitted bacterial disease **syphilis**.

Søren Sørenson (1879–1963), a Danish chemist, was studying protein—a substance that is sensitive to the effects of acids and alkali. Sørenson worked on a way of quantifying acidity and alkalinity and, as a result, **devised the pH scale**. This scale rated substances according to whether they were acid (1–6), alkaline (8–14), or neutral (7).

The previous year, **Louis Blériot** (1872–1936), a French engineer, had witnessed the public manned flights of Wilbur Wright. Inspired into action, he set his sights on the English Channel. French inventor Jean Pierre Blanchard had crossed it in a balloon (see 1785) but in July, Bleriot became the **first person to cross the English Channel in a manned, powered plane**. The *Blériot XI* left Calais at sunrise on July 25 and landed in Dover 36 minutes later. As well as receiving international acclaim, Blériot won a £1,000 prize from London's *Daily Mail* newspaper.

In America, paleontologist **Charles Walcott** (1850–1927) made an **important discovery**.

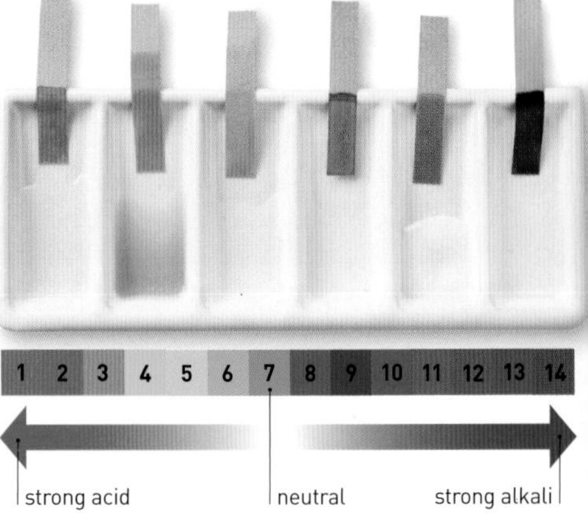
1 2 3 4 5 6 7 8 9 10 11 12 13 14
strong acid
neutral
strong alkali

Measuring pH value
Indicator paper contains a chemical that reacts with acid or alkali to produce a colour to match a scale according to the subject's "strength."

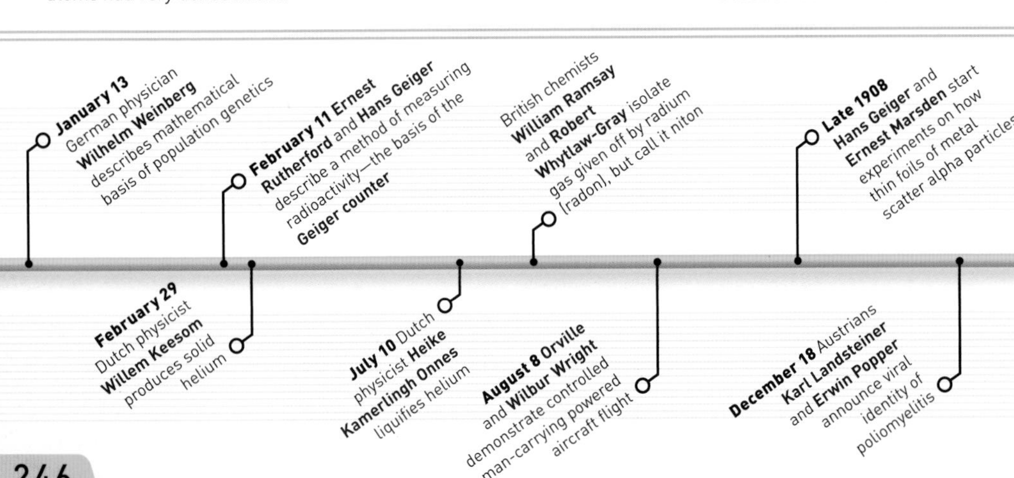
January 13 German physician **Wilhelm Weinberg** describes mathematical basis of population genetics
February 29 Dutch physicist **Willem Keesom** produces solid helium
February 11 Ernest Rutherford and **Hans Geiger** describe a method of measuring radioactivity—the basis of the **Geiger counter**
July 10 Dutch physicist **Heike Kamerlingh Onnes** liquifies helium
British chemists **William Ramsay** and **Robert Whytlaw-Gray** isolate gas given off by radium (radon), but call it niton
August 8 Orville and **Wilbur Wright** demonstrate controlled man-carrying powered aircraft flight
Late 1908 Hans Geiger and **Ernest Marsden** start experiments on how thin foils of metal scatter alpha particles
December 18 Austrians **Karl Landsteiner** and **Erwin Popper** announce viral identity of poliomyelitis

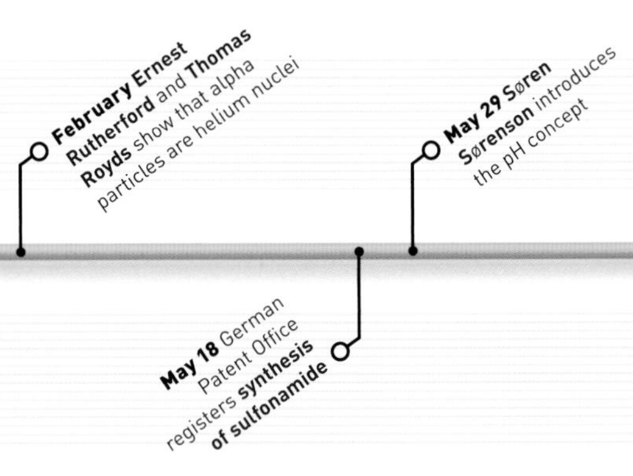
February Ernest Rutherford and **Thomas Royds** show that alpha particles are helium nuclei
May 18 German Patent Office registers **synthesis of sulfonamide**
May 29 Søren Sørenson introduces the pH concept

65 THOUSAND THE NUMBER OF SPECIMENS **UNEARTHED** BY **WALCOTT** AT **BURGESS SHALE**

motor cortex controls coordinated muscle movements

premotor cortex creates the intention to move

prefrontal cortex is involved in determining personality and thought.

Broca's area is associated with the production of language

primary auditory cortex receives and analyzes nerve impulses from the ears

auditory cortex integrates auditory data with memories and other senses

somatosensory cortex receives and analyses nerve impulses from touch receptors

sensory cortex processes sensory information

visual cortex integrates visual data with memories and other senses

primary visual cortex receives and analyses nerve impulses from the eyes

the Wernicke's area is associated with language

BRAIN FUNCTIONS

This "map" of the most complex part of the brain came out of the painstaking work of Korbinian Brodmann. The surface of the cerebral hemispheres—called the cerebral cortex—is divided into sensory, motor, and association areas. Sensory areas receive signals from the rest of the body and motor areas control the dispatch of signals to muscles. Association areas are those involved in complicated, higher processing, such as decision-making and language.

Coming to the end of fieldwork at **Burgess Shale** in the Canadian Rocky Mountains, Walcott found a large deposit of fossils. Subsequent investigation revealed it to be one of **the oldest and best-preserved fossil sites**, with specimens that dated back 500 million years. Walcott attempted to classify the animals into known groups, but scientists have since discovered that many of them belonged to ancient evolutionary dead ends.

German neuroscientist **Korbinian Brodmann** (1868–1918) had been working on the fine structure of the part of the brain called the cerebral cortex, which controls higher functions—for example, decision-making and emotion. Using microscopic studies, Brodmann managed **to identify the different functional regions of this part of the brain** that could be linked to processes demonstrated experimentally by other scientists. Brodmann's cerebral map formed the basis for modern understanding of higher brain function.

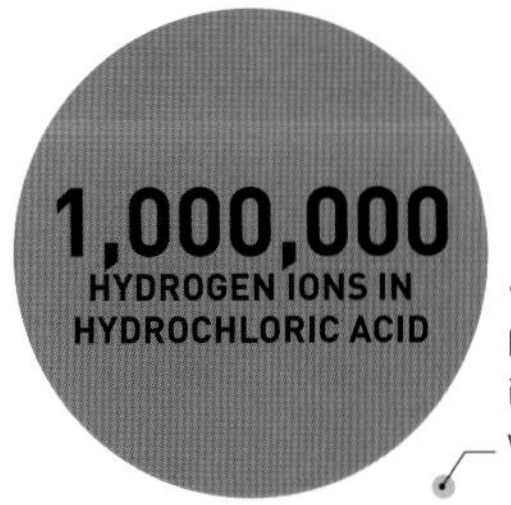
1,000,000
HYDROGEN IONS IN HYDROCHLORIC ACID
1 hydrogen ion in water

Comparing pH value
Acidity is determined by hydrogen ions. The concentration of hydrogen ions in hydrochloric acid is million times greater than that of water.

The distinctive corkscrew shaped bacteria that cause syphilis were finally conquered by an arsenic-based drug initially known as 606.

ON MAY 20, HALLEY'S COMET came closer to Earth than at any time since 1835. *The New York Times* had warned of an imminent apocalypse as astronomers had described the **comet's poisonous cyanide-containing tail**. However, the event passed without disaster.

The Danish and American astronomers, Ejnar Hertzsprung (1873–1967) and Henry Russell (1877–1957), published the **Hertzsprung–Russell, or H-R, diagram they devised to classify star types**. The H-R diagram is a scatter graph that plots the relationship between **temperature, luminosity, and size of stars** and distinguishes types of stars as clusters within the chart—white dwarfs, main sequence, super-giants, and red giants. It remains a standard astronomical tool today.

In April, Paul Ehrlich announced the **completion of his arsenic-based drug, 606**, to treat syphilis. By November, the German drug company Hoescshst AG had begun marketing it as **Salvarsan**. Demand for this new drug grew quickly as the treatment was more effective than any previous medication. However, the toxic dangers of its arsenic component remained a concern and, 30 years later, Salvarsan would be replaced by antibiotics (see 1940).

American physicist **Robert W. Wood** (1868–1955)—an expert in optics—was the **first to use infrared (IR) and ultraviolet (UV) radiation to produce photographs**, and published the first examples. Wood pioneered this type of photography and the technology would lead to the modern-day black lights that emit UV radiation and minimal visible light.

PAUL ERLICH (1854–1915)

Starting his career in the medical study of blood, Paul Ehrlich developed stains for revealing cells—including bacteria. From these studies, he pursued the "magic bullet"—a drug that could target a specific infectious organism—initiating the concept of chemotherapy. He also worked on immunization and proposed a theory that explained immune response.

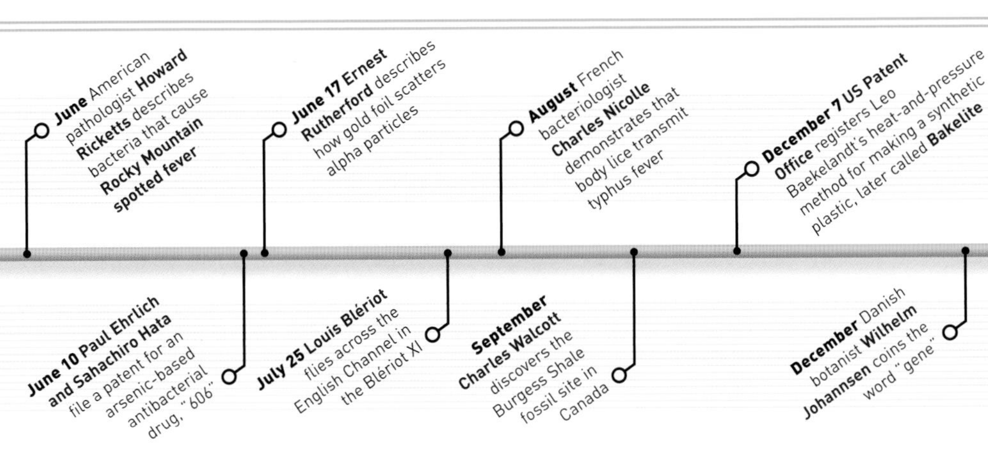
June American pathologist Howard Ricketts describes bacteria that cause Rocky Mountain spotted fever
June 10 Paul Ehrlich and Sahachiro Hata file a patent for an arsenic-based antibacterial drug, "606"
June 17 Ernest Rutherford describes how gold foil scatters alpha particles
July 25 Louis Blériot flies across the English Channel in the Blériot XI
August French bacteriologist Charles Nicolle demonstrates that body lice transmit typhus fever
September Charles Walcott discovers the Burgess Shale fossil site in Canada
December 7 US Patent Office registers Leo Baekelandt's heat-and-pressure method for making a synthetic plastic, later called Bakelite
December Danish botanist Wilhelm Johannsen coins the word "gene"

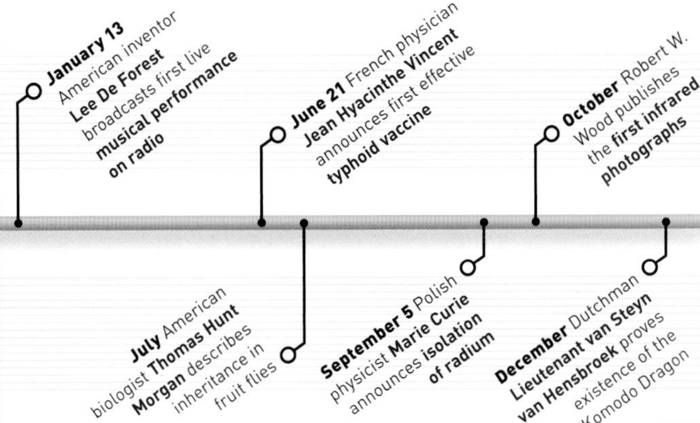
January 13 American inventor Lee De Forest broadcasts first live musical performance on radio
June 21 French physician Jean Hyacinthe Vincent announces first effective typhoid vaccine
July American biologist Thomas Hunt Morgan describes inheritance in fruit flies
September 5 Polish physicist Marie Curie announces isolation of radium
October Robert W. Wood publishes the first infrared photographs
December Dutchman Lieutenant van Steyn van Hensbroek proves existence of the Komodo Dragon

The delegate list of the first Solvay Conference included many important scientists including Ernest Rutherford, Max Planck, Albert Einstein, and Marie Curie.

> **“** WE WOULD LIKE TO HOPE THAT **OUR CONFERENCE** WILL… **EXERT** AN IMPORTANT **INFLUENCE** ON THE **DEVELOPMENT OF PHYSICS. ”**
>
> **Walther Nernst, Principal instigator of the Solvay Conference,** 1911

NEW ZEALAND-BORN PHYSICIST Ernest Rutherford was gathering evidence for a **new theory of atomic structure**. Experiments had indicated that atoms had a dense core, so the plum pudding model was wrong (see 1904). Rutherford proposed that atoms contained a **central dense mass with a positive charge that was surrounded by electrons**. In 1912, he called this central area the **nucleus**, and noted that the value of its charge was related to its atomic mass. Dutch physicist **Antonius Van den Broek** (1870–1926) suggested that the charge value was equal to an element's atomic number, or position in the periodic table. **Henry Moseley** (see 1913) later proved Van den Broek right.

In 1908, Danish physicist **Heike Kamerlingh Onnes** (1853–1926) had liquified helium and was now using it as a coolant to study the electrical properties of frozen mercury. Onnes found that at −452°F (−269°C) mercury's electrical resistance dropped to zero—he had **discovered superconductivity**.

In October, **physicists gathered at the first Solvay Conference**, founded by Belgian industrialist Ernest Solvay. This was the first opportunity **to debate the new field of quantum physics**.

At Columbia University, American biologist Thomas Hunt Morgan (1866–1945) was **experimenting with heredity in fruit flies**. Following on from the work of Gregor Mendel (see 1866) and Hugo de Vries (see 1900), Morgan studied mutations in the insects that caused obvious characteristics, such as eye color, to help him trace patterns of inheritance. In 1911, Morgan discovered that the **genes were on the chromosomes that determined sex**.

In Antarctica, Robert Scott's (1866–1912) **British Terra Nova expedition set out for the South Pole**. In what was later described as "the worst journey in the world," they succeeded in part of their mission and found a rookery of Emperor Penguins in midwinter. But in December, a Norwegian expedition, led by **Roald Amundsen** (1872–1928), **reached the South Pole first**. The British Polar party all died on their return journey.

German chemist **Philip Monnartz** described a way of **producing steel to improve its corrosion resistance.** The result—**stainless steel**—would be patented in 1912 and in 1913, British engineer Harry Brearley (1871–1948) would cast the first commercial stainless steel in Sheffield, England.

ERNEST RUTHERFORD
(1871–1937)

Born in New Zealand, Ernest Rutherford worked at McGill University, Canada, moving to the University of Manchester, England, in 1907. In 1919, he was appointed director of Cambridge's Cavendish Laboratory. Rutherford established the field of nuclear physics with his model of the atom, and explained that radioactivity was caused by atoms decaying into different forms.

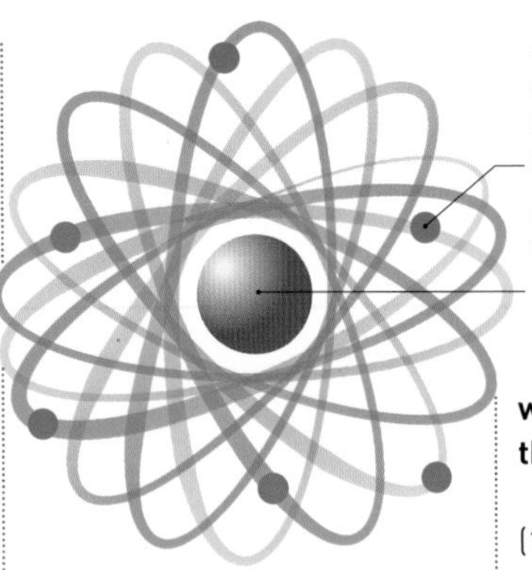
electrons circle round the central nucleus
dense nucleus with a positive charge

Rutherford's atomic model
Rutherford proposed that an atom had a dense nucleus and that the electrons orbited the space around it.

MORGAN'S FRUIT FLY EXPERIMENT

Thomas Hunt Morgan's breeding experiments revealed that the inheritance of a characteristic is linked to gender because its gene occurs on one of the sex chromosomes—usually the X. This can be seen in the eye color of fruit flies because the red-eye variant (allele), is dominant. As a result, in the first generation, a red-eyed female and a white-eyed male would produce only red-eyed offspring, but some will carry the white-eye allele. In the next generation, if a red-eyed female has a white-eye trait it will show up in her sons.

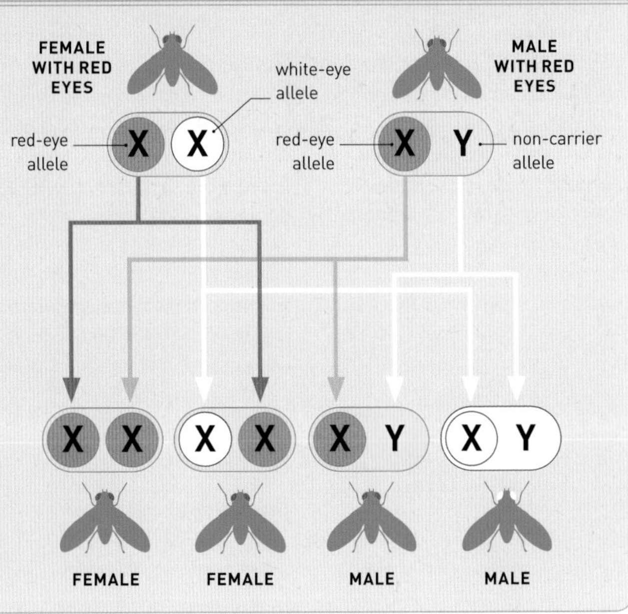
FEMALE WITH RED EYES
white-eye allele
MALE WITH RED EYES
red-eye allele
X X
red-eye allele
X Y
non-carrier allele
X X
X X
X Y
X Y
FEMALE
FEMALE
MALE
MALE

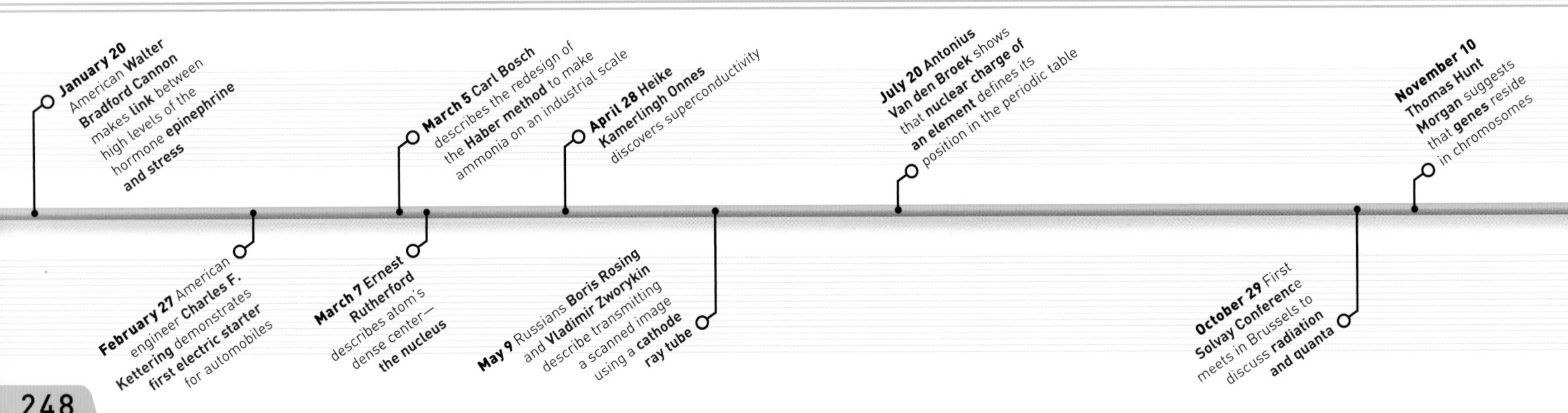
January 20 American Walter Bradford Cannon makes link between high levels of the hormone epinephrine and stress
February 27 American engineer Charles F. Kettering demonstrates first electric starter for automobiles
March 5 Carl Bosch describes the redesign of the Haber method to make ammonia on an industrial scale
March 7 Ernest Rutherford describes atom's dense center—the nucleus
April 28 Heike Kamerlingh Onnes discovers superconductivity
May 9 Russians Boris Rosing and Vladimir Zworykin describe transmitting a scanned image using a cathode ray tube
July 20 Antonius Van den Broek shows that nuclear charge of an element defines its position in the periodic table
October 29 First Solvay Conference meets in Brussels to discuss radiation and quanta
November 10 Thomas Hunt Morgan suggests that genes reside in chromosomes

1912

The archaeologist Charles Dawson (left) convinced paleontologist Arthur Smith-Woodward (center) that he had found the ape–human missing link.

ON FEBRUARY 14, British archaeologist **Charles Dawson** (1864–1916) uncovered **primate skull fragments and a jaw at** the **Piltdown** gravel pit, England. They were reported to be the missing link between humans and apes and attracted much interest from experts at the British Museum of Natural History. More than 40 years later, in 1953, **Piltdown Man was shown to be a hoax**—it was in fact fragments of a modern human and an orang-utan.

This year also saw the dawn of a new way of analyzing chemical structures. German physicists **Walter Friedrich** (1874–1958) **and Paul Knipping** (1883–1935) showed that when X-rays were scattered through a crystal, the resulting diffraction pattern recorded on a photographic plate could be used to **identify the positions of the crystal's atoms**. Later, German physicist **Max von Laue** (1879–1960) **explained the theory** for this scattering behavior and used it to show that X-rays were **not particles, but waves with very short wavelengths**. X-ray diffraction would be developed as a mainstay of analytical chemistry.

British biochemist **Frederick Hopkins** (1861–1947) was experimenting with the diets of animals and announced that, along with carbohydrates, proteins, and fats, a healthy diet needed **accessory food factors**. Polish-born biochemist **Casimir Funk** (1884–1967) proposed a name for them—**vitamines**, from "vital" and "amine." He thought that they belonged to a class of chemicals called amines, but when it was shown that this was not the case, the name was changed to **vitamin**.

186

MILES/SEC THE SPEED THE **ANDROMEDA GALAXY** MOVES AWAY FROM THE **SUN**

American astronomer **Vesto Slipher** (1875–1969), using a technique of light analysis called spectroscopy (see 1860), observed a shift toward **red wavelengths for light** coming from the **Andromeda galaxy**. This **redshift** indicated that the galaxy was moving away. Slipher's studies provided early evidence for the idea that **the Universe was expanding**.

1913

" IT IS PERHAPS… INDELICATE TO ASK… MOTHER EARTH HER AGE, BUT SCIENCE ACKNOWLEDGES NO SHAME AND… HAS… ATTEMPTED TO WREST FROM HER A SECRET WHICH IS PROVERBIALLY WELL GUARDED. "

Arthur Holmes, British geologist, from *The Age of the Earth,* 1913

BRITISH PHYSICIST FREDERICK SODDY had worked with Ernest Rutherford on the nature of radioactivity and identification of decay products. He found that these products had different atomic masses, but held positions in the periodic table that appeared to be occupied by known elements. Soddy concluded that they must be varieties of the elements, and in 1913 he called them **isotopes**.

Danish physicist **Niels Bohr** (1885–1962) **modified Rutherford's nuclear model of the atom** to take into account new ideas from quantum physics. Bohr proposed that electrons orbiting the atom's nucleus occupied various shells or energy levels called orbitals. The electrons in the outer shell determined the way elements reacted chemically.

At the BASF (*Badische Anilinund Soda-Fabrik*) chemical plant in Germany, chemist Carl Bosch oversaw the **first industrial production of ammonia** for use in agricultural fertilizers. Bosch had modified a chemical process first described by Fritz Haber (see 1905)—now known as **the Haber–Bosch process**. The factory would reach full production capacity in 1914.

In December, British physicist **Henry Moseley** (1887–1915) found that elements emit characteristic X-rays of wavelengths determined by the numbered position in the periodic table. His discovery supported Van den Broek's idea that the position had physical meaning in the form of the **number of positive charges in an element's atomic nucleus**—its atomic number.

British geologist **Arthur Holmes** (1890–1965) had adopted a radiometric technique for calculating **the age of rocks** pioneered by Bertram Boltwood (see 1907). Holmes published his results in *The Age of the Earth*, but—at 1.6 million years—his calculation was still more than 4,000 times shorter than modern results.

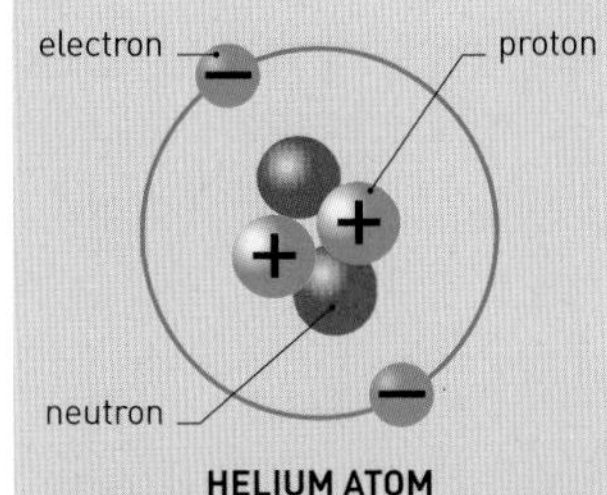

electron
proton
neutron

HELIUM ATOM

ATOMIC NUMBER

Henry Moseley proved that the vaguely defined concept of atomic number related to a measurable physical property of the atomic nucleus, later found to be the number of protons. For a neutral atom, this equals the number of encircling electrons. The nucleus is made up of protons and neutrons (neutral particles).

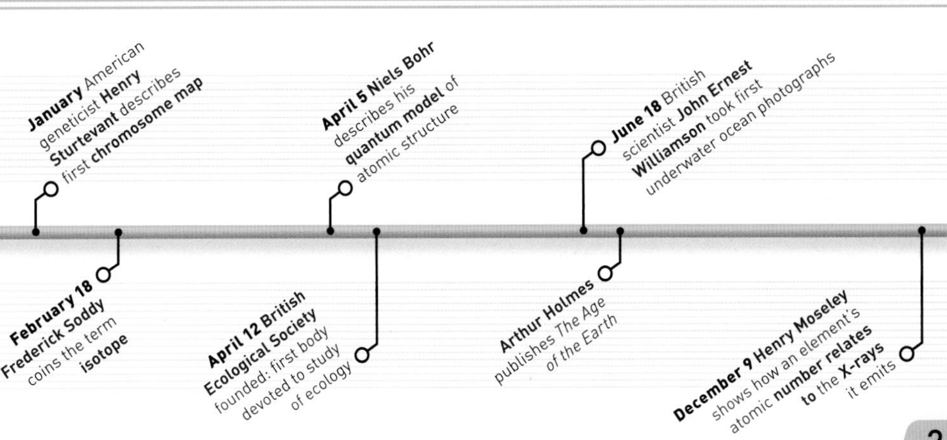

electron
nucleus
first shell
second, outer shell

Bohr's atomic model
Bohr's model suggested electron "shells" around the nucleus. As electrons move between the shells, the atom emits or absorbs electromagnetic radiation.

January 6 German geophysicist **Alfred Wegener** describes his theory that modern continents have **split** from an **ancient supercontinent**

April 21 Walter Friedrich and **Paul Knipping** explain that X-rays diffracted through a crystal can show positions of atoms

June 8 Max von Laue shows that **X-rays are waves**, not particles

July 15 Frederick Hopkins states that a healthy diet needs **accessory food factors**

September 17 Vesto Slipher demonstrates Andromeda Galaxy is moving

November 11 William Lawrence Bragg explains how crystals diffract X-rays

January American geneticist **Henry Sturtevant** describes first **chromosome map**

February 18 Frederick Soddy coins the term **isotope**

April 5 Niels Bohr describes his **quantum model** of atomic structure

April 12 British Ecological Society founded: first body devoted to study of ecology

June 18 British scientist **John Ernest Williamson** took first underwater ocean photographs

Arthur Holmes publishes *The Age of the Earth*

December 9 Henry Moseley shows how an element's atomic **number relates to the X-rays** it emits

UNDERSTANDING ATOMIC STRUCTURE

PARTICLES AND FORCES COMBINE TO MAKE UP ATOMS, THE BUILDING BLOCKS OF MATTER

Solids, liquids, and gases—the three main types of matter—are made up of particles called atoms, each so small that several million would fit side by side on the period at the end of this sentence. All atoms are composed of just three types of particles—protons, neutrons, and electrons.

ERNEST RUTHERFORD
New Zealand-born physicist Ernest Rutherford discovered the atomic nucleus in 1911 and the proton in 1920.

In the 19th century scientists became certain of the existence of atoms. The assumption was that atoms were the smallest parts of matter—the word "atom" means indivisible. However, in 1897 the discovery of the electron suggested that atoms have inner structure. Electrons carry negative electric charge, but atoms are neutral overall so a positive charge also had to exist. Scientists used alpha particles, which are produced by radioactive substances, to probe the atom more deeply. In 1911, Ernest Rutherford found that some of the positively charged alpha particles he fired at a thin metal foil bounced back, and proved that this could happen only if each atom has a positive charge concentrated at its center—he had discovered the atomic nucleus.

ATOMS AND ELEMENTS

Except in hydrogen (see right), an atom's nucleus consists of protons, which are positively charged, and neutrons, which have no charge, held together by an attractive force called the strong nuclear force. Different elements have different numbers of protons in the nucleus (see below). Negatively charged electrons orbit the nucleus and are prevented from escaping by electric attraction to protons. There are equal numbers of protons and electrons, so atoms have no overall charge.

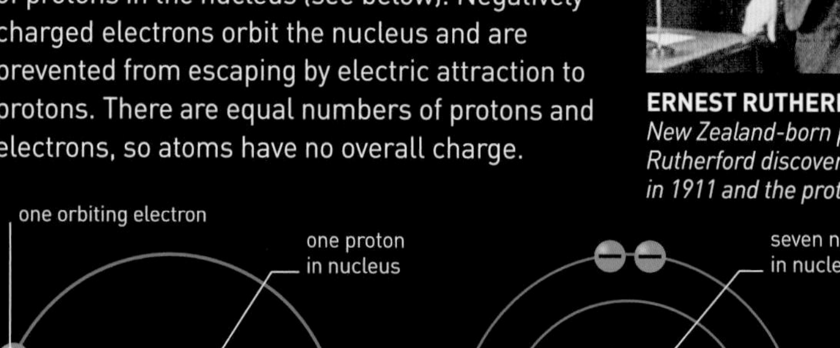

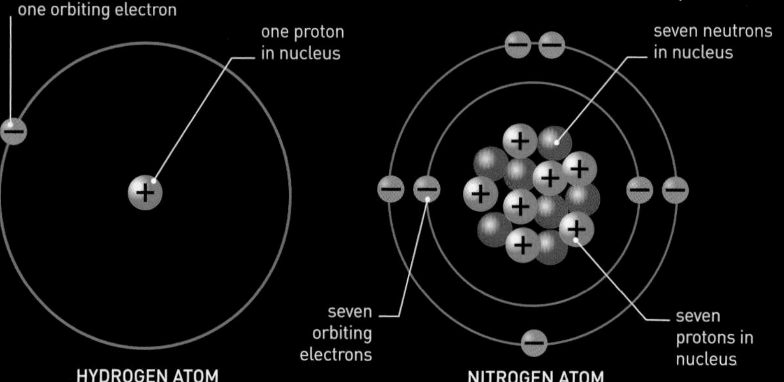

one orbiting electron
one proton in nucleus
HYDROGEN ATOM
seven neutrons in nucleus
seven orbiting electrons
seven protons in nucleus
NITROGEN ATOM
orbital is a region where there is a high probability of finding electrons

ATOMIC SIZES

The nucleus accounts for nearly all the mass of an atom but only a tiny proportion of its volume—it is an atom's electrons that define this. Electrons occupy orbits at specific distances from the nucleus, and only certain numbers of electrons can occupy each orbit. As a result, atoms with more electrons have more orbits, at increasing distances from the nucleus, producing a larger atomic radius.

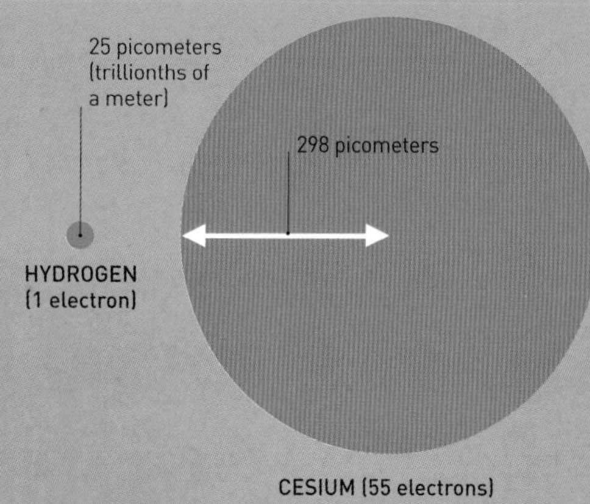

25 picometers (trillionths of a meter)
298 picometers
HYDROGEN (1 electron)
CESIUM (55 electrons)

ATOMIC RADIUS
Hydrogen, the lightest element, has only one electron orbiting its nucleus, so it has the smallest atoms. A cesium atom has 55 electrons orbiting its nucleus and is about 12 times wider than a hydrogen atom.

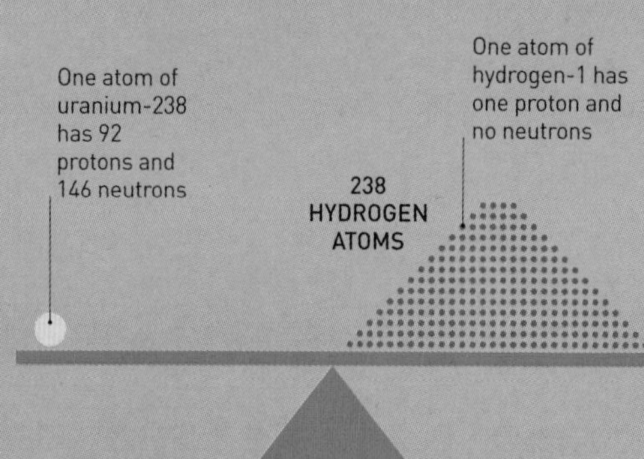

One atom of uranium-238 has 92 protons and 146 neutrons
One atom of hydrogen-1 has one proton and no neutrons
238 HYDROGEN ATOMS

ATOMIC MASS
Protons and neutrons have identical mass; electrons have negligible mass. So the atomic mass is simply the total number of protons and neutrons. Different versions (isotopes) of an element vary in the number of neutrons.

> **“ PROTONS** GIVE AN ATOM ITS **IDENTITY, ELECTRONS** GIVE IT ITS **PERSONALITY. ”**

American author Bill Bryson,
A Short History of Nearly Everything, 2003

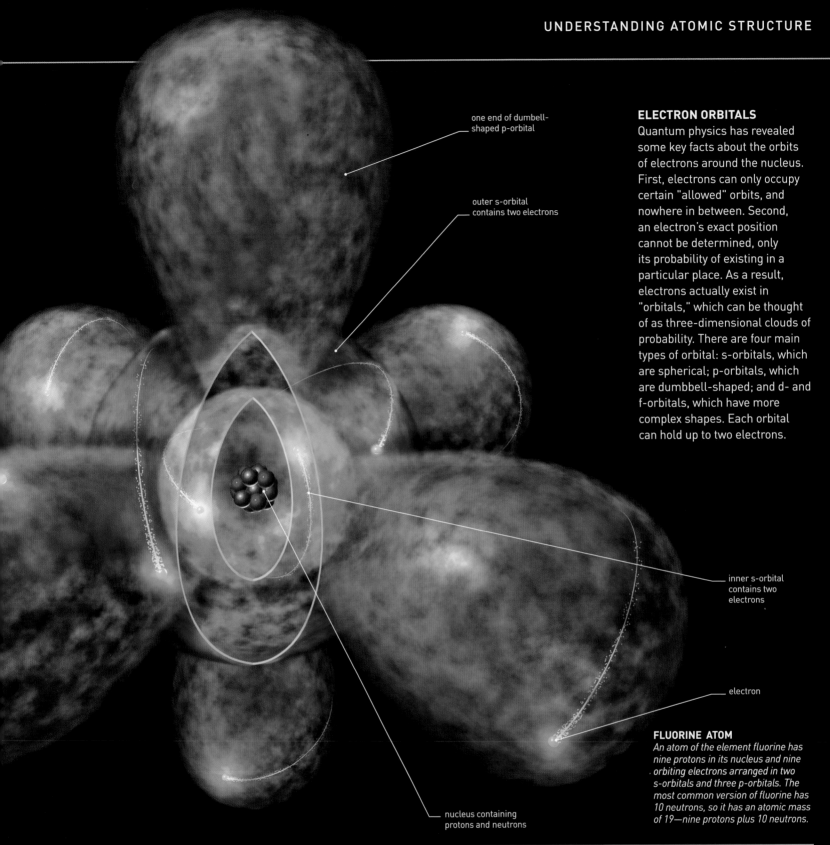

one end of dumbell-shaped p-orbital

outer s-orbital contains two electrons

inner s-orbital contains two electrons

electron

nucleus containing protons and neutrons

ELECTRON ORBITALS

Quantum physics has revealed some key facts about the orbits of electrons around the nucleus. First, electrons can only occupy certain "allowed" orbits, and nowhere in between. Second, an electron's exact position cannot be determined, only its probability of existing in a particular place. As a result, electrons actually exist in "orbitals," which can be thought of as three-dimensional clouds of probability. There are four main types of orbital: s-orbitals, which are spherical; p-orbitals, which are dumbbell-shaped; and d- and f-orbitals, which have more complex shapes. Each orbital can hold up to two electrons.

FLUORINE ATOM
An atom of the element fluorine has nine protons in its nucleus and nine orbiting electrons arranged in two s-orbitals and three p-orbitals. The most common version of fluorine has 10 neutrons, so it has an atomic mass of 19—nine protons plus 10 neutrons.

INCANDESCENCE AND LUMINESCENCE

There are two ways in which matter emits light: incandescence and luminescence. Incandescence is the production of light from hot matter—things glow red hot or white hot, for example. In contrast, luminescence does not require heat. Fluorescence, glowing when illuminated by ultraviolet radiation, and phosphorescence, glowing in the dark after being illuminated, are two familiar examples. Light produced by luminescence is the result of energy lost by electrons "falling" to a lower energy level, closer to an atom's nucleus.

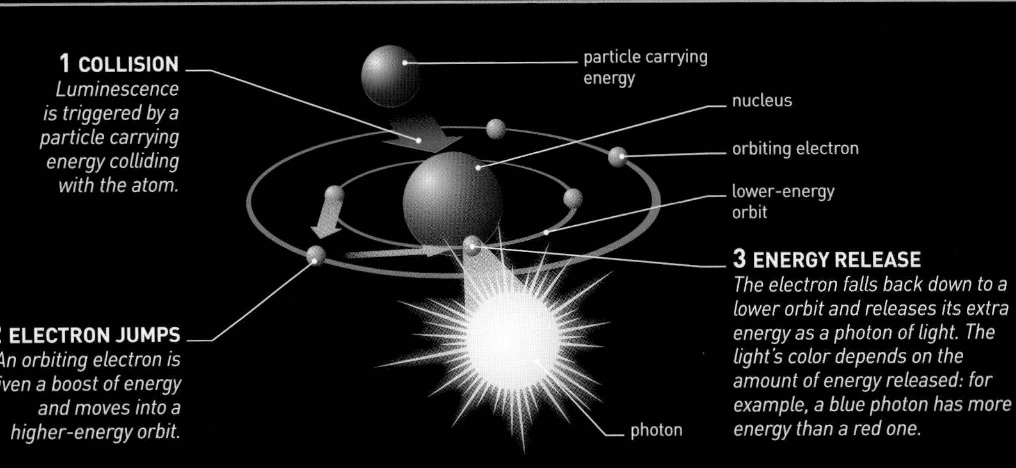
particle carrying energy
nucleus
orbiting electron
lower-energy orbit
photon

1 COLLISION
Luminescence is triggered by a particle carrying energy colliding with the atom.

2 ELECTRON JUMPS
An orbiting electron is given a boost of energy and moves into a higher-energy orbit.

3 ENERGY RELEASE
The electron falls back down to a lower orbit and releases its extra energy as a photon of light. The light's color depends on the amount of energy released: for example, a blue photon has more energy than a red one.

> **“** THE **LUMINESCENT** PROPERTIES OF **NEON...** CONSTITUTE SOURCES OF **LIGHT** OF **GREAT BRILLIANCY. ”**
>
> **George Claude**, *Vacuum discharge-tube for lighting purposes*, US Patent Office, 1915

A gas-discharge light tube works by applying voltage across the low-pressure gas to create ions (charged particles), which causes a colored, neon, light to be emitted.

IN 1914, BELGIAN PHYSICIAN ALBERT HUSTIN (1882–1967) found he could **stop blood clotting** by adding a reagent called sodium citrate. Previously, the danger of coagulation meant that transfusions had to be performed directly from donor to recipient. In March, Hustin performed the **first safe nondirect transfusion** using citrate-treated blood from storage.

In 1915, two German scientists would change the way we viewed our world. Geologist **Alfred Wegener** expanded on a **theory of continental drift** that he first proposed in 1912 (see below). Physicist **Albert Einstein's theory of general relativity** was more radical still. His special relativity theory (see 1905) had asserted that measurements of distances and time change for different frames of reference, even though the speed of light stays constant everywhere. The frames of reference referred to situations where the observers making the measurements were moving at different fixed speeds. Einstein's new theory of general relativity accommodated acceleration and deceleration effects too. Acceleration is equivalent to gravity and an outcome was that the effect of gravity was to bend light. Furthermore, he argued that **strong gravitational fields** should distort time and appearance; in other words, they would **warp a space–time continuum**. Physicists consider space and time to be related: space makes up the three (everyday) dimensions, and time is the fourth dimension.

In the same year, French engineer George Claude received his patent for a new type of lighting: **the neon discharge tube**.

ALFRED WEGENER (1880–1930)

Born in Germany, Alfred Wegener began his career as a meteorologist, and he participated in several Arctic expeditions to investigate climate. He is best known for his theory of continental drift. Wegener collected geological evidence but could not explain how the movement occurred; during his lifetime the theory was not taken seriously.

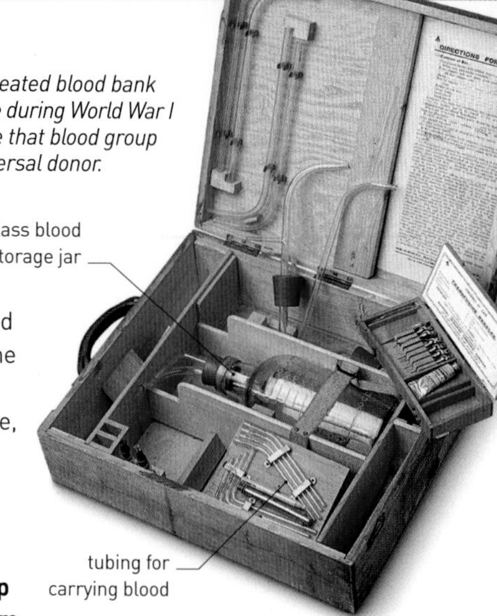
glass blood storage jar
tubing for carrying blood

Blood transfusion kit
The first anticoagulant-treated blood bank kits—used on the frontline during World War I—relied on the knowledge that blood group O could be used as a universal donor.

CONTINENTAL DRIFT

Alfred Wegener collected evidence for his theory that modern landmasses originated from a prehistoric super-continent that fragmented millions of years ago. He saw not only that transatlantic coastlines seemed to match, but also that geological formations (including fossils) were similar on either side. This suggested that the coastlines were once joined.

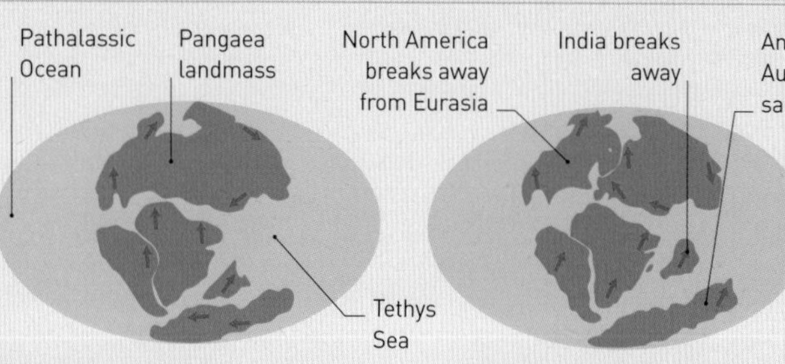
Pathalassic Ocean
Pangaea landmass
Tethys Sea
North America breaks away from Eurasia
India breaks away
Antartica and Australia are the same continent
South America separates from Africa

200 million years ago
A single supercontinent called Pangaea was formed when ancient landmasses came together. As tectonic plates continued to move, it began to fragment.

130 million years ago
By this era, there were several separated continents, including India (drifting northward) and an Australia-Antarctic landmass.

70 million years ago
For millions of years, South America was isolated as the Atlantic Ocean opened up to separate it from Africa.

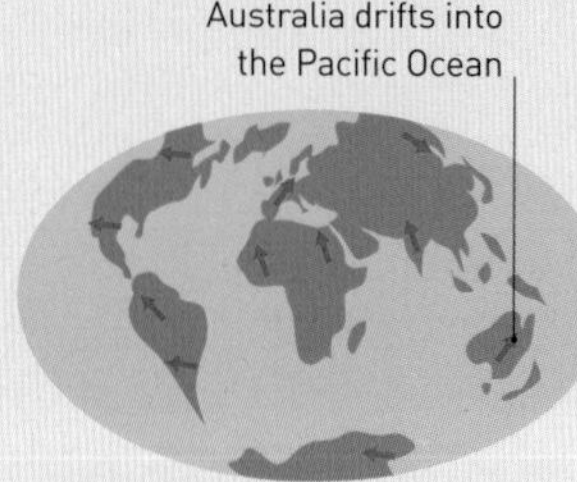
Australia drifts into the Pacific Ocean

Present day
India collided with Asia, creating the Himalayas. South America was joined to North America across a narrow Central American land bridge.

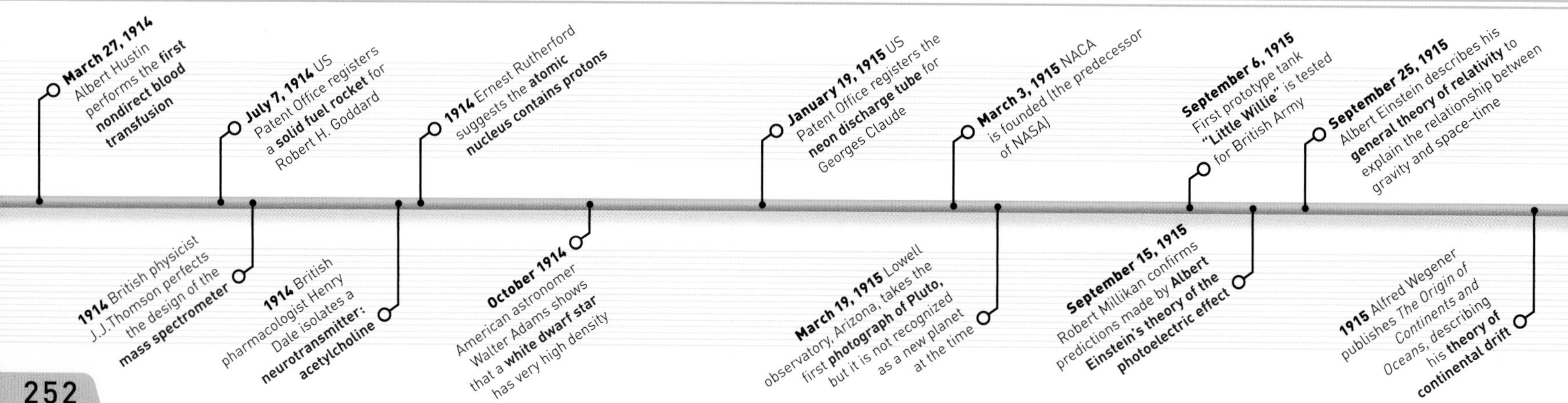
March 27, 1914 Albert Hustin performs the **first nondirect blood transfusion**
1914 British physicist J.J. Thomson perfects the design of the **mass spectrometer**
July 7, 1914 US Patent Office registers a **solid fuel rocket** for Robert H. Goddard
1914 British pharmacologist Henry Dale isolates a **neurotransmitter: acetylcholine**
1914 Ernest Rutherford suggests the **atomic nucleus contains protons**
October 1914 American astronomer Walter Adams shows that a **white dwarf star** has very high density
January 19, 1915 US Patent Office registers the **neon discharge tube** for Georges Claude
March 3, 1915 NACA is founded (the predecessor of NASA)
March 19, 1915 Lowell observatory, Arizona, takes the first **photograph of Pluto**, but it is not recognized as a new planet at the time
September 6, 1915 First prototype tank **"Little Willie"** is tested for British Army
September 15, 1915 Robert Millikan confirms predictions made by **Albert Einstein's theory of the photoelectric effect**
September 25, 1915 Albert Einstein describes his **general theory of relativity** to explain the relationship between gravity and space–time
1915 Alfred Wegener publishes *The Origin of Continents and Oceans*, describing his **theory of continental drift**

300 THOUSAND LIGHT YEARS

SHAPLEY'S ESTIMATE OF THE DIAMETER OF THE MILKY WAY

American astronomer Harlow Shapley's first measurement of the Milky Way galaxy was an overestimate. It is 100,000 light-years across and contains more than 100 billion stars.

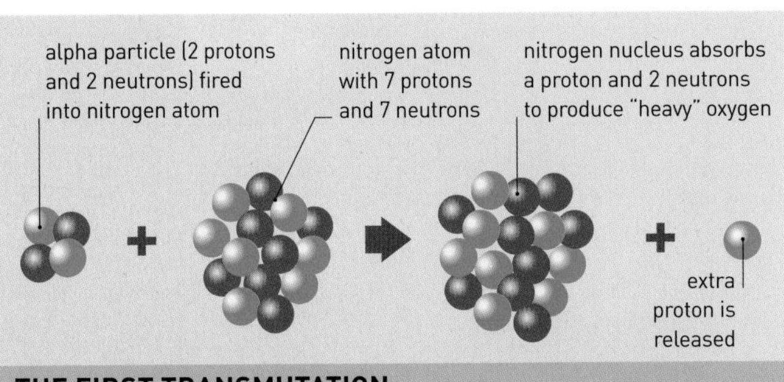
alpha particle (2 protons and 2 neutrons) fired into nitrogen atom
nitrogen atom with 7 protons and 7 neutrons
nitrogen nucleus absorbs a proton and 2 neutrons to produce "heavy" oxygen
extra proton is released

THE FIRST TRANSMUTATION

British physicist Ernest Rutherford was the first person to cause one element to change into another—a phenomenon called transmutation. He showed that it was possible to change the size of an atom's nucleus by bombarding it with other particles. Later scientists exploited this process to release huge amounts of energy.

IN 1916, AMERICAN CHEMIST GILBERT LEWIS (1875–1946) suggested that when atoms bonded together to make bigger molecules, they **shared their outer electrons**. In 1919, his colleague Irving Langmuir (1881–1957) would expand the idea and call them covalent bonds.

Swedish geologist Lennart von Post (1884–1951) explained a new way of studying geological deposits. Different **plant species produce distinctly shaped pollen grains**, and von Post was able to identify pollen to make conclusions about the historic vegetation of peat beds. The study of pollen (palynology) is still an important analytical study used for example in forensics.

American astronomer Harlow Shapley (1885–1972) started a **photographic survey of star clusters**. His systematic measurements would show that clusters surround our galaxy like a halo, and are not centered on Earth.

Pollen grain *The distinctive sculpturing of pollen grains makes it possible to identify the source plant species by microscopy.*

In 1917, at the Mount Wilson Observatory in California, the assembly of the **Hooker reflecting telescope**—the largest in the world at the time – was finished. To this day, its 100 in (254 cm) mirror is the largest solid-glass mirror ever made. The telescope was used to make the first measurements of stars other than the Sun.

In November, New-Zealand born British physicist **Ernest Rutherford** achieved a significant breakthrough in atomic physics: the first ever man-made transmutation of elements. He succeeded in **turning nitrogen atoms into a form of oxygen**, by firing them with alpha particles.

When the US entered World War I in April 1917, British-born medical officer **Oswald Robertson** (1886–1966) introduced a **blood bank system** to the frontline. The Royal Army Medical Corps called it the most important medical advance of the war. Treating blood with **anticoagulant made blood transfusions faster and safer**. British physician **Arthur Cushny** (1866–1926) published the **first major study of kidney physiology**. He correctly deduced that the organ filtered blood and reabsorbed nutrients to produced waste urine.

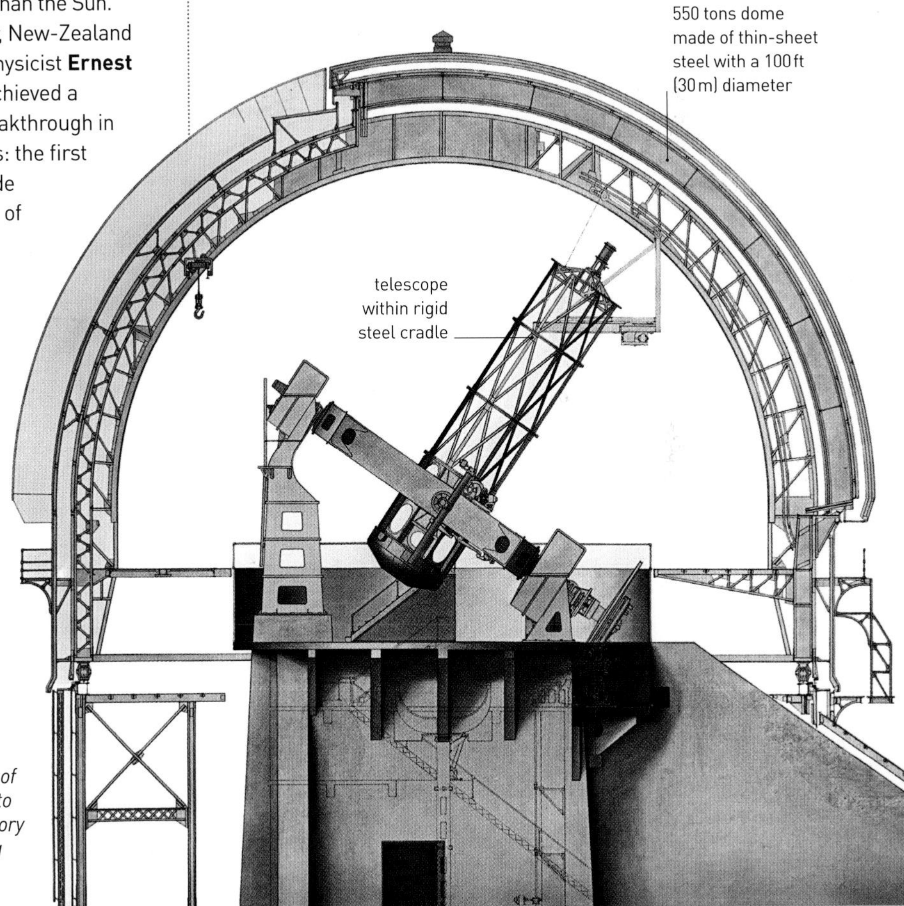
550 tons dome made of thin-sheet steel with a 100 ft (30 m) diameter
telescope within rigid steel cradle

Hooker telescope *This telescope was used to make the first measurements of star sizes, and to develop the theory of an expanding universe.*

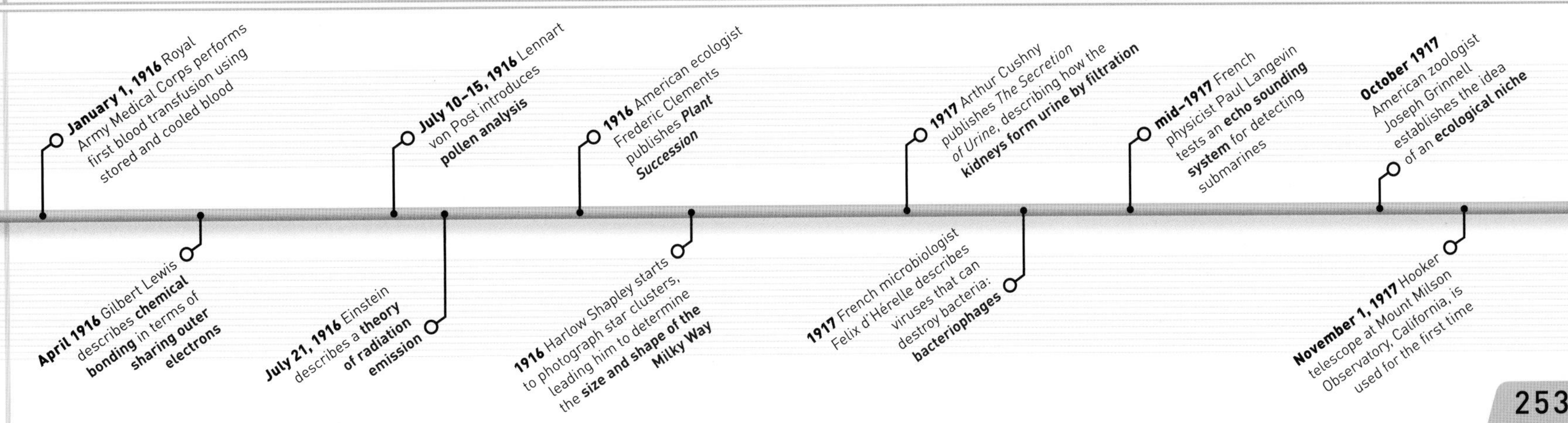
January 1, 1916 Royal Army Medical Corps performs first blood transfusion using stored and cooled blood
April 1916 Gilbert Lewis describes **chemical bonding** in terms of **sharing outer electrons**
July 10–15, 1916 Lennart von Post introduces **pollen analysis**
July 21, 1916 Einstein describes a **theory of radiation emission**
1916 American ecologist Frederic Clements publishes ***Plant Succession***
1916 Harlow Shapley starts to photograph star clusters, leading him to determine the **size and shape of the Milky Way**
1917 Arthur Cushny publishes *The Secretion of Urine*, describing how the **kidneys form urine by filtration**
1917 French microbiologist Felix d'Hérelle describes viruses that can destroy bacteria: **bacteriophages**
mid-1917 French physicist Paul Langevin tests an **echo sounding system** for detecting submarines
October 1917 American zoologist Joseph Grinnell establishes the idea of an **ecological niche**
November 1, 1917 Hooker telescope at Mount Milson Observatory, California, is used for the first time

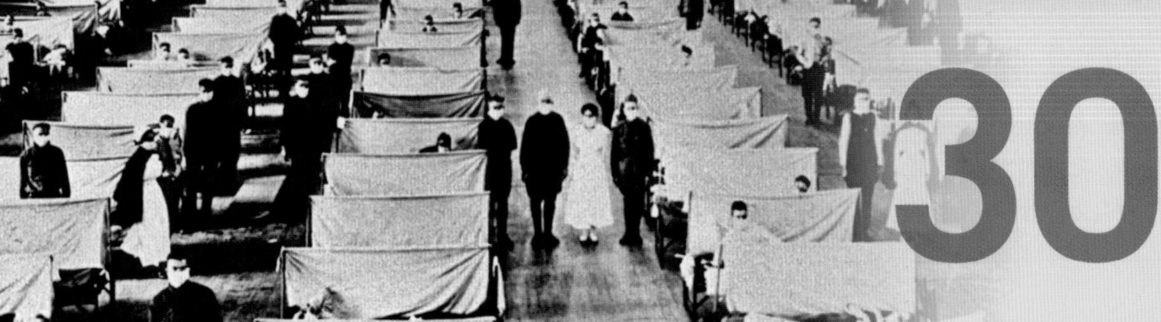

There were so many victims of the Spanish flu epidemic that several makeshift hospitals—such as this gymnasium in the US—had to be created.

30 MILLION

THE NUMBER OF **PEOPLE** WHO **DIED** OF **SPANISH FLU** IN **SIX MONTHS** OF **1918**

CONFIRMING THE GENERAL THEORY OF RELATIVITY

In his theory of general relativity, Einstein had predicted that the strong gravitational field of the Sun would warp space-time. This distortion would deflect light coming from the stars, causing their apparent position to differ from their real position. Independent observations by Arthur Eddington and Andrew Crommelin confirmed this theory. During a solar eclipse, they accurately measured the positions of stars close to the Sun. They compared their results with the stars' apparent positions and noticed a slight difference between the two.

IN THE WAKE OF WORLD WAR I, a natural global disaster killed more people than the great plague of the 1300s. **Spanish flu** spread to all the continents, and medical science seemed powerless to stop it. Modern research suggests that it originated as a mutated virus and many people lacked the immunity to fight it.

In February 1918, French physicist Paul Langevin (1872–1946) demonstrated a system to detect the sound of **underwater submarines**—a technology that would have important implications in wartime. Called the **ASDICS** device (Aided Sonar Detection Integration and Classification System), it was the forerunner to **sonar** (see pp.292–93).

In Germany, electrical engineer Albert Scherbius (1878–1929) invented a **cipher machine**, which he called **Enigma**. It worked by a series of rotating wheels. In 1918, Scherbius offered it to the German Navy. They began using it in February 1926, and the German Army followed shortly afterward.

In 1917, New Zealand-born physicist Ernest Rutherford had found that firing alpha particles from radioactive elements at nitrogen atoms yielded small elementary particles. He realized that these were equivalent to **positive hydrogen ions** and later called them **protons**. It would be shown that they are responsible for the positive charge in the nucleus of atoms.

During a solar eclipse in 1919, British astronomer Arthur Eddington (1882–1944) and French astronomer Andrew Crommelin (1865–1939) observed that **rays of starlight were bent as they passed close to the Sun**. This proved **Einstein's general theory of relativity** (see pp.244–45).

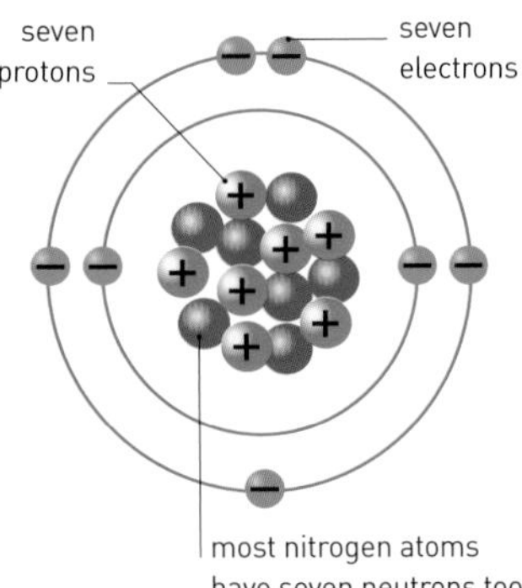
seven protons
seven electrons
most nitrogen atoms have seven neutrons too

Balanced charge
The number of protons in a non-ionized atom equals the number of electrons; this means that an atom, overall, has no charge. Neutrons are particles that have no charge.

Zur Beachtung!
metal cover to fit over cylinders
motor cylinders
keys to type messages
plugboard setting altered regularly to change cipher
Klappe schliessen
ENIGMA

Enigma cipher machine
These machines were used during World War II for decrypting and encrypting confidential messages.

January 8, 1918 American astronomer Harlow Shapley publishes his estimate of the **size and shape of the Milky Way** and deduces that it is not centered on Earth

January 1918 Spanish flu pandemic first identified in Kansas

February 1918 Paul Langevin demonstrates **ASDICS**—the forerunner to modern sonar systems

April 18, 1918 Arthur Scherbius offers his **Enigma cipher machine** to the German Navy

June 1918 British physicist Ernest Rutherford explains that **hydrogen ions** are **elementary particles**

May 29, 1919 Arthur Eddington and Andrew Crommelin test **Einstein's general theory of relativity**

June 1, 1919 American chemist Irving Langmuir describes the **arrangement of electrons** in atoms and molecules, and introduces the term **covalent bond**

November 30, 1919 Spanish flu pandemic is declared at an **end**

December 1, 1919 American chemist Phoebus Levens describes **nucleotide composition of DNA**

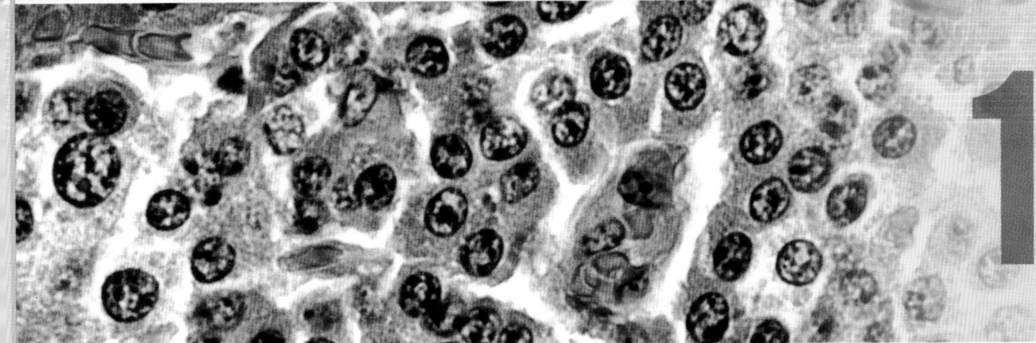

These clusters of cells, called Islets of Langerhans, are the source of the hormone insulin in the pancreas.

1 MILLION THE NUMBER OF ISLETS OF LANGERHANS IN THE PANCREAS OF AN AVERAGE ADULT

AMERICAN PHYSICIST ROBERT GODDARD (1882–1945) had recently published a treatise promoting the idea of rocket space travel. On January 13, 1920, the *New York Times* ridiculed Goddard's views. A few years later he successfully launched a liquid-fueled rocket (see 1926).

In 1920, British physician Edward Mellanby (1884–1955) showed that a plain oatmeal diet caused a bone-softening deformity called **rickets**. He also found that the effects were eliminated by supplementing the diet with **cod-liver oil**. He deduced that cod liver must contain a factor that was needed for essential growth. In 1922, this factor was identified as **Vitamin D**.

BCG vaccine
Consisting of a harmless strain of bovine TB bacteria (orange), the BCG vaccine stimulates the body's immune system.

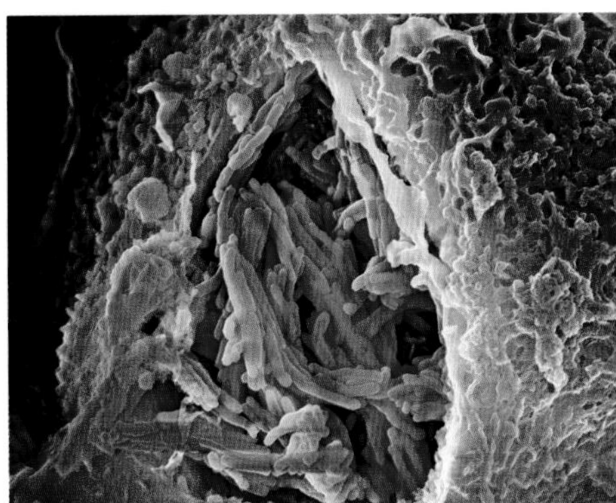

Later research revealed that exposure to the Sun's rays stimulates the formation of Vitamin D in the body naturally.

In the same year, Arthur Eddington suggested that a **star's energy** comes from the conversion of hydrogen to helium by nuclear fusion. This was later supported by the findings of British astronomer Cecilia Payne-Gaposchkin (1900–79), who showed that these elements predominated in a star.

In Germany, botanist Hans Winkler (1877–1945) combined the terms "gene" (the particle of inheritance) and "chromosome" (solidified thread of genetic material that becomes visible during cell division) to coin a new scientific term—**genome**. He used it in his book on biological reproduction to refer to a single set of genetic material.

In July 1921, the **BCG (Bacille Calmette–Guérin) vaccine** against tuberculosis (TB) was ready for its first human trial—15 years after Albert Calmette and Camille Guérin started developing it (see 1906). French physicians Benjamin Weill-Hallé (1875–1958) and Raymond Turpin (1895–1988) administered the vaccine orally to a newborn infant whose mother had died of TB. It was the start of a successful vaccination program that would extend worldwide.

In August, Canadian biologists Frederick Banting (1891–1941) and Charles Best (1899–1978) **isolated insulin**. It was known that the pancreas produced a secretion that regulated sugar levels in the body; the active ingredient had been called "insuline." However, the pancreas also produced digestive enzymes, and the destructive influence of these had always interfered with the extraction of insulin. Banting developed a way of tying off the digestive ducts so that insulin could be retrieved.

In 1921, the **Nobel Committee for Physics** decided that none of the nominations satisfied the prize criteria, and carried it over to the next year. In 1922, **Albert Einstein** received the Nobel Prize for 1921, nearly two decades after his "miracle year," 1905, when he had published several ground-breaking theories. Einstein was awarded the prize for his services to Theoretical Physics and especially for his discovery of the law of the **photoelectric effect**.

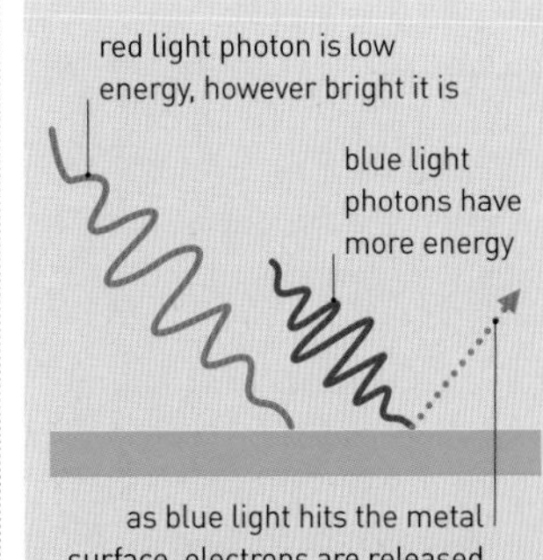

red light photon is low energy, however bright it is
blue light photons have more energy
as blue light hits the metal surface, electrons are released

PHOTOELECTRIC EFFECT

The emission of electrons from a metallic surface when light shines on it is called the photoelectric effect. Albert Einstein explained it in terms of light particles with distinct quantities of energy called photons. He said that photons of red light (long wavelength) do not have enough energy to allow electrons to be emitted, but photons of blue light (short wavelength) do.

BLOOD SUGAR REGULATION

The pancreas releases insulin when blood sugar concentration rises, such as after a meal. This prompts the liver to convert sugar to stored carbohydrate, causing the blood sugar to drop. When this happens, another hormone, glucagon, is released, prompting the liver to break down stored carbohydrate and release more sugar. The combined effect of the two hormones regulates the level of sugar in the blood.

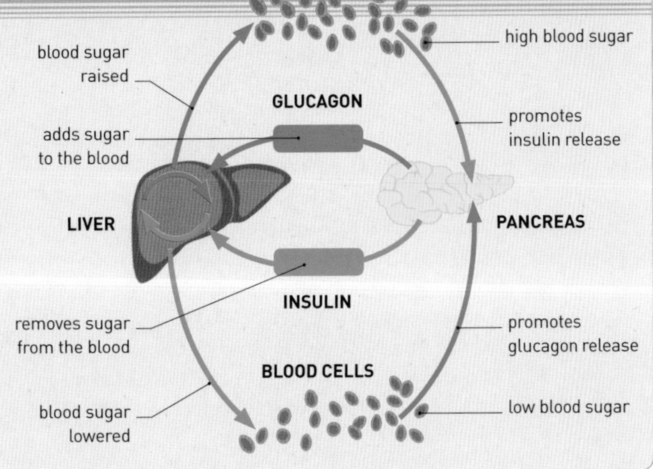

blood sugar raised
high blood sugar
GLUCAGON
adds sugar to the blood
promotes insulin release
LIVER
PANCREAS
INSULIN
removes sugar from the blood
promotes glucagon release
BLOOD CELLS
blood sugar lowered
low blood sugar

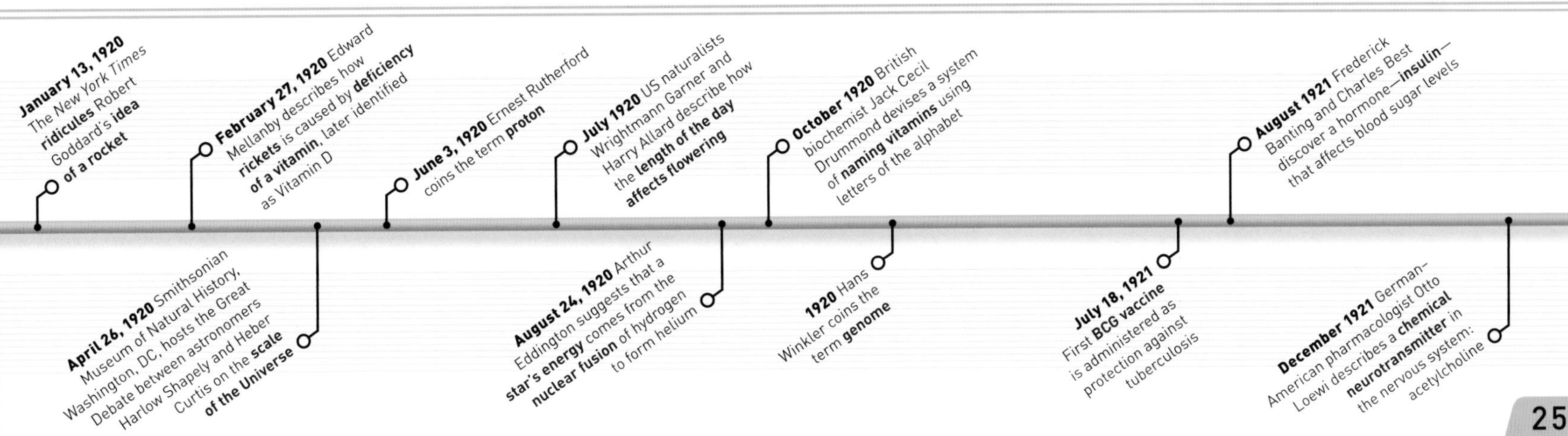

January 13, 1920 The New York Times ridicules Robert Goddard's idea of a rocket
April 26, 1920 Smithsonian Museum of Natural History, Washington, DC, hosts the Great Debate between astronomers Harlow Shapely and Heber Curtis on the scale of the Universe
February 27, 1920 Edward Mellanby describes how rickets is caused by deficiency of a vitamin, later identified as Vitamin D
June 3, 1920 Ernest Rutherford coins the term proton
August 24, 1920 Arthur Eddington suggests that a star's energy comes from the nuclear fusion of hydrogen to form helium
July 1920 US naturalists Wrightmann Garner and Harry Allard describe how the length of the day affects flowering
October 1920 British biochemist Jack Cecil Drummond devises a system of naming vitamins using letters of the alphabet
1920 Hans Winkler coins the term genome
July 18, 1921 First BCG vaccine is administered as protection against tuberculosis
August 1921 Frederick Banting and Charles Best discover a hormone—insulin—that affects blood sugar levels
December 1921 German-American pharmacologist Otto Loewi describes a chemical neurotransmitter in the nervous system: acetylcholine

1922

> **“** WE CAN… CONSIDER THAT **FIRST** PIECE OF ORGANIC **SLIME…** AS BEING THE FIRST **ORGANISM. ”**
>
> **A. Oparin,** from *Origin of Life*, 1924

RUSSIAN BIOCHEMIST ALEXANDER OPARIN (1894–1980) proposed a theory regarding the **origins of life**. He claimed that the first **living things had evolved from nonliving matter**. He also suggested that Earth's earliest atmosphere was highly reduced—that is, gases had an abundance of bonded hydrogen atoms, just like molecules in organisms. The diversity of these gases provided the other elements needed as building blocks for these molecules: methane as a source of carbon, ammonia for nitrogen, and water vapor for oxygen. Oparin suggested that the gases reacted to form a "soup" of organic molecules, from which the first living cells evolved. In 1953, American chemist Stanley Miller (1930–2007) would show that organic molecules could be made from nonorganic ones.

Russian physicist Alexander Friedmann (1888–1925) was studying the curvature of space. One of his models suggested that the **radius of the universe** was constantly **increasing** with time. The idea of an expanding universe would be independently proposed by Belgian astronomer Georges Lemaître (see 1927); and a few years later, the American astronomer Edwin Hubble would collect evidence to suggest that galaxies were indeed receding (see 1929–30).

1923

Edwin Hubble observes the stars through the enormous Hooker Telescope in California.

> **“** THE HISTORY OF **ASTRONOMY** IS A HISTORY OF **RECEDING HORIZONS. ”**
>
> **Edwin Hubble, American astronomer,** from *The Realm of the Nebulae*, 1936

IN OCTOBER, EDWIN HUBBLE suggested that certain **stars of the Andromeda nebula** were much **farther away** than previously supposed—even beyond the Milky Way. His findings were revolutionary, because at the time scientists thought that the Milky Way was the extent of the entire universe.

At about the same time, American chemists **Gilbert Lewis** (1875–1946) and **Merle Randall** (1888–1950) published *Thermodynamics and the Free Energy of Chemical Substances*, which explained **chemical reactions in terms of energy**. In the 19th century, American physicist Willard Gibbs (1839–1903) had shown how the energy associated with chemical reactions could be quantified. Different substances have varied amounts of chemical energy associated with them depending on their chemical composition; this affects their tendency to react. When a chemical reaction occurs, substances change from one form to another—and so does their chemical energy. According to the **laws of thermodynamics**, the total energy stays the same. So if products have less chemical energy than reactants, then some energy, such as heat, must be released. In their book, Lewis and Merle introduced "free energy" values for different substances, which would enable scientists to calculate energy changes associated with reactions.

Chemists also redefined acids and bases. In 1884, 19th-century Swedish chemist **Svante Arrhenius** had shown that **acids and bases** could be recognized by a prevalence of hydrogen (H+) and hydroxyl (OH-) ions respectively. Now, Danish chemist **Johannes Brønsted** (1879–1947) and British chemist **Martin Lowry** (1874–1936) independently defined them in terms of hydrogen ions alone: an acid donates hydrogen ions and a base accepts them. Gilbert Lewis took this one step further and explained them in terms of electrons—the negative subatomic particles that determine the chemical properties of all substances.

HYDROCHLORIC ACID

chlorine atom

hydrogen atom

hydrogen ion

hydrogen ion from hydrochloric acid acts as an electron "accepter"

SODIUM HYDROXIDE

oxygen atom

sodium atom

hydrogen atom

hydroxyl ion

the hydroxyl ion from sodium hydroxide acts as an electron donor

WATER MOLECULE

ACID-BASE THEORY

This theory defines acids and bases in terms of electrons. Acids are electron acceptors: that is, they release positively charged hydrogen ions (protons), which readily combine with negative electrons. Bases are electron donors—they provide the electrons that the hydrogen ions accept. Many bases—called alkalis—release electron-rich hydroxyl ions, which then react with hydrogen ions to form water.

1924

THE YEAR BEGAN with another important development in the field of astronomy. In 1920, British astronomer **Arthur Eddington** had proposed that a star's energy came from the nuclear fusion of hydrogen to form helium, a theory later supported by fellow British

May Alexander Oparin suggests that **life** evolved from **nonliving things**

Alexander Friedmann proposes a solution to Einstein's field equation that predicts that the **universe is expanding**

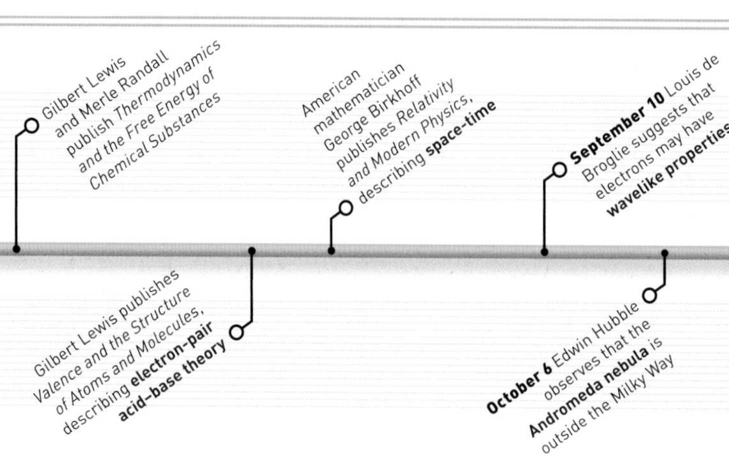

Gilbert Lewis and Merle Randall publish *Thermodynamics and the Free Energy of Chemical Substances*

Gilbert Lewis publishes *Valence and the Structure of Atoms and Molecules*, describing **electron-pair acid-base theory**

American mathematician George Birkhoff publishes *Relativity and Modern Physics*, describing **space-time**

September 10 Louis de Broglie suggests that electrons may have **wavelike properties**

October 6 Edwin Hubble observes that the **Andromeda nebula** is outside the Milky Way

February 14 **IBM is founded** in New York

2.5 MILLION THE NUMBER OF LIGHT YEARS **ANDROMEDA** IS FROM EARTH

Andromeda lies far beyond our own Milky Way. It has long arms spiraling out from a bulging center.

astronomer Cecilia Payne, later, Cecilia Payne-Gaposchkin (1900–79). In March, Eddington published his analysis of the **relationship** between a **star's mass and its luminosity** (brightness). Although a star's brightness increases with its mass, the link between the two is not a simple one, because bigger stars have disproportionately greater luminosities. However, with Eddington's mathematical equation, it became possible to calculate a star's mass from its measured luminosity. This had important implications for understanding the life cycle of stars, as the more massive stars have shorter lifespans.

In November, French physicist **Louis de Broglie** (1892–1987) published an idea that would revolutionize scientists' perception of atoms. He suggested that matter—like light—could have both **particle- and wavelike properties**. He devised a way to calculate the theoretical wavelength of a particle, such as an electron. He found that the wavelength values diminished to almost negligible levels for particles above the size of an atom, but for subatomic particles they were more significant. A few years later, experimental evidence would suggest that de Broglie's idea was right: electrons diffracted in wavelike ways, just like light.

At the end of 1924, more than a year after he had observed that certain stars were farther away than the other stars in the Milky Way, Edwin Hubble announced that **Andromeda was not a spiral nebula**, as was previously thought, but an entire **galaxy**, equivalent to Milky Way. Andromeda stars can be more than 20 times the distance of the farthest stars of the Milky Way. The revelation settled the Great Debate of 1920—held at the Smithsonian Museum of Natural History, Washington, DC, between American astronomers Harlow Shapley and Heber Curtis (1872–1942). Shapley had argued that the Milky Way defined the extent of the universe, but Eddington's results supported Curtis's view that a multigalactic universe was much bigger.

Luminous star
As a star slowly expodes, nuclear reactions at its core release energy as light. Eddington's equation made it possible to determine a star's mass based on its luminosity—in general, the brighter a star burns, the greater its mass.

WAVE-PARTICLE DUALITY

Electrons fired through a thin sheet of graphite onto a luminescent screen form a pattern of rings. These kinds of rings ordinarily arise when diffracted (scattered) waves, such as those of light, interfere with one another. This indicates that electrons, conventionally regarded as subatomic particles, must have wavelike properties too.

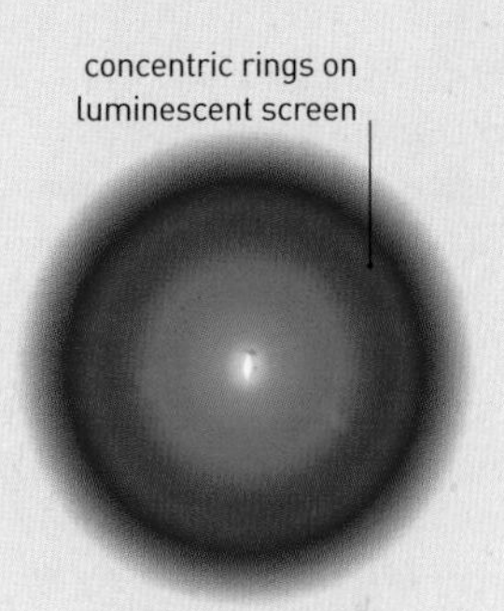
concentric rings on luminescent screen

ARTHUR EDDINGTON (1882–1944)

Born into a Quaker family, Arthur Eddington studied at Cambridge University, UK, before pursuing a career in astronomy. He tested Albert Einstein's general theory of relativity during a solar eclipse in West Africa, and formulated theories concerning the life of stars. He was awarded a knighthood in 1930, followed by the Order of Merit in 1938.

March 5 Arthur Eddington explains the relationship between the **mass and luminosity of stars**

British physiologist Charles Sherrington discovers the **stretch reflex**

American physicist Wolfgang Pauli starts to develop his **Exclusion Principle**—that it is impossible for two particles to have the same quantum state

Indian and German physicists Satyendra Bose and Albert Einstein explain the behaviour of **fundamental force-carrying particles** that are later called bosons

October 1 British mathematician John Lennard-Jones describes the interactions between **nonbonding atoms**

November 25 Louis de Broglie describes his **particle-wave theory** of matter

December 30 Edwin Hubble reports that Andromeda and the Milky Way are both galaxies, suggesting that the **universe is bigger** than previously thought

1925

“ THE BEST THAT MOST OF US CAN HOPE TO ACHIEVE IN **PHYSICS** IS SIMPLY TO **MISUNDERSTAND AT A DEEPER LEVEL. ”**

Wolfgang Pauli, Austrian physicist, in a letter to Jagdish Mehra in Berkeley, California, 1958

SCOTTISH ENGINEER JOHN LOGIE BAIRD (1888–1946) had been experimenting with transmitting moving images and now, using a semimechanical device, he achieved it. In March, **television was demonstrated** in a London department store. At first images were silhouettes, but by October, he was **transmitting pictures with grayscale** made up of 30 vertical lines running at five pictures per second.

In May, **John Scopes** (1900–70), of Tennessee, was arrested for teaching Darwin's theory of evolution, which was prohibited in the state's schools. He was tried, found guilty, and fined $100.

Quantum physics had shown how the electrons of an atom existed in fixed energy levels, but Austrian physicist **Wolfgang Pauli** (1900–58) went further with his **Exclusion Principle** saying that no two particles could occupy the same quantum state at the same time. Later, two categories of particles would be distinguished—matter-associated particles (fermions) that obeyed Pauli's principle, and force-associated particles (bosons) that did not (see 1974).

The German Atlantic Meteor Expedition began to survey the ocean's floor and **discovered an unbroken ridge running from north to south**.

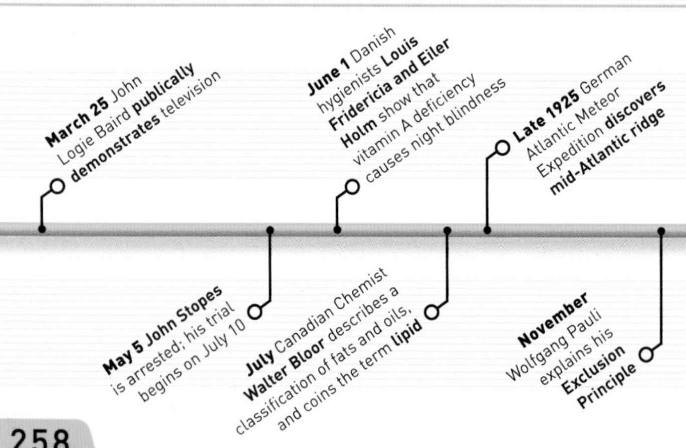
Atlantic Ridge

Mid-Atlantic ridge
The complementary shapes of Atlantic coastlines suggest that they were once joined together. This ridge shows where new seafloor has pushed them part.

1926

Two ventriloquist's dolls became the first TV personalities when John Logie Baird demonstrated his television system to scientists of the Royal Institution and members of the press.

IN JANUARY, JOHN LOGIE BAIRD was back showing off his improved television. This time it was to members of the Royal Institution in London, UK. He now achieved **clearer images with tonal gradation of grays**. Although the images were still blurred, Baird had increased the picture frequency, so **the movements were smoother**.

Austrian physicist **Erwin Schrodinger** (1887–1961) examined earlier suggestions that matter was both particle-like and wavelike in nature. Schrodinger calculated a particle's distribution of energy in space—its wave function—and urged that the wave idea alone was critical to understanding reality. His work would be the **basis for quantum mechanics**.

Six years after the *New York Times* had ridiculed his suggestion that **a rocket was a possibility,** American physicist Robert Goddard (1882–1945) launched one. Using his specially developed liquid fuel, **Goddard's rocket reached a height of 40ft (12m)** in less than three seconds.

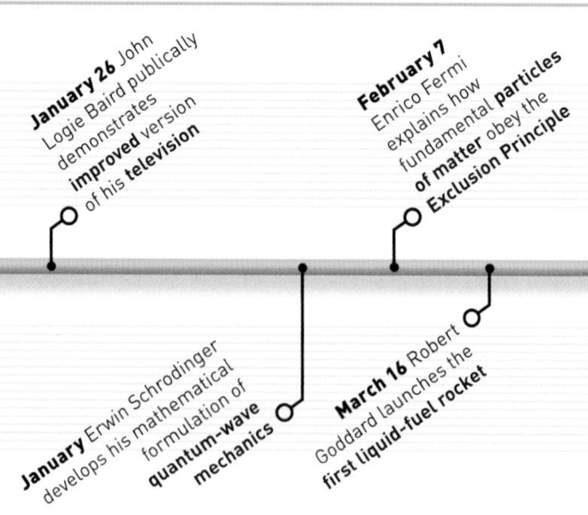

The first rocket
Robert Goddard launched the first liquid-fuel rocket on his aunt's farm in Massachusetts. Three people witnessed the event: his wife, his machinist, and a fellow physicist.

In August, American chemist **James Sumner** (1887–1955) made a breakthrough in biochemistry. By grinding up jack beans, he had **isolated crystals of the enzyme urease**. Enzymes are substances found in living tissues that speed up the chemical reactions of metabolism. When Sumner analyzed his crystals, he found that they were protein. In this way, he had **demonstrated that biological enzymes were proteins**.

In December, another American chemist, Gilbert Lewis (1875–1946), **coined the term photon** for a unit of radiant energy. It came to be used to describe a particle of light energy.

PVC, polyvinyl chloride (often called vinyl), **made its first**

QUANTUM THEORY

The introduction of quantum theory was a pivotal point in early 20th century physics that began with Max Planck's explanation of black body radiation (see 1900). The theory is based on the idea that energy exists in discrete parcels, or quanta (from the Latin *quantus*, meaning "how much"). These parcels equate to the fixed quantities of energy absorbed or emitted by atoms. This is realized as specific wavelengths of radiation, each associated with particular quanta of energy: red light wavelength is transmitted by low-energy parcels and blue light by high-energy ones. Danish physicist Niels Bohr explained this in terms of the atom's electrons moving between energy levels, or orbits.

March 25 John Logie Baird **publically demonstrates** television

May 5 John Stopes is arrested; his trial begins on July 10

June 1 Danish hygienists **Louis Fridericia and Eiler Holm** show that vitamin A deficiency causes night blindness

July Canadian Chemist **Walter Bloor** describes a classification of fats and oils, and coins the term **lipid**

Late 1925 German Atlantic Meteor Expedition **discovers mid-Atlantic ridge**

November Wolfgang Pauli explains his **Exclusion Principle**

January 26 John Logie Baird publically demonstrates **improved** version of his **television**

January Erwin Schrodinger develops his mathematical formulation of **quantum-wave mechanics**

February 7 Enrico Fermi explains how fundamental **particles of matter** obey the **Exclusion Principle**

March 16 Robert Goddard launches the **first liquid-fuel rocket**

August 26 British physicist **Paul Dirac** independently gives Fermi's explanation of fundamental particles

August James Sumner describes isolating the enzyme urease and shows that **enzymes are proteins**

“ I WAS… ABLE TO TRANSMIT THE LIVING IMAGE, AND IT WAS THE FIRST TIME IT HAD BEEN DONE. BUT HOW TO CONVINCE THE… SCEPTICAL SCIENTIFIC WORLD? ”

John Logie Baird, Scottish engineer, in *The Times*, January 28, 1926

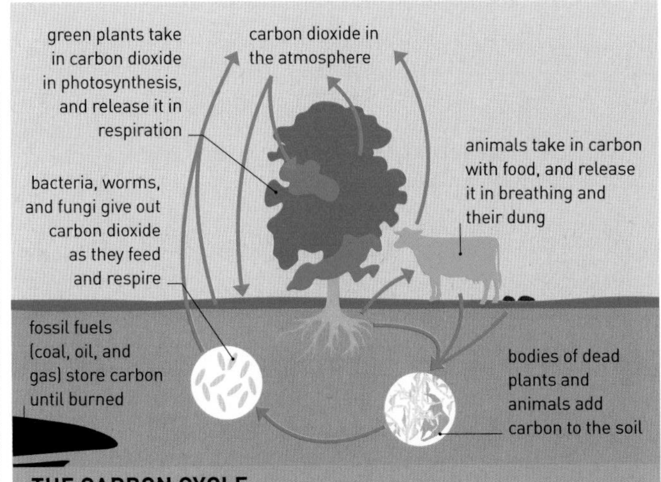

green plants take in carbon dioxide in photosynthesis, and release it in respiration
carbon dioxide in the atmosphere
animals take in carbon with food, and release it in breathing and their dung
bacteria, worms, and fungi give out carbon dioxide as they feed and respire
fossil fuels (coal, oil, and gas) store carbon until burned
bodies of dead plants and animals add carbon to the soil

THE CARBON CYCLE

At the heart of Vladimir Vernadsky's book *The Biosphere* are his ideas about nature's constant recycling of matter. Elemental atoms react and recombine in different ways. Carbon atoms in complex organic materials of living things—animals, plants, and soil bacteria—are respired into the atmosphere as carbon dioxide, before reacting to make plant sugars in photosynthesis.

appearance in its modern form. Chemists had been making this chemical polymer since the 1800s, but American chemist **Waldo Semon** (1898–1999) found a way of making it malleable and less brittle. It would become one of the most widely used plastics.

Russian geochemist **Vladimir Vernadsky** (1863–1945) had been working on theories that unified geology and biology, Earth and life. He published his thoughts in a book called *The Biosphere*. In it he described some of the main ideas behind a concept today referred to as the **ecosystem: a system where living things interact with nonliving matter**. Verdansky recognized that all life on Earth **relies upon solar energy**, and that particles of matter **undergo a process of recycling**.

1927

A computer-generated simulated image of a particle collision shows matter thought to have been produced microseconds after the Big Bang.

THE YEAR STARTED WITH A MILESTONE in communication technology. On January 7, a collaboration between the American Telephone and Telegraph Company and Britain's General Post Office **opened the first transatlantic telephone service**. On its first day 31 calls were made between London and New York.

A further development came in the field of quantum physics. Erwin Schrodinger had laid the foundations of quantum mechanics with his description of the wavelike characteristics of particles. Now, German physicist **Werner Heisenberg** (1901–76) reasoned that the wave function of a particle could not be localized to a specific point in space and have a definable wavelength. Heisenberg developed this as his **Uncertainty Principle**. Its consequences are extraordinary: the more accurately a particle's position is measured, the less accurately it is possible to determine its movement, and *vice versa*. Later, the Copenhagen Interpretation stated that it is impossible to experimentally measure wavelike and particle-like properties at the same time.

While Baird worked in London on his television, the **American Bell Telephone Company** was also developing the technology. In April, Bell had a breakthrough when the company sent the first **long-distance TV transmission** using the semimechanical television from Washington to New York. Five months later, American inventor **Philo Farnsworth** (1906–71) introduced a way of scanning and transmitting electronically. Russian-American inventor **Vladimir Zworykin** (1888–1982) was working on similar technology at the same time, but it was Farnsworth who made it a reality.

In April, Belgian astronomer **Georges Lemaître** (1894–1966) published a scientific paper containing a revolutionary theory: that the **Universe is expanding**.

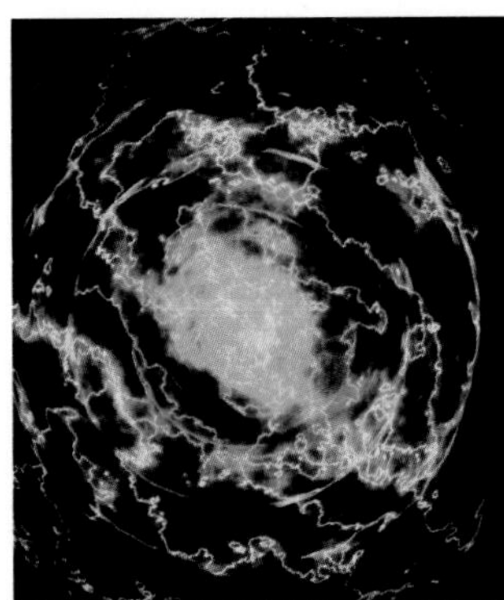

Electron clouds
Modern quantum physics has revised the idea that atoms—such as this helium atom—have electrons in fixed orbits. A more realistic interpretation sees electrons existing as clouds of probability.

When Lemaître presented his conclusions to the British Association for the Advancement of Science in 1931, he elaborated upon his theory, suggesting the Universe had originated from a primeval atom. His "exploding cosmic egg" model anticipated the work of American astronomer Edwin Hubble (see 1929) and was the forerunner of the Big Bang theory.

WERNER HEISENBERG
(1901–76)

Heisenberg studied physics at the universities of Munich and Göttingen, where he met Niels Bohr in 1922. In 1925, he developed a mathematical way of understanding quantum physics called matrix mechanics. He derived his Uncertainty Principle before working on the German nuclear energy project in World War II.

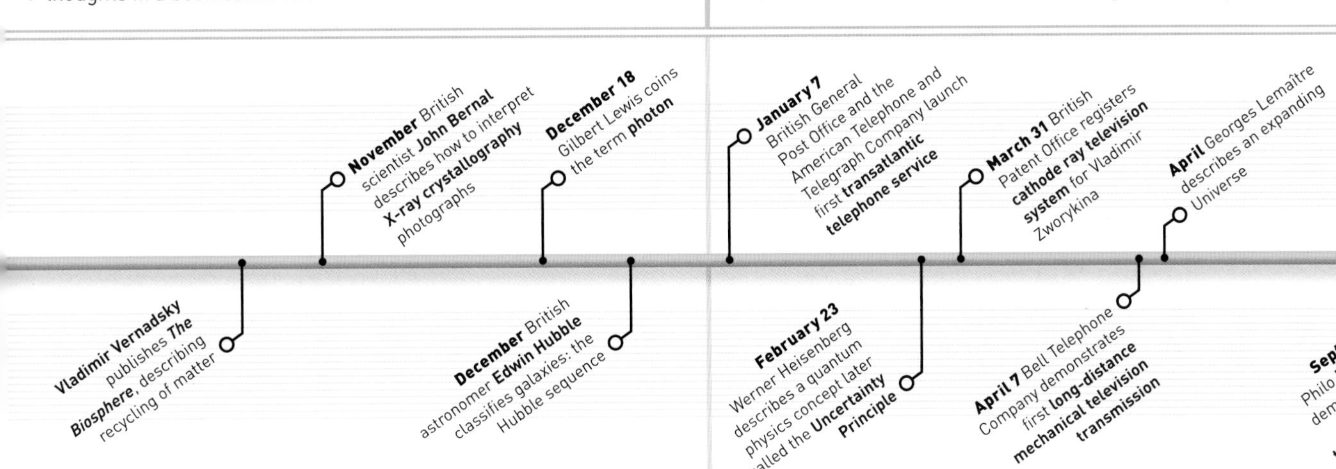

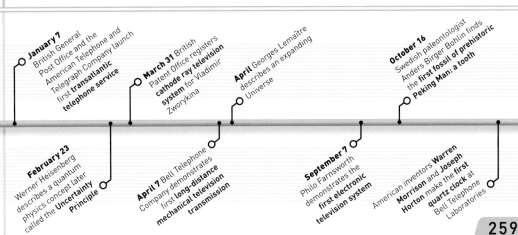

Vladimir Vernadsky publishes The Biosphere, describing recycling of matter
November British scientist John Bernal describes how to interpret X-ray crystallography photographs
December British astronomer Edwin Hubble classifies galaxies: the Hubble sequence
December 18 Gilbert Lewis coins the term photon
January 7 British General Post Office and the American Telephone and Telegraph Company launch first transatlantic telephone service
February 23 Werner Heisenberg describes a quantum physics concept later called the Uncertainty Principle
March 31 British Patent Office registers cathode ray television system for Vladimir Zworykina
April 7 Bell Telephone Company demonstrates first long-distance mechanical television transmission
April Georges Lemaître describes an expanding Universe
September 7 Philo Farnsworth demonstrates the first electronic television system
October 16 Swedish paleontologist Anders Birger Bohlin finds the first fossil of prehistoric Peking Man: a tooth
American inventors Warren Morrison and Joseph Horton make the first quartz clock at Bell Telephone Laboratories

Bakelite radio
Bakelite is a good insulator, which is why it is used for electrical appliances such as this 1950s Tesla Talisman radio. It is also tough and shiny, and can be dyed and molded into domestic appliances, including telephones and kitchenware.

single loudspeaker

hard molded bakelite shell

1862
Parkesine
Alexander Parkes develops the first plastic—parkesine. It is used to make the first cheap buttons.

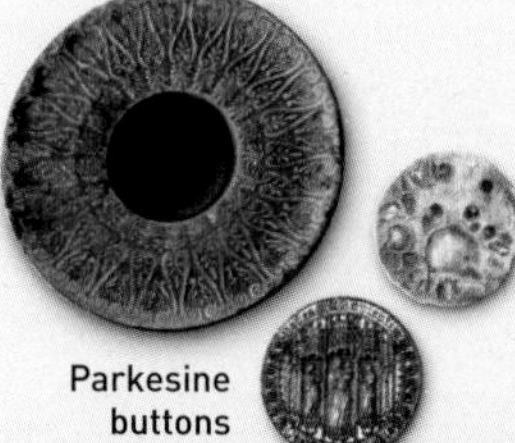
Parkesine buttons

1887
Celluloid
American John Hyatt and Englishman Daniel Spill both develop a material called celluloid that is similar to parkesine. It is used to make flexible film for photographs to replace glass plates. This is a crucial step for movie-making.

Celluloid film

1909
Bakelite
American chemist Leo Baekeland first develops bakelite by treating phenol resin made from coal tar with formaldehyde. It is the first entirely synthetic plastic. Not only can it be molded, like earlier plastics, but once it sets it is hard and heatproof.

Golf ball

1872
PVC
This extemely tough plastic is first developed in 1872 by German chemist Eugen Baumann. It is thought to be useless until the 1920s.

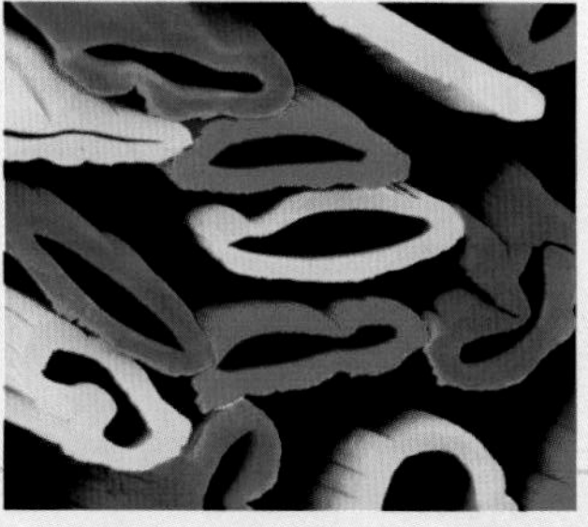
Viscose fibers

1894
Viscose rayon
Two English chemists produce a synthetic material called viscose (rayon) by reconstituting wood fibers in sodium hydroxide and spinning them into thread.

Sweet wrappers

1912
Cellophane
Cellophane, a thin, transparent sheet made of processed cellulose, is first developed. It provides an airtight wrapping and is useful for packaging food.

THE STORY OF PLASTICS

BY THE END OF THE 20TH CENTURY THE AGE OF PLASTICS HAD ARRIVED, TRANSFORMING MANY INDUSTRIES AND THE HOME

Plastic is one of the most remarkable of all man-made materials, used in everything from spaceships and computers to bottles and artificial body parts. What gives plastic its special quality is the shape of its molecules. Most plastics are made from long organic molecules known as polymers.

In the mid-19th century, people knew that cellulose (the woody substance in plants) could be made into a brittle substance called cellulose nitrate. In 1862, British chemist Alexander Parkes added camphor to it, producing a tough but moldable plastic called parkesine. In 1869, American inventor John Hyatt created a similar substance called celluloid, which was used to make photographic film by Kodak in 1889. Today there are thousands of synthetic plastics, each with their own properties and uses. Many are still based on hydrocarbons (oil or natural gas), but in recent decades carbon fibers and other materials have been added to create superlight, superstrong plastics such as Kevlar and CNRP.

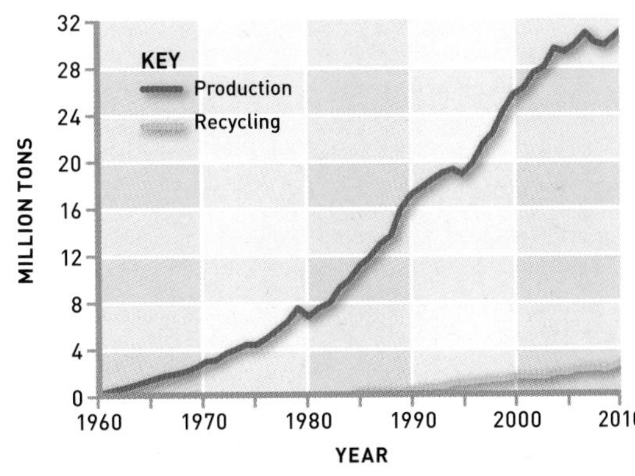
KEY
Production
Recycling
MILLION TONS
0
4
8
12
16
20
24
28
32
1960
1970
1980
1990
2000
2010
YEAR

Production and recycling of plastics
In recent years, there have been concerted attempts to recycle plastics, made easier by the establishment of recycling centers and collection services. However, the slight rise in recycled plastic lags far behind the soaring production of new plastics, as this graph clearly shows.

RECYCLING

Plastics are widely used because they are durable and tough, but they are not biodegradable. Once disposed of, they linger in the environment for a very long time. The vast amount of waste plastic now in the oceans—maybe hundreds of millions of tons—is damaging marine wildlife. It is important to reduce plastic use and recycle as much as possible. Not all plastics can be recycled easily, and it takes energy to heat the plastic to reform it, so recycling rates are low.

> “I THOUGHT I SHOULD MAKE SOMETHING **REALLY SOFT** INSTEAD THAT COULD BE **MOLDED** INTO **DIFFERENT SHAPES.**”

Leo Baekeland, Belgian chemist, on inventing bakelite

1926
Vinyl
American chemist Waldo Semon exposes PVC to heat and a range of chemicals to produce vinyl. It is used to make objects ranging from shoes to shampoo bottles.

1933
Polyethylene
British chemists Eric Fawcett and Reginald Gibson create practical polyethylene in 1933, although it was first made in 1898. It is tough, soft, and flexible, and now the most widely used of all plastics.

Polyethylene crop tunnel

1935
Nylon
American chemist Wallace Carothers invents nylon, the first thermoplastic—it is liquid when hot and sets hard when it cools. Best known for its use in stockings, it has many other uses.

Nylon toothbrush

1936
Polystyrene
Styrol is the oily substance in the resin of Turkish sweetgum trees. In 1936 German chemical company I G Farben use it to produce polystyrene.

1937
Teflon
PTFE, or Teflon, is invented by American chemist Roy Plunket. It is not made from hydrocarbons but from fluorine joined to carbon, and is often used in frying pans.

Teflon frying pan

1954
Polypropylene
This rugged plastic resists many solvents and acids. It has uses ranging from wrapping to bottles for medical chemicals.

Polyproylene rope

1966
Kevlar
American chemist Stephanie Kwolek spins heat-resistant fibers from liquid hydrocarbons. These fibers can be woven together to make materials such as Kevlar.

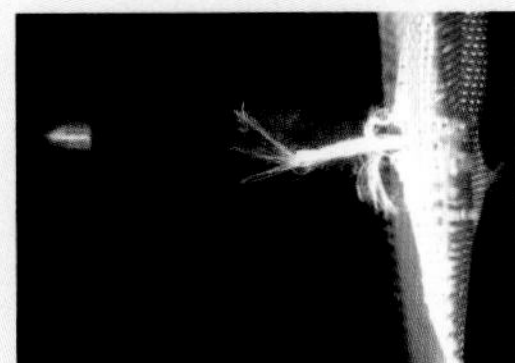

Bulletproof Kevlar

1991
CNRP
Japanese physicist Sumio Iijima rolls carbon molecules into nanotubes. These can reinforce plastic to make strong, light CNRP.

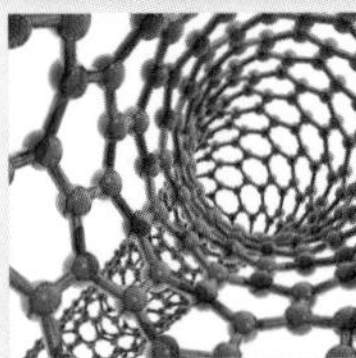

Carbon nanotube molecule

" MY... PART IN THE STORY WAS THAT I SAW **SOMETHING UNUSUAL** AND **APPRECIATED** SOMETHING OF **ITS IMPORTANCE** SO... I SET TO WORK ON IT. **"**

Alexander Fleming, Scottish biologist, address to the University of Edinburgh, 1952

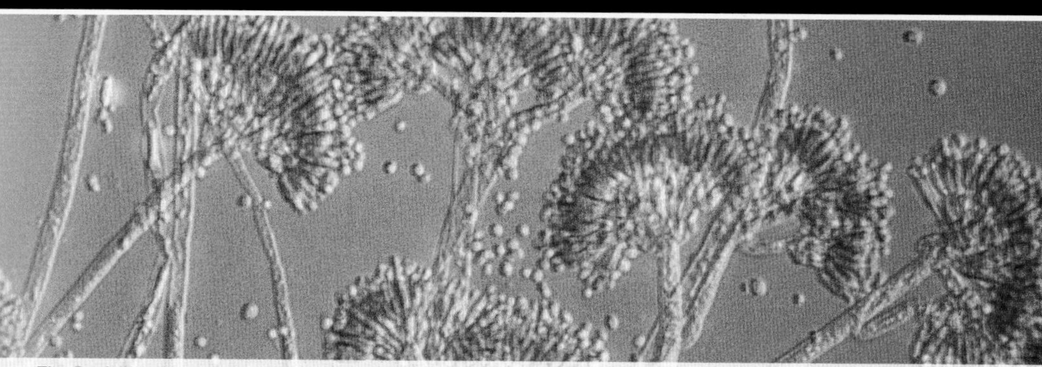

The *Penicillium* mold is a common fungus, spread by spores from the capsules seen here; some species produce the penicillin seen by Fleming.

THE YEAR BEGAN WITH EXPERIMENTAL CONFIRMATION that characteristics and inheritance were determined by a chemical substance. British physician **Frederick Griffith** (1879–1941) studied strains of pneumonia bacteria, some of which could cause disease, while others were harmless. Griffith's work indicated that a **transforming factor** could move from the harmful bacteria to others and make them harmful too (see panel, below). This factor would later be identified as DNA (see 1943–44).

Another advance came in physics. Austrian physicist Erwin Schrödinger had described the wave function of a particle (see 1926). Now British physicist **Paul Dirac** described a new form of Schrödinger's wave equation for electrons. As a result, he predicted a new class of matter called anti-electrons, which had positive, instead of negative, charges. Dirac's work was the first **modern theory of antimatter**. The anti-electrons would later be discovered and renamed positrons (see 1932–33). It would also become evident that corresponding antimatter particles existed for most subatomic particles.

In September the previous year, Britain's **John Logie Baird** had sent his technical assistant to New York to prepare for the biggest test of his television system so far: **a transatlantic transmission**. After several false starts, success came at midnight, London time, on February 8, 1928. Baird himself briefly appeared as a fuzzy image on screen in America.

In the wake of World War I, Scottish biologist **Alexander Fleming** (1881–1955) was developing ways of fighting infections that went beyond the routine use of antiseptics, which were often ineffective for serious wounds. As part of his research, Fleming had been growing cultures of infectious bacteria. On September 3, in his laboratory at St. Mary's Hospital, London, Fleming noticed that one of his cultures had become contaminated: **a mold had spread on the culture dish**. But significantly, just around the spreading mold there was a region that was clear of bacteria.

" IF YOU ARE **RECEPTIVE** AND HUMBLE, MATHEMATICS **WILL LEAD YOU... "**

Paul Dirac, British physicist, November 27, 1975

Clearly, the mold was producing something that had **killed the bacteria**. Fleming took samples of the mold, cultivated it, and identified it as *Penicillium*. The following year he called the active ingredient of his **anti-bacterial mold juice penicillin**. Despite his efforts, Fleming was unable to isolate penicillin in its raw chemical form, but he did preserve his culture of mold. In the following decades penicillin would not only be purified then manufactured, but also lauded as one of the **first effective antibiotic cures** of bacterial infection (see 1940–41).

PAUL DIRAC (1902–84)

British physicist Paul Dirac held the Lucasian Chair of Mathematics at Cambridge University, UK, from 1932 to 1969. He advanced quantum physics by applying it to Einstein's theory of relativity (see pp.244–45). His work on the quantum wave equations predicted the existence of antimatter. In 1993 he shared the Nobel Prize for Physics with Erwin Schrödinger.

A medical breakthrough of a different kind came in Sydney, Australia. In 1926, physicians **Mark Lidwell** (1878–1969) and **Edgar Booth** (1893–1963) had devised a portable plug-in **artificial heart pacemaker**. In 1928 they used it to **revive a stillborn infant**.

BACTERIAL TRANSFORMATION

Pneumococcus bacteria exist in harmless (rough) and harmful (smooth) forms. Frederick Griffith found that bacteria can exchange a chemical that changes these characteristics. Heat-killed harmful bacteria fail to cause an infection on their own, but if mixed with living harmless bacteria, a substance moves from dead bacteria to living ones, changing them into harmful forms that kill the mouse and appear in its blood.

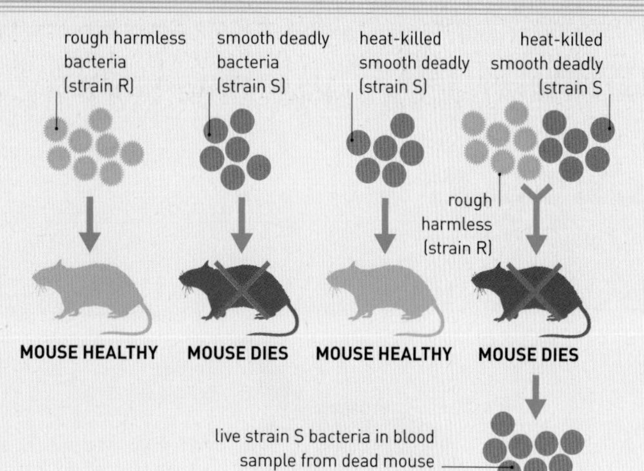
rough harmless bacteria (strain R)
smooth deadly bacteria (strain S)
heat-killed smooth deadly (strain S)
heat-killed smooth deadly (strain S
rough harmless (strain R)
MOUSE HEALTHY
MOUSE DIES
MOUSE HEALTHY
MOUSE DIES
live strain S bacteria in blood sample from dead mouse

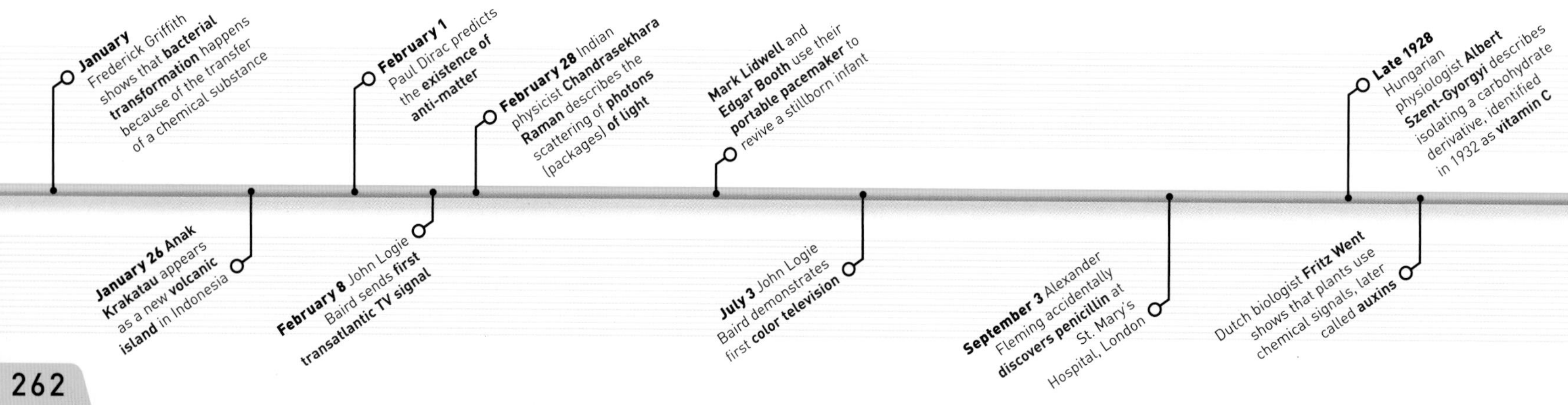
January Frederick Griffith shows that bacterial transformation happens because of the transfer of a chemical substance
January 26 Anak Krakatau appears as a new volcanic island in Indonesia
February 1 Paul Dirac predicts the existence of anti-matter
February 8 John Logie Baird sends first transatlantic TV signal
February 28 Indian physicist Chandrasekhara Raman describes the scattering of photons (packages) of light
Mark Lidwell and Edgar Booth use their portable pacemaker to revive a stillborn infant
July 3 John Logie Baird demonstrates first color television
September 3 Alexander Fleming accidentally discovers penicillin at St. Mary's Hospital, London
Late 1928 Hungarian physiologist Albert Szent-Györgyi describes isolating a carbohydrate derivative, identified in 1932 as vitamin C
Dutch biologist Fritz Went shows that plants use chemical signals, later called auxins

46 BILLION THE SIZE OF THE UNIVERSE VISIBLE FROM EARTH IN LIGHT-YEARS

This image taken by the Hubble Space Telescope reveals some of the most distant galaxies in the Universe. They are billions of light-years beyond the foreground stars.

BY 1920 ASTRONOMERS HAD REALIZED that the Universe was not centered on our Milky Way, but rather that the **Milky Way was a single galaxy** among many others. As the decade passed, another still more extraordinary revelation came to light. It appeared that the **Universe** did not have a fixed finite size. Astronomers had discovered that the light spectrum of stars carried wavelengths that were "stretching out" toward the red end of the spectrum—which they **called redshift** (see panel, right). This redshift indicated that these stars were moving away. In other words, **the Universe is expanding**. In 1929, US astronomer **Edwin Hubble** (1889–1953) explained this relationship and noted that the most **distant galaxies were receding at the fastest rate**. This later became known as **Hubble's Law**.

Television technology took another step forward in June 1929 when the **Bell Laboratory** in the US demonstrated the **first color image** transmission. The subjects they used were chosen for maximum impact: a woman with flowers and the US flag.

In April 1930, US chemist **Elmer Bolton** (1886–1968) made one of the first synthetic rubbers. A derivative of acetylene, it was later called **neoprene**. More corrosion-resistant than natural rubber, neoprene was suitable for use in extreme conditions, such as for hoses and fire-resistant coatings.

Photographs taken at the Lowell Observatory in Arizona in March 1917 had recorded the faint image of a body that became known as "X." In February 1930, American astronomer **Clyde Tombaugh** (1906–97) confirmed that **X was a planet**, and in May, it was given its name: **Pluto**.

Two years after Alexander Fleming made his accidental discovery of penicillin, this new antibiotic had its first curative use. British physician **Cecil George Paine** (1905–94), a former student of Fleming, had taken samples of the mold to his workplace in Sheffield, UK. In August 1930, he successfully used its filtrate to **treat eye infections**.

Paul Dirac (see panel, opposite) used quantum physics theory to correctly predict the existence of antimatter. In 1930, he published ***Principles of Quantum Mechanics***, which for decades would be the standard textbook.

REDSHIFT AND BLUESHIFT

The movement of a light-emitting object, such as a galaxy, affects its visible light spectrum. If an object is moving away, its wavelengths appear stretched out. As longer wavelengths belong to red light, this is described as a redshift. Movement toward the observer squashes up the wavelengths, giving a blueshift. Galaxies exhibit a redshift, so they are moving away.

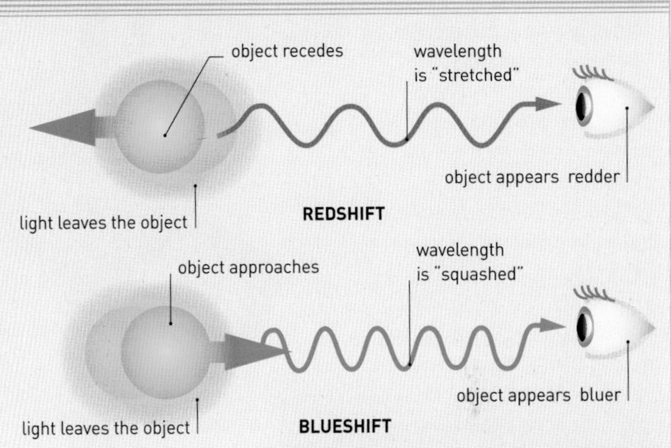

object recedes
wavelength is "stretched"
object appears redder
light leaves the object
REDSHIFT
object approaches
wavelength is "squashed"
object appears bluer
light leaves the object
BLUESHIFT

First color television demonstration
Open doors reveal the working parts. Light is directed at the operator through a scanning disk (center) onto photoelectric cells (left).

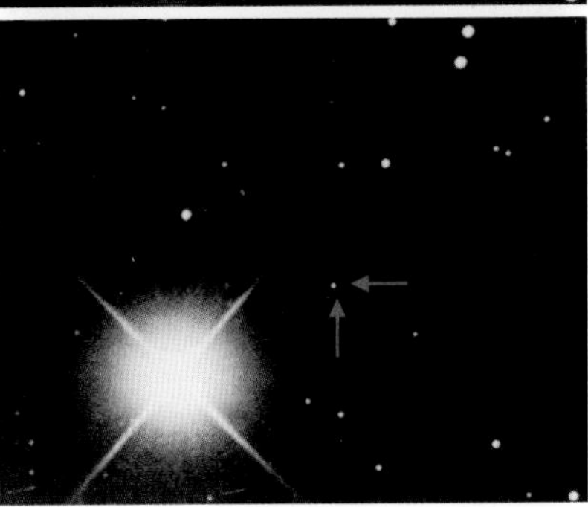

Existence of Pluto
These two photographs of the sky taken in 1930 on different nights show a "body" (arrowed) that has changed position, which indicates that it is closer than surrounding stars. The body was a planet and was given the name Pluto.

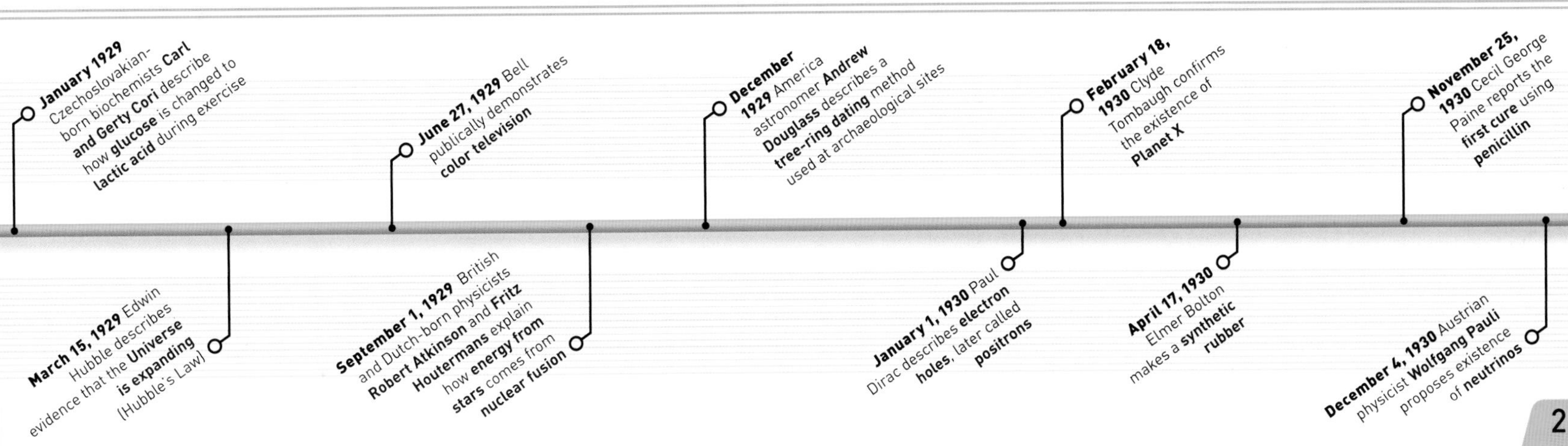

January 1929 Czechoslovakian-born biochemists Carl and Gerty Cori describe how glucose is changed to lactic acid during exercise
March 15, 1929 Edwin Hubble describes evidence that the Universe is expanding (Hubble's Law)
June 27, 1929 Bell publicly demonstrates color television
September 1, 1929 British and Dutch-born physicists Robert Atkinson and Fritz Houtermans explain how energy from stars comes from nuclear fusion
December 1929 America astronomer Andrew Douglass describes a tree-ring dating method used at archaeological sites
January 1, 1930 Paul Dirac describes electron holes, later called positrons
February 18, 1930 Clyde Tombaugh confirms the existence of Planet X
April 17, 1930 Elmer Bolton makes a synthetic rubber
November 25, 1930 Cecil George Paine reports the first cure using penicillin
December 4, 1930 Austrian physicist Wolfgang Pauli proposes existence of neutrinos

1931

❝ IT IS A **MIRACLE** THAT… **DIFFICULTIES HAVE BEEN SOLVED** TO AN EXTENT THAT SO **MANY SCIENTIFIC DISCIPLINES…** CAN **REAP ITS BENEFITS. ❞**

Ernst Ruska, Nobel lecture on electron microscopy, December 8, 1986

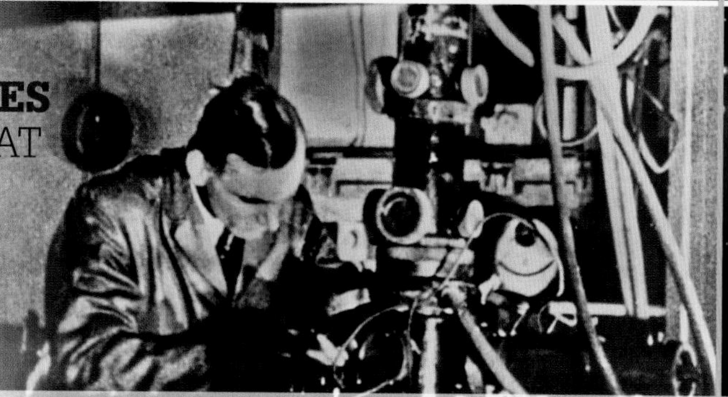

Ernst Ruska worked with the German company Siemens and Halske to produce the first commercial electron microscope in 1939.

THE 1930S WITNESSED THE BIRTH OF TECHNOLOGIES that would help reveal the secrets of the microscopic world. Physicists had devised ways of firing charged particles at very high speeds so that they could study resulting collision products. They did this with accelerators that shot particle beams through evacuated tubes and used electric fields to keep them going. In 1931, American physicist **Ernest Lawrence** (1901–58) invented a new kind of **particle accelerator** with a spiral center called a cyclotron, designed to shoot hydrogen ions. His first model had a diameter of 5 in (12.5 cm) and energized ions with **80,000 electron volts** (energy acquired by an electron as it accelerates through the potential of one volt). Lawrence's team went on to make ever more powerful cyclotrons. By 1946, his laboratory at Berkeley, California, had built a 173 in- (440 cm-) device powered by **100 million electron volts**.

❝ IT **MAY** BRING TO LIGHT… A **DEEPER** KNOWLEDGE OF THE **STRUCTURE** OF MATTER. **❞**

Ernest Lawrence, Nobel Prize speech, November 29, 1940

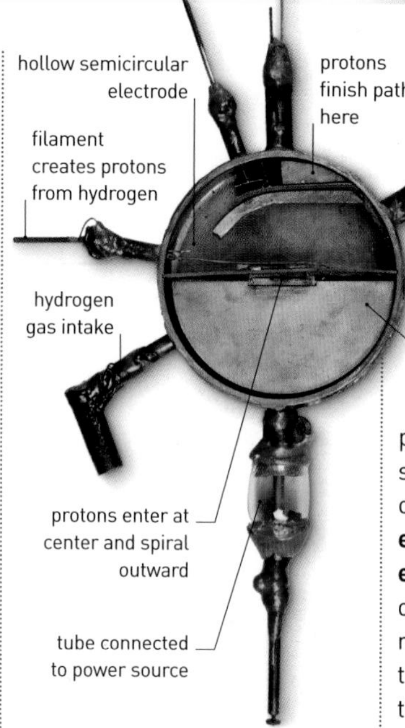

hollow semicircular electrode
protons finish path here
filament creates protons from hydrogen
hydrogen gas intake
hollow semicircular electrode
protons enter at center and spiral outward
tube connected to power source

The first working cyclotron
Protons enter the cyclotron and, as voltage switches between electrodes, they are drawn from one to the other, speeding up each time they cross over.

The particles targeted by physicists are **billions of times smaller** than the smallest **objects that can be seen** with light microscopes. Until 1930, all microscopes magnified images using conventional optics that refract (bend) light rays using lenses. However, the wavelength of light, which is measured in 10 thousandths of a millimeter, restricts the amount of detail that can be seen in objects that approach this size. German physicist **Ernst Ruska** (1906–88) found a radical way to solve this problem using radiation with a smaller wavelength than that of visible light—**beams of electrons**. He used powerful **electromagnetic fields** instead of glass lenses to bend the radiation—the stronger the field, the **greater the magnification** that could be achieved. Ruska's **electron microscope** allowed extremely large magnifications, and scientists could examine tiny molecules, possibly even atoms.

Meanwhile, chemistry was making advances. German chemist **Eric Hückel** (1896–1980) was using quantum physics (the physics of ultra-small particles) to build more realistic ideas about the nature of chemical bonds and proposed a theory that described them in terms of the **behavior of electrons** within molecules.

In November, American chemist **Harold Clayton Urey** (1893–1981) made a discovery about hydrogen atoms—the smallest and lightest of all elements. He found a **heavier variety** (isotope) **of hydrogen**. A hydrogen nucleus contains a single proton, but this heavier isotope (later called **deuterium**) had a nucleus with an extra new subatomic particle that would not be identified until 1932.

DEUTERIUM, OR HEAVY HYDROGEN

Heavy hydrogen (deuterium) accounts for fewer than one in 6,000 of Earth's hydrogen atoms. All hydrogen atoms have a single proton (positively charged particle), so have an atomic number of 1 (see panel, 1913). Normal hydrogen has only one particle in its nucleus. Deuterium has an additional particle—a neutron (particle with no charge)—in its nucleus.

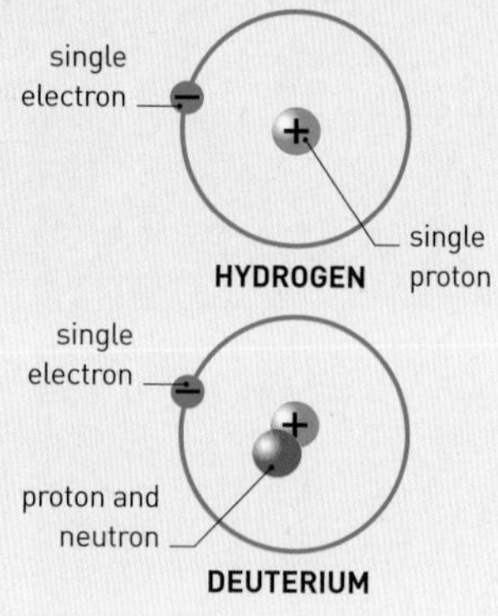

single electron
single proton
HYDROGEN
single electron
proton and neutron
DEUTERIUM

1932–33

Between these bright galaxies lies invisible dark matter.

ERNEST RUTHERFORD HAD PREDICTED the existence of a second proton-sized subatomic particle (see 1919). In 1932, British physicist **James Chadwick** (1891–1974) examined a type of radiation that could knock protons from atoms and found Rutherford's **missing particle: the neutron**. Atomic capabilities advanced in April, when British and Irish physicists **John Cockcroft** (1897–1967) and **Ernest Walton** (1903–95) split lithium atoms **into helium atoms** (see panel, opposite).

British physicist Paul Dirac had predicted the existence of anti-particles of electrons (see 1928). Confirmation came in August 1932 when American physicist **Carl Anderson** (1905–91) studied trails of charged particles in a detector called a cloud chamber and found that some electron-like particles were **positive instead of negatively charged**.

600,000
THE NUMBER OF **VOLTS** NEEDED TO **SPLIT THE ATOM**

January 2 Ernest Lawrence uses the **first cyclotron**

March 3 Ernst Ruska and German engineer **Max Knoll** achieve the first electron optical magnification with the **prototype** of the **electron microscope**

August American physicist **Karl Guthe Jansky** discovers **radio waves** coming from the **Milky Way galaxy**

August 1 American botanists **Harriet Creighton and Barbara McClintock** explains **genetic recombination** by shuffling of chromosome fragments

French physicist **Irène Joliot-Curie** demonstrates radiation that she thinks is **gamma rays**, but **James Chadwick** later shows it to be **neutrons**

November 26 Harold Clayton Urey discovers **heavy hydrogen**, later called **deuterium**

January 1, 1932 German physicist **Werner Heisenberg** begins the **proton- neutron model** of the nucleus and uses it to explain isotopes

January 1932 Dutch-American biologist **Cornelius van Niel** suggests that plants use water as an **electron source** for **photosynthesis**

22.5 PERCENT THE PROPORTION OF THE UNIVERSE THAT IS DARK MATTER

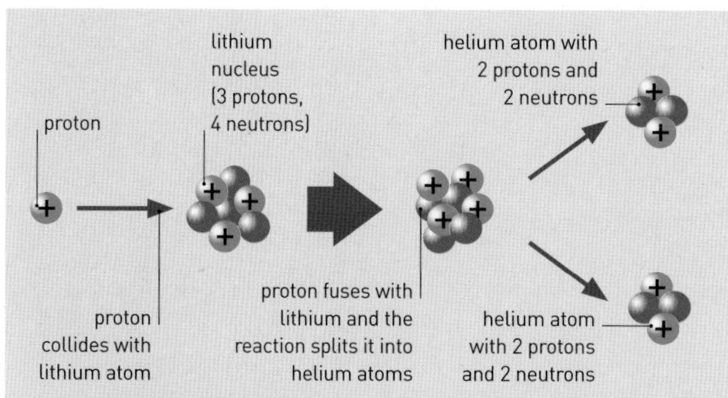
proton
lithium nucleus (3 protons, 4 neutrons)
helium atom with 2 protons and 2 neutrons
proton collides with lithium atom
proton fuses with lithium and the reaction splits it into helium atoms
helium atom with 2 protons and 2 neutrons

SPLITTING THE ATOM

Transmutation of elements was first achieved by changing nitrogen to oxygen (see 1917). In 1932, John Cockcroft and Ernest Walton used a similar technique to split lithium. Lithium atoms are the lightest of any metal and contain just three protons. When an extra proton collides with an atom, a nuclear reaction occurs: the four-neutron, four-proton total splits into two helium atoms.

In 1932, **Hans Krebs** (1900–81), a German-born biochemist, was looking at how the body processes waste nitrogen. Excess amino acids (the building blocks of protein) are recycled into carbohydrate. Krebs discovered the **cycle of chemical reactions** used by **liver cells** to process the **nitrogen content** into a compound called **urea**, which is excreted by the body.

Dutch and Swiss astronomers **Jan Oort** (1900–81) and **Fritz Zwicky** (1898–1974) discovered an **anomaly in galaxies**: they were much larger than the quantity of star material suggested. This indicated the existence of a type of matter not previously detected. It became known as **dark matter**—because it neither emits nor absorbs light.

In July 1933, Polish-born chemist **Tadeus Reichstein** (1897–1996) became the first person to make a vitamin by artificial means when he made ascorbic acid (**vitamin C**).

Positron–electron tracks
Charged particles appear to shoot upwards and shower down, forming tracks. In a magnetic field, negative electrons coil one way and positive particles (positrons) the other.

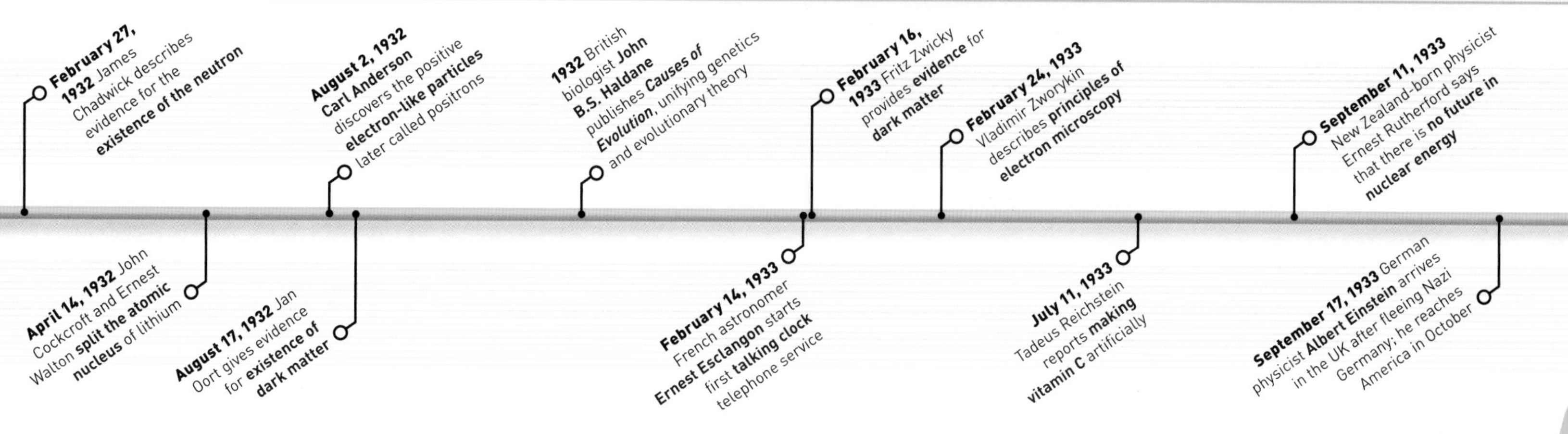
February 27, 1932 James Chadwick describes evidence for the **existence of the neutron**
April 14, 1932 John Cockcroft and Ernest Walton **split the atomic nucleus** of lithium
August 2, 1932 Carl Anderson discovers the positive **electron-like particles** later called positrons
August 17, 1932 Jan Oort gives evidence for **existence of dark matter**
1932 British biologist **John B.S. Haldane** publishes ***Causes of Evolution***, unifying genetics and evolutionary theory
February 14, 1933 French astronomer **Ernest Esclangon** starts **first talking clock** telephone service
February 16, 1933 Fritz Zwicky provides **evidence** for **dark matter**
February 24, 1933 Vladimir Zworykin describes **principles of electron microscopy**
July 11, 1933 Tadeus Reichstein reports **making vitamin C** artificially
September 11, 1933 New Zealand-born physicist Ernest Rutherford says that there is **no future in nuclear energy**
September 17, 1933 German physicist **Albert Einstein** arrives in the UK after fleeing Nazi Germany; he reaches America in October

UNDERSTANDING RADIOACTIVITY

THE DISCOVERY THAT SOME ELEMENTS ARE RADIOACTIVE TRIGGERED A REVOLUTION IN PHYSICS

In 1896, French physicist Henri Becquerel found that compounds containing uranium produce invisible radiation. Within months, Polish scientist Marie Curie showed that the rays emanate from the uranium atoms themselves, and in 1898 she coined a term to describe the phenomenon: radioactivity.

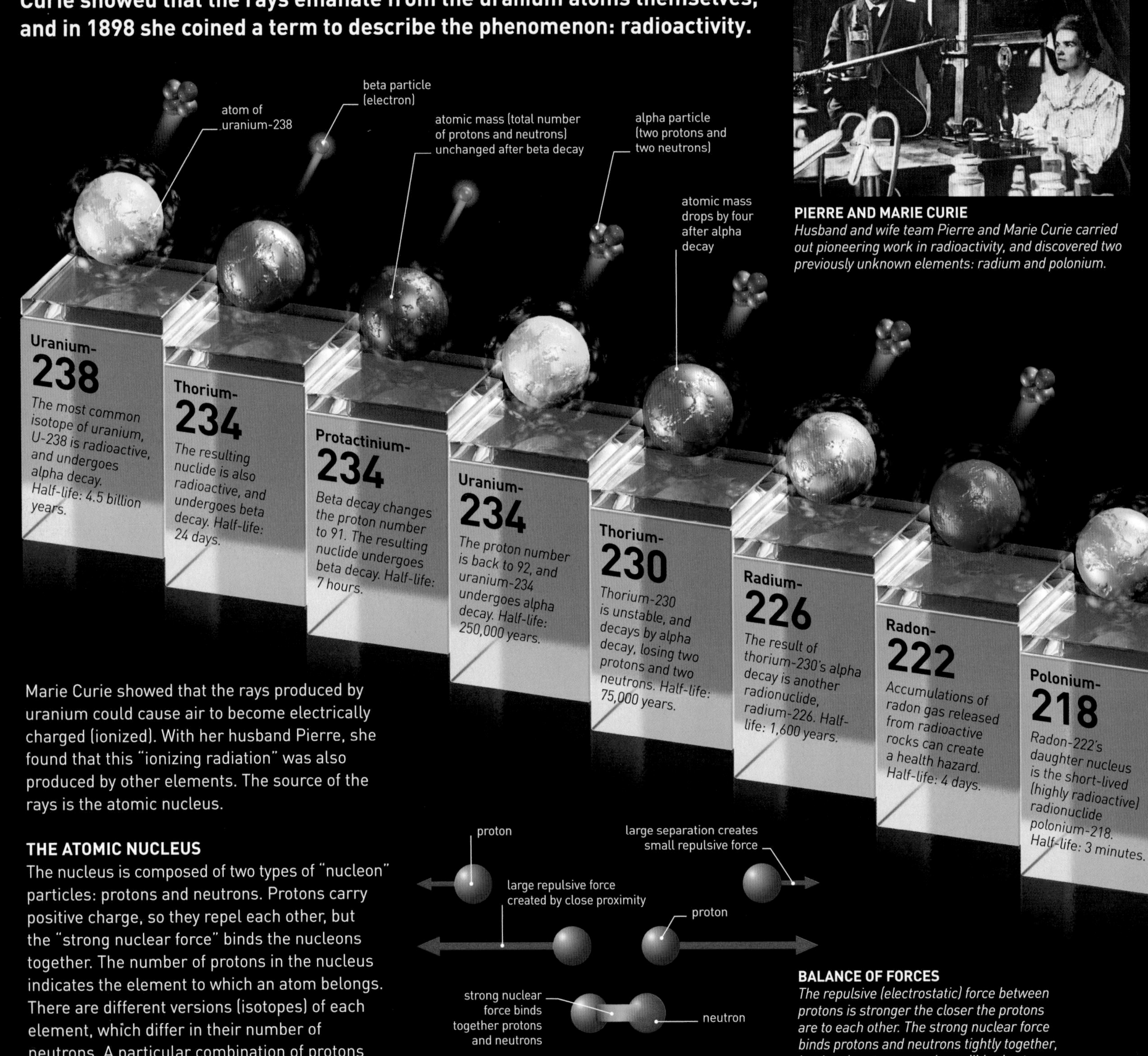

atom of uranium-238
beta particle (electron)
atomic mass (total number of protons and neutrons) unchanged after beta decay
alpha particle (two protons and two neutrons)
atomic mass drops by four after alpha decay
Uranium-238
The most common isotope of uranium, U-238 is radioactive, and undergoes alpha decay. Half-life: 4.5 billion years.
Thorium-234
The resulting nuclide is also radioactive, and undergoes beta decay. Half-life: 24 days.
Protactinium-234
Beta decay changes the proton number to 91. The resulting nuclide undergoes beta decay. Half-life: 7 hours.
Uranium-234
The proton number is back to 92, and uranium-234 undergoes alpha decay. Half-life: 250,000 years.
Thorium-230
Thorium-230 is unstable, and decays by alpha decay, losing two protons and two neutrons. Half-life: 75,000 years.
Radium-226
The result of thorium-230's alpha decay is another radionuclide, radium-226. Half-life: 1,600 years.
Radon-222
Accumulations of radon gas released from radioactive rocks can create a health hazard. Half-life: 4 days.
Polonium-218
Radon-222's daughter nucleus is the short-lived (highly radioactive) radionuclide polonium-218. Half-life: 3 minutes.
proton
large separation creates small repulsive force
large repulsive force created by close proximity
proton
strong nuclear force binds together protons and neutrons
neutron
strong nuclear force is stronger than repulsion
proton

PIERRE AND MARIE CURIE
Husband and wife team Pierre and Marie Curie carried out pioneering work in radioactivity, and discovered two previously unknown elements: radium and polonium.

Marie Curie showed that the rays produced by uranium could cause air to become electrically charged (ionized). With her husband Pierre, she found that this "ionizing radiation" was also produced by other elements. The source of the rays is the atomic nucleus.

THE ATOMIC NUCLEUS

The nucleus is composed of two types of "nucleon" particles: protons and neutrons. Protons carry positive charge, so they repel each other, but the "strong nuclear force" binds the nucleons together. The number of protons in the nucleus indicates the element to which an atom belongs. There are different versions (isotopes) of each element, which differ in their number of neutrons. A particular combination of protons and neutrons is called a nuclide—or, if unstable, a radionuclide.

BALANCE OF FORCES
The repulsive (electrostatic) force between protons is stronger the closer the protons are to each other. The strong nuclear force binds protons and neutrons tightly together, but it only acts over an incredibly short range. This is the main reason why larger nuclei tend to be unstable.

RADIOACTIVE DECAY

At some point, an unstable nucleus will disintegrate, or "decay." The two most common types of decay are called alpha decay and beta decay (see below). In each case, the nucleus always has excess energy to lose, and that energy is carried away by very short-wavelength, high-energy gamma radiation. The probability that a particular atom will decay is fixed, but there is no way of telling when it will happen. However, in a sample containing a large number of atoms of the same radioactive element, it always takes exactly the same amount of time for half the atoms to decay; that period is known as the element's half-life.

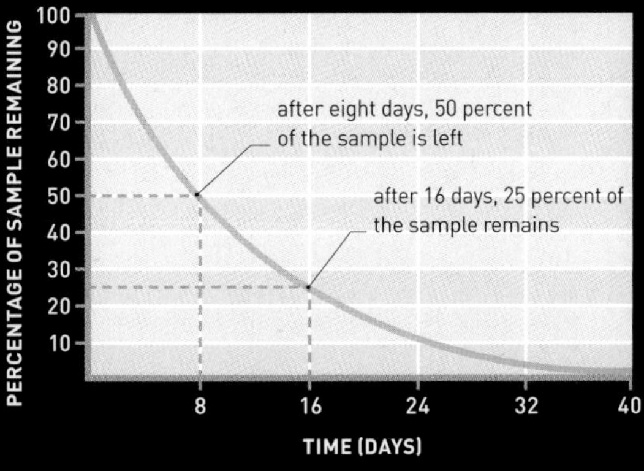
PERCENTAGE OF SAMPLE REMAINING
100
90
80
70
60
50
40
30
20
10
8
16
24
32
40
TIME (DAYS)
after eight days, 50 percent of the sample is left
after 16 days, 25 percent of the sample remains

HALF-LIFE

This graph shows the decay curve of a sample of a radionuclide with an eight-day half-life. Every eight days, the number of atoms of that radionuclide halves.

EFFECTS OF RADIOACTIVITY

Radioactivity generates heat, and this is harnessed by the radioisotope thermal generators that power unmanned space satellites and probes. Radioactive substances pose a threat to health because their ionizing effect can damage chemical bonds in compounds essential to life. This can cause a general malaise, called radiation sickness. Damage to DNA inside cells can cause mutations that can result in cancers. Ironically perhaps, radioactive substances are also used in medicine—particularly in radiation therapy to treat cancer.

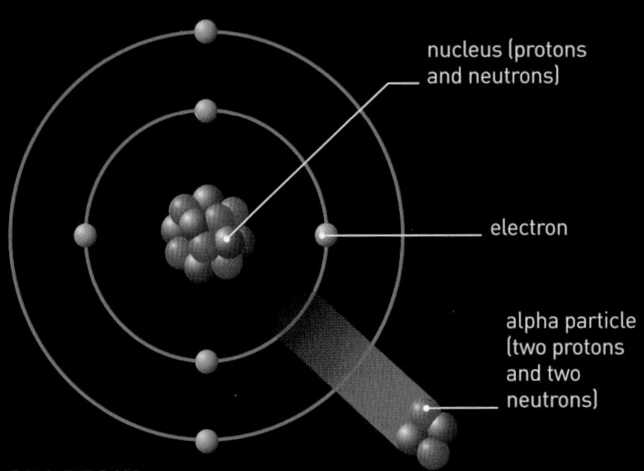
nucleus (protons and neutrons)
electron
alpha particle (two protons and two neutrons)

ALPHA DECAY

An unstable nucleus jettisons a particle made up of two protons and two neutrons: an alpha particle. This makes the nucleus smaller, which sometimes results in greater stability.

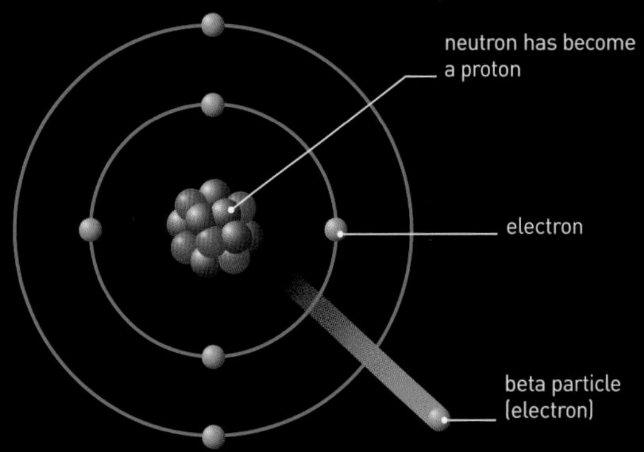
neutron has become a proton
electron
beta particle (electron)

BETA DECAY

Inside some unstable nuclei, a neutron may spontaneously become a proton, emitting an electron at the same time. In this case, the electron is known as a beta particle.

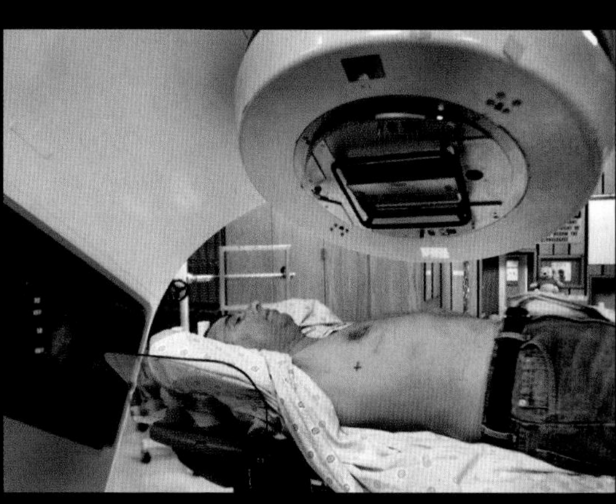

RADIATION THERAPY

During radiation therapy, radiation is used to kill cancer cells. The radiation must be carefully controlled because it can damage healthy cells as well as cancerous ones.

RADIOACTIVE DECAY CHAIN

Elements are defined by the number of protons in the nuclei of their atoms. Decay changes the proton number, so the "daughter" nucleus is of a different element. Often the daughter nucleus is also radioactive, and sometimes the process continues in a decay chain like the one shown here.

Lead-214
Polonium-218 emits an alpha particle, and its daughter nucleus, lead-214 is also radioactive. Half-life: 27 minutes.
Bismuth-214
The result of lead-214's beta decay, bismuth-214, also undergoes beta decay. Half-life: 20 minutes.
Polonium-214
Polonium-214 is extremely unstable, so has a very short half-life. Half-life: 0.0002 seconds.
Lead-210
The radionuclide lead-210 formed by alpha decay of polonium-214 is itself unstable. Half-life: 22 years.
Bismuth-210
Bismuth-210 undergoes beta decay to produce polonium-210. Half-life: 5 days.
Polonium-210
This substance has been used as a radioactive poison in assassinations. Half-life: 138 days.
Lead-206
The end of the long decay chain is a stable nuclide, lead-206.
stable nuclide

5.6

OUNCES OF **POTASSIUM** ARE **CONTAINED** IN THE **HUMAN BODY** AND AROUND **4,400** ATOMS **DECAY** PER SECOND

1934

The husband-and-wife team Irène and Frédéric Joliot-Curie continued the work of Marie Curie—Irène's mother—after she died in 1934.

FRENCH CHEMISTS Irène (1897–1956) and Frédéric Joliot-Curie (1900–58) demonstrated that it was possible to induce **artificial radioactivity**. They created nuclear reactions by making high-energy particles collide with the nuclei of atoms. In February, they published their findings on making radioactive isotopes (variants) of phosphorus and nitrogen by firing alpha particles at nonradioactive targets.

Meanwhile, Italian physicist **Enrico Fermi** (1901–54) decided to use recently discovered subatomic particles called **neutrons** (see 1932–33), instead of alpha particles, in nuclear reactions. He worked his way up the periodic table to see how different elements captured the neutrons. At the time nobody thought that neutrons would have enough energy to split a heavier atom. But then, in January 1934, Fermi succeeded in splitting uranium by bombarding it with neutrons. Fermi thought he had made the first **transuranium element** (an elements with an atomic number above 92, uranium), but he had in fact achieved **atomic fission** (see 1938).

German astronomer **Walter Baade** (1893–1960) and his Swiss counterpart **Fritz Zwicky** suggested that the smallest, densest stars that could be detected were made of neutrons. It was later confirmed that a **neutron star** is the remnant of a star that has exploded and is toward the end of its life.

While experimenting with ultrasound, German scientists **H. Frenzel** and **H. Schultes** noticed that high-frequency sound affected photographic plates—indicating light production. Known as **sonoluminescence**, this occurs when sound creates underwater bubbles that focus its energy a trillion times, creating flashes of light and heat. In the natural world, pistol shrimps use this process to stun their prey.

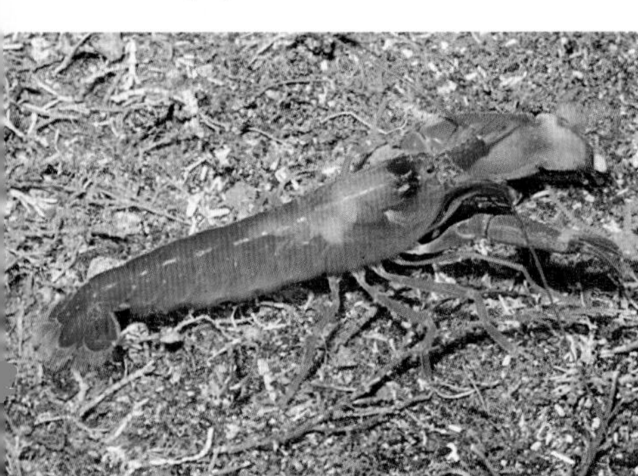

Pistol shrimp
The snapping sound of this shrimp's enlarged pincer produces enough energy in a fraction of a second to release heat and light as well as stun its prey.

1935

In the rain forest's ecosystem, the living organisms—vegetation and animals—interact with the nonliving air and soil through the release and absorption of carbon dioxide and the exchange of nutrients.

INSTRUMENTS TO MEASURE EARTHQUAKES date back to antiquity, but by the early 1900s seismometers were being used to detect movements in Earth's crust. A sophisticated system of levers generated a paper trace showing the magnitude of an earthquake. Working in earthquake-prone California, American physicist **Charles Richter** (1900–85) developed a scale that would make it easier to compare the amount of energy released by different earthquakes. Initially devised just for local use, the **Richter scale** was soon applied throughout the world.

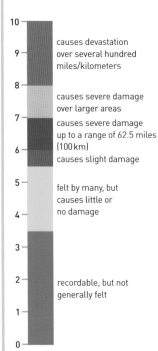

Richter scale	Effect
8–10	causes devastation over several hundred miles/kilometers
7–8	causes severe damage over larger areas
6–7	causes severe damage up to a range of 62.5 miles (100 km)
5–6	causes slight damage
3.5–5	felt by many, but causes little or no damage
0–3.5	recordable, but not generally felt

Richter scale
By linking measurable earthquake movements to environmental effects, Richter created a scale that could be understood by everybody.

480 MEGATONS

THE **ENERGY RELEASED IN AN EARTHQUAKE** MEASURING **9 ON** THE **RICHTER SCALE**

In February, German bacteriologist **Gerhard Domagk** (1895–1964), described a chemical dye with powerful antibacterial properties. He reported on clinical trials suggesting that the dye—later sold under the name Prontosil—could be used as a drug to cure common, dangerous infections. Later in the year, Italian pharmacologist **Daniel Bovet** (1907–92) discovered the chemical basis for Prontosil's effectiveness. He found its active ingredient to be a sulfur-based compound called **sulfonamide**. Sulfonamides were the most important **antibacterial drugs** until the introduction of penicillin (see 1940–41).

During 1935, American chemist **Wallace Carothers** (1896–1937) led a team of scientists at DuPont chemical laboratories in the development of new polymers—long molecules made by bonding smaller molecular building blocks into chains. The team had already manufactured an artificial silk called **polyester**, but was now working with a different kind of building block. The resulting polymer, called polyamide 6-6, could be drawn out into tough filaments. It was later called **nylon** (see 1937).

In July, a scientific paper in the journal *Ecology* introduced the concept of an **ecosystem**. Its author, British botanist **Arthur Tansley** (1871–1955), combined two key themes: American botanist Frederic Clements's 1916 concept of vegetation as a community of different species that changes over time, and Russian scientist Vladimir Vernadsky's treatise on cycles of matter (see 1926). Tansley's ecosystem was an ecological structure in which living organisms interacted with each other as well as with the

January Enrico Fermi unwittingly achieves **atomic fission**

February 10 Description of artificial induction of **radioactivity** is published

May It is suggested that **super novae** are a transition of normal stars to **neutron stars**

July 4 Hungarian physicist Leó Szilárd files a patent for the means of producing a **nuclear chain reaction**—the nuclear bomb

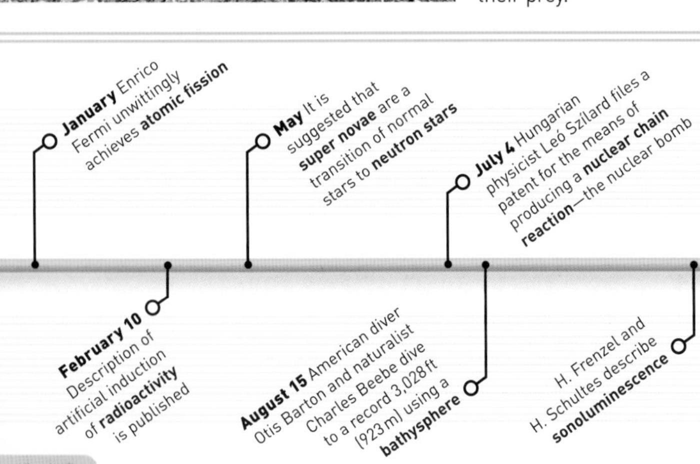

August 15 American diver Otis Barton and naturalist Charles Beebe dive to a record 3,028 ft (923 m) using a **bathysphere**
H. Frenzel and H. Schultes describe **sonoluminescence**

January Charles Richter invents a scale for the magnitude of **earthquakes**

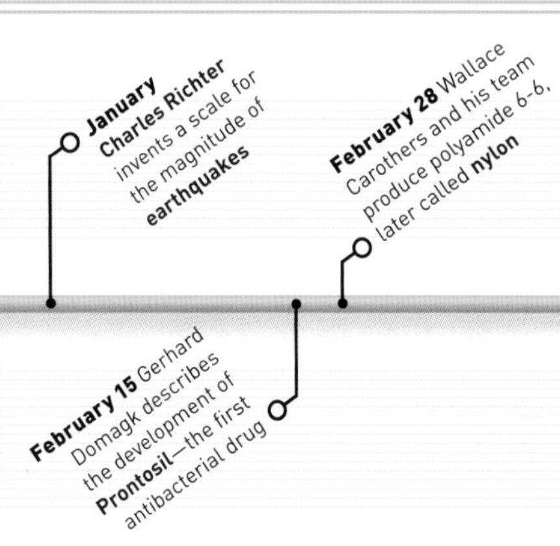

February 15 Gerhard Domagk describes the development of **Prontosil**—the first antibacterial drug

February 28 Wallace Carothers and his team produce polyamide 6-6, later called **nylon**

July Arthur Tansley introduces the concept of the **ecosystem**

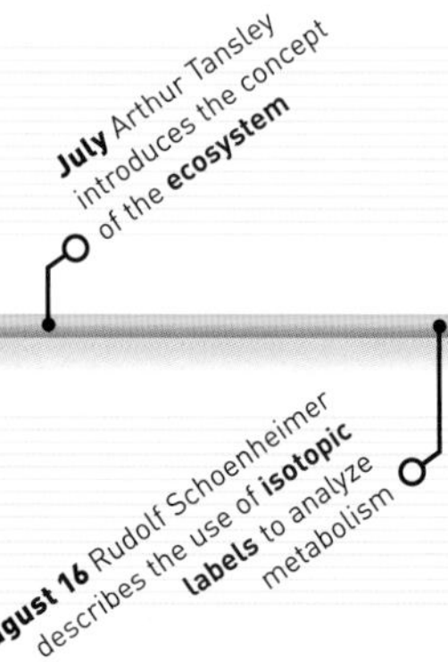

August 16 Rudolf Schoenheimer describes the use of **isotopic labels** to analyze metabolism

THOUGH ORGANISMS… CLAIM OUR PRIMARY INTEREST… WE CANNOT SEPARATE THEM FROM THEIR SPECIAL ENVIRONMENTS, WITH WHICH THEY **FORM ONE PHYSICAL SYSTEM.**

Arthur Tansley, from *Ecology Journal,* 1935

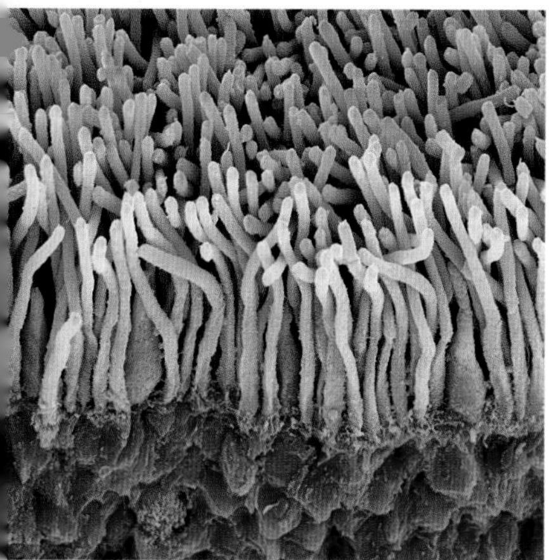

Retina rods and cones
Cells on the retina contain pigments for light-absorption; rods (sepia) have one kind, but cones (green) collectively have three kinds of pigment for detecting different colors.

nonliving environment that surrounded them.

Rudolf Schoenheimer (1898–1941) was a German biochemist studying **metabolism**—the complex pattern of chemical reactions in the body. In order to understand these patterns, he found a way of tagging, or marking, substances in the body with detectable isotopes (variants) of elements. By tracing the pathways of the isotopes, he was able to work out the sequences of chemical reactions taking place. **Isotopic labeling** would become the standard method for studying the biochemistry of metabolism.

Danish ophthalmologist **Gustav Østerberg** published a study on the cellular makeup of the retina at the back of the eye. In the 19th century, German anatomist Max Schultze had identified the layers of the retina and drawn detailed illustrations of structures later called rods and cones (see 1866). Østerberg recorded the **first accurate count of the rods and cones**. It was later shown that cones have high sensitivity in low light intensity, but cannot detect color. Color-sensitive cones only work in high light intensity and are concentrated in an area of the retina called the fovea, which collects light from a point of focus at the center of the field of vision to form an image.

7 MILLION CONES IN THE **HUMAN EYE**

130 MILLION RODS IN THE **HUMAN EYE**

1936

At the impressionable stage immediately after hatching, a brood of goslings assumed Konrad Lorenz to be their surrogate mother and followed him everywhere.

IN 1936, DUTCH BIOLOGIST NIKOLAAS TINBERGEN (1907–88) met Austrian biologist **Konrad Lorenz** (1903–89) at a symposium, and the two men spent many months discussing aspects of **animal behavior**. Their collaboration marked the **foundation of ethology**—the modern science of animal behavior. They distinguished between innate behavior that was inherited and learned behavior that became modified through experience. Lorenz famously demonstrated how **goslings** can become attached (imprinted) to humans just after hatching; Tinbergen studied the innate **courtship behavior** of the **stickleback fish**.

The world's **first practical helicopter**—the Focke-Wulf Fw61—took its maiden flight in June. Built by German aviator **Heinrich Focke** (1890–1979), it flew using twin rotary blades extending to the left and right of the fuselage. Its first flight lasted just 28 seconds—but it was far easier to control than previous versions.

In July, Hungarian biologist **Hans Selye** (1907–82) was the first person to describe the **scientific basis for physiological stress**. In the 19th century, French physiologist Claude Bernard had proposed that the living body maintained a steady state by processes of internal regulation. By the early 1900s it was understood how aspects of the nervous system could make the body respond to changes in circumstances—for example, initiating the "fight or flight" response to danger. Selye further explained how hormonal changes were associated with stress, and that these changes could affect the function of the body's immune system.

In September, the last surviving **thylacine** died in Hobart Zoo in Tasmania, after being exposed to extreme weather conditions when a keeper inadvertently left it out of its shelter one night. Commonly known as the Tasmanian wolf or tiger, the thylacine was the **largest carnivorous marsupial**. Relentless hunting had driven the wild population to extinction some years earlier.

IT IS A GOOD **MORNING EXERCISE** FOR A RESEARCH SCIENTIST TO **DISCARD A PET HYPOTHESIS** EVERY DAY BEFORE BREAKFAST. IT KEEPS HIM YOUNG.

Konrad Lorenz, from *On Aggression,* 1966

The last thylacine
Named Benjamin, this was the last thylacine in existence. A predator of kangaroos and wallabies, the species earned an exaggerated reputation for attacking livestock.

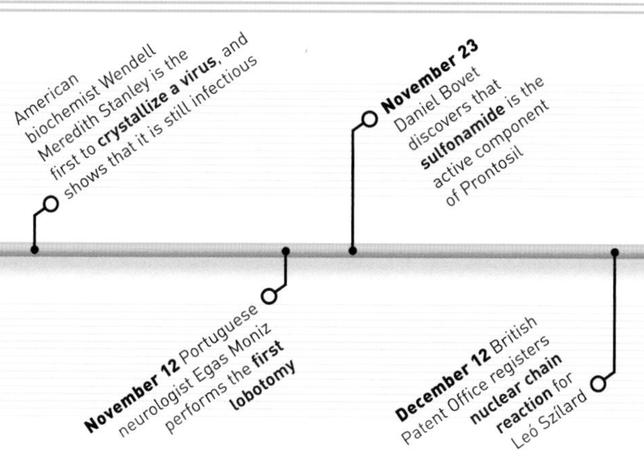
American biochemist Wendell Meredith Stanley is the first to **crystallize a virus**, and shows that it is still infectious
November 12 Portuguese neurologist Egas Moniz performs the **first lobotomy**
November 23 Daniel Bovet discovers that **sulfonamide** is the active component of Prontosil
December 12 British Patent Office registers **nuclear chain reaction** for Leó Szilard

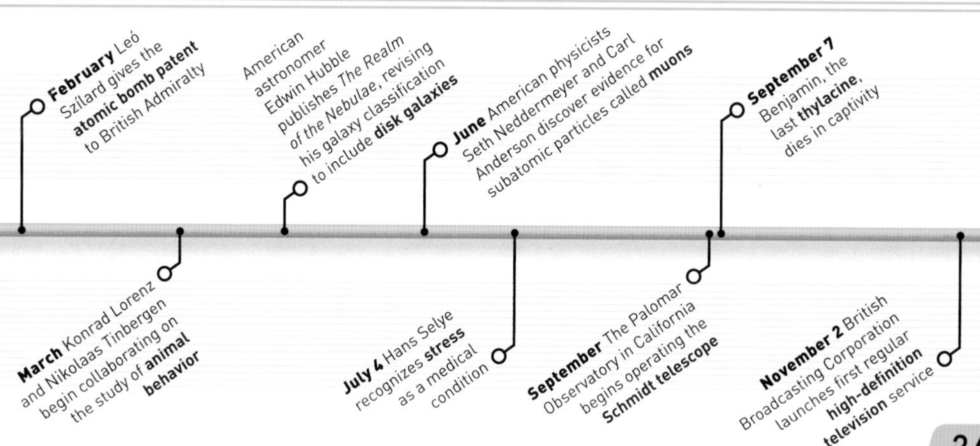
February Leó Szilard gives the **atomic bomb patent** to British Admiralty
March Konrad Lorenz and Nikolaas Tinbergen begin collaborating on the study of **animal behavior**
American astronomer Edwin Hubble publishes *The Realm of the Nebulae*, revising his galaxy classification to include **disk galaxies**
June American physicists Seth Neddermeyer and Carl Anderson discover evidence for subatomic particles called **muons**
July 4 Hans Selye recognizes **stress** as a medical condition
September The Palomar Observatory in California begins operating the **Schmidt telescope**
September 7 Benjamin, the last **thylacine**, dies in captivity
November 2 British Broadcasting Corporation launches first regular **high-definition television** service

1937

German-born, Professor Hans Krebs emigrated to England in 1933. He is shown at work in the laboratory at Sheffield University where he was based from 1935 to 1954 and carried out much of his research.

IN JANUARY, ITALIAN physicists **Carlo Perrier** (1866–1948) and **Emilio Sègre** (1905–89) reported on a **new artificial element** formed by atomic reaction. They had made **technetium** in a radioactive-contaminated part of a cyclotron (see 1931).

In February, American chemist **Wallace Carothers** patented his **new chemical polymer**—polyamide 6-6. By 1938, the DuPont Company had called this polymer **nylon** and was using it commercially to make toothbrush bristles.

> **" A SPECIES** IS A **STAGE** IN **PROGRESS,** NOT A **STATIC UNIT. "**
>
> **Theodosius Dobzhansky**, in *Genetics and the Origin of Species*, 1937

In 1927, French-American engineer **Eugene Houdry** (1892–1937) had invented a way of **cracking petroleum from crude oil** using a silica-alumina based catalyst. Sun Oil started up the first petroleum cracking unit using the invention in March 1937.

In September, American electrical engineer **Grote Reber** (1911–2002) built the **first radio telescope**.

German-born biochemist **Hans Krebs** found that **citric acid** could keep cells alive and showed that it was a key intermediary in a process that provided cells with energy. The process became known as the **citric acid, or Krebs, cycle**.

Ukrainian biologist **Theodosius Dobzhansky** (1900–75) published *Genetics and the Origin of Species*. In it he showed how Darwin's natural selection (see 1859) could be explained in terms of the genetic makeup of populations and helped lay the foundations of **modern evolutionary biology**.

The first radio telescope
American Grote Reber built his radio telescope in his back yard. He used it to confirm that radio signals could be detected from space.

1938

The coelacanth lives in the deep ocean waters and although it was known to local fishermen, paleontologists knew it only as a fossil.

AMERICAN ZOOLOGIST DONALD GRIFFIN (1915–2003) was studying the migratory behavior of bats, but could not understand how they managed to navigate in the dark. In the 18th century, Italian biologist **Lazzaro Spallanzani** (1729–1799) had demonstrated that bats could fly around objects when deprived of sight—but not if their ears were blocked. Working with the American neurologist **Robert Galambos** (1914–2010), Griffin used a special ultrasound microphone to discover that **bats emit high-pitched sounds** that are beyond the range of human hearing. Griffin's theory was that the animals were using a form of sonar, listening for echoes of their sounds as they bounced off obstacles or prey. At first, his theory was derided, but by 1944 the accepted phenomenon was called **echolocation**.

The German chemist **Otto Hahn** (1879–1968) was experimenting with chemical transmutation—how one element could change into another through a nuclear reaction. Transmutation had been first demonstrated 20 years earlier by New Zealand-born physicist **Ernest Rutherford** (see 1916–17). But scientists believed that there were limits to what could be achieved, and that smashing the heaviest atoms, such as those of uranium, could not make much lighter atoms. However, toward the end of the year, Hahn reported that he had achieved just that—he **split uranium to produce barium**, a process known as **nuclear fission**. The Italian physicist Enrico Fermi had achieved a similar process—although he thought he had synthesized a new element (see 1934). Later, Fermi would oversee the first fission chain reaction (see 1942).

In December, **Marjorie Courtenay-Latimer**

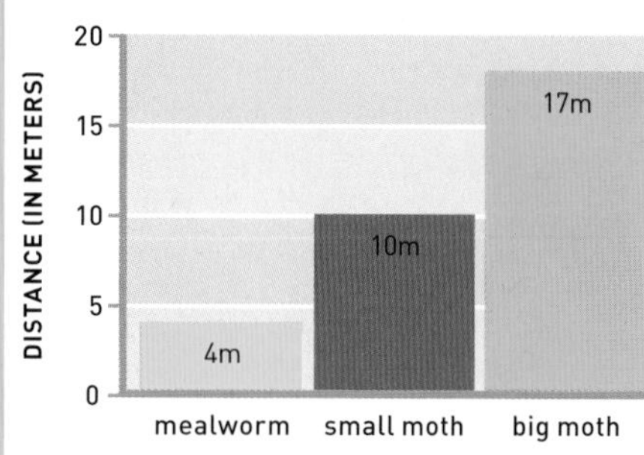

DISTANCE (IN METERS)
20
15
10
5
0
4m
10m
17m
mealworm
small moth
big moth
PREY

Echolocation in bats
Bats hunt for food by listening for echoes of their high-pitched calls bouncing off their prey. The larger the prey, the greater is the maximum detection distance.

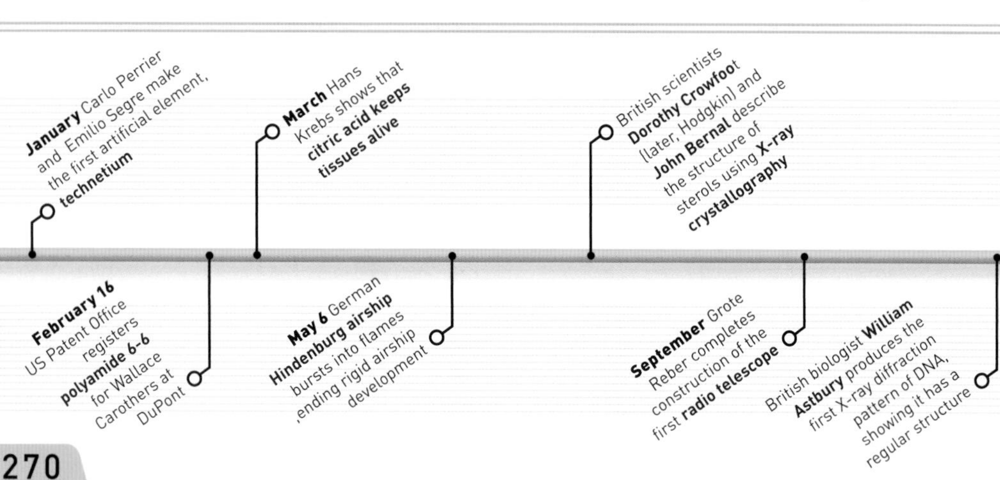

January Carlo Perrier and Emilio Segre make the first artificial element, **technetium**
February 16 US Patent Office registers **polyamide 6-6** for Wallace Carothers at DuPont
March Hans Krebs shows that **citric acid keeps tissues alive**
May 6 German **Hindenburg airship** bursts into flames, ending rigid airship development
British scientists **Dorothy Crowfoot** (later, Hodgkin) and **John Bernal** describe the structure of sterols using **X-ray crystallography**
September Grote Reber completes construction of the first **radio telescope**
British biologist **William Astbury** produces the first X-ray diffraction pattern of DNA, showing it has a regular structure

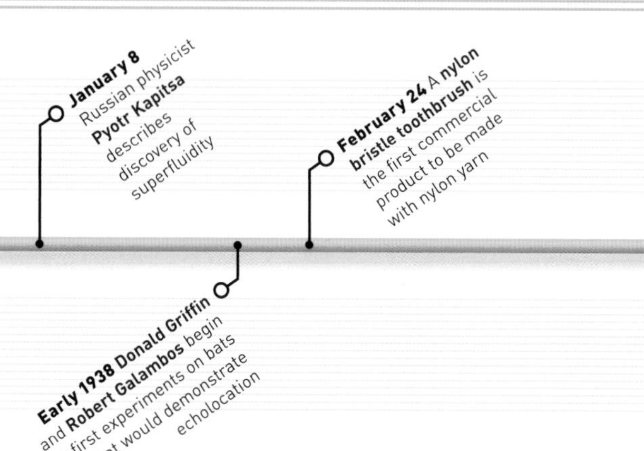

January 8 Russian physicist **Pyotr Kapitsa** describes discovery of superfluidity
Early 1938 Donald Griffin and **Robert Galambos** begin the first experiments on bats that would demonstrate echolocation
February 24 A **nylon bristle toothbrush** is the first commercial product to be made with nylon yarn

uranium-235 nucleus
smaller product nucleus (barium)
neutron
nucleus splits into two fragments (fission)
burst of energy from fission releases smaller product nucleus and neutrons
shower of excess neutrons
more uranium-235 undergo fission in chain reaction

NUCLEAR FISSION CHAIN REACTION

Nuclear fission releases atomic energy—which is maximized when the process is done in a way that initiates a chain reaction. Uranium-235 is favored as it is the most abundant fissionable isotope and has plenty of critical neutrons. When the uranium nucleus (fissionable nucleus) is bombarded with an external source of neutrons, the uranium nuclei split into smaller fragments. More neutrons are emitted as by-products, which in turn split other uranium nuclei in the chain reaction.

(1907–2004), curator at a museum in East London, South Africa, was asked to **collect a specimen** from the local fishing docks. There she found an unusual fish that she could not identify and, in the absence of any technical support, reluctantly had it stuffed. South African zoologist **James Smith** (1897–1968) later **identified it as a coelacanth**—a fish until then known only from fossils and **thought to have become extinct** with the dinosaurs. The coelacanth belongs to an **ancient group of lobe-finned fishes** related to the ancestors of the first vertebrates that evolved to live on land. Initially, the modern coelacanth was known only in deep waters of the western Indian Ocean, but a second species was found in Indonesian waters in 1997.

1939

Made by German company Heinkel in 1939, the first jet aircraft—the He178—was the prototype for later models built for combat at the end of World War II.

IN APRIL, GROTE REBER DISCOVERED a new kind of astronomical object with his radio telescope—**a radio galaxy called Cygnus A**. This object remains one of the strongest sources of radio signals that has ever been detected.

Six years after Hungarian-born physicist Léo Szilárd (1898–1964) conceived the idea of releasing the atomic bomb. Einstein had prompted what would become the **Manhattan Project**: the Allied program of research and development that would eventually unleash the only **nuclear weapons so far used in wartime combat**.

Although many scientists left Germany in the months leading up to World War II, others stayed described **the behavior of electrons** in forming ionic bonds (different atoms bonding by gaining and losing electrons) and covalent bonds (atoms bonding by sharing electrons). In particular, Pauling developed the idea that shared electrons are not in a fixed position, but **orbit around both nuclei associated with the bond**.

> **“ SCIENCE** IS THE **SEARCH FOR TRUTH**—IT IS **NOT A GAME** IN WHICH ONE TRIES TO **BEAT HIS OPPONENT** OR DO **HARM** TO **OTHERS.** ”
>
> Linus Pauling, American chemist and peace activist, in *Liberation*, 1958

energy in a nuclear chain reaction, there was **growing concern among scientists** that Nazi Germany would **develop an atomic bomb**. Italian physicist Enrico Fermi and German chemist Otto Hahn had already demonstrated that controlled nuclear fission by a chain reaction was possible. In August, German-born physicist **Albert Einstein**, by now in exile in America and working at Princeton University, wrote to **President Roosevelt** expressing the scientists' concerns. He later urged the US president to enter the race to be the first to make behind to continue their work in advancing science and technology. The first **jet engine was designed** and patented by German physicist **Hans von Ohain** (1911–98). On August 27, the first aircraft to fly under turbojet power—the **Heinkel He178**—had its **maiden flight**. After the war, von Ohain was one of many German scientists recruited to help advance scientific research in America as part of **Operation Paperclip**.

American chemist **Linus Pauling** published his most celebrated book: *The Nature of the Chemical Bond*. In it he

LINUS PAULING
(1901–94)

American chemist and peace activist, Linus Pauling is the only person to have received two unshared Nobel Prizes. His work spanned quantum mechanics in chemistry and the structures of complex biological molecules, such as proteins. After World War II he campaigned against the further use of nuclear weapons.

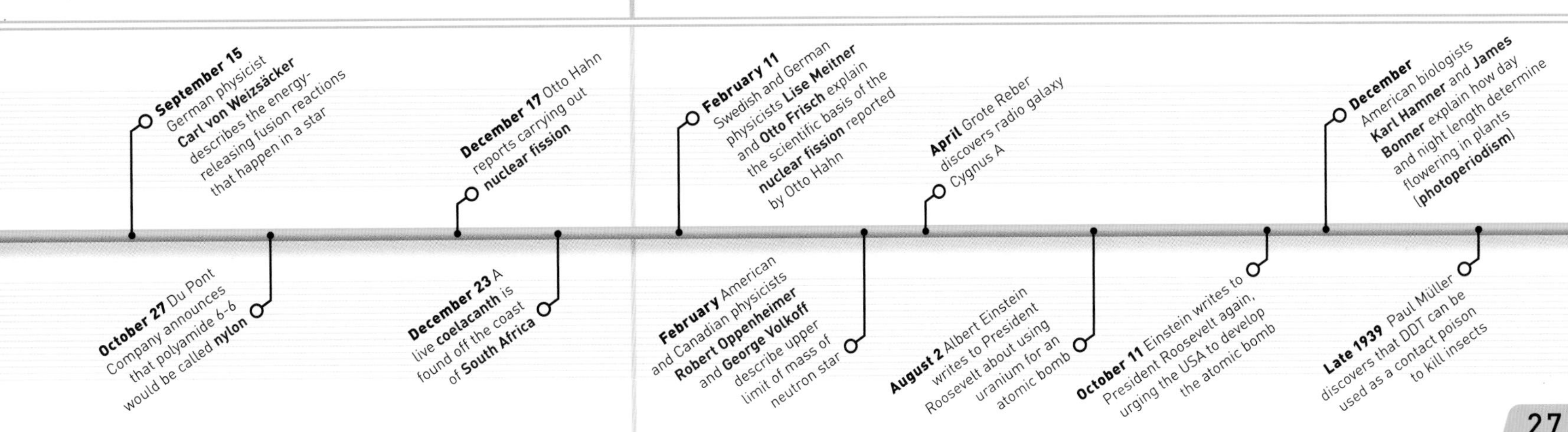

September 15 German physicist **Carl von Weizsäcker** describes the energy-releasing fusion reactions that happen in a star
October 27 Du Pont Company announces that polyamide 6-6 would be called **nylon**
December 17 Otto Hahn reports carrying out **nuclear fission**
December 23 A live **coelacanth** is found off the coast of **South Africa**
February 11 Swedish and German physicists **Lise Meitner** and **Otto Frisch** explain the scientific basis of the **nuclear fission** reported by Otto Hahn
February American and Canadian physicists **Robert Oppenheimer** and **George Volkoff** describe upper limit of mass of neutron star
April Grote Reber discovers radio galaxy Cygnus A
August 2 Albert Einstein writes to President Roosevelt about using uranium for an atomic bomb
October 11 Einstein writes to President Roosevelt again, urging the USA to develop the atomic bomb
December American biologists **Karl Hamner** and **James Bonner** explain how day and night length determine flowering in plants (**photoperiodism**)
Late 1939 Paul Müller discovers that DDT can be used as a contact poison to kill insects

2.2 POUNDS

THE AMOUNT OF PLUTONIUM NEEDED FOR **AN EXPLOSION** EQUAL TO THAT PRODUCED BY **20,000 TONS OF CHEMICAL EXPLOSIVE**

The production method of the glowing radioactive element plutonium was initially kept secret because it was key to the manufacture of early atomic bombs.

IN THE FIRST PART OF THE 20TH CENTURY, geologists used data from studying **seismic waves** to conclude that Earth had a distinct core (see 1906). By studying the way waves were transmitted during earthquakes, they deduced that the core was made up of different materials from the rest of the planet. In 1940, Canadian geologist **Reginald Aldworth Daly** (1871–1957) published ***Strength and Structure of the Earth***, in which he identified a multi-layered arrangement around Earth's core. He stated that a rigid but brittle envelope overlaid a much thicker semisolid layer, which was hotter toward the core. Today, it is now known that **Earth's layers differ chemically and physically**, with a silicon-rich rocky surround and a core that is roughly 80 percent iron (see below). The rigid surround includes a low-density surface crust, with a mantle layer made up of higher-density rock beneath it.

Layers of outer Earth
Earth's iron core is surrounded by a thick mantle covered in crust. The hard rock of the uppermost mantle and crust forms the movable plates that account for drifting continents.

solid inner core
liquid outer core
continental crust
lithosphere is rigid surface rock composed of low-density crust floating on a surface layer of high-density mantle
rigid lower mantle called the mesosphere
semifluid upper mantle called the asthenosphere on which the lithosphere's rigid plates move
oceanic crust

In 1940, at University of California, Berkeley, scientists successfully made the **first transuranium elements**—those with an atomic number higher than that of uranium (92). By using a particle accelerator, scientists successfully made elements 93 and 94. These new elements were named **neptunium** and **plutonium**, respectively, after Neptune and Pluto—the two outer planets of the Solar System. Publication of the discovery of these elements was delayed until after World War II as it was found that plutonium had the potential to be used as fuel for an atomic bomb.

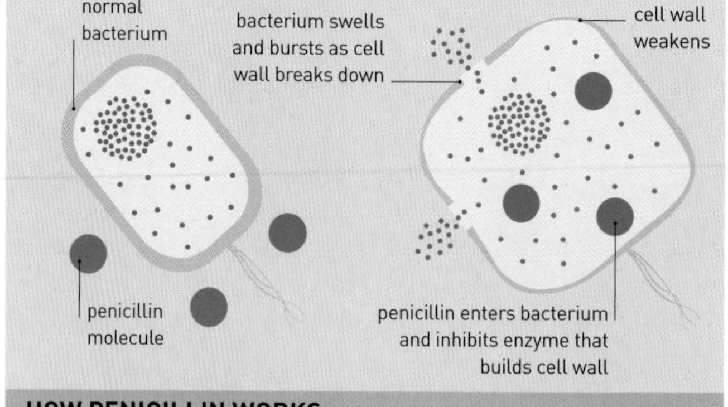
normal bacterium
bacterium swells and bursts as cell wall breaks down
cell wall weakens
penicillin molecule
penicillin enters bacterium and inhibits enzyme that builds cell wall

HOW PENICILLIN WORKS

Penicillin is an antibiotic—a member of a group of chemicals that either prevent the growth of bacteria or kill them altogether. These chemicals attack targets that are present only in bacterial cells and not in infected tissue. Penicillin inhibits the process that many bacteria use to build their cell walls. As a result, the cell walls weaken, making the bacteria absorb water and burst.

More than 30 years after Scottish bacteriologist Alexander Fleming had identified the antibacterial properties of the *Penicillium* mold (see 1928), Australian biologist **Howard Walter Florey** (1898–1968) and British biochemists **Ernst Boris Chain** (1906–79) and **Norman Heatley** (1911–2004) not only demonstrated that its antibiotic secretion, **penicillin, could be used to cure infection** (see panel, above), but that it could be produced and isolated in useful amounts as well. Medical trials began in January 1941, and techniques for the **mass production of the antibiotic** would be developed by the time World War II was at its height.

At the same time, German-born American biochemist **Fritz Albert Lipmann** (1899–1986) had been studying the **chemistry of metabolism** and made a breakthrough in understanding the way living cells process their energy. In 1941, he reported that a phosphate-rich substance called **adenosine triphosphate (ATP)** was the chemical key to the process. While ATP had been discovered more than 10 years earlier, biologists were only now able to recognize its function. By burning calorific nutrients, such as carbohydrates and fats, living cells harness their energy in the phosphate bonds of a pool of ATP molecules. When energy is required—such as for growth or movement—ATP is broken down, unlocking the energy of its phosphate bonds to do the work (see panel, opposite).

While working at the Dow Chemical Company, American engineer **Ray McIntire** (1918–96) had been asked to develop an insulating material as part

> “I WAS A **28-YEAR-OLD KID** AND I DIDN'T STOP TO RUMINATE ABOUT IT.”
>
> **Glenn Theodore Seaborg, American scientist,** on being part of the team that discovered plutonium, 1947

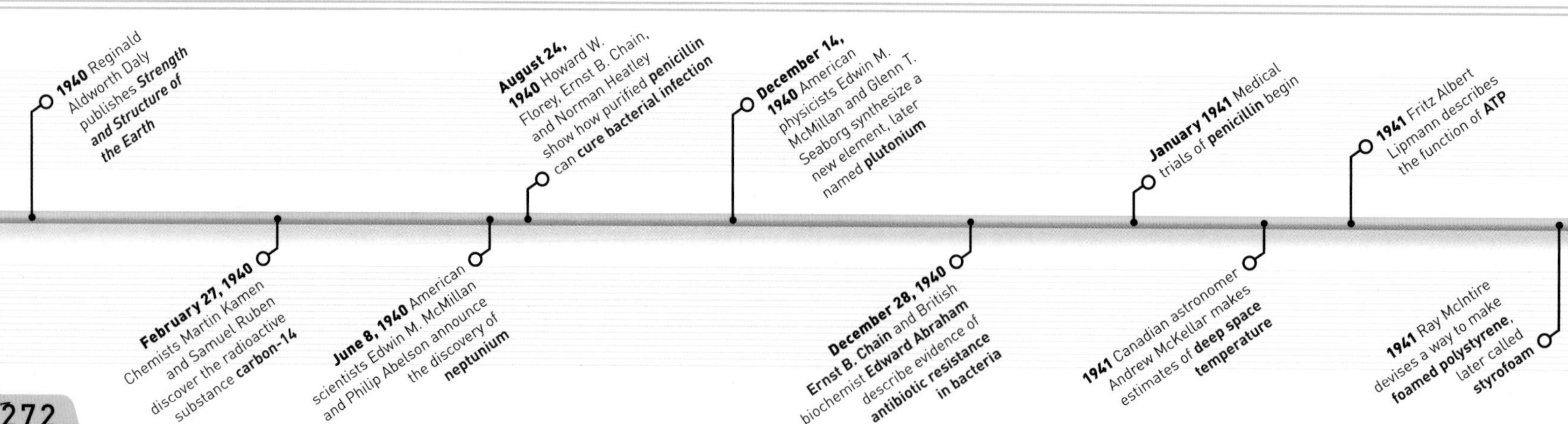
1940 Reginald Aldworth Daly publishes ***Strength and Structure of the Earth***
February 27, 1940 Chemists Martin Kamen and Samuel Ruben discover the radioactive substance **carbon-14**
June 8, 1940 American scientists Edwin M. McMillan and Philip Abelson announce the discovery of **neptunium**
August 24, 1940 Howard W. Florey, Ernst B. Chain, and Norman Heatley show how purified **penicillin** can **cure bacterial infection**
December 14, 1940 American physicists Edwin M. McMillan and Glenn T. Seaborg synthesize a new element, later named **plutonium**
December 28, 1940 Ernst B. Chain and British biochemist **Edward Abraham** describe evidence of **antibiotic resistance in bacteria**
January 1941 Medical trials of **penicillin** begin
1941 Canadian astronomer Andrew McKellar makes estimates of **deep space temperature**
1941 Fritz Albert Lipmann describes the function of **ATP**
1941 Ray McIntire devises a way to make **foamed polystyrene**, later called **styrofoam**

1942

In the 1940s, aircraft were used to spray DDT over the widest possible areas. Its polluting effects lasted for many years.

2 MILLION TONS THE AMOUNT OF DDT USED THROUGHOUT THE WORLD SINCE 1940

of the war effort. He used a material called polystyrene—a type of plastic originally made from tree resin—and chemically treated it to produce countless bubbles in the manufacturing process. The resulting foam polystyrene, trademarked as **styrofoam**, was inexpensive and lightweight.

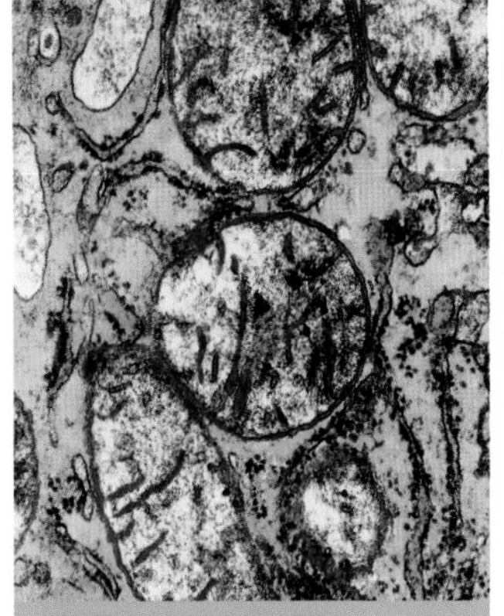

THE ROLE OF ATP IN METABOLISM

Plant and animal cells contain powerhouses called mitochondria (sepia). They are packed with membrane folds that carry the molecular machinery needed to make a chemical called adenosine triphosphate (ATP) from high-energy foods. Energy from ATP is released to order to drive cellular activity, such as building DNA and protein.

IN 1939, SWISS CHEMIST PAUL HERMANN MÜLLER (1899–1965) had discovered that a chlorine-containing chemical called **DDT** (dichlorodiphenyltrichloethane) was lethal when it came in contact with insects, and could possibly be used to control insect pests. In September, the US received the **first stocks of DDT to begin using it on a wide scale**. During World War II, DDT was used widely to control lice-born typhus as well as malaria-carrying mosquitos. In the post-war years, DDT application increased further as agriculturalists began using it to kill crop-eating pests. However, by the 1960s, the world had realized that DDT poisoned the environment by accumulating in food chains (see 1962), and the chemical that had earlier earned Müller a Noble Prize was **banned from most countries.**

Two American biologists, German-born **Max Delbrück** (1906–81) and Italian-born **Salvador Luria** (1912–91), embarked on a collaborative study of **bacteriophages** (also known as phages)—viruses that infect and kill bacteria. They investigated the process whereby some bacteria mutate to become genetically resistant to infection, and in 1943, explained that this happened spontaneously, and was not induced by the environment.

Scientists had already demonstrated the practical possibility of a self-sustained nuclear reaction in a controlled setting (see 1938). In December, Italian-American physicist **Enrico Fermi** oversaw the operation of the world's **first nuclear reactor**—Chicago pile-1—built beneath a football stand at the University of Chicago. The reactor bombarded a sample of uranium with neutrons to **trigger a chain reaction**, which was controlled with neutron-absorbing cadmium rods. It ran successfully for four and a half minutes before Fermi stopped the process. This event marked a critical step in the Manhattan Project—the US government-led Allied research and development initiative that produced the first atomic bombs during World War II.

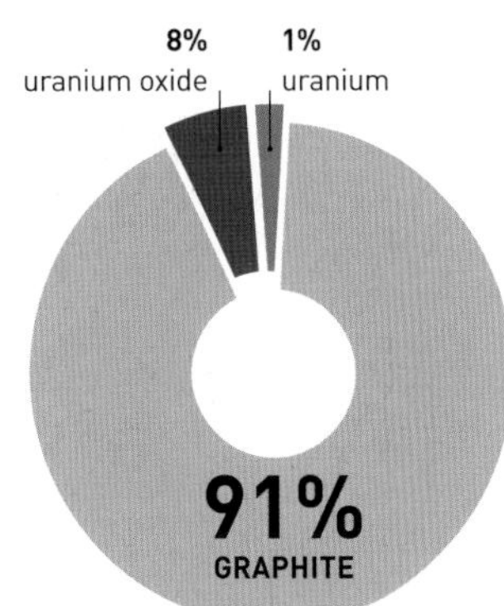
8%
uranium oxide
1%
uranium
91%
GRAPHITE

Enrico Fermi's nuclear reactor
Neutron-releasing uranium pellets were at the heart of the nuclear chain reaction in Fermi's reactor. Graphite blocks slowed the neutrons.

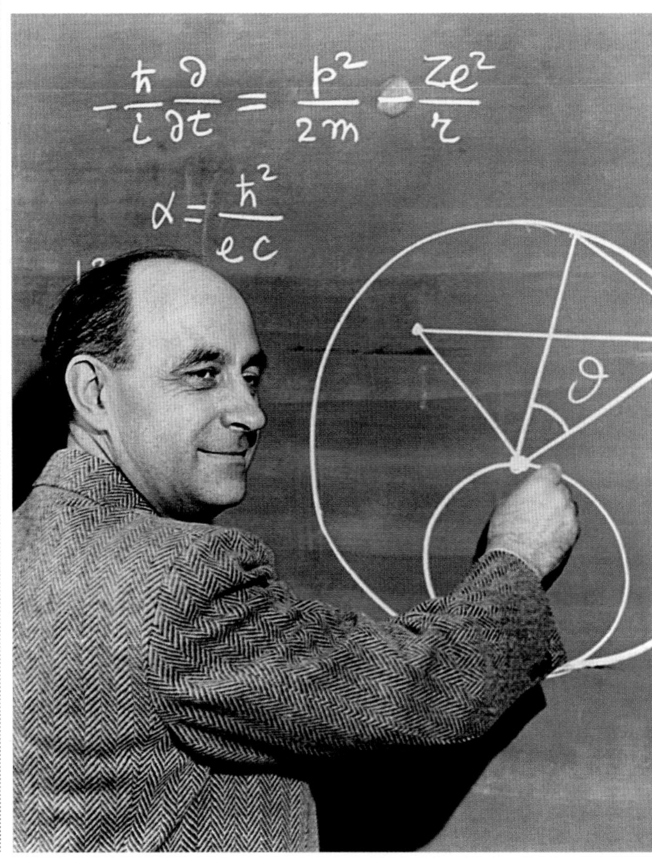

Enrico Fermi
In 1938, Enrico Fermi moved to the US with his Jewish wife to escape Italy's anti-Semitic policies. He received the Nobel Prize for Physics the same year.

> "THE **ITALIAN NAVIGATOR** HAS LANDED IN THE **NEW WORLD.**"
>
> **Arthur Compton, Director of the Metallurgical Laboratory, University of Chicago**, in a coded message referring to Enrico Fermi's success, 1942

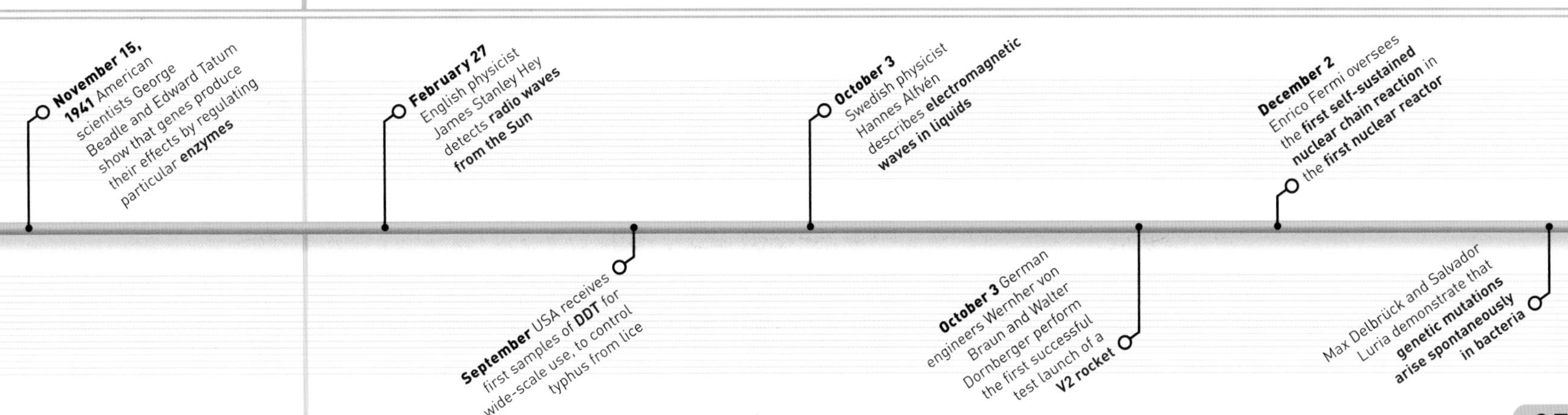
November 15, 1941 American scientists George Beadle and Edward Tatum show that genes produce their effects by regulating particular enzymes
February 27 English physicist James Stanley Hey detects radio waves from the Sun
September USA receives first samples of DDT for wide-scale use, to control typhus from lice
October 3 Swedish physicist Hannes Alfvén describes electromagnetic waves in liquids
October 3 German engineers Wernher von Braun and Walter Dornberger perform the first successful test launch of a V2 rocket
December 2 Enrico Fermi oversees the first self-sustained nuclear chain reaction in the first nuclear reactor
Max Delbrück and Salvador Luria demonstrate that genetic mutations arise spontaneously in bacteria

1943–44

These are the control panels of the first electronic programmable computer—Colossus—which was used to decode German messages during World War II.

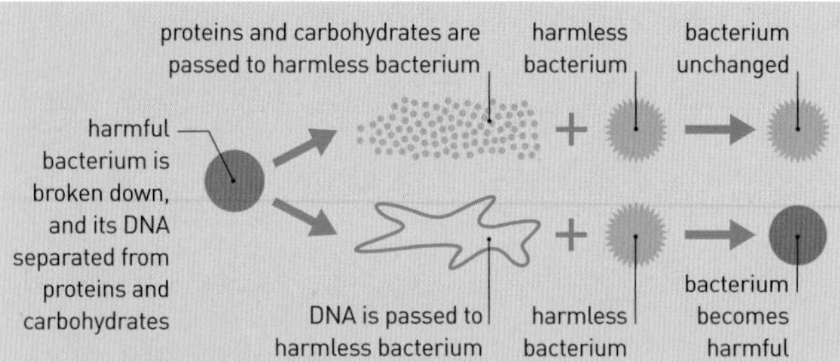

proteins and carbohydrates are passed to harmless bacterium
harmless bacterium
bacterium unchanged
harmful bacterium is broken down, and its DNA separated from proteins and carbohydrates
DNA is passed to harmless bacterium
harmless bacterium
bacterium becomes harmful

DISCOVERING THE NATURE OF DNA

Oswald Avery sought to identify the chemical substance that could transform harmless bacteria into harmful bacteria. Avery killed harmful bacteria and broke them down into their various chemical components (protein and carbohydrates and DNA). He added each component in turn to harmless bacteria, and he found that only the bacteria's DNA could cause a change.

DURING WORLD WAR II, the pioneering code-breaking work of British mathematician **Alan Turing** (1912–54) triggered an outburst of computer technology. One of the **earliest electronic digital computers** was built in Britain in 1943: **Colossus**.

French engineer **Émile Gagnan** (1900–79) modified a gas-regulator valve for a new use. Working with French biologist **Jacques Cousteau** (1910–97), he adapted the device so that it could be used to control the air supply in an aqualung. This invention would **revolutionize underwater exploration**.

A major step forward in unlocking the science of genes and inheritance was made in 1944. Canadian-born physician **Oswald Avery** (1877–79) wanted to identify the transforming principle first identified by Frederick Griffith (see 1928). Avery performed similar experiments to Griffith, but he analyzed the genetic factor in more detail. He demonstrated that if proteins and their enzymes were removed from bacteria, another chemical present—DNA—could still cause transformation. This showed that **genes are made of DNA**, not protein, as was previously thought.

> “ WHY DIDN'T **AVERY** GET THE **NOBEL PRIZE?** BECAUSE MOST **PEOPLE DIDN'T TAKE HIM SERIOUSLY.** ”
>
> **James Watson, American geneticist,** from *Nature*, April 1983

Quinine, a naturally occurring substance found in the the bark of the South American cinchona tree, had long been valued for its antimalarial properties. Supply had become difficult during World War II, but in May 1944, American chemists **Robert Woodward** (1917–79) and **William Doering** (1917–2011) announced that they had successfully manufactured it.

Austrian physician **Hans Asperger** (1906–80) had been studying mental disorders in children and **formalized the diagnosis of autism**. He examined a group of autistic children who, because of the nature of their way of learning, he called little professors. Their condition later became known as **Asperger's syndrome**.

1945

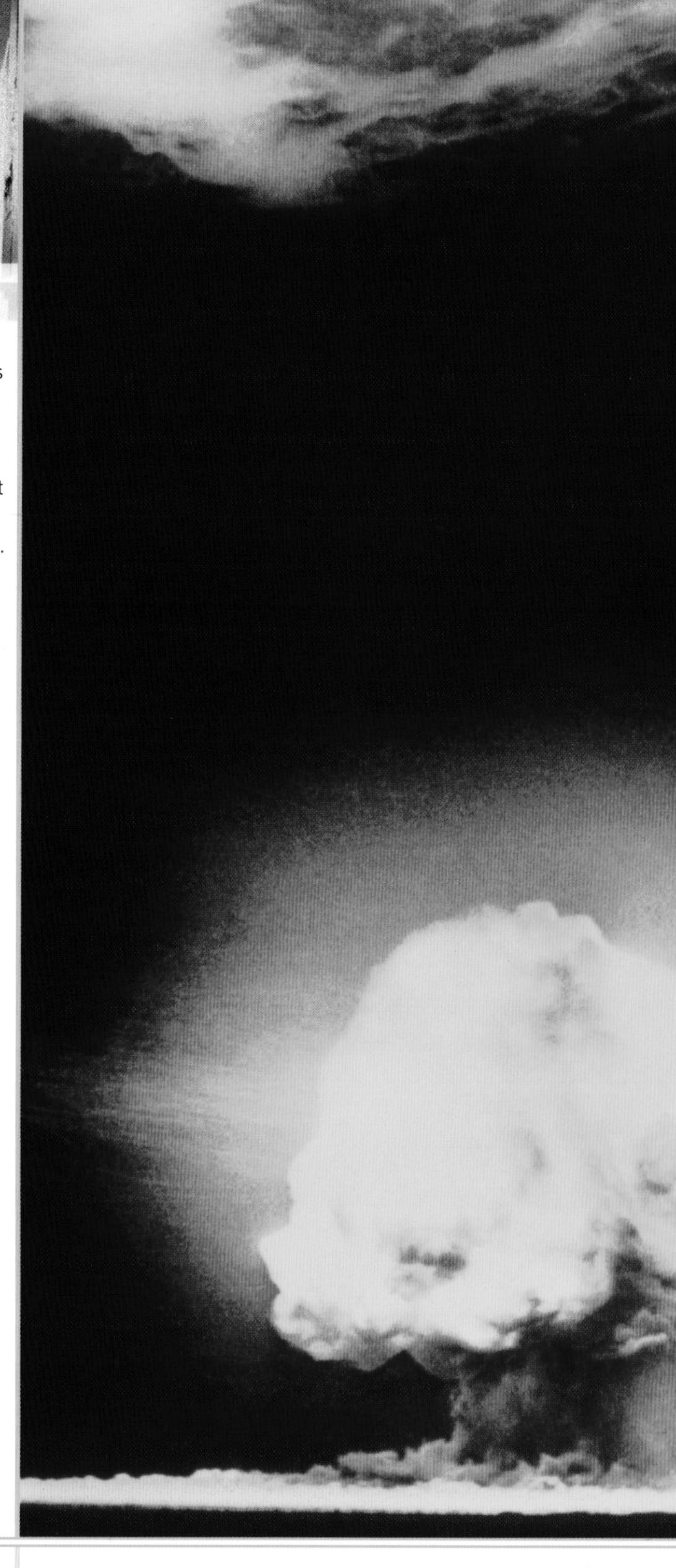

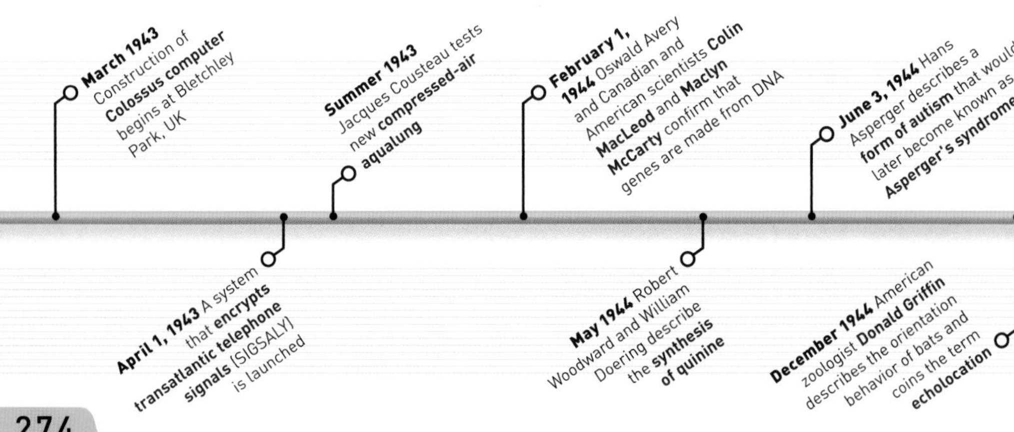

March 1943 Construction of **Colossus computer** begins at Bletchley Park, UK
April 1, 1943 A system that **encrypts transatlantic telephone signals** (SIGSALY) is launched
Summer 1943 Jacques Cousteau tests new **compressed-air aqualung**
February 1, 1944 Oswald Avery and Canadian and American scientists **Colin MacLeod** and **Maclyn McCarty** confirm that genes are made from DNA
May 1944 Robert Woodward and William Doering describe the **synthesis of quinine**
June 3, 1944 Hans Asperger describes a **form of autism** that would later become known as **Asperger's syndrome**
December 1944 American zoologist **Donald Griffin** describes the orientation behavior of bats and coins the term **echolocation**

February Dorothy Hodgkin describes the central chemical **structure of penicillin**

February 16 American physician Raymond L. Libby describes a method of delivering **penicillin by mouth**

March American chemists **Jacob Marinsky, Lawrence Glendenin**, and **Charles Coryell** isolate a new element—**romethium**

This polarized light micrograph shows crystals of the vitamin folic acid (folacin), which is needed during pregnancy for normal development of the baby in the womb.

AS WORLD WAR II ENTERED ITS SIXTH YEAR, so did the **Manhattan Project**—the Allies' atomic bomb research project (see 1939). The idea of an artificial nuclear chain reaction had been developed by Hungarian-born physicist **Léo Szilárd** (1898–1964), but it did not become a reality until 1942. Nuclear reactions are processes that change the nuclei of atoms, either through **fission** (splitting them) or **fusion** (combining them). Both release energy, but only **fission could be achieved artificially**. Fission happens naturally when radioactive elements decay, but it can be **induced by bombarding elements with neutrons** (see 1938). If elements capable of sustaining a nuclear fission chain reaction (fissile material), such as uranium or plutonium, are condensed to a critical mass, the neutrons emitted by the radioactivity trigger instantaneous decay and release massive amounts of energy. On July 16, the US Army exploded the **Manhatten Project's first atomic bomb** in the New Mexico desert—the Trinity Test—witnessed by fewer than 300 people. Three weeks later, two atomic bomb attacks on Japan (on **Hiroshima**, and **Nagasaki**), helped bring World War II to an end, but hundreds of thousands of lives were lost.

The world's first atomic bomb
The first bomb was exploded by the US Army in the early hours of July 16, 1945, at Alamogordo, in the New Mexico desert. It had the energy equivalent of 20,000 tons of TNT.

> “WHEN YOU SEE SOMETHING THAT IS **TECHNICALLY SWEET,** YOU GO AHEAD AND DO IT. THAT IS THE WAY IT WAS WITH THE **ATOMIC BOMB.**”
>
> **J. Robert Oppenheimer, American physicist,** testifying in court, 1954

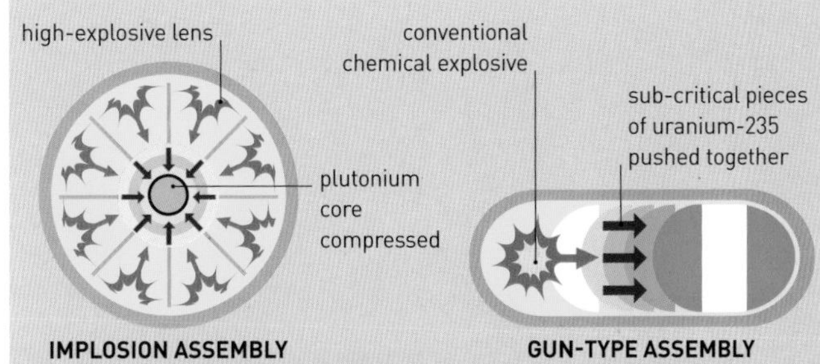

high-explosive lens
conventional chemical explosive
sub-critical pieces of uranium-235 pushed together
plutonium core compressed
IMPLOSION ASSEMBLY
GUN-TYPE ASSEMBLY

BUILDING ATOMIC BOMBS

Two methods have been used. In the implosion method (used for the Trinity Test and the Nagasaki bomb), explosives compress a central core of fissile material. In the gun-type assembly (used on Hiroshima), elements were pushed together. The naturally emitted neutrons strike neighboring atoms, causing a chain reaction of fission events (see 1938), which releases huge amounts of energy.

Scientists had already developed X-ray crystallography, a technique that could be used to work out positions of a crystal's atoms (see 1912), and it proved especially useful for studying structures of complex biological molecules. By July, British chemist **Dorothy Hodgkin** had helped resolve the complex **structures** of both **penicillin and cholesterol**.

In August, chemical company American Cyanamid announced that it had made **folic acid**—the vitamin needed for healthy growth in the developing fetus.

DOROTHY HODGKIN
(1910–94)

Born in Egypt, Dorothy Hodgkin studied chemistry at Oxford University, UK, and earned a PhD in the study of biological substances called sterols. She studied the three-dimensional structure of complex molecules such as cholesterol, penicillin, and insulin. She was awarded the 1964 Nobel Prize for her research into vitamin B12.

British writer **Arthur C. Clarke** (1917–2008) was looking to the future. Among his many predictions was the idea that **geostationary satellites,** (satellites that would sit at a fixed point in space relative to Earth's position and rotation period) could be **used for telecommunications**. This became a reality less than 20 years later.

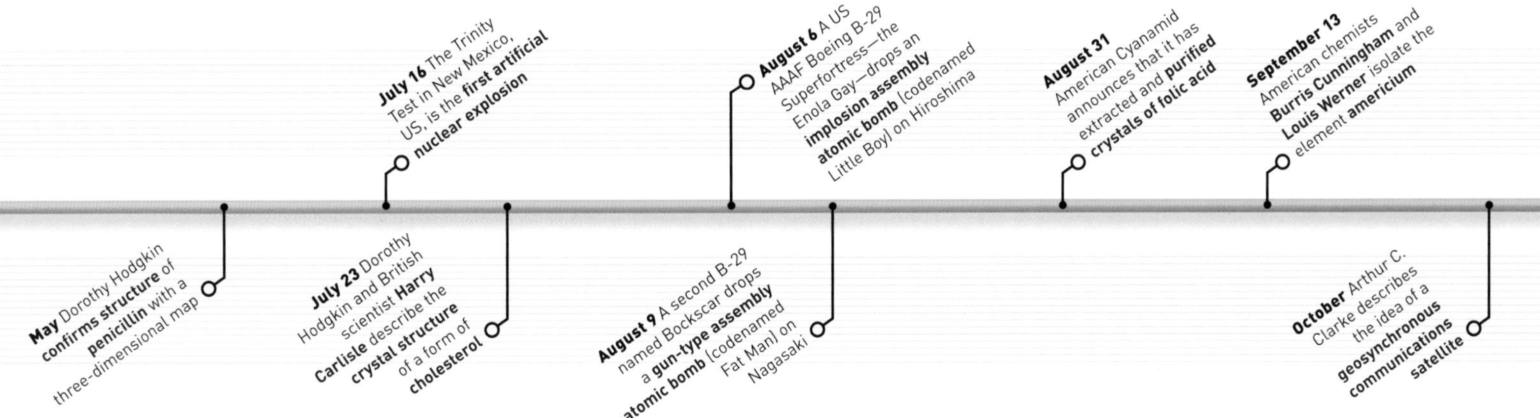

May Dorothy Hodgkin **confirms structure of penicillin** with a three-dimensional map
July 16 The Trinity Test in New Mexico, US, is the **first artificial nuclear explosion**
July 23 Dorothy Hodgkin and British scientist **Harry Carlisle** describe the **crystal structure** of a form of **cholesterol**
August 6 A US AAAF Boeing B-29 Superfortress—the Enola Gay—drops an **implosion assembly atomic bomb** (codenamed Little Boy) on Hiroshima
August 9 A second B-29 named Bockscar drops a **gun-type assembly atomic bomb** (codenamed Fat Man) on Nagasaki
August 31 American Cyanamid announces that it has extracted and **purified crystals of folic acid**
September 13 American chemists **Burris Cunningham** and **Louis Werner** isolate the element **americium**
October Arthur C. Clarke describes the idea of a **geosynchronous communications satellite**

6

THE INFORMATION AGE
1946–2013

As electronic devices repeatedly shrank in size and cost, digital power expanded exponentially, ushering in a new era of information technology, global communication, and exploration into the remotest parts of the Universe.

1946

238,000 MILES
THE DISTANCE FROM EARTH TO THE MOON, AS SHOWN BY PROJECT DIANA

The distance between Earth and the Moon was measured by bouncing a radar signal off the Moon.

THE VERY FIRST CONTACT WITH A SPACE BODY was made on January 10, when the US Army Signal Corps successfully detected the **echo of radar** reflected from the Moon—just 2.5 seconds after it was sent. **Project Diana** was initially established to examine whether long-range radar could detect incoming missiles, but it marked the start of the **US space program** as well. In addition to determining the distance to the Moon, it demonstrated that signals made on Earth could be used to communicate in space.

The same month saw the publication of studies of a phenomenon that would revolutionize medical imaging: **nuclear magnetic resonance (NMR)**. Swiss physicists **Felix Bloch** (1905–83) and American physicist **Edward Purcell** (1912–97) showed that when a sample is exposed to an intense magnetic field, certain nuclei within the sample resonate at characteristic frequencies. This technique was subsequently used to study the structures of molecules, and was later modified to produce images of bigger internal structures, such as those of the living body in **MRI (magnetic resonance imaging)**.

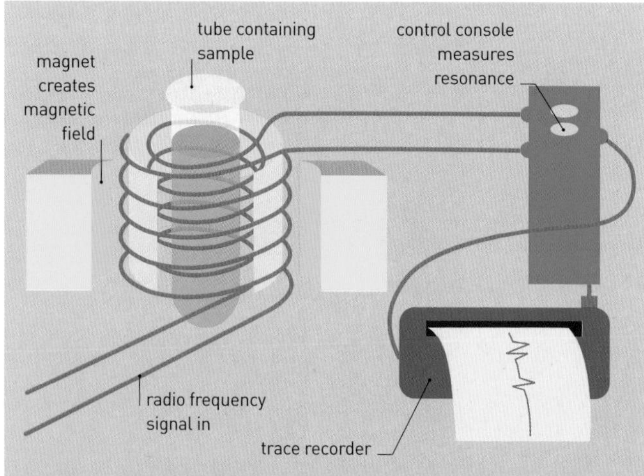
magnet creates magnetic field
tube containing sample
control console measures resonance
radio frequency signal in
trace recorder

NUCLEAR MAGNETIC RESONANCE

The magnetic resonance of atomic nuclei can be used to detect the chemical components of a substance. The atomic nuclei in a test sample that has been placed in a strong magnetic field absorb and emit electromagnetic radiation at characteristic frequencies. Scientists can use this to establish the kinds of atoms present in the sample and determine its chemical structure.

American biologists **Edward Tatum** (1909–75) and **Joshua Lederberg** (1925–2008) found that bacteria had a sexual process similar to that of more complex organisms. When they combined different strains, a few **bacteria** developed new abilities that the original strains were incapable of doing individually. They discovered that bacterial cells were binding together and exchanging genetic material in a process known as conjugation, thereby sharing chemical capabilities. This process has important implications, such as the spread of **antibiotic resistance**.

In July, the US undertook the **first post-war nuclear test** using the same sort of bomb as dropped on Nagasaki in 1945. Called **Test Able**, it was conducted on the Bikini Atoll in the Pacific Ocean, and was designed to see the effects on 78 experimental ships anchored in the lagoon—some carrying living test animals. More than 60 nuclear tests would later be carried out in the Bikini Atoll.

1947

The Bell X-1 manned airplane was launched from the bomb bay of a Boeing B-29 to reach a record-breaking altitude and break the sound barrier.

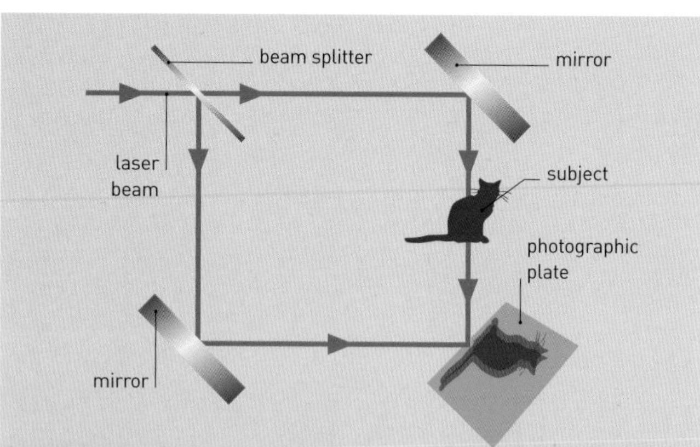
beam splitter
mirror
laser beam
subject
photographic plate
mirror

HOLOGRAMS

Hologram technology involves directing a laser beam onto a photographic plate using mirrors. The beam is split so that part of it bounces off the subject before hitting the plate, while the other part hits the plate directly. This produces an interference pattern, which is photographed. A 3-D image is made by shining a laser onto the negative at the same angle as the original beam.

IN 1947, AMERICAN PHYSICIST LUIS ALVAREZ (see 1980) oversaw the construction of the first **proton linear accelerator** in Berkley, California. In the same year, American **physicist William Hansen** (1909–49) produced the **first electron linear accelerator** at Stanford University, California. Particle acceleration is used to study the building blocks of matter, and it has practical applications, such as in the treatment of cancer.

In September at the American Chemical Society, Jacob Marinsky (1918–2005) announced the **discovery of radioactive element number 61,** named **promethium** after the Greek god Prometheus. Promethium is a rare earth element, which closed the gap in the periodic table.

In October, US Air Force pilot Charles "Chuck" Yeager (b.1923) became the **first person to break the sound barrier**. He flew **Bell X-1** and achieved a speed of Mach 1.07 or 815 mph (1,311 kmph).

Hungarian–British **physicist Dennis Gabor** (1900–79) filed a patent in December for a

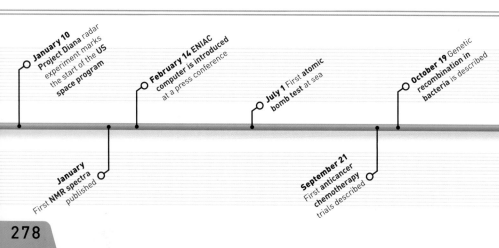

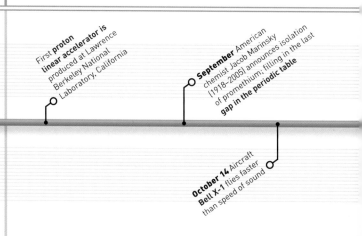

January 10 Project Diana radar experiment marks the start of the **US space program**

January First **NMR spectra** published

February 14 ENIAC computer is introduced at a press conference

July 1 First **atomic bomb test** at sea

September 21 First **anticancer chemotherapy** trials described

October 19 Genetic **recombination in bacteria** is described

First **proton linear accelerator is** produced at Lawrence Berkeley National Laboratory, California

September American chemist Jacob Marinsky (1918–2005) announces isolation of promethium, filling in the last **gap in the periodic table**

October 14 Aircraft **Bell X-1** flies faster than speed of sound

200 INCHES THE DIAMETER OF THE PRIMARY MIRROR OF THE HALE TELESCOPE

For 45 years, the Hale telescope was the world's biggest effective telescope. It is still in use today, collecting data on around 290 nights every year.

Proton linear accelerator
The first device to accelerate protons in a straight line was a 40 ft (12 m) construction. It helped advance research and understanding of fundamental particles of matter.

theoretical technique for producing 3D-style images, or holograms. He described how to produce an image of an object that changed orientation depending on the viewing angle. However, practical application of his theory became possible only with the development of the first working lasers in 1960.

American engineer **Percy Spencer** (1894–1970) was involved in **radar design** with a company called Raytheon. He accidentally discovered that the microwaves produced by a vacuum tube called a magnetron—a core component of radar—heated food. By confining the food "target" in a metal box, Spencer invented the **first microwave oven**—patented in 1945. Raytheon sold the first commercial model in 1947.

IN MARCH, American physicists **Julian Schwinger** (1918–94) and **Richard Feynman** (1918–88) introduced a new field of science: **quantum electrodynamics**. It described how electrically charged particles interacted with packets of electromagnetic radiation, called photons. The conference included the first presentations of **"Feynman diagrams"** to show interactions between subatomic particles, with time along one axis and space along another.

In April, Russian physicist **George Gamow** (1904–68) and American cosmologist **Ralph Alpher** (1921–2007) proposed that elements produced when the Universe formed (see pp.344–45) occurred in fixed proportions. Hydrogen and helium are still the most abundant elements in the Universe today.

In June, Caltech's (California Institute of Technology) Palomar Observatory was completed. Its dome houses the **Hale telescope**, named after American **astronomer George Hale**. This telescope—first used by astronomer Edwin Hubble—has aided in the discovery of quasars and stars of distant galaxies.

Takahe
Since their rediscovery in 1948, the New Zealand takahe have been moved to predator-free islands as part of a conservation program.

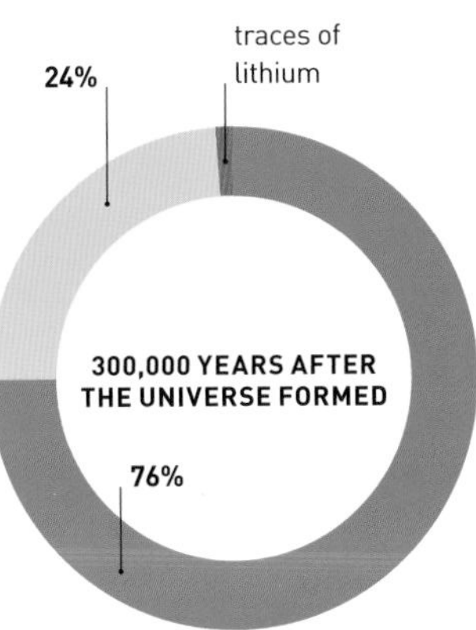

24%
traces of lithium
300,000 YEARS AFTER THE UNIVERSE FORMED
76%

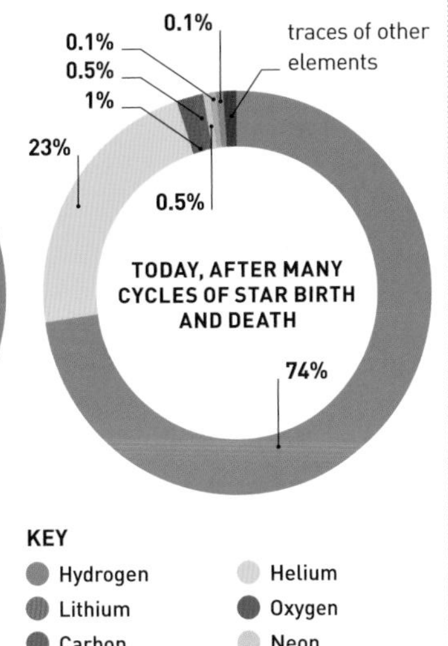

0.1%
0.1%
0.5%
1%
traces of other elements
23%
0.5%
TODAY, AFTER MANY CYCLES OF STAR BIRTH AND DEATH
74%
KEY
Hydrogen
Lithium
Carbon
Iron
Helium
Oxygen
Neon
Nitrogen

Composition of the Universe
The first elements formed after the Universe formed were the lightest. Cosmologists theorize that the heavier elements were formed inside stars by fusion of atoms.

In the same month, Albert 1 became the **first monkey astronaut** to leave for space onboard a V2 rocket. However, he died of suffocation 39 miles (63 km) into the ascent, before reaching the Karman line that marks the beginning of space at 62 miles (100 km).

Austrian-American chemist **Erwin Chargaff** reported on a study of the **makeup of DNA**—a substance that had recently been shown to be the chemical of heredity. Five years before DNA structure was revealed as a double helix, Chargaff's analysis showed that DNA components called bases occurred in fixed proportions. The proportions of base adenine matched that of thymine, while the proportion of guanine matched cytosine. The double helix model would show that these matches were due to the bases pairing up. Later work revealed that the base sequence along the double helix was the basis for inherited information.

ERWIN CHARGAFF
(1905–2002)

In studying the chemistry of biological molecules, Austrian-born chemist Erwin Chargaff made breakthroughs in a range of topics, such as how blood clots. He moved to the USA when Nazi influence spread through Europe. Chargaff discovered that DNA varied between species, but the proportions of key components were fixed.

On November 20, English ornithologist **Geoffrey Orbell** (1908–2007) made a remarkable discovery when he saw **takahes** in the mountains of South Island, New Zealand. These flightless relatives of rails and moorhens had been thought to be extinct for the previous 50 years.

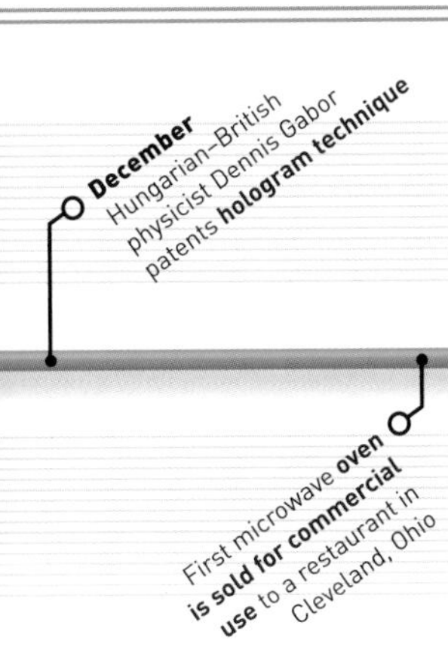

December Hungarian-British physicist Dennis Gabor patents **hologram technique**
First microwave **oven is sold for commercial use** to a restaurant in Cleveland, Ohio

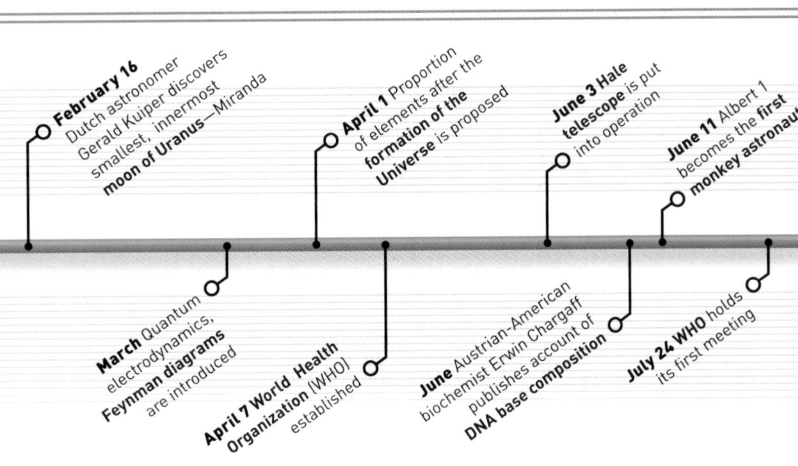

February 16 Dutch astronomer Gerald Kuiper discovers smallest, innermost **moon of Uranus—Miranda**
March Quantum electrodynamics, **Feynman diagrams** are introduced
April 1 Proportion of elements after the **formation of the Universe** is proposed
April 7 World Health Organization (WHO) established
June 3 Hale telescope is put into operation
June Austrian-American biochemist Erwin Chargaff publishes account of **DNA base composition**
June 11 Albert 1 becomes the **first monkey astronaut**
July 24 WHO holds its first meeting

November 20 Takahe (bird) rediscovered in New Zealand

1949

EDSAC was one of the first recognizably modern computers in terms of design, but it nearly filled a room and needed 60 in- (152 cm-) long tubes of mercury to help with memory storage.

ON MARCH 28, BRITISH ASTRONOMER FRED HOYLE coined the term "**Big Bang**" on BBC radio, while explaining the Steady State Theory—his view that the Universe had an infinite past and an infinite future. He disagreed with the theory that all matter was created "in one big bang at a particular time in the remote past." In doing so, he unwittingly gave the name to an idea that would later gain almost universal acceptance.

Monkey astronaut
The V2 rocket carrying a rhesus monkey went beyond the Karman Line—the boundary between the Earth's atmosphere and outer space.

Dutch-American astronomer **Gerald Kuiper** had established that the atmosphere of Mars was made of carbon dioxide and Saturn's rings were made of ice. In May, he **discovered Nereid**—the outermost moon of Neptune. Kuiper later attributed Nereid's eccentric orbit to its origin in an hypothesized ring of icy bodies beyond Neptune. The existence of this ring—called the **Kuiper Belt**—was confirmed in 1992.

On May 6, **EDSAC** (Electronic Delay Storage Automatic Calculator), a new computer at Cambridge University, UK, ran its first program. EDSAC could work through around 700 operations per second. It became the **first computer to routinely help scientists** with **complex calculations**.

In the US, rocket scientists successfully launched the **first mammal into space** (see 1948). Albert II—a rhesus monkey—went beyond the Earth's atmosphere when his V2 rocket reached a height of 81 miles (130.6 km). Albert II survived the flight, but he died on return to Earth due to parachute failure.

FRED HOYLE (1915–2001)

British astronomer Fred Hoyle was one of the great scientific thinkers of the 20th century, and stimulated widespread interest in cosmology with his support of the Steady State Theory. This idea has been supplanted by the Big Bang Theory (see pp.344–45). However, Hoyle's Stellar Nucleosynthesis Theory prevails.

3 THOUSAND
THE **NUMBER** OF **VACUUM TUBES** USED IN **EDSAC**

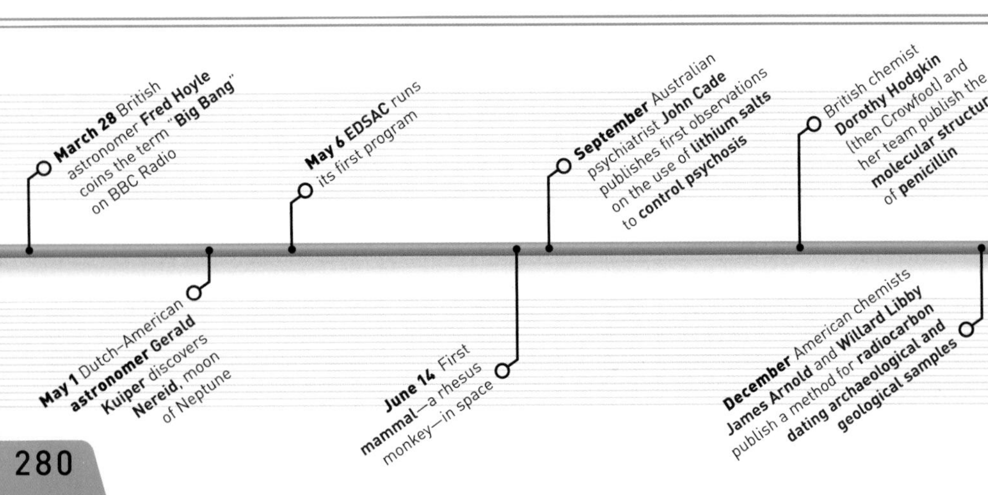
March 28 British astronomer Fred Hoyle coins the term "Big Bang" on BBC Radio
May 1 Dutch-American astronomer Gerald Kuiper discovers Nereid, moon of Neptune
May 6 EDSAC runs its first program
June 14 First mammal—a rhesus monkey—in space
September Australian psychiatrist John Cade publishes first observations on the use of lithium salts to control psychosis
December American chemists James Arnold and Willard Libby publish a method for radiocarbon dating archaeological and geological samples
British chemist Dorothy Hodgkin (then Crowfoot) and her team publish the molecular structure of penicillin

1950

Deployment of the myxoma virus was the first biological control method used for a mammal pest, dramatically reducing the number of rabbits in Australia.

THE YEAR 1950 SAW RAPID ADVANCEMENT in nuclear technology. On January 31, US President **Harry Truman** announced—largely in response to the Soviet Union's detonation of an atomic bomb in August 1949—that he had **authorized** the **development of a hydrogen bomb**. Its design would form the basis for all future thermonuclear weapons. A month later, at the University of California, nuclear chemist **Stanley Thompson** (1912–76) and his team created **californium**—the 98th element of the periodic table. Despite its instability, californium is still the heaviest element that does not quickly decay—and unlike most other ultraheavy elements, it can be made in quantities visible to the naked eye.

Computing science was racing forward too. The US had **ENIAC** (Electronic Numerical Integrator And Computer), which became the **first computer** to be used in **predicting weather**. It started the first 24-hour weather forecast service on March 5. In October, British computer scientist **Alan Turing** proposed a **test for artificial intelligence**.

> **" A COMPUTER** WOULD DESERVE TO BE **CALLED INTELLIGENT** IF IT COULD **DECEIVE A HUMAN** INTO **BELIEVING** THAT IT **WAS HUMAN. "**
>
> **Alan M. Turing, British computer scientist,** 1950

Rabbits had become a national problem for Australia. A century and a half before, European settlers had brought rabbits with them for food. But in a land without predators, the rabbit population exploded, wreaking havoc on crops and native wildlife. When shooting, poisoning, and containment with fences failed to control them, **Frank Fenner** (1914–2010), a microbiologist at the Australian National University, oversaw the release of **myxoma—**a deadly virus that caused **myxomatosis**. It **curbed** the **rabbit plague**, and although rabbits were not

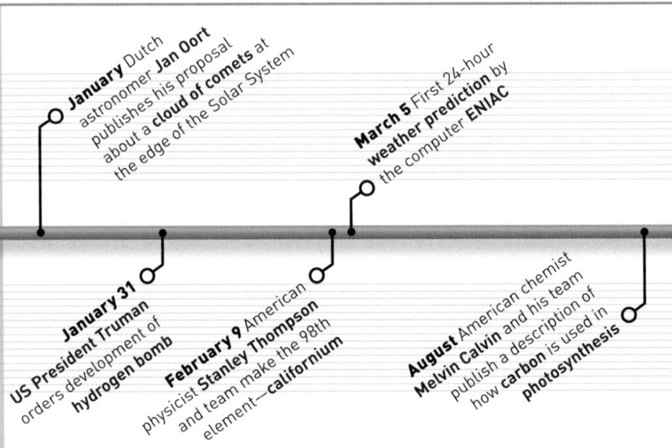
January Dutch astronomer Jan Oort publishes his proposal about a cloud of comets at the edge of the Solar System
January 31 US President Truman orders development of hydrogen bomb
February 9 American physicist Stanley Thompson and team make the 98th element—californium
March 5 First 24-hour weather prediction by the computer ENIAC
August American chemist Melvin Calvin and his team publish a description of how carbon is used in photosynthesis

600,000,000

PEAK NUMBER OF **RABBITS** IN **AUSTRALIA BEFORE** INTRODUCTION OF **MYXOMA**

Science influenced fashion at the 1951 Festival of Britain, where fabric and wallpaper designs based on X-ray crystallography (see 1945) were exhibited.

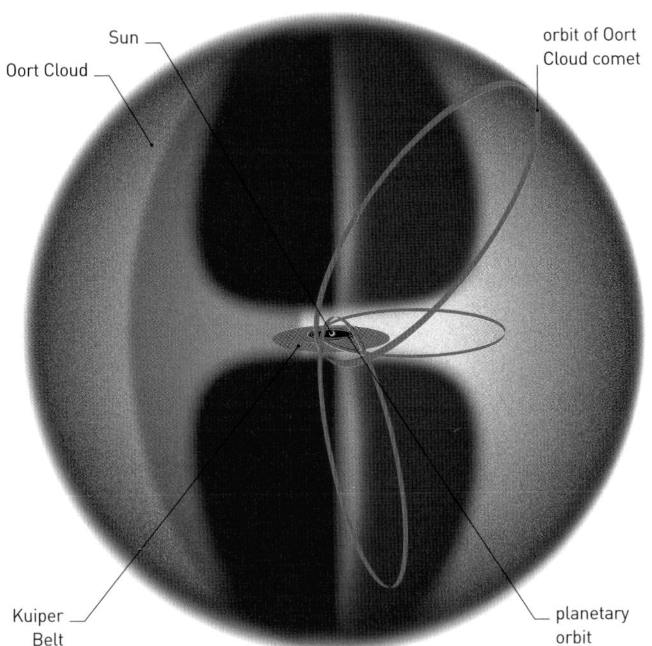
Sun
Oort Cloud
orbit of Oort Cloud comet
Kuiper Belt
planetary orbit

Oort Cloud
Made up of billions of comets, the Oort Cloud marks the hypothetical outer boundary of the Solar System. Orbits of Oort Cloud comets are more than a thousand times bigger than planetary orbits.

eradicated, their numbers never recovered to pre-1950s levels. In the 1970s, Fenner went on to use his skills in disease control and played a crucial role in the World Health Organization's successful **global elimination of human smallpox**.

The start of a new decade also saw an effort to bring into focus the most mysterious part of Earth's surface: the **ocean floor**. Geologists such as **Marie Tharp** (1920–2006) and **Bruce Heezen** (1924–77) had used photographic methods to locate sunken aircraft from World War II. They went on to **map the underwater seascape**—discovering submerged mountain ranges along the way.

In this year, Dutch astronomer and physicist **Jan Oort** (1900–92) suggested that **comets came from a cloudlike reservoir** at the edge of the Solar System. Modern astronomers believe that Oort was right.

IN 1951, the *Lancet*, a medical journal, published an article by British physician **Richard Asher** (1912–69) about a new mental disorder in which sufferers sought attention by fabricating medical ailments. Asher called it the **Münchhausen syndrome**—after the 18th-century German baron who invented wild stories about his life that he claimed were true.

The same year also saw an important breakthrough in the field of molecular biology. Unraveling the molecular structure of complex biological substances such as proteins had been one of the challenges of analytical chemistry. British biochemist **Fred Sanger** (b.1918) studied one particular protein, **insulin**, which consisted of interlocked chains of smaller variable components called amino acids. By chemically splitting these chains, Sanger was able to determine the **types of amino acids**, and even the sequence in which they were linked together. Sanger was the first scientist to show that this **sequence was the same for all insulin** molecules, and that **different kinds of proteins** have **unique amino acid sequences**. It would take scientists more than a decade to fully appreciate the implications of Sanger's findings: that in the living body, individual genes in the DNA hold the instructions for assembling particular proteins by linking amino acids together in the correct order.

On July 4, American inventor **William Shockley** (1910–89), working at Bell Telephone Laboratories in New Jersey, announced the invention of the **junction transistor**. Shockley and his team had made their first transistor in 1947, but it was the improved design of 1951 that would become the standard component of electronic devices for the next 30 years.

Early transistors
Transistors revolutionized the world of electronic devices and circuitry. Their ability to act as switches also proved to be especially valuable in the growing field of computer technology.

POINT-CONTACT TRANSISTOR
JUNCTION TRANSISTOR

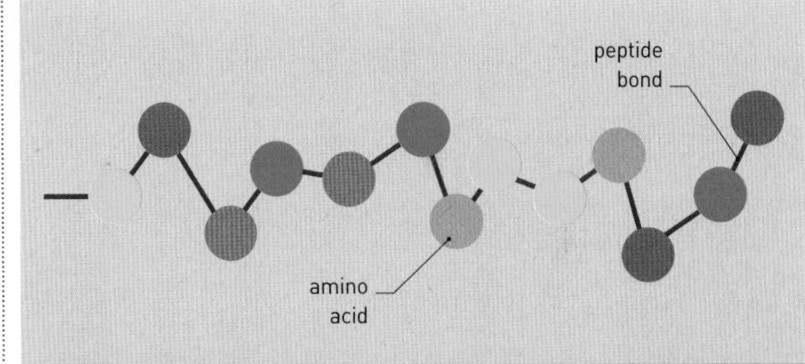
peptide bond
amino acid

AMINO ACID CHAINS

Protein molecules perform critical roles in the bodies of living things, such as driving metabolism and helping cells absorb nutrients. In 1951, Fred Sanger found that the chainlike molecule of a certain kind of protein—insulin—consisted of a unique sequence of amino acids. He also found that different kinds of proteins have different sequences—this determines how each chain folds into a shape for a particular purpose.

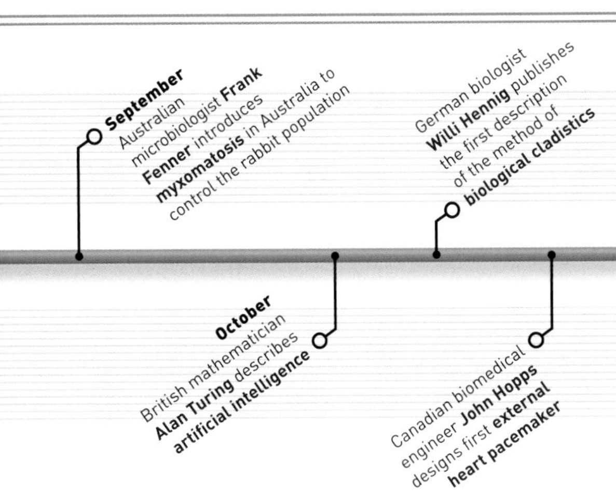
September Australian microbiologist **Frank Fenner** introduces **myxomatosis** in Australia to control the rabbit population
October British mathematician **Alan Turing** describes **artificial intelligence**
German biologist **Willi Hennig** publishes the first description of the method of **biological cladistics**
Canadian biomedical engineer **John Hopps** designs first **external heart pacemaker**

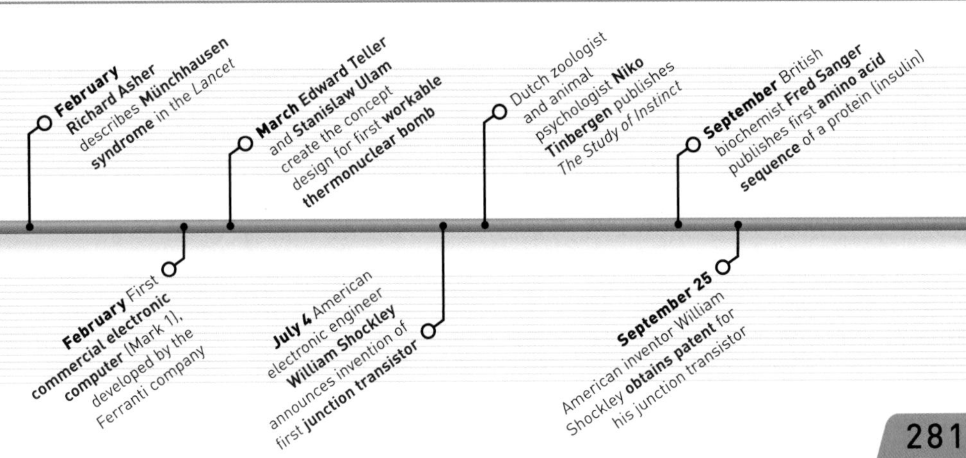
February Richard Asher describes **Münchhausen syndrome** in the *Lancet*
February First **commercial electronic computer** (Mark 1), developed by the Ferranti company
March **Edward Teller** and **Stanislaw Ulam** create the concept design for first **workable thermonuclear bomb**
July 4 American electronic engineer **William Shockley** announces invention of first **junction transistor**
Dutch zoologist and animal psychologist **Niko Tinbergen** publishes *The Study of Instinct*
September British biochemist **Fred Sanger** publishes first **amino acid sequence** of a protein (insulin)
September 25 American inventor William Shockley **obtains patent** for his junction transistor

1952

The hydrogen bomb test on the Enewetak Islands in the Pacific was the first thermonuclear explosion that combined atomic fusion and fission. Its mushroom cloud rose 10 miles (17km) into the air.

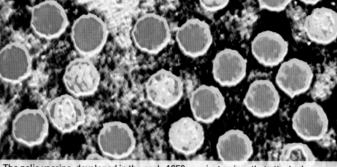

THE YEAR 1952 saw some important breakthroughs. American microbiologist **Alfred Hershey** and geneticist **Martha Chase** worked together at New York's Cold Spring Harbor Laboratory to tackle a key question in biology: was **life's genetic material** made from protein or DNA? Some scientists thought only protein was sufficiently complex to suit the task. Hershey and Chase examined the genetic material that phages—viruses that infect bacteria—inject into host cells. They discovered that the injected cells contained phosphorus—an element found in DNA, but not in protein. This indicated that genes were made of DNA.

On the other side of the world, **Alan Walsh**, a British physicist living in Australia, pioneered a technique that revolutionized analytical chemistry. He explored the practical applications of the fact that atoms of different elements absorb radiation of highly specific wavelengths. Walsh developed the **atomic absorption spectrometer**, which measured this radiation and detected the tiniest levels of elements in a mixture. The spectrometer, patented the following year, later became a standard tool in forensic science and other fields that demanded high-precision chemical analysis.

12 MEGATONS OF TNT
THE **SIZE** OF THE 1952 **IVY MIKE EXPLOSION**

In the US, **Walton Lillehei** and **John Lewis** performed the first **open-heart surgery** in September. They induced hypothermia (cooling below normal body temperature) in their patient, giving them 10 minutes to correct a congenital heart defect. Just over a week later, American surgeon **Charles Hufnagel** implanted the first **artificial heart valve** in a patient with rheumatic fever, prolonging her life by nearly a decade. Plans were also made for the first **separation of conjoined twins**. The Brodie brothers, fused at the head, had been born the year before. **Oscar Sugar** and his team separated the twins and saved one of them.

On November 1, the first **hydrogen bomb** test—codenamed **Ivy Mike**—was conducted by the US on the Enewetak Islands in the northwest Pacific. It produced a 3 mile (5km) fireball and obliterated a small island. Previous thermonuclear bombs had used atomic fission (see 1938), but Ivy Mike showed that an explosion could come, at least in part, from fusion (see 1988–89) too.

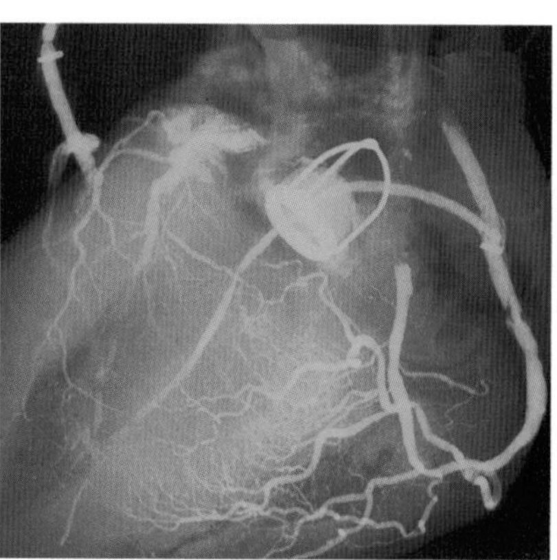

Artificial heart valve
The first artificial heart valve was a caged-ball design. Blood leaves the heart as the ball is pushed against the cage. When the heart relaxes, the ball falls back to seal the valve.

1953

The polio vaccine, developed in the early 1950s against a virus that attacked the central nervous system, saved thousands from a lifetime of paralysis.

IN 1953, SCIENTISTS gained insight into how **inheritance and reproduction** worked at a **chemical level**. English scientists **James Watson** and **Francis Crick** (see pp.284–85) believed that DNA was the key. While the chemical makeup of DNA was partially known, the physical arrangement of its components had until then been a mystery.

By 1953, a new technique—**X-ray crystallography**—was being used to produce 3-D images of the structure of complicated biological molecules. At Kings College, London, a team that included Rosalind Franklin and Maurice Wilkins used it to study DNA. Franklin perfected a technique of preparing samples of DNA that yielded especially clear results. The early indications were that DNA was shaped like a helix. In spite of Franklin's reluctance to draw premature conclusions, Watson and Crick began building a **helical model of DNA** based on the evidence, and published their results in the science journal *Nature*. They depicted DNA as two molecular chains entwined around one another in a double helix. This model of DNA was ground-breaking, suggesting a way living things could reproduce their genetic material.

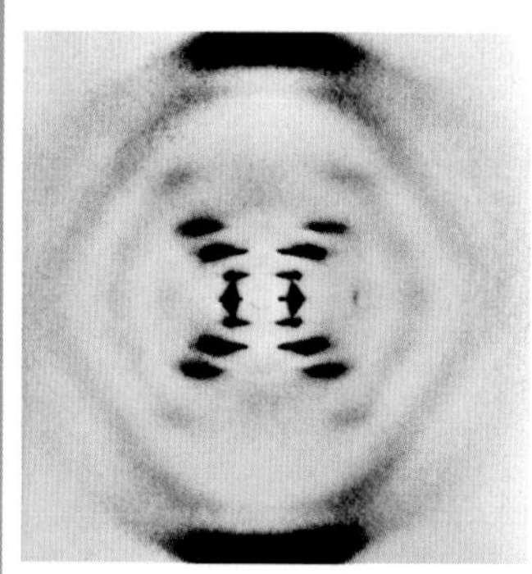

X-ray diffraction image of DNA
Franklin's remarkable, distinct X-ray diffraction image with its striking "X" pattern clearly indicated that DNA was shaped like a double helix.

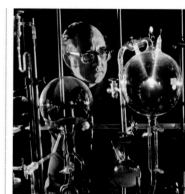

A month later, another journal, *Science*, reported on an experiment conducted the year before. American researchers **Stanley Miller** and **Harold Urey** had tried to re-create the **origin of life in a laboratory**. They had heated a mixture of ammonia, water, methane, and hydrogen in a sealed flask, and sparked it with electrodes to simulate lightning. Within two weeks the mixture was

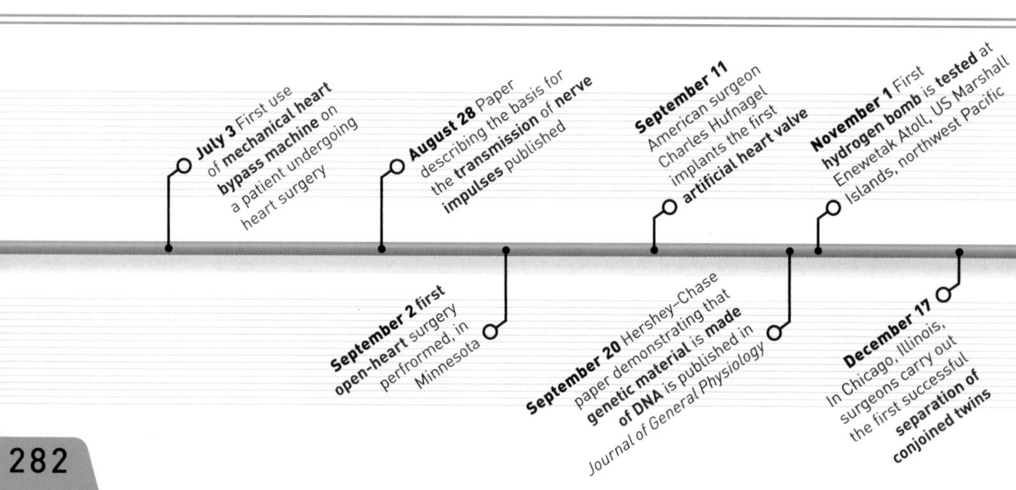

July 3 First use of **mechanical heart bypass machine** on a patient undergoing heart surgery
August 28 Paper describing the basis for the **transmission of nerve impulses** published
September 11 American surgeon Charles Hufnagel implants the first **artificial heart valve**
November 1 First **hydrogen bomb** is **tested** at Enewetak Atoll, US Marshall Islands, northwest Pacific
September 2 first open-heart surgery performed, in Minnesota
September 20 Hershey–Chase paper demonstrating that **genetic material** is **made of DNA** is published in *Journal of General Physiology*
December 17 In Chicago, Illinois, surgeons carry out the first successful **separation of conjoined twins**

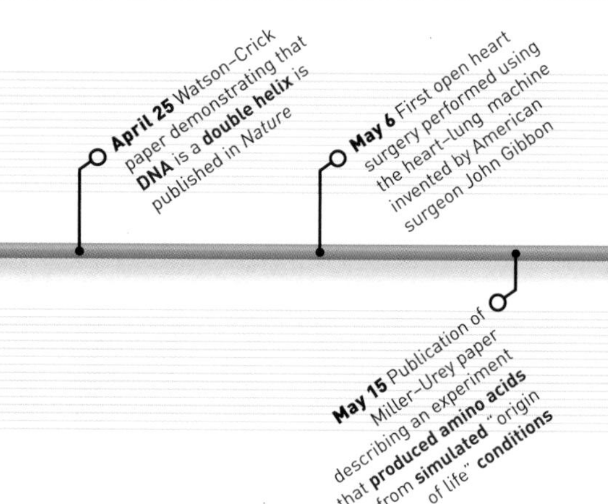

April 25 Watson–Crick paper demonstrating that **DNA** is a **double helix** is published in *Nature*
May 6 First open heart surgery performed using the heart–lung machine invented by American surgeon John Gibbon
May 15 Publication of Miller–Urey paper describing an experiment that **produced amino acids** from **simulated** "origin of life" **conditions**

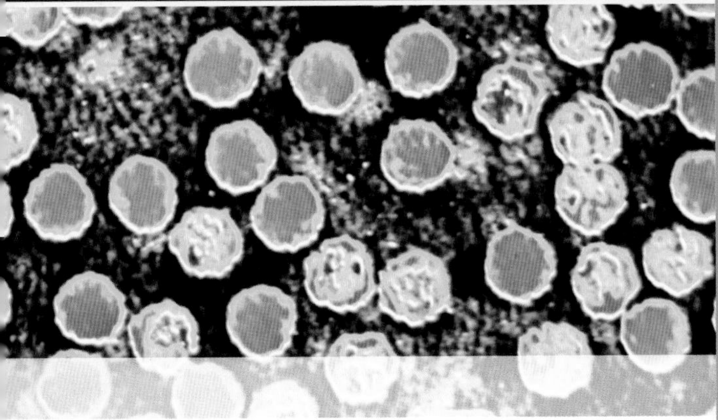

The Boeing 367-80—the prototype for the Boeing 707, which was the most successful of the early passenger jetliners—was donated to the Smithsonian Air and Space Museum in Washington, D.C. after being used to test design innovations.

Life in a flask
Stanley Miller and Harold Urey recreated the dawn of life in a flask: their "primordial soup" produced amino acids within two weeks.

found to contain amino acids—the building blocks of proteins. This showed that the building blocks of life could be made from the simplest of substances.

In November, **Jonas Salk**, an American virologist, announced a breakthrough of a more humanitarian nature. He developed a **vaccine against poliomyelitis** based on a dead form of the polio virus. It was a far safer version of a live vaccine that had previously been tried and found unsuccessful earlier.

Just as one discovery was being championed, another was being demolished. A **hominoid skull** in the Natural History Museum, London, supposedly excavated in Piltdown, East Sussex, in 1912 and acclaimed as a valuable link in the story of human evolution, was **exposed as a hoax**. The anatomists and paleontologists Kenneth Oakley, Wilfred Le Gros Clark, and Joseph Weiner declared that the skull actually consisted of a medieval human cranium, the teeth of a fossil chimpanzee, and the jaw of an orangutan. To this day, no one knows who perpetrated the hoax and duped the academic world for so long.

ROSALIND FRANKLIN (1920–58)

Trained in chemistry, Rosalind Franklin applied her skills to study the structure of biological molecules. She produced an X-ray diffraction image of DNA—"Photograph 51"—that showed a cross pattern. This suggested that DNA was helical in shape. It became a key piece of evidence in Watson and Crick's double helix model. Franklin died in 1958, and she did not share the Nobel Prize awarded for this achievement.

ENCOURAGED BY EARLY RESULTS, the US conducted a nationwide field trial—the largest medical field trial ever—of Salk's **polio vaccine**. On February 23, a **mass vaccination** program involving 1.8 million schoolchildren began. In 1955, a license was issued for its routine use. Salk's vaccine went on to protect children from polio across the world and heralded the World Health Organization's international campaign to eradicate polio.

In May, the American aerospace company **Boeing** rolled out a new type of jet aircraft. The **367-80** was the prototype for the 707 passenger aircraft that came into use in the 1960s and 1970s. Until then, civil aviation had mostly been centered on aircraft that were propeller-driven, but the 367-80 demonstrated that **jet propulsion** was the way forward.

From 1954, scientists worldwide were able to apply a standard unit of measurement for temperature, following the General Conference on Weight and Measures held in France. The conference had been established to oversee what would be known as the **International System of Units (SI)**. In 1954, "kelvin" (named after British physicist Lord Kelvin) was deemed to be the SI unit for temperature.

At the same time, **Nikolay Basov** and **Alexander Prokhorov** at the USSR Academy of Sciences published their description of **maser** (microwave amplification by stimulated emission of radiation)—a system for concentrating beams of radiation. Maser came to be used in atomic clocks and helped amplify tiny signals in long-distance television broadcasts. Since then, researchers have explored its potential for other uses, such as in medical body scanners.

The year ended with the first successful **kidney transplant**, carried out by **Joseph Murray** in Boston, Massachusetts.

273.16 K
THE TEMPERATURE AT WHICH **WATER, ICE, AND VAPOR** CAN **COEXIST**

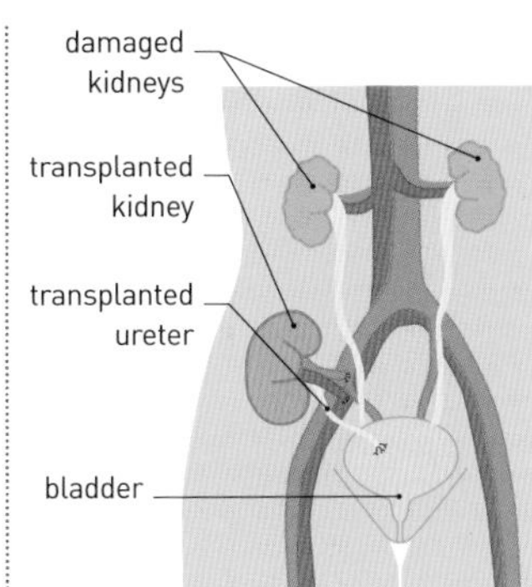

damaged kidneys
transplanted kidney
transplanted ureter
bladder

Kidney transplant
In a kidney transplant, the failed kidney is usually left in place. The donated kidney is implanted lower down the body and connected to a different part of the blood system.

YEAR	NUMBER OF POLIO CASES IN US
1948	27,726
1949	42,033
1950	33,300
1951	28,386
1952	57,879
1953	35,592
1954	38,476
1955	28,985
1956	15,140

Polio vaccinations
The use of Salk's polio vaccine for mass vaccination of children across the US brought down the number of polio cases by the thousands.

November 13 *New York Times* reports that **Jonas Salk guarantees** his antipolio **vaccine**

November 30 *Time* magazine reports that the **Piltdown Man skull** is a **hoax**

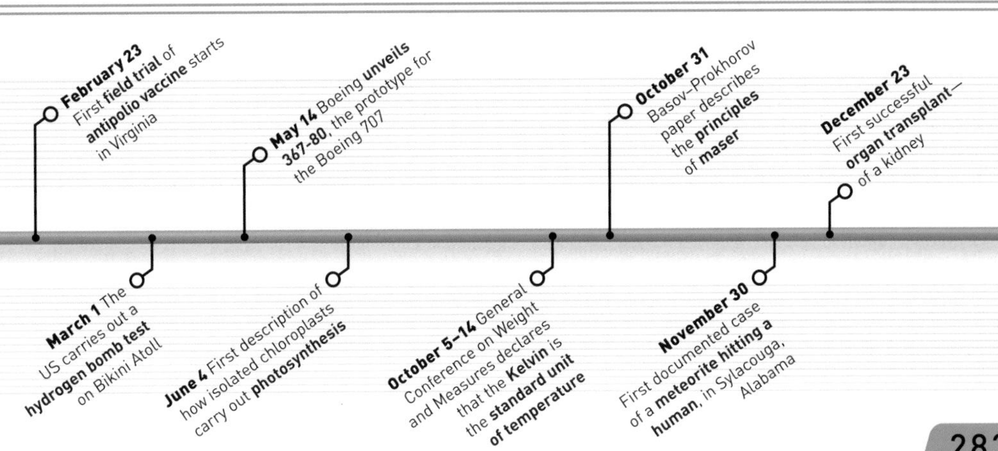

February 23 First **field trial** of **antipolio vaccine** starts in Virginia
March 1 The US carries out a **hydrogen bomb test** on Bikini Atoll
May 14 Boeing **unveils 367-80**, the prototype for the Boeing 707
June 4 First description of how isolated chloroplasts carry out **photosynthesis**
October 5–14 General Conference on Weight and Measures declares that the **kelvin** is the **standard unit of temperature**
October 31 Basov–Prokhorov paper describes the **principles of maser**
November 30 First documented case of a **meteorite hitting a human**, in Sylacouga, Alabama
December 23 First successful **organ transplant**—of a kidney

UNDERSTANDING DNA

A SELF-COPYING MOLECULE CALLED DNA IS THE CHEMICAL CODE OF LIFE ITSELF

The characteristics of living things are produced by chemical processes that happen in every cell. Twentieth-century science traced these processes to their source—a molecule that not only carries genetic information but also has the remarkable capacity to copy itself. It is called DNA.

By the turn of the 20th century, scientists had discovered that inherited characteristics come from particles passed down through generations. However, they did not understand the composition of these units of genetic material, which we now know as genes. In 1919, Lithuanian biochemist Phoebus Levene dismissed nucleic acid—a material present in the nucleus of every cell—as too simple to be involved directly in inheritance. But in the decades that followed, experiments proved that a form of nucleic acid is, in fact, the substance of genes.

By the 1950s, advances in analytical techniques meant that the best-known form of nucleic acid, DNA (deoxyribonucleic acid), could be examined in ways never before possible. A method called X-ray crystallography even promised to reveal its three-dimensional shape.

BASE PAIRS

In 1953, results from these new methods convinced American biologist James Watson and British biophysicist Francis Crick that DNA had a helical structure. Evidence indicated two coiled chains (a double helix) with variable components (bases) that hold the chains together. Four varieties of bases were always present in certain proportions. Watson and Crick deduced that this was because they were bonded in fixed ways: adenine with thymine and guanine with cytosine. This would be key to understanding not only how DNA carried inherited information but also how this information replicated at reproduction.

JAMES WATSON AND FRANCIS CRICK
To test their double helix theory, Watson and Crick built a model of the structure to check that the chemical pieces would fit.

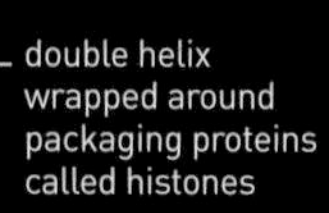
double helix wrapped around packaging proteins called histones
DNA molecule coiled into a double helix
thymine
adenine
DNA backbone, consisting of deoxyribose (a form of sugar) and phosphates

CHROMOSOMES

When a cell divides, to prevent entanglement each DNA molecule is bundled into condensed structures called chromosomes. The number of chromosomes per cell varies from species to species.

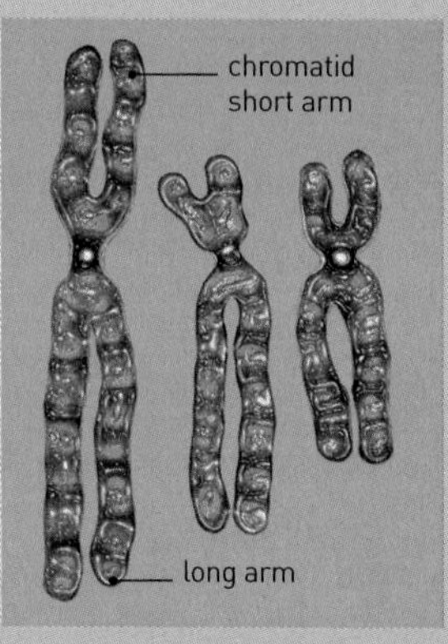
chromatid short arm
long arm

Chromosome micrograph
Because DNA replication happens before cell division, every chromosome appears with duplicated "chromatids."

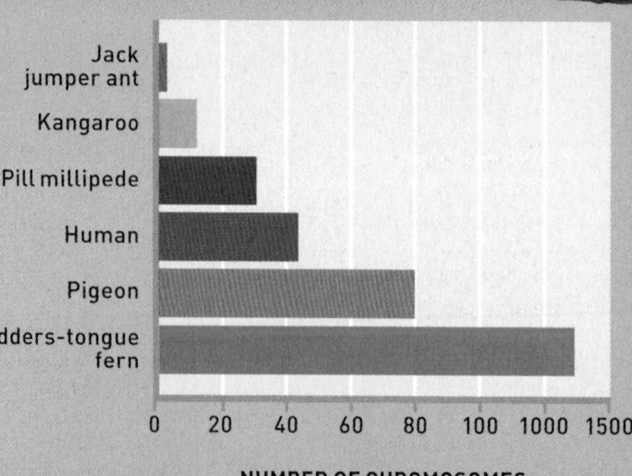
Jack jumper ant
Kangaroo
Pill millipede
Human
Pigeon
Adders-tongue fern
0
20
40
60
80
100
1000
1500
NUMBER OF CHROMOSOMES

DNA REPLICATION

Just before a cell divides, it replicates its entire DNA. Each DNA molecule "unzips," and the paired bases separate. Because of the strict base-pair ruling, the base sequence along one strand determines the sequence along the other: they are complementary. Free DNA building blocks are linked to make new complementary strands along each existing strand "template." This creates material for two new, genetically identical double helices. At cell division, one double helix goes to one cell, and one goes to the other.

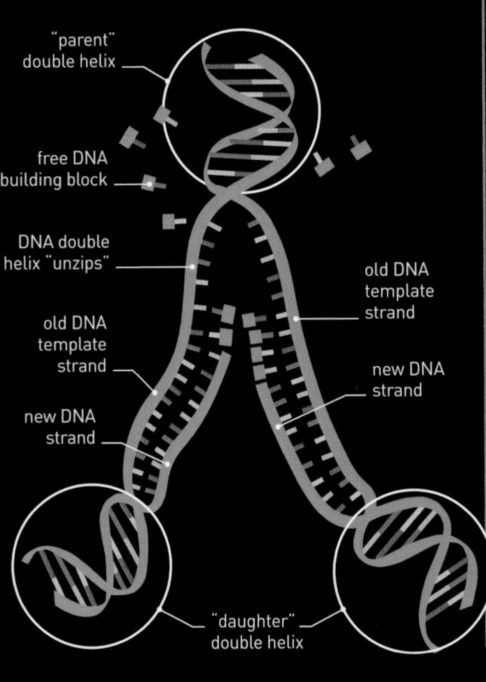

"parent" double helix
free DNA building block
DNA double helix "unzips"
old DNA template strand
old DNA template strand
new DNA strand
new DNA strand
"daughter" double helix

MAKING PROTEINS

A gene is a section of DNA containing instructions to assemble a protein molecule to carry out a specific task, such as making a pigment. In this way, genes determine an organism's characteristics. Before a protein is assembled, the gene's base sequence must be "copied" in the cell nucleus, a process called transcription. This copy is then sent to the cytoplasm. In another process, called translation, the base sequence information is used to assemble the protein.

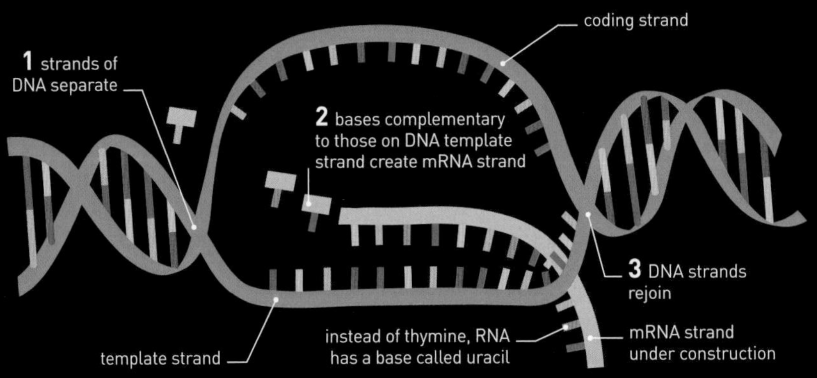

1 strands of DNA separate

coding strand

2 bases complementary to those on DNA template strand create mRNA strand

3 DNA strands rejoin

template strand

instead of thymine, RNA has a base called uracil

mRNA strand under construction

TRANSCRIPTION

Inside the nucleus, part of a DNA double helix unravels to expose the coding region of a gene, ready for "copying." This involves making a strand of nucleic acid called RNA (ribonucleic acid) by bonding together free RNA building blocks.

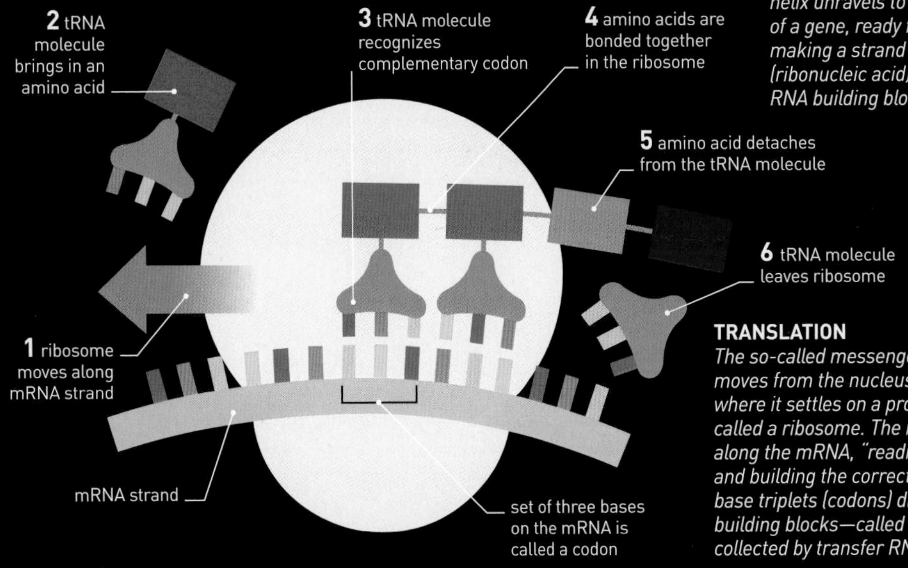

2 tRNA molecule brings in an amino acid

3 tRNA molecule recognizes complementary codon

4 amino acids are bonded together in the ribosome

5 amino acid detaches from the tRNA molecule

6 tRNA molecule leaves ribosome

1 ribosome moves along mRNA strand

mRNA strand

set of three bases on the mRNA is called a codon

TRANSLATION

The so-called messenger RNA (mRNA) moves from the nucleus to the cytoplasm, where it settles on a protein-making granule called a ribosome. The ribosome moves along the mRNA, "reading" its base sequence and building the correct protein. Specific base triplets (codons) dictate specific protein building blocks—called amino acids—collected by transfer RNA (tRNA) molecules.

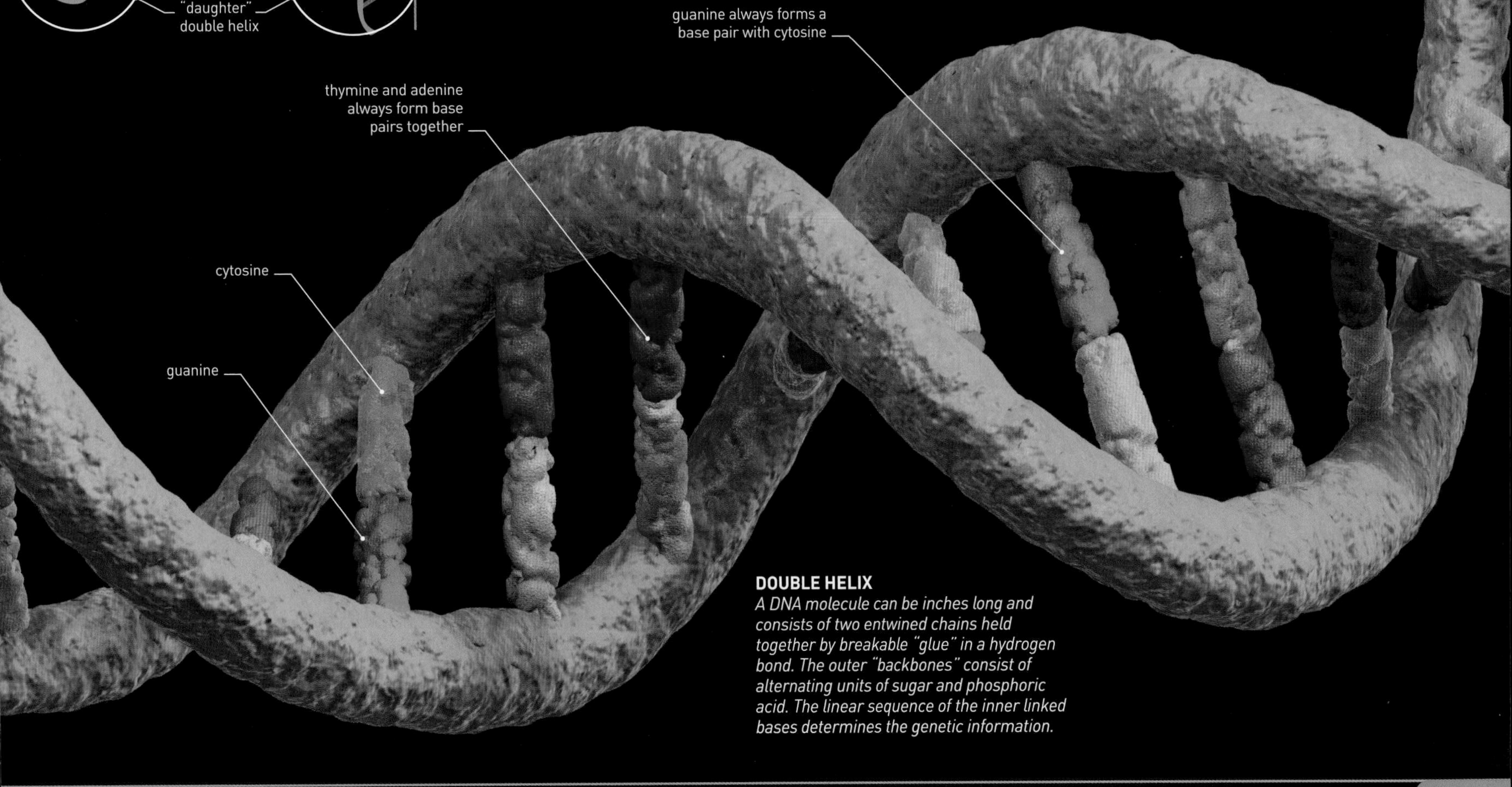

guanine always forms a base pair with cytosine

thymine and adenine always form base pairs together

cytosine

guanine

DOUBLE HELIX

A DNA molecule can be inches long and consists of two entwined chains held together by breakable "glue" in a hydrogen bond. The outer "backbones" consist of alternating units of sugar and phosphoric acid. The linear sequence of the inner linked bases determines the genetic information.

1955

The first radiometric dating of rock layers, such as these in Bryce Canyon, USA, put estimates of Earth's age at only about 2 billion years, far below the actual age of the planet.

IN 1955, AMERICAN GEOCHEMIST CLAIR PATTERSON (1922–95) studied the atoms of rocks to estimate **the age of Earth**. He focused his analysis on meteorites, which are considered remnants of an age when the Solar System—and therefore Earth—was first made. Patterson **isolated lead from samples of rock** and studied the relative proportions of lead isotopes—variants of the metal. **Radioactive atoms**, such as those of uranium, decay at a known rate (see p.267), and thereby provide information about the age of samples (see below). Some of these atoms decay into specific lead isotopes,

4.55 BILLION YEARS THE AGE OF EARTH

which build up over time. Patterson calculated Earth's age as 4.55 billion years—older, and more accurate than prevailing estimates. This changed the way scientists viewed our world.

British physicist **Louis Essen** (1908–97), working at the UK's National Physics Laboratory, designed the **first atomic clock**. Its time-keeping was based on radiation emitted by atoms of cesium metal. It would make **time measurement more precise** than ever—the new atomic clock would gain or lose only a second in 300 years. Modern atomic clocks are accurate to 1 second in 6 million years.

In December, in the laboratory of Swedish geneticist Albert Levan at the Institute of Genetics in Lund, Sweden, visiting Javanese-born American biologist **Joe Hin Tjio** (1919–2001) made a discovery that corrected a 50-year error in the field of **genetics**: that human cells contained 48 chromosomes. He showed that there are in fact **46 chromosomes in a normal human body cell**. Tjio used an improved microscopic technique to squash cells into single layers, rather than relying on slicing thin sections of tissue. He also used a technique to spread out the tiny chromosomes in a sample, so that they separated easily and clearly, without fragmenting. Tjio continued his career in the US, where he became a significant figure in developing the new branch of biology called **cytogenetics** (cell genetics).

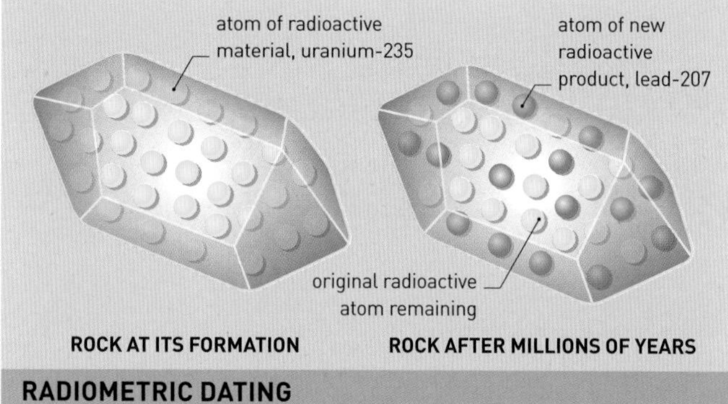

atom of radioactive material, uranium-235
atom of new radioactive product, lead-207
original radioactive atom remaining
ROCK AT ITS FORMATION
ROCK AFTER MILLIONS OF YEARS

RADIOMETRIC DATING

Radiometric dating is a technique used to date rocks, minerals, or fossils using natural rates of radioactive decay. Its origins are in the work done by physicists in the early 1900s. Radioactive material decays at a known rate. By working out the ratio of a radioactive material (for example, uranium) to its decay products (lead), the age of the rock can be calculated.

1956

2,244 MILES THE LENGTH OF TRANSATLANTIC TELEPHONE CABLE

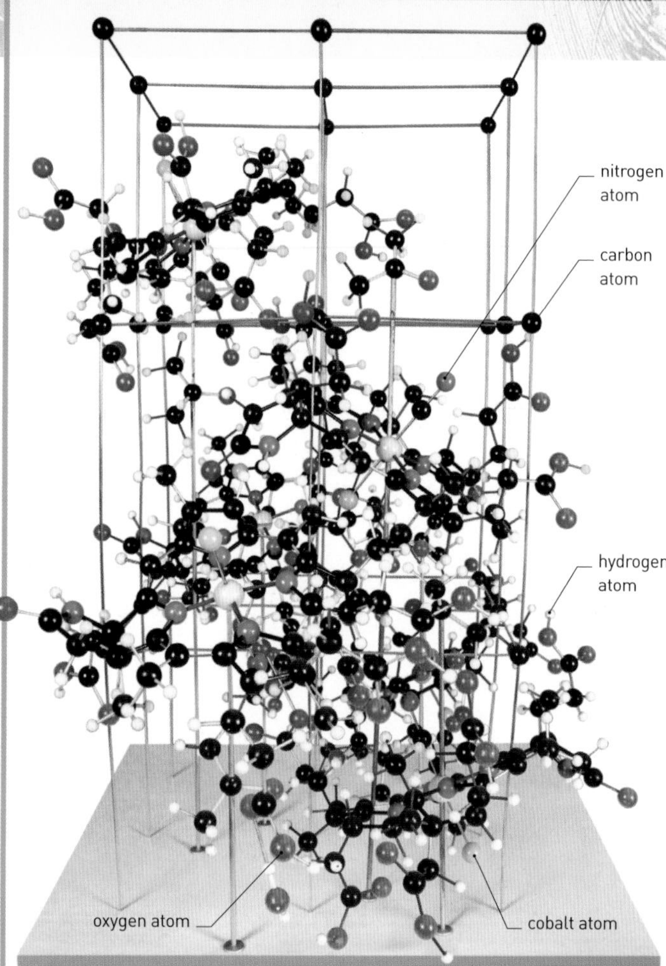

nitrogen atom
carbon atom
hydrogen atom
oxygen atom
cobalt atom

Structure of vitamin B12
A model of a vitamin B12 molecule reveals the complexity of it structure. This was first revealed by Dorothy Hodgkin, through a method called X-ray crystallography (see 1945).

A NEW DISEASE affecting a fishing community in Minamata, Japan, was described in May 1956. Called **Minamata disease**, it was characterized by progressive paralysis reminiscent of many nervous system disorders. It was not until November that the disease was traced to **heavy metal poisoning**. In the years that followed, the disease was linked specifically to mercury that had been leaking into the

American geochemist Clair Patterson uses radiometric dating to recalculate **age of Earth**

April 12 US Food and Drug Administration **approves Salk's polio vaccine**

August 24 British physicist Louis Essen launches the first workable **atomic clock**

December 22 Javanese-born biologist Joe Hin Tjio discovers that **human cells contain 46, not 48, chromosomes**

May 1 Minamata disease described

July 14 Vitamin **B12 structure** is published

September 25 Launch of **transatlantic submarine telephone** cable

The transatlantic telephone cable was laid from a ship in coiled batches.

"... THE FIRST **ARTIFICIAL EARTH SATELLITE** HAS BEEN BUILT."

Telegraph Agency of the Soviet Union (TASS), published in *Pravda* newspaper, October 5, 1957

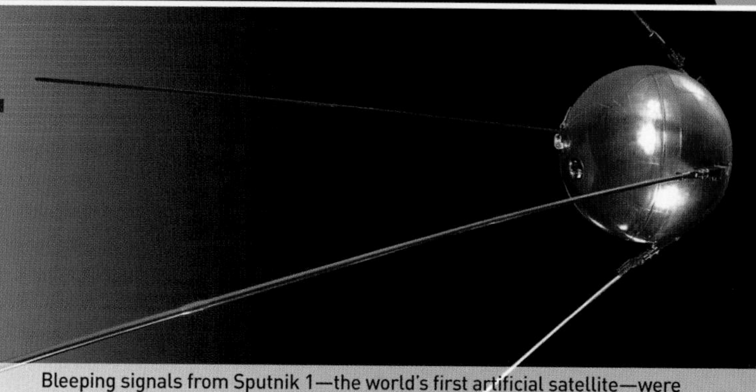
Bleeping signals from Sputnik 1—the world's first artificial satellite—were heard by people on their radio sets around the world.

fishing waters from a chemical factory. The mercury had accumulated in fish and shellfish, and had poisoned the people who ate them. This incident was one of the first well-documented cases of the effects of toxic chemicals polluting food chains.

In July, British chemist **Dorothy Hodgkin** (see 1945) published a study of the **structure of vitamin B12**—a vitamin needed to prevent pernicious anemia. Using X-ray crystallography, she found that the vitamin contained a ringlike structure, porphyrin, that surrounded a central atom of the element cobalt.

The **world's first underwater telephone cable** became operational on September 25. It ran between the US and Europe under the Atlantic Ocean. In the first 24 hours of operation, the service hosted 588 London to US telephone calls, which were clearer than any previous transatlantic communication.

In September, IBM launched the 305 RAMAC (Random Access Memory Accounting Machine)—the **first computer with a hard disk drive** and random access memory. RAMAC weighed over a ton and stored 5 MB of data on a stack of 50 large disks.

22
THE **NUMBER OF DAYS** THAT **SPUTNIK 1** TRANSMITTED **SIGNALS** BACK TO EARTH

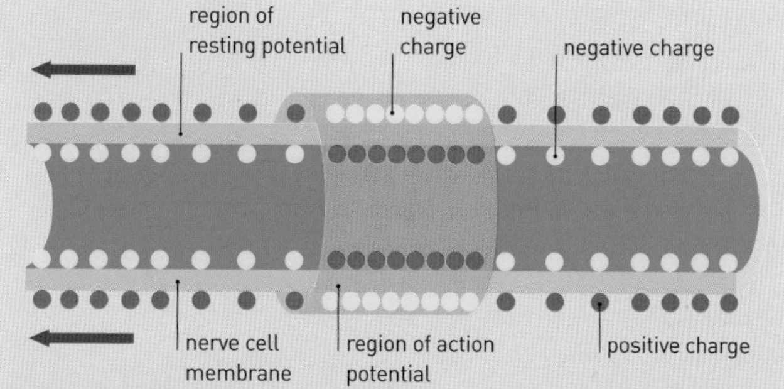
region of resting potential
negative charge
negative charge
nerve cell membrane
region of action potential
positive charge

CREATING A NERVE IMPULSE

Membranes around nerve fibres are electrically charged because they contain protein "pumps" that push ions (charged particles) of sodium out and pull potassium ions in. This makes positive charge accumulate on the outer surface: a so-called resting potential. When stimulated, protein channels open in the membrane, making positive charge leak in. This reverses the resting potential to create an action potential. The region of action potential fires down the nerve fiber membrane as a nerve impulse.

THE HAMILTON ELECTRIC 500, launched in January, was the **first ever wristwatch that did not need winding**. Although its battery life was relatively short, the innovation proved to be very popular.

Danish chemist **Jens Skou** (b.1918) published the basis of how the **nervous system** works. He discovered molecules in the cell membranes of crab nerves that pump ions to "charge up" membrane surfaces, using up cellular energy in the process. This makes the molecules excitable so that they can carry an impulse (see panel, below) when stimulated.

On October 4, Russia launched Sputnik 1—the first artificial satellite to orbit Earth.

At the start of the nuclear arms race, and prompted by US President Eisenhower's "atoms for peace" speech in 1953, the United Nations set up the **International Atomic Energy Agency** (IAEA) in July 1957 to control and develop **atomic energy**. Thirteen years later, the IAEA would oversee the Treaty on Non-Proliferation of Nuclear Weapons.

In the same year, aviation company **Boeing** launched **the first commercial jet airliner**, the 707. It marked the dawn of a new age of air travel powered by turbine engines that made aeroplanes fly higher and faster than before.

In November, American physicist **Gordon Gould** (1920–2005) suggested a method of amplifying light into an intense beam and coined the term for it as **laser** (Light Amplification by Stimulated Emission of Radiation). An operational laser, however, would not be produced until 1960.

British scientist **James Lovelock** (b.1919) invented the **electron capture detector (ECD)** as an ultrasensitive way of analyzing gas mixtures. The device emits electrons (negatively charged particles), which are then absorbed by certain gases to produce a signal. It helped environmentalists detect tiny quantities of atmospheric pollutants, such as the pesticide DDT and chlorofluorocarbons (CFCs)—ozone-depleting (see 1973) compounds used in refrigeration and as propellants in aerosol cans.

James Lovelock holds ECD
The electron capture detector (ECD) picked up the tiniest amounts of electron-binding chemicals in the atmosphere, such as the chlorine in pesticides and other compouds.

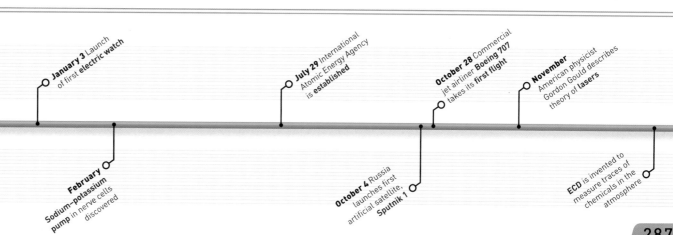

September IBM launches first computer with a **hard disk drive**

November 4 Minamata disease in Japan traced to **heavy metal poisoning**

January 3 Launch of first **electric watch**

February **Sodium–potassium pump** in nerve cells discovered

July 29 International Atomic Energy Agency is **established**

October 4 Russia launches first artificial satellite, **Sputnik 1**

October 28 Commercial jet airliner **Boeing 707** takes its **first flight**

November American physicist Gordon Gould describes theory of **lasers**

ECD is invented to measure traces of chemicals in the atmosphere

1958

New satellite technology helped explain how Earth's magnetic field channels energetic solar particles into a collision course with the atmosphere to produce the aurora borealis, or Northern Lights.

THE SOVIET UNION'S SPUTNIK 1—the first artificial satellite—burned up on January 4, after spending three months in orbit. The US joined the space race with the **launch of Explorer 1** later that month. For the month or so of their battery-powered lives, these satellites sent back important information about airspace—a term that refers to the realm of the atmosphere and the accessible space above it. Sputnik 1 measured the **density of the upper atmosphere**, and Explorer 1 found how Earth's **magnetosphere** (see panel, below) could **deflect harmful radiation** before it reached the ground.

In California, **Matthew Meselson** (b.1930) and **Franklin Stahl** (b.1929) were unlocking the secrets of DNA in what would come to be described as "the most beautiful experiment in biology." They traced what happened to components of DNA when it replicated, and found that **a double helix molecule** unravelled into two strands, each of which was **a genetic "template" for making more DNA**. After replication, each new double helix contained one old strand and one new one. This method of semiconservative replication had been proposed by Watson and Crick in 1953, when they developed the double helix model of DNA.

In the same year, a practical expression of this replication came with the work of British botanists **Frederick Steward** (1904–93) and **John Gurdon** (b.1933). Both teased cells from mature organisms—Gurdon from a tadpole and Steward from a carrot—and grew clones from them as new genetically identical plants. It was the **first cloning using material** from differentiated **"body" tissues**.

The year also saw a breakthrough in electronic engineering. American electrical engineers **Jack Kilby** (1923–2005) and **Robert Noyce** (1927–90) simultaneously came up with the idea of condensing all the components of an electronic circuit into a single plate of silicon, thereby inventing the **microchip**.

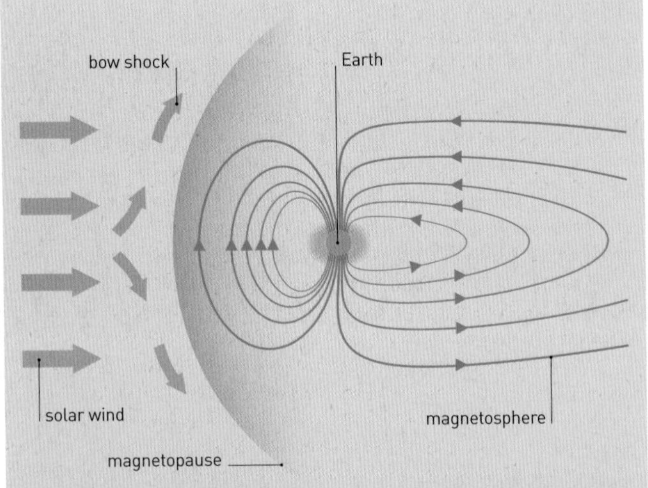
bow shock
Earth
solar wind
magnetosphere
magnetopause

THE MAGNETOSPHERE

Earth is enveloped by the magnetosphere, a magnetic "blanket" that results from magnetism deep inside the planet. The magnetosphere deflects harmful high-energy particles from the Sun through a bow shock, or shock wave created at the magnetopause—an abrupt boundary between incoming solar wind and the magnetosphere. Without the bow shock, the solar wind would destroy life on Earth.

1959

Hidden from the gaze of astronomers since antiquity, the far side of the Moon was finally revealed when USSR's Luna 3 probe photographed it from orbit.

TWO MONKEYS, ABLE AND BAKER, went up in the US missile **Jupiter AM-18**, and became the first primates to survive a space flight. Meanwhile, the **USSR** launched three **lunar probes**—Luna 1 and 3 achieved "flyby." Luna 3 also took the first photograph of the Moon's far side.

In Cambridge University, UK, molecular biologist **Max Perutz** (1914–2002) was studying **hemoglobin**—the red oxygen-carrying pigment of blood—using techniques that had cracked the structure of DNA. He found that it had four protein chains, each with oxygen-grabbing iron.

Australopithecine skull
Dated at 1.75 million years old, this skull belongs to an australopithecine, also known as "nutcracker man" because of his large cheek teeth.

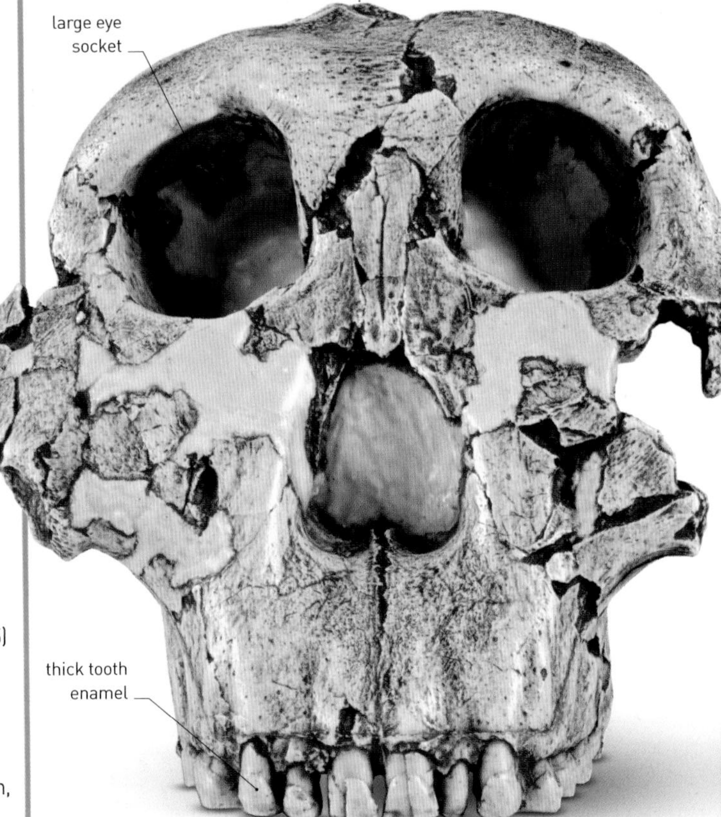
large eye socket
thick tooth enamel

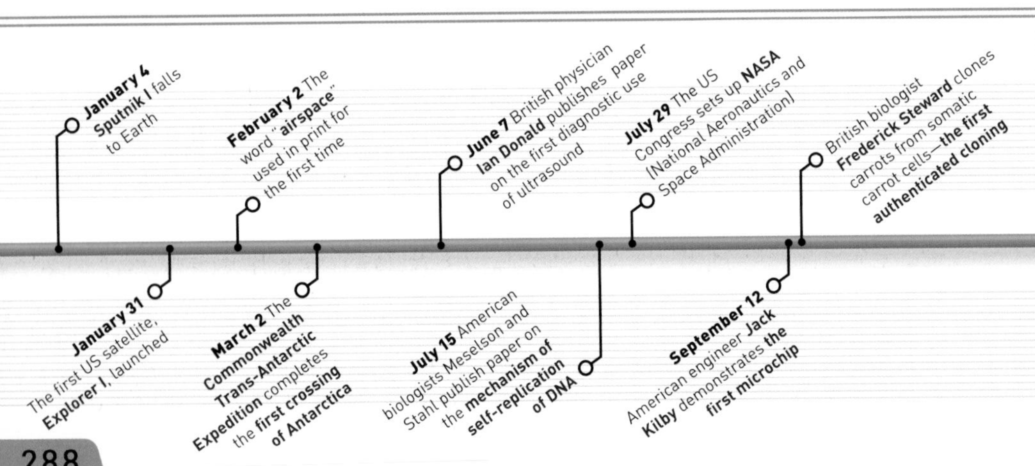
January 4 Sputnik I falls to Earth
January 31 The first US satellite, **Explorer I**, launched
February 2 The word "**airspace**" used in print for the first time
March 2 The **Commonwealth Trans-Antarctic Expedition** completes the **first crossing of Antarctica**
June 7 British physician **Ian Donald** publishes paper on the first diagnostic use of ultrasound
July 15 American biologists Meselson and Stahl publish paper on the **mechanism of self-replication of DNA**
July 29 The US Congress sets up **NASA** (National Aeronautics and Space Administration)
September 12 American engineer **Jack Kilby** demonstrates **the first microchip**
British biologist **Frederick Steward** clones carrots from somatic carrot cells—**the first authenticated cloning**

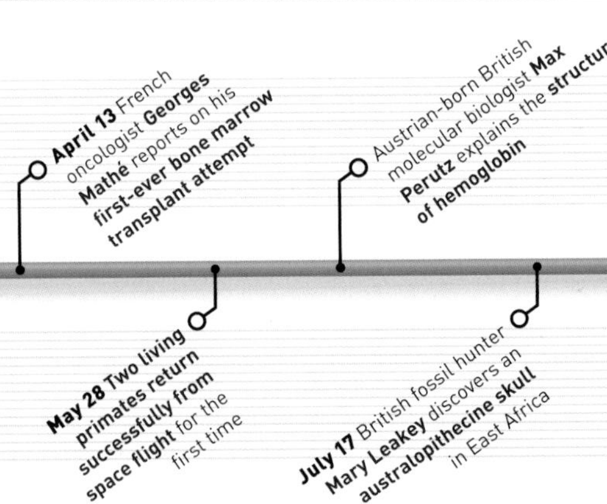
April 13 French oncologist **Georges Mathé** reports on his **first-ever bone marrow transplant attempt**
May 28 Two living **primates return successfully from space flight** for the first time
Austrian-born British molecular biologist **Max Perutz** explains the **structure of hemoglobin**
July 17 British fossil hunter **Mary Leakey** discovers an **australopithecine skull** in East Africa

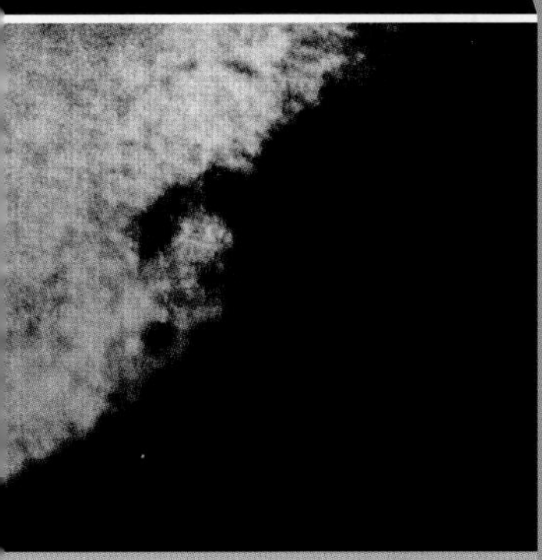

35,797 FEET

THE **DEPTH** REACHED BY THE ***TRIESTE***

The submersible *Trieste* was designed to withstand the immense pressures of the deepest part of the Marianas Trench. Until 2012, it was the only manned vessel to get there.

LOUIS LEAKEY (1903–72)

British anthropologist Louis Leakey was instrumental in advancing the understanding of human evolution. Together with his wife, Mary, he spent much of his career studying fossils in East Africa. He showed that humankind originated in Africa, and later in his career helped inspire the work of primatologists such as Jane Goodall and Dian Fossey.

Anthropologists **Louis and Mary Leakey** had been excavating prehistoric stone tools at Olduvai Gorge in East Africa for two decades. In July 1959, Mary found a prehistoric skull. It belonged to an **australopithecine**, an ancestor of modern humans who was later thought to be the **first "apeman" to use stone tools**.

JUST AFTER MIDDAY ON JANUARY 23, the US Navy submersible ***Trieste*** touched down at **the deepest spot of the world's oceans: Challenger Deep** in the Marianas Trench of the western Pacific Ocean. The submersible's onboard team of **Don Walsh** (b.1931) and **Jacques Piccard** (1922–2008) spent 20 minutes there, and saw that some animals were adapted to survive even at these great depths.

Harry Hess (1906–69), a geologist with the US Navy, had studied the ocean depths during World War II. In 1960, he suggested that **the entire seafloor was moving**. He later stated that the molten magma spewing out of Earth's crust from underwater ridges was cooling, expanding, and pushing the oceanic plates on either side. Today, Hess's theory is accepted by geologists. It is believed that as new mantle forms at a ridge, old mantle plunges back into Earth elsewhere. This process is responsible for **continental drift**, a theory first suggested by Alfred Wegener (see 1914–15) nearly 50 years before.

In April, NASA launched the **first successful weather satellite, TIROS-1** (Television Infrared Observation Satellite Program-1). For 78 days, television cameras aboard the satellite took thousands of images of cloud cover and other aspects of atmospheric conditions from aerospace.

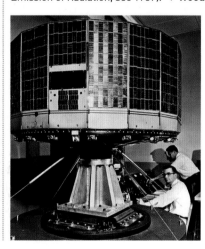

In August, American physicist **Theodore Maiman** (1927–2007) demonstrated a new way of producing a concentrated "pencil-beam" of light known as a **laser** (Light Amplification by Stimulated Emission of Radiation, see 1957). His method involved using a rod of synthetic ruby to produce a series of laser pulses. This technology was later modified to produce a continuous beam that today has applications ranging from eye surgery to compact disc players and supermarket scanners.

American chemist **Robert Woodward** (1917–79) had spent the last decade studying the chemical structures of complex biological substances, such as cholesterol and quinine. He showed that the rules of structural chemistry could be used to produce these substances in the laboratory. In 1960, he succeeded in **artificially creating chlorophyll II**—one of the main components of the green plant pigment that traps light energy for photosynthesis.

First weather satellite
TIROS-1 carried television cameras and photographed Earth's weather patterns from a height of at least 435 miles (700 km).

In October, the **11th General Conference on Weights and Measures** published a series of unit standards. Known as **Le Système International d'Unités** (SI units), these were adopted by scientists and technologists.

Also in October, British surgeon **Michael Woodruff** (1911–2001) performed the **UK's first kidney transplant operation**. It was performed between identical twins to minimize the risk of rejection. Both donor and recipient survived the operation and went on to live many years.

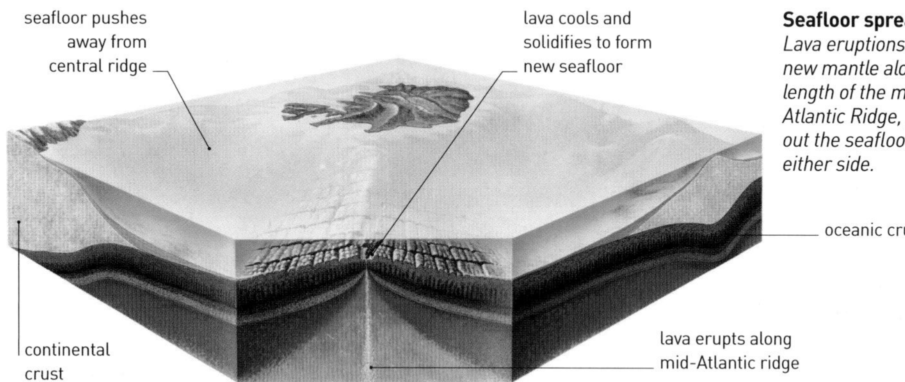

seafloor pushes away from central ridge
lava cools and solidifies to form new seafloor
oceanic crust
continental crust
lava erupts along mid-Atlantic ridge

Seafloor spreading
Lava eruptions create new mantle along the length of the mid-Atlantic Ridge, pushing out the seafloor on either side.

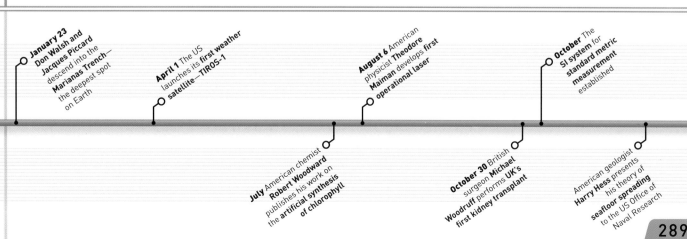

August Chinese-American biologist **Min Chueh Chang** successfully performs **in vitro fertilization** in **rabbits**

October 7 The Soviet **Lunar 3** probe **photographs the far side of the Moon**

January 23 **Don Walsh and Jacques Piccard** descend into the **Marianas Trench**—the deepest spot on Earth

April 1 The US launches its **first weather satellite—TIROS-1**

July American chemist **Robert Woodward** publishes his work on the **artificial synthesis of chlorophyll**

August 6 American physicist **Theodore Maiman** develops **first operational laser**

October 30 British surgeon **Michael Woodruff** performs **UK's first kidney transplant**

October The **SI system** for **standard metric measurement** established

American geologist **Harry Hess** presents his theory of **seafloor spreading** to the US Office of Naval Research

1961

1962

108 MINUTES THE DURATION OF **YURI GAGARIN'S** SPACE FLIGHT

Soviet pilot Yuri Gagarin was selected from 20 candidates to be the first man in space.

DDT was developed as a contact poison to control insect pests and was claimed to be harmless to people and the environment.

IN FEBRUARY, PHYSICISTS at the University of California Berkeley succeeded in producing **atoms of a new heavy element** by bombarding a sample of the element californium with nuclei of boron atoms. They called the element **lawrencium**, after American physicist **Ernest Lawrence**, the inventor of the cyclotron—a particle accelerator. With an atomic number of 103, lawrencium was the last—and the heaviest—of a group of radioactive metals called actinides.

On April 12, Soviet pilot **Yuri Gagarin** (1934–68) became the **first cosmonaut** when he **traveled to outer space** aboard Vostok 1. He orbited Earth once before returning safely and was awarded his country's highest honor—the title of "Hero of the Soviet Union." He went on to train new cosmonauts in Russia.

The USSR launched its **Venera program** to gather **information about Venus**. The first probe, Venera 1, is thought to have passed within 62,137 miles (100,000 km) of Venus. It was the first man-made object to fly by another planet.

In April, prompted by the plight of wildlife in Africa and elsewhere, a team including biologist **Julian Huxley** (1887–1975) and ornithologist **Peter Scott** (1909–89) proposed the establishment of an international organization for wildlife conservation. They created the International Secretariat of the **World Wildlife Fund (WWF)**—now known as the World Wide Fund for Nature—in Switzerland. The WWF went on to set up offices worldwide and harnessed the expertise of scientists to protect endangered species.

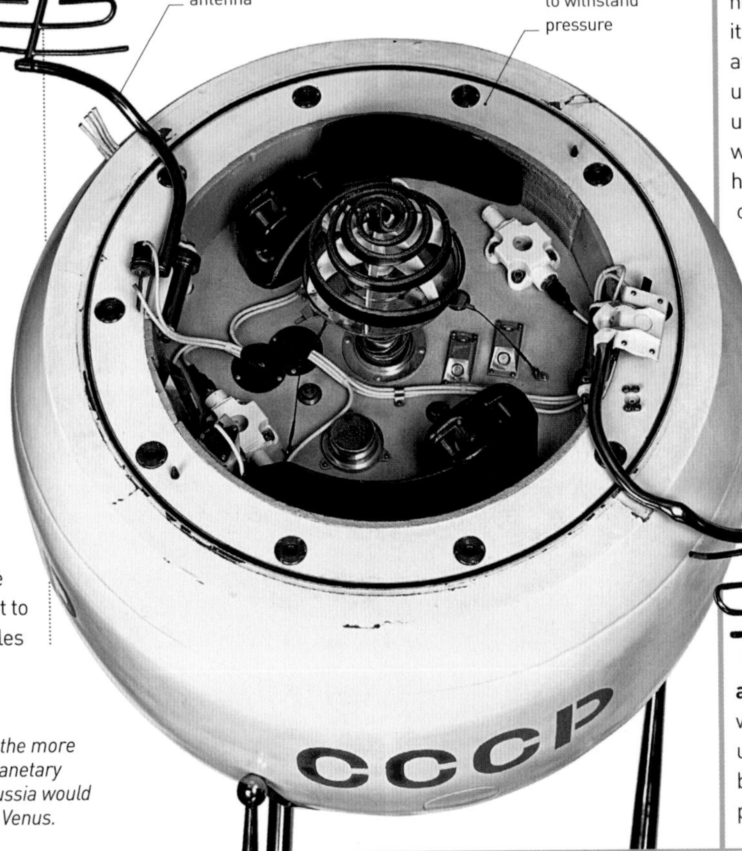

antenna
toughened body to withstand pressure
CCCP

Soviet Venera space probe
The Venera probes were among the more sophisticated of the first interplanetary space probes. Over the years, Russia would succeed in landing 10 probes on Venus.

IN JUNE, *The New Yorker* magazine began to serialize a book by American marine biologist **Rachel Carson**, titled ***Silent Spring***. It proclaimed that human activity, particularly the **use of pesticides** such as **DDT** (dichlorodiphenyltrichloroethane), threatened the environment with damage and destruction. Carson explained how the intensive techniques used to satisfy humankind's demand for food and its drive to eliminate pests were affecting the environment on an unprecedented scale. Widespread use of pesticides was harming wildlife, and would ultimately harm humans too. Carson was concerned about DDT. Developed as a contact poison to control the spread of insect-borne diseases during World War II, DDT was later adopted as an agricultural pesticide. However, it accumulated in food chains, killing wildlife. Carson's book, coming a year after the inauguration of the WWF, served as an alarm call. It created a **new environmental awareness**, especially in the US, where environmental concerns ultimately precipitated a national ban on DDT and other highly potent pesticides.

In July, the multinational **communications satellite Telstar** was sent into space aboard a NASA rocket. The first of its kind, Telstar made it possible for **live television signals** to be transmitted across the Atlantic. The first pictures were seen on television screens on July 10. Telstar became the prototype for later, more efficient, models.

RACHEL CARSON
(1907–64)

Trained in marine biology, Rachel Carson achieved acclaim as a writer of popular natural history books before she became famous for her book, *Silent Spring*. Her work led to the establishment of the US Environmental Protection Agency. She was posthumously awarded the Presidential Medal of Freedom in 1980.

February 14 **Lawrencium** is synthesized

April 12 Soviet pilot **Yuri Gagarin**, the first human in space, **orbits Earth**

April 29 World Wildlife Fund **(WWF)** founded

May 19 **Venera 1** passes **close to Venus**

May 25 The US Congress announces **launch of Apollo** program to land a man on the Moon

November 9 First edition of **International Code of Zoological Nomenclature** is designed to standardize naming of animal species

February 20 **John Glenn** becomes first **American to orbit Earth**

June The New Yorker begins **serialization of** Rachel Carson's ***Silent Spring***

July 10 **Telstar**, the first active communications satellite, is launched

> **“** BUT **MAN** IS PART OF NATURE, AND HIS **WAR AGAINST NATURE** IS INEVITABLY A **WAR AGAINST HIMSELF. ”**

Rachel Carson, marine biologist, from *Silent Spring*, 1962

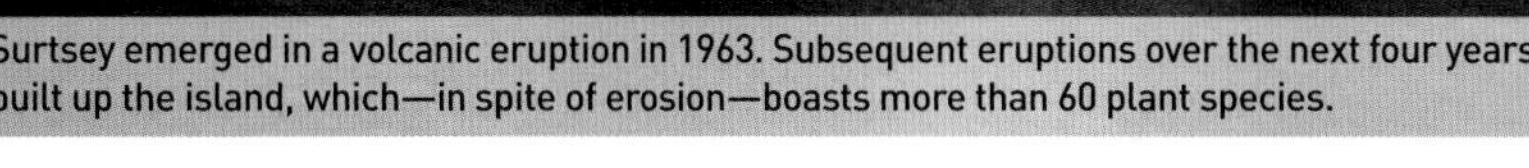

Surtsey emerged in a volcanic eruption in 1963. Subsequent eruptions over the next four years built up the island, which—in spite of erosion—boasts more than 60 plant species.

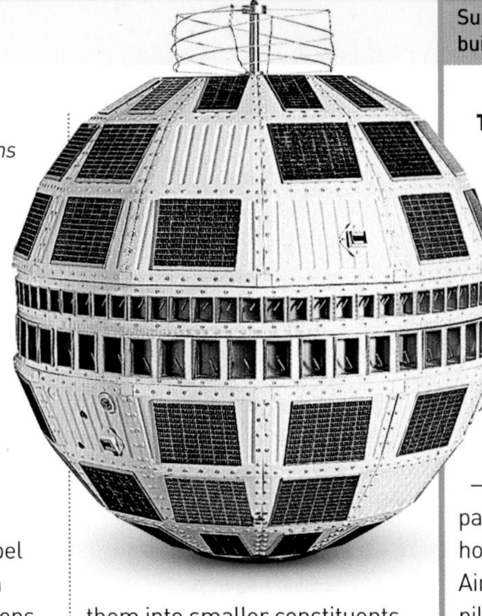

Telstar
The world's first communications satellite, Telstar transmitted its signals intermittently from 1962 to early 1963. Although it has ceased to communicate, it remains in orbit to this day.

American biologist **Gerald Edelman** (b.1929) and English biochemist **Rodney Porter** (1917–85) independently made a breakthrough that would eventually win them the Nobel Prize. They were working on **antibodies**—natural secretions that help the immune system fight infection by targeting and "neutralizing" harmful foreign particles called antigens. Edelman and Porter analyzed antibodies by chemically splitting them into smaller constituents. They found that each Y-shaped **antibody molecule was made up of protein** chains. Their work helped unravel the chemical structure of antibodies. Later work would show how the human body produces different kinds of antibodies to target different kinds of antigens, so that the immune system can attack specific infections.

The structure of antibodies
An antibody molecule is made up of two "light" and two "heavy" protein chains, held together by tight bonds in a Y-shaped structure.

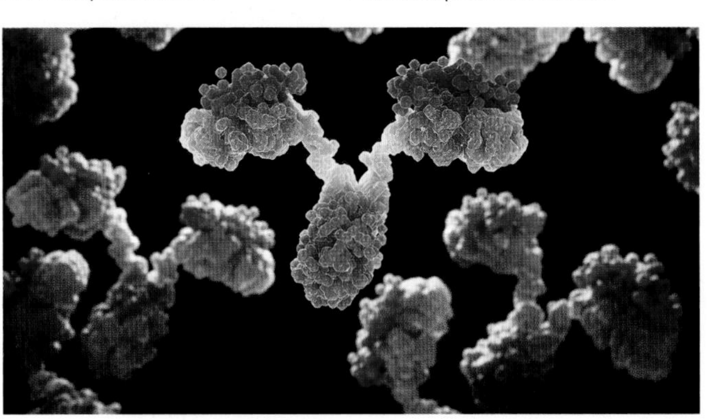

THE BELL COMPANY developed the **first push-button telephone for public use**. Scientists also saw breakthroughs in science on a grander and more fundamental scale.

Soviet cosmonaut **Valentina Tereshkova** (b.1937) became the **first woman**—and the first civilian—**to fly into space**. An amateur parachutist, she became an honorary inductee into the Soviet Air Force before she trained to pilot Vostok 6. The mission helped Russian scientists understand how the female body reacted to time in space.

Scientists also reached a fundamental turning point in **understanding the nature of matter**. Experiments conducted in the 1950s had shown that conventional subatomic particles, such as protons and neutrons, could explode to create even smaller fundamental entities. However, no one could be sure what these were. In 1963, American physicists **Murray Gell-Mann** (b.1929) and **George Zweig** (b.1937) independently proposed a **"quark model" of matter**, suggesting that a variety of different entities called quarks come together in combinations to make subatomic particles. Over the next few years, experiments in particle physics indicated that this quark model was essentially correct.

American mathematician **Edward Lorenz** (1917–2008) was revising the way we look at systems, such as weather, on a much larger scale. He suggested that a small, seemingly insignificant change in one place can have major repercussions in the long term. By alluding to the effect of tiny flapping wings leading to hurricane-scale devastation, he came up with an evocative name for his idea: **the butterfly effect**. With this, he laid the foundation for **chaos theory**.

48

THE NUMBER OF TIMES **TERESHKOVA ORBITED EARTH**

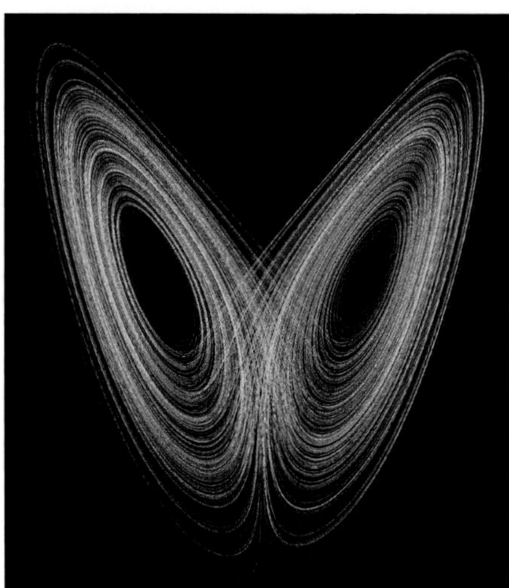

The butterfly effect
A "Lorenz attractor" is a butterfly-shaped graphical plot based on mathematical equations that describe a chaotic system.

In November, change on a massive scale was seen in the geographic realm, when an island was born near Iceland. An underwater volcano on the mid-Atlantic ridge erupted to push **Surtsey** out of the water. This gave scientists a rare chance to **study Earth's active geology** first-hand. In subsequent years, scientists were able to see how life colonized the island in a process of biological succession to form a new ecosystem.

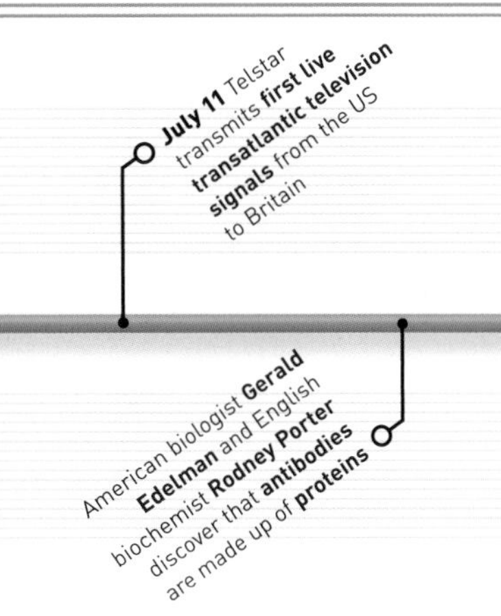

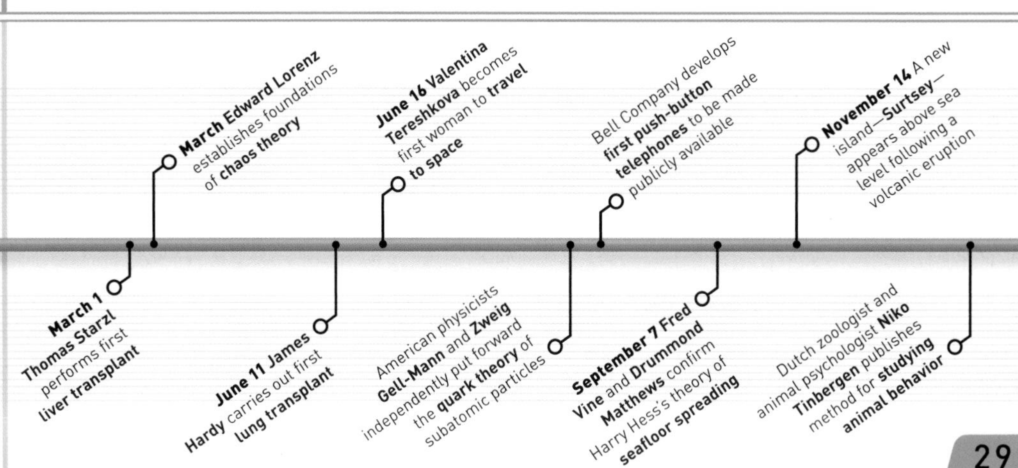

July 11 Telstar transmits **first live transatlantic television signals** from the US to Britain

American biologist **Gerald Edelman** and English biochemist **Rodney Porter** discover that **antibodies** are made up of **proteins**

British doctor **James Black** invents **beta-blockers**—a class of synthetic drugs used to treat heart disease

March 1 Thomas Starzl performs first **liver transplant**

March Edward Lorenz establishes foundations of **chaos theory**

June 11 James Hardy carries out first **lung transplant**

June 16 Valentina Tereshkova becomes first woman to **travel to space**

American physicists **Gell-Mann** and **Zweig** independently put forward the **quark theory** of subatomic particles

Bell Company develops **first push-button telephones** to be made publicly available

September 7 Fred Vine and **Drummond Matthews** confirm Harry Hess's theory of **seafloor spreading**

November 14 A new island—**Surtsey**—appears above sea level following a volcanic eruption

Dutch zoologist and animal psychologist **Niko Tinbergen** publishes method for **studying animal behavior**

THE STORY OF OCEANOGRAPHY

ONE OF THE WORLD'S MOST UNEXPLORED REALMS HAS GRADUALLY BEEN REVEALING ITS SECRETS

For a long time, oceans have been the least understood features on Earth. However, knowledge of marine life and the topography of the ocean floor has gradually accumulated and, in recent years, new exploration techniques have led to several discoveries.

The earliest records of sea exploration date back 3,000 years to the Phoenicians, who made charts to navigate and used weights to plumb the oceans' depths. Ancient Greek philosopher Aristotle was one of the first to speculate about marine life, and other Greeks developed instruments to help ships locate their position when far from shore.

However, the open oceans remained unexplored by Westerners until the 1400s, when Christopher Columbus sailed westward into the Atlantic in the hope of finding land on the far side. This paved the way for further voyages of exploration, such as Ferdinand Magellan's circumnavigation, which finally revealed the extent of the world's oceans and allowed mapmakers to chart their shapes.

Scientific attempts to study beneath the surface of the oceans began in the 19th century. Early surveys were conducted using sounding chains and sample nets. The introduction of sonar after World War II helped in mapping the ocean floor. Recently, improved sonar, satellite techniques, and an array of submersibles have helped add to our knowledge of marine life and ocean currents as well as the oceans' geography.

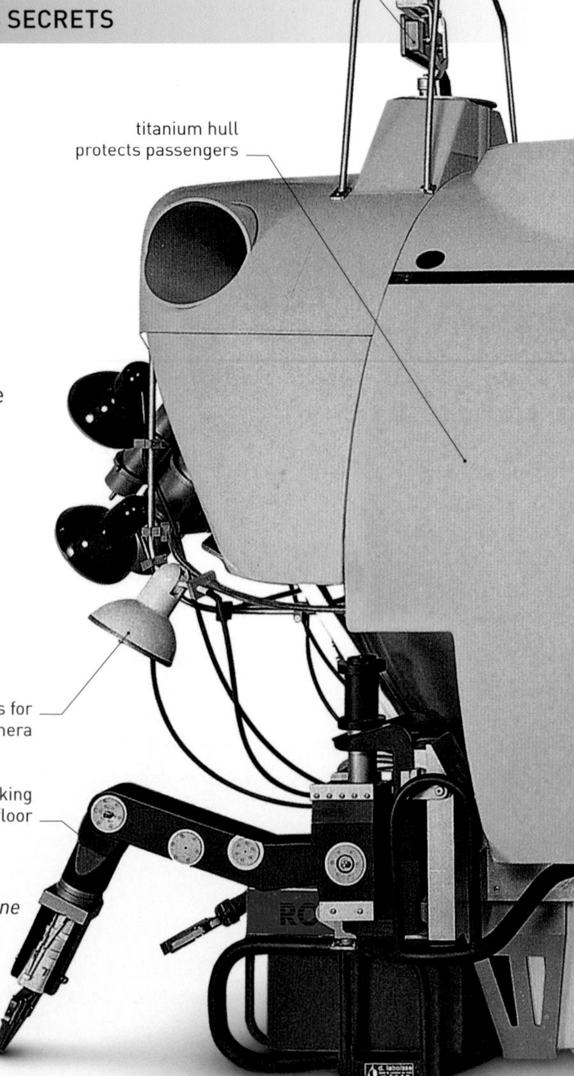
sonar equipment
titanium hull protects passengers
lights for video camera
manipulator arm for picking objects off seafloor

Nautile
French miniature submarine Nautile *is only 26 ft (8 m) long. However, its tough hull allows it to reach a depth of 3¾ miles (6 km), and external robotic arms, video cameras, and floodlights enable detailed exploration.*

SONAR

The first maps of the ocean floor were produced by sonar, a technology developed during World War II to detect submarines by picking up reflections of sound waves from underwater objects. It can also be used to detect schools of fish. Nowadays, new systems, such as side-scan sonar, are being used together with GPS to survey vast areas quickly.

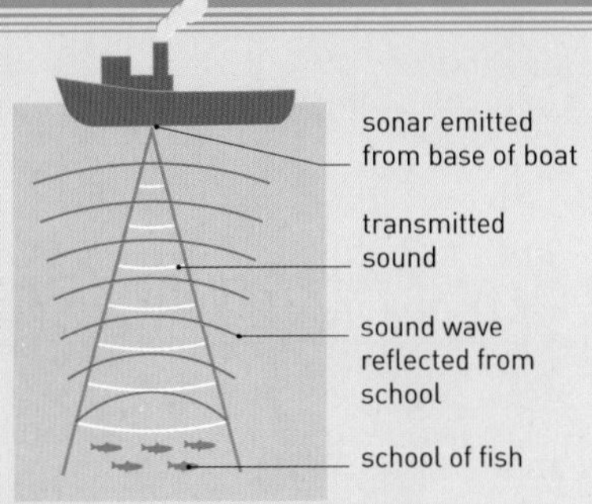
sonar emitted from base of boat
transmitted sound
sound wave reflected from school
school of fish

1200–250 BCE
Phoenician traders
The first seafarers, the Phoenicians plumb the seabed to find channels. They develop the first coins to facilitate trade.

Ancient Phoenician coin

500–200 BCE
Greek marine science
Aristotle identifies many marine species, such as crustaceans, mollusks, echinoderms, and fish.

Ancient Grecian bowl showing a sailing boat

c.80 BCE
Antikythera mechanism
The Greeks develop instruments, such as the clockwork Antikythera mechanism, to plot the movement of the heavens and to navigate at sea.

1492
Columbus' voyage
Italian navigator Christopher Columbus's voyage to the Americas shows that it is possible to cross the Atlantic and even sail around the world.

Christopher Columbus

1519–22
Strait of Magellan
Portuguese explorer Ferdinand Magellan is the first to sail from the Atlantic to the Pacific Ocean, and discovers the Strait of Magellan on the way.

Map of the Strait of Magellan

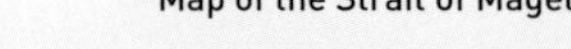

1769–71
Captain Cook's *Endeavour*
British naval captain James Cook makes voyages to the Southern Oceans and is the first European to reach New Zealand and Australia.

Endeavour

1842
Matthew Maury
Considered to be the "father of oceanography," US naval officer Maury compiles sea charts of the world's oceans.

“HOW **INAPPROPRIATE** TO CALL THIS PLANET EARTH WHEN IT IS **QUITE CLEARLY OCEAN.**”

Arthur C. Clarke, British writer, 1917–2008

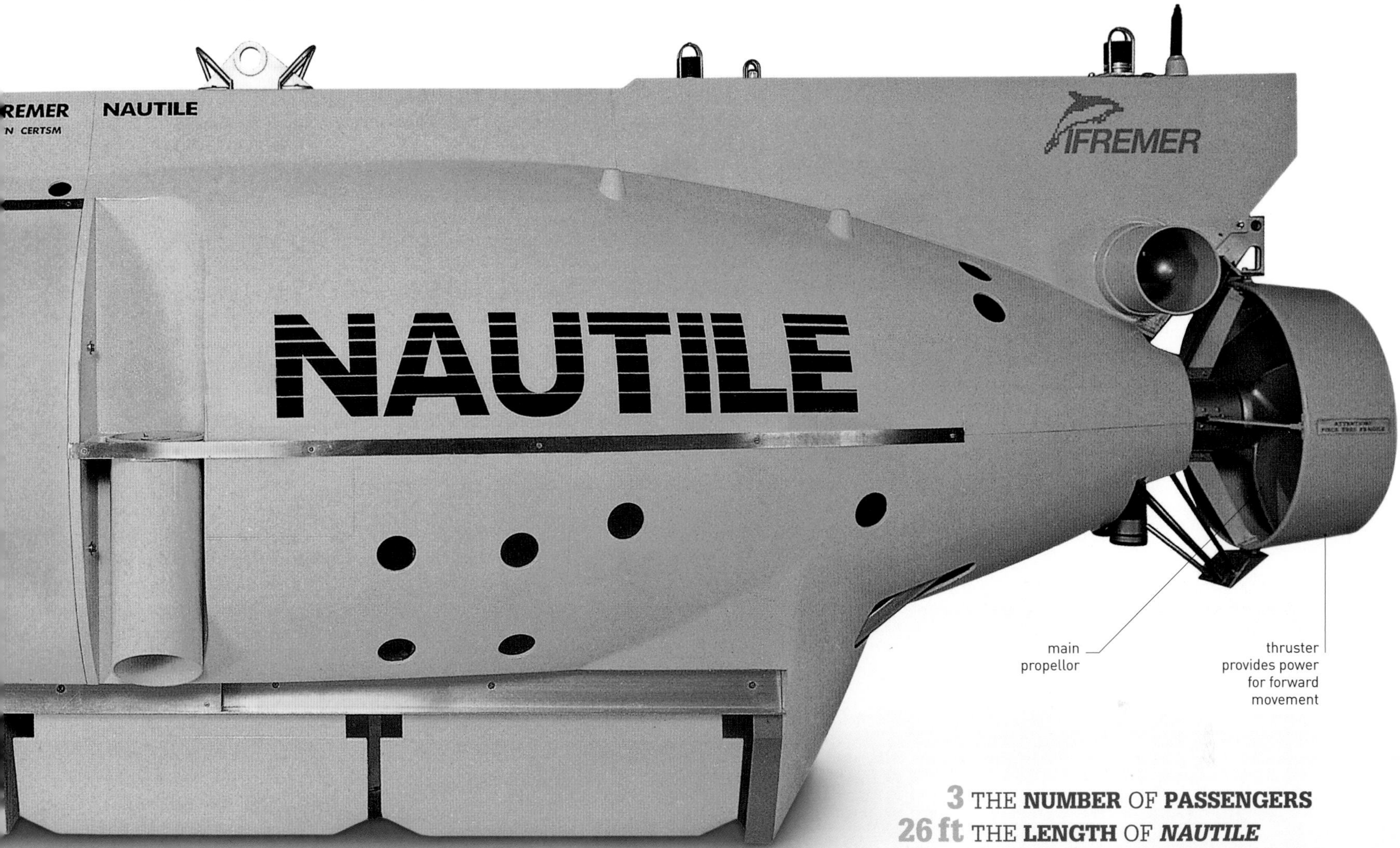

main propellor

thruster provides power for forward movement

3 THE **NUMBER** OF **PASSENGERS**
26 ft THE **LENGTH** OF ***NAUTILE***
4.6 miles THE **RANGE** OF ***NAUTILE***

1872–76
HMS *Challenger*
On its voyage around the world, the HMS *Challenger* collects a huge amount of data about the oceans.

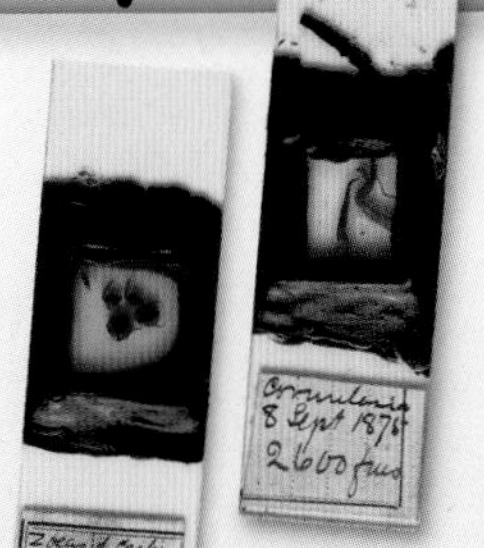

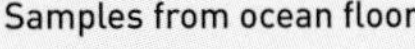

Samples from ocean floor

1956
Mid-ocean ridge
Marie Tharp and Bruce Heezen, American oceanographers, discover the mid-ocean ridge—an undersea ridge running down the Atlantic seabed.

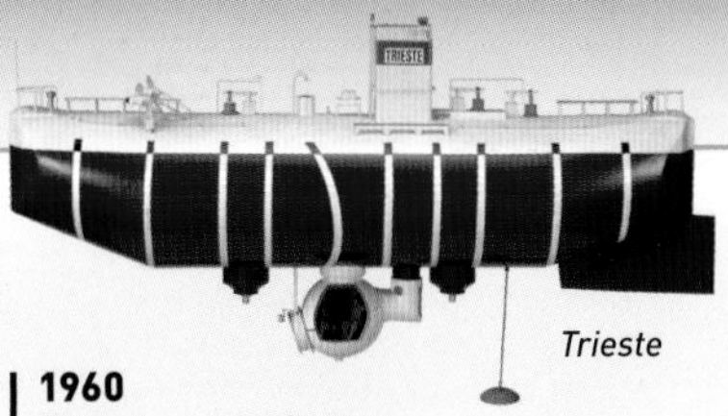

Trieste

1960
Descent to the bottom
The bathyscaphe (diving vessel) *Trieste* dives 35,797 ft (10,911 m) down the Pacific's Marianas Trench, to make the first descent to the deepest part of the ocean.

1968
Deep sea drilling
Rock samples taken from the mid-ocean ridge show magnetic striping confirming that the ocean floor is actively spreading.

1977
Ocean floor
Marie Tharp and Bruce Heezen make the first accurate relief map of all the world's ocean floors, mainly using data recorded by sonar.

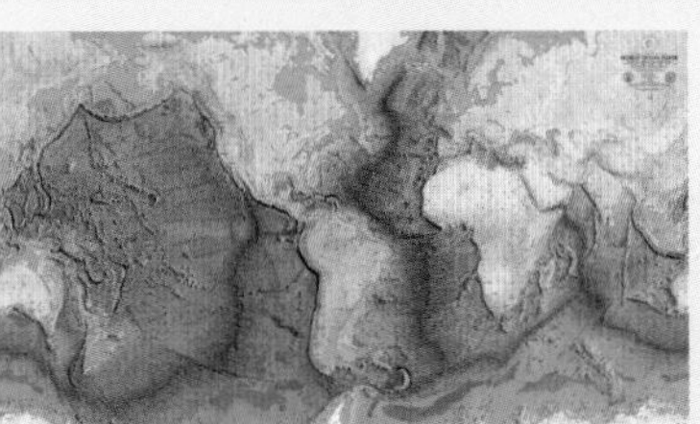

Map of ocean floor

1984
Nautile
The bathyscaphe *Nautile* is used to film the wreck of the RMS *Titanic* and search for the flight data recorder from Air France Flight 447, which crashed into the Atlantic Ocean in 2009.

2000–10
Marine census
A Census of Marine Life that catalogued the diversity of life in oceans worldwide is completed in 2010. This octopod is one of the many strange discoveries.

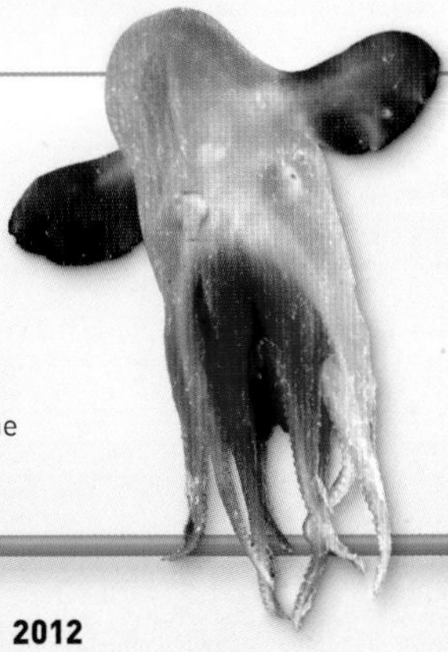

Dumbo octopod

2012
Deepsea Challenger
Canadian filmmaker James Cameron goes to the bottom of the Marianas Trench in the submersible *Deepsea Challenger* and makes a film about life there.

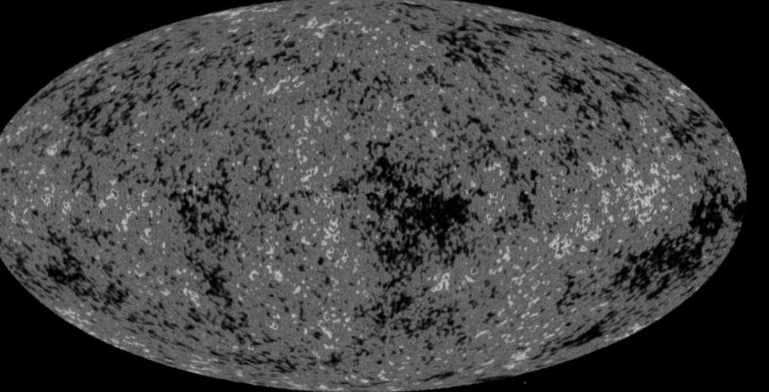

A map of microwave radiation emitted after the Big Bang defines an expanding universe that is 13.7 billion years old.

> **"** WELL BOYS, **WE'VE BEEN SCOOPED. "**
>
> **Robert Dicke, American physicist,** on the accidental detection of microwave radiation by Arno Penzias and Robert Wilson, 1965

AMERICAN PHYSICISTS Arno **Penzias** (b.1933) and **Robert Wilson** (b.1936) were studying radio waves from satellites. Despite removing all known sources of interference, their antenna continued to pick up background noise. What they were hearing, by accident, was the **cosmic microwave background radiation** (CMB) left over from the formation of the universe—evidence of the Big Bang (see p.344).

For more than a century, physicists had hypothesized the existence of astronomical objects with such compacted mass that not even light could escape from them. In 1964, these massive objects got a name: **black holes**. In June, a rocket discovered the strongest source of X-rays near Earth, **Cygnus X-1**—later shown to be a black hole. These holes are now known to be formed when massive stars collapse.

British physician **Robert Macfarlane (1907–87)**, and American scientists **Oscar Ratnoff** (1916–2008) and **Earl Davie** (b.1927) independently showed how proteins solidified in blood when exposed to air due to chemical reactions that involved different clotting factors. American physiologist **Judith Pool** (1919–75) isolated the chemical factor that was eventually used in treating hemophiliacs, people with impaired **blood clotting**.

American chemist **Jerome Horwitz** (1919–2012) made a drug called **azidothymidine (AZT)** that was a modified component of DNA. By injecting it into tumors, he hoped it would confuse cancer cells and stop them from dividing. AZT became an effective antiviral treatment for AIDS.

rotating wheel for positioning aperture
aluminum earlike aperture
cab containing receiver to measure incoming signals

Holmdel Horn Antenna
Classified as a National Historic Landmark, this radio telescope at the Bell Laboratories in New Jersey, US, was the first detector of background radiation left over from the Big Bang.

CCCP

Alexey Leonov's historical spacewalk lasted for 12 minutes and 9 seconds.

IN MARCH, SOVIET COSMONAUT ALEXEY LEONOV (b.1934) became the **first person to walk in space**. Tethered to his spacecraft, Voskhod 2, Leonov spent over 10 minutes in extravehicular activity (EVA). He nearly failed to get back inside the spacecraft because his suit had swelled up in the vacuum of space.

In the late 1950s, astronomers found a celestial object that gave off brilliant light. First detected by their radio waves, these objects were called **quasars** (for quasi-stellar radio sources). In 1965, however, American astronomer Allan Sandage (1926–2010) found the first radio-quiet quasar, which had weak radio emissions, but could emit other types of radiation. It took another 20 years for astronomers to identify a **quasar as the core of a galaxy with a black hole at its center**.

Until 1965, biologists thought that human cells could divide continuously. But in March, American biologist **Leonard Hayflick** (b.1928) published evidence that cultures of human cells only went through about 50 rounds of division before stopping altogether. Ten years later, it was found that this happened because each division corroded the ends of chromosomes

January 18 American journalist Ann Ewing first uses the term **"black hole"**

May American physicist Arno Penzias and Robert Wilson discover **cosmic microwave background radiation (CMB)**

May Blood clotting is described in detail

June 16 US Naval Research Laboratory rocket flight discover **Cygnus X-1** black hole

Azidothymidine (AZT) antiviral is developed to **treat cancer** and is later used for **HIV treatment**

> **“** THE **SPACESUIT** STARTED **BEHAVING** ABSOLUTELY **DIFFERENT** FROM WHAT IT DID **ON THE GROUND. ”**
>
> **Alexey Leonov, Soviet cosmonaut,** 1965

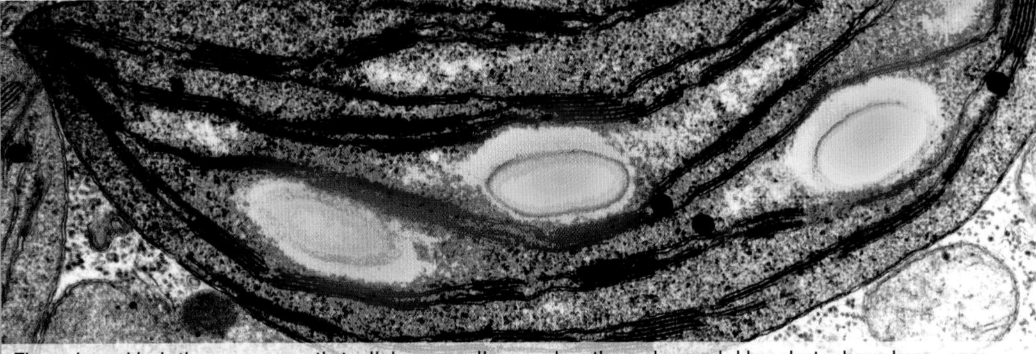

The endosymbiosis theory proposes that cellular organelles—such as the nucleus and chloroplasts shown here—once lived as independent organisms. Evidence for this comes in the form of their own functioning DNA.

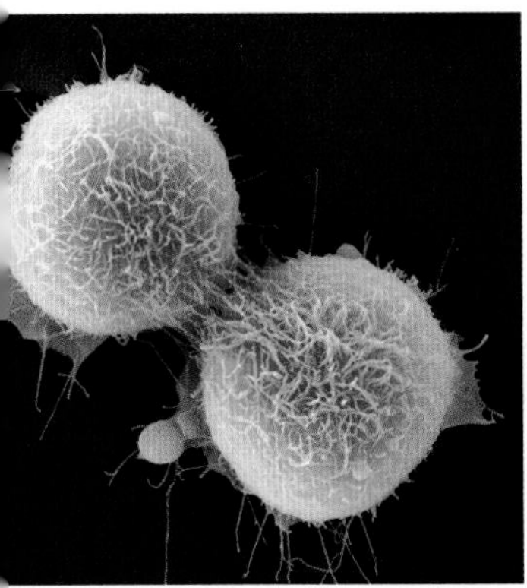

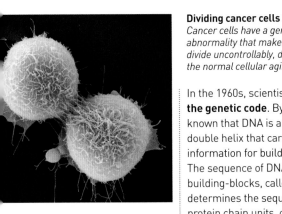

Dividing cancer cells
Cancer cells have a genetic abnormality that makes them divide uncontrollably, deviating from the normal cellular aging process.

until the cells were no longer viable. Hayflick's discovery had important implications in the **biology of cancer**. Cancer cells are abnormal in that their cell division is unchecked—so continuous division produces a tumor. A modern strategy in cancer treatment involves exposing the affected cells to drugs that encourage the natural corrosion of chromosomes.

In the 1960s, scientists **cracked the genetic code**. By 1961 it was known that DNA is a two-chain double helix that carries information for building proteins. The sequence of DNA chain building-blocks, called bases, determines the sequence of protein chain units, called amino acids. Between 1961 and 1966, cell biologists worked out the triplet combinations of bases that are encoded for all 20 different kinds of amino acids. The final link came in 1965, when American biochemist **Robert Holley** (1922–93) unraveled the **structure of tRNA** (transfer RNA)—the molecule that provided a physical link between DNA base code and assembling protein.

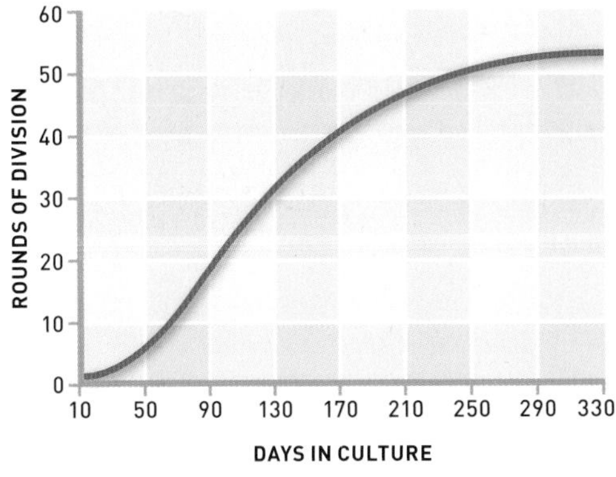
ROUNDS OF DIVISION
60
50
40
30
20
10
0
10
50
90
130
170
210
250
290
330
DAYS IN CULTURE

Hayflick's graph of cell proliferation
Normal cells nurtured in the laboratory soon divide to produce a living culture—but after about 50 rounds of division, reproduction stops.

IN JUNE, after being rejected by more than a dozen scientific journals, a young scientist finally managed to publish a theory that would eventually revolutionize our understanding of the early history of life on Earth. American biologist **Lynn Margulis** (then married to Carl Sagan, see 1970) proposed that components of cells—such as the nucleus and chloroplasts—originally had independent lives. She suggested that millions of years ago bacteria-like life forms engulfed one another to form **the first complex cells, called eukaryotes**. Today, all animals, plants, and many microbes are made up of eukaryotic cells. Her **endosymbiotic theory** was initially resisted by most other scientists.

In July, Japanese husband and wife biologists **Kimishige** (b.1925) **and Teruko Ishizaka** (b.1926), working in the field of immunology, reported that they had discovered a **new class of antibodies**. These substances, called **immunoglobulin E (IgE)**, play a central role in making people oversensitive to certain triggers called allergens. Although they help in defending the body against certain kinds of parasites, IgE antibodies can also make it overproduce chemicals such as **histamine**, which triggers the massive inflammation associated with allergic reactions.

Meanwhile, in a Mt. Hotham Resort ski hut in Victoria, Australia, the discovery of a mouse-sized animal was causing a sensation among zoologists. This **mountain pygmy possum**—the only marsupial adapted to the snow-capped habitats of the Australian Alps—was previously known only from fossils and thought to be extinct for more than half a century. In 1966, the first living specimen was found.

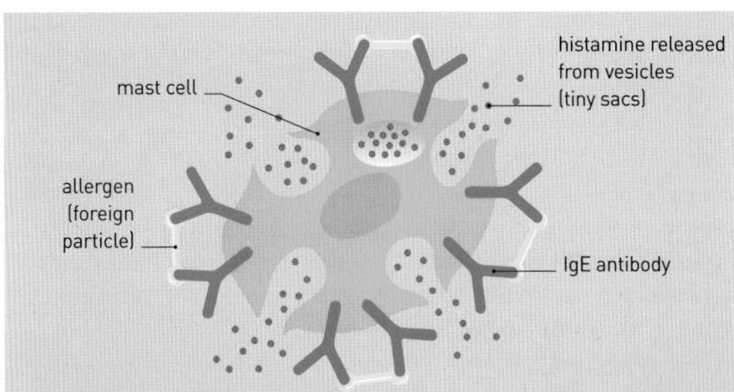
mast cell
histamine released from vesicles (tiny sacs)
allergen (foreign particle)
IgE antibody

ALLERGIC RESPONSE

IgE antibodies are the basis for allergic responses. When first exposed to an allergen (allergy-causing particle, such as pollen), white blood cells release IgE antibodies, which then bind to mast cells. When these IgE molecules bind to a second exposure of the same allergens, it makes the mast cell release histamine. This triggers the body's allergic response symptoms.

Mountain pygmy possum
The mouse-sized alpine possum was discovered as a fossil in 1896, but a living animal was found in Australia in 1966.

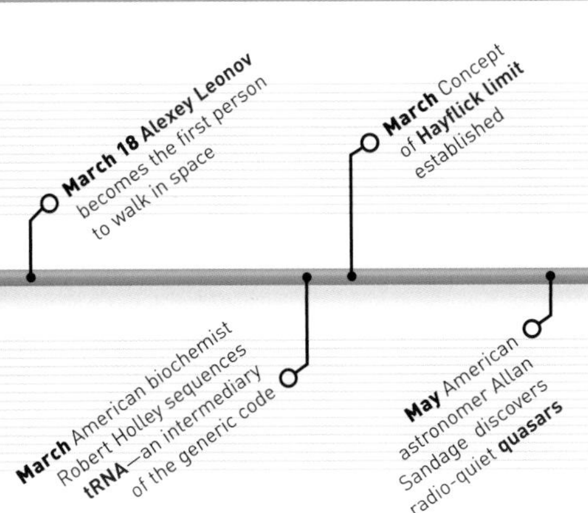
March 18 Alexey Leonov becomes the first person to walk in space
March American biochemist Robert Holley sequences **tRNA**—an intermediary of the generic code
March Concept of **Hayflick limit** established
May American astronomer Allan Sandage discovers radio-quiet **quasars**

June 8 Endosymbiotic theory of **eukaryotic cell origins** established

July Japanese biologists Teruka and Kimishige Ishizaka discover **antibody class IgE**

First living example of a **mountain pygmy possum** discovered in ski resort on Mt. Hotham, Australia

1967

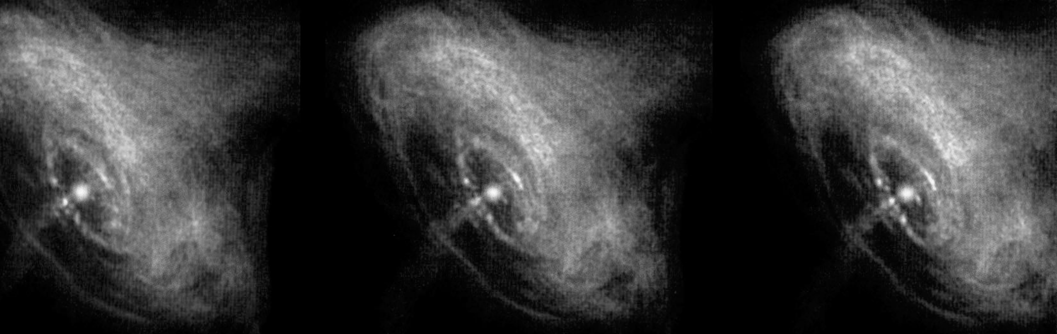

First detected by their radio signals, pulsars are now known to emit other forms of radiation, such as visible light and, as shown here, X-rays.

ALTHOUGH IT WAS PRACTICALLY IGNORED at the time, American physicist **Steven Weinberg's** (b.1933) 1967 study of the forces of nature eventually became one of the most quoted scientific papers. In it, Weinberg explained how electromagnetism and weak nuclear force were just different aspects of a **single unified set of "electroweak" forces**. He also proposed that breaking the symmetry of these forces provided particles with a fundamental property—their mass. For his work, Weinberg went on to win the 1979 Nobel Prize in Physics, with Pakistani nuclear physicist Abdus Salam (1926–96) and American theoretical physicist Sheldon Glashow (b.1932).

In November, British astronomers **Antony Hewish** (b.1924) and **Jocelyn Bell Burnell** observed pulses of radio signals coming from a fixed position in the sky. They whimsically called them **LGM-1 (Little Green Men-1)**. It was later found that the signals were coming from the radiation beam of a rotating neutron star (a dense, compact star thought to be composed mainly of neutrons), and each pulse corresponded to a single rotation. In 1968, these stars were termed pulsars.

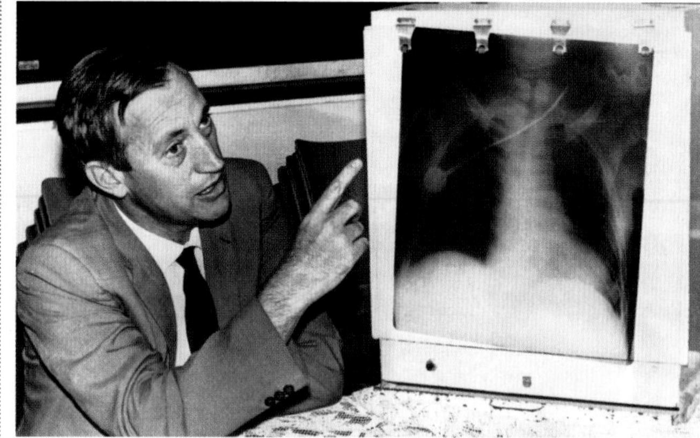

First heart transplant
South African surgeon Christiaan Barnard shows a chest X-ray image of 54-year-old Louis Washansky, the first person to undergo a successful heart transplant.

In December, surgeon **Christiaan Barnard** (1922–2001) performed **the world's first successful heart transplant** at Groote Schuur Hospital in Cape Town, South Africa. The patient, with diabetes and incurable heart disease, received a heart from a young road-accident victim. The recipient survived for just over two weeks, ultimately dying of pneumonia. Barnard was nonetheless celebrated for his achievement, and went on to perform other similar operations. His longest-surviving recipient went on to live for 23 years.

JOCELYN BELL BURNELL (b.1943)

British astrophysicist Jocelyn Bell Burnell came to prominence as a postgraduate, when she discovered radio pulsars with her thesis supervisor, Antony Hewish. Controversially, Bell did not share Hewish's 1974 Nobel Prize for this work. More recently, she served for two years as president of London's Institute of Physics.

1968

This image shows tracks made by neutrinos—the most abundant subatomic particles in the universe—caught in a nanosecond inside a bubble chamber.

THE STANFORD LINEAR ACCELERATOR CENTER in California houses the longest linear accelerator—2 miles (3.2 km) long. In 1968, it provided the **first evidence for the existence of fundamental particles called quarks** (see 1963) by bombarding and shattering subatomic particles.

A solution to the **solar neutrino problem** was proposed in 1968. Neutrinos—subatomic particles with no charge and negligible mass—are generated by atomic reactions in the Sun. However, fewer neutrinos were recorded striking Earth than had been estimated. Italian physicist **Bruno Pontecorvo** (1913–1993) accounted for this discrepancy by proposing that neutrinos had appreciable mass, which allowed them to change types. Many went undetected by neutrino detectors, which monitored only one type.

American physician **Henry Nadler** (b.1936) reported on the **first prenatal diagnosis of Down syndrome using amniocentesis** (see panel, below). His observations were confirmed by direct diagnosis on the fetus.

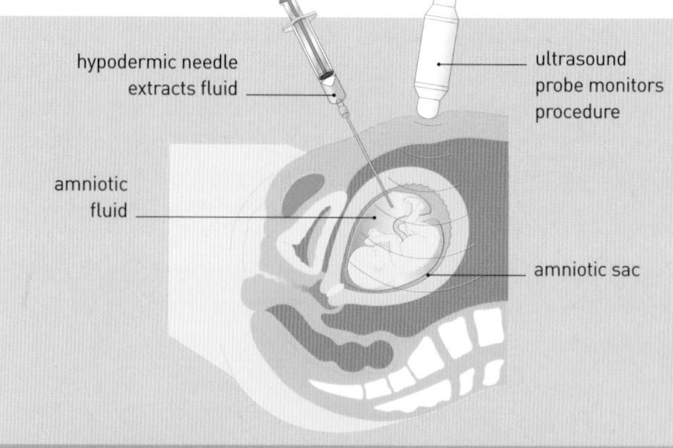

hypodermic needle extracts fluid
ultrasound probe monitors procedure
amniotic fluid
amniotic sac

AMNIOCENTESIS

The amniotic fluid, which surrounds a fetus, contains cells that come from the unborn child. Amniocentesis is a method by which a sample of this fluid is extracted to test for abnormality. In 1952, British obstetrician Douglas Bevis (1919–94) discovered how it could be used as a diagnostic tool. By the 1960s, scientists could use it to detect chromosome abnormalities, including Down syndrome.

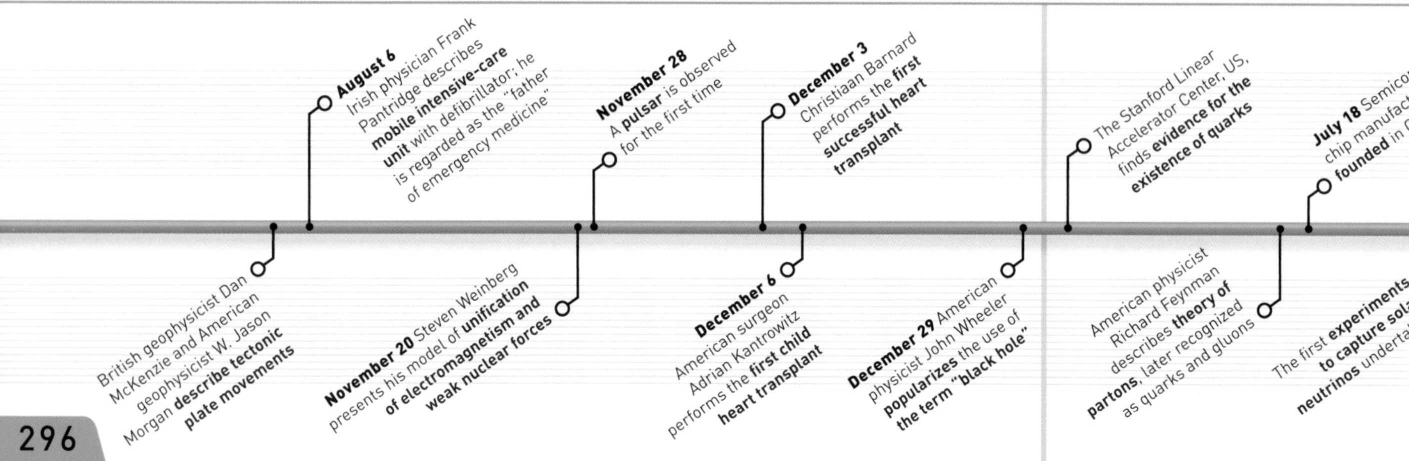

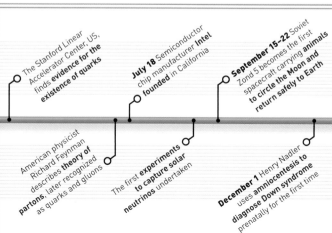

British geophysicist Dan McKenzie and American geophysicist W. Jason Morgan **describe tectonic plate movements**
August 6 Irish physician Frank Pantridge describes **mobile intensive-care unit** with defibrillator; he is regarded as the "father of emergency medicine"
November 20 Steven Weinberg presents his model of **unification of electromagnetism and weak nuclear forces**
November 28 A **pulsar** is observed for the first time
December 3 Christiaan Barnard performs the **first successful heart transplant**
December 6 American surgeon Adrian Kantrowitz performs the **first child heart transplant**
December 29 American physicist John Wheeler **popularizes the term "black hole"**
The Stanford Linear Accelerator Center, US, finds **evidence for the existence of quarks**
American physicist Richard Feynman describes **theory of partons**, later recognized as quarks and gluons
July 18 Semiconductor chip manufacturer **Intel founded** in California
The first **experiments to capture solar neutrinos** undertaken
September 15–22 Soviet Zond 5 becomes the first spacecraft carrying **animals to circle the Moon and return safely to Earth**
December 1 Henry Nadler uses **amniocentesis to diagnose Down syndrome** prenatally for the first time

1969

“ ONE **SMALL STEP** FOR [A] **MAN,** ONE **GIANT LEAP** FOR **MANKIND.** ”

Neil Armstrong, American astronaut, 1969

IN THE WAKE OF Christiaan Barnard’s ground-breaking heart transplant (see 1967), American surgeon **Denton Cooley** (b.1920) **implanted the first artificial heart** on April 4. At times when natural hearts were unavailable and surgical intervention urgent, early artificial hearts could give the patient time until a donor could be found. The first patient to receive an artificial heart survived long enough to get a donor.

In an event that was televised live across the world, American astronauts **Neil Armstrong** (1930–2012) and **Buzz Aldrin** (b.1930) became **the first humans to set foot on the Moon** on July 21 Coordinated Universal Time (UTC). Their spacecraft, Apollo 11, had touched down on the Moon’s Sea of Tranquillity, a large plain, the day before. Armstrong and Aldrin spent two and a half hours on the surface of the Moon, collecting samples of lunar rock. They left behind instruments, including a series of reflectors for laser-ranging experiments, which would help **determine the distance between Earth and the Moon** to a degree of accuracy never before possible. The astronauts returned to Earth on July 24, splashing down in a module in the Pacific Ocean. The entire mission took just over eight days to complete.

British biochemist **Dorothy Hodgkin** (see 1945) specialized in working out the structures of complex biological molecules. After her success with steroids, penicillin, and vitamins, she moved on to a **much more complicated substance—insulin**, a protein hormone. Fred Sanger had determined the sequential building blocks of insulin in 1951. Ten years later, Hodgkin worked out the **3-D structure of insulin** by using X-ray diffraction techniques that had earlier been applied to DNA and other complex structures.

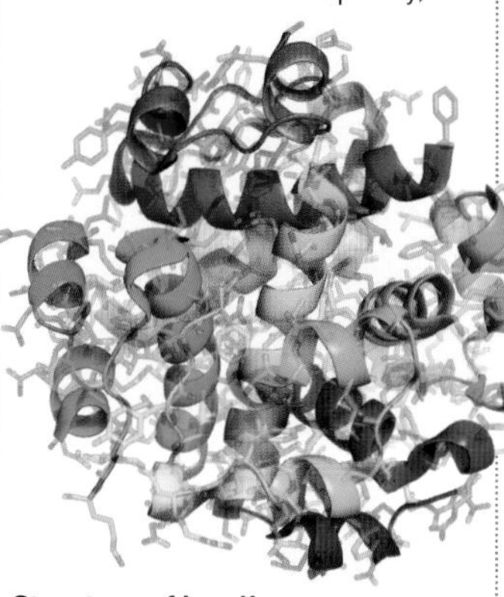

Structure of insulin
Dorothy Hodgkin showed how the building blocks of insulin were spatially arranged to form a fixed complex shape.

First man on the Moon
Neil Armstrong was the first person to step on the Moon, followed soon afterward by Buzz Aldrin. Television images of the event were sent back to at least 600 million people on Earth.

- **January Peptide hormones described** in the hypothalamus of the brain
- **March** Norwegian-American meteorologist Jacob Bjerknes describes **pattern of global climate oscillation**, exemplified by El Niño
- **March 31** Formaldehyde discovered between stars—the first large **organic molecule discovered in interstellar space**
- **April 4** Denton Cooley implants the **first temporary artificial heart**
- **May 16–17, 1969 USSR’s Venera 5 and 6** land on Venus
- **July 20** US spacecraft **Apollo 11** lands on the Moon
- **July 21** Apollo 11 astronauts Neil Armstrong and Buzz Aldrin become the **first humans to walk on the surface of the Moon**
- **July** American paleontologist John Ostrom points out the **birdlike characteristics** of *Deinonychus*, a **bipedal dinosaur**
- **August** American microbiologists Thomas Brock and Hudson Freeze describe **discovery of bacteria living in hot springs**
- **November 1** Dorothy Crowfoot (later Hodgkin) publishes **3-D structure of insulin**

1949
Animals in space
The first astronauts were animals. American rocket scientists send Rhesus monkey Albert II into space in 1949. Soviet dog Laika became the first animal to orbit Earth in 1957.

SHQIPERIA
3 NËNDOR 1957
LAIKA
1
Laika

1957
Man-made satellite
On October 4, USSR launches the first satellite into orbit around Earth, Sputnik 1. Today there are over 500 working satellites.

1959
Luna 2 and 3
The Soviet lunar probes Luna 2 and 3 are the first craft to successfully reach the Moon. Luna 3 captures the first images of the far side of the Moon.

Luna 2

1961
First person in space
On April 12, Soviet cosmonaut Yuri Gagarin becomes the first person in space. He completes one orbit of the Earth in Vostok 1. In May, Alan Shephard (1923–98) becomes the first American in space.

Yuri Gargarin

1962
Mission to Venus
On December 14, the Mariner 2 becomes the first spacecraft to fly past another planet, revealing Venus as a hothouse planet.

1963
First female cosmonaut
On June 7, Soviet cosmonaut Valentina Tereshkova becomes the first woman in space. A crater on the far side of the Moon is named after her. The first US female astronaut was Sally Ride, 20 years later.

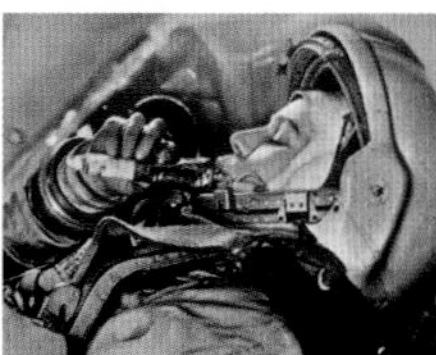
Tereshkova

1965
First spacewalks
On March 18, Soviet Aleksei Leonov becomes the first astronaut to venture outside his craft. American astronaut Edward White completes a spacewalk in June.

Edward White

1966
Landing on the Moon
On February 3, Soviet probe Luna 9 becomes the first spacecraft to successfully land on the Moon. On May 30, the US Surveyor 1, made the second soft landing.

THE STORY OF SPACE EXPLORATION

FROM THE FIRST ARTIFICIAL SATELLITE TO REACHING OTHER WORLDS AND THE EDGE OF THE SOLAR SYSTEM

The launch of the Soviet Union's Sputnik 1 in October 1957 is usually taken to mark the start of space exploration, even though some earlier flights had left Earth's atmosphere. It was the beginning of a series of adventures that has taken astronauts to the Moon and probes to distant planets.

Sputnik 1 was followed a month later by the voyage of the Soviet dog Laika, which became the first animal to orbit Earth. The first human to travel in space was Yuri Gagarin, who orbited Earth in the Soviet spacecraft Vostok 1 in April 1961. These Soviet successes threw down the challenge to the US to step up its space exploration programme. In 1965, the American Mariner 4 sent back the first close-up pictures of another planet, Mars. In 1966, a Soviet probe, Luna 9, made a soft-landing on the Moon and sent back the first pictures from the surface. Three years later, American astronauts Neil Armstrong and Buzz Aldrin set foot on the Moon. Their first moments there were broadcast live to people around the world. Astronauts returned to the Moon several times in the 1970s, but most voyages of exploration since have been by robot craft, which have now traveled to every planet in the Solar System and some even beyond that (see panel, right).

hatch through which astronauts entered and exited module

Apollo 11 command module
This was part of the spacecraft that carried Neil Armstrong, Buzz Aldrin, and Michael Collins on their historic mission to the Moon in 1969.

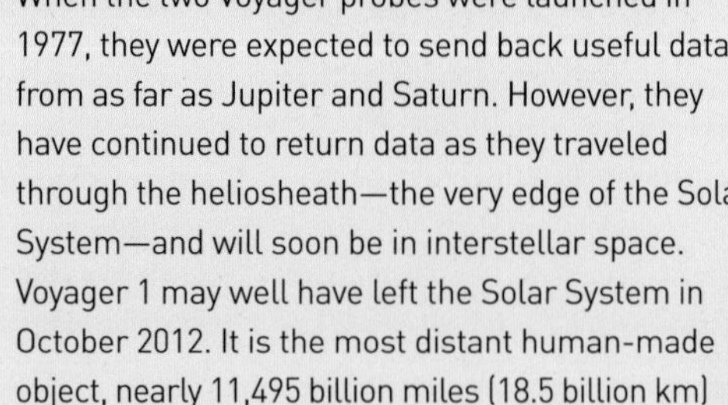

VOYAGER 1 AND 2

When the two Voyager probes were launched in 1977, they were expected to send back useful data from as far as Jupiter and Saturn. However, they have continued to return data as they traveled through the heliosheath—the very edge of the Solar System—and will soon be in interstellar space. Voyager 1 may well have left the Solar System in October 2012. It is the most distant human-made object, nearly 11,495 billion miles (18.5 billion km) away in March 2013. Voyager 2 is not far behind, over 9,320 billion miles (15 billion km) away.

> “IF OUR LONG-TERM **SURVIVAL IS AT STAKE,** WE HAVE A BASIC RESPONSIBILITY TO OUR SPECIES TO **VENTURE TO OTHER WORLDS.**”
>
> **Carl Sagan, American cosmologist,** 1934–96

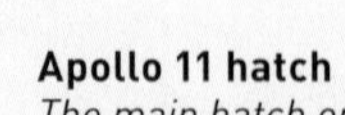

Apollo 11 hatch
The main hatch on the Apollo Command Module had to provide a perfect seal to protect the crew. It was redesigned to open outward after an accident in 1967 in which astronauts were trapped inside the capsule as it caught fire.

1969
Man on the Moon
On July 21, Neil Armstrong becomes the first human to set foot on the Moon. As he steps on to the Moon's surface, he says, "One small step for (a) man, one giant leap for mankind."

Lunar rover

1971
First Lunar Roving Vehicle
American astronauts on the Apollo 15, 16, and 17 missions explore the Moon with an electric car called the Lunar Roving Vehicle.

1971
First space station
On April 19, the first space station, the Salyut 1, is launched by the Soviet Union. Later in the year, the first crew of three stays on board for 23 days.

1973
Missions to Mars
In 1973, Soviet probe Mars 2 reaches Mars—one part of it sent back pictures from orbit, the other crashed while attempting to land. In 1975, the American Viking 1 successfully landed and sent back data for six years.

1981
Space Shuttle
The first spacecraft were designed to be used only once. The Space Shuttle is the first reusable craft, able to land after a mission like an aeroplane. The USSR built the similar, though much less successful, Buran.

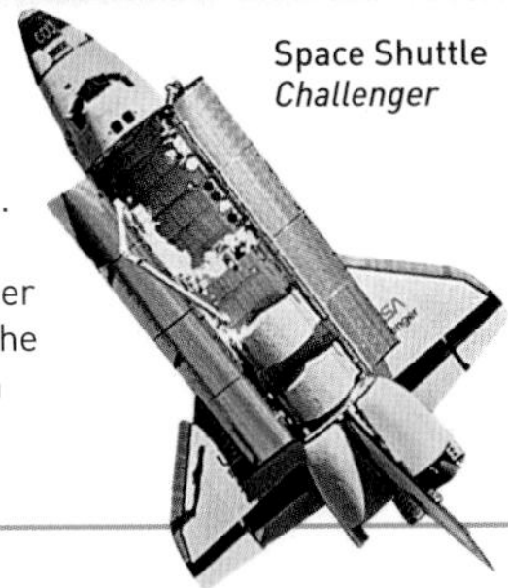

Space Shuttle *Challenger*

1990
Space telescope
Orbiting telescopes give astronomers a view of space not influenced by Earth's atmosphere. Launched in 1990, the Hubble Space Telescope (HST) is the most well-known space telescope.

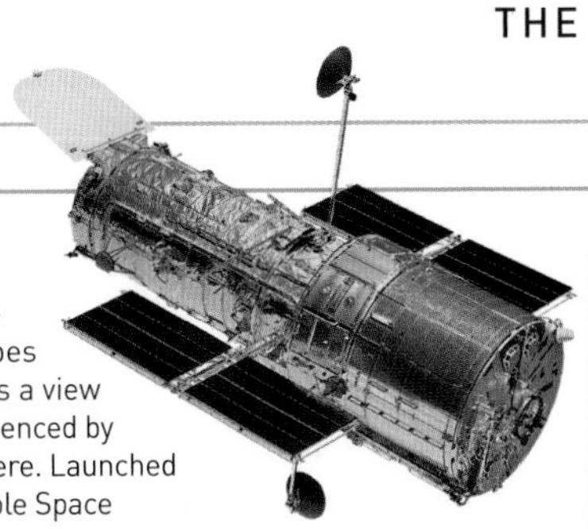

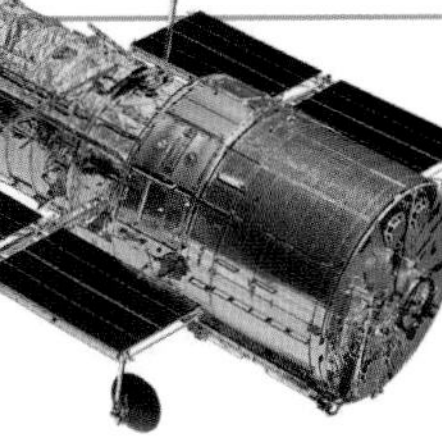
Hubble Space Telescope

1995
Galileo
NASA's Galileo probe becomes the first spacecraft to orbit the Solar System's largest planet, Jupiter. It beams back many images of Jupiter's moons and also the impact of comet Shoemaker Levy 9.

1998
ISS
The International Space Station is a collaboration between 16 countries. It is launched in 1998 and assembled bit by bit over 14 years.

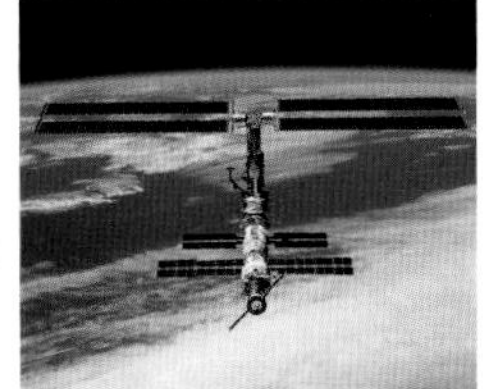
ISS

2012
Mars Curiosity
NASA's Curiosity Rover, a car-sized robot exploration vehicle, lands in Gale Crater on Mars on August 6. The latest in a series of similar missions, it studies rocks and climate, and searches for signs of water and microbial life.

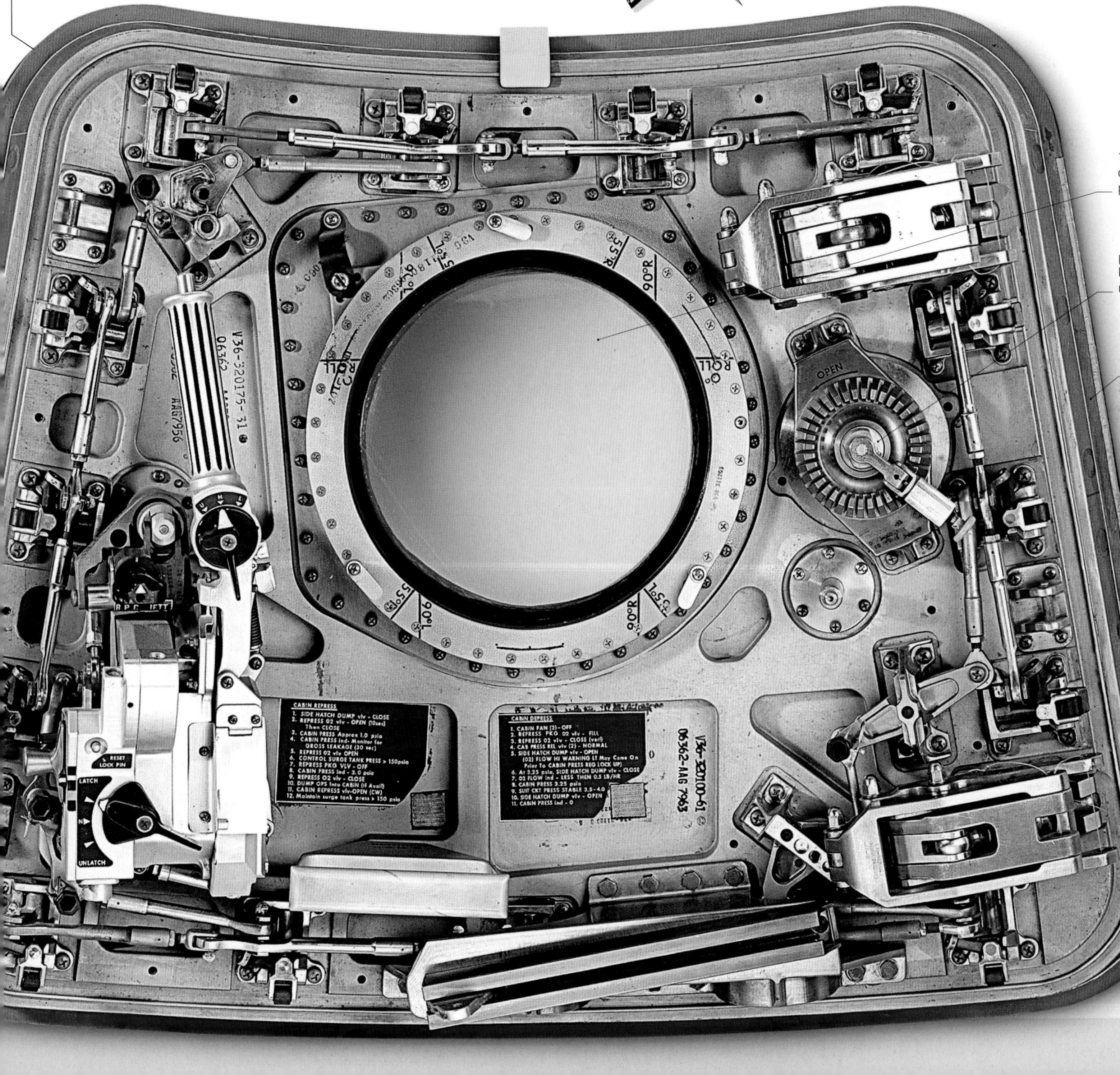

11 in (27 cm) diameter porthole

pressurized locking mechanism

cabin seal

1970

452 THE PASSENGER **CAPACITY** OF THE **BOEING 747-100**

The first Boeing 747-100—*Clipper Victor*—went into commercial service on January 22, on a PanAm flight from New York to London.

THE AGE OF WIDE-BODY PASSENGER JET TRAVEL began when the **first Boeing 747 jumbo jet made its maiden commercial voyage** in January of 1970. The idea of the airliner was conceived in the mid-1960s, partly to ease congestion at busy airports. By 2013, more than 1,500 Boeing 747s had been built.

In February, **Japan** became the fourth country—after the USSR, the US, and France—to **send a rocket into space**, when its national space development agency (NASDA) launched the experimental satellite called Ōsumi. **China** became the fifth country to do so, in April, with the successful launch of its experimental satellite, **Dong Fang Hong 1**.

After the successes of its first three manned missions to the Moon, **NASA's Apollo program** suffered a setback in April 1970, **during the Apollo 13 mission**. Around 55 hours after lift off—more than 200,000 miles (320,000 km) from Earth—one of the spacecraft's two oxygen cylinders exploded. Oxygen was crucial not only to make the air in the spacecraft breathable, but also to generate electrical power and to make drinking water in the fuel cells. The mission was aborted, and the drama of the journey back to Earth was played out on radio and television. The spacecraft splashed down in the South Pacific Ocean, five days after the incident.

Later this year, NASA successfully launched Uhuru, the **first dedicated orbiting X-ray observatory**. X-ray astronomy is only possible at high altitude—ideally, in orbit—because the atmosphere absorbs most of the X-ray radiation from space.

The first Earth Day—an annual celebration of the world's natural environment and a call to action for environmentalists to protect it—was observed this year. It was first proposed in 1969 by American peace activist John McConnell (1915–2012), as an event that would take place on the spring equinox (Northern Hemisphere) each year—around March 21. The **first Earth Day** took place on April 22, and is seen as a major event in the history of environmentalism. It was celebrated with events across the US, but since 1990, Earth Day events have been held worldwide.

Rescue mission
Three Apollo 13 astronauts are lifted aboard a helicopter in a rescue net, after their Lunar Module returned them safely to Earth on April 17.

LYNN MARGULIS (1938–2011)

American biologist Lynn Margulis is best known for her theory of complex cell evolution, which she first published in 1966 when working at Boston University. *Origin of Eukaryotic Cells* (1970), expanded on her endosymbiotic theory, but earned criticism within the scientific community. It took 30 years before sufficient evidence led to the theory's acceptance.

In May, American biologist **Lynn Margulis** (see panel, above) published a book expanding on her **endosymbiotic theory of the origin of eukaryotic cells** (cells with complex structures contained within a membrane), which she first proposed in 1966. Eukaryotic cells contain structures called organelles, with specific functions (see pp.194–95); for example, plant cells contain organelles called chloroplasts, in which photosynthesis (see 1787–88) takes place. Margulis' idea was that organelles were once simple cells in their own right, and that eukaryotic cells evolved as a symbiosis of these subunits.

It was a year of important breakthroughs in genetics too. Genetic information is carried in two similar compounds inside cells: DNA and RNA (see pp.284–85). DNA stores genetic information, while RNA transfers this information and is involved in building protein molecules. Until 1970, biologists believed that information could flow in only one direction: from DNA to RNA. In June, American

> **“HOUSTON,** WE'VE HAD A **PROBLEM.”**
>
> **John L. "Jack" Swigert, US astronaut,** Apollo 13 mission, 1970

January 22 **Jumbo jet** enters commercial service

February 11 Japan launches first **satellite**, Ōsumi

April 17 NASA's **Apollo 13 returns** to Earth after the mission is aborted

April 24 China launches **first satellite**, Dong Fang Hong 1

April 22 First **Earth Day** is celebrated

June 19 Two cosmonauts set **endurance record for human space flight**

June 27 Howard Temin and American biologist David Baltimore independently discover **reverse transcriptase**

CLEAN AIR

Earth Day protests like this one in New York appealed for changes in government policy that would promote a sustainable way of life.

“ **WE DECLARE** THAT THE PROPER USE OF **SCIENCE** IS **NOT TO CONQUER NATURE** BUT TO **LIVE IN IT.** ”

Professor Barry Commoner, American biologist, Earth Day, 1970

geneticist Howard Temin (1934–94) and American biologist David Baltimore (b.1938) independently discovered that **some viruses carry information in RNA**, which is then passed to **DNA**. These retroviruses include the human immunodeficiency virus (HIV), which causes AIDS (see 1982). The enzyme involved in the process is called **reverse transcriptase**. For their work, Temin and Baltimore jointly won the 1975 Nobel Prize in Physiology or Medicine.

In July, **Type II restriction enzymes** that cut DNA into fragments at specific points were discovered. Restriction enzymes play a central role in modern genetic technologies, including DNA profiling (see 1984), which is used in fields such as paternity testing, criminal investigation, and ecological studies. American microbiologists **Hamilton Smith** (b.1931) and **Daniel Nathans** (1928–99), along with Swiss microbiologist **Werner Arber** (b.1929), all shared the 1978 Nobel Prize in Physiology or Medicine for this discovery.

The USSR had a particularly successful year in space exploration. In June, two of its cosmonauts set an **endurance record for space flight**. Their Soyuz 9 mission lasted 17 days, 16 hours, and 59 minutes. In September, the unmanned lunar probe **Luna 16** drilled into the Moon's surface and returned samples to Earth. In November, the unmanned Luna 17 mission put the first rover—the Lunakhod—on the lunar surface. The Lunokhod, spent nearly 11 months analyzing the lunar soil and sending back photographs. In December, **Venera 7** became the first spacecraft **to make a soft landing on another planet**, when it touched down on Venus. Engineers back on Earth detected no signals from the probe after its landing and assumed it had been destroyed. Later analysis of their recordings revealed 23 minutes of data, which indicated that the surface temperature was extremely hot at 887°F (475°C).

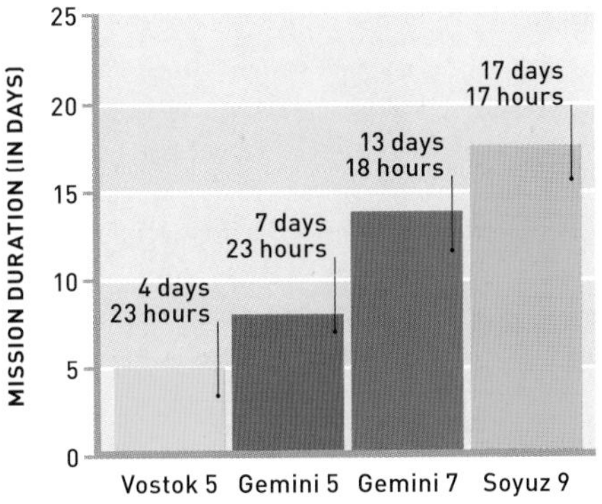

Space missions
Endurance records for manned space flight steadily improved between 1963 and 1970. The Soyuz 9 record was broken in 1971, when cosmonauts on board Soyuz 11 remained in orbit for 24 days.

Mission	Mission duration (in days)
Vostok 5	4 days 23 hours
Gemini 5	7 days 23 hours
Gemini 7	13 days 18 hours
Soyuz 9	17 days 17 hours

Lunokhod 1
The first lunar rover landed on the Moon's surface on November 17, aboard the USSR's Luna 17 spacecraft.

directional antenna

omnidirectional antenna

television camera

drive wheels

July 28 Isolation of the first Type II **restriction enzyme**

August 7 Corning Glass scientist Don Keck makes **optical fibers** a practical reality

September 20 Soviet **Luna 16** drills into the Moon's surface

November 17 Soviet **Lunakhod rover** begins 11 months of lunar exploration

December 12 The US launches **Uhuru**, the first **orbiting X-ray observatory**

December 15 Soviet **Venera 7** lands on Venus

1971

NASA's Apollo Lunar Roving Vehicle (LRV) is driven on the surface of the Moon. During the Apollo 15 mission in July, the first LRV traveled a total of nearly 17 miles (28 km).

THE FIRST ALL-ELECTRONIC CALCULATORS appeared in the 1950s, but they were large and expensive. In the 1960s, calculators became smaller and cheaper, thanks to the use of integrated circuits, or "chips"—tiny but complex electronic circuits etched onto a single piece of semiconductor material. In 1968, adding-machine company Busicom was designing a portable calculator, and approached two chip manufacturers: Intel and Mostek. Intel developed a chip with the entire central processing unit of a computer in its tiny circuits. It was the **world's first commercially available "microprocessor"** (see panel, right), the Intel 4004. Busicom chose not to use the Intel 4004, because it would have made the cost of producing the calculator prohibitively high. It chose a simpler chip, made by Mostek. The **first truly pocket-sized calculator**, the LE-120A HANDY, went on sale in January 1971.

In May, US computer company IBM announced a new convenient data storage device—**the floppy disk**. The floppy played a key role in enabling the development and popularity of **personal computers** because users could store electronic documents, carry them from computer to computer, and send them through the mail. Floppy disks were thin plastic disks coated with magnetic particles, and were attached to a spindle inside a rigid plastic housing. The first floppy disks were 8 in (20.3 cm) in diameter, and **held 80 kB (kilobytes) of information**.

Another important first in 1971 was e-mail (electronic mail). For about a decade, different users logged into the same computer were able to leave messages for each other. With the creation of the US military's ARPANET network in 1969, organizations **could send information across large distances**. American computer programmer Ray Tomlinson designed an e-mail system for ARPANET and **sent the first true e-mail**.

In space, the USSR's **Salyut 1 became the first ever orbiting space station**, remaining in orbit for 175 days. The cosmonauts of the Soyuz 11 mission were the first to enter the craft, in June. They spent three weeks inside, but all three were killed as they were leaving; they remain the only people to have died outside Earth's atmosphere.

In July, astronauts from NASA's Apollo 15 mission drove around the lunar surface in the first **Lunar Roving Vehicle**, or Rover.

In medical science, the first commercially available CT (computed tomography) scanner was released—and the first scans of the human brain produced. American inventor Raymond Damadian published the results of his experiments into **magnetic resonance imaging** (MRI, see 1977). British pharmacologist John Vane published his research that **explained how the painkiller aspirin works**—by blocking the production of prostaglandins, compounds that play a central role in the mechanism of pain and the body's inflammatory response.

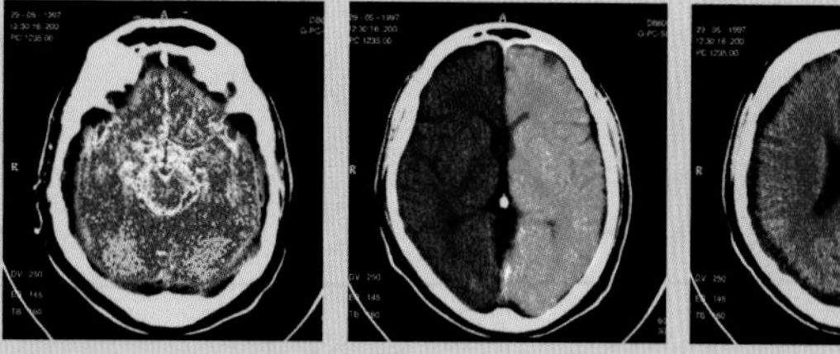

COLORED CT SCANS OF THE HUMAN BRAIN

CT SCANNING

In computed tomography (CT), an X-ray source and a detector are placed on opposite sides of a rotating drum. A person lies inside the drum and an X-ray beam passes through him or her. The signal in the detector depends on the average density of the material through which the beam passes. As the drum rotates, a computer combines multiple scans, to produce a 2-D "slice" through the person.

BUSICOM HANDY-LE

First pocket calculator
Busicom's HANDY LE, had a red LED display. It featured a fixed decimal point, with a choice of four, two, or no decimal places. It cost $395.

> **“** I THOUGHT ABOUT OTHER **SYMBOLS,** BUT @ DIDN'T APPEAR IN ANY **NAMES, SO IT WORKED. ”**
>
> **Ray Tomlinson, US computer programmer and inventor of e-mail,** 1998

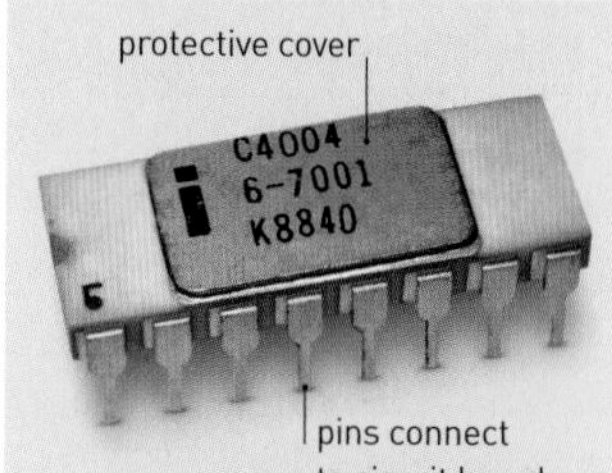
protective cover
C4004
6-7001
K8840
pins connect to circuit board

FIRST MICROPROCESSOR

A microprocessor is a single-chip central processing unit (CPU). Beneath the cover is a single flat crystal of silicon onto which thousands, millions, or, more recently, billions of transistors and other components are etched. Microprocessors are one of the defining components of the microelectronics industry.

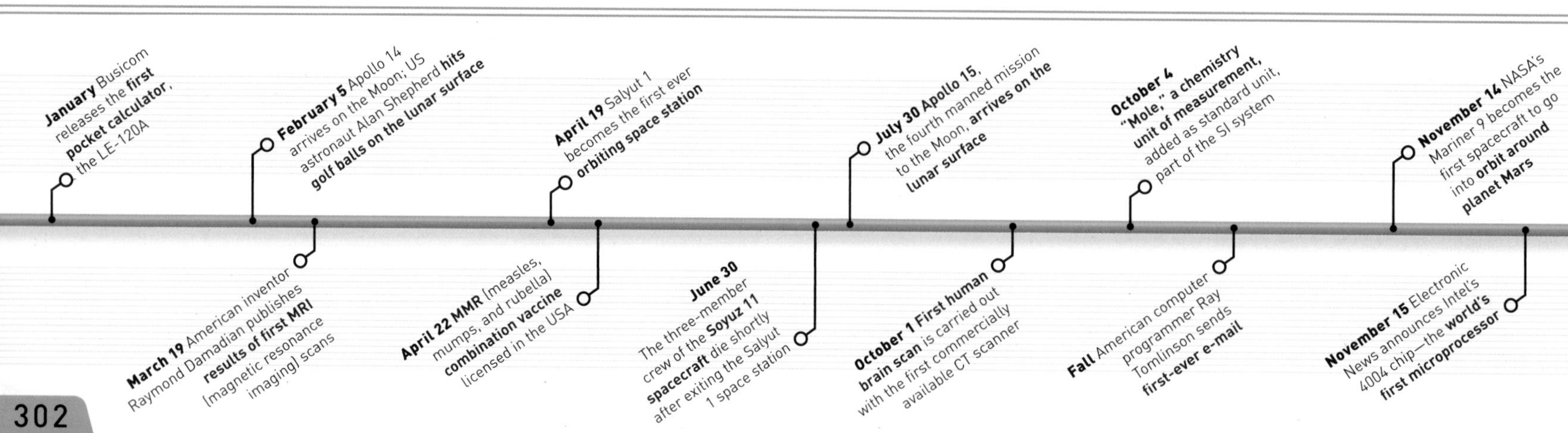
January Busicom releases the **first pocket calculator**, the LE-120A
February 5 Apollo 14 arrives on the Moon; US astronaut Alan Shepherd **hits golf balls on the lunar surface**
March 19 American inventor Raymond Damadian publishes **results of first MRI** (magnetic resonance imaging) scans
April 19 Salyut 1 becomes the first ever **orbiting space station**
April 22 MMR (measles, mumps, and rubella) **combination vaccine** licensed in the USA
June 30 The three-member crew of the **Soyuz 11 spacecraft** die shortly after exiting the Salyut 1 space station
July 30 Apollo 15, the fourth manned mission to the Moon, **arrives on the lunar surface**
October 1 First human brain scan is carried out with the first commercially available CT scanner
October 4 **"Mole," a chemistry unit of measurement,** added as standard unit, part of the SI system
Fall American computer programmer Ray Tomlinson sends **first-ever e-mail**
November 14 NASA's Mariner 9 becomes the first spacecraft to go into **orbit around planet Mars**
November 15 Electronic News announces Intel's 4004 chip—the **world's first microprocessor**

1972

Apollo 17 astronauts captured this image—the celebrated "Blue Marble" photograph—on December 7. Seeing our planet from space helped raise concern for the environment.

THE LIMITS TO GROWTH

A simple graph (right) illustrates that while population grows geometrically—the more people there are, the faster it grows—a finite resource, such as food supply, can only increase at a steady rate. This inevitably leads to crisis at some point, showing that continued growth cannot be sustained indefinitely.

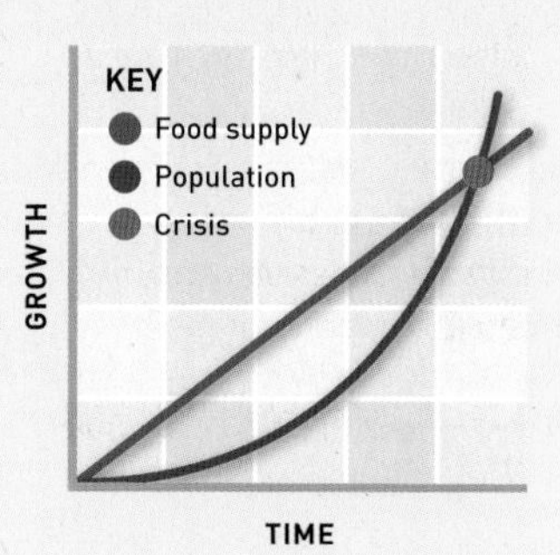

KEY
Food supply
Population
Crisis
GROWTH
TIME

THE MICROELECTRONICS INDUSTRY produced a number of advances in 1972. American company Magnavox introduced the Odyssey, the world's **first home video game console**, which plugged into televisions. The popular Pong video game—a two-dimensional on-screen version of table tennis—was released by Atari in November.

The **first digital watch**, the Hamilton Watch Company's Pulsar, went on sale in the fall. The company had announced the watch's development in 1970 and begun producing it in 1971.

Computer users had been able to send information across the telephone network since the mid-1960s, but in December, Vadic Corporation **introduced the first practical modem**, the VA3400. It could send data at 1,200 bits per second.

In January, the international think tank The Club of Rome published *The Limits to Growth*. It set out the results of a computer simulation investigating the possible effects of **unbridled industrial development and population growth**. The report had many critics at the time, but was nonetheless very important for **raising public awareness of environmental concerns**, such as the affects of economic growth on finite natural resources.

In May, a team led by American geneticist **Walter Fiers** revealed that they had worked out the **sequence of nucleotide bases along the entire length of a gene**—the first time this had been accomplished. Also in May, Japanese-born evolutionary biologist **Susumu Ohno introduced the term "junk DNA."** Geneticists knew that mutations introduced when DNA replicates inside cells impose a limit on the number of genes a genome can carry. That upper limit is around 30,000. Since the amount of DNA in each human cell could carry 3 million genes, the **great majority of our genome must have no function**—hence the term "junk DNA." Most geneticists prefer to use the term "noncoding DNA": the DNA in genes carry codes used to build proteins inside cells, but other DNA may yet be found to have other functions.

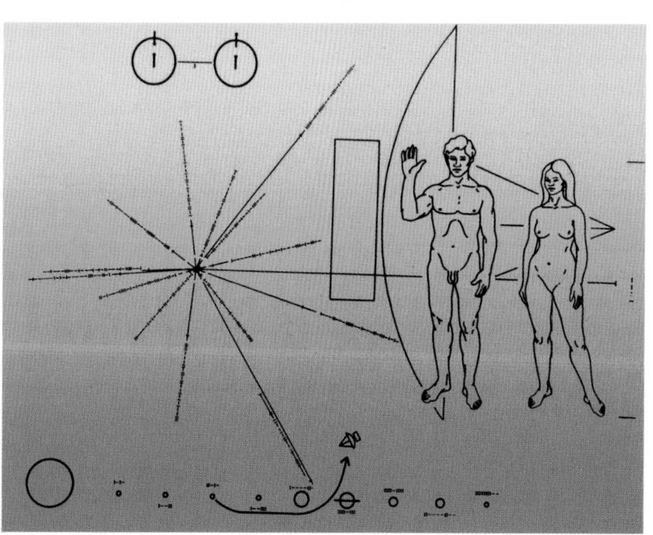

Information plaque about Earth
Pioneer 10 carries this plaque, which aimed to give extraterrestrials an idea of the human form and planet Earth's place in the Solar System.

> **"** IT WAS JUST A **'EUREKA'.** YOU KNOW, I LOOK AT THE **TV SET** AND SAY TO MYSELF, **'WHAT CAN I DO WITH THIS?' "**
>
> **Ralph Baer, inventor of Magnavox Odyssey video game console,** 2007

In October, a team of American molecular biochemists led by **Paul Berg** reported that they had **combined sequences of DNA from different organisms**. This process of making "recombinant DNA" is central to genetic engineering.

In July, NASA launched the Earth Resources Technology Satellite—the first of the USA's Landsat satellites. NASA launched seven satellites to **gather information about Earth from space**—providing data on land use, geology, oceans, lakes and rivers, and pollution. The last two Apollo missions, 16 and 17, took another six astronauts to the Moon. Three further Apollo missions had been planned, but were canceled, partly because the budget was cut and partly so that NASA could concentrate on developing an orbiting space station. In March, NASA launched its Pioneer 10 probe, bound for Jupiter (which it flew past in 1974).

245
POUNDS
THE WEIGHT OF **MOON ROCKS COLLECTED** BY **APOLLO 17** ASTRONAUTS

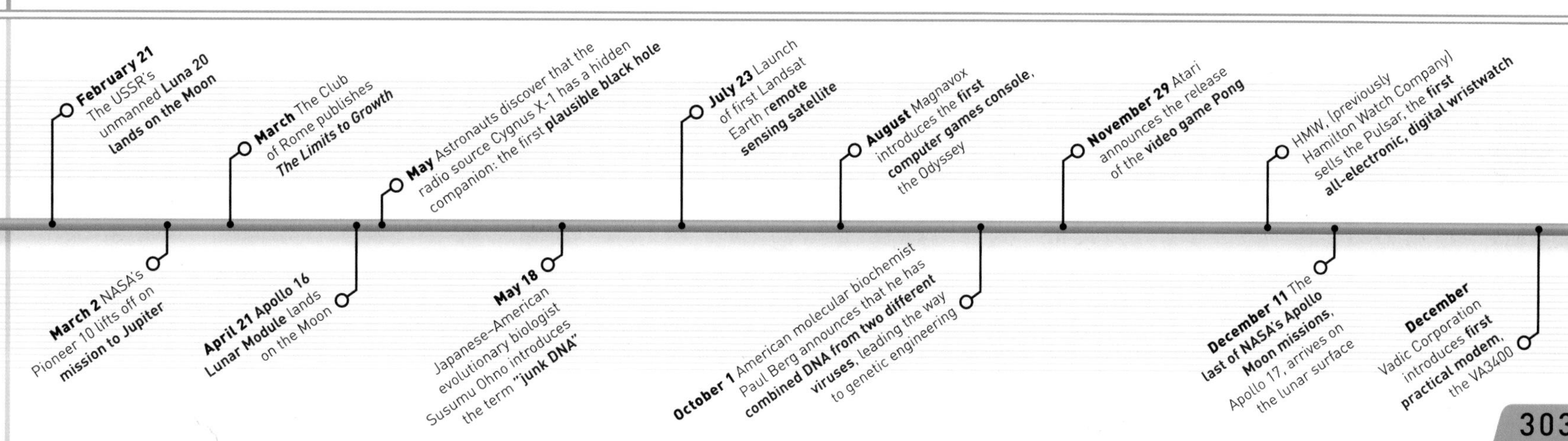

February 21 The USSR's unmanned Luna 20 lands on the Moon
March 2 NASA's Pioneer 10 lifts off on mission to Jupiter
March The Club of Rome publishes The Limits to Growth
April 21 Apollo 16 Lunar Module lands on the Moon
May Astronauts discover that the radio source Cygnus X-1 has a hidden companion: the first plausible black hole
May 18 Japanese-American evolutionary biologist Susumu Ohno introduces the term "junk DNA"
July 23 Launch of first Landsat Earth remote sensing satellite
August Magnavox introduces the first computer games console, the Odyssey
October 1 American molecular biochemist Paul Berg announces that he has combined DNA from two different viruses, leading the way to genetic engineering
November 29 Atari announces the release of the video game Pong
HMW (previously Hamilton Watch Company) sells the Pulsar, the first all-electronic, digital wristwatch
December 11 The last of NASA's Apollo Moon missions, Apollo 17, arrives on the lunar surface
December Vadic Corporation introduces first practical modem, the VA3400

2,249 THE NUMBER OF **DAYS SKYLAB** WAS IN **ORBIT AROUND EARTH**

THIS YEAR SAW THE LAUNCH of the USSR's second space station, Salyut 2, in April, while a month later the US launched Skylab. This space station remained in orbit until 1977.

In April, in New York, Motorola researcher Martin Cooper (b.1928) made the first telephone call from a **truly mobile, handheld phone**—albeit one that weighed 2.5lb (1.1kg). Around the same time, engineers at the European Organization for Nuclear Research (CERN), on the Swiss–French border, were developing the **world's first touch screen**, which was put to use on computers in a sophisticated control room.

Later in the year, American computer scientists **Vint Cerf** (b.1943) and **Robert Kahn** (b.1938) drafted the **internet protocol suite**—a set of networking standards that enable different interconnected computer networks to communicate. The most important components, Transfer Control Protocol and the Internet Protocol (TCP/IP), are the **basis of nearly all traffic across the internet today**. At Xerox's Palo Alto Research Center (PARC) in California, computer scientists produced the Alto—the first computer to have a **graphical user interface (GUI) and a mouse**.

In November, American geneticists Herbert Boyer (b.1936) and Stanley Cohen (b.1935) announced that they had created the **first-ever transgenic organism**, heralding the dawn of genetic engineering. Boyer and Cohen inserted **an antibiotic-resistant gene** from one bacterium into the genome of another, endowing the recipient bacterium with the **donor's resistance** against antibiotics.

Also in November, British inventor Stephen Salter (b.1938) applied for a patent for an alternative energy device, known as **Salter's duck**, which could **extract energy from water waves**. This device could be used to generate electricity.

Skylab in orbit
Skylab housed a laboratory, in which astronauts carried out scientific and medical experiments, and carried a telescope for studying the Sun.

Computer touch screen
Danish electronics engineer Bent Stumpe holds one of the first touch screens, which he developed with British engineer Frank Beck.

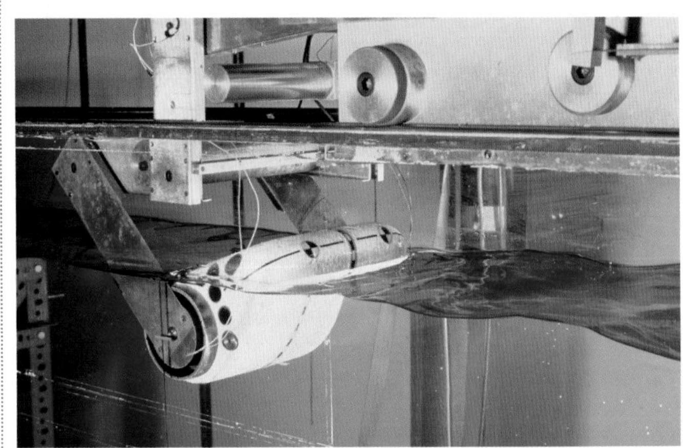

Salter's duck
A prototype of the duck generates electric power from waves in a laboratory test. The water behind the duck is flat because its energy has been extracted.

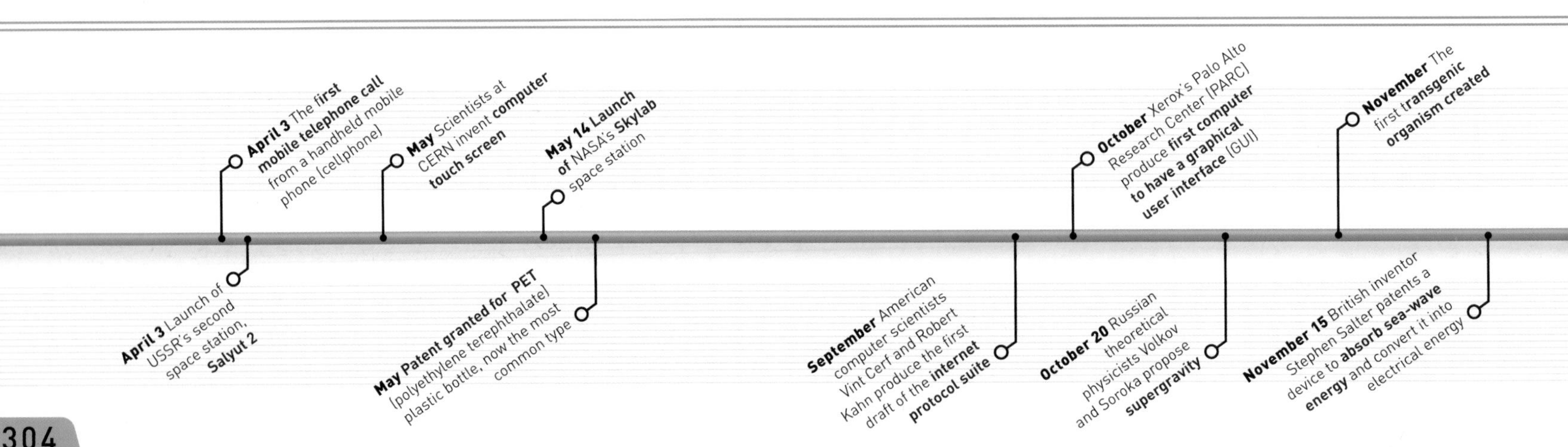
April 3 Launch of USSR's second space station, **Salyut 2**
April 3 The first **mobile telephone call** from a handheld mobile phone (cellphone)
May Scientists at CERN invent **computer touch screen**
May Patent granted for **PET** (polyethylene terephthalate) plastic bottle, now the most common type
May 14 Launch of NASA's **Skylab** space station
September American computer scientists Vint Cerf and Robert Kahn produce the first draft of the **internet protocol suite**
October Xerox's Palo Alto Research Center (PARC) produce **first computer to have a graphical user interface** (GUI)
October 20 Russian theoretical physicists Volkov and Soroka propose **supergravity**
November The first **transgenic organism created**
November 15 British inventor Stephen Salter patents a device to **absorb sea-wave energy** and convert it into electrical energy

The first extreme-loving organisms were discovered in Colorado, in Yellowstone National Park's sulfur-rich hot springs. In 1974, such organisms were classified as extremophiles.

STEPHEN HAWKING (b.1942)

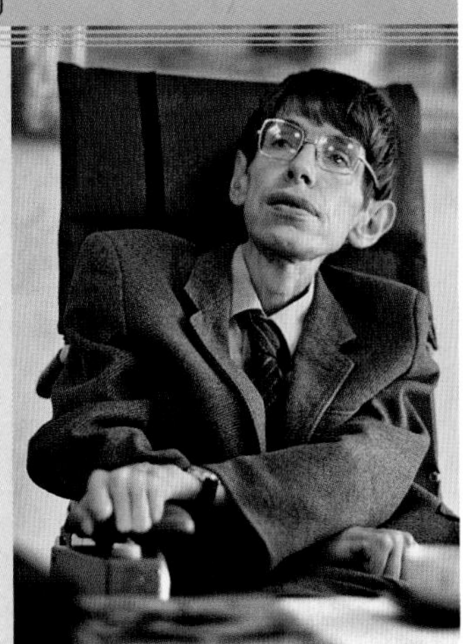

Perhaps best known to the public for *A Brief History of Time* (1988), British theoretical physicist Stephen Hawking has made pioneering contributions to our understanding of black holes, cosmology, and quantum physics. He is almost completely paralyzed, a result of the degenerative condition known as motor neuron disease, with which he was diagnosed when he was just 21.

IN JULY, AMERICAN MICROBIOLOGIST Robert MacElroy coined the term "extremophiles" to refer to **organisms that thrive in extremes**—of acidity, pressure, or temperature, for example. This classification sparked interest from astrobiologists, since such organisms might be found in hostile environments on other planets, and from evolutionary biologists, as extremophiles are mostly primitive organisms that evolved early in Earth's history.

German geneticist **Rudolf Jaenisch** (b.1942) and American embryologist **Beatrice Mintz** (b.1921) reported that they had **created the first transgenic animal**, by introducing virus DNA into a mouse embryo. The pace of development in genetics caused concern in the general public and the scientific community about the possible environmental effects of genetic engineering.

More concern about the environmental effects of scientific progress arose after scientists published their findings about the dangers of **chlorofluorocarbons** (CFCs). These man-made compounds were widely used as refrigerants and propellents in aerosol cans. The researchers warned that **CFCs could deplete the ozone layer**, which protects the planet from harmful ultraviolet radiation.

In November, American anthropologist **Donald Johanson** (b.1943) discovered fossil fragments of a hominid skeleton in the Afar Triangle, Ethiopia. The fossils were from a **3.2 million year-old *Australopithecus afarensis***—a species of bipedal (walked on two legs) hominid. The find was named "Lucy"—after the Beatles' song *Lucy in the Sky with Diamonds*, which was often played at the camp where the bones were found—and was by far the oldest such find at the time.

British theoretical physicist **Stephen Hawking** published a **revolutionary idea about black holes**. One of the findings of quantum physics is that empty space is constantly thronging with pairs of virtual particles, which exist only fleetingly, before annihilating each other. Hawking uncovered a strange possibility: for any virtual pairs created at the event horizon (boundary) of a black hole, one member would disappear inside, while the other would travel out into space, the result being that black holes "radiate" particles. At first controversial, **Hawking radiation** is now part of mainstream physics, and the search is on to detect it.

Lucy
A reconstructed skull of an Australopithecus afarensis, *a species of early bipedal apes.*

Also in physics, a rapid succession of discoveries known as the **November Revolution**—which included the discoveries of the charm quark and the particle called J or psi—gave an enormous boost to the emerging understanding of **subatomic particles**, known as the Standard Model of particle physics (see panel, below).

STANDARD MODEL OF PARTICLE PHYSICS

The "Standard Model," developed in the 1960s and 1970s, is still the best-fit explanation of fundamental particles—the most basic building blocks of matter and force. The model's success rests largely on its predictions of the existence of particles that have since been observed, including the charm quark in 1974. According to the theory, there are two families of fundamental particles: the fermions, which make matter, and the bosons, which carry force. The fermions are further divided into quarks—which normally occur in tightly bound twos or threes to make composite particles such as the proton and the neutron—and leptons, which occur singly and include the electron.

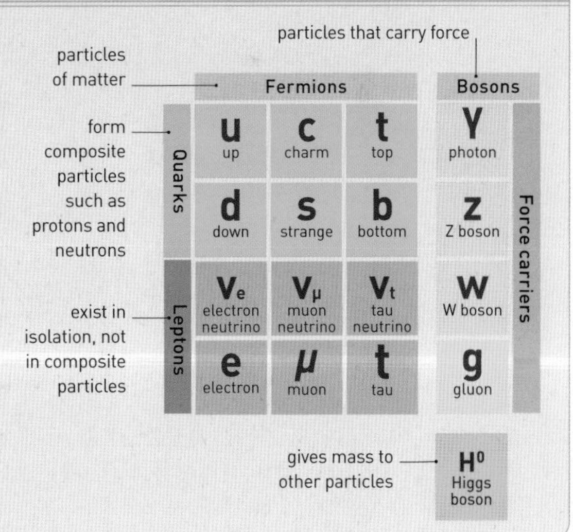

particles that carry force
particles of matter
Fermions
Bosons
form composite particles such as protons and neutrons
Quarks
u up
c charm
t top
γ photon
d down
s strange
b bottom
Z Z boson
exist in isolation, not in composite particles
Leptons
νe electron neutrino
νμ muon neutrino
νt tau neutrino
W W boson
e electron
μ muon
t tau
g gluon
Force carriers
gives mass to other particles
H0 Higgs boson

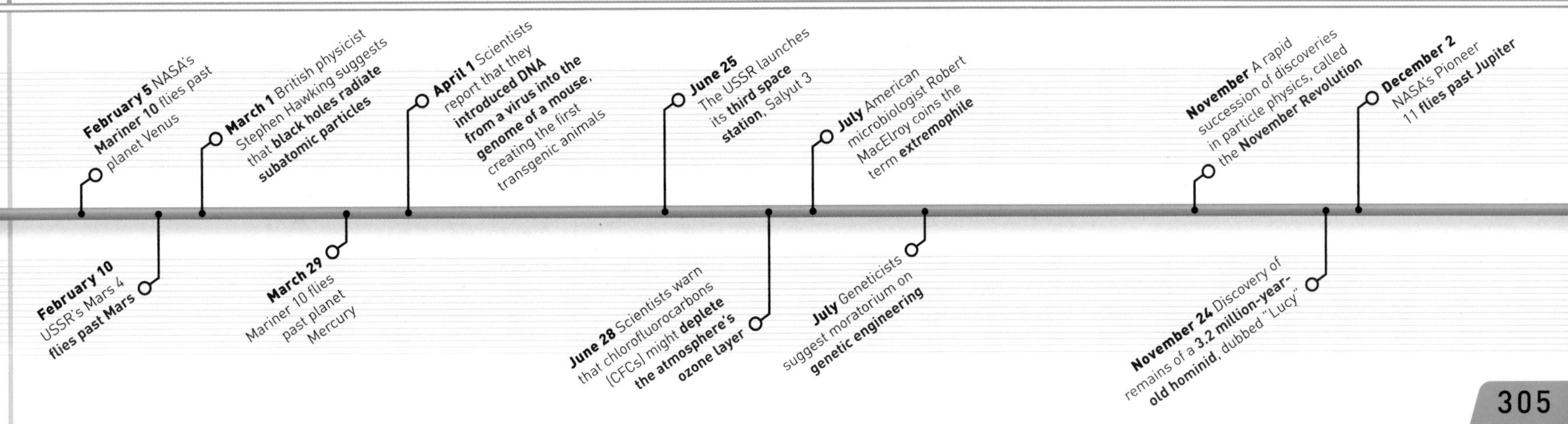

February 5 NASA's Mariner 10 flies past planet Venus
February 10 USSR's Mars 4 flies past Mars
March 1 British physicist Stephen Hawking suggests that black holes radiate subatomic particles
March 29 Mariner 10 flies past planet Mercury
April 1 Scientists report that they introduced DNA from a virus into the genome of a mouse, creating the first transgenic animals
June 25 The USSR launches its third space station, Salyut 3
June 28 Scientists warn that chlorofluorocarbons (CFCs) might deplete the atmosphere's ozone layer
July American microbiologist Robert MacElroy coins the term extremophile
July Geneticists suggest moratorium on genetic engineering
November A rapid succession of discoveries in particle physics, called the November Revolution
November 24 Discovery of remains of a 3.2 million-year-old hominid, dubbed "Lucy"
December 2 NASA's Pioneer 11 flies past Jupiter

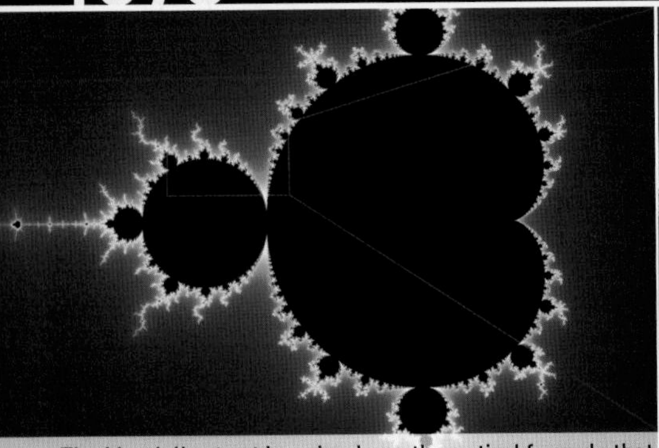

The Mandelbrot set is a simple mathematical formula that, when visualized using computer graphics, reveals intricate beauty at different scales.

53 MINUTES THE LENGTH OF TIME THE **VENERA 9 LANDER** OPERATED ON THE **SURFACE** OF **VENUS**

IN JUNE 1975, Japanese electronics company Sony released the first home video recorder format, **Betamax**. **Video cassette recorders** (VCRs) allowed people to record television programs and watch prerecorded films rented or bought from video stores. Although VCRs had been available since the 1960s, they were expensive and very few households owned one. Betamax was inexpensive enough to be purchased by many families for home use. The following year, another Japanese company, JVC, came out with a rival format—**VHS** (Video Home System). Over the next decade, the two systems competed in a "format war"; by the end of the 1980s, Betamax VCRs and cassettes represented only about 5 percent of the home video market, although they remained the standard for broadcasters and professional video editors until the rise of digital video production.

In July, millions of people around the world watched a historic moment live on their televisions: the docking of a **Russian Soyuz spacecraft** with an **American Apollo** module. The two craft were together in orbit for more than 40 hours, during which the crews carried out joint experiments, exchanged gifts, and un-docked and re-docked their craft several times.

Farther out in space, the Russian unmanned space probe Venera 9 became the first spacecraft to **orbit Venus**. A spherical entry pod detached from the orbiter and opened to release a lander probe, which descended to the planet's surface. It was the first probe to **send images back from** the **surface of another planet**; the orbiter acted as a relay station between Venus and astronomers on Earth. The lander relayed data and images for 53 minutes, after which radio contact was lost.

In November, Polish–French mathematician **Benoit Mandelbrot** (1924–2010) coined the word **fractal** in his book *Les Objets Fractals: Forme, Hasard, et Dimension* (*Fractals: Form, Probability, and Dimension*). The mathematics of fractals provided a way of bringing apparently **rough, irregular, and chaotic phenomena** into the domain of mathematics. It allowed mathematicians to understand and model intricate natural forms

mass spectrometer

helical antenna

air brake

Venera 9 lander
This is a model of the Venera 9 lander probe that carried out tests and sent panoramic photographs from Venus.

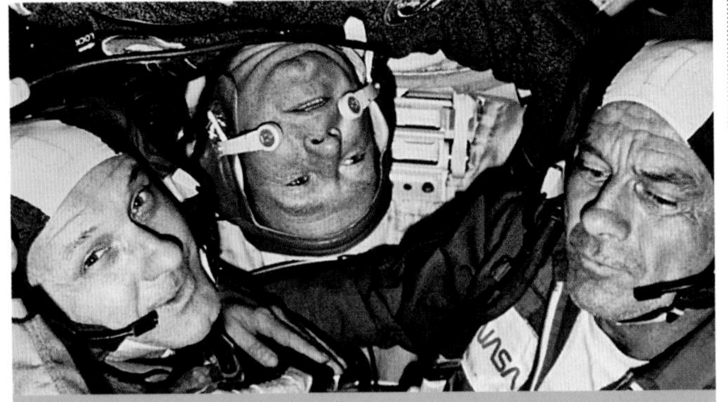

COLLABORATION IN SPACE

Since the 1950s, the US and USSR had been engaged in a "space race," each superpower trying to assert its technological superiority. The Russians were the first to put a satellite into orbit, launch a person into space, and send probes to the Moon. Not far behind, the Americans achieved perhaps the biggest coup, by landing astronauts on the Moon's surface. The docking of the two nations' craft in orbit—the Apollo-Soyuz Test Project—was an important statement of peace and collaboration between the US and USSR, a reflection of the easing of tensions between them.

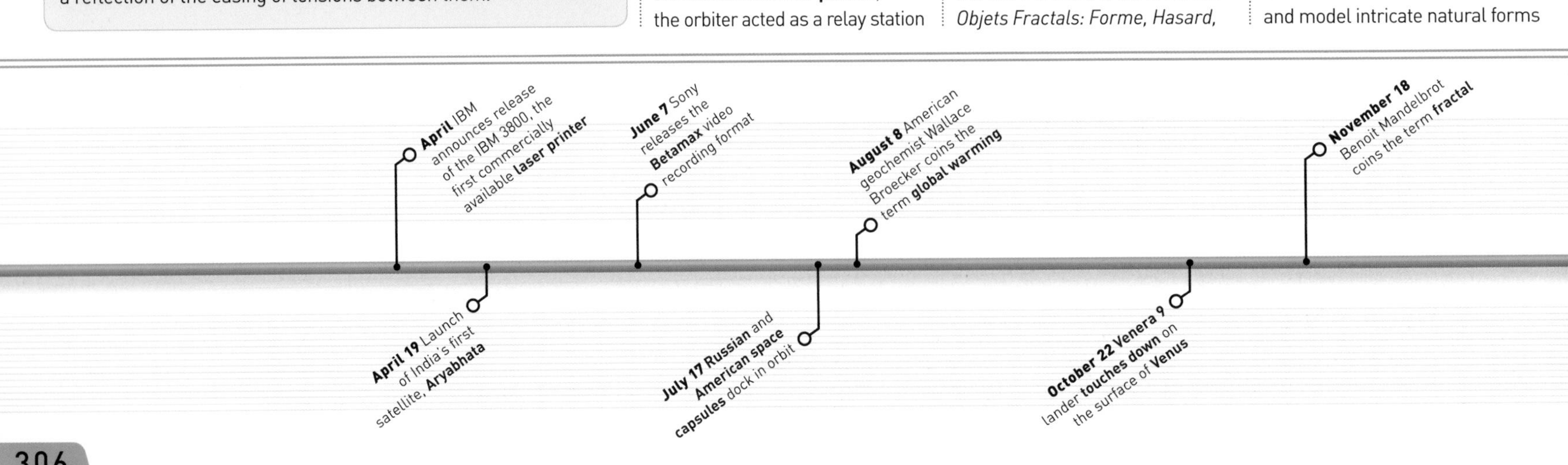
April IBM announces release of the IBM 3800, the first commercially available **laser printer**
April 19 Launch of India's first satellite, **Aryabhata**
June 7 Sony releases the **Betamax** video recording format
July 17 Russian and **American space capsules** dock in orbit
August 8 American geochemist Wallace Broecker coins the term **global warming**
October 22 Venera 9 lander **touches down** on the surface of **Venus**
November 18 Benoit Mandelbrot coins the term **fractal**

1976

1,354

MILES PER HOUR

THE **TOP SPEED** OF **CONCORDE**

Concorde could travel at more than twice the speed of sound.

such as mountains, clouds, snowflakes, plants, and lightning bolts as simple repeating patterns on different scales. Mandelbrot's work also fed into the emerging field of **computer graphics**, enabling computer game designers and artists to create stunning **virtual worlds** with realism. With its strong links to another emerging discipline, chaos theory, fractal mathematics helped scientists understand **unpredictable systems**, such as stock markets and earthquakes.

In December, US engineer **Steven Sasson** (b.1950) took the first digital photograph with a device he had invented, which proved to be a prototype of the **digital camera**. Sasson's images had a resolution of just 0.01 megapixels, and took 23 seconds to be stored on magnetic tape.

Also in December, US physicist **Martin Perl** (b.1927) announced the discovery of a subatomic particle that he called **tau lepton**. It was further evidence in favor of the Standard Model of fundamental particles (see 1974). In 1995, Perl was awarded the Nobel Prize in Physics for his discovery.

gamma-ray detector

motherboard
cassette board connector

Apple 1
The first Apple computer came as a single circuit board to which customers would connect a keyboard and a video display, and a cassette player to load programs.

THE YEAR DAWNED with the maiden flight of **Concorde**, the first supersonic jet airliner. It was developed jointly by British Aerospace (formerly British Aircraft Corporation) and the French company Aérospatiale. After more than 10 years of development and testing, Concorde's **first commercial flights** took place on January 21, one from London to Bahrain and the other from Paris to Rio. Concorde remained in service until 2003. Only one other supersonic airliner has ever gone into service: the Russian Tupolev Tu-144, in 1977. However, it was grounded in 1978 after a crash.

In March, British evolutionary biologist **Richard Dawkins** published *The Selfish Gene*. In it, he suggested that **evolution** was best understood at the **level of genes**, and not whole organisms. The gene-centric theory developed in the 1960s, and it accounts for certain behaviors that are otherwise hard to explain; for example, the altruism that organisms show toward those most closely related (and therefore likely to carry many of the same genes). Dawkins' use of the word "selfish," although figurative, captured the imagination of the general public and scientists alike. Many consider it to be a landmark in the development of evolutionary biology.

In April, American computer scientists **Steve Jobs** (1955–2011) and **Steve Wozniak** (b.1950), and electronics expert Ronald Wayne (b.1934) founded a new company called Apple Computers. Their first product, simply called the Apple Computer (later known as **Apple 1**), came as a single circuit board. It had 8 kilobytes of RAM, and sold for $666.66.

A few months later, US computer company Cray Research delivered a ground-breaking **supercomputer, Cray-1**, to the Los Alamos National Laboratory in New Mexico. Cray-1 was the brainchild of Seymour Cray (1925–96), who had been working since the mid-1960s on making processing units work in parallel, to increase computer power.

Surface of Mars
The Viking landers, which spent a combined total of 10 years studying the Martian surface, sent back many color pictures to Earth.

In July and September, NASA's unmanned space probes **Viking 1** and **Viking 2** orbited Mars and dropped lander probes on its surface. The landers sent back high resolution images, and also carried out **chemical analyses of the soil**, partly with the aim of finding clues that life had once existed on the planet; however, no such evidence was found.

RICHARD DAWKINS
(b.1941)

British evolutionary biologist Richard Dawkins was born in Kenya, and graduated with a degree in zoology from Oxford University in 1962. For most of the 1960s, he was a researcher in ethology (the study of animal behavior). He has written a number of influential and hugely popular books explaining evolution, criticizing creationism, and promoting atheism.

December Martin Perl announces the discovery of the **tau lepton**

December Steven Sasson takes the **first digital pictures**

January 21 First commercial flight of **Concorde**

March 16 Publication of Richard Dawkins' ***The Selfish Gene***

April 8 Belgian molecular biologist Walter Fiers announces sequencing the entire **genome of a virus**

June 18 Launch of NASA's **Gravity Probe A**

July 20 **Viking 1** lands on Mars

July The **Apple Computer** goes on sale; around 200 computers are sold

July The first appearance of the **Ebola virus**, the cause of an infectious disease that has killed thousands

September Shugart Associates introduce the first 5¼in (13.3cm) **floppy disk drive**

September 3 **Viking 2** lands on Mars

September 9 JVC releases the **VHS** video system

September 21 Cray Research delivers **Cray-1**, the world's most successful supercomputer

447 SECONDS

THE LENGTH OF TIME THE **GOSSAMER CONDOR** FLEW TO WIN THE **KREMER PRIZE**

The Gossamer Condor was the first human-powered aircraft to achieve sustained flight.

" IN A WAY, **I'M PROUD** THAT WHAT I WAS HAS HELPED OTHER **PEOPLE HAVE CHILDREN.** "

Louise Brown, first baby conceived through IVF, 1998

IN FEBRUARY, a team of oceanographers and geologists set out to explore the Galápagos Ridge in the East Pacific Ocean, searching for **hydrothermal vents**—cracks in the ocean floor where seawater meets hot magma. They found that the mineral-rich water gushing out of the hydrothermal vents supported a rich and diverse community of organisms never seen before. Later in the year, American paleobotanist **Elso Barghoorn** (1915–84) and his Ph.D. student Andrew Knoll discovered 3.4 billion-year-old fossils of simple single-celled organisms in rocks in South Africa. Their find pushed back the **date of the earliest known life** by several hundred million years.

In the same year, the science of **genomics**—analysis of the sequence of nucleotide bases (genetic material) along the length of an organism's DNA—took two major steps forward. First, in February, British biochemist **Frederick Sanger** determined the sequence of the 5,000 or so bases in the genome of a simple virus. Then, in December, Sanger published details of a new, rapid method of **genome sequencing**. The Sanger method remained the basis of genome sequencing until automated techniques became available in the early 2000s.

In August, **Jerry Ehman**, an American astronomer, detected a remarkably strong and consistent **radio signal** that seemed to have come from an **extraterrestrial source**. The signal's strength, frequency, and consistency suggested that it was broadcast on purpose by intelligent extraterrestrial beings. Ehman used Ohio State University's "Big Ear" radio telescope as part of the **Search for Extraterrestrial Intelligence (SETI)** project. The signal has never been detected again.

Also in August, the **Gossamer Condor** became the first **human-powered aircraft** to achieve sustained, controlled flight, completing a figure of eight around two markers set 0.4 mile (0.8 km) apart, at Minter Field, California. It had a mass of just 70 lb (32 kg) and was driven by pedal power. The aircraft's designer, Paul MacCready, was awarded the **Kremer Prize** (£50,000), which was set up in 1959 by British industrialist Henry Kremer to reward innovations in human-powered flight.

HYDROTHERMAL VENTS

A common feature around mid-ocean ridges, hydrothermal vents are fissures in the ocean floor from which chemically rich water emerges. They can either be hot "black smokers," which form rocky, chimneylike structures, or cool "white smokers." Some biologists believe white smokers to be the site of the origin of life.

MARY LEAKEY (1913–96)

Born in London, Mary Nicol married archaeologist and anthropologist Louis Leakey. The couple spent many years in Africa, inspired by the idea that human origins lay in that continent. In 1959, Mary found the 1.75 million-year-old remains of a hominid in Tanzania. Her find contributed to later understanding of human evolution.

THE US AIR FORCE LAUNCHED the first of 24 NAVSTAR (Navigation System using Timing And Ranging) satellites for its **Global Positioning System (GPS)** in February. The satellites acted as orbiting radio beacons. Each one carried an accurate atomic clock and broadcast constant signals announcing its position and the exact time. Ground-based receivers could detect signals from at least four satellites from any point on Earth. A receiver worked out its exact location by triangulating the signals from the satellites. In 1983, US President Ronald Reagan announced that the system would be available for public use once the first group of satellites was complete. A second group of satellites, the first of which was launched in 1989, enhanced accuracy.

Also in February, a team led by British paleontologist Mary Leakey revealed a set of **prehistoric footprints** that were made by bipedal hominids an estimated **3.4 million years ago**. The 80 ft- (24 m-) long set of footprints, discovered at **Laetoli**, in Tanzania, were made by three individuals walking in volcanic ash. Light rainfall soon cemented the footprints, which were then covered in another layer of ash and preserved until erosion revealed them to Leakey and her team. The oldest hominid footprints known before the discovery at Laetoli were made by Neanderthals just 80,000 years ago.

In July, the first human baby conceived outside a woman's body was born. Louise Brown was conceived through **in vitro fertilization (IVF)**—a technique that was the result of nearly a decade of work by British embryologist Robert Edwards and surgeon Patrick Steptoe.

February 17 Scientists discover **hydrothermal vents** near the Galápagos Ridge, in the East Pacific

February British biochemist Frederick Sanger produces the first DNA-based complete **genome sequence**

April 22 The first transmission of **telephone signals** via **optical fibers**

July 3 US researcher Raymond Damadian conducts the first full body **MRI scan** on a human being

August 15 The Big Ear radio telescope detects an unexplained **signal from outer space**

August 23 The Gossamer Condor becomes the first **human-powered aircraft** to achieve sustained, controlled flight

October 28 Elso Barghoorn and Andrew Knoll announce their discovery of **3.4 billion-year-old fossils** of bacteria

December Frederick Sanger announces a new, quicker method of **sequencing DNA**

February 22 Launch of the first **NAVSTAR satellite**, part of the Global Positioning System (GPS) network

February 24 Mary Leakey reports discovery of **fossilized footprints** made by bipedal hominids 3.6 million years ago

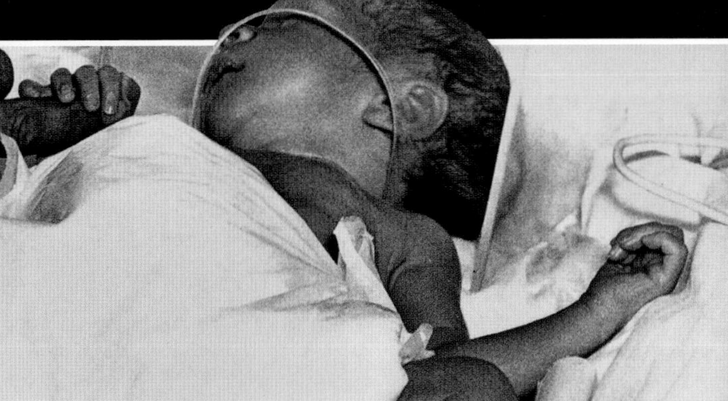

Louise Brown, the first baby conceived through IVF, holds her father's hand shortly after her birth at Oldham General Hospital, UK.

Newspaper reports of the birth of the first "test tube baby" caused a flurry of both praise for the achievement and condemnation for "meddling with nature." Since the birth of Louise Brown, several million babies have been conceived through IVF.

In another technological substitute for a natural process, US company Genetech announced that it had managed to **genetically engineer** *E.coli* bacteria to produce the hormone **insulin**. People suffering from diabetes (Type I) cannot produce enough of this hormone, and at the time, their only source of life-saving supplementary insulin was from pigs and cattle. The US Food and Drug Administration (FDA) approved Genetech's product in 1982, making it the first genetically engineered product to gain approval.

Ancient footprints
The footprints at Laetoli, preserved in hardened volcanic ash, were made by three individuals apparently on their way to a watering hole.

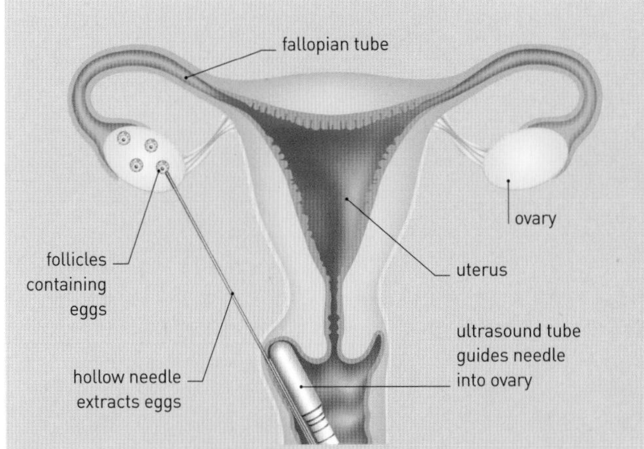
fallopian tube
ovary
follicles containing eggs
uterus
ultrasound tube guides needle into ovary
hollow needle extracts eggs

IN VITRO FERTILIZATION

In this assisted fertilization technique, a woman's ovaries are stimulated (using drugs) to produce several eggs. The eggs are extracted, and combined with sperm in the laboratory. One or two of the embryos produced are then implanted into the woman's womb to develop into fetuses. Since first introduced, IVF has become more sophisticated and there are many ways to achieve egg fertilization.

June 19 A team led by US physicist David Wineland announces that they have **trapped** a cloud of **magnesium atoms** using a laser—it would prove significant in the development of quantum computing

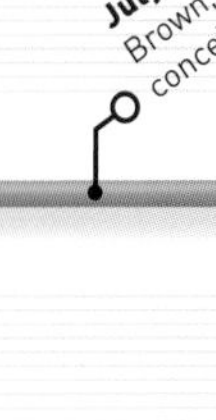

July 25 Birth of Louise Brown, the first baby conceived through **IVF**

September 6 US company Genetech announces that it has **genetically engineered bacteria** to produce **insulin**

November 13 NASA launches the **Einstein Observatory**—the first dedicated X-ray telescope

140,000

THE NUMBER OF **PEOPLE EVACUATED** FROM THE AREA AROUND **THREE MILE ISLAND** AFTER THE **DISASTER**

An aerial view of the Three Mile Island nuclear power plant, near Harrisburg, Pennsylvania, which is considered the site of the US's worst nuclear accident.

> “ FOR EACH OF OUR **ACTIONS** THERE ARE ONLY **CONSEQUENCES.** ”
>
> **James Lovelock, British biologist,** in *Gaia: A New Look at Life on Earth,* 1979

AS THE DECADE DREW TO A CLOSE, progress in space exploration continued. NASA's **two Voyager space probes** reached their first target, Jupiter, two years after leaving Earth. Voyager 1 made its closest approach in March; Voyager 2 in July. The probes sent back exceptionally clear photographs and data to reveal **new information about Jupiter** and its moons, including the discovery of volcanoes on the innermost moon, Io. In September, **NASA's Pioneer 11** became the first space probe to fly past Saturn, another gas giant. Pioneer 11 came as close as 13,000 miles (21,000 km) to the tops of the planet's deep, dense clouds. At the end of the year, the European Space Agency launched the first of its highly successful **Ariane rockets**, from French Guyana.

In May, astronomers based at Kitt Peak National Observatory, Arizona, discovered two quasars close to each other in the sky; a quasar (quasi-stellar radio source) is the energetic centre of a distant galaxy. The two quasars appeared so similar that the astronomers concluded they were two views of the same object. This was the **first example of gravitational lensing** (see panel, left): the light from the quasar was bent, as if by a lens, due to the distortion of space-time by a massive galaxy cluster that lies between the quasar and Earth.

On March 28, the US suffered its **worst-ever nuclear power plant accident**. In the early hours of the morning, one of the reactors at the **Three Mile Island nuclear power plant**, Pennsylvania, went into partial meltdown after a valve failed to operate correctly. A large amount of pressurized water circulating the reactor core then leaked out into a containment building. Although the reactor was stabilized by the evening, concern over rising radiation levels caused local officials to advise children and pregnant women within an 5 mile- (8 km-) radius to evacuate; two days after the accident, the radius was extended to 20 miles (32 km). Radiation levels around the power plant did not rise significantly, and nuclear

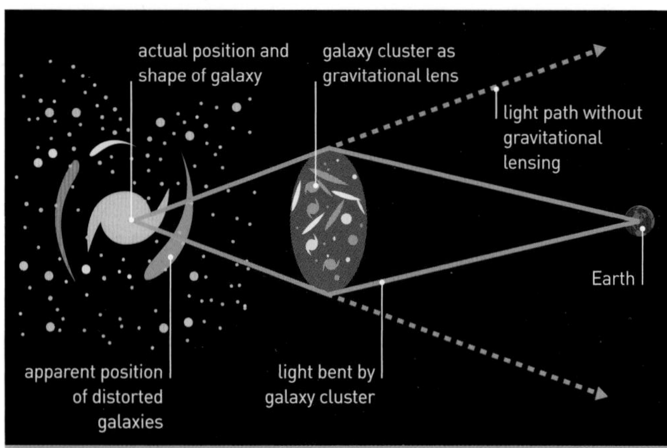

actual position and shape of galaxy
galaxy cluster as gravitational lens
light path without gravitational lensing
Earth
apparent position of distorted galaxies
light bent by galaxy cluster

GRAVITATIONAL LENSING

The gravitational field of matter present in space acts as a lens, causing light to be deflected. As a result, people viewing a celestial object from Earth may see multiple distorted images of it. This phenomenon, known as gravitational lensing, was first proposed in 1936 by German–American physicist Albert Einstein. The first example of gravitational lensing was discovered in 1979, when astronomers observed two views of the same quasar.

fission in nuclear fuel heats water filling reactor

control rods slide down between fuel rods to slow down the rate of fission

steam used to spin turbines and drive electricity generators

hot water produces steam in the steam generator

nuclear fuel rods

concrete dome

steam is condensed to water and returned to steam generator

coolant pump circulates water around reactor

NUCLEAR REACTOR

A nuclear reactor generates electric power by using nuclear fission to heat water. This in turn produces steam, which is used to drive huge turbines. A reactor core contains fuel rods made of radioactive material, and control rods, which limit the rate of fission. There are two separate cooling circuits, in which water circulates to carry heat away from the reactor core.

- **January** Bell Laboratories releases **Version 7 of the UNIX** computer operating system
- **March 5** Space probe Voyager 1 makes its **closest approach to Jupiter**
- **March 28 Meltdown** at the Three Mile Island **nuclear power plant**
- **May 31** Astronomers publish first evidence of **gravitational lensing**
- **June 12** Human-powered airplane **Gossamer Albatross crosses the English Channel**, winning the Kremer Prize for the second time
- **July 1 Sony introduces the Walkman** portable audio cassette player
- **July 9** Space probe **Voyager 2** makes its **closest approach to Jupiter**
- **September 1** NASA's **Pioneer 11** becomes the **first spacecraft to fly past Saturn**

Erosion has revealed the 65-million-year-old Cretaceous–Paleogene boundary—which corresponds to the extinction of the dinosaurs—in these rocks in the Hoodoo Formation, Alberta, Canada.

scientists concluded that neither people nor the environment had come to any harm.

Sony, a Japanese company, introduced a revolutionary new product in July: **the Walkman**. It was the first truly **personal portable audio device**, allowing people to carry their music with them on audio cassettes and listen to it on headphones.

Earlier in the year, the US company Bell Laboratories released the **seventh, and most important, version of UNIX**—a computer operating system that is the direct ancestor of the popular modern operating systems Mac OS X and Linux.

In October, British biologist **James Lovelock** published *Gaia: A New Look at Life on Earth*. The book set out **the Gaia hypothesis**. This states that Earth can be considered a single, self-regulating, living organism. Lovelock asserted that Earth's living things interact with the physical environment, keeping the oceans and atmosphere favorable for life to continue. Lovelock had been developing the idea and gathering relevant evidence since the early 1960s. The hypothesis influenced how many people tend to think about the environment, and the **interconnectedness and interdependence** of all living things in the world.

LUIS ALVAREZ (1911–88)

American physicist Luis Walter Alvarez's hypothesis about the cause of the extinction of dinosaurs came near the end of a long and illustrious career in physics. He made many contributions to radar technology during World War II, but his speciality was subatomic particles. In 1968, he won the Nobel Prize in Physics.

IN JANUARY, AMERICAN THEORETICAL PHYSICIST Alan Guth (b.1947) proposed a **refinement to the Big Bang theory** (see pp.344–45), which states that the Universe began in an incredibly tiny, hot, dense state several billion years ago, and has been expanding ever since. He suggested that the **universe underwent "cosmic inflation,"** expanding by a factor of a trillion trillion trillion trillion trillion trillion trillion trillion (1 followed by 72 zeroes) in a tiny fraction of a second. Guth's proposition solved many existing problems with the Big Bang theory. Evidence in support of Guth's theory has since come from astronomers and particle physicists, and although some mysteries remain, it is now almost certain that cosmic inflation did occur.

In June, **Luis Alvarez** (see panel, left) put forward a theory to explain the **extinction of dinosaurs** 65 million years ago. Geologists had already noticed a distinct transition in rock layers dated to the time of extinction. Alvarez analyzed the rocks at this Cretaceous–Paleogene (K–Pg) boundary and found high levels of iridium, an element more common in asteroids. He suggested that **an asteroid hit Earth**, throwing up dust that blocked out the Sun for thousands of years, and leaving a still-visible crater. His hypothesis remained controversial until 1990, when a huge 65-million-year-old crater was discovered off the Yucatán peninsula, Mexico.

In the following month, pharmacologists announced the isolation of a compound called mevinolin from the mold (fungus) *Aspergillus terreus*. Mevinolin was shown to inhibit **the production of cholesterol**, which is associated with increased risk of heart disease. It was renamed lovastatin, and became the **first cholesterol-reducing "statin" drug** ever sold. In 1987, the US Food and Drug Administration (FDA) approved it, under the brand name Mevacor. Lovastatin has since been discovered in other fungi, such as oyster mushrooms.

Mushroom
*The cholesterol-reducing compound lovastatin accounts for around 2 percent of the dry weight of edible oyster mushrooms (*Pleurotus ostreatus*), shown here.*

Earlier in the year, the thirty-third World Health Assembly declared that **smallpox had been eradicated globally**.

Space probe Voyager 1 made its **closest approach to Saturn** in November, flying within 77,000 miles (124,000 km) of the planet's cloud tops.

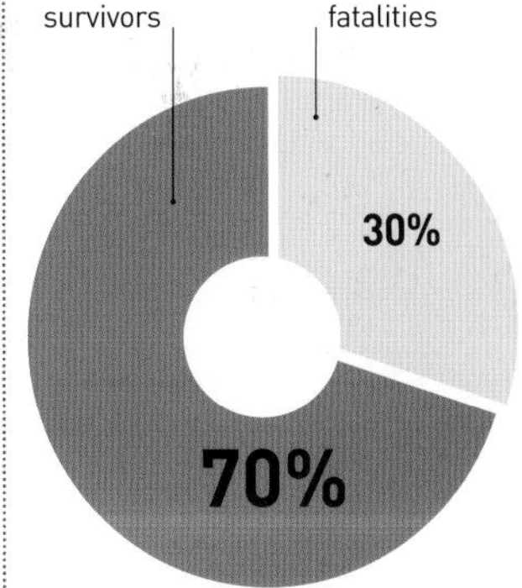
survivors
fatalities
30%
70%

Deaths from smallpox
Until its global eradication in 1980, smallpox had claimed millions of human lives, killing about one-third of all those afflicted by it.

84 MINUTES
THE TIME IT TOOK **VOYAGER'S RADIO SIGNALS** FROM **SATURN** TO REACH ASTRONOMERS **ON EARTH**

October 4 British biologist James Lovelock publishes ***Gaia: A New Look at Life on Earth***

December 24 The European Space Agency and Arianespace launch the **first Ariane rocket**

January 23 American theoretical physicist Alan Guth publishes his **Cosmic Inflation theory**

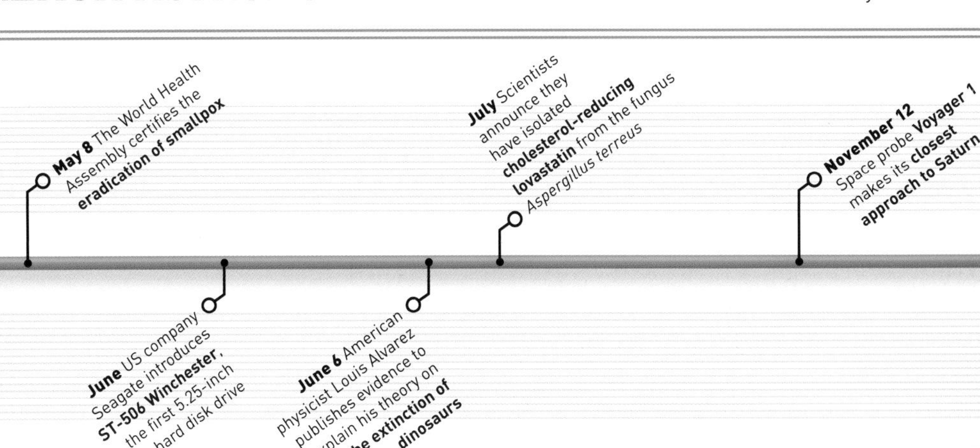

May 8 The World Health Assembly certifies the **eradication of smallpox**

June US company Seagate introduces **ST-506 Winchester**, the first 5.25-inch hard disk drive

June 6 American physicist Louis Alvarez publishes evidence to explain his theory on **the extinction of dinosaurs**

July Scientists announce they have isolated **cholesterol-reducing lovastatin** from the fungus *Aspergillus terreus*

November 12 Space probe **Voyager 1** makes its **closest approach to Saturn**

25 TON 10 TON
USA

April The **first artificial skin** for burn victims tested

April 3 The launch of the Osborne 1, the **first portable computer**

April 12 NASA's first reusable spacecraft, the **Space Shuttle *Columbia***, goes into orbit for the first time

April 27 Xerox introduces the **Xerox "Star,"** the first commercial computer with a **mouse, icons, and menus**

May American physicist Richard Feynman lays down the foundations for **development of quantum computers**

July 9 British scientists Martin Evans and Matthew Kaufman report that they have cultured **pluripotent cells from mouse embryos**

“ WHAT WE DO IS **MAKE A SCAFFOLDING** ON WHICH THE **CELLS GROW.** ”

John Burke, American trauma surgeon, in *New York Times*, 1981

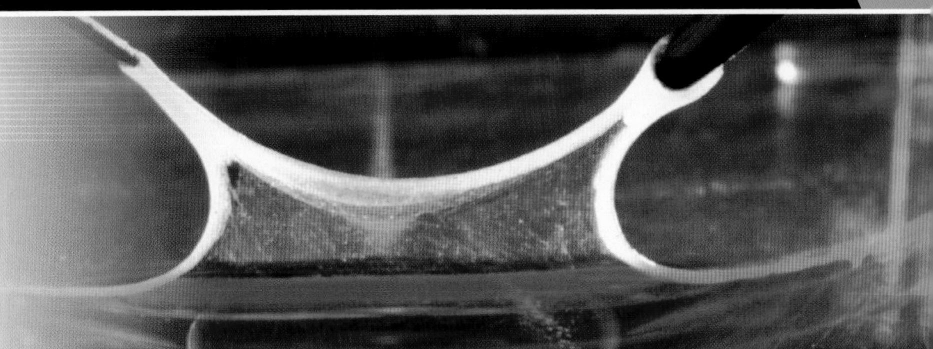
Artificial human skin, shown here, can be created by culturing human skin cells and growing them around a collagen matrix.

IN APRIL, NASA'S REUSABLE SPACECRAFT, the Space Shuttle, made its first full, orbital test flight. It was the first of 28 flights made by Space Shuttle *Columbia*, which tragically broke up on reentry in 2003. The Shuttle's main engines lifted the craft into orbit, with extra fuel provided by an external fuel tank; solid-fuel booster rockets provided extra thrust. Once empty, the boosters and the fuel tank detached from the Shuttle. The boosters were recovered and reused in future missions. NASA used **five space shuttles**, which made a total of 135 successful missions, lifting into orbit satellites as well as components for the **International Space Station**.

American company Osborne Computer Corporation released the **first successful portable computer**, the Osborne 1. This was followed by Japanese company Epson's **first laptop computer**, the HX-20, which weighed just 3.5 lb (1.6 kg) and worked on rechargeable batteries. Xerox launched the 8010 Information System ("Star"), a workstation computer significant as the first commercially available computer with files displayed graphically as icons inside "window" folders, and an on-screen pointer controlled by a mouse. The **most important development in computer technology** this year was the release of the IBM 5150, usually referred to as the IBM-PC. The computer's operating system was PC-DOS, a version of MS-DOS (Microsoft Disk Operating System). Microsoft's enormously successful Windows software (see 1985) was built around MS-DOS. The IBM-PC had a huge influence on the **development of the personal computer**. Its success led other companies to make clones—called IBM compatibles, which have dominated the personal computer market ever since. Companies could make IBM compatibles using off-the-shelf hardware, but the operating system had to be licensed from Microsoft—making that company extremely successful.

IBM-PC
One configuration of the IBM 5150 had two disk drives. There was no mouse, because the only interaction with the operating system was through typed commands.

In April, Swiss physicist **Heinrich Rohrer** and German physicist **Gerd Binnig** succeeded in building the first **scanning tunneling microscope** (STM, see panel below)—an instrument with which scientists can produce accurate images of the atoms that make up solid surfaces. Also in April, American trauma surgeon **John Burke** and Greek-born American chemical engineer **Ioannis Yanas** developed the **first successful artificial skin** for patients with severe burns. Their skin was made with collagen from sharks and cows, sealed with a layer of silicone rubber. The collagen formed a scaffold onto which the body could generate its own collagen, and build up new skin—and the silicone layer could then be removed.

Two separate teams, one in the UK and one in the US, developed the technology to isolate and culture **embryonic stem (ES) cells** from mouse embryos. ES cells are pluripotent—they have the potential to develop into any kind of cell. They can also replicate indefinitely. Human ES cells—first cultured in 1998—hold **great promise for future medical treatments**. For example, ES cells could be used to generate tissues for transplant, or implanted into the body, could help repair damage done by disease, injury, or aging.

Preparing for launch
NASA's Space Shuttle Columbia *sits on the launch pad at Kennedy Space Center in Florida, US. It is attached to the huge external liquid fuel tank and two slimmer solid-fuel boosters.*

256 THE MAXIMUM **NUMBER OF KILOBYTES** OF RAM MEMORY IN THE **IBM-PC**

SCANNING TUNNELING MICROSCOPE

At the heart of a scanning tunneling microscope (STM) is an extremely sharp metal tip that scans across a surface at very close range, measuring the electric current created by electrons "tunneling" across the gap between the surface and the tip of the probe. The color-enhanced STM image to the right shows a clump of gold atoms (yellow and brown) on a graphite surface (carbon atoms, green).

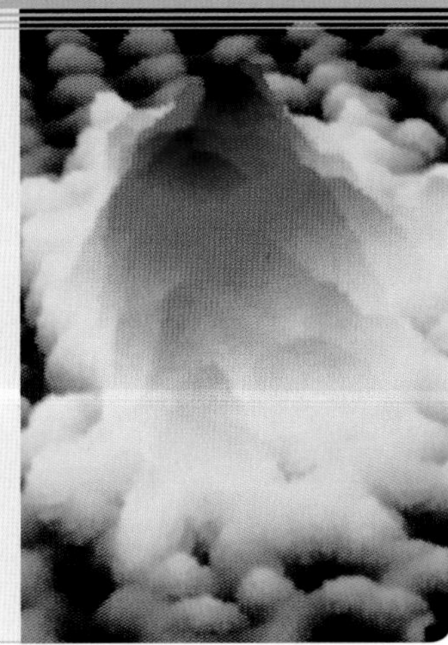

August Microsoft releases first version of **MS-DOS operating system**, a branded version called PC-DOS given away with the IBM-PC

August 12 IBM launches the **first successful personal computer**, the IBM-PC

August 26 Space probe **Voyager 2** makes closest approach to Saturn and **sends back stunning images**

November Launch of the Epson HX-20, the world's **first lightweight laptop computer**

December 4 Vaccine for **hepatitis-B virus** first approved in the US

December American molecular biologist Gail Martin reports that she had cultured pluripotent cells from mouse embryos, and she coins the term **embryonic stem cells**

10 BILLION DOLLARS THE COST OF DAMAGE CAUSED BY EL NINO IN 1982–83

WIDE LOAD
In 1982, El Niño caused sustained high rainfall, resulting in the raised water level in the San Lorenzo River, California.

AN IMPORTANT DEVELOPMENT IN DIGITAL SOUND reproduction, the **compact disk (CD)**, became available in October this year.

The CD, developed jointly by Philips and Sony, is a polycarbonate disk with a thin layer of aluminum inside. The aluminum is pitted with **millions of tiny indentations**, arranged as a spiral track more than 3 miles (5km) long. The pits represent binary digits, which in turn represent the original sound. A laser inside the CD player reflects light off the pits as the disk spins, and a microprocessor reconstructs the original sound wave from the pattern of reflected light. Within a few years, **CDs became the most popular format** for buying recorded music. The compact disk was adapted as a medium for read-only computer data storage (CD-ROM), and later, as writeable data disks (CDRs).

Also in October, American synthesizer pioneer **Robert Moog** announced the **Musical Instrument Digital Interface** (MIDI), a new way of recording and playing back musical performances. MIDI consists of simple messages that relate to the notes played; the messages can be produced by playing MIDI instruments such as keyboards, or by manipulating software. MIDI messages trigger sound samples, giving musicians great flexibility in composition, recording, and performance.

In July, a recently identified disease was given its name: **acquired immunodeficiency syndrome** (AIDS). The disease had claimed many lives in the gay community in New York and California. However, it had become clear that, while the disease spread easily among homosexual men via sexual contact, it was by no means confined to the gay community. The **rapid spread of the disease** led to major health campaigns, encouraging people to use condoms during intercourse, and intravenous drug users to avoid sharing needles.

Widespread extreme weather alerted the general public to the **existence of the El Niño phenomenon**—in which the Pacific Ocean remains warmer for longer than usual because of a change in the world's trade winds (the prevailing easterly winds found in the tropics). El Niño events happen occasionally, and typically last for several months. This year's El Niño, **one of the most devastating on record**, began in July and lasted into the next year. It decimated fish stocks in Peru, brought drought and bush fires to Australia and parts of Africa, and extreme rainfall in California and Peru—causing an estimated 2,000 deaths.

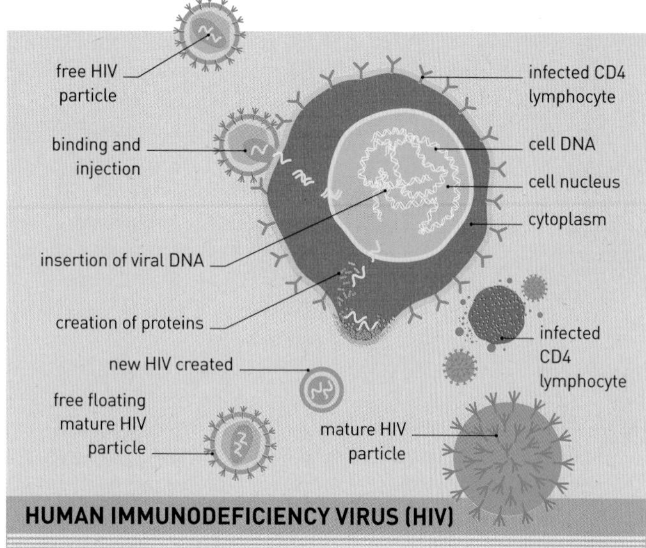
free HIV particle
binding and injection
insertion of viral DNA
creation of proteins
new HIV created
free floating mature HIV particle
infected CD4 lymphocyte
cell DNA
cell nucleus
cytoplasm
infected CD4 lymphocyte
mature HIV particle

HUMAN IMMUNODEFICIENCY VIRUS (HIV)

The cause of AIDS was identified as a virus in 1984, and that virus was given the name human immunodeficiency virus (HIV) in 1986. HIV is transmitted in bodily fluids and infects cells crucial to the body's immune system—in particular, CD4 cells—and uses them to reproduce. The cells may die as a result or be destroyed by other immune system cells.

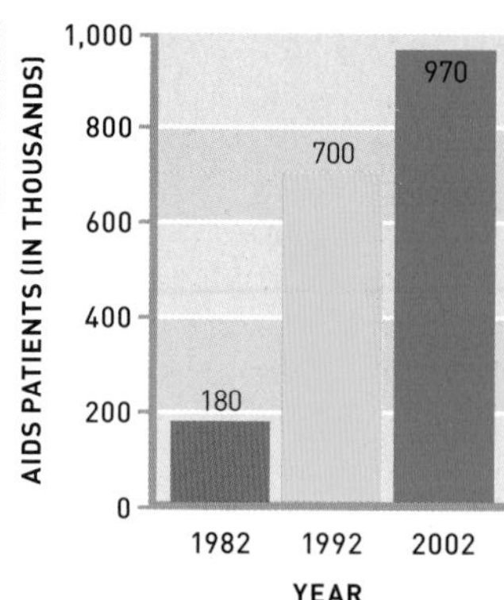
AIDS PATIENTS (IN THOUSANDS)
1,000
800
600
400
200
0
180
700
970
1982
1992
2002
YEAR

Living with AIDS
This graph shows the number of people living with AIDS in the USA. Worldwide, 30 million people were living with the disease by 2002.

In May, Taiwanese biologist **Chiaho Shih** (b.1950) and American biologist **Robert Weinberg** (b.1942) reported that they had isolated a human oncogene, which is a **genetic cause of cancer**.

In December, American surgeon William DeVries (b.1943) performed surgery on retired dentist Barney Clark to implant the **first permanent artificial human heart**, the Jarvik-7. The artificial heart, designed by American inventor Robert Jarvik (b.1946), kept Clark alive for 112 days.

Artificial heart
A Jarvik-7 artificial heart like this was given to Barney Clark at the University of Utah. Clark lived for 112 days after the operation.

January US computer company Commodore introduces the **hugely successful Commodore 64** personal computer

February 1 Intel introduces the first microprocessor with more than **one hundred thousand transistors** on it

April 9 American neurologist Stanley Prusiner defines the term "prions" to describe the **newly discovered infectious particles** that cause scrapie, BSE, and CJD

May 1 Isolation of first human oncogene (cancer-causing gene) announced

July Onset of weather phenomenon known as **El Niño**

July 27 At a meeting in Washington DC, the disease now known as **AIDS gains its name**

October 1 Sony launches the world's **first commercially released compact disk player**, the Sony CDP-101

1983

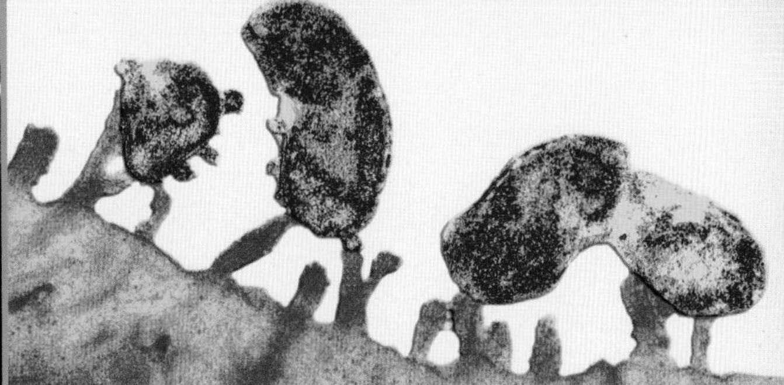

> **“I DRANK** IT DOWN **IN ONE GO** AND THEN **FASTED** FOR THE REST OF THE DAY. A FEW **STOMACH GURGLES** OCCURRED, **WAS IT THE BACTERIA** OR WAS I **JUST HUNGRY?”**

Barry Marshall, Australian physician, from *Nobel Lecture*, December 8, 2005

Rod-shaped, helical bacteria, known as *Helicobacter pylori*, attached to the stomach lining. These bacteria cause stomach ulcers.

Orchestra for sale?

FAIRLIGHT CMI

Full orchestra
A MIDI controller keyboard came with a standard set of MIDI instruments, but MIDI messages can play any sounds.

THIS YEAR, THREE IMPORTANT DEVELOPMENTS in consumer electronics reflected advances made exactly 10 years earlier. The first day of the year was the deadline for all computers connected to the worldwide network known as ARPANET to be using the internet protocol, TCP/IP (see 1973). Many computer historians consider that day as the **switching on of the modern internet**. Until then, some computers were still using a different protocol to communicate, but TCP/IP has formed the **basis of all internet traffic ever since**. The first computer to have a graphical user interface (GUI) was the Alto, developed at Xerox's research center in California (see 1973). Ten years later, Apple Computer Inc. launched the **first personal computer with a GUI**.

In October, American businessman David Meilahn made the first commercial mobile phone call, over a cell-based wireless network—10 years after the first call on a prototype device (see 1973).

Apple Lisa
The Apple Lisa, the first personal computer with a graphical user interface, and the team that designed it.

Physicists at the **European Organization for Nuclear Research** (CERN) **discovered three particles** whose existence gave more weight to the Standard Model (see 1974). The W^+, W^-, and Z bosons carry the "weak interaction," which is involved in radioactive decay (see pp.266–67). The existence of these particles had been **predicted in 1968**, as a result of a unified theory of the weak interaction and the electromagnetic force. Their discovery was only possible because of CERN's powerful new particle accelerator, the **Super Proton Synchrotron**, which had been operational since 1976.

In October, the 17th General Conference on Weights and Measures defined the meter as equal to the **length of the path traveled by light in a vacuum** during a time interval of one 299,792,458ths of a second.

At the University of Western Australia, in Perth, physician **Barry Marshall** (b.1951) and pathologist **Robin Warren** (b.1937) identified the most **common cause of stomach ulcers**—a common condition that **can lead to death** by gastrointestinal bleeding or stomach cancer. Their work began when Warren became curious about large numbers of a new species of bacteria he had found in patients' stomachs. At the time, no one expected bacteria to be present inside the stomach, because of the strong acid there. After finding the bacteria in patients with stomach ulcers, Warren and Marshall hypothesized that these organisms were infecting the lining of the stomach and the duodenum (part of the small intestine), leading to inflammation and causing the ulcers. Marshall realized he needed to carry out a definitive **test of the hypothesis on a human subject**—and he decided to use himself as the subject. After making sure the bacteria were not present in his stomach, he prepared a culture, mixed it with chicken broth, and drank it. Marshall's stomach lining became inflamed, and the hypothesis was soon proven. The link between the new species of bacteria—later named *Helicobacter pylori*—and stomach ulcers meant that most ulcers can be treated using antibiotics, saving hundreds of thousands of lives. In recognition of their discovery, Marshall and Warren were awarded the 2005 Nobel Prize in Physiology or Medicine.

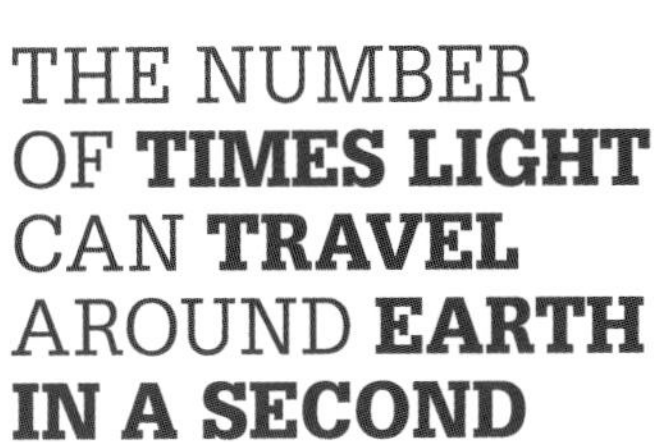

7.5 THE NUMBER OF **TIMES LIGHT** CAN **TRAVEL** AROUND **EARTH IN A SECOND**

openings connect to major arteries and veins
polyester shell

Mobile phone
The world's first commercially available mobile phone, Motorola's DynaTAC 8000x. Early mobile phones like this were affectionately known as "bricks."

October Introduction of **MIDI** (Musical Instrument Digital Interface)

December 2 American Barney Clark becomes the first person to be given a **permanent artificial heart**

January 1 The date on which the modern **internet is "turned on"**

January 19 Apple releases its Lisa computer, the **first home computer** with a **graphical user interface** (GUI)

January Particle physicists at CERN discover the first particles **responsible for the weak nuclear force**, the two W bosons

May Particle physicists at **CERN discover** the other particle responsible for the weak nuclear force, the **Z boson**

June First description of a **new species of bacteria**, *Helicobacter pylori*, which survives in the stomach and causes stomach ulcers

October 13 **First wireless call** made, on a Motorola DynaTAC 8000x, the world's first commercial portable mobile phone

October 21 The **meter defined** as equal to the length of the **path traveled by light in a vacuum** during a set time interval

Cuneiform inscription
c.3200 BCE
The earliest written language was cuneiform—a set of symbols inscribed on wet clay with a stylus.

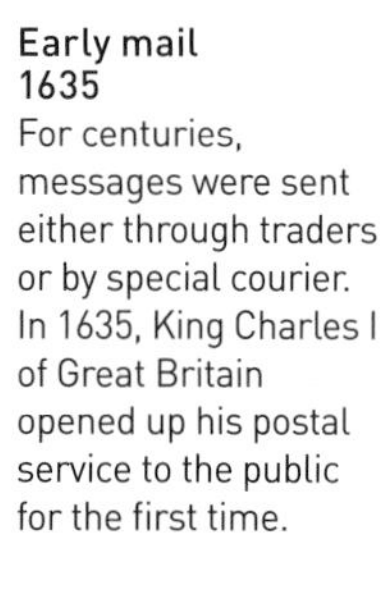

Early mail
1635
For centuries, messages were sent either through traders or by special courier. In 1635, King Charles I of Great Britain opened up his postal service to the public for the first time.

POSTAGE
ONE PENNY

Message carrier
Early 20th century
The idea of using specially trained pigeons to carry mail originated in Persia. Messages would be placed inside metal carriers and attached to pigeons' legs.

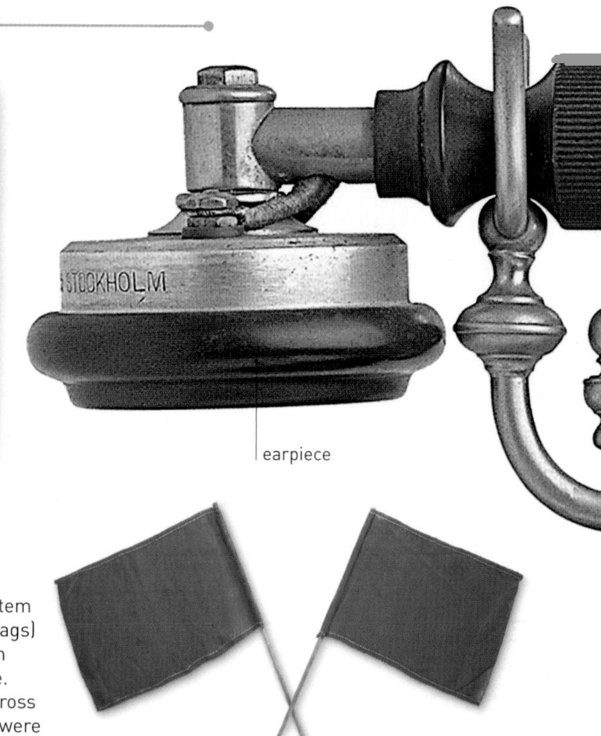
STOCKHOLM
earpiece

Semaphore flags
1792
The first semaphore system (visual signaling using flags) was developed by French engineer Claude Chappe. Signals could be sent across a network of towers but were limited by the weather.

COMMUNICATION

TECHNOLOGY FOR TALKING AND SENDING MESSAGES OVER LONG DISTANCES HAS SHAPED THE MODERN WORLD

The ability to communicate complex ideas through language is a key trait separating humans from animals. In recent years, technological advances have enabled us to send messages faster than a person can travel.

Most prehistoric cultures communicated solely through the spoken word, relying on an oral tradition to pass on or record important information. The appearance of writing in the 4th century BCE transformed human society forever. However, written messages still had to be delivered by hand. It was only in the 19th century that the harnessing of electricity paved the way for modern forms of instantaneous communication.

Bell's electric telephone
1876
Alexander Graham Bell's telephone transmitted sound using electrical signals. As the signal reached the receiver, a metallic disk would vibrate, producing soundwaves.

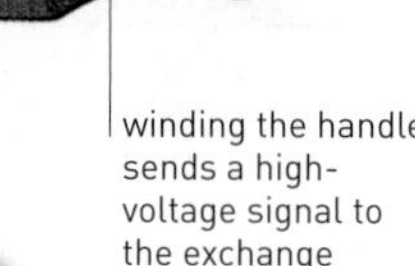
winding the handle sends a high-voltage signal to the exchange

tickertape records received messages for decoding
key to send electrical pulses

Morse telegraph
1836
American inventor Samuel Morse designed a telegraph that could send signals over long distances along a single wire. His colleague Alfred Vail devised a code that used short and long pulses ("dots" and "dashes") to represent letters of the alphabet.

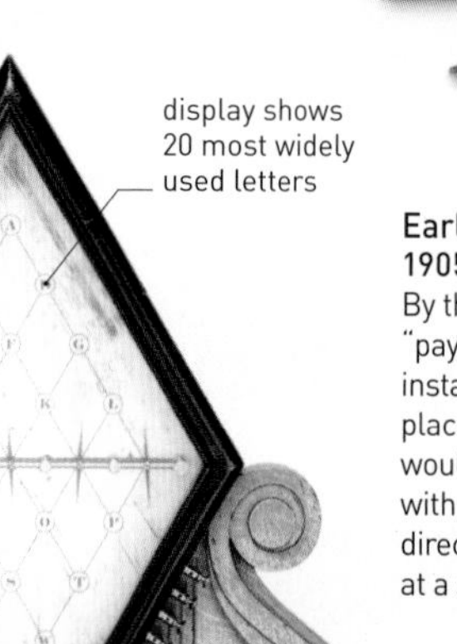
display shows 20 most widely used letters

Cooke and Wheatstone telegraph
1837
This telegraph used electricity to send signals. Messages were composed using the five needles across the middle of the grid; deflecting any two would point to specific letters.

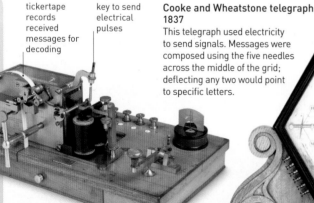

Early payphone
1905
By the 20th century, "payphones" were installed in public places. Several phones would be connected, with calls being directed by operators at a single exchange.

integrated speaker and mouthpiece unit

Ericsson table phone
1890
This design was one of the first to have an integrated speaker and mouthpiece unit. Winding the handle alerted the operator at a telephone exchange to "open the line" for a call.

bell rings to indicate incoming signal from exchange

Mobile telephone
1983
Cellphones use radio waves to make wireless telephone calls via local antennae that form a cellular telephone network. The Motorola DynaTAC 8000x was the first truly hand-held cellular phone.

iPhone
2012
Digital technology has changed cellphones beyond all recognition. The Apple iPhone, launched in 2007, brought together computer and phone technology.

Webcam
2000s
The development of webcams has enabled video calling over the Internet, and much communication has now been transferred to computers.

photoelectric sensors convert image on paper into electric signals

keyboard

Facsimilie telegraph
1956
Pictures were transmitted via the telegraph system as early as 1865, but the first device to use telephone lines for this was patented by Xerox in 1964. Once popular, faxes have now largely been superseded by e-mail.

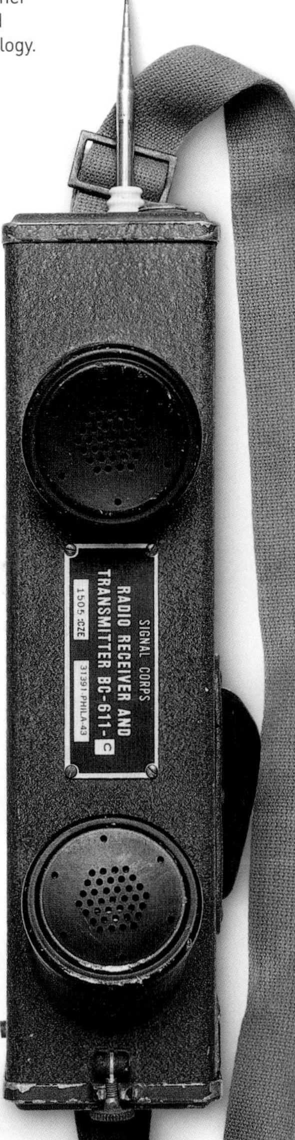
SIGNAL CORPS
RADIO RECEIVER AND
TRANSMITTER BC-611-C

Walkie talkie
1940
Compact short-distance wireless "telephones" developed rapidly during World War II. Signals were typically sent using AM (amplitude modulated) radio waves.

numbered dial

Rotary dial phone
1931
Popular during the mid-20th century, rotary telephones used a numbered dial to send a series of electrical pulses along the line. Switches at the exchange connected calls automatically.

512×342 PIXELS

THE **RESOLUTION OF THE SCREEN** OF THE **FIRST MACINTOSH** COMPUTER

A worker on Apple's assembly line in Fremont, California checks and cleans a Macintosh computer.

FOLLOWING ON FROM THEIR SUCCESS with the Apple Lisa computer the previous year, Apple Computer Inc. launched a **groundbreaking new personal computer**: the Macintosh. Easy to use, with a modern design, and a high-profile advertising campaign, the Macintosh was aimed at breaking the growing dominance of IBM-compatible computers (see 1981). The following year, Microsoft would launch the Windows operating system, which gave users of IBM-compatibles a graphical user interface, strengthening that dominance.

In February, American astronauts **Bruce McCandless** and **Robert Stewart** made the **first ever untethered spacewalk**. They were strapped into Manned Maneuvering Units (MMUs), which could move and orient in space thanks to 24 small retro rockets that emitted jets of nitrogen gas. McCandless ventured nearly 328ft (100m) away from the spacecraft.

US secretary of Health and Human Services, Margaret Heckler, announced that American virologist **Robert Gallo** (b.1937) had **discovered the probable cause of AIDS** (see 1982). Gallo had been collaborating with a team in France, led by French virologist Luc Montagnier (b.1932), who had also discovered a new virus that seemed to be related to AIDS. In June, Gallo and Montagnier announced that the two new viruses were one and the same; the virus eventually gained its name, human immunodeficiency virus (HIV), in 1986.

> **“ BEGINNING** WITH A **SINGLE MOLECULE** OF... DNA, THE PCR CAN GENERATE **100 BILLION SIMILAR MOLECULES** IN AN AFTERNOON. **”**
>
> **Kary Mullis, American biochemist,** in *Scientific American*, 1990

This year, two independent teams of geneticists reported the same discovery—the genetic code in the DNA of the fruit fly (*Drosophila melanogaster*) that **control the development** of the insect's **major anatomical features**. These so-called homeobox sequences code for proteins that **switch other genes on or off** during the insect's embryonic stage. Homeobox genes have since been found in nearly all types of living organism, from yeasts to humans.

In September, British geneticist **Alec Jeffreys** (b.1950) developed DNA profiling—a tool that can be used to **identify individuals from samples containing their**

Untethered space walk
American astronaut Bruce McCandless II became the world's first human satellite, when he performed the untethered spacewalk, in February.

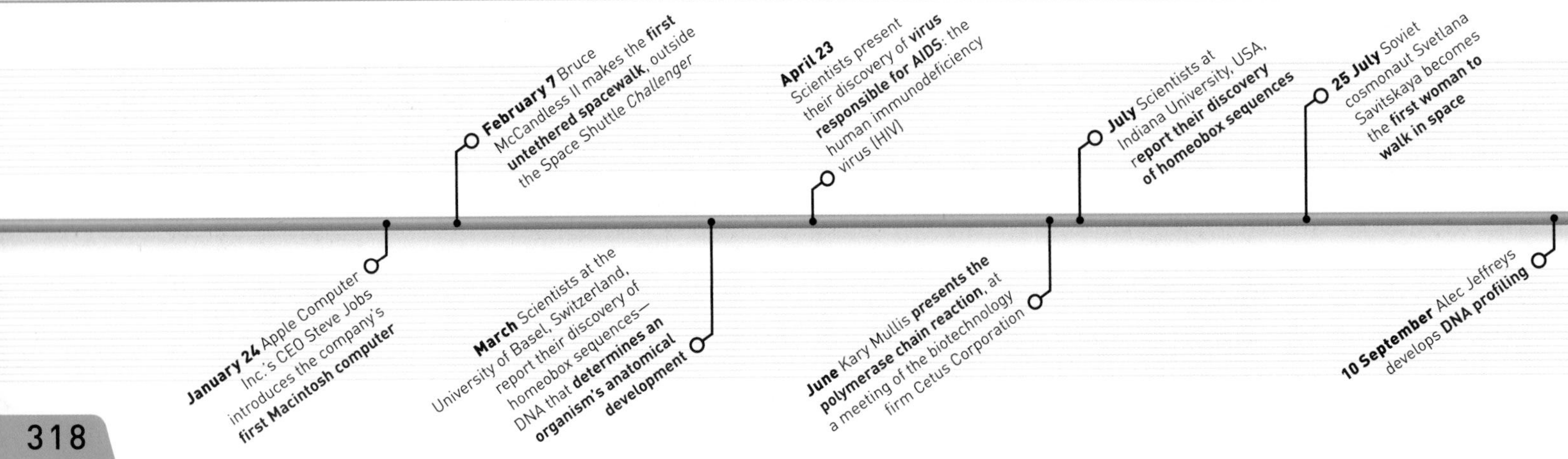

January 24 Apple Computer Inc.'s CEO Steve Jobs introduces the company's **first Macintosh computer**

February 7 Bruce McCandless II makes the **first untethered spacewalk**, outside the Space Shuttle *Challenger*

March Scientists at the University of Basel, Switzerland, report their discovery of homeobox sequences—DNA that **determines an organism's anatomical development**

April 23 Scientists present their discovery of **virus responsible for AIDS**: the human immunodeficiency virus (HIV)

June Kary Mullis **presents the polymerase chain reaction**, at a meeting of the biotechnology firm Cetus Corporation

July Scientists at Indiana University, USA, **report their discovery of homeobox sequences**

25 July Soviet cosmonaut Svetlana Savitskaya becomes the **first woman to walk in space**

10 September Alec Jeffreys develops **DNA profiling**

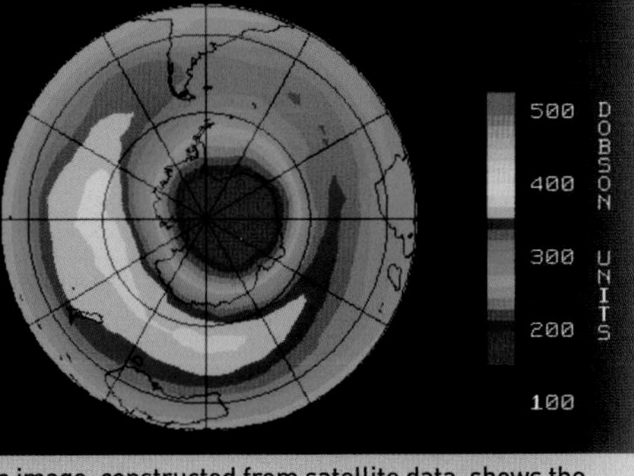
500
400
300
200
100
DOBSON UNITS
An image, constructed from satellite data, shows the ozone "hole" over Antarctica, in October 1985.

66 THE PERCENTAGE BY WHICH **OZONE** IS **DEPLETED** EACH SPRING OVER **ANTARCTICA**

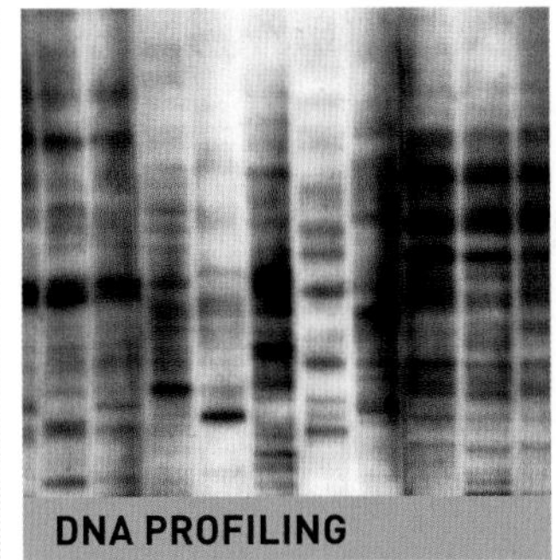

DNA PROFILING

Although DNA differs minimally between any two individuals, some sections of the genome do vary. In DNA profiling, these sections can be cut and then be arranged, in a gel, according to length. Photographs of these fragments resemble barcodes and can be used to identify an individual with a high degree of certainty.

DNA, such as blood or saliva. He later improved the sensitivity of his new procedure, by employing a technique called the **polymerase chain reaction** (PCR), which was first reported in June by American biochemist **Kary Mullis** (b.1944). This process enables molecular biochemists to **multiply small sections of DNA** almost indefinitely. PCR is a crucial element in many important DNA technologies, including DNA sequencing, cloning, and profiling.

UNTIL THE 1980s, scientists had made very little progress in **understanding Alzheimer's disease**—an age-related disorder affecting nerve cells in the brain first described in 1906 by German psychiatrist Alois Alzheimer. One phenomenon that seemed to be common in sufferers' brains was the **build-up of proteins that form "plaques"** in the spaces between neurons. This year, a team headed by Australian neuropathologist Colin Masters (b.1947) published the **first clear analysis of the protein** present in these plaques. The protein, called A-Beta amyloid, had first been described by American pathologist George Glenner (1927–95) just a year earlier. Within two years, another protein, called tubulin associated unit (tau), would also be implicated in the disease (see 1986–87).

In May, scientists from the British Antarctic Survey (BAS) announced they had discovered a **downward trend in the concentration of ozone** above the Antarctic. Most atmospheric ozone is found in a layer between 12 and 18 miles (20 and 30 km) above ground, and is greatest above the poles. The ozone layer plays a vital role in **protecting life on Earth,** blocking out potentially harmful ultraviolet radiation. Within two years, the main cause of the depletion of atmospheric ozone was confirmed: synthetic compounds called chlorofluorocarbons (CFCs), which were widely used in refrigerators and as propellants in aerosol cans (see 1986–87).

Alzheimer's protein
A beta-amyloid protein molecule shows how the twisted structure of these molecules helps make them clump together, forming plaques.

twisted structure

The **element carbon is extremely versatile**, forming chains and rings in countless compounds. Even when pure, this versatility is apparent: two of its best known allotropes (forms) are diamond, in which carbon atoms are arranged in tetrahedrons (see 1797), and graphite, in which they form flat planes of hexagons. In September of this year, chemists at Sussex University, UK, and Rice University, Texas, found **evidence of a stable allotrope of carbon** consisting of molecules with 60 atoms. The scientists quickly worked out the molecules' structure: the carbon atoms are joined in hexagons and pentagons. The structure is similar to that of a geodesic dome designed by American architect Richard Buckminster Fuller, so the new allotrope was named **buckminsterfullerene**. The new allotrope had been predicted by other chemists, and it has since been found occurring naturally. It spawned interest in an **important new class of materials**, called fullerenes (see 1990–91).

Buckyball
A buckminsterfullerene, also known as a buckyball, has a structure of alternating pentagons and hexagons, like the sections of a soccer ball.

carbon atom

60 THE NUMBER OF **CARBON ATOMS IN** EACH MOLECULE OF **BUCKMINSTERFULLERENE**

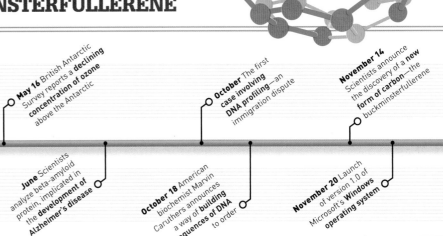

May 16 British Antarctic Survey reports a **declining concentration of ozone** above the Antarctic

June Scientists analyze beta-amyloid protein, implicated in the **development of Alzheimer's disease**

October The first **case involving DNA profiling**—an immigration dispute

October 18 American biochemist Marvin Caruthers announces a way of **building sequences of DNA** to order

November 14 Scientists announce the discovery of a **new form of carbon**—the buckminsterfullerene

November 20 Launch of version 1.0 of Microsoft's **Windows operating system**

“CHERNOBYL... SHOWED US A WORLD IN WHICH THE VERY EARTH AND AIR AND WATER FROM WHICH WE DRAW SUSTENANCE LIES POISONED.”

Satya Das, Canadian author, in speech at the University of Alberta, 1986

Repairs are carried out on the Chernobyl nuclear power plant, after what most consider as the world's worst ever nuclear accident.

TWO DISASTERS DOMINATED technology news in 1986. In January, NASA's Space Shuttle *Challenger* broke apart as a result of a **catastrophic explosion** soon after lift-off. All seven astronauts on board were killed—including a civilian, schoolteacher Christine McAuliffe—and the Space Shuttle program was **halted for more than two years**. Three months later, the **Chernobyl nuclear reactor in Ukraine, USSR, exploded** after a power surge during a routine test. Two workers died immediately, and a further 28 in the following few weeks. The area around the reactor was highly contaminated by radioactive material. Around five percent of the obliterated reactor core was carried high into the air after the explosion and the resulting fires spread contamination over a wide area far beyond Ukraine. The accident was attributed to design flaws and inadequate personnel training.

Above Earth's atmosphere, a Russian Proton-K rocket lifted the first element of the USSR's **modular space station Mir** into orbit. Over the next decade, another six modules were attached, and a wide range of scientific experiments were conducted in them—including research into the effects of long periods spent in space. Mir was more or less permanently inhabited, by successive crews of astronauts and

73

THE NUMBER OF **SECONDS AFTER LIFT OFF** BEFORE SPACE SHUTTLE CHALLENGER **EXPLODED**

Space shuttle tragedy
NASA's space shuttle Challenger *disintegrates after an explosion shortly after lift-off. All seven astronauts on board the shuttle died.*

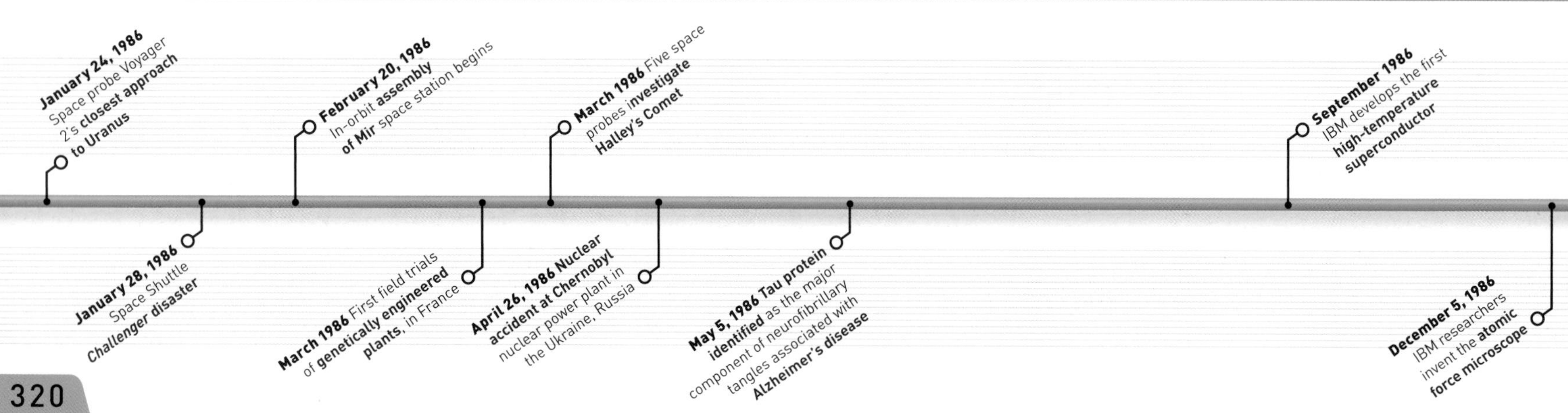

January 24, 1986 Space probe Voyager 2's **closest approach to Uranus**

January 28, 1986 Space Shuttle **Challenger disaster**

February 20, 1986 In-orbit **assembly of Mir** space station begins

March 1986 First field trials of **genetically engineered plants**, in France

March 1986 Five space probes **investigate Halley's Comet**

April 26, 1986 Nuclear accident at Chernobyl nuclear power plant in the Ukraine, Russia

May 5, 1986 Tau protein identified as the major component of neurofibrillary tangles associated with **Alzheimer's disease**

September 1986 IBM develops the first **high-temperature superconductor**

December 5, 1986 IBM researchers invent the **atomic force microscope**

This image shows a field of rapeseed that has been genetically modified to be resistant to herbicides. In some countries, as much as 90 percent of all rapeseed is genetically modified.

Mir Space Station
The Soviet space station Mir in orbit about 223 miles (360 km) above the Pacific Ocean. This photograph was taken in 1995, from NASA's space shuttle Discovery.

cosmonauts from 12 different countries, for most of its 15-year lifetime.

Also up in space, five probes **flew close to Halley's comet**, which was venturing in toward the inner Solar System. The European Space Agency's Giotto probe passed within 373 miles (600 km) of the centre of the comet, capturing the **first ever images of a comet's nucleus**.

Researchers attempting to unravel **the mystery of Alzheimer's disease** discovered the nature of the neurofibrillary tangles (NFTs) found inside neurones in the brains of people with the disease. They found that NFTs are made of Tau (tubule associated unit) proteins, which stabilize microtubules—vital for maintaining cell structure. This was the **second major breakthrough** in understanding Alzheimer's disease since the characterization of beta-amyloid protein (see 1985) that makes up plaques between neurons in Alzheimer's sufferers.

The first field trials of **genetically modified (GM) plants** began in France and the USA: tobacco plants with altered DNA that gave them resistance to herbicides. Since 1986, many GM crops have been produced, including cotton, potatoes, and rapeseed. From the outset, the production of **GM crops** has been **controversial**, with concerns about the unpredictable consequences of GM organisms on the environment and some people simply opposed to "tampering" with nature.

41.6 MPH THE AVERAGE SPEED OF THE **SUNRAYCER SOLAR CAR**

In February 1987, astronomers in the Southern Hemisphere observed a **new source of light in the sky**, as bright as many of the stars visible to the naked eye. It turned out to be a supernova—the explosive death throes of a giant star—and was labeled SN 1987a. The explosion happened in a nearby galaxy, the Large Magellanic Cloud, nearly 170,000 light-years away. It was the first naked-eye supernova for more than 300 years.

The discovery of the ozone hole over Antarctica (see 1985) led to more research into the effects of chlorofluorocarbons (CFCs) on the ozone layer. The United Nations proposed an international treaty to **limit the production of CFCs**, and the Montreal Protocol was opened for signature in September 1987, and came into force in 1989.

In Berlin, German ophthalmic surgeon **Theo Seiler** (b.1949) carried out the **first laser eye surgery** on a human patient.

Laser eye surgery
An ophthalmologist prepares to carry out laser eye surgery by first measuring the curvature of the patient's eye.

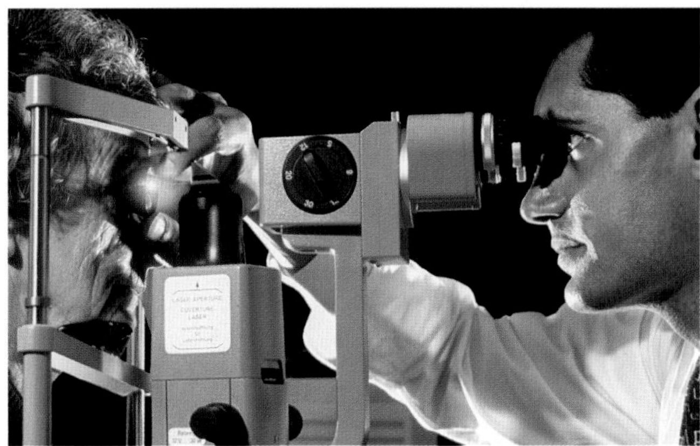

Corrective eye surgery began in the 1970s, with the introduction of radial keratotomy, in which radial lines are cut into the cornea, changing its shape. In 1983, American ophthalmologist **Stephen Trokel** (b.1934) developed a way of **changing the cornea's shape** by burning away corneal tissue with an ultraviolet laser. It was this method, called photo-refractive keratectomy (PRK), that Seiler was using. A more sophisticated procedure, called LASIK (laser in-situ keratomileusis) was patented in 1989 and available commercially from 1991.

In November, the first World Solar Challenge—a race held to promote research into **solar technology for cars**—was held in Australia. The race was won by General Motors' Sunraycer.

PENTAX WORLD SOLAR CHALLENGE

Solar car
General Motors' car Sunraycer beat the competition in the first ever Solar Challenge, a race for solar powered vehicles across Australia at the height of summer.

February 23, 1987 Supernova SN1987a, bright enough to be **observed with the naked eye**

March 1987 Scientists observe that brain cells in Alzheimer's patients **have defective Tau protein**

March 1987 The first **licensed AIDS treatment**, AZT, is approved in the US

October 1987 Chinese-born American bioengineer Bert Fung **coins the term "tissue engineering"**

November 1987 Sunraycer wins the first ever **World Solar Challenge** for solar-powered cars

November 1987 World's first **laser eye surgery** is performed on a human being

A culture of the bacterium *Clostridium botulinum*, which produces the botulinum toxin. The toxin was approved for medical use in the USA in 1988.

IN JUNE 1988, JAMES HANSEN (b.1941) the director of NASA's Goddard Institute for Space Studies, in New York, reported to the US Senate Committee on Energy and Natural Resources that the average **global temperature was increasing** above what would be expected by normal climate variation. He noted that the world's temperature was greater than it had ever been since systematic recording had begun about a hundred years earlier. He said raised temperatures would probably cause an **increase in heat waves** and other **extreme weather events**. Importantly, he suggested that the main cause of the warming was the greenhouse effect (see pp.326–27), which was being enhanced by the enormous amounts of **carbon dioxide released into the atmosphere** by the burning of fossil fuels.

Climate scientists were already well aware of global warming, and the possible challenges the world might face if the warming continued. In 1986, the World Meteorological Organization and the United Nations Environment Program had set up a body to examine the phenomenon: the **Advisory Group on Greenhouse Gases**. This small group was superseded by the UN's **Intergovernmental Panel on Climate Change** (IPCC), which was formed in late 1988. The IPCC's first assessment report would be published two years later (see 1990).

> “...**GLOBAL WARMING** HAS REACHED A LEVEL SUCH THAT WE CAN ASCRIBE... **A CAUSE AND EFFECT RELATIONSHIP** BETWEEN THE **GREENHOUSE EFFECT** AND THE **OBSERVED WARMING.**”
>
> **James Hansen, American climate scientist,** testifying before the US Senate Committee on Energy and Natural Resources, 1988

In November, Dutch computer scientist Piet Beertema (b.1943) initiated a connection to the **US National Science Foundation Network** (NSFnet). The NSFnet was a nationwide set of interconnected computer networks for academics. It formed the **backbone of the early Internet**. Other organizations in the Netherlands and across Europe became connected soon afterward. Fortunately for computer scientists in the Netherlands, their connection to NSFnet was made two weeks after the **first internet worm** was released. The Morris worm—written by student Robert Morris at Cornell University, Ithaca, New York—**infected several thousand computers** across NSFnet, slowing them down. The cost in terms of downtime and the effort involved in removing the worm from infected computers is not known, but is estimated to be around $1 million. Morris was convicted under the US Computer Fraud and Abuse Act.

In February 1989, the first in a new phase of satellites was launched to **modernize the Global Positioning System** (GPS, see 1978). Over the next decade, 18 new satellites were placed in orbit. Since GPS was initially a US military enterprise, access to highly accurate GPS signals was restricted to the armed services—mainly to ensure that military enemies would not be able to benefit. By the end of the 1990s, this restriction was lifted, and the general public was given access to the full service, **opening up a new market in navigational devices**—including in-car "satellite navigation" devices and GPS-enabled mobile phones. **Russia has a similar satellite navigation system**—the Globalnaya Navigatsionnaya Sputnikovaya Sistema (GLONASS), whose roots also lie in the 1970s. It became fully operational in 1993. Since the late 2000s, many satellite navigation devices use both GPS and GLONASS.

In December, botulinum toxin, **better known as "botox,"** was approved in the US for problems associated with eye muscles. The

HOW SATELLITE NAVIGATION WORKS

Satellite navigation (sat-nav) devices pick up signals sent out by satellites in orbit around Earth. Each satellite carries a very accurate atomic clock. From the difference between the time signals that are received from three or more satellites, a sat-nav device can calculate how far it is from each satellite. Using those distances, the device can work out its geographic position with great accuracy.

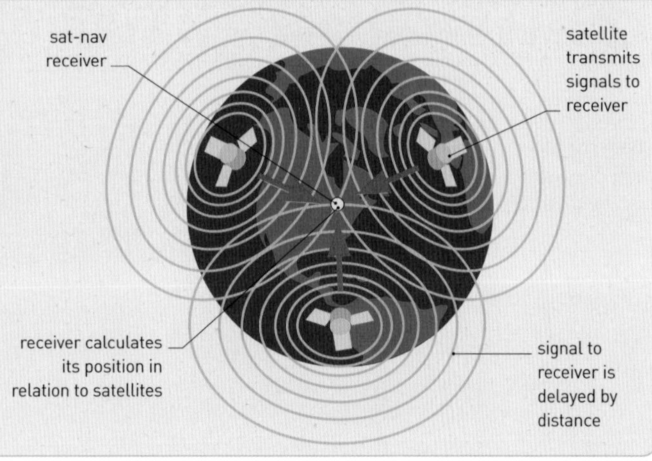
sat-nav receiver
satellite transmits signals to receiver
receiver calculates its position in relation to satellites
signal to receiver is delayed by distance

April 1, 1988 Publication of Stephen Hawking's hugely popular book ***A Brief History of Time***

May 16, 1988 US Surgeon General declares **nicotine as addictive as heroin** or cocaine

June 23, 1988 James Hansen raises awareness of the term **"global warming"**

November 2, 1988 The Morris Worm becomes the first computer virus to **infect computers across the Internet**

November 1988 The Netherlands becomes the **second country connected** to the Internet

December 6, 1988 United Nations establishes the **Intergovernmental Panel on Climate Change** (IPCC)

8 THE **TYPICAL NUMBER** OF **CELLS IN AN EMBRYO** BEFORE REMOVING A CELL FOR **GENETIC DIAGNOSIS**

A single cell is removed from an embryo, the first stage of preimplantation genetic diagnosis. The embryo can continue to develop unharmed—the removed cell is subjected to genetic testing.

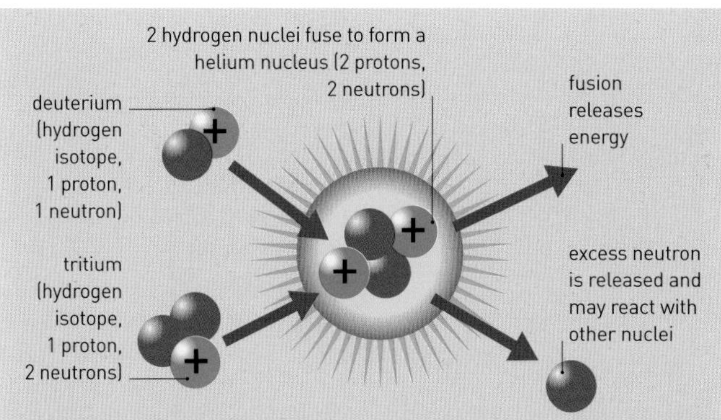
2 hydrogen nuclei fuse to form a helium nucleus (2 protons, 2 neutrons)
deuterium (hydrogen isotope, 1 proton, 1 neutron)
fusion releases energy
tritium (hydrogen isotope, 1 proton, 2 neutrons)
excess neutron is released and may react with other nuclei

NUCLEAR FUSION

Fusion is the process of joining together atomic nuclei, most commonly deuterium and tritium (heavy hydrogen) nuclei fusing to form a nucleus of helium. This releases a burst of energy and is the power source deep inside stars. Fusion has been achieved in experiments but, so far, the amount of energy input to create the heat and pressure is much greater than the energy released.

toxin, produced by bacteria in the genus *Clostridium*, can be lethal even in tiny amounts. Upon injection, it causes **paralysis of facial muscles** for around three months. In the same year, American plastic surgeon Richard Clark reported that injections of botox had removed unwanted wrinkling above one eye in a patient who had paralysis in the facial nerve on one side. The fact that botox **injections can reduce wrinkles**—one of the most visible signs of aging—was of great interest to many customers of cosmetic surgeons. Some began receiving the treatment for cosmetic reasons only, illegally at first; botox injections were approved for cosmetic use in the US in 2002, and shortly afterward in other countries.

In the 11 years since the birth of the first baby conceived by IVF (see 1978), the technology of assisted reproduction advanced greatly. The fact that embryos are created outside the body (and typically implanted after three days' growth) opened up the possibility of **genetic testing on embryos** from parents with certain genetic diseases. Since a cycle of IVF creates several embryos, any that carried the genes giving rise to the disease would be discarded. A team headed by British IVF doctors Alan Handyside (b.1951) and Robert Winston (b.1940) carried out the first human **preimplantation genetic diagnosis** (PGD). The procedure was controversial, with some disability rights groups claiming it was a high-tech version of eugenics.

There was also controversy surrounding a claim made by American physicist **Stanley Pons** (b.1943) and British physicist **Martin Fleischmann** (1927–2012), who were at the University of Utah. The pair announced they had conducted an experiment in which more energy was produced than could be explained by chemical reactions alone. They claimed that the extra energy had come from nuclear fusion—normally only possible at **extremely high temperatures and pressures** (see panel, left). There was great interest in this **"cold fusion,"** but many scientists were sceptical, and no one could repeat the experiment with the same result—leading the scientific community to conclude Pons' and Fleischmann's **claim was almost certainly incorrect**.

In August, NASA's Voyager 2 made its closest approach to Neptune, capturing the **first detailed images** of the planet. It was Voyager 2's last visit to any planet or moon before it headed toward the outer reaches of the Solar System.

Neptune
NASA's Voyager 2 spacecraft captured images of the gas giant Neptune's clouds, whose blue color is due to the presence of methane.

12 YEARS THE **TIME VOYAGER 2** TOOK TO **REACH** THE **PLANET NEPTUNE**

February 14, 1989 The first of a new set (Block II) of **GPS satellites launched**

March 1989 British computer scientist Tim Berners-Lee writes the **blueprint for the World Wide Web**

March 1, 1989 Discovery of giant magnetoresistance effect, which allows data to be **stored at high densities**

March 23, 1989 Researchers announce that they had **achieved cold fusion**—a claim later shown to be false

April 14, 1989 American oncologist Bert Vogelstein discovers the function of p53, the **tumor-suppressing protein**

May 11, 1989 Intel announces its 80486 chip, the first to have **a million transistors**

August 25, 1989 Space probe Voyager 2's **closest approach to Neptune**, capturing stunning images of the planet

October 1989 Alan Handyside carries out the first preimplantation genetic diagnosis, **screening IVF embryos** for genetic diseases before implantation

110 MILES THE DIAMETER OF THE CHICXULUB CRATER

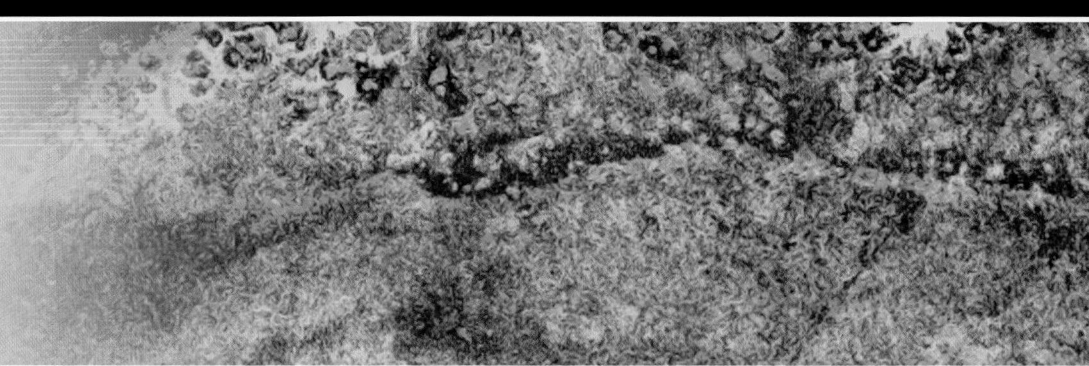

This radar image shows a small portion of the Chicxulub crater, Mexico. The crater was first discovered nearly 20 years before geologists realized it was caused by the object that probably contributed to the demise of the dinosaurs.

THE INTERGOVERNMENTAL PANEL ON CLIMATE CHANGE (see 1988) published its first assessment report in 1990. The report suggested that the average **global temperature was increasing** by about 32°F (0.3°C) per year due largely to man-made emissions of greenhouse gases, such as carbon dioxide (see pp.326–27). It set out some of the **potential impacts of global warming**, such as rising sea levels and threats to biodiversity. Further assessment reports have maintained and refined the scientific conclusions.

TIM BERNERS-LEE
(b.1955)

British computer scientist Tim Berners-Lee earned a physics degree from Oxford University. While working at the European Organization for Nuclear Research (CERN) he developed the concept of the Web. In 1994, Berners-Lee founded the World Wide Web Consortium (W3C), the international organization that develops Web standards.

British computer scientist **Tim Berners-Lee** created the **world's first Web browser**—called WorldWideWeb. At the time, the internet was growing rapidly, but was mainly used by academics, communicating via typed commands on bulletin boards—systems that allowed users to exchange software and post messages. There were several different operating systems in use, and few common programs or document formats. Berners-Lee devised a computer language—**hypertext markup language** (html)—that could be used on any computer, to create pages of information. Crucially, these pages could contain links to pages on other, specially programmed, internet-linked computers called servers. The result would be a "web" of information—hence the name of the software, and eventually the World Wide Web itself. Berners-Lee created the first web server at CERN, in Switzerland, where he was working at the time.

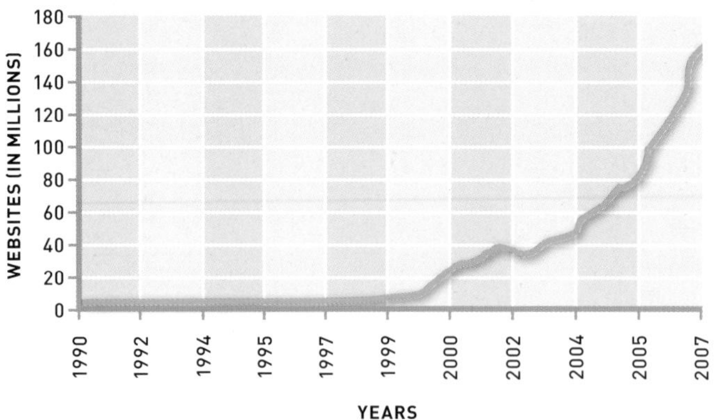

WEBSITES (IN MILLIONS)
180
160
140
120
100
80
60
40
20
0
1990
1992
1994
1995
1997
1999
2000
2002
2004
2005
2007
YEARS

Rise in number of websites
The number of websites rose steeply after the internet became commonplace among businesses and home users.

In April 1990, NASA's Space Shuttle *Discovery* carried the **Hubble Space Telescope** (HST) into a low-Earth orbit. The telescope, named after the American astronomer Edwin Hubble (see 1923), carries a range of instruments that can detect infrared, ultraviolet, and visible light. The Hubble Space Telescope has produced stunning images of a wide range of objects in space, and has provided huge amounts of information for astronomers, astrophysicists, and cosmologists.

Japanese biological physicist **Seiji Ogawa** (b.1934) developed an extension of **magnetic resonance imaging** (MRI) that could differentiate between oxygenated and deoxygenated blood. Ogawa realized that this could **reveal which regions of a brain are most active.** His technique forms the basis of functional MRI (fMRI), which is used to measure brain activity. He produced fMRI images of rats in 1990; the first human fMRI imaging was carried out in 1992.

In June 1990, researchers in the US began the world's first clinical **trial to use gene therapy**. A gene was inserted into white blood cells, and the resulting transgenic cells were injected into a girl suffering from a severe immune disorder.

Hydrogen fueled car
In 1991, the Japanese firm Mazda revealed its first HR-X concept car, which has an internal combustion engine that burns hydrogen.

February 14, 1990 The Voyager 1 space probe turns around to **take the "Family Portrait"** of all the planets except Mercury and Mars

April 1990 Launch of the **Human Genome Project**

April 24, 1990 Hubble Space Telescope launched

June 1990 US company General Instruments introduces first digital **high-definition television** system

September 10, 1990 The **first internet search engine**, called Archie, released

September 14, 1990 The first gene therapy carried out

September 15, 1990 NASA's Magellan spacecraft began **detailed 3-D mapping** of the surface of Venus

October 26, 1990 The IPCC published its **First Assessment Report** on climate change

December 1990 Technology underlying fMRI, which allows **brain processes to be imaged** in real time, is developed

December 25, 1990 The world's **first web browser**, called WorldWideWeb, is created

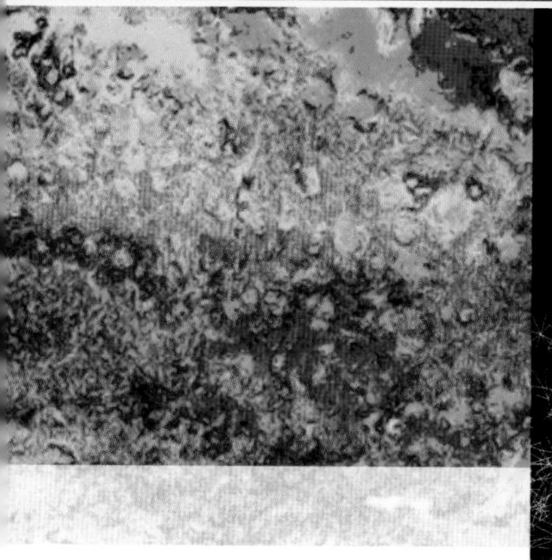

In 1991, Canadian geophysicist **Alan Hildebrand** (b.1955) announced that the **Chicxulub crater**, centered on the coast of the Yucatán Peninsula, Mexico, was almost certainly created by the impact that American physicist Luis Alvarez had hypothesized as the cause of **the extinction of the dinosaurs** (see 1980). The ages of the rocks and the size of the object that would have made the crater tallied well with Alvarez's hypothesis.

In November, Japanese physicist **Sumio Iijima** (b.1939) published his research into **nanoscale tubes of pure carbon**, known as carbon nanotubes. Although these tubes had been observed before, Iijima's work was inspired by the growing interest in carbon allotropes called fullerenes (see 1985) and helped develop it further.

American inventor **Roger Billings** (b.1948) revealed the first **electric vehicle powered by a hydrogen fuel cell**. In the same year, Japanese firm Mazda announced a concept hydrogen car with an internal combustion engine that burns hydrogen.

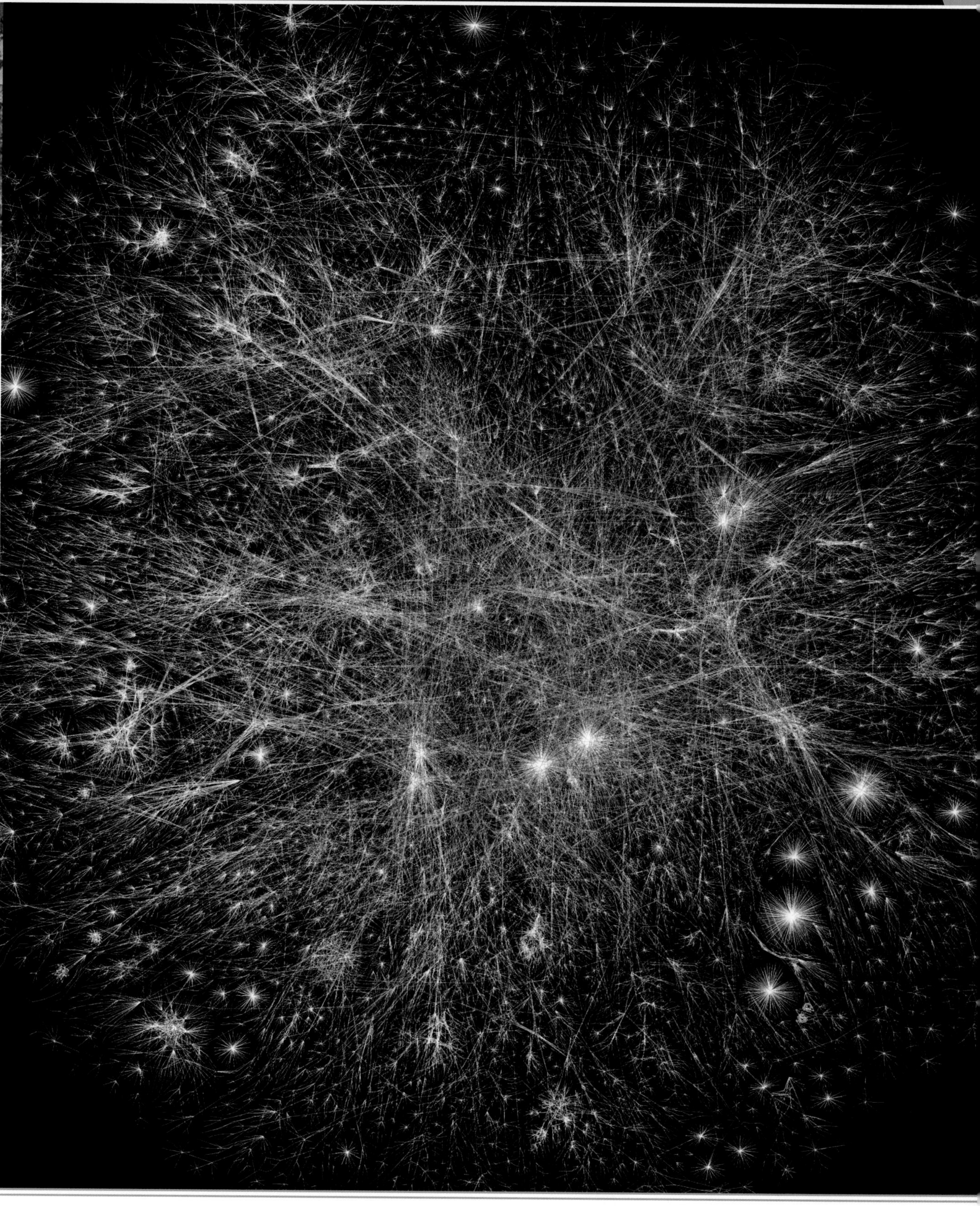

Global connections
This image represents a map of computer networks in the early 2000s. The World Wide Web is the sum of interconnected information that is stored on servers throughout this complex interconnected network.

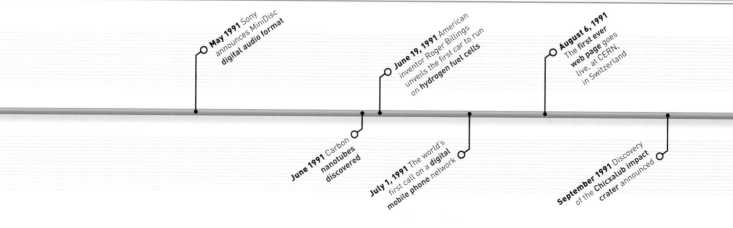

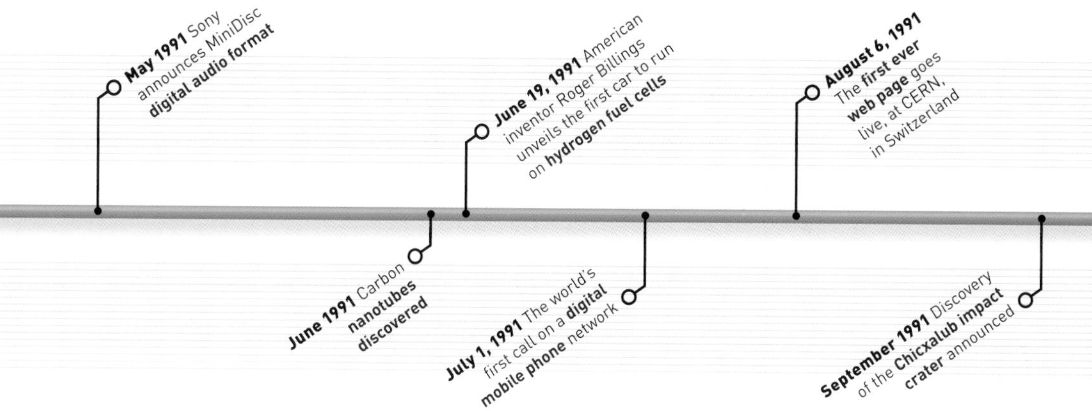
May 1991 Sony announces MiniDisc **digital audio format**
June 1991 Carbon **nanotubes discovered**
June 19, 1991 American inventor Roger Billings unveils the first car to run on **hydrogen fuel cells**
July 1, 1991 The world's first call on a **digital mobile phone** network
August 6, 1991 The **first ever web page** goes live, at CERN, in Switzerland
September 1991 Discovery of the **Chicxulub impact crater** announced

UNDERSTANDING GLOBAL WARMING

HUMAN ACTIVITY IS ENHANCING THE EARTH'S GREENHOUSE EFFECT AND WARMING ITS ATMOSPHERE

Over the past 200 years, Earth's average temperature has risen rapidly against a prevailing trend of cooling. Evidence suggests that this global warming is not the result of natural climate variation but is due largely to human activity—and it could have disastrous consequences.

Energy from the Sun reaches Earth in the form of electromagnetic radiation—mostly visible light and infrared and ultraviolet radiation. Some of the energy is absorbed, with the rest reflected back into space. The absorbed energy heats the planet—and since any warm object emits infrared radiation, Earth also loses heat to space. At a certain temperature, the planet radiates energy at the same rate as it absorbs it.

GREENHOUSE EFFECT

With no atmosphere, Earth's equilibrium temperature would be around -1°F (-18°C). However, the atmosphere absorbs some of the incoming and outgoing radiation, and it warms up. The warmed atmosphere produces its own infrared radiation, some of which is absorbed by the surface below. As a result, the equilibrium temperature is higher—around 59°F (14°C). This phenomenon is known as the greenhouse effect, since a greenhouse also "traps" heat, making it warmer than it would otherwise be.

GREENHOUSE GASES

The first experimental evidence for the greenhouse effect came from Irish physicist John Tyndall. In the 1850s, Tyndall measured how much infrared radiation various gases absorb. The strongest "greenhouse gas" is water vapor, but methane, carbon dioxide, and ozone also contribute significantly. In the 20th century, climate scientists found that the concentration of carbon dioxide is increasing, enhancing the greenhouse effect and raising the equilibrium

JOHN TYNDALL
Irish physicist John Tyndall (1820–93) studied magnetism and atmospheric physics. He was a great popularizer of science.

> GLOBAL WARMING MUST BE SEEN AS AN **ECONOMIC** AND **SECURITY** THREAT.
>
> **Kofi Annan,** ***former United Nations Secretary General,*** 2009

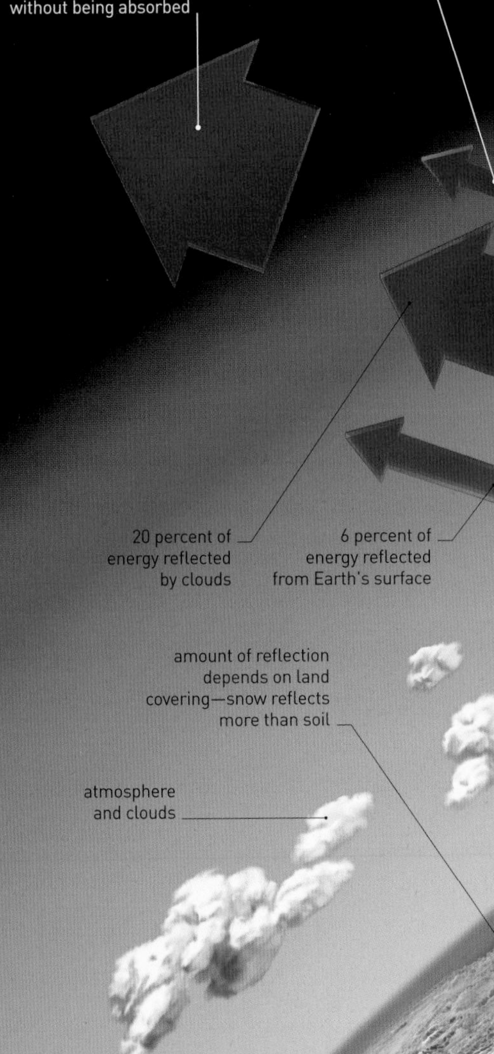

total of 30 percent of incoming energy (52 PW) reflected without being absorbed
4 percent of energy reflected by atmosphere
20 percent of energy reflected by clouds
6 percent of energy reflected from Earth's surface
amount of reflection depends on land covering—snow reflects more than soil
atmosphere and clouds

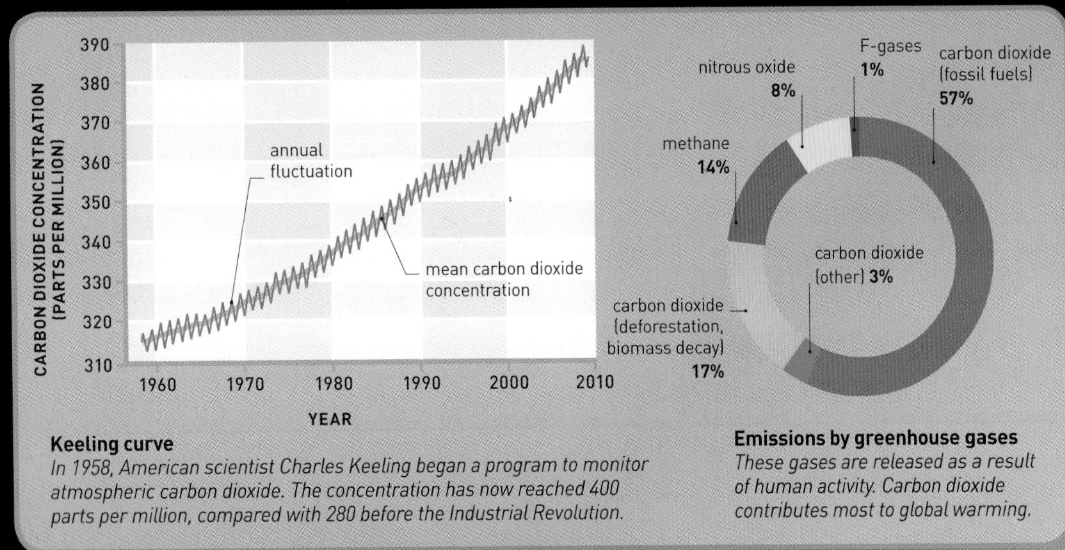

CARBON DIOXIDE CONCENTRATION (PARTS PER MILLION)
390
380
370
360
350
340
330
320
310
1960
1970
1980
1990
2000
2010
YEAR
annual fluctuation
mean carbon dioxide concentration
nitrous oxide 8%
F-gases 1%
carbon dioxide (fossil fuels) 57%
methane 14%
carbon dioxide (other) 3%
carbon dioxide (deforestation, biomass decay) 17%

Keeling curve
In 1958, American scientist Charles Keeling began a program to monitor atmospheric carbon dioxide. The concentration has now reached 400 parts per million, compared with 280 before the Industrial Revolution.

Emissions by greenhouse gases
These gases are released as a result of human activity. Carbon dioxide contributes most to global warming.

temperature. The increase in carbon dioxide concentration is largely the result of burning fossil fuels in vehicles and power stations.

EFFECTS OF GLOBAL WARMING

There is consensus among climate scientists that global warming is anthropogenic (caused by human activity). International agreements such as the Kyoto Protocol represent efforts to curb greenhouse gas emissions. Rising temperatures will have drastic consequences—meltwater will raise the sea level and increase flooding, while extreme weather events are likely to increase.

THE PREDICTED **RISE** IN SEA **LEVEL** BY THE END OF THE **CENTURY**

EXTREME WEATHER

Global warming appears to be increasing the frequency and severity of hurricanes, as the warmer atmosphere injects more energy and moisture into the climate system.

EARTH'S ENERGY BUDGET

Earth receives 174 PW (petawatts), which means it receives energy at 174 trillion joules per second. About 30 percent reflects back into space. The rest warms the surface and atmosphere, which produce energy as infrared radiation. Overall, the system is balanced—but the balance shifts as the concentration of greenhouse gases increases.

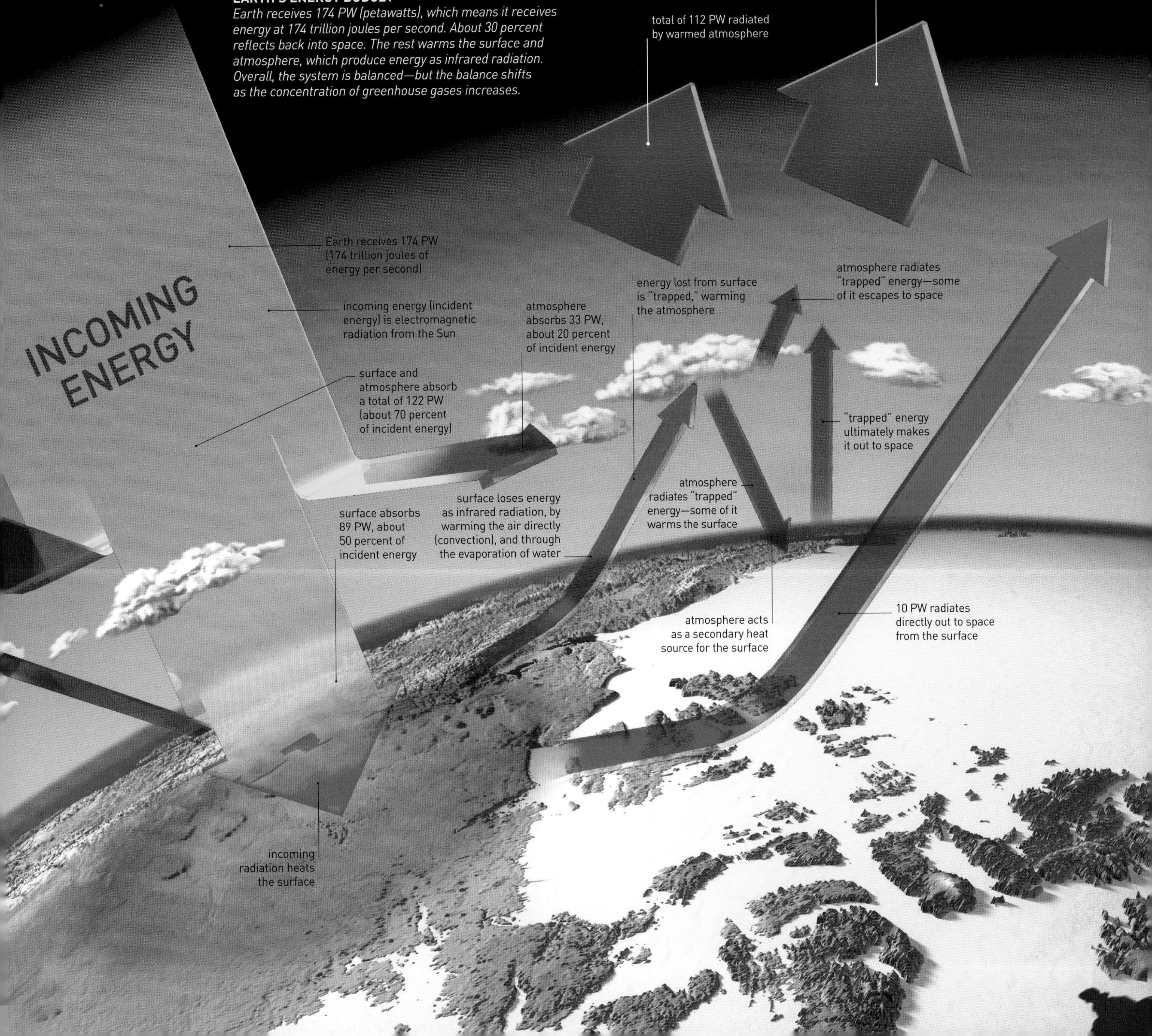

INCOMING ENERGY

Earth receives 174 PW (174 trillion joules of energy per second)

incoming energy (incident energy) is electromagnetic radiation from the Sun

surface and atmosphere absorb a total of 122 PW (about 70 percent of incident energy)

atmosphere absorbs 33 PW, about 20 percent of incident energy

surface absorbs 89 PW, about 50 percent of incident energy

incoming radiation heats the surface

surface loses energy as infrared radiation, by warming the air directly (convection), and through the evaporation of water

energy lost from surface is "trapped," warming the atmosphere

total of 112 PW radiated by warmed atmosphere

overall, 122 PW radiated to space from the atmosphere and surface (equal to the total absorbed)

atmosphere radiates "trapped" energy—some of it escapes to space

"trapped" energy ultimately makes it out to space

atmosphere radiates "trapped" energy—some of it warms the surface

atmosphere acts as a secondary heat source for the surface

10 PW radiates directly out to space from the surface

1 CELL

THE NUMBER OF **SPERM CELLS INJECTED DIRECTLY** INTO AN EGG DURING AN **INTRACYTOPLASMIC SPERM INJECTION**

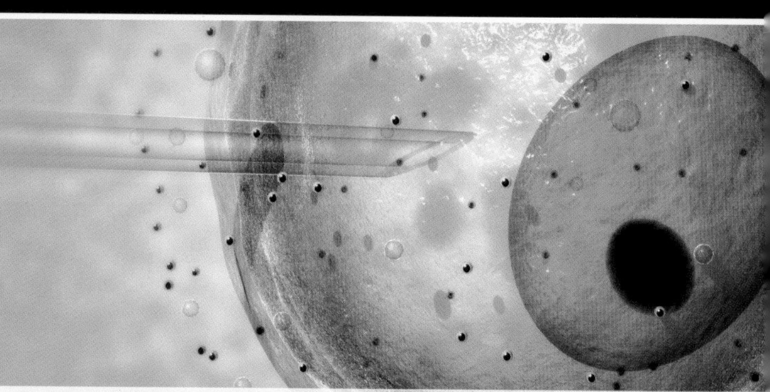

During intracytoplasmic sperm injection (ISCI), a single sperm cell is placed directly inside an ovum via an extremely fine glass needle.

ASTRONOMERS HAD ASSUMED for decades that there existed **extrasolar planets**—planets outside our Solar System. They had found several possible candidates, but **no definitive proof**. Confirmation came at last in 1992, with the detection of a planet orbiting a pulsar—which is a rotating neutron star that emits beams of radio waves. Within three years, astronomers had detected a planet in orbit around an ordinary star—one that is in the main part of its lifetime (see 1995).

“ MERRY CHRISTMAS. ”

Neil Papworth, British engineer, first SMS, 1992

Data collected by the **Cosmic Background Explorer** satellite (COBE) led to the most significant development in cosmology since the discovery of cosmic background radiation (CMB, see 1964). COBE carried out an extremely sensitive all-sky survey of the CMB, and **detected slight variations in the radiation**, which correspond to slight variations in temperature in the early Universe. These temperature variations in turn represent variations in density. The density variations are important—if the density had been perfectly uniform, matter would never have clumped together to form stars, galaxies, and galaxy clusters.

In June 1992, representatives of governments and non-governmental organizations (NGOs) attended the **UN Conference on Environment and Development** in Rio de Janeiro, Brazil, commonly referred to as the **Earth Summit**. The main goal of the conference was to discuss issues concerning the **sustainable use of Earth's natural resources** in an increasingly industrialized and highly populated world. Two major conventions came out of the summit. The first was the **Convention on Biological Diversity** in 1993 (see opposite); the second convention was the United Nations Framework Convention on Climate Change (UNFCCC). An ambitious initiative aimed at combatting climate change—called the Kyoto Protocol (see 1997)—arose from the UNFCCC, and came into force in 2005.

hole in ozone above Antarctica

Ozone hole
At the end of September 1992, atmospheric scientists reported that the ozone hole over the Antarctic (see 1985) had grown by 15 percent in a single year.

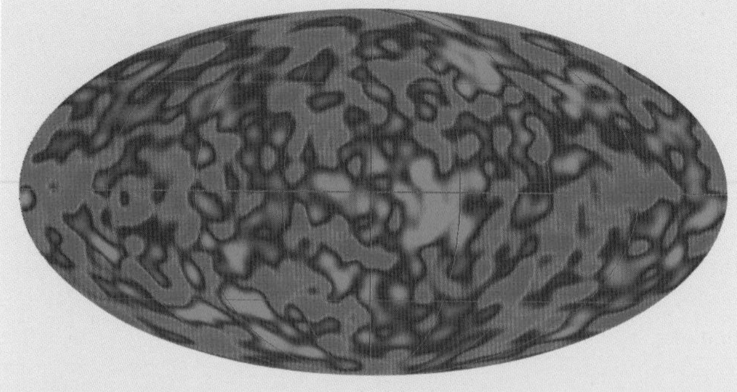

COSMIC BACKGROUND RADIATION

The cosmic background radiation (CMB) is the heat radiation that filled the Universe 380,000 years after the Big Bang (see pp.342–43), and it provides a record of the temperature of the Universe at that time. The CMB is remarkably isotropic—the same in every direction. Slight variations, or anisotropies, in the CMB are shown in the false-colour map above—red areas are slightly warmer than blue.

At the end of September 1992, atmospheric scientists reported that the **ozone hole over the Antarctic** (see 1985) had grown by 15 percent in a single year—and was **the size of North America**.

In vitro fertilization (see 1978) had been developed mainly to overcome female infertility; in the early 1990s, an extension of the technology—**intracytoplasmic sperm injection** (ICSI)—was developed mainly to address male infertility. This revolutionary technique involves injecting a single sperm into an ovum (egg cell) to overcome problems arising from low sperm count or low sperm motility (ability to move toward the ovum). It was developed by Italian fertility specialist **Gianpiero Palermo** and Belgian doctor **André van Steirteghem** at the Vrije Universiteit Brussel in Belgium. The first births resulting from the technique were confirmed in 1992.

Small, simple **liquid crystal displays** (LCDs) had been available since the 1970s and were incorporated into consumer electronics devices such as calculators, digital watches, and video cassette recorders. In 1992, the

hydraulic suspension
carbon fiber lower leg

Prosthetic knee
This microprocessor-controlled knee prosthesis automatically adjusts to the wearer's gait.

January 9, 1992 Astronomers discover the first confirmed **planets outside our Solar System**

April 23, 1992 Scientists find the **seeds of galaxy formation** in the early Universe, in data from the Cosmic Background Explorer (COBE)

June 3, 1992 The **Earth Summit in Rio de Janeiro**, Brazil, opens, bringing together 172 countries to discuss sustainability

September 29, 1992 NASA reports that the ozone hole over Antarctica **grew rapidly during the year**

1992 Hitachi develops the first practical, **high-resolution LCDs** (liquid crystal displays)

December 3, 1992 The **first SMS message** sent over a GSM network

1992 First birth of a child **conceived by direct injection** of sperm into an egg

1993 Blatchford releases the Intelligent Prosthesis, the first commercially available **microprocessor-controlled prosthetic knee**

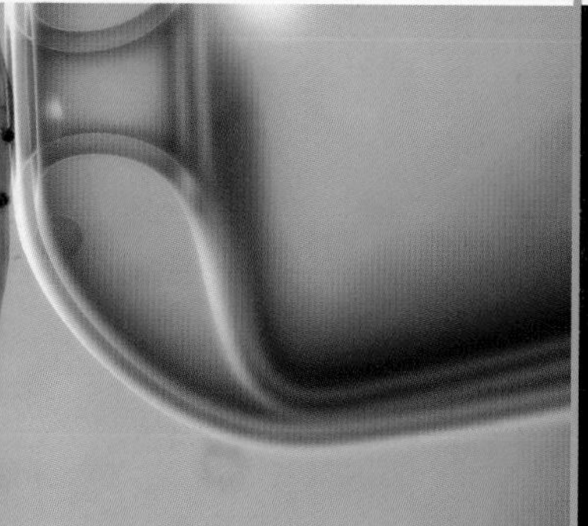

These fragments of comet Shoemaker-Levy 9 were photographed by the Hubble Space Telescope two months before their collisions with Jupiter. Shoemaker-Levy 9 was the first comet ever observed in orbit around anything other than the Sun.

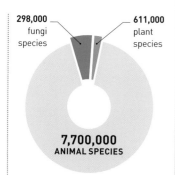

298,000 fungi species

611,000 plant species

7,700,000 ANIMAL SPECIES

Biodiversity
More than eight million plant, animal, and fungal species are known—there are many more undiscovered.

Japanese company Hitachi developed a new technology, which was called in-plane switching, making it possible to use large LCDs as television screens. By 2007, LCD televisions were outselling those with cathode ray tube screens.

In 1993, the **UN Convention on Biological Diversity** came into force. This treaty, which opened for signatures at the Earth Summit, aims to **conserve biodiversity** and encourage the sharing of profits made utilizing traditional knowledge.

The British company Blatchford unveiled their first prosthesis that was **controlled by a microchip**. This knee prosthesis could automatically adjust to the wearer's gait and became commercially available in 1993.

EIGHT YEARS AFTER THE FIRST FIELD TRIALS of genetically modified crops (see 1986), the first **commercially available genetically modified food** became available this year, when American company Calgene's Flavr Savr tomato was **approved for sale** by the US Food and Drug Administration. Calgene had inserted a gene into the genome of the tomato that hindered the production of an enzyme normally involved in breaking down of cell walls and softening of the fruit. The result was a tomato that could **stay firm and fresh for longer**. The Flavr Savr was discontinued in 1997, after initial commercial success waned.

Astronomers around the world had their telescopes fixed on Jupiter in July, when 21 mountain-sized fragments of **comet Shoemaker-Levy 9 plunged into the atmosphere** of the planet. The comet had been captured by Jupiter's gravitational influence, and had probably been in orbit around the planet for more than 20 years. It was discovered in 1993, by American astronomers **Eugene Shoemaker** (1928–97), **Carolyn Shoemaker** (b.1929), and Canadian astronomer **David Levy** (b.1948). The comet had broken into fragments after passing close to the huge planet some time in 1992, and the pieces formed a chain about 6,000 miles (10,000 km) long. The impacts produced explosions that left scars in the atmosphere. The **Galileo space probe** (see 1995), en route to Jupiter, was well placed to gather data and images of the collisions.

In December, a team led by American medical researcher Jeffrey Friedman (b.1954) reported the discovery of a hormone that is **closely involved with appetite** and therefore with obesity. The hormone, named leptin from the Greek word leptos, meaning thin, **acts on the brain's hunger center** in the hypothalamus (see panel, below). The discovery was made after Friedman's team studied mice with a mutant gene that gave them a voracious appetite. The mutant obese mice were first discovered in 1950. Injecting leptin into the blood of the obese mice caused the mice to **eat much less and lose weight rapidly**. Medical researchers hoped that leptin might form the basis of a cure for morbid obesity in humans, but it remains an elusive hope.

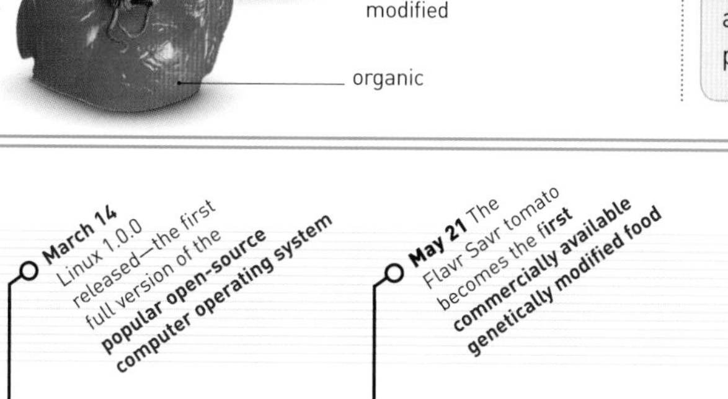
genetically modified
organic

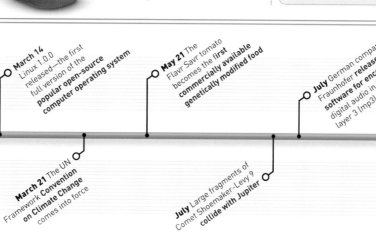

GM food
Firm, fresh-looking genetically modified Flavr Savr tomatoes (right) and three ordinary tomatoes, which have begun softening. All six are at the same stage of ripening.

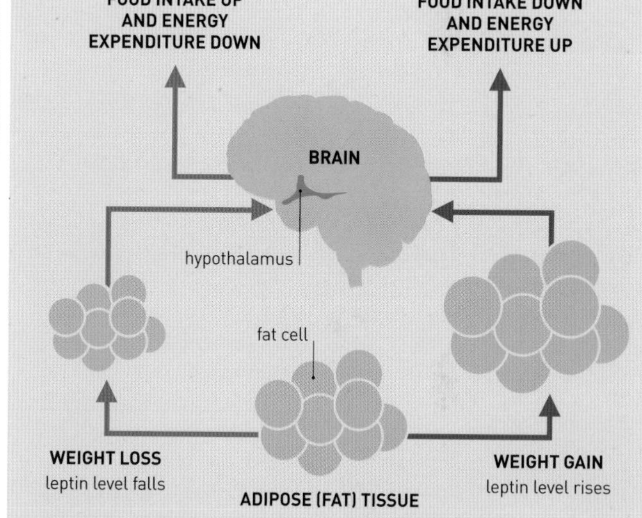
FOOD INTAKE UP AND ENERGY EXPENDITURE DOWN
FOOD INTAKE DOWN AND ENERGY EXPENDITURE UP
BRAIN
hypothalamus
fat cell
WEIGHT LOSS
leptin level falls
ADIPOSE (FAT) TISSUE
WEIGHT GAIN
leptin level rises

LEPTIN AND APPETITE

The energy-regulating hormone leptin is produced by adipocytes (fat cells). Leptin levels are controlled by an area of the brain called the hypothalamus. A person gains weight when the adipocytes in their body fat store more fat. In this state, adipocytes produce more leptin, and the hypothalamus reduces appetite, leading to weight loss. If person's weight reduces, leptin level falls, and appetite increases.

December 29, 1993 The Convention on Biological Diversity comes into force

March 14 Linux 1.0.0 released—the first full version of the **popular open-source computer operating system**

March 21 The UN Framework **Convention on Climate Change** comes into force

May 21 The Flavr Savr tomato becomes the **first commercially available genetically modified food**

July Large fragments of Comet Shoemaker-Levy 9 **collide with Jupiter**

July German company Fraunhofer **release software for encoding** digital audio in the mpeg layer 3 (mp3) format

December 1 Discovery of leptin, a **hormone released by fat cells** that controls appetite

1995–96

7,000

THE DISTANCE IN **LIGHT-YEARS** TO THE **EAGLE NEBULA**

IN APRIL 1995, NASA AMALGAMATED 32 individual Hubble Space Telescope photographs to produce an image known as **The Pillars of Creation**. It shows immense clouds of interstellar gas and dust several light-years long, shaped by intense ultraviolet radiation from nearby stars, in which **new stars are forming**.

Also in space, NASA's Galileo spacecraft dropped a probe into Jupiter's atmosphere. Analysis of **meteorite ALH84001**, which originated from Mars and was discovered in Antarctica in 1984, revealed tiny structures that resembled fossilized bacteria. The discovery prompted speculation that these were the first **definitive signs of extraterrestrial life**. However, further analysis has since shown that the structures are almost certainly not of biological origin.

Wireless endoscopy
The PillCam is a swallowable capsule that contains a tiny camera, a flashing light, and a radio transmitter to send images to a receiver.

Pillars of Creation
In this dramatic composite image, areas of star birth in the Eagle Nebula can be seen in the tiny globules at the tops of the pillars.

In October 1995, astronomers detected a planet orbiting the star 51 Pegasi, by measuring the wobble in the star's motion caused by the planet's presence. It was the **first confirmed planet around an ordinary star** other than the Sun.

Israeli inventor **Gavriel Iddan patented the PillCam** in 1995. It is a small capsule that a patient swallows, and which then passes through the digestive tract taking pictures and sending them wirelessly to a receiver. The PillCam gave gastroenterologists a new, safe, low-cost window on processes and problems in the large section of the digestive system **where endoscopes cannot reach**.

There were three important advances in modern physics in 1995. At the time, theoretical physicists had developed five **separate superstring theories**, which propose that the particles of matter and force are actually **tiny vibrating one-dimensional objects** (strings). Each theory assumed the existence of several dimensions in addition to the three space dimensions with which we are familiar in everyday life. The extra dimensions are

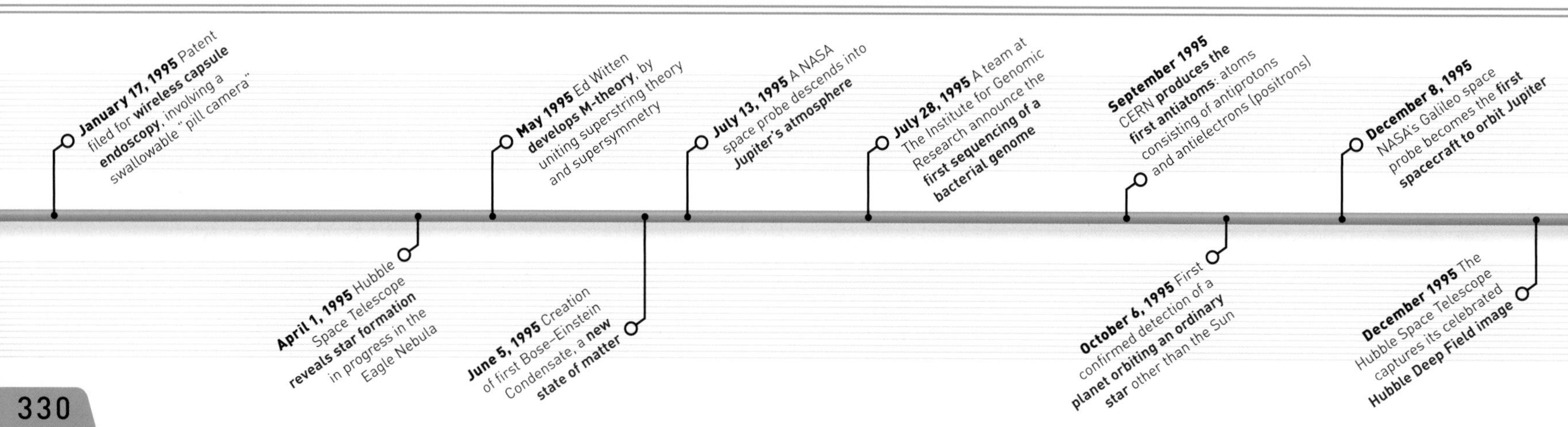

January 17, 1995 Patent filed for **wireless capsule endoscopy**, involving a swallowable "pill camera"
April 1, 1995 Hubble Space Telescope **reveals star formation** in progress in the Eagle Nebula
May 1995 Ed Witten **develops M-theory**, by uniting superstring theory and supersymmetry
June 5, 1995 Creation of first Bose-Einstein Condensate, a **new state of matter**
July 13, 1995 A NASA space probe descends into **Jupiter's atmosphere**
July 28, 1995 A team at The Institute for Genomic Research announce the **first sequencing of a bacterial genome**
September 1995 CERN **produces the first antiatoms**: atoms consisting of antiprotons and antielectrons (positrons)
October 6, 1995 First confirmed detection of a **planet orbiting an ordinary star** other than the Sun
December 8, 1995 NASA's Galileo space probe becomes the **first spacecraft to orbit Jupiter**
December 1995 The Hubble Space Telescope captures its celebrated **Hubble Deep Field image**

71 THE NUMBER OF STATES THAT SIGNED THE TREATY TO BAN NUCLEAR MISSILE TESTS

Delegates at the United Nations observe the result of their voting on the Comprehensive Nuclear-Test-Ban Treaty—an outright ban on all nuclear explosions.

rolled up tightly and cannot be perceived directly. All the theories had inconsistencies, but at a conference at the University of Southern California, American theoretical physicist **Ed Witten** (b.1951) proposed a way of combining them into a **single, super-theory**, which became known as M-theory. To date, it is the most complete "theory of everything," explaining the existence of the particles in the Standard Model (see 1974), but it is hard to test its validity.

In June 1995, physicists at the University of Colorado **created an unusual state of matter** called a Bose–Einstein condensate (BEC), predicted in the 1920s, in which several particles at a temperature just above absolute zero (see 1847–48) attain exactly the same quantum state, and act as a single system.

In September, physicists at the European Organization for Nuclear Research (CERN), on the border of Switzerland and France, **created antiatoms**—composed of antiprotons and positrons (antielectrons). Antimatter particles are created naturally, in cosmic ray collisions, for example, but when an antiparticle meets a particle, both are annihilated. Modern physics has not explained why matter, rather than antimatter, dominates the Universe.

CLONING

The process that created Dolly the sheep began with the removal of the nucleus (enucleation) of a cell taken from an adult sheep. The nucleus, containing the animal's DNA, was inserted into an enucleated egg cell from another sheep. The egg was fertilized and then grew into a sheep with a genome identical to that of the original sheep.

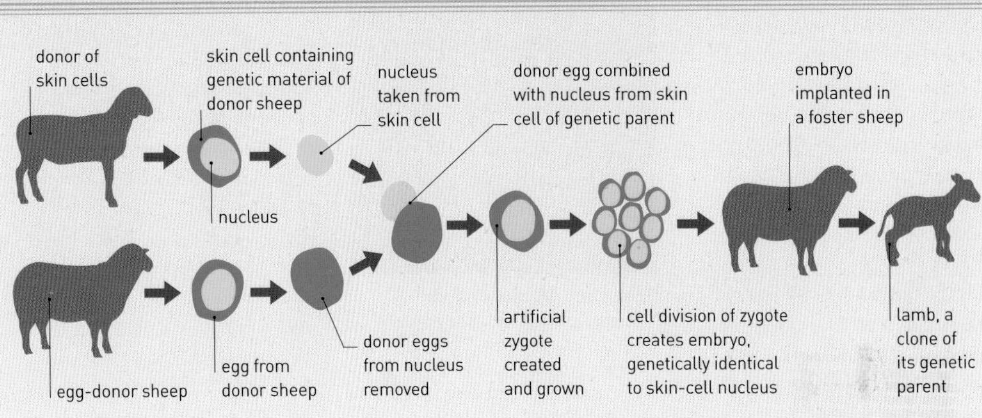
donor of skin cells
skin cell containing genetic material of donor sheep
nucleus taken from skin cell
donor egg combined with nucleus from skin cell of genetic parent
embryo implanted in a foster sheep
nucleus
egg-donor sheep
egg from donor sheep
donor eggs from nucleus removed
artificial zygote created and grown
cell division of zygote creates embryo, genetically identical to skin-cell nucleus
lamb, a clone of its genetic parent

In 1995, molecular biologists at the The **Institute for Genomic Research** in Maryland completed the first genome sequence of a bacterium.

In 1996, a global collaboration among biologists around the world resulted in the **first complete sequencing of the genome of a eukaryote** (an organism whose cells have a nucleus). Another significant breakthrough in genome science and technology was the **creation of a cloned sheep**, which was **named Dolly**, at the Roslin Institute, Scotland. Clones of many animals, including mammals, had been carried out before, but Dolly was the result of transferring DNA from a cell taken from an adult sheep into an egg—a procedure known as **somatic cell nuclear transfer** (see panel, above).

> "LIKE AN **ICE CREAM CONE,** WITH A NEWLY UNCOVERED STAR PLAYING… **THE CHERRY ON TOP.**"
>
> **Jeff Hester, US physicist,** 1995

In September, the UN adopted the Comprehensive Nuclear-Test-Ban Treaty, which places a ban on all nuclear explosions. It has still not been fully ratified. In technology, the **Universal Serial Bus** (USB) connection was launched in 1996, and the first commercial DVD players became available in Japan. In November 1996, Japanese inventor **Shuji Nakamura** (b.1954) invented a continuous, low-power **blue light LED** (light emitting diode) **laser**. Since blue light has a shorter wavelength than the red light used in DVD players, Nakamura's invention allows much more information to be carried on DVD-like disks.

SHUJI NAKAMURA (b.1954)

Born in Ikata, Japan, Shuji Nakamura studied electronic engineering at the University of Tokushima. He made the first practical LEDs (light emitting diodes) to use gallium nitride, resulting in brighter LEDs, and the first LEDs to produce blue light. Nakamura's development of the blue LED laser represents a milestone in consumer electronics.

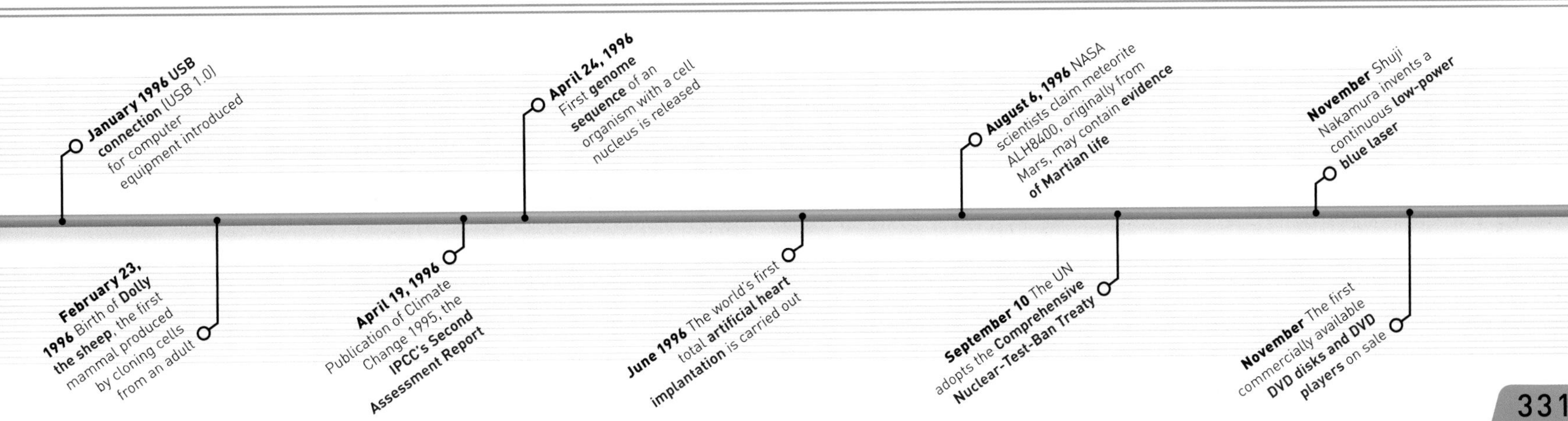
January 1996 USB connection (USB 1.0) for computer equipment introduced
February 23, 1996 Birth of **Dolly the sheep**, the first mammal produced by cloning cells from an adult
April 19, 1996 Publication of Climate Change 1995, the **IPCC's Second Assessment Report**
April 24, 1996 First **genome sequence** of an organism with a cell nucleus is released
June 1996 The world's first total **artificial heart implantation** is carried out
August 6, 1996 NASA scientists claim meteorite ALH8400, originally from Mars, may contain **evidence of Martian life**
September 10 The UN adopts the **Comprehensive Nuclear-Test-Ban Treaty**
November Shuji Nakamura invents a continuous **low-power blue laser**
November The first commercially available **DVD disks and DVD players** on sale

Journalists in Hong Kong wear face masks to reduce the risk of airborne infection by the H5N1 virus, commonly known as bird flu.

6 THE NUMBER OF **FATALITIES** IN **THE 18** PEOPLE **INFECTED WITH BIRD FLU** IN 1997

CHESS-PLAYING COMPUTER PROGRAMS first appeared in the late 1950s. Increases in computing power led to more powerful programs. In 1997, for the first time, a **computer won a match** against the reigning world chess champion. **IBM's Deep Blue** computer won two of the six games of the match, with three draws, while its human opponent, Russian **grandmaster Garry Kasparov**, won only one game.

The popular Internet search engine **Google**—originally called BackRub—**got its name** this year. This new name was derived from the word googol—a mathematical term for the number represented by 1 followed by one hundred zeroes. The creators of Google—American computer scientist **Larry Page** (b.1973) and Russian-born computer scientist **Sergey Brin** (b.1973)—were still developing the search engine at Stanford University, California. They incorporated the company, Google Inc., in 1998.

A particularly **virulent influenza virus** known as H5N1, which had affected birds since the 1950s, **crossed the species barrier** into humans. An outbreak of the disease it causes, nicknamed bird flu, killed six people in Hong Kong. Health authorities were concerned that the disease could become a pandemic and issued hygiene advice to international travelers and to people working in the poultry trade. Although **other bird flu outbreaks** have occurred since then, fears of a pandemic have not been realized.

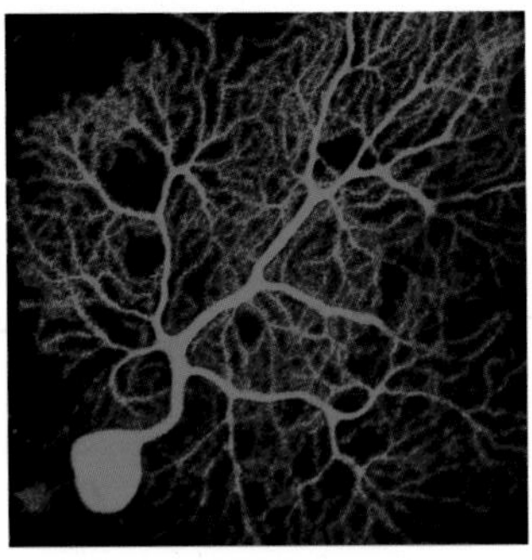

Green fluorescent protein
A micrograph shows a cell taken from a mouse's brain. The cell is producing green fluorescent protein (GFP), a substance used to track gene expression.

At a conference in December in Kyoto, Japan, the United Nations reached an **agreement called the Kyoto Protocol** that relates to the UN Framework Convention on Climate Change (UNFCCC, see 1992). The countries that signed and ratified the agreement were committed to reducing their emissions of greenhouse gases—most importantly, carbon dioxide released by burning fossil fuels (see pp.326–27). Each participating nation had a target: its emissions for the period 2008–12 had to be reduced by a certain percentage compared with the emissions in a base year (in most cases, 1990). The targets did not take into account emissions from aviation and international shipping. At a conference at Doha, Qatar, in December 2012, parties to the **UNFCCC agreed to new targets** for the second assessment period—2013 to 2020.

A team led by Japanese geneticist **Masaru Okabe** hit the headlines when they produced **genetically modified mice** that produce a green glow under ultraviolet light. The glow is produced by a compound called **green fluorescent protein** (GFP), which exists naturally in certain jellyfish. The gene that codes for GFP was first sequenced in 1994, and the protein is now an important tool in molecular biology. By inserting the GFP-coding gene from the jellyfish into a part of the genomes of other organisms, researchers can tell when and whether that part of the **genome is being activated**. Okabe injected the gene into mouse embryos with the hope of tracking the development of the mice's sperm cells, but instead, the protein was produced in nearly every type of cell in the mice's bodies.

> **“ DEEP BLUE** WAS ONLY **INTELLIGENT** THE WAY YOUR PROGRAMMABLE ALARM CLOCK IS INTELLIGENT. NOT THAT **LOSING TO A $10 MILLION ALARM CLOCK** MADE ME FEEL ANY BETTER. **”**
>
> **Garry Kasparov, Russian chess grandmaster,** 1997

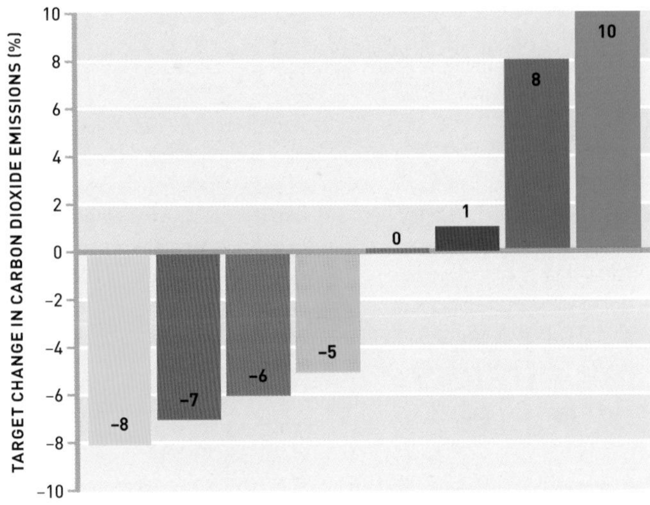

TARGET CHANGE IN CARBON DIOXIDE EMISSIONS (%)
10
8
6
4
2
0
-2
-4
-6
-8
-10
-8
-7
-6
-5
0
1
8
10
KEY
Austria, Belgium, Bulgaria, Czech Republic, Denmark, Estonia, Finland, France, Germany, Ireland, Italy, Latvia, Liechtenstein, Lithuania, Luxembourg, Monaco, Netherlands, Portugal, Romania, Slovakia, Slovenia, Spain, Sweden, Switzerland, United Kingdom of Great Britain and Northern Ireland
United States of America
Canada, Hungary, Japan, Poland
Croatia
New Zealand, Russian Federation, Ukraine
Norway
Australia
Iceland

Kyoto Protocol emission targets
Emissions targets are shown for the first assessment period (2008–12) of the Kyoto Protocol. Most countries had to reduce their emissions by the end of this period; some had leeway to increase them. The US signed the protocol but did not ratify it.

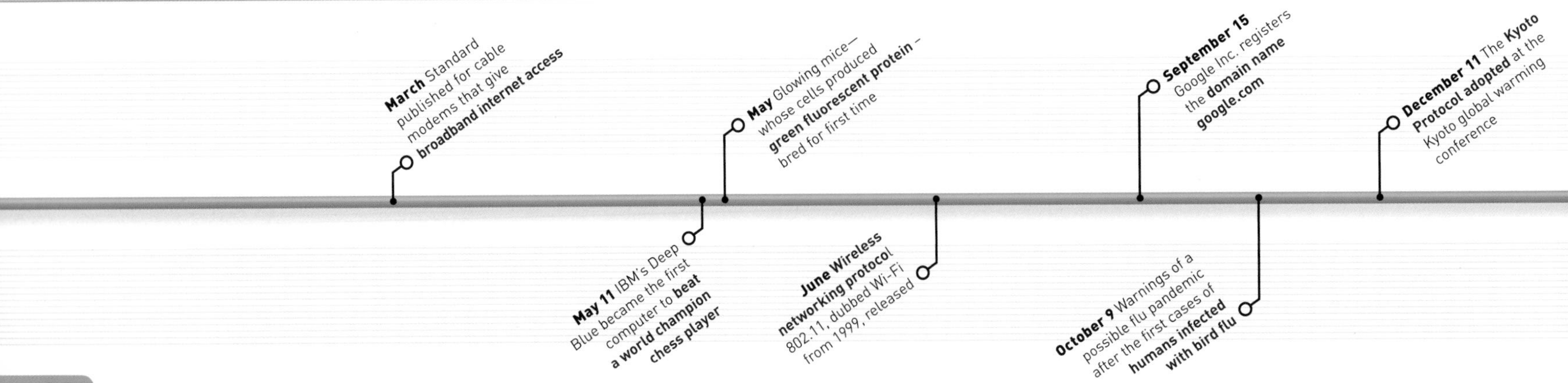

March Standard published for cable modems that give **broadband internet access**
May 11 IBM's Deep Blue became the first computer to **beat a world champion chess player**
May Glowing mice—whose cells produced **green fluorescent protein**—bred for first time
June Wireless networking protocol 802.11, dubbed Wi-Fi from 1999, released
September 15 Google Inc. registers the **domain name google.com**
October 9 Warnings of a possible flu pandemic after the first cases of **humans infected with bird flu**
December 11 The **Kyoto Protocol adopted** at the Kyoto global warming conference

> "THERE IS AN **INCREDIBLE AMOUNT OF MATTER** BETWEEN US AND THE CENTER OF THE MILKY WAY TO **OBSCURE OUR VIEW.**"
>
> **Terry Oswalt, National Science Foundation program manager for Stellar Astronomy and Astrophysics,** 1998

This color-enhanced X-ray image shows the region around Sagittarius A*, the supermassive black hole at the galactic center that produces regular X-ray flares as a result of vaporizing asteroids and other matter.

INTERNET USE BEGAN TO GROW steeply this year. A new, **high speed internet connection** through telephone lines became available in 1998. With asymmetric digital subscriber line (ADSL), users could receive information from the internet at 8 megabits (8 million bits) per second. The previous year, another **broadband technology** had its debut for home users: the cable modem, which connected to the internet via existing coaxial cables that also delivered television signals. These new technologies allowed users to easily **download larger files**, such as mp3 music files, more easily. The world's first **portable mp3 player**, the MPMan F10, was released this year, by SaeHan Information Systems from South Korean.

In September, American astronomer **Andrea Ghez** (b.1965) reported that she had detected the presence of a supermassive **black hole at the center of our galaxy**. Astronomers have since found evidence that supermassive black holes are present in most, if not all, galactic centers.

Astrophysicists studying supernovas in distant galaxies came to the conclusion that the **expansion of the Universe is accelerating**; this was the first concrete evidence of the existence of a cosmological constant—some kind of repulsive agent, also known as **dark energy** (see pp.344–45), that drives the expansion of space-time ever faster. Dark energy probably accounts for around three-quarters of the mass-energy in the Universe.

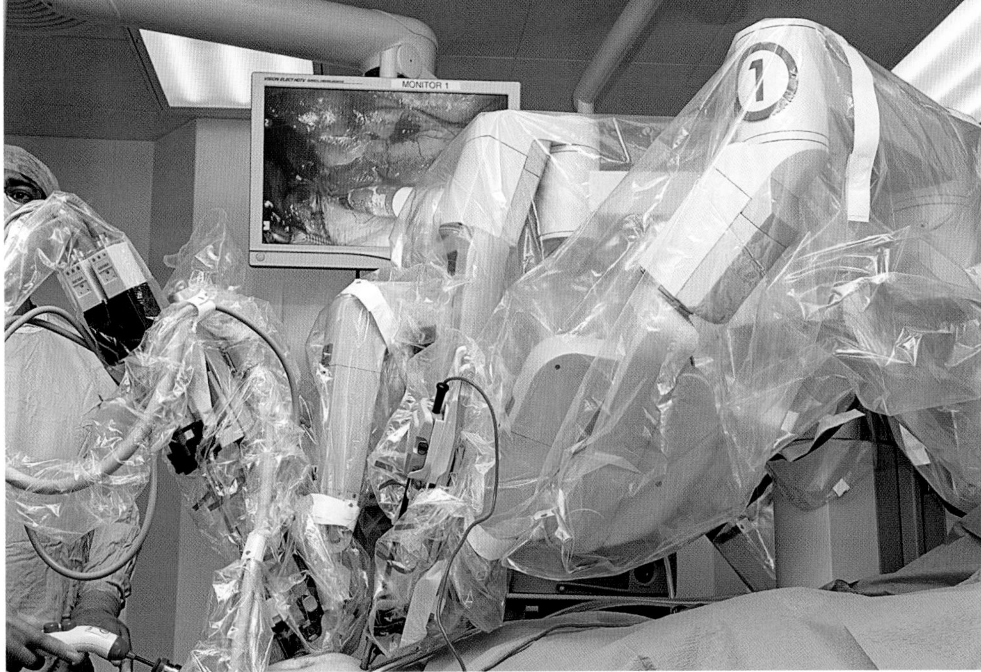
MONITOR 1

Robotic surgery
The da Vinci Surgical System has four robot arms, which are controlled by the surgeon. One of the arms carries a high definition, 3-D vision system.

In November, a Russian Proton rocket launched the first module of the **International Space Station** (ISS), into orbit. The ISS has been inhabited continuously since November 2000. The project involves **five space agencies** representing a total of 16 countries.

A team led by American cell biologist **James Thomson** (b.1958) created a culture of **human embryonic stem cells** (hESCs). These cells can develop into any kind of tissue, and have the potential of creating donor-matched organs for transplantation. The cultures, or lines, began with cells **harvested from human embryos**—a fact that raised significant ethical concerns, since the technique led to the destruction of human embryos. Thomson later managed to create induced stem cells—similar to hESCs, but created from **reprogrammed adult cells**, not taken from embryos (see 2007).

In another milestone in genomics (see 1977 and 1995), a nematode worm became the **first multicellular organism** to have its **genome sequenced**. Within two years, the draft sequence of the human genome was complete (see 2000).

In May, German surgeon **Friedrich-Wilhelm Mohr** (b.1951) performed the first **robotically assisted heart surgery**. In robotic surgery, the surgeon benefits from the assistance of computer-controlled instruments, which do not suffer from vibration or fatigue. With information and images relayed via the internet, an expert surgeon can carry out **operations in distant locations** using surgical robots remotely.

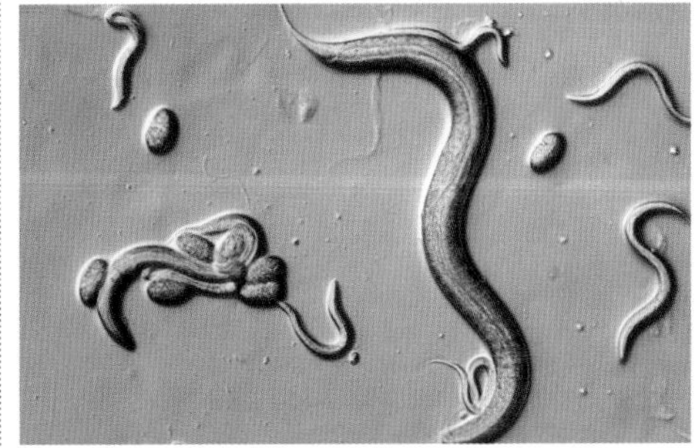

Worm genome
Caenorhabditis elegans *was the first multicellular organism to have its genome sequenced. This nematode worm lives in soil and grows to about 1/32 in (1 mm) in length.*

March Release of the **first consumer mp3 player**, the MPMan F10

April First experimental demonstration of a **quantum computer**

May First robotically assisted heart bypass

September Astrophysicists announce that the **expansion of the Universe is accelerating**

September 8 Astronomers conclude that there is a supermassive **black hole at the center of our galaxy**

October ADSL (asymmetric digital subscriber line) technology introduced for the first time in the USA

November 6 James Thomson creates human **embryonic stem cell lines (cultures)**

20 November First component of the **International Space Station** launched into orbit

December 11 First genome sequencing for a multicellular organism reported

THE STORY OF ROBOTICS

ROBOTS HAVE DEVELOPED FROM BASIC MECHANICAL TOOLS AND TOYS TO PLAYING AN IMPORTANT ROLE IN MODERN SOCIETY

Robots are built in a wide range of forms, and there is no single definition of a robotic device that encompasses all its aspects. However, the vast majority are electromechanical machines that can perform tasks and manipulate their environment in accordance with a set of preestablished instructions.

Robots vary hugely in form and function. Physically, they range from a jointed arm with tools at the end to the human-shaped "android" beloved of science fiction enthusiasts. Operationally, they are just as diverse and include machines whose function is defined by their form—such as ancient Egyptian water clocks, or clepsydra, and Jacquard looms that could weave textile patterns based on instructions stored on punched cards. There are also robots that are versatile devices capable of performing a wide range of tasks and reacting to external stimuli, such as robot space probes.

MODERN ROBOTS

The word "robot" was first used by Czech writer Karel Čapek in his 1920 science fiction play *R.U.R.* (*Rossum's Universal Robots*), and derives from the Czech term for forced labor—indeed, most robots do perform their tasks by obeying instructions either embedded in their design, or passed on to them by a governing computerized controller in the form of software. Such robots excel at carrying out repetitive, complex, or detailed tasks at high speed and without fatigue. More versatile robots, such as the da Vinci surgical robot, often function under the direct command of a human operator—a technique known as telepresence, which may use cameras or more complex technology to relay the robot's "view" of a hostile, dangerous, or complex environment. Increasing numbers of robots are being equipped with artificial intelligence, which allows them to make decisions of their own.

ARTIFICIAL INTELLIGENCE

For many robotics enthusiasts, the ultimate goal of computing is to develop a form of intelligence capable of mimicking human behavior. Most forms of artificial intelligence (AI) used in robots so far have involved the machine recognizing and reacting to scenarios according to a set of preprogrammed rules that its computer can understand. AI research has made its greatest advances in applications such as chess computers.

> **" ROBOTICS** HAS BECOME A SUFFICIENTLY **WELL DEVELOPED TECHNOLOGY** TO WARRANT ARTICLES AND BOOKS ON ITS HISTORY AND **I HAVE WATCHED THIS IN AMAZEMENT,** AND IN SOME DISBELIEF, **BECAUSE I INVENTED IT.** NO, **NOT THE TECHNOLOGY;** THE **WORD. "**

Isaac Asimov, US writer, from *Counting the Eons*, 1983

c.250 BCE
Early robotics
Automated machinery, such as Egyptian water clocks, are considered by some to be an early form of robotics.

Clesibius's clepsydra

c.1206
Al-Jazari's automata
One of Arabic inventor al-Jazari's many achievements is a band of automata that can be commanded to play different pieces of music.

Al-Jazari's band

1801
Programmable loom
The Jacquard loom, a programmable device that can weave textile patterns automatically, is developed.

Jacquard loom

19th century
Mechanical toy
From the European Renaissance onward, robotic figures, called automata, are collected as playthings and curios by the wealthy.

Drum automaton

1942
Asimov's laws
In his book *I, Robot*, US writer Isaac Asimov devises three laws to govern his fictional robots. These go on to influence real-life robot builders.

1949
Educational robot
The highly popular Turtle design developed by William Grey Walter is equipped with a variety of sensors that allow it to react to its environment.

Turtle robot

1961
Industrial robot
The first commercial manufacturing robot, a robotic arm called Unimate, enters service at a General Motors plant in the US.

Robotic arm

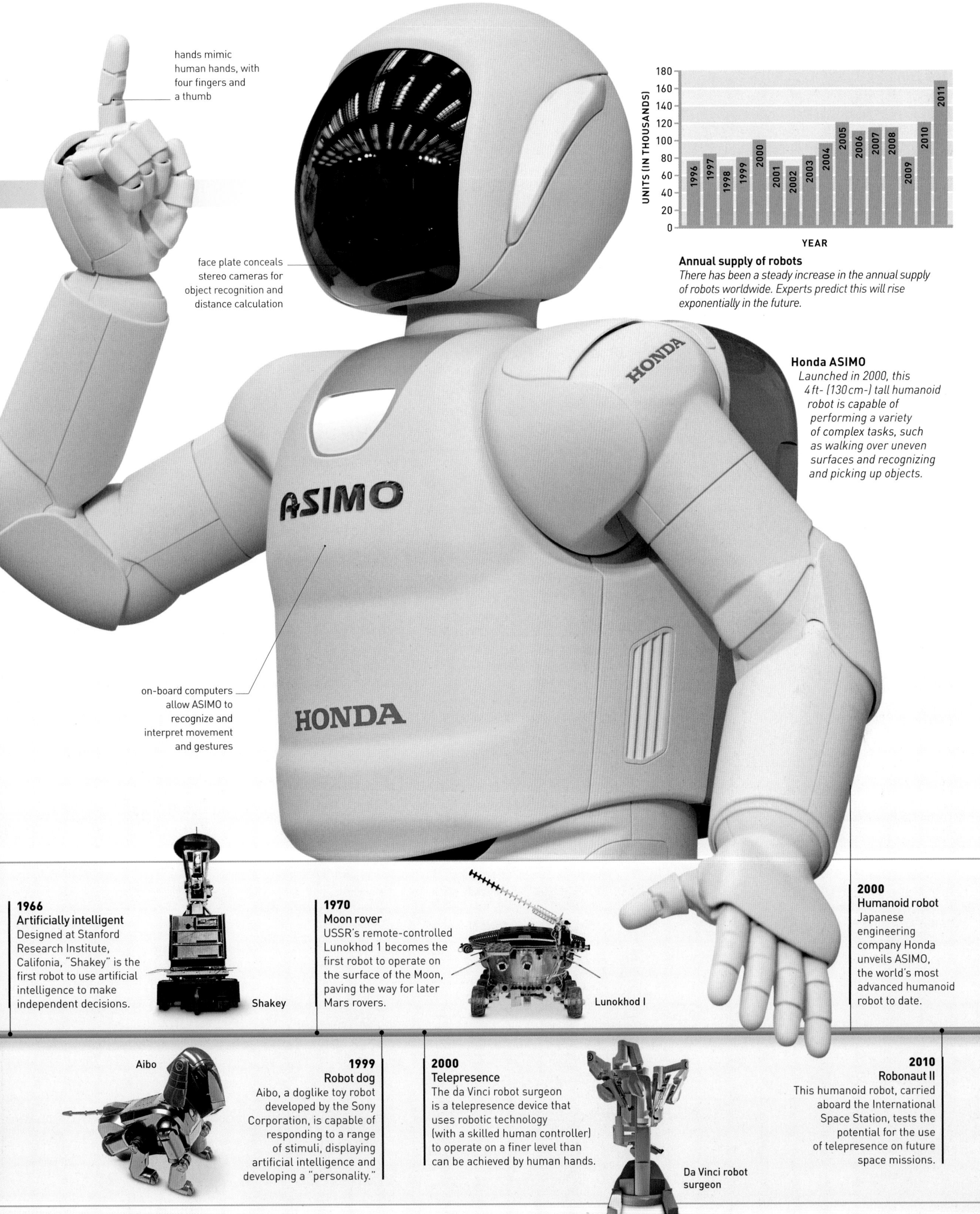
hands mimic human hands, with four fingers and a thumb
face plate conceals stereo cameras for object recognition and distance calculation
on-board computers allow ASIMO to recognize and interpret movement and gestures
HONDA
ASIMO
HONDA
UNITS (IN THOUSANDS)
180
160
140
120
100
80
60
40
20
0
1996
1997
1998
1999
2000
2001
2002
2003
2004
2005
2006
2007
2008
2009
2010
2011
YEAR
Shakey
Lunokhod I
Aibo
Da Vinci robot surgeon

Annual supply of robots
There has been a steady increase in the annual supply of robots worldwide. Experts predict this will rise exponentially in the future.

Honda ASIMO
Launched in 2000, this 4 ft- (130 cm-) tall humanoid robot is capable of performing a variety of complex tasks, such as walking over uneven surfaces and recognizing and picking up objects.

1966
Artificially intelligent
Designed at Stanford Research Institute, Califonia, "Shakey" is the first robot to use artificial intelligence to make independent decisions.

1970
Moon rover
USSR's remote-controlled Lunokhod 1 becomes the first robot to operate on the surface of the Moon, paving the way for later Mars rovers.

2000
Humanoid robot
Japanese engineering company Honda unveils ASIMO, the world's most advanced humanoid robot to date.

1999
Robot dog
Aibo, a doglike toy robot developed by the Sony Corporation, is capable of responding to a range of stimuli, displaying artificial intelligence and developing a "personality."

2000
Telepresence
The da Vinci robot surgeon is a telepresence device that uses robotic technology (with a skilled human controller) to operate on a finer level than can be achieved by human hands.

2010
Robonaut II
This humanoid robot, carried aboard the International Space Station, tests the potential for the use of telepresence on future space missions.

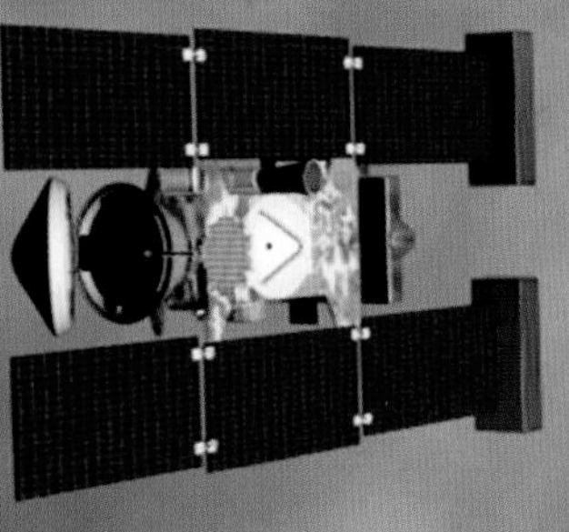

> **“** IN THIS **FIVE-YEAR JOURNEY** WE REACHED BACK IN TIME TO **COLLECT PARTICLES** THAT HAVEN'T BEEN CHANGED IN **4.6 BILLION YEARS. ”**
>
> **Tom Duxbury, Project Manager,** Stardust project, 2004

NASA's Stardust probe flew past comet Wild 2 and a retractable aerogel panel collected the dust from the comet's coma (dust cloud).

IN FEBRUARY, NASA's STARDUST PROBE BEGAN an unprecedented mission: to **collect samples** from the dust cloud (coma) surrounding a **comet's nucleus**. As it traveled, it also collected interstellar dust. When Stardust flew past Earth in 2006, it released a sample-return capsule, which returned to Earth.

Scientists **created three identical goats** by transferring the contents of embryo cells into empty egg cells. The embryonic cells **had an extra gene**, which caused the goats to produce a blood clotting factor that could be harvested from their milk and **used in human medicines**.

The increasing number of digital gadgets around the home encouraged the **development of wireless data connections**. The most widely used wireless networking protocol—IEEE 802.11 (see 1997)—was given a user-friendly name this year: Wi-Fi. For one-to-one connections between various types of device, a **new protocol was released**: Bluetooth. Developed by Swedish company Ericsson in 1994, the most popular use for Bluetooth is sending digital audio signals to wireless headphones or speakers. American climatologists Michael Mann (b.1965) and Raymond Bradley (b.1948) and American dendrochronologist (studies tree rings) Malcolm Hughes (b.1943) constructed a graph of the **estimated average global temperature** over the past thousand years. The data were derived from meteorological readings, with older estimates based on historical records and tree rings. The graphs showed a gradual cooling, with a **rapid rise in temperature** corresponding to the rise in global population and industrialization. The **graph's shape** reminded American climate scientist Jerry Mahlman (1940–2012) of a **hockey stick**. To climate scientists, the graph is a symbol of the **effect of human activity** on the climate—and a stark reminder of the need to reduce carbon dioxide emissions. However, many who were opposed to the idea that human activity is causing global warming disputed the accuracy of the graph.

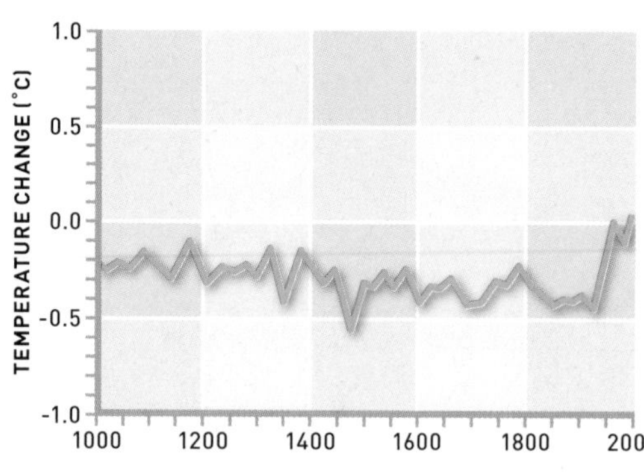

1.0
0.5
0.0
-0.5
-1.0
TEMPERATURE CHANGE (°C)
1000
1200
1400
1600
1800
2000
YEAR

Hockey-stick graph
A simplified version (without error bars) of the controversial graph that shows estimated average global temperature over the past thousand years.

Researchers studying growth hormones **discovered a protein**, which they called ghrelin, that is secreted by cells in the stomach lining. Ghrelin acts on the **appetite center in the brain** (see leptin, 1994). The release of ghrelin into the bloodstream is determined by stretch receptors in the stomach lining: when the stomach is full, less ghrelin is produced, but as the stomach empties between meals, more ghrelin is produced, **leading to the sensation of hunger**.

Also this year, British physicist **John Pendry** (b.1943) described a new class of materials with **unusual properties**, called metamaterials. One metamaterial in particular held the promise of invisibility (see panel, left).

3 THE NUMBER OF **IDENTICAL GOATS** PRODUCED BY **CLONING** AN **ADULT** ANIMAL

METAMATERIALS

Metamaterials are man-made materials that have properties not found in naturally occurring materials. One example (right) is this material with tiny metal coils embedded within it, which can divert microwaves around itself. This makes the material invisible to the microwaves. A metamaterial that does the same to visible light could act as an invisibility cloak.

January 22 Italian cell biologist Angelo Vescovi announces the creation of **blood cells from brain stem cells** in mice

February 7 Launch of NASA's Stardust mission to **asteroid 5535 Annefrank and comet Wild 2**

April 23 Publication of the controversial graph showing the recent **rise in the average global temperature**

April 27 Scientists **produce three clones** of an adult goat

July 31 NASA's Lunar Prospector is **deliberately crashed into the lunar north pole**, revealing water ice in the collision debris

November Physicists publish details of the first metamaterials, opening up the **possibility of invisibility cloaking**

December 1 Ericsson releases the **wireless networking protocol Bluetooth**, which they invented in 1994

December 9 Appetite-regulating hormone ghrelin discovered

IN THE MONTHS LEADING UP TO THE TURN OF THE MILLENNIUM, people were warned of the possibility that **electronic devices and systems could fail**, as computer operating systems used two digits to represent years and internal clocks would revert to the year 1900. In a world in which people's livelihoods and even their safety increasingly rely upon computerized systems, this possibility—known as the **Y2K problem or the Millennium Bug**—led to widespread panic. Software engineers worked hard to ensure that the fears would turn out to be unfounded, and indeed only a small number of systems were affected.

Fruit fly
The fruit fly has been used as a model organism in genetic research since 1910, and so was a natural choice to have its genome sequenced.

January 14 Creation of vitamin A-rich **transgenic rice** announced

1.8 BILLION

THE NUMBER OF **SEEDS CONSERVED** IN THE MILLENNIUM SEED BANK

Seeds are shown after they have been collected and sorted as part of the Millennium Seed Bank Partnership. The project aims to store seeds of 25 percent of the world's plants by 2020.

The cells inside a developing embryo have the potential to **develop into any kind of cell** (see 1981). There are similar stem cells in other tissues, but they only differentiate (develop) into a limited number of different cell types. In 1999, a team led by Italian researcher **Angelo Vescovi** (b.1962) took stem cells from mouse brains, injected them into mouse blood, and found they developed into various types of blood cell. This year, the same team took stem cells from mouse brains and found they developed into muscle cells just by being in contact with muscle cells in laboratory glassware. These experiments proved that **non-embryonic stem cells** are more flexible than was believed—a boon to stem cell research since it avoids the harvesting of embryonic stem cells, which involves the destruction of the embryo.

After a huge international effort, scientists involved in the **Human Genome Project** announced the completion of the draft sequence of the **entire human genome**. Thale cress (*Arabidopsis thaliana*), a model organism widely used in genetic experiments, became the first plant to have its genome sequence completed. The genome of another model organism, the fruit fly (*Drosophila melanogaster*), was also sequenced this year.

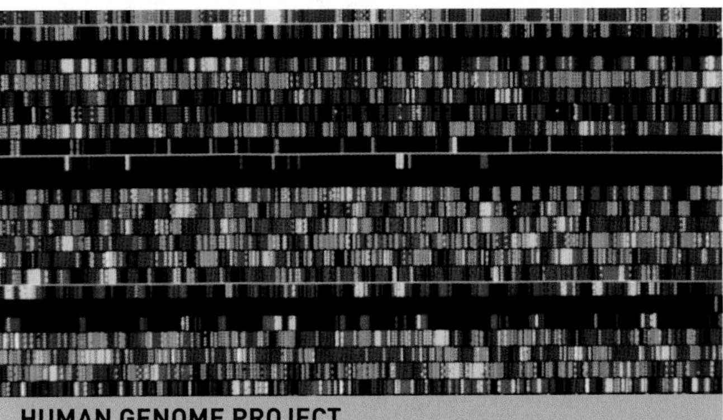

HUMAN GENOME PROJECT

Officially launched in 1990, the human genome project was an international effort to determine all 3 billion DNA base pairs along the length of the human genome and identify all the 25,000 or so genes the genome contains. This knowledge holds great benefits for medical science as well as studies of human evolution and genetics in general. The project was declared complete in 2003.

German biotechnologist **Ingo Potrykus** (b.1933) and German cell biologist **Peter Beyer** (b.1952) announced that they had created a **genetically modified (GM) variety of rice**. The new rice plant produced beta carotene, the precursor to vitamin A, in the edible grain. Vitamin A deficiency is commonplace in many developing countries, causing an estimated two million deaths every year, and is a major, preventable cause of blindness. The rice earned the nickname "golden rice," because of the color of the beta carotene. Biotechnology companies agreed to give the rice seeds free to farmers who made less than $10,000 per year, and the project was supported by various humanitarian organizations. However, the project has **attracted resistance** from anti-GM protesters (and anti-capitalists), who fear that multinational food companies will have an economic hold on poor farmers. Controversy delayed field trials and approval, but in 2013, golden rice seeds were given to farmers in the Philippines—with several other nations considering following suit.

In November, plant conservationists at the Royal Botanic Gardens, Kew, UK, launched the **Millennium Seed Bank Partnership**—a project involving more than 50 countries that aims to collect and preserve the seeds of tens of thousands of plants. Seeds are sorted, cleaned and dried, then stored in cold, dry conditions in large underground freezers. Climate change and changes in land use could put **many plant species at risk of extinction**, and protecting the seeds means that extinct species could be reintroduced.

In November, engineers at the Japanese car manufacturer Honda revealed the first model of their **humanoid robot ASIMO** (Advanced Step in Innovative Mobility), a popular humanlike robot that can talk, walk, and run.

ASIMO
HONDA

ASIMO
The first version of Japanese car manufacturer Honda's humanoid ASIMO robot was just 4 ft (1.2 m) tall and weighed 106 lb (48 kg).

> “ WE ARE HERE TODAY TO CELEBRATE A **MILESTONE** ALONG A TRULY **UNPRECEDENTED VOYAGE,** THIS ONE **INTO OURSELVES.** ”
>
> **Francis Collins, American geneticist**, June 26, 2000

March 24 Scientists complete the sequencing of the **genome of the fruit fly**

June 26 Rough draft of the entire human genome announced by **the Human Genome Project**

October Angelo Vescovi reports that he has turned mouse **stem cells into muscle cells**

November 20 Opening of the **Millennium Seed Bank**, at Kew Gardens in the UK

November 20 Honda **debuts ASIMO**, a popular humanlike robot

December 14 The sequencing of the **genome of the model organism** thale cress is completed

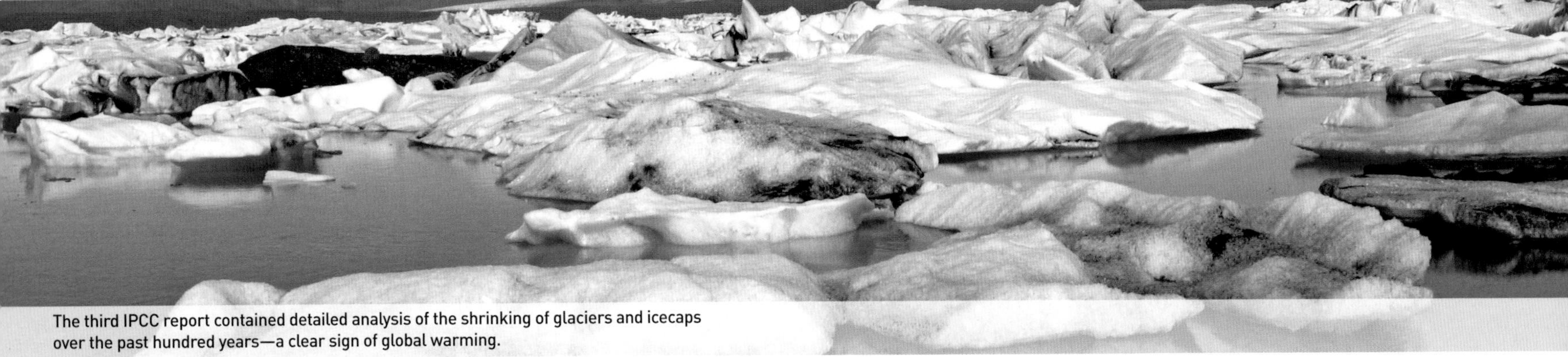
The third IPCC report contained detailed analysis of the shrinking of glaciers and icecaps over the past hundred years—a clear sign of global warming.

THE INTERGOVERNMENTAL PANEL ON CLIMATE CHANGE or IPCC (see 1988) published its third major report into climate change in 2001. The document supported and extended the conclusions of the organization's previous reports (see 1990, 1995) and provided more detailed projections of climate change for the decades ahead. The report included the hockey stick graph, produced by climate scientists two years earlier (see 1999).

Water on Mars
This polar icecap is made of water ice covered with a layer of carbon dioxide ice. NASA's Mars Odyssey also found evidence of huge amounts of underground water.

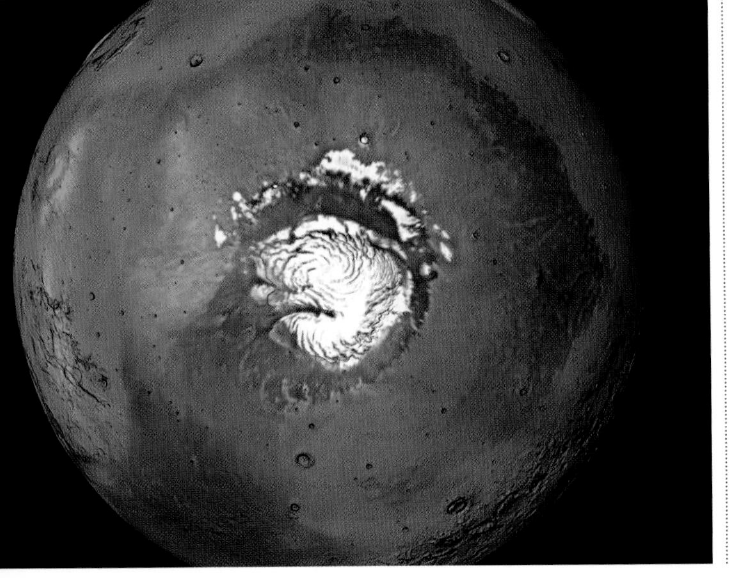

One of the early indications of the emerging Web 2.0—in which users create content on the world wide web—was the launch of the online encyclopedia Wikipedia. Supported by the non-profit-making Wikimedia Foundation, Wikipedia can be written and edited by anyone. In 2007, it became the largest encyclopedia ever written, with more than two million articles.

In October, Apple Inc. launched its **portable digital music player**, the iPod. Similar products already existed, but sleek design and an intuitive user interface, coupled with Apple's music library and iTunes software, made the iPod a landmark consumer product.

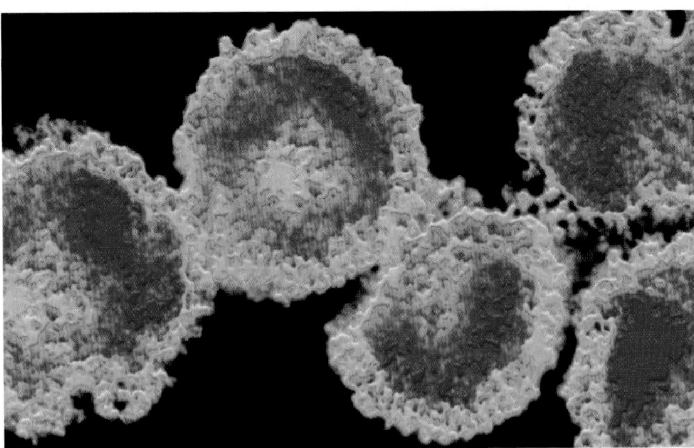

SARS virus
This electron microscope image shows the corona virus—the virus found to be the cause of the mysterious respiratory illness known as SARS.

Researchers at the Sudbury Neutrino Observatory, Canada, detected definitive evidence of **neutrino oscillation**. Neutrinos are fundamental particles produced in nuclear reactions. There are three types, or flavors, of neutrino: muon, tau, and electron. According to the Standard Model (see 1974), neutrinos should have no mass. In the 1960s, American physicist Raymond Davis had studied solar neutrinos (see 1968), but detected one-third as many as was predicted. Davis' experiment could only detect electron neutrinos. One way to explain the deficit was that neutrinos change flavor, or oscillate, and this is what the Sudbury experiment found. Neutrino oscillation is only possible if neutrinos have mass—something the Standard Model has yet to account for.

In 2002, NASA's **Mars Odyssey space probe** detected huge amounts of water on Mars, just below the surface in the planet's Arctic region. Much of the water was locked in claylike minerals. NASA's Phoenix lander probe visited that region in 2008 and confirmed Odyssey's observations.

In Guandong Province, China, doctors became aware of an outbreak of a pneumonialike illness that gave people trouble breathing and, in some cases, led to their death. Treatment with antibiotics was ineffective, and researchers found none of the bacteria or viruses known to cause pneumonia. The number of cases increased rapidly and it became clear that **SARS (severe acute respiratory syndrome)** was highly infective. The disease began to spread and risked becoming a major pandemic. A combination of quarantines and the screening of airline passengers curbed the spread of the disease, and the last known case occurred in May 2004. In all, there were 8,273 reported cases, with 775 deaths—most of them in China and Hong Kong.

> “ **IMAGINE** A WORLD IN WHICH **EVERY SINGLE PERSON…** IS GIVEN **FREE ACCESS** TO… **ALL HUMAN KNOWLEDGE.** THAT'S WHAT WE'RE DOING. ”
>
> **Jimmy Wales, founder of *Wikipedia*,** 2004

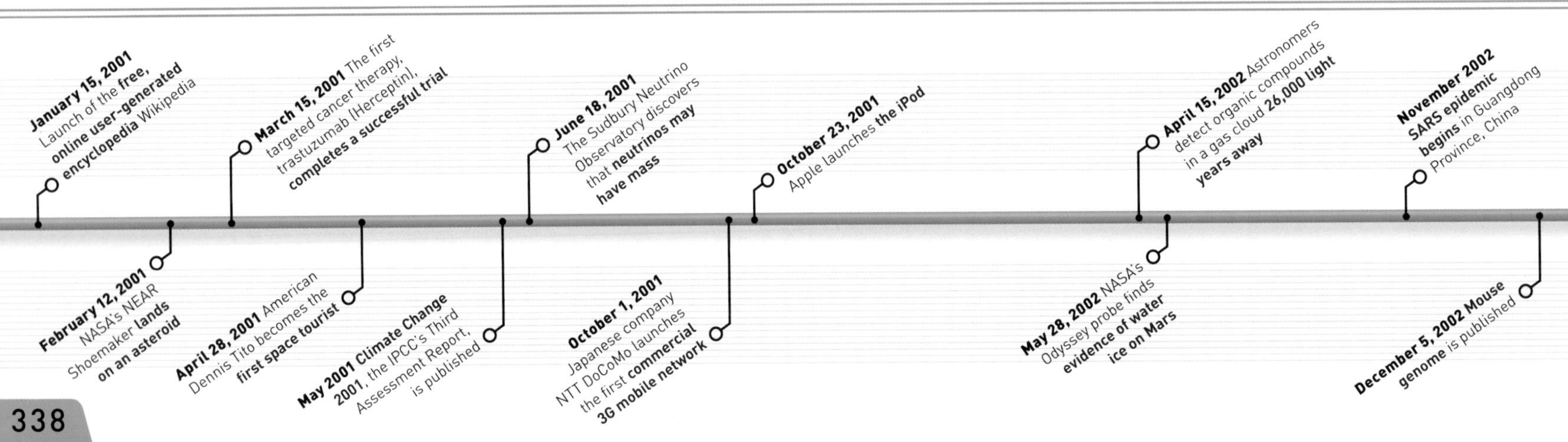
January 15, 2001 Launch of the free, online user-generated encyclopedia Wikipedia
February 12, 2001 NASA's NEAR Shoemaker lands on an asteroid
March 15, 2001 The first targeted cancer therapy, trastuzumab (Herceptin), completes a successful trial
April 28, 2001 American Dennis Tito becomes the first space tourist
May 2001 Climate Change 2001, the IPCC's Third Assessment Report, is published
June 18, 2001 The Sudbury Neutrino Observatory discovers that neutrinos may have mass
October 1, 2001 Japanese company NTT DoCoMo launches the first commercial 3G mobile network
October 23, 2001 Apple launches the iPod
April 15, 2002 Astronomers detect organic compounds in a gas cloud 26,000 light years away
May 28, 2002 NASA's Odyssey probe finds evidence of water ice on Mars
November 2002 SARS epidemic begins in Guangdong Province, China
December 5, 2002 Mouse genome is published

2003

13.75 BILLION YEARS
THE **AGE** OF OUR **UNIVERSE**

This super computer simulation of the structure of the Universe from 2005, created at the Max Planck Institute in Germany, shows the distribution of dark matter in part of the Universe.

IN THIS YEAR, THE CHINESE NATIONAL SPACE administration launched its first manned space mission, Shenzhou 5. Astronaut Yang Liwei spent 21 hours and 23 minutes in space, orbiting Earth 14 times.

In February, the US **space shuttle *Columbia* disintegrated while it was reentering** Earth's atmosphere. The incident resulted in the halting of the space shuttle program for two years—almost as long as the gap in the program caused by the loss of the *Challenger* (see 1986).

The same month, cosmologists and astrophysicists published the first year's observations of the **Wilkinson Microwave Anisotropy Probe (WMAP)**. One of the aims of the program was to provide a more detailed map of the cosmic background radiation than the COBE satellite did (see 1992). Variations in the cosmic background radiation (CMB) reflect variations in the density of the early Universe, which in turn caused matter to clump together, forming galaxies and galaxy clusters. The map produced by WMAP also refined an emerging model of the Universe, the Lambda-Cold Dark Matter model (see p.345). The lambda part is the **cosmological constant**—a repulsive force that is accelerating the Universe's expansion. Evidence for the existence of the cosmological constant—or dark energy—came from studies of supernovas in other galaxies (see 1998). Cold dark matter is matter that neither produces nor interacts with electromagnetic radiation. The presence of cold dark matter can be inferred only by its gravitational interaction with ordinary matter. WMAP's observations also enabled scientists to put the Universe's age, with a high degree of accuracy, at **13.7 billion years** (it has since been updated to closer to 13.82 billion years).

Three years after the draft version of the human genome was published, the International Human Genome Sequencing Consortium finally **completed the full sequencing of all 3 billion DNA base pairs** that make up the human genome. A draft of the chimpanzee genome, which is nearly 99 percent identical to the human genome, was also published in 2003. The chimpanzee is the closest living relative to humans, and comparisons between the two genomes give biologists an unprecedented opportunity to study primate evolution.

Yang Lewei in space
Chinese astronaut, Yang Liwei, is shown aboard the Shenzhou 5 capsule. China became only the third nation to send humans into space, joining the US and Russia.

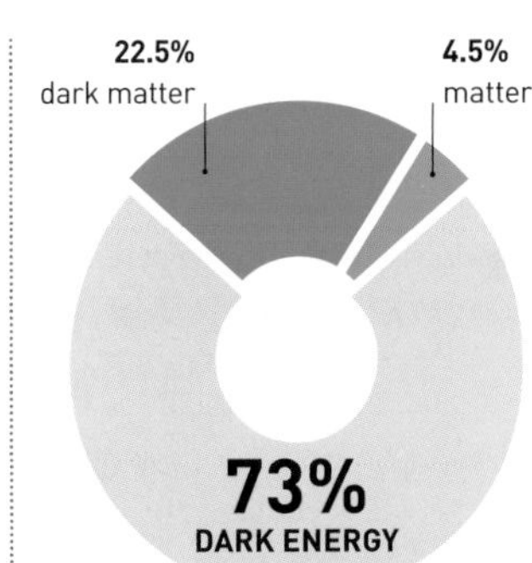
22.5%
dark matter
4.5%
matter
73%
DARK ENERGY

Composition of the Universe
According to WMAP observations, ordinary matter accounts for only a small percentage of the Universe's total mass-energy.

> “I'M FEELING VERY GOOD IN **SPACE,** AND IT **LOOKS EXTREMELY SPLENDID** AROUND HERE.”

Yang Liwei, Chinese astronaut, on the telephone to his wife from space, 2003

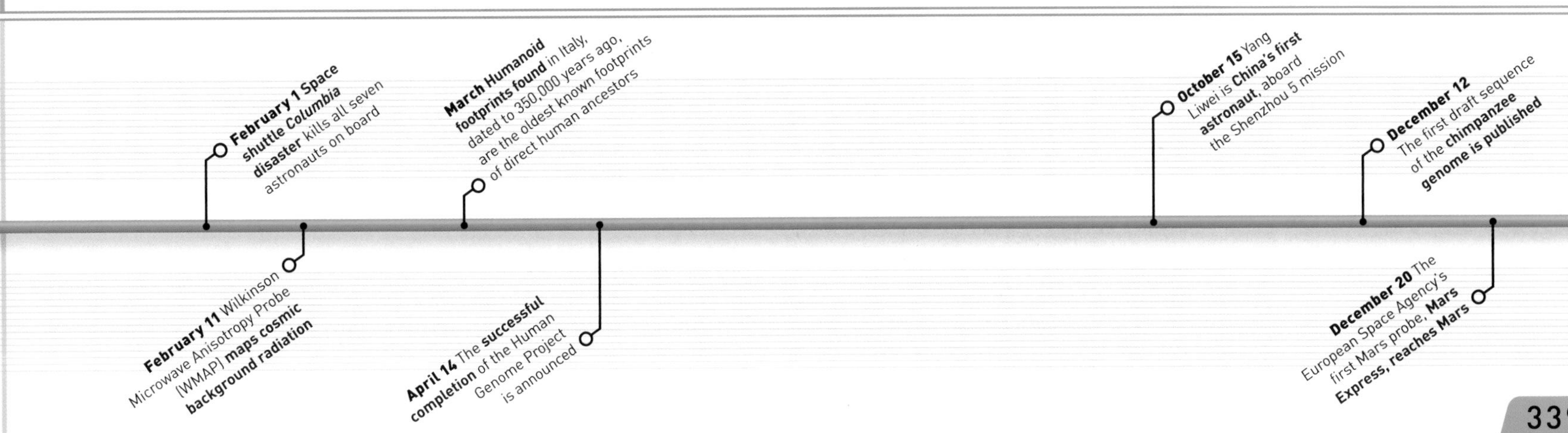
February 1 Space shuttle *Columbia* disaster kills all seven astronauts on board
February 11 Wilkinson Microwave Anisotropy Probe (WMAP) **maps cosmic background radiation**
March Humanoid footprints found in Italy, dated to 350,000 years ago, are the oldest known footprints of direct human ancestors
April 14 The **successful completion** of the Human Genome Project is announced
October 15 Yang Liwei is **China's first astronaut**, aboard the Shenzhou 5 mission
December 12 The first draft sequence of the **chimpanzee genome is published**
December 20 The European Space Agency's first Mars probe, **Mars Express, reaches Mars**

10,000

THE **NUMBER OF GALAXIES** VISIBLE IN THE **HUBBLE ULTRA DEEP FIELD**

Part of the Hubble Ultra Deep Field, showing some of the farthest galaxies ever seen. The entire image contains about 10,000 individual galaxies, some of which are seen by light that left them 13 billion years ago.

IN MARCH, NASA RELEASED A REMARKABLE IMAGE of a tiny part of the sky taken by the Hubble Space Telescope (HST): the **Hubble Ultra Deep Field**, a follow-up to the Hubble Deep Field of 1995. This new image was the result of 800 long exposures—with a total exposure time of nearly 280 hours—captured by an instrument called the Advanced Camera for Surveys, which was installed on the HST in 2002. The image showed thousands of faint galaxies, many of which are so far away that their light would have left them billions of years ago, when the Universe was very young. Astronomers are still studying the image, which provides new **information about galaxy formation**. In 2012, the HST produced an even more detailed view of the same part of the sky. Called the eXtreme Deep Field, the 2012 image revealed about 5,500 galaxies more than shown in the Ultra Deep Field.

In October, Australian palaeoanthropologists unveiled the partial **remains of an unusual humanlike skeleton**: a 3 ft- (1 m-) tall adult with a small skull that would have housed a very small brain. Australian and Indonesian archaeologists discovered the specimen in a cave on the Island of Flores, east of Bali, Indonesia, together with sophisticated tools and the remains of animals. The skeletal remains were around 18,000 years old, and the researchers confirmed them as being from a **new species in the genus *Homo***, which includes our closest ancestors, such as *Homo erectus*, as well as our own species, *Homo sapiens*. The researchers called the **new species *Homo floriensis***.

cranium has a volume about a quarter that of a modern human

maxilla (upper jaw)

mandible (lower jaw)

***Homo floriensis* skull**
This skull of the recently extinct, newly discovered species of human, Homo floriensis, *is from a female who would have been about 3 ft (1 m) tall. The small, human features led to this species being nicknamed the "Hobbit."*

GRAPHENE

Graphene takes its name from graphite, which is composed of layers of carbon atoms. In graphite, the layers move over each other easily, which is why it is slippery, but they are extremely robust because the atoms form strong hexagonal bonds, as shown in this color-enhanced electron micrograph; graphene is simply a single layer of graphite.

> **THIS CANNOT BE A PECULIAR MODERN HUMAN.**
>
> **Chris Stringer, British anthropologist,** 2004

Remains of a total of seven individuals were found altogether. The indigenous people of Flores have ancient but detailed legends about a race of small, hairy people who murmured in their own language, and so it is likely that *Homo floriensis* lived side-by-side with modern humans.

Also in October, scientists at Manchester University, UK, succeeded in producing **samples of graphene** (see panel, above), a form of pure carbon not previously produced in bulk. The new material was just one atom thick but remarkably strong, transparent, and was later found to conduct electricity at room temperature.

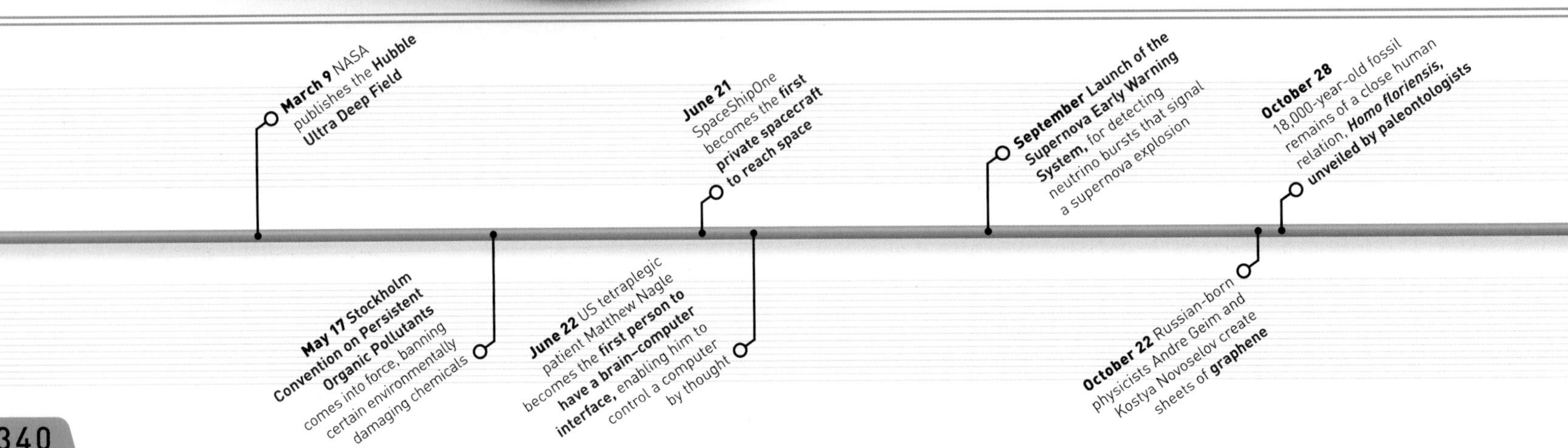

March 9 NASA publishes the **Hubble Ultra Deep Field**
May 17 Stockholm Convention on Persistent Organic Pollutants comes into force, banning certain environmentally damaging chemicals
June 21 SpaceShipOne becomes the **first private spacecraft to reach space**
June 22 US tetraplegic patient Matthew Nagle becomes the **first person to have a brain-computer interface**, enabling him to control a computer by thought
September Launch of the Supernova Early Warning System, for detecting neutrino bursts that signal a supernova explosion
October 22 Russian-born physicists Andre Geim and Kostya Novoselov create sheets of **graphene**
October 28 18,000-year-old fossil remains of a close human relation, ***Homo floriensis*, unveiled by paleontologists**

“A NEW PAGE OF **AERONAUTICAL HISTORY** HAS BEEN **WRITTEN.**”

Jacques Chirac, French politician, 2005

The Airbus A380 takes off from its production site in Toulouse, France, on its maiden flight on 27 April 2005. The world's largest passenger airliner, the A380 can carry up to 853 people.

IN THE FIRST-EVER LANDING OF A SPACECRAFT in the outer Solar System—and only the second landing on any moon—the European Space Agency's (ESA's) Huygens probe parachuted on to **the surface of Saturn's largest moon, Titan**, in January. The probe was part of the joint NASA–ESA Cassini-Huygens mission, and the spacecraft that carried Huygens to Titan, Cassini, remained in orbit around Titan, gathering data and images and relaying signals from Huygens back to astronomers on Earth. Titan had long been of interest to astrobiologists, as its **atmosphere was known to contain organic compounds**—carbon-rich compounds that could form the basis of life (see panel, below). The Huygens probe captured more than 300 images of Titan's surface during its descent. Images of the surface taken after landing show rock-shaped objects on "sandy" ground—although these objects and the sand are both probably mostly frozen water.

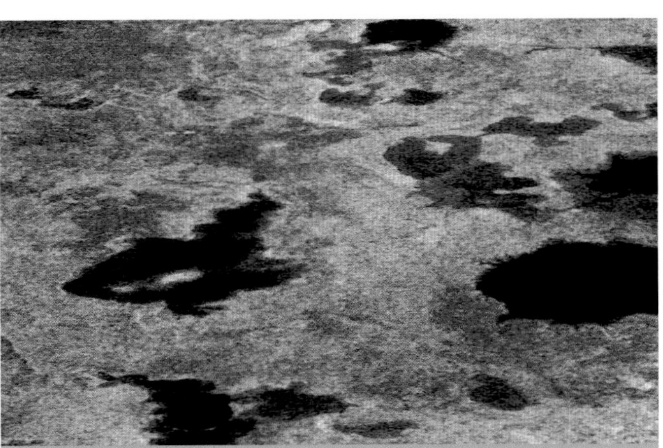

LIFE ON TITAN?

Titan's atmosphere has clouds that produce rain of the carbon compound methane, and the space probe Cassini found lakes of methane and ethane (the blue areas in this color-enhanced radar image) around Titan's poles. In 2012, NASA also found evidence of a huge subsurface ocean of water, which led to speculation that simple life forms may exist on this seemingly inhospitable world.

In the following month, an international team of astronomers released their analysis of a **huge, galaxy-sized mass of hydrogen** known as **VIRGOHI21**, which lies about 50 million light years away and had been discovered by radio astronomers in 2004. The motion of the galaxy suggests that it **may be composed mostly of cold, dark matter**—a strange form of matter that does not interact with light or other types of electromagnetic radiation but does have a gravitational effect (see 2003). Current astrophysical theories suggest that all galaxies contain dark matter—it is the only way to account for the motion and distribution of the galaxies' stars. However, VIRGOHI21 was the first strong candidate for a dark matter galaxy, a galaxy dominated by dark matter.

In March, US paleontologist **Mary Schweitzer** reported finding **68-million-year-old soft tissue,** which she had extracted from the fossilized femur (thigh bone) of a ***Tyrannosaurus rex*** dinosaur by dissolving away the mineral parts of the fossil. This method is commonly used on living bone as a way of extracting soft tissue. The same procedure carried out on a fossil would normally dissolve the whole specimen, since most fossils are completely mineralized. However, with this specimen the process revealed a soft, elastic bone matrix made mostly of what seemed like collagen. Viewing this matrix under a microscope, Schweitzer found what appeared to be **blood vessels and bone cells**. She saw similar structures in soft tissue extracted from the femur of a modern ostrich, a close relative of *T. rex*. Repeating the procedure on other dinosaur fossils, Schweitzer found two other samples of soft tissue. Many scientists were sceptical of the analysis, suggesting that the organic matter was the result of contamination, but more detailed examination in 2008 and 2011 seemed to support Schweitzer's interpretation.

In April, a new aircraft made its maiden flight: the **Airbus A380,** which replaced the Boeing 747 (see 1970) as the **world's largest commercial passenger aircraft**. Designed and built by European corporation Airbus, the wide-bodied A380 has two decks and is 239 ft (73 m) long. It began commercial service in 2007.

3,200

MILES

THE **DIAMETER** OF **TITAN,** SATURN'S LARGEST MOON

Facing the press
The first person to receive a partial face transplant, Isabelle Dinoire gave a press conference in February 2006, only three months after her pioneering operation.

Toward the end of the year, in November, 38-year-old Frenchwoman Isabelle Dinoire became the **first person to have a partial face transplant.** She received skin, blood vessels, and muscles after her face had been badly damaged when she was mauled by a dog six months earlier. The first full face transplant was performed in Spain in 2010.

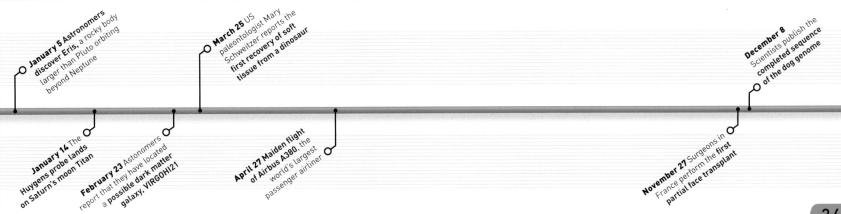

January 5 Astronomers discover Eris, a rocky body larger than Pluto orbiting beyond Neptune

January 14 The **Huygens probe lands on Saturn's moon Titan**

February 23 Astronomers report that they have located a **possible dark matter galaxy, VIRGOHI21**

March 25 US paleontologist Mary Schweitzer reports the **first recovery of soft tissue from a dinosaur**

April 27 Maiden flight of Airbus A380, the world's largest passenger airliner

November 27 Surgeons in France perform the **first partial face transplant**

December 8 Scientists publish the **completed sequence of the dog genome**

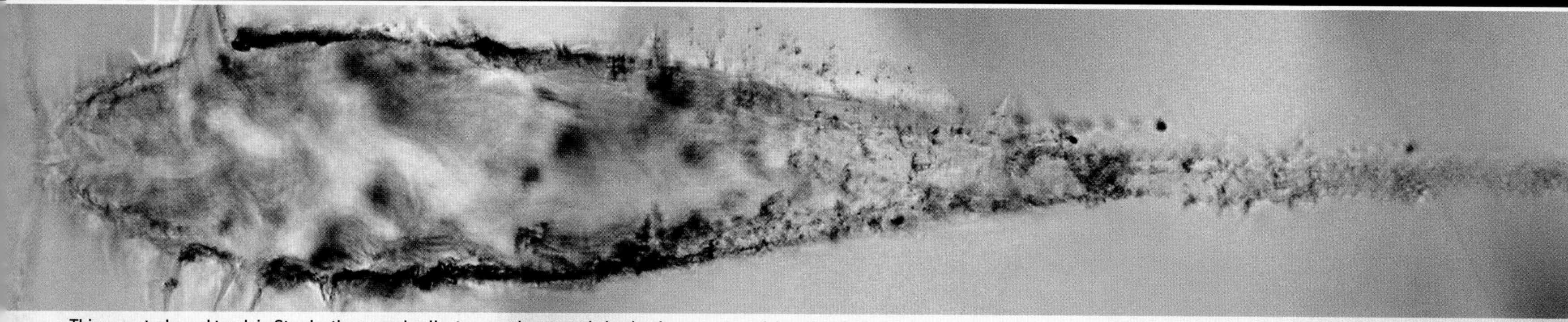

This carrot-shaped track in Stardust's aerogel collector panel was made by the tiny comet particle—the black speck to the right of this image—which entered the aerogel at a relative speed of several miles per second.

AT THE BEGINNING OF 2006, the Solar System officially had nine planets. The outermost, **Pluto**, stood out as being different from the others. For example, while all the other planets' orbits lie more or less in the same plane, Pluto's orbit is inclined at a fairly steep angle to it; Pluto's orbit is also very eccentric and passes inside Neptune's. When, in the 1970s, an object similar to Pluto was discovered, **Pluto's classification as a planet was in doubt**. In 2005, astronomers detected a rocky object outside the orbit of Neptune with a mass greater than Pluto's. After much deliberation, the International Astronomical Union decided in 2006 to designate Pluto as a dwarf planet, **one of many similar asteroid-like objects** in the Kuiper Belt (see 1949).

After seven years in space, **NASA's Stardust probe** released its sample return capsule, which parachuted to Earth and landed safely in the desert in Utah. The capsule contained about a **million specks of dust** collected from Stardust's encounter with Comet Wild 2 (see 1999).

In January, shortly after it became the **world's largest neutrino observatory**, the IceCube Neutrino Observatory detected its first neutrino. Trillions of neutrinos pass through the detector every second, but these elusive particles interact only very weakly with matter, making them very difficult to detect. Every so often, one will interact with the nucleus of an atom, resulting in a **tiny flash of light**. The observatory has strings of detectors in the rock and ice under the ground in Antarctica that pick up the tiny flashes of light. Neutrinos originating from high-energy phenomena, such as supernovas and mysterious gamma-ray bursts, provide astrophysicists with a **unique window on the Universe**. Construction of the observatory began in 2005, and was completed in 2010.

Just 10 years after the launch of DVDs (see 1995–96), this year saw the introduction of **two rival formats for video disks** designed to carry high definition video. HD DVD and BluRay can carry around **10 times as much data as DVDs**. Each had different electronics and entertainment companies bent on promoting it. Japanese company Toshiba was the main protagonist for HD DVDs, Sony for BluRay. Within just two years, **BluRay had won this format war**, largely because it garnered more support, but also partly as a result of the inclusion of a BluRay player in Sony's popular PlayStation 3 games console, which was released in November 2006.

In August, doctors in Australia administered for the first time a **vaccine for the human papilloma virus** (HPV). There are many different strains of HPV; several types are transmitted during sexual contact, and are the most common cause of genital warts and cervical cancer. Medical researchers suggested that the vaccine should be given to girls and boys routinely, in an effort to **reduce incidence of cervical and other cancers**. This idea was controversial, with some groups claiming it would encourage underage sex. Nevertheless, the vaccine is now routinely given in many countries.

In the journal *Nature*, a team of paleoanthropologists (who study hominid history through fossil evidence) described an

8 THE **NUMBER** OF **PLANETS** IN OUR **SOLAR SYSTEM** AFTER **AUGUST 24, 2006**

164 ft (50 m) below surface

IceCube lab

each detector string has 60 sensors

sensors pick up light flashes produced by neutrinos interacting with atomic nuclei

8,038 ft (2,450 m) below surface

9,252 ft (2,820 m) below surface

bedrock

Neutrino observatory
The IceCube Neutrino Observatory consists of deep holes drilled into the Antarctic rock and ice. Strings of detectors pick up tiny flashes as neutrinos interact with atomic nuclei.

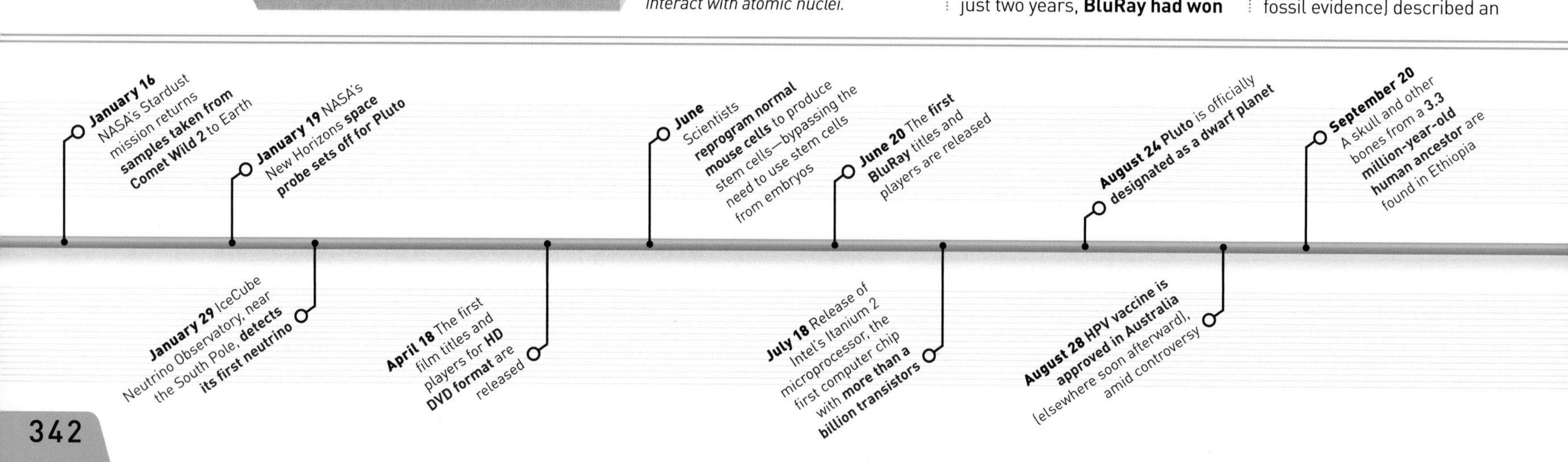
January 16 NASA's Stardust mission returns **samples taken from Comet Wild 2** to Earth
January 19 NASA's New Horizons **space probe sets off for Pluto**
January 29 IceCube Neutrino Observatory, near the South Pole, **detects its first neutrino**
April 18 The first film titles and players for **HD DVD format** are released
June Scientists **reprogram normal mouse cells** to produce stem cells—bypassing the need to use stem cells from embryos
June 20 The **first BluRay** titles and players are released
July 18 Release of Intel's Itanium 2 microprocessor, the first computer chip with **more than a billion transistors**
August 24 Pluto is officially **designated as a dwarf planet**
August 28 HPV vaccine is approved in Australia (elsewhere soon afterward), amid controversy
September 20 A skull and other bones from a **3.3 million-year-old human ancestor** are found in Ethiopia

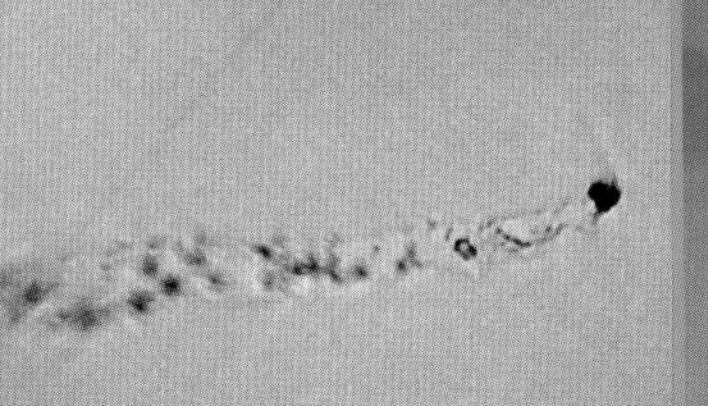

26 GIGAWATT THE GLOBAL **WINDPOWER CAPACITY** IN 2007

The IPCC stressed the need to increase the proportion of electricity generated by renewable sources, such as wind power, to mitigate climate change.

Skull of Selam
This well-preserved skull is that of a human ancestor nicknamed Lucy's Child, who would have walked upright on two legs.

important find: a partial skeleton of a **bipedal human ancestor** who lived and died around **3.3 million years ago**. The fossilized remains were of a young female about three years of age, and had been discovered in 2000, in Dikika, Ethiopia—close to where the skeleton nicknamed Lucy was found (see 1974). The individual, from the same species as Lucy (*Australopithicus afarensis*), earned the nickname Lucy's Child—although she predates Lucy by around 120,000 years.

IN ITS FOURTH ASSESSMENT REPORT, the Intergovernmental Panel on Climate Change (IPCC) further refined and extended its analysis of **global climate change** presented in its earlier reports (see 1990, 1995, 2001), while reiterating the same conclusions with still more certainty: that human activities—particularly carbon dioxide emissions from burning fossil fuels—are producing a greenhouse effect (see pp.326–27), causing a **rise in average global temperature**.

While mobile phones were becoming commonplace in developed nations, most devices could do little more than make calls and send SMS (text) messages. American company Apple Inc. **revolutionized the mobile phone industry** with the introduction of the iPhone. In addition to calling and texting, users could browse the web, download and use a huge range of apps (applications), listen to music, watch and record videos, and take pictures. The iPhone was an enormous success, and other manufacturers soon created smartphones with similar capabilities.

Stem cell research promises many benefits in medicine—for example, stem cells could regenerate brain cells to cure dementia. In an example of this kind of potential, researchers in the UK **grew heart tissue** from bone marrow stem cells in 2007. One stumbling block in stem cell research was the fact that only embryos contain cells that can develop into any kind of cell. Many people found that idea objectionable, because the embryos were destroyed in the process. However, in 2007 two teams of scientists independently managed to **reprogram ordinary cells** (see panel, below) to make them into stem cells—a feat previously achieved with cells taken from mice. Also this year, a team used DNA bases to **synthesize a copy of a bacterial chromosome**.

Apple iPhone
The iPhone from Apple introduced a convenient touch interface, with gestures such as swiping and pinching, to mobile phones.

CELL PROGRAMMING

A major breakthrough in stem cell research involves cells called fibroblasts, which are found in skin and connective tissue. Fibroblasts are responsible for producing collagen and other proteins that repair the skin. Turning these cells into pluripotent stem cells, which have the potential to develop into any kind of cell, involves adding compounds that switch on certain genes. This causes the cells to revert to the state in which all cells begin.

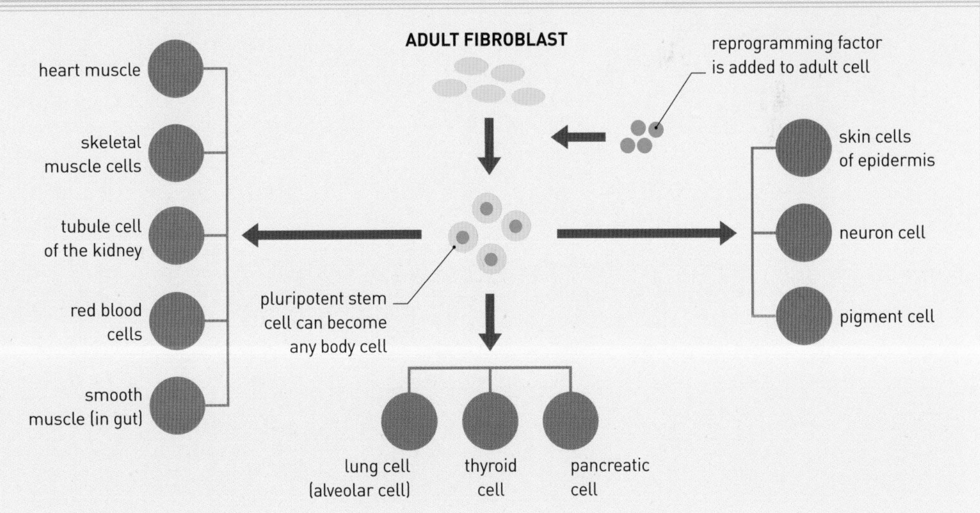
ADULT FIBROBLAST
reprogramming factor is added to adult cell
pluripotent stem cell can become any body cell
heart muscle
skeletal muscle cells
tubule cell of the kidney
red blood cells
smooth muscle (in gut)
skin cells of epidermis
neuron cell
pigment cell
lung cell (alveolar cell)
thyroid cell
pancreatic cell

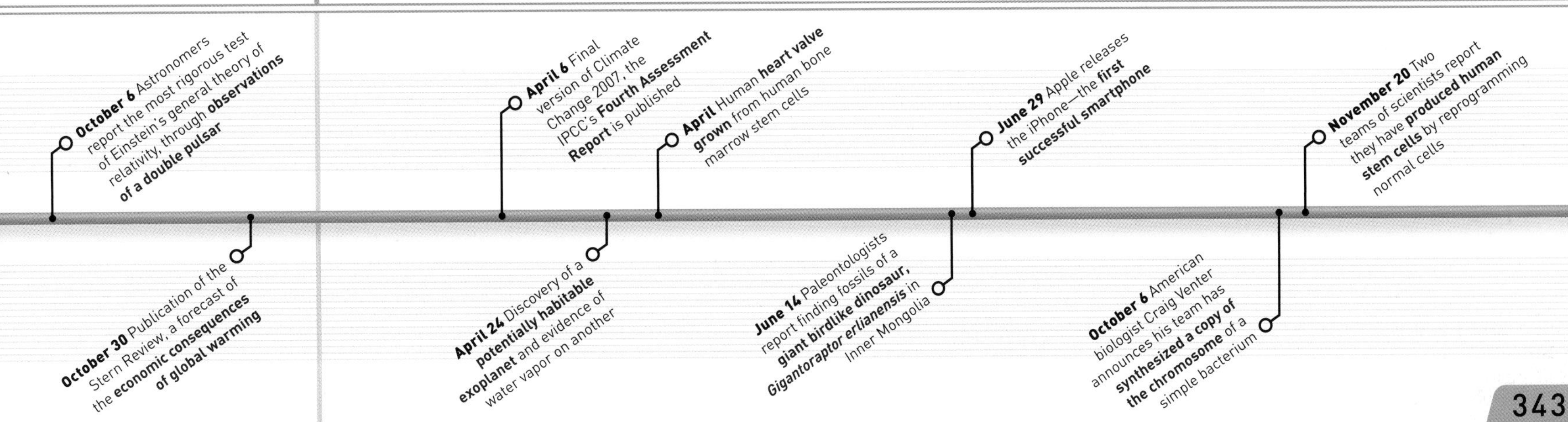
October 6 Astronomers report the most rigorous test of Einstein's general theory of relativity, through **observations of a double pulsar**
October 30 Publication of the Stern Review, a forecast of the **economic consequences of global warming**
April 6 Final version of Climate Change 2007, the IPCC's **Fourth Assessment Report** is published
April 24 Discovery of a **potentially habitable exoplanet** and evidence of water vapor on another
April Human **heart valve grown** from human bone marrow stem cells
June 14 Paleontologists report finding fossils of a **giant birdlike dinosaur, *Gigantoraptor erlianensis*** in Inner Mongolia
June 29 Apple releases the iPhone—the **first successful smartphone**
October 6 American biologist Craig Venter announces his team has **synthesized a copy of the chromosome** of a simple bacterium
November 20 Two teams of scientists report they have **produced human stem cells** by reprogramming normal cells

UNDERSTANDING COSMOLOGY

EARLY 20TH-CENTURY ASTRONOMERS DEVELOPED A THEORY TO EXPLAIN THE ORIGIN OF THE UNIVERSE

Cosmology is the scientific study of the Universe as a whole—the term is derived from the Greek word for Universe, cosmos. Cosmologists are interested in how the Universe began, how it works (particularly on the largest scales), how it will develop in the future, and if and how it will end.

It was only in the 1920s that astronomers discovered other galaxies outside our own. Belgian astronomer and priest Georges Lemaître (1894–1966) applied the equations of Einstein's General Relativity to the Universe as a whole; the results suggested that the Universe might be expanding. He proposed that if that were true, it must long ago have been very small, very dense, and very hot—a state he called the primordial atom.

IN THE BEGINNING

In 1929, American astronomer Edwin Hubble discovered that galaxies are moving away in every direction, suggesting that the Universe is indeed expanding—and that Lemaître's theory might be correct. British astronomer Fred Hoyle rejected these ideas, but coined the term Big Bang in 1950 to help explain them. Big Bang theory remains a very likely explanation of how the Universe began.

GEORGES LEMAITRE
After a strict Jesuit upbringing, Lemaître studied civil engineering, then physics and mathematics. He was ordained as a Jesuit priest in 1923.

THE ORIGIN OF THE UNIVERSE
As the Universe expanded after the Big Bang, it also cooled. Some of its energy turned into fundamental particles, and the basic forces of nature came into existence.

DIAMETER	3×10^{-26}ft/10^{-26}m	33ft/10m	62miles (10^5m/100km)
TEMPERATURE	10^{27}k (1,800 trillion trillion °F/100 trillion trillion °C)	10^{27}k (1,800 trillion trillion °F/100 trillion trillion °C)	10^{22}k (18 billion °F/100 billion billion °C)
	Cosmic inflation *At the moment of the the Big Bang, the entire Universe was much smaller than an atomic nucleus. Within a tiny fraction of a second, it underwent an inconceivably rapid expansion called cosmic inflation.*	**Particle soup** *When cosmic inflation ended, still well within the first second, the Universe was tiny and hot. Energy created pairs of particles and antiparticles, which pop into existence fleetingly before annihilating each other.*	**Separation of forces** *Originally, what we know as electromagnetism, gravity, and the weak and strong nuclear forces were unified as a single force. After cosmic inflation, the unified force separated, giving rise to the laws of nature we know today.*
TIME	A hundred-billionth of a yoctosecond (10^{-35}k seconds)	A hundred-millionth of a yoctosecond (10^{-32}k seconds)	1 yoctosecond (10^{-24}k seconds)

all time, space, and energy begins as a point of unimaginable density

cosmic inflation—a dramatic expansion of the Universe

early Universe contains a soup of fundamental particles, such as quarks, gluons, and force-carrying bosons

protons and neutrons—compound particles made of quarks held together by gluon fields—formed nuclei of the lightest elements

The first compelling piece of evidence in favour of the Big Bang theory was the discovery of the cosmic background radiation. This radiation was produced around 300,000 years after the Big Bang, and shows that the Universe was much smaller and hotter than it is now. The Universe's expansion has stretched the radiation, so that it is now mostly long-wavelength microwave radiation. Expanding 3-D space is best visualized as the growing surface of an inflating sphere. The relatively recent increase in the rate of expansion is caused by a poorly understood form of mass-energy called dark energy (see right and 1998). Measurements of the expansion and other key parameters suggest that the Big Bang occurred 13.8 billion years ago.

Edwin Hubble found that the further away a galaxy is, the faster it is receding. The mathematical relationship between a galaxy's distance and its speed is called Hubble's Law. It is best explained by the fact that space itself is expanding.

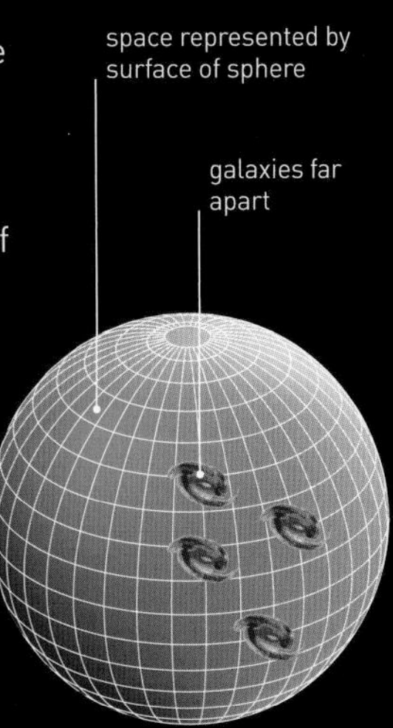

galaxies close together in early Universe
space between galaxies expanding
space represented by surface of sphere
galaxies far apart

The Universe appears to be dominated by dark matter and dark energy—forms of matter and energy that have not been directly observed but whose existence is inferred from their gravitational influence and effect on the Universe's rate of expansion. If this is indeed the case, the fate of the Universe depends on whether the mutual gravitational attraction of observable and dark matter in the Universe is enough to slow and even reverse the expansion driven by dark energy.

Three possible scenarios for the fate of the Universe are shown here. The gravitational influence of matter and dark matter could slow the expansion of the Universe, until it reaches a maximum. It is more likely that expansion will reverse or continue forever.

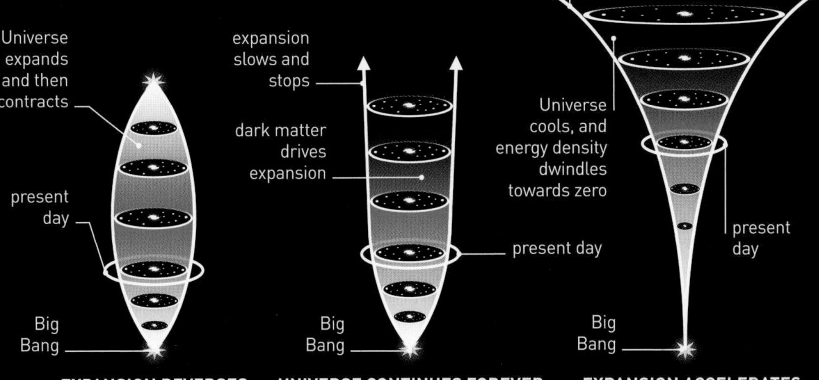

Universe expands and then contracts
present day
Big Bang
EXPANSION REVERSES
expansion slows and stops
dark matter drives expansion
present day
Big Bang
UNIVERSE CONTINUES FOREVER
expansion accelerates
Universe cools, and energy density dwindles towards zero
present day
Big Bang
EXPANSION ACCELERATES

60 billion miles (100 billion km)	1,000 light-years	100 million light-years
10^{13}k (18 trillion°F/ 10 trillion°C)	10^{8}k (180 million°F/100 million°C)	3,000k (4,900°F/2,700°C)
Protons and neutrons *The Universe continued to expand and cool, though more slowly, and fundamental particles called quarks combined into compound particles called baryons. The most important are protons and neutrons—these later formed the nuclei of atoms.*	**Opaque era** *For the next 300,000 years, the temperature was too high for atoms to form. Charged particles such as protons and electrons constantly produced and absorbed photons (particles of electromagnetic radiation), making the Universe opaque.*	**Matter era** *As the Universe cooled, electrons began settling down into orbits around the atomic nuclei. The cosmic background radiation (see above) dates from this time. The Universe has continued to expand ever since.*
1 microsecond	200 seconds	300,000 years
10^{-6} seconds (1 millionth of a second)		

27 MILES THE APPROXIMATE **CIRCUMFERENCE** OF THE **LHC**

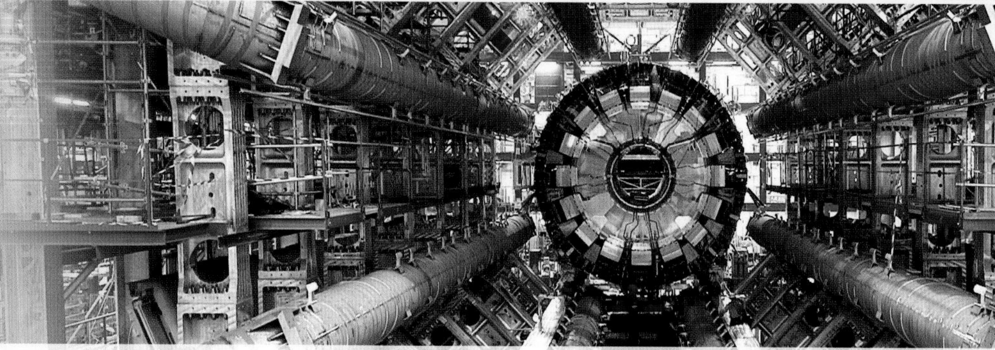
The ATLAS detector in the Large Hadron Collider is surrounded by eight huge electromagnets, which deflect the particles produced during collisions to that their mass and charge can be determined.

IN FEBRUARY 2008, A SECURE SEED BANK WAS OPENED ON the island of Spitsbergen in the Norwegian archipelago of Svalbard in the Arctic Ocean. Seed banks (see 2000) store seeds that could boost the populations of important plant species should climate change, wars, or natural disasters threaten their survival. However, seed banks themselves can also be vulnerable: in 2004, war claimed a seed bank at Abu Ghraib in Iraq, while floods destroyed one in the Philippines in 2006 after a typhoon hit. The Svalbard Global Seed Vault has the **capacity to hold up to 4.5 million seeds** and is embedded in an icy mountain in a remote location, providing more secure storage.

Icy seed storage
Situated on the Norwegian island of Spitsbergen, the Svalbard Global Seed Vault can hold up to 4.5 million seeds at freezing 0°F (–18°C).

The world's largest and most powerful particle accelerator, the **Large Hadron Collider (LHC), was turned on for the first time** in September 2008. The LHC is situated in a huge, circular, underground tunnel that straddles the border between France and Switzerland, at the European Organization for Nuclear Research (CERN). When the LHC is in operation, beams of protons (or for some experiments, ions) circle at extremely high speeds and are made to collide inside detectors. The **energy of the collision creates new particles**, and the detectors record the tracks of these particles as they hurtle out in all directions. These tracks are then analyzed by a network of powerful computers to look for evidence of specific types of particle—in particular, the **Higgs boson**. This particle is associated with the Higgs field, which, according to theory, is responsible for giving particles mass. Scientists have since revealed convincing evidence for the existence of the Higgs boson, and therefore the Higgs field (see 2011–12).

Several hundred extrasolar planets (planets outside our Solar System) had been found since the first example was detected (see 1992–93). In March 2009, NASA launched the **Kepler space observatory to find Earth-sized planets** around relatively nearby stars and to help astronomers estimate what proportion of stars have Earth-like planets. By 2012 it had discovered over 2,000 candidates.

In May 2009, Norwegian palaeontologist Jørn Hurum (b.1967) unveiled a remarkable specimen: an almost complete, fossilized skeleton of a **previously unknown species of lemur-like animal** that lived 47 million years ago. The specimen was claimed to be a transitional fossil—a **missing link** between lower primates, such as lemurs, and higher primates such as monkeys, apes, and humans. The fossil was

Darwinius masillae
Nicknamed "Ida," this fossil of a female is the only known example of Darwinius masillae, *quite possibly a direct human ancestor that lived 47 million years ago.*

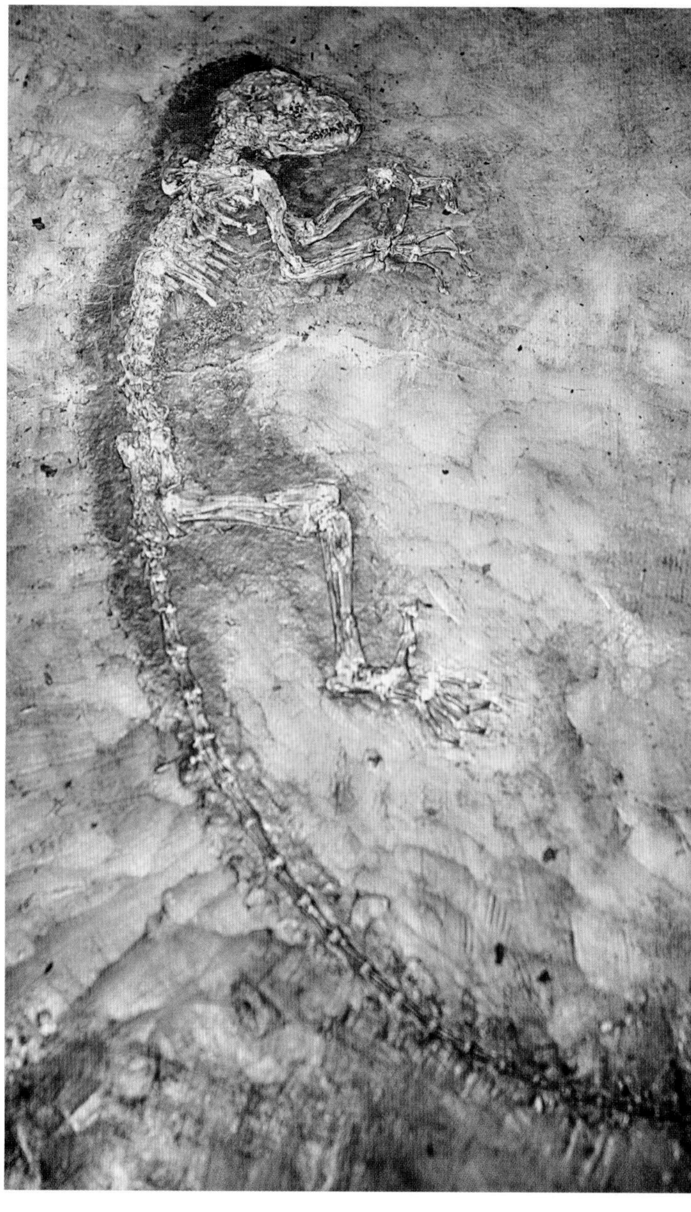

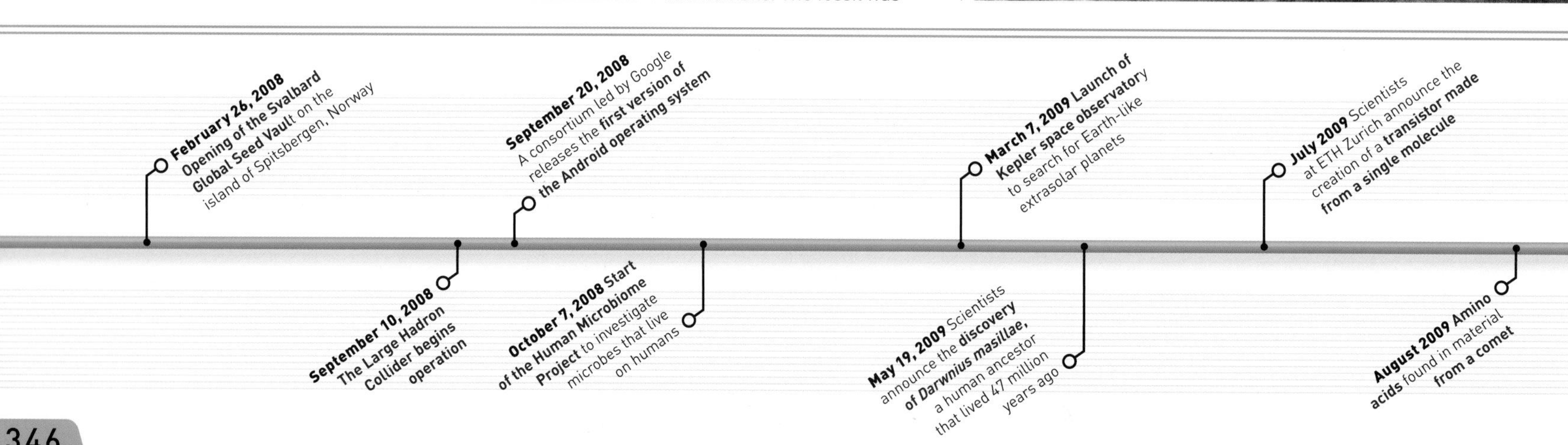
February 26, 2008 Opening of the **Svalbard Global Seed Vault** on the island of Spitsbergen, Norway
September 10, 2008 The Large Hadron Collider begins operation
September 20, 2008 A consortium led by Google releases the **first version of the Android operating system**
October 7, 2008 Start of the Human Microbiome Project to investigate microbes that live on humans
March 7, 2009 Launch of **Kepler space observatory** to search for Earth-like extrasolar planets
May 19, 2009 Scientists announce the **discovery of *Darwinius masillae*,** a human ancestor that lived 47 million years ago
July 2009 Scientists at ETH Zurich announce the creation of a **transistor made from a single molecule**
August 2009 Amino acids found in material **from a comet**

> **WE ARE ON** THE **VERGE** OF **PERPETUAL FLIGHT.**

Bertrand Piccard, Swiss co-founder of the Solar Impulse project, 2010

SOLARIMPULSE

The Solar Impulse used 2,200 sq ft (200 sq m) of solar panels to generate electricity to power the electric motors that drove its propellers.

originally found in 1983, in a disused quarry in Germany; Hurum came across it in 2006 and was intrigued by features that made it stand apart from the lower primates—including humanlike nails and opposable big toes. The species was named *Darwinius masillae*, in honor of British naturalist and pioneer of the theory of evolution Charles Darwin (see 1859). Later that year, in October, scientists announced that the bones of a 4.4-million-year-old hominid fossil, ***Ardipithecus ramidus***, were **the oldest hominid remains ever discovered**.

Meanwhile, two projects began to increase knowledge of our own species. In October 2008, the **Human Microbiome Project** was launched. Spearheaded by the US National Institutes of Health, this was an initiative to investigate the **microbes that colonize different parts of the body** and establish the roles of these organisms in health and disease. A year later, the **Human Epigenome Project**, an international project to map the epigenome (factors that affect which genes are switched on and when) published its **first map of the human epigenome**.

IN 2010, IBM AND INTEL BEGAN MANUFACTURING CHIPS with 32-nanometer transistors. Semiconductor manufacturers had been cramming ever more transistors on to a single chip since the invention of integrated circuits (see 1958). Advances like these helped increase the power and portability and decrease the cost of various electronic devices, and the first commercially successful tablet computer, **Apple's iPad**, was released in 2010. It was soon followed by tablet computers running the Android operating system, which had been developed by a consortium headed by Google, and had its first release in 2008. Android was already a popular operating system for smartphones.

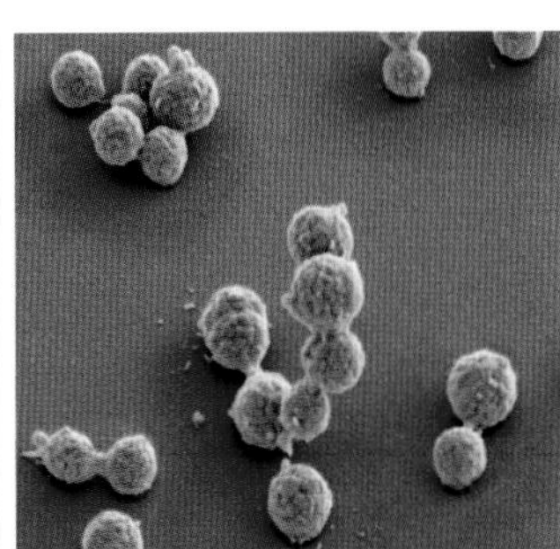

Synthetic mycoplasma bacteria
This colored micrograph shows the first organisms with a synthetic genome. Nicknamed "Synthia," the new species is officially called Mycoplasma mycoides *JCVI-syn1.0.*

There is a limit to how small conventional transistors can be, so researchers were looking for alternatives to silicon as the basis of future electronics. In February 2010, IBM researchers created the **first reliable, fast-switching transistor** from the material graphene (see 2004). In July 2009, scientists at ETH Zurich, Switzerland, had created a transistor made from a single molecule, and in May 2010 a team at the University of New South Wales, Australia, made a transistor with just seven atoms, using phosphorus.

In May, a team led by US biologist **Craig Venter** (see panel, right) created the **first synthetic life form**. Using a technique called oligonucleotide synthesis—in which DNA sequences stored in a computer are pieced together to order—Venter and his co-workers had already created a complete virus genome (2003) and a synthetic chromosome (2007). In 2010, they synthesized a copy of the **genome of a bacterium called *Mycoplasma mycoides***, made a few changes (including adding a kind of "watermark"), then inserted it into another bacterial cell that had had its DNA removed. The new genome functioned as required, manufacturing proteins and causing the cell to reproduce. Biologists hope to learn a great deal about living systems by building them themselves, rather than just taking them apart. Synthetic life also holds the promise of designing **new life forms that could be beneficial**—for example, by helping clean up oil spills or produce biofuels.

In July, a solar-powered aircraft called **Solar Impulse made a continuous 26-hour flight**. Electricity generated by solar panels was stored in batteries during the day, allowing the aircraft to fly through the night. The project was the brainchild of Swiss balloonist Bertrand Piccard (b.1958) and Swiss entrepreneur André Borschberg (b.1958), and it aimed to encourage development of renewable energy. Later in the year, US company SpaceX launched Dragon, the **first commercially owned spacecraft to go into orbit**. In 2012, it completed the first of a number of scheduled missions delivering supplies to the International Space Station.

CRAIG VENTER (b.1946)

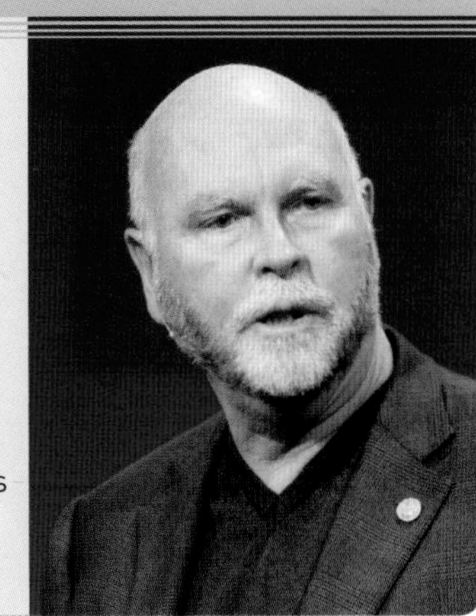

US biologist Craig Venter is a pioneer in genome studies and synthetic biology. In 1998, he helped form Celera Genomics, a private company that helped speed up sequencing of the human genome (see 2000). In 2006, he founded the J. Craig Venter Institute in California, a centre of expertise in genomics that produced the first synthetic life form.

1 MILLION
THE NUMBER OF **DNA BASE PAIRS** USED TO MAKE THE **FIRST ORGANISM** WITH A **SYNTHETIC GENOME**

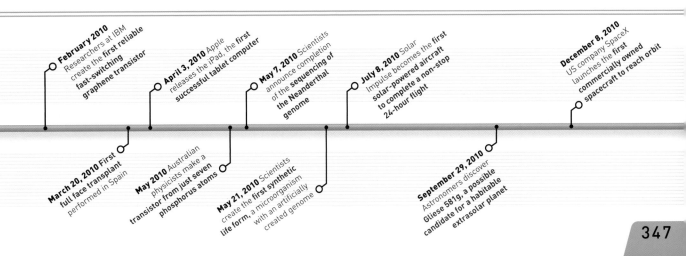

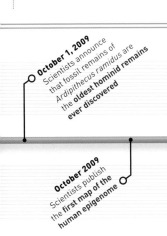

October 1, 2009 Scientists announce that fossil remains of *Ardipithecus ramidus* are the **oldest hominid remains ever discovered**

October 2009 Scientists publish the **first map of the human epigenome**

February 2010 Researchers at IBM create the **first reliable fast-switching graphene transistor**

March 20, 2010 First full face transplant performed in Spain

April 3, 2010 Apple releases the iPad, the **first successful tablet computer**

May 2010 Australian physicists make a **transistor from just seven phosphorus atoms**

May 7, 2010 Scientists announce completion of the **sequencing of the Neanderthal genome**

May 21, 2010 Scientists create the **first synthetic life form**, a microorganism with an artificially created genome

July 8, 2010 Solar Impulse becomes the **first solar-powered aircraft to complete a non-stop 24-hour flight**

September 29, 2010 Astronomers discover **Gliese 581g, a possible candidate for a habitable extrasolar planet**

December 8, 2010 US company SpaceX launches the **first commercially owned spacecraft to reach orbit**

This image shows the "Rocknest," an area of the Martian surface where Curiosity found mysterious bright objects and where it carried out its first X-ray diffraction experiments.

IN MAY 2011, NASA ANNOUNCED RESULTS FROM ITS GRAVITY PROBE B, which was designed to test Albert Einstein's general theory of relativity, originally published in 1916. Since its launch in 2004, the probe had been measuring the curvature of space-time in the vicinity of Earth (see pp.244–45), and the extent to which Earth's rotation drags space-time around with it. The results of both tests provided the **best confirmation to date of Einstein's general theory of relativity**.

In June, a team of geophysicists presented the **first map of the terrain beneath the Antarctic ice sheet**. Part of a long-term project to examine the geology of Antarctica, the map was produced from data provided by several instruments, including ice-penetrating radar. It revealed various geological features in glacial landscapes and also gave valuable information about the icecap's formation, which began about 30 million years ago. In August, a team of Australian and British scientists found **microfossils in rocks 3.4 billion years old**, pushing back the date of the earliest known life on Earth by a few million years. The primitive cells' metabolisms were based on sulfur rather than oxygen.

The following year, 2012, was a momentous one for science as physicists at CERN announced a result of experiments at the Large Hadron Collider (LHC, see 2008). The LHC aims to recreate energies and conditions that existed a tiny fraction of a second after the Big Bang (see pp.344–45), when the Universe formed. On July 4, 2012, physicists at CERN announced that they had found **compelling evidence for the existence of the Higgs boson**. This particle is a mainstay of the Standard Model of particle physics (see 1974) and is associated with the Higgs field. According to a theory developed by British physicist Peter Higgs and others in the 1960s, the Higgs field exists throughout space, and interaction with the Higgs field is **responsible for giving fundamental particles**, such as quarks and leptons, **their mass**.

In August 2012, NASA's **Curiosity rover landed in a large crater (the Gale crater) on Mars**, beginning the most comprehensive mission to the planet so far. The landing took place autonomously, during a radio black-out that NASA engineers called "seven minutes of terror." In the last stage of descent, four rockets fired to slow the craft to almost a hover and the rover itself was lowered to the surface on cables to prevent kicking up dust that could damage its instruments.

Soon after landing, the rover began transmitting stunning, **high-resolution panoramas and close-ups from Mars**. As well as cameras, Curiosity also carries a number of instruments to collect and analyze regolith (soil) and rock samples. A laser vaporizes any rock samples of interest and a spectroscope records the spectrum of the light emitted by the vapor to determine the rock's makeup. Other instruments work out the crystal structure of minerals using X-ray diffraction, and an environmental monitoring station records temperature, wind speed, atmospheric pressure, and humidity. By the end of 2012, Curiosity had traveled more than 1,650 ft (500 m) and had **analyzed regolith from more than 30 different locations**.

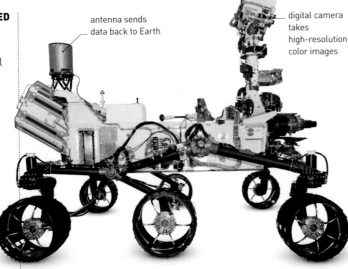

antenna sends data back to Earth

digital camera takes high-resolution color images

Roving over Mars
NASA's Curiosity rover is about the size of a family car. It carries an array of instruments, including ones designed to detect chemical compounds that could support life.

1,982 lb THE **MASS** OF **CURIOSITY**
9.5 ft THE **LENGTH** OF **CURIOSITY**
7.2 ft THE **HEIGHT** OF **CURIOSITY**

PETER HIGGS (b.1929)

Theoretical physicist Peter Higgs was born in Newcastle-upon-Tyne, UK. In the early 1960s, he developed a theoretical mechanism (which was also proposed by several other physicists at around the same time) to explain why particles have mass. In 1964, he predicted the existence of a particle associated with that mechanism: the Higgs boson.

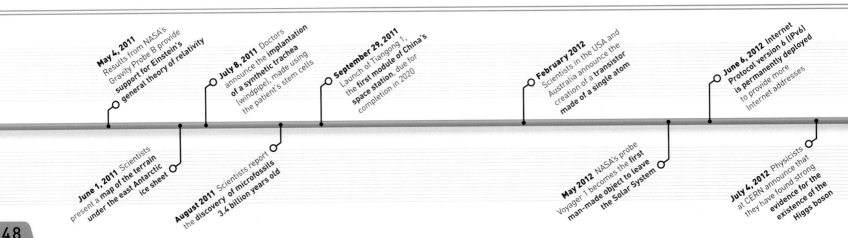

May 4, 2011 Results from NASA's Gravity Probe B provide **support for Einstein's general theory of relativity**

June 1, 2011 Scientists present a **map of the terrain under the east Antarctic ice sheet**

July 8, 2011 Doctors announce the **implantation of a synthetic trachea** (windpipe), made using the patient's stem cells

August 2011 Scientists report the discovery **of microfossils 3.4 billion years old**

September 29, 2011 Launch of Tiangong 1, the **first module of China's space station**, due for completion in 2020

February 2012 Scientists in the USA and Australia announce the creation of a **transistor made of a single atom**

May 2012 NASA's probe Voyager 1 becomes the **first man-made object to leave the Solar System**

June 6, 2012 Internet **Protocol version 6 (IPv6) is permanently deployed** to provide more Internet addresses

July 4, 2012 Physicists at CERN announce that they have found strong **evidence for the existence of the Higgs boson**

> **“ AS A LAYMAN I WOULD NOW SAY ‘I THINK WE HAVE IT’. ”**
>
> **Rolf Heuer, CERN Director General,** announcing the discovery of the Higgs boson, July 4, 2012

In September 2012, NASA published the **eXtreme Deep Field**, the most detailed image of deep space obtained to date (see 2004). Back on Mars, in February 2013, Curiosity had moved on to analyze subsurface rocks after drilling a 2.5in (6.4cm) hole in the Martian surface.

Regenerative medicine took two important steps forward in early 2013: scientists in the US successfully implanted a fully functioning laboratory-grown kidney into a rat, while a team in Bolivia **injected stem cells** into rats' brains soon after the rats had strokes and restored full brain function.

In China, a new strain of "bird flu" (see 1997), called H7N9, infected humans for the first time, causing concern of an epidemic.

Meanwhile, new calculations of the **age of the Universe** were being made. In March, data from the **European Space Agency's Planck satellite** refined cosmologists' estimate of the age of the Universe to 13.82 billion years old—about 100 million years older than previously thought.

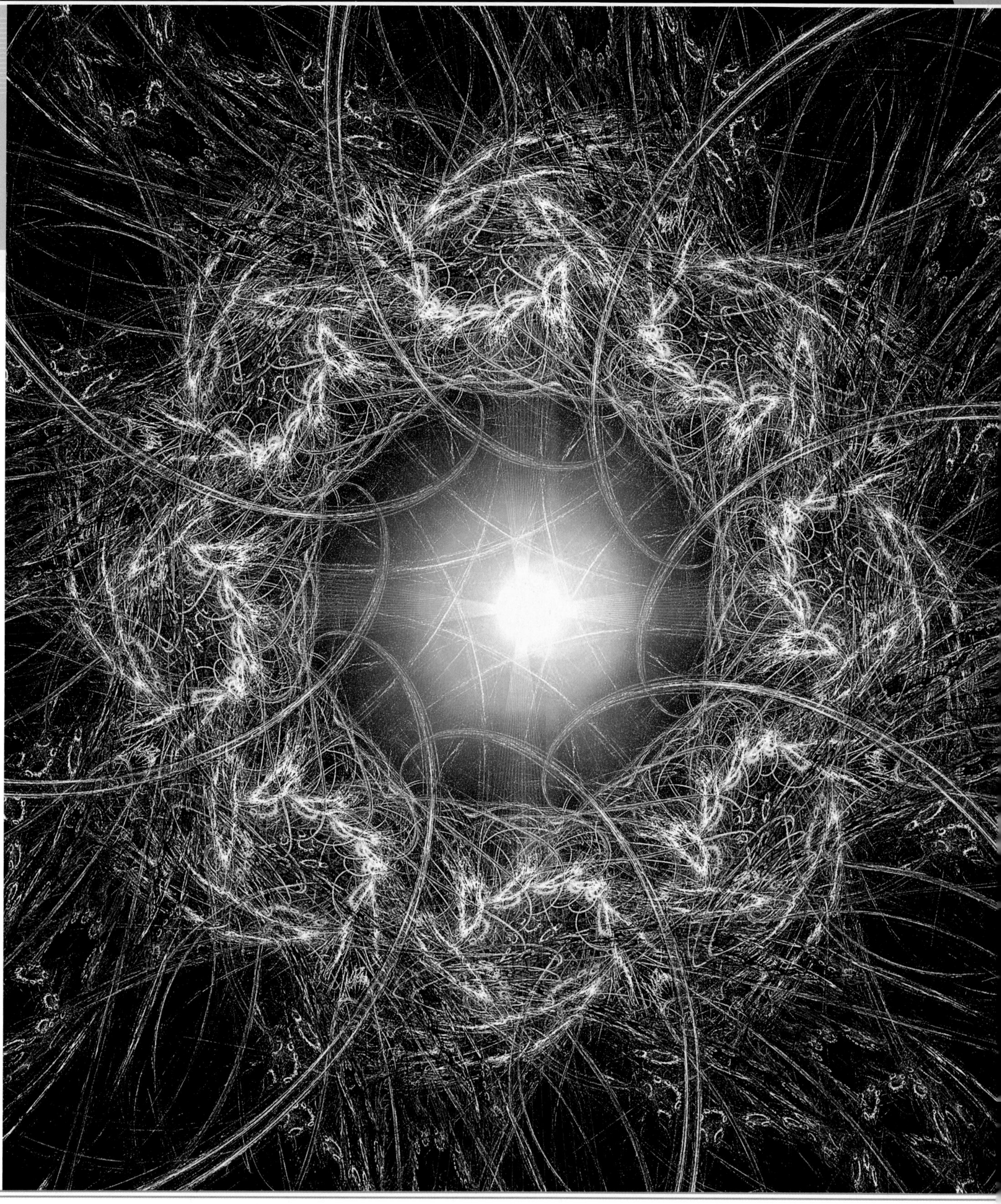

Searching for the Higgs boson
This computer-generated image shows tracks from particle collisions inside the Large Hadron Collider. Analysis of such tracks provided evidence for the Higgs boson.

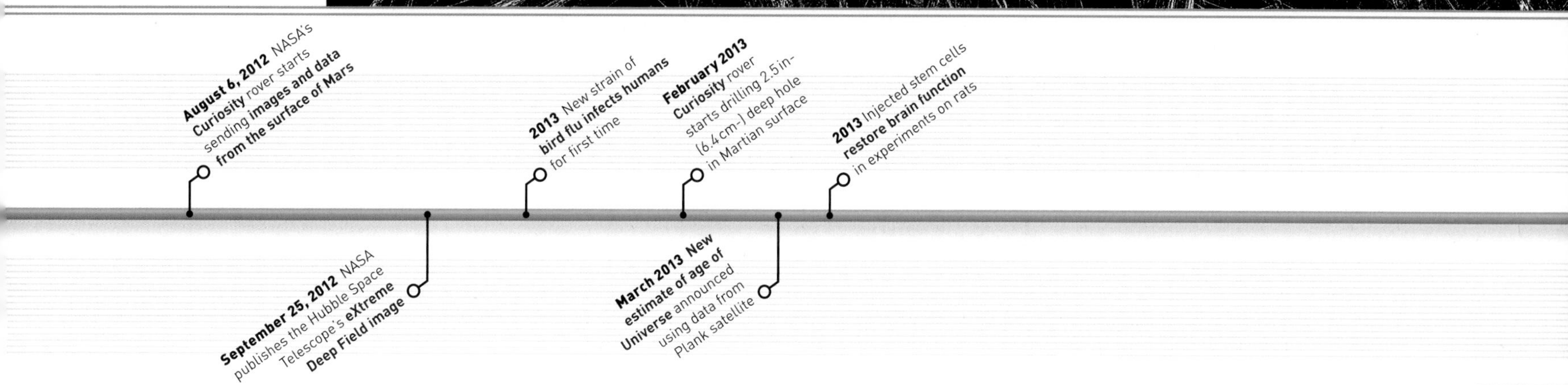
August 6, 2012 NASA's **Curiosity** rover starts sending **images and data from the surface of Mars**
September 25, 2012 NASA publishes the Hubble Space Telescope's **eXtreme Deep Field image**
2013 New strain of **bird flu infects humans** for first time
February 2013 Curiosity rover starts drilling 2.5in- (6.4cm-) deep hole in Martian surface
March 2013 New estimate of age of Universe announced using data from Plank satellite
2013 Injected stem cells **restore brain function** in experiments on rats

7

REFERENCE

MEASUREMENTS AND UNITS

BASE SI UNITS

The SI (Système International d'Unités) is the modern form of the metric system of measurements, and it is used by most countries. It consists of series of units of measurement built around seven interdependant base units of individual physical qualities. All other physical qualities are obtained from these units.

UNIT	SYMBOL	DEFINITION
Meter	m	This unit is the length of a path traveled by light in a vacuum during a time interval of 1/299,792,458 of a second.
Kilogram	kg	The unit of mass, equal to the mass of the international prototype kilogram.
Second	s	The second is the duration of 9,192,631,770 periods of the radiation corresponding to the transition between the two hyperfine levels of the ground state of the cesium-133 atom.
Ampere	A	The constant electric current which, if maintained in two straight parallel conductors of infinite length and negligible cross section, and placed 1 meter apart in a vacuum, would produce between these conductors a force equal to 2×10^{-7} newtons per meter.
Kelvin	K	The unit of thermodynamic temperature, this is the fraction 1/273.16 of the thermodynamic temperature of the triple point of water.
Candela	cd	The luminous intensity, in a given direction, of a source that emits monochromatic radiation of frequency 540×10^{12} hertz and that has a radiant intensity in that direction of 1/683 watt per steradian.
Mole	mol	The mole amount of substance that contains as many elementary units as there are carbon atoms in 0.012 kilograms of carbon-12.

SI PREFIXES

SI prefixes and symbols are used to indicate decimal multiples and submultiples of SI units to avoid having to write either very large or extremely small numeric values, from 10^{18} to 10^{-18}. When the number is written, the prefix attaches directly to the name of the unit—for example "nanosecond." Similarly, a prefix symbol can also be attached to the symbol for the unit—for example "ns."

FACTOR	PREFIX	SYMBOL
10^{18}	exa-	E
10^{15}	peta-	P
10^{12}	tera-	T
10^{9}	giga-	G
10^{6}	mega-	M
10^{3}	kilo-	k
10^{2}	hecto-	h
10^{1}	deca-	da

FACTOR	PREFIX	SYMBOL
10^{-1}	deci-	d
10^{-2}	centi-	c
10^{-3}	milli-	m
10^{-6}	micro-	µ
10^{-9}	nano-	n
10^{-12}	pico-	p
10^{-15}	femto-	f
10^{-18}	atto-	a

SUPPLEMENTARY AND DERIVED SI UNITS

The SI is an evolving system in which units are created and definitions are modified as the technology and precision of measurement improves. In addition to the seven base SI units, there are two supplementary units and many other units derived from the SI base units.

SUPPLEMENTARY UNITS	SYMBOL	DEFINITION
Radian	rad	The unit of measurement of angle; it is the angle subtended at the center of a circle by an arc equal in length to the circle's radius.
Steradian	sr	The unit of measurement of solid angle; it is the solid angle subtended at the center of a circle by a spherical cap equal in area to the square of the circle's radius.

DERIVED UNITS	SYMBOL	DEFINITION
Hertz	Hz	The unit of frequency; 1 hertz has a periodic interval of 1 second.
Newton	N	A unit of force equal to the force that imparts an acceleration of 1 m/sec/sec to a mass of 1 kilogram.
Pascal	Pa	A unit of pressure equal to 1 newton per square meter.
Joule	J	A unit of energy exerted by a force of 1 newton acting to move an object through a distance of 1 meter.
Watt	W	A unit of power equal to 1 joule per second; it is also the power dissipated by a current of 1 ampere flowing across a resistance of 1 ohm.
Coulomb	C	A unit of electrical charge equal to the amount of charge transferred by a current of 1 ampere in 1 second.
Volt	V	A unit of potential equal to the potential difference between two points on a conductor carrying a current of 1 ampere when the power dissipated between the two points is 1 watt; it is equivalent to the potential difference across a resistance of 1 ohm when 1 ampere of current flows through it.
Farad	F	A unit of capacitance of a capacitor that has an equal and opposite charge of 1 coulomb on each plate and a voltage difference of 1 volt between the plates.
Ohm	Ω	A unit of electrical resistance equal to the resistance between two points on a conductor when a potential difference of 1 volt between them produces a current of 1 ampere.
Siemens	S	A unit of electrical conductance, the reciprocal of 1 ohm: 1 divided by 1 ohm.

SUPPLEMENTARY UNITS	SYMBOL	DEFINITION
Weber	Wb	A unit of magnetic flux that produces an electromagnetic force of 1 volt when the flux is reduced to zero at a uniform rate of 1 second.
Tesla	T	A unit of magnetic flux density, the equivalent of 1 weber of magnetic flux per meter squared.
Henry	H	A unit of inductance in which an induced electromotive force of 1 volt is produced when the current is varied at the rate of 1 ampere per second.
Degree Celsius	°C	A unit of temperature on the centigrade scale, with the ice point at 0°C and boiling point at 100°C.
Lumen	lm	A unit of luminous flux equal to the amount of light given out through a solid angle of 1 steradian by a point source of 1 candela intensity radiating uniformly in all directions.
Lux	lx	A unit of illumination that describes the illumination given by 1 lumen over an area of 1 square meter.
Becquerel	Bq	A unit of radioactivity. One becquerel descibes the activity of radioactive material in which one nucleus decays per second.
Gray	Gy	A unit of an absorbed dose of ionizing radiation and the energy it gives off. It is equal to the absorption of 1 joule per kilogram of irradiated material.
Sievert	Sv	A unit of the dose equivalent needed for protection against ionizing radiation.
Katal	kat	A unit that measures the activity, or property, of catalysts such as enzymes. For example, 1 katal of trypsin is the amount needed to break a mole of peptide bonds in 1 second.

COMMON PHYSICAL PROPERTIES

Scientists use a number of symbols when defining processes using mathematical formulae. The following table shows some of the most common physical properties and the symbols that are used to represent them. The units by which the different properties are measured are indicated by their SI units together with the relevant symbols.

PHYSICAL PROPERTY	SYMBOL	SI UNIT	SI UNIT SYMBOL
Acceleration, deceleration	a	meter/second^2 kilometer/hour/second	m s^{-2} km h^{-1} s^{-1}
Angular velocity	ω	radian/second	rad s^{-1}
Density	ρ	kilogram/meter^3 kilogram/milliliter	kg m^{-3} kg ml^{-1}
Electric charge	Q, q	coulomb	C
Electric current	I, i	ampere (coulomb/ second)	A (C s^{-1})
Electrical energy	–	megajoule kilowatt-hour	MJ kWh
Electrical power	P	watt (joule/second)	W (J s^{-1})
Electromotive force (EMF)	E	volt (watt/ampere)	V (W A^{-1})
Electrical conductance	S	siemen (ohm^{-1})	A V^{-1}
Electrical resistance	R	ohm (volt/ampere)	Ω (V A^{-1})
Frequency	*f*	hertz (cycles/second)	Hz (s^{-1})
Force	F	newton (kilogram meter/ second^2)	N (kg m s^{-2})
Gravitational intensity, field strength	–	newton/kilogram	N kg^{-1}
Magnetic field strength	H	ampere/meter	A m^{-1}
Magnetic flux	Φ	weber	Wb
Magnetic flux density	B	tesla (weber/meter^2)	T (Wb m^{-2})
Mass	m	kilogram	kg
Mechanical power	P	watt (joule/second)	W (J s^{-1})
Moment of inertia	I	kilogram meter^2	kg m^2
Momentum	p	kilogram meter/second	kg m s^{-1}
Pressure	P	pascal (newton/meter^2)	Pa (N m^{-2})
Quantity of substance	n	mole	mol
Specific heat capacity	C or c	joule/kilogram/kelvin	J kg^{-1} K^{-1}
Specific latent heats of fusion, vaporization	L	joule/kilogram	J kg^{-1}
Torque	τ	newton meter	N m
Velocity, speed	u, v	meter/second kilometer/hour	m s^{-1} km h^{-1}
Volume	V	meter^3 milliliter	m^3 ml
Wavelength	λ	meter	m
Weight	W	newton	N
Work, energy	W	joule (newton meter)	J (N m)

SI CONVERSION FACTORS

This table lists units of measurements from non-SI systems and the factors needed to convert them to SI units. The conversion factor in the column headed "SI equivalent" can be used to convert from the non-SI unit to the SI unit named. The reverse conversion can be made using the conversion factor in the column headed reciprocal.

UNIT	SYMBOL	QUANTITY	SI EQUIVALENT	SI UNIT	RECIPROCAL
Acre		area	0.405	hm^2	2.471
Ångström	Å	length	0.1	nm	10
Astronomical unit	AU	length	0.150	Tm	6.684
Atomic mass unit	amu	mass	1.661×10^{-27}	kg	6.022×10^{26}
Bar	bar	pressure	0.1	MPa	10
Barrel (US) = 42 US gal	bbl	volume	0.159	m^3	6.290
Calorie	cal	energy	4.187	J	0.239
Cubic foot	cu ft	volume	0.028	m^3	35.315
Cubic inch	cu in	volume	16.387	cm^3	0.061
Cubic yard	cu yd	volume	0.765	m^3	1.308
Curie	Ci	activity of radionuclide	37	GBq	0.027
Degree Celsius	°C	temperature	1	K	1
Degree Fahrenheit	°F	temperature	0.556	K	1.8
Electronvolt	eV	energy	0.160	aJ	6.241
Erg	erg	energy	0.1	μJ	10
Fathom (6 ft)		length	1.829	m	0.547
Fermi	fm	length	1	fm	1
Foot	ft	length	30.48	cm	0.033
Foot per second	$ft\ s^{-1}$	velocity	0.305 1.097	$m\ s^{-1}$ $km\ h^{-1}$	3.281 0.911
Gallon (UK)	gal	volume	4.546	dm^3	0.220
Gallon (US) = 231 cu in	gal	volume	3.785	dm^3	0.264
Gauss	Gs, G	magnetic flux density	100	μT	0.01
Grain	gr	mass	1	g	15.432
Hectare	ha	area	0.746	hm^2	1
Horsepower	hp	power	2.54	kW	1.341
Inch	in	length	9.807	cm	0.394
Kilogram-force	kgf	force	1.852	N	0.102
Knot	kn	velocity	9.461×10^{15}	$km\ h^{-1}$	0.540
Light-year	ly	length	1	m	1.057×10^{-16}
Liter	l	volume	1,193.3	dm^3	1
Mach number	Ma	velocity	10	$km\ h^{-1}$	8.380×10^{-4}
Maxwell	Mx	magnetic flux	1	nWb	0.1
Micron	μ	length	1.852	μm	1
Mile (nautical)		length	1.609	km	0.540
Mile (statute)		length	1.609	km	0.621
Miles per hour (mph)	$mile\ h^{-1}$	velocity	2.91×10^{-4}	$km\ h^{-1}$	0.621
Ounce (avoirdupois)	oz	mass	31.103	g	0.035
Ounce (troy) = 480 gr		mass	30,857	g	0.032
Parsec	pc	length	10	Tm	0.0000324
Pint (UK)	pt	volume	0.1	dm^3	1.760
Pound	lb	mass	4.448	kg	2.205
Pound force	lbf	force	6.895	N	0.225
Pound force/in		pressure	0.138	kPa	0.145
Pounds per square inch	psi	pressure	0.01	kPa	0.145
Röntgen	R	exposure	0.258	$mC\ kg^{-1}$	3.876
Second = (1/60')	"	plane angle	4.85×10^{-6}	mrad	2.063×10^5
Solar mass	M	mass	1.989×10^{30}	kg	5.028×10^{-31}
Square foot	sq ft	area	9.290	dm^2	0.108
Square inch	sq in	area	6.452	cm^2	0.155
Square mile (statute)	sq mi	area	2.590	km^2	0.386
Square yard	sq yd	area	0.836	m^2	1.196
Stere	st	volume	1	m^3	1
Therm = 105 btu		energy	0.105	GJ	9.478
Ton = 2240 lb		mass	1.016	Mg	0.984
Ton-force	tonf	force	9.964	kN	0.100
Ton-force/ sq in		pressure	15.444	MPa	0.065
Tonne	t	mass	1	Mg	1

PHYSICS

NEWTON'S LAWS

British physicist Isaac Newton proposed a series of laws of motion, which he published in his masterpiece *Principia* (see 1698–99). Newton's laws describe how objects move, remain at rest, or interact with other objects and forces. Newton also formulated the law of universal gravity to describe the force of attraction that acts between physical bodies.

LAW	DESCRIPTION
First law of motion	Unless disturbed by a force, an object will either stay still or travel at a constant speed in a straight line.
Second law of motion	The force acting on a body is equal to its rate of change of momentum, which is the product of its mass and its acceleration.
Third law of motion	For every action there is an equal and opposite reaction.

LAW OF UNIVERSAL GRAVITATION

$$F = \frac{Gm_1m_2}{r^2}$$

F = force
G = universal gravitational constant
m_1, m_2 = masses
r^2 = square of distances between masses

MECHANICS OF FORCES

A force is a push or pull that makes an object move in a straight line or turn. Forces can act alone or together, and they can be harnessed to make machines work more effectively. The relationship between the properties of moving objects—such as time, distance, direction and speed—can be described using equations.

MOTION FORMULAE

QUANTITY	DESCRIPTION	FORMULA
Speed	$\frac{\text{distance}}{\text{time}}$	$S = \frac{d}{t}$
Time	$\frac{\text{distance}}{\text{speed}}$	$t = \frac{d}{S}$
Distance	speed × time	d = St
Velocity	$\frac{\text{displacement (distance in a given direction)}}{\text{time}}$	$v = \frac{s}{t}$
Acceleration	$\frac{\text{change in velocity}}{\text{time taken for change}}$	$a = \frac{(v-u)}{t}$
Resultant force	mass × acceleration	F = ma
Momentum	mass × velocity	p = mv

EQUATIONS OF MOTION UNDER CONSTANT ACCELERATION

These four equations are used to express constant acceleration in different ways:

$$s = \frac{(u + v)t}{2}$$

$$v = u + at$$

$$v^2 = u^2 + 2as$$

$$s = ut + \tfrac{1}{2}at^2$$

s = displacement
u = original velocity
v = final velocity
a = acceleration
t = time taken

HOOKE'S LAW

$$F_s = -kx$$

F_s = force of spring
k = spring constant (indication of spring's stiffness)
x = extension

TURNING FORCES

FORCE	DESCRIPTION	FORMULA	KEY
Moment of inertia	The equivalent of mass for an object rotating about an axis	$I = mr^2$	I = moment of inertia m = mass r^2 = square of distance from axis
Angular velocity	The velocity of an object rotating about an axis	$\omega = \frac{\Delta\theta}{\Delta t}$	ω = angular velocity Δθ = angular displacement Δt = change in time
Angular momentum	The momentum of an object rotating about an axis	$L = I\omega$	L = angular momentum I = moment of inertia ω = angular velocity

LAWS OF THERMODYNAMICS

Thermodynamics is the study of the interrelationship between heat, work, and internal energy. The laws of thermodynamics describe what happens when a thermodynamic system goes though an energy change. Energy cannot be created or destroyed (as stated in the first law), but it can be converted into other forms.

LAW	DESCRIPTION
First law	Energy can be neither created nor destroyed.
Second law	The total entropy of an isolated system increases over time.
Third law	There is a theoretical minimum temperature at which the motion of the particles of matter would cease.
Zeroth law	If two bodies (distinct systems) are each in thermal equilibrium with a third body, these two bodies will also be in thermal equilibrium with each other.

TEMPERATURE SCALES

Heat is a form of kinetic energy. The amount of energy contained in an object—its temperature—can be measured using one of three scales: kelvin (one of the SI base units), Celsius (an SI derived scale), and Fahrenheit (first proposed by the Dutch-German-Polish physicist Daniel Gabriel Fahrenheit (see 1724). Absolute 0 kelvin is the point at which atoms in a substance have no heat energy and do not vibrate at all.

	KELVIN	CELSIUS	FAHRENHEIT
	373K	100°C	212°F
	300K	27°C	81°F
	273K	0°C	32°F
	255K	-18°C	0°F
	200K	-73°C	-99°F
	100K	-173°C	-279°F
Absolute zero	0K	-273°C	-460°F

GAS LAWS

Gas is a state of matter with relatively low density and viscosity, variable pressure and temperature, and the ability to diffuse readily and distribute uniformly throughout any container. The gas laws described below relate movements of molecules within a gas to its volume, pressure, and temperature, and state how each measure responds when the others change. Most of the laws are named after the person who discovered them.

LAW	DESCRIPTION	FORMULA	KEY
Avogadro's law	At a constant temperature and pressure, volume is proportional to the number of molecules. Equal volumes of different gases, in identical conditions of temperature and pressure, will contain equal numbers of molecules.	$V \propto n$	V = volume n = number of molecules $\propto$ = proportional to
Boyle's law	For a given mass of gas at a constant temperature, volume is inversely proportional to pressure. So, for example, if volume doubles, pressure halves.	PV = constant	P = pressure V = volume
Charles's law	For a given mass of gas at a constant pressure, volume is directly proportional to absolute temperature (measured in kelvin).	$\frac{V}{T}$ constant	V = volume T = (absolute) temperature
Gay-Lussac's law	For a given mass of gas at a constant volume, pressure is directly proportional to absolute temperature (measured in kelvin).	$\frac{P}{T}$ constant	P = pressure T = (absolute) temperature
Ideal gas law	An ideal gas is a hypothetical gas in which particles may collide but do not have attractive forces between them. The ideal gas law is a good approximation of the behavior of many gases in various conditions.	$PV = nRT$	P = total pressure n = number of moles R = universal gas constant T = absolute temperature
Dalton's law of partial pressures	This law describes mixtures of two or more gases. It states that the total pressure of a gaseous mixture equals sum of the partial pressures of individual component gases. The law is used to work out gas mixtures breathed by scuba divers.	$P = \Sigma p$ or $P_{total} = p_1 + p_2 + p_3$...	P = total pressure Σp = sum of individual partial pressures

PRESSURE AND DENSITY

Pressure is the force applied to an object divided by the area over which the force is applied, and it varies according to the density of the body exerting the force. Pressure and density can be described using a series of equations.

LAW	DESCRIPTION	FORMULA
Pressure	$\frac{\text{force}}{\text{area}}$	$P = \frac{F}{A}$
Density	$\frac{\text{mass}}{\text{volume}}$	$\rho = \frac{m}{V}$
Volume	$\frac{\text{mass}}{\text{density}}$	$V = \frac{m}{\rho}$
Mass	volume × density	$m = V\rho$

EINSTEIN'S THEORIES OF RELATIVITY

In 1905 and 1915, German physicist Albert Einstein published his ground-breaking theories that challenged the previously accepted theory of gravitation.

THEORY	PROPOSITION
Special theory of relativity	1) All physical laws are the same in all frames of reference in uniform motion with respect to one another. 2) The speed of light is a constant, regardless of the motions of the light source and the observer.
General theory of relativity	Spacetime is curved; strong gravity causes distortions of time and mass, and large objects (such as stars) warp spacetime around them.

MAXWELL'S EQUATIONS

This is a series of equations, or laws, formulated by Scottish physicist James Clerk Maxwell (see 1855) that provide a full description of how electromagnetic waves behave. The equations show how electromagnetic fields are produced and how the rates of change in the fields are related to their sources.

LAW	STATEMENT	APPLICATION
Gauss's law for electricity	The electric flux through any closed surface is proportional to the total charge contained within that surface.	Used to calculate electric fields around charged objects.
Gauss's law for magnetism	For a magnetic dipole (one of a pair of equal and oppositely magnetized poles) with any closed surface, the magnetic flux drawn inward toward the south pole will equal the flux directed outward from the north pole; the net flux will always be zero.	Describes sources of magnetic fields and shows that they will always be closed loops.
Faraday's law of induction	The induced electromotive force (EMF) around any closed loop equals the negative of the rate of change of the magnetic flux through the area enclosed by the loop.	Describes how a changing magnetic field can generate an electric field; this is the operating principle for electric generators, inductors, and transformers.
Ampère's law with Maxwell's correction	In static electric field, the line integral of the magnetic field around any closed loop is proportional to the electric current flowing through it.	Used to calculate magnetic fields. Shows that magnetic current can be generated by electric current and by changing electric fields.

ELECTRICITY AND CIRCUIT LAWS

Electricty can be made to travel in a flow, or current. Electric current can be generated by a source of electromotive force (EMF), such as a battery cell, and directed around wire loops called circuits to power electrical devices. There are two main ways in which flow is driven around a circuit: an alternating current (AC), which flows to and fro; and direct current, which flows in one direction only. Particular laws govern the way that a current flows around a circuit.

LAW	DESCRIPTION	FORMULA	KEY
Coulomb's law	The force of attraction or repulsion between two charged particles is directly proportional to the product of the charges and inversely proportional to the distance between them.	$F = \frac{k q_1 q_2}{r^2}$	k = Coulomb's constant q_1 and q_2 = electric (point) charges r^2 = square of distance
Ohm's law	This law expresses the relationships between voltage, resistance, and current, which can be expressed in several ways.	$I = \frac{V}{R}$ $R = \frac{V}{I}$ $V = I R$	R = resistance I = current V = potential difference (voltage)
Kirchhoff's current law	The sum of the electric currents entering any junction in a circuit is equal to the sum of those leaving the junction.	$\sum I = 0$	$\sum$ = summation symbol I = current
Kirchhoff's voltage law	The sum of the voltage changes around the path of any closed loop is zero.	$\sum V = 0$	$\sum$ = summation symbol V = potential difference (voltage)

SUBATOMIC PARTICLES

These are the basic building blocks of matter. Physicists distinguish between elementary particles (with no substructure) and composite particles (composed of smaller structures). Elementary particles are the fundamental constituents of everything in the Universe. Scientists think that every particle is twinned with an opposing antiparticle, which if paired can annihilate each other, producing packages of light (photons).

ELEMENTARY PARTICLES	COMPOSITE PARTICLES (HADRONS)
Quarks Particles that make protons and neutrons. There are six "flavors": up, down, charm, strange, top, and bottom. **Leptons** A group of six particles, comprising electron, muon, and tau particles and their associated neutrinos (the electron-neutrino, muon-neutrino, and tau-neutrino).	**Baryons** Particles made up of three quarks. The best known are the proton (two up and one down quarks) and the neutron (one up and two down quarks)
Gauge bosons Particles associated with the four fundamental forces (see left). No gauge boson has yet been found for gravity, although the existence of a "graviton" has been hypothesized.	**Mesons** Particles made from a quark with an antiquark. There are many types, including the positive pion (up quark with down antiquark) and the negative kaon (strange quark with up antiquark).

THE FOUR FUNDAMENTAL FORCES

All matter in the Universe is subject to four basic forces: gravity, electromagnetism, and the strong and weak nuclear forces. Each one is associated with a subatomic "messenger" particle. Physicists theorize that the forces were once unified in a single force that split in the first fraction of a second after the Big Bang. Matter particles affected by a particular force produce and absorb specific force carriers.

PARTICLE	FORCE	RELATIVE STRENGTH	RANGE IN (M)
Graviton	Gravity	10^{-41}	Infinite
Photon	Electromagnetic	1	Infinite force
Gluon	Strong nuclear force	25	10^{-15}
W, Z bosons	Weak nuclear force	0.8	10^{-18}

COMMON EQUATIONS

The following equations are commonly used in physics.

QUANTITY	STATEMENT	FORMULA
Kinetic energy	½ mass × velocity²	$E_k = ½mv^2$
Weight	mass × gravitational field strength	$W = mg$
Power	$\frac{\text{work done}}{\text{time taken}}$ or $\frac{\text{energy transferred}}{\text{time taken}}$	$P = \frac{W}{t}$
Speed	$\frac{\text{distance moved}}{\text{time taken}}$	$s = \frac{d}{t}$
Velocity	$\frac{\text{displacement}}{\text{time taken}}$	$v = \frac{s}{t}$
Acceleration	$\frac{\text{change in velocity}}{\text{time taken for this change}}$	$a = \frac{(v-u)}{t}$
Resultant force	mass × acceleration	$F = ma$
Momentum	mass × velocity	$m v$
Refractive index	$\frac{\text{speed of light in a vacuum}}{\text{speed of light in a medium}}$	$n = \frac{c}{v}$
Wave speed	frequency × wavelength	$v = f \lambda$
Electric charge	current × time taken	$q = I t$
Potential difference (voltage)	current × resistance or $\frac{\text{energy transferred}}{\text{charge}}$	$V = I R$ $V = \frac{W}{q}$
Resistance	$\frac{\text{voltage}}{\text{current}}$	$R = \frac{V}{I}$
Electrical energy	potential difference (or voltage) × current × time taken	$E = V I t$
Work done	force × distance moved in the direction of the force	$W = Fs$
Efficiency	$\frac{\text{work output}}{\text{work input}} \times 100\%$	$\frac{W_o}{W_i} \times 100\%$

CHEMISTRY

THE PERIODIC TABLE

The modern periodic table contains 118 known elements, of which 90 occur naturally on Earth. The elements are grouped according to their atomic structure. They are positioned on the chart in ascending order of their atomic number (the number of protons in an element's nucleus) and grouped within the chart according to the arrangement of the outer electrons. This way a chemist can predict the likely characteristics of an element from its position on the table.

KEY

- Alkali metals
- Transition metals
- Other metals
- Alkali earth metals
- Rare earth metals
- Metalloids
- Other nonmetals
- Halogens
- Noble gases
- Unknown

PERIOD / GROUP	1	2	3	4	5	6	7	8	9	10	11	12	13	14	15	16	17	18
1	1 1 H Hydrogen																	2 4 He Helium
2	3 7 Li Lithium	4 9 Be Beryllium											5 11 B Boron	6 12 C Carbon	7 14 N Nitrogen	8 16 O Oxygen	9 19 F Fluorine	10 20 Ne Neon
3	11 23 Na Sodium	12 24 Mg Magnesium											13 27 Al Aluminum	14 28 Si Silicon	15 31 P Phosphorus	16 32 S Sulfur	17 35 Cl Chlorine	18 40 Ar Argon
4	19 39 K Potassium	20 40 Ca Calcium	21 45 Sc Scandium	22 48 Ti Titanium	23 51 V Vanadium	24 52 Cr Chromium	25 55 Mn Manganese	26 56 Fe Iron	27 59 Co Cobalt	28 59 Ni Nickel	29 64 Cu Copper	30 65 Zn Zinc	31 70 Ga Gallium	32 73 Ge Germanium	33 75 As Arsenic	34 79 Se Selenium	35 80 Br Bromine	36 84 Kr Krypton
5	37 85 Rb Rubidium	38 88 Sr Strontium	39 89 Y Yttrium	40 91 Zr Zirconium	41 93 Nb Niobium	42 96 Mo Molybdenum	43 98 Tc Technetium	44 101 Ru Ruthenium	45 103 Rh Rhodium	46 106 Pd Palladium	47 108 Ag Silver	48 112 Cd Cadmium	49 115 In Indium	50 119 Sn Tin	51 122 Sb Antimony	52 128 Te Tellurium	53 127 I Iodine	54 131 Xe Xenon
6	55 133 Cs Cesium	56 137 Ba Barium		72 178 Hf Hafnium	73 181 Ta Tantalum	74 184 W Tungsten	75 186 Re Rhenium	76 190 Os Osmium	77 192 Ir Iridium	78 195 Pt Platinum	79 197 Au Gold	80 201 Hg Mercury	81 204 Tl Thallium	82 207 Pb Lead	83 209 Bi Bismuth	84 209 Po Polonium	85 210 At Astatine	86 220 Rn Radon
7	87 223 Fr Francium	88 226 Ra Radium		104 261 Rf Rutherfordium	105 262 Db Dubnium	106 263 Sg Seaborgium	107 264 Bh Bohrium	108 265 Hs Hassium	109 268 Mt Meitnerium	110 281 Ds Darmstadtium	111 273 Rg Roentgenium	112 285 Cn Copernicium	113 284 Uut Ununtrium	114 289 Fl Flerovium	115 288 Uup Ununpentium	116 292 Lv Livermorium	117 292 Uus Ununseptium	118 294 Uuo Ununoctium

LANTHANOIDS	57 139 La Lanthanum	58 140 Ce Cerium	59 141 Pr Praseodymium	60 144 Nd Neodymium	61 145 Pm Promethium	62 150 Sm Samarium	63 152 Eu Europium	64 157 Gd Gadolinium	65 159 Tb Terbium	66 163 Dy Dysprosium	67 165 Ho Holmium	68 167 Er Erbium	69 169 Tm Thulium	70 173 Yb Ytterbium	71 175 Lu Lutetium
ACTINOIDS	89 227 Ac Actinium	90 232 Th Thorium	91 231 Pa Protactinium	92 238 U Uranium	93 237 Np Neptunium	94 244 Pu Plutonium	95 243 Am Americium	96 247 Cm Curium	97 247 Bk Berkelium	98 251 Cf Californium	99 252 Es Einsteinium	100 257 Fm Fermium	101 258 Md Mendelevium	102 259 No Nobelium	103 262 Lr Lawrencium

INDIVIDUAL ENTRY

Each block on the table describes an individual element. The relative atomic mass is the average number of protons and neutrons in the nucleus and is given in the table above as a rounded figure.

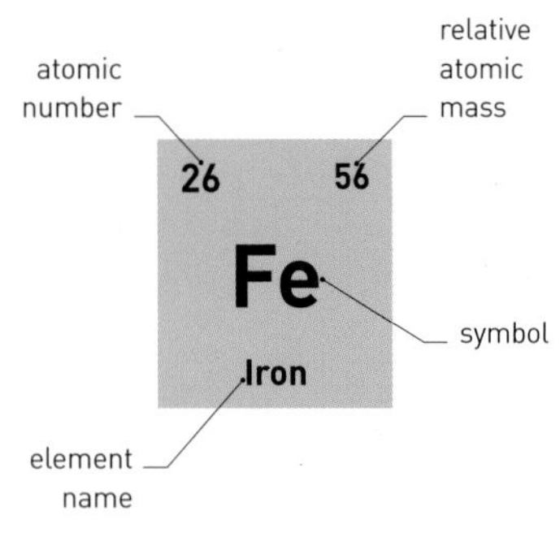
atomic number
relative atomic mass
26
56
Fe
Iron
symbol
element name

BUILDING BLOCKS OF THE TABLE

The periodic table is divided in three different directions: groups form the vertical columns; periods, the horizontal rows; and series are indicated by blocks of color.

The overall groupings put elements that demonstrate close family resemblances together. Reactive metallic elements are on the left, progressing across the chart through less reactive metals, metalloids, and non metals, to barely reactive gases on the far right.

7

Periods
Elements that have the same number of electron shells are grouped horizontally in rows. Periods 6 and 7 are too long to fit on the table so are positioned below the chart.

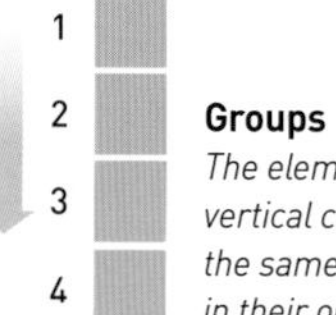
1
2
3
4

Groups
The elements in each of the vertical columns all have the same number of electrons in their outer shells.

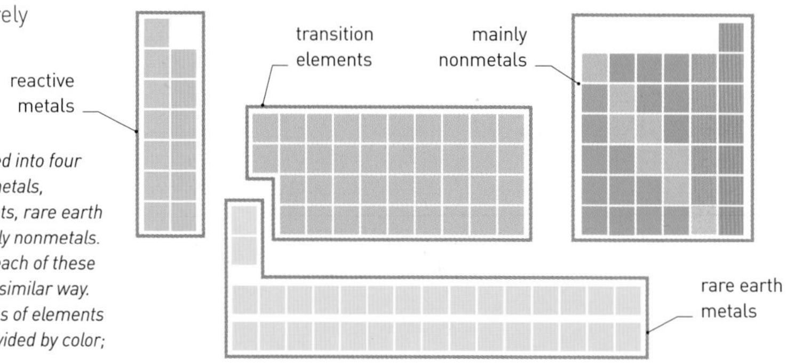

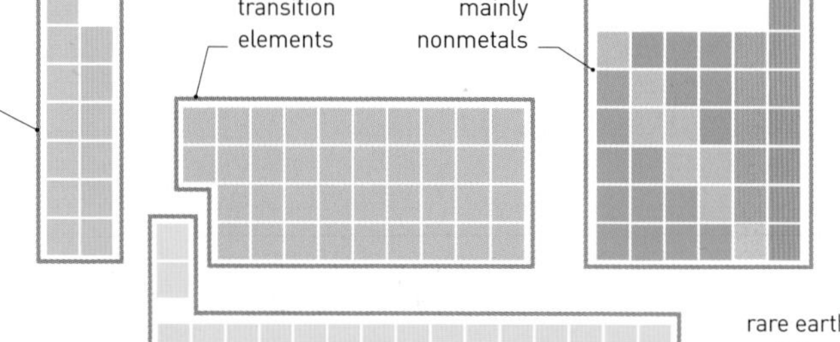
reactive metals
transition elements
mainly nonmetals
rare earth metals

Series
The table is divided into four series: reactive metals, transition elements, rare earth metals, and mainly nonmetals. The elements in each of these groups react in a similar way. The different types of elements are further subdivided by color; see key above.

THE ELEMENTS

The chart below lays out essential information for each of the known elements. It is arranged by atomic number—the number of protons in the nucleus of an element's atoms. For each element, the table includes the commonly used symbol, the relative atomic mass (in this chart given to the nearest two decimal places), and the valency (the number of chemical bonds formed by the atoms) for each given element.

ATOMIC NUMBER	ELEMENT	SYMBOL	ATOMIC MASS	MELTING POINT °C	MELTING POINT °F	BOILING POINT °C	BOILING POINT °F	VALENCY
1	Hydrogen	H	1.00	-259	-434	-253	-423	1
2	Helium	He	4.00	-272	-458	-269	-452	0
3	Lithium	Li	6.94	179	354	1,340	2,440	1
4	Beryllium	Be	9.01	1,283	2,341	2,990	5,400	2
5	Boron	B	10.81	2,300	4,170	3,660	6,620	3
6	Carbon	C	12.01	3,500	6,332	4,827	8,721	2,4
7	Nitrogen	N	14.01	-210	-346	-196	-321	3,5
8	Oxygen	O	16.00	-219	-362	-183	-297	2
9	Fluorine	F	19.00	-220	-364	-188	-306	1
10	Neon	Ne	20.18	-249	-416	-246	-410	0
11	Sodium	Na	22.99	98	208	890	1,634	1
12	Magnesium	Mg	24.31	650	1,202	1,105	2,021	2
13	Aluminum	Al	26.98	660	1,220	2,467	4,473	3
14	Silicon	Si	28.09	1,420	2,588	2,355	4,271	4
15	Phosphorus	P	30.97	44	111	280	536	3,5
16	Sulfur	S	32.07	113	235	445	832	2,4,6
17	Chlorine	Cl	35.45	-101	-150	-34	-29	1,3,5,7
18	Argon	Ar	39.95	-189	-308	-186	-303	0
19	Potassium	K	39.10	64	147	754	1,389	1
20	Calcium	Ca	40.08	848	1,558	1,487	2,709	2
21	Scandium	Sc	44.96	1,541	2,806	2,831	5,128	3
22	Titanium	Ti	47.87	1,677	3,051	3,277	5,931	3,4
23	Vanadium	V	50.94	1,917	3,483	3,377	6,111	2,3,4,5
24	Chromium	Cr	52.00	1,903	3,457	2,642	4,788	2,3,6
25	Manganese	Mn	54.94	1,244	2,271	2,041	3,706	2,3,4,6,7
26	Iron	Fe	55.85	1,539	2,802	2,750	4,980	2,3
27	Cobalt	Co	58.93	1,495	2,723	2,877	5,211	2,3
28	Nickel	Ni	58.69	1,455	2,651	2,730	4,950	2,3
29	Copper	Cu	63.55	1,083	1,981	2,582	4,680	1,2
30	Zinc	Zn	65.41	420	788	907	1,665	2
31	Gallium	Ga	69.72	30	86	2,403	4,357	2,3
32	Germanium	Ge	72.63	937	1,719	2,355	4,271	4
33	Arsenic	As	74.92	817	1,503	613	1,135	3,5
34	Selenium	Se	78.96	217	423	685	1,265	2,4,6
35	Bromine	Br	79.90	-7	19	59	138	1,3,5,7
36	Krypton	Kr	83.80	-157	-251	-152	-242	0
37	Rubidium	Rb	85.47	39	102	688	1,270	1
38	Strontium	Sr	87.62	769	1,416	1,384	2,523	2
39	Yttrium	Y	88.91	1,522	2,772	3,338	6,040	3
40	Zirconium	Zr	91.22	1,852	3,366	4,377	7,911	4
41	Niobium	Nb	92.91	2,467	4,473	4,742	8,568	3,5
42	Molybdenum	Mo	95.96	2,610	4,730	5,560	10,040	2,3,4,5,6
43	Technetium	Tc	97.91	2,172	3,942	4,877	8,811	2,3,4,6,7
44	Ruthenium	Ru	101.07	2,310	4,190	3,900	7,052	3,4,6,8
45	Rhodium	Rh	102.91	1,966	3,571	3,727	6,741	3,4
46	Palladium	Pd	106.42	1,554	2,829	2,970	5,378	2,4
47	Silver	Ag	107.87	962	1,764	2,212	4,014	1
48	Cadmium	Cd	112.41	321	610	767	1,413	2
49	Indium	In	114.82	156	313	2,028	3680	1,3
50	Tin	Sn	118.71	232	450	2,270	4118	2,4
51	Antimony	Sb	121.76	631	1,168	1,635	2975	3,5
52	Tellurium	Te	127.60	450	842	990	1814	2,4,6
53	Iodine	I	126.90	114	237	184	363	1,3,5,7
54	Xenon	Xe	131.29	-112	-170	-107	-161	0
55	Cesium	Cs	132.91	29	84	671	1,240	1
56	Barium	Ba	137.33	725	1,337	1,640	2,984	2
57	Lanthanum	La	138.91	921	1,690	3,457	6,255	3
58	Cerium	Ce	140.12	799	1,470	3,426	6,199	3,4
59	Praseodymium	Pr	140.91	931	1,708	3,512	6,354	3
60	Neodymium	Nd	144.24	1,021	1,870	3,068	5,554	3
61	Promethium	Pm	144.91	1,168	2,134	2,700	4,892	3
62	Samarium	Sm	150.36	1,077	1,971	1,791	3,256	2,3
63	Europium	Eu	151.96	822	1,512	1,597	2,907	2,3
64	Gadolinium	Gd	157.25	1,313	2,395	3,266	5,911	3
65	Terbium	Tb	158.93	1,356	2,473	3,123	5,653	3
66	Dysprosium	Dy	162.50	1,412	2,574	2,562	4,644	3
67	Holmium	Ho	164.93	1,474	2,685	2,695	4,883	3
68	Erbium	Er	167.26	1,529	2,784	2,863	5,185	3
69	Thulium	Tm	168.93	1,545	2,813	1,947	3,537	2,3
70	Ytterbium	Yb	173.04	819	1,506	1,194	2,181	2,3
71	Lutetium	Lu	174.97	1,663	3,025	3,395	6,143	3
72	Hafnium	Hf	178.49	2,227	4,041	4,602	8,316	4
73	Tantalum	Ta	180.95	2,996	5,425	5,427	9,801	3,5
74	Tungsten	W	183.84	3,410	6,170	5,660	10,220	2,4,5,6
75	Rhenium	Re	186.21	3,180	5,756	5,627	10,161	1,4,7
76	Osmium	Os	190.23	3,045	5,510	5,090	9,190	2,3,4,6,8
77	Iridium	Ir	192.22	2,410	4,370	4,130	7,466	3,4
78	Platinum	Pt	195.08	1,772	3,222	3,827	6,921	2,4
79	Gold	Au	196.97	1,064	1,947	2,807	5,080	1,3
80	Mercury	Hg	200.59	-39	-38	357	675	1,2
81	Thallium	Tl	204.38	303	577	1,457	2,655	1,3
82	Lead	Pb	207.20	328	622	1,744	3,171	2,4
83	Bismuth	Bi	208.98	271	520	1,560	2,840	3,5
84	Polonium	Po	208.98	254	489	962	1,764	2,3,4
85	Astatine	At	209.99	300	572	370	698	1,3,5,7
86	Radon	Rn	222.02	-71	-96	-62	-80	0
87	Francium	Fr	223.02	27	81	677	1,251	1
88	Radium	Ra	226.02	700	1,292	1,200	2,190	2
89	Actinium	Ac	227.03	1,050	1,922	3,200	5,792	3
90	Thorium	Th	232.04	1,750	3,182	4,787	8,649	4
91	Protactinium	Pa	231.04	1,597	2,907	4,027	7,281	4,5
92	Uranium	U	238.03	1,132	2,070	3,818	6,904	3,4,5,6
93	Neptunium	Np	237.05	637	1,179	4,090	7,394	2,3,4,5,6
94	Plutonium	Pu	244.06	640	1,184	3,230	5,850	2,3,4,5,6
95	Americium	Am	243.06	994	1,821	2,607	4,724	2,3,4,5,6
96	Curium	Cm	247.07	1,340	2,444	3,190	5,774	2,3,4
97	Berkelium	Bk	247.07	1,050	1,922	710	1,310	2,3,4
98	Californium	Cf	251.08	900	1,652	1,470	2,678	2,3,4
99	Einsteinium	Es	252.08	860	1,580	996	1,825	2,3
100	Fermium	Fm	257.10	unknown		unknown		2,3
101	Mendelevium	Md	258.10	unknown		unknown		2,3
102	Nobelium	No	259.10	unknown		unknown		2,3
103	Lawrencium	Lr	262.11	unknown		unknown		3
104	Rutherfordium	Rf	261.11	unknown		unknown		unknown
105	Dubnium	Db	262.11	unknown		unknown		unknown
106	Seaborgium	Sg	263.12	unknown		unknown		unknown
107	Bohrium	Bh	264.13	unknown		unknown		unknown
108	Hassium	Hs	265.13.	unknown		unknown		unknown
109	Meitnerium	Mt	268.14	unknown		unknown		unknown
110	Darmstadtium	Ds	281.16	unknown		unknown		unknown
111	Roentgenium	Rg	273.15	unknown		unknown		unknown
112	Copernicium	Cn	[285]	unknown		unknown		unknown
113	Ununtrium	Uut	[284]	unknown		unknown		unknown
114	Flerovium	Fl	[289]	unknown		unknown		unknown
115	Ununpentium	Uup	[288]	unknown		unknown		unknown
116	Livermorium	Lv	[293]	unknown		unknown		unknown
117	Ununseptium	Uus	[292]	unknown		unknown		unknown
118	Ununoctium	Uuo	[294]	unknown		unknown		unknown

BIOLOGY

TAXONOMIC RANKS

Biologists classify organisms into groups according to their characteristics as a way of illustrating their evolutionary relationships. All life has descended from a single common ancestor, which lived billions of years ago, and has diversified through evolution. The most distantly related organisms belong to groups that are ranked as domains. These are subdivided into groups of successively lower ranks to contain organisms that are progressively more closely related. At any rank, modern biologists strive to define groups that are monophyletic: ones that contain all organisms descended from a single point of ancestry. At the lowest rank, species, organisms are so closely related that they can usually interbreed. In the charts opposite and on pp.362–63, dotted lines are used to define informal assemblages of taxonomic ranks. Although these are not natural evolutionary groups, they often provide a convenient and useful way to refer to collections of organisms.

DOMAIN	KINGDOM	PHYLUM	CLASS	ORDER	FAMILY	GENUS	SPECIES
Introduced in the 1990s in response to discoveries that were made in cellular biology, the domain rank defines groups that represent the most ancient divisions of life on Earth. These emerged as distinct groups around 4 billion years ago and include bacteria and the more complex multicelled life forms.	Domains are divided into different kingdoms that include familiar groups of organisms (animals, plants, and fungi)—as well as other groups (including a range of single-celled organisms and algae) that evolved around a billion years ago.	Kingdoms are divided into phyla (singular: phylum). Organisms united in a phylum share a particular body plan; for animals and plants, this includes groups that originated 0.5–1 billion years ago, when the seas were teeming with life and land was being colonized for the first time.	Phyla are divided into classes. Organisms within a class are defined by body structure and life history. For example, land vertebrates are split into classes that include amphibians, reptiles, birds, and mammals. Plant classes include monocotyledon and dicotyledon flowering plants.	Classes are divided into orders. Animal orders are based largely on body structure, but plant orders are also defined by the chemicals made in their tissues. For example, the mammals are split into orders that include primates, rodents, and bats. Plant orders include Ranunculales (buttercups and relatives) and Lamiales (mints and relatives).	Orders are divided into families. Names of plant, algal, and fungal families conventionally end in "-aceae" (for example, Liliaceae for the lily family); names of animal families end in "-idae" (for example, Sciuridae for the squirrel family).	Families are divided into genera (singular: genus). The name of each genus is the first part of the scientific name of a species. For example, big cats are united in the genus *Panthera* and include the species *Panthera leo* (lion) and *Panthera tigris* (tiger). The generic and species names are always given in italics.	The species group is the only taxonomic rank that can be defined in biological terms. It is often taken to be a group of organisms that can interbreed. But in practice, most species—including those that cannot reproduce sexually or where reproductive biology is unknown—are defined by their physical characteristics.

DOMAINS AND KINGDOMS

Biologists used to divide all life into plants and animals, but the roots of biological diversity are now known to be far more complex. The earliest life diversified as single-celled organisms. Two lineages of these organisms, out of countless others, later gave rise to plants and animals. This means that most groups at the ranks of domain and kingdom include single-celled organisms. The most basic differences evolved when the earliest life split into three domains: simple single-celled Bacteria and Archaeae; and the more complex Eukaryotes, whose genetic material became packaged into a cellular nucleus. Many groups of eukaryotic organisms retained their single-celled nature, but others formed the multicelled bodies that became fungi, plants, and animals.

LIFE ON EARTH

BACTERIA
BACTERIA

These are single-celled organisms, most of which have a cell wall made from a tough material called murein. Their genetic material (DNA) is not packaged into a nucleus. Other cellular organelles (membrane-bound structures, such as mitochondria) are also absent.

8,000+ SPECIES

ARCHAEA
ARCHAEA

These are single-celled organisms with genetic material (DNA) reinforced by packaging proteins called histones, but they do not have a nucleus or other organelles. Many are adapted to survive in harsh environments, such as hot, acidic pools.

2,000+ SPECIES

EUKARYOTES
EUKARYA

These are single- or multicelled organisms whose genetic material is packaged into a nucleus and reinforced with histones. During cell division, this material solidifies into structures called chromosomes. The cells also have other organelles, such as mitochondria and chloroplasts.

2 MILLION+ SPECIES

FUNGI
FUNGI

Single- or multicelled organisms, fungi reproduce by spores and absorb food by decomposing dead material or behaving as parasites. Multicelled forms are usually made up of a network of microscopic fibers called a mycelium.

70,000+ SPECIES

PLANTS
PLANTAE

Multicelled organisms that make food through the light-absorbing process of photosynthesis, plants grow into a branching, often leafy, body. The most primitive forms reproduce by free-swimming sperm and drifting spores; so-called higher plants form seeds.

290,000+ SPECIES

ANIMALS
ANIMALIA

These are multicelled organisms that obtain food by eating other organisms or dead material. Fast response times and movement are possible because animals have a nervous system that transmits electrical signals to contracting muscles.

1.6 MILLION+ SPECIES

ALL OTHER KINGDOMS

The remaining eukaryotes were formerly classified as a single kingdom (called Protoctista). It is now known that they do not form a natural evolutionary group, but instead comprise at least seven different kingdoms and include a wide range of single- and multicelled organisms. Many, such as algae and seaweeds, photosynthesize like plants. Some are animal-like predators or parasites, such as amoebas. Others, including slime molds, grow like fungi.

70,000+ SPECIES

FUNGI

A distinction can be made between simple fungi (the microsporidians and chytrids) and "higher" fungi, which have an extensive mycelium (a network of microscopic filaments). The common feature of chytrids—also called water molds—is that they produce spores that swim by beating flagella (microscopic hairs). Some kinds of sac or club fungi form lichens: partnerships with certain algae or bacteria to supplement nutrition by photosynthesis.

FUNGI
FUNGI
70,000+ SPECIES

TYPICAL CHYTRIDS
CHYTRIDIOMYCOTA
Decomposers in soil and water, or parasites in animals (including one that infects amphibians).
700+ SPECIES

GUT CHYTRIDS
NEOCALLIMASTIGOMYCOTA
Fungi that inhabit gut of herbivorous vertebrates (for example, cows) and help digest plant fiber.
20+ SPECIES

MICROSPORIDIANS
MICROSPORIDIA
Tiny, single-celled microbes that live as parasites inside cells of animals. Many infect insects.
1,200+ SPECIES

BLASTOCLADIAL CHRYTRIDS
BLASTOCLADIOMYCOTA
Decomposers in soil, or parasites of plants or invertebrate animals.
180+ SPECIES

MOSTLY PARASITIC SINGLE- OR MULTICELLED FUNGI THAT LACK AN EXTENSIVE MYCELIUM

PIN MOLDS
ZYGOMYCOTA
Fungi with threads of mycelium with no cross-walls between adjacent cellular nuclei.
1,100+ SPECIES

SAC FUNGI
ASCOMYCOTA
Fungi with spores forming in minute pods. Includes forms with mycelium and single-celled yeasts.
33,000+ SPECIES

GLOMEROMYCOTES
GLOMEROMYCOTA
Soil-living fungi found on roots of simple plants, where they exchange nutrients.
230+ SPECIES

CLUB FUNGI
BASIDIOMYCOTA
Mushrooms and toadstools whose spores form on tiny, clublike structures, often in fruiting bodies.
32,000+ SPECIES

FUNGI THAT MOSTLY GROW BY MULTICELLED MYCELIUM

PLANTS

Plant classification is based largely on their reproduction and life cycle, which alternates between sperm- and egg-producing generations (gametophyte) and spore-producing generations (sporophyte). Gametophytes of simple plants, such as mosses, need moist habitats to thrive, while in seed plants both generation stages happen inside reproductive shoots (such as cones or flowers), so that the life cycle can succeed in drier conditions.

PLANTS
PLANTAE
290,000+ SPECIES

LIVERWORTS
MARCHANTIOPHYTA
Simple flat or leafy body produces eggs and free-swimming sperm. Spores on umbrella-like shoots.
8,000+ SPECIES

HORNWORTS
ANTHOCEROTOPHYTA
Flat body produces eggs and free-swimming sperm. Spores on erect, hornlike structures.
100+ SPECIES

MOSSES
BRYOPHYTA
Leafy or tufted body produces eggs and free-swimming sperm. Spores in upright capsules.
12,000+ SPECIES

FERNS AND HORSETAILS
PTERIDOPHYTA
Fronds (ferns) or bottle-brushlike whorls (horsetails) produce spores. Tiny egg-producing stage.
12,000+ SPECIES

LYCOPODS
LYCOPODIOPHYTA
Leafy, upright body producing spores. The egg-producing stage may be underground.
1,200+ SPECIES

CYCADS
CYCADOPHYTA
Palmlike trees, confined to the tropics, that produce seeds in cones.
300+ SPECIES

CONIFERS
PINOPHYTA
Mostly trees that produce cones. Many are needle-leaved to cope with cold or drought.
630+ SPECIES

GNETOPHYTES
GNETOPHYTA
Mostly tropical woody plants producing seeds in cones. Transport vessels more open than in conifers.
70+ SPECIES

GINKGO
GINKGOPHYTA
Tree that produces seeds without cones or true fruit. Naturally confined to China.
1 SPECIES

FLOWERING PLANTS
MAGNOLIOPHYTA
Plants that produce seeds in fruit that develops from a flower; includes tiny herbs and big trees.
260,000+ SPECIES

WATER LILIES AND RELATIVES
NYMPHAEA AND OTHER GENERA
Soft-bodied aquatic plants, growing either submerged or with floating leaves.
100+ SPECIES

AMBORELLA
AMBORELLA
Shrubs with tiny flowers and without open transport vessels. Confined to New Caledonia.
1 SPECIES

STAR ANISE AND RELATIVES
ILLICIUM AND OTHER GENERA
Trees, shrubs, and climbing plants that produce berrylike fruit. From North America and Indo-Pacific.
100+ SPECIES

BASAL FLOWERING PLANTS

MONOCOTYLEDONS
MONOCOTYLEDONEAE
Plants with mostly straplike leaves with parallel veins. Pollen grains have one opening and a single seed-leaf.
58,000+ SPECIES

MAGNOLIAS AND RELATIVES
MAGNOLIIDAE
Mostly woody plants, superficially like dicotyledons, but pollen grains have a single opening.
7,100+ SPECIES

ADVANCED DICOTYLEDONS
EUDICOTYLEDONEAE
Plants with variable leaves and net-like vein pattern. Pollen grains have three openings and two seed-leaves.
190,000+ SPECIES

ANIMALS

Animals are grouped into phyla according to the internal arrangement of their organs and body cavities. The simplest bodies have a single opening to the gut and even lack blood circulatory systems. In more advanced animals, the body has organs for respiration and excretion, as well as a sophisticated brain. Most animals reproduce sexually by producing eggs and swimming sperm, but some are asexual. More than 90 percent of animals in over 30 phyla are invertebrates—animals that lack a backbone. All vertebrates (including humans) belong to the single phylum Chordata, but even this phylum includes some invertebrates.

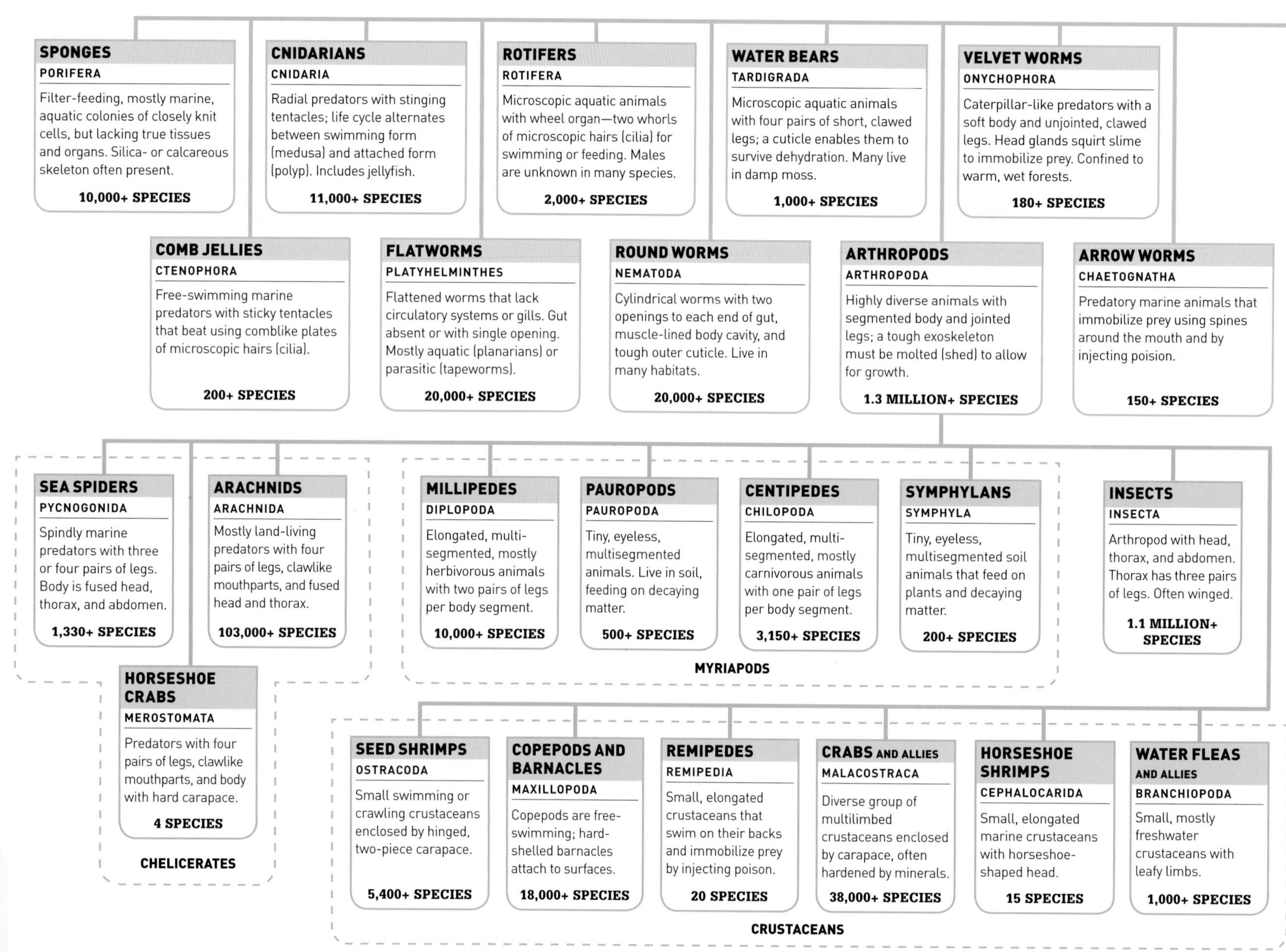

SPONGES
PORIFERA
Filter-feeding, mostly marine, aquatic colonies of closely knit cells, but lacking true tissues and organs. Silica- or calcareous skeleton often present.
10,000+ SPECIES
CNIDARIANS
CNIDARIA
Radial predators with stinging tentacles; life cycle alternates between swimming form (medusa) and attached form (polyp). Includes jellyfish.
11,000+ SPECIES
ROTIFERS
ROTIFERA
Microscopic aquatic animals with wheel organ—two whorls of microscopic hairs (cilia) for swimming or feeding. Males are unknown in many species.
2,000+ SPECIES
WATER BEARS
TARDIGRADA
Microscopic aquatic animals with four pairs of short, clawed legs; a cuticle enables them to survive dehydration. Many live in damp moss.
1,000+ SPECIES
VELVET WORMS
ONYCHOPHORA
Caterpillar-like predators with a soft body and unjointed, clawed legs. Head glands squirt slime to immobilize prey. Confined to warm, wet forests.
180+ SPECIES
COMB JELLIES
CTENOPHORA
Free-swimming marine predators with sticky tentacles that beat using comblike plates of microscopic hairs (cilia).
200+ SPECIES
FLATWORMS
PLATYHELMINTHES
Flattened worms that lack circulatory systems or gills. Gut absent or with single opening. Mostly aquatic (planarians) or parasitic (tapeworms).
20,000+ SPECIES
ROUND WORMS
NEMATODA
Cylindrical worms with two openings to each end of gut, muscle-lined body cavity, and tough outer cuticle. Live in many habitats.
20,000+ SPECIES
ARTHROPODS
ARTHROPODA
Highly diverse animals with segmented body and jointed legs; a tough exoskeleton must be molted (shed) to allow for growth.
1.3 MILLION+ SPECIES
ARROW WORMS
CHAETOGNATHA
Predatory marine animals that immobilize prey using spines around the mouth and by injecting poision.
150+ SPECIES
SEA SPIDERS
PYCNOGONIDA
Spindly marine predators with three or four pairs of legs. Body is fused head, thorax, and abdomen.
1,330+ SPECIES
ARACHNIDS
ARACHNIDA
Mostly land-living predators with four pairs of legs, clawlike mouthparts, and fused head and thorax.
103,000+ SPECIES
HORSESHOE CRABS
MEROSTOMATA
Predators with four pairs of legs, clawlike mouthparts, and body with hard carapace.
4 SPECIES
CHELICERATES
MILLIPEDES
DIPLOPODA
Elongated, multi-segmented, mostly herbivorous animals with two pairs of legs per body segment.
10,000+ SPECIES
PAUROPODS
PAUROPODA
Tiny, eyeless, multisegmented animals. Live in soil, feeding on decaying matter.
500+ SPECIES
CENTIPEDES
CHILOPODA
Elongated, multi-segmented, mostly carnivorous animals with one pair of legs per body segment.
3,150+ SPECIES
SYMPHYLANS
SYMPHYLA
Tiny, eyeless, multisegmented soil animals that feed on plants and decaying matter.
200+ SPECIES
MYRIAPODS
INSECTS
INSECTA
Arthropod with head, thorax, and abdomen. Thorax has three pairs of legs. Often winged.
1.1 MILLION+ SPECIES
SEED SHRIMPS
OSTRACODA
Small swimming or crawling crustaceans enclosed by hinged, two-piece carapace.
5,400+ SPECIES
COPEPODS AND BARNACLES
MAXILLOPODA
Copepods are free-swimming; hard-shelled barnacles attach to surfaces.
18,000+ SPECIES
REMIPEDES
REMIPEDIA
Small, elongated crustaceans that swim on their backs and immobilize prey by injecting poison.
20 SPECIES
CRABS AND ALLIES
MALACOSTRACA
Diverse group of multilimbed crustaceans enclosed by carapace, often hardened by minerals.
38,000+ SPECIES
HORSESHOE SHRIMPS
CEPHALOCARIDA
Small, elongated marine crustaceans with horseshoe-shaped head.
15 SPECIES
WATER FLEAS AND ALLIES
BRANCHIOPODA
Small, mostly freshwater crustaceans with leafy limbs.
1,000+ SPECIES
CRUSTACEANS

HUMAN ANCESTORS

Humans—members of the genus Homo—appeared on Earth within the last 2 million years, having descended from a group of apelike ancestors that included the genera *Paranthropus* and *Australopithecus*. Several different species of *Homo*, such as *Homo neanderthalensis,* have lived in recent times, but only a single species—*Homo sapiens*—survives today. All humans and apes belong to the family Hominidae in the mammal order Primates.

AUSTRALOPITHECUS ANAMENSIS (4.2–3.9 MYA)
SAHELANTHROPUS TCHADENSIS (7–6 MYA)
ARDIPITHECUS KADABBA (5.8–5.2 MYA)
ORRORIN TUGENENSIS (6.2–5.6 MYA)
ARDIPITHECUS RAMIDUS (4.5–4.3 MYA)
6 MYA
5 MYA
4 MY

ANIMALS

ANIMALIA

1.6 MILLION+ SPECIES

ENTOPROCT MOSS ANIMALS

ENTOPROCTA

Colonial, filter-feeding marine animals, similar to ectoprocts, but with both mouth and anus of gut within tentacle crown.

150+ SPECIES

SEGMENTED WORMS

ANNELIDA

Animals with segmented body, an internal body cavity, blood vessels, and sets of muscles for swimming and burrowing. Includes earthworms.

21,000+ SPECIES

MOLLUSKS

MOLLUSCA

Diverse animals with a soft body carried on a muscular foot and overhung by a fleshy mantle that makes a shell. Includes mussels and snails.

110,000+ SPECIES

ACORN WORMS AND ALLIES

HEMICHORDATA AND OTHER FAMILIES

Burrowing, marine filter-feeders. Like vertebrates, these animals have a dorsal nerve cord.

130+ SPECIES

+ 18 MORE MINOR PHYLA

MINOR PHYLA

Around half of animal phyla include fewer than a hundred species. Some of these minor phyla—almost all known only from the marine habitat—have been discovered within the last two decades.

1,000+ SPECIES

ECTOPROCT MOSS ANIMALS

ECTOPROCTA

Colonial, filter-feeding, mostly marine animals with tiny "crowns" of tentacles. Anus emerges outside the crown.

6,000+ SPECIES

RIBBON WORMS

NEMERTEA

Predatory marine animals with muscular, barbed proboscis for harpooning prey and injecting poison. Includes some of the longest animals.

1,400+ SPECIES

LAMPSHELLS

BRACHIOPODA

Marine animals superficially similar to molluscs. A two-valved shell attaches to rock. Filter-feeds using tentacles with beating cilia.

400+ SPECIES

ECHINODERMA

ECHINODERMATA

Five-part radial marine animals with spiny skin and numerous tiny tube feet. Transport system relies on circulating water. Includes starfishes.

7,000+ SPECIES

CHORDATES

CHORDATA

Animals with body supported by a stiff rod (notochord) that develops into a spine in adult vertebrates becoming part of a cartilaginous or bony skeleton.

70,000+ SPECIES

INVERTEBRATE CHORDATES

PELAGIC TUNICATES

THALIACEA

Drifting filter-feeders that siphon water through their body. Can form big colonies.

80 SPECIES

SEA SQUIRTS

ASCIDIACEA

Saclike filter feeders attached to rocks that siphon water through their body. Larval form has notochord.

2,900+ SPECIES

LANCELETS

LEPTOCARDII

Small filter-feeders that resemble larvae of fish. Have gill slits and body muscle blocks for swimming.

30 SPECIES

VERTEBRATE CHORDATES

JAWLESS FISHES

CYCLOSTOMATA

Fishes with a simple skull and incomplete cartilaginous spine. Suckerlike mouth ringed by "teeth".

130 SPECIES

RAY-FINNED FISHES

ACTINOPTERYGII

Marine fishes with bony skeletons; fins are supported by jointed rods.

31,000+ SPECIES

AMPHIBIANS

AMPHIBIA

Mostly four-legged vertebrates (caecilians are legless) that breathe using lungs and moist skin.

6,640+ SPECIES

BIRDS

AVES

Two-legged vertebrates with wings and feathered skin. All lay hard-shelled eggs.

10,200+ SPECIES

CARTILAGINOUS FISHES

CHONDRICHTHYES

Mostly predatory fishes with a skeleton made from cartilage. Includes sharks.

1,200 SPECIES

LOBE-FINNED FISHES

SARCOPTERGYII

Fishes with fins supported by a strong muscular base; some can shuffle on land.

8 SPECIES

REPTILES

REPTILIA

Mostly four-limbed (snakes and some lizards are legless) with scaly skin. Most lay hard-shelled eggs.

9,400+ SPECIES

MAMMALS

MAMMALIA

Four-limbed, warm-blooded vertebrates with hairy skin. Most give birth to live young.

5,400+ SPECIES

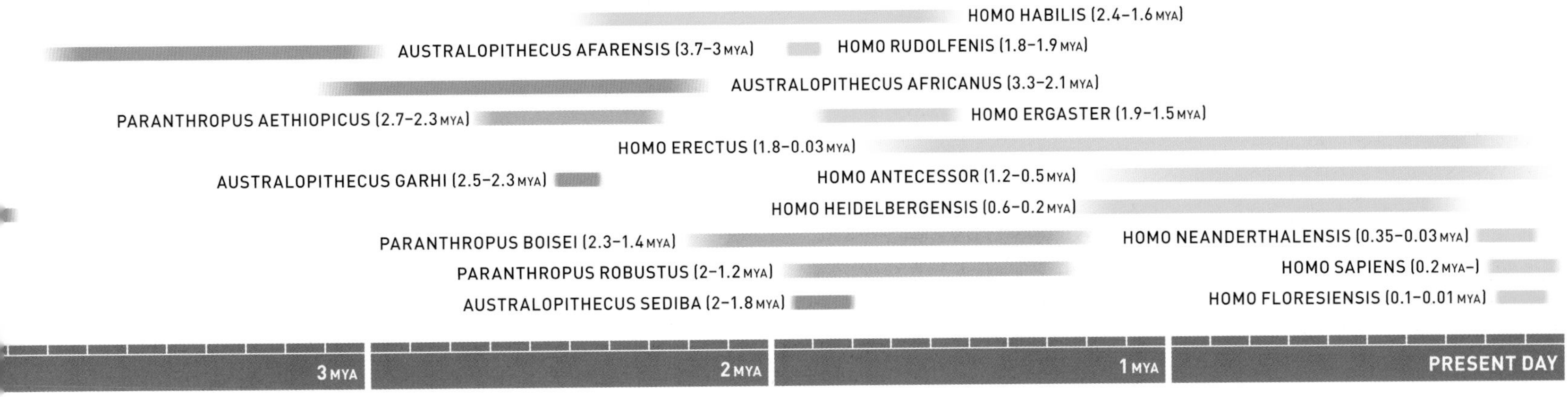

HOMO HABILIS (2.4–1.6 MYA)
AUSTRALOPITHECUS AFARENSIS (3.7–3 MYA)
HOMO RUDOLFENIS (1.8–1.9 MYA)
AUSTRALOPITHECUS AFRICANUS (3.3–2.1 MYA)
PARANTHROPUS AETHIOPICUS (2.7–2.3 MYA)
HOMO ERGASTER (1.9–1.5 MYA)
HOMO ERECTUS (1.8–0.03 MYA)
AUSTRALOPITHECUS GARHI (2.5–2.3 MYA)
HOMO ANTECESSOR (1.2–0.5 MYA)
HOMO HEIDELBERGENSIS (0.6–0.2 MYA)
PARANTHROPUS BOISEI (2.3–1.4 MYA)
HOMO NEANDERTHALENSIS (0.35–0.03 MYA)
PARANTHROPUS ROBUSTUS (2–1.2 MYA)
HOMO SAPIENS (0.2 MYA–)
AUSTRALOPITHECUS SEDIBA (2–1.8 MYA)
HOMO FLORESIENSIS (0.1–0.01 MYA)

3 MYA | 2 MYA | 1 MYA | PRESENT DAY

ASTRONOMY AND SPACE

THE PLANETS OF THE SOLAR SYSTEM

The Solar System consists of our local star, the Sun, and a large number of objects that orbit around it, including eight planets. In the inner region, nearest the Sun, there are four rocky planets: Mercury, Venus, Earth and Mars. The four outer planets are known as the gas giants: Jupiter, Saturn, Uranus, and Neptune.

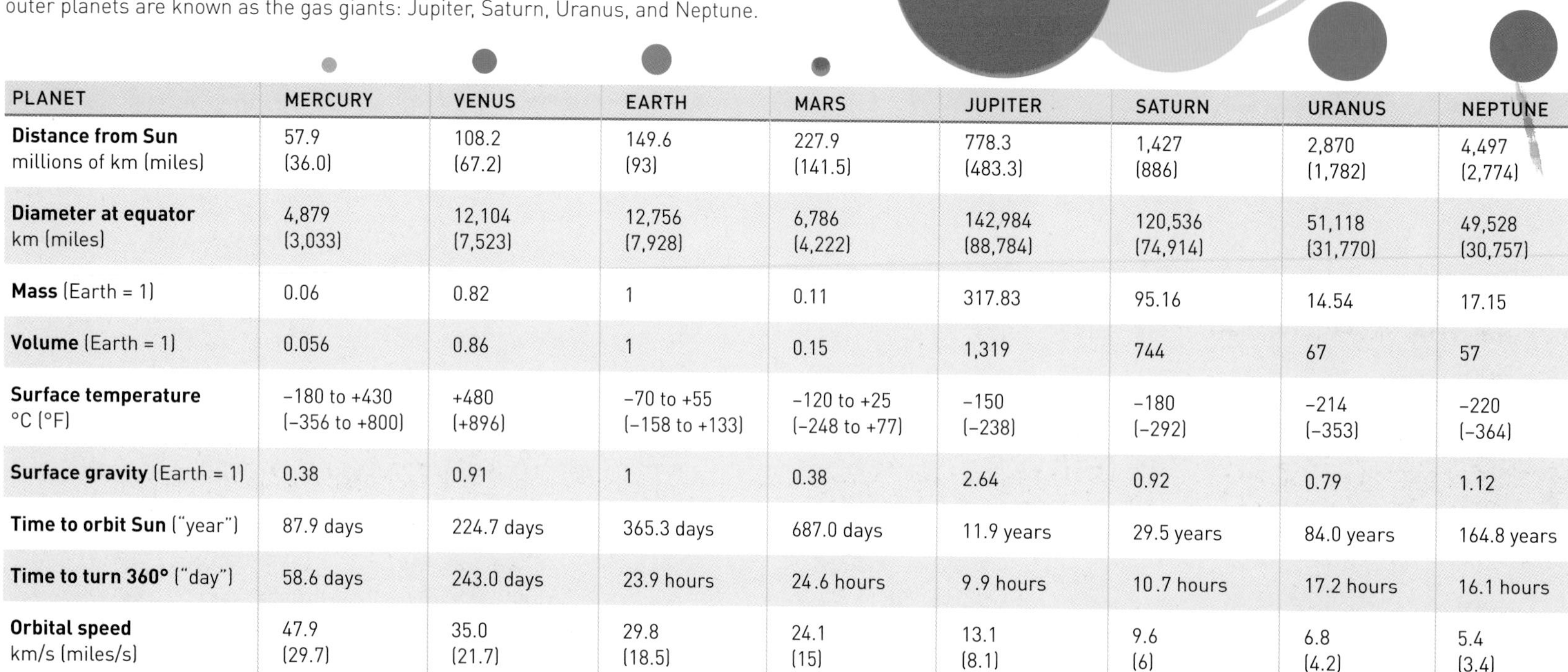

PLANET	MERCURY	VENUS	EARTH	MARS	JUPITER	SATURN	URANUS	NEPTUNE
Distance from Sun millions of km (miles)	57.9 (36.0)	108.2 (67.2)	149.6 (93)	227.9 (141.5)	778.3 (483.3)	1,427 (886)	2,870 (1,782)	4,497 (2,774)
Diameter at equator km (miles)	4,879 (3,033)	12,104 (7,523)	12,756 (7,928)	6,786 (4,222)	142,984 (88,784)	120,536 (74,914)	51,118 (31,770)	49,528 (30,757)
Mass (Earth = 1)	0.06	0.82	1	0.11	317.83	95.16	14.54	17.15
Volume (Earth = 1)	0.056	0.86	1	0.15	1,319	744	67	57
Surface temperature °C (°F)	−180 to +430 (−356 to +800)	+480 (+896)	−70 to +55 (−158 to +133)	−120 to +25 (−248 to +77)	−150 (−238)	−180 (−292)	−214 (−353)	−220 (−364)
Surface gravity (Earth = 1)	0.38	0.91	1	0.38	2.64	0.92	0.79	1.12
Time to orbit Sun ("year")	87.9 days	224.7 days	365.3 days	687.0 days	11.9 years	29.5 years	84.0 years	164.8 years
Time to turn 360° ("day")	58.6 days	243.0 days	23.9 hours	24.6 hours	9.9 hours	10.7 hours	17.2 hours	16.1 hours
Orbital speed km/s (miles/s)	47.9 (29.7)	35.0 (21.7)	29.8 (18.5)	24.1 (15)	13.1 (8.1)	9.6 (6)	6.8 (4.2)	5.4 (3.4)
Number of observed moons	0	0	1	2	64	62	27	13

KEPLER'S LAWS OF PLANETARY MOTION

These laws, first formulated by 17th-century astronomer Johannes Kepler (1571–1630), show how the planets move around the Sun. The laws demonstrate that the planets travel in eliptical, not circular, orbits (see pp.100–101) and that the farther they are from the Sun, the slower the orbiting speed.

LAW	DESCRIPTION
First law	This law, sometimes called the law of ellipses, states that planets move around the Sun in a regular oval-shaped path, known as an ellipse, with the Sun at one focus. An ellipse has two focuses along its major axis. On any particular ellipse, the total distance from one focus to any point on the ellipse and back to the other focus is always the same.
Second law	Also known as the law of equal areas, this law describes how the speed of a planet changes as it orbits the Sun. A line drawn from the center of the Sun to the center of a planet will sweep out equal areas in equal intervals of time. Therefore, a planet moves faster when it is near the Sun and slower when it is farther away from it.
Third law	Also known as the law of harmonies, the third law describes the mathematical relationship between the distances of the planets from the Sun and their orbital periods. It states that the square of each planet's orbital period (the time it takes to travel one orbit of the Sun) is directly proportional to the cube of its average distance from the Sun. This law allows orbital period and distance to be calculated for each of the planets.

SPECTRAL CLASSIFICATION OF STARS

Light from a star can be split into a band of its component wavelengths called a spectrum. The positions of dark absorption lines and bright emission lines on this spectrum indicate the chemical components in the star's atmosphere. Based on their spectra, stars are divided into seven main classes—O, B, A, F, G, K, M.

TYPE	COLOUR	PROMINENT SPECTRAL LINES	AVERAGE TEMPERATURE	EXAMPLE OF STAR
O	Blue	He^+, He, H, O^{2+}, N^{2+}, C^{2+}, Si^{3+}	45,000°C (80,000°F)	Regor
B	Blueish white	He, H, C^+, O^+, N^+ Fe^{2+}, Mg^{2+}	30,000°C (55,000°F)	Rigel
A	White	H, ionized metals	12,000°C (22,000°F)	Sirius
F	Yellowish white	H, Ca^+, Ti^+, Fe^+	8,000°C (14,000°F)	Procyon
G	Yellow	H, Ca^+, Ti^+, Mg, H, some molecular bands	6,500°C (12,000°F)	The Sun
K	Orange	Ca^+, H, molecular bands	5,000°C (9,000°F)	Aldebaran
M	Red	TiO, Ca, molecular bands	3,500°C (6,500°F)	Betelgeuse

THE MAGNITUDES OF STARS

Astronomers measure the luminosity, or brightness, of stars in units called magnitudes. The chart below is a modern scale that describes stars in terms of intensities of brightness as seen from Earth. The smaller the magnitude number, the brighter the star. The very brightest stars have negative magnitude values. Each step in the scale represents an increase or decrease in brightness of 2.5 times, so five magnitude steps correspond to an increase or decrease in brightness by a factor of about 100. Astronomers can now measure differences in brightness as small as one-hundredth of a magnitude. For the purposes of comparison, the scale shown here includes the planet Venus, which sometimes appears in the sky as a far brighter object than any star.

MAGNITUDE SCALE

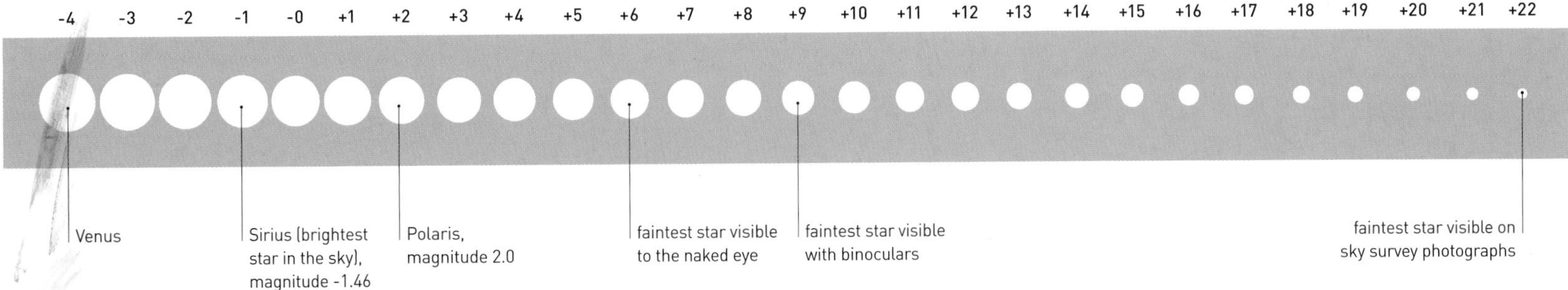
-4 -3 -2 -1 -0 +1 +2 +3 +4 +5 +6 +7 +8 +9 +10 +11 +12 +13 +14 +15 +16 +17 +18 +19 +20 +21 +22
Venus
Sirius (brightest star in the sky), magnitude -1.46
Polaris, magnitude 2.0
faintest star visible to the naked eye
faintest star visible with binoculars
faintest star visible on sky survey photographs

THE HERTZSPRUNG-RUSSELL DIAGRAM

The Hertzsprung-Russell (H-R) diagram, devised by Swedish and American astronomers Ejnar Hertsprung and Henry Norris Russell (see 1910), plots stars on a chart according to their intrinsic values: luminosity, surface temperature, magnitude, and spectral type. The chart shows that most stars obey a simple relationship between luminosity and temperature (brighter at higher temperatures) and it is one the most useful diagrams in astronomy. It also reveals that the majority of stars lie on a diagonal called the main sequence that links faint red dwarf stars with the rarer and very bright blue giants. Stars can only be seen at one stage in their incredibly long lives, and so during a human life any star will appear at only one point on the diagram. However, as hydrogen fuel in their cores is exhausted and they near the end of their lives, most stars move off the main sequence band, shifting to a new position on the diagram that is dictated by their mass.

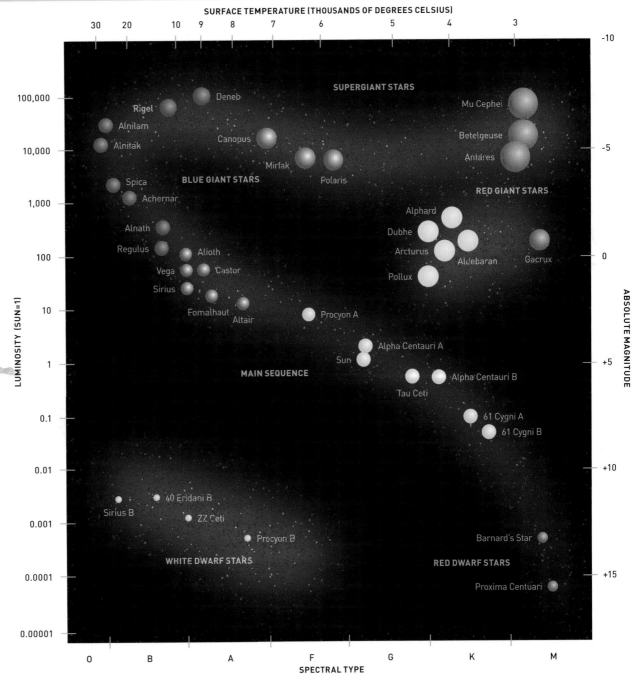

SURFACE TEMPERATURE (THOUSANDS OF DEGREES CELSIUS)
30 20 10 9 8 7 6 5 4 3
LUMINOSITY (SUN=1)
100,000 10,000 1,000 100 10 1 0.1 0.01 0.001 0.0001 0.00001
ABSOLUTE MAGNITUDE
-10 -5 0 +5 +10 +15
SUPERGIANT STARS
Deneb
Rigel
Alnilam
Alnitak
Canopus
Mirfak
Polaris
Mu Cephei
Betelgeuse
Antares
BLUE GIANT STARS
Spica
Achernar
RED GIANT STARS
Alphard
Dubhe
Arcturus
Aldebaran
Gacrux
Pollux
Alnath
Regulus
Alioth
Vega
Castor
Sirius
Fomalhaut
Altair
Procyon A
Alpha Centauri A
Sun
MAIN SEQUENCE
Tau Ceti
Alpha Centauri B
61 Cygni A
61 Cygni B
Sirius B
40 Eridani B
ZZ Ceti
Procyon B
WHITE DWARF STARS
Barnard's Star
RED DWARF STARS
Proxima Centuari
O B A F G K M
SPECTRAL TYPE

EARTH SCIENCE

THE GEOLOGICAL TIMESCALE

This timescale provides scientists with an internationally recognized chronology of Earth's history over 4 billion years. The history of Earth is divided into a hierarchical system of named units: the largest are called eons, followed in order of size by eras, periods, epochs, and ages (the latter are not included on the chart below). The timescale allows geologists to go anywhere in the world, examine the rock strata, identify the fossils within them, and give them an approximate age as they know that they are all referring to the same events, strata, and time periods.

The timescale has been developed by examining the history of global changes in ocean and atmospheric chemistry preserved in sedimentary rocks, as well as several other lines of evidence. Lithostratigraphy looks at sedimentary rock types and sequences. Biostratigraphy examines fossils—fossils in the same layer can be matched up across the world. Chronostratigraphy, or radiometric dating, calculates when certain minerals were crystalized, while magnetostratigraphy is a tool that uses the record of the changing polarity of Earth's magnetic field.

	PRECAMBRIAN						
EON	ARCHEAN				PROTEROZOIC		
ERA	EOARCHEAN	PALEOARCHEAN	MESOARCHEAN	NEOARCHEAN	PALEOPROTEROZOIC	MESOPROTEROZOIC	NEOPROTEROZOIC

millions of years ago (MYA): 4,000.0 | 3,600.0 | 3,200.0 | 2,800.0 | 2,500.0 | 1,600.0 | 1,000.0 | 541.0

EON	PHANEROZOIC															
ERA	PALEOZOIC									MESOZOIC						
PERIOD	Carboniferous						Permian			Triassic			Jurassic			
EPOCH	Mississippian			Pennsylvanian			Cisuralian	Guadalupian	Lopingian	Lower	Middle	Upper	Lower	Middle	Upper
	Lower	Middle	Upper	Lower	Middle	Upper									

millions of years ago (MYA): 358.9 | 346.7 | 330.9 | 323.2 | 315.2 | 307.0 | 298.8 | 272.3 | 259.9 | 252.2 | 247.2 | 237.0 | 201.3 | 174.1 | 163.5

MINERAL CLASSIFICATION

Most minerals are solid, naturally occurring inorganic materials with well-defined chemical compositions and characteristic crystal structures. More than 4,000 are known, although only about 100 are abundant. Minerals are classified according to their chemical composition, and are commonly divided into the groups listed below.

GROUP	APPROXIMATE MINERALS	EXAMPLES
Sulfides	600	Pyrite, galena
Silicates	500	Olivine, quartz, feldspar, garnet
Oxides and hydroxides	400	Chromite, aematite
Phosphates and vanadates	400	Apatite, carnotite
Sulfates	300	Anhydrite, barite, gypsum
Carbonates	200	Calcite, aragonite, dolomite
Halides	140	Fluorite, halite, sylvite
Borates and nitrates	125	Borax, colemanite, kernite, nitratine
Molybdates and tungstates	42	Wulfenite, wolframite
Native elements	20	Gold, platinum, copper, sulfur, carbon

EARTH'S ROCK TYPES

Rocks are naturally occurring assemblages of minerals. All of Earth's rocks can be categorized as one of the three main types: igneous, sedimentary, and metamorphic. Within each type, geologists recognize many different rocks. Much of this rocky materials is also recycled over geological time.

TYPE	DESCRIPTION
Igneous	Rocks formed by cooling and crystallizing lava or magma. They range from quick-cooled, fined-grained volcanic lava to coarse-grained rocks that have cooled more slowly.
Sedimentary	Rocks formed by the deposition of material on Earth's surface. Weathering and erosion of rock transports sediment to inland areas, where it is laid down in layers. Plant and animal fossils are found in sedimentary rock.
Metamorphic	When igneous or sedimentary rock is subjected to high temperature and pressure, it is pushed into Earth's crust, which causes it to flow and recrystallize as metamorphic rock.

EON	PHANEROZOIC													
ERA	PALEOZOIC													
PERIOD	Cambrian				Ordovician			Silurian				Devonian		
EPOCH	Terrenuvian	Series 2	Series 3	Furongian	Lower	Middle	Upper	Llandovery	Wenlock	Ludlow	Pridoli	Lower	Middle	Upper

millions of years ago (MYA): 541.0 | 521.0 | 509.0 | 497.0 | 485.4 | 470.0 | 458.4 | 443.4 | 433.4 | 427.4 | 423.0 | 419.2 | 393.3 | 382.7 | 358.9

EON	PHANEROZOIC								
ERA	MESOZOIC		CENOZOIC						
PERIOD	Cretaceous		Paleogene			Neogene		Quaternary	
EPOCH	Lower	Upper	Paleocene	Eocene	Oligocene	Miocene	Pliocene	Pleistocene	Holocene

millions of years ago (MYA): 145.0 | 100.5 | 66.0 | 56.0 | 33.9 | 23.0 | 5.3 | 2.6 | 0.01

TECTONIC PLATES

Earth's lithosphere (its crust and uppermost mantle) is divided into nine major tectonic plates, about six or seven medium-sized plates, and numerous much smaller plates called microplates.

The boundaries between the plates are of three different types: divergent, where the plates have moved apart; convergent, where they have moved together; and transform, where plates slide past one another along fault planes. The movement of divergent and convergent plates has shifted continents, open and closed oceans, and formed mountains.

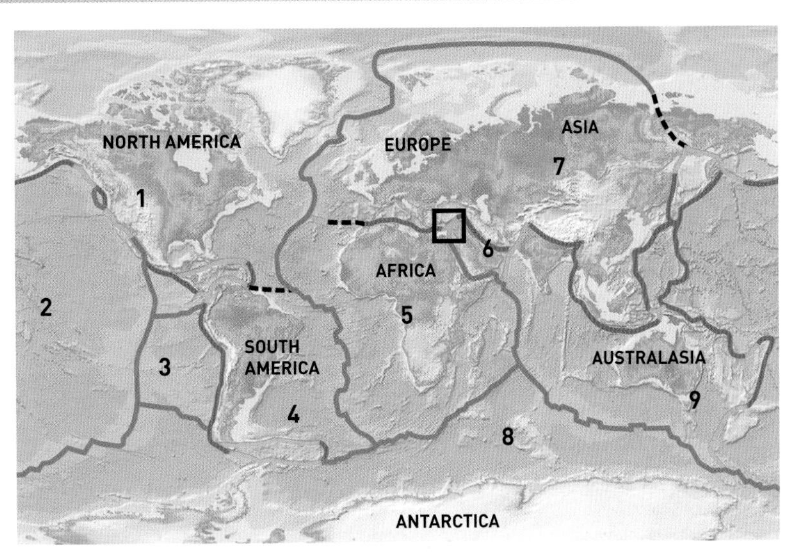
NORTH AMERICA
EUROPE
ASIA
AFRICA
SOUTH AMERICA
AUSTRALASIA
ANTARCTICA
1
2
3
4
5
6
7
8
9

MAJOR TECTONIC PLATES

1 North American Plate
2 Pacific Plate
3 Nazca Plate
4 South American Plate
5 African Plate
6 Arabian Plate
7 Eurasian Plate
8 Antarctic Plate
9 Indo-Australian Plate

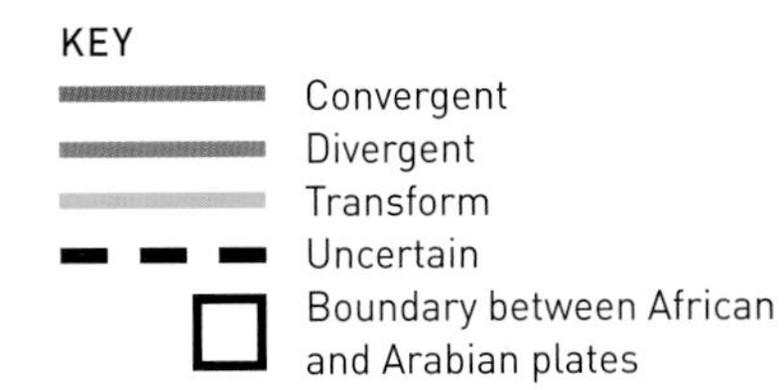
KEY
Convergent
Divergent
Transform
Uncertain
Boundary between African and Arabian plates

WHO'S WHO

The selection of people included in this Who's Who reflects the main experimenters, philosophers, and scientists represented in the book. Cross references have been included for scientists with biography panels within the main timeline pages.

Alhazen (965–1040) Arab mathematician, astronomer, and physicist widely considered to be the father of modern optics. Arguing that light enters the eye rather than being emitted from it, his influential treatise *Optics* described the laws of reflection and refraction, as well as the anatomy of the human eye. He also tried to develop realistic cosmological models.

Al-Khwarizmi (*c.*780–*c.*850) Persian mathematician, geographer, and astronomer responsible for introducing Hindu–Arabic numerals and algebra to the West. Working in Baghdad's translation and research center, the House of Wisdom, he produced two mathematical textbooks and updated Ptolemy's *Geography*, presenting coordinates for places around the world. The word "algorithm" is derived from the Latin pronunciation of Al-Khwarizmi's name.

Al-Kindi (Abu Yusuf Ya'qub Ibn 'Ishaqal-Kindi), Arab philosopher (*c.*801–*c.*873) See p.46

Al-Razi (Rhazes), Arab philosopher (*c.*865–*c.*925) See p.48

Alvarez, Luis Walter, American physicist (1911–88) See p.311

Ampère, André-Marie, French mathematician and physicist (1775–1836) See p.181

Ångström, Anders Jonas (1814–74) Swedish physicist and father of spectroscopy who discovered that hot gas emits and absorbs light at the same wavelengths at which it absorbs light when cooler. Ångström wrote on heat, magnetism, optics, and the solar spectrum and was the first to examine the spectrum of the aurora borealis. The Ångström unit (Å) for measuring atomic distances is 10^{-10} m.

Anning, Mary, British fossil hunter (1799–1847) See p.176

Archimedes (*c.*290–*c.*212 BCE) Greek inventor, philosopher, and mathematician, who stated that any object immersed in a liquid will experience an upward force equal to the weight of the liquid displaced. Archimedes wrote works on arithmetic, geometry, and mechanics, and constructed siege machines to defend Syracuse against the Romans. He is also credited with the creation of the Archimedes screw water pump.

Aristarchus of Samos (*c.*310–230 BCE) Greek astronomer who first suggested that Earth revolves around the Sun. Aristarchus's treatise *On the Sizes and Distances of the Sun and Moon* incorrectly calculated the Sun as 20 times as far from Earth as the Moon, and 20 times the size of the Moon, but his pioneering methods paved the way for future astronomical studies.

Aristotle, Greek philosopher (384–322 BCE) See p.29

Arkwright, Richard (1732–92) English textile industrialist whose water frame invention enabled the automated spinning of cotton threads. Arkwright installed his water frames in specially built factories, an early example of mass production and the Industrial Revolution.

Arrhenius, Svante August (1859–1927) Swedish physicist and chemist who was awarded the Nobel Prize in Chemistry for his electrolytic theory of dissociation. He was also the first to recognize that carbon dioxide in the atmosphere could create a greenhouse effect on Earth's surface. The lunar crater Arrhenius is named after him.

Avicenna (see Ibn Sina)

Avogadro, Amedeo (1776–1856) Italian mathematical physicist whose law states that equal volumes of gases contain equal numbers of molecules when at the same temperature and pressure. As a tribute, the number of elementary particles in a mole was called Avogadro's constant.

Babbage, Charles (1791–1871) English mathematician and inventor regarded in Britain as the pioneer of modern computers. Babbage devoted his life to building two mechanical calculating machines, including his Analytical Engine, designed to perform arithmetic using punched cards as its memory source. Neither machine was successfully completed.

Bacon, Francis, English philosopher (1561–1626) See p.98

Bacon, Roger, English scholar (*c.*1220–92) See p.60

Baekeland, Leo Hendrik (1863–1944) Belgium-born American chemist who invented Velox, the first photographic paper that could be developed under artificial light. In 1899, Baekeland sold his Velox rights to American innovator George Eastman for $1 million and used the proceeds to develop his most famous invention, Bakelite—the first synthetic plastic that could be poured into molds to harden in different shapes.

Baird, John Logie (1888–1946) Scottish engineer, inventor, and television pioneer. Baird first televised objects in 1924, moving objects in 1926, and produced the first color transmission in 1928. When the BBC began broadcasting in 1936, Baird's mechanical scanning system competed with Guglielmo Marconi's EMI electronic system, which the corporation adopted exclusively from 1937.

Banks, Joseph (1743–1820) English botanist, naturalist, president of the Royal Society, and often called Australia's first scientist. Banks traveled around the world on Captain Cook's HMS *Endeavour* and introduced many new plants to the West. Geographical features and plants have been named after him. He helped establish the Royal Botanic Gardens at Kew, London, and persuaded the government to invest in scientific exploration.

Bardeen, John (1908–91) American physicist awarded the Nobel Prize in Physics twice: in 1956 for coinventing the transistor; and in 1972 for developing the theory of superconductivity. The transistor paved the way for modern electronics, while superconductivity was used in medical advances such as MRI (magnetic resonance imaging). Bardeen was professor of electrical engineering and physics at Illinois University from 1951 to 1975.

Barnard, Christiaan Neethling (1922–2001) South African surgeon who performed the first human heart transplant. Barnard introduced open heart surgery, performing heart transplants on dogs and designing an artificial heart valve. In 1967 he performed the world's first human heart transplant on grocer Louis Washkansky, who later died from pneumonia.

Bassi, Laura, Italian physicist (1711–78) See p.137

Becquerel, Antoine-Henri (1852–1908) French physicist who shared the 1903 Nobel Prize in Physics with Marie and Pierre Curie for discovering radioactivity. He discovered radioactivity accidentally by experimenting with phosphorescence and uranium salts. This led to the isolation of radium and paved the way for modern nuclear physics.

Bell, Alexander Graham, American inventor (1847–1922) See p.217

Bell Burnell, Jocelyn, British astrophysicist (1943–) See p.296

Benz, Karl (1844–1929) German inventor who, together with Gottlieb Daimler, created the first gasoline-powered motor vehicle. Benz patented his three-wheeled, four-stroke cylinder Motorwagen in 1886 and produced the first four-wheel automobile in 1893. This laid the foundation for the motor industry and in 1899 Benz & Co. began producing the world's first racing cars.

Berg, Paul (1926–) American biochemist and cowinner of the 1980 Nobel Prize in Chemistry for developing recombinant DNA techniques for splicing and recombining DNA from different organisms, which led to modern genetic engineering.

Berners-Lee, Tim, British computer scientist (1955–) See p.324

Bernoulli, Daniel (1700–82) Swiss physicist and mathematician, who proposed that the pressure in a fluid decreases as the speed of its flow increases—Bernoulli's principle. Bernoulli's 1738 work *Hydrodynamica* was very important for kinetic theory of gases and fluids, and proposed practical applications of watermills, water propellers, and water pumps. Bernoulli also investigated medicine, biology, astronomy, and oceanography.

Bernoulli, Johann, Swiss mathematician (1667–1748) See p.124

Berzelius, Jöns Jakob (1779–1848) Swedish chemist considered to be the founding father of modern chemistry. Berzelius is noted for formulating his electrochemical theory, producing a list of atomic weights, and developing modern chemical symbols. A professor of medicine and member of the Royal Swedish Academy of Sciences, Berzelius discovered and isolated several elements, developed analytical techniques, and investigated isomerism and catalysis.

Bessemer, Henry (1813–98) English engineer who introduced the Bessemer process for creating the first inexpensive steel by blowing air through molten iron. The son of a metallurgist, Bessemer manufactured gold paint powder, invented a sugarcane crushing machine, and developed a cast-iron cannon for the Crimean War.

Biot, Jean-Baptiste (1774–1862) French physicist who established the existence of meteorites and made the first scientific balloon flight for scientific purposes. He won a Royal Society award for his study of light polarization and helped develop saccharimetry, a technique for analyzing sugar solutions. Working with fellow physicist Félix Savart, he formulated the Biot–Savart law, a fundamental component of modern electromagnetic theory.

Bjerknes, Vilhelm (1862–1951) Norwegian meteorologist and physicist who helped found modern weather forecasting. As a professor at Stockholm University, Sweden, Bjerknes studied hydrodynamics and thermodynamics and their relation to atmospheric motion. This led to the theory of air masses, an essential component of modern-day weather forecasting. He later founded the Geophysical Institute and Weather Service of Bergen in Norway.

Black, Joseph (1728–99) Scottish chemist and physician famous for discovering that fixed air (carbon dioxide) is present as a distinct gas in the atmosphere. He also discovered latent heat by showing that when ice melts, it takes up heat without changing temperature.

Bode, Johann Elert (1747–1826) German astronomer responsible for Bode's law (Titus–Bode rule) predicting the relative spacing between the Sun and its planets.

Bohr, Niels (1885–1962) Danish physicist awarded the 1922 Nobel Prize in Physics for using quantum theory to explain atomic structure. Bohr's 1913 model of the atom describes a central atomic nucleus with electrons in orbit around it. Bohr joined the Manhattan Project during World War II, but later advocated the peaceful use of nuclear energy.

Bonnet, Charles, Swiss naturalist and philosopher (1720–93) See p.148

Boole, George (1815–64) English mathematician who pioneered Boolean algebra—symbolic logic and the rules that govern it. His ideas proved vital for modern computer science.

Bosch, Carl (1874–1940) German chemist whose Haber–Bosch process for the high pressure synthesis of ammonia won him the 1931 Nobel Prize in Chemistry, and is today the standard industrial procedure for nitrogen fixation.

Bose, Satyendranath (1894–1974) Indian mathematician and physicist who collaborated with Albert Einstein in the study of quantum mechanics. Together, they developed Bose-Einstein statistics for studying the behavior of bosons (particles with integral spin values named after Bose), important for lasers and superfluid helium.

Boyle, Robert, English chemist, physicist, and inventor (1627–91) See p.111

Brahe, Tycho, Danish astronomer (1546–1601) See p.87

Bramah, Joseph (1748–1814) English locksmith noted for inventing the hydraulic press, an improved water closet, a machine for printing bank notes, and a wood planing machine. He also built a pick-proof lock—a model of which was left as a challenge in his store window and remained unpicked for 67 years, despite numerous attempts.

Brewster, David (1781–1868) Scottish physicist best known for his work in optics, including polarization, reflection, refraction, and light absorption. The invention of the kaleidoscope and an improved stereoscope popularized Brewster's name and his portrait was displayed on cigar boxes.

Broca, Paul (1824–80) French surgeon who discovered the part of the frontal lobe responsible for articulate speech—now known as Broca's area. Broca found that lesions in this area of the brain caused aphasia, which impairs the ability to form articulate words. His studies of the brain helped establish physical anthropology.

Brunel, Isambard Kingdom (1806–59) English engineer whose bridges, railroad lines, and steamships revolutionized modern engineering. Brunel helped his father build the first tunnel under the Thames River, designed the Clifton Suspension Bridge across the Avon River, and constructed the Great Western Railroad from London to Cornwall. He also built three steamships, including the *Great Western*—the first regular transatlantic passenger ship.

Buffon, Georges (1707–88) French naturalist and mathematician best known for his 36-volume *Natural History* (1749–1788). After studying law, medicine, and mathematics, Buffon devoted himself to the study of natural history and was an early proponent of evolution.

Carnot, Nicolas Leonard Sadi (1796–1832) French physicist and military engineer often considered the father of thermodynamics. Carnot's 1824 *Reflections on the Driving Power of Fire* presented the Carnot cycle, which is now considered the most efficient heat engine allowed by physical laws. Carnot's work was largely ignored until after his death, when he was credited with introducing the second law of thermodynamics.

Carson, Rachel Louise, American marine biologist (1907–64) See p.290

Cassini, Giovanni Domenico (1625–1712) Italian-born French astronomer who discovered a dark gap between two of Saturn's rings, now called Cassini's division. He also discovered four of Saturn's moons and Jupiter's Great Red Spot. Cassini was the first to regard zodiacal light as a cosmic rather than a meteorological phenomenon.

Cauchy, Augustin-Louis, Baron (1789–1857) French mathematician, writer, and a pioneer of analysis. In five textbooks and over 800 research articles, he presented innovative research on infinitesimal calculus, probability, mathematical physics, and other subjects.

Cavendish, Henry (1731–1810) English physicist and chemist and noted for his study of "inflammable air" (hydrogen). A wealthy recluse, Cavendish devoted his life to conducting scientific experiments on a wide range of topics, including chemistry, electricity, and a celebrated experiment to calculate the weight of Earth.

Celsius, Anders, Swedish astronomer (1701–44) See p.140

Chadwick, James (1891–1974) English physicist awarded the 1935 Nobel Prize in Physics for discovering the neutron, a particle without electric charge in the nucleus of an atom. He joined the Manhattan Project and was knighted in 1945.

Chandrasekhar, Subrahmanyan (1910–95) Indian-born American astrophysicist awarded the 1983 Nobel Prize in Physics for showing that white dwarf stars can exist only up to a maximum mass, the Chandrasekhar limit, of about 1.44 times that of the Sun. Initially rejected, this later helped the understanding of neutron stars, supernovas, and black holes.

Chambers, Robert (1802–71) Scottish publisher and anonymous writer of the enormously controversial *Vestiges of the Natural History of Creation* (1844). Only acknowledged posthumously as the book's author, Chambers wrote several other historical, literary, and geological titles and published the *Edinburgh Journal* and *Chambers' Encyclopaedia*.

Chappe, Claude (1763–1805) French engineer who invented a mechanical semaphore system to connect the French mainland and bring news of Napoleon Bonaparte's campaign. In 1772, Chappe and his brother Ince successfully delivered their first message between Paris and Lille. By 1774, 513 semaphore towers spanned France and parts of Europe. Chappe's system was superseded by the electric telegraph in 1846.

Chargaff, Erwin, Austrian biochemist (1905–2002) See p.279

Châtelet, Émilie du (1706–49) French physicist and mathematician noted for her translation of Isaac Newton's *Principia Mathematica*, still the only complete one. Living with her lover Voltaire, she wrote several important books on science, philosophy, and religion.

Cherenkov, Pavel Alekseyevich (1904–90) Russian physicist who shared the 1958 Nobel Prize in Physics with Igor Tamm and Ilya Frank for discovering Cherenkov radiation. Cherenkov observed that electrons emit a blue glow when traveling through a medium such as water at speeds faster than light in that medium. Based on this effect, the Cherenkov detector is used in experimental nuclear and particle physics.

Cohen, Stanley Norman (1935–) American geneticist, microbiologist, and early pioneer of genetic engineering. From 1972, Cohen collaborated with Stanford University colleagues Herbert Boyer and Paul Berg to combine and transplant genes. This led to the first genetic engineering experiment, which transferred frog ribosomal RNA into bacteria cells.

Cope, Edward Drinker (1840–97) Pioneering American paleontologist whose discovery of over 1,000 extinct Tertiary Period vertebrates helped define modern paleontology. Cope wrote over 1,200 papers on his finds, which included extinct fish and dinosaurs. From 1877, Cope competed with rival Othniel Marsh in the "bone wars" to discover the greatest number of fossils, damaging the reputation and finances of both men.

Copernicus, Nicolaus, Polish astronomer (1473–1543) See p.76

Coriolis, Gaspard-Gustave de (1792–1843) French engineer and mathematician best known for the Coriolis force, which affects movement across a rotating body, such as air masses around Earth. Coriolis dedicated his life to studying applied mechanics, friction, and hydraulics, and introduced the terms "work" and "kinetic energy" into scientific parlance.

Crick, Francis (1916–2004) British biophysicist and neuroscientist who determined the structure of deoxyribonucleic acid, or DNA, with colleague James Watson. Their discovery confirmed that DNA contained life's hereditary information and earned Crick, Watson, and biophysicist Maurice Wilkins the 1962 Nobel Prize in Physiology or Medicine.

Crookes, William (1832–1919) British chemist and physicist noted for pioneering vacuum tubes and discovering the element thallium. After inheriting a fortune, he devoted himself to scientific research, inventing the Crookes tube to investigate cathode rays, founding the journal *Chemical News*, and inventing the radiometer to convert light radiation into rotary motion.

Curie, Marie, Polish-French physicist and chemist (1867–1934) See p.233

Cuvier, Georges (1769–1832) French zoologist who established comparative anatomy and paleontology by comparing fossils with living animals. His studies proved that whole species of creatures had become extinct. He attributed mass extinctions to extreme catastrophic events, a theory known as catastrophism.

Da Vinci, Leonardo, Italian artist, architect, botanist, mathematician, and engineer (1452–1519) See p.71

Daguerre, Louis (1787–1851) French painter and physicist who perfected a process of creating permanent photographs on thin copper sheets, called daguerreotypes.

Dalton, John, British chemist and physicist (1766–1844) See p.172

Darwin, Charles, British naturalist (1809–82) See p.206

Darwin, Erasmus (1731–1802) English physician, poet, and inventor, and grandfather of naturalist Charles Darwin. Darwin was a prominent figure best known for his scientific poetry, freethinking ideas, and mechanical inventions. His *Zoonomia* outlined his radical theories on evolution.

Davy, Humphry (1778–1829) English chemist and pioneer of electrochemistry, noted for using electrolysis to isolate and discover several chemical elements, including sodium, potassium, barium, and magnesium. He also invented the Davy gas safety lamp for miners and was made a baronet in 1818.

Dawkins, Richard, British zoologist and evolutionary biologist (1941–) See p.307

Delbrück, Max (1906–81) German-born US biophysicist and pioneer of molecular biology. Trained in physics, Delbrück began working on chemistry after fleeing Nazi Germany for America in 1937. He was awarded the 1969 Nobel Prize in Physiology or Medicine with Alfred Day Hershey and Salvador Luria for their work on bacteriophages—viruses that infect bacteria and then replicate.

Descartes, René (1596–1650) French mathematician known as the father of modern philosophy. His principle "I think, therefore I am" summarizes his determination to build only on knowledge that is certain. He also founded Cartesian geometry and contributed to optics.

Diesel, Rudolf (1858–1913) German engineer famous for inventing the diesel engine, a four-stroke, vertical cylinder compression engine that made him rich. Later, he disappeared from the deck of a channel steamer, and was presumed drowned.

Diophantus of Alexandria (*c.*200–*c.*284) Greek mathematician based in Alexandria and reputed as a founding father of algebra. His only surviving work is *Arithmetica*, the earliest known treatise on algebra, which greatly influenced Islamic scholars and also the French mathematician Pierre de Fermat, who helped found modern number theory.

Dirac, Paul, British theoretical physicist (1902–84) See p.262

Dollond, John (1706–61) English optician and manufacturer of astronomical instruments. Born to Huguenot silk weavers, Dollond is best known for reducing color distortion with achromatic lenses. He also invented the heliometer, a telescope used to measure the distances between stars.

Doppler, Christian Johann (1803–53) Austrian mathematician and physicist best known for the Doppler effect, which describes how the perceived frequency of light and sound waves produced by a moving source depends on the position of the observer. In 1850, he became professor of experimental physics at the University of Vienna.

Duchenne, Guillaume (1806–75) French neurologist who studied nervous and muscular disorders and developed electrotherapy to treat diseased nerves and atrophied muscles. Duchenne became the first to use deep tissue biopsy, clinical photography, and nerve conduction tests.

Eddington, Arthur, British astronomer, mathematician, and astrophysicist (1882–1944) See p.257

Edison, Thomas Alva, American inventor (1847–1931) See p.221

Ehrlich, Paul, German bacteriologist (1854–1915) See p.247

Einstein, Albert, German-born American physicist (1879–1955) See p.242

Eratosthenes (*c.*276–*c.*194 BCE) Greek mathematician and astronomer who first calculated the circumference of Earth. He was chief librarian at Alexandria in Egypt, where he measured the tilt of Earth's axis and calculated its circumference to be 250,000 stadia. Although the value of stadia is uncertain, his estimate was within the current range. He also created a calendar that included leap years, and created a system of latitude and longitude.

Euclid (*c.*330–*c.*260 BCE) Prominent Greek mathematician, often considered the father of geometry. A teacher at the mathematical school in Alexandria, Euclid is best known for his 13-volume treatise on geometry, *Elements*. It is considered the most important mathematical textbook of antiquity and was still in general use until the 19th century.

Euler, Leonhard, Swiss mathematician (1703–83) See p.152

Fabricius, Hieronymus, Italian surgeon (1537–1619) See p.93

Fahrenheit, Gabriel Daniel (1686–1736) German physicist and engineer who invented alcohol and mercury thermometers. Fahrenheit worked as a glassblower and chemistry lecturer in the Netherlands, where he also manufactured barometers, altimeters, and thermometers. In addition to developing the Fahrenheit scale, he discovered that water can remain a liquid below its freezing point.

Falloppio, Gabriele, Italian anatomist (1523–62) See p.83

Faraday, Michael, English chemist and physicist (1791–1867) See p.192

Fermat, Pierre de, French mathematician (1601–65) See p.104

Fermi, Enrico (1901–54) Italian physicist best known for developing atomic energy. A professor of theoretical physics at the University of Rome, Fermi was awarded the 1938 Nobel Prize in Physics for his work in induced radioactivity. A leading figure in the US's Manhattan Project to build an atomic bomb, he later designed the country's first nuclear reactor.

Feynman, Richard (1918–88) American physicist and co-winner of the 1965 Nobel Prize in Physics for developing quantum electrodynamics—the theory of the interaction between light and matter. He also created pictorial representations (Feynman diagrams) of interacting particles, provided an explanation of the physics of supercooled liquid helium, and contributed to the Manhattan Project.

Fibonacci, Leonardo, Italian mathematician (*c.*1170–*c.*1250) See p.59

Flamsteed, John (1646–1719) The first Astronomer Royal of England and who helped establish the Greenwich Observatory in London. Educated at Cambridge University and ordained a clergyman, Flamsteed is noted for his 1725 *Historia Coelestis Britannica*, which cataloged 3,000 stars. His observational data helped Isaac Newton verify his gravitational theory.

Fleming, Alexander (1881–1955) Scottish bacteriologist and co-winner of the 1945 Nobel Prize in Physiology or Medicine for discovering penicillin. He is also noted for discovering the antiseptic properties of the enzyme lysozyme, and for being the first to use antityphoid vaccines on humans.

Florey, Howard Walter (1898–1968) Australian pathologist who collaborated with Ernst Boris Chain to purify, isolate, and produce penicillin for medical use, for which both scientists were awarded the 1945 Nobel Prize in Physiology or Medicine. The manufacture of penicillin began in 1943, and saved the lives of countless war casualties.

Fossey, Dian (1932–85) American zoologist famous for her 18-year-long study of mountain gorillas in Rwanda. Anthropologist Louis Leakey convinced Fossey to undertake the study, which she began in 1967. She lived reclusively among gorillas and became a leading authority on their behavior. Fossey was murdered in 1985, after ensuring worldwide media coverage of the issue of gorilla poaching.

Foucault, Jean Bernard Leon (1819–68) French physicist famous for measuring the speed of light and showing that it travels more slowly through water than air. Foucault is also noted for inventing the gyroscope and using a giant pendulum to demonstrate that Earth rotates on its axis.

Fourier, Joseph, French mathematician (1768–1830) See p.183

Franklin, Benjamin, American inventor and scientist (1706–90) See p.143

Franklin, Rosalind, British chemist and biophysicist (1920–58) See p.283

Fraunhofer, Joseph von (1787–1826) German physicist who discovered the dark lines of the Sun's spectrum, now known as Fraunhofer lines, which later helped reveal the chemical composition of the Sun's atmosphere. To observe the lines, Fraunhofer designed and constructed achromatic lenses of high magnitude. He is considered the founder of the German optical industry.

Fresnel, Augustin Jean, French engineer (1788–1827) See p.179

Freud, Sigmund (1856–1939) Austrian neurologist and founder of psychoanalysis. Freud's methods advocated dialogue and "free association" to interpret childhood dreams, recollections, and infantile sexuality. Always controversial, Freud's ideas gained importance after World War I, especially in the US, but in 1933, Hitler banned psychoanalysis and Freud fled to England.

Gabor, Dennis (1900–79) Hungarian-British engineer and physicist, awarded the 1971 Nobel Prize in Physics for inventing holography, a method of 3-D photography. Originally a research engineer in Berlin, Gabor moved to London in 1933, where he worked on optics, oscilloscopes, and television. Holograms could not be produced until lasers were invented.

Galen, Claudius, Roman physician, surgeon, and philosopher (*c.*130–*c.*210) See p.37

Galilei, Galileo, Italian natural philosopher, astronomer, and mathematician (1564–1642) See p.97

Galvani, Luigi (1737–98) Italian physiologist who discovered that he could make the muscles of a dead frog twitch by applying two pieces of metal to nerve endings in its leg. This showed that nervous messages are carried by what was called "animal electricity," later shown to be the same as the electricity produced by a battery. Galvanization, or rust prevention, is named after Galvani.

Gamow, George (1904–68) Russian-born American nuclear physicist and cosmologist who helped develop the Big Bang theory of creation. He also correctly proposed that patterns within DNA form a genetic code. Gamow authored popular science books, including the notable *Mr. Tompkins* series.

Gassendi, Pierre (1592–1655) French priest, mathematician, and philosopher who tried to reconcile an atomic theory of matter based on Epicureanism with Christian doctrine. Gassendi took the harmony of nature as proof of the existence of God and is noted for his 1642 *Objections to Descartes' Meditations*. He was the first to observe the planetary transit of Mercury in 1631.

Gauss, Carl Friedrich, German mathematician and physicist (1777–1855) See p.163

Gay-Lussac, Joseph-Louis (1778–1850) French chemist and physicist noted for his investigations into gases. An assistant to chemist Berthollet, Gay-Lussac conducted experiments on gases, vapors, temperature, and terrestrial magnetism, sometimes from an ascending hot-air balloon. He discovered the law of combining volumes of gases as well as the element boron.

Geiger, Hans (1882–1945) German physicist who developed the Geiger counter for detecting and measuring radioactivity. Working under Ernest Rutherford at Manchester University, Geiger and Ernest Marsden undertook an experiment to show that an atom has a nucleus. He later worked with his student Walther Müller to improve the sensitivity of his Geiger counter.

Gilbert, William (1544–1603) English physicist and royal physician often regarded as the father of magnetic studies. Held in high esteem by his contemporaries, Gilbert was the first to establish the magnetic nature of Earth, and to use the terms: electric attraction, electric force, and magnetic pole.

Goddard, Robert H. (1882–1945) American physicist and inventor who created the first liquid-fueled rocket. A professor at Clark University, Goddard wrote *A Method of Reaching Extreme Altitudes* (1919)—considered a classic treatise on 20th-century rocket science. He developed three-axis control, gyroscopes, and steerable thrust for rockets, and successfully launched 34 rockets between 1926 and 1941.

Goeppert-Mayer, Marie (1906–72) German-born American theoretical physicist awarded the 1963 Nobel Prize in Physics for proposing the shell theory of nuclear structure. Goeppert-Mayer is also noted for her work in quantum electrodynamics and spectroscopy, and for researching organic molecules with her husband, American chemist Joseph Mayer. She also worked on the separation of uranium isotopes for the Manhattan Project.

Golgi, Camillo (1843–1926) Italian biologist, pathologist, and co-winner of the 1906 Nobel Prize in Physiology or Medicine for his investigations into the central nervous system. Golgi's development of a silver nitrate nerve-tissue staining technique called the "black reaction," allowed him to discover a connecting nerve cell, known as the Golgi cell.

Goodall, Jane (1934–) English ethologist best known for her 45-year study on the chimpanzees of Gombe Stream National Park, Tanzania. A one-time assistant to anthropologist Louis Leakey, Goodall established her Gombe Stream camp in 1960. She discovered that chimpanzees are omnivores, capable tool-makers, and have highly complex social behaviors.

Gould, Stephen Jay (1941–2002) American paleontologist and evolutionary biologist best known for creating the theory of punctuated equilibrium with Niles Eldredge. This theory proposes that evolution undergoes periods of relative stability, punctuated by short bursts of change. A Harvard University professor and popularizer of evolutionary theory, Gould campaigned against creationism and argued that science and religion be kept as two distinct fields.

Greene, Brian (1963–) American physicist, mathematician, and advocate of string theory, which tries to reconcile relativity and quantum theory, and proposes that minuscule strands of energy are responsible for creating every particle and force in the Universe. A well-known popularizer of science, his best-selling books include Pulitzer Prize finalist *The Elegant Universe*.

Guericke, Otto von (1602–86) German physicist, engineer, philosopher, and mayor of Magdeburg. Guericke invented the air pump, with which he was able to investigate atmospheric pressure and the properties of the vacuum, which he demonstrated to Emperor Ferdinand III. In 1663, Guericke produced static electricity by rubbing a spinning sulfur globe.

Gutenberg, Johannes, German inventor (*c.*1395–*c.*1468) See p.69

Guth, Alan (1947–) American theoretical physicist, cosmologist, and creator of the inflationary universe theory, which states that a rapid period of inflation during the Big Bang caused the Universe to expand exponentially—from microscopic to cosmic.

Haber, Fritz (1868–1934) German chemist awarded the 1918 Nobel Prize in Chemistry for synthesizing ammonia—an essential component of explosives and fertilizers. Together with Carl Bosch, Haber developed a process for the mass production of ammonia for use in fertilizer, a method still widely used today. Known as the father of chemical warfare, Haber also developed poisonous gases for use in World War I.

Hadley, George (1685–1768) English physicist and meteorologist whose theory of the trade winds explained why Northern Hemisphere winds blow from the north, and Southern Hemisphere winds blow from the southeast. Now known as Hadley's principle, it remained unacknowledged from 1735 until its rediscovery by John Dalton in 1793.

Haeckel, Ernst (1834–1919) German zoologist and Darwinist who was the first to map a genealogical tree relating all forms of life. A professor at Jena University, Germany, Haeckel studied marine organisms, described and named thousands of new animal species, and created the now discarded recapitulation theory, summarized as "ontogeny recapitulates phylogeny" (evolution can be seen in embryonic development).

Hahn, Otto (1879–1968) German chemist and pioneer of radioactivity and radiochemistry. Hahn's first great breakthrough came in 1917 when he and colleague Lise Meitner discovered the radioactive element protactinium. This was followed by the discovery of nuclear fission in 1938, for which he won the 1944 Nobel Prize in Chemistry. Hahn later became an outspoken opponent of nuclear weapons.

Hales, Stephen (1677–1761) English botanist and clergyman whose pioneering research on plant and animal physiology was described in his *Vegetable Staticks*. The first to note the upward flow of sap and measure vapor emission in plants, he also measured blood pressure and blood output from the heart. His inventions included an artificial ventilator and pneumatic trough.

Halley, Edmond (1656–1742) English astronomer and mathematician who calculated the orbit of the eponymous Halley's Comet and its subsequent 1758 date of return to Earth. Later Astronomer Royal, Halley published influential papers on magnetic variation, the trade winds and monsoons. He was also responsible for the publication of Isaac Newton's *Principia*.

Harrison, John (1693–1776) English carpenter and clockmaker who invented the marine chronometer, which enabled sailors to establish their position at sea. Harrison designed and built four chronometers in response to a £20,000 government prize offered in 1714 for a way of accurately finding longitude at sea. Despite the great accuracy of his chronometers, Harrison was not paid in full until 1773.

Harvey, William, English physician (1578–1657) See p.103

Hawking, Stephen, British theoretical physicist (1942–) See p.305

Heisenberg, Werner, German physicist (1901–76) See p.259

Henry, Joseph (1797–1878) American physicist who discovered the phenomenon of self-inductance—a defining principle of electronic circuitry. His many contributions included constructing the first electromagnetic motor, developing the telegraph with Samuel Morse, and introducing an early weather forecasting system.

Herschel, Caroline (1750–1848) German-born British astronomer who had a long collaboration with her brother, William Herschel. Planning to be an opera singer, she moved to her brother's house in England at age 22, and is noted for discovering three nebulae and eight comets, as well as for completing their star catalog.

Herschel, William (1738–1822) German-born British astronomer noted for discovering Uranus in 1781. Originally a music teacher, Herschel took up astronomy and specialized in making very large telescopes. He developed a theory of nebulae and star evolution, observed and cataloged many stars, and showed that the Solar System moves through space.

Hertz, Heinrich, German physicist (1857–94) See p.224

Hertzsprung, Ejnar (1873–1967) Danish astronomer best known for his 1913 Hertzsprung–Russell diagram, a star classification system still used today. The diagram, developed with Henry Norris Russell, plotted the brightness of stars against their spectral types. He also researched open star clusters and variable stars, and developed a method for positioning double-stars.

Hevelius, Johannes (1611–87) Polish astronomer and early lunar topographer, best known for his detailed map of the Moon's surface. A city councillor of Gdansk, Hevelius built an observatory on top of his house to investigate the night sky. He cataloged over 1,500 stars, discovered several constellations, and named many lunar features.

Higgs, Peter, British physicist (1929–) See p.348

Hipparchus (*c.*170–*c.*120 BCE) Greek astronomer and mathematician often considered the founder of trigonometry. Hipparchus's contributions to astronomy include a study of solar eclipses, discovery of the precession of the equinoxes, and a description of the Sun and Moon's orbits and their distances from Earth.

Hippocrates (*c.*460–*c.*377 BCE) Greek physician widely regarded as the father of medicine. A medical pragmatist, Hippocrates based his practice on studies of the body and the symptoms and treatments of illness. He was the first to describe many diseases and coin terms such as "acute," "chronic," and "relapse." Hippocrates' code of ethics for his medical students is today known as the Hippocratic oath sworn by all doctors.

Hodgkin, Dorothy, British chemist (1910–94) See p.275

Hooke, Robert (1635–1703) English inventor and natural philosopher who, after being Robert Boyle's assistant, became Curator of Experiments at the newly established Royal Society in London. He worked on theoretical astronomy, as well as inventing a compound (two-lens) microscope to study microscopic life—producing the Royal Society's first publication: *Micrographia*. He was also the first to record biological cells. Given the range of his scientific contribution, he is widely hailed as England's Leonardo.

Hopper, Grace (1906–92) American mathematician and pioneer of computer programming and technology. A rear admiral in the US Navy, Hopper was one of the first programmers of the Harvard Mark I and helped develop UNIVAC I, the first commercial electronic computer. She also contributed to the COBOL computer language, and introduced the term "bug." US Navy's missile-destroyer, the *Hopper*, is named after her.

Hoyle, Fred, British mathematician and astronomer (1915–2001) See p.280

Hubble, Edwin (1889–1953) American astronomer considered the founder of extragalactic astronomy for discovering that the Universe is expanding. While working at the Mount Wilson Observatory, California, Hubble established that previously thought nebulae of the Milky Way were in fact different galaxies receding away from our own. The rate at which the Universe is expanding is known as the Hubble constant.

Humboldt, Alexander von (1769–1859) German naturalist, explorer, and pioneer of biogeography best known for investigating the geography and flora and fauna of Latin America with French botanist Aimé Bonpland. An enthusiastic popularizer of science, Humbolt spent 25 years writing *Cosmos*, an account of the structure of the Universe—four volumes of which were published during his lifetime.

Hutton, James, British geologist (1726–97) See p.157

Huxley, Thomas Henry (1825–95) English biologist, surgeon, and champion of Darwinism. His studies on comparative anatomy led him to conclude that birds evolved from dinosaurs. He debated evolutionary theory against Samuel Wilberforce in 1860, earning him the nickname "Darwin's bulldog." Huxley also declared himself agnostic—a term he coined.

Huygens, Christiaan (1629–95) Dutch physicist, mathematician, and astronomer known for the Huygens–Fresnel principle, which states that light is made up of waves. Huygens discovered the rings of Saturn and its fourth moon Titan, and also invented the pendulum clock and other time-keeping innovations.

Ibn Sina (Avicenna), Persian physician (980–1037) See p.50

Ingenhousz, Jan, Dutch physician (1730–99) See p.155

Isidore of Seville, Saint, Spanish theologian (*c.*560–636 CE) See p.42

Jeans, James (1877–1946) English physicist, mathematician, and astronomer. A great popularizer of astronomy, Jeans investigated spiral nebulae, multiple star systems, and giant and dwarf stars. He was also the first to hypothesize a continuous creation of matter throughout the Universe. Among his best-known books is the 1929 *The Universe Around Us*.

Jenner, Edward (1749–1823) English physician who developed a vaccine for smallpox. Learning from dairymaids, Jenner observed that a person infected with the cowpox disease would not succumb to the deadly smallpox virus. Within five years, his cowpox inoculation was in widespread use. Smallpox was declared eradicated in 1980.

Joule, James Prescott (1818–89) English physicist who provided the foundation for the theory of conservation of energy, which states that energy can change form, but cannot be created or destroyed. Joule showed that heat is energy and helped establish the mechanical equivalent of heat.

Kamerlingh Onnes, Heike (1853–1926) Dutch physicist awarded the 1913 Nobel Prize in Physics for his research on low-temperature physics and for discovering liquid helium. His work in cryogenics led him to discover superconductivity.

Kant, Immanuel (1724–1804) German philosopher whose theories on knowledge, ethics, and esthetics profoundly influenced subsequent philosophical thought. Kant attempted to reconcile the theories of rationalism (we know only what our minds can construct) and empiricism (we know only what our senses reveal) by asking "what can we know?"

Kekulé, Friedrich August (1829–96) German chemist and founder of structural theory in organic chemistry. A professor at Ghent and Bonn universities, Kekulé showed that carbon atoms can link together to form chains, which led to his later discovery of benzene's six-carbon cyclical structure.

Kepler, Johannes, German astronomer (1571–1630) See p.95

Khayyam, Omar, Persian mathematician and astronomer (1048–1131) See p.53

Koch, Robert (1843–1910) German physician awarded the 1905 Nobel Prize in Physiology or Medicine for isolating the tuberculosis bacillus. Considered one of the founders of microbiology and bacteriology, Koch also discovered the bacteria responsible for anthrax and cholera. His postulates establish four criteria for investigating the relationship between a causative microbe and a disease.

Krebs, Hans (1900–81) German-British physician and biochemist who discovered the citric acid cycle in living organisms. The discovery of this metabolic cycle, also known as the Krebs cycle, won him and Fritz Lipmann the 1953 Nobel Prize in Physiology or Medicine. Krebs also discovered the urea cycle, during which mammals convert ammonia into urea.

Lamarck, Jean-Baptiste, French biologist (1744–1829) See p.169

Laplace, Pierre-Simon (1749–1827) French astronomer and mathematician noted for his research into the stability of the Solar System and often called "the French Newton." Using calculus, Laplace reformed astronomical mathematics in his five-volume *Celestial Mechanics*, and introduced determinism into Newtonianism. Several operators and transforms are named after him.

Laue, Max von (1879–1960) German physicist awarded the 1914 Nobel Prize in Physics for studying the diffraction of X-rays in crystals. This proved important for X-ray crystallography, solid-state physics, and modern electronics. Director of the Max Planck Institute and the Institute for Theoretical Physics, Laue also researched into superconductivity, quantum theory, and optics.

Lavoisier, Antoine Laurent, French chemist (1743–94) See p.160

Lawrence, Ernest O. (1901–58) American physicist awarded the 1939 Nobel Prize in Physics for inventing the cyclotron, which accelerates particles to study subatomic interactions. Lawrence used his cyclotron to produce radioactive iodine, phosphorus, and other isotopes for medical use. A professor at University of California, Berkeley, Lawrence later contributed to the Manhattan Project. The element lawrencium is named after him.

Leakey, Louis, British archaeologist and anthropologist (1903–72) See p.289

Leakey, Mary, British archaeologist and paleontologist (1913–96) See p.308

Leavitt, Henrietta Swan (1868–1921) American astronomer who discovered the relationship between the brightness and time span of Cepheid variable stars. Leavitt worked at the Harvard College Observatory, where she examined the luminosity of stars from photographic plates, and observed that Cepheid variable stars showed a regular pattern of brightness. Her work proved crucial for measuring the distance between Earth and other galaxies.

Lee, Tsung-Dao (1926–) Chinese-born American physicist and co-winner of the 1957 Nobel Prize in Physics for discovering violations of the law of parity conservation, which led to important developments in particle physics. Lee created a solvable model of quantum field theory, called the Lee model, and helped study the violations of time-reversal invariance.

Leeuwenhoek, Anton van (1632–1723) Dutch microscopist often considered the father of microbiology. Originally employed in the textile trade, Leeuwenhoek built and used microscopes, becoming the first to observe single-celled organisms, including bacteria and protozoa, as well as muscle fibres and blood flow in capillaries.

Leibniz, Gottfried von (1646–1716) German philosopher and mathematician who made major contributions to physics, metaphysics, optics, logic, statistics, mechanics, and technology. Leibniz developed calculus independently of Isaac Newton, built a calculating machine, and refined the binary system, which forms the foundation of digital technology. He published no major philosophical treatises.

Lenard, Philipp (1862–1947) German physicist awarded the 1905 Nobel Prize in Physics for his research on cathode rays. A professor at four German universities, Lenard supported Nazi doctrine and denounced "Jewish" science, including Einstein's theory of relativity.

Liebig, Justus von (1803–73) German chemist whose pioneering work in the fields of organic chemistry, biochemistry, and agriculture contributed to the establishment of the fertilizer industry. Appointed professor at Giessen University at 21 years old, Liebig was the first to establish the laboratory-based teaching methodology that spread to the US and the rest of Europe.

Lind, James (1716–94) Scottish physician who tried to eradicate scurvy from the British navy by introducing citrus juice to the shipboard diet. Although the Navy was slow to adopt his ideas, he also introduced fumigation below decks, better hygiene for sailors, and the distillation of seawater into drinking water.

Linnaeus, Carolus (Carl von Linné), Swedish naturalist (1707–78) See p.139

Lippershey, Hans (c.1570–c.1619) Dutch eyeglass-maker commonly credited with inventing the telescope. In 1608, Lippershey sold his invention to the Dutch government for use in warfare. Later astronomers, notably Galileo, recognized the telescope's importance for science. A planet and lunar crater were named after Lippershey.

Lister, Joseph (1827–1912) British surgeon and founder of antiseptic medicine. A professor and president of the Royal Society, Lister pioneered the principle of bacteria prevention during surgery, using carbolic acid to sterilize surgical instruments, and keeping postoperative wounds clean.

Lockyer, Joseph (1836–1920) English astronomer known for discovering the element helium in the Sun's atmosphere and naming it. Originally a civil servant, Lockyer observed solar prominences in the Sun's chromosphere, devised the spectroscopic observation of sunspots, and founded the periodical *Nature*.

Lodge, Oliver (1851–1940) English physicist noted for his pioneering work in wireless telegraphy. Lodge is best known for improving detector devices for transcribing Morse code radio waves onto paper. A keen promoter of spiritualism, Lodge also received patents for several wireless inventions.

Lomonosov, Mikhail, Russian chemist, physicist, geographer, and astronomer (1711–65) See p.145

Lonsdale, Kathleen (1903–71) Irish crystallographer who developed X-ray techniques to investigate chemical structures. Lonsdale established the hexagonal shape of carbon atoms in benzene, and determined the structure of hexachlorobenzene. In 1945, Lonsdale became the first woman to be elected a Royal Society fellow.

Lord Kelvin (see Thomson, William)

Lorentz, Hendrik Antoon (1853–1928) Dutch physicist who shared the 1902 Nobel Prize in Physics with Pieter Zeeman for their work on electromagnetic radiation. The first to describe the force of charged particles within an electromagnetic field, Lorentz's analyzed how events may be perceived at different times in different frames of reference, and developed the transformation equations that underpinned Einstein's theory of relativity.

Lorenz, Konrad (1903–89) Austrian founder of ethology and co-winner of the 1973 Nobel Prize in Physiology or Medicine for research on animal behavior. Lorenz is noted for studying imprinting in birds, and also animal aggression, which he argued is motivated purely by survival.

Lovelace, Ada, British mathematician (1815–52) See p.197

Lovelock, James (1919–) English chemist best known for his 1979 Gaia hypothesis, which proposes that Earth is a living organism "maintained and regulated by life on the surface." An ardent environmentalist, Lovelock invented the electron capture detector to reveal chlorofluorocarbons in the atmosphere.

Lyell, Charles (1797–1875) Scottish geologist who proposed that the geological features of Earth's surface were shaped by processes still operating at the same rate as in the past. His theory of uniformitarianism, presented in *Principles of Geology* (1803–33), was vital for Charles Darwin's theories because it provided a greatly expanded time frame for Earth's history.

Malpighi, Marcello (1628–94) Italian physician and biologist who founded the science of microscopic anatomy through his study of plant and animal tissue. Personal physician to Pope Innocent XII and a pioneer of brain anatomy, Malpighi named capillaries, contributed to embryology, discovered taste buds, and investigated the anatomy of frog lungs.

Malthus, Thomas Robert (1766–1834) English economist, clergyman, and philosopher who argued that natural growth in human population will always outstrip the food supply. To preserve humanity, Malthus proposed strict limits on reproduction or that overpopulation be left to be checked by war or famine. His theory, known as Malthusianism, greatly influenced social, political, and economic thought.

Mandelbrot, Benoit (1924–2010) Polish-born French-American mathematician who introduced the Mandelbrot set and fractal geometry, which shows how visual complexity can be created from simple shapes. A professor at Yale University, he examined many phenomena which, like a rocky coastline, seem equally rough or jagged however close or far away you get.

Marconi, Guglielmo (1874–1937) Italian physicist and inventor of the wireless telegraph. Marconi sent the first wireless signal across the English channel in 1896 and across the Atlantic Ocean in 1902. He shared the 1909 Nobel Prize in Physics with Ferdinand Braun and helped develop shortwave wireless communication.

Margulis, Lynn, American biologist (1938–2011) See p.300

Maudslay, Henry (1771–1831) English inventor and engineer considered a founding father of the machine-tool industry. Originally a locksmith's apprentice, Maudslay invented many important machines during the Industrial Revolution, such as the metal lathe, marine engines, and methods for desalinating seawater and printing calico cloth.

Maxwell, James Clerk, British physicist (1831–79) See p.209

Mayer, Julius Robert von (1814–78) German physicist, physician, and early founder of thermodynamics. Mayer was the first to determine the mechanical equivalent of heat, although this was credited to James Joule. He also described oxidation as the primary energy source for living creatures.

Mendel, Gregor (1822–84) Austrian monk and botanist whose plant experiments laid the foundation for modern genetics. Experimenting with garden peas, Mendel discovered that the characteristics of an individual are controlled by hereditary factors, now called genes. The significance of Mendel's findings was not recognized until the early 20th century.

Mendeleev, Dmitri, Russian chemist (1834–1907) See p.211

Mercator, Gerardus, Flemish cartographer (1512–94) See p.73

Michell, John (1724–93) English clergyman, astronomer, and pioneer of seismology. In 1760, Michell proposed that earthquakes were wave motions in Earth's crust, and in 1790, he created a torsion balance to measure Earth's density.

Michelson, Albert Abraham (1852–1931) Polish-American physicist who accurately measured the speed of light. His experiments with Edward Morley to detect the drift of an ether were important for the understanding of Einstein's relativity theory. He was awarded the Nobel Prize in Physics in 1907, the first American to receive a scientific Nobel prize.

Millikan, Robert (1868–1953) American physicist awarded the 1923 Nobel Prize in Physics for measuring the electrical charge of the electron with his oil-drop experiments. Millikan also confirmed Einstein's photoelectric equation and conducted studies on the nature of cosmic rays, X-rays, and electric constants.

Mitchell, Maria (1818–89) First American woman to work as a professional astronomer, and discoverer of a comet named after her. Mitchell became director of the Vassar Female College's Observatory in 1865, and she also founded the Association for the Advancement of Women.

Montagu, Lady Mary Wortley, English writer (1689–1762) See p.131

Morgan, Thomas Hunt (1866–1945) American geneticist and biologist whose research on the Drosophila fruit fly helped establish genetics. Morgan's experiments showed that genes are arranged on chromosomes and are responsible for hereditary traits. He received the Nobel Prize in Physiology or Medicine in 1933.

Moseley, Henry (1887–1915) English physicist who used X-ray spectroscopy to prove the theory of atomic numbers. Working under Ernest Rutherford at Manchester University, Moseley confirmed physically the atomic numbers for elements that had been derived chemically. Like Mendeleev, his research enabled him to predict elements for gaps in the periodic table. Moseley was killed in World War I.

Murchison, Roderick Impey (1792–1871) Scottish geologist best known for establishing the Silurian, Permian, and Devonian geological time periods. Murchison's findings were regarded as the crowning achievement of 19th-century geology and saw him elected president of the Geological Society in 1831.

Muybridge, Eadweard (1830–1904) English photographer known for his pioneering work in photographing motion. An establishd landscape photographer, Muybridge captured images of horses in full gallop by using up to 24 cameras and fast shutter speeds. He presented apparently moving images of animals using his zoopraxiscope.

Nakamura, Shuji, Japanese electronic engineer and inventor (1954–) See p.331

Napier, John (1550–1617) Scottish mathematician who invented logarithms. The 8th Laird of Merchistoun, Napier introduced the use of the decimal point in fractions, developed logarithms for mathematical calculations, and devised a set of calculating rods called Napier's bones. He also designed secret weapons to defend Scotland against a perceived Catholic attack.

Newcomen, Thomas (1663–1729) English engineer and inventor of the first practical steam engine. Developed in conjunction with Thomas Savery, the Newcomen engine was originally used to pump water from a coal mine in Dudley, Staffordshire. Over the next 75 years, hundreds of Newcomen engines greatly increased coal production in England and contributed significantly to industrialization.

Newton, Isaac, English physicist and mathematician (1642–1727) See p.118

Nightingale, Florence (1820–1910) English nurse who reformed hospitals and founded modern nursing. Dubbed "the lady with the lamp" for her nursing work during the Crimean War, Nightingale founded the nurses training school at London's St Thomas' Hospital in 1861. She also helped improve public health in India and introduced new statistical techniques.

Nobel, Alfred (1833–96) Swedish chemist who invented dynamite, a less sensitive form of nitroglycerine, and founded the Nobel Prizes. He willed the majority of his fortune to creating the Nobel Prize, an annual award for achievements in physics, chemistry, medicine, literature, and peace.

Noether, Emmy (1882–1935) German mathematician and pioneering leader of abstract algebra. Appointed a lecturer at Göttingen University in 1919, she won acclaim for her research on noncommutative algebras and the general theory of ideals in rings. She emigrated to the US to escape the Nazis.

Ockham, William of, German philosopher (c.1285–c.1349) See p.65

Ørsted, Hans Christian (1777–1851) Danish chemist and physicist who showed that electricity and magnetism are related by observing the needle of a magnetic compass move when close to a wire carrying an electric current. The unit of magnetic induction was named after Ørsted.

Ohm, Georg Simon (1789–1854) German physicist and discoverer of Ohm's law, which uses the concept of resistance to formulate the relationship of current and voltage. Ohm's law was so badly received at the time that he resigned as professor. Its value was recognized later.

Olbers, Heinrich Wilhelm (1758–1840) German astronomer and physician who carried out theoretical work on comets, and discovered two asteroids and five comets. Olbers' paradox, which asks why the sky is dark at night, remained unanswered during his lifetime.

Oppenheimer, Robert (1904–67) American physicist best known as the father of the atomic bomb. Oppenheimer investigated subatomic particles, before becoming director of the Manhattan Project in 1941 under General Groves. Although he won the Presidential Medal of Merit in 1946, Oppenheimer was accused of communism in 1953. He received the Enrico Fermi Award in 1963 as a gesture of reconciliation.

Otto, Nikolaus August (1832–91) German engineer who invented the four-stroke internal combustion engine. Otto's prize-winning engines offered a practical alternative to steam power. Described theoretically by the Otto cycle, four-stroke engines were used by Karl Benz and Gottlieb Daimler in the first motorcars.

Oughtred, William (1574–1660) English mathematician and teacher who invented the slide rule. His popular and influential textbook *The Key to Mathematics* (1631), introduced symbols such as "x" for multiplication.

Owen, Richard (1804–92) English anatomist and paleontologist who coined the word Dinosauria or "terrible reptile." Owen published several texts on dinosaurs, classifying them differently from other reptiles. Owen helped establish London's Museum of Natural History, and although he believed in evolution, was an outspoken opponent of Darwin's theory.

Papin, Denis (1647–c.1712) French physicist and inventor whose steam digester led to the development of steam engines. Papin also invented a steam safety valve, a condensing pump, and a paddle-wheel boat.

Paracelsus (1493–1541) Swiss physician and alchemist who established the use of chemistry in medicine. Traveling and practicing medicine across Europe, Paracelsus introduced laudanum, sulfur, lead, and mercury as medicinal remedies and gave a clinical description of syphilis. An outspoken opponent of university medicine, he gained huge influence by writing and speaking in German.

Pascal, Blaise, French mathematician and physicist (1623–62) See p.107

Pasteur, Louis, French chemist, biologist, and microbiologist (1822–95) See p.214

Pauli, Wolfgang (1900–58) Austrian-born American theoretical physicist awarded the 1945 Nobel Prize in Physics for his Pauli exclusion principle, which states that no two electrons in an atom can exist in the same quantum state simultaneously. Pauli also devised an atomic model of the thermal properties of metal, and was the first to propose the existence of neutrinos.

Pauling, Linus Carl, American chemist (1901–94) See p.271

Pavlov, Ivan Petrovich (1849–1936) Russian physiologist whose experiments on dogs led to his discovery of the conditioned reflex. Pavlov showed that dogs salivate in anticipation of food, not just at the sight of it. He was awarded the 1904 Nobel Prize in Physiology or Medicine and summarized his work on behaviorism in the 1926 book, *Lectures on Conditioned Reflexes*.

Perkin, William (1838–1907) English chemist celebrated for creating the first synthetic dye, aniline purple, which became extremely fashionable. While synthesizing quinine, Perkin came across a bluish dye now called aniline purple, which he patented and manufactured, enabling him to retire at 35.

Petit, Alexis Therese (1791–1820) French physicist who discovered the Dulong–Petit Law with Pierre Dulong. This states that for all solid elements, the specific heat multiplied by the atomic weight is a constant. He also designed a thermometer to measure the dilation coefficients of metals.

Planck, Max, German physicist (1858–1947) See p.236

Plato, Greek philosopher (424 BCE–348 BCE) See p.25

Poincaré, Henri, French mathematician (1854–1912) See p.227

Priestley, Joseph (1733–1804) English chemist and clergyman who discovered several gases including the one later identified as oxygen. Learning the theories of electricity from Benjamin Franklin, Priestley began his own electrical experiments and presented his findings in the popular 1767 work *The History and Present State of Electricity*. He then experimented with gases and made important discoveries, although he believed in the phlogiston theory that was later discarded.

Proust, Joseph-Louis (1754–1826) French chemist best known for formulating the law of definite proportions (Proust's law), which states that in any compound, the elements are present in a fixed proportion by weight.

Ptolemy (Claudius Ptolemaeus) (c.100–c.170 CE) Greek astronomer and geographer whose Ptolemaic system placed Earth at the center of the Universe and incorporated complex epicycles. Based in Alexandria, Ptolemy also made a map of the world and wrote the encyclopedia *Almagest*.

Pythagoras (580–500 BCE) Greek philosopher and mathematician whose teachings contributed to mathematics and rational philosophy. Pythagoras taught that nature and the world could be interpreted through numbers, and greatly influenced Plato and Aristotle. He is credited with discovering the chief musical intervals and the Pythagorean theorem of geometry.

Raman, Chandrasekhara Venkata (1888–1970) Indian physicist awarded the 1930 Nobel Prize in Physics for work on the scattering of light, called the Raman effect. This shows that when light passes through transparent material, a small proportion of the deflected light changes in wavelength (that is, in energy).

Ramón y Cajal, Santiago, Spanish histologist and neuroscientist (1852–1934) See p.229

Ramsay, William (1852–1916) Scottish chemist awarded the 1904 Nobel Prize in Chemistry for discovering the inert gases argon, neon, xenon, and krypton. He also discovered the rare gas radon and isolated helium from liquid air.

Ray, John (1627–1705) English naturalist and botanist whose contributions helped found modern taxonomy. A fellow of Cambridge's Trinity College, Ray lost his position during the Restoration and began studying botany and zoology across Europe. He set out his classification of plants in *Historia Plantarum*, which established the species as the basic unit of taxonomy.

Réaumur, René, French physicist and entomologist (1683–1757) See p.134

Rhazes (see Al-Razi)

Richter, Charles (1900–85) American physicist, seismologist, and developer of the logarithmic Richter scale, which records the magnitude of an earthquake at its epicenter. Richter also devised a map of the most earthquake-prone areas in America.

Rømer, Ole Christensen (1644–1710) Danish astronomer who established that light travels at a finite speed. Rømer calculated the speed of light to be 140,000 miles (225,000 km) per second, around 47,000 miles (75,000 km) per second slower than modern estimates. Rømer also invented a temperature scale and introduced the first Danish system for weights and measures.

Röntgen, Wilhelm (1845–1923) German physicist and recipient of the first Nobel Prize in Physics in 1901 for discovering X-rays. A professor of physics, Röntgen researched elasticity, capillarity, polarized light, and the specific heat of gases. His 1895 discovery of X-rays was enormously important for medicine and modern physics.

Rumford, Benjamin Thompson (1753–1814) American-born British physicist, inventor, soldier, and administrator, best known for his work on heat. Rumford correctly theorized that heat was produced by the motion of particles, rather than being a liquid form of matter as thought previously. He helped to found the Royal Institution of London in 1799.

Russell, Henry Norris (1877–1957) American astronomer who helped establish the modern science of theoretical astrophysics. Russell is known for discovering the relationship between a star's magnitude and its spectral type, which he presented in the 1910 Hertzsprung–Russell diagram. He also theorized an abundance of hydrogen in stellar atmospheres—now considered a fundamental tenet of modern cosmology.

Rutherford, Ernest, New Zealand-born chemist and physicist (1871–1937) See p.248

Salam, Abdus (1926–96) Pakistani nuclear physicist and co-winner of the 1979 Nobel Prize in Physics for formulating the electroweak theory, which unifies the weak nuclear force and electromagnetic interactions of elementary particles. Salam was a professor of theoretical physics in London, and the first Muslim scientist to win a Nobel prize.

Salk, Jonas Edward (1914–95) American physician and medical researcher who discovered the first effective vaccine for polio. After working on an influenza vaccine at Michigan University, Salk began human trials of his polio vaccine in 1952. In 1955, the vaccine was released for use in America, virtually eradicating polio.

Sanger, Frederick (1918–) English biochemist and the only person to be awarded the Nobel Prize in Chemistry twice. Sanger's 1958 Prize was for his research on the structure of proteins, in particular insulin. His 1980 Prize was for his method of sequencing DNA molecules, which was used to develop the first fully sequenced DNA-based genome.

Schrödinger, Erwin (1887–1961) Austrian theoretical physicist and co-recipient of the 1933 Nobel Prize in Physics for his contributions to quantum mechanics. Schrödinger is best known for his equations of wave mechanics, but his book *What is Life?* (1948) greatly influenced molecular biology.

Schwann, Theodor (1810–82) German physiologist who founded histology by proposing that all organisms are composed of cells. He discovered the digestive enzyme pepsin and the cells that surround nerve axons. He helped disprove the theory of spontaneous generation, and coined the term "metabolism."

Semmelweis, Ignaz (1818–65) Hungarian physician who pioneered the use of antisepsis to prevent deaths caused by puerperal fever. Although he showed that the high mortality rates associated with childbirth could be reduced by physicians washing their hands in chlorinated lime, this practice was not introduced until many years later.

Servetus, Michael, Spanish physician (c.1511–53) See p.80

Shen Kuo (1031–95) Polymathic Chinese scholar who discovered magnetic declination and described the first magnetic needle compass. Shen's finding is one of the many recorded in his famous book, *Brush Talks from Dream Brook*. He also described movable type, formulated a geological hypothesis about fossils, and undertook an ambitious project of mapping the stars.

Shockley, William Bradford (1910–89) American physicist who shared the 1956 Nobel Prize in Physics with John Bardeen and Walter Brattain for inventing the transistor. Professor of Engineering at Stanford University, Shockley commercialized his transistor, which led to the development of California's Silicon Valley. He later caused controversy by advocating eugenics and proposing sterilization for those with low IQs.

Shoujing, Guo (1231–1316) Chinese engineer astronomer, and mathematician, known for his Shoushi Calendar, which accurately presented 365 days in a year. Shoujing also invented an astrological compass, built hydraulic clocks, engineered the Kunming Lake reservoir in Beijing and developed spherical trigonometry.

Siemens, Werner von (1816–92) German electrical engineer remembered for his role in developing the telegraph industry. Inventor of the electric dynamo and an electroplating process, Siemens laid the first telegraph line in Germany and co-founded the telegraph company that is now called Siemens AG. The unit of electrical conductance bears his name.

Smith, William (1769–1839) English geologist and engineer who founded the science of stratigraphy. While working as a canal-site surveyor across Britain, Smith studied regional rock strata and the fossils within each layer, which enabled him to establish geological time periods. Smith produced the first geological map of England and Wales, in 1815.

Snell, Willebrord (1580–1626) Dutch physicist and mathematician credited with discovering the law of refraction. In 1617, he presented a method of measuring Earth by triangulation, and in 1621, developed his law of refraction.

Snow, John (1813–58) English physician and pioneer of modern epidemiology. Snow is best known for showing cholera to be a waterborne disease—a theory he published in 1839 and confirmed in 1854 through his investigation of London's Broad Street pump outbreak. He also promoted gaseous anesthesia after administering chloroform to Queen Victoria.

Somerville, Mary (1780–1872) Scottish astronomer, geographer, and popularizer of science. With little formal education, Somerville won acclaim for her 1831 translation of Pierre-Simon Laplace's *The Mechanism of the Heavens*. Celebrated as the "Queen of the Sciences" for her numerous and wide-ranging books, in 1835, she and Caroline Herschel became the first women members of the Royal Astronomical Society.

Sørensen, Søren Peder Lauritz (1868–1939) Danish biochemist famed for introducing the pH scale for expressing hydrogen ion concentration as a measure of acidity. Sørensen also contributed to the chemical technology of Denmark's spirits and explosives industries.

Spallanzani, Lazzaro (1729–99) Italian biologist and physiologist noted for his experimental research on animal reproduction and bodily functions. Spallanzani discredited the theory of spontaneous generation and showed that living cells use oxygen and give off carbon dioxide. He also proved that mammal reproduction requires semen and an ovum, and was the first to artificially inseminate a dog.

Spitzer, Lyman (1914–97) American theoretical physicist and astronomer who made significant contributions to the study of interstellar matter, plasma physics, and the dynamics of star clusters. Spitzer's 1946 proposal of a space telescope led to the development of the Hubble telescope, and he helped design the ultraviolet astronomy satellite, Copernicus.

Stahl, Georg (1660–1734) German physician and chemist who founded the phlogiston theory of combustion, which states that all substances that burn contain a substance called phlogiston. Although later discarded, this theory was accepted for decades because it was very useful, especially in the mining industry.

Swammerdam, Jan (1637–80) Dutch microscopist who helped found the fields of comparative anatomy and entomology. After designing a dissecting microscope, he recorded his observations of the structure, classification of insects, and metamorphosis. The first to describe red blood cells, he also discovered Swammerdam valves in the lymphatic vessels.

Swan, Joseph (1828–1914) English physicist and chemist whose early incandescent light bulb predated that of Thomas Edison. While an assistant at a chemical manufacturing firm, Swan made important contributions to photography, and later legal disputes between Swan and Edison over the light bulb led to the Edison & Swan United Electric Light Company partnership.

Talbot, William Henry Fox (1800–77) English chemist and photography pioneer. Talbot is best known for his calotype photographic process that produced negatives from which prints could be taken. Talbot had 12 patents to his name and published over 50 articles on mathematics, astronomy, and physics. His book, *The Pencil of Nature*, was the first to feature photographic illustrations.

Tansley, Arthur (1871–1955) English ecologist and conservationist who coined the term ecosystem. Tansley advocated the study of plants within their natural communities—an approach central to modern ecology. Tansley's best-known book was the 1939 *The British Islands and their Vegetation*.

Tesla, Nikola (1856–1943) Serbian engineer and pioneering inventor in the fields of electricity and radio transmission. Tesla emigrated to America in 1884 where he worked with Thomas Edison and sold patents to George Westinghouse. He invented the Tesla coil transformer, the induction motor, and discovered the rotating magnetic field. The unit of magnetic induction bears his name.

Thomson, Joseph John (1856–1940) English physicist who discovered the electron. A professor of experimental physics at Cambridge University, Thomson also developed the mathematical theory of electricity and magnetism, discovered the natural radioactivity of potassium, and invented the mass spectrometer. He received the 1906 Nobel Prize in Physics for his study of conduction of electricity through gases.

Thomson, William (Lord Kelvin) (1824–1907) Scottish physicist and thermodynamic pioneer known for determining the correct value of absolute zero. Appointed a professor at Glasgow University at 22, Thomson developed the second law of thermodynamics and the electromagnetic theory of light, determined the value of absolute zero, and helped lay the first transatlantic telegraph cables. Deeply religious, he used his estimate of Earth's age to argue against evolution by natural selection.

Trevithick, Richard (1771–1833) English engineer and inventor of the first successful steam-powered railroad locomotive. Trevithick's engines were first used to power stationary mill and mine machinery, but he invented the first self-propelled road vehicle in 1801, and the railroad locomotive in 1803.

Tull, Jethro, English agronomist and inventor (1674–1741) See p.126

Turing, Alan (1912–54) English mathematician widely regarded as the father of computer science. During World War II, Turing developed the "Bombe," a prototype for electronic computers, which helped to crack the German enigma code. His development of the theoretical Turing machine, the Automatic Computing Engine, and the Ferranti Mark I paved the way for modern computing.

Venter, Craig, American biologist (1946–) See p.347

Vesalius, Andreas (1514–64) Flemish physician and founder of modern anatomy at Padua University. Vesalius's dissections of human cadavers informed his *The Seven Books on the Structure of the Human Body*, which included many detailed illustrations of internal human anatomy. He revolutionized anatomical teaching by insisting on the importance of close observation and the use of human cadavers.

Villasante, Manuel Losada (1929–) Spanish biologist and biochemist noted for his research on the photosynthetic assimilation of nitrogen. Losada's work focuses on biochemical and biological systems that can transform solar energy into chemical energy.

Virchow, Rudolf Carl (1821–1902) German physician and founding figure of modern pathology. Virchow popularized the expression "every cell is derived from a cell" and showed that disease occurs as a result of changes in normal cells. A pioneer of social medicine, he advocated the advancement of public health.

Volta, Alessandro (1745–1827) Italian physicist and inventor of the voltaic pile—the first electric battery to produce an electric current. A professor of physics at Pavia University, Italy, Volta also invented the electrophorus—a device that produced a static electric charge, and was the first to isolate methane gas. The unit of electrical potential—the volt—bears his name.

Vries, Hugo de (1848–1935) Dutch botanist known for his research into the nature of mutations in plant breeding. While a professor at Amsterdam University, Vries coined the terms "mutation," "isotonic," and "pangene" (later shortened to gene), and rediscovered Gregor Mendel's laws of heredity, while being unaware of Mendel's work at the time.

Waals, Johannes Diderik van der (1837–1923) Dutch physicist awarded the 1910 Nobel Prize in Physics for his equation of state for gases and liquids. The equation explained why real fluids do not obey the ideal gas laws at high pressures. His work led to the liquefaction of hydrogen and helium, and the study of temperatures near absolute zero.

Wallace, Alfred Russel, British naturalist (1823–1913) See p.203

Warburg, Otto Heinrich (1883–1970) German biochemist and physician awarded the 1931 Nobel Prize in Physiology or Medicine for his research on cancerous tumors and the respiration of cells. Director of the Kaiser Wilhelm Institute, Warburg discovered the nature and mode of action of the respiratory yellow enzyme, outlining his research in *The Metabolism of Tumors* (1931).

Watson, James Dewey (1928–) American geneticist and co-discoverer of the double helix structure of deoxyribonucleic acid (DNA). Watson shared the 1962 Nobel Prize in Physiology or Medicine with Francis Crick and Maurice Wilkins for the discovery. Watson became Director of the Cold Spring Harbor Laboratory, and took a leading role in the Human Genome Project.

Watt, James, British engineer and inventor (1736–1819) See p.150

Wegener, Alfred, German geophysicist and meteorologist (1880–1930) See p.252

Weinberg, Steven (1933–) American physicist and co-recipient of the 1979 Nobel Prize in Physics for his work in formulating the electroweak theory. Weinberg's theory, published in his 1967 article *A Model of Leptons* explains that electromagnetic and weak forces are indistinguishable at extremely high temperatures, such as those occurring during the Big Bang.

Weismann, August (1834–1914) German biologist and founder of the modern science of genetics. Weismann is known for his germ plasm theory, which states that all living things are born with a special and stable hereditary substance. A supporter of Darwin, Weismann opposed the idea of the inheritance of acquired characters.

White, Gilbert (1720–93) English naturalist, clergyman and author of *The Natural History and Antiquities of Selborne*. White's observation journals on his garden have attained English classic status.

Wöhler, Friedrich (1800–82) German chemist who became the first to synthesize an organic compound, urea, from an inorganic substance, ammonium cyanate. Professor of chemistry at Göttingen University, Germany, Wöhler also discovered calcium carbide, isolated the elements silicon and beryllium, and developed a way of preparing metallic aluminum.

Yalow, Rosalyn Sussman (1921–2011) American medical physicist and co-recipient of the 1977 Nobel Prize in Physiology or Medicine for developing radioimmunoassay. This is a method of measuring minute substances in the blood, such as hormones, enzymes, and vitamins. Director of the Berson Research Laboratory, Yalow was the second woman to win a Nobel Prize in this field.

Yukawa, Hideki (1907–81) Japanese physicist awarded the 1949 Nobel Prize in Physics for his theory of elementary particles. Yukawa's prediction of the existence of the meson, a subatomic particle hundred times heavier than the electron, would inform later research on nuclear and high-energy physics. He joined other scientists in signing the Russell–Einstein Manifesto for nuclear disarmament in 1955.

Zhang Heng (78–139) Chinese geographer, mathematician, and astronomer who invented a device to record any earthquake within 310 miles (500 km) by causing a ball to drop from a model dragon into a frog's mouth below, thus making a sound. Zhang also calculated the value of pi, created a comprehensive star map, and explained solar and lunar eclipses.

GLOSSARY

Terms defined elsewhere in the glossary are in *italics*.

aberration Any of various defects that may occur in an image formed by a lens or mirror.

absolute scale Also called Kelvin scale, a temperature scale starting at *absolute zero*.

absolute zero The lowest possible temperature (0 K, –459.67°F, or –273.15°C), when there is no random energy of movement in atoms and molecules.

absorption (1) The taking up of one substance by another. (2) The capture of electromagnetic radiation by matter.

acceleration The rate of change of velocity.

acid A compound containing hydrogen that splits up in water to give reactive hydrogen ions.

acoustics (1) The study of sound. (2) The properties of a particular space, such as a concert hall, in terms of how sound travels around it.

active transport In biology, any transport of substances across cell membranes that requires energy input.

acupuncture A medical treatment originating in China, which involves fine needles being inserted into the skin at particular points.

adaptation Any inherited aspect of an organism's structure or behavior that helps fit it to its environment; also, the evolutionary process giving rise to such features.

ADP Adenosine diphosphate, a compound formed when *ATP* releases energy.

adrenal gland Either of two glands situated one on top of each kidney.

alchemy A medieval science that tried, among other things, to change different metals into gold.

algae (singular: **alga**) Simple water-living organisms that produce their food by photosynthesis. They include single-celled forms as well as large seaweeds.

algebra A branch of mathematics that involves performing calculations using letters and other general symbols.

algorithm Set of rules by which calculations can be performed automatically, especially by a machine, to get a particular result.

alkali A *base* that dissolves in water.

alkaline A solution with a pH greater than 7.

allotropes Different forms of the same element. For example, graphite and diamond are allotropes of carbon.

alloy A metal that is a mixture of more than one element, either all metallic elements or with non-metals mixed in.

alternating current An electric current whose direction reverses at regular intervals.

alternator An electric generator that produces an alternating current.

amino acid Any of a group of small molecules that are the building blocks of proteins. They also have various other roles in the body.

amniocentesis Obtaining a sample of the fluid surrounding a baby in the womb by passing a hollow needle through the mother's body wall under anesthetic.

amp (ampere) The SI unit of electric current.

amplitude The size of a vibration or the height of a wave.

anaphase The stage of *mitosis* or *meiosis* where *chromosomes* or chromatids separate from one another.

anatomy The study of the internal structure of living things.

anesthesia The production of pain relief by loss of sensation or consciousness. Drugs that achieve this are called anesthetics.

angle of incidence The angle between a light ray hitting a surface and an imaginary line perpendicular to that surface.

angle of reflection The angle between a light ray reflected from a surface and an imaginary line perpendicular to that surface.

anion A negatively charged *ion*.

anode A positive *electrode*.

anther A pollen-producing structure of a flower, which together with its supporting stalk (filament) makes up a stamen.

antibiotic A drug used to kill or inhibit the growth of bacteria that cause infections.

antibodies Proteins produced by the body that identify and attack foreign particles, such as invading bacteria.

antigen Anything that stimulates the body to produce antibodies, such as the outer coat of an invading microorganism.

antiparticle A version of a subatomic particle that is opposite in electrical charge to the normal version.

aphasia The inability to produce and/or understand speech.

area The size of a two-dimensional surface.

arithmetical progression A sequence of numbers each differing from the previous one by the same amount.

armillary sphere A metal model in the form of an open sphere, representing the apparent movements of the Sun, stars, planets, etc. as seen from Earth.

artery A blood vessel leading away from the heart. See also *vascular circulation*.

asteroid A small rocky body orbiting the Sun. See also *meteoroid*.

asthenosphere The relatively soft upper layer of Earth's mantle, below the *lithosphere*.

astrolabe A historical astronomical instrument used by astronomers and sailors for locating the Sun, Moon, planets, and stars.

astronomical unit A unit of distance used in astronomy equal to the distance between Earth and the Moon.

astronomy The scientific study of space and the Universe beyond Earth's atmosphere.

atmosphere (1) The gases that surround the Sun, Earth, and some planets. (2) A measurement of pressure.

atmospheric pressure The normal pressure of the air, especially near the ground.

atom The smallest part of an element that has the chemical properties of that element. An atom consists of a nucleus of protons and neutrons surrounded by orbiting electrons.

atomic mass Also called atomic weight, a measure of the relative amount of material in different kinds of atoms. (Hydrogen atoms have the smallest atomic mass.)

atomic number The number of protons in the nucleus of an *atom*. All atoms of the same element have the same atomic number.

atomic theory Any theory that states that matter is made up of atoms.

atomic weight See *atomic mass*.

ATP Adenosine triphosphate, an important energy-carrying *molecule* in all living cells.

aurora borealis A display of lights in the night sky in Arctic regions, caused by the impact of electrically charged particles from the Sun with Earth's atmosphere.

axis (1) The imaginary line about which a body, such as a planet, rotates. (2) A reference line on a graph.

axle A structural rod upon which a wheel or wheels revolve, or a rod which revolves along with the wheels.

background radiation See *cosmic background radiation*.

bacteriophage A virus that attacks bacteria, called phage for short.

bacterium (plural: **bacteria**) A member of a kingdom of single-celled microscopic life forms whose cells lack nuclei. See also *prokaryotic cell*.

barometer An instrument for measuring air pressure.

base (1) A substance that reacts with an *acid* to form a salt. (Soluble bases are called alkalis.) (2) Any of the four similar molecules found repeatedly in molecules of DNA, whose order "spells out" the genes of living things. (3) In mathematics, the specific number that forms the basis of how numbers are conventionally written down. For example, in everyday life 10 means "ten" (= one ten and no units) but in the binary system 10 means two (= one two and no units); this latter system is also called "base 2".

base pair A matching pair of complementary bases opposite each other on the strands of a DNA molecule.

battery Originally, two or more voltaic cells connected together; now often just means a single voltaic cell.

beta decay A form of radioactive decay in which beta particles (fast-moving electrons or positrons) are given out.

Big Bang The moment, estimated at about 13.8 billion years ago, when the present Universe is thought to have begun by explosion and expansion from a tiny point.

binomial system The standard system of giving a two-part scientific name to each biological species. For example, the human species is named *Homo sapiens*.

biodegradable Able to be broken down by natural biological processes.

biomass (1) The amount of living material of a specific kind or in a given area. (2) Non-fossil plant material, such as wood, usable as fuel.

biosphere The surface regions of Earth where living things are found.

bit In computing, a fundamental unit of information having just two possible values, as either of the binary digits 0 or 1.

black body A theoretical object that can absorb all electromagnetic radiation, and can also emit radiation of all wavelengths, dependent only on its temperature.

black hole A super-dense body of matter with gravity so intense that not even light can escape from it.

blastocyst A hollow ball of cells that is an early stage in the formation of an embryo.

blood type Also called blood group, any of several categories into which blood can be classified, defined by differences in the surface chemistry of red blood cells.

blood vessel An *artery*, *vein*, or *capillary*. See also *vascular circulation*.

boiling point The temperature at which a particular liquid changes into a gas.

bond A binding connection between atoms.

botany The study of plants.

Brownian motion The random movement of tiny particles in a liquid or a gas, caused by molecules colliding with them.

buoyancy The tendency of an object to rise upward in a fluid (liquid or gas) when the object is less dense than the fluid.

byte A unit of information storage and transmission in computing and telecommunications. A kilobyte is a thousand bytes, a megabyte is a million bytes, and a gigabyte is a billion bytes.

calculus (1) A branch of mathematics based on calculations involving tiny infinitesimal changes. It comprises differential calculus, which is concerned with rates of change, and integral calculus, which can be used to calculate areas, volumes, etc. (2) A medical name for a hard mass formed in the body, such as a kidney stone.

calendar round cycle In the Maya civilization, a cycle of 52 years after which the two separate Mayan calendar systems become aligned with one other.

calx A powdery or crumbly substance left when a mineral or metal has been burnt.

capacitor A device used to store electric charge temporarily.

capillaries The tiny blood vessels that supply tissues and connect the arteries and veins. See also *vascular circulation*.

carbon A chemical element (symbol C, atomic number 6) which forms more compounds than any other element, including the important chemicals of life.

carbon cycle The cycling of carbon through the living and non-living parts of Earth and its atmosphere.

cartography The science and practice of mapmaking.

catalyst A substance that speeds up a chemical reaction without being changed itself at the end of the reaction.

cathode A negative *electrode*, from which electrons flow or are emitted.

cathode ray tube A kind of vacuum tube incorporating a fluorescent screen, best known from its use in televisions and monitors before the era of flat screens.

cauterization Destroying tissues by applying heat: used in medicine, especially in the past, to remove small growths or stop bleeding.

celestial body A natural body in space, such as a planet or star.

celestial sphere Imaginary sphere on which the stars seem to lie when seen from Earth.

cell (1) The "unit of life": a tiny structure composed of genes, surrounding fluid that carries out chemical reactions, and an enclosing membrane. See also *eukaryotic cell*, *prokaryotic cell*. (2) See *voltaic cell*.

cell division The process by which one cell splits to produce two daughter cells.

Celsius scale The temperature scale in which, under normal conditions, water freezes at 0° and boils at 100°.

centrifuge A device used to separate substances of different densities by spinning them at high speed.

cerebellum Part of the brain near the back of the skull. Its primary role is to control the detailed coordination of movements.

cerebrum The largest part of the brain in mammals, and in humans responsible for most conscious thought and activity.

chain reaction A chemical or nuclear reaction in which the product of one step triggers the next step, which in turn triggers the next step, and so on.

chaos theory A mathematical theory for analyzing complex systems whose behavior is very dependent on initial conditions, for example weather systems.

charged particle A small particle which has a net positive or negative electrical charge.

chlorophyll The green pigment found in plants that absorbs light to provide the energy for photosynthesis.

chloroplasts The chlorophyll-containing structures in the cells of plants and *algae* where photosynthesis occurs.

chromatid One of two identical strands of a chromosome. During cell division, the strands part and become separate *chromosomes*.

chromosomes Structures within living cells that contain copies of the genes of an organism. Each chromosome consists of a single long DNA molecule combined with various proteins. For example, humans have 23 pairs of chromosomes, with a complete set present in nearly every cell of the body.

circuit See *electric circuit*.

circulation See *vascular circulation*.

circumference The distance round the perimeter of an object or shape.

climate The average weather conditions of a region over a long period.

clone An identical copy or set of copies. Depending on the context, it can refer to: a copied DNA molecule; a set of identical descendants of a given cell; an animal bred artificially using the cell nucleus of an adult.

cloud chamber An early form of apparatus for detecting subatomic particles.

codon A sequence of three adjacent *bases* that forms part of the genetic code. Most codons represent the code for adding a specific *amino acid* to a protein being synthesized in the cell.

cohesion The force of attraction between two particles of the same substance.

coke A solid fuel, mostly carbon, obtained by heating coal in the absence of air.

combustion A chemical reaction (burning) in which a substance combines with oxygen, producing heat energy.

comet Any of millions of bodies in the outer Solar System consisting of a mixture of rocky particles and ice. A comet becomes apparent when it orbits close to the Sun, its evaporating ice and dust particles producing a visible tail.

companion planting Growing plants of different crops together for mutual benefit.

compass Any of various devices to indicate the direction of north and south.

compound A *molecule* or chemical substance made up of atoms of two or more elements bonded together.

concave Curving or projecting inward.

concentric spheres Hollow spheres, arranged one outside the other, with the same center.

conductor A structure or material that conducts electricity and/or heat easily.

cones Light-sensitive cells in the retina of the eye of humans and some other animals that make it possible to see colors.

conic section Any of several mathematically important curves and shapes, produced by intersecting a cone with a plane surface.

conjugation In bacteria, the transfer of genetic material by direct cell-to-cell contact.

conjunctiva Mucous membrane covering the inside of the eyelids and the front of the eye.

conservation of energy The principle that energy can neither be created nor destroyed, but only changed from one form to another.

constellation A named pattern of stars that astronomers use when referring to different regions of the sky.

continental drift The moving of Earth's continents with respect to each other over millions of years, resulting from plate tectonic activity.

convection The transfer of heat through a fluid by currents within the fluid.

convergent boundary The line along which two tectonic plates that are moving toward each other meet. See *plate tectonics*.

convergent evolution The phenomenon by which unrelated species evolve similar features as a result of *adaptation* to similar environments or ecological niches.

convex Curving or projecting outward.

Coriolis effect The deflection of winds and ocean currents by the rotation of Earth.

corona The outer atmosphere of the Sun or another star.

cosine (cos) The ratio of the adjacent side of an angle to the hypotenuse in a right-angled triangle. Also, the mathematical function describing how this ratio varies when the angle changes as the hypotenuse sweeps round a circle.

cosmic background radiation Microwave radiation coming from all directions of outer space that represents a relic of the Big Bang.

cosmic rays High-energy particles bombarding Earth from space.

cosmological principle The principle that the Solar System and Earth are not positioned in any special or central place in the Universe.

cosmology The study of the Universe on the largest scale.

coulomb The SI unit of electric charge.

covalent bond A chemical bond formed by atoms sharing one or more electrons.

cross-fertilization Fertilization of a plant by a different member of the same species (in contrast to self-fertilization).

crystal A solid whose constituent atoms, ions, or molecules are arranged in a regular, repeating geometrical pattern.

cubic equation A mathematical equation containing at least one variable number multiplied by itself twice (for example, x×x×x, also written x^3), but no variables multiplied more times than this.

cumulus clouds Rounded fluffy clouds that form when moisture-containing air rises.

cuneiform A form of writing using wedge-shaped impressions made in clay, characteristic of some ancient civilizations.

curvature of space The idea derived from relativity that, viewed on a large scale, space itself is curved, not in three straight dimensions as common sense would suggest.

curve In mathematics, a plot on a graph of one quantity against another, or a line that represents a particular geometrical shape.

dark energy A little-understood theoretical phenomenon proposed to explain why the expansion of the Universe is accelerating.

dark matter Matter that cannot be detected by conventional means but which must exist in galaxies to explain their gravitational properties. Its properties mean that it cannot be made of atoms as we understand them.

dead reckoning Navigating by estimating speed and direction only, without using other checks such as astronomical observations.

decibel The standard unit for measuring sound intensity.

decomposition (1) The decay of organic material. (2) In chemistry, a reaction that breaks a larger *molecule* into smaller ones.

detector In electronics, the circuit in a radio receiver that separates out the sound signal from a radio wave.

differential calculus See *calculus*.

differentiation The type of calculation carried out in differential calculus.

diffraction The bending of waves around obstacles or the spreading out of waves when they pass through a narrow aperture.

diffusion The spreading of one substance through another by random motions of its atoms or molecules.

digital A term referring to the storage and transmission of information (such as sound or video information) using patterns of discrete units, such as the 0 and 1 values of the binary system.

digital sound Sound recorded digitally.

diode An electronic component that lets electricity flow in one direction only.

dioptre A unit of refractive power of a lens. See *refraction*.

diploid cell A cell containing two copies of each chromosome.

DNA Short for deoxyribonucleic acid, the large *molecule* that carries the genetic information in all living things, except for some viruses which use *RNA*.

double helix A double spiral: the term is used particularly to refer to the two intertwining strands of a DNA molecule.

driving mechanism A mechanism that transmits mechanical movement and power.

dwarf planet An astronomical object (which includes the former planet Pluto) that is big enough to have become rounded by its own gravity, but not big enough to have cleared surrounding space of objects.

dye A substance that colors a material.

dynamics Branch of physics, which studies the movement of objects when under the influence of forces.

dynamo A generator that produces direct current.

eclipse The temporary hiding of one astronomical object behind another, especially the Sun behind the Moon when seen from Earth (a solar eclipse), or Earth between the Moon and the Sun, when Earth's shadow falls on the Moon (a lunar eclipse).

ecliptic The curved path, representing the plane of the Solar System, through which the Sun and planets appear to move through the skies over the course of a year.

ecology The study of the relationships between organisms and their environment.

ecosystem A community of living things considered together with the interactions between them and their physical environment.

egg (1) A female sex cell (gamete), also called an ovum. (2) A structure that protects the growing embryo in birds and other animals.

elasticity The tendency of a substance to "bounce back" to its original shape or volume when an applied force is removed.

electric charge A basic property of many subatomic particles that makes them interact electromagnetically. Charge can be either positive or negative.

electric circuit A complete loop of conducting material that carries an electric current and connects electrical devices such as switches and light bulbs.

electric current The flow of electrical energy.

electric motor A device in which electrical power is converted into rotational mechanical power.

electrical resistance Resistance to the flow of electricity, usually resulting in heat being given off.

electrode An electrical terminal, which can conduct electricity into or out of a system.

electrolysis Chemical change or breakdown in an electrolyte caused by electrical current being passed through it.

electrolyte A substance that conducts electricity when molten or in solution.

electromagnetic induction See *induction*, senses 2 and 3.

electromagnetic radiation Waves of energy that are in the form of electric and magnetic fields vibrating at right angles to each other.

electromagnetic rotation Mechanical rotation produced by electromagnetic means.

electromagnetic spectrum The complete range of electromagnetic radiation, including (from the highest to the lowest frequency and energy): gamma rays, X-rays, ultraviolet radiation, visible light, infrared radiation, microwaves, and radio waves.

electromagnetism The physics of the electromagnetic field created by the interaction of electricity and magnetism.

electromotive force (emf) The potential difference of a battery or generator, which "pushes" an electric current around a *circuit*.

electron A tiny subatomic particle with a negative electric charge. Electrons are leptons, with about a thousandth of the mass of a proton or neutron. They orbit the nuclei of atoms in a cloud, and their movement is responsible for electric currents in circuits.

electron micrograph Permanent magnified image of an object obtained using an electron microscope.

electron microscope A microscope that uses a beam of electrons instead of light to obtain a magnified image of an object.

electron shell One of the layers in which electrons orbit around the nucleus of an atom.

electron volt A small unit of energy, convenient when discussing the energies of subatomic particles.

electrophoresis A technique for analyzing and separating large molecules and small particles by using the different speeds at which they move through a medium when electricity is supplied.

electroscope An instrument demonstrating the presence of electric charge.

electrostatic Relating to stationary electric charges.

electrostatic field The field of force surrounding a stationary electrically charged object.

element In chemistry, a substance whose atoms are all of the same kind (that is, all have the same number of protons in their nuclei).

ellipse A flat symmetrical oval shape or outline, like a flattened circle.

embryo An early stage in the development of a new individual (animal or plant). In humans, an embryo more than eight weeks old is called a fetus.

emission lines Bright lines in the spectrum of light emitted by a body, usually indicating the presence of particular elements.

endangered species A species of living thing that is at risk of becoming extinct.

endocrinology The study of hormones and of the endocrine glands that produce them.

endoscope Any of various instruments for directly viewing inside the body.

energy Traditionally described as the capacity to do work, energy is difficult to define fully but can be regarded as the agent that can produce change in the Universe.

entanglement In quantum physics, the linking of two particles as one, so that when the particles move apart, a change in one instantly causes a change in the other.

environment The surroundings of living things, sometimes including the living things themselves.

enzyme A catalyst in living things that increases the speed of a particular biochemical reaction. There are thousands of different kinds, nearly all of them proteins.

equinox Either of the twice-yearly moments when the Sun crosses the celestial equator and the day-length of the Northern and Southern Hemispheres is momentarily equal.

escapement Mechanism in a clock or watch that allows its motive power to be released in a way that provides exactly timed motion.

eukaryotic cell The typical cell of an animal or plant, in which the genes are contained in a nucleus. See also *prokaryotic cell*.

evolution The gradual process by which living things develop and change over a long period of time.

evolutionary biology The study of evolution and its related areas of biology.

exoplanet A planet that orbits a star other than the Sun; also known as an extrasolar planet.

exothermic Of a chemical reaction, resulting in the release of heat.

Fahrenheit scale The temperature scale named after Gabriel Fahrenheit, in which, under normal conditions, water freezes at 32° and boils at 212°.

faience Pottery decorated with glazes.

fermion Any of the group of subatomic particles associated with matter, such as electrons, quarks, and protons, rather than those that carry force, such as bosons.

Ferrel cell The circulation of the atmosphere in midlatitudes that brings westerly winds at surface level and returning easterly winds at high level.

fertilization The joining of two gametes as the first stage of producing a new organism.

fetus The unborn, developing offspring of a mammal. In humans, it covers pregnancy after the first eight weeks.

field of force A condition produced in the space around (for example) a magnet (magnetic field) or an electric charge (electric field), which can be diagrammed as curved lines showing the direction in which any nearby object influenced by these forces will tend to be moved.

filament See *anther*.

fissile A term referring to certain atomic nuclei, such as one type of uranium, which are capable of being split into two roughly equal parts when bombarded with neutrons.

flintlock mechanism A mechanism for discharging a gun by using the spark obtained from a flint striking metal.

fluid A substance that can flow, including solids, liquids, or plasmas.

FM Frequency modulation. The transmission of a signal by changing the frequency of the carrier wave, such as a radio wave.

formula (plural: **formulas** or **formulae**) (1) In chemistry, a set of symbols that represent the makeup of a substance. (2) A set of mathematical symbols expressing a rule, principle, or method for finding an answer.

fossil The long-preserved remains of a living thing, especially when it has been mineralized (turned to stone).

Fraunhofer lines Dark lines in the spectrum of the Sun and other stars. They show where particular chemical elements in a star's outer layers are absorbing light from the stars.

freezing point The temperature at which a given liquid freezes. Freezing point also depends on pressure.

friction A force that resists or stops the movement of objects in contact with each other. The friction between an object and a fluid, such as air or water, is known as drag.

fundamental particle Also called an elementary particle, any subatomic particle such as electrons, which it is believed does not consist of simpler particles. Electrons are an example, but not protons or neutrons since they are made of quarks.

fuse (1) A safety device used in electrical circuits, such as a thin wire which melts if too much current passes through it. (2) A cord or other device that can be ignited or activated and used to set off an explosive.

Gaia hypothesis The concept that all the living things and physical components of Earth interact to form a complex self-regulating system, like a huge organism.

galaxy A huge grouping of stars, dust, and gas, all loosely held together by gravity. Our galaxy is called the Milky Way.

gamete A sex cell, such as a sperm or egg cell. Gametes have half the number of *chromosomes* of most other cells (see *haploid cell*) so that, when they join together in fertilization, the normal number of chromosomes is restored.

gametocytes Cells that represent an early stage in the production of gametes.

ganglion Concentration of nerve cell bodies, more so outside the central nervous system.

Geiger counter An instrument used to detect and measure radioactivity.

gene The basic unit of inheritance in living things, a segment of DNA (or RNA in some viruses) that typically codes for making a particular protein and also incorporates features allowing it to be switched on and off.

gene map A plot of the sequence of genes along an entire strand of DNA.

gene sequencing Finding out the order of *bases* in the DNA of particular genes.

generator A device that converts mechanical energy into electrical energy.

genetic code The code by which sequences of DNA "spell out" the recipe for making a particular protein. See also *codon*.

genetic drift Change in the overall genetic composition of a population as a result of random events rather than natural selection.

genetic engineering Techniques of artificially modifying an organism's characteristics by manipulating its genetic material.

genetic fingerprinting The analysis of a DNA sample to identify who it belongs to.

genome The complete set of genes for an organism.

genotype The genetic makeup of an organism.

geological period One of the time divisions (such as the Jurassic period) into which Earth's history is divided.

geostationary Term applied to a satellite orbiting Earth at the same rate as Earth rotates, so staying above one particular point on the surface.

geothermal Relating to the internal heat of Earth, or energy obtained from it.

germ theory The theory that infectious living agents (germs) cause many diseases.

glaciation Coverage of land areas by glaciers and ice caps.

gluons Particles within protons and neutrons that hold their component quarks together.

gravitational force The force of gravity. It is regarded as one of the four fundamental forces in the Universe.

gravitational lensing Phenomenon where the gravity of a large astronomical object can bend light coming from another behind it, sometimes resulting in several images of the more distant object being visible.

gravity The tendency of every body possessing mass to attract every other body.

greenhouse effect Heating effect due to heat rays from the ground being absorbed by some of the gases in the atmosphere.

greenhouse gases Gases that lead to the greenhouse effect, including water vapor, carbon dioxide, and methane.

habitat The environment where a particular living thing occurs naturally.

Hadley cell The circulation pattern of the atmosphere in the tropics, which brings north- and southeasterly trade winds toward the equator at surface level and returns westerly winds at high level.

half-life (1) The time taken for the radioactive emission of any particular radioactive material to drop to half of its initial value. (2) The time taken for a drug in the body to reduce to half of its original concentration.

haploid cell A cell containing only one copy of each chromosome.

hemoglobin An iron-containing protein that is the carrier of oxygen in the blood.

heredity The passing on of characteristics from one generation to the next.

hertz (Hz) The SI unit of frequency. One hertz is one cycle per second.

histology The study of tissues of the body.

homeobox A sequence of DNA that is part of the genes controlling body development in animals, plants, and other organisms.

hominid Any member of the primate family *Hominoidea*, including humans.

horology The science of clockmaking and of measuring time.

H-R diagram Hertzsprung–Russell diagram. A chart showing how stars evolve from one type to another over time.

html Hypertext markup language. The main computer language used on websites.

http Hypertext transfer protocol. The call and response system used to link websites to the *Internet*.

Hubble's law This law states that a galaxy's distance is proportional to the speed at which it is moving away from us; the more distant the galaxy, the faster it is receding. This shows that the Universe is expanding.

Human Genome Project The worldwide science project completed in 2003 to map the entire sequence of genes in human DNA.

hydraulic pressure The pressure created by a fluid, for example, when pushed through a pipe.

hydraulics The study and phenomena of liquids flowing through pipes, especially when used as a source of power.

hydrocarbon A chemical compound made up of carbon and hydrogen only.

hydrogen The lightest, most abundant chemical, which makes up about 75 percent of the total mass of elements in the Universe.

hydrostatics The branch of physics that studies the pressure and equilibrium of liquids at rest.

imaginary number Any number that is a multiple of the square root of –1, which does not exist as a "normal" number.

imaging Any method of producing images, especially when done indirectly by analyzing X-rays, magnetic responses of materials, etc.

immune system The body's natural defense mechanisms, which react to foreign material such as microorganisms, with effects such as inflammation and antibody production.

immunization The priming of the body's immune system, by *inoculation*, to fight against future infection.

indeterminate equation A mathematical equation which has more than one solution.

induction Any of the processes by which (1) an object becomes electrically charged when near another charged object; (2) a magnetizable object becomes magnetized when in the presence of an electric field, including one produced by an electric current; (3) an electric current is produced in a *circuit* by a varying magnetic field.

inertia The tendency of an object to remain at rest or to keep moving in a straight line until a force acts on it.

infrared radiation A type of electromagnetic radiation with a wavelength just longer than that of visible light but shorter than that of microwaves. It is often experienced as heat.

inheritance The pattern or manner in which genetic characteristics are passed on.

inhibitor In chemistry and biology, a substance that prevents or hinders a reaction or physiological response.

inoculation The deliberate introduction of disease-causing organisms into the body in a mild or harmless form to stimulate the production of antibodies that will provide future protection against the disease.

inorganic chemistry Branch of chemistry that deals with all chemicals except the large number of organic compounds (those that contain carbon–hydrogen bonds).

insulator A material that stops or reduces the flow of electricity, heat, or sound.

integrated circuit A tiny electric circuit made of components built into the surface of a silicon chip.

interference The disturbance of signals where two or more waves meet.

interferometry Techniques for analyzing the interference patterns of waves.

Internet The electronic information network linking computers around the world.

interstellar space The space between stars, where density of matter is typically very low.

ion An atom or molecule that has lost or gained one or more electrons to become electrically charged.

ionic bond A chemical bond formed when one or more electrons have been transferred from one atom to another, creating two ions of opposite charge that attract each other.

ionosphere The part of Earth's atmosphere that reflects radio waves. It lies within the thermosphere.

irrational number Any number that cannot be expressed as one whole number divided by another.

isomer A chemical compound with the same formula but a different structure to another compound.

isotope A version of a chemical element in which the atoms have different numbers of neutrons in their nuclei compared with other atoms of that element.

IVF In-vitro fertilization. The techniques (informally called the "test-tube baby" method) of arranging for a sperm to fertilize an egg outside the body before implanting the early embryo back in the womb.

joule The SI unit of work or energy.

karyotype Chromosomal characterization of a species or individual in terms of the number, size, and structure of each chromosome; also, a diagram that shows this.

Kelvin scale See *absolute scale*.

kinetic energy The energy an object has because of its movement.

Kyoto Protocol An international agreement on climate change that sets industrialized countries binding targets for reducing greenhouse gas emissions.

Lamarckism The theory that evolution depends on inheritance of characteristics acquired during an organism's life.

laser A device used to produce an intense narrow beam of light in which the light rays are parallel.

latent heat The heat absorbed or given out, without change of temperature, when a substance changes between a liquid and a gas, or between a solid and a liquid.

lathe A machine designed to spin objects around while cutting them into shape.

latitude A measure of distance from the equator (the poles are at 90° latitude and the equator is at 0°). Lines of latitude are imaginary lines drawn around Earth, parallel with the equator.

lens A transparent object shaped to refract light so that it produces or causes to be produced a sharp image.

lepton A family of fundamental particles, including the electron, that are not affected by the strong nuclear force, unlike quarks and particles made up of quarks.

leucocyte A white blood cell.

Leyden jar *Capacitor*, invented in the 18th century, capable of delivering electric shocks.

line of force One of the imaginary lines in a field of force.

linear equation A mathematical equation that contains no variable number multiplied by itself (for example, there is no x^2, x^3, etc.). Linear equations result in straight lines when plotted out on graphs.

lithosphere The rigid outer layer of Earth, consisting of the crust and the uppermost layer of the underlying mantle.

lock and key Term applied to a situation (for example involving biological molecules) where two parts have to match and interact like a lock and key to effect a change.

lodestone A naturally magnetic piece of the iron-containing mineral magnetite.

logarithm In mathematics, the *power* to which a *base*, such as 10, must be raised to yield a given number.

long count An indefinitely long calendar with a starting point of several thousand years ago, used by the Maya and other Mesoamericans.

longitude A measure of position on Earth, measured by degrees east or west from an imaginary line (the prime meridian) running from the north to the south pole via Greenwich in London. All other lines of longitude also run from the north to the south pole.

longitudinal wave A wave whose to-and-fro movement takes place along the line that the wave is traveling, not at right angles to it. Sound waves are an example.

low frequency Involving a relatively small number of vibrations in a given time period.

luminosity The amount of light given out by an object, such as a star.

lunar eclipse See *eclipse*.

lymphatic system A network of tubes and small organs that drains a fluid called lymph from the body's tissues into the bloodstream.

lymphocytes Types of white blood cell that play specialist roles in the immune system.

magnetic dip The downward-pointing angle at which a freely moving *compass* points, representing the fact that Earth's magnetic poles lie below the surface.

magnetic poles (1) The two regions of a magnet where magnetic effect is strongest. (2) The two variable points on Earth where Earth's magnetic field is strongest and toward which a *compass* needle points.

magnetism The invisible force of attraction or repulsion produced by a magnetic field.

magnetosphere The magnetic field around a star or planet.

mass The amount of matter in an object.

matter Anything that has mass and occupies space.

megabyte See *byte*.

megalith A large stone, especially one deliberately set in position as a marker or monument in prehistoric times.

meiosis A specialized type of cell division (strictly speaking, of nuclear division) that takes place in two stages, and in which haploid sex cells are produced.

Mercator's projection A way of representing the Earth's surface on a flat map so that longitude and latitude lines are at right angles to each other.

merozoite A stage in the life cycle of some microscopic parasites.

mesopause The boundary between the mesosphere and the thermosphere, about 50 miles (80 km) above Earth's surface.

mesosphere (1) The layer of the atmosphere above the *stratosphere*. (2) The layer of the Earth's mantle below the *asthenosphere*.

metabolism The sum total of all the chemical reactions taking place in a living organism.

metal A substance typically having a combination of properties including shiny appearance, ability to be bent into shape, and high conductivity to heat and electricity. Most chemical elements are metals, and there are also thousands of metallic *alloys*.

metaphase The stage of *mitosis* and *meiosis* preceding *anaphase*, during which the chromosomes are aligned along the middle of the cell.

meteoroid Any rocky body, smaller than an asteroid, moving freely through space in the Solar System. If one falls to Earth and is not completely burnt up, it is called a meteorite.

microorganism A tiny organism which can be seen only with the aid of a microscope.

microscope An instrument that produces magnified images of very small objects.

mid-ocean ridge A ridge down the middle of the ocean floor, created by volcanic material erupting from the gap between oceanic plates. See also *plate tectonics*.

mitosis The division of the nucleus during normal cell division, in which each "daughter" nucleus has the same number of chromosomes as the parent cell.

mode (1) A particular pattern of vibrations. (2) In statistics, the value that occurs most frequently in a set of data.

model organism A specimen studied by scientists with a view to developing knowledge that can be applied more generally for understanding other organisms.

modular arithmetic Sometimes called clock arithmetic, a method of counting where one starts again at the beginning after a set point is reached.

modulation Transmitting information by superimposing an extra pattern upon a radio wave (called a carrier wave) or other waves.

molecule Smallest free unit of an element or compound, made up of at least two atoms.

momentum A quantity equal to an object's mass multiplied by its velocity.

Monocotyledons (Monocots) A major subgroup of flowering plants (including grasses, orchids, spring bulbs, palms, etc.) originally identified because they have only one cotyledon (seed-leaf) in their seeds.

motor nerve A nerve that transmits impulses from the central nervous system to operate a muscle or control a gland.

MRI Magnetic resonance imaging. A noninvasive form of medical imaging.

multiple A number is described as "a multiple of x" when x is multiplied by 2, 3, 4, or any other whole number.

mutation A random change in a chromosome of a cell, either to a particular gene or on a larger scale.

myelin Fatty material wrapped around some neurons, which speeds up the transmission of signals.

myofibril Any of the tiny structures in a muscle that enable it to contract.

nano- A prefix meaning a billionth (thousand-millionth).

nanometer A billionth of a meter.

natural selection The process whereby the inheritable characteristics that increase one's chances of survival and reproduction are passed on to the next generation.

Neanderthal A member of an extinct species closely related to modern humans.

nebula Originally, a term for any distant cloudlike object visible beyond Earth's atmosphere. It now applies specifically to huge clouds of dust and gas which are often the locations for new stars to be formed.

negative number A number less than zero.

nephron One of the million or so purification and filtration units in the kidney.

nerve A cablelike structure transmitting information and control instructions in the body. A typical nerve consists of strands of many separate nerve cells (*neurons*) running parallel to, but insulated from each other.

nervous system The network of nerve cells (including the brain) that controls the body.

neuron A nerve cell.

neutrino A tiny, almost massless, uncharged subatomic particle abundant in the Universe but rarely interacting with other matter.

neutron A subatomic particle found in all atomic nuclei except the normal form of hydrogen. It is similar in size to a proton but has no electric charge.

neutron star A small but extremely dense star made mostly of neutrons, formed by the gravitational collapse of a giant star.

newton The SI unit of force.

nitrogen A chemical element that as a gas makes up most of Earth's atmosphere, and combined into compounds is essential to living things.

noble gases Gases such as helium and neon that have a complete complement of electrons in their outer shell and are very unreactive.

nomenclature A body or system of names.

nuclear fission See *fission*.

nuclear fusion A reaction in which the nuclei of light atoms such as hydrogen fuse to form a heavier nucleus, releasing energy.

nuclear reaction A change in the nucleus of an atom.

nucleolus A small, dense, round body inside the nucleus of a cell.

nucleus (1) Central part of an atom, made up of protons and neutrons. (2) The structure in eukaryote cells that contains *chromosomes*.

nutrients Substances that are used by living organisms for growth, maintenance, and reproduction.

observatory A building or institution where astronomers study space.

ohm The SI unit of electrical resistance.

Oort cloud Huge spherical region containing comets that is thought to exist toward the outer boundary of the Solar System.

opiates Drugs related to opium. They are strong painkillers but have many side-effects.

optical fibers Thin glass fibers along which light travels, used in communication.

optics The study of the behavior of light and how it is affected by devices such as lenses, mirrors, etc.

orbit The path of any body that is circling around another one.

orbital period The time a celestial body takes to complete one *orbit* around another.

organ A group of tissues, usually grouped together in a discrete structure, that has a special function, such as the brain.

organic Adjective that can refer to (1) Any compound containing carbon, with the exception of some simple molecules such as carbon dioxide, (2) Food produced without the use of artificial fertilizers or pesticides.

Orrery A mechanical model of the Solar System, showing the relative positions and orbits of the planets and their moons.

oscillation A regular movement back and forth.

oscillator A circuit or instrument that produces an alternating current of known frequency.

oscilloscope An instrument that shows electrical signals on a screen.

osmosis The movement of water through a semipermeable membrane from a weak solution to a more concentrated one.

ovary (1) The structure in animals that produces female sex cells (gametes). (2) A specialized region of the female part of a flower that contains the ovules.

ovule A structure within a flower that develops into a seed after fertilization.

ovum An egg cell.

oxidation Originally, a reaction where a substance combines with oxygen; now used for any reaction where a substance loses electrons. Its opposite is reduction.

oxide A compound of oxygen with one other element.

oxidizing agent A compound that can cause oxidation.

oxygen A reactive gas that makes up 21 percent of Earth's atmosphere and is essential to life.

ozone A very reactive form of oxygen that has three atoms in each *molecule* instead of two.

P wave Primary wave. A fast-moving earthquake wave that alternately stretches and squeezes rocks as it moves.

palynology The study of living and fossil pollen grains and spores.

pancreas A gland close to the stomach that secretes digestive enzymes and also hormones that regulate glucose levels.

parallax The apparent movement of objects against each other when an observer moves position, such as nearby trees against distant hills. Astronomers use the same principle to measure the distance to nearby stars.

parallel circuit A circuit in which there are at least two independent paths to get back to the source.

particle In physics, usually short for subatomic particle.

particle accelerator A giant machine in which subatomic particles are accelerated along or around a tunnel by electromagnets, and smashed together at very high speeds.

particle physics The branch of physics that deals with subatomic particles.

pasteurization The heating of food to destroy disease-causing bacteria.

pathogen A disease-causing microorganism.

peptide A *molecule* similar to a protein in structure but typically smaller.

pericardium A tough double-layered membrane surrounding the heart.

periodic table A table of the chemical elements in order of their atomic numbers, whose vertical columns bring together elements with similar properties.

perpetuum mobile Also called perpetual motion, the theoretically impossible notion that a machine can be made to run forever without energy input although work is extracted from it.

petal One of a set of structures surrounding the sexual organs of most flowers, usually shaped and colored for attracting pollinating animals.

pH A measure of acidity or alkalinity of a solution. A pH of 7 is neutral, below 7 is *acid*, and above 7 is alkaline.

pharmacology The study of drugs and how they act in the body.

phlogiston theory A now-disproved 18th-century theory that all burning involved giving off a substance called "phlogiston."

photoelectric effect The emission of electrons from the surfaces of some objects when light hits them.

photon The particle that makes up light and other electromagnetic radiation.

photoperiodism Any situation where life-processes in living things are affected by the length of daylight they are exposed to.

photosynthesis The process by which plants and algae make food from water and carbon dioxide, using energy from the Sun.

physiology The study of body processes. Also a term for the body processes.

pi The ratio of the circumference of a circle to its diameter, approximately 22 divided by 7, or about 3.14159.

piezoelectric effect The production of electricity by applying mechanical stress to certain crystals, such as quartz.

pistil A female organ of a flower.

piston A close-fitting sliding disk or short solid cylinder attached to a rod that is pushed up and down in the cylinder of an engine to provide power.

pitch (1) The property of a sound that makes it high or low. (2) The angle of an aeroplane wing, propeller blade, etc.

Planck constant Symbol h, the ratio of the energy in one photon of electromagnetic radiation to its frequency. It is a fundamental constant in quantum physics.

plane figure A two-dimensional shape.

planet A large spherical or almost spherical body that orbits a star. See also *dwarf planet*.

plankton Plants, animals, or other life forms living in open water that cannot swim strongly and so drift with the currents. Most are small or microscopic.

plant A member of one of the major kingdoms of living things, making their own food using photosynthesis. They include everything from trees and flowers to ferns and mosses, but not most algae (see *alga*).

plasmid A normally circular strand of DNA in bacteria or protozoa.

plate tectonics The theory and phenomena of Earth's *lithosphere* being divided into huge rigid plates that move with respect to each other. Some plates include continents or parts of continents, while others comprise only deep ocean floor.

platelets Irregular disk-shaped microscopic structures in the blood which function to coagulate the blood and stop bleeding.

pluripotent Term applied to a stem cell that can give rise to any of several other cell types.

pneumatics Branch of physics studying the mechanical properties of air and other gases.

polarized light Light in which the wave vibrations occur only in one plane.

pollen tube The tube that a germinating pollen grain forms as it grows down the female part of a flower to fertilize an ovule.

pollination The deposition of pollen on a flower so that its ovules can be fertilized and it can set seed.

polymer A long, thin *molecule* made of many identical or very similar small molecules joined together; also, a substance made of such molecules.

population In biology, a group of individuals of the same species, especially when able to interbreed with each other.

positron Positively charged counterpart of an electron, sometimes called antielectron.

potential difference The electrical equivalent of pressure. A high potential difference is like a high pressure forcing electricity round a *circuit*. Also called voltage.

potential energy Stored energy that a body has because of its position or internal state.

power (1) The rate of change of energy. (2) The number of times a number is multiplied by itself: for example, x×x×x, or x^3, is also called "x to the power of 3".

precipitate Tiny solid particles formed in a liquid as a result of a chemical reaction.

predator An animal that feeds by attacking and eating other animals (its prey), especially those that are relatively large in relation to its own size.

preservation The process of keeping something in its original state, or free from harm, erosion, or decay.

pressure Continual physical force pushing against an object, especially considered as force per unit area.

prey See *predator*.

prime number Any positive whole number that cannot be divided to give another whole number except by dividing by itself and 1.

prism (1) Any solid geometric form with sides that form a parallelogram. (2) A prism-shaped block of glass, especially one with triangular sides, that is used to split white light into the colors of the spectrum.

probability The likelihood of an event happening, normally expressed as a value between 0 and 1.

product In mathematics, the result of one number being multiplied by another.

prokaryotic cell The cell of microscopic organisms such as bacteria: it is smaller than a *eukaryotic cell* and does not have a separate nucleus.

prosthetic An artificial body part.

protein Any of thousands of different types of large molecules made by the body and coded for by genes. See also *amino acid*.

proton A particle in the nucleus of an atom that has a positive electric charge.

pulsar A neutron star which can be detected because its rapid rotation causes pulses of radiation to be beamed outward.

quadrant A navigational instrument.

quadratic equation Mathematical equation containing at least one variable number that is multiplied by itself once (for example x×x, also written x^2), but none which have been multiplied more times than this.

quantum electrodynamics The quantum physics theory that deals with the interactions between electrons, positrons, and photons.

quantum physics The branch of science that deals with subatomic particles and energy interactions in terms of minute discrete energy packets called quanta.

quantum theory The theory that light and other electromagnetic radiation is made up of a stream of photons, each carrying a certain amount of energy.

quark One of a group of fundamental particles that do not exist separately but make up protons, neutrons, and some other subatomic particles.

quasar An immensely powerful source of radiation occurring beyond our own galaxy. Quasars are thought to be the central regions of other galaxies that are producing far more radiation than the Milky Way's center does.

radar A way of detecting objects by sending out radio waves and collecting their returning "echoes."

radiation Any stream of fast-moving particles or waves.

radio waves Invisible waves at the low-frequency end of the electromagnetic spectrum, whose wavelength can range from kilometers to centimeters (microwaves).

radioactive Emitting high-energy subatomic particles or radiation as part of radioactive decay.

radioactive decay The process whereby unstable nuclei emit high-energy particles, or radiation, as they break up or transform.

radioactive tracers Substances that contain radioactive atoms to allow easier detection and measurement.

radioactivity Phenomena involving radioactive decay.

radiometric dating The process of finding a rock's absolute age by detecting the stage of radioactive decay of particular isotopes in it.

RAM Random access memory. Computer memory chips where information can be stored and retrieved.

rarefaction The opposite of compression, especially in a gas, where the density becomes lowered.

ratio The proportional relation between two numbers.

reactant A substance taking part in a chemical reaction.

reaction (1) A force same in magnitude, but opposite in direction, to another force. Every force has a reaction. (2) Any change that alters the chemical properties of a substance or forms a new substance.

red giant A star that is nearing the end of its life and has expanded to a giant size and become reddish in color.

red shift The tendency for wavelengths of light to be shifted toward the red end of the spectrum when their source is receding rapidly from the observer. The effect also occurs with other wavelengths of electromagnetic radiation.

reduction Originally, a reaction where a substance loses oxygen; now used more generally for any reaction where a substance gains electrons. Its opposite is oxidation.

reflex An automatic reaction to something.

refraction The bending of light rays as they enter a different medium at an angle, such as from air to water.

refractive index The ratio of the speed of light in one medium to the speed of light in a second medium.

relativity The description of space and time, energy and matter according to the theories of Albert Einstein, which depend on the constancy of the speed of light in a vacuum.

repoussé The ancient art of decorating metal by hammering on the back of the piece.

reproduction Process of creating offspring.

resistance See *electrical resistance*.

resistor An electrical device or component that resists current flow.

resonance Situation when the vibrations of an object become large because it is being made to vibrate at its "natural" frequency.

respiration (1) Breathing. (2) Also called cellular respiration, the biochemical processes within cells that break down food molecules to provide energy, usually by combining the food molecules with oxygen.

retrograde motion Motion that is the opposite of another motion, such as a satellite orbiting a planet in the opposite direction to the planet's own rotation. Retrograde motion can only be apparent as when a planet seems to move backward against the stars because Earth is overtaking it on its orbit round the Sun.

retrovirus An RNA virus—such as HIV—which reproduces itself by inserting a DNA copy of its genes into the host cell.

Richter scale A scale for measuring the size of an earthquake in terms of the amount of energy released.

right angle An angle created by lines meeting perpendicularly, forming an equal angle on either side.

RNA Ribonucleic acid. A *molecule* similar to DNA with various roles in cells, including acting as an intermediary between DNA and the rest of the cell.

robot (1) An intelligent humanlike machine (mainly in fictional contexts). (2) A machine, especially a programmable one, that can carry out a complex series of movements.

ruminant An animal such as a cow or deer that chews the cud.

S wave Secondary wave. An earthquake wave that travels through the ground as lateral or horizontal waves.

satellite An object that orbits a planet. There are natural satellites, such as a moon, and artificial satellites, such as a craft used to retransmit radio signals.

secretion The release of specific substances by the cells of living things.

sedimentary rock Formed when fragments of material settle on the floor of a sea or lake and are cemented together over time.

sedimentation The geological process in which loose material is laid down on ocean and river beds, and by wind and moving ice.

seismic wave A wave that travels through the ground, such as from an earthquake.

seismograph A device for measuring and recording earthquake waves.

seismology The study of earthquakes.

seismometer A device for measuring earthquake waves. The term is now interchangeable with *seismograph*, since modern seismometers record their measurements as well.

selective breeding The process of choosing particular domestic animals for breeding to encourage the development of desired traits over the generations.

semiconductor A substance whose resistance is intermediate between a conductor and an insulator. The properties of semiconductor devices can be controlled and altered very exactly, making them vital in modern electronics.

sensory nerve A nerve that transmits information about the environment (touch, taste, etc.) to the central nervous system.

sepal One of a set of petal-like or leaflike structures usually found around the rim or base of a flower outside the petals.

sex cell See *gamete*.

sextant A navigational instrument designed to measure the altitude of an object, such as the Sun above the horizon at noon.

sexual reproduction Reproduction that involves the fusion of two gametes (sex cells) to produce a new individual.

SI unit A unit in the international system of measure based on the meter, kilogram, second, ampere, kelvin, candela, and mole.

silicon A semi-metallic element related to carbon, which is a constituent of many of Earth's rocks.

sine (sin) The ratio of the opposite side of an angle to the hypotenuse in a right-angled triangle. Also, the mathematical function describing how this ratio varies when the angle changes as the hypotenuse sweeps round a circle.

skeleton The frame of bone and cartilage in vertebrates that supports the body and protects its organs, or any structure in other animals that serves similar functions.

slide rule A ruler with a central sliding bar designed to make quick calculations using logarithms.

smelting Extracting a metal from its ore.

software The programs used by a computer.

solar constant The amount of heat energy from the Sun received per unit area of Earth's surface.

solar eclipse See *eclipse*.

solar flare A sudden burst of radiation from the Sun.

Solar System The system consisting of the Sun, the planets, and other objects orbiting the Sun, and the surrounding regions of space in which the Sun's influence is discernable.

solenoid A cylindrical coil of wire that becomes a magnet when an electric current is passed through it.

solstice Each of the two times in the year, one at midsummer and one at midwinter, when the Sun reaches its highest or lowest point in the sky at noon.

solubility The ability of a *solute* to dissolve.

solute The substance that dissolves in a solvent to form a solution.

solution A liquid in which individual atoms, molecules, or ions of another substance (as distinct from small solid particles) are evenly dispersed.

solvent A substance, especially a liquid, that can dissolve other substances.

somatic nuclear transfer A laboratory technique for creating a fertilized ovum using a somatic cell (ordinary body cell) to create a clone of the organism.

sonar A means of detecting objects and navigating under water by sending out sound waves and receiving their echoes.

space probe An unpiloted vehicle (other than an Earth satellite) that is designed to explore space.

space station A human-occupied orbiting structure for the purposes of carrying out experiments, observations, etc.

space-time The three dimensions of space combined with time in a single continuum.

species A particular kind of living thing, often defined in terms of the ability of individuals of that species to mate and produce fully fertile offspring (although this definition does not work for all cases).

specific heat capacity The quantity of heat required to raise the temperature of a unit mass of a given substance by one degree.

spectroscope A machine that measures and analyses *spectra*.

spectroscopy The study and measurement of *spectra*.

spectrum (plural: **spectra**) Originally, light separated by refraction so that its different wavelengths (colors) are spread out in sequence. The term is now also applied to other electromagnetic radiation, and also to refer to characteristic patterns of radiation given off by particular sources.

speed The rate at which something is moving. See also *velocity*.

sperm A male sex cell (*gamete*) that can move to locate a female cell. All animals and some lower plants produce sperm.

spherical trigonometry Trigonometry modified to apply to the surface of a sphere rather than a flat surface.

sporozoite A stage in the life cycle of some microscopic parasites.

square root A number which when multiplied by itself yields a given number.

stade (1) An ancient Greek unit of measurement. (2) A period of geological time when glaciers have stopped retreating.

stamen The male organ of a flower. See also *anther*.

standard model The principle theoretical framework of particle physics, combining theories on how three of the four fundamental forces interact (electromagnetism and the strong and weak nuclear forces) with 12 basic particles (six quarks and six leptons).

star A huge luminous ball of ionized gas (plasma) whose energy emissions are powered by nuclear reactions in its core.

static electricity Phenomena involving nonmoving electric charges on objects.

stem cell A type of cell in the body that is able to divide and grow into other more specialized cells.

sterilization (1) Giving special treatment to equipment to kill life forms such as harmful bacteria. (2) Rendering an animal infertile by performing an operation, using radiation, etc.

stethoscope A diagnostic instrument used to listen to sounds within the body, especially the chest.

stigma The top of the female part of a flower (pistil). It is usually sticky to receive pollen.

stratigraphy The study of rock layers.

stratopause The boundary between the *stratosphere* and *mesosphere*.

stratosphere The part of Earth's atmosphere between the *troposphere* and *mesosphere*.

stratus cloud Usually low cloud and in the form of flat sheets, often bringing light rain.

style The stalk that supports a stigma.

subatomic particle A particle smaller than an atom or its nucleus, such as a proton, neutron, or electron.

subduction boundary A boundary between two tectonic plates in the deep ocean where one plate is being pushed beneath the other.

submersible A usually small vessel for underwater exploration.

substance Any kind of matter.

sunspot A region of the Sun's surface whose temperature is temporarily lowered and which therefore appears darker than its surroundings in images.

superconductivity Phenomenon in which some substances lose all electrical resistance close to *absolute zero*.

supernova An enormous explosion taking place at the end of the life of a very large star.

supersonic Faster than the speed of sound.

surface tension The effect that makes a liquid seem as though it has an elastic "skin," caused by cohesion between the surface molecules.

suspension A mixture of tiny solid particles or globules of liquid in a surrounding medium.

switch A device that turns an electric current on or off, or more generally changes something from one state to another.

synapse A junction between two nerve cells, or between a nerve cell and a muscle or gland cell.

synthesis The combining of separate parts or different theories to make a whole.

taxonomy The classification of living things, and also the principles behind classification.

tectonic Relating to the structure of Earth's crust and its movements. See also *plate tectonics*.

telophase The final stage of *mitosis* and of each part of *meiosis*, in which a nuclear membrane forms around each set of separated *chromosomes*.

temperature A measure of how hot or cold something is.

theodolite A surveying instrument that measures angles using a rotating telescope.

theorem A mathematical rule or statement, especially a truth that is not self-evident but which can be proved by reasoning.

thermal (1) Relating to heat (adjective). (2) A current of rising hot air in the atmosphere.

thermodynamics The branch of physics dealing with the relationship between heat and other forms of energy.

thermoelectric effect Any of various effects that can occur when there are temperature differences in an electric circuit.

thermosphere The layer of Earth's atmosphere above the *mesosphere*.

three dimensions (3-D) Length, breadth, and depth.

tissue Living material made up of broadly similar kinds of cells and performing a particular function: for example, nerve tissue, muscle tissue.

topography The study of landforms.

torque A twisting force.

trace elements Elements that are needed in only minute amounts by living things.

trade wind A wind that blows from the southeast or northeast all the year round in equatorial regions.

transformer A device that increases voltage while decreasing current, or vice versa. It works only with alternating currents.

transfusion The transfer of blood from a donor to a recipient.

transistor A semiconductor electronic device that acts as a switch, amplifier, or rectifier.

translocation (1) A situation where part of one chromosome has moved to another location, either on the same chromosome or a different one. (2) Movement of materials through a plant.

transmission The conveying of something from one place to another.

transmutation (1) The evolutionary change of one species into another. (2) The conversion of one kind of atom to another through a nuclear reaction.

transparent Allowing light or other radiation to pass through.

transpiration The loss of water from the surfaces of a plant, especially from its leaves.

transplant A tissue or organ taken from one part of the body to be placed in another part, or in another individual.

transverse wave A wave whose to-and-fro movement takes place at right angles to the line that the wave is traveling. Light waves are an example.

triangulation A method of surveying by measuring angles and distances using the mathematical properties of triangles.

trigonometry The branch of mathematics dealing with calculations involving the sides and angles of triangles.

tropical Relating to the warm regions of Earth that lie between the Tropic of Cancer (23.5° north of the equator) and the Tropic of Capricorn (23.5° south of the equator), or to climates typical of those regions.

tropopause In Earth's atmosphere, the boundary between the *troposphere* and *stratosphere*.

troposphere The lowest layer of Earth's atmosphere, starting from ground level, where most weather events take place.

tsunami A water wave, sometimes of huge size, generated by an earthquake, underwater landslide, or other major disturbance.

turbine A rotating wheel driven by water, or moving gas or air, in order to provide power.

ultrasound Sound with a frequency above that which the human ear can detect.

ultraviolet A type of electromagnetic radiation with wavelengths shorter than visible light.

uncertainty principle The quantum physics principle that it is impossible to measure both the position and momentum of objects at the subatomic level exactly, because the observation of one changes the other.

Universe Traditionally, a term for the totality of everything that exists. Now often used for everything created as a result of the Big Bang, which leaves open the possibility of other universes existing.

urine A fluid that most animals discharge to get rid of waste products and excess water.

vaccine A special preparation of substances that stimulate an immune response, used for inoculation.

vacuum A space in which there is no matter. (In real-life examples a vacuum is only approximate.)

vacuum tube A sealed tube, usually glass, from which most air has been removed and which contains electrodes. Application of electricity causes a beam of electrons to be emitted from the negative electrode (cathode). It is a general term and includes various devices used or formerly used in electronics. Also called an electron tube.

valency The number of chemical bonds that an atom can make with another atom.

valve A device or structure that restricts the flow of fluid or electricity to one direction. In electrical contexts it refers to a kind of *vacuum tube*.

Van der Waals bond A relatively weak kind of chemical bond.

vascular circulation The circulation of blood through blood vessels (arteries, capillaries, and veins) and back to the heart.

vein A blood vessel leading back toward the heart. See also *vascular circulation*.

velocity *Speed* in a particular direction.

Vernier scale A small movable, graduated scale added to a larger scale for increased accuracy in precision measuring instruments, named after Pierre Vernier.

virus (1) Tiny parasitic noncellular life forms consisting mainly of genes with a protective coat that are able to hijack living cells to make copies of themselves. (2) A piece of computer software that can spread itself through computer systems in a manner similar to a biological virus.

visible light Electromagnetic radiation with wavelengths that can be detected as light by the eye.

vitamin Any of various organic compounds needed in small amounts in the diet to preserve health.

viviparous Giving birth to live young as distinct from eggs.

voltage See *potential difference*.

voltaic cell A device that converts the energy of a chemical reaction directly into electricity; colloquially, a battery. See also *battery*.

volume (1) The amount of space something takes up. (2) The loudness of a sound.

vulgar fraction A fraction expressed as one number divided by another rather than using decimal points.

wave A regular oscillation in intensity or concentration. Waves typically travel in a particular direction or directions and transmit energy in that direction.

wavelength The distance between each successive peak or crest of a series of waves.

weak force The force in atomic nuclei responsible for beta decay. It is so called in comparison with the strong force.

weak interaction Another name for the weak force.

white dwarf A small, faint, very dense star, thought to represent the final stage of evolution of stars below a certain mass.

World Wide Web A vast network on the *Internet* for gathering and exchanging data and documents through hypertext links.

worm gear A gear arrangement in which one of the gears is a cylinder with grooves cut in it.

X-ray diffraction A technique of beaming X-rays at materials to obtain information about their internal structure by analyzing the diffraction patterns produced; also called X-ray crystallography.

X-rays A type of high-energy, high-frequency electromagnetic radiation.

INDEX

Page numbers in **bold** indicate main treatements of a topic.

A

B

E

F

G

K

L

M

R

S

T

U

V

W

X

Z

ACKNOWLEDGMENTS

Dorling Kindersley would like to thank the following people: Irene Lyford and Steve Setford for proofreading; Nikki Sims and Kathryn Hennessy for additional editorial assistance; Lili Bryant and Rhiannon Carroll for assistance with photography.

Smithonian Institution: Deborah Warner, Roger Sherman (**National Museum of American History**); Robert F. van der Linden, Paul Ceruzzi, Andrew K. Johnston, Hunter Hollins (**National Air and Space Museum**); Alex Nagel (**Freer Gallery of Art and the Arthur M. Sackler Gallery**); Michael Brett-Surman, Salima Ikram (**National Museum of Natural History**).

New photography: The Whipple Museum of the History of Science, Cambridge, UK. The Whipple Museum holds an internationally important collection of scientific instruments and models, dating from the Middle Ages to the present. DK is thankful for the help offered by Professor Liba Taub and Dr. Claire Wallace for assistnce with photography at the museum.

Leeds University Museum for the History of Science: Thanks to Dr. Emily Winterburn, curator at the Leeds University Museum for the History of Science, Technology and Medicine, Professor Denis Greig for help with objects in the physics department of the University and Dr. Claire Jones, museum director.

DK India would like to thank Rupa Rao for editorial assistance; Priyabrata Roy Chowdhury, Parul Gambhir, Supriya Mahajan, Ankita Mukherjee, Neha Sharma, Shefali Upadhyay for design assistance; Neeraj Bhatia and Arvind Kumar for repro assistance

The publisher would like to thank the following for their kind permission to reproduce their photographs:

(Key: a-above; b-below/bottom; c-center; f-far; l-left; r-right; t-top)

12 Alamy Images: The Art Gallery Collection (cr). SWNS.com Ltd: Pedro Saura (t). **13** Corbis: Sakamoto Photo Research Laboratory (c). Dorling Kindersley: Museum of London (cra, crb). Dreamstime.com: Joe Gough (t). **14** The Bridgeman Art Library: Heini Schneebeli (br). Corbis: EPA (t); Michel Gounot / Godong (cl). **15** Corbis: Minden Pictures / Jim Brandenburg (t). Dorling Kindersley: Museum of London (cl). Getty Images: DEA Picture Library (br). **16** Corbis: Roger Wood (br). Dorling Kindersley: The Trustees of the British Museum (bl); Museum of London (c); University Museum of Archaeology and Athropology, Cambridge (tl, ca, cb, crb, bc). Getty Images: De Agostini (tc); SSPL (tr). **17** Alamy Images: UK Alan King (cla). Dorling Kindersley: The Trustees of the British Museum (tr, c, cr, bc, bl); Courtesy of Museo Tumbas Reales de Sipan (br); Courtesy of the Board of Trustees of the Royal Armouries (tc). **18** Corbis: Nik Wheeler (t); Visuals Unlimited (c). Dorling Kindersley: The Trustees of the British Museum (b). **19** Alamy Images: The Art Archive (c). Dorling Kindersley: University Museum of Archaeology and Athropology, Cambridge (t). **20–21** Photo SCALA, Florence: DeAgostini Picture Library (c). 20 Alamy Images: World History Archive (bl). Corbis: Werner Forman (br, crb). **21** Dorling Kindersley: National Motor Museum, Beaulieu (bc). Getty Images: Universal Images Group (clb). Science Museum / Science & Society Picture Library: National Railway Museum (cb). Fu Xinian: "Zhongguo meishu quanji, huihua bian 3: Liang Song huihua, shang" (Beijing: Wenwu chubanshe, 1988), pl. 19, p. 34, Collection of the National Palace Museum, Beijing (bl). **22** ChinaFotoPress: akg / Erich Lessing (t). from S. Percy Smith, "Hawaiki, The Original Home of the Maori; with a sketch of Polynesian History", Whitcombe & Tombs Ltd, 1904: (cr). **23** Getty Images: De Agostini (l); Werner Forman (t). **24** Alamy Images: The Art Gallery Collection (cr). Getty Images: Universal Images Group (t). **25** Alamy Images: imagebroker (t); The Art Archive (br). Getty Images: Bridgeman Art Library (c). **26–27** Getty Images: SSPL (cb). Museum of the History of Science, University of Oxford: (t). **26** Fotolia: Pedrosala (br); Sculpies (clb). Getty Images: British Library / Robana (bl). Science Photo Library: (cl). **27** Fotolia: Andreas Nilsson (crb). Getty Images: SSPL (bc, tr). **28** Getty Images: De Agostini (cl). University of Pennsylvania Museum: (t). **29** Alamy Images: Mary Evans Picture Library (cr). Getty Images: De Agostini (bl). **30** Alamy Images: Hans-Joachim Schneider (bl). Getty Images: Universal Images Group (t). Wikipedia: (cr). **31** Alamy Images: The Art Gallery Collection (t). **32** Corbis: Araldo de Luca (cr). Getty Images: De Agostini (br). from "Vitruvius Teutsch" (German edition by Walther Ryff), Peter Flotner, 1548, p645: (t). **33** Alamy Images: Interfoto (t). Corbis: Sygma (c). Dorling Kindersley: The Science Museum, London (br). **34** Science Photo Library: Sheila Terry (cl). **36** Alamy Images: The Art Gallery Collection (t). Dorling Kindersley: The Science Museum, London (b). **37** Alamy Images: Ian Macpherson Europe (t). Corbis: Bettmann (clb); (cl). **38** Getty Images: De Agostini (cr). Pedro Szekely: (br). **38–39** Corbis: (t). **39** Getty Images: Universal Images Group (r). **40** Corbis: Heritage Images (clb). Getty Images: (t). **41** Getty Images: De Agostini (t); (cr, clb). **42** Alamy Images: Patrick Forget / Sagaphoto.com (cr); Sonia Halliday (t). Getty Images: Alireza Firouzi (clb). **43** Dorling Kindersley: National Maritime Museum, London (cl). Science Photo Library: (t). **46** ChinaFotoPress: De Agostini Picture Library (crb). Getty Images: De Agostini (tl). **46–47** Walter Callens: (t). **47** Getty Images: Hulton Archive (cb); Universal Images Group (tr). **48** Alamy Images: Art Directors & TRIP (l). Science Photo Library: Sheila Terry (tr). 49 Fotolia: Charles Taylor (cla). Getty Images: DEA / M Seemuller (tr). **50** Alamy Images: The Art Gallery Collection (tl). Getty Images: De Agostini (br). Photos.com: (bl). **51** Alamy Images: ImagesClick, Inc. (t). Dorling Kindersley: Courtesy of the Board of Trustees of the Royal Armouries (cr). **52–53** Getty Images: De Agostini. **53** Alamy Images: Art Directors & TRIP (cr). ChinaFotoPress: R. u. S. Michaud (tr). **54** Alamy Images: Keystone Pictures USA (tr). Corbis: (cr). NASA: ESA, M. Livio and the Hubble 20th Anniversary Team (STScI) (c). **56** Alamy Images: The Art Archive (c); The Art Gallery Collection (crb). Corbis: Alfredo Dagli Orti / The Art Archive (tl). Getty Images: SSPL (clb). **57** Science Photo Library: Sheila Terry (t). Wikipedia: Drawing from treatise "On the Construction of Clocks and their Use", Ridhwan al-Saati, 1203 C.E. (cl). **58** Getty Images: De Agostini (l). **59** Corbis: Stefano Bianchetti (cr). NASA: (c). Science Photo Library: Mehau Kulyk (tl). **60** ChinaFotoPress: British Library (r). Getty Images: De Agostini (t). Photos.com: (clb). **61** Corbis: Heritage Images (cb). Flickr.com: http://www.flickr.com/photos/takwing/5066964602 (cla). Getty Images: De Agostini (tr). **62** Getty Images: De Agostini (br); SSPL (clb); Feng Wei Photography (bl). Mark Heers (http://www.travel-wonders.com): courtesy Parc Leonardo da Vinci au Château du Clos Lucé (t). **63** Dorling Kindersley: The Science Museum, London (bl). Getty Images: SSPL (cb). Science Photo Library: David Parker (crb). **64** Getty Images: (cr); SSPL (bl). **64–65** Getty Images: (t). **65** Alamy Images: Photos 12 (bl). Jeff Moore (jeff@jmal.co.uk): (cr). **66** Corbis: (bl). Getty Images: Hulton Archive (tl). **67** Corbis: Christel Gerstenberg (cr). Getty Images: Bridgeman Art Library (tl); SSPL (c). **68** Corbis: Sandro Vannini (tl). Fotolia: Zechal (tr). **69** Alamy Images: Interfoto (tr). Fotolia: Georgios Kollidas (cl). Photo SCALA, Florence: White Images (r). **70** Corbis: Baldwin H. & Kathryn C. Ward (r). Dorling Kindersley: Glasgow City Council (Museums) (tl). **71** Corbis: Bettmann (tr). Getty Images: De Agostini (cl); Universal Images Group (cb). **72** Alamy Images: Antiquarian Images (tl). Getty Images: SSPL (cb); Universal Images Group (cr). **72–73** Dorling Kindersley: National Museum of Wales (t). **73** Getty Images: Bridgeman Art Library (c); De Agostini (crb). National Library of Medicine: Jacopo Berengario da Carpi, Isagogae breues, perlucidae ac uberrimae, in anatomiam humani corporis a communi medicorum academia usitatam, Bologna, Beneditcus Hector 1523 (clb). **76** Corbis: Bettmann (cl). Getty Images: Universal Images Group (crb). Science Photo Library: SOHO-EIT / NASA / ESA (tl). **77** Corbis: Stefano Bianchetti. **78** Alamy Images: Lebrecht Music & Arts Photo Library (bc). The Bridgeman Art Library: The Royal Collection © 2011 Her Majesty Queen Elizabeth II (t). Dorling Kindersley: The Trustees of the British Museum (clb). Getty Images: Universal Images Group (bl); (br). **79** Corbis: Bettmann (br). Dorling Kindersley: The Science Museum, London (bl); University College, London (bc). Getty Images: AFP (cra); CMSP (crb); SSPL (cb, cl). Wikipedia: Andreas Vesalius, De Humani Corporis Fabrica, 1543, page 178 (clb). **80** Corbis: Heritage Images (c); David Lees (tl). Dorling Kindersley: Natural History Museum (crb). **81** Getty Images: De Agostini (tl). Science Photo Library: (tc). **82–83** Getty Images: Hulton Archive (t). **82** Dorling Kindersley: Natural History Museum (cb). National Library of Medicine: IHM / Realdo Colombo, De Re Anatomica, 1559, title page (crb). David Nicholls: courtesy St Mary's Tenby (t). Project Gutenberg (www.gutenberg.org): Georgius Agricola, De re metallica, 1556 figure 305 (cl). **83** Corbis: Sygma (crb). Getty Images: Universal Images Group (tr). Science Photo Library: Science Source (cl). **84** Dorling Kindersley: The Trustees of the British Museum (tl); Judith Miller / Branksome Antiques (r); The Science Museum, London (cb, br). Dreamstime.com: Erierika (bl). Paul Marienfeld GmbH & Co. KG: (fbl). **85** Dorling Kindersley: The Trustees of the British Museum (cr); National Maritime Museum, London (tl). Fotolia: Coprid (cra). Getty Images: Ryoichi Utsumi (bc). Schuler Scientific (www.schulersci.com): (crb). **86** Fotolia: Jenny Thompson (tl). Science Photo Library: Science Source (clb). **86–87** Corbis: Mike Agliolo (t). **87** ChinaFotoPress: Massimiliano Pezzolini (cb). Getty Images: Universal Images Group (cr). NASA: ESA, D. Lennon and E. Sabbi (ESA / STScI), J. Anderson, S. E. de Mink, R. van der Marel, T. Sohn, and N. Walborn (STScI), N. Bastian (Excellence Cluster, Munich), L. Bedin (INAF, Padua), E. Bressert (ESO), P. Crowther (University of Sheffield), A. de Koter (University of Amsterdam), C. Evans (UKATC / STFC, Edinburgh), A. Herrero (IAC, Tenerife), N. Langer (AifA, Bonn), I. Platais (JHU), and H. Sana (University of Amsterdam) (t). **88** Corbis: Heritage Images (tl). Getty Images: Bridgeman Art Library (c). **89** Alamy Images: Matthew Johnston (crb). Getty Images: Bridgeman Art Library (clb). Science Photo Library: Sheila Terry (c). **90** Dorling Kindersley: The Trustees of the British Museum (tr); The Science Museum, London (br, bl). SuperStock: Marka (tl). **91** Photos.com: Lyudmyla Nesterenko (tr). **92** Alamy Images: Interfoto (tr). Getty Images: SSPL (c). Science Photo Library: (tl). **93** Corbis: Bettmann (tr). Dorling Kindersley: Natural History Museum (br). Getty Images: Bridgeman Art Library (bl); SSPL (tl); Universal Images Group (r). **94** Corbis: EPA (clb). **94–95** Science Photo Library: Sheila Terry (t). **95** Getty Images: (clb). Science Photo Library: New York Public Library (tr). **96** Corbis: Bettmann (clb). **96–97** NASA: ESA, M. Robberto (STScI / ESA) and the Hubble Space Telescope Orion Treasury Project Team (t). **97** Dorling Kindersley: The Science Museum, London (cb). Getty Images: Hulton Archive (cr); SSPL (cl). **98–99** Alamy Images: Charistoone-Images (t). **98** Science Photo Library: Maria Platt-Evans (crb). **99** Alamy Images: Pictorial Press (tr). **100** Corbis: Bettmann (c). **102** Science Photo Library: Sidney Moulds (tl). SuperStock: Science Faction / Jay Pasachoff (cl). **102–103** Getty Images: SSPL (t). **103** Alamy Images: World History Archive (cl); The Art Gallery Collection (tr). **104** Corbis: Lebrecht Music & Arts (ca). Getty Images: Hulton Archive (tl). NASA: SDO, AIA (tr). **105** Corbis: David Lees (cr). Dorling Kindersley: The Science Museum,

London (cb). Science Photo Library: AMI Images (tl). **106** Corbis: Bettmann (cr). Library Of Congress, Washington, D.C.: Image No 3c10457u / Louis Figuier, Les merveilles de la science ou description populaire des inventions modernes, Vol 1 p33 fig 18, Paris 1867 (tl). **107** Alamy Images: Science Photo Library (cb). Corbis: Bettmann (tl). Getty Images: Bridgeman Art Library (cr); SSPL (tr). **108** Alamy Images: The Art Archive (bc); Stefano Cavoretto (clb). Getty Images: De Agostini (bl); Diane Macdonald (cr); SSPL (br). 108–109 Getty Images: James Strachan. **109** Getty Images: (tr); SSPL (bc, cb, bl). 110 Alamy Images: North Wind Picture Archives (cl). NASA: JPL (tl). **110–111** Getty Images: Hulton Archive (t). **111** Dorling Kindersley: NASA (tr). Getty Images: (cr); SSPL (cl). **112** Corbis: Heritage Images (tl). NASA: Steve Lee University of Colorado, Jim Bell Cornell UniversityNASA, Steve Lee University of Colorado, Jim Bell Cornell University (ca). Science Photo Library: (clb). **112–113** Dorling Kindersley: The Science Museum, London (c). **113** Corbis: The Gallery Collection (c). Getty Images: Universal Images Group (tr). Science Photo Library: (tl). **114** Dorling Kindersley: Judith Miller / Branksome Antiques (bl); The Science Museum, London (tl). Getty Images: SSPL (tc). **115** Dorling Kindersley: Natural History Museum (tr). Getty Images: Doyeol Ahn (ca); SSPL (bl, br). www.antique-microscopes.com: (clb). Science Photo Library: Andy Crump (bc). **116** Corbis: Blue Lantern Studio (crb); Michael Jenner (clb). **117** Getty Images: Bridgeman Art Library (l). NASA: JPL / Space Science Institute (tr). Science Photo Library: Mehau Kulyk (ca). **118** Corbis: Bettmann (clb). Getty Images: Bridgeman Art Library (tc, cr); SSPL (c). 118–119 NASA: (t). **119** Florida Center for Instructional Technology: (cl). **120** Getty Images: FPG (cl). **121** Alamy Images: NASA (cr). Dreamstime.com: Jiawangkun (tl). **122** Corbis: Bettmann (c). Getty Images: Photodisc / Siede Preis (l). **123** Getty Images: SSPL (cr). **124** Corbis: Science Faction / David Scharf (tl). Getty Images: SSPL (clb). **124–125** Corbis: Shah Rogers Photography / Fiona Rogers (t). **125** Getty Images: SSPL (cr). **126** Alamy Images: The Art Archive (tl). Corbis: Bettmann (tr). Getty Images: Time & Life Pictures (clb). Science Photo Library: (c). **127** Getty Images: Bridgeman Art Library (cr); Digital Vision (tl); SSPL (clb). **128** Getty Images: SSPL (clb); Universal Images Group (tl). **128–129** Corbis: Minden Pictures / Flip Nicklin (t). **129** Alamy Images: Prisma Archivo (cr). Corbis: Minden Pictures / Scott Leslie (clb). Getty Images: SSPL (tr). **130** Canada-France-Hawaii Telescope: Coelum / Jean-Charles Cuillandre / Giovanni Anselmi (tl). 130–131 Corbis: Science Faction / David Scharf (t). **131** Getty Images: Universal Images Group (clb). Science Photo Library: Science Source / NYPL (c); Paul D. Stewart (tr). **132** Dorling Kindersley: The Trustees of the British Museum (tl); National Maritime Museum, London (c, bl); Judith Miller / Branksome Antiques (bc). **133** Alamy Images: Ian Dagnall (br). Dorling Kindersley: By kind permission of The Trustees of the Imperial War Museum, London (tr); National Maritime Museum, London (cla, cl, tl); The Science Museum, London (bl, cr). **134** Alamy Images: James Jackson (l); Interfoto (c). Dorling Kindersley: Natural History Museum (crb). NASA: (tl). **134–135** Alamy Images: Mikhail Yurenkov (t). **135** Corbis: Bettmann (crb). Getty Images: SSPL (tr). **136** Getty Images: SSPL (tl). **137** Corbis: The Gallery Collection (tr). Getty Images: Bridgeman Art Library (crb); Field Museum Library (cl). Photos.com: (tl). Science Photo Library: Sheila Terry (cr). **138** Fotolia: Mikhail Olykainen (tl). **138–138** Getty Images: Dan Rosenholm (t). **138–139** Wikipedia: (c). **139** Dorling Kindersley: Linnean Society of London (cr). **140** Alamy Images: Mary Evans Picture Library (t). Getty Images: (c). **141** Corbis: (cl). Dorling Kindersley: The Science Museum, London (crb). Getty Images: SSPL (tl, tr). **142** Corbis: Bettmann (clb); Christie's Images (crb). Science Photo Library: US National Library of Medicine (tl). **142–143** NASA: ESA / The Hubble Heritage Team (STScl / AURA) (t). **143** Corbis: Bettmann (cr). Getty Images: Hulton Archive (tr). Photos.com: (c). 144 Alamy Images: Pictorial Press Ltd. (tl). Dorling Kindersley: The Science Museum, London (cl). **144–145** Corbis: Bettmann (b). NASA: JPL-Caltech / UCLA (t). **145** Corbis: The State Hermitage Museum, St. Petersburg, Russia (ca). Science Photo Library: Royal Astronomical Society (br). **146** Dorling Kindersley: Judith Miller / Lawrence's Fine Art Auctioneers (bl). **147** Dorling Kindersley: The Science Museum, London (cla). **148** Corbis: Michael Nicholson (crb). Dorling Kindersley: National Maritime Museum, London (cl). **149** Getty Images: SSPL (crb). Photos.com: (tc). **150** Corbis: Bettmann (cr). Dorling Kindersley: The Science Museum, London (cl). NASA: (t). **151** Corbis: Historical Picture Archive (cr). Dorling Kindersley: The Science Museum, London (cl). Getty Images: Universal Images Group (tr). **152** Alamy Images: Interfoto (tc). Getty Images: (clb). **152–153** Getty Images: Universal Images Group (t). **153** Dorling Kindersley: Courtesy of the National Trust (tr). Getty Images: SSPL (c). **154** Getty Images: Chris Hepburn (t); SSPL (clb, cr). **155** Corbis: Hulton-Deutsch Collection (cl). Science Photo Library: (crb); Science Source (tr). **156–157** Alamy Images: Ivy Close Images (t). **156** Corbis: Bettmann (r). Science Museum / Science & Society Picture Library: (cl). **157** The Bridgeman Art Library: Christie's Images (clb). **160** Alamy Images: North Wind Picture Archives (cr). Corbis: (l). **161** Corbis: Design Pics (tl). Getty Images: SSPL (tr); Universal Images Group (cr). 162 Alamy Images: incamerastock (ca). Getty Images: SSPL (crb). **162–163** Corbis: Photononstop / Gérard Labriet (t). **163** Corbis: Bettmann (ca). Getty Images: Hulton Archive (crb). NASA: JPL-Caltech / UIUC (tr). **164** Dorling Kindersley: Hunterian Museum (University of Glasgow) (cla); The Oxford University Museum of Natural History (tl); Natural History Museum (cb); Courtesy of The Oxford Museum of Natural History (tr). **164–165** Dorling Kindersley: Oxford University Museum of Natural History (c). **165** Dorling Kindersley: National Museum of Wales (crb); Courtesy of The Oxford Museum of Natural History (tl, br); Natural History Museum (tc); Senckenberg Nature Museum, Frankfurt (tr). **166** Alamy Images: Pictorial Press Ltd. (tl). Dorling Kindersley: The Science Museum, London (cr). **166–167** Dorling Kindersley: The Trustees of the British Museum (t). **167** Corbis: Michael Maslan Historic Photographs (t). Dorling Kindersley: The Science Museum, London (r). **168** Library Of Congress, Washington, D.C.: LC-USZ62-110377 (tr). Science Photo Library: Sheila Terry (cl). **169** Getty Images: Leemage (cl). NASA: Tod Strohmayer (GSFC) / Dana Berry (Chandra X-Ray Observatory) (tr). **170** Dorling Kindersley: The Science Museum, London (clb, br). Fotolia: Volker Witt (cl). Getty Images: Bridgeman Art Library (bl); SSPL (bc, cb). **170–171** Getty Images: SSPL (t). **171** Alamy Images: Marc Tielemans (bc). Getty Images: SSPL (crb). Library Of Congress, Washington, D.C.: LC-USZ62-110411 (clb). Courtesy of Mazda: (br). Matthias Serfling: (cb). **172** Corbis: Design Pics / Robert Bartow (t). Getty Images: SSPL (clb); Time & Life Pictures (crb). National Library of Medicine: Hanaoka Seishu's Surgical Casebook, Japan, c1825 (cra). **173** Corbis: Blue Lantern Studio (tl). Getty Images: SSPL (cr); Hulton Archive (tr). **176** Corbis: Theo Allofs (tl). Getty Images: Bridgeman Art Library (c); Universal Images Group (tr). **177** Getty Images: Bridgeman Art Library (ca); SSPL (tc). **178** British Geological Survey: (cr). Getty Images: SSPL (cb). NASA: Earth Observatory (tr). **179** Getty Images: SSPL (tr); Hulton Archive (cr). . Michael Morgan (www.flickr.com/photos/morgamic/) : (cl). **180** Alamy Images: Gavin Thorn (c). Corbis: Bettmann (tl). **180–181** Corbis: Stapleton Collection (t). **181** Alamy Images: The Art Archive (cr). **182** Corbis: Bettmann (tr). **183** Science Photo Library: Paul D. Stewart (cr). SuperStock: Science Faction (t). Wikipedia: "Portraits et Histoire des Hommes Utiles, Collection de Cinquante Portraits," Societe Montyon et Franklin, 1839-1840. (http://web.mit.edu/2.51/www/fourier.jpg) (cl). **184** Corbis: ZUMA Press / Aristidis Vafeiadakis (cla). Dorling Kindersley: The Science Museum, London (tc). Getty Images: SSPL (tr, cra). Photos.com: (cr). **185** Dorling Kindersley: The Science Museum, London (b). Getty Images: SSPL (tr, cra); Hulton Archive (crb). **186** Alamy Images: Hemis (tl). Corbis: (tr). Getty Images: SSPL (cl). Wikipedia: J.B. Perrin, "Mouvement brownien et réalité moléculaire," Ann. de Chimie et de Physique (VIII) 18, 5-114, (1909) (cr). **187** Alamy Images: Mary Evans Picture Library (t). **188** Getty Images: SSPL (tl, c). **188–189** Getty Images: Grant Dixon (t). **189** Science Photo Library: Dr David Furness, Keele University (tr). **190** Corbis: Bettmann (tl). Getty Images: SSPL (tr, c). **191** Alamy Images: Mary Evans Picture Library (tr). Corbis: Stefano Bianchetti (cb). **192** Getty Images: SSPL (cla); Hulton Archive (clb). **193** Getty Images: SSPL (cl); Hulton Archive (tr). **194** Library Of Congress, Washington, D.C.: LC-USZ62-48656 (cra). Science Photo Library: Dr Torsten Wittmann (cb). **195** Science Photo Library: Don Fawcett (br). **196** Getty Images: SSPL (tr); Hulton Archive (clb). **196–197** Dorling Kindersley: Courtesy of the Senckenberg Nature Museum, Frankfurt (c). **197** Getty Images: SSPL (crb, ca). NASA: H.E. Bond and E. Nelan (Space Telescope Science Institute, Baltimore, Md.); M. Barstow and M. Burleigh (University of Leicester, U.K.); and J.B. Holberg (University of Arizona) (tr). **198–199** NASA: (t). **198** Eon Images: (tl). Getty Images: SSPL (cr, clb). **199** Alamy Images: Everett Collection Historical (tr). **200** Alamy Images: The Art Archive (cr). Science Photo Library: NASA (tl). **201** Getty Images: SSPL (tl). Science Photo Library: James Cavallini (tr). **202** Corbis: Bettmann (tl). **202–203** Getty Images: Michael Mellinger (t). **203** Alamy Images: VintageMedStock (cr). Getty Images: (tr); Hulton Archive (cl). **205** Alamy Images: Natural History Museum, London (tr). **206** Corbis: adoc-photos (cra). Getty Images: Jessie Reeder (t). **207** Getty Images: SSPL (t, c, clb). **208–209** Corbis: (t). **209** Alamy Images: Everett Collection Historical (tr). Corbis: Stefano Bianchetti (cla). **210** Getty Images: AFP (cr); Steve Gschmeissner / SPL (ca); SSPL (clb). **210–211** Getty Images: De Agostini (t). **211** Dorling Kindersley: The Science Museum, London (cr). Getty Images: SSPL (cl). **212** Dorling Kindersley: Army Medical Services Museum (c, bc, br); The Science Museum, London (t); Old Operating Theatre Museum, London (cb). Getty Images: SSPL (crb). **213** Dorling Kindersley: Gettysburg National Military Park (cr); The Science Museum, London (bl, tl, cra); Collection of Jean-Pierre Verney (tr). Getty Images: Joseph Clark (br). **214** Corbis: Image Source (t). Photos.com: (cr). Science Photo Library: Paul D. Stewart (clb). **215** Corbis: Visuals Unlimited (t). Getty Images: Time & Life Pictures (cr). Linde AG: (crb). **216** Corbis: Visuals Unlimited (c). Getty Images: SSPL (tc). **216–217** Getty Images: Time & Life Pictures (t). **217** Dorling Kindersley: The Science Museum, London (cb). Getty Images: Time & Life Pictures (cl). **218** Dorling Kindersley: Judith Miller / Hamptons (tl). Getty Images: Photodisc / C Squared Studios (cla); SSPL (tc). Library Of Congress, Washington, D.C.: LC-DIG-cwpbh-04044 (bl). **218–219** Getty Images: SSPL (c). **219** Alamy Images: D. Hurst (cr). Dorling Kindersley: Judith Miller / Manic Attic (ca); The Science Museum, London (cla). Getty Images: SSPL (tl, tc); Time & Life Pictures (tr). **220** Corbis: Hulton-Deutsch Collection (t). Dorling Kindersley: The Science Museum, London (cl, c). **221** Getty Images: SSPL (c); Universal Images Group (tr); Hulton Archive (crb). Library Of Congress, Washington, D.C.: LC-DIG-det-4a25922 (cl). **222** Corbis: (tr). Dorling Kindersley: Natural History Museum (cr). Science Photo Library: Biology Media (clb). **223** Corbis: Visuals Unlimited (t). Getty Images: Universal Images Group (cra). **224** Alamy Images: Bilwissedition Ltd. & Co. KG (cra). Library Of Congress, Washington, D.C.: LC-DIG-ppmsca-17974 (t). **225** Getty Images: De Agostini (tr); SSPL (c). **226** Corbis: Bettmann (cla). Photos.com: (tr). **227** Alamy Images: Mary Evans Picture Library (cr). Getty Images: National Geographic (clb); Hulton Archive (c). Science Photo Library: CNRI (tr). **228** Corbis: John Springer Collection (clb). Getty Images: Time & Life Pictures (tr). **229** Corbis: Visuals Unlimited (t). Wikipedia: http://en.wikipedia.org/wiki/File:Cajal-Restored.jpg (c). **232** Corbis: Bettmann (clb); Minden Pictures / Tim Fitzharris (tl). **232–233** Getty Images: SSPL (cb). **233** Corbis: Hulton-Deutsch Collection (tl, crb). Science Photo Library: London School of Hygiene & Tropical Medicine (tr). **234** Fotolia: cbpix (bl). **235** Corbis: Roger Ressmeyer (c). Fotolia: Monkey Business (bc). Science Photo Library: Martin Bond (br);

Tony McConnell (tr). **236** Corbis: DPA (tl). Getty Images: SSPL (cb). **236–237** Alamy Images: Mary Evans Picture Library (t). **237** Corbis: Bettmann (bl). Science Photo Library: C. Powell, P. Fowler & D. Perkins (tr). **238** Corbis: NASA (tl). Dr James LaFountain, Dept of Biological Sciences, University at Buffalo, The State University of New York and Dr Rudolf Oldenbourg, Marine Biological Laboratory, Woods Hole, Mass: (cl). **238–239** Getty Images: SSPL (t). **239** Corbis: Bettmann (cl, tr). Dorling Kindersley: The Science Museum, London (crb). **240** Corbis: Image Source (bl). **241** Dorling Kindersley: Old Flying Machine Company (tl). **242** Getty Images: Roger Viollet (clb). Science Photo Library: Du Cane Medical Imaging Ltd (cr). **242–243** Corbis: Bettmann (t). **243** Corbis: Bettmann (tr). Getty Images: Hulton Archive (crb). **244** © CERN : Maximilien Brice (bc). Corbis: Bettmann (tr). **245** Corbis: Science Faction / William Radcliffe (cr). **246** Corbis: Ocean (tr). Getty Images: Hulton Archive (tl). **247** Corbis: Visuals Unlimited (tr). Getty Images: SSPL (cr). **248** Corbis: Bettmann (cl). Getty Images: SSPL (tl). **249** Science Photo Library: Natural History Museum, London (tl). **250** Getty Images: SSPL (cra). **252** Corbis: Bettmann (c). Getty Images: SSPL (t, cr). **253** Dorling Kindersley: EMU Unit of the Natural History Museum, London (clb). Getty Images: SSPL (crb). NASA: ESA / The Hubble Heritage Team (STScl / AURA) (t). **254** Alamy Images: Everett Collection Historical (tl). Corbis: National Geographic Society / US Naval Observatory (ca). Getty Images: SSPL (r). **255** Corbis: Visuals Unlimited (t). Science Photo Library: (ca). **256** Alamy Images: Pictorial Press Ltd. (tc). **256–257** NASA: ESA; Z. Levay and R. van der Marel, STScl; T. Hallas; and A. Mellinger (t); Yves Grosdidier (University of Montreal and Observatoire de (b). **257** Corbis: Hulton-Deutsch Collection (crb). Science Photo Library: Andrew Lambert Photography (cra). **258** American Institute of Physics, Emilio Segre Visual Archives: Esther C. Goddard (cr). Corbis: Hulton-Deutsch Collection (t). Science Photo Library: NOAA (cl). **259** Corbis: Bettmann (cr). Science Photo Library: ArSciMed (cb); Cern (t). **260** Alamy Images: Interfoto (t, cb). Getty Images: SSPL (clb). Photos.com: Lisa Ison (bl). Science Photo Library: (bc). **261** Corbis: Science Photo Library / Laguna Design (br). Fotolia: Albert Lozano-Nieto (bc). Getty Images: AFP (cra). Science Photo Library: Edward Kinsman (crb). **262** Corbis: Bettmann (cr); Clouds Hill Imaging Ltd (tl). **262–263** NASA: ESA / The Hubble Heritage Team (STScl / AURA) (t). **263** Courtesy of AT&T Archives and History Center: (clb). Science Museum / Science & Society Picture Library: (crb). **264** Alamy Images: Interfoto (tc). Lawrence Berkeley National Laboratory: Roy Kaltschmidt (c). **264–265** NASA: ESA, CFHT, CXO, M.J. Jee (University of California, Davis), and A. Mahdavi (San Francisco State University) (t). **265** Science Photo Library: Goronwy Tudor Jones, University of Birmingham (r). **266** Corbis: Bettmann (cr). **267** Getty Images: Boston Globe (cr). **268** Corbis: Bettmann (tl). FLPA: Steve Trewhella (clb). **269** Corbis: John Carnemolla (crb); Science Photo Library / Photo Quest Ltd (cl). Getty Images: Time & Life Pictures (tr). **270** Corbis: Bettmann (tl). NRAO : AUI (cl). **270–271** Science Photo Library: Peter Scoones (t). **271** Getty Images: (cr). Science Photo Library: (tr). **272–273** Science Photo Library: US Department of Energy (t). **273** Corbis: Bettmann (tc); (crb). Getty Images: Ron Boardman (cl). **274** Getty Images: SSPL (tl). **274–275** Science Photo Library: Los Alamos National Laboratory (c). **275** Corbis: Bettmann (cr). Science Photo Library: Pasieka (tr). **278** Fotolia: David Woods (tl). **278–279** NASA: Dryden Flight Research Center (t). **279** American Institute of Physics, Emilio Segre Visual Archives: Photo courtesy of Berkeley Lab (cl). Corbis: Roger Ressmeyer (tr). Science Photo Library: (cr). **280** Corbis: John Carnemolla (tr); Hulton-Deutsch Collection (ca). Getty Images: SSPL (tl). National Air and Space Museum, Smithsonian Institution: (cl). **281** Getty Images: SSPL (tr, cra). **282** Corbis: (tl). Science Photo Library: (cb, crb). **282–283** Corbis: Roger Ressmeyer (ca). Science Photo Library: A.B. Dowsett (t). **283** Corbis: Reuters (tr). National Library of Medicine: From the personal collection of Jenifer Glynn, photo Vittorio Luzzati (clb). **284** Science Photo Library: A. Barrington Brown (cra); James Cavallini (bc). **286** Getty Images: SSPL (cr). **287** Corbis: Bettmann (tl). Science Photo Library: Anthony Howarth (cr). **288–289** Alamy Images: ITAR-TASS Photo Agency (t). **288** Alamy Images: Chad Ehlers (tl). Dorling Kindersley: Courtesy of The Oxford Museum of Natural History (cr). **289** Corbis: Jonathan Blair (cl). Getty Images: National Geographic (tr). NASA: (cb). **290–291** Corbis: Bettmann (t). **290** Getty Images: Time & Life Pictures (cr). Science Photo Library: Ria Novosti (tl). **291** Alamy Images: FLPA (tr); RGB Ventures LLC dba SuperStock (clb). Dorling Kindersley: The Science Museum, London (ca). Wikipedia: Wikimol (http://en. wikipedia.org/wiki/File:Lorenz_ system_r28_s10_b2-6666.png) (cr). **292** Corbis: Michael Brennan (br); Historical Picture Archive (crb). Getty Images: Universal Images Group (clb); (bl). Library Of Congress, Washington, D.C.: LC-DIG-highsm-02106 / Carol M. Highsmith (bc). **292–293** Courtesy of IFREMER, Paris (t). **293** David Shale Deep Sea Images: (crb). Marie Tharp: (br). **294** Alamy Images: Everett Collection Historical (c). NASA: WMAP Science Team (tl). **294–295** Science Photo Library: Ria Novosti (t). **295** Dr. Linda Broome: (cb). Getty Images: Visuals Unlimited (tr). Science Photo Library: Steve Gschmeissner (cl). **296** Corbis: Bettmann (ca); Science Faction / Dan McCoy (tr). Science Photo Library: Robin Scagell (clb). **297** Getty Images: Oxford Scientific / Laguna Design (l). NASA: (r). **298** Alamy Images: Index Stock / NASA (cra). Corbis: Hulton-Deutsch Collection (tr). Getty Images: (cla); Hulton Archive (ca); SSPL (cb). NASA: (tl, crb). **299** NASA: (tc, tr). National Air and Space Museum, Smithsonian Institution: (SI-99-15152) (b). **300** Flickr.com: Aero Icarus (t). Getty Images: Nancy R. Schiff / Archive Photos (cr). NASA: (cl). **301** Corbis: Sygma (t). Getty Images: Gamma-Keystone (c). **302** Corbis: Ocean (ca). Yann Mousseaux: (www.flickr.com/ photos/ymousseaux/4756622927/) (cr). NASA: Apollo 17 / Arizona State University (M. Furst) (t). www.vintagecalculators.com: (clb). **303** NASA: Johnson Space Center (t); (cb). **304** © CERN : (cra). Dorling Kindersley: NASA (cl). Science Photo Library: Martin Bond (crb). **305** Corbis: Michael S Yamashita (cl). Dorling Kindersley: Courtesy of The Oxford Museum of Natural History (cr). Fotolia: dotypicture (t). **306** Dr. Wolfgang Beyer: (tl, tc, tr). Corbis: Bettmann (cl). Dorling Kindersley: Courtesy of ESA (cr). **307** Dr. Wolfgang Beyer: (tl). Corbis: Richard Baker (tr); Kim Kulish (c). Dorling Kindersley: NASA (cb). Getty Images: Time & Life Pictures (cr). **308** Corbis: Bettmann (tl, cr). NOAA: WHOI, UCSB, Univ. S. Carolina (cl). **309** Alamy Images: Images of Africa Photobank (r); Trinity Mirror / Mirrorpix (tl). **310–311** Corbis: Bettmann (t). **311** Alamy Images: Michelle Gilders (tr). Corbis: Bettmann (cl). Fotolia: Yves Roland (cra). **312** NASA: KSC. **313** Getty Images: SSPL (cl). Science Photo Library: Philippe Plailly (crb); J.C. Revy (tr). **314–315** Corbis: Frans Lanting (t). Getty Images: AFP (cb). **315** Corbis: Roger Ressmeyer (ca); Visuals Unlimited (tc). Dorling Kindersley: Judith Miller / Graham Cooley (crb). Fairlight CMI: advertisement from Keyboard Magazine, February 1982 page 15 (cl). **316–317** Dorling Kindersley: The Science Museum, London (c). **316** Dorling Kindersley: National Museums of Scotland (cla); University Museum of Archaeology and Athropology, Cambridge (tl); The Science Museum, London (c); National Maritime Museum, London (cra). Getty Images: SSPL (bc). **317** Dorling Kindersley: Imperial War Museum, London (cr). Fotolia: Volodymyr Krasyuk (ftr). Getty Images: (tr); SSPL (crb). **318–319** Corbis: Sygma (t). **318** NASA: (c). **319** Getty Images: SSPL (cl). Science Photo Library: NASA / GSFC (tc); Pasieka (c). **320** Corbis: (clb). Getty Images: AFP (t); Time & Life Pictures (cr). **321** Corbis: Firefly Productions (cra). Fotolia: (t). Used with permission, GM Media Archives: (crb). **322** Getty Images: CMSP / Wood (t). **323** Advanced Cell Technology: (tr). NASA: JPL (cr). **324** © CERN : (cl). Corbis: Car Culture (cb). **324–325** NASA: JPL / Caltech (t). **325** The Opte project: http://blyon. com/blyon-cdn/opte/maps/ static/1069646562. LGL.2D.4000x4000.png (r). **326** Corbis: (cra). **327** Corbis: NASA (tc). **328–329** Corbis: (t). **328** Chas. A. Blatchford & Sons Ltd: (r). NASA: (ca). **329** NASA: Dr. Hal Weaver and T. Ed Smith (STScl) (tr). Science Photo Library: (bc). **330** Getty Images: Alan Crawford (clb). NASA: ESA, STScl, J. Hester and P. Scowen (Arizona State University) (cr). **331** Corbis: Reuters (crb). Getty Images: AFP (t). **332** Corbis: Visuals Unlimited (ca). Press Association Images: AP (t). **333** Corbis: Visuals Unlimited (crb). NASA: CXC / MIT / F.K. Baganoff et al. (t). Science Photo Library: Dr P. Marazzi (c). **334** Getty Images: AFP (tr); Time & Life Pictures (clb, crb); De Agostini (bl, bc); Bloomberg (br). Photos. com: (cb). **335** Alamy Images: Aflo Co. Ltd. (t). Corbis: ABK / BSIP (bc). Getty Images: Time & Life Pictures (clb). **336** Isaac Ehrenberg and Dylan Erb: (clb). Getty Images: Visuals Unlimited (crb). NASA: JPL (tl). **336–337** Corbis: Frans Lanting (t). **337** Science Photo Library: James King-Holmes (c). **338** Getty Images: Visuals Unlimited (ca). NASA: JPL / GSFC (clb). **339** Getty Images: AFP (bl). Max Planck Institute for Astrophysics: (t). **340** Dorling Kindersley: Swedish Museum of Natural History (cl). Getty Images: MCT (cra). NASA: ESA, R. Ellis (Caltech), and the UDF 2012 Team (t). **341** Corbis: Reuters (t). Getty Images: Bloomberg (cra). NASA: JPL-Caltech / ASI (clb). **342** NASA: JPL / University of Washington (t). **343** Corbis: Reuters (cl). Fotolia: Sergiy Serdyuk (t). **344** Corbis: Bettmann (cr). **346** Corbis: Reuters (cr). Global Crop Diversity Trust (www.croptrust.org): Cary Fowler (clb). **346–347** © CERN : Maximilien Brice (t). **347** Corbis: Visuals Unlimited (clb). Getty Images: Bloomberg (tr, cra). **348** Getty Images: (crb). NASA: JPL-Caltech / MSSS (t); JPL / Caltech (c). **349** Science Photo Library: (r)

Jacket images: Front and back: Alamy Images: Dennis Hallinan
All other images © Dorling Kindersley
For further information see: www. dkimages.com

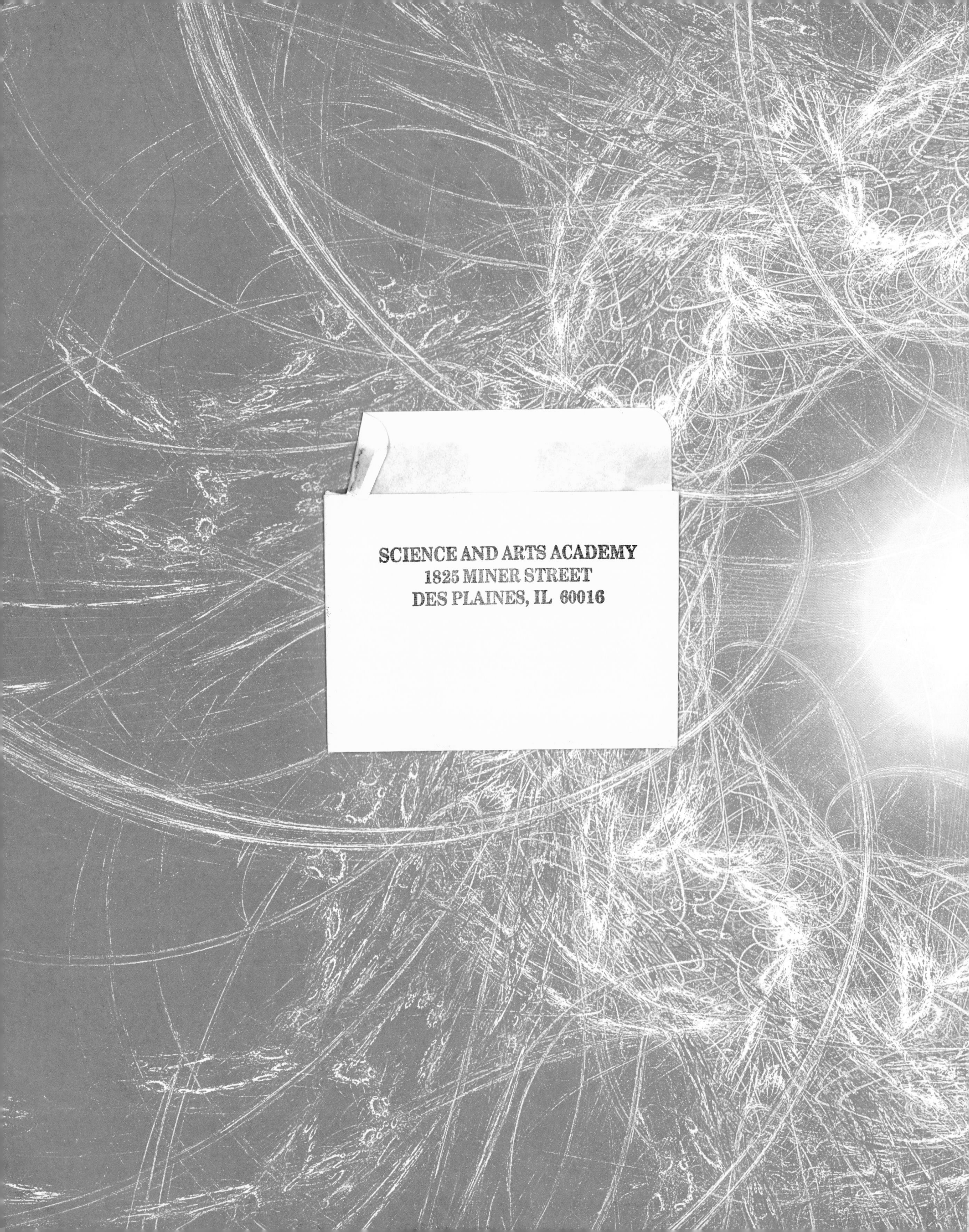
SCIENCE AND ARTS ACADEMY
1825 MINER STREET
DES PLAINES, IL 60016